INDEX to RECORD AND TAPE REVIEWS

A Classical Music Buying Guide

1977

ANTOINETTE O. MALEADY

CHULAINN PRESS

San Anselmo, California 1978

ISBN 0-917600-03-7
Library of Congress Catalog Card No. 72-3355
Manufactured in the United States of America

CHULAINN PRESS
Post Office Box 770
San Anselmo, California 94960

To Mary

CONTENTS

INTRODUCTION

This Index brings together in one volume a listing of all classical music recordings reviewed in 1977 in the major reviewing media of the United States, England and Canada.

The Index has four sections. Section I is a straight listing by composer. Section II, "Music in Collections", lists records or tapes with several composers on one disc or tape. Section II main entries are arranged alphabetically by name of the manufacturer, and then serially by manufacturer's number within each main entry. The work of any composer in each of these collections appears in Section I under the name of that composer, with a reference to the Section II entry. Section III lists, alphabetically by title, anonymous works, with a reference to their location in Section I or Section II. Section IV is a Performer Index to recordings listed in Sections I and II, with reference by citation number to its location within the section. Each citation gives, if available, the disc label and number; variant labels and numbers; tape cassette, cartridge or reel numbers; quadraphonic disc or tape numbers; reissues; location of the reviews and the reviewers evaluation of the recording. The main entry for each recording is in upper case letters. Tapes or discs reviewed in 1977 that were also reviewed in 1976, 1975, 1974, 1973, 1972 and/or 1971 have all the reviews brought together in the 1977 Index.

A key is provided for understanding the entries of the four sections. The entries are fictitious entries to show the various possibilities of form.

Section I

MACONCHY, Elizabeth

1219 Ariadne, soprano and orchestra. WALTON: Songs: Daphne; Old Sir
[entry Faulk; A song for the Lord Mayor's table; Through gilded
no.] trellises. Heather Harper, s [Heather Harper, soprano]; Paul Hamburger, pno [Paul Hamburger, piano]; ECO [English Chamber Orchestra]; Raymond Leppard [conductor]. Columbia M 30443 [disc number] (2) [number of discs in set] Tape (c) MT 30443 [cassette number] (ct) MA 30443 [cartridge number] (r) L 30443 [reel number] (Q) MQ 30443 [quadraphonic disc number] Tape (ct) MAQ 30443 [quadraphonic tape cartridge number]. (also CBS 72941) [recording also available on CBS label]

++Gr 9-75 p1025 [evaluation excellent; review in Gramophone, September 1975, page 1025]

/NR 2-76 p4 [evaluation fair, review in New Records February 1976 page 4]

-St 7-77 p55 tape [evaluation poor, review in Stereo Review, July 1977, page 55]

+NYT 7-23-77 pD38 [evaluation good, review in New York Times, July 23, 1977 page D38]

SECTION II

LONDON

OS 36578 (also Decca SXL 3315) [label number. Also available on Decca label] (Reissue from OS 3716) [reissue from London OS 3716]

2395 [entry no.] Baroque flute sonatas. BLAVET: Sonata, flute, no. 2, F major. GUALTIER: Suite, G minor. HANDEL: Sonata, flute, op. 1, no. 5, G major. LOEILLET: Sonata, flute, F major. VINCI: Sonata, flute, D major. Jean-Pierre Rampal, flt [Jean-Pierre Rampal, flute]; Raymond Leppard, hpd [Raymond Leppard, harpsichord] Claude Viala, vlc [Claude Viala, violoncello]; AMF [Academy of St. Martin-in-the-Fields]; Raymond Leppard [conductor]

+-MQ 4-77 p53 [evaluation mixed, review in Musical Quarterly, April 1977, page 53]

+STL 1-15-77 p30 [evaluation good, review in Sunday Times, London, January 15, 1977, page 30]

Section III

Kyrie trope: Orbis factor. cf TALLIS: Missa salve intemerata. [Anonymous work, appears in Section I under entry for Tallis]

Deo gracias Anglia, Agincourt carol. cf BASF 25 22286-1. [Anonymous work, appears in Section II under entry BASF 25 22286-1]

Section IV

Baker, Janet, mezzo-soprano (contralto) 231, 494, 939, 1376, 1776, 2513 [performer, listed in reviews as mezzo-soprano or contralto. Citation numbers in Section I and II where artist has performed]

ABBREVIATIONS

Periodicals Indexed

AR	American Recorder
ARG	American Record Guide
ARSC	Assocation for Recorded Sound Collections Journal
Audio	Audio
CJ	Choral Journal
CL	Clavier
FF	Fanfare
Gr	Gramophone
GTR	Guitar Review
Ha	Harpers
HF	High Fidelity
HFN	Hi-Fi News & Record Review
HPD	Harpsichord
IN	The Instrumentalist
LJ	Library Journal/School Library Journal Previews
MJ	Music Journal
MM	Music & Musicians
MQ	Musical Quarterly
MR	Music Review
MT	Musical Times
MU	Music/American Guild of Organists
NC	Nineteenth Century Music
NR	New Records
NYT	New York Times
OC	Opera Canada
ON	Opera News
Op	Opera, London
OR	Opera Review
PNM	Perspectives of New Music
PRO	Pro Musica
RR	Records and Recording
SFC	San Francisco Examiner and Chronicle
SR	Saturday Review/World
St	Stereo Review
ST	Strad
STL	Sunday Times, London
Te	Tempo

Performers

Orchestral

AMF	Academy of St. Martin-in-the-Fields

BBO	Berlin Bach Orchestra
BeSO	Berlin Symphony Orchestra
BPhO	Berlin Philharmonic Orchestra
BPO	Boston "Pops" Orchestra
Brno PO	Brno State Philharmonic Orchestra
BRSO	Berlin Radio Symphony Orchestra
BSO	Boston Symphony Orchestra
CnSO	Cincinnati Symphony Orchestra
CO	Cleveland Orchestra
COA	Concertgebouw Orchestra, Amsterdam
CPhO	Czech Philharmonic Orchestra
CSO	Chicago Symphony Orchestra
DBS	Deutsche Bach Solisten
ECO	English Chamber Orchestra
FK	Frankfurt Kantorei
HCO	Hungarian Chamber Orchestra
HRT	Hungarian Radio and Television
HSO	Hungarian State Symphony Orchestra
LAPO	Los Angeles Philharmonic Orchestra
LOL	Little Orchestra, London
LPO	London Philharmonic Orchestra
LSO	London Symphony Orchestra
MB	Munich Bach Orchestra
MPAC	Munich Pro Arte Chamber Orchestra
MPO	Moscow Philharmonic Orchestra
MRSO	Moscow Radio Symphony Orchestra
NPhO	New Philharmonia Orchestra
NSL	New Symphony, London
NWE	Netherlands Wind Ensemble
NYP	New York Philharmonic Orchestra
ORTF	O.R.T.F. Philharmonic Orchestra
OSCCP	Orchestre de la Societe des Concerts du Conservatoire de Paris
OSR	l'Orchestre de la Suisse Romande
PCO	Prague Chamber Orchestra
PH	Philharmonia Hungarica
PhO	Philharmonia Orchestra, London
PO	Philadelphia Orchestra
PSO	Prague Symphony Orchestra
ROHO	Royal Opera House Orchestra, Covent Garden
RPO	Royal Philharmonic Orchestra
SBC	Stuttgart Bach Collegium
SCO	Stuttgart Chamber Orchestra
SDR	Stuttgart S.D.R. Symphony Orchestra
SPO	Stuttgart Philharmonic Orchestra
SSO	Sydney Symphony Orchestra
VCM	Vienna Concentus Musicus
VPM	Vienna Pro Musica
VPO	Vienna Philharmonic Orchestra
VSO	Vienna Symphony Orchestra
VSOO	Vienna State Opera Orchestra

<u>Instrumental</u>

acc	accordion	cld	clavichord
bal	balalaika	clt	clarinet
bs	bs	cor	cornet

bsn	bassoon	dr	drums
c	celesta	Eh	English horn
cimb	cimbalon	flt	flute
Fr hn	French horn	pic	piccolo
gtr	guitar	pno	piano
harm	harmonica	rec	recorder
hn	horn	sax	saxophone
hp	harp	sit	sitar
hpd	harpsichord	tpt	trumpet
lt	lute	trom	trombone
mand	mandolin	vib	vibraphone
mar	marimba	v	viol
ob	oboe	vla	viola
ond	ondes martenot	vlc	violoncello
org	organ	vln	violin
perc	percussion	z	zither

Vocal

bar	baritone	ms	mezzo-soprano
bs	bass	s	soprano
con	contralto	t	tenor
c-t	countertenor		

Qualitative Evaluation of Recordings

++	excellent or very good	/	fair
+	good	-	poor
+-	mixed	*	no evaluation

Section I

COMPOSERS

ABBOTT, Alan
Alla caccia. cf Musical Heritage Society MHS 3547.

ABEL, Karl
Sonata, G major. cf BIS LP 22.

ACHRON
In a little cottage. cf Olympic OLY 105.

ADAM
O holy night. cf London OS 26437.
O holy night. cf Philips 6392 023.

ADAM, Adolphe-Charles
Cantique Noel. cf Decca SXL 6781.
Le chalet: Vallons de l'Helvetie. cf Rubini GV 39.
Le chalet: Vive le vin, l'amor, et le tabac. cf Club 99-107.

1 Giselle: Pas de deux, Act 1; Grand pas de deux and finale (arr. Busser). MINKUS: Don Quixote: Pas de deux. TCHAIKOVSKY: The nutcracker, op. 71: Pas de deux, Act 2. Sleeping beauty, op. 66: Blue bird pas de deux. Swan Lake, op. 20: Pas de deux, Act 3. LSO, OSR, OSCCP; Richard Bonynge, Ernest Ansermet, Jean Martinon. Decca SPA 487.
+HFN 1-77 p119 ++RR 12-76 p62

Variations on a theme of Mozart's "Ah, vous dirai-je Maman". cf BIS LP 45.

ADAM DE LA HALLE
De ma dame vient. cf DG 2710 019.
Fines amouretes ai. cf Nonesuch H 71326.
J'os bien a m'amie parler. cf DG 2710 019.
The play of Robin and Marion, excerpts. cf Enigma VAR 1020.
Songs: De cueur pensieu; En mai, quant rosier. cf Turnabout 37086.
Tant con je vivrai. cf Nonesuch H 71326.

ADAMS, Stephen
Songs: The holy city. cf Argo ZFB 95/6.
Songs: The holy city. cf L'Oiseau-Lyre DSLO 20.

ADASKIN
Algonquin symphony. cf Citadel CT 6011.

ADDISON, John

2 Concerto, trumpet, strings and percussion. ARNOLD: Concerto, 2 violins and strings, op. 77. CROSSE: Some marches on a ground, op. 28. SEIBER: Concertino, clarinet and strings. Leon Rapier, tpt; Paul Kling, Peter McHugh, vln; James Livingston, clt; Louisville Orchestra; Jorge Mester. RCA GL 2-5018.
+Gr 10-76 p577 +RR 11-76 p49
+-HFN 12-76 p136 +St 1-77 p777

3 Divertimento, op. 9. DODGSON: Sonata, brass. BENNETT: Commedia

IV. GARDNER: Theme and variations, op. 7. Philip Jones Brass Ensemble. Argo ZRG 813.
++Gr 12-75 p1066 ++NR 8-76 p6
++HF 4-77 p114 ++RR 12-75 p36
HFN 12-75 p156

ADLER, Samuel
4 Xenia, a dialogue for organ and percussion. COOPER: Variants, organ. HARRISON: Concerto, organ and percussion. David Craighead, org; Gordon Stout, perc; Los Angeles Percussion Ensemble; William Kraft. Crystal S 858.
+NR 11-77 p14

ADRIAENSSEN, Emanuel
Branle Englese. cf DG 2533 302.
Branle simple de Poictou. cf DG 2533 302.
Courante. cf DG 2533 302.
Fantasia. cf DG 2533 302.
Io vo gridando. cf DG 2533 323.
Madonna mia pieta. cf DG 2533 323.

ADSON, John
Ayres (2). cf Golden Crest CRS 4148.

AGAPKIN
March of the Pechera Regiment. cf HMV CSD 3782.
Slav girl's farewell. cf HMV CSD 3782.
AGE OF GOLD. cf CBS 61781

AGRICOLA, Alexander
Comme femme. cf HMV SLS 5049.
Oublier veul. cf Argo ZRG 823.
Songs: Christum wir sollen loben schon; Gelobet seist du Jesu Christ. cf ABC L 67002.

AICHINGER, Gregor
Jubilate Deo. cf Golden Crest CRS 4148.

AITKEN
Maire my girl. cf Philips 6392 023.

AITKEN, Hugh
5 Cantatas, nos. 1, 3-4. Fantasy, piano. Charles Bressler, t; Jean Hakes, s; Karl Kraber, flt; Melvin Kaplan, ob; Helen Kwalwasser, vln; Ynez Lynch, vla; Fortunato Arico, vlc; Julius Levin, double bas; Gary Kirkpatrick, pno. CRI SD 365.
+-ARG 7-77 p11 +NR 10-77 p11
*MJ 10-77 p29
Fantasy, piano. cf Cantata, nos. 1, 3-4.
Montages, solo bassoon. cf Crystal S 351.

ALABIEFF, Alexander
The nightingale. cf Columbia 34294.
The Russian nightingale. cf BIS LP 45.

ALAIN, Jehan
Choral Dorien, op. 47. cf Organ works (Odeon C 065-12170).
Climat, op. 53. cf Organ works (Odeon C 065-12170)
Danses, op. 81. cf Organ works (Odeon C 069-12798).
Danses a Agni Yavishta, op. 52 (2). cf Organ works (Cathedral CRMS 858).
Danses a Agni Yavishta, op. 52. cf Organ works (Odeon C 069-12798).
Fantasias, nos. 1 and 2, opp. 51 and 73. cf Organ works (Odeon 069-12798)
Fantasia, no. 1, op. 51. cf Organ works (Cathedral CRMS 858)
Fantasia, no. 2, op. 73. cf Organ works (Cathedral CRMS 858)
Grave, op. 32. cf Organ works (Odeon C 065-12170).

Le jardin suspendu, op. 50. cf Organ works (Odeon C 065-12170)
Lamento, op. 12. cf Organ works (Odeon C 065-12170).
Litanies, op. 79. cf Organ works (Cathedral CRMS 838).
Litanies, op. 79. cf Organ works (Odeon C 065-12170).
Litanies, op. 79. cf Organ works (Odeon C 069-12798).
Litanies, op. 79. cf Decca SDD 499.
Litanies, op. 79. cf Decca PFS 4416.

6 Organ works: Danses a Agni Yavishta, op. 52 (2). Fantasia, no. 1, op. 51. Fantasia, no. 2, op. 73. Litanies, op. 79. Postlude, op. 21. LANGLAIS: Suite breve: Jeux; Cantilene; Plainte; Dialogue sur les mixtures. Nicholas Kynaston, org. Cathedral CRMS 858.
+HFN 6-77 p131

7 Organ works: Choral Dorien, op. 47. Climat, op. 53. Le jardin suspendu, op. 50. Grave, op. 32. Lamento, op. 12. Litanies, op. 79. Suite, op. 48: Choral. Variations sur Lucis Creator, op. 28. Variations sur un theme de Jannequin, op. 78. Andre Merineau, org. Odeon C 065-12170.
+-MU 4-77 p13

8 Organ works: Danses, op. 81 (3). Danses a Agni Yavishta, op. 52 (2). Fantasias, nos. 1 and 2, opp. 51 and 73. Litanies, op. 79. Jean-Louis Gil, org. Odeon C 069-12798.
+MU 4-77 p13

Postlude, op. 21. cf Organ work (Cathedral CRMS 858).
Suite, op. 48: Choral. cf Organ works (Odeon C 065-12170).
Suite pour orgue. cf Calliope CAL 1922/4.
Variations sur Lucis Creator, op. 28. cf Organ works (Odeon C 065-12170).

9 Variations sur un theme de Jannequin, op. 78. DUPRE: Prelude and fugue, op. 7, no. 2, F minor. Andre Isoir, org. Calliope CAL 1924.
+Gr 3-77 p1454

Variations sur un theme de Jannequin, op. 78. cf Organ works (Odeon C 065-12170).
Variations sur un theme de Jannequin, op. 78. cf (Calliope 1922/4).

ALBENIZ, Isaac

Bajo la palmera, op. 212. cf RCA ARL 1-0456.
Cantos de Espana, op. 232: Cordoba. cf Angel S 36094.
Capricho catalan. cf RCA ARL 1-1323.
Espana, op. 165. cf Discopaedia MB 1012.

10 Espana, op. 165, no. 2: Tango (arr. Kreisler). FALLA: La vida breve: Spanish dance. GRANADOS: Danse espagnole, op. 37 (arr. Kreisler). KREISLER: Caprice Viennois. Concerto, violin, C major (after Vivaldi). Liebesfreud. Liebesleid. Recitative and scherzo-caprice, solo violin, op. 6. Schon Rosmarin. Syncopation. Tambourin Chinois, op. 3. Erich Gruenberg, vln; NPhO; Elgar Howarth. Decca PFS 4423.
-Gr 12-77 p1150

11 Espana, op. 165, no. 2: Tango (arr. Godowsky). CHOPIN: Ballade, no. 3, op. 47, A flat major. Etudes, op. 10, no. 4, C sharp minor; op. 25, no. 7, C sharp minor; op. 25, no. 10, B minor. SCHUBERT: Moment musical, op. 94, no. 3, D 780, F minor (arr. Godowsky). SCHUMANN: Symphonic etudes, op. 13. Shura Cherkassky, pno. L'Oiseau-Lyre DSLO 15.
+-Gr 11-76 p843 +-RR 12-76 p86
+-HFN 1-77 p104

Espana, op. 165, no. 2: Tango. cf Angel S 36094.

Espana, op. 165, no. 2: Tango. cf Discopaedia MB 1015.
Espana, op. 165, no. 3: Malaguena. cf HMV ASD 3346.
12 Iberia. TURINA: Danzas fantasticas. Blanca Uribe, pno. Orion ORS 75202/3 (2).
+-HF 6-77 p79 +NR 11-76 p14
13 Iberia. Aldo Ciccolini, pno. Seraphim SIB 6091 (2).
+ARG 3-77 p10 +SR 2-19-77 p42
+-HF 6-77 p79 +St 8-77 p110
+NR 7-77 p13
14 Iberia, Bks 1 and 2. Michel Block, pno. Connoisseur Society CS 2120.
+ARG 10-77 p13 +NYT 7-31-77 pD13
++HF 6-77 p79 ++SFC 4-3-77 p43
+-MJ 5-77 p32 +St 8-77 p110
+NR 7-77 p13
15 Iberia, Bks 3 and 4. Navarra. Michel Block, pno. Connoissuer Society CS 2121.
+ARG 10-77 p13 +NYT 7-31-77 pD13
+-MJ 5-77 p32 ++SFC 4-3-77 p43
+NR 7-77 p13 +St 8-77 p110
Iberia: Evocacion. cf RCA ARL 1-0456.
Iberia: Fete dieu a Seville. cf CBS 73589.
16 Iberia: Triana; El corpus en Sevilla (orch. Arbos). Navarra (compl. Severac, orch Arbos). FALLA: Love the magician: Ritual fire dance. The three-cornered hat: Suites, nos. 1 and 2. GRANADOS: Danzas espanolas, op. 37: Andaluza (orch. de Grignon). RPO; Artur Rodzinski. Classics for Pleasure CFP 40261. (Reissue from HMV SLP 1688)
-Gr 6-77 p48 +-RR 6-77 p48
+-HFN 6-77 p137
Improvisations cf International Piano Archives IPA 109.
Navarra. cf Iberia, Bks 3 and 4.
Navarra. cf Iberia: Triana; El corpus en Sevilla.
Recuerdos de viaje, op. 71: Rumores de la caleta. cf London 7015.
Sonata, guitar. cf RCA RK 1-1735.
Suite espanola, no. 1: Granada. cf Classics for Pleasure 40012.
Suite espanola, no. 1: Granada. cf RCA RL 2-5099.
Suite espanola, no. 3: Sevillanas. cf RCA GL 4-2125.
Suite espanola, no. 5: Asturias. cf Philips 6833 159.
Zambra granadina. cf London CS 7015.

ALBENIZ, Melchior
Sonata, guitar, D major. cf Swedish Society SLT 33205.

ALBINONI, Tommaso
17 Adagio. GABRIELI, G.: Canzoni, nos. 5, 8. GABRIELI, A.: La bataille de Marignan. VIVALDI: Concerto, piccolo and strings, C major. Le Grande Ecurie et la Chambre du Roy; Jean-Claude Malgoire. Odyssey Y 34605.
+NR 8-77 p5
Adagio, G major. cf VIVALDI: The four seasons, op. 8, nos. 1-4.
Adagio, G minor. cf DG 2548 219.
Adagio, G minor. cf HMV ESD 7011.
18 Adagio, G minor. CORELLI: Concerto grosso, op. 6, no. 8, G minor. MOZART: Serenade, no. 13, K 525, G major. PACHELBEL: Canon. Toulouse Chamber Orchestra; Louis Auriacombe. Seraphim S 60271.
+NR 3-77 p3
19 Adagio, organ and strings, G minor (Giazotto). Concerto, oboe, op. 9, no. 2, D minor. BONONCINI: Sinfonia da chiesa a quattro,

op. 5, no. 1. PACHELBEL: Canon, D major (arr. Lam). PURCELL: Chacony, G minor. Alastair Ross, org; Sarah Barrington, ob; Richard Hickox Orchestra; Richard Hickox. Argo ZRG 866 Tape (c) KZRC 866.
+-Gr 7-77 p188
+-HFN 6-77 p116
+HFN 9-77 p155 tape
+RR 6-77 p38
+RR 9-77 p98 tape

20 Adagio, organ and strings, G minor (arr. Giazotto). GABRIELI, G.: Canzoni, nos. 5, 8. JANNEQUIN: La bataille de Marignan (arr. A. Gabrieli). VIVALDI: Concerto, piccolo and strings, C major. La Grand Ecurie et la Chambre du Roy; Jean-Claude Veilhan, rec; Jean-Claude Malgoire. CBS 76535.
-Gr 5-77 p1704
+HFN 5-77 p116
+RR 5-77 p42

21 Adagio, organ and strings, G minor (arr. Giazotto). VIVALDI: The four seasons, op. 8, nos. 1-4. Giuliano Badini, vln; Sinfonia di Siena. Saga 5433 Tape (c) CA 5443.
+-HFN 11-77 p187 tape
/-RR 2-77 p40
/-RR 11-77 p118 tape

Concerto, op. 5, no. 5, A major. cf Supraphon 110 1890.

22 Concerti, op. 7 (12). I Solisti Veneti; Claudio Scimone. Erato STU 70883 (2).
+Gr 3-77 p1386
+HFN 5-77 p116
-MT 7-77 p564
++RR 3-77 p41

23 Concerti, op. 7 (12). Hans Werner Watzig, ob; Berlin Chamber Orchestra; Vittorio Negri. Philips 6747 138 (2).
+-Gr 6-77 p32
+HF 10-77 p99
+HFN 6-77 p116
++MJ 11-77 p30
++NR 10-77 p6
+RR 5-77 p41
++SFC 9-11-77 p42

Concerto, oboe, op. 9, no. 2, D minor. cf Adagio, organ and strings, G minor.

Concerto, oboe, op. 9, no. 2, D minor. cf HMV ASD 3318.

24 Concerto, trumpet, op. 7, no. 3, B flat major. Concerto, trumpet, op. 7, no. 6, D major (arr. Paumgartner). HANDEL: Concerto, trumpet, D minor (arr. Thilde). HERTEL: Concerto, trumpet, E flat major. TELEMANN: Concerto, trumpet, D major (ed. Tottcher/Grebe). Maurice Andre, tpt; ECO; Charles Mackerras. HMV SQ ASD 3394.
++Gr 11-77 p854
+HFN 12-77 p164
++RR 11-77 p59

Concerto, trumpet, op. 7, no. 6, D major. cf Concerto, trumpet, op. 7, no. 3, B flat major.

Concerto, trumpet and 6 clarinets, A major. cf Philips 6581 018.

25 Concerto, 2 trumpets, C major. BACH: Cantata, no. 78: Air, 2 trumpets. Cantata, no. 51: Chorale. Sonata, flute and harpsichord, S 1031, E flat major: Siciliano. Toccata, adagio and fugue, S 564, C major: Adagio. ZIPOLI: Suite, F major. Maurice Andre, Jean Pirot, Lionel Andre, tpt; Jean-Francois Paillard Chamber Orchestra; Jean-Francois Paillard. Musical Heritage Society MHS 3614.
++FF 9-77 p71

Sonata a 5, op. 5, no. 9, E minor. cf DG 2548 219.

ALBRIGHT, William

Pneuma. cf Marc MC 5355.

ALCOCK, John

Voluntary, D major. cf Cambridge CRS 2540.

ALCOCK, Walter
Introduction and passacaglia. cf Guild GRSP 7011.
ALDER
Bi-Centenary USA march. cf Rediffusion 15-56.
ALFONSO X, El Sabio
Rosa das rosas. cf CRD CRD 1019.
ALFORD, Harry
Colonel Bogey on parade. cf Grosvenor GRS 1055.
On the quarter deck. cf HMV SQ ASD 3341.
ALFVEN, Hugo
26 Festspel, op. 25. Swedish rhapsody, no. 1, op. 19. Swedish rhapsody, no. 3, op. 48. Stockholm Philharmonic Orchestra; Kungl Hovkapellet; Stig Westerberg, Hugo Alfven. Swedish Society SLT 33145.
+RR 7-76 p42 ++St 4-77 p114
27 En Skargardssagen (A legend of the Skerries), op. 20. Symphony, no. 5, op. 55, A minor: 1st movement. Swedish Radio Symphony Orchestra; Stig Westerberg. Swedish Society SLT 33174.
+-RR 5-76 p33 +St 1-77 p116
Songs: Sa tag mitt hjarta; Jag langtar dig. cf RCA LSC 9884.
Swedish rhapsody, no. 1, op. 19. cf Festspel, op. 25.
Swedish rhapsody, no. 3, op. 48. cf Festspel, op. 25.
28 Symphony, no. 1, op. 7, F minor. Swedish Radio Symphony Orchestra; Stig Westerberg. Swedish Society SLT 33213.
++Gr 3-76 p1457 +St 1-77 p116
++HFN 1-76 p101
29 Symphony, no. 2, op. 11, D major. Stockholm Philharmonic Orchestra; Leif Segerstam. Swedish Society SLT 33211.
+HFN 10-76 p163 +St 1-77 p116
+RR 10-76 p38
30 Symphony, no. 3, op. 23, E major. Stockholm Philharmonic Orchestra; Nils Grevillius. Swedish Society SLT 33161.
+-RR 5-76 p33 +St 1-77 p116
31 Symphony, no. 4, op. 39, C minor. Sven Erik Vikstrom, t; Gunilla af Malmborg, s; Stockholm Philharmonic Orchestra; Nils Grevillius. Swedish Society SLT 33186.
+-RR 2-75 p29 +St 1-77 p116
+-RR 5-76 p33
Symphony, no. 5, op. 55, A minor: 1st movement. cf En Skargardssagen, op. 20.
ALISON, Richard
Dolorosa pavan. cf L'Oiseau-Lyre 12BB 203/6.
Go from my windoe. cf RCA RL 2-5110.
Lady Francis Sidney's almayne. cf RCA RL 2-5110.
Sharp pavin. cf Saga 5438.
ALLDAHL, Per Gunnar
Stem the blood flow. cf BIS LP 32.
ALMEIDA, Laurindo
32 Jazz-tune at the mission. Late last night (Feather). GNATALLI: Sonata, guitar and violoncello. LENNON-MCCARTNEY: Yesterday (arr. Almeida). Laurindo Almeida, gtr; Frederick Seykora, vlc; Assisting artists. Crystal Clear CCS 8001.
+ARG 8-77 p35
Late last night. cf Jazz-tune at the mission.
ALMOROX (15th century Spain)
O dichoso. cf Enigma VAR 1024.

ALTHEN
Songs: Land du valsignade. cf RCA LSC 9884.
ALWYN, William
33 Symphony, no. 1, D major. LPO; Kenneth Alwyn. Lyrita SRCS 86. (also HNH 4040)
+FF 11-77 p4 ++HFN 4-77 p132
++Gr 4-77 p1541 ++RR 4-77 p40
34 Symphony, no. 4. Symphony, no. 5. LPO; William Alwyn. Musical Heritage Society MHS 3574.
+ARG 9-77 p4 +FF 9-77 p3
Symphony, no. 5. cf Symphony, no. 4.
ALONSO (16th century Spain)
La tricotea Samartin. cf Nonesuch H 71326.
AMBROSIAN CHANTS. cf DG 2533 284.
AMERICA SINGS. cf Vox SVBX 5353.
AMERICA SINGS: The founding years, 1620-1800. cf Vox SVBX 5350.
AMERICAN ORGAN MUSIC OF THREE CENTURIES. cf Orion ORS 76255.
AMMERBACH, Elias
Wer das Tochterlein haben will. cf ABC ABCL 67008.
AMON, Johann
35 Quartet, horn, op. 20, no. 1. BEETHOVEN: Sonata, horn and piano, op. 17, F major. SCHUBERT: Auf dem Strome, op. 119, D 943. SCHUMANN: Adagio and allegro, op. 70, A flat major. Albert Linder, hn; Ingemar Bergfelt, pno; Marta Schele, s; Vaclav Rabl, vln; Henry Stenstrom, vla; Goran Holmstrand, vlc. BIS LP 47.
+HFN 12-77 p168 +-RR 11-77 p85
ANCLIFFE, Charles
Nights of gladness. cf HMV SRS 5197.
Nights of gladness. cf Starline SRS 197.
ANDERSON
Fiddle-faddle. cf Rediffusion 15-57.
Sleigh ride. cf HMV ESD 7024.
ANDERSON, Leroy
Bugler's holiday. cf Golden Crest CRS 4148.
ANDERSON, Robert
Canticle of praise: Te deum. cf Delos FY 025.
ANDRE PREVIN'S MUSIC NIGHT, volume 2. cf HMV SQ ASD 3338.
ANDRIESSEN, Hendrik
Theme and variations. cf Vista VPS 1035.
ANGLES, Rafael
Aria, D minor. cf RCA RL 2-5099.
d'ANGLETERRA, Gallot
36 Carillon, G major. CAMPIAN: Pieces, D major. VISEE: Suite, D minor. WEISS: Chaconne, E flat major. Suite, C minor. David Rhodes, gtr, lt. Titanic TI 5.
+NR 7-76 p14 +-St 8-77 p59
ANIK
Changed. cf Orion ORS 77271.
ANNAPOLIS SOUNDS. cf Richardson RRS 3.
ANTES, John
37 Chorales (3). CRUSE: Sonatas, trombones (3). HUS: Communion hymn. ANON.: The liturgical year; Moravian chorale cycle; Moravian funeral chorales. Los Angeles Philharmonic Trombone Ensemble; Jeffrey Reynolds. Crystal S 222.
+HF 10-77 p128 +St 8-77 p124
+NR 9-77 p7

How beautiful upon the mountains. cf Vox SVBX 5350.
ANTHEIL, George
38 Ballet mecanique. A jazz symphony. Sonatas, violin, nos. 1 and 2. Vera Beths, vln; Reinbert de Leeuw, pno; Nederlands Blazers Ensemble; Reinbert de Leeuw. Harlekijn Holland 2925 524.
*IN 10-77 p24
A jazz symphony. cf Ballet mecanique.
Sonata, piano, no. 2. cf Vox SVBX 5303.
39 Sonata, trumpet. COPE: Bright angel. GEORGE: Tuckets and sennets, trumpet and piano. Marice Stith, tpt; Stuart Raleigh, pno. Redwood ES 5.
+NR 7-77 p15
Sonatas, violin, nos. 1 and 2. cf Ballet mecanique.
APKALNS, Longins
40 Burial songs. Ligo songs. KALNINS: The long night. Longins Apkalns, t; Karlis Bauers-Zemgalis, bs-bar; South West German Radio Orchestra; Longins Apkalns, H. J. Dahmen. Kaibala 30C01.
+NR 6-77 p13
Ligo songs. cf Burial songs.
APPLEBY, Thomas
41 Magnificat. DAVY: Ah mine heart. MASON: Quales sumus; Vae nobis miserere. PRESTON: Beatus Laurentius. Magdalen College Choir; Ian Crabbe, org; Bernard Rose. Argo ZRG 846.
+Gr 6-76 p77 *NR 2-77 p9
+HFN 5-76 p103 +RR 4-76 p67
+MT 2-77 p833
ARAJO
Batalha de sexto tono. cf Harmonia Mundi HM 759.
d'ARAUXO, Francisco
Tiento de medio registro de baxon de sexto tono. cf Harmonia Mundi HM 759.
ARBAN, Jean-Baptiste
Etude characteristique. cf Argo ZRG 851.
ARDITI, Luigi
L'Orologio. cf Decca SDD 507.
Parla. cf Court Opera Classics CO 342.
Parla. cf Rubini GV 26.
Se saran rose. cf HMV RLS 719.
AREL, Bulent
42 Music for a sacred service: Prelude and postlude. DAVIDOVSKY: Electronic study, no. 2. GABURO: For Harry. Lemon drops. USSACHEVSKY: Linear contrasts. Metamorphosis. Tape realized at the Columbia-Princeton Electronic Music Center or University of Illinois Electronic Music Studio and Composers home studio. CRI SD 356.
+MJ 3-77 p46 +NR 5-77 p15
ARENSKY, Anton
43 Concerto, violin, op. 54, A minor. RIMSKY-KORSAKOV: Fantasy on Russian themes, op. 33. WIENIAWSKI: Polonaise, violin. Aaron Rosand, vln; Luxembourg Radio Orchestra; Louis de Froment. Turnabout QTV 34629.
+MJ 1-77 p27 +NR 12-76 p7
44 Trio, op. 32, D minor. RAVEL: Trio, A minor. Temple University Trio. Golden Crest CRSQ 4149.
/-ARG 4-77 p25
45 Variations on a theme by Tchaikovsky, op. 35a (ed. Maganini). TCHAIKOVSKY: Serenade, strings, op. 48, C major. LSO; John

Barbirolli. HMV SXLP 30239 Tape (c) TC SXLP 30239. (Reissue from ASD 646)
+Gr 5-77 p1689 +HFN 8-77 p99 tape
+-HFN 5-77 p137 +RR 5-77 p60
ARIAS AND SONGS OF THE ITALIAN BAROQUE. cf HNH 4008.

ARMA, Paul
Evolutions pour basson seul. cf Gasparo GS 103.

ARNE, Thomas
The British Grenadiers. cf HMV SQ ASD 3341.
Concerto, harpsichord, no. 5, G minor. cf Argo D69D3.
Rule Britannia. cf BBC REH 290.
Rule Britannia. cf CRD Britannia BR 1077.
Rule Britannia. cf HMV SQ ASD 3341.
Songs: The lass with the delicate air. cf Abbey LPB 778.
Songs: Where the bee sucks. cf HMV ESD 7002.
Songs: Blow, blow thou winter wind; Come away death. cf National Trust NT 002.

ARNELL, Richard
Baroque prelude and fantasia. cf Organ works (RCA RL 2-5075).
Classical sonata. cf Organ works (RCA RL 2-5075).
A fugal flourish. cf Organ works (RCA RL 2-5075).
46 Organ works: Baroque prelude and fantasia. Classical sonata. A fugal flourish. Sonata, no. 2. Three related peices. Variations on "Ein feste Burg". Nicholas Jackson, org. RCA RL 2-5075.
+Gr 6-77 p70 +RR 7-77 p72
+-HFN 9-77 p133 +ST 8-77 p343
Sonata, no. 2. cf Organ worsk (RCA RL 2-5075).
Three related pieces. cf Organ works (RCA RL 2-5075).
Variations on "Ein feste Burg". cf Organ works (RCA RL 2-5075).

ARNOLD, Malcolm
Concerto, 2 violins and strings, op. 77. cf ADDISON: Concerto, trumpet, strings and percussion.
Cornish dances, op. 91. cf Symphony, no. 5, op. 74.
English dances, nos. 1-8. cf Symphony, no. 2, op. 40.
Fantasy, bassoon. cf Gasparo GS 103.
Fantasy, trombone. cf Argo ZRG 851.
Little suite, no. 2. cf Rediffusion 15-56.
Peterloo overture, op. 97. cf Symphony, no. 5, op. 74.
47 Serenade, guitar and strings. CASTELNUOVO-TEDESCO: Concerto, guitar, no. 1, op. 99, D major. DODGSON: Concerto, guitar, no. 2. John Williams, gtr; ECO; Charles Groves. CBS 76634.
+-Gr 12-77 p1067
48 Symphony, no. 2, op. 40. English dances, nos. 1-8. Bournemouth Symphony Orchestra; Charles Groves. HMV SQ ASD 3353 Tape (c) TC ASD 3353.
+Audio 12-77 p48 +HFN 8-77 p76
+Gr 6-77 p32 ++RR 8-77 p25
/Gr 8-77 p346 tape +ST 8-77 p343
49 Symphony, no. 5, op. 74. Cornish dances, op. 91 (4). Peterloo overture, op. 97. Birmingham (City) Symphony Orchestra; Malcolm Arnold. HMV ASD 2878.
+Audio 12-77 p48 ++HFN 5-73 p978
+-Gr 5-73 p2041 +RR 5-73 p39

ARRIAGA Y BALZOLA, Juan
50 Los esclavos felices: Overture. Symphony, D minor. ECO; Jesus Lopez Cobos. Pye NEL 2016. (also HNH 4001)

+Gr 6-75 p31 +RR 5-75 p26
+HF 9-77 p108 ++St 9-77 p90
++NR 8-77 p5

51 Symphony, D major. SCHMIDT: Variations on a Hussar's song. NPhO; Hans Bauer. HMV CSD 3679. (also Odeon CSD 3769)
+FF 9-77 p48 +-MT 10-76 p831
+-Gr 4-76 p1588 ++NR 1-77 p3
++HFN 4-76 p99 +RR 4-76 p35
+MM 2-77 p35

Symphony, D minor. cf Los esclavos felices: Overture.
THE ART OF BENIAMINO GIGLI. cf Seraphim S 60280.
THE ART OF COURTLY LOVE. cf HMV SLS 863.
THE ART OF LOTTE LEHMANN. cf Seraphim IB 6105.
THE ART OF SONIA ESSIN. cf Orion ORS 77271.
THE ART OF THE ALTO SAXOPHONE. cf Citadel CT 6012.
THE ART OF THE NETHERLANDS. cf HMV SLS 5049.

ARTHUR
Today. cf HMV RLS 715.
ARTHUR FIEDLER: A LEGENDARY PERFORMER. cf RCA CRL 1-2064.

ASENCIO
Dipso. cf RCA ARL 1-0864.

ATTAINGNANT, Pierre
52 Basse dance. Tant que vivray. BALLARD: Branles de village. Courante. Entree de Luth, nos. 1-3. BESARD: Air de cour. Allemande. Ballet. Branle. Branle gay (20). Chorea rustica. Gagliarda. Gagliarda vulgo dolorata. Gesamtzeit mit Pausen. Guillemette. Pass'e mezo. Volte. LE ROY: Basse dance. Branle gay. Destre amoureux. Haulberroys. Passemeze. Konrad Ragossnig, Renaissance lute. DG 2533 304.
+Gr 6-76 p70 +RR 7-76 p69
++HFN 5-76 p103 +St 8-77 p59
+NR 7-76 p14

Basse dance. cf Enigma VAR 1024.
La Brosse: Tripla, Tourdion. cf DG 2723 051.
Content desir basse danse. cf L'Oiseau-Lyre 12BB 203/6.
Galliard. cf Enigma VAR 1024.
La gatta. cf DG 2723 051.
La Magdalena. cf DG 2723 051.
Pavan. cf Enigma VAR 1024.
Tant que vivray. cf Basse dance.

ATTWOOD, Thomas
Turn thee again, O Lord, at the last. cf Guild GRS 7008.

AUBER, Daniel
Manon Lescaut: L'eclat de rire. cf Odyssey Y 33130.
Manon Lescaut: C'est h'histoire amoureuse. cf Rubini GV 26.

AULIN
Idyll. cf Discopaedia MOB 1020.
Polska. cf Discopaedia MOB 1020.

AURIC, Georges
Adieu, New York. cf Supraphon 111 1721/2.
THE AVANT GARDE WOODWIND QUINTET IN THE U.S.A. cf Vox SVBX 5307.

AZZAIOLO, Filippo
Quando le sera; Sentomi la formicula. cf L'Oiseau-Lyre 12BB 203/6.

BABBIT, Milton
53 Phenomena, soprano and piano. Phenomena, soprano and tape. Post partitions. Reflections, piano and tape. BASSETT: Music for saxophone and piano. SMITH: Fancies for clarinet alone.

WUORINEN: Variations, bassoon, harp and timpani. Gordon Gottlieb, timpani; Susan Jolles, hp; Donald MacCourt, bsn; Robert Miller, Ellen Wechler, Jerry Kuderna, pno; Donald Sinta, sax; William O. Smith, clt; Lynn Webber, s. New World Records NW 209.
+ARG 12-77 p49

Phenomena, soprano and tape. cf Phenomena, soprano and piano.
Post partitions. cf Phenomena, soprano and piano.
Reflections, piano and tape. cf Phenomena, soprano and piano.

BABIN, Victor

54 David and Goliath. BLOCH: Suite, solo violoncello, no. 1. ELWELL: Blue symphony. SHEPHERD: Sonata, violin and piano. Cleveland Institute of Music Faculty. Golden Crest GCCL 201 (2).
+ARG 2-77 p49

BACARISSE, Salvador

55 Concertino, guitar and orchestra, op. 72, A minor. RODRIGO: Concierto de Aranjuez, guitar and orchestra. Manuel Cubedo, gtr; Barcelona Symphony Orchestra; Jose Ferrer. Pye GSGC 15030. (Reissue from Contour 2870 365)
+-Gr 1-77 p1139 -RR 1-77 p57
-HFN 1-77 p119

Concertino, guitar and orchestra, op. 72, A minor. cf HALFFTER: Concerto, guitar.

BACH, Carl Philipp Emanuel

Concerto, orchestra, D major: 2nd movement. cf Decca DDS 507/1-2.
Concerto, violoncello, B flat major. cf Bruno Walter Society IGI 323.

56 Duets (4). BACH, J.C.: Sonata, 2 harpsichords, op. 15, no. 5, G major. BACH, J.S.: Canons (14). Variations, harpsichord, S 988. BACH, W.F.: Concerto, 2 harpsichords, F major. Rolf Junghanns, Bradford Tracy, hpd. Toccata 53622.
++MQ 10-77 p563

57 Fantasias, W 58/6, E flat; W 59/5, F major. Rondos, W 56/1, C major; W 57/1, E major; W 61/4, D minor. Sonatas, fortepiano, W 55/1, C major; W 57/4, D minor. Huguette Dreyfus, fortepiano. DG 2533 327.
+Gr 1-77 p1160 +MT 4-77 p307
+HFN 1-77 p101 +RR 12-76 p76

Fantasias, SQ 59/6, Wq 61/6, C minor. cf Fantasia, WQ 254, C minor.

58 Fantasia, Wq 254, C minor. Fantasias, Wq 59/6, Wq 61/6, C minor. BACH, J.S.: Chromatic fantasia and fugue, S 903, D minor. BACH, W.F.: Fantasia, D minor. MOZART: Fantasia, K 397, D minor. Colin Tilney, cld. DG 2533 326.
+ARG 3-77 p39 ++SFC 9-11-77 p42
++HF 5-77 p100 ++St 6-77 p91
+-NR 7-77 p14

59 Magnificat. Felicity Palmer, s; Helen Watts, con; Robert Tear, t; Stephen Roberts, bs; King's College Chapel Choir; AMF: Philip Ledger. Argo ZRG 853 Tape (c) KZRC 853.
+Gr 3-77 p1437 +NR 8-77 p8
+Gr 5-77 p1738 tape +-RR 4-77 p80
+-HFN 3-77 p97 +-RR 5-77 p90 tape
+HFN 5-77 p138 tape +SFC 12-18-77 p53

Minuet and trio (2). cf DG 2723 051.
Polonaises (5). cf DG 2723 051.

Rondos, W 56/1, C major; W 57/1, E major; W 61/4, D minor. cf Fantasias, W 58/6, E flat; W 59/5, F major.
Sonatas, fortepiano, W 55/1, C major; W 57/4, D minor. cf Fantasias, W 58/6, E flat; W 59/5, F major.
Sonatas, harpsichord, G major, E minor. cf Sonata, harpsichord, A minor: Poco adagio.
60 Sonata, harpsichord, A minor: Poco adagio. Sonatas, harpsichord, G major, E minor. Sonata, harpsichord, no. 6, A major. GIUSTINI: Sonata, no. 8, A major. James Bonn, fortepiano, hpd; David Hart, flt. Pleiades P 105.
+CL 12-77 p6 +HF 7-77 p114
+Gr 9-77 p504 +NR 12-77 p15
Sonata, harpsichord, no. 6, A major. cf Sonata, harpsichord, A minor: Poco adagio.
Sonata, organ, no. 5, D major. cf Crescent ARS 109.
Sonata, saxophone, A minor. cf Coronet LPS 3036.
Trio. cf DG 2723 051.
61 Trios, flute, violin and harpsichord, C major, B minor, B flat major. Eugenia Zukerman, flt; Pinchas Zukerman, vln; Samuel Sanders, hpd; Timothy Eddy, vlc. Columbia M 34216.
+ARG 3-77 p51 +NR 5-77 p9
+-HF 4-77 p92 -St 8-77 p110
+MU 3-77 p14

BACH, Johann Christian
62 Concerto, harp, op. 1, no. 6, D major. HANDEL: Royal fireworks music. HAYDN: Symphony, no. 101, D major. Annie Challan, hp; BPhO, VSO; Lorin Maazel, Wolfgang Sawallisch. Philips Tape (c) 7328 612.
+-HFN 7-77 p127 tape +-RR 7-77 p98 tape
Concerto, violoncello, C minor. cf Bruno Walter Society IGI 323.
Duets, A major, F major. cf Sonatas, flute and piano, op. 18, nos. 1-4.
Rondo, B flat major. cf Kaibala 20B01.
63 Sinfonia, op. 6, no. 6, C minor. HAYDN, M.: Symphony, G major (with introduction by Mozart, K 444). MOZART: Cassation (serenade), no. 1, K 62a, D major (with March, K 62, D major). St. Paul Chamber Orchestra; Dennis Russell Davies. Nonesuch H 73123.
+Gr 11-76 p775 +MQ 7-77 p446
+HF 8-76 p91 +-NR 8-76 p3
+HFN 11-76 p155 +RR 11-76 p69
+IN 12-76 p20 +SFC 7-18-76 p31
64 Sinfonia, op. 9, no. 2, E flat major. Sinfonias, op. 18, nos. 1-6. Sinfonia concertante, F major. NWE; David Zinman. Philips 6780 025 (2).
++ARG 11-77 p19 +MJ 11-77 p30
+-Gr 5-77 p1689 +NR 11-77 p3
+HF 11-77 p103 +RR 5-77 p45
+HFN 5-77 p116 ++SFC 9-11-77 p42
Sinfonias, op. 18, nos. 1-6. cf Sinfonia, op. 9, no. 2, E flat major.
65 Sinfonias, op. 18, nos. 2, 4, 6. TELEMANN: Don Quichotte. Stuttgart Chamber Orchestra; Karl Munchinger. Decca SXL 6755. (also London 6988)
+-Gr 10-76 p577 +NR 7-77 p6
+-HF 11-77 p103 +-RR 9-76 p43
+-HFN 10-76 p163 ++St 8-77 p110

Sinfonia concertante, F major. cf Sinfonia, op. 9, no. 2, E flat major.

66 Sonatas, flute and fortepiano, op. 16, nos. 1-6. Ingrid Haebler, fortepiano; Kurt Redel, flt. Philips 6580 214. (Reissue from 6500 121)
+Gr 12-77 p1098 +RR 11-77 p84
++HFN 11-77 p185

67 Sonatas, flute and piano, op. 18, nos. 1-4. Duets, A major, F major. Nicholas McGegan, flt; Christopher Hogwood, Colin Tilney, fortepiano. L'Oiseau-Lyre DSLO 516.
+Gr 10-76 p616 ++NR 9-77 p7
+HFN 11-76 p155 ++RR 10-76 p85
+MJ 9-77 p35 ++SFC 9-11-77 p42
+MT 7-77 p564

Sonata, harpsichord, op. 15, no. 3, D major. cf Crescent BUR 1001.

Sonata, 2 harpsichords, op. 15, no. 5, G major. cf BACH, C.P.E.: Duets (4).

Symphony, op. 3, no. 1, D major. cf Philips 6580 114.

BACH, Johann Christoph

Aria and 15 variations, A minor. cf Toccata 53623.

BACH, Johann Christoph Friedrich

68 Sonata, fortepiano, flute and violoncello, D major. RICHTER: Sonata, harpsichord, flute and violoncello. STANLEY: Sonata, flute, G major. Swiss Baroque Soloists. Amon Ra SAR 4.
+RR 5-77 p61

BACH, Johann Sebastian

Allabreve, S 589, D major. cf Organ works (DG 2722 014).

Anna Magdalena notebook, S 508, excerpts. cf RCA ARL 1-1323.

Anna Magdalena notebook, S 508: Bist du bei mir. cf Works, selections (Angel Q S 37229).

Ave Maria. cf Decca SXL 6781.

Ave Maria. cf HMV RLS 719.

Ave Maria. cf London OS 26437.

69 Brandenburg concerti, nos. 1-6, S 1046-1051. Leonhardt Consort; Gustav Leonhardt. ABC AB 67020/2 (2).
+HF 12-77 p77 ++St 12-77 p134
++SFC 12-18-77 p53

70 Brandenburg concerti, nos. 1-6, S 1046-1051. Philomusica of London; Thurston Dart, hpd. Decca DAP 557/8 (2). (Reissue from L'Oiseau-Lyre SOL 60005/6)
+Gr 7-77 p167 +-RR 5-77 p42
+-HFN 5-77 p137

71 Brandenburg concerti, nos. 1-6, S 1046-1051. Suites, orchestra, S 1066-9. Munich Bach Orchestra; Karl Richter. DG Tape (c) 3376 003.
-Gr 1-77 p1185 tape +HF 11-75 p149 tape

72 Brandenburg concerti, nos. 1-6, S 1046-1051. Northern Sinfonia Orchestra; George Malcolm. Enigma VAR 1041/2.
-HFN 11-77 p164 +-RR 9-77 p46

73 Brandenburg concerti, nos. 1-6, S 1046-1051. Stuttgart Chamber Orchestra; Karl Munchinger. London STS 15366/7 (2).
+-HF 6-77 p80 ++SFC 2-13-77 p33
+NR 7-77 p6 -St 5-77 p102

74 Brandenburg concerti, nos. 1-6, S 1046-1051 (trans. Reger). Martin Berkofsky, David Hagan, pno. Musical Heritage Society MHS 3522/3 (2).
+ARG 10-77 p14 +NYT 6-5-77 p19

75 Brandenburg concerti, nos. 1-6, S 1046-1051. Virtuosi of England; Arthur Davison. Vanguard SRV 313/4 (2). (also Classics for Pleasure CFP 40010/1 Tape (c) TC CFP 40010/1)
-Gr 10-72 p679
+-Gr 2-77 p1325 tape
+HF 9-75 p81
+-HFN 9-72 p1659
+HFN 12-76 p153 tape
+NR 7-75 p3
+RR 10-72 p53
++St 7-76 p106

Brandenburg concerto, no. 3, S 1048, G major. cf Philips 6580 114
Brandenburg concerto, no. 3, S 1048, G major: Allegro. cf RCA FRL 1-3504.
Canons (14). cf BACH, C.P.E.: Duets.

76 Canon, 6 part, S 1076. Concerto, 3 harpsichords, S 1063, D minor. Concerto, 3 harpsichords, S 1064, C maor. Concerto, 4 harpsichords, S 1065, A minor. Jorg Ewald Dahler, Christine Daxelhofer, Ernst Gerber, Marc Meystre, hpd; Ilse Mathieu, Catrin Demenga, vln; Erika Crafoord, vla; Johannes Buhler, vlc; Edgar Kremsa, bs. Claves D 614.
+-HFN 5-77 p116
+-RR 6-77 p43

77 Canons on the First 8 bass notes of the "Goldberg variations" aria, S 1087. REGER: Sonata, clarinet and piano, no. 3, op. 107, B flat major. Various instrumentalists; David Singer, clt; Rudolf Serkin, pno. Marlboro Recording Society MRS 12.
+ARG 10-77 p14
+HF 5-77 p84
+-MJ 4-77 p50
+SR 7-9-77 p48

78 Cantatas, nos. 1, 4, 6, 12, 23, 67, 87, 92, 104, 108, 126, 158, 182. Edith Mathis, s; Anna Reynolds, Hertha Topper, ms; Peter Schreier, Ernst Hafliger, t; Dietrich Fischer-Dieskau, bar; Theo Adam, bs; Munich Bach Orchestra and Choir; Karl Richter. DG Archive 2722 022 (6).
+-Gr 4-77 p1578
+-HFN 4-77 p133
+-RR 4-77 p79

79 Cantatas, nos. 1, 51, 56, 76, 93, 106. Edith Mathis, s; Peter Schreier, Ernst Hafliger, t; Dietrich Fischer-Dieskau, bar; Theo Adam, bs; Munich Bach Orchestra and Choir; Karl Richter. DG Tape (c) 3376 009.
+Gr 1-77 p1183 tape

80 Cantata, no. 4, Christ lag in Todesbanden. Cantata, no. 71, Gott ist mein Konig. Kathrin Graf, s; Hanna Schwarz, Helrun Gardow, alto; Aldo Baldin, Adalbert Kraus, Alexander Senger, t; Niklaus Tueller, Wolfgang Schone, bs; Bach Collegium Stuttgart; Gachinger Kantorei; Helmuth Rilling. Claudius Verlag CLV 71926.
+CJ 9-77 p17

Cantata, no. 6, Bleib uns: Hochgelobter Gottessohn. cf Works, selections (Angel (Q) S 37229).

81 Cantatas, nos. 10-11, 24, 30, 34, 39, 44, 68, 76, 93, 129, 135, 175. Edith Mathis, s; Anna Reynolds, ms; Peter Schreier, t; Dietrich Fischer-Dieskau, bar; Kurt Moll, bs; Munich Bach Orchestra and Choir; Karl Richter. DG 2722 025 (6).
+-Gr 10-77 p672
+HFN 11-77 p185
+-RR 9-77 p85

82 Cantata, no. 10, Meine Seele erhebt den Herren. Cantata, no. 24, Ein ungefarbt Gemute. Cantata, no. 135, Ach Herr, mich armen Sunder. Edith Mathis, s; Anna Reynolds, con; Peter Schreier, t; Kurt Moll, Dietrich Fischer-Dieskau, bs; Munich Bach Orchestra and Choir; Karl Richter. DG Archive 2533 329.
++ARG 9-77 p17
+Gr 12-76 p1027
+HFN 1-77 p101
++NR 8-77 p8
+-RR 12-76 p88

83 Cantata, no. 11, Lobet Gott in seinen Reichen. Cantata, no. 44, Sie werden euch in den Bann tun. Edith Mathis, s; Anna Reynolds, con; Peter Schreier, t; Dietrich Fischer-Dieskau, bar; Munich Bach Orchestra and Choir; Karl Richter. DG Archive 2533 355 Tape (c) 3310 355. (Reissue from 2722 018)
+-Gr 6-77 p76 +-RR 5-77 p78
+-Gr 7-77 p225 tape +RR 9-77 p98 tape
++HFN 5-77 p137

Cantata, no. 11, Lobet Gott: Ach, bleibe doch. cf Works, selections (Angel (Q) S 37229).

84 Cantata, no. 12, Weinen, Klagen, Sorgen, Zagen. Cantata, no. 102, Herr, deine Augen sehen nach dem Glauben. Helen Watts, Eva Randova, alto; Adalbert Kraus, Kurt Equiluz, t; Wolfgang Schone, bs; Bach Collegium Stuttgart; Gachinger Kantorei; Helmuth Rilling. Claudius Verlag CLV 71911.
+CJ 9-77 p19

85 Cantata, no. 18, Gleich wie der Regen und Schnee vom Himmel fallt. Cantata, no. 196, Der Herr denket an uns. Doris Soffel, Eva Csapo, s; Gabriele Schnaut, alto; Aldo Baldin, Adalbert Kraus, t; Wolfgang Schone, Niklaus Tueller, bs; Bach Collegium Stuttgart; Gachinger Kantorei; Helmut Rilling. Claudius Verlag CLV 71927.
+CJ 9-77 p17

86 Cantata, no. 18: Sinfonia. Cantata, no. 21: Sinfonia. Cantata, no. 29: Sinfonia. Cantata, no. 31: Sonata. Cantata, no. 35: Sinfonia I and II. Cantata, no. 42: Sinfonia. Cantata, no. 49: Sinfonia. Cantata, no. 248: Sinfonia. VCM: Nikolaus Harnoncourt. Telefunken AF 6-41970.
++RR 7-76 p42 ++SFC 9-11-77 p42

Cantata, no. 21: Sinfonia. cf Cantata, no. 18: Sinfonia.

Cantata, no. 21: Sinfonia. cf Decca DDS 507/1-2.

87 Cantata, no. 23, Du wahrer Gott und Davids Sohn. Cantata, no. 87, Bisher habt ihr nichts gebeten in meinem Namen. Edith Mathis, s; Anna Reynolds, con; Peter Schreier, t; Dietrich Fischer-Dieskau, bs; Munich Bach Orchestra and Choir; Karl Richter. DG Archive 2533 313.
-Gr 12-76 p1027 ++NR 2-77 p9
-HF 4-77 p92 -RR 12-76 p88
+-HFN 1-77 p101 +St 5-77 p110

Cantata, no. 24, Ein ungefarbt Gemute. cf Cantata, no. 10, Meine Seele erhebt den Herren.

Cantata, no. 29: Sinfonia. cf Cantata, no. 18: Sinfonia.

Cantata, no. 29: Sinfonia. cf Works, selections (Columbia M 34272).

Cantata, no. 29: Sinfonia. cf Rediffusion Royale 2015.

Cantata, no. 29, Wir danken dir, Gott: Sinfonia. cf Works, selections (DG 2530 647).

88 Cantata, no. 30, Freue dich, erloste Schar. Edith Mathis, s; Anna Reynolds, ms; Peter Schreier, t; Dietrich Fischer-Dieskau, bar; Munich Bach Orchestra and Choir; Karl Richter. DG 2533 330 Tape (c) 3310 330.
-Gr 9-77 p460 -RR 8-77 p73
+HFN 8-77 p95 +SFC 12-11-77 p60

Cantata, no. 31: Sonata. cf Cantata, no. 18: Sinfonia.

Cantata, no. 31: Sonata. cf Works, selections (Columbia M 34272).

Cantata, no. 33: Aria. cf RCA FRL 1-3504.

89 Cantata, no. 34, O ewiges Feuer, O Ursprung der Liebe. Cantata, no. 68, Also hat Gott die Welt geliebt. Cantata, no. 175, Er

rufet seinen Schafen mit Namen. Edith Mathis, s; Anna Reynolds, con; Peter Schreier, t; Dietrich Fischer-Dieskau, bs; Munich Bach Orchestra and Choir; Karl Richter. DG Archive 2533 306. (Reissue from 1711 019)
+Gr 12-76 p1028 ++NR 2-77 p9
+HF 4-77 p92 -RR 12-76 p88
+-HFN 1-77 p102 +St 5-77 p110

Cantata, no. 34, O Ewiger Feuer: Wohl euch ihr auser wahlten Seelen. cf Works, selections (Angel (Q) S 37229).

Cantata, no. 35: Sinfonias I and II. cf Cantata, no. 18: Sinfonia.

Cantata, no. 35: Sinfonias I and II. cf Works, selections (Columbia M 34272).

Cantata, no. 42: Sinfonia. cf Cantata, no. 18: Sinfonia.

Cantata, no. 44, Sie werden euch in den Bann tun. cf Cantata, no. 11, Lobet Gott in seinen Reichen.

Cantata, no. 49: Sinfonia. cf Cantata, no. 18: Sinfonia.

Cantata, no. 49: Sinfonia. cf Works, selections (Columbia M 34272).

90 Cantatas, nos. 51-52, 54-56. Marianne Kweksilber, s; Seppi Kronwitter, boy soprano; Paul Esswood, c-t; Kurt Equiluz, t; Michael Schopper, bs; Hanover Boys Choir; Leonhardt Consort; Gustav Leonhardt. Telefunken 6-35304. (also SKW 14)
+Gr 12-76 p1028 ++NR 9-76 p8
+-HF 8-76 p79 +-RR 11-76 p92
+HFN 12-76 p136 +-St 5-77 p110
++MJ 10-76 p24

Cantata, no. 51: Chorale. cf ALBINONI: Concerto, 2 trumpets, C major.

91 Cantata, no. 54, Widerstehe doch der Sunde. Cantata, no. 172, Erschallet, ihr Lieder erklinget, ihr Saiten. Eva Csapo, s; Julia Hamari, Doris Soffel, alto; Adalbert Kraus, t; Wolfgang Schone, bs; Frankfurt Kantorei, Bach Collegium Stuttgart; Helmuth Rilling. Claudius Verlag CLV 71929.
+CJ 9-77 p17

92 Cantatas, nos. 57-64. Peter Jelosits, Seppi Kronwitter, boy sopranos; Paul Esswood, c-t; Kurt Equiluz, t; Ruud van der Meer, bs; Tolzer Boys Choir; VCM; Nikolaus Harnoncourt. Telefunken 6-35305/6 (4) (also SK 15/6)
++ARG 7-77 p14 -MJ 7-77 p70
++Gr 4-77 p1581 +-MT 9-77 p733
++HF 8-77 p77 +-NR 9-77 p9
++HFN 5-77 p117 +-RR 4-77 p78
+MJ 2-77 p30

93 Cantata, no. 61, Nun komm, der Heiden Heiland. Cantata, no. 110, Unser Mund sei voll Lachens. Helen Donath, Kathrin Graf, s; Helrun Gardow, alto; Adalbert Kraus, Aldo Baldin, t; Wolfgang Schone, bs; Bach Collegium Stuttgart; Gachinger Kantorei; Helmuth Rilling. Claudius Verlag CLV 71920.
+CJ 9-77 p17

94 Cantatas, nos. 65-68. Peter Jelosits, treble; Kurt Equiluz, t; Max van Egmond, Ruud van der Meer, bs; Paul Esswood, c-t; Tolzer Boys Choir, Hanover Boys Choir, Ghent Collegium Vocale; VCM, Leonhardt Consort; Nicolaus Harnoncourt, Gustav Leonhardt. Telefunken EX 6-35335 (2).
+-Gr 10-77 p672 +-RR 10-77 p80
+HFN 10-77 p164

Cantata, no. 68, Also hat Gott die Welt geliebt. cf Cantata, no. 34, O ewiges Feuer, O Ursprung der Liebe.

95 Cantata, no. 69, Lobe den Herrn, meine Seele. Cantata, no. 120, Gott man lobet dich in der Stille. Helen Donath, s; Gachinger Kantorei; Bach Collegium, Stuttgart; Helmuth Rilling. Musical Heritage Society Tape MCH 2119.
++HF 1-77 p151 tape

Cantata, no. 71, Gott ist mein Konig. cf Cantata, no. 4, Christ lag in Todesbanden.

Cantata, no. 78: Air, 2 trumpets. cf ALBINONI: Concerto, 2 trumpets, C major.

Cantata, no. 87, Bisher habt ihr nichts gebeten in meinem Namen. cf Cantata, no. 23, Du wahrer Gott und Davids Sohn.

96 Cantata, no. 92, Ich hab in Gottes Herz und Sinn. Cantata, no. 126, Erhalt uns, Herr, bei deinem Wort. Edith Mathis, s; Anna Reynolds, ms; Peter Schreier, t; Dietrich Fischer-Dieskau, Theo Adam, bs; Munich Bach Orchestra and Choir; Karl Richter. DG Archive 2533 312.
++Gr 7-76 p199
++HF 4-77 p92
+HFN 8-76 p75
++NR 2-77 p9
+RR 8-76 p72
+St 5-77 p110

Cantata, no. 95, Content am I to leave thee. cf Abbey LPB 779.

Cantata, no. 102, Herr, deine Augen sehen nach dem Glauben. cf Cantata, no. 12, Weinen, Klagen, Sorgen, Zagen.

97 Cantata, no. 106, Gottes Zeit ist die allerbeste Zeit (Actus tragicus). Cantata, no. 131, Aus der Tiefe rufe ich, Herr, zu dir. Eva Csapo, s; Hanna Schwarz, alto; Adalbert Kraus, t; Wolfgang Schone, bs; Bach Collegium, Stuttgart; Gachinger Kantorei; Helmuth Rilling. Claudius Verlag CLV 71925.
+CJ 9-77 p17

Cantata, no. 110, Unser Mund sei voll Lachens. cf Cantata, no. 61, Nun komm, der Heiden Heiland.

98 Cantata, no. 114, Ach lieben Christen, seid getrost. Cantata, no. 179, Siehe zu, dass deine Gottesfrucht nicht Heuchelei sei. Kathrin Graf, Gabriele Schnaut, s; Else Paaske, alto; Kurt Equiluz, t; Wolfgang Schone, bs; Frankfurt Kantorei, Gachinger Kantorei; Bach Collegium; Helmuth Rilling. Musical Heritage Society MHS 3207.
+-St 5-77 p110

Cantata, no. 118, O Jesu Christ mein Lebens Licht. cf Magnificat, S 243, D major.

Cantata, no. 120, Gott, man lobet dich in der Stille. cf Cantata, no. 69, Lobe den Herrn, meine Seele.

Cantata, no. 126, Erhalt uns, Herr, bei deinem Wort. cf Cantata, no. 92, Ich hab in Gottes Herz und Sinn.

Cantata, no. 129, Gelobet sei der Herr. cf Works, selections (Angel (Q) S 37229).

99 Cantata, no. 131, Aus der Tiefe rufe ich, Herr. HANDEL: Sing unto God, anthem. Wendy Eathorne, s; Paul Esswood, c-t; Neil Jenkins, t; John Noble, bar; London Bach Society Chorus; Steinitz Bach Players; Paul Steinitz. HMV CSD 3741. (also Nonesuch H71294 Tape (r) NST 71294)
+Gr 11-73 p962
++HF 10-74 p87
+HF 7-75 p102 tape
++HFN 12-73 p106
+HFN 7-77 p125
+NR 8-74 p7
+RR 11-73 p80
+RR 4-77 p82

Cantata, no. 131, Aus der Tiefe rufe ich, Herr, zu dir. cf Cantata, no. 106, Gottes Zeit ist die allerbeste Zeit.

Cantata, no. 135, Ach Herr, mich armen Sunder. cf Cantata, no. 10, Meine Seele erhebt den Herren.

Cantata, no. 140, Wachet auf, ruft uns die Stimme. cf Works, selections (DG 2530 647).
Cantata, no. 140, Zion hears her watchmen's voices. cf Abbey 779.
100 Cantata, no. 143, Lobe den Herrn, meine Seele. Cantata, no. 182, Himmelskonig, sei willkommen. Eva Csapo, s; Doris Soffel, alto; Adalbert Kraus, Aldo Baldin, t; Wolfgang Schone, Philippe Huttenlocher, bs; Frankfurt Kantorei; Bach Collegium Stuttgart; Helmuth Rilling. Claudius Verlag CLV 71928.
+CJ 9-77 p17
101 Cantata, no. 146, Wir mussen durch viel Trubsal in das Reich Gottes eingehen. Helen Donath, s; Marga Hoffgen, alto; Kurt Equiluz, t; Hanns-Friedrich Kunz, bs; Bach Collegium Stuttgart; Gachinger Kantorei; Helm. Rilling. Claudius Verlag CLV 71917.
+CJ 9-77 p17
Cantata, no. 146: Sinfonia. cf Works, selections (Columbia M 34272).
Cantata, no. 147: Chorale. cf Works, selections (Columbia M 34272).
Cantata, no. 147, Jesu bleibt meine Freude. cf Works, selections (DG 2530 647).
102 Cantata, no. 147, Jesu, joy of man's desiring. Cantata, no. 208, Sheep may safely graze. Cantata, Shepherds Christmas music. CORELLI: Concerto grosso, op. 6, no. 8, G minor. VIVALDI: L'Estro armonico, op. 3, no. 11, D minor. Igor Kipnis, hpd; Leopold Stokowski and His Orchestra. Vanguard SRV 363SD.
*ARG 3-77 p51 ++NR 3-77 p3
+MU 3-77 p14
Cantata, no. 147, Jesu joy of man's desiring. cf Abbey LPB 779.
Cantata, no. 147, Jesu joy of man's desiring. cf Decca PFS 4387.
Cantata, no. 147, Jesu joy of man's desiring. cf HMV ASD 3318.
Cantata, no. 147, Jesu joy of man's desiring. cf HMV SQ ASD 3375.
Cantata, no. 147, Jesu joy of man's desiring. cf Inner City 1006.
Cantata, no. 147, Jesu joy of man's desiring. cf St. John the Divine Cathedral Church.
Cantata, no. 147, Jesu joy of man's desiring. cf Works, selections (DG 2584 001).
Cantata, no. 161, Komm, du susse Todesstunde. cf Works, selections (Angel (Q) S 37229).
Cantata, no. 169: Sinfonia. cf Works, selections (Columbia 34272).
Cantata, no. 172, Erschallet, ihr Lieder erklinget, ihr Saiten. cf Cantata, no. 54, Widerstehe doch der Sunde.
Cantata, no. 175, Er rufet seinen Schafen mit Namen. cf Cantata, no. 34, O ewiges Feuer, O Ursprung der Liebe.
Cantata, no. 179, Siehe zu, dass deine Gottesfurcht nicht Heuchelei sei. cf Cantata, no. 114, Ach lieben Christen, seid getrost.
Cantata, no. 182, Himmelskonig, sei willkommen. cf Cantata, no. 143, Lobe den Herrn, meine Seele.
Cantata, no. 190, Lobe, Zion, deinen Gott. cf Works, selections (Angel (Q) S 37229).
Cantata, no. 196, Der Herr denket an uns. cf Cantata, no. 18, Gleich wie der Regen und Schnee vom Himmel fallt.
103 Cantata, no. 202, Weichet nur betrubte Schatten. MOZART: Exsultate jubilate, K 165. Helen Donath, s; Kosice State Philharmonic Chamber Orchestra; Klaus Donath. Opus 9112 0373.
+Gr 5-77 p1717 +-RR 4-77 p78
+HFN 5-77 p119
Cantata, no. 208, Sheep may safely graze. cf Cantata, no. 147, Jesu, joy of man's desiring.

Cantata, no. 208, Sheep may safely graze (arr. Walton). cf Works, selections (DG 2584 001).

104 Cantata, no. 208, Was mir behagt, ist nur die muntre. Helen Donath, Elisabeth Speiser, s; Wilfrid Jochims, t; Jakob Stampfli, bs; Gedachtniskirche Figuralchor; Bach Collegium, Stuttgart; Helmuth Rilling. Musical Heritage Society MHF 3297.
+St 5-77 p110

105 Cantata, no. 209: Sinfonia. Concerto, E minor (reconstr. Radeke, based on S 1059 & S 35). Suite, orchestra, S 1067, B minor. Jean-Pierre Rampal, flt; Jean-Francois Paillard Chamber Orchestra; Jean-Francois Paillard. RCA FRL 5693 Tape (c) FRK 1-5693 (ct) FRS 1-5693.
++HF 3-77 p123 tape ++NR 11-76 p5
+MJ 2-77 p30

Cantata, no. 211, Schweigt stille, plaudert nicht. cf Golden Age GAR 1001.

Cantata, no. 248: Sinfonia. cf Cantata, no. 18: Sinfonia.

Cantata, Shepherds Christmas music. cf Cantata, no. 147, Jesu, joy of man's desiring.

Canzona, S 588, D minor. cf Organ works (DG 2722 014).

106 Capriccio, S 992, B flat major. Concerto, harpsichord, S 971, F major. Duets, S 802-805. Partita, harpsichord, no. 7, S 831, B minor. Isolde Ahlgrimm, hpd. Philips 6580 142.
+-Gr 11-77 p860 +RR 10-77 p73
+HFN 10-77 p148

Chaconne. cf CBS Tape (c) 40-72728.

Chaconne, solo violin, D minor. cf Finnlevy SFLP 8569.

Chorale preludes (13). cf BRUCKNER: Symphony, no. 9, D minor.

107 Chorale preludes: Valet will ich dir geben;Allein Gott in der Hoh. HANFF: Chorale preludes: Ach Gott vom Himmel sieh darein; Auf meinem lieben Gott; Ein feste Burg ist unser Gott; Erbarm dich mein lieber Gott; Helft mir Gottes Gute preisen; War Gott nicht mit uns diese Zeit. PACHELBEL: Was Gott tut, das ist wohlgetan. WALTHER: Jesu meine Freude, partita. Erich Stoffers, org. Pelca PSR 40583.
+-NR 1-77 p14

108 Chorale preludes (Orgelbuchlein), S 599-644. Trio sonatas (4). Helmut Walcha, org. DG Tape (c) 3376 005.
+Gr 1-77 p1183 tape

109 Chorale preludes (Orgelbuchlein), S 599-644. Robert Kobler, org. Telefunken DX 6-35318 (2).
+-HFN 9-77 p133 +-RR 9-77 p79

Chorale preludes: Heut triumphiret, S 630; Ich ruf zu dir, S 639. cf St. John the Divine Cathedral Church.

Choral preludes, S 636, S 711, S 729, S 731. cf BIS LP 27.

Chorale preludes: Ich ruf zu dir, Herr Jesu Christ, S 639; Nun komm der Heiden Heiland, S 659; Herzlich tut mich verlangen (Befiehl du deine Wege), S 727; In dulci jubilo. cf Works, selections (DG 2530 647).

110 Chorale preludes (Schubler), S 645-50. Pastorale, S 590, F major. Prelude and fugue, S 544, B minor. David Lumsden, org. Abbey LPB 760.
+Gr 9-77 p456 +MT 11-77 p921
+HFN 8-77 p76 +RR 7-77 p76

111 Chorale preludes (Schubler), S 645-50. MOZART: Fantasia, K 594, F minor. Fantasia, K 608, F minor. Simon Preston, org. Argo ZK 13 Tape (c) KZKC 13. (Reissue from Argo ZRK 5419)

+Gr 8-77 p318 +HFN 11-77 p187 tape
+Gr 9-77 p508 tape +-RR 7-77 p76
+HFN 7-77 p125

Chorale prelude, S 645, Wachet auf, ruft uns die Stimme. cf Decca PFS 4416.

Chorale prelude, S 650, Kommst du nun, Jesu. cf Pelca PSR 40571.

112 Chorale preludes, S 651-668. Peter Hurford, org; Alban Singers. Argo ZRG 843/4 (2).
+ARG 12-76 p10 +MU 9-77 p14
+MT 6-77 p485 ++St 2-77 p112

Chorale preludes, S 659, Nun komm, den Heiden Heiland. cf TR 1006.

Chorale preludes, S 680, Wir glauben all an einen Gott. cf Vista VPS 1060.

Chorale prelude, S 731, Liebster Jesu. cf RC Records RCR 101.

Chorale prelude, S 734, Rejoice, beloved Christians. cf International Piano Library IPL 104.

113 Chorale prelude, S 767, O Gott du frommer Gott. BULL: Revenant. BUXTEHUDE: Chorale prelude: Wie schon leuchtet der Morgenstern, Trauermusic. GRIGNY: Veni creator. Marion Rowlatt, s; Peter Hurford, org. Argo ZRG 835.
+Gr 10-76 p633 +MT 9-77 p739
+HF 10-77 p126 +NR 9-77 p14
+HFN 10-76 p167 +-RR 10-76 p86

Chorale prelude, S 767, O Gott, du frommer Gott. cf Advent 4007.

114 Christmas oratorio, S 248. Hans Buchhierl, treble; Andreas Stein, alto; Theo Altmeyer, t; Barry McDaniel, bar; Tolzer Boys Choir; Collegium Aureum; Gerhard Schmidt-Gaden. BASF 59 21749-3 (3).
+Gr 3-76 p418 +-MM 1-77 p31
+-HFN 1-76 p102 +RR 1-76 p54

115 Christmas oratorio, S 248. Gundula Janowitz, s; Hertha Topper, con; Christa Ludwig, ms; Dietrich Fischer-Dieskau, bar; Keith Engen, bs; Munich Bach Orchestra and Choir; Karl Richter. DG Tape (c) 3376 007.
+-Gr 1-77 p1183 tape

116 Christmas oratorio, S 248. Elly Ameling, s; Janet Baker, ms; Robert Tear, t; Dietrich Fischer-Dieskau, bar; King's College Chapel Choir; AMF; Philip Ledger. HMV SQ SLS 5098 (3). (also Angel S 3840)
+ARG 12-77 p6 +NR 12-77 p9
+-Gr 10-77 p671 +RR 11-77 p98
+Gr 12-77 p1146 tape +SFC 12-18-77 p53
++HFN 11-77 p164 +St 12-77 p83

Christmas oratorio, S 248: Bereite dich, Zion. cf Works, selections (Angel (Q) S 37229).

Christmas oratorio, S 248: Sinfonia pastorale. cf Decca DDS 507.

117 Chromatic fantasia and fugue, S 903, D minor. Concerto, harpsichord, S 971, F major. French suite, no. 5, S 816, G major. Toccata, S 912, D major. George Malcolm, hpd. Decca ECA 788 Tape (c) KECC 788. (Reissue from SXL 2259)
+Gr 10-76 p622 +-RR 10-76 p82
+HFN 11-76 p171 +-RR 1-77 p87 tape
+HFN 12-76 p155 tape -RR 11-77 p118 tape

118 Chromatic fantasia and fugue, S 903, D minor. Fantasia, S 922, A minor. Toccatas, S 912, D major; S 914, E minor; S 916, G major. Blandine Verlet, hpd. Philips 6833 184.
++ARG 11-76 p13 +NR 12-76 p15
++MJ 11-76 p60 ++St 2-77 p112

Chromatic fantasia and fugue, S 903, D minor. cf BACH, C.P.E.: Fantasia, Wq 254, C minor.

119 Clavierubung, Pt 3. Fritz Heimann, org. Telefunken AJ 6-41977.
-RR 2-77 p74

Concerto, E minor (reconstr. Radeke, based on S 1059 & S 35). cf Cantata, no. 209: Sinfonia.

Concerto, flute and strings, G minor (arr. from S 1056). cf Argo D69D3.

120 Concerto, harpsichord, S 971, F major (Italian). Partita, harpsichord, no. 7, S 831, B minor. Igor Kipnis, hpd. Angel S 36096. (also HMV HQS 1392)
+ARG 12-76 p12 +NR 12-76 p15
+Gr 11-77 p860 ++RR 10-77 p74
++HF 1-77 p110 +-SFC 10-31-76 p35
+-HFN 10-77 p148 +St 2-77 p112
+MJ 4-76 p31

121 Concerto, harpsichord, S 971, F major. Fantasia, S 906, C minor. Inventions, 2 part, S 772-786. Suite, lute (harpsichord), S 997, C minor. Lionel Party, hpd. Desmar DSM 1008.
+MJ 3-76 p25 ++St 2-77 p112
+RR 4-76 p62

Concerto, harpsichord, S 971, F major. cf Capriccio, S 992, B flat major.

Concerto, harpsichord, S 971, F major. cf Chromatic fantasia and fugue, S 903, D major.

122 Concerti, harpsichord, nos. 1, 4-5, S 1052, 1055, 1056. Concerti, 3 harpsichords, S 1063-64. Concerto, 4 harpsichords, S 1065, A minor. Concerti, violin and strings, S 1041-42. Concerto, 2 violins and strings, S 1043, D minor. Karl Richter, hpd; Eduard Melkus, vln; Munich Bach Orchestra; Eduard Melkus. DG Tape (c) 3376 006.
+-Gr 1-77 p1183 tape +HF 9-76 p84

123 Concerto, harpsichord, no. 1, S 1052, D minor. Concerto, harpsichord, no. 2, S 1053, E major. George Malcolm, hpd; Stuttgart Chamber Orchestra; Karl Munchinger. Decca JB 9 Tape (c) KJBC 9. (Reissue from SXL 6101)
+Gr 9-77 p417 +HFN 10-77 p169 tape
++Gr 11-77 p899 tape ++RR 9-77 p46
+HFN 9-77 p151 ++RR 12-77 p90 tape

124 Concerto, harpsichord, no. 1, S 1052, D minor. Concerto, 2 harpsichords, S 1061, C major. Anatole Vedernikov, Sviatoslav Richter, pno; Moscow Chamber Orchestra; Rudolf Barshai. Everest 3415.
+-NR 11-77 p6

Concerto, harpsichord, no. 2, S 1053, E major. cf Concerto, harpsichord, no. 1, S 1052, D minor.

125 Concerto, harpsichord, no. 5, S 1056, F minor. FRANCK: Symphonic variations. SHOSTAKOVICH: Concerto, piano and trumpet, no. 1, op. 35, C minor. Maria Grinberg, pno; MRSO; Gennady Rozhdestvensky. Westminster WGS 8325.
+CL 10-77 p8 +NR 3-77 p7
+-HF 8-77 p82 +SFC 2-13-77 p33

Concerto, harpsichord, no. 5, S 1056, F minor: Largo. cf Works, selections (DG 2530 647).

126 Concerti, 2 harpsichords, S 1060-1062. Zoltan Kocsis, Andras Schiff, pno; Liszt Academy Orchestra; Albert Simon. Hungaroton SLPX 11751.
-ARG 10-77 p13

Concerto, 2 harpsichords, S 1061, C major. cf Concerto, harpsichord, no. 1, S 1052, D minor.
Concerto, 3 harpsichords, S 1063, D minor. cf Canon, 6 part, S 1076.
Concerti, 3 harpsichords, S 1063-64. cf Concerti, harpsichord, nos. 1, 4-5, S 1052, 1055, 1056.
Concerto, 3 harpsichords, S 1064, C major. cf Canon, 6 part, S 1076.
Concerto, 4 harpsichords, S 1065, A minor. cf Canon, 6 part, S 1076.
Concerto, 4 harpsichords, S 1065, A minor. cf Concerti, harpsichord, nos. 1, 4-5, S 1052, 1055, 1056.
Concerto, oboe, violin and strings, D minor (arr. from S 1060). cf Argo D69D3.

127 Concerto, oboe d'amore, A major. STRAUSS, R.: Concerto, oboe, D major. VAUGHAN WILLIAMS: Concerto, oboe, A minor. Leon Goossens, ob, ob d'amore; PhO; Alceo Galliera, Walter Susskind. World Records SH 243. (Reissues from HMV CLP 1656, Columbia DX 1444/6)
+Gr 10-77 p648 +RR 10-77 p59
+HFN 10-77 p148

Concerto, organ, no. 2, S 593, A minor. cf Organ works (Classics for Pleasure CFP 40241).

128 Concerto, organ, no. 5, S 596, D minor (after Vivaldi). Preludes and fugues, S 553-60 (Little). E. Power Biggs, org. Columbia M 33975.
++NR 5-76 p13 +St 12-76 p72
+SFC 1-2-77 p34

Concerti, violin and strings, S 1041-42. cf Concerti, harpsichord, S 1052, 1055, 1056.

129 Concerto, violin and strings, S 1041, A minor. Concerto, violin and strings, S 1042, E major. Concerto, violin, oboe and strings, S 1060, C minor. Isaac Stern, vln; LSO, NYP; Harold Gomberg, ob; Isaac Stern, Leonard Bernstein. CBS 61573 Tape (c) 40-61573. (Reissue from SBRG 72531)
+Gr 9-77 p417 +-RR 10-77 p32
+-HFN 9-77 p151 +-RR 12-77 p90 tape
+-HFN 10-77 p169 tape

130 Concerto, violin and strings, S 1041, A minor. Concerto, 2 violins and strings, S 1043, D minor. MENDELSSOHN: Concerto, violin, op. 64, E minor. Stoika Milanova, Georgi Badev, vln; Sofia Soloists Chamber Ensemble, TVR Symphony Orchestra; Vassil Kazandjiev, Vassil Stefanov. Monitor MCS 2156.
+FF 11-77 p4

131 Concerto, violin and strings, S 1041, A minor. Concerto, violin and strings, S 1042, E major. HAYDN: Concerto, violin, no. 1, C major. Arthur Grumiaux, vln; ECO; Raymond Leppard. Philips Fontana 6530 004. (Reissue from SAL 3489)
++Gr 12-77 p1067 +RR 11-77 p45
+HFN 11-77 p183

132 Concerto, violin and strings, S 1041, A minor. Concerto, violin and strings, S 1042, E major. Concerto, 2 violins and strings, S 1043, D minor. Suite, orchestra, S 1068, D major: Air. Henryk Szeryng, Maurice Hasson, vln; AMF: Neville Marriner. Philips 9500 226 Tape (c) 7300 537.
+-Gr 8-77 p291 +NR 11-77 p6
+-HFN 7-77 p104 +-RR 9-77 p46

133 Concerto, violin and strings, S 1042, E major. Concerto, 2 violins and strings, S 1043, D minor. Igor Oistrakh, David Oistrakh, vln; Moscow Chamber Orchestra; Rudolf Barshai. Everest SDBR 3410.
+-NR 11-77 p6

134 Concerto, violin and strings, S 1042, E major. MOZART: Concerto, violin, no. 5, K 219, A major. Adolf Busch, vln; Busch Chamber Orchestra; Adolf Busch. Rococo 2099. (Reissues from Columbia M 530, 78s)
+-NR 2-77 p5

Concerto, violin and strings, S 1042, E major. cf Concerto, violin and strings, S 1041, A minor (CBS 61573).

Concerto, violin and strings, S 1042, E major. cf Concerto, violin and strings, S 1041, A minor (Philips 6530 004).

Concerto, violin and strings, S 1042, E major. cf Concerto, violin and strings, S 1041, A minor (Philips 9500 226).

135 Concerto, 2 violins and strings, S 1043, D minor. Sonata, solo flute, S 1013, A minor. Suite, orchestra, S 1067, B minor. Okko Kamu, Leif Segerstam, vln; Gunilla von Bahr, flt; Stockholm National Museum Chamber Orchestra; Claude Genetay. BIS LP 21.
+Gr 3-77 p1456

136 Concerto, 2 violins and strings, S 1043, D minor. BEETHOVEN: Symphony, no. 1, op. 21, C major. MOZART: Serenade, no. 13, K 525, G major. Louis Zimmerman, Ferdinand Hellmann, vln; NYP, COA; Willem Mengelberg. Bruno Walter Society RR 501.
+-NR 7-77 p3

Concerto, 2 violins and strings, S 1043, D minor. cf Concerti, harpsichords, nos. 1, 4-5, S 1052, 1055, 1056.

Concerto, 2 violins and strings, S 1043, D minor. cf Concerto, violin and strings, S 1041, D minor (Monitor MCS 2156).

Concerto, 2 violins and strings, S 1043, D minor. cf Concerto, violin and strings, S 1041, A minor (Philips 9500 226).

Concerto, 2 violins and strings, S 1043, D minor. cf Concerto, violin and strings, S 1042, E major.

Concerto, 2 violins and strings, S 1043, D minor. cf CBS SQ 79200.

Concerto, violin, oboe and strings, S 1060, C minor. cf Concerto, violin and strings, S 1041, A minor.

Duets, S 802-805. cf Capriccio, S 992, B flat major.

Easter oratorio, S 249: Saget, saget, mir geschwinde. cf Works, selections (Angel (Q) S 37229).

Easter oratorio, S 249: Sinfonia. cf Works, selections (DG 2584 001).

137 English suites, nos. 1-6, S 806-11. Glenn Gould, pno. Columbia M2 34578 (2).
+-HF 10-77 p100 ++NR 12-77 p14
+MJ 11-77 p30 ++NYT 6-5-77 pD19

138 English suites, nos. 1-6, S 806-11. French suites, nos. 1-6, S 812-17. Gustav Leonhardt, hpd. Philips 6709 500 (5). (also Philips 6575 070)
+Gr 10-77 p661 +-RR 10-77 p73
++HFN 9-77 p133

139 English suites, nos. 1-6, S 806-11. Joseph Payne, hpd. RCA RL 2-5053 (3).
+-Gr 8-77 p318 +RR 9-77 p79
+HFN 10-77 p148

Fantasia, S 562, C minor. cf Organ works (Angel S 37264/5).

Fantasia, S 562, C minor. cf Organ works (DG 2722 014).
Fantasia, S 562, C minor. cf Wealden WS 131.
Fantasia, S 563, B minor. cf Organ works (Angel S 37264/5).
Fantasia, S 570, C major. cf Organ works (Angel S 37264/5).
140 Fantasia, S 572, G major. Passacaglia and fugue, S 582, C minor. Trio sonata, no. 1, S 525, E flat major. Toccata, adagio and fugue, S 564, C major. Gillian Weir, org. Argo ZK 10.
++Gr 4-77 p1570 +RR 4-77 p68
+-HFN 4-77 p132
Fantasia, S 572, G major. cf Organ works (Angel S 37264/5).
Fantasia, S 572, G major. cf Organ works (DG 2722 014).
Fantasia, S 573, C major. cf Crescent ARS 109.
141 Fantasia, S 906, C minor. Toccata and fugue, S 910, F sharp minor. MOZART: Fantasia, K 396, C minor. Sonata, piano, no. 17, K 576, D major. Variations on a minuet by Duport, K 573. Grant Johannesen, pno. Golden Crest CRSQ 4142.
++ARG 3-77 p11 +HF 9-76 p105
Fantasia, S 906, C minor. cf Concerto, harpsichord, S 971, F major.
Fantasia, S 922, A minor. cf Chromatic fantasia and fugue, S 903, D minor.
142 Fantasia and fugue, S 537, C minor (orch. Elgar). ELGAR: Concerto, violin, op. 61, B minor. Yehudi Menuhin, vln; LSO, Royal Albert Hall Orchestra; Edward Elgar. HMV HLM 7107. (Reissues from HMV DB 1751/6, RLS 708)
+Gr 4-77 p1548
Fantasia and fugue, S 537, C minor. cf Organ works (Angel S 37264/5
Fantasia and fugue, S 537, C minor. cf Organ works (DG 2722 014).
Fantasia and fugue, S 537, C minor. cf ELGAR: Concerto, violin, op. 61, B minor.
Fantasia and fugue, S 540, F major. cf Organ works (Angel S 37264/5
143 Fantasia and fugue, S 542, G major. Passacaglia and fugue, S 582, C minor. Prelude and fugue, S 548, E minor. Toccata and fugue, S 565, D minor. Karl Richter, org. Decca SPA 459. (Reissue from SXL 2219)
+-Gr 12-76 p1015 +-RR 12-76 p75
+HFN 1-77 p121
Fantasia and fugue, S 542, G minor. cf Organ works (Angel S 37264/5)
Fantasia and fugue, S 542, G minor. cf Organ works (Classics for Pleasure CFP 40241).
Fantasia and fugue, S 542, G minor. cf Organ works (DG 2722 014).
Fantasia and fugue, S 565, D minor. cf Organ works (Angel S 37264/5
French suites, nos. 1-6, S 812-17. cf English suites, nos. 1-6, S 806-11.
French suite, no. 5, S 816, G major. cf Chromatic fantasia and fugue, S 903, D minor.
Fugue, A minor. cf Kaibala 20B01.
Fugue, S 574, C minor. cf Organ works (DG 2722 014).
Fugue, S 577, G major. cf Organ works (Classics for Pleasure CFP 40241).
Fugue, S 577, G major (arr. Holst). cf Works, selections (DG 2584 001).
Fugue, S 578, G minor. cf Organ works (DG 2722 014).
Fugue, S 578, G minor, excerpt. cf TR 1006.
Fugue, S 579, B minor. cf Organ works (DG 2722 014).
Fugue, S 1080, D minor. cf Organ works (DG 2722 014).
144 Fugue, lute, S 1000, G minor. Prelude, S 999, C minor. Suite,

lute, S 995, G minor. Suite, S 1006a, E major. Narciso Yepes, lt. DG Archive 2533 351. (Reissue from 2708 030)
+-Gr 5-77 p1708 +RR 5-77 p67
+HFN 5-77 p137

145 Fugue, lute, S 1000, G minor. Prelude, S 999, C minor. Suite, lute, S 996, E minor, excerpt. Suite, lute, S 1006a, E major, excerpt. Suite, lute, S 1007, A major. Walter Gerwig, lt. Oryx Bach 1202. (Reissie from Nonesuch H 71137)
+-Gr 3-73 p1711 +St 8-77 p59
+-HFN 2-73 p332

Fugue, lute, S 1001, G minor. cf Works, selections (DG 2535 248).

Gavotte, nos. 1 and 2. cf London CS 7015.

Inventions, 2 part, S 772-786. cf Concerto, harpsichord, S 971, F major.

Inventions, 2 part, S 775, D minor. cf Atlantic SD 2-7000.

146 Inventions, 3 part, S 787-801. Trio sonata, no. 1, S 525, E flat major. Trio sonata, no. 3, S 527, D minor. Emilia Csanky, ob; Kalman Berkes, clt; Gyorgy Hortobagyi, bsn. Hungaroton SLPX 11814.
+HFN 1-77 p102 +RR 2-77 p66
+NR 5-77 p9

147 Die Kunst der Fuge (The art of the fugue), S 1080. AMF; Neville Marriner. Philips 6747 172 (2) Tape (c) 7699 007.
+Audio 2-77 p81 ++HFN 9-75 p92
+Gr 10-75 p601 +-MT 1-76 p39
+-Gr 12-76 p1066 tape ++NR 12-75 p4
+-HF 1-76 p76 ++RR 9-75 p26
+-HF 7-77 p125 tape +RR 11-76 p108 tape

Die Kunst der Fuge, S 1080: Contrapuncti (2). cf University of Iowa Press unnumbered.

The art of the fugue, S 1080: Contrapunctus VII. cf Crystal S 206.

148 Magnificat, S 243, D major. VIVALDI: Magnificat. Felicity Palmer, s; Helen Watts, con; Robert Tear, t; Stephen Roberts, bs; King's College Choir; AMF; Philip Ledger. Argo ZRG 854.
+-Gr 9-77 p460 +RR 9-77 p89
+-HFN 9-77 p133

149 Magnificat, S 243, D major. Cantata, no. 118, O Jesu Christ mein Lebens Licht. Shelagh Molyneux, s; Paul Esswood, c-t; Ian Partridge, t; John Noble, bs; London Bach Society Chorus; Steinitz Bach Players; Paul Steinitz. Unicorn UNS 248.
+-Gr 9-77 p460 +RR 6-77 p20
+HFN 8-77 p76

Magnificat, S 243, D major: Et exultavit. cf Works, selections (Angel (Q) S 37229).

150 Mass, S 232, B minor. Arleen Auger, Julia Hamari, s; Adalbert Kraus, t; Wolfgang Schone, Siegmund Nimsgern, bs; Gachinger Kantorei; Stuttgart Bach Collegium; Helmuth Rilling. CBS 79307 (3).
+-Gr 12-77 p1118 +RR 12-77 p73
+-HFN 12-77 p164

151 Mass, S 232, B minor. Vienna Boys Choir; VCM; Nikolaus Harnoncourt. Telefunken 6-35019 Tape (c) MH 35019.
+Gr 10-77 p705 tape +RR 9-77 p97 tape

Minuet and badinerie. cf RCA Tape (c) RK 2-5030.

152 Motets: Jesu, meine Freude, S 227; Furchte dich nicht, S 229; Singet dem Herrn, S 225. Regensburg Domspatzen, Vienna Capella Academica; Hubert Gumz, org; Laurenzius Strehl, violone; Hans-

Martin Schneidt. DG Archive 2533 349 Tape (c) 3310 349. (Reissue from 2708 031)
+-Gr 6-77 p76 ++HFN 5-77 p137
-Gr 7-77 p225 tape +RR 5-77 p78

153 Motets: Jesu, meine Freude, S 227. BRAHMS: Liebeslieder, op. 52: Waltzes. South Church Choral Society. (Available from South Congregational Church, 90 Main St., New Britain, Conn).
+MJ 8-77 p10

154 A musical offering, S 1079. Leonhardt Consort. ABC L 67007.
-ARG 9-77 p16 +-PRO 5-77 p25
+-HF 3-77 p108 +SFC 12-12-76 p55
+-MJ 2-77 p30 +-St 5-77 p54
+NR 3-77 p14

155 A musical offering, S 1079. Yorkshire Sinfonia Members; Manoug Parikian, vln, cond. Enigma VAR 1044.
+Gr 10-77 p653 ++RR 10-77 p32
+HFN 10-77 p148

156 A musical offering, S 1079. Suttgart Chamber Orchestra; Karl Munchinger. London CS 7045 Tape (c) CS 5-7045. (also Decca SXL 6824 Tape (c) KSXC 6824)
+Audio 12-77 p50 +-HFN 5-77 p116
+Gr 5-77 p1689 +HFN 8-77 p99 tape
+Gr 7-77 p225 tape +-RR 5-77 p42
+HF 10-77 p101 +RR 12-77 p90 tape

157 Organ works: Fantasias, S 562, C minor; S 563, B minor; S 570, C major; S 572, G major. Fantasias and fugues, S 537, C minor; S 542, G minor. Toccatas and fugues, S 538, D minor; S 540, F major; S 565, D minor. Toccata, adagio and fugue, S 564, C major. Lionel Rogg, org. Angel S 37264/5 (2).
+Ha 6-77 p88 +NR 6-77 p14
+HF 9-77 p98 +-St 8-77 p110

158 Organ works: Concerto, organ, no. 2, S 593, A minor (after Vivaldi). Fantasia and fugue, S 542, G minor. Fugue, S 577, G major. Prelude and fugue, S 544, B minor. Toccata and fugue, S 540, F major. Nicolas Kynaston, org. Classics for Pleasure CFP 40241.
+Gr 11-76 p837 -RR 2-77 p74
+-HFN 8-76 p75

159 Organ works: Allebreve, S 589, D major. Canzona, S 588, D minor. Fantasias, S 562, C minor; S 572, G major. Fantasias and fugues, S 537, C minor; S 542, G minor. Fugues, S 574, C minor; S 578, G minor; S 579, B minor; S 1080, D minor. Passacaglia and fugue, S 582, C minor. Pastorale, S 590, F major. Preludes and fugues, S 531-536, S 539, S 541, S 543-552. Trio sonatas, nos. 1-6, S 525-530. Toccata, adagio and fugue, S 546, C major. Toccatas and fugues, S 538, D minor; S 540, F major; S 565, D minor. Helmut Walcha, org. DG 2722 014 (8) Tape (c) 3376 004. (Reissue from 2722 002/1)
*Gr 8-75 p347 +-HFN 7-75 p72
+-Gr 1-77 p1183 +HF 6-76 p97 tape

160 Partita, flute, S 1013, A minor. Sonatas, flute and harpsichord, S 1030, S 1032, S 1034, S 1035 (alternative instrumentations and early versions). Frans Bruggen, flt; Gustav Leonhardt, hpd. ABC AB 67015/2 (2).
-FF 11-77 p4

161 Partita, flute, S 1013, A minor. Sonatas, flute and harpsichord, S 1031-S 1033, S 1020. Alexander Murray, flt; Martha Goldstein,

hpd. Pandora PAN 104/5 (2).
-FF 11-77 p4
Partita, flute, S 1013, A minor. cf Sonatas, flute and harpsichord, S 1030-5 (CRD CRD 1014/5).
Partita, flute, S 1013, A minor. cf Sonatas, flute and harpsichord, S 1030-5 (Vanguard VSD 71215/6).
162 Partita, harpsichord, no. 1, S 825, B flat major. Partita, harpsichord, no. 2, S 826, C minor. Igor Kipnis, hpd. Angel S 36097.
++ARG 9-77 p16 ++SFC 6-26-77 p46
+-HF 10-77 p100 ++St 9-77 p124
++NR 7-77 p14
163 Partita, harpsichord, no. 1, S 825, B flat major. Partita, harpsichord, S 826, C minor. Partita, harpsichord, no. 3, S 827, A minor. Alexis Weissenberg, pno. Connoisseur Society CS 2117.
++ARG 5-77 p12 +NR 7-77 p12
+-HF 10-77 p100 +-St 9-77 p124
+MJ 5-77 p32
Partita, harpsichord, no. 2, S 826, C minor. cf Partita, harpsichord, no. 1, S 825, B flat major (Angel 36097).
Partita, harpsichord, no. 2, S 826, C minor. cf Partita, harpsichord, no. 1, S 825, B flat major (Connoisseur Society CS 2117).
Partita, harpsichord, no. 3, S 827, A minor. cf Partita, harpsino. 1, S 825, B flat major.
164 Partita, harpsichord, no. 4, S 828, D major. Partita, harpsichord, no. 7, S 831, B minor. Alexis Weissenberg, pno. Connoisseur Society CS 2118.
++ARG 5-77 p12 +NR 7-77 p12
+-HF 10-77 p100 +-St 8-77 p124
+MJ 5-77 p32
Partita, harpsichord, no. 7, S 831, B minor. cf Capriccio, S 992, B flat major.
Partita, harpsichord, no. 7, S 831, B minor. cf Concerto, harpsichord, S 971, F major.
Partita, harpsichord, no. 7, S 831, B minor. cf Partita, harpsichord, no. 4, S 828, D major.
165 Partitas, violin, nos. 1-3, S 1002, S 1004, S 1006. Michael Goldstein, vln. Polydor 066 6556.
++RR 5-77 p67
166 Partita, violin, no. 2, S 1004, D minor: Chaconne. Prelude, fugue and allegro, S 998, E flat major. Suite, solo violoncello, no. 3, S 1009, C major. Diego Blanco, gtr. Swedish Society SLT 33226.
+-HFN 7-77 p105 +-RR 7-77 p72
Partita, violin, no. 1, S 1002, B minor: Bourree. cf Discopaedia MOB 1019.
Partita, violin, no. 1, S 1002, B minor: Sarabande. cf Works, selections (DG 2535 248).
Partita, violin, no. 2, S 1004, D minor: Chaconne. cf Works, selections (DG 2534 248).
Partita, violin, no. 2, S 1004, D minor: Chaconne (arr. Busoni). cf International Piano Library IPL 104.
Partita, violin, no. 3, S 1006, E major. cf Discopaedia MOB 1020.
Partita, violin, no. 3, S 1006, E major: Preludio. cf Discopaedia MB 1012.
Partita, violin, no. 3, S 1006, E major: Preludio (Soldat). cf Discopaedia MOB 1019.

Partita, violin, no. 3, S 1006, E major: Praeludium, Gavotte. cf Works, selections (DG 2584 001).

Partita, violin, no. 3, S 1006, E major: Prelude, gavotte, gigue. cf Angel S 37219.

167 Passacaglia and fugue, S 582, C minor. Pastorale, S 590, F major. Prelude and fugue, S 548, E minor. Trio sonata, no. 1, S 525, E flat major. Hans Fagius, org. BIS LP 63.
+-RR 3-77 p74 ++St 6-77 p126

168 Passacaglia and fugue, S 582, C minor. Prelude and fugue, S 552, E flat major. Toccata and fugue, S 565, D minor. Toccata, adagio and fugue, S 564, C major. Anthony Newman, org. Turnabout QTV 34656 Tape (c) CT 2137.
-NR 6-77 p15

Passacaglia and fugue, S 582, C minor. cf Fantasia, S 572, G major.

Passacaglia and fugue, S 582, C minor. cf Fantasia and fugue, S 542, G minor.

Passacaglia and fugue, S 582, C minor. cf Organ works (DG 2722 014).

Passacaglia and fugue, S 582, C minor. cf Argo ZRG 806.

Passacaglia and fugue, S 582, C minor. cf Wealden WS 159.

Pastorale, S 590, F major. cf Chorale preludes, S 645-50.

Pastorale, S 590, F major. cf Organ works (DG 2722 014).

Pastorale, S 590, F major. cf Passacaglia and fugue, S 582, C minor.

Prelude, A minor. cf London CS 7015.

Prelude, E major. cf Discopaedia MOB 1020.

Prelude, S 927, F major. cf Works, selections (Telefunken 6-42091).

Prelude, S 929, G minor. cf Works, selections (Telefunken 6-42091).

Prelude, S 937, E major. cf Works, selections (Telefunken 6-42091).

Prelude, S 939, C major. cf Works, selections (Telefunken 6-42091).

Prelude, S 940, D minor. cf Works, selections (Telefunken 6-42091).

Prelude, S 999, C minor. cf Fugue, lute, S 1000, G minor (DG 2533 351).

Prelude, S 999, C minor. cf Fugue, lute, S 1000, G minor (Oryx 1202).

Prelude, S 999, C minor. cf Works, selections (DG 2535 248).

169 Preludes and fugues (48). Anthony Newman, org. Vox QSVBX 5479/80.
+-HF 9-77 p98 -NR 9-77 p14
++MU 4-77 p15

Preludes and fugues, S 531-536, S 539, S 541, S 543-552. cf Organ works (DG 2722 014).

Prelude and fugue, S 532, D major. cf TR 1006.

Prelude and fugue, S 532, D major. cf Vista VPS 1030.

Prelude and fugue, S 535, G minor. cf Works, selections (DG 2584 001).

170 Prelude and fugue, S 544, B minor. Prelude and fugue, S 546, C minor. Trio sonata, no. 5, S 529, C major. Trio sonata, no. 6, S 530, G major. Charles Benbow, org. Philips 6581 019.
+Gr 8-77 p318 +-RR 7-77 p76
+HFN 7-77 p104

Prelude and fugue, S 544, B minor. cf Chorale preludes, S 645-50.

Prelude and fugue, S 544, B minor. cf Organ works (Classics for Pleasure CFP 40241).

Prelude and fugue, S 545, C major. cf International Piano Library IPL 104.

Prelude and fugue, S 545, C major. cf Pelca PSR 40599.

Prelude and fugue, S 546, C minor. cf Prelude and fugue, S 544, B minor.
Prelude and fugue, S 548, E minor. cf Fantasia and fugue, S 542, G minor.
Prelude and fugue, S 548, E minor. cf Passacaglia and fugue, S 582, C minor.
Prelude and fugue, S 550, G major. cf TR 1006.
Prelude and fugue, S 552, E flat major. cf Passacaglia and fugue, S 582, C minor.
Preludes and fugues, S 553-69 (Little). cf Concerto, organ, no. 5, S 596, D minor.
Prelude and fugue, S 554, D minor. cf Westminster WG 1012.

171 Prelude, bouree and gigue (arr. Kynaston). DIEMENTE: Diary, Pt II. IBERT: Concertino da camera. KYNASTON: Dawn and jubilation. Trent Kynaston, David Fish, alto saxophone; Steven Hesla, pno. Coronet LPS 3035.
+-NR 1-77 p15

172 Prelude, fugue and allegro, S 998, E flat major. KELLNER: Fantasy, C major. Fantasy, A minor. WEISS: Sonata, lute, A minor. Eugen Dombois, lt. Philips 6565 005. (also ABCL 67006)
+Gr 9-74 p554 +PRO 5-77 p25
+HF 3-77 p108 +RR 10-74 p18
+NR 7-77 p15

Prelude, fugue and allegro, S 998, E flat major. cf Partita, violin, no. 2, S 1004, D minor: Chaconne.
Prelude, fugue and allegro, S 998, E flat major. cf Works, selections (DG 2535 248).
Prelude, fugue and allegro, S 998, E flat major. cf CBS 72526.
Presto, A minor. cf London CS 7015.
St. John Passion, S 245. cf St. Matthew Passion, S 244.
St. John Passion, S 245: Es ist vollbracht. cf Works, selections (Angel (Q) S 37229).

173 St. John Passion, S 245: Herr, unser Herrscher; Simon Petrus aber folgete Jesu nach; Ich folge dir gleichfalls mit freudigen Schritten; Und Hannas sandte ihn gebunden; Bist du nicht seiner Junger einer; Er leugnete aber und sprach; Ach, mein Sinn, wo willt du endlich hin; Petrus, der nicht denkt zuruck; Da sprach Pilatus zu ihm; Nicht diesen, diesen nicht; Barrabas aber war ein Morder; Betrachte, mein Seel mit angstlichem Vergnugen; Und die Kriegsknechte flochten eine Krone; Sei gegrusset, lieber Judenkonig; Und gaben ihm Backenstreiche; Kreuzige; Pilatus sprach zu ihnen; Wir haben ein Gesetz, und nach dem Gesetz; Da Pilatus das Wert horete, furchtet er sich; Durch dein Gefangnis, Gottes Sohn; Die Juden aber Schreien; Lassest du diesen los so bist du des Kaisers Freund nicht; Da Pilatus das Wort horete, fuhrete er Jesum; Weg mit dem kreuzige ihn; Spricht Pilatus zu ihnen; Wir haben Keinen Konig; De uber antwortete er ihn, dass er gekreuziget wurde; All da kreuzigten sie ihn; Schreibe nicht; Pilatus antwortete; In meines Herzens Grunde, dein Nam; Es ist vollbracht, o Trost, Der Held aus Juda siegt; Ruht wohl ihr heiligen Gebeine. Elly Ameling, Marianne Koehnlein-Goebel, s; Julia Hamari, con; Dieter Ellenbeck, Wolfgang Isenhardt, Werner Hollweg, t; Walter Berry, Hans Georg Ahrens, Manfred Ackerman, Herman Prey, bs; Stuttgart Hymnus Boys Choir; Stuttgart Chamber Orchestra; Karl Munchinger. Decca SXL 6778 Tape (c) KSXC 6778. (Reissue from SET 590/2)
+Gr 4-77 p1578 +HFN 5-77 p138 tape

+-Gr 7-77 p225 tape +RR 4-77 p79
+HFN 4-77 p151 ++RR 7-77 p99 tape

174 St. Matthew Passion, S 244. Elsie Suddaby, s; Kathleen Ferrier, con; Eric Greene, t; Bruce Boyce, Gordon Clinton, Henry Cummings, William Parsons, bs; Bach Choir; Thornton Lofthouse, hpd; Osborne Peasgood, org; Jacques Orchestra; Reginald Jacques. Decca D42D3 (3). (Reissue from K 1673/9, AK 2001/21)
+-Gr 4-77 p1578 +-RR 4-77 p79
+-HFN 5-77 p138

175 St. Matthew Passion, S 244. St. John Passion, S 245. Irmgard Seefried, Antonia Fahberg, Evelyn Lear, s; Hertha Topper, con; Ernst Hafliger, t; Dietrich Fischer-Dieskau, bar; Keith Engen, Max Proebstl, Hermann Prey, bs; Munich Bach Orchestra and Choir; Munich Boys Choir; Karl Richter. DG Archive 2722 010 (7) Tape (c) 3376 001/2. (Reissue from SAPM 198009/12, SAPM 198329/30)
+Gr 11-74 p935 +HF 11-75 p149 tape
+-Gr 1-77 p1183 tape +-RR 12-74 p19

176 St. Matthew Passion, S 244. Adele Stolte, Erika Wustman, Eva Hassbecker, s; Annelies Burmeister, Gerda Schriever, con; Peter Schreier, Hans-Joachim Rotzsch, t; Theo Adam, bs-bar; Johannes Kunzel, Siegfried Vogel Hans Martin Nau, Hermann Christian Polster, Gunther Leib, bs; Dresden Kreuzchor, St. Thomas Choir; Leipzig Gewandhaus Orchestra; Rudolf Mauersberger. RCA LRL 4-5098 (4).
+-Gr 2-76 p1363 -MM 1-77 p31
+-HFN 3-76 p87 +-RR 2-76 p54

177 Sonata, flute, E flat major. BRAHMS: Four serious songs, op. 121. RUSSEL: Suite concertante. ZINDARS: Trigon. San Francisco Symphony Orchestra; Naomi Cooley, piano; Floyd Cooley, tuba. Avant AV 1020.
+SFC 3-6-77 p34

Sonata, solo flute, S 1013, A minor. cf Concerto, 2 violins and strings, S 1043, D minor.

178 Sonatas, flute and harpsichord, S 1030-1035. Partita, flute, S 1013, A minor. Stephen Preston, flt; Trevor Pinnock, hpd; Jordi Saval, vla da gamba. CRD CRD 1014/5 (2) Tape (c) CRD 4014/5.
+Gr 8-75 p336 +RR 12-77 p90 tape
+Gr 10-77 p705 tape ++St 4-76 p110
+HFN 11-77 p187 tape +STL 11-2-75 p38
+RR 8-75 p54

179 Sonatas, flute and harpsichord, S 1030-1035. Christopher Taylor, flt; Leslie Pearson, hpd; Dennis Vigay, vlc. Gale GMFD 2-76-002-3 (2).
+HF 10-77 p112 +HFN 9-77 p131

180 Sonatas, flute and harpsichord, S 1030-1035. Partita, flute, S 1013, A minor. Paula Robison, flt; Kenneth Cooper, hpd; Timothy Eddy, vlc. Vanguard VSD 71215/6 (2).
+-ARG 3-77 p10 +NR 5-77 p9
++HF 3-77 p95 +St 5-77 p102
+MJ 2-77 p30

181 Sonatas, flute and harpsichord, S 1030, S 1032, S 1034-5. Sonatas, 2 flutes and harpsichord, S 1038-9. Leopold Stastny, Frans Bruggen, flt; Alice Harnoncourt, vln; Nikolaus Harnoncourt, vlc; Herbert Tachezi, hpd. Telefunken EX 6-35339 (2).
+Gr 2-77 p1296 ++RR 12-76 p75
+HF 10-77 p99 ++St 12-77 p132
+HFN 5-77 p117

Sonatas, flute and harpsichord, S 1030, S 1032, S 1034, S 1035. cf Partita, flute, S 1013, A minor.
Sonatas, flute and harpsichord, S 1031-1033, S 1020. cf Partita, flute, S 1013, A minor.
Sonata, flute and harpsichord, S 1031, E flat major: Siciliano. cf Works, selections (DG 2530 647).
Sonata, flute and harpsichord, S 1031, E flat major: Siciliano. cf ALBINONI: Concerto, 2 trumpets, C major.
Sonatas, 2 flutes and harpsichord, S 1038-9. cf Sonatas, flute and harpsichord, S 1030, S 1032, S 1034-5.

182 Sonatas, viola da gamba and harpsichord, nos. 1-3, S 1027-29. TELEMANN: Sonata, viola da gamba and harpsichord, A minor. Eva Heinitz, vla da gamba; Malcolm Hamilton, hpd. Delos DELS 15341.
+NR 9-77 p7 +-St 3-75 p96

Sonata, violin, no. 1, S 1001, G minor: Adagio. cf Discopaedia MOB 1019.
Sonata, violin, no. 2, S 1003, A minor. cf Discopaedia MB 1016.
Sonata, violin, no. 2, S 1003, A minor: Andante. cf HMV RLS 723.
Sonata, violin, no. 3, S 1005, C major: Adagio. cf Works, selections (DG 2584 001).

183 Sonatas, violin and harpsichord, nos. 1-6, S 1014-1019. Jaime Laredo, vln; Glenn Gould, pno. Columbia M2 34226 (2). (also CBS 79209)
+-ARG 3-77 p12 +MU 3-77 p14
-Gr 6-77 p68 ++NR 1-77 p7
+-HF 3-77 p90 +-RR 4-77 p68
+-HFN 4-77 p132 +-St 4-77 p114
+MJ 4-77 p50

184 Sonatas, violin and harpsichord, nos. 1-6, S 1014-1019. Eduard Melkus, vln; Huguette Dreyfus, hpd. DG Archive 2708 032 (2) Tape (c) 3375 002.
+Gr 8-76 p311 +NR 8-76 p5
++Gr 11-76 p887 tape ++RR 9-76 p79
+-HF 9-76 p85 ++SFC 1-2-77 p34
+HFN 9-76 p112 ++St 9-76 p116

185 Sonatas, violin and harpsichord, nos. 1-6, S 1014-1019. Alice Harnoncourt, vln; Nikolaus Harnoncourt, vla da gamba; Herbert Tachezi, hpd. Telefunken EX 6-35310.
+-Gr 9-76 p436 ++NR 1-77 p7
+HF 3-77 p90 ++RR 10-76 p84
+HFN 9-76 p112 +-St 4-77 p114

Sonatas, violin and harpsichord, nos. 1-6, S 1014-1019. cf Sonata, violin and harpsichord, S 1021, G major.

186 Sonata, violin and harpsichord, S 1021, G major. Sonata, violin and harpsichord, S 1023, E minor. Sonatas, violin and harpsichord, S 1014-1019. Endre Granat, vln; Edith Kilbuck, hpd. Orion ORS 79213/5 (3).
+-HF 3-77 p90 +-St 4-77 p114
++NR 1-77 p7

Sonata, violin and harpsichord, S 1023, E minor. cf Sonata, violin and harpsichord, S 1021, G major.

187 Sonatas and partitas, solo violin, S 1001-6. Yehudi Menuhin, vln. Angel SC 3817 (3). (also HMV SLS 5045)
+-Gr 4-76 p1617 ++NR 7-76 p12
+-HF 10-76 p102 -NYT 7-24-77 pD11
+-HFN 6-76 p80 +-RR 7-76 p69
+-MT 9-76 p744 +-St 11-76 p136

188 Sonatas and partitas, solo violin, S 1001-6. Joseph Szigeti, vln. Bach Guild HM 54/6 (3).
+-NYT 7-24-77 pD11
189 Sonatas and partitas, solo violin, S 1001-6. Nathan Milstein, vln. DG 2709 047 (3) Tape (c) 3371 030.
++Audio 6-76 p100 ++RR 4-75 p47
++Gr 4-75 p1830 ++SFC 10-26-75 p24
++Gr 11-77 p900 tape ++St 2-76 p104
++HF 1-76 p81 ++STL 5/4/75 p37
++NR 12-75 p15
190 Sonatas and partitas, solo violin, S 1001-6. Salvatore Accardo, vln. Philips 6703 076 (3).
+Audio 12-77 p48 ++HFN 7-77 p104
+Gr 8-77 p317 ++RR 7-77 p77
191 Sonatas and partitas, solo violin, S 1001-6. Sandor Vegh, vln. Telefunken EK 6-35344 (3).
+Gr 8-77 p317 +-RR 7-77 p78
+-HFN 7-77 p104
Songs (choral works) Dir, dir, Jehova, will ich singen, S 452; Gibt dich zufrieden und sei stille, S 511; Kanon zu Stimmen, S 1073, S 1075, S 1077, S 1078; Nicht so traurig, nicht so sehr, S 384; O Herzensangst, O Bangigkeit und Zagen, S 400; Quodlibet, S 524; So oft ich meine Tobacks-Pfeife, S 515a; Vergiss mein nicht, mein allerliebster Gott, S 505; Was betrubst du dich, mein Herze, S 423; Wer nur den lieben Gott lasst walten, S 434, S 690. cf Works, selections (Telefunken 6-42091).
Songs: Mein Jesu, was fur Seelenweh, S 487. cf Seraphim SIB 6094.
Songs (Motet): Singet dem Herrn ein neues Lied, S 225. cf Pelca PSR 40607.
Suite, no. 3: Aria. cf HMV RLS 723.
Suite, lute, S 995, G minor. cf Fugue, lute, S 1000, G minor.
Suite, lute, S 995, G minor. cf Philips 6833 159.
Suite, lute, S 996, E minor, excerpt. cf Fugue, lute, S 1000, G minor.
Suite, lute, S 996, E minor: Bourree. cf Westminster WG 1012.
Suite, lute, S 996, E minor: Sarabande. cf Works, selections (DG 2535 248).
192 Suite, lute, S 997, C minor. PURCELL: Fairy Queen. VISEE: Suite, G major. WEISS: Tombeau sur la mort de M. Comte de Logy. Carlos Bonell, gtr. Enigma VAR 1050.
+HFN 12-77 p171 +RR 12-77 p62
Suite, lute (harpsichord), S 997, C minor. cf Concerto, harpsichord, S 971, F major.
Suite, lute, S 1006a, E major. cf Fugue, lute, S 1000, G minor.
Suite, lute, S 1006a, E major, excerpt. cf Fugue, lute, S 1000, G minor.
Suite, lute, S 1007, A major. cf Fugue, lute, S 1000, G minor.
193 Suites orchestra, S 1066-9. AMF; William Bennett, flt; Thurston Dart, hpd; Neville Marriner. Argo ZRG 687/8 Tape (c) KRZC 687/8.
+AR 8-72 p96 +HFN 3-72 p513 tape
+Gr 3-72 p1607 -NR 2-72 p2
+HF 3-72 p76 +NYT 12-12-71 pD31
+HF 4-77 p121 tape ++SFC 12-5-71 p38
+HFN 9-71 p1630 ++St 4-72 p75
194 Suites, orchestra, S 1066-9. Stuttgart Chamber Orchestra; Karl

Munchinger. Decca DPA 589/90. (Reissue from SXL 2300/1).
/-Gr 12-77 p1067 +-RR 11-77 p45
+-HFN 11-77 p183

195 Suites, orchestra, S 1066-69. Lorant Kovacs, flt; Liszt Chamber Orchestra; Frigyes Sandor. Hungaroton SPLX 11787/8 (2).
+-FF 11-77 p5

196 Suites, orchestra, S 1066-69. Saar Chamber Orchestra; Karl Ristenpart. Sine Qua Non Tape (c) 147.
+HF 9-77 p119 tape

Suites, orchestra, S 1066-9. cf Brandenburg concerti, nos. 1-6, S 1046-51 (DG Tape (c) 3376 003).

Suite, orchestra, S 1067, B minor. cf Cantata, no. 209: Sinfonia.

Suite, orchestra, S 1067, B minor. cf Concerto, 2 violins and strings, S 1043, D minor.

Suite, orchestra, S 1067, B minor: Minuet; Badinerie. cf RCA LRL 1-5094.

197 Suite, orchestra, S 1068, D major. COUPERIN, F.: Les gouts reunis, no. 8, G major (arr. Oubradous). Pieces en concert (arr. Bazelaire). Roderick Brydon, hpd; John Wilbraham, tpt; Paul Tortelier, vlc; Scottish Chamber Orchestra; Paul Tortelier. HMV ASD 3321 Tape (c) TC ASD 3321.
+-Gr 4-77 p1570 +-HFN 8-77 p99 tape
+-HFN 5-77 p117 +-RR 5-77 p45

Suite, orchestra, S 1068, D major. cf Bruno Walter Society RR 443.

Suite, orchestra, S 1068, D major: Air. cf Concerto, violin and strings, S 1041, A minor.

Suite, orchestra, S 1068, D major: Air on the G string (Soldat). cf Discopaedia MOB 1019.

Suite, orchestra, S 1068, D major: Air on the G string. cf Discopaedia 1020.

Suite, orchestra, S 1068, D major: Air on the G string. cf Rediffusion Royale 2015.

198 Suite, solo violoncello, nos. 1-6, S 1007-1012. Henri Honegger, vlc. Telefunken EX 6-35345 (3).
+-Gr 9-77 p455 +-RR 7-77 p77
+-HFN 7-77 p104

199 Suite, solo violoncello, no. 1, S 1007, G major. PENDERECKI: Capriccio per Siegfried Palm. SESSIONS: Pieces, violoncello (6). Roy Christensen, vlc. Gasparo GS 102.
+Audio 5-77 p98 +NR 3-77 p14

Suite, solo violoncello, no. 1, S 1007, G major: 3 movements. cf RCA ARL 1-0864.

200 Suite, solo violoncello, no. 3, S 1009, C major. KODALY: Sonata, solo violoncello, op. 8. Frans Helmerson, vlc. BIS LP 25. (also HNH 4021)
+-ARG 11-77 p6 +RR 8-76 p67
++Gr 6-76 p62 ++St 8-76 p92

Suite, solo violoncello, no. 3, S 1009, C major. cf Partita, violin, no. 2, S 1004, D minor: Chaconne.

201 Suite, solo violoncello, no. 5, S 1011, C minor. BRITTEN: Suite, violoncello, no. 1, op. 72. Frans Helmerson, vlc. BIS LP 5. (also HNH 4016)
+-ARG 10-77 p26 +HF 9-77 p108
+ARG 11-77 p6 ++HFN 11-75 p148
+Gr 10-76 p622 ++RR 11-75 p68

Toccata, C major: Adagio. cf HMV SQ ASD 3283.

Toccata, D minor, excerpt. cf TR 1006.

Toccata, S 912, D major. cf Chromatic fantasia and fugue, S 903, D minor (Decca ECS 788).
Toccata, S 912, D major. cf Chromatic fantasia and fugue, S 903, D minor (Philips 6833 184).
Toccata, S 914, E minor. cf Chromatic fantasia and fugue, S 903, D minor.
Toccata, S 916, C major. cf Chromatic fantasia and fugue, S 903, D minor.
Toccata, adagio and fugue, S 564, C major. cf Fantasia, S 572, G major.
Toccata, adagio and fugue, S 564, C major. cf Organ works (Angel S 37264/5).
Toccata, adagio and fugue, S 564, C major. cf Organ works (DG 2722 014).
Toccata, adagio and fugue, S 564, C major. cf Passacaglia and fugue, S 582, C minor.
Toccata, adagio and fugue, S 564, C major: Adagio. cf ALBINONI: Concerto, 2 trumpets, C major.
Toccata and fugue, S 538, D minor. cf Organ works (Angel S 37264/5)
Toccata and fugue, S 538, D minor. cf Organ works (DG 2722 014).
Toccata and fugue, S 540, F major. cf Organ works (Angel S 37264/5)
Toccata and fugue, S 540, F major. cf Organ works (Classics for Pleasure CFP 40241).
Toccata and fugue, S 540, F major. cf Organ works (DG 2722 014).
Toccata and fugue, S 565, D minor. cf Fantasia and fugue, S 542, G minor (Decca SPA 459).
Toccata and fugue, S 565, D minor. cf Organ works (Angel S 37264/5)
Toccata and fugue, S 565, D minor. cf Organ works (DG 2722 014).
Toccata and fugue, S 565, D minor. cf Passacaglia and fugue, S 582, C minor.
Toccata and fugue, S 565, D minor. cf Works, selections (DG 2584 001).
Toccata and fugue, S 565, D minor. cf Advent ASP 4007.
Toccata and fugue, S 565, D minor. cf Decca SDD 499.
Toccata and fugue, S 565, D minor. cf Decca PFS 4416.
Toccata and fugue, S 565, D minor. cf Odeon HQS 1356.
Toccata and fugue, S 565, D minor. cf Seraphim SIB 6094.
Toccata and fugue, S 910, F sharp minor. cf Fantasia, S 906, C minor.
Trio sonatas (4). cf Chorale preludes (Orgelbuchlein), S 599-644.
Trio sonatas, nos. 1-6, S 525-530. cf Organ works (DG 2722 014).
Trio sonata, no. 1, S 525, E flat major. cf Fantasia, S 572, G major.
Trio sonata, no. 1, S 525, E flat major. cf Inventions, 3 part, S 787-801.
Trio sonata, no. 1, S 525, E flat major. cf Passacaglia and fugue, S 582, C minor.
Trio sonata, no. 3, S 527, D minor. cf Inventions, 3 part, S 787-801.
Trio sonata, no. 5, S 529, C major. cf Prelude and fugue, S 544, B minor.
Trio sonata, no. 5, S 529, C major. cf Abbey LPB 752.
Trio sonata, no. 6, S 530, G major. cf Prelude and fugue, S 544, B minor.

202 Variations, harpsichord, S 988 (Goldberg). Rosalyn Tureck, pno. Everest 3397.
+ARG 3-77 p12 +-NR 4-77 p14

203 Variations, harpsichord, S 988. Sylvia Marlow, hpd. RCA AGL 1-2447.
+-NR 12-77 p14

Variations, harpsichord, S 988. cf BACH, C.P.E.: Duets.

204 The well-tempered clavier, S 846-8, 850-3, 860-2, 866-7. Wilhelm Kempff, pno. DG 2530 807.
+-Gr 4-77 p1570 -RR 4-77 p68
+-HFN 4-77 p132

The well-tempered clavier, Dk I: Prelude and fugue, C sharp major. cf International Piano Library IPL 112.

205 Works, selections: Anna Magdalena notebook, S 508: Bist du bei mir. Cantata, no. 6, Bleib ei uns: Hochgelobter Gottessohn. Cantata, no. 11, Lobet Gott: Ach, bleibe doch. Cantata, no. 34, O Ewiger Feuer: Wohl euch ihr auser wahlten Seelen. Cantata, no. 129, Gelobet sei der Herr. Cantata, no. 161, Komm, du susse Toddestunde. Cantata, no. 190, Lobe, Zion, deinen Gott. Christmas oratorio, S 248: Bereite dich, Zion. Easter oratorio, S 249: Saget, saget, mir geschwinde. Magnificat, S 243, D major: Et exultavit. St. John Passion, S 245: Es ist vollbracht. Janet Baker, ms; AMF; Neville Marriner. Angel (Q) S 37229. (also HMV Tape (c) TC ASD 3265)
++ARG 8-77 p12 +NR 8-77 p11
+Gr 3-77 p1458 tape ++RR 2-77 p96 tape
+HF 7-77 p98 +-St 8-77 p124
+HFN 1-77 p123 tape

206 Works, selections: Cantata, no. 49: Sinfonia. Cantata, no. 35: Sinfonia I and II. Cantata, no. 49: Sinfonia. Cantata, no. 31: Sonata. Cantata, no. 146: Sinfonia. Cantata, no. 147: Chorale. Cantata, no. 169: Sinfonia. E Power Biggs, org; Leipzig Gewandhaus Orchestra; Hans-Joachim Rotzsch. Columbia (Q) 34272.
+ARG 4-77 p12 +MU 5-77 p17
+MJ 4-77 p50 +-NR 11-77 p14

207 Works, selections: Cantata, no. 29, Wir danken dir Gott: Sinfonia. Cantata, no. 140, Wachet auf, ruft uns die Stimme. Cantata, no. 147, Jesu bleibt meine Freude. Chrale preludes: Herzlich tut mich verlangen (Befiehl du deine Wege), S 727; Ich ruf zu dir, Herr Jesu Christ, S 639; In dulci jubilo; Nun komm der Heiden Heiland, S 659. Concerto, harpsichord, no. 5, S 1056, F minor: Largo. Sonata, flute and harpsichord, S 1031, E flat major: Siciliano. GLUCK: Orfeo ed Euridice: Ballet music. HANDEL: Minuet, G minor. (Piano transcriptions by Kempff) Wilhelm Kempff, pno. DG 2530 647.
++ARG 4-77 p28 +NR 12-77 p14
-Gr 5-76 p1788 +RR 5-76 p63
+HF 9-77 p114 +St 5-77 p120
+-HFN 5-76 p92

208 Works, selections: Fugue, lute, S 1001, G minor. Prelude, S 999, C minor. Partita, violin, no. 1, S 1002, B minor: Sarabande. Partita, violin, no. 2, S 1004, D minor: Chaconne. Prelude, fugue and allegro, S 998, E flat major. Suite, lute, S 996, E minor: Sarabande. Narciso Yepes, gtr. DG 2535 248 Tape (c) 3335 248.
+Gr 12-77 p1146 tape -RR 11-77 p91
+HFN 11-77 p183

209 Works, selections: Cantata, no. 147, Jesu, joy of man's desiring. Cantata, no. 208, Sheep may safely graze (arr. Walton). Pre-

lude and fugue, S 535, G minor (arr. Cailliet). Fugue, S 577, G major (arr. Holst). Easter oratorio, S 249: Sinfonia (arr. Whittaker). Partita, violin, no. 3, S 1006, E major: Praeludium, Gavotte (arr. Bachrich). Sonata, violin, no. 3, S 1005, C major: Adagio (arr. Bachrich). Toccata and fugue, S 565, D minor (arr. Stokowski). BPO; Arthur Fiedler. DG 2584 001.
+-ARG 7-77 p12 +-RR 6-76 p33
+Gr 5-76 p1807 +SFC 1-24-77 p40
+-HFN 6-76 p80 +St 11-77 p160

210 Works, selections: Prelude, S 927, F major. Prelude, S 929, G minor. Prelude, S 937, E major. Prelude, S 939, C major. Prelude, S 940, D minor. Songs (choral works): Dir, dir, Jehova, will ich singen, S 452; Gibt dich zufrieden und sei stille, S 511; Kanon zu Stimmen, S 1073, S 1075, S 1077, S 1078; Nicht so traurig, nicht so sehr, S 384; O Herzensangst, O Bangigkeit und Zagen, S 400; Quodlibet, S 524; So oft ich meine Tobacks-Pfeife, S 515a; Vergiss mein nicht, mein allerliebster Gott, S 505; Was betrubst du dich, mein Herze, S 423; Wer nur den lieben Gott lasst walten, S 434, S 690. Agnes Giebel, s; Marie Luise Gilles, con; Bert van t'Hoff, t; Peter Christoph Runge, bs; Leonhardt Consort; Gustav Leonhardt, hpd, org; Anner Bylsma, vlc. Telefunken 6-42091.
+NR 10-77 p11

BACH, Wilhelm Friedemann
Concerto, 2 harpsichords, F major. cf BACH, C.P.E.: Duets.
Fantasia, D minor. cf BACH, C.P.E.: Fantasia, Wq 254, C minor.

BACHELET, Alfred
Chere nuit. cf Columbia 34294.
Chere nuit. cf Odyssey Y 33793.

BACHILLER
The widowes mite. cf RCA RL 2-5110.

BACK, Sven-Eric
211 ...for Eliza. HAMBRAEUS: Shogaku. LIGETI: Volumina. RAXACH: The looking glass. Karl Erik Welin, org. Caprice CAP 1108.
+RR 5-77 p71

BACKER GRONDAHL, Agathe
Ballade, op. 36, no. 5. cf Piano works (Norwegian Polydor NKF 30008).
Concert study, op. 11, no. 1, B flat minor. cf Piano works (Norwegian Polydor NKF 30008)
Concert study, op. 58, no. 2, G minor. cf Piano works (Norwegian Polydor NKF 30008).
In the blue mountains. cf Piano works (Norwegian Polydor NKF 30008).

212 Piano works: Ballade, op. 36, no. 5. Concert study, op. 11, no. 1, B flat minor. Concert study, op. 58, no. 2, G minor. In the blue mountains. Serenade, op. 15, no. 1. Song of the roses, op. 39, no. 4. Summer ballad, op. 45, no. 3. Liv Glaser, pno. Norwegian Polydor NKF 30008.
+HFN 3-77 p109 +-RR 1-77 p68

Serenade, op. 15, no. 1. cf Piano works (Norwegian Polydor NKF 30008).
Song of the roses, op. 39, no. 4. cf Piano works (Norwegian Polydor NKF 30008).
Summer ballad, op. 45, no. 3. cf Piano works (Norwegian Polydor NKF 30008).

BACON, Ernst
The Burr frolic. cf Orion ORS 76247.
BADARZEWSKA-BARANOWSKA, Tekla
The maiden's prayer. cf Decca SPA 473.
BAGPIPE MARCHES AND MUSIC OF SCOTLAND. cf Olympic 6118.
BAIRSTOW, Edward
213 Pange lingua. Vexilla Regis. Songs: As Moses lifted up the serpent; Let my prayer come up; Our Father in the heavens. JACKSON: Allelui, laudate pueri; Evening hymn; Impromptu. NARES: Introduction and fugue, F major. The souls of the righteous. BBC Northern Singers; Francis Jackson, org; Stephen Wilkinson. Abbey LPB 737.
+Gr 10-77 p685 +MT 12-77 p1017
+HFN 11-77 p177 +-RR 10-77 p84
Songs: As Moses lifted up the serpent; Let my prayer come up; Our Father in the heavens. cf Pange lingua.
Vexilla Regis. cf Pange lingua.
BAKER, David
214 Le chat qui peche (The fishing cat). GOULD: Symphonette, no. 2. Linda Anderson, s; Jamey Aebersold, alto and tenor sax; Dan Haerle, pno; John Clayton, bs; Charlie Craig, drums; Louisville Orchestra; Jorge Mester. Louisville LS 751.
+-HF 3-77 p95 +-St 7-76 p106
+MJ 7-76 p57
BAKER, George
Far-West toccata. cf Delos FY 025.
BAKFARK, Balint
Fantasias (4). cf DG 2533 294.
Fantasia (after "D'amour me plains" by Roger). cf Hungaroton SLPX 11721.
215 Transcriptions: Aspice Domine (Gombert); Benedicta es (Pieton); Fantasies (anon.); Hierusalem Luge (Richafort): Martin Menoit (Jannequin); Ultimi mei sospiri (Verdelot). Daniel Benko, lt. Hungaroton SLPX 11817.
++ARG 6-77 p10 +RR 5-77 p67
+NR 8-77 p15
BALADA, Leonardo
Analogias. cf DG 2530 802.
BALAKIREV, Mily
216 Concerto, piano, no. 1, op. 1, F sharp minor. MEDTNER: Concerto, piano, no. 1, op. 33, C minor. Igor Zhukov, pno; MSRO; Alexander Dmitriev. HMV ASD 3339.
+-Gr 5-77 p1689 +MT 8-77 p650
+-HFN 4-77 p133 +-RR 5-77 p46
+MM 10-77 p45
217 Concerto, piano, no. 2, E flat major. LIAPUNOV: Rhapsody on Ukrainian themes, op. 28. Michael Ponti, pno; Westpahlian Symphony Orchestra; Siegfried Landau. Turnabout (Q) QTVS 34645 Tape (c) KTVC 34645.
+Gr 8-77 p291 +-MT 12-77 p1014
+-Gr 10-77 p698 tape +-NR 6-77 p5
+-HFN 8-77 p76 +-RR 8-77 p43
+-HFN 10-77 p169 tape +RR 12-77 p91 tape
+MJ 5-77 p32 +St 6-77 p126
218 King Lear: Overture. BORODIN: Symphony, no. 3, A minor. MUSSORGSKY: A night on the bare mountain. RIMSKY-KORSAKOV: The legend of Sadko, op. 5. LPO; David Lloyd-Jones. Philips 5680 053

Tape (c) 7317 032.
++Gr 6-72 p62 +HFN 8-77 p99 tape
+HFN 6-72 p1103 +RR 10-77 p98 tape

219 Russia. BORODIN: Symphony, no. 2, B minor. RIMSKY-KORSAKOV: Skazka, op. 29. Bournemouth Symphony Orchestra; Anshel Brusilow. HMV SQ ASD 3193 Tape (c) TC ASD 3193.
+Gr 7-76 p167 +-RR 7-76 p48
+-Gr 11-76 p887 tape +RR 1-77 p88 tape
+HFN 7-76 p83 +STL 7-4-76 p36
+HFN 11-76 p175 tape

220 Symphony, no. 1, C major. RACHMANINOFF: Caprice bohemien, op. 12. USSR Symphony Orchestra; Yevgeny Svetlanov. HMV ASD 3315.
+Gr 3-77 p1386 +MM 10-77 p45
+-HFN 3-77 p97 +RR 3-77 p41

221 Symphony, no. 1, C major. USSR Symphony Orchestra; Yevgeny Svetlanov. Melodiya/Angel SR 40272.
+NR 4-77 p3 +-St 5-77 p102

222 Thirty Russian national pieces, piano, 4 hands. Viktoria Postnikova, Gennady Rozhdestvensky, pno. Melodiya 33S10-07565/6.
+-ARG 11-77 p36

BALAZS
Songs: Salute to tomorrow; Toast. cf Hungaroton SLPX 11823.

BALFE, Michael
223 Satanella. Opera Integra, Addison Orchestra. Rare Recorded Editions SSRE 173/4 (2).
+-HFN 12-77 p164

BALL, Eric
Morning rhapsody. cf DECCA SB 328.
224 Sinfonietta, brass band. ELGAR: Severn suite, op. 87. FLETCHER: Epic symphony. RUBBRA: Variations on "The shining river", op. 101. Black Dyke Mills Band; Roy Newsome, Peter Parkes. RCA RL 2-5078 Tape (c) RK 2-5078.
+Gr 6-77 p64 ++RR 6-77 p48
+Gr 9-77 p508 tape ++RR 10-77 p96 tape
+-HFN 8-77 p76 +ST 8-77 p343

BALLARD, Robert
Allemande. cf BIS LP 22.
Branles de village. cf ATTAINGNANT: Basse dance.
Courante. cf ATTAINGNANT: Basse dance.
Courante. cf BIS LP 22.
Entree de Luth, nos. 1-3. cf ATTAINGNANT: Basse dance.
Prelude. cf BIS LP 22.
Recontins. cf BIS LP 22.

BALLIF, Claude
225 Sonates pour orgue, op. 14. Louis Robbilliard, org. Arion ARN 236003 (2).
+MJ 1-77 p27 +RR 6-74 p63

BANKS, Don
226 Concerto, horn. MUSGRAVE: Concerto, clarinet. SEARLE: Aubade, op. 28. Barry Tuckwell, hn; Gervase de Peyer, clt; NPhO, LSO; Norman Del Mar. Argo ZRG 726.
++Gr 8-75 p315 ++NR 2-76 p4
+HF 6-76 p90 ++RR 9-75 p40
+-HFN 11-75 p149 +SR 1-24-76 p53
+MQ 10-77 p572 +-Te 3-76 p31
++MT 3-76 p235

BARATI, George

Triple exposure, solo violoncello. cf Gasparo GS 103.

BARBER, Samuel

Adagio, strings. cf Angel S 37409.

Adagio, strings. cf HMV (SQ) ASD 3338.

Adagio, strings. cf Seraphim SIB 6094.

227 Concerto, piano, op. 38. COPLAND: Concerto, piano. Abbott Ruskin, pno; MIT Symphony Orchestra; David Epstein. Turnabout QTVS 34683.

+NYT 11-27-77 pD15

228 Essays, orchestra, no. 1, op. 12. Essays, orchestra, no. 2, op. 17. Night flight, op. 19a. Symphony, no. 1, op. 9. LSO; David Measham. Unicorn RHS 342.

++Gr 9-76 p409 +HFN 9-76 p117
-HF 3-77 p95 +RR 8-76 p27

Essays, orchestra, no. 2, op. 17. cf Essays, orchestra, no. 1, op. 12.

229 Excursions. MACDOWELL: Sonata, piano, no. 4, op. 59. WALKER: Sonata, piano, no. 3. Leon Bates, pno. Orion ORS 76237.

+HF 3-77 p115 +St 8-77 p116
++NR 8-77 p14

230 Hermit songs, op. 29. COPLAND: Poems of Emily Dickinson (12). Sandra Browne, ms; Michael Isador, pno. Enigma VAR 1029.

+-Gr 10-77 p677 +RR 7-77 p85
+HFN 10-77 p149

Die Natali, op. 37: Chorale prelude on "Silent night". cf Vista VPS 1038.

Night flight, op. 19a. cf Essays, orchestra, no. 1, op. 12.

Nocturne, op. 33. cf Vox SVBX 5303.

231 Quartet, strings, op. 11, B minor. IVES: Quartet, strings, no. 2. Scherzo, string quartet. Cleveland Quartet. RCA ARL 1-1599 (Q) ARD 1-1599.

++HF 10-76 p110 ++SFC 7-3-77 p34
-MJ 11-76 p44 ++St 11-76 p136
+NR 9-76 p6

Quartet, strings, op. 11, B minor: Adagio. cf Argo ZRG 845.

Sonata, piano, op. 26: Fuga. cf Vox SVBX 5303.

232 Sonata, violoncello and piano, op. 6, C minor. DIAMOND: Sonata, violoncello and piano. Harry Clark, vlc; Sandra Schuldmann, pno. Musical Heritage Society MHS 3378.

+-HF 3-77 p95

Songs: Sure on this shining night. cf New World Records NW 243.

A stopwatch and an ordinance map. cf Vox SVBX 5353.

Summer music, woodwind quintet, op. 31. cf Vox SVBX 5307.

333 Symphony, no. 1, op. 9, in one movement. MAYER: Octagon, piano and orchestra. William Masselos, pno; Milwaukee Symphony Orchestra; Kenneth Schermerhorn. Turnabout (Q) QTVS 34564.

+-HF 3-77 p95 +SFC 6-22-75 p26
++MJ 1-75 p49 ++St 5-75 p96
++NR 1-75 p5

Symphony, no. 1, op. 9. cf Essays, orchestra, no. 1, op. 12.

234 Vanessa. Eleanor Stever, s; Rosalind Elias, Regina Resnik, ms; Nicolai Gedda, t; Giorgio Tozzi, bs; George Cehanovsky, bar; Metropolitan Opera Orchestra and Chorus; Dimitri Mitropoulos. RCA ARL 2-2094. (Reissue)

++MJ 7-77 p68 +SFC 7-3-77 p34
+NR 5-77 p11

Variations on a Shapenote hymn, op. 34. cf Orion ORS 76255.
Wondrous love: Variations on a Shapenote hymn, op. 34. cf Vista VPS 1038.
BARBERIIS (16th century)
Madonna, qual certezza. cf L'Oiseau-Lyre 12BB 203/6.
BARBERIS
Munasterio e Santo Chiara. cf London SR 33221.
BARBETTA, Giulio
Moresca detta le Canarie. cf DG 2533 173.
BARBIERI, Francisco
Songs: El barberillo de lavapies; Cancion de paloma; Jugar con fuego; Romanza de la Duqesa. cf London OS 26435.
BARBIREAU, Jacques
Songs: En frolyk weson. cf Argo ZK 24.
Songs: Ein frohlich wesen. cf HMV SLS 5049.
Songs: En frolyk weson. cf Saga 5444.
BARDOS, Lajos
Songs: Shepherd's pipe tune; Hymn to the sun. cf Hungaroton SLPX 11823.
BARIE
Toccata, op. 7. cf Calliope CAL 1922/4.
BARNARD, Charlotte
Come back to Erin. cf HMV RLS 719.
BAROQUE MASTERPIECES, TRUMPET AND ORGAN. cf Nonesuch H 71279.
BARRAUD, Henri
235 Symphonic concertante, trumpet and orchestra. MILHAUD: Saudades do Brasil. ORTF; Roger Delmotte, tpt; Manuel Rosenthal. Barclay Inedits 995 034.
+Gr 1-74 p1371 +St 3-77 p138
-RR 12-73 p70
BARRET
March of the cobblers. cf EMI TWOX 1058.
BARRIOS, Agustin
236 Las abejas. La catedral. Danza paraguaya. LAURO: Venezuelan waltzes (4). PONCE: Folie d'Espagne: Theme and variations with fugue. SOJO: Pieces from Venezuela (5). Diego Blanco, gtr. BIS LP 33.
+Gr 3-77 p1454 -RR 1-77 p73
Aconquija. cf Guitar works (CBS 76662).
Aire de zambal. cf Guitar works (CBS 76662).
La catedral. cf Las abejas.
La catedral. cf Guitar works (CBS 76662).
Choro de saudade. cf Guitar works (CBS 76662).
Cueca. cf Guitar works (CBS 76662).
Danza paraguaya. cf Las abejas.
Danza paraguaya. cf L'Oiseau-Lyre SOL 349.
Danza paraguaya. cf Saga 5412.
Estudio. cf Guitar works (CBS 76662).
237 Guitar works: Aconquija. Aire de zambal. La catedral. Cueca. Estudio. Choro de saudade. Una limosna el amor de Dios. Mazurka apasionata. Madrigal. Minuet. Preludio. Maxixa. Sueno en la foresta. Valse, no. 3. Villancico de Navidad. John Williams, gtr. CBS 76662.
+Gr 12-77 p1106 +RR 12-77 p61
++HFN 12-77 p164
Una limosna el amor de Dios. cf Guitar works (CBS 76662).
Madrigal. cf Guitar works (CBS 76662).

Maxixa. cf Guitar works (CBS 76662).
Mazurka apasionata. cf Guitar works (CBS 76662).
Minuet. cf Guitar works (CBS 76662).
Preludio. cf Guitar works (CBS 76662).
Sueno en la foresta. cf Guitar works (CBS 76662).
Valse, no. 3. cf Guitar works (CBS 76662).
Villancico de Navidad. cf Guitar works (CBS 76662).

BARSANTI, Francesco
Sonata, recorder, C minor. cf BIS LP 48.

BARTA, Lubor
238 Quintet, winds, no. 2. FELD: Quintet, winds, no. 2. FLOSMAN: Sonata, wind quintet and piano. KALABIS: Little chamber music, wind quintet, op. 27. Prague Wind Quintet. Supraphon 111 1426.
+Gr 1-77 p1159 +NR 7-76 p5
+HFN 10-77 p165 +RR 8-76 p62

BARTLETT, Homer
Grande polka de concert. cf New World Records NW 257.

BARTLETT, John
Of all the birds that I do know. cf BG HM 56/8.
What thing is love. cf Enigma VAR 1023.

BARTOK, Bela
239 Bluebeard's castle, op. 11. Tatiana Troyanos, ms; Siegmund Nimsgern, bs; BBC Symphony Orchestra; Pierre Boulez. CBS 76518 Tape (c) 40-76518. (also Columbia M 34217)
++Gr 9-76 p466 ++NR 2-77 p11
+HF 2-77 p96 +-ON 2-12-77 p41
++HFN 9-76 p114 ++RR 9-76 p26
++HFN 11-76 p175 tape ++RR 1-77 p87 tape
+MJ 4-77 p50 ++St 2-77 p81
++MT 3-77 p214 +ST 1-77 p775

240 Concerto, orchestra. DELLA JOIO: Meditations on Ecclesiastes. HAYDN: Symphony, no. 96, D major. WAGNER: Die Meistersinger von Nurnberg: Prelude. Boston University Symphony Orchestra; Joseph Silverstein. Boston University BU 101.
++IN 12-77 p26

241 Concerto, orchestra. Dance suite. LSO; Georg Solti. Decca Tape (c) KSXC 6212. (also London 6784)
++HFN 2-77 p135 tape ++RR 8-74 p82 tape

242 Concerto, orchestra. Dance suite. Hungarian Radio and Television Orchestra; Gyorgy Lehel. DG Tape (c) 3335 202.
++Gr 1-77 p1183 +RR 1-77 p87 tape

243 Concerto, orchestra. Hungarian sketches. Israel Philharmonic Orchestra; Zubin Mehta. London CS 6949 Tape (c) 5-6949. (also Decca SXL 6730 Tape (c) KSXC 6730)
++ARG 12-76 p13 +HFN 1-77 p123 tape
+Gr 12-76 p989 +MJ 1-77 p26
+Gr 3-77 p1458 tape /NR 12-76 p3
/HF 12-76 p98 +RR 12-76 p51
+HFN 12-76 p136 +-RR 1-77 p87 tape

244 Concerto, orchestra. Dance suite. COA; Bernard Haitink. Philips Tape (c) 7317 116.
+HFN 2-77 p135 tape

245 Concerto, orchestra. CPhO; Karel Ancerl. Supraphon Tape (c) 045 0515.
+HFN 2-77 p135 tape ++RR 8-74 p82 tape
+-HFN 2-77 p135 tape

246 Concerti, piano, nos. 1-3. Rhapsody, op. 1. Pascal Roge, pno;

LSO; Walter Weller. Decca SXL 6815/6 Tape (c) KSXC 6815/6.
+-Gr 2-77 p1270 +-HFN 5-77 p138 tape
+-Gr 4-77 p1541 ++MT 7-77 p564
+Gr 5-77 p1738 tape +-RR 2-77 p40
+HFN 2-77 p112 +-RR 3-77 p41
+-HFN 4-77 p133 +RR 4-77 p91 tape

247 Concerto, piano, no. 1, A major. Rhapsody, op. 1. Pascal Roge, pno; LSO; Walter Weller. Decca Tape (c) KSXC 6815.
+HFN 4-77 p155 tape

248 Concerto, piano, no. 1, A major. Concerto, piano, no. 3, E major. Stephen Bishop-Kovacevich, pno; LSO; Colin Davis. Philips 9500 043.
+Gr 7-76 p167 +-HFN 7-76 p84
+Gr 3-77 p1458 tape +NR 10-76 p6
+-HF 1-77 p110 ++RR 7-76 p47

Concerto, piano, no. 3, E major. cf Concerto, piano, no. 1, A major.

249 Contrasts. Bela Bartok, pno; Joseph Szigeti, vln; Benny Goodman, clt. Odyssey 32160220E.
+St 5-77 p72

Dance suite. cf Concerto, orchestra (Decca Tape (c) KSXC 6212).
Dance suite. cf Concerto, orchestra (DG 3335 202).
Dance suite. cf Concerto, orchestra (Philips 7317 116).

250 Etudes, op. 18 (3). BUSONI: Short pieces for the cultivation of polyphonic playing (6). MESSIAEN: Etudes de rythme (4). STRAVINSKY: Etudes, op. 7 (4). Paul Jacobs, pno. Nonesuch H 71334.
++ARG 4-77 p29 ++NR 8-77 p14
++Gr 3-77 p1430 ++RR 1-77 p68
++HF 5-77 p99 ++SFC 6-12-77 p41
+HFN 2-77 p112 ++St 6-77 p136

251 Hungarian folk songs (trans. Szigeti). HINDEMITH: Sonata, violin and piano, E major. STRAVINSKY: Pulcinella: Suite, excerpts. YSAYE: Mazurka, op. 11, no. 3, B minor. Sonata, violin and piano, no. 3, op. 27. Igor Oistrakh, vln; Natalia Zertsalova, pno. Westminster WGS 8326.
++NR 4-77 p5

Hungarian sketches. cf Concerto, orchestra.

252 The miraculous Mandarin, op. 19: Ballet suite. Music, strings. percussion and celesta. BSO; Seiji Ozawa. DG 2530 887.
+-Gr 12-77 p1067 ++SFC 12-18-77 p42
++RR 12-77 p43

253 The miraculous Mandarin, op. 19: Suite. The wooden prince, op. 13: Suite. Minnesota Orchestra; Stanislaw Skrowaczewski. Candidi QCE 31097.
+ARG 12-77 p20 ++NR 10-77 p4
+HF 11-77 p104 ++SFC 8-14-77 p50

Music, strings, percussion and celesta. cf The miraculous Mandarin, op. 19: Ballet suite.

254 Quartets, strings, nos. 1-6. Vegh Quartet. Telefunken 6-35023.
+-NYT 10-23-77 pD15

255 Quartets, strings, nos. 1-6. New Hungarian Quartet. Vox SVBX 593 (3).
+ARG 4-77 p12 ++NR 4-77 p6
+HF 10-77 p101

Rhapsody, op. 1. cf Concerti, piano, nos. 1-3.
Rhapsody, op. 1. cf Concerto, piano, no. 1, A major.

Rumanian folk dances (6). cf Kelsey Records DEL 7601.

256 Sonata, 2 pianos and percussion. BRAHMS: Variations on a theme by Haydn, op. 56a. Jeffry and Ronald Marlow, pno; Kivnick, Woodhull, perc. Devon DV 751.
+-NR 9-77 p15

257 Sonata, solo violin (ed. Menuhin). BLOCH: Suite, nos. 1 and 2. Yehudi Menuhin, vln. HMV ASD 3368.
+-Gr 9-77 p456 +RR 9-77 p80
+HFN 8-77 p76

258 Sonatas, violin and piano, nos. 1 and 2. Gidon Kremer, vln; Yuri Smirnov, pno. Hungaroton SLPX 11655.
+Gr 1-75 p1361 +SFC 10-16-77 p47
+NR 5-75 p9 +-St 7-75 p94

259 Songs: Hungarian folksongs (1907-17). Songs, op. 15 (5). Songs, op. 16 (5). Village scenes (5). Elizabeth Suderburg, s; Bela Siki, pno. Turnabout TV 34592.
++NR 11-75 p10 +St 5-77 p102

260 Songs, op. 15 (5). Songs, op. 16 (5). Eszter Kovacs, s; Adam Fellegi, pno. Hungaroton SLPX 11603.
+-Gr 8-76 p324 +-RR 6-76 p77
+-HF 8-76 p79 ++SFC 8-14-77 p50
+NR 6-76 p14

261 Songs, op. 16 (5). KILPINEN: Songs of death, op. 62. Songs of love, opp. 60 and 61. Rolf Leanderson, bar; Helene Leanderson, pno. BIS LP 43.
+HFN 11-77 p165 +RR 11-77 p104

Songs: Letter to the home folks. cf Hungaroton SLPX 11823.

262 Suite, op. 14. PROKOFIEV: Sonata, piano, no. 2, op. 14, D minor. SAEVERUD: Ballade of revolt, op. 22, no. 5. Rondo amoroso, op. 14, no. 7. VILLA-LOBOS: Festa no sertao (Country festival). Impressoes seresteiras (Impressions of a serenade). Inger Wikstrom, pno. Swedish Society SLT 33196.
+-Gr 1-77 p1162 +-RR 1-77 p68

263 The wooden prince, op. 13. NYP; Pierre Boulez. Columbia M 34514 Tape (c) MT 34514. (also CBS SQ 76625)
++ARG 7-77 p16 +NR 6-77 p3
++Gr 10-77 p618 ++RR 10-77 p37
+HF 11-77 p104 +SFC 8-14-77 p50
+HFN 10-77 p149 ++St 10-77 p86

The wooden prince, op. 13: Suite. cf The miraculous Mandarin, op. 19: Suite.

BARTOLOZZI

Collage. cf CBS 61453.

BARTON, Andrew

264 The disappointment, or The force of credulity. Ruth Denison, s; Elain Bonazzi, ms; Arden Hopkin, Milford Fargo, Tonio di Paolo, John Maloy, t; William Sharp, Joseph Bias, Richard Hudson, Richard Reif, bar; Eastman Philharmonia Chamber Ensemble; Samuel Adler. Turnabout TVS 34650.
+HF 6-77 p80 +NYT 7-3-77 pD11
+NR 3-77 p10 +-St 5-77 p103

BASART, Robert

265 Fantasy. BERRY: Trio, piano. GOOSSEN: Clausulae. Temple music. Alfio Pignotti, vln; Margaret Moores, vlc; Dady Mehta, pno; Janet Ketchum, flt; Nathan Schwartz, pno; Steinerius Duo. CRI SD 371.
*MJ 10-77 p29 +NR 9-77 p6

BASHFORD, Rodney
A Purcell suite: Rondo. cf Decca SB 715.
A Windsor flourish. cf Decca SB 715.
BASHMAKOV, Leonid
Fantasia per flauti. cf BIS LP 50.
BASSA, Jose
Minuet. cf London CS 7046.
BASSANO, Jerome
Dic nobis Maria. cf Argo ZRG 859.
BASSETT, Leslie
Music for saxophone and piano. cf BABBITT: Phenomena, soprano and piano.
266 Suite, solo trombone. MILHAUD: Four seasons: Concertino d'hiver. PERSICHETTI: Serenade, no. 6, op. 44. PERGOLESI: Sinfonia, F major (trans. Sauer). Ralph Sauer, trom; Alan deVeritch, vla; Ronald Leonard, vlc; Paul Pitman, pno. Crystal S 381.
++NR 9-77 p6
BATCHELAR, Daniel
Mounsiers almaine. cf CBS Tape (c) 40-72728.
Pavan and galliard. cf L'Oiseau-Lyre DSLO 510.
BATTEN, Adrian
O sing joyfully. cf Guild GRS 7008.
BATTISHILL, Jonathan
O Lord, look down from heaven. cf Guild GRS 7008.
BAX, Arnold
Burlesque. cf Saga 5445.
267 Coronation march (arr. O'Brien). BLISS: Processional (arr. O'Brien). ELGAR: Imperial march, op. 32. Pomp and circumstance march, op. 39, no. 1, D major (arr. Farrell). WALTON: Orb and sceptre (arr. McKie). Crown imperial (arr. Murrill). Christopher Herrick, Timothy Farrell, org. Vista VPS 1055.
+-Gr 10-77 p712 +-RR 8-77 p44
Coronation march. cf CRD Britannia BR 1077.
Fanfare for the wedding of Princess Elizabeth, 1948. cf Decca SPA 500.
The happy forest. cf Symphony, no. 3.
268 Symphony, no. 3. The happy forest. LSO; Edward Downes. RCA GL 4-2247. (Reissue from SB 6806)
+Gr 7-77 p167 ++RR 8-77 p44
+HFN 9-77 p151
269 Symphony, no. 7, A flat major. LPO; Raymond Leppard. Lyrita SRCS 83. (also HNH 4010)
+Gr 11-75 p801 ++RR 9-75 p28
+HF 9-77 p108 +SFC 1-24-77 p40
+HFN 11-75 p149 ++St 10-77 p136
++NR 9-77 p1
What is it like to be young and fair. cf RCA GL 2-5062.
BAYCO
Royal Windsor march. cf Lismor LILP 5078.
BEACH, Amy
270 Concerto, pianoforte, op. 45. MASON: Prelude and fugue, op. 20. Mary Louise Boehm, pno; Westphalian Symphony Orchestra; Siegfried Landau. Turnabout QTVS 34665.
+ARG 10-77 p17 +-NR 8-77 p6
+HF 12-77 p77 +NYT 7-3-77 pD11
+MJ 5-77 p32 +SFC 8-28-77 p46

271 Sonata, violin and piano, op. 34, A minor. FOOTE: Sonata, violin and piano, op. 20, G minor. Joseph Silverstein, vln; Gilbert Kalish, pno. New World Records NW 268.
+ARG 10-77 p28 +St 12-77 p132
+NYT 7-3-77 pD11
The year's at the spring, op. 44, no. 1. cf Decca SDD 507.
BEACH, Perry
272 Then said Isaiah. KANTOR: Playthings of the wind. STRAVINSKY: Anthem; Ave Maria; Pater noster. Los Angèles Camerata. H. Vincent Mitzelfelt. Crystal S 890.
+NR 10-77 p9
BEATLES
Eleanor Rigby; Penny Lane. cf Golden Crest CRS 4148.
BEAUVARLET-CHARPENTIER (18th century France)
Fugue, G minor. cf L'Oiseau-Lyre SOL 343.
BEDFORD, David
A garland for Dr. K. cf Universal Edition UE 15043.
273 Instructions for angels. David Bedford, keyboards, synthesisers, church organ, perc; Mike Oldfield, gtr; Mike Ratledge, synthesiser; Mick Baines, Jeff Bryant, Robin Davies, hn; Jim Douglas, ob; Diana Coulson, Sophie Dickson, vocals; Jennifer Angel, Julia Foguel, Celia Hutchison, Mary Rose Lillingston, Fiona Lofts, flt; Leicestershire Schools Symphony Orchestra; Leicestershire Schools Recorder Group; Eric Pinkett, Jane Ward. Virgin V 2090.
+-RR 11-77 p46
274 The odyssey. David Bedford, Mike Oldfield; Queen's College Choir and others. Virgin V 2070.
+Gr 2-77 p1270 +-Te 3-77 p27
275 The rime of the ancient mariner. David Bedford, keyboards, recorders, percussion; Mike Oldfield, gtr; Robert Powell, narrator; Queen's College Girls Choir, London. Virgin V 2038 Tape (c) TCV 2038.
-Gr 2-76 p1331 +RR 12-75 p71
+-HFN 11-75 p150 +-Te 3-77 p27
276 Star's end. Mike Oldfield, gtr; Chris Cutler, perc; RPO; Vernon Handley. Virgin V 2020 Tape (c) TCV 2020 (ct) TC 2020.
+-Gr 3-75 p1639 +-Te 3-77 p27
*RR 1-75 p26
BEETHOVEN, Ludwig van
277 Adagio and allegro, music box. DANZI: Quintet, op. 68, no. 3, D minor. HAYDN: Pieces, flute clock (7). MOZART: Andante, K 616, F major. Fantasy, organ, K 608, F minor. Danzi Quintet. ABC AB 67016.
+FF 11-77 p59 ++SFC 9-25-77 p50
278 Ah perfido, op. 65. Egmont, op. 84: Die Trommel geruhret; Freudvoll und Liedvoll. Songs: No non turbati. SCHUBERT: Alfonso und Estrella, D 732: Konnt ich ewig hier verweilen. Lazarus, D 689: So schlummert auf Rosen. Rosamunde, op. 26, D 797: Der Vollmond strahlt. Songs: Zogernd Leise, D 920. Janet Baker, ms; ECO and Choir; Raymond Leppard. Philips 9500 307.
+-Gr 9-77 p469 +RR 9-77 p86
+HFN 9-77 p137
Ah perfido, op. 65. cf Works, selections (DG 2721 138).
Ah perfido, op. 65. cf Philips 6767 001.
279 An die ferne Geliebte, op. 98. BRAHMS: Vier ernste Gesange, op. 121. SCHUMANN: Songs: Die beiden Grenadiere, op. 49, no. 1.

Belsazar, op. 57. Dichterliebe, op. 48: Ich grolle nicht. Liederkreis, op. 24: Schone Wiege meiner Leiden. Die Lotosblume, op. 25, no. 7. Widmung, op. 25, no. 1. Norman Bailey, bar; John Constable, pno. Saga 5450.
++Gr 10-77 p686 +-RR 9-77 p85
+HFN 11-77 p167

An die ferne Geliebte, op. 98. cf Works, selections (DG 2721 138).

280 Andante favori, F major. Sonata, piano, no. 21, op. 53, C major. Sonata, piano, no. 31, op. 110, A flat major. Alfred Brendel, pno. Philips 6500 762 Tape (c) 7300 351.
+Gr 11-75 p859 ++MJ 10-76 p32
+HF 8-76 p80 +NR 6-76 p11
++HF 5-77 p101 tape +-RR 11-75 p74
+-HFN 10-75 p137 ++St 12-76 p134
++HFN 7-75 p90 tape

Andante favori, F major. cf Piano works (DG 2548 266).

Bagatelle, A minor. cf Inner City IC 1006.

Bagatelle, no. 25, A minor. cf Piano works (DG 2548 266).

Bagatelle, no. 25, A minor. cf RCA GL 4-2125.

Batatelles, op. 126, nos. 1, 4, 6. cf Sonata, piano, no. 29, op. 106, B flat major.

Christ on the Mount of Olives, op. 85. cf Works, selections (DG 2721 138).

281 Concerti, piano, nos. 1-5. Vladimir Ashkenazy, pno; CSO; Georg Solti. Decca Tape (c) K44K43.
+Gr 5-77 p1738 tape +RR 4-77 p90 tape
+HFN 5-77 p138 tape

282 Concerti, piano, nos. 1-5. Concerto, violin, op. 61, D major. Concerto, violin, violoncello and piano, op. 56, C major. Romances, nos. 1 and 2, opp. 40, 50. Wilhelm Kempff, pno; Geza Anda, pno; Christian Ferras, Wolfgang Schneiderhan, vln; Pierre Fournier, vlc; BPhO, Berlin Radio Symphony Orchestra, David Orchestra; Ferdinand Leitner, Herbert von Karajan, Ferenc Fricsay, Eugene Goossens. DG 2721 128 (6).
++Gr 3-77 p1415 +RR 4-77 p42

283 Concerti, piano, nos. 1-5. Concerto, violin, op. 61, D major. Concerto, violin, violoncello and piano, op. 56, C major. Romances, nos. 1 and 2, opp. 40, 50. Wilhelm Kempff, Geza Anda, pno; Wolfgang Schneiderhan, David Oistrakh, vln; Pierre Fournier, vlc; BPhO; Ferenc Fricsay. DG 2740 131.
+Gr 3-77 p1415

284 Concerti, piano, nos. 1-5. Vladimir Ashkenazy, pno; CSO; Georg Solti. London CSA 2404 (4) Tape (c) D 10270 (r) W 480270. (also Decca SXLG 6594/7 Tape (c) KSXCG 7019/21)
++Gr 9-73 p477 +HFN 12-73 p2621 tape
+HF 10-73 p98 +RR 9-73 p57
++HF 5-74 p122 tape +-RR 2-74 p71 tape
++HF 1-76 p111 tape ++SFC 8-12-73 p32
++HF 6-77 p103 tape ++St 11-73 p74

285 Concerti, piano, nos. 1-5. Glenn Gould, pno; Columbia Symphony Orchestra, NYP, American Symphony; Vladimir Golschmann, Leonard Bernstein, Leopold Stokowski. Odyssey Y4 34640 (4).
+NR 7-77 p6

286 Concerti, piano, nos. 1-5. Fantasia, op. 80, C minor. Alfred Brendel, pno; LPO and Chorus; Bernard Haitink. Philips 6767 002 (5) Tape (c) 7699 061.

+Gr 11-77 p824 +-RR 10-77 p37
++HFN 10-77 p149 +RR 12-77 p89 tape
+MJ 11-77 p25 +SFC 10-16-77 p47
+NR 12-77 p3

287 Concerto, piano, no. 1, op. 15, C major. Fantasia, op. 77, G minor. MOZART: Fantasia, K 475, C minor. Edwin Fischer, pno; BPhO Members. Bruno Walter Society RR 450.
+NR 5-77 p7

288 Concerto, piano, no. 1, op. 15, C major. Sonata, piano, no. 8, op. 13, C minor. Vladimir Ashkenazy, pno; CSO; Georg Solti. Decca SXL 6651 Tape (c) KSXC 6651. (Reissue from SXLF 6594/7, SXL 6706) (also London CS 6853. Reissue from CSA 2404)
+Gr 2-76 p1332 +NR 7-77 p7
+HFN 2-76 p115 +RR 2-76 p25
++HFN 4-76 p125 tape +-RR 5-76 p77 tape
++MJ 3-77 p68

289 Concerto, piano, no. 1, op. 15, C major. Variations (32). Anika Szegedi, pno; Budapest Philharmonic Orchestra; Andras Korodi. Hungaroton SLPX 11793.
+-NR 7-77 p6 ++SFC 5-23-77 p42

290 Concerto, piano, no. 1, op. 15, C major. WEBER: Euryanthe: Overture. Adrian Aeschbacher, pno; Lucerne Festival Orchestra, RAI Torino Orchestra; Wilhelm Furtwangler. Rococo 2106.
+-NR 2-77 p5

Concerto, piano, no. 1, op. 15, C major. cf EMI Italiana C 153 52425/31.

291 Concerto, piano, no. 2, op. 19, B flat major. Concerto, piano, no. 4, op. 58, G major. John Lill, pno; Scottish National Symphony Orchestra; Alexander Gibson. Classics for Pleasure CFP 40271.
+HFN 12-77 p165

292 Concerto, piano, no. 2, op. 19, B flat major. Sonata, piano, no. 21, op. 53, C major. Vladimir Ashkenazy, pno; CSO; Georg Szell. Decca SXL 6652 Tape (c) KSXC 6652. (Reissues from SXL 6594/7, SXL 6706) (also London CS 6854)
+ARG 3-77 p51 +NR 3-77 p7
+Gr 1-76 p1188 +RR 2-76 p24
+HFN 12-75 p171

293 Concerto, piano, no. 2, op. 19, B flat major. Sonata, piano, no. 1, op. 2, no. 1, F minor. Sonata, piano, no. 20, op. 49, no. 2, G major. Claudio Arrau, pno; Orchestra; Bernard Haitink. Philips Tape (c) 7317 145.
+HFN 10-77 p169 tape

294 Concerto, piano, no. 3, op. 37, C minor. John Lill, pno; Scottish National Symphony Orchestra; Alexander Gibson. Classics for Pleasure CFP 40259.
++Gr 6-77 p32 +-RR 6-77 p43
+HFN 6-77 p117

295 Concerto, piano, no. 3, op. 37, C minor. MOZART: Rondos, piano, K 382, D major; K 386, A major. Annie Fischer, pno; Bavarian State Orchestra; Ferenc Fricsay. DG 2548 238. (Reissue from SLPM 138087)
++Gr 8-76 p278 +RR 3-77 p43
+HFN 8-76 p93

296 Concerto, piano, no. 3, op. 37, C minor. Sonata, piano, no. 26, op. 81, E flat major. Vladimir Ashkenazy, pno; CSO; Georg Solti. London CS 6855. (also Decca SXL 6653)

+MJ 4-77 p33 +RR 1-76 p30
+-NR 12-76 p6

297 Concerto, piano, no. 3, op. 37, C minor. LISZT: Concerto, piano, no. 1, G 124, E flat major. Claudio Arrau, pno; PO; Eugene Ormandy. Odyssey 34601. (Reissues from ML 4302, ML 4665)
++ARG 7-77 p16

298 Concerto, piano, no. 4, op. 58, G major. SCHUBERT: Impromptu, op. 142, no. 3, D 935, B flat major. Clifford Curzon, pno; VPO; Hans Knappertsbusch. Decca ECS 752. (Reissues from LXT 2948, 2781)
+ARG 6-77 p50 -HFN 5-76 p115
+-Gr 8-76 p278 +-RR 8-76 p28

299 Concerto, piano, no. 4, op. 58, G major. Maurizio Pollini, pno; VPO; Karl Bohm. DG 2530 791 Tape (c) 3300 791.
++Gr 3-77 p1391 ++NR 7-77 p7
++Gr 4-77 p1603 tape +NYT 3-13-77 pD22
++HF 9-77 p99 +RR 3-77 p43
+-HFN 3-77 p99 ++SFC 4-17-77 p42
-MJ 7-77 p68 ++St 7-77 p112

300 Concerto, piano, no. 4, op. 58, G major. Leonore overture, no. 3, op. 72. Symphony, no. 5, op. 67, C minor. Claudio Arrau, pno; Bavarian Radio Symphony Orchestra; Leonard Bernstein. DG 2721 153 (2).
+ARG 5-77 p12 ++NR 5-77 p3
+-Gr 3-77 p1391 +RR 3-77 p42
+HFN 3-77 p97 +-St 7-77 p112
+MJ 7-77 p68

301 Concerto, piano, no. 4, op. 58, G major. Sonata, piano, no. 27, op. 90, E minor. Ivan Moravec, pno; Vienna Musikverein Orchestra; Martin Turnovsky. Rediffusion Legend LGD 009 Tape (c) LGD 009.
+-Gr 8-77 p291 +-HFN 8-77 p93
+-HF 2-77 p95 tape +RR 11-77 p46

Concerto, piano, no. 4, op. 58, G major. cf Concerto, piano, no. 2, op. 19, B flat major.

Concerto, piano, no. 4, op. 58, G major. cf EMI Italiana C 153 52425/31.

302 Concerto, piano, no. 5, op. 73, E flat major. John Lill, pno; Scottish National Orchestra; Alexander Gibson. Classics for Pleasure CFP 40087 Tape (c) TC CFP 40087.
+Gr 10-74 p681 +RR 11-77 p118 tape
+RR 10-74 p34

303 Concerto, piano, no. 5, op. 73, E flat major. Friedrich Gulda, pno; VPO; Horst Stein. Decca JB 18. (Reissue from SDDE 304/7)
+Gr 12-77 p1068

304 Concerto, piano, no. 5, op. 73, E flat major. Emil Gilels, pno; CO; Georg Szell. HMV SXLP 30223 Tape (c) TC SXLP 30223. (Reissue from World Records SM 156/60)
+Gr 11-76 p775 +HFN 12-76 p155
+Gr 11-76 p887 tape +-RR 2-77 p40
+-HFN 11-76 p171

305 Concerto, piano, no. 5, op. 73, E flat major. Rudolf Serkin, pno; NYP; Bruno Walter. Odyssey Y 34607.
+St 11-77 p134

306 Concerto, piano, no. 5, op. 73, E flat major. Alfred Brendel, pno; LPO; Bernard Haitink. Philips 9500 243 Tape (c) 7300 542.
+-MJ 7-77 p68 +St 11-77 p134
+-NR 7-77 p6

307 Concerto, piano, no. 5, op. 73, E flat major. Rudolf Firkusny, pny; RPO; Rudolf Kempe. RCA GL 2-5014 Tape (c) GK 2-5014. (Previously issued by Reader's Digest)
+-Gr 11-76 p775 +-HFN 12-76 p153 tape
+Gr 2-77 p1325 tape +-RR 11-76 p50
+-HFN 12-76 p137 +-RR 12-76 p104 tape
Concerto, piano, no. 5, op. 73, E flat major. cf Works, selections (Decca D77D5).
Concerto, piano, no. 5, op. 73, E flat major. cf Decca D62D4.
Concerto, piano, no. 5, op. 73, E flat major. cf EMI Italiana C 153 52425/31.

308 Concerto, violin, op. 61, D major. BRAHMS: Concerto, violin, op. 77, D major. BRUCH: Concerto, violin, no. 1, op. 26, G minor. MENDELSSOHN: Concerto, violin, op. 64, E minor. Zino Francescatti, Isaac Stern, vln; Columbia Symphony Orchestra, NYP, PO; Bruno Walter, Georg Szell, Thomas Schippers, Eugene Ormandy. CBS 78309 (3).
+-Gr 12-77 p1097 +RR 11-77 p47
+-HFN 11-77 p183

309 Concerto, violin, op. 61, D major. Leonid Kogan, vln; OSCCP; Constantin Silvestri. Connoisseur Society CS 2132.
+MJ 7-77 p68 +NR 6-77 p5

310 Concerto, violin, op. 61, D major. Pinchas Zukerman, vln; CSO; Daniel Barenboim. DG 2530 903 Tape (c) 3300 903.
+Gr 11-77 p824 +HFN 12-77 p165
+-Gr 12-77 p1142 tape ++RR 12-77 p44

311 Concerto, violin, op. 61, D major. Wolfgang Schneiderhan, vln; BPhO; Wilhelm Furtwangler. DG 2535 809. (Reissue from KL 27/31)
+Gr 5-77 p1690 +RR 6-77 p44
+HFN 6-77 p137

312 Concerto, violin, op. 61, D major. Igor Oistrakh, vln; VSO; David Oistrakh. RCA GL 2-5005 Tape (c) GK 2-5005.
++Gr 10-76 p578 +-RR 10-76 p40
+-HFN 12-76 p137 +-RR 12-76 p104 tape
+-HFN 12-76 p153 tape +ST 1-77 p777

313 Concerto, violin, op. 61, D major. Josef Suk, vln; NPhO; Adrian Boult. Vanguard SRV 353SD.
+ARG 11-76 p13 ++NR 11-76 p4
+HF 10-76 p94 +-SFC 9-12-76 p31
+MJ 1-77 p27 ++St 1-77 p112
Concerto, violin, op. 61, D major. cf Concerti, piano, nos. 1-5 (DG 2721 128).
Concerto, violin, op. 61, C major. cf Concerti, piano, nos. 1-5 (DG 2740 131).

314 Concerto, violin, violoncello and piano, op. 56, C major. Franzjosef Maier, vln; Anner Bylsma, vlc; Paul Badura-Skoda, pno; Collegium Aureum. BASF BAC 3097. (also BASF Harmonia Mundi 20-22063-3)
+ARG 6-77 p51 +MT 10-75 p885
+-Gr 7-75 p174 +-RR 7-75 p22
+-HFN 6-75 p85
Concerto, violin, violoncello and piano, op. 56, C major. cf Concerti, piano, nos. 1-5 (DG 2721 128).
Concerto, violin, violoncello and piano, op. 56, C major. cf Concerti, piano, nos. 1-5 (DG 2740 131).
Consecration of the house, op. 124: Overture. cf Overtures (DG 2721 137).

Consecration of the house, op. 124: Overture. cf Overtures (Philips 6780 031).
Consecration of the house, op. 124: Overture. cf Symphony, no. 2, op. 36, D major.
Consecration of the house, op. 124: Overture. cf Works, selections (HMV SXDW 3032).
Contradances (Country dances), G 141. cf DG 2533 182.
Coriolan overture, op. 62. cf Overtures (DG 2721 137).
Coriolan overture, op. 62. cf Overtures (Philips 6780 031).
Coriolan overture, op. 62. cf Symphony, no. 1, op. 21, C major.
Coriolan overture, op. 62. cf Symphony, no. 3, op. 55, E flat major.
Coriolan overture, op. 62. cf Symphony, no. 6, op. 68, F major.
Coriolan overture, op. 62. cf Symphony, no. 7, op. 92, A major.
Coriolan overture, op. 62. cf Works, selections (HMV SXDW 3032).
Coriolan overture, op. 62. cf Works, selections (Vanguard 359/62).
Coriolan overture, op. 62. cf Bruno Walter Society RR 443.
Ecossaises (arr. Busoni). cf International Piano Library IPL 104.
Ecossaises. cf International Piano Library IPL 117.

315 Egmont, op. 84: Overture. MOZART: Symphony, no. 40, K 550, G minor. STRAUSS, R.: Don Juan, op. 20. BPhO; Bruno Walter. Bruno Walter Society BWS 726.
+NR 1-77 p4

316 Egmont, op. 84: Overture. HOLST: The planets, op. 32: Jupiter.* LISZT: Les preludes, no. 3, G 97.* RIMSKY-KORSAKOV: Scheherazade, op. 35: Festival of Bagdad; The sea; The shipwreck.* LPO; Bernard Haitink. Philips 6833 227. (*Reissues from 6500 072, SAL 3750, 6500 410)
+-Gr 10-77 p711

Egmont, op. 84: Overture. cf Overtures (Philips 6780 031).
Egmont, op. 84: Overture. cf Symphonies, nos. 1-9 (Seraphim 6093).
Egmont, op. 84: Overture. cf Symphonies, nos. 1-3, 5-8.
Egmont, op. 84: Overture. cf Symphony, no. 2, op. 36, D major.
Egmont, op. 84: Overture. cf Symphony, no. 5, op. 67, C minor (DG 2535 810).
Egmont, op. 84: Overture. cf Symphony, no. 5, op. 67, C minor (Hungaroton LPX 11457).
Egmont, op. 84: Overture. cf Symphony, no. 6, op. 68, F major.
Egmont, op. 84: Overture. cf Symphony, no. 7, op. 92, A major (Decca 4342).
Egmont, op. 84: Overture. cf Works, selections (Decca D77D5).
Egmont, op. 84: Overture. cf Works, selections (HMV SXDW 3032).
Egmont, op. 84: Overture. cf Works, selections (Vanguard 359/62).
Egmont, op. 84: Overture and incidental music. cf Overtures (DG 2721 137).
Egmont, op. 84: Die Trommel geruhret; Freudvoll und Leidvoll. cf Ah perfido, op. 65.
Egmont, op. 84: Die Trommel geruhret; Freudvoll und Leidvoll. cf Philips 6767 001.
Fantasia, op. 77, G minor. cf Concerto, piano, no. 1, op. 15, C major.
Fantasia, op. 80, C minor. cf Concerti, piano, nos. 1-5.
Fantasia, op. 80, C minor. cf Works, selections (DG 2721 138).

317 Fidelio. Gwyneth Jones, Edith Mathis, s; James King, Peter Schreier, Eberhard Buchner, t; Gunter Leib, bar; Theo Adam, bs-bar; Martti Talvela, Franz Crass, bs; Leipzig Radio Chorus; Dresden State Opera Orchestra and Chorus; Karl Bohm. DG 2721 136 (3).

+Gr 3-77 p1416 +RR 4-77 p34

Fidelio, op. 72: Abscheulicher wo wilst du hin. cf BASF 22-22645-3.
Fidelio, op. 72: Gott, welch Dunkel hier. cf Decca ECA 812.
Fidelio, op. 72: O welche Lust. cf Decca SXL 6826.
Fidelio, op. 72: Overture. cf Overtures (DG 2721 137).
Fidelio, op. 72: Overture. cf Overtures (Philips 6780 031).
Fidelio, op. 72: Overture. cf Symphonies, nos. 1-9 (Philips 6767 003).
Fidelio, op. 72: Overture. cf Symphony, no. 5, op. 67, C minor (DG 2548 255).
Fidelio, op. 72: Overture. cf Symphony, no. 5, op. 67, C minor (Philips 6580 145).
Fidelio, op. 72: Overture. cf Symphony, no. 7, op. 92, A major.
Fidelio, op. 72: Overture. cf Works, selections (Vanguard 359/62).
Folksongs arrangements (29). cf Works, selections (DG 2721 138).

318 Fugue, op. 137, D major. Septet, op. 20, E flat major. Vienna Philharmonic Chamber Ensemble. DG 2530 799.
+Gr 4-77 p1568 ++RR 4-77 p62
++HFN 4-77 p133

Gellert Lieder, op. 48. cf Works, selections (DG 2721 138).

319 German dances, WoO 8 (12). March, WoO 24, D major. Minuet, WoO 142. The ruins of Athens, op. 113: Turkish march. MOZART: March, K 408, no. 2, D major. March, K 408, no. 3, C major. Contradanses, K 609 (5). Minuets, K 463 (2). Contradanse, K 535. German dances, K 509 (6). Rotterdam Philharmonic Orchestra; Edo de Waart. Philips 9500 080 Tape (c) 7300 479.
+Gr 1-77 p1154 +NR 3-77 p6
+HF 10-77 p129 tape +RR 1-77 p45
+HFN 1-77 p102 ++SFC 1-16-77 p43
+MJ 2-77 p30 ++St 1-77 p112

Die Geschopfe des Prometheus (The creatures of Prometheus), op. 43: Ballet music. cf Works, selections (HMV SXDW 3032).
The creatures of Prometheus, op. 43: Overture. cf Overtures (DG 2721 137).
The creatures of Prometheus, op. 43: Overture. cf Overtures (Philips 6780 031).
The creatures of Prometheus, op. 43: Overture. cf Symphonies, nos. 1-9 (Seraphim SIH 6093).
The creatures of Prometheus, op. 43: Overture. cf Classics for Pleasure CFP 40236.
The creatures of Prometheus, op. 43: Overture. cf Decca SXL 6782.
The creatures of Prometheus, op. 43: Overture. cf HMV ESD 7011.

320 Grosse Fuge, op. 133, B flat major. Quartets, strings, op. 18, nos. 1-16. Quartet, F major (after Sonata, op. 14, no. 1). Quintet, strings, op. 29, C major. Amadeus Quartet; Cecil Aronowitz, vla. DG 2721 130 (11).
+Gr 3-77 p1416 +-RR 4-77 p61

Grosse Fuge, op. 133, B flat major. cf Quartets, strings, nos. 12-16 (DG 2740 168).
Grosse Fuge, op. 133, B flat major. cf Quartets, strings, nos. 12-16 (Philips 6707 031).
Grosse Fuge, op. 133, B flat major. cf Symphony, no. 3, op. 55, E flat major.
Grosse Fuge, op. 133, B flat major. cf Symphony, no. 4, op. 60, B flat major (DG 2535 813).
Grosse Fuge, op. 133, B flat major. cf Symphony, no. 4, op. 60,

B flat major (Philips 9500 033).
King Stephen, op. 117: Overture. cf Overtures (DG 2721 137).
King Stephen, op. 117: Overture. cf Overtures (Philips 6780 031).
King Stephen, op. 117: Overture. cf Symphony, no. 4, op. 60, B flat major.
King Stephen, op. 117: Overture. cf Works, selections (HMV SXDW 3032).

321 Leonore. Edda Moser, Helen Donath, s; Richard Cassilly, Eberhard Buchner, Rainer Goldberg, t; Karl Ridderbusch, Theo Adam, Hermann-Christian Polster, Siegfried Lorenz, bs; Leipzig Radio Chorus; Dresden Staatskapelle Orchestra; Herbert Blomstedt. HMV SQ SLS 999 (3).
+Gr 8-77 p338 +RR 9-77 p20
++HFN 9-77 p137

Leonore overtures, nos. 1-3. cf Ovetures (DG 2721 137).
Leonore overtures, nos. 1-3. cf Overtures (Philips 6780 031).
Leonore overtures, nos. 1-3. cf Works, selections (HMV SXDW 3032).
Leonore overtures, nos. 1 and 3, opp. 138 and 72. cf Symphony, no. 4, op. 60, B flat major.
Leonore overture, no. 2, op. 72. cf Symphony, no. 2, op. 36, D major.
Leonore overture, no. 2, op. 72. cf HMV RLS 717.
Leonore overture, no. 3, op. 72. cf Concerto, piano, no. 4, op. 58, G major.
Leonore overture, no. 3, op. 72. cf Symphonies, nos. 1-9 (Seraphim SIH 6093).
Leonore overture, no. 3, op. 72. cf Symphonies, nos. 1-3, 5-8.
Leonore overture, no. 3, op. 72. cf Symphony, no. 2, op. 36, D major.
Leonore overture, no. 3, op. 72. cf Symphony, no. 5, op. 67, C minor.
Leonore overture, no. 3, op. 72. cf Symphony, no. 7, op. 92, A major.
Leonore overture, no. 3, op. 72. cf Symphony, no. 8, op. 93, F major.
Leonore overture, no. 3, op. 72. cf Symphony, no. 9, op. 125, D minor.
Leonore overture, no. 3, op. 72. cf Works, selections (Decca D77D5).
Leonore overture, no. 3, op. 72. cf Works, selections (Vanguard 359/62).
Leonore overture, no. 3, op. 72. cf CBS SQ 79200.
Leonore overture, no. 3, op. 72. cf Pye PCNHX 6.
March, B flat major. cf Works, selections (London STS 15387).
March, B flat major. cf Works, selections (Philips 9500 087).
March, WoO 24, D major. cf German dances, WoO 8.

322 Mass, op. 86, C major. Felicity Palmer, s; Helen Watts, con; Robert Tear, t; Christopher Keyte, bs; Stephen Cleobury, org; St. John's College Chapel Choir; AMF; George Guest. Argo ZRG 739.
++Gr 5-74 p2049 -RR 5-74 p62
+NR 9-77 p10

323 Mass, op. 86, C major. Mass, op. 123, D major. Gundula Janowitz, s; Christa Ludwig, Julia Hamari, con; Fritz Wunderlich, Horst Laubenthal, t; Walter Berry, Ernst Schramm, bs; Vienna Singverein; BPhO, Munich Bach Orchestra and Choir; Herbert von Karajan, Karl Richter. DG 2721 135.

+Gr 3-77 p1416 +-RR 4-77 p80

324 Mass, op. 86, C major. Rodina Choir; Sofia Philharmonic Orchestra; Konstantin Iliev. Monitor MCS 2154.

+-ARG 9-77 p20 +-NR 9-77 p10
+-MJ 7-77 p65 +-ON 3-5-77 p33

325 Mass, op. 123, D major. Heather Harper, s; Janet Baker, ms; Robert Tear, t; Hans Sotin, bs; New Philharmonic Chorus; LPO; Carlo Maria Giulini. Angel SB 3836 (2). (also HMV SQ SLS 989 Tape (c) TC SLS 989)

+-ARG 12-76 p13 +NR 2-77 p10
+Gr 7-76 p200 -ON 3-5-77 p33
-Gr 10-76 p658 tape +RR 8-76 p72
+-HF 12-76 p98 +RR 10-76 p104 tape
+-HFN 9-76 p118 +St 2-77 p113
+HFN 10-76 p185 tape +STL 7-4-76 p36

Mass, op. 123, D major. cf Mass, op. 86, C major.
Meerestille und Gluckliche Fahrt (Calm sea and prosperous voyage), op. 112. cf Works, selections (DG 2721 138).
Minuet, G major. cf Inner City IC 1006.

326 Minuets, WoO 7 (12). Minuets, WoO 10 (6). Minuets, 2 violins and double bass, WoO 9 (6). PH; Hans Ludwig Hirsch Telefunken AW 6-41935.

+-Gr 3-76 p1458 +NR 3-77 p5
+MJ 2-77 p30 ++SFC 1-16-77 p43

Minuets, WoO 10 (6). cf Minuets, WoO 7.
Minuet, WoO 142. cf German dances, WoO 8 (12).
Minuets, 2 violins and double bass, WoO 9(6). cf Minuets, WoO 7.
Modlinger Tanz (dances), nos. 1-8. cf DG 2723 051.
Modlinger dances: Waltzes, nos. 3, 10-11. cf Saga 5421.
Namensfeier (Name day), op. 115: Overture. cf Overtures (DG 2721 137).
Namensfeier, op. 115: Overture. cf Overtures (Philips 6780 031).
Octet, op. 103, E flat major. cf Works, selections (DG 2721 129).
Octet, op. 103, E flat major. cf Works, selections (London STS 15387).
Octet, op. 103, E flat major. cf Works, selections (Philips 9500 087).
Ode to joy. cf Inner City IC 1006.

327 Overtures: Coriolan, op. 62. Consecration of the house, op. 124. The creatures of Prometheus, op. 43. Fidelio, op. 72. King Stephen, op. 117. Leonore, nos. 1-3. Name day, op. 115. The ruins of Athens, op. 113. Egmont, op. 84: Overture and incidental music. Gundula Janowitz, s; Erich Schellow, speaker; BPhO; Herbert von Karajan. DG 2721 137(3)

+Gr 3-77 p1416 +-RR 4-77 p42

328 Overtures: Consecration of the house, op. 124. Coriolon, op. 62. The creatures of Prometheus, op. 43. Egmont, op. 84. Fidelio, op. 72. King Stephen, op. 117. Leonore, nos. 1-3. Namensfeier, op. 115. The ruins of Athens, op. 113. Leipzig Gewandhaus Orchestra; Kurt Masur. Philips 6780 031 (2). (Reissue from 6747 135)

+Gr 4-77 p1541 +RR 3-77 p42
+-HFN 4-77 p151

329 Piano works: Andante favori, F major. Bagatelle, no. 25, A minor. Rondo, op. 51, no. 1, C major. Rondo, op. 51, no. 2, G major. Rondo a capriccio, op. 129, G major. Sonata, piano, no. 8, op. 13, C minor: 2nd movement. Sonata, piano, no. 14, op. 27, no.

2, C sharp minor: 1st movement. Sonata, piano, no. 20, op. 49, no. 2, G major. Wilhelm Kempff, pno. DG 2548 266 Tape (c) 3348 266. (Reissues from SLPM 138934, SKL 905)
+Gr 8-77 p323

Quartet, F major (after Sonata, op. 14, no. 1). cf Grosse Fuge, op. 133, B flat major.

330 Quartets, piano, nos. 1-3. Trio, clarinet. Trios, piano, nos. 1-6. Trio, piano, E flat major. Variations, opp. 44, 121a. Wilhelm Kempff, pno; Henryk Szeryng, vln; Pierre Fournier, vlc; Karl Leister, clt; Christoph Eschenbach, pno; Amadeus Quartet Members. DG 2721 132 (6).
+Gr 3-77 p1416 +RR 4-77 p61

331 Quartets, strings, nos. 1-6, op. 18. Juilliard Quartet. CBS 77362 (3). (also Columbia M3 30084)
+Gr 12-76 p1010 +-RR 12-76 p68
+HFN 1-77 p121

Quartets, strings, op. 18, nos. 1-16. cf Grosse Fuge, op. 133, B flat major.

332 Quartet, strings, no. 1, op. 18, no. 1, F major. Quartets, strings, no. 2, op. 18, no. 2, G major. Gabrieli Quartet. Decca SDD 478 Tape (c) KSDC 478. (also London STS 15398)
+Gr 4-76 p1623 ++NR 8-77 p7
+Gr 6-76 p102 tape +-RR 3-76 p52
+-HF 11-77 p105 ++STL 5-9-76 p38
+HFN 3-76 p89

Quartet, strings, no. 2, op. 18, no. 2, G major. cf Quartet, strings, no. 1, op. 18, no. 1, F major.

333 Quartet, strings, no. 3, op. 18, no. 3, D major. Quartet, strings, no. 4, op. 18, no. 4, C minor. Fine Arts Quartet. Concert Disc 210.
++St 5-77 p72

Quartet, strings, no. 4, op. 18, no. 4, C minor. cf Quartet, strings, no. 3, op. 18, no. 3, D major.

334 Quartets, strings, nos. 7-9, op. 59. Quartetto Italiano. Philips 6747 139 (2).
+Gr 12-75 p1062 ++NR 2-76 p6
+-HF 3-76 p79 +RR 11-75 p58
++HFN 12-75 p148 +St 11-75 p64
+MJ 1-76 p24 ++St 5-77 p72

335 Quartets, strings, nos. 12-16. Grosse Fuge, op. 133, B flat major. LaSalle Quartet. DG 2740 168 (4).
+HFN 12-77 p165 +RR 12-77 p58

336 Quartets, strings, nos. 12-16. Grosse Fuge, op. 133, B flat major. Quartetto Italiano. Philips 6707 031 (4).
++ARG 11-77 p11

337 Quartet, strings, no. 15, op. 132, A minor. LaSalle Quartet. DG 2530 728.
+-ARG 9-77 p20 +HFN 1-77 p113
+-Gr 3-77 p1416 +-RR 2-77 p67
+-HF 7-77 p98 +-St 8-77 p111

Quintet, E flat major. cf Works, selections (Philips 9500 087).

Quintet, oboe, 3 horns and bassoon, B flat major, excerpt. cf Works, selections (London STS 15387).

338 Quintet, piano, op. 16, E flat major. POULENC: Sextet, piano. Gothenburg Wind Quintet; Eva Knardahl, pno. BIS LP 61.
+HFN 11-77 p165 +RR 12-77 p58

339 Quintet, piano, op. 16, E flat major. Trio, piano, op. 36, D major

(after Symphony, no. 2). Diana Steiner, vln; Frances Steiner, vlc; David Berfield, Frank Glazer, pno; New York Woodwind Quintet. Orion ORS 76224.
+IN 10-77 p26 +NR 5-77 p8
+MJ 7-77 p68

Quintet, piano, op. 16, E flat major. cf Works, selections (DG 2721 129).

Quintet, strings, op. 29, C major. cf Grosse Fuge, op. 133, B flat major.

340 Romances, nos. 1 and 2, opp. 40, 50. SIBELIUS: Concerto, violin, op. 47, D minor. Pinchas Zukerman, vln; LPO; Daniel Barenboim. DG 2530 552 Tape (c) 3300 496.
+-Gr 10-75 p633 +NR 2-76 p4
-HF 7-76 p92 +-NYT 1-18-76 pD1
+-HF 1-77 p151 tape +RR 10-75 p55
+HFN 11-75 p151 ++RR 10-76 p106 tape
+-MJ 5-76 p29 +-St 6-76 p108

Romances, nos. 1 and 2, opp. 40, 50. cf Concerti, piano, nos. 1-5.

Romances, nos. 1 and 2, opp. 40, 50. cf Concerti, piano, nos. 1-5 (DG 2721 128).

Romance, no. 2, op. 50, F major (Soldat). cf Discopaedia MOB 1019.

Rondino, E flat major. cf Works, selections (London STS 15387).

Rondino, E flat major. cf Works, selections (Philips 9500 087).

Rondino, E flat major. cf Telefunken EK 6-35334.

Rondo. cf Works, selections (DG 2721 133).

Rondo, op. 51, no. 1, C major. cf Piano works (DG 2548 266).

Rondo, op. 51, no. 2, G major. cf Piano works (DG 2548 266).

Rondo a capriccio, op. 129, G major. cf Piano works (DG 2548 266).

Die Ruinen von Athens (The ruins of Athens), op. 113: Overture. cf Overtures (DG 2721 137).

The ruins of Athens, op. 113: Overture. cf Overtures (Philips 6780 031).

The ruins of Athens, op. 113: Overture. cf Symphony, no. 1, op. 21, C major.

Die Ruinen von Athens, op. 113: Overture. cf HMV RLS 717.

The ruins of Athens, op. 113: Turkish march. cf German dances, WoO8.

The ruins of Athens, op. 113: Turkish march. cf International Piano Library 103.

The ruins of Athens, op. 113: Turkish march. cf International Piano Library IPL 5001/2.

341 Septet, op. 20, E flat major. Vienna Octet. Decca SDD 200 Tape (c) KSDC 200. (also London STS 15361)
+ARG 2-77 p22 +RR 7-75 p69 tape
+HFN 7-75 p90 tape

342 Septet, op. 20, E flat major. Melos Ensemble. L'Oiseau-Lyre 60015.
++St 5-77 p74

Septet, op. 20, E flat major. cf Fugue, op. 137, D major.

Septet, op. 20, E flat major. cf Works, selections (DG 2721 129).

Serenade, op. 25, D major. cf Works, selections (DG 2721 129).

Sextet, op. 71, E flat maor. cf Works, selections (DG 2721 129).

Sextet, op. 71, E flat major. cf Works, selections (London STS 15387).

Sextet, op. 71, E flat major. cf Works, selections (Philips 9500 087).

Sextet, op. 81b, E flat major. cf Works, selections (DG 2721 129).

343 Sonata, flute and piano, B flat major. SCHUBERT: Introduction and variations, op. 160, D 802, E minor. Sid Zeitlin, flt; Layton James, Ginger Smith, pno. Coronet LPS 3037.
+-NR 1-77 p15

344 Sonata, flute and piano, B flat major. Variations on "Air russe", op. 107, no. 7. Variations on "Air de la petite Russie", op. 107, no. 3. MOZART: Sonata, flute and harpsichord, no. 4, K 13, F major. SCHUBERT: Introduction and variations, op. 160, D 802, E minor. Richard Adeney, flt; Ian Brown, pno and hpd. Enigma VAR 1029.
+Gr 8-77 p313 +-RR 7-77 p79

345 Sonata, horn and piano, op. 17, F major. DANZI: Sonata, horn, op. 28, E flat major. SAINT-SAENS: Romance, op. 67. SCHUMANN: Adagio and allegro, op. 270, A flat major. Barry Tuckwell, hn; Vladimir Ashkenazy, pno. Decca SXL 6717. (also London CS 6938).
+Gr 8-75 p342 +NR 4-77 p15
++HF 4-77 p118 +RR 8-75 p54
+HFN 8-75 p84 ++St 3-77 p149

Sonata, horn and piano, op. 17, F major. cf Works, selections (DG 2721 129).

Sonata, horn and piano, op. 17, F major. cf AMON: Quartet, op. 20, no. 1.

346 Sonatas, mandolin (4). GIANNEO: Suite Argentine. HASSE: Concerto, mandolin, G major. Jacob Thomas, mand; Gunter Krieger, hpd; Heidelberg Chamber Orchestra. CMS/Oryx EXP 40.
+NR 7-77 p15 +St 7-77 p126

347 Sonatas, piano, nos. 1, 3 7-8, 13-14, 17-18, 21-23, 26-32. Solomon, pno. HMV RLS 722 (7). (Reissues from ALP 1573, C 3847/9, ALP 1062, 1900, BPL 1051, ALP 1303, 1160, 1546, 1272, 294, 1141).
+-Gr 11-76 p838 +RR 1-77 p69
+-HFN 11-76 p155

348 Sonata, piano, no. 1, op. 2, no. 1, F minor. Sonata, piano, no. 7, op. 10, no. 3, D major. Sviatoslav Richter, pno. Angel S 37266. (also HMV SQ ASD 3364)
++ARG 10-77 p22 +-RR 9-77 p80
+Gr 9-77 p456 +NR 12-77 p13
+HF 10-77 p101 +St 12-77 p133
+HFN 9-77 p136

Sonata, piano, no. 1, op. 2, no. 1, F minor. cf Concerto, piano, no. 2, op. 19, B flat major.

349 Sonata, piano, no. 2, op. 2, no. 2, A major. Sonata, piano, no. 3, op. 2, no. 3, C major. Vladimir Ashkenazy, pno. London CS 7028 Tape (c) CS 5-7028. (also Decca SXL 6808 Tape (c) KSXC 6808)
+-Gr 2-77 p1302 +HFN 2-77 p113
+Gr 4-77 p1603 tape +NR 8-77 p14
+Ha 9-77 p108 ++RR 2-77 p75
++HF 6-77 p81

350 Sonata, piano, no, 3, op. 2, no. 3, C major. MEDTNER: Sonata, piano, op. 22, G minor. Emil Gilels, pno. Westminster 8273.
+CL 10-77 p6

Sonata, piano, no. 3, op. 2, no. 3, C major. cf Sonata, piano, no. 2, op. 2, no. 2, A major.

351 Sonata, piano, no. 6, op. 10, no. 2, F major. Sonata, piano, no. 27, op. 90, E minor. Sonata, piano, no. 30, op. 109, E major. Alfred Brendel, pno. Philips 9500 076.

++ARG 7-77 p16 +MJ 5-77 p32
+Gr 11-76 p837 +NR 6-77 p14
+HFN 11-76 p156 +-RR 11-76 p88

352 Sonata, piano, no. 7, op. 10, no. 3, D major. Sonata, piano, no. 9, op. 14, no. 1, E major. Sonata, piano, no. 10, op. 14, no. 2, G major. Alfred Brendel, pno. Turnabout TV 34118 Tape (c) KTVC 34118.
+Gr 8-77 p346 tape

Sonata, piano, no. 7, op. 10, no. 3, D major. Sonata, piano, no. 1, op. 2, no. 1, F minor.

353 Sonatas, piano, nos. 8, 13, 23. Bruno-Leonardo Gelber, pno. Concoisseur Society CSQ 2113.
+MJ 12-76 p28 +St 2-77 p113
+NR 2-77 p14

354 Sonatas, piano, nos. 8, 14, 21. Rudolf Firkusny, pno. Decca PFS 4341. (also London SPC 21080)
-Gr 12-75 p1075 /-RR 11-75 p73
-HFN 11-75 p151 +St 2-77 p113
+-NR 2-77 p14

355 Sonatas, piano, nos. 8, 14, 23. Vladimir Horowitz, pno. Columbia M 34509 Tape (c) MT 34509. (Reissues)
+NR 8-77 p14

356 Sonatas, piano, nos. 8, 18, 19. Alfred Brendel, pno. Philips 9500 077 Tape (c) 7300 478.
+-ARG 9-77 p18 +-HFN 2-77 p135 tape
+Gr 3-77 p1458 tape ++NR 8-77 p14
+-HF 11-77 p105

357 Sonatas, piano, nos. 8, 21, 26. Vladimir Ashkenazy, pno. Decca SXL 6706 Tape (c) KSCX 6706. (also London CS 6921)
+Gr 5-75 p1992 ++NR 11-76 p12
+-HF 10-76 p103 +-RR 5-75 p52
++HFN 6-75 p87 ++St 2-77 p113
+HFN 7-75 p90 tape

358 Sonata, piano, no. 8, op. 13, C minor. BRAHMS: Pieces, piano, op. 117, no. 2, B flat minor. CHOPIN: Ballade, no. 4, op. 52, F minor. Prelude, op. 28, no. 24, D minor. DEBUSSY: Suite bergamasque: Clair de lune. Preludes, Bk 2: Ondine. Ivan Moravec, pno. Connoisseur CS 2129.
+NR 6-77 p14

359 Sonata, piano, no. 8, op. 13, C minor. Sonata, piano, no. 14, op. 27, no. 2, C sharp minor. Sonata, piano, no. 23, op. 57, F minor. John Lill, pno. Enigma VAR 1001.
+Gr 12-76 p1015 +RR 12-76 p76
-HFN 1-77 p102

360 Sonata, piano, no. 8, op. 13, C minor. Sonata, piano, no. 14, op. 27, no. 2, C sharp minor. Sonata, piano, no. 23, op. 57, F minor. Solomon, pno. Seraphim M 60286.
+ARG 12-77 p20 ++NR 12-77 p13
++FF 9-77 p4

Sonata, piano, no. 8, op. 13, C minor. cf Concerto, piano, no. 1, op. 15, C major.

Sonata, piano, no. 8, op. 13, C minor. cf Works, selections (Decca D77D5).

Sonata, piano, no. 8, op. 13, C minor: Adagio cantabile. cf Decca SPA 473.

Sonata, piano, no. 8, op. 13, C minor: 2nd movement. cf Piano works (DG 2548 266).

361 Sonatas, piano, nos. 9, 10, 28. Alfred Brendel, pno. Philips 9500 041 Tape (c) 7300 475.
+-ARG 12-76 p18 +MJ 1-77 p38
+Gr 10-76 p622 ++NR 2-77 p14
+-Gr 10-76 p658 tape +-RR 10-76 p86
+-HFN 10-76 p165

Sonata, piano, no. 9, op. 14, no. 1, E major. cf Sonata, piano, no. 7, op. 10, no. 3, D major.

Sonata, piano, no. 10, op. 14, no. 2, G major. cf Sonata, piano, no. 7, op. 10, no. 3, D major.

362 Sonata, piano, no. 12, op. 26, A flat major. Sonata, piano, no. 16, op. 31, no. 1, G major. Emil Gilels, pno. DG 2530 654 Tape (c) 3300 654.
+-Gr 6-76 p62 ++NR 5-77 p13
+-Gr 10-76 p658 tape +RR 6-76 p67
+HFN 6-76 p81 +St 5-77 p103

363 Sonata, piano, no. 12, op. 26, A flat major. Sonata, piano, no. 30, op. 109, E major. Frantisek Rauch, pno. Rediffusion Tape (c) LGC 010.
+-RR 2-77 p96 tape

364 Sonata, piano, no. 12, op. 26, A flat major. SCHUBERT: Sonata, piano, D 537, A minor. Arturo Benedette Michelangeli, pno. Rococo 2117.
-NR 5-77 p14

365 Sonata, piano, no. 13, op. 27, no. 1, E flat major. Sonata, piano, no. 17, op. 31, no. 2, D minor. Sonata, piano, no. 19, op. 49, no. 1, G minor. John Lill, pno. Enigma VAR 1002.
+-Gr 5-77 p1713 +-RR 5-77 p68

366 Sonata, piano, no. 14, op. 27, no. 2, C sharp minor. BRAHMS: Pieces, piano, op. 118, no. 2, A major. CHOPIN: Nocturne, op. 27, no. 2, D flat major. Mazurka, op. 63, no. 3, C sharp minor. DEBUSSY: Children's corner suite: Golliwog's cake walk; Snow is dancing. Preludes, Bk 2: Feux d'artifice. Ivan Moravec, pno. Connoisseur Society CS 2123.
+NR 6-77 p14 +SFC 8-21-77 p46

Sonata, piano, no. 14, op. 27, no. 2, C sharp minor. cf Sonata, piano, no. 8, op. 13, C minor (Enigma VAR 1001).

Sonata, piano, no. 14, op. 27, no. 2, C sharp minor. cf Sonata, piano, no. 8, op. 13, C minor (Seraphim M 60286).

Sonata, piano, no. 14, op. 27, no. 2, C sharp minor. cf Works selections (Decca D77D5).

Sonata, piano, no. 14, op. 27, no. 2, C sharp minor: 1st movement. cf Piano works (DG 2548 266).

367 Sonata, piano, no. 16, op. 31, no. 1, G major. Sonata, piano, no. 18, op. 31, no. 3, E flat major. Sonata, piano, no. 20, op. 49, no. 2, G major. John Lill, pno. Enigma VAR 1003 Tape (c) TC VAR 1003.
+-Gr 7-77 p201 +-RR 7-77 p78
+-HFN 7-77 p105 +RR 8-77 p85

Sonata, piano, no. 16, op. 31, no. 1, G major. cf Sonata, piano, no. 12, op. 26, A flat major.

368 Sonatas, piano, nos. 17, 28-32. Alfred Brendel, pno. Philips 6747 312 (3).
+-RR 2-77 p75

Sonata, piano, no. 17, op. 31, no. 2, D minor. cf Sonata, piano, no. 13, op. 27, no. 1, E flat major.

369 Sonata, piano, no. 18, op. 31, no. 3, E flat major. Sonata, piano,

no. 23, op. 57, F minor. Lazar Berman, pno. CBS 76533 Tape (c) 40-76533. (also Columbia M 34218 Tape (c) MT 34218 (ct) MA 34218)
/Gr 12-76 p1016
+-HF 3-77 p96
+-HF 5-77 p101 tape
+-HFN 1-77 p102
+-HFN 3-77 p119 tape
+MJ 3-77 p74
+-MT 10-77 p825
++NR 5-77 p13
+-RR 12-76 p76
+SR 2-19-77 p42
+-St 1-77 p113

370 Sonata, piano, no. 18, op. 31, no. 3, E flat major. SCHUMANN: Fantasiestucke, op. 12. Artur Rubinstein, pno. RCA ARL 1-2397 Tape (c) ARK 1-2397 (ct) ARS 1-2397.
+NR 11-77 p14
+-NYT 9-25-77 pD19

Sonata, piano, no. 18, op. 31, no. 3, E flat major. cf Sonata, piano, no. 16, op. 31, no. 1, G major.

Sonata, piano, no. 19, op. 49, no. 1, G minor. cf Sonata, piano, no. 13, op. 27, no. 1, E flat major.

Sonata, piano, no. 20, op. 49, no. 2, G major. cf Concerto, piano, no. 2, op. 19, B flat major.

Sonata, piano, no. 20, op. 49, no. 2, G major. cf Piano works (DG 2548 266).

Sonata, piano, no. 20, op. 49, no. 2, G major. cf Sonata, piano, no. 16, op. 31, G major.

Sonata, piano, no. 20, op. 49, no. 2, G major: Tempo di minuetto. cf Supraphon 113 1323.

371 Sonata, piano, no. 21, op. 53, C major. Sonata, piano, no. 23, op. 57, F minor. Peter Frankl, pno. Gale GMFD 1-76-005.
+HF 10-77 p112
+HFN 9-77 p132

372 Sonata, piano, no. 21, op. 53, C major. Variations and fugue, op. 35, E flat major. Emanuel Ax, pno. RCA ARL 1-2083 Tape (c) ARK 1-2083 (ct) ARS 1-2083.
+-HF 6-77 p82
+-MJ 5-77 p32
+-NR 5-77 p13
++St 5-77 p103

Sonata, piano, no. 21, op. 53, C major. cf Andante favori, F major.

Sonata, piano, no. 21, op. 53, C major. cf Concerto, piano, no. 2, op. 19, B flat major.

Sonata, piano, no. 21, op. 53, C major. cf International Piano Archives IPA 5007/8.

373 Sonata, piano, no. 22, op. 54, F major. HAYDN: Sonata, piano, no. 12, A major. LARSSON: Concertino, piano, op. 45, no. 12. MOZART: Fantasia, K 396, C minor. Hans Palsson, pno; Musica Sveciae; Sven Verde. BIS LP 36.
+Gr 9-77 p500
+-RR 11-77 p90
+-St 6-77 p132

374 Sonata, piano, no. 23, op. 57, F minor. Sonata, piano, no. 31, op. 110, A flat major. Michael Studer, pno. Claves P 612.
+-HFN 5-77 p119
/RR 7-77 p78

375 Sonata, piano, no. 23, op. 57, F minor. LISZT: Sonata, piano, G 178, B minor. Lazar Berman, pno. Saga 5430 Tape (c) CS 5430. (Reissue from SIX 5019)
+-ARG 11-76 p14
+-Gr 6-76 p69
+-Gr 2-77 p1322 tape
+-HF 3-77 p46
-RR 5-76 p66
+-St 9-76 p117

376 Sonata, piano, no. 23, op. 57, F minor. Sonata, piano, no. 26, op. 81, E flat major. Sonata, piano, no. 27, op. 90, E minor. Alfred Brendel, pno. Turnabout TV 34116. Tape (c) KTVC 34116.
+Gr 8-77 p346 tape

Sonata, piano, no. 23, op. 57, F minor. cf Sonata, piano, no. 8, op. 13, C minor (Enigma VAR 1001).
Sonata, piano, no. 23, op. 57, F minor. cf Sonata, piano, no. 8, op. 13, C minor (Seraphim M 60286).
Sonata, piano, no. 23, op. 57, F minor. cf Sonata, piano, no. 18, op. 31, no. 3, E flat major.
Sonata, piano, no. 23, op. 57, F minor. cf Sonata, piano, no. 21, no. 53, C major.
Sonata, piano, no. 23, op. 57, F minor. cf Works selections (Decca D77D5).

377 Sonata, piano, no. 26, op. 81a, E flat major. GRIEG: Ballade, op. 24, G minor. SCHUMANN: Carnaval, op. 9. Leopold Godowsky, pno. International Piano Library IPL 105.
++NR 5-73 p13 +RR 3-77 p26

Sonata, piano, no. 26, op. 81a, E flat major. cf Concerto, piano, no. 3, op. 37, C minor.
Sonata, piano, no. 26, op. 81a, E flat major. cf Sonata, piano, no. 23, op. 57, F minor.

378 Sonatas, piano, nos. 27-32. Alfred Brendel, pno. Philips 6747 312. (Reissues from 9500 076, 9500 041, 6500 139, 6500 762, 6500 138)
+Gr 3-77 p1422 ++MT 7-77 p564
++HFN 3-77 p117

Sonata, piano, no. 27, op. 90, E minor. cf Concerto, piano, no. 4, op. 58, G major.
Sonata, piano, no. 27, op. 90, E minor. cf Sonata, piano, no. 6, op. 10, no. 2, F major.
Sonata, piano, no. 27, op. 90, E minor. cf Sonata, piano, no. 23, op. 57, F minor.

379 Sonata, piano, no. 28, op. 101, A major. Sonata, piano, no. 30, op. 109, E major. Vladimir Ashkenazy, pno. London CS 7029 Tape (c) CS 5-7029. (also Decca SXL 6809 Tape (c) KSXC 6809)
+Gr 7-77 p291 ++NR 8-77 p14
+HF 11-77 p105 +RR 7-77 p78
++HFN 7-77 p107 ++SFC 8-21-77 p46
+HFN 8-77 p99 tape

380 Sonata, piano, no. 29, op. 106, B flat major. Bagatelles, op. 126, nos. 1, 4, 6. Sviatoslav Richter, pno. Rococo 2110.
+-NR 5-77 p14

381 Sonata, piano, no. 30, op. 109, E major. Sonata, piano, no. 31, op. 110, A flat major. Maurizio Pollini, pno. DG 2530 645 Tape (c) 3300 645.
++Gr 5-76 p1782 +-NYT 3-13-77 pD22
+-HF 1-77 p110 +-RR 5-76 p64
++HFN 5-76 p95 +-RR 9-76 p93 tape
+-MJ 12-76 p28 ++SFC 9-12-76 p31
+NR 10-76 p14 ++St 12-76 p134

Sonata, piano, no. 30, op. 109, E major. cf Sonata, piano, no. 6, op. 10, no. 2, F major.
Sonata, piano, no. 30, op. 109, E major. cf Sonata, piano, no. 12, op. 26, A flat major.
Sonata, piano, no. 30, op. 109, E major. cf Sonata, piano, no. 28, op. 101, A major.
Sonata, piano, no. 31, op. 110, A flat major. cf Andante favori, F major.
Sonata, piano, no. 31, op. 110, A flat major. cf Sonata, piano, no. 23, op. 57, F minor.

Sonata, piano, no. 31, op. 110, A flat major. cf Sonata, piano, no. 30, op. 109, E major.

382 Sonatas, violin and piano, nos. 1-10. Denes Kovacs, vln; Mihaly Bacher, pno. Hungaroton SLPX 11700/4 (5).
+- Gr 5-77 p1704 ++RR 2-77 p75
+-HFN 2-77 p112

383 Sonatas, violin and piano, nos. 1-10. Jascha Heifetz, vln; Emanuel Bay, Brooks Smith, pno. RCA RL 4-2004 (5). (Reissues from HMV ALP 1422, 1423, 1424, 1425, 1426, RCA RB 16277)
+Gr 9-77 p444 +RR 9-77 p80
+HFN 10-77 p149

Sonatas, violin and piano, nos. 1-10. cf Works, selections (DG 2721 133).

384 Sonata, violin and piano, no. 1, op. 12, no. 1, D major. Sonata, violin and piano, no. 10, op. 96, G major. Itzhak Perlman, vln; Vladimir Ashkenazy, pno. Decca SXL 6790 Tape (c) KSXC 6790.
+Gr 7-77 p192 ++HFN 9-77 p155 tape
+-Gr 10-77 p705 tape ++RR 7-77 p79
++HFN 7-77 p107

385 Sonata, violin and piano, no. 2, op. 12, A major. Sonata, violin and piano, no. 9, op. 47, A major. Itzhak Perlman, vln; Vladimir Ashkenazy, pno. Decca SXL 6632 Tape (c) KSXC 6632. (also London CS 6845)
+-Gr 2-75 p1504 ++NR 8-75 p8
++Gr 8-77 p346 tape +RR 2-75 p49
+HF 7-75 p71 ++SFC 7-13-75 p21
++HFN 9-77 p155 tape +St 8-75 p95

386 Sonata, violin and piano, no. 2, op. 12, A major. Sonata, violin and piano, no. 4, op. 23, A minor. Arthur Grumiaux, vln; Claudio Arrau, pno. Philips 9500 263.
+Gr 8-77 p313 ++RR 9-77 p81
++HFN 7-77 p107

387 Sonata, violin and piano, no. 3, op. 12, E flat major. Sonata, violin and piano, no. 8, op. 30, G major. Itzhak Perlman, vln; Vladimir Ashkenazy, pno. Decca SXL 6789 Tape (c) KSXC 6789.
+Gr 12-76 p1010 ++HFN 1-77 p102
++Gr 3-77 p1458 tape +RR 12-76 p77

Sonata, violin and piano, no. 3, op. 12, E flat major. cf RCA CRM 6-2264.

388 Sonata, violin and piano, no. 4, op. 23, A minor. Sonata, violin and piano, no. 5, op. 24, F major. Oleg Kagaan, vln; Sviatoslav Richter, pno. HMV ASD 3295.
+Gr 1-77 p1159 +RR 12-76 p77
++HFN 12-76 p137

Sonata, violin and piano, no. 4, op. 23, A minor. cf Sonata, violin and piano, no. 2, op. 12, A major.

389 Sonata, violin and piano, no. 5, op. 24, F major. MOZART: Sonata, violin and piano, no. 32, K 454, B flat major. Kaja Danczowska, vln; Ewa Bukojemska, pno. Muza SX 1102.
+-NR 5-77 p8

Sonata, violin and piano, no. 5, op. 24, F major. cf Sonata, violin and piano, no. 4, op. 23, A minor.

Sonata, violin and piano, no. 5, op. 24, F major. cf Works, selections (Decca D77D5).

390 Sonata, violin and piano, no. 6, op. 30, no. 1, A major. Sonata, violin and piano, no. 7, op. 30, no. 2, C minor. Itzhak Perl-

man, vln; Vladimir Ashkenazy, pno. Decca SXL 6791. (also London CS 7014)
++Gr 12-77 p1098

391 Sonata, violin and piano, no. 7, op. 30, no. 2, C minor. Sonata, violin and piano, no. 8, op. 30, no. 3, G major. Arthur Grumiaux, vln; Claudio Arrau, pno. Philips 9500 220.
+-Gr 3-77 p1416 +RR 3-77 p75
+-HFN 3-77 p99

Sonata, violin and piano, no. 7, op. 30, no. 2, C minor. cf Sonata, violin and piano, no. 6, op. 30, no. 1, A major.

Sonata, violin and piano, no. 8, op. 30, no. 3, G major. cf Sonata, violin and piano, no. 3, op. 12, E flat major.

Sonata, violin and piano, no. 8, op. 30, no. 3, G major. cf Sonata, violin and piano, no. 7, op. 30, no. 2, C minor.

Sonata, violin and piano, no. 8, op. 30, no. 3, G major. cf RCA CRM 6-2264.

392 Sonata, violin and piano, no. 9, op. 47, A major. MOZART: Sonata, violin and piano, no. 32, K 454, B flat major. Georg Kulenkampff, vln; Georg Solti, pno. London R 23214.
+NR 5-77 p8

Sonata, violin and piano, no. 9, op. 47, A major. cf Sonata, violin and piano, no. 2, op. 12, A major.

Sonata, violin and piano, no. 10, op. 96, G major. cf Sonata, violin and piano, no. 1, op. 12, no. 1, D major.

393 Sonatas, violoncello and piano, nos. 1-5. Variations on Mozart's "Bei Mannern" (7). Variations on Mozart's "Ein Madchen oder Weibchen" (12). Variations on a theme from Handel's "Judas Maccabeus" (12). Jacqueline du Pre, vlc; Daniel Barenboim, pno. Angel SCB 3823 (3). (also HMV SLS 5042)
+-Gr 5-76 p1775 +NR 6-76 p8
-HF 10-76 p95 +-NYT 7-24-77 pD11
+HFN 5-76 p92 +RR 3-76 p20
+HFN 10-76 p165 ++St 9-76 p117
+MT 6-76 p495 +STL 6-6-76 p37

394 Sonatas, violoncello, nos. 1-5. Daniel Shafran, vlc; Anton Ginzburg, pno. Odyssey/Melodiya Y2 34645 (2).
-ARG 10-77 p19 +NR 5-77 p9
+MJ 7-77 p68 +St 7-77 p126

395 Sonatas, violoncello and piano, nos. 1-5. Lynn Harrell, vlc; James Levine, pno. RCA ARL 2-2241 (2) Tape (c) ARS 2-2241.
+-ARG 10-77 p19 +NR 8-77 p7
+-Gr 11-77 p854 ++NYT 7-24-77 pD11
+HF 12-77 p78 +-RR 11-77 p85
+-MJ 10-77 p27 +St 11-77 p134

Sonatas, violoncello, nos. 1-5. cf Works, selections (DG 2721 133).

396 Songs (Folksongs): Behold my love, op. 108, no. 9; Come fill fill my good fellow, op. 108, no. 13; Duncan Gray; The elfin fairies; Faithful Johnie, op. 108, no. 20; He promised me at parting; The Highlander's lament; The Highland watch, op. 108, no. 22; The miller of the dee; Music, love and wine, op. 108, no. 1; O sweet were the hours, op. 108, no. 3; Oh had my fate been join'd with thine, op. 10, no. 12; The pulse of an Irishman; Put round the bright wine. Edith Mathis, s; Alexander Young, t; Dietrich Fischer-Dieskau, bar; RIAS Chamber Choir; Andreas Rohn, vln; Georg Donderer, vlc; Karl Engel, pno. DG 2535 241 Tape (c) 3335 241. (Reissue from 2720 017)
+Gr 9-77 p467 +HFN 10-77 p169 tape
+Gr 9-77 p508 tape

397 Songs (Folksong settings): La Biondina in Gondoletta; By the side of the Shannon; Charlie is my darling; God save the King; A health to the brave; I am bowed down with years; The miller of dee; O sanctissima; My Harry was a gallant gay; Sir Johnie Cope; Since all thy vows; The soldier. Accademia Monteverdiana; Denis Stevens. Nonesuch H 71340.
+-FF 11-77 p7 +NR 10-77 p11
+HF 11-77 p106

398 Songs: Egmont, op. 84: Freudvoll und Leidvoll. Gellert songs, op. 48. Ich denke dein; Des Kriegers Abschied; Die laute Klage; Opferlied, op. 121b. Goethe songs, op. 83: Wonne der Wehmut; Sehnsucht; Mit einem gemalten Bande. Songs, op. 52: Urians Reise um die Welt; Mailied; Die Liebe; Marmotte; Das Blumchen Wunderhold. Songs, op. 75: Neue Liebe, neues Leben; Flohlied sus Faust. Peter Schreier, t; Walter Olbertz, Gisela Franke, pno. Telefunken 6-41997.
+HF 11-77 p106 ++St 12-77 p132

399 Songs: An die ferne Geliebte, op. 98; An die Geliebte; An die Hoffnung, opp. 32 and 94; Das Gluck der Freunchschaft, op. 88; Der Jungling in der Fremde; Klage; Die Liebende; Ruf vom Berge; Selbstgesprach. Peter Schreier, t; Walter Olbertz, pno. Telefunken AP 6-42082.
++Gr 12-77 p1121 +-HFN 11-77 p167

400 Songs (Folksong settings, arr.): Irish songs: The return to Ulster; Sweet power of song; Farewell bliss and farewell Nancy. Scottish songs, op. 108: The sweetest lad was Jamie; The shepherd's song Glencoe; Auld lang syne. Welsh songs: Sion the son of Evan; Love without hope; O let the night my blushes hide. Accademia Monteverdiana; Denis Stevens. Vanguard SRV 356.
+HF 11-77 p106

Songs (Lieder) (41). cf Works, selections (DG 2721 138).

Songs: Andenken; Klarchen's song, no. 1; Die Trammel geruhret; Wonne der Wehmut. cf Bruno Walter Society BWS 729.

Songs: In questa tomba oscura. cf Rubini GV 57.

Songs: No non turbati. cf Ah perfido, op. 65.

Songs: No non turbati. cf Philips 6767 001.

401 Symphonies, nos. 1-9. NYP; Leonard Bernstein. CBS 61901/7 (7) Tape (c) 40-61901/7.
+Gr 11-77 p823 +-RR 12-77 p43
+-HFN 11-77 p183 +-RR 12-77 p97 tape
+-HFN 12-77 p187 tape

402 Symphonies, nos. 1-9. Gwyneth Jones, s; Tatiana Troyanos, con; Jess Thomas, t; Karl Ridderbusch, bs; Vienna State Opera Chorus; VPO; Karl Bohm. DG 2721 154 (8).
+Gr 3-77 p1415 +-RR 4-77 p42

403 Symphonies, nos. 1-9. Anna Tomowa-Sintow, s; Agnes Baltsa, ms; Peter Schreier, t; Jose van Dam, bar; Vienna Singverein; BPhO; Herbert von Karajan. DG 2740 172 (8) Tape (c) 3378 070.
++Gr 10-77 p617 +-RR 10-77 p37
+-Gr 12-77 p1142 tape ++SFC 12-4-77 p56
+HFN 10-77 p165

404 Symphonies, nos. 1-9. Hannelore Bode, s; Helen Watts, con; Horst Laubenthal, t; Benjamin Luxon, bs; LPO and Chorus; Bernard Haitink. Philips 6747 307 (7) Tape (c) 7699 037.
+-ARG 2-77 p23 +MJ 2-77 p30
+-Gr 1-77 p1139 +NR 2-77 p1
+Gr 10-77 p698 tape +RR 1-77 p45

+-HF 7-77 p99 +-RR 10-77 p92 tape
++HFN 1-77 p100 +-SFC 1-16-77 p43
+HFN 10-77 p169 tape +St 4-77 p120

405 Symphonies, nos. 1-9. Fidelio, op. 72: Overture. To van der Sluys, s; Suze Luger, con; Louis van Tulder, t; Willem Ravelli, bar; Amsterdam Tonkunstchor, Konigliche Oratorienvereinigung; COA; Willem Mengelberg. Philips 6767 003 (3). (Reissues from Pearl HE 301, Philips GL 5806, 5689)
+-Gr 9-77 p417 +-MJ 11-77 p25
+-HF 12-77 p78 +-NR 12-77 p1
+-HFN 9-77 p136 +-RR 9-77 p47

406 Symphonies, nos. 1-9. Overtures: Egmont, op. 84. Leonore, no. 3, op. 72. The creatures of Prometheus, op. 43. Ursula Koszut, s; Brigitte Fassbaender, ms; Nicolai Gedda, t; Donald McIntyre, bar; Munich Motet Choir; Munich Philharmonic Orchestra and Chorus; Rudolf Kempe. Seraphim SIH 6093 (8).
+-HF 8-76 p73 +-St 4-77 p120
+NR 6-76 p1

407 Symphonies, nos. 1-3, 5-8. Egmont, op. 84: Overture. Leonore overture, no. 3, op. 72. Joan Sutherland, s; Marilyn Horne, con; James King, t; Martti Talvela, bs; Vienna State Opera Chorus; VPO; Hans Schmidt-Isserstedt. Decca JB 1-5 (5) Tape KJBC 1-5. (Reissues from SXL 6233, 6329, 6437, 6447, 6396)
+Gr 9-77 p418 +HFN 9-77 p151
+Gr 11-77 p899 tape +-RR 9-77 p48

408 Symphony, no. 1, op. 21, C major. Symphony, no. 2, op. 36, D major. VPO; Hans Schmidt-Isserstedt. Decca Tape (c) KJBC 3.
+HFN 12-77 p187 tape +-RR 11-77 p118 tape

409 Symphony, no. 1, op. 21, C major. Symphony, no. 8, op. 93, F major. CSO; Georg Solti. Decca SXL 6760 Tape (c) KSXC 6760. (Reissue from 11BB 188/96)
-Gr 9-76 p409 ++HFN 10-76 p181
+Gr 4-77 p1603 tape +-RR 10-76 p39

410 Symphony, no. 1, op. 21, C major. Symphony, no. 4, op. 60, B flat major. Coriolan overture, op. 62. The ruins of Athens, op. 113: Overture. BRAHMS: Variations on a theme by Haydn, op. 56a. Casals Orchestra, LSO; Pablo Casals. Electrola Da Capo 187-03 039/40 (2).
+-ARSC vol 9, no. 1, 1977 p87

411 Symphony, no. 1, op. 21, C major. Symphony, no. 8, op. 93, F major. HSO; Janos Ferencscik. Hungaroton SLPX 11890.
+FF 9-77 p5 +-NR 8-77 p3
+-HFN 5-77 p119 /RR 5-77 p46

412 Symphony, no. 1, op. 21, C major. Symphony, no. 8, op. 83, F major. COA; Eugen Jochum. Philips 6580 148. (Reissue from AXS 90001/9)
+-Gr 2-77 p1270 +-RR 2-77 p43
+-HFN 3-77 p117

413 Symphony, no. 1, op. 21, C major. Symphony, no. 8, op. 93, F major. CPhO; Jean Meylan. Rediffusion Legend Tape (c) LGC 005.
/-RR 2-77 p95

414 Symphony, no. 1, op. 21, C major. Symphony, no. 4, op. 60, B flat major. BBC Symphony Orchestra; Arturo Toscanini. World Records SH 134.
+FF 11-77 p7

Symphony, no. 1, op. 21, C major. cf BACH: Concerto, 2 violins and strings, S 1043, D minor.

415 Symphony, no. 2, op. 36, D major. Egmont, op. 84: Overture. CSO; Georg Solti. Decca SXL 6761. (Reissue from 11BB 188/196)
+Gr 11-76 p989 +-RR 1-77 p45
++HFN 1-77 p121
416 Symphony, no. 2, op. 36, D major. Symphony, no. 4, op. 60, B flat major. CSO; Georg Solti. Decca Tape (c) KBB 2-7042.
+Gr 4-77 p1603 tape +-RR 3-77 p98 tape
+-HFN 4-77 p155 tape
417 Symphony, no. 2, op. 36, D major. Symphony, no. 8, op. 93, F major. HSO; Janos Ferencsik. Hungaroton SLPX 11891.
-ARG 6-77 p12 +-HFN 5-77 p119
+FF 9-77 p5 +-NR 8-77 p3
+HF 10-77 p101 /-RR 5-77 p47
418 Symphony, no. 2, op. 36, D major. Leonore overture, no. 2, op. 72. COA; Eugen Jochum. Philips 6580 175. (Reissue from AXS 9000)
+Gr 11-77 p823 +RR 10-77 p39
+HFN 10-77 p165
419 Symphony, no. 2, op. 36, D major. Leonore overture, no. 3, op. 72. BBC Symphony Orchestra, LSO; Colin Davis. Philips 9500 160 Tape (c) 7300 525.
+Audio 12-77 p50 +-HFN 9-77 p155 tape
-Gr 5-77 p1690 +NR 11-77 p3
+-Gr 9-77 p508 tape +-RR 5-77 p46
+HF 11-77 p106 +-RR 10-77 p92 tape
+-HFN 6-77 p116
420 Symphony, no. 2, op. 36, D major. Consecration of the house, op. 124: Overture. CPhO; Janos Ferencsik. Rediffusion Legend LGD 021.
+-Gr 11-77 p823 +-RR 10-77 p39
+-HFN 10-77 p165
Symphony, no. 2, op. 36, D major. cf Symphony, no. 1, op. 21, C major.
Symphonies, nos. 3, 5-7. cf Works, selections (Vanguard 359/62).
421 Symphony, no. 3, op. 55, E flat major. Grosse Fuge, op. 133, B flat major. BPhO, VPO; Wilhelm Furtwangler. Bruno Walter Society RR 520.
+-NR 7-77 p3
422 Symphony, no. 3, op. 55, E flat major. VPO; Hans Schmidt-Isserstedt. Decca JB 6. (Reissue)
++HFN 12-77 p185 -RR 12-77 p44
423 Symphony, no. 3, op. 55, E flat major. BERLIOZ: Roman carnival, op. 9. SIBELIUS: Kuolema, op. 44: Valse triste. En saga, op. 9. WAGNER: Die Walkure: Ride of the Valkyries. LPO; Victor de Sabata. Decca 6BB 236/7 (2). (Reissues from K 1507/13, K 1552, K 1504/6, K 1562). (also London 26022)
+Gr 10-76 p612 +HFN 11-76 p155
+-HF 9-77 p94 +RR 11-76 p50
424 Symphony, no. 3, op. 55, E flat major. Scottish National Symphony Orchestra; Carlos Paita. Decca SXL 4367. (also London SPC 21152 Tape (c) 5-21152)
+-ARG 12-76 p13 +-RR 8-76 p28
+-HFN 8-76 p76 +-St 5-77 p103
/NR 3-77 p5 +STL 8-8-76 p29
425 Symphony, no. 3, op. 55, E flat major. CSO; Georg Solti. Decca SXL 6829 Tape (c) KSXC 16829. (Reissue from 11BB 188/96)

+Gr 5-77 p1690 +HFN 8-77 p99 tape
+-Gr 9-77 p508 tape +RR 5-77 p47
++HFN 5-77 p137 +-RR 8-77 p86 tape

426 Symphony, no. 3, op. 55, E flat major. Halle Orchestra; James Loughran. Enigma VAR 1033 Tape (c) TC VAR 1033.
+Gr 2-77 p1270 ++RR 2-77 p43
+Gr 5-77 p1738 tape +-RR 8-77 p85 tape
+-HFN 3-77 p97

427 Symphony, no. 3, op. 55, E flat major. CSO; Georg Solti. London CS 7049.
+NR 8-77 p4

428 Symphony, no. 3, op. 55, E flat major. VPO; Erich Kleiber. London 23202.
+NR 5-77 p4 ++SR 5-28-77 p42
+NYT 1-16-77 pD13

429 Symphony, no. 3, op. 55, E flat major. Coriolan overture, op. 62. LPO; Bernard Haitink. Philips 6500 986 Tape (c) 7300 459.
+-Gr 9-77 p508 tape +RR 11-77 p47
+HFN 11-77 p183 -RR 8-77 p86 tape
+HFN 9-77 p155 tape

430 Symphony, no. 3, op. 55, E flat major (Liszt). Roger Woodward, pno. RCA RL 2-5090.
+Gr 8-77 p318 -RR 8-77 p70
+-HFN 10-77 p149

431 Symphony, no. 3, op. 55, E flat major. London Sinfonia; Wyn Morris. Rediffusion Symphonica SYM 5.
+-Gr 12-77 p1067

432 Symphony, no. 3, op. 55, E flat major. CPhO; Lovro von Matacic. Rediffusion Legend LGD 013 Tape (c) LGC 013. (Reissue from Parliament PLPS 129)
+Gr 9-77 p418 +RR 2-77 p95 tape
+-HFN 8-77 p93 +-RR 7-77 p45

433 Symphony, no. 3, op. 55, E flat major. VPO; Erich Kleiber. Richmond R 23202. (Reissue from B 19051)
+-ARG 12-76 p13 +-HF 9-77 p94

Symphony, no. 3, op. 55, E flat major. cf HMV RLS 717.

434 Symphony, no. 4, op. 60, B flat major. PFITZNER: Symphony, op. 46, C major. VPO; Wilhelm Furtwangler. Bruno Walter Society RR 437.
+-NR 5-77 p4

435 Symphony, no. 4, op. 60, B flat major. Consecration of the house overture, op. 124. VPO; Hans Schmidt-Isserstedt. Decca JB 7. (Reissue)
+HFN 12-77 p185 +-RR 12-77 p44

436 Symphony, no. 4, op. 60, B flat major.* WEBER: Oberon: Overture. CSO; Georg Solti. Decca SXL 6830. (*Reissue from 11BB 188/60) (also London CS 7050)
+Gr 8-77 p292 +-RR 7-77 p45
++HFN 7-77 p125 ++SFC 8-21-77 p46

437 Symphony, no. 4, op. 60, B flat major. RPO; Antal Dorati. DG 2535 218 Tape (c) 3335 218.
+-Gr 7-77 p167 +-HFN 7-77 p105
+Gr 9-77 p346

438 Symphony, no. 4, op. 60, B flat major. Grosse Fuge, op. 133, B flat major. BPhO; Wilhelm Furtwangler. DG 2535 813. (Reissues from LPM 18742, 18859)
+Gr 9-77 p418 +-RR 6-77 p43
+HFN 9-77 p136 +-RR 8-77 p44

439 Symphony, no. 4, op. 60, B flat major. HSO; Janos Ferencsik. Hungaroton SLPX 11894.
-ARG 6-77 p12
+FF 9-77 p5
+HF 10-77 p101
+-HFN 5-77 p119
+-NR 8-77 p3
-RR 5-77 p47

440 Symphony, no. 4, op. 60, B flat major. Symphony, no. 5, op. 67, C minor. CO; Georg Szell. Odyssey 34600.
+SFC 3-6-77 p34

441 Symphony, no. 4, op. 60, B flat major. Leonore overtures, nos. 1 and 3, opp. 138 and 72. COA; Eugen Jochum. Philips 6580 146 Tape (c) 7317 161. (Reissue from AXS 9000)
+Gr 4-77 p1542
+-HFN 3-77 p117
+-HFN 5-77 p138 tape
+-RR 3-77 p43

442 Symphony, no. 4, op. 60, B flat major. Grosse Fuge, op. 133, B flat major. AMF; Neville Marriner. Philips 9500 033 Tape (c) 7300 456.
++Audio 3-77 p91
+-Gr 4-76 p1591
+HF 12-76 p147 tape
+HFN 3-76 p88
+HFN 7-76 p104 tape
+-MJ 2-77 p30
+NR 12-76 p3
++RR 3-76 p35
+SFC 9-19-76 p33
++St 12-76 p134

443 Symphony, no. 4, op. 60, B flat major. King Stephen, op. 117: Overture. CPhO; Janos Ferencsik. Rediffusion Legend LGD 003 Tape (c) LGC 003.
+-Gr 7-77 p167
+-HFN 6-77 p137
+RR 2-77 p95 tape
+RR 6-77 p43

Symphony, no. 4, op. 60, B flat major. cf Symphony, no. 1, op. 21, C major (World Records SH 134).

Symphony, no. 4, op. 60, B flat major. cf Symphony, no. 1, op. 21, C major (Electrola 187 039).

Symphony, no. 4, op. 60, B flat major. cf Symphony, no. 2, op. 36, D major.

Symphonies, nos. 5-6, 9. cf Works, selections (Decca D77D5).

444 Symphony, no. 5, op. 67, C minor. SIBELIUS: En saga, op. 9. VPO; Wilhelm Furtwangler. Bruno Walter Society RR 507.
+-NR 7-77 p3

445 Symphony, no. 5, op. 67, C minor. WAGNER: Siegfried Idyll. Rehearsal sequences. Columbia Symphony Orchestra; Bruno Walter. CBS 61772/3 (2).
+Gr 3-77 p1391
+-HFN 3-77 p99

446 Symphony, no. 5, op. 67, C minor (Rehearsa 1st and 2nd movements and performance). WAGNER: Siegfried Idyll (Rehearsal and performance). Columbia Symphony Orchestra; Bruno Walter. CBS 79001.
+RR 3-77 p43

447 Symphony, no. 5, op. 67, C minor. Symphony, no. 8, op. 93, F major. BPhO; Andre Cluytens. Classics for Pleasure CFP 40007 Tape (c) TC CFP 40007. (Reissue)
+Gr 2-77 p1325 tape
+-HFN 12-76 p153 tape

448 Symphony, no. 5, op. 67, C minor. Symphony, no. 8, op. 93, F major. VPO; Hans Schmidt-Isserstedt. Decca Tape (c) KJBC 5.
+HFN 10-77 p169 tape
+RR 10-77 p92 tape

449 Symphony, no. 5, op. 67, C major. Symphony, no. 8, op. 93, F major. VPO; Hans Schmidt-Isserstedt. Decca SXL 6369 Tape (c) KJBC 4. (Reissue)
+HFN 10-77 p169 tape

450 Symphony, no. 5, op. 67, C minor. VPO; Carlos Kleiber. DG 2530

516 Tape (c) 3300 472.

++Audio 3-76 p65 +NR 12-75 p3
++Gr 6-75 p32 ++NYT 8-10-75 pD14
+Gr 3-77 p1458 tape ++RR 6-75 p32
++HF 11-75 p91 +RR 7-75 p68 tape
+HF 9-76 p84 tape ++RR 10-75 p96 tape
++HFN 6-75 p85 +SR 11-1-75 p45
++MT 10-75 p885 ++St 12-75 p81

451 Symphony, no. 5, op. 67, C minor. Egmont, op. 84: Overture. BPhO; Wilhelm Furtwangler. DG 2535 810. (Reissue from LPM 18724)

+-Gr 5-77 p1690 +-RR 5-77 p47
+HFN 6-77 p137

452 Symphony, no. 5, op. 67, C minor. Fidelio, op. 72: Overture. Bavarian Radio Symphony Orchestra; Eugen Jochum. DG 2548 255 Tape (c) 3348 255. (Reissue from SLPM 138024)

+Gr 1-77 p1139 +-HFN 1-77 p119
+Gr 2-77 p1325 tape +RR 1-77 p46

453 Symphony, no. 5, op. 67, C minor. VPO; Wilhelm Furtwangler. EMI 3C 053 00771.

+-RR 8-77 p40

454 Symphony, no. 5, op. 67, C minor. Egmont, op. 84: Overture. HSO; Janos Ferencsik. Hungaroton LPX 11457.

+FF 9-77 p5

455 Symphony, no. 5, op. 67, C minor. Leonore overture, no. 3, op. 72. CSO; Georg Solti. London CS 6930 Tape (c) 5-6930. (also Decca SXL 6762 Tape (c) KSXC 6762)

+-ARG 12-76 p18 ++HFN 1-77 p121
+Gr 3-77 p1458 tape +NR 3-77 p5

456 Symphony, no. 5, op. 67, C minor. MOZART: Symphony, no. 40, K 550, G minor. COA, LPO; Erich Kleiber. London R 23232. (Reissues from LL 912, Decca/London originals, c1949)

+-HF 9-77 p96 +NR 6-77 p4

457 Symphony, no. 5, op. 67, C minor. Fidelio, op. 72: Overture. COA; Eugen Jochum. Philips 6580 145. (Reissue from AXS 9000)

+-Gr 1-77 p1139 +HFN 1-77 p119
+Gr 1-77 p46

458 Symphony, no. 5, op. 67, C minor. SCHUBERT: Symphony, no. 8, D 759, B minor. BSO; Seiji Ozawa. RCA GL 2-5002.

+-Gr 10-76 p578 +-RR 11-76 p51
+-HFN 12-76 p137 +ST 1-77 p779

Symphony, no. 5, op. 67, C minor. cf Concerto, piano, no. 4, op. 58, G major.

Symphony, no. 5, op. 67, C minor. cf Symphony, no. 4, op. 60, B flat major.

459 Symphony, no. 6, op. 68, F major. Egmont, op. 84: Overture. VPO; Hans Schmidt-Isserstedt. Decca Tape (c) KJBC 2.

+HFN 10-77 p169 tape +-RR 10-77 p92 tape

460 Symphony, no. 6, op. 68, F major. CSO; Georg Solti. Decca SXL 6763 Tape (c) KSXC 16763. (Reissue from 11BB 188/96) (also London CS 6931)

++Gr 4-76 p1591 +NR 7-77 p5
+Gr 3-77 p1458 tape +-RR 3-76 p35
+HFN 3-76 p109 +-RR 8-76 p83 tape
+HFN 7-76 p104 tape

461 Symphony, no. 6, op. 68, F major. RPO; Antal Dorati. DG 2535 219 Tape (c) 3335 219.

+HFN 10-77 p169 tape

462 Symphony, no. 6, op. 68, F major. Coriolan overture, op. 62. VPO; Karl Bohm. DG Tape (c) 3300 476.
+Gr 3-77 p1458 tape ++RR 10-75 p96 tape
463 Symphony, no. 6, op. 68, F major. Halle Orchestra; James Loughran. Enigma VAR 1036.
+-HFN 12-77 p165 +RR 12-77 p44
464 Symphony, no. 6, op. 68, F major. Munich Philharmonic Orchestra; Rudolf Kempe. HMV SQ ESD 7004. (Reissue from Q4 SLS 892)
+Gr 2-77 p1273
465 Symphony, no. 6, op. 68, F major. Hungarian State Symphony Orchestra; Janos Ferencsik. Hungaroton SLPX 11790.
+ARG 12-76 p15 ++NR 1-77 p2
++HF 3-77 p96
466 Symphony, no. 6, op. 68, F major. BBC Symphony Orchestra; Colin Davis. Philips 6500 463 Tape (c) 7300 361.
++ARG 11-76 p14 +-MJ 2-77 p30
+Gr 5-76 p1749 -NR 12-76 p3
+-HF 6-77 p82 -RR 5-76 p35
+-HFN 2-76 p116 tape +SFC 9-19-76 p33
+-HFN 5-76 p92
467 Symphony, no. 6, op. 68, F major. LSO; Antal Dorati. Philips Fontana 6531 009 Tape (c) 7328 007
-Gr 2-77 p1326 +HFN 7-77 p127 tape
+-Gr 7-77 p225 tape +-RR 7-77 p98 tape
468 Symphony, no. 6, op. 68, F major. COA; Eugen Jochum. Philips 6580 139. (Reissues)
+HFN 12-77 p185 +-RR 12-77 p44
469 Symphony, no. 6, op. 68, F major. BPhO; Wilhelm Furtwangler. Rococo 2077. (also Unicorn WF 59)
+-NR 9-75 p6 +STL 1-9-77 p35
470 Symphony, no. 6, op. 68, F major. Orchestra dell'Augusteo; Victor de Sabata. World Records SH 235. (Reissue from HMV DB 6473/7) (also EMI Odeon SH 235)
+ARSC, vol 9, no. 1 1977, p80 +HF 9-77 p94
+HFN 10-76 p165
+Gr 12-76 p989 +RR 11-76 p50
471 Symphony, no. 7, op. 92, A major. WAGNER: Tristan und Isolde: Prelude, Act 1: Liebestod. Stockholm Philharmonic Orchestra; Wilhelm Furtwangler. Bruno Walter Society RR 505.
+NR 5-77 p4
472 Symphony, no. 7, op. 92, A major. Leonore overture, no. 3, op. 72. VPO; Hans Schmidt-Isserstedt. Decca Tape (c) KJBC 4.
+-RR 10-77 p92 tape
473 Symphony, no. 7, op. 92, A major. Egmont, op. 84: Overture. NPhO; Leopold Stokowski. Decca PFS 4342. (also London 21139 Tape (c) 5-21139)
+-ARG 6-77 p11 +NR 3-77 p5
+Gr 2-76 p1331 +-MM 6-76 p43
+-HFN 3-76 p88 +St 2-77 p114
+MJ 7-77 p68
474 Symphony, no. 7, op. 92, A major. VPO; Carlos Kleiber. DG 2530 706 Tape (c) 3300 706.
++ARG 6-77 p11 +NR 3-77 p6
+-Gr 9-76 p409 ++RR 9-76 p45
+Gr 4-77 p1603 tape +RR 10-76 p105 tape
+Ha 6-77 p88 ++SFC 1-16-77 p43
++HFN 10-76 p163 +-St 2-77 p114

475 Symphony, no. 7, op. 92, A major. Bavarian Radio Symphony Orchestra; Rafael Kubelik. DG 2535 252 Tape (c) 3335 252.
+Gr 11-77 p824 +-HFN 11-77 p165
+-Gr 12-77 p1146 tape +-RR 12-77 p94 tape

476 Symphony, no. 7, op. 92, A major. Halle Orchestra; James Loughran. Enigma VAR 1037 Tape (c) TC VAR 1037.
+Gr 7-77 p167 +-RR 5-77 p48
+-HFN 7-77 p105 +RR 8-77 p85 tape

477 Symphony, no. 7, op. 92, A major. Coriolan overture, op. 62. CSO; Georg Solti. London CS 6932.
+-NR 8-77 p4

478 Symphony, no. 7, op. 92, A major. COA; Eugene Jochum. Philips 6580 176. (Reissue from AXS 9000)
+-Gr 8-77 p292 +-RR 6-77 p43
+HFN 6-77 p137

479 Symphony, no. 7, op. 92, A major. Fidelio, op. 72: Overture. COA; Eugene Jochum. Philips Tape (c) 7317 180.
+-HFN 10-77 p169 tape -RR 10-77 p92 tape

480 Symphony, no. 7, op. 92, A major. LSO; Colin Davis. Philips 9500 219 Tape (c) 7300 562.
+ARG 6-77 p11 /MJ 7-77 p68
+Audio 9-77 p46 -NR 7-77 p5
+-Gr 3-77 p1391 +-RR 3-77 p44
+HF 6-77 p82 +-RR 12-77 p94 tape
+-HFN 3-77 p97 ++SFC 3-6-77 p34

481 Symphony, no. 8, op. 93, F major. Symphony, no. 9, op. 125, D minor. Carole Farley, s; Alfreda Hodgson, con; Stuart Burrows, t; Norman Bailey, bs; Brighton Festival Chorus; RPO; Antal Dorati. DG 2726 073 (2).
+-Gr 12-76 p989 +-RR 1-77 p46
+-HFN 12-76 p137

482 Symphony, no. 8, op. 93, F major. Symphony, no. 9, op. 125, D minor. Luise Helletsgruber, s; Rosette Anday, alto; Georg Maikl, t; Richard Mayr, bs; Vienna State Opera Chorus; VPO; Felix Weingartner. Turnabout THS 65076/7 (2). (Reissues from European Columbia originals, 1935-36)
+HF 9-77 p94 +NR 7-77 p3

483 Symphony, no. 8, op. 93, F major. Leonore overture, no. 3, op. 72. (Also rehearsal of the overture) Stockholm Philharmonic Orchestra; Wilhelm Furtwangler. Unicorn WFS 5.
+SFC 3-16-75 p25 +STL 1-9-77 p35

Symphony, no. 8, op. 93, F major. cf Symphony, no. 1, op. 21, C major (Decca SXL 6760).

Symphony, no. 8, op. 93, F major. cf Symphony, no. 1, op. 21, C major (Hungaroton SLPX 11890).

Symphony, no. 8, op. 93, F major. cf Symphony, no. 1, op. 21, C major (Philips 6580 148).

Symphony, no. 8, op. 93, F major. cf Symphony, no. 1, op. 21, C major (Rediffusion Legend LGC 005).

Symphony, no. 8, op. 93, F major. cf Symphony, no. 2, op. 36, D major.

Symphony, no. 8, op. 93, F major. cf Symphony, no. 5, op. 67, C minor (Classics for Pleasure CFP 40007).

Symphony, no. 8, op. 93, F major. cf Symphony, no. 5, op. 67, C minor (Decca KJBC 5).

Symphony, no. 8, op. 93, F major. cf Symphony, no. 5, op. 67, C minor (Decca SXL 6396)

484 Symphony, no. 9, op. 125, D minor. Leonore overture, no. 3, op. 72. STRAUSS, R.: Don Juan, op. 20. Irmgard Seefried, s; Rosette Anday, con; Anton Dermota, t; Paul Schoffler, bar; VPO; Wilhelm Furtwangler. Bruno Walter Society RR 460 (2).
+-NR 7-77 p3 +-NR 5-77 p4
485 Symphony, no. 9, op. 125, D minor. PO; Eugene Ormandy. CBS 61747 Tape (c) 40-61747.
+-HFN 1-77 p123 tape
486 Symphony, no. 9, op. 125, D minor. Joan Sutherland, Marilyn Horne, s; James King, t; Martti Talvela, bs; VPO; Hans Schmidt-Isserstedt. Decca Tape (c) KJBC 1. (Reissue from KSXC 6233)
+-HFN 12-77 p187 tape +RR 11-77 p118 tape
487 Symphony, no. 9, op. 125, D minor. Soloists; Prague Philharmonic Choir; CPhO; Vaclav Neumann. Denon PCM OB 7333/4 ND (2).
+HF 10-77 p112
488 Symphony, no. 9, op. 125, D minor. Kerstin Lindberg, s; Else Jena, ms; Erik Sjoberg, t; Holger Bryding, bar; Danish State Radio Orchestra and Chorus; Fritz Busch. DG 2535 814. (also Nordisk Polyphon 0008633)
+Gr 8-77 p292 +-RR 8-77 p49
+HFN 9-77 p136
489 Symphony, no. 9, op. 125, D minor. VPO; Erich Kleiber. London R 23201. (Reissue from LL 632/3, Richmond B 19083)
+HF 9-77 p94 +SFC 9-19-77 p33
+NYT 1-16-77 pD13
490 Symphony, no. 9, op. 125, D minor. PFITZNER: Symphony, op. 46, C major. VPO; Wilhelm Furtwangler. Rococo 2109 (2).
+-NR 2-77 p3
Symphony, no. 9, op. 125, D minor. cf Symphony, no. 8, op. 93, F major (DG 2726 073).
Symphony, no. 9, op. 125, D minor. cf Symphony, no. 8, op. 93, F major (Turnabout THS 65076/7)
491 Trio, bassoon, flute and piano. LARSSON: Concertino, bassoon, op. 45, no. 4. Concertino, flute, op. 45, no. 1. Gunilla von Bahr, flt; Knut Sonstevold, bsn; Lucia Negro, pno; Musica Sveciae; Sven Verde. BIS LP 40.
+Gr 9-77 p500 +-RR 3-77 p64
Trio, bassoon, flute and piano, G major. cf Works, selections (DG 2721 129).
Trio, clarinet. cf Quartets, piano, nos. 1-3.
Trios, piano, E flat major, B flat major. cf Quartets, piano, nos. 1-3.
Trios, piano, nos. 1-6. cf Quartets, piano, nos. 1-3.
492 Trio, piano, no. 1, op. 1, no. 1, E flat major. Trio, piano, no. 3, op. 1, no. 3, C minor. George Malcolm Piano Trio. Crescent ARS 108.
+-Gr 5-77 p1704 +RR 5-77 p62
+-HFN 5-77 p119
Trio, piano, no, 3, op. 1, no. 3, C minor. cf Trio, piano, no. 1, op. 1, E flat major.
493 Trio, piano, no. 6, op. 97, B flat major. Isaac Stern, vln; Leonard Rose, vlc; Eugene Istomin, pno. Columbia MS 6819.
++St 5-77 p71
494 Trio, piano, no. 6, op. 97, B flat major. Suk Trio. Rediffusion Legend LGD 002 Tape (c) LGC 002.
+Gr 7-77 p192 +RR 2-77 p96 tape
+-HFN 6-77 p137 +RR 6-77 p72

Trio, piano, op. 36, D major (after Symphony, no. 2). cf Quintet, piano, op. 16, E flat major.
495 Trios, strings, opp. 3 and 9. Trio, strings, op. 8, D major. Italian String trio. DG 2721 131 (3).
+Gr 3-77 p1416 +RR 4-77 p61
Trio, strings, op. 3, no. 1, E flat major. cf RCA CRM 6-2264.
496 Trio, strings, op. 8, D major. KODALY: Duo, op. 7. Jascha Heifetz, vln; Gregor Piatigorsky, vlc; William Primrose, vla. RCA LSC 2550.
++St 5-77 p71 ++ST 1-77 p775
Trio, strings, op. 8, D major. cf Trios, strings, opp. 3 and 9.
Trio, strings, op. 9, no. 1, G major. cf RCA CRM 6-2264.
Trio, strings, op. 9, no. 3, C minor. cf RCA CRM 6-2264.
Variations. cf Works, selections (DG 2721 133).
Variations (32). cf Concerto, piano, no. 1, op. 15, C major.
Variations, opp. 44 and 121a. cf Quartets, piano, nos. 1-3.
Variations, opp. 46 and 66. cf Works, selections (DG 2721 133).
Variations and fugue, op. 35, E flat major. cf Sonata, piano, no. 21, op. 53, C major.
497 Variations on a theme by Diabelli, op. 120 (33). Alfred Brendel, pno. Philips 9500 381.
+-Gr 11-77 p860 +RR 11-77 p84
+-HFN 11-77 p165
498 Variations on a theme by Diabelli, op. 120. Sviatoslav Richter, pno. Rococo 2098.
+-NR 5-77 p14
499 Variations on a theme by Diabelli, op. 120. Daniel Barenboim, pno. Westminster WG 1007.
+-RR 2-77 p76
Variations on a theme from Handel's "Judas Maccabeus" (12). cf Sonatas, violoncello and piano, nos. 1-5.
Variations on "Air de la petite Russie", op. 107, no. 3. cf Sonata, flute, B flat major.
Variations on "Air russe", op. 107, no. 7. cf Sonata, flute, B flat major.
Variations on Mozart's "Bei Mannern" (7). cf Sonatas, violoncello and piano, nos. 1-5.
Variations on Mozart's "Bei Mannern" (7). cf HMV RLS 723.
Variations on Mozart's "Ein Madchen oder Weibchen" (12). cf Sonatas, violoncello and piano, nos. 1-5.
500 Works, selections: Concerto, piano, no. 5, op. 73, E flat major. Egmont, op. 84: Overture. Leonore overture, no. 3, op. 72. Sonatas, piano, nos. 8, 14, 23. Sonata, violin and piano, no. 5, op. 24, F major. Symphonies, nos. 5-6, 9. Joan Sutherland, s; Marilyn Horne, con; James King, t; Martti Talvela, bs; Vladimir Ashkenazy, Wilhelm Backhaus, pno; Itzhak Perlman, vln; VPO, CSO; Vienna State Opera Chorus; Hans Schmidt-Isserstedt, Georg Solti. Decca D77D5 (5). (Reissues from SXL 6396, 6233, 5736, 2190, 2241, 11BB 188/96, SXLG 6594/7, SWL 8016, SXLP 6684)
+Gr 12-77 p1068 +-RR 12-77 p44
+-HFN 12-77 p185
501 Works, selections: Octet, op. 103, E flat major. Quintet, piano, op. 16, E flat major. Septet, op. 20, E flat major. Serenade, op. 25, D major. Sextet, op. 71, E flat major. Sextet, op. 81b, E flat major. Sonata, horn and piano, op. 17, F major. Trio, bassoon, flute and piano, G major. Jorg Demus, Aloys

Kontarsky, pno; Karlheinz Zoller, flt; Klaus Thunemann, bsn; Siegbert Uberschaer, vla; Thomas Brandis, vln; Gerd Siefert, hn; BPhO Members, Drolc Quartet. DG 2721 129 (4).
+Gr 3-77 p1415 +-RR 4-77 p61

502 Works, selections: Rondo. Sonatas, violin and piano, nos. 1-10. Sonatas, violoncello, nos. 1-5. Variations. Variations, opp. 46 and 66. Yehudi Menuhin, vln; Pierre Fournier, vlc; Wilhelm Kempff, pno. DG 2721 133 (8).
++Gr 3-77 p1416 +-RR 4-77 p61

503 Works, selections: Ah perfido, op. 65. An die ferne Geliebte, op. 98. Christ on the Mount of Olives, op. 85. Calm sea and prosperous voyage, op. 112. Fantasia, op. 80, C minor. Gellert Lieder, op. 48. Folksongs arrangements (29). Lieder (41). Edith Mathis, Birgit Nilsson, Elizabeth Harwood, s; Julia Hamari, con; Alexander Young, James King, t; Franz Crass, bs; Dietrich Fischer-Dieskau, bar; RIAS Chamber Choir, Vienna Singverein; Georg Donderer, vlc; Andreas Rohn, vln; Karl Engel, Jorg Demus, pno; VSO; Ferdinand Leitner, Bernhard Klee. DG 2721 138.
+Gr 3-77 p1416 +RR 4-77 p80

504 Works, selections: Consecration of the house, op. 124: Overture. The creatures of Prometheus, op. 43: Ballet music. Coriolan overture, op. 62. Egmont, op. 84: Overture. King Stephen, op. 117: Overture. Leonore overtures, nos. 1-3. PhO, NPhO; Otto Klemperer. HMV SXDW 3032 (2) Tape (c) TC2 SXDW 3032. (Reissues from Columbia 33CX 1575, SAX 2373, 2354, 2542, 2331, 2451/3)
+-Gr 6-77 p41 +HFN 7-77 p107
-Gr 8-77 p346 tape ++RR 7-77 p45

505 Works, selections: March, woodwinds, B flat major. Octet, op. 103, E flat major. Quintet, oboe, 3 horns and bassoon, B flat major, excerpt. Rondino, E flat major. Sextet, op. 71, E flat major. London Wind Soloists. London STS 15387.
+MJ 7-77 p68

506 Works, selections: March, B flat major. Octet, op. 103, E flat major. Quintet, E flat major. Rondino, E flat major. Sextet, op. 71, E flat major. NWE. Philips 9500 087.
+Gr 6-77 p68 +RR 6-77 p72
+HFN 6-77 p117 ++SFC 8-21-77 p46
+-MT 10-77 p825 +St 11-77 p134
++NR 10-77 p7

507 Works, selections: Symphonies, nos. 3, 5-7. Egmont, op. 84: Overture. Coriolan overture, op. 62. Leonore overture, no. 3, op. 72. Fidelio, op. 72: Overture. Philharmonia Promenade Orchestra; Adrian Boult. Vanguard 359/62 (4).
++Audio 11-77 p128 +SFC 6-19-77 p46
+-FF 11-77 p7

BEHRENS, Jack

508 The feast of life. GILBERT: Poems VI and VII. ROBB: Dialogue. SHIELDS: Wildcat songs. Jack Behrens, pno; John Donald Robb, gtr, pno; Paul Dunkel, pic; Humbert Lucarelli, ob; David Gilbert, flt; Stepanie Turash, s. Opus One 13.
++ARG 2-77 p51

BELLINI, Vicenzo

Adelson e Salvini: Ecco, signor, la sposa. cf Philips 9500 203.
Almen se non poss'io. cf L'Oiseau-Lyre SOL 345.
Beatrice di Tenda: Deh, so un urna. cf Decca D65D3.

509 I Capuleti ed i Montecchi: Eccomi in lieta vesta...Oh quante volte. ROSSINI: Il barbiere di Siviglia: Una voce poco fa, Contro un cor che accende amore...Cara immagine ridente. L'Assedio di Corinto: Cielo che diverro...Si ferite, il chieggo, il merto... Dah soggiorno degli estinti...No non piu spero, oh Dio. VERDI: La traviata: E strano...Ah fors'e lui...Follie, Follie... Sempre libera. Beverly Sills, s; orchestral accompaniments. Angel S 37255.
+-NR 6-77 p10

510 Concerto, oboe, E flat major. MOLIQUE: Concertino, oboe, G minor. MOSCHELES: Concertante, flute and oboe, F major. RIETZ: Konzertstuck, oboe, op. 33, F minor. Heinz Holliger, ob; Aurele Nicolet, flt; Frankfurt Radio Symphony Orchestra; Eliahu Inbal. Philips 9500 070 Tape (c) 7300 515.
+ARG 11-77 p43 +MJ 9-77 p35
+FF 11-77 p57 +NR 12-77 p4
+-Gr 1-77 p1140 ++RR 1-77 p51
++HF 10-77 p97 +RR 10-77 p92 tape
+HFN 1-77 p113 +St 10-77 p148
+HFN 7-77 p127 tape

Little overture. cf Westminster WGS 8338.

511 Norma. Joan Sutherland, Marilyn Horne, s; Yvonne Minton, con; John Alexander, Joseph Ward, t; Richard Cross, bs; LSO and Chorus; Richard Bonynge. Decca Tape (c) K21K32 (2).
+-Gr 7-77 p225 tape +RR 10-77 p92 tape
+HFN 7-77 p127 tape

Norma: Casta diva. cf Decca D65D3.
Norma: Casta diva. cf HMV SLS 5104.
Norma: Casta diva. cf Rubini GV 57.
Norma: Dormono entrambi...Teneri, teneri figlia, In mia man alfin tu sei. cf Court Opera Classics CO 347.
Norma: Ite sul colle. cf VERDI: Simon Boccanegra.
Norma: Sediziose voci...Casta diva. cf HMV SLS 5057.
Norma: Sgombra e la sacra selva...Deh proteggi me, o Dio. cf Club 99-106.
Il pirata: Oh, s'io potessi...Col sorriso d'innocenza. cf HMV SLS 5057.

512 I puritani. Joan Sutherland, s; Anita Caminada, ms; Luciano Pavarotti, Renato Cazzaniga, t; Piero Cappuccilli, bar; Nicolai Ghiaurov, Gian Carlo Luccardi, bs; ROHO Chorus; LSO; Richard Bonynge. London OSA 13111 (3) Tape (c) OSA 5-13111. (also Decca SET 587/9 Tape (c) K25K32)
+Gr 7-75 p237 +ON 4-19-75 p54
++HF 6-75 p84 ++ON 6-75 p10
+HF 7-77 p125 tape +RR 7-75 p16
+HFN 7-75 p76 +RR 4-77 p90 tape
+HFN 6-77 p137 ++SFC 2-23-75 p22
+HFN 5-77 p138 tape ++St 5-75 p73
+MT 2-76 p139 +STL 10-5-75 p36
+NYT 4-6-75 pD18

513 I puritani: Ah, per sempre io ti perdei; A te, o cara; E la vergine adorata..son vergin vezzosa; Cinta di fiori; O rendetemi la speme...Qui la voce sua soave; Torno il riso...Vien, diletto; Riccardo Riccardo...Suoni la tromba; Qual suon...A una fonte afflitto e solo; Dunque m'ami, mio Arturo si...Vieni fra questa braccia; Credeasi misera. Joan Sutherland, s; Anita Caminada, ms; Gian Carlo Luccardi, Luciano Pavarotti, t; Piero

Cappuccilli, bar; Nicolai Ghiaurov, bs; ROHO Chorus; LSO; Richard Bonynge. Decca SET 619. (Reissue from SET 587/9)
+Gr 5-77 p1724 +RR 5-77 p34

514 I puritani: Qui la voce...Vien diletto. La sonnambula: Ah, non credea mirarti...Ah, non giunge. DONIZETTI: Lucia di Lammermoor: Il dolce suono...Ardon gl'incensi. ROSSINI: The barber of Seville: Una voce poco fa. VERDI: Rigoletto: Caro nome. La traviata: Ah, fors'e lui...Sempre libera. Anna Moffo, s; PhO; Colin Davis. Seraphim S 60281.
+NR 6-77 p8 +St 11-77 p162

I puritani: O rendetemi la speme...Qui la voce. cf HMV SLS 5057.

I puritani: O rendetemi la speme...Qui la voce...Vien diletto. cf Angel S 37446.

I puritani: Qui la voce. cf Decca D65D3.

I puritani: Son vergin vizzosa. cf HMV SLS 5014.

515 Songs: Dolente immagine; Il fervido desidereo; Vaga luna, che inargenti. CHOPIN: Songs: Dumka; Chants polonaise, op. 74, no. 1: The maiden's wish; no. 5: The warrior; no. 13: Spring. DONIZETTI: Songs: La corrispondenza amorosa; A mezzanotte; La mere et l'enfant. ROSSINI: Songs: La danza; L'orgia; La promessa. Leyla Gencer, s; Marcello Guerrini, pno. Cetra LPO 2003. (also LPS 69003)
+-Gr 7-77 p221 -HF 9-77 p105

516 Songs: Bella nice; Il fervido desiderio; Malinconia ninfa gentile; Vanne o rosa fortunata. DONIZETTI: Songs: A mezzanotte; Me voglio fa na casa; La zingara. ROSSINI: Songs: La fioraia fiorentina; Mi lagnero tacendo; L'invito; La promesa. VERDI: Songs: Perduta ho la pace; Ad una stella; Stornello; Lo spazzacamino. Anna Moffo, s; Giorgio Favaretto, pno. Westminster WG 1014.
+-RR 4-77 p80

517 La sonnambula: Perdona o mia diletta...Prendi l'anel ti dono. DONIZETTI: Linda di Chamounix: Linda, Linda...Da quel di che t'incontrai. VERDI: Aida: La fatal pietra...O terra, addio. Otello: Gia nella notte densa. La traviata: Libiamo ne lieti calici; Un di felice; Signora...Che t'accadde...Parigi, O cara. Joan Sutherland, Jacquelyn Fugelle, s; Elizabeth Connell, ms; Luciano Pavarotti, t; London Opera Chorus; National Philharmonic Orchestra; Richard Bonynge. Decca SXL 6828. (also London OS 26499)
+Gr 12-77 p1138 +RR 12-77 p36
+HFN 12-77 p175

La sonnambula: Ah non credea mirarti. cf HMV SLS 5104.

La sonnambula: Ah non credea mirarti...Ah, non giunge. cf I puritani: Qui la voce...Vien diletto.

La sonnambula: Care compagne...Come per me sereno. cf HMV SLS 5057.

La sonnambula: Come per me sereno. cf Decca D65D3.

BELKNAP

The seasons. cf Vox SVBX 5350.

BELTON

Down the mall. cf DJM Records DJM 22062.

BEMBERG, Herman

Songs: Les anges pleurent; Un ange est venu; Chant hindou; Chant venetien; Elaine: L'amour est pur; Nymphs et Sylvains; Sur le lac. cf HMV RLS 719.

BENATSKY, Ralph
Im Weissen Rossl: Mein Liebeslied muss ein Walzer sein. cf HMV ESD 7043.
BENDA, Franz (Frantisek)
518 Concerto, flute and strings, E minor. RICHTER: Concerto, flute and strings, D major. Vladislav Brunner, Jr., flt; Alexander Cattarino, hpd; Slovak Chamber Orchestra; Bohdan Warchal. Opus 9111 0375.
+Gr 5-77 p1695 +RR 3-77 p58
BENDA, Jiri Antonin (Georg)
519 Concerti, harpsichord and strings, F minor, B minor, G major. Josef Hala, hpd; String Quintet. Supraphon 111 2138.
++ARG 10-77 p22 +NR 12-77 p4
+FF 9-77 p8 +RR 6-77 p72
+-Gr 9-77 p418 +St 10-77 p134
+HFN 7-77 p107
Sonatina, D major. cf RCA ARL 1-0864.
Sonatina, D minor. cf RCA ARL 1-0864.
BENEDICT, Julius
La capinera. cf BIS LP 45.
Greetings to America. cf Swedish SLT 33209.
BEN-HAIM, Paul
Pieces, piano, op. 34: Toccata. cf Decca PFS 4387.
BENJAMIN, Arthur
Fanfares: For a state occasion; For a brilliant occasion; For a gala occasion. cf RCA RL 2-5081.
Fanfare for a festive occasion. cf RCA RL 2-5081.
Jamaican rumba. cf CBS 61039.
BENNET, John
Songs: All creatures now. cf Argo ZK 25.
Songs: All creatures now are merry. cf DG 2533 347.
BENNETT, Richard Rodney
Alba. cf Vista VPS 1034.
520 Calendar. GOEHR: Choruses (2). MAXWELL DAVIES: Leopardi fragments. WILLIAMSON: Symphony, voices. Mary Thomas, s; Rosemary Phillips, Pauline Stevens, con; Geoffrey Shaw, bar; John Alldis Choir; Melos Ensemble; John Carewe. Argo ZRG 758. (Reissue from HMV ASD 640)
+Gr 5-75 p2000 +MT 9-75 p797
-HFN 5-75 p125 +NR 10-75 p6
+MQ 10-77 p572 *RR 4-75 p59
Commedia IV. cf ADDISON: Divertimento, op. 9.
Fanfare, brass quintet. cf Argo ZRG 851.
Heare us, O heare us, Lord. cf Abbey LPB 783.
Impromptu. cf Universal Edition UE 15043.
521 Telegram. LUYTENS: Bagatelles, op. 49 (5). Intermezzi (5). Piano e forte, op. 43. Plenum, op. 87. WATKINS: Synthesis. WILLIAMSON: Ritual of admiration. Richard Deering, pno. Pearl SHE 537.
+Gr 5-77 p1714 +RR 7-77 p82
Tom O'Bedlams song. cf Argo ZK 28/9.
BENNETT, Robert Russell
522 The fun and faith of William Billings, American. BILLINGS: Hymns (3). SCHUMAN: New England triptych. University of Maryland Chorus; National Philharmonic Orchestra; Antal Dorati. London OS 26442.
+-SFC 7-3-77 p34 +St 6-77 p126

BENNETT, William Sterndale
Psalm, no. 71. cf Vista VPS 1037.
Songs: The carol singers. cf Argo ZFB 95/6.
BENTZON, Nils
523 Quartet, strings, no. 8, op. 228. ROSENBERG: Quartet, strings, no. 12. Copenhagen Quartet. Caprice CAP 1100.
+HFN 11-77 p167 +RR 11-77 p74
BERG
Herdegossen. cf Swedish SLT 33209.
BERG, Alban
524 Lyric suite. SCHOENBERG: Verklarte Nacht, op. 4. NYP; Pierre Boulez. CBS 76305 Tape (c) 40-76305.
+-Gr 9-77 p423 +RR 8-77 p62
++HFN 9-77 p137 +RR 11-77 p121
+HFN 10-77 p169 tape
BERGER
Amoureuse. cf Angel (Q) S 37304.
BERGER, Arthur
Quartet, C major. cf Vox SVBX 5307.
BERGER, Jean
I lift up my eyes. cf Columbia M 34134.
BERGHMANS
La femme a barbe. cf Boston Brass BB 1001.
BERIO, Luciano
525 Allelujah II. Concerto, 2 pianos. Nones. Bruno Canino, Antonio Ballista, pno; LSO, BBC Symphony Orchestra; Pierre Boulez, Luciano Berio. RCA ARL 1-1674.
+ARG 4-77 p14 +MT 12-77 p1014
+Gr 8-77 p292 ++NR 3-77 p7
+HFN 10-77 p152 +-RR 9-77 p57
-MJ 10-77 p29 ++SFC 12-19-76 p50
526 A-Ronne. Cries of London. Swingle II; Luciano Berio. Decca HEAD 15.
++ARG 8-77 p650 ++HFN 12-76 p137
+Gr 1-77 p1162 ++RR 12-76 p89
Children's play, wind quartet, op. zoo. cf Vox SVBX 5307.
527 Chimens IV. Concertino. Linea. Points on the curve to find. Anthony di Bonaventura, Katia and Marielle Labeque, pno; Anthony Pay, clt; Nona Liddell, vln; Heinz Holliger, ob; Jean-Pierre Dronet, vibraphone; Sylvio Gualda, marimba; London Sinfonietta. RCA ARL 1-2291.
++FF 11-77 p9
Concertino. cf Chimens IV.
Concerto, 2 pianos. cf Allelujah II.
Cries of London. cf A-Ronne.
Linea. cf Chemens IV.
The modification and instrumentation of a famous hornpipe as merry and altogether sincere homage to Uncle Alfred. cf Universal Edition UE 15043.
Nones. cf Allelujah II.
O King. cf Delos DEL 25406.
Points on the curve to find. cf Chimens IV.
528 Rounds. BOULEZ: Sonata, piano, no. 1. Sonata, piano, no. 3. Trope. HOLLIGER: Elis. MESSIAEN: Canteyodiaya. Klara Kormendi, pno. Hungaroton SLPX 11771.
+ARG 5-77 p48 ++SFC 6-12-77 p41
+NR 8-77 p14 +St 6-77 p126

Sequenza IV. cf CP 3-5.

BERKELEY, Lennox

529 Concerto, guitar. RODRIGO: Concierto de Aranjuez, guitar and orchestra. Julian Bream, gtr; Monteverdi Orchestra; John Eliot Gardiner. RCA ARL 1-1181 Tape (c) ARK 1-1181 (ct) ARS 1-1181.
+Gr 11-75 p807 ++RR 12-75 p60
+HF 2-76 p130 tape +-RR 1-77 p91 tape
++HFN 12-75 p149 +SR 11-29-75 p50
+MM 8-76 p35 +St 1-77 p72
+NR 12-75 p14

530 Duo, violoncello and piano. DALBY: Variations, violoncello and piano. FRICKER: Sonata, violoncello and piano, op. 28. MCCABE: Partita, solo violoncello. Julian Lloyd Webber, vlc; John McCabe, pno. L'Oiseau-Lyre DSLO 18.
+Gr 4-77 p1569 +MT 9-77 p736
+-HFN 4-77 p133 +RR 3-77 p78

The Lord is my shepherd. cf Abbey LPB 770.

Sonata, op. 51. cf RCA RK 1-1735.

Sonatina. cf RCA Tape (c) RK 2-5030.

Spring at the hour. cf RCA GL 2-5062.

BERLINSKI, Herman

The burning bush. cf Delos FY 025.

BERLIOZ, Hector

Beatrice et Benedict: Dieu, Que viens-je d'entendre...Il m'en souvient. cf CBS 76522.

Beatrice et Benedict: Overture. cf Overtures (Angel 37170).

Benvenuto Cellini, op. 23. cf Overtures (Angel S 37170).

La carnival romain (Roman carnival), op. 9. cf Overtures (Angel S 37170).

Roman carnival, op. 9. cf BEETHOVEN: Symphony, no. 3, op. 55, E flat major.

Roman carnival, op. 9. cf Decca SXL 6782.

Roman carnival, op. 9. cf Pye PCNHX 6.

531 Le Corsaire, op. 21. MENDELSSOHN: A midsummer night's dream, op. 21: Overture. The fair Melusina, op. 32. ROSSINI: Cambiale di matrimonio: Overture. La gazza ladra. RPO; Thomas Beecham. Quintessence PMC 7004.
+HF 9-77 p108 +St 11-77 p147

532 Le Corsaire, op. 21. The damnation of Faust, op. 24: Rakoczy march. BIZET: L'Arlesienne: Suite, no. 2: Farandole. FALLA: The three-cornered hat: Jota. TCHAIKOVSKY: Eugene Onegin, op. 24: Polonaise. CO; Lorin Maazel. Telarc 5020.
+ARG 7-77 p42 +MJ 9-77 p34
+Gr 6-77 p68 +-St 10-77 p144
+HF 7-77 p122

Le Corsaire, op. 21. cf Overtures (Angel S 37170).

The damnation of Faust, op. 24, excerpts. cf GOUNOD: Faust.

La damnation de Faust, op. 24: D'amour l'ardente flamme. cf CBS 76522.

The damnation of Faust, op. 24: Hungarian march (Rakoczy). cf DG 2548 148.

The damnation of Faust, op. 24: Rakoczy march. cf Le Corsaire, op. 21.

The damnation of Faust, op. 24: Rakoczy march. cf Decca SB 713.

La damnation de Faust, op. 24: Mephisto's serenade. cf International Piano Library IPL 101.

La damnation de Faust, op. 24: Serenade. cf Rubini GV 39.

533 L'Enfance du Christ, op. 25. Janet Baker, ms; Eric Tappy, Philip Langridge, t; Thomas Allen, Raimund Herincx, bar; Joseph Rouleau, Jules Bastin, bs; John Alldis Choir; LSO; Colin Davis. Philips 6700 106 (2) Tape (c) 7699 058.
+Gr 10-77 p677 ++RR 10-77 p81
+HFN 10-77 p152 ++SFC 10-9-77 p40
+NR 12-77 p8 ++St 12-77 p133
+-NYT 10-2-77 pD19

Les Francs Juges, op. 3. cf Overtures (Angel S 37170).

Les Francs Juges, op. 3. cf Symphonie fantastique, op. 14.

534 Harold in Italy, op. 16. Pinchas Zukerman, vla; Orchestre de Paris; Daniel Barenboim. CBS 76593. (also Columbia M 34541)
+Gr 6-77 p41 +NYT 10-2-77 pD19
+HFN 7-77 p108 +RR 7-77 p45
+MT 11-77 p921 -SFC 10-9-77 p40
+NR 12-77 p3

535 Harold in Italy, op. 16. Donald McInnes, vla; French National Orchestra; Leonard Bernstein. HMV SQ ASD 3389. (also Angel S 37413)
+-Gr 11-77 p829 -NYT 10-2-77 pD19
+-HFN 12-77 p165 +RR 12-77 p45

536 Harold in Italy, op. 16. Nobuko Imai, vla; LSO; Colin Davis. Philips 9500 026 Tape (c) 7300 441.
+Gr 3-76 p1463 ++NR 6-76 p4
+-HF 10-76 p103 +RR 2-76 p26
+-HF 2-77 p118 tape ++SFC 5-23-76 p36
+HFN 2-76 p92 ++St 9-76 p118
+MT 6-77 p485

Irlande, op. 2: La belle voyageuse. cf Songs (Philips 6500 009).

537 Lelio, op. 14b. Jean Topard, speaker; Nicolai Gedda, Charles Burles, t; Jean van Gorp, bar; ORTF Chorus; Jean Martinon. Angel S 37139.
+-ARG 11-77 p12 -NYT 10-2-77 pD19
+-HF 10-77 p101 +SFC 8-21-77 p46
+-NR 10-77 p11 +St 11-77 p134

538 La mort de Cleopatre. Les nuits d'ete, op. 7. Yvonne Minton, ms; Stuart Burrows, t; BBC Symphony Orchestra; Pierre Boulez. CBS 76576.
+-FF 9-77 p8 +-HFN 8-77 p77
+-Gr 8-77 p328 +-RR 8-77 p74

539 La mort de Cleopatre: Lyric scene. Les Troyens: Act 5, scenes 2 and 3. Janet Baker, ms; Bernadette Greevy, con; Gwynne Howell, bar; Keith Erwen, t; Ambrosian Opera Chorus; LSO; Alexander Gibson. HMV SXLP 30248 Tape (c) TC SXLP 30248. (Reissues from ASD 2516)
++Gr 9-77 p470 +HFN 11-77 p185
+Gr 11-77 p900 tape ++RR 11-77 p36

540 Les nuits d'ete, op. 7. RAVEL: Sheherazade. Regine Crespin, s; OSR; Ernest Ansermet. Decca JB 15 Tape (c) KJBC 15. (Reissue from SXL 6081)
+Gr 10-77 p706 tape +HFN 10-77 p169 tape
+Gr 9-77 p467 +-RR 9-77 p87
+HFN 9-77 p153 +RR 12-77 p94 tape

Les nuits d'ete, op. 7. cf La mort de Cleopatre.

Les nuits d'ete, op. 7. cf Songs (Philips 6500 009).

Les nuits d'ete, op. 7: Le spectre de la rose; Absence. cf HMV RLS 716.

541 Overtures: Beatrice et Benedict. Benvenuto Cellini, op. 23. La carnival romain, op. 9. Le Corsaire, op. 21. Les Francs Juges, op. 3. LSO; Andre Previn. Angel (Q) S 37170. (also HMV SQ ASD 3212 Tape (c) TC ASD 3212)
+Gr 6-76 p38 +RR 6-76 p36
+-Gr 10-76 p658 tape ++RR 7-76 p47
+-HF 9-76 p89 +RR 2-77 p96 tape
++HFN 6-76 p81 -SFC 5-23-76 p36
++HFN 11-76 p175 ++St 10-76 p120
+NR 6-76 p175
542 Requiem (Grande messe des morts), op. 5. Stuart Burrows, t; French National Radio Orchestra and Chorus; Leonard Bernstein. CBS 79205 (2). (also Columbia M2 34202 Tape (c) MT 34202)
+Gr 11-76 p844 +MU 2-77 p8
+-HF 4-77 p93 +NR 12-76 p9
+-HF 11-77 p138 tape ++RR 11-76 p94
++HFN 11-76 p153 ++SFC 10-24-76 p35
+MJ 1-77 p29 +St 3-77 p126
543 Requiem, op. 5. Peter Schreier, t; Bavarian Radio Orchestra and Chorus; Charles Munch. DG 2726 050 (2). (Reissue from SLPM 139264/5)
+Gr 3-77 p1437 +RR 3-77 p86
+HFN 8-77 p95
544 Requiem, op. 5. Robert Tear, t; Birmingham City Symphony Orchestra and Chorus; Louis Fremaux. HMV SLS 982 (2) Tape (c) TC SLS 982. (also Angel (Q) SB 3814)
+Gr 9-75 p499 /NR 11-76 p10
+Gr 12-75 p1121 tape ++RR 10-75 p81
/-HF 4-77 p93 +RR 3-76 p76 tape
+HFN 10-75 p138 +St 3-77 p126
+-HFN 12-75 p173 tape ++STL 9-7-75 p37
+-MT 2-76 p131
Requiem, op. 5: Sanctus. cf Decca SXL 6781.
Requiem, op. 5: Sanctus. cf London OS 26437.
545 Romeo et Juliette, op. 17. Julia Hamari, ms; Jean Dupouy, t; Jose van Dam, bs-bar; New England Conservatory Chorus; BSO; Seiji Ozawa. DG 2707 089 (2) Tape (c) 3370 011.
+-ARG 5-77 p15 +NYT 10-2-77 pD19
++Audio 8-77 p93 +ON 4-16-77 p37
+-Gr 11-76 p844 +-OR 6/7-77 p28
-HF 6-77 p82 +-RR 11-76 p94
+-HFN 11-76 p158 ++SFC 2-20-77 p39
+-MJ 5-77 p32 ++St 5-77 p104
++NR 6-77 p8
546 Romeo et Juliette, op. 17. Regina Resnik, ms; Andre Turp, t; David Ward, bs; LSO and Chorus; Pierre Monteux. Westminster WGD 2002 (2).
+-RR 1-77 p78
547 Romeo and Juliet, op. 17: Orchestral excerpts. Vienna State Opera Chorus; VPO; Lorin Maazel. Decca SXL 6800 Tape (c) KSXC 6800. (Reissue from SET 570/1)
+-Gr 3-77 p1392 +-HFN 5-77 p138 tape
+Gr 8-77 p346 tape +RR 3-77 p44
+-HFN 3-77 p117 +RR 5-77 p90 tape
548 Romeo and Juliet, op. 17: Love scene. TCHAIKOVSKY: Romeo and Juliet: Fantasy overture.* COA, LSO; Antal Dorati. Philips 6585 026 Tape (c) 7300 045. (*Reissue from Mercury AMS 16116)

+Gr 4-77 p1542 +RR 3-77 p47
+-HFN 3-77 p117 +-RR 8-77 p87 tape
+HFN 7-77 p127

Romeo and Juliet, op. 17: Queen Mab scherzo. cf RCA CRM 5-1900.

549 Romeo and Juliet, op. 17: Romeo's reverie and fete of the Capulets; Love scene; Queen Mab scherzo. TCHAIKOVSKY: Romeo and Juliet: Fantasy overture. RPO, LPO; Antal Dorati, Adrian Boult. Quintessence PMC 7045.
+-ARG 12-77 p43

550 Romeo et Juliette, op. 17: Suite. Les Troyens: Royal hunt and storm. Orchestre de Paris; Daniel Barenboim. CBS 76524 Tape (c) 40-76524.
-Gr 10-76 p584 +-HFN 1-77 p123 tape
-Gr 8-77 p346 tape +-RR 11-76 p51
+-HFN 11-76 p156 +-RR 5-77 p90 tape

551 Songs: La captive, op. 12; Le chasseur danois, op. 19, no. 6; La jeune patre breton, op. 13, no. 4; Les nuits d'ete, op. 7; Irlande, op. 2: La belle voyageuse; Zaide, op. 19, no. 1. Sheila Armstrong, Josephine Veasey, s; Frank Patterson, t; John Shirley-Quirk, bar; LSO; Colin Davis Philips 6500 009.
/HF 2-71 p74 +NR 2-71 p10
+MJ 3-71 p78 +St 3-71 p88
+MJ 5-77 p33

Songs: Irlande, op. 2: La belle voyageuse, Le coucher du soleil, L'origine de la harpe. cf BIZET: Songs (Saga 5388).

552 Symphonie fantastique, op. 14. Les Francs Juges, op. 3. CSO; Georg Solti. Decca Tape (c) KSXC 16571.
+RR 1-77 p87 tape

553 Symphonie fantastique, op. 14. BPhO; Herbert von Karajan. DG 2530 597 Tape (c) 3300 498.
++Gr 3-76 p1463 +-MJ 1-76 p39
++Gr 10-76 p658 tape +-NR 1-76 p5
+HF 1-76 p83 +-RR 3-76 p36
+HF 9-76 p84 tape +RR 1-77 p87 tape
+-HFN 3-76 p89 +-SFC 1-4-76 p27
+HFN 6-76 p105 tape ++STL 2-8-76 p36

554 Symphonie fantastique, op. 14. Lamoureux Orchestra; Igor Markevitch. DG 2548 172 Tape (c) 3348 172. (Reissue from 138712)
/Gr 11-75 p840 +HFN 2-76 p117
+Gr 2-77 p1325 +-RR 11-75 p41
++HFN 10-75 p152 +RR 9-76 p92 tape

555 Symphonie fantastique, op. 14. Lamoureux Orchestra; Igor Markevitch. DG Tape (c) 3318 034.
+-RR 1-77 p87 tape

556 Symphonie fantastique, op. 14. ORTF; Jean Martinon. HMV (Q) Q4ASD 2945. (also Angel (Q) S 31738 Tape (c) 4XS 37138)
++Gr 6-74 p40 ++NR 10-76 p4
+HF 1-77 p111 +-RR 2-74 p24
+MJ 11-76 p45 ++St 12-76 p135

557 Symphonie fantastique, op. 14. French National Radio Orchestra; Jean Martinon. HMV SQ ASD 3263 Tape (c) TC ASD 3263. (Reissue from Q4 ASD 2945)
++Gr 10-76 p584 +HFN 2-77 p135 tape
+Gr 1-77 p1178 tape +-RR 10-76 p40
+-HFN 11-76 p170 +RR 3-77 p98 tape

558 Symphonie fantastique, op. 14. French National Orchestra; Leonard Bernstein. HMV SQ ASD 3397. (also Angel S 37414)
++Gr 11-77 p829 +NYT 10-2-77 pD19
-HFN 12-77 p165 +-RR 12-77 p46
+-NR 12-77 p8 ++SFC 10-9-77 p40

559 Symphonie fantastique, op. 14. Budapest Symphony Orchestra; Charles Munch. Hungaroton SLPX 11842.
+HF 10-77 p102 -RR 12-76 p52
+HFN 12-76 p139 +-St 4-77 p115
+-NR 1-77 p3

560 Symphonie fantastique, op. 14. Halle Orchestra; John Barbirolli. Pye GSGC 15010 Tape (c) ZCCCB 15010. (Reissue from GSGC 14005).
/Gr 8-75 p335 +RR 8-75 p29
/HFN 8-75 p87 +-RR 1-77 p87

561 Symphonie fantastique, op. 14. Sydney Symphony Orchstra; Willem van Otterloo. RCA GL 2-5012.
+Gr 10-76 p584 +RR 10-76 p40
+-HFN 12-76 p139 +ST 1-77 p779

562 Symphonie fantastique, op. 14. CPhO; Carlo Zecchi. Rediffusion Legend Tape (c) LGC 001.
+RR 2-77 p96 tape

563 Symphonie fantastique, op. 14. North German Radio Orchestra; Pierre Monteux. Turnabout TVS 34616 Tape Vox (c) CT 2107.
+-HF 10-77 p102

564 Symphonie fantastique, op. 14. Leningrad Philharmonic Orchestra; Arvid Jansons. Westminster WGS 8350.
-NR 12-77 p3 +SFC 10-9-77 p40

565 Symphonie funebre et triomphale, op. 15. Musique des Gardiens de la Paix; Desire Dondeyne. Calliope CAL 1859 Tape (c) CAL 4859.
+Gr 3-77 p1392 ++RR 2-77 p43
+-HFN 2-77 p113

566 Te deum, op. 22. Jean Dupouy, t; Jean Guillou, org; Choeur d'Enfants de Paris; Maitrise de la Resurrections; Orchestre de Paris and Chorus; Daniel Barenboim. CBS SQ 76578. (also Columbia M 34536 Tape (c) MT 34536)
+-ARG 12-77 p21 +-NR 12-77 p8
+FF 11-77 p10 *NYT 10-2-77 pD19
+-Gr 8-77 p328 +-RR 9-77 p87
+HFN 8-77 p77 +-St 12-77 p136
-HF 12-77 p80

Les Troyens: Act 5, scenes 2 and 3. cf La mort de Cleopatre: Lyric scene.

567 Les Troyens: Royal hunt and storm. DEBUSSY: Prelude a l'apres-midi d'un faune. STRAVINSKY: The firebird: Suite. LPO; John Pritchard. Pye GSGC 15002 Tape (c) ZCCCB 15002. (Reissue from Virtuoso TPLS 13032/3)
*Gr 8-75 p335 +-RR 8-75 p44
/HFN 9-75 p109 /-RR 2-77 p97 tape

Les Troyens: Royal hunt and storm. cf Romeo et Juliette, op. 17: Suite.

Les Troyens: Trojan march. cf HMV RLS 717.

BERMUDO

Cantus del primero por mi bequadro. cf Harmonia Mundi HM 759.

Conditor alme siderum. cf Harmonia Mundi HM 759.

BERNARD DE CLUNY

Pantheon abluiter. cf DG 2710 019.

BERNARD DE VENTADORN
Can vei a lauzeta mover. cf Telefunken 6-41126.
BERNIA (16th century)
Toccata chromatica. cf Saga 5438.
BERNSTEIN
Fanfare for Bima. cf Crystal S 204.
The lark: Choruses. cf Vox SVBX 5353.
BERNSTEIN, Charles
568 Poeme transcendental. Rhapsody Israelien. Trio, strings. Yoshiko Nakura, vln; Milton Thomas, vla; Charles Brennard, vlc. Laurel LR 105.
+ARG 7-77 p17 +NR 2-77 p6
-MJ 10-77 p29
Rhapsody Israelien. cf Poeme transcendental.
Trio, strings. cf Poeme transcendental.
BERNSTEIN, Leonard
569 Candide: Overture. COPLAND: Appalachian spring. GERSHWIN: An American in Paris. LAPO; Zubin Mehta. Decca SXL 6811 Tape (c) KSXC 6811. (also London CS 7031)
+Gr 9-76 p410 ++NR 7-77 p4
+HFN 9-76 p119 ++RR 10-76 p40
++HFN 12-76 p155 ++RR 2-77 p96 tape
Candide: Overture. cf Mass, excerpts.
Candide: Overture. cf London CSA 2246.
Fancy free: Galop; Waltz; Danzon. cf CBS 61780.
570 Mass, excerpts. Candide: Overture. BPO; Arthur Fiedler. DG 2584 002.
+ARG 11-77 p14 +-SR 9-3-77 p42
+NR 10-77 p2
571 Sonata, clarinet and piano. VANHAL: Sonata, clarinet, B flat major. VAUGHAN WILLIAMS: Studies in English folksong (6). WAGNER (attrib.): Adagio. Jerome Bunke, clt; Hidemitsu Hayashi, pno. Musical Heritage Society MHS 1887.
+HF 2-77 p108 ++St 3-75 p111
++MJ 3-75 p26
West side story, excerpts. cf MENC 76-11.
572 West side story: Symphonic dances. GERSHWIN: Concerto, piano, F major. Roberto Szidon, pno; San Franciso Symphony Orchestra, LPO; Seiji Ozawa, Edward Downes. DG 2535 210. (Reissues from 2530 309, 2330 055)
+Gr 8-77 p349 +-RR 6-77 p48
+HFN 7-77 p125
BERRY, Wallace
Trio, piano. cf BASART: Fantasy.
BERTE, Heinrich
573 Lilac time (Schubert). BRAHMS: Waltzes, op. 39 (orch. Darvas). SCHUBERT: Wiener Tanze. HSO, HRT Orchestra; Tamas Brody, Gyorgy Lehel. Qualiton SLPX 16584.
/NR 6-77 p12
BERTOLI, Giovanni
Sonata prima. cf Cambridge CRS 2826.
BERWALD, Franz
Alvalek. cf Works, selections (HMV SLS 5096)
Bajadarfesten. cf Works, selections (HMV SLS 5096).
Concerto, piano, D major. cf Works, selections (HMV SQ SLS 5014).
Concerto, piano, D major. cf Works, selections (HMV SLS 5096).
Concerto, violin, op. 2, C sharp minor. cf Works, selections (HMV SQ SLS 5014).

Concerto, violin, op. 2, C sharp minor. cf Works, selections (HMV SLS 5096).
Estrella di Soria: Overture. cf Works, selections (HMV SLS 5096).
Estrella di Soria: Tragic overture. cf Works, selections (HMV SQ SLS 5014).
Festival of the Bayaderes. cf Works, selections (HMV SQ SLS 5014).
Kapplopning. cf Works, selections (HMV SLS 5096).
Memories of the Norwegian Alps. cf Works, selections (HMV SLS 5014
Play of the elves. cf Works, selections (HMV SQ SLS 5014).
574 Quartet, piano, E flat major. RIMSKY-KORSAKOV: Quintet, piano, B flat major. Eva Knardahl, pno; Goteborg Wind Quintet. BIS LP 44.
+HFN 11-77 p167 +RR 11-77 p74
The Queen of Golconda: Overture. cf Works, selections (HMV SQ SLS 5014).
The Queen of Golconda: Overture. cf Works, selections (HMV 5096).
Reminiscences from the Norwegian mountains. cf Works, selections (HMV SLS 5096).
Serious and light fancies. cf Works, selections (HMV SQ SLS 5014).
Serious and joyful fancies. cf Works selections (HMV SLS 5096).
Symphonies, nos. 1-4. cf Works, selections (HMV SQ SLS 5014).
Symphonies, nos. 1-4. cf Works, selections (HMV SLS 5096).
Tournament. cf Works, selections (HMV SQ SLS 5014).
575 Works, selections: Concerto, piano, D major. Concerto, violin, op. 2, C sharp minor. Festival of the Bayaderes. Estrella di Soria: Tragic overture. Play of the elves. The Queen of Golconda: Overture. Tournament. Memories of the Norwegian Alps. Serious and light fancies. Symphonies, nos. 1-4. Marian Migdal, pno; Arve Tellefsen, vln; RPO; Ulf Bjorlin. HMV SQ SLS 5014 (4).
+-Gr 10-77 p623 ++HFN 11-77 p167
576 Works, selections: Alvalek. Concerto, piano, D major. Concerto, violin, op. 2, C sharp minor. Bajadarfesten. Estrella di Soria: Overture. Kapplopning. The Queen of Golconda: Overture. Reminiscences from the Norwegian mountains. Serious and joyful fancies. Symphonies, nos. 1-4. Marian Migdal, pno; Arve Tellefsen, vln; RPO; Ulf Bjorlin. HMV SLS 5096 (4).
+-RR 11-77 p47

BESARD, Jean
Air de cour. cf ATTAINGNANT: Basse dance.
Allemande. cf ATTAINGNANT: Basse dance.
Ballet. cf ATTAINGNANT: Basse dance.
Branle. cf ATTAINGNANT: Basse dance.
Branle gay (2). cf ATTAINGNANT: Basse dance.
Branle gay. cf DG 2723 051.
Chorea rustica. cf ATTAINGNANT: Basse dance.
Gagliarda. cf ATTAINGNANT: Basse dance.
Gagliarda vulgo dolorata. cf ATTAINGNANT: Basse dance.
Gesamtzeit mit Pausen. cf ATTAINGNANT: Basse dance.
Guillemette. cf ATTAINGNANT: Basse dance.
Pass'e mezo. cf ATTAINGNANT: Basse dance.
Volte. cf ATTAINGNANT: Basse dance.

BESLEY
Little fairy songs: The fairy children; Canterbury bells. cf Decca SDD 507.

BESTOR, Charles
577 Sonata, piano. BREHM: Variations, piano. PLESKOW: Pentimento.

WATTS: Sonata, piano. Dwight Peltzer, pno. Serenus SRS 12069.
+NR 12-77 p12

BIBER, Carl Heinrich
Sonata, trumpet, no. 4, C major. cf HMV ASD 3318.
Suite, 2 clarino trumpets. cf Philips 6500 926.

BILLI
E canta il grillo. cf Club 99 CL 99-96.
E canta il grillo. cf RUBINI RS 301.

BILLINGS, William
Hymns (3). cf BENNETT, R.R.: The fun and faith of William Billings, American.
Jargon. cf Folkways FTS 32378.
Judea. cf Folkways FTS 32378.
Songs: The hart panteth; Consonance; Thus saith the high, the loft one. cf Vox SVBX 5350.
Songs: Thus saith the high. cf Richardson RRS 3.

BINCHOIS, Gilles de
Amoreux suy. cf HMV SLS 863.
Bien puist. cf HMV SLS 863.
Files a marier. cf Enigma VAR 1024.
Files a marier. cf HMV SLS 863.
Je ne fai toujours. cf HMV SLS 863.
Jeloymors. cf HMV SLS 863.
Votre tres doulz regart. cf HMV SLS 863.

BINDER
Sabbath at the concluding meal. cf Olympic OLY 105.

BINGE, Ronald
Elizabethan serenade. cf HMV SRS 5197.
Elizabethan serenade. cf Starline SRS 197.

BIRD, Arthur
Carnival scene. cf Louisville LS 753/4.

BIRTWISTLE, Harrison
Chronometer. cf The triumph of time.
Four interludes for a tragedy. cf L'Oiseau-Lyre DSLO 17.

578 Grimethorpe aria. HENZE: Ragtimes and habaneras. HOWARTH: Fireworks, brass band. TAKEMITSU: Garden rain (arr. Howarth). Grimethorpe Colliery Band, Besses o' th' Barn Band; Elgar Howarth. Decca HEAD 14.
+Gr 8-77 p313 +RR 8-77 p55
+-HFN 10-77 p156

Some petals from the garland. cf Universal Edition UE 15043.

579 Tragoedia. CROSSE: Concerto da camera. WOOD: Pieces, piano, op. 6 (3). Manoug Parikian, vln; Susan McGaw, pno; Melos Ensemble; Edward Downes, Lawrence Foster. Argo ZRG 759. (Reissues from HMV ASD 2333)
+Gr 5-76 p1776 +NR 6-76 p8
+HFN 4-76 p123 +-RR 12-75 p72
+MQ 10-77 p570

580 The triumph of time. Chronometer. BBC Symphony Orchestra; Pierre Boulez. Argo ZRG 790.
+Gr 7-75 p174 ++NR 1-76 p4
++HF 3-76 p80 +RR 6-75 p32
+-HFN 10-75 p137 +ST 2-76 p741
+MQ 10-77 p572 +Te 12-75 p43
+MT 7-76 p580

BISHOP, Henry
Grand march, E major. cf Saga 5417.

My pretty Jane. cf Philips 6392 023.
Songs: Bid me discourse; Home, sweet home; Lo, hear the gentle lark. cf HMV RLS 719.
Songs: Home, sweet home. cf Prelude PRS 2505.
Songs: Home, sweet, home; Lo, here the gentle lark; Pretty mocking bird. cf Columbia 34294.
Songs: Lo, here the gentle lark. cf BIS LP 45.

BIXIO

Parlami d'amore Mariu. cf London SR 33221.
The sun and the golden mountains. cf Club 99-105.

BIZET, Georges

Agnus Dei. cf Decca SXL 6781.
Agnus Dei. cf London OS 26437.
L'Arlesienne: Minuetto, no. 1. cf Angel S 37219.

581 L'Arlesienne: Suites, nos. 1 and 2. Carmen: Suites, nos. 1 and 2. National Philharmonic Orchestra; Leopold Stokowski. Columbia M 34503 Tape (c) MT 34503 (ct) MA 34503. (also CBS 76587 Tape (c) 40-76587)
+Audio 9-77 p100 ++NR 5-77 p5
+Gr 5-77 p1695 +RR 5-77 p48
+Gr 8-77 p346 tape +-RR 11-77 p118 tape
+-HFN 6-77 p117 +-St 5-77 p104
+HFN 8-77 p99 tape

L'Arlesienne: Suites, nos. 1 and 2. cf Works, selections (Decca DPA 559/60).

582 L'Arlesienne: Suites, nos. 1 and 2, excerpts. Carmen: Suites, nos. 1 and 2. CSO, Orchestra; Jean Martinon, Morton Gould. Quintessence PMC 7024.
+St 11-77 p146

L'Arlesienne: Suite, no. 1: Adagietto. cf Citadel CT 6013.
L'Arlesienne: Suite, no. 1: Minuet. cf RACHMANINOFF: Concerto, piano, no. 2, op. 18, C minor.
L'Arlesienne: Suite, no. 2: Farandole. cf BERLIOZ: Le Corsaire, op. 21.
L'Arlesienne: Suite, no. 2: Farandole. cf Angel S 37250.
L'Arlesienne: Suite, no. 2: Farandole. cf Seraphim S 60277.

583 Carmen. Anna Moffo, Helen Donath, Arleen Auger, s; Jane Berbie, ms; Franco Corelli, Karl-Ernst Mercker, t; Piero Cappuccilli, Barry McDaniel, Jean-Christopher Benoit, bar; Jose van Dam, bs; German Opera Orchestra; Lorin Maazel. Cetra LPS 3276 (3). (also Eurodisc XG 80489)
+Gr 7-77 p216

584 Carmen. Victoria de los Angeles, Janine Micheau, Denise Monteil, Marcelle Croisier, Monique Linval, s; Michel Hamel, Nicolai Gedda, t; Ernest Blanc, Jean-Christopher Benoit, Bernard Plantey, bar; Xavier Depraz, bs; Les Petits Chanteurs de Versailles; French National Radio Orchestra and Chorus; Thomas Beecham. HVM SLS 5021 (3) Tape (c) TC SLS 5021. (Reissue from ASD 331/3) (also Angel S 3613)
++Gr 2-76 p1374 +RR 1-76 p24
+Gr 1-77 p1183 tape +RR 4-76 p80 tape
+-HFN 2-76 p92

585 Carmen. Kiri Te Kanawa, Norma Burrowes, s; Tatiana Troyanos, Jane Berbie, ms; Placido Domingo, Michel Senechal, t; Jose van Dam, Michel Roux, Thomas Allen, bar; Pierre Thau, bs; John Alldis Choir; LPO; Georg Solti. London OSA 13115 (3) Tape (c) 5-13115. (also Decca D11D3 Tape (c) K11K33)

++ARG 12-76 p19 +NR 1-77 p12
+Gr 10-76 p640 +-OC 6-77 p45
+-Gr 1-77 p1183 tape +-ON 1-1-77 p48
/HF 12-76 p89 +-OR 6/7-77 p29
+HFN 10-76 p161 +RR 11-76 p34
++HFN 1-77 p123 tape +RR 3-77 p98 tape
+MJ 2-77 p30 +St 12-76 p97
+-MT 1-77 p42

586 Carmen. Leontyne Price, Mirella Freni, s; Franco Corelli, t; Robert Merrill, bar; VPO; Herbert von Karajan. RCA Tape (c) RK 40004 (2).
+-RR 3-77 p98 tape

587 Carmen: Mais nous ne voyons pas la Carmencita; L'amour est un oiseau rebelle; Pres des remparts de Seville; Les tringles des sistres tintaient; Vivat vivat le torero; Votre toast; Nous avons en tete une affair; La fleur que tu m'avais jetee; Non tu ne m'aimes pas; Melons Coupons...En vain pour eviter; Je dis que rien ne m'epouvante; C'est toi, C'est moi. Kiri Te Kanawa, Norma Burrowes, s; Tatiana Troyanos, Jane Berbie, ms; Placido Domingo, Michel Senechal, t; Jose van Dam, Michel Roux, Thomas Allen, bar; Pierre Thau, bs; John Alldis Choir; LPO; Georg Solti. Decca SET 621. (Reissue from D11D3) (also London OS 26504)
+-Gr 11-77 p885 +RR 11-77 p36
+-HFN 11-77 p185

588 Carmen: Prelude; Avec la garde montante; L'amour est un oiseau rebelle; Pres de remparts de Seville; Entre'acte; Les tringles des sistres tintaient; Votre toast; Nous avons en tete une affaire; La fleur que tu m'avais jetee; Melons, Coupons; Entre' act; Finale. Colette Boky, s; Marilyn Horne, Marcia Baldwin, ms; James McCracken, Russell Christopher, Andrea Velis, t; Tom Krause, bar; Donald Gramm, bs; Manhattan Opera Children's Chorus; Metropolitan Opera Orchestra and Chorus; Leonard Bernstein. DG 2530 534 Tape (c) 3300 478. (Reissue from 2740 101)
+Gr 10-75 p676 ++HFN 9-75 p109
+Gr 1-77 p1183 tape +RR 8-75 p22

589 Carmen: Prelude; Entrance of Carmen and habanera; Seguidilla and duet; Gypsy song; Toreador's song; Eh bien; Flower song; Card scene; Micaela's aria; Duet and final chorus. Victoria de los Angels, Janine Micheau, Denise Monteil, Monique Linval, s; Nicolai Gedda, Michel Hamel, t; Ernest Blanc, Jean-Christopher Benoit, bar; French Radio Orchestra and Chorus; Thomas Beecham. HMV ESD 7047 Tape (c) TC ESD 7047. (Reissue from ASD 331/3)
+Gr 11-77 p885 +-HFN 11-77 p185
+Gr 12-77 p1149 tape +RR 11-77 p36

Carmen: Arias. cf Free Lance FLPS 675.
Carmen: L'amour est un oiseau rebelle. cf HMV SLS 5104.
Carmen: L'amour est un oiseau rebelle; Pres des remparts de Seville; Les tringles des sistres tintaient. cf Seraphim M 60291.
Carmen: Eccola...Ella vien...Habanera, Seguidilla, Chanson boheme, Vieni lassu sulla montagna, Air des cartes, Aragonaise, Su tu m'ami, Sei tu, Son io. cf Club 99-103.
Carmen: La fleur que tu m'avais jetee. cf HMV ASD 3302.
Carmen: La fleur que tu m'avais jetee. cf RCA CRM 1-1749.
Carmen: La fleur que tu m'avais jetee. cf RCA TVM 1-7201.
Carmen: Flower song. cf HMV HLP 7109.

Carmen: Habanera; Chanson boheme; Seguidilla. cf Club 99-96
Carmen: Habanera; Chanson boheme; Seguidilla. cf Rubini RS 301.
Carmen: Habanera; Seguidilla; Toreador song; Flower song; Card trio. cf Works, selections (Decca DPA 559/60).
Carmen: Je dis que rien ne m'epouvante. cf Advent S 5023.
Carmen: March of the smugglers; Gypsy dance. cf CBS 30091.
Carmen: Micaela's aria. cf Decca D65D3.
Carmen: Seguidilla. cf Decca SPA 450.
Carmen: Suites, nos. 1 and 2. cf L'Arlesienne: Suites, nos. 1 and 2.
Carmen: Suites, nos. 1 and 2. cf L'Arlesienne: Suites, nos. 1 and 2, excerpts.
Carmen: Toreador song. cf Golden Age GAR 1001.
Carmen fantasy, op. 25 (arr. Busoni). cf International Piano Library IPL 104.
Carmen Fantasy, op. 25. cf RCA JRL 1-2315.
The fair maid of Perth: Serenade. cf BBC REC 267.
The fair maid of Perth: Serenade. cf Decca SPA 491.
The fair maid of Perth: Sweet echo, come tune thy lay. cf HMV HLM 7066.

590 Jeux d'enfants, op. 22. La jolie fille de Perth. Symphony, C major. OSR; Ernest Ansermet. Decca ECS 801. (Reissue from SXL 2275)
+Gr 7-77 p168 +-RR 6-77 p44
+-HFN 7-77 p125

Jeux d'enfants, op. 22. cf Symphony, C major.
Jeux d'enfants, op. 22. cf Works, selections (Decca DPA 559/60).
La jolie fille de Perth. cf Jeux d'enfants, op. 22.
La jolie fille de Perth. cf Symphony, C major.
La jolie fille de Perth: Quand la flamme de l'amour. cf Philips 6580 174.
La jolie fille de Perth: Suite. cf Works, selections (Decca DPA 559/60).
Little duet. cf Gasparo GS 103.
Pastorale. cf HMV RLS 719.
Les pecheurs de perles: Je crois entendre encore. cf HMV RLS 715.
Les pecheurs de perles: Je crois entendre. cf HMV ASD 3302.
I pescatori di perle: Mi par di udir encor; Del templo al limitar. cf RCA TVM 1-7203.
Les pecheurs de perles: L'orage s'est calme...O nadir, tendre ami de mon jeune age. cf Philips 6580 174.

591 Songs: Chanson d'Avril; Ouvre ton coeur; Pastorale; Vieille chanson. MASSENET: Songs: Crepuscule; Premiere danse; Le poete et le fantome; Ouvre tes yeux bleus; Le saistu; Serenade. SAINT-SAENS: Songs: L'Attente; Aimons-nous; Pourquoi rester seulette. Joan Patenaude, s; Mikael Eliasen, pno. Musical Heritage Society MHS 3433.
+OC 6-77 p45 +St 11-77 p162

592 Songs: Adieux de l'hotesse arabe; Chanson d'Avril; La chanson de la rose; Vous ne priez pas. BERLIOZ: Songs: Irlande, op. 2; La belle voyageuse, Le coucher du soleil, L'Origine de la harpe. DEBUSSY: Songs: Proses lyriques; Noel des enfants qui n'ont plus de maison. Jill Gomez, s; John Constable, pno. Saga 5388.
+ARG 9-77 p48 +RR 2-75 p58
++Gr 1-75 p1380 +-STL 1-12-75 p36

593 Symphony, C major. Jeux d'enfants, op. 22. La jolie fille de Perth. ORTF; Jean Martinon. DG 2535 238 Tape (c) 3335 238.
+-HFN 10-77 p169 tape

594 Symphony, C major. TCHAIKOVSKY: Francesca da Rimini, op. 32. RPO; Charles Munch. Quintessence PMC 7048.
++ARG 12-77 p43 +SFC 10-30-77 p44
Symphony, C major. cf Jeux d'enfants, op. 22.
Symphony, C major. cf Works, selections (Decca DPA 559/60).
595 Works, selections: L'Arlesienne: Suites, nos. 1 and 2. Carmen: Habanera; Seguidilla; Toreador song; Flower song; Card trio. Jeux d'enfants, op. 22. La jolie fille de Perth: Suite. Symphony, C major. Georgette Spanellys, s; Regina Resnik, Yvonne Minton, con; Mario del Monaco, t; Tom Krause, Claude Cales, bar; Robert Geay, bs; OSR; Ernest Ansermet. Decca DPA 559/60 Tape (c) KDPC 559/60.
+Gr 2-77 p1326 tape +-HFN 12-76 p155 tape
+-HFN 12-76 p152 -RR 12-76 p52

BLACKWOOD, Easley
596 Sonata, violin, no. 2. SESSIONS: Sonata, violin. Paul Zukovsky, vln; Easley Blackwood, pno. CP 1.
+MJ 10-77 p28 +NR 11-77 p8

BLANK, Allan
597 Songs (2). SMIT: At the corner of the sky. Songs of wonder. WILSON: Sometimes. Jan deGaetani, ms; Martha Hanneman, s; William Brown, t; Arthur Weisberg, bsn; Leo Smit, pno; Henrik Svitzer, flt; Nora Post, ob; Leo Smit, speaker; St. Paul's Cathedral Men and Boys Choir; Electronic Music Studio, Univ. of California, Berkeley; Frederick Burgomaster. CRI SD 370.
+NR 10-77 p10

BLISS, Arthur
Antiphonal fanfare, 3 brass choirs. cf Decca SPA 500.
598 A colour symphony. Things to come: Prelude; Ballet for children; The world in ruins; The building of the new world; Attack on the moon gun; Epilogue (ed. arr. Palmer). RPO; Charles Groves. HMV SQ ASD 3416.
+Gr 11-77 p829
Concerto, piano. cf HMV SLS 5080.
599 Concerto, violoncello. Miracle in the Gorbals: Ballet suite. Arto Noras, vlc; Bournemouth Symphony Orchestra; Paavo Berglund. HMV (SQ) ASD 3342 Tape (c) TC ASD 3342.
+Gr 6-77 p41 +MT 12-77 p1014
/Gr 8-77 p346 tape +RR 7-77 p48
+HFN 7-77 p108
Fanfares (3). cf HMV HQS 1376.
Fanfare. cf Guild GRSP 701.
Fanfare for a coming of age. cf RCA RL 2-5081.
Fanfare for a dignified occasion. cf RCA RL 2-5081.
Fanfare for heroes. cf RCA RL 2-5081.
Fanfare for the bride. cf RCA RL 2-5081.
Fanfare for the Lord Mayor of London. cf RCA FL 2-5081.
Fanfare, homage to Shakespeare. cf RCA RL 2-5081.
Gala fanfare. cf Lismor LILP 5078.
A garland for coronation morning. cf RCA GL 2-5062.
Interlude. cf RCA RL 2-5081.
600 Kenilworth. ELGAR: Severn suite, op. 87. HOLST: Moorside suite. IRELAND: Comedy overture. Grimethorpe Colliery Band; Elgar Howarth. Decca SXL 6820 Tape (c) KSXC 6820.
+Gr 5-77 p1703 +RR 5-77 p54
+Gr 9-77 p508 tape +RR 11-77 p120 tape
+-HFN 5-77 p120

Miracle in the Gorbals: Ballet suite. cf Concerto, violoncello.
Processional. cf BAX: Coronation march.
Processional. cf HMV SQ ASD 3341.
Royal fanfare. cf RCA RL 2-5081.
Royal fanfare, no. 1: Sovereign's fanfare. cf RCA RL 2-5081.
Royal fanfares, nos. 5, 6. cf RCA RL 2-5081.
Things to come: Epilogue. cf Lismor LILP 5078.
Things to come: Prelude; Ballet for children; The world in ruins; The building of the new world; Attack on the moon gun; Epilogue. cf A colour symphony.
Welcome the Queen. cf Decca SPA 500.
Welcome the Queen. cf Lismor LILP 5078.

BLITHEMAN, William
Eterne rerum. cf ABC ABCL 67008.

BLITZSTEIN, Marc
601 The airborne symphony. Orson Welles, narrator; Andrea Velis, t; David Watson, bar; Choral Art Society; NYP; Leonard Bernstein. Columbia M 34136.
+HF 1-77 p111 +ON 11-76 p98
+NR 9-76 p3 +-St 11-76 p136

BLOCH, Ernest
602 America, an epic rhapsody. Symphony of the Air; Leopold Stokowski. Vanguard SRV 545.
+-Audio 5-77 p101
Baal Shem: Ningun. cf Discopaedia MB 1013.
Baal Shem: Ningun. cf Enigma VAR 1025.
603 Concerto grosso, no. 1. SCHOENBERG: Pieces, op. 16 (5). CSO: Rafael Kubelik. Mercury SRI 75036.
++SFC 10-2-77 p44
Jewish life: 3 pieces. cf Orion ORS 75181.
Meditation hebraique. cf Orion ORS 75181.
604 Nocturnes (3). HOPKINS: Diferencias sobre una tema original. PISTON: Trio, piano. SCHWANTNER: Autumn canticles. Western Arts Trio. Laurel LR 104.
++ARG 8-77 p36 ++NR 10-76 p6
+MJ 10-77 p29 +SFC 12-19-76 p50
605 Sacred service. Utah Symphony Orchestra; Maurice Abravanel. Angel S 37305.
+NYT 11-27-77 pD15
606 Schelomo. SCHUMANN: Concerto, violoncello, op. 129, A minor. Mstislav Rostropovich, vlc; Paris Symphony Orchestra; Leonard Bernstein. Angel (Q) S 37256 Tape (c) 4XS 37256. (also HMV ASD 3334 Tape (c) TC ASD 3334)
++ARG 9-77 p21 +-HFN 6-77 p139 tape
+-Gr 3-77 p1392 +-NYT 7-24-77 pD11
+-HF 11-77 p106 +-RR 4-77 p56
+-HF 11-77 p138 tape +-St 8-77 p111
+-HFN 4-77 p133
607 Schelomo. Suite, viola and orchestra. Laszlo Varga, vlc; Milton Katims, vla; Westphalian Symphony Orchestra; Seattle Symphony Orchestra; Siegfried Landau, Henry Siegl. Turnabout TVS 34622 Tape (c) KTVC 34622.
+Gr 8-77 p292 +NR 12-76 p5
+-Gr 10-77 p698 tape +-RR 8-77 p50
+-HFN 8-77 p77 +SFC 10-76 p32
+-MJ 5-77 p32 +-St 2-77 p114
Suites, nos. 1 and 2. cf BARTOK: Sonata, solo violin.

Suite, viola and orchestra. cf Schelomo.
Suite, solo violoncello, no. 1. cf BABIN: David and Goliath.
608 Suite hebraique. HINDEMITH: Der Schwanendreher. MARTIN: Sonata da chiese, viola d'amore and strings. Marcus Thompson, vla d'amore, vla; MIT Symphony Orchestra; David Epstein. Turnabout QTV 34687.
+NR 11-77 p9
609 Suite symphonique. Symphony, trombone and orchestra. Howard Prince, trom; Portland Junior Symphony Orchestra; Jacob Avshalomov. CRI SD 351.
+ARG 12-76 p22 +NR 11-76 p3
+-Gr 9-77 p500 -RR 6-77 p44
+HF 12-76 p99 +St 2-77 p114
+MJ 11-76 p46
Symphony, trombone and orchestra. cf Suite symphonique.

BLOMDAHL, Karl-Birger
610 Symphony, no. 2. PETTERSSON: Symphony, no. 10. Stockholm Philharmonic Orchestra, Swedish Radio Symphony Orchestra; Antal Dorati. HMV 4E 061 35142.
+ARG 8-77 p23 +-Gr 12-75 p1061

BLONDEL de NESLE
Quant je plus. cf Enigma VAR 1020.

BLOW, John
611 Amphion Angelicus: Songs. Ode on the death of Mr. Henry Purcell. Nobuko Yamamoto, Nelly van der Spek, s; Rene Jacobs, James Bowman, c-t; Marius van Altena, t; Max van Egmond, bs; Instrumental Ensemble; Gustav Leonhard. ABC ABCL 67004.
+HF 3-77 p108 +ON 3-12-77 p40
+-MJ 2-77 p30 +SFC 12-12-76 p55
+-NR 3-77 p12 ++St 5-77 p54
Fugue, F major: Vers. cf Philips 6500 926.
God spake sometime in visions. cf Decca SPA 500.
Ode on the death of Mr. Henry Purcell. cf Amphion Angelicus: Songs.
612 Sing unto the Lord, O ye saints, anthem. HUMFREY: O give thanks; Hear, O heav'ns; By the waters of Babylon. LOCKE: The king shall rejoice; When the son of man shall come in his glory. James Bowman, c-t; Robert Tear, t; Christopher Keyte, bs; St. John's College Choir, Matheson Consort, Philomusica, London; George Guest. Argo ZRG 855.
+Gr 3-77 p1443 +MT 10-77 p826
+-HFN 3-77 p105 ++RR 3-77 p92

BLUMENFELD, Felix
Etude for left hand, op. 36. cf Pye PCNH 9.

BOCCANEGRA
Teresita. cf Club 99-105.

BOCCHERINI, Luigi
Concerto, violoncello, B flat major. cf HMV RLS 723.
Concerto, violoncello, D major. cf Bruno Walter Society IGI 323.
Minuet. cf Discopaedia MOB 1020.
Minuet. cf International Piano Library IPL 101.
613 Quartets, strings, op. 32. Esterhazy Quartet. Telefunken EK 6-35337 (2).
++Gr 12-77 p1098 +RR 11-77 p79
+HFN 11-77 p167
Quartet, strings, op. 61, no. 1, D major: Adagio. cf Decca 507.
614 Quintet, guitar. MOZART: Sonata, violoncello and bassoon, K 292.

RAVEL: Chansons madecasses. Marlboro MRS 10.
+MJ 4-77 p50
Sonata, violoncello, no. 6, A major: Adagio and allegro. cf HMV RLS 723.
615 Symphonies, op. 12, nos. 1-6. NPhO; Raymond Leppard. Philips 6703 034 (3).
+Gr 9-77 p423 ++RR 9-77 p57
+HFN 9-77 p137
616 Symphony, op. 12, no. 4, D minor. Symphony, op. 41, C minor. I Solisti Veneti; Claudio Scimone. Erato STU 70828.
-Gr 7-74 p199 +RR 5-74 p28
+-Gr 3-77 p1395 ++RR 3-77 p47
+-HFN 4-77 p135
Symphony, op. 41, C minor. cf Symphony, op. 12, no. 4, D minor.

BOELLMANN, Leon
Suite Gothique: Toccata. cf Calliope CAL 1922/4.

BOELY, Alexandre
Andante, G minor. cf Organ works (Telefunken DX 6-35293).
Bin ich gleich von dir Gewichen. cf Organ works (Telefunken DX 6-35293).
Dialogue de hautbois et cromorne. cf Organ works (Telefunken DX 6-35293).
Duo. cf Organ works (Telefunken DX 6-35293).
Fantasie et fugue. cf Organ works (Telefunken DX 6-35293).
Fugue a deux sujets. cf Organ works (Telefunken DX 6-35293).
Fugue sur le Kyrie. cf Organ works (Telefunken DX 6-35293).
Grands jeux. cf Organ works (Telefunken DX 6-35293).
617 Organ works: Andante, G minor. Bin ich gleich von dir Gewichen. Dialogue de hautbois et cromorne. Duo. Fantasie et fugue. Fugue sur le Kyrie. Fugue a deux sujets. Grands jeux. Quatuor. Quel etonnement vient saisir mon ame. Tierce en taille. Trio. Le vermeil du soleil. Verset sur le Christe. Verset sur le Kyrie. Voici la premiere entree. BOYVIN: Livre d'orgue, no. 1. Jean Boyer, Jean-Albert Villard, org. Telefunken DX 6-35293 (2).
+Gr 2-77 p1302
Quatuor. cf Organ works (Telefunken DX 6-35293).
Quel etonnement vient saisir mon ame. cf Organ works (Telefunken DX 6-35293).
Tierce en taille. cf Organ works (Telefunken DX 6-35293).
Trio. cf Organ works (Telefunken DX 6-35293).
Le vermeil du soleil. cf Organ works (Telefunken DX 6-35293).
Verset sur le Christe. cf Organ works (Telefunken DX 6-35293).
Verset sur le Kyrie. cf Organ works (Telefunken DX 6-35293).
Voici la premiere entree. cf Organ works (Telefunken DX 6-35293).

BOGAR, Istvan
Three movements, brass quartet. cf Hungaroton SLPX 11811.

BOHAC, Josef
618 Suita drammatica. KALABIS: Symphony, no. 4, op. 34. CPhO; Vaclav Neumann, Zdenek Kosler. Supraphon 110 1784.
/-NR 3-77 p2

BOHM, Carl
Still as the night. cf Prelude PRS 2505.

BOHM, Theobald
619 Fantasy on a theme by Schubert. HOLMES: Petites pieces (3). KRUGER: Suite (sonata). REINECKE: Ballade, op. 288. Arthur Hoberman, flt; Neil Stannard, pno. Orion ORS 76257.
+ARG 8-77 p37 +NR 6-77 p7

BOISMORTIER, Joseph Bodin de
620 Concerto, 5 recorders without bass, D minor. HEINICHEN: Concerto, 4 recorders and strings. SCARLATTI. A.: Concerto, recorder and strings, A minor. TELEMANN: Concerto, recorder and strings, F major. Concerto, 2 recorders and strings, B minor. Clas Pehrsson, rec; Musica Dolce; Drottingholm Baroque Ensemble. BIS LP 8.
+RR 9-77 p78 ++St 11-76 p166
Concerto, 5 recorders, D minor. cf BIS LP 57.
Rondeau, A minor. cf Cambridge CRS 2826.
621 Sonata, trumpet, G minor (arr. Bilgram). HANDEL: Sonata, flute, F major (arr. Alain). TELEMANN: L'amour. L'armement. La douceur. L'esperance. La generosite. La gaillardise. La Grace. La Majeste. Le rojoussance (arr. Alain). La Vaillance. Maurice Andre, tpt; Hedwig Bilgram, org. RCA FRL 1-7021 Tape (c) FRK 1-7021 (ct) FRS 1-7021.
++NR 11-77 p14
BOITO, Arrigo
Mefistofele: L'altra notte. cf Club 99-100.
Mefistofele: L'altra notte. cf Seraphim 60274.
Mefistofele: Dai campi dai prati. cf RCA TVM 1-7203
Mefistofele: Dai campi dai prati. cf Rubini GV 43.
Mefistofele: Lontano; L'altra notte. cf Club 99-96.
Mefistofele: Lontano, lontano. cf Rubini RS 301.
BOLCOM, William
622 Commedia. Open house. Paul Sperry, t; Saint Paul Chamber Orchestra; Dennis Russell Davies. Nonesuch H 71324.
+NR 7-76 p9 +ON 6-76 p52
+NYT 4-11-76 pD23 ++St 11-76 p136
+NYT 11-27-77 pD15
Open house. cf Commedia.
BOLLING, Claude
623 Concerto, guitar and piano. Alexandre Lagoya, gtr; Claude Bolling, pno; Michel Gaudry, bs; Marcel Sabiani, drum. RCA FRL 1-0149 Tape (c) ARK 1-0149 (ct) FRS 1-0149.
+-ARG 12-76 p23 +-NR 12-76 p8
+HF 7-77 p125 tape +-St 1-77 p112
624 Suite, flute and jazz piano. Jean-Pierre Rampal, flt; Claude Bolling, pno; Max Hediguer, double bs; Marcel Sabiani, drums. Columbia M 33233 Tape (c) MT 33233.
+HF 7-77 p125 tape ++St 2-76 p99
+NR 10-75 p4
BOLOGNA, Jacopo da
Fenice fu. cf Argo D40D3.
BOLTON, C.
Bobby's tune. cf Grosvenor GRS 1043.
BOND
Sonata, violoncello cf Laurel Protone LP 13.
BOND, Carrie Jacobs
Songs: A perfect day. cf Argo ZFB 95/6.
BONINCONTRO
Kom ater. cf Rubini GV 43.
BONNAL, Joseph
La vallee de Behorleguy. cf Calliope CAL 1922/4.
BONNEAU
Caprice en forme de valse. cf Citadel CT 6012.

BONONCINI, Giovanni
Sinfonia da chiesa a quattro, op. 5, no. 1. cf ALBINONI: Adagio, organ and strings, G minor.
BOOTH, Margaret
Britons awake. cf CRD Britannian BR 1077.
Elizabeth. cf CRD Britannia BR 1077.
Salute to the Prince of Wales. cf CRD Britannia BR 1077.
BORLET (14th century)
He tres doulz roussignol. cf HMV SLS 863.
Ma tredol rosignol. cf HMV SLS 863.
BORODIN, Alexander
625 In the Steppes of Central Asia. GLINKA: Russlan and Ludmilla: Overture. RIMSKY-KORSAKOV: The golden cockerel: Suite. May night: Overture. Lamoureux Orchestra; Igor Markevitch. DG 2548 247. (Reissue from SLPEM 136225, SLPEM 133006)
-Gr 2-77 p1326 +-RR 1-77 p57
+HFN 1-77 p119
626 In the Steppes of Central Asia. Prince Igor: Polovtsian dances. MUSSORGSKY: Night on the bare mountain. RIMSKY-KORSAKOV: Russian Easter festival overture, op. 26. Orchestras; Igor Markevitch, Willem van Otterloo, Jean Fournet. Philips Fontana 6530 022.
/Gr 2-77 p1326
627 In the Steppes of Central Asia. GLIERE: The red poppy, op. 70: Sailors dance. MUSSORGSKY: Khovanschina: Prelude, Act 1; Dance of the Persian maidens; Entr'acte, Act 4. A night on the bare mountain (Stokowski). RIMSKY-KORSAKOV: Russian Easter festival overture, op. 36. Nicola Moscona, bs; Symphony Orchestra; Leopold Stokowski. Quintessence PMC 7026.
+-ARG 12-77 p42
628 In the Steppes of Central Asia. GLINKA: Russlan and Ludmilla: Overture. TCHAIKOVSKY: Capriccio italien, op. 45. Overture, the year 1812, op. 49. CPhO; Karel Ancerl. Rediffusion Legend LGD 008 Tape (c) LGG 008.
/Gr 5-77 p1749 *RR 2-77 p96 tape
+HFN 6-77 p117 -RR 6-77 p70
In the Steppes of Central Asia. cf Works, selections (RCA RL 2-5098).
In the Steppes of Central Asia. cf CBS 61781.
Mlada: Final dance. cf Works, selections (RCA RL 2-5098).
Nocturne, string orchestra. cf Works, selections (RCA RL 2-5098).
Petite suite. cf Works, selections (RCA RL 2-5098).
Prince Igor: Dance of the Polovtsian maidens. cf CBS 61781.
Prince Igor: Igor's aria. cf Golden Age GAR 1001.
Prince Igor: Overture; March; Polovtsian dances. cf Works, selections (RCA RL 2-5098).
Prince Igor: Polovtsian dances. cf In the Steppes of Central Asia.
Prince Igor: Overture; Polovtsian dances. cf Symphonies, nos. 1-3.
629 Prince Igor: Polovtsian dances. RIMSKY-KORSAKOV: Scheherazade, op. 35. OSR; Ernest Ansermet. Decca SDD 496. (Reissue from SXL 2268)
+-Gr 4-77 p1556 +RR 1-77 p57
+HFN 1-77 p119
630 Prince Igor: Polovtsian dances. MUSSORGSKY: A night on the bare mountain. RIMSKY-KORSAKOV: Russian Easter festival overture, op. 36. Capriccio espagnol, op. 34. CSO; Daniel Barenboim. DG 2536 379 Tape (c) 3336 379.

+-Gr 11-77 p909 ++HFN 12-77 p167
+-Gr 12-77 p1146 +RR 12-77 p46

631 Prince Igor: Polovtsian dances (orch. Rimsky-Korsakov/Glazunov). MUSSORGSKY: A night on the bare mountain (orch. Rimsky-Korsakov). RIMSKY-KORSAKOV: Capriccio espagnol, op. 34. Russian Easter festival overture, op. 36. Orchestre de Paris: Gennady Rozhdestvensky. HMV ESD 7006 Tape (c) TC ESD 7006. (Reissue from Columbia TWO 395)
++Gr 9-76 p435 +RR 9-76 p46
+HFN 2-77 p135 tape

Prince Igor: Polovtsian dances. cf DG 2535 254.
Prince Igor: Vladimir's cavatina. cf HMV RLS 715.
Prince Igor: Vladimir's recitative and cavatina. cf RACHMANINOFF: Francesca da Rimini, op. 25.

632 Quartet, strings, no. 2, D major. SHOSTAKOVICH: Quartet, strings, no. 8, op. 110, C minor. Borodin Quartet. London STS 15046. (also Decca ECS 795. Reissue from SXL 6036)
+Gr 3-77 p1421 +RR 3-77 p69
+HFN 3-77 p117 +St 5-77 p73

633 Quartet, strings, no. 2, D major. DVORAK: Quartet, strings, no. 12, op. 96, F major. Quartetto, Italiano. Philips 802814.
+St 5-77 p73

Quartet, strings,no. 2, D major: Nocturne. cf RCA CRL 3-2026.

634 Symphonies, nos. 1-3. Prince Igor: Overture; Polovtsian dances. Toronto Mendelssohn Choir; Toronto Symphony Orchestra; Andrew Davis. CBS 79214 (2). (also Columbia M2 34587)
+Gr 11-77 p830 +RR 11-77 p48
+-HFN 11-77 p168

Symphonies, nos. 1-3. cf Works, selections (RCA RL 2-5098).
Symphony, no. 2, B minor. cf BALAKIREV: Russia.
Symphony, no. 3, A minor. cf BALAKIREV: King Lear: Overture.

635 Works, selections: In the Steppes of Central Asia. Mlada: Final dance. Nocturne, string orchestra (arr. Sargent). Petite suite. Prince Igor: Overture; March; Polovtsian dances. Symphonies, nos. 1-3. National Philharmonic Orchestra; John Alldis Choir; Loris Tjeknavorian. RCA RL 2-5098 (3).
+-Gr 8-77 p295 +RR 8-77 p50
+HFN 9-77 p139

BORRONO DA MILANO, Pietro
Casteliono book: Pieces (3). cf Saga 5438.

BOSSI, Marco
Divertimento in forma de giga. cf Vista VPS 1035.

BOTTEGARI
Mi stare pone Totesche. cf L'Oiseau-Lyre 12BB 203/6.

BOUGHTON, Rutland
The immortal hour: Faery song. cf Abbey LPB 778.
The immortal hour: Faery song. cf HMV HLP 7109.

BOUIN (17th century)
La Montauban. cf DG 2723 051.

BOULEZ, Pierre
Pour le Dr. Kalmus. cf Universal Edition UE 15043.
Sonata, piano, no. 1. cf BERIO: Rounds.
Sonata, piano, no. 1. cf CP 3-5.
Sonata, piano, no. 3. cf BERIO: Rounds.
Trope. cf BERIO: Rounds.

BOURGAULT-DUCOUDRAY, Louis
Thamara: Reve de Noureddin. cf Rubini GV 38.

BOURGEOIS
Dark'ning night the land doth cover. cf Abbey LPB 779.
BOUTRY
Capriccio. cf Boston Brass BB 1001.
BOWEN, York
Sonata, op. 35. cf Saga 5445.
BOWLES, Paul
Songs: Once a lady was here; Song of an old woman. cf New World Records NW 243.
BOYCE, William
Concerto, 2 violins, violoncello and strings, B minor. cf Concerto grosso, strings, E minor.
636 Concerto grosso, strings, E minor. Concerto 2 violins, violoncello and strings, B minor. WOODCOCK: Concerto, flute and strings, D major. Concerto, oboe and strings, E flat major. ANON.: Concerto, 2 oboes and strings, F major. William Bennett, flt; Neil Black, ob; Thames Chamber Orchestra; Michael Dobson. CRD 1031.
+ARG 12-77 p48
+Gr 3-77 p1415
+HF 9-77 p113
+HFN 2-77 p112
+NR 11-77 p7
+RR 2-77 p40
+SFC 9-11-77 p42
Heart of oak. cf HMV ESD 7002.
I have surely built Thee an house. cf Guild GRS 7008.
637 Songs (choral): By the waters of Babylon; I have surely built Thee an house; O where shall wisdom be found; Turn Thee unto me, O Lord. Voluntaries, organ, nos. 1-2, 4-10. Arthur Wills, org; Ely Cathedral Choir; Arthur Wills. Saga 5440 Tape (c) CA 5440.
+Gr 3-77 p1437
+-HFN 3-77 p99
+HFN 11-77 p187 tape
+RR 3-77 p88
++RR 11-77 p119 tape
Trio sonata, D major. cf National Trust NT 002.
Voluntaries, organ, D major, G minor. cf Cambridge CRS 2540.
Voluntaries, organ, nos. 1-2, 4-10. cf Songs (Saga 5440).
Voluntary, trumpet and organ, no. 1, D major. cf Nonesuch H 71279.
BOYVIN, Jacques
Livre d'orgue, no. 1. cf BOELY: Organ works (Telefunken DX 6-35293)
BOZAY, Attila
Formazioni, op. 16. cf Works, selections (Hungaroton SLPX 11742).
Improvisations, zither solo. cf Works, selections (Hungaroton SLPX 11742).
Intervalli, op. 15. cf Works, selections (Hungaroton SLPX 11742).
Lux perpetua, op. 17. cf Works, selections (Hungaroton SLPX 11742).
Movements, oboe and piano, op. 18 (2). cf Works, selections (Hungaroton SLPX 11742).
Sorozat Kamaraegyuttestre, op. 19. cf Works, selections (Hungaroton SLPX 11742).
638 Works, selections: Formazioni, op. 16. Improvisations, zither solo. Intervalli, op. 15. Lux perpetua, op. 17. Movements, oboe and piano, op. 18 (2). Sorozat Kamaraegyuttestre, op. 19. Klara Kormendi, pno; Laszlo Mezo, vlc; Heinz Holliger, ob; Attila Bozay, zither; Hungarian Radio and TV Chamber Orchestra and Chorus; Chamber Ensemble; Ferenc Sapszon, Peter Eotvos. Hungaroton SLPX 11742.
*NR 2-77 p15

BOZZA, Eugene
Aria. cf Citadel CT 6012.
Ballade, trombone. cf Pandora PAN 2001.
Sonatina, brass quintet. cf Crystal S 204.
Sonatina, brass quintet. cf University of Iowa unnumbered.
Sonatina, flute and bassoon. cf Crystal S 351.
BRADE, William
Pieces. cf Golden Crest CRS 4148.
BRADY, Victor
Easy and hold on. cf Inner City IC 1006.
Rosebud. cf Inner City IC 1006.
BRAGA
Batalha de 6 tom. cf Pelca PSR 40571.
BRAHE, May
Songs: Bless this house. cf Argo ZFB 95/6.
Songs: Bless this house. cf HMV MLF 118.
Songs: Bless this house. cf L'Oiseau-Lyre DSLO 20.
BRAHMS, Johannes

639 Academic festival overture, op. 80. Tragic overture, op. 81. Variations on a theme by Haydn, op. 56a. Columbia Symphony Orchestra; Bruno Walter. CBS 61784 Tape (c) 40-61784 (Reissues from Philips SABL 182, 183, 185)
+-Gr 8-77 p296 +-HFN 8-77 p93
+-HFN 9-77 p155 tape +-RR 10-77 p93 tape

640 Academic festival overture, op. 80. Hungarian dances, nos. 5 and 6. Serenade, no. 2, op. 16, A major. NYP; Leonard Bernstein. CBS 61789 Tape (c) 40-61789. (Reissues from 73197, 61259, 30018, SBRG 72524)
-Gr 7-77 p168 +-HFN 8-77 p99 tape
+-HFN 8-77 p93

Academic festival overture, op. 80. cf Symphonies, nos. 1-4 (Decca D39D4).
Academic festival overture, op. 80. cf Symphonies, nos. 1-4 (HMV SQ SLS 5093).
Academic festival overture, op. 80. cf Symphonies, nos. 1-4 (London 2405).
Academic festival overture, op. 80. cf Works, selections (Vanguard VCS 10117/20).
Academic festival overture, op. 80. cf Decca SXL 6782.
Academic festival overture, op. 80. cf International Piano Library IPL 5001/2.
Academic festival overture, op. 80. cf Pye Golden Hour GH 643.

641 Alto rhapsody, op. 53. German requiem, op. 45. Nanie, op. 82. Agnes Giebel, s; Helen Watts, con; Hermann Prey, bar; OSR; Lausanne Pro Arte Choruses; Ernest Ansermet. Decca DPA 583/4 (2). (Reissue from SET 333/4)
-Gr 5-77 p1718 +-RR 5-77 p79
+HFN 5-77 p137

642 Alto rhapsody, op. 53. Nanie, op. 82. Song of destiny, op. 54. Songs with viola, op. 91 (2). Helen Watts, con; Ambrosian Chorus; Cecil Aronowitz, vla; Geoffrey Parsons, pno; NPhO, OSR and Chorus, Lausanne Pro Arte Chorus; Claudio Abbado, Ernest Ansermet. Decca ECS 798. (Reissues from SXL 6386, SET 333/4, SOL 268)
++Gr 3-77 p1437 ++RR 3-77 p88
+HFN 3-77 p118

643 Alto rhapsody, op. 53.* STRAUSS, R.: Liebeshymnus, op. 32, no. 3; Muttertandelei, op. 43, no. 2; Das Rosenband, op. 36, no. 1; Ruhe, meine Seele, op. 27, no. 1. WAGNER: Wesendonk Lieder. Janet Baker, ms; LPO; John Alldis Choir; Adrian Boult. HMV SQ ASD 3260 Tape (c) GC ASD 3260. (*Reissue from ASD 2749) (also Angel S 37199)
+ARG 11-77 p44 +MT 12-76 p1007
+Gr 9-76 p454 +NR 7-77 p11
+-HF 10-77 p124 +RR 9-76 p87
++HFN 10-76 p166 +-St 11-77 p158
++HFN 12-76 p155

Alto rhapsody, op. 53. cf German requiem, op. 45.

644 Ballades, op. 10. Fantasias, op. 116. Emil Gilels, pno. DG 2530 655 Tape (c) 3300 655.
+ARG 2-77 p25 +HFN 7-76 p85
+Gr 7-76 p193 +-NR 12-76 p14
+-Gr 10-76 p658 ++RR 6-76 p68
+HF 1-77 p112 +-St 1-77 p112

Ballades, op. 10. cf Piano works (Telefunken FX 6-35303).

Ballade, no. 1, op. 10, D minor. cf Piano works (Vanguard VSD 71213).

645 Chorale preludes, op. 122 (11). O Traurigkeit, O Herzlied, chorale prelude. Prelude and fugue, A minor. Haid Mardirosian, org. Musical Heritage Society 1751 Tape (r) MHS BBC 1751.
-HF 8-77 p96 tape

Choralvorspiele, op. 122, nos. 5, 8, 11. cf BIS LP 27.

646 Concerto, piano, no. 1, op. 15, D minor. TCHAIKOVSKY: Concerto, piano, no. 1, op. 23, B flat minor. Vladimir Horowitz, pno; COA, Philharmonic Orchestra; Bruno Walter. Bruno Walter Society BWS 728.
+-NR 5-77 p7

647 Concerto, piano, no. 1, op. 15, D minor. Bruno-Leonardo Gelber, pno; Munich Philharmonic Orchestra; Franz Paul Decker. Connoisseur Society CS 2102 Tape (c) Advent E 1052. (Reissue from Odeon SMC 91337)
+-ARG 11-76 p17 +NR 9-76 p6
+HF 2-77 p97 ++SFC 10-10-76 p32
+HF 9-77 p119 tape +St 11-76 p138
+MJ 5-77 p32

648 Concerto, piano, no. 1, op. 15, D minor. Artur Rubinstein, pno; Israel Philharmonic Orchestra; Zubin Mehta. Decca SXL 6797 Tape (c) KSXC 6797. (also London 7018 Tape (c) 5-7018)
-ARG 11-76 p17 +HFN 11-76 p175 tape
+-Gr 10-76 p589 +-NR 11-76 p4
+-HF 2-77 p97 +RR 9-76 p46
+HFN 10-76 p166 +St 1-77 p113

649 Concerto, piano, no. 1, op. 15, D minor. Artur Rubinstein, pno; CSO; Fritz Reiner. RCA ARL 1-2044 Tape (c) ARK 1-2044 (ct) ARS 1-2044. (Reissue from LM 1831)
++ARG 9-77 p23 ++NR 6-77 p5
+HF 6-77 p84 ++SFC 3-27-77 p41
+MJ 5-77 p32 ++St 6-77 p127

650 Concerto, piano, no. 1, op. 15, D minor. Roger Woodward, pno; NPhO; Kurt Masur. RCA RL 2-5031.
+Gr 2-77 p1273 +-RR 3-77 p47
+-HFN 3-77 p101

651 Concerto, piano, no. 1, op. 15, D minor. Claudio Arrau, pno; PhO; Carlo Maria Giulini. Seraphim S 60264. (Reissue from Angel S 35892)
+HF 2-77 p97 +MJ 5-77 p32
652 Concerto, piano, no. 1, op. 15, D minor. Rudolf Kerer, pno; MRSO; Gennady Rozhdestvensky. Westminster WG 8345.
+-ARG 9-77 p23 +-NR 6-77 p4
+CL 10-77 p8
Concerto, piano, no. 1, op. 15, D minor. cf HMV SLS 5094.
653 Concerto, piano, no. 2, op. 83, B flat major. Bruno-Leonardo Gelber, pno; RPO; Rudolf Kempe. Connoisseur Society CSQ 2088 Tape (c) Advent E 1053.
+HF 6-76 p77 +-NR 3-76 p4
+HF 9-77 p119 tape ++SFC 11-16-75 p32
+-MJ 4-76 p30 +St 7-76 p108
654 Concerto, piano, no. 2, op. 83, B flat major. Julius Katchen, pno; LSO; Janos Ferencsik. Decca SPA 458. (Reissue from SXL 2236)
+-Gr 9-77 p424 -HFN 8-77 p93
655 Concerto, piano, no. 2, op. 83, B flat major. Cecile Ousset, pno; Leipzig Gewandhaus Orchestra; Kurt Masur. Decca SDD 522.
+Gr 2-77 p1273 +-RR 3-77 p47
+HFN 3-77 p99
656 Concerto, piano, no. 2, op. 83, B flat major. Maurizio Pollini, pno; VPO; Claudio Abbado. DG 2530 790 Tape (c) 3300 790.
+-Gr 8-77 p295 +HFN 10-77 p169 tape
+HFN 8-77 p77 ++RR 8-77 p51
657 Concerto, piano, no. 2, op. 83, B flat major. Solomon, pno; PhO; Issay Dobrowen. Turnabout THS 65071. (Reissue from HMV originals)
+ARG 8-77 p13 ++NR 6-77 p4
+HF 8-77 p78 ++St 10-77 p134
Concerto, piano, no. 2, op. 83, B flat major. cf Decca D62D4.
Concerto, piano, no. 2, op. 83, B flat major. cf HMV SLS 5094.
658 Concerto, violin, op. 77, D major. MENDELSSOHN: Concerto, violin, op. 63, E minor. Gioconda de Vito, vln; RAI Torino Orchestra; Wilhelm Furtwangler. Bruno Walter Society RR 510.
+-NR 6-77 p4
659 Concerto, violin, op. 77, D major. Nathan Milstein, vln; VPO; Eugen Jochum. DG 2530 592 Tape (c) 3300 592.
+Audio 11-76 p111 +-NR 4-76 p3
+-Gr 12-75 p1031 +RR 12-75 p22
+-HF 5-76 p80 +RR 4-76 p81 tape
+-HF 1-77 p151 tape ++SFC 2-29-76 p25
+HFN 3-76 p113 tape +St 6-76 p101
++MJ 5-76 p29 +STL 1-11-76 p36
660 Concerto, violin, op. 77, D major. Andrei Korsakov, vln; Belgian Radio Symphony Orchestra; Rene Defossez. DG 2548 263 Tape (c) 3348 263.
+-Gr 7-77 p173 +-RR 7-77 p49
+HFN 7-77 p108
661 Concerto, violin, op. 77, D major. Gidon Kremer, vln; BPhO; Herbert von Karajan. HMV SQ ASD 3261. (also Angel 37226)
-Gr 10-76 p590 +-RR 9-76 p46
+-HF 2-77 p98 -SFC 12-26-76 p34
-HFN 10-76 p166 +-SR 2-19-77 p42
+-MJ 1-77 p27 +-St 3-77 p126
+NR 2-77 p5

662 Concerto, violin, op. 77, D major. Itzhak Perlman, vln; CSO; Carlo Maria Giulini. HMV SQ ASD 3385 Tape (c) TC ASD 3385.
++Gr 11-77 p830 +-HFN 11-77 p168
+-Gr 12-77 p1142 ++RR 11-77 p49

663 Concerto, violin, op. 77, D major. Hermann Krebbers, vln; COA; Bernard Haitink. Philips 6580 087 Tape (c) 7317 149.
+Gr 3-77 p1458 tape +-RR 3-75 p24
+HFN 5-77 p139 tape

664 Concerto, violin, op. 77, D major. Nathan Milstein, vln; PhO; Anatole Fistoulari. Seraphim S 60265.
+-NR 6-76 p6 +St 3-77 p126

Concerto, violin, op. 77, D major. cf BEETHOVEN: Concerto, violin, op. 61, D major.

Concerto, violin, op. 77, D major: Adagio. cf Discopaedia MB 1015.

665 Concerto, violin and violoncello, op. 102, A minor. Variations and fugue on a theme by Handel, op. 24, B flat major (orch. Rubbra). Isaac Stern, vln; Leonard Rose, vlc; PO; Eugene Ormandy. CBS 61785 Tape (c) 40-61785. (Reissue from SBRG 72295)
+Gr 7-77 p168 +-HFN 8-77 p93
+-HFN 10-77 p169 tape +-RR 12-77 p94 tape

666 Concerto, violin and violoncello, op. 102, A minor. DVORAK: Slavonic dance, op. 46, no. 3, A flat major. Slavonic dance, op. 72, no. 2, E minor. David Oistrakh, vln; Mstislav Rostropovich, vlc; CO; Georg Szell. HMV ASD 3312 Tape (c) TC ASD 3312. (Reissues from SLS 786, 2653)
++Gr 5-77 p1695 +HFN 8-77 p99 tape
-Gr 8-77 p346 tape ++RR 5-77 p48
+HFN 6-77 p137

667 Concerto, violin and violoncello, op. 102, A minor. SAINT-SAENS: Concerto, violoncello, no. 1, op. 33, A minor. Nathan Milstein, vln; Gregor Piatigorsky, vlc; Robin Hood Dell Orchestra, RCA Symphony Orchestra; Fritz Reiner. RCA AVM 1-2020.
+-ARG 5-77 p16 +MJ 10-77 p27

Concerto, violin and violoncello, op. 102, A minor. cf HMV RLS 723.

668 Ein deutsches Requiem (German requiem), op. 45. Tragic overture, op. 81. Variations on a theme by Haydn, op. 56a. Anna Tomowa-Sintow, s; Jose van Dam, bs-bar; Vienna Singverein; BPhO; Herbert von Karajan. Angel SB 3838 (2) Tape (c) 4X2S 3838. (also HMV SLS 996 Tape (c) TC SLS 996)
+-ARG 10-77 p24 +NR 6-77 p8
+-Gr 6-77 p76 +RR 6-77 p86
+-Gr 8-77 p346 tape +-SFC 5-15-77 p50
+-HF 9-77 p99 +-St 12-77 p138
+-HFN 6-77 p117

669 German requiem, op. 45. Alto rhapsody, op. 53. Ileana Cotrubas, s; Yvonne Minton, con; Hermann Prey, bar; Ambrosian Singers; NPhO and Chorus; Lorin Maazel. CBS 79211 (2). (also Columbia M 34583)
+-Gr 11-77 p868 +-RR 11-77 p99
+HFN 11-77 p168 ++SFC 10-16-77 p47
+NR 12-77 p7

670 German requiem, op. 45. Variations on a theme by Haydn, op. 56a (St. Antoni). Gundula Janowitz, s; Eberhard Wachter, bar; Vienna Singverein; BPhO; Herbert von Karajan. DG 2726 078 (2). (Reissue from SKL 138/9)
+-Gr 6-77 p76

German requiem, op. 45. cf Alto rhapsody, op. 53.
German requiem, op. 45: How lovely is thy dwelling place. cf Abbey LPB 779.
Es ist ein Ros entsprungen, op. 122, no. 8. cf Odeon HQS 1356.
F-A-E sonata: Scherzo. cf Enigma VAR 1025.

671 Fantasias, op. 116. Pieces, piano, op. 117. Rhapsody, op. 79, no. 1, B minor. Ludwig Olshansky, pno. Monitor MCS 2152.
+NR 8-77 p13 -St 10-77 p134

Fantasias, op. 116. cf Ballades, op. 10.

672 Fantasias, op. 151 (4). Waltzes, op. 39 (16). Variations on a theme by Schumann, op. 23. Viktoria Postnikova, Gennady Rozhdestvensky, pno. Eurodisc-Melodiya 89152 KK.
+ARG 4-77 p10

673 Fugue, A flat minor. FRANCK: Chorale, no. 3, A minor. LISZT: Variations on Bach's "Weinen, Klagen, Sorgen, Zagen", G 673. REGER: Toccata and fugue, op. 59. Timothy Day, org. Wealden WS 163.
+Gr 8-77 p327

Gigues, op. posth., A minor, B minor. cf Piano works (Telefunken FX 6-35303).
Hungarian dance, F minor. cf HMV ASD 3346.

674 Hungarian dances (21). Michel Beroff, Jean-Philippe Collard, pno. Connoisseur Society CS 2083 (Q) CSQ 2083.
+HF 6-76 p77 ++SFC 10-12-75 p22
-MJ 5-77 p32 ++St 5-76 p78
+NR 3-76 p12

675 Hungarian dances, nos. 1-21. Alfons and Aloys Kontarsky, pno. DG 2530 710 Tape (c) 3300 710.
++ARG 6-77 p15 +NR 6-77 p14
++Gr 10-76 p622 ++RR 9-76 p80
+HFN 10-76 p166 +-RR 9-77 p98 tape
+-MJ 5-77 p32 /SFC 3-27-77 p41

676 Hungarian dances, nos. 1-21. Michel Beroff, Jean-Philippe Collard, pno. HMV SQ HQS 1380 Tape (c) TC HQS 1380.
+-Gr 7-77 p192 +-RR 9-77 p81
+HFN 7-77 p108 +-RR 10-77 p93 tape

677 Hungarian dances, nos. 1-7, 10-12, 15, 17-21. LSO; Antal Dorati. Philips 6582 017. (Some reissues from Mercury AMS 16005)
+Gr 5-77 p1749 +HFN 5-77 p137

Hungarian dances, nos. 1 and 2. cf Discopaedia MOB 1019.
Hungarian dances, nos. 1, 3, 19. cf Symphony, no. 3, op. 90, F major.
Hungarian dances, nos. 1, 7, 17. cf Discopaedia MB 1013.
Hungarian dance, no. 1, G minor. cf Discopaedia MOB 1018.
Hungarian dance, no. 1, G minor. cf International Piano Library IPA 117.
Hungarian dance, no. 2, D minor. cf Discopaedia MB 1014.
Hungarian dance, no. 5, G minor. cf CBS 61039.
Hungarian dance, no. 5, G minor. cf Connoisseur Society CS 2131.
Hungarian dances, nos. 5 and 6. cf Academic festival overture, op. 80.
Hungarian dance, no. 6. cf CBS 61780.
Hungarian dance, no. 6. cf Discopaedia MOB 1020.
Liebeslieder, op. 52: Waltzes. cf BACH: Motet: Jesu, meine Freude, S 227.
Lullaby. cf Decca SPA 491.
Lullaby. cf Kaibala 40D03.

Nanie, op. 82. cf Alto rhapsody, op. 53 (Decca DPA 583).
Nanie, op. 82. cf Alto rhapsody, op. 53 (Decca ECS 798).
"O Traurigkeit, O Herzlied", chorale prelude. cf Chorale preludes, op. 122.

678 Piano works: Ballade, no.1, op. 10, D minor. Pieces, piano, op. 76, nos. 2 and 6. Pieces, piano, op. 117, no. 2, B flat major. Pieces, piano, op. 118, nos. 1-6. Rhapsodies, op. 79 (2). Bruce Hungerford, pno. Vanguard VSD 71213.
+ARG 11-76 p15 +NR 11-76 p13
+-HF 1-77 p112 ++SFC 10-31-76 p35
+-MJ 11-76 p45

679 Piano works: Ballades, op. 10. Gigues, op. posth., A minor, B minor. Sarabandes, B minor, A major. Scherzo, op. 50, E flat minor. Sonatas, nos. 1-3. Variations and fugue on a theme by Handel, op. 24, B flat major. Variations on a Hungarian song, op. 21, no. 2. Variations on a theme by Schumann, op. 9, F sharp minor. Variations on an original theme, op. 21, no. 1, D major. Detlef Kraus, pno. Telefunken FX 6-35303 (4).
+-Gr 4-77 p1570 +-RR 10-77 p75
+-HFN 5-77 p120

Pieces, piano, op. 76, nos. 2 and 6. cf Piano works (Vanguard VSD 71213).

680 Pieces, piano, op. 116: Fantasias. Pieces, op. 117: Intermezzi. Rhapsody, op. 79, no. 1, B minor. Ludwig Olshansky, pno. Monitor MCS 2152.
+-FF 11-77 p10 +NR 8-77 p13
-MJ 7-77 p70 -St 10-77 p134

681 Pieces, piano, opp. 117, 118 (6), 119 (4). Dmitri Alexeev, pno. Angel S 37290. (also HMV HQS 1370)
+-ARG 9-77 p24 -MT 9-77 p733
+-Gr 3-77 p1422 +NR 8-77 p13
-HF 7-77 p100 +-RR 5-77 p68
+-HFN 5-77 p119 +-St 10-77 p134
+-MJ 9-77 p34

682 Pieces, piano, op. 117 (3). Trio, piano, no. 1, op. 8, B major. Gyula Kiss, pno; Eszler Perenyi, vln; Miklos Perenyi, vlc. Hungaroton SLPX 11796.
++ARG 4-77 p15 +RR 3-77 p69
+NR 11-76 p6 ++SFC 10-10-76 p32

683 Pieces, piano, op. 117 (3). Pieces, piano, op. 119 (4). Variations and fugue on a theme by Handel, op. 24, B flat major. Van Cliburn, pno. RCA ARL 1-2280 Tape (c) ARK 1-2280 (ct) ARS 1-2280.
+-ARG 9-77 p24 ++SFC 5-15-77 p50
+-MJ 7-77 p70 +St 10-77 p134
+-NR 8-77 p13

684 Pieces, piano, op. 117 (3). Sonata, piano, no. 3, op. 5, F minor. Josef Palinecek, pno. Rediffusion Legend Tape (c) LGC 011.
+RR 2-77 p96 tape

Pieces, piano, op. 117. cf Fantasias, op. 116.
Pieces, piano, op. 117: Intermezzi. cf Pieces, piano, op. 116: Fantasias.
Pieces, piano, op. 117, no. 1, E flat major. cf Sonata, piano, no. 3, op. 5, F minor.
Pieces, piano, op. 117, no. 2, B flat minor. cf Piano works (Vanguard VSD 71213).

Pieces, piano, op. 117, no. 2, B flat minor. cf BEETHOVEN: Sonata, piano, no. 8, op. 13, C minor.
Pieces, piano, op. 118, nos. 1-6. cf Piano works (Vanguard VSD 71213).
Pieces, piano, op. 118, no. 2, A major. cf BEETHOVEN: Sonata, piano, no. 14, op. 27, no. 2, C sharp minor.
Pieces, piano, op. 118, no. 3, G minor. cf Decca SPA 519.
Pieces, piano, op. 119 (4). cf Pieces, piano, op. 117 (3).
Pieces, piano, op. 119, no. 3, C major. cf Sonata, piano, no. 3, op. 5, F minor.
Prelude and fugue, A minor. cf Chorale preludes, op. 122.
Quartet, piano, no. 1, op. 25, G minor. cf Works, selections (Hungaroton SLPX 11596/600).
Quartet, piano, no. 2, op. 26, A major. cf Works, selections (Hungaroton SLPX 11596/600).

685 Quartet, piano, no. 3, op. 60, C minor. SCHUMANN: Quartet, piano and strings, op. 47, E flat major. Pro Arte Piano Quartet. L'Oiseau-Lyre S 320.
+St 5-77 p73

Quartet, piano, no. 3, op. 60, C minor. cf Works, selections Hungaroton SLPX 11596/600).

686 Quartets, strings, nos. 1-3. Quintets, strings, nos. 1-2. Sextets, strings, nos. 1-2. Bartok Quartet. Hungaroton SLPX 11591/5 (5).
/-Gr 1-77 p1159 +-RR 2-77 p67

687 Quartets, strings, nos. 1-3. SCHUMANN: Quartets, strings, nos. 1-3. Quartetto Italiano. Philips 6703 029 (3).
+-Gr 5-77 p1704 /NYT 6-17-73 pD28
+HF 4-73 p77 +-RR 5-77 p66
+HFN 5-77 p120 +SFC 2-11-73 p31
++NR 2-73 p6

688 Quartet, strings, no. 3, op. 67, B flat major. SCHUMANN: Quartet, strings, op. 41, no. 1, A minor. Musikverein Quartet. Decca SDD 510 Tape (c) KSDC 510.
+-Gr 9-77 p444 +-RR 9-77 p77
+-HFN 9-77 p139

689 Quintet, clarinet, op. 115, B minor. Gervase de Peyer, clt; Melos Ensemble. Angel S 36280.
++St 5-77 p74

690 Quintet, clarinet, op. 115, B minor. Richard Stoltzman, clt; Cleveland Quartet. RCA ARL 1-1993 Tape (c) ARK 1-1993 (ct) ARS 1-1993.
++ARG 4-77 p19 +MJ 4-77 p50
+Gr 8-77 p313 +NR 4-77 p6
+-HF 4-77 p94 +RR 8-77 p66
+HF 5-77 p101 tape ++St 5-77 p104
+-HFN 10-77 p152

Quintet, clarinet, op. 115, B minor. cf Works, selections (Hungaroton SLPX 11596/600).

691 Quintet, piano, op. 34, F minor. Sviatoslav Richter, pno; Borodin Quartet. Saga 5448.
+-Gr 7-77 p192 +-RR 11-77 p79
+-HFN 6-77 p137

Quintet, piano, op. 34, F minor. cf Works, selections (Hungaroton SLPX 11596/600).
Quintets, strings, nos. 1-2. cf Quartets, strings, nos. 1-3.

692 Quintet, strings, no. 1, op. 88, F major. Quintet, strings, no. 2,

op. 111, G major. Sextet, strings, no. 1, op. 18, B major. Sextet, strings, no. 2, op. 36, G major. Amadeus Quartet; Cecil Aronowitz, vla; William Pleeth, vlc. DG 2733 011 (3). (Reissue from 104973/87)
+-Gr 4-77 p1568 +-MT 9-77 p733
+HFN 4-77 p151 +RR 8-77 p65

Quintet, strings, no. 2, op. 111, G major. cf Quintet, strings, no. 1, op. 88, F major.
Rhapsodies, op. 79 (2). cf Piano works (Vanguard VSD 71213).
Rhapsody, op. 79, no. 1, B minor. cf Fantasias, op. 116.
Rhapsody, op. 79, no. 1, B minor. cf Pieces, piano, op. 116: Fantasias.
Sarabandes, B minor, A major. cf Piano works (Telefunken FX 35303).
Scherzo, op. 50, E flat minor. cf Piano works (Telefunken FX 6-35303).

693 Serenade, no. 1, op. 11, D major. COA; Bernard Haitink. Philips 9500 322 Tape (c) 7300 584.
++SFC 11-20-77 p54

694 Serenade, no. 2, op. 16, A major. MOZART: Sonatas, violin and piano, K 12 and K 13, A major and F major. Pina Carmirelli, vln; Peter Serkin, pno; Marlboro Festival Orchestra; Pablo Casals. Marlboro MRS 1.
/LJ 1-15-71 p179 ++MJ 4-77 p50

Serenade, no. 2, op. 26, A major. cf Academic festival overture, op. 80.
Sextets, strings, nos. 1-2. cf Quartets, strings, nos. 1-3.

695 Sextet, strings, no. 1, op. 18, B flat major. SCHUBERT: Quartet, strings, no. 12, D 703, C minor. Alberni Quartet; Roger Best, vla; Moray Welsh, vlc. CRD CRD 1034.
+-Gr 8-77 p314 +-RR 8-77 p65
+-HFN 9-77 p139 ++SFC 11-13-77 p50

Sextet, strings, no. 1, op. 18, B major. cf Quintet, strings, no. 1, op. 88, F major.
Sextet, strings, no. 2, op. 36, G major. cf Quintet, strings, no. 1, op. 88, F major.

696 Sonata, clarinet, no. 1, op. 120, no. 1, F minor. Songs (arr.): Wo bist du meine Koenigin, op. 32, no. 9; Dein blaues Auge, op. 59, no. 8; Die Mainacht, op. 43, no. 2. WEBER: Duo concertant, op. 48, E flat major. David Pino, clt; Frances Mitchum Webb, pno. Orion ORS 77266.
+-NR 10-77 p8

697 Sonata, clarinet, no. 1, op. 120, no. 1, F minor. Trio, clarinet, violoncello and piano, op. 114, A minor. Klaus Storck, vlc; Jost Michaels, clt; Detlef Kraus, pno. Oryx 3C325.
+NR 5-77 p8

698 Sonata, clarinet, no. 2, op. 120, no. 2, E flat major. WEBER: Duo concertant, op. 48, E flat major. Frealon Bibbins, clt; Roslyn Frantz, pno. Cambridge CRS 2828.
+NR 4-77 p6 +SFC 3-6-77 p34

Sonatas, piano, nos. 1-3. cf Piano works (Telefunken FX 6-35303).

699 Sonata, piano, no. 2, op. 2, F sharp minor. Variations on a theme by Paganini, op. 35. Claudio Arrau, pno. Philips 9500 066.
+-ARG 11-76 p18 +MJ 12-76 p28
++GR 6-76 p69 ++MT 9-77 p733
+HF 12-76 p99 +NR 12-76 p14
++HFN 8-76 p97 +RR 5-76 p64

700 Sonata, piano, no. 3, op. 5, F minor. Pieces, piano, op. 117, no. 1, E flat major; op. 119, no. 3, C major. Clifford Curzon, pno. London STS 15272. (Reissue from CS 6341) (also Decca SDD 498. Reissue from SXL 6041)
+Gr 12-76 p1016 +RR 5-77 p68
++HF 6-75 p87 ++St 7-75 p95
+HFN 1-77 p121

Sonata, piano, no. 3, op. 5, F minor. cf Pieces, piano, op. 117 (3).

Sonata, piano, no. 3, op. 5, F minor. cf International Piano Library IPA 112.

701 Sonata, violin and piano, no. 1, op. 78, G major. Sonata, violin and piano, no. 3, op. 108, D minor. Miklos Szenthelyi, vln; Andras Schiff, pno. Hungaroton SLPX 11731.
+-HF 3-77 p97 -RR 1-76 p47
+-NR 3-76 p5

702 Sonata, violin and piano, no. 1, op. 78, G major. Trio, horn, violin and piano, op. 40, E flat major. Arthur Grumiaux, vln; Francis Orval, hn; Gyorgy Sebok, pno. Philips 9500 161.
++ARG 4-77 p15 ++MJ 2-77 p30
+-Gr 1-77 p1159 +-NR 2-77 p8
+HF 4-77 p94 +-RR 3-77 p75
++HFN 1-77 p103

703 Sonata, violin and piano, no. 1, op. 78, G major. Sonata, violin and piano, no. 2, op. 100, A major. Sonata, violin and piano, no. 3, op. 108, D minor. Georg Kulenkampff, vln; Georg Solti, pno. Richmond R 23213. (Reissues)
+NR 10-77 p8 +-St 12-77 p139
+SFC 10-17-77 p47

704 Sonata, violin and piano, no. 2, op. 100, A major. Sonata, violin and piano, no. 3, op. 108, D minor. Pinchas Zukerman, vln; Daniel Barenboim, pno. DG 2530 806. (Reissue from 2709 058)
+Gr 9-77 p444 +-RR 8-77 p70
+HFN 9-77 p151

705 Sonata, violin and piano, no. 2, op. 100, A major. SCHUMANN: Sonata, violin, no. 1, op. 105, A minor. Miklos Szenthelyi, vln; Andras Schiff, pno. Hungaroton SLPX 11819.
+-NR 12-77 p5

706 Sonata, violin and piano, no. 2, op. 100, A major. Sonata, violin and piano, no. 3, op. 108, D minor. Arthur Grumiaux, vln; Gyrogy Sebok, pno. Philips 9500 108.
++ARG 4-77 p15 ++HFN 11-76 p156
+Gr 11-76 p829 ++NR 2-77 p8
+HF 4-77 p94 ++RR 11-76 p88

Sonata, violin and piano, no. 2, op. 100, A major. cf Sonata, violin and piano, no. 1, op. 78, G major.

707 Sonata, violin and piano, no. 3, op. 108, D minor. RAVEL: Tzigane. TCHAIKOVSKY: Souvenir d'un lieu cher, op. 42: Meditation. Valse scherzo, op. 34. Jela Spitkova, vln; Pavol Kovac, pno. Opus 9111 0363.
+-Gr 8-77 p314 +-HFN 6-77 p117

Sonata, violin and piano, no. 3, op. 108, D minor. cf Sonata, violin and piano, no. 1, op. 78, G major (hungaroton SLPX 11731).

Sonata, violin and piano, no. 3, op. 108, D minor. cf Sonata, violin and piano, no. 1, op. 78, G major (London R 23213).

Sonata, violin and piano, no. 3, op. 108, D minor. cf Sonata, violin and piano, no. 2, op. 100, A major (DG 2530 806).

Sonata, violin and piano, no. 3, op. 108, D minor. cf Sonata, violin and piano, no. 2, op. 100, A major (Philips 9500 108).

Sonata, violin and piano, no. 3, op. 108, D minor. cf Discopaedia MOB 1018.

708 Sonata, violoncello and piano, no. 1, op. 38, E minor. Sonata, violoncello and piano, no. 2, op. 99, F major. Eckart Sellheim, pno; Friedrich-Jurgen Sellheim, vlc. CBS 76639.
+-Gr 12-77 p1098

709 Sonata, violoncello and piano, no. 1, op. 38, E minor. Sonata, violoncello and piano, no. 2, op. 99, F major. Gabor Rejto, vlc; Adolph Baller, pno. Orion ORS 77260.
+-NR 10-77 p8

710 Sonata, violoncello and piano, no. 1, op. 38, E minor. Sonata, violoncello and piano, no. 2, op. 99, F major. Gregor Piatigorsky, vlc; Artur Rubinstein, pno. RCA ARL 1-2085 Tape (c) ARK 1-2085 (ct) ARS 1-2085.
+-ARG 6-77 p14 ++MT 10-77 p825
+Gr 7-77 p192 +NR 5-77 p8
+HF 6-77 p84 +-RR 7-77 p79
+HFN 9-77 p139 +-SFC 5-15-77 p50
+MJ 5-77 p32

711 Sonata, violoncello and piano, no. 2, op. 99, F major. ELGAR: Concerto, violoncello, op. 85, E minor. Pablo Casals, vlc; Mieczysalw Horszowski, pno; BBC Symphony Orchestra; Adrian Boult. HMV HLM 7110 (Reissues from DB 3059/62, 2EA 10641/7, DB 6338/41)
+Gr 6-77 p48 +-MT 10-77 p826
+HFN 7-77 p108 +RR 7-77 p50

Sonata, violoncello and piano, no. 2, op. 99, F major. cf Sonata, violoncello and piano, no. 1, op. 38, E minor (CBS 76639).

Sonata, violoncello and piano, no. 2, op. 99, F major. cf Sonata, violoncello and piano, no. 1, op. 38, E minor (Orion ORS 77260).

Sonata, violoncello and piano, no. 2, op. 99, F major. cf Sonata, violoncello and piano, no. 1, op. 38, E minor (RCA 1-2085).

712 Songs: Feldeinsamkeit, op. 86, no. 2; Immer lieser wird mein Schlummer, op. 105, no. 2; Liebestreu, op. 3, no. 1; Madchenspricht, op. 107, no. 3; Die Mainacht, op. 43, no. 2; Ruhe, Sussliebchen im Schatten, op. 33, no. 9; Sapphische Ode, op. 94, no. 4; Standchen, op. 106, no. 1; Der Tod das ist die kuhle Nacht, op. 96, no. 1; Von ewiger Liebe, op. 43, no. 1; Zigeunerlieder, op. 103. Christa Ludwig, s; Leonard Bernstein, pno. Columbia M 34535.
++FF 11-77 p11 +NR 10-77 p12

713 Songs (German folk songs): Des Abends kann ich nicht schlafen gehn; Ach, englische Schaferin; Ach konnt diesen Abend; Da unten im Tale; Dort in den Weiden steht ein Haus; Es ging ein Maidlein zarte; Es ritt eine Ritter; Es reit Herr und auch sein Knecht; Es steht ein Lind; Es war eine schone Judin; Es wohnet ein Fiedler; Feinsliebchen, du sollst mir nicht barfuss gehn; Gunhilde lebt gar stille und fromm; In stiller Nacht; Maria ging aus wandern; Och Mod'r ich will en Ding han; Der Reiter spreitet seinen Mantel aus; Schwesterlein; So wunsch ich ihr ein Nacht; Soll sich der Mond nicht heller scheinen; Die Sonne scheint nicht mehr; Wach auf, mein Hort. Edith Mathis, s; Peter Schreier, t; Karl Engel, pno. DG 2536 279. (Reissue from 2709 057)
+Gr 5-77 p1718 +RR 5-77 p79
++HFN 4-77 p135

714 Songs: Auf dem See; Alte Liebe; Est steht ein Lind; Der Jager; Das Madchen; Schwesterlein; Sommerfadin; Der Tod das ist die kuhle Nacht, op. 96, no. 1; Wir wandelten. SCHUMANN: Frauenliebe und Leben, op. 42. Sheila Armstrong, s; Martin Jones, pno. Gale GMFD 7-86-006.
+-HFN 9-77 p132

715 Songs: Ave Maria, op. 12; Psalm 13, op. 27; Sacred choruses, op. 37 (3); Songs, op. 17 (4); Songs and romances, op. 44 (12). Zoltan Kodaly Women's Choir; Ilona Andor. Hungaroton SLPX 11691.
++HF 8-76 p83 +SFC 7-31-77 p40
+NR 6-76 p12

716 Songs: Das Madchen spricht, op. 107, no. 3; Die Mainacht, op. 43, no. 2; Nachtigall, op. 97, no. 1; Von ewiger Liebe, op. 43, no. 1. SCHUBERT: Songs: Die abgebluhte Linde, D 514; Heimliches Lieben, D 922; Minnelied, D 429; Der Musensohn, D 764. SCHUMANN: Frauenliebe und Leben, op. 42. Janet Baker, ms; Martin Isepp, pno. Saga 5277 Tape (c) CA 5277.
+-Gr 2-77 p1322 tape +HFN 10-76 p185 tape
+HF 11-76 p124

Songs: Des Abends kann ich nicht schlafen gehn; Ach, englische schaferin; Ach konnt ich diesen Abend; All mein Gedanken, Gar lieblich hat sich gesellet; Erlaube mir, feins Madchen; Es ging ein Maidlein zarte; Fahr wohl, o Voglein; Feinsliebchen sollst nicht barfuss gehn; Guten Abend, mein tausinger Schatz; Mein Madel hat einen Rosenmund; Schwesterlein; Die Sonne scheint nicht mehr; Wach auf mein Herzens schone; Wie komm ich denn zur Tur herein; Wiegenlied. cf Telefunken DT 6-48085.

Songs: An eine Aeolscharfe, op. 19, no. 5; Meine Liebe ist grun, op. 63, no. 5; O wusst ich doch den Weg zuruck, op. 63; Der Tod, das ist die kuhle Nacht, op. 96, no. 1; Verzagen, op. 72, no. 4. cf Prelude PRS 2505.

Songs: Bei dir sind meine Gedanken; Gute Nacht; Madchenlied; Unbewegte Laue Luft. cf Orion ORS 77271.

Songs: Geistliches Wiegenlied, op. 91, no. 2. cf Argo ZRG 871.

Songs: Die Mainacht, op. 43, no. 2. cf Seraphim IB 6105.

Songs: Serenata inutile. cf Decca SDD 507.

Songs: Wiegenlied. cf Bruno Walter Society BWS 729.

Songs: Wo bist du meine Koenigin, op. 32, no. 9; Dein blaues Auge, op. 59, no. 8; Die Mainacht, op. 43, no. 2. cf Sonata, clarinet, no. 1, op. 120, no. 1, F minor.

Songs of destiny, op. 54. cf Alto rhapsody, op. 53.

Songs with viola, op. 91 (2). cf Alto rhapsody, op. 53.

717 Symphonies, nos. 1-4. Variations on a theme by Haydn, op. 56a. Academic festival overture, op. 80. CO; Lorin Maazel. Decca D39D4 (4). (No. 1 reissue from SXL 6783)
+-Gr 9-77 p423 +-RR 10-77 p40
+-HFN 9-77 p139

718 Symphonies, nos. 1-4. VPO; Karl Bohm. DG 2740 154 (4). (also DG 2711 017 Tape (c) 3371 023)
+-ARG 2-77 p26 +HFN 10-76 p165
+-Gr 10-76 p589 +NR 3-77 p5
-Gr 12-76 p1066 tape +RR 10-76 p45
+-HF 3-77 p97 ++RR 10-76 p105 tape

719 Symphonies, nos. 1-4. Academic festival overture, op. 80. Tragic overture, op. 81. LPO; Eugen Jochum. HMV SQ SLS 5093 (4) Tape (c) TC SLS 5093.

+Gr 10-77 p623 +-HFN 12-77 p167
+-Gr 12-77 p1142 tape +RR 11-77 p49

720 Symphonies, nos. 1-4. Academic festival overture, op. 80. Tragic overture, op. 81. Variations on a theme by Haydn, op. 56a. CO; Lorin Naazel. London 2405 Tape (c) 5-2405.
+SFC 10-23-77 p45

721 Symphonies, nos. 1-4. COA; Bernard Haitink. Philips 6747 325 Tape (c) 7699 011.
+-HF 7-77 p125 tape

Symphonies, nos. 1-4. cf Works, selections (Vanguard VCS 10117/20).

722 Symphony, no. 1, op. 68, C minor. Halle Orchestra; James Loughran. Classics for Pleasure CFP 40096 Tape (c) TC CFP 40096.
+Gr 3-75 p1648 +RR 3-75 p24
+Gr 2-77 p1325 tape +-RR 11-77 p117 tape
++HFN 12-76 p153 tape

723 Symphony, no. 1, op. 68, C minor. COA; Eduard van Beinum. Decca ECS 793. (Reissue from LXT 2675)
+-Gr 4-77 p1542 +-RR 7-77 p48
+-HFN 3-77 p117

724 Symphony, no. 1, op. 68, C minor. CO; Lorin Maazel. Decca SXL 6783 Tape (c) KSXC 6783. (also London 7007 Tape (c) 5-7007)
+-Gr 11-76 p776 +NR 1-77 p5
-HF 5-77 p81 +-RR 10-76 p45
+-HFN 10-76 p165 ++SFC 12-5-76 p58
+-MJ 2-77 p31

725 Symphony, no. 1, op. 68, C minor. BPhO; Wilhelm Furtwangler. DG 2530 744. (also DG 2535 162)
+ARG 11-76 p16 +-NR 10-76 p2
+Gr 5-76 p1772 +RR 5-76 p23
+-HF 11-76 p105 ++STL 1-9-77 p35

726 Symphony, no. 1, op. 68, C minor. PhO; Otto Klemperer. HMV SXLP 30217 Tape (c) TC SXLP 30217. (Reissue from Columbia SAX 2262)
++Gr 12-76 p990 +-HFN 1-77 p123 tape
+Gr 1-77 p1178 tape +RR 12-76 p53

727 Symphony, no. 1, op. 68, C minor. LSO; Jascha Horenstein. Quintessence PMC 7028.
+HF 9-77 p108 +St 11-77 p146

728 Symphony, no. 2, op. 73, D major. Halle Orchestra; James Loughran. Classics for Pleasure CFP 40219 Tape (c) TC CFP 40219.
+Gr 11-75 p808 -HFN 12-76 p153 tape
+Gr 2-77 p1325 tape +RR 12-75 p45
+HFN 11-75 p152

729 Symphony, no. 2, op. 73, D major. Tragic overture, op. 81. PhO; Otto Klemperer. HMV SXLP 30238 Tape (c) SXLP 30238. (Reissue from Columbia SAX 2362)
+Gr 8-77 p295

730 Symphony, no. 2, op. 73, D major. Slovak Philharmonic Orchestra; Ludovic Rajter. Opus 9110 0364.
-Gr 7-77 p168 /HFN 3-77 p99

731 Symphony, no. 2, op. 73, D major. Danish State Radio Orchestra; Jascha Horenstein. Unicorn UNS 236.
+Gr 11-76 p776 -NR 8-77 p5
-HF 4-77 p96

Symphony, no. 2, op. 73, D major. cf HMV RLS 717.

732 Symphony, no. 3, op. 90, F major. Hungarian dances, nos. 1, 3, 19. Halle Orchestra; James Loughran. Classics for Pleasure CFP 40237 Tape (c) TC CFP 40237.

+Gr 7-76 p168 +-HFN 12-76 p153 tape
+-Gr 2-77 p1325 tape +RR 2-76 p26
+-HFN 3-76 p91

733 Symphony, no. 3, op. 90, F major. CSO; James Levine. RCA ARL 1-2097 Tape (c) ARK 1-2097 (ct) ARS 1-2097.
-NR 11-77 p3 -NYT 12-4-77 pD18

734 Symphony, no. 4, op. 98, F minor. Munich Philharmonic Orchestra; Rudolf Kempe. BASF BAC 3064. (also 20-22394-9)
+FF 11-77 p11 +-HFN 10-75 p139
+Gr 11-75 p808 +RR 10-75 p38

735 Symphony, no. 4, op. 98, E minor. Halle Orchestra; James Loughran. Classics for Pleasure CFP 40084 Tape (c) TC CFP 40084.
++Gr 10-74 p682 +RR 10-74 p38
+Gr 2-77 p1325 +RR 11-77 p117 tape

736 Symphony, no. 4, op. 98, E minor. BPhO; Victor de Sabata. DG 2535 812. (Reissue from Polydor LY 6171/6)
+Gr 8-77 p295 +HFN 9-77 p136

737 Symphony, no. 4, op. 98, E minor. COA; Willem Mengelberg. Past Masters 5.
+FF 11-77 p11

738 Symphony, no. 4, op. 98, E minor. RPO; Fritz Reiner. RCA AGL 1-1961 Tape (c) GK 1-1961. (Reissue)
+Gr 7-77 p168 ++NR 1-77 p5
+-HFN 8-77 p99 tape +-RR 10-77 p93 tape
+-MJ 2-77 p31

739 Symphony, no. 4, op. 98, E minor. CPhO; Dietrich Fischer-Dieskau. Supraphon SQ 27971.
-ARG 9-77 p51

740 Symphony, no. 4, op. 98, E minor. RPO; Fritz Reiner. RCA AGL 1-1961 Tape (c) GK 1-1961.
+ARG 3-77 p12 +HFN 7-77 p108
+HF 4-77 p96 +SFC 11-21-76 p35

741 Symphony, no. 4, op. 98, E minor. BPhO; Wilhelm Furtwangler. Turnabout TV 4476. (also Unicorn WFS 8)
-HF 7-74 p113 +STL 1-9-77 p35

742 Tragic overture, op. 81. MENDELSSOHN: The fair Melusine, op. 32. SCHUMANN: Manfred overture, op. 115. WAGNER: Faust overture. Prague Symphony Orchestra; Dean Dixon. Rediffusion Legend LGD 014 Tape (c) LGC 014.
+-HFN 10-77 p165 +-RR 2-77 p96 tape
+-RR 10-77 p58

743 Tragic overture, op. 81. SCHUBERT: Symphony, no. 8, D 759, B minor. WEBER: Overon: Overture. USSR Symphony Orchestra; Paul Kletzki. Westminster WG 8344.
+ARG 8-77 p40 +NR 8-77 p3

Tragic overture, op. 81. cf Academic festival overture, op. 80.
Tragic overture, op. 81. cf Ein deutsches Requiem, op. 45.
Tragic overture, op. 81. cf Symphonies, nos. 1-4 (HMV SQ SLS 5093).
Tragic overture, op. 81. cf Symphonies, nos. 1-4 (London 2405).
Tragic overture, op. 81. cf Symphony, no. 2, op. 73, D major.
Tragic overture, op. 81. cf Works, selections (Vanguard VCS 10117).
Tragic overture, op. 81. cf Bruno Walter Society RR 443.

744 Trio, clarinet, violoncello and piano, op. 114, A minor. WEBER: Duo concertant, op. 48, E flat major. Keith Puddy, clt; Ian Brown, pno; Gabrieli Ensemble. Enigma VAR 1021.
+-Gr 1-77 p1159 +-RR 3-77 p69
+-HFN 1-77 p103

Trio, clarinet, violoncello and piano, op. 114, A minor. cf Sonata, clarinet, no. 1, op. 120, no. 1, F minor.

745 Trio, horn, violin and piano, op. 40, E flat major. SCHUMANN: Quintet, piano, op. 44, E flat major. Michael Tree, vln; Myron Bloom, hn; Rudolf Serkin, pno. Columbia MS 7266.
++St 5-77 p72

Trio, horn, violin and piano, op. 40, E flat major. cf Sonata, violin and piano, no. 1, op. 78, G major.

Trio, piano, no. 1, op. 8, B major. cf Pieces, piano, op. 117.

746 Variations and fugue on a theme by Handel, op. 24, B flat major. Variations on a theme by Paganini, op. 35, Bks I and II. Garrick Ohlsson, pno. Angel (Q) S 37249. (also (HMV SQ HQS 1379)
+FF 11-77 p12 +HFN 8-77 p77
+Gr 8-77 p323 +NR 12-77 p12
+-HF 11-77 p108 +-St 11-77 p135

Variations and fugue on a theme by Handel, op. 24, B flat major. cf Concerto, violin and violoncello, op. 102, A minor.

747 Variations and fugue on a theme by Handel, op. 24, B flat major. CHOPIN: Mazurka, op. 17, no. 4, A minor. HANDEL: Air with variations, B flat major. Lincoln Mayorga, pno. Sheffield LAB 4.
+MJ 7-77 p70

Variations and fugue on a theme by Handel, op. 24, B flat major. cf Piano works (Telefunken FX 6-35303).

Variations and fugue on a theme by Handel, op. 24, B flat major. cf Pieces, piano, op. 117.

Variations and fugue on a theme by Handel, op. 24, B flat major. cf Variations on an original theme, op. 21.

Variations on a Hungarian song, op. 21, no. 2. cf Piano works (Telefunken FX 6-35303).

Variations on a theme by Haydn, op. 56a. cf Academic festival overture, op. 80.

Variations on a theme by Haydn, op. 56a. cf Ein deutsches Requiem, op. 45 (Angel 3838).

Variations on a theme by Haydn, op. 56a. cf German requiem, op. 45 (DG 2726 078).

Variations on a theme by Haydn, op. 56a. cf Symphonies, nos. 1-4 (Decca D39D4).

Variations on a theme by Haydn, op. 56a. cf Symphonies, nos. 1-4 (London 2405).

Variations on a theme by Haydn, op. 56a. cf Works, selections (Vanguard VCS 10117/20).

Variations on a theme by Haydn, op. 56a. cf BARTOK: Sonata, 2 pianos and percussion.

Variations on a theme by Haydn, op. 56a. cf BEETHOVEN: Symphony, no. 1, op. 21, C major.

Variations on a theme by Haydn, op. 56a. cf ELGAR: Enigma variations, op. 36.

Variations on a theme by Paganini, op. 35. cf Sonata, piano, no. 2, op. 2, F sharp minor.

Variations on a theme by Paganini, op. 35, Bks I and II. cf Variations and fugue on a theme by Handel, op. 24, B flat major.

Variations on a theme by Schumann, op. 9, F sharp minor. cf Piano works (Telefunken FX 6-35303).

Variations on a theme by Schumann, op. 23. cf Fantasias, op. 151.

748 Variations on an original theme, op. 21. Variations and fugue on

a theme by Handel, op. 24, B flat major. Hans Petermandl, pno. Supraphon 111 1614.
-FF 9-77 p9

Variations on an original theme, op. 21, no. 1, D major. cf Piano works (Telefunken FX 6-35303).

Vier ernste Gesange (Four serious songs), op. 121. cf BACH: Sonata, flute, E flat major.

Vier ernste Gesange, op. 121. cf BEETHOVEN: An die ferne Geliebte, op. 98.

Waltzes, op. 39. cf BERTE: Lilac time.

Waltzes, op. 39 (16). cf Fantasias, op. 151.

Waltz, op. 39, no. 2, E major. cf International Piano Library IPA 117.

Waltz, op. 39, no. 15, A flat major. cf International Piano Library IPA 117.

Waltzes, op. 39, nos. 15 and 16, A flat major, C sharp minor. cf International Piano Library IPA 112.

749 Works, selections: Quartet, piano, no. 1, op. 25, G minor. Quartet, piano, no. 2, op. 26, A major. Quartet, piano, no. 3, op. 60, C minor. Quintet, piano, op. 34, F minor. Quintet, clarinet, op. 115, B minor. Bartok Quartet; Csilla Szabo, Istvan Lantos, Dezso Ranki, Sandor Falvai, pno; Bela Kovacs, clt. Hungaroton SLPX 11596/600 (5).
+-Gr 11-77 p854

750 Works, selections: Academic festival oveture, op. 80. Symphonies, nos. 1-4. Tragic overture, op. 81. Variations on a theme by Haydn, op. 56a. Utah Symphony Orchestra; Maurice Abravanel. Vanguard VCS 10117/20 (4).
+ARG 12-77 p22 +-NYT 12-4-77 pD18
+HF 10-77 p102 ++St 12-77 p138

BRANDL

Der Liebe Augustin: Du alter Stefansturm. cf Discopaedia MB 1012.

BRAXTON, Anthony

751 P-JOS..4K-D (MIX). EISLER: Sonata, piano, no. 3, RZEWSKI: Variations on "No place to go but around". Frederick Rzewski, pno. Finnadar SR 9011.
+-HF 9-77 p115 +-NR 8-77 p15
+-MJ 10-77 p28 ++St 6-77 p136

BRAYSSING

Fantasia. cf Lyrichord LLST 7299.

BREHM, Alvin

Variations, piano. cf BESTOR: Sonata, piano.

BREMNER, James

Trumpet tune. cf Vista VPS 1038.

BRESNICK, Martin

752 B's garlands. EDWARDS: Exchange misere. HELLERMANN: On the edge of a node. JONES: Piece mouvante. Tison Street, vln; Frances Uitti, vlc; William Hellermann, gtr; Angelo Persichilli, flt, pic; Michele Incenzo, clt; Massimo Coen, vln, vla; David Saperstein, Noriko Hiraga, pno. CRI SD 336.
+-NR 10-77 p6

BRIAN, Havergal

Festival fanfare. cf RCA RL 2-5081.

753 Symphonies, nos. 6, 16. LPO; Myer Fredman. Lyrita SRCS 67. (also Musical Heritage Society MHS 3426)
++ARG 6-77 p16 ++RR 5-75 p28
+Gr 5-75 p1962 ++Te 12-75 p45
++HFN 5-75 p125

BRIDGE, Frank

First book of organ pieces. cf Vista VPS 1050.

754 Idylls, string quartet (3). Noveletten, string quartet (3). BRITTEN: Phantasy quartet, oboe, violin, viola and violoncello, op. 2. Quartet, strings, D major. Janet Craxton, ob; Gabrieli Quartet. Decca SDD 497.

++FF 11-77 p13 ++MM 10-77 p44
+Gr 6-77 p69 +RR 6-77 p73
+-HFN 6-77 p117 +ST 8-77 p343

Noveletten, string quartet (3). cf Idylls, string quartet.

755 Phantasm rhapsody. MOERAN: Rhapsody, F sharp major. Peter Wallfisch, John McCabe, pno; LPO, NPhO; Nicholas Braithwaite. HNH 4042. (also Lyrita SRCS 9)

+Gr 4-77 p1542 ++MM 10-77 p44
++HF 12-77 p80 ++RR 4-77 p41
+HFN 4-77 p135 ++St 11-77 p135

756 Prelude and minuet. REUBKE: Sonata on the 94th psalm. Raymond Humphrey, org. Wealden WS 151.

+Gr 8-77 p327

Sketches (3). cf Saga 5445.

757 Sonata, violin and piano. Sonata, violoncello and piano. Levon Chilingirian, vln; Rohan de Saram, vlc; Clifford Benson, Druvi de Saram, pno. Pearl SHE 541.

+Gr 6-77 p68 -MM 10-77 p44
+-HFN 6-77 p117 +RR 11-77 p85

Sonata, violoncello and piano. cf Sonata, violin and piano.

Songs: Goldenhair; Journey's end; Tis but a week; When you are old; So perverse. cf Argo ZK 28/9.

BRIGHT, Houston

Rainsong. cf Columbia M 34134.

BAIRSTOW, George

Dream land, op. 59. cf New World Records NW 257.

BRITTEN, Benjamin

758 A charm of lullabies, op. 41. Folksong arrangements: The Ash grove; The bonny Earl o' Moray; Come you not from Newcastle; O can ye sew cushions; O waly, waly; Oliver Cromwell; The Salley gardens; There's none to soothe; Sweet Polly Oliver; The trees they grow so high. Bernadette Greevy, con; Paul Hamburger, pno. London STS 15166.

+HF 4-77 p97 ++SFC 10-2-77 p44
+NR 6-76 p14 +St 9-76 p120

Concerto, piano, op. 13, D major. cf HMV SLS 5080.

759 Concerto, violin, op. 15, D minor. Serenade, op. 31. Ian Partridge, t; Rodney Friend, vln; Nicholas Busch, hn; LPO; John Pritchard. Classics for Pleasure CFP 40250.

++ARG 6-77 p51 +-HFN 12-76 p139
+-Gr 2-77 p1274 +-RR 1-77 p51

760 Concerto, violin, op. 15, D minor. RUBBRA: Improvisation, op. 89. Sidney Harth, Paul Kling, vln; Louisville Orchestra; Robert Whitney. RCA GL 2-5096.

+Gr 11-77 p830 +RR 11-77 p68

761 Diversions, op. 21. Sinfonia da Requiem, op. 20. Julius Katchen, pno; LSO, Danish State Radio Orchestra; Benjamin Britten. Decca ECS 799. (Reissue from LXT 2981)

+Gr 6-77 p42 +RR 6-77 p45
+HFN 7-77 p125

Fanfare for St. Edmonsbury. cf BIS LP 59.

Folksong arrangements: The Ash grove; The bonny Earl o' Moray; Come you not from Newcastle; O can ye sew cushions; O waly, waly; Oliver Cromwell; The Salley gardens; There's none to soothe; Sweet Polly Oliver; The trees they grow so high. cf A charm of lullabies, op. 41.

Hymn to St. Cecilia, op. 27. cf RCA RL 2-5112.

A hymn to the virgin. cf HMV CSD 3774.

Missa brevis, op. 63, D major. cf Songs (Argo ZK 19).

762 Nocturnal, op. 70. Songs from the Chinese, op. 58. CASTELNUOVO-TEDESCO: Sonata, op. 77, D major. Songs, op. 207 (6). Marta Schele, s; Josef Holecek, gtr. BIS LP 31.
+Gr 3-77 p1454 ++St 3-77 p126
+-RR 1-77 p78

763 Noye's fludde. Trevor Anthony, speaker; Caroline Clack, Marie-Therese Pinto, Eileen O'Donovan, Patricia Garrod, Margaret Hawes, Kathleen Petch, Gillian Saunders, girl sopranos; David Pinto, Darien Angadi, Stephen Alexander, boy trebles; Sheila Rex, con; Owen Brannigan, bs-bar; Children's Chorus; English Opera Group Orchestra, Suffolk Children's Orchestra; Norman Del Mar. Argo ZK 1. (Reissue from ZNF 1)
+Audio 6-77 p134 ++HFN 1-77 p121
+Gr 12-76 p1045 +RR 1-77 p34

764 Peter Grimes, op. 33. Irish Kells, s; Lauris Elms, ms; Peter Pears, Raymond Nilsson, John Lanigan, t; James Pease, Geraint Evans, bar; David Kelly, Owen Brannigan, bs; ROHO and Chorus; Benjamin Britten. Decca Tape (c) K71K33 (3). (also London OSA 1305)
++Gr 9-77 p508 tape +NYT 10-9-77 pD21
+HFN 9-77 p155 tape ++RR 9-77 p97 tape

765 Peter Grimes: Four sea interludes and passacaglia, op. 33. Suite on English folk tunes. NYP; Leonard Bernstein. CBS 76640. (also Columbia M 34529 Tape (c) MT 34529)
+-FF 9-77 p10 +NYT 8-7-77 pD13
+-Gr 10-77 p624 +RR 10-77 p40
+-HFN 10-77 p152

766 Peter Grimes: Four sea interludes, op. 33a. Young person's guide to the orchestra (Variations and fugue on a theme by Purcell), op. 34. PH; Carlo Maria Giulini. HMV SXLP 30240 Tape (c) TC SXLP 30240. (Reissue from Columbia SAX 2555)
+Gr 7-77 p173 +RR 7-77 p49
+HFN 8-77 p93

767 Peter Grimes: Four sea interludes and passacaglia, op. 33 a and b. Sinfonia da Requiem, op. 20. LSO; Andre Previn. Angel S 37142. (also HMV SQ ASD 3154)
+Gr 3-76 p1463 ++NR 3-76 p2
+HF 1-77 p112 +-RR 3-76 p37
+HFN 3-76 p91 ++SFC 2-22-76 p29
+MJ 4-76 p31 +St 7-76 p109

768 Phaedra. Prelude and fugue, op. 29. Songs: Sacred and profane, op. 91: A shepherd's carol; Sweet was the song; The sycamore tree; A wealden trio. Janet Baker, s; ECO, Wilbye Consort; Stuart Bedford. Decca SXL 6847.
++Gr 7-77 p207 +RR 7-77 p84
++HFN 7-77 p109 +RR 10-77 p93 tape
++HFN 9-77 p155 tape +ST 8-77 p345

Phaedra, op. 93. cf Works, selections (Decca SXL 6847).

Phantasy quartet, oboe, violin, viola and violoncello, op. 2. cf Idylls, string quartet.

Prelude and fugue. cf Songs (Argo ZK 19).
Prelude and fugue, op. 29. cf Works, selections (Decca SXL 6847).
Prelude and fugue, op. 29. cf Phaedra.
Prelude and fugue on a theme by Vittoria. cf HMV SQS 1376.
Quartet, strings, D major. cf Idylls, string quartet.
Sacred and profane, op. 91, nos. 1-8. cf Works, selections (Decca SXL 6847).

769 St. Nicolas, op. 42. Hamline University/Twin Cities Orchestra; William Jones. Musical Heritage Tape (c) MHC 5484.
+-HF 11-77 p138 tape

770 St. Nicolas, op. 42. Robert Tear, t; Bruce Russell, c-t; King's College Chapel Choir, Cambridge; AMF; David Willcocks. Seraphim S 60296.
+NR 12-77 p8 +SFC 12-25-77 p42

Serenade, op. 31. cf Concerto, violin, op. 15, D minor.
A shepherd's carol. cf Works, selections (Decca SXL 6847).
Sinfonia da Requiem, op. 20. cf Diversions, op. 21.
Sinfonia da Requiem, op. 20. cf Peter Grimes: Four sea interludes and passacaglia, op. 33 a and b.

771 Songs: Hymn to the virgin; Hymn to St. Peter, op. 56a; Hymn to St. Columba; Hymn to St. Cecilia, op. 17; Antiphon, op. 56b; Te Deum, C major. David Lumsden, Patrick Russill, Paul Trepte, org; New College Choir, Oxford. Abbey LPB 753.
++Gr 7-77 p207 ++MT 11-77 p921
+HFN 7-77 p109 +RR 8-77 p75

772 Songs (church music): Hymn to St. Cecilia, op. 17; Hymn to St. Peter; Hymn to the virgin; Festival Te Deum, op. 32; Jubilate Deo. Missa brevis, op. 63, D major. Prelude and fugue. St. Johns College Chapel Choir, LSO Chorus; Simon Preston. Argo ZK 19. (Reissues)
++HFN 12-77 p185 ++RR 12-77 p74

773 Songs (folksong arrangements): The ash grove; La belle est au jardin d'amour; Come ye not from Newcastle; Early one morning; Il est quelqu'un sur terre; Fileuse; The foggy, foggy dew; How sweet the answer; I will give my love an apple; The last rose of summer; Little Sir William; Master Kilby; The minstrel boy; O waly, waly; Oft in the stilly night; The plough boy; Le roi s'en va-t'en chasse; Sailor boy; The Salley gardens; The soldier and the sailor; Sweet Polly Oliver; There's none to soothe. Peter Pears, t; Osian Ellis, hp. Decca SXL 6793.
+Gr 6-77 p83 +RR 6-77 p88
+HFN 6-77 p119 +St 8-77 p343

Songs: The birds. cf Abbey LPB 778.
Songs: Sacred and profane, op. 91; A shepherd's carol; Sweet was the song; The sycamore tree; A wealden trio. cf Phaedra.
Songs from the Chinese, op. 58. cf Nocturnal, op. 70.
Suite on English folk tunes. cf Peter Grimes: Four sea interludes and passacaglia, op. 33.
Sweet was the song. cf Works, selections (Decca SXL 6847).
The sycamore tree. cf Works, selections (Decca SXL 6847).
Te Deum, C major. cf Guild GRSP 701.

774 Variations on a theme by Frank Bridge, op. 10. BUTTERWORTH: The banks of green willow. English idylls (2). A Shropshire lad. AMF; Neville Marriner. Argo ZRG 860 Tape (c) KZRC 860.
+-ARG 9-77 p26 +-MT 2-77 p133
+-Gr 11-76 p776 +NR 10-77 p2
++HFN 11-76 p157 +-RR 11-76 p52

++HFN 1-77 p123 tape ++RR 1-77 p88 tape
+MM 5-77 p37 +-St 7-77 p113

Variations on a theme by Frank Bridge, op. 10. cf Argo D26D4.

775 War requiem, op. 66. Galina Vishnevskaya, s; Peter Pears, t; Dietrich Fischer-Dieskau, bar; Simon Preston, org; Bach Choir, Highgate School Choir; LSO and Chorus, Melos Ensemble; Benjamin Britten. London A 1225 Tape (c) OSA 5-1255. (also Decca Tape (c) K27K22)

++CJ 4-76 p30 +HF 4-77 p64
+Gr 4-77 p1603 tape +HFN 3-77 p119 tape
++HF 6-77 p103 tape ++RR 5-77 p90 tape

A Wealden trio: Christmas song of the women. cf Works, selections (Decca SXL 6847).

Winter words, op. 53. cf Enigma VAR 1027.

776 Works, selections: Prelude and fugue, op. 29. Phaedra, op. 93. Sacred and profane, op. 91, nos. 1-8. A shepherd's carol. The sycamore tree. Sweet was the song. A Wealden trio: Christmas song of the women. Janet Baker, Elaine Barry, Rosemary Hardy, s; Margaret Cable, ms; Nigel Rogers, t; Geoffrey Shaw, bs; Wilbye Consort, ECO; Benjamin Britten, Steuart Bedford, Peter Pears. Decca SXL 6847 Tape (c) KSXC 6847.

++Gr 7-77 p207 +RR 7-77 p84
++HFN 7-77 p109 +RR 10-77 p93 tape
++HFN 9-77 p155 tape +ST 8-77 p345

777 Young person's guide to the orchestra, op. 34 (Variations and fugue on a theme by Purcell). PROKOFIEV: Peter and the wolf, op. 67. Richard Baker, narrator; NPhO; Raymond Leppard. Classics for Pleasure CFP 185 Tape (c) TC CFP 185.

+Gr 1-72 p1208 ++HFN 2-72 p305
+Gr 2-77 p1325 tape +HFN 12-76 p153 tape

Young person's guide to the orchestra, op. 34. cf Peter Grimes: Four sea interludes, op. 33a.

BRIXI, Frantisek Xavier

778 Mass. MICHNA Z OSTRADOVIC: Christmas music (6). Czech Singers Chorus, Prague Madrigal Singers; PSO, Reicha Wind Quintet, CPhO, Members; Josef Veselka, Miroslav Venhoda. Musical Heritage Society MHS 3456.

+-MU 10-77 p8

BROUWER, Leo

Parabola. cf DG 2530 802.

BROWN

Love is where you find it. cf HMV MLF 118.

BROWN, Rayner

779 Concertino, piano and band. DAVIS: Though men call us free. SCHMIDT: Vendor's call. Delcina Stevenson, s; Sharon Davis, pno; David Atkins, clt; Band; Robert Henderson. WIM WIMR 13.

+-NR 10-77 p16

BROWNE

El cumbanchero. cf Transatlantic XTRA 1169.

Stabat iuxta Christi crucem. cf Coimbra CCO 44.

BROWNSON

Salisbury. cf Vox SVBX 5350.

BRUCH, Max

780 Concerto, violin, no. 1, op. 26, G minor. MENDELSSOHN: Concerto, violin, op. 64, E minor. Nathan Milstein, vln; PhO; Leon Barzin. HMV SXLP 30245. (Reissue from Capitol SP 8518)

+Gr 10-77 p624 +RR 11-77 p61
+-HFN 12-77 p185

781 Concerto, violin, no. 1, op. 26, G minor. GLAZUNOV: Concerto, violin, op. 82, A minor. Erica Morini, vln; BRSO; Ferenc Fricsay. DG 2548 170 Tape (c) 3348 170. (Reissue from 138044)
+Gr 11-75 p840 +HFN 10-75 p152
+Gr 3-77 p1458 tape +RR 11-75 p42
782 Concerto, violin, no. 1, op. 26, G minor. MENDELSSOHN: Concerto, violin, op. 64, E minor. Itzhak Perlman, vln; LSO; Andre Previn. HMV Tape (c) TC ASD 2926. (also Angel S 36963 Tape (c) 4XS 36963 (ct) 8XS 36983)
+Gr 3-77 p1458 tape +RR 4-77 p92 tape
Concerto, violin, no. 1, op. 26, G minor. cf BEETHOVEN: Concerto, violin, op. 61, D major.
Concerto, violin, no. 1, op. 26, G minor. cf HMV SLS 5068.
Concerto, violin, no. 1, op. 26, G minor: Adagio. cf Discopaedia MB 1015.
783 Concerto, violin, no. 2, op. 44, D minor. Scottish fantasia, op. 46. Itzhak Perlman, vln; NPhO; Jesus Lopez-Cobos. Angel (Q) S 37210. (also HMV ASD 3310 Tape (c) TC ASD 3310)
+Gr 6-77 p42 +NYT 7-24-77 pD11
+HF 11-77 p108 +RR 7-77 p49
+HFN 7-77 p109 ++SFC 5-15-77 p50
+HFN 8-77 p99 +-St 9-77 p124
+-NR 8-77 p6
784 Kol Nidrei, op. 47. MEDINS: Suite concertante. STRAUSS, R.: Sonata, violoncello and piano, op. 6. Ingus Naruns, vlc; Anatol Berzkalns, pno. Kaibala 60F01.
++Audio 8-76 p75 ++NR 3-77 p14
Kol Nidrei, op. 47. cf Orion ORS 75181.
785 Pieces, piano, violoncello and clarinet, op. 83. GLINKA: Trio pathetique, D minor. Montagnana Trio. Celos DEL 25433.
+NR 12-77 p6
Scottish fantasia, op. 46. cf Concerto, violin, no. 2, op. 44, D minor.
786 Symphony, no. 2, op. 36, F minor. RIETZ: Concert overture, op. 7. Louisville Orchestra; Jorge Mester. RCA GL 2-5017.
+Gr 10-76 p590 /RR 12-76 p53
+-HFN 12-76 p139 +ST 1-77 p777
Waldpsalm, op. 38, no. 1. cf Telefunken DT 6-48085.

BRUCKNER, Anton

Ecce sacerdos magnus. cf Abbey LPB 776.
787 Mass, no. 2, E minor. Kenneth Jewel Chorale; Interlochen Arts Academy Wind Ensemble. (Available from Kenneth Jewel Chorale, 10805 Elgin, Huntington Woods, Mich)
+MU 6-77 p8
788 Mass, no. 2, E minor. Te deum. Judith Blegen, s; Margarita Lilowa, ms; Claes Haaken Ahnsjo, t; Peter Meven, bs; Vienna State Opera Chorus; VPO; Zubin Mehta. Decca SXL 6837. (also London 26506)
-Gr 12-77 p1211 +-RR 12-77 p74
+-HFN 12-77 p167
Songs: Ave Maria; Graduale; Virga Jesse. cf Argo ZRG 871.
789 Symphony, no. 1, C minor. VSO; Volkmar Andreae. Amadeo AVRS 19016.
+-HFN 12-77 p167 +RR 11-77 p50
790 Symphony, no. 3, D minor. VPO; Karl Bohm. London CS 6717. (also Decca SXL 6505 Tape (c) KSXC 6505)

+Gr 10-71 p629 -MJ 10-72 p58
+HF 12-71 p90 ++RR 9-77 p98 tape
+HFN 10-71 p1854 -SFC 4-2-72 p35
++HFN 8-77 p99 tape +St 2-72 p80
+NR 12-71 p7

791 Symphony, no. 3, D minor. MRSO; Gennady Rozhdestvensky. Westminster WHS 8327.
++ARG 4-77 p22 ++NR 5-77 p6
+HF 10-77 p104

792 Symphony, no. 4, E flat major. BPhO; Herbert von Karajan. DG 2530 674 Tape (c) 3300 674.
+ARG 6-77 p18 ++RR 10-76 p46
+Gr 10-76 p590 ++RR 1-77 p88 tape
-HF 7-77 p100 +SFC 5-29-77 p42
+-HFN 10-76 p166 ++St 7-77 p112
-MJ 9-77 p35 ++STL 10-10-76 p37
+NR 5-77 p2

793 Symphony, no. 4, E flat major (ed. Haas). Symphony, no. 7, E major (ed. Haas). Leipzig Gewandhaus Orchestra; Kurt Masur. Eurodisc 27913 XGK.
+FF 11-77 p14 +HF 11-77 p93

794 Symphony, no. 4, E flat major. Leipzig Gewandhaus Orchestra; Kurt Masur. RCA RL 2-5106.
+-Gr 11-77 p830 +-RR 11-77 p50

795 Symphony, no. 5, B flat major (orig. version). Symphony, no. 8, C minor: 4th movement. VPO, Berlin State Opera Orchestra; Wilhelm Furtwangler, Herbert von Karajan. Bruno Walter Society RR 508 (2).
+NR 8-77 p4

796 Symphony, no. 6, A major. VPO; Horst Stein. Decca SXL 6682 Tape (c) KSXC 6682. (also London 6880)
+-Gr 4-75 p1802 ++NR 2-76 p2
+HF 2-76 p94 +-RR 4-75 p26
+HFN 9-77 p155 tape +RR 10-77 p93 tape
+MJ 2-76 p47

797 Symphony, no. 6, A major. BSO; William Steinberg. RCA GL 2-5009.
-Gr 11-76 p781 +-RR 10-76 p48
+-HFN 12-76 p140 +ST 1-77 p779

798 Symphony, no. 7, E major (orig. version). BPhO; Wilhelm Furtwangler. DG 2535 161.
+-Gr 5-76 p1772 +STL 1-9-77 p35
+RR 5-76 p22

799 Symphony, no. 7, E major. WAGNER: Siegfried Idyll. BPhO; Herbert von Karajan. DG 2707 102 Tape (c) 3370 023.
+NYT 12-4-77 pD18 ++SFC 10-16-77 p47

800 Symphony, no. 7, E major. Symphony, no. 8, C minor (ed. Nowak). VPO; Karl Bohm. DG 2709 068 (3) Tape (c) 3371 027.
+-Gr 6-77 p42 +-NYT 12-4-77 pD18
+HF 11-77 p93 +-RR 6-77 p45
+-HFN 6-77 p119 +-RR 7-77 p99
++NR 10-77 p1 +St 12-77 p139

801 Symphony, no. 7, E major. WAGNER: Parsifal: Prelude, Act 1. Tristan und Isolde: Prelude, Act 1. BPhO; Herbert von Karajan. HMV SLS 5086 (2) Tape (c) TC SLS 5086. (Reissues from SLS 811, 963, ASD 3160)
+Gr 6-77 p42 +HFN 8-77 p93
-Gr 8-77 p346 tape +RR 6-77 p45

802 Symphony, no. 7, E major. MAHLER: Lieder eines fahrenden Gesellen. Dietrich Fischer-Dieskau, bar; BPhO, VPO; Wilhelm Furtwangler. Rococo 2105 (2).
+NR 2-77 p3

803 Symphony, no. 7, E major. BPhO; Jascha Horenstein. Unicorn UN 111. (Reissue from Polydor 66802/8)
+-Gr 12-76 p990 +NR 11-77 p5
+HF 1-77 p112 +St 1-77 p113

Symphony, no. 7, E major. Symphony, no. 4, E flat major.

804 Symphony, no. 8, C minor. BPhO; Herbert von Karajan. DG 2707 085 (2).
++ARG 4-77 p6 ++NR 4-77 p2
+Gr 5-76 p1750 ++RR 8-76 p29
+-HF 3-77 p97 ++SFC 12-26-76 p34
+HFN 8-76 p77 ++St 3-77 p129
+MJ 2-77 p31 +STL 7-4-76 p36
+-MT 9-77 p733

805 Symphony, no. 8, C minor. BPhO; Herbert von Karajan. HMV SXDW 3024 (2). (Reissue from Columbia 33 CX 1586/7)
+Gr 5-76 p1750 +RR 8-76 p30
+MT 9-77 p733

Symphony, no. 8, C minor. cf Symphony, no. 7, E major.

Symphony, no. 8, C minor: 4th movement. cf Symphony, no. 5, B flat major.

806 Symphony, no. 9, D minor. NYP; Leonard Bernstein. Columbia M 30828. (also CBS 61646 Tape (c) 40-61646)
+-ARG 5-72 p407 -NR 12-71 p6
-Gr 3-77 p1395 -RR 3-77 p47
/HF 1-72 p82 -RR 5-77 p90 tape
+-HFN 3-77 p117 -SFC 1-21-73 p41
+-HFN 5-77 p139 tape +-St 8-72 p74

807 Symphony, no. 9, D minor. BPhO; Herbert von Karajan. DG 2530 828 Tape (c) 3300 828.
++Gr 6-77 p42 +-RR 6-77 p46
++HFN 6-77 p119 +SFC 10-16-77 p46
+NYT 12-4-77 pD18

808 Symphony, no. 9, D minor. CSO; Carlo Maria Giulini. HMV SQ ASD 3382. (also Angel S 37287)
+-Gr 12-77 p1068 +NYT 12-4-77 pD18
++HFN 12-77 p167 +RR 12-77 p46

809 Symphony, no. 9, D minor. COA; Eduard van Beinum. Philips Fontana 6530 058. (Reissue)
+HFN 10-77 p165 +RR 12-77 p1068
+RR 10-77 p40

810 Symphony, no. 9, D minor. BACH: Chorale preludes (13) (arr. Mytsalski). MRSO; Gennady Rozhdestvensky. Westminster WGS 8347 (2).
-HF 10-77 p104 +NYT 12-4-77 pD18
+-NR 8-71 p4 +SFC 5-29-77 p42

811 Te deum. MOZART: Mass, no. 16, K 317, C major. Anna Tomowa-Sintow, s; Agnes Baltsa, ms; Peter Schreier, Werner Krenn, t; Jose van Dam, bs-bar; Rudolf Scholz, org; Vienna Singverein; BPhO; Herbert von Karajan. DG 2530 704 Tape (c) 3300 704.
++ARG 9-77 p20 +NR 9-77 p9
+-Gr 2-77 p1308 +RR 2-77 p87
+-Gr 5-77 p173 tape +RR 3-77 p99 tape
+HF 10-77 p105 ++SFC 7-31-77 p40
+HFN 2-77 p113 ++St 10-77 p141

812 Te deum. VERDI: Pezzi sacri: Te deum. Uta Spreckelsen, s; Heidrun Ankersen, ms; Adalbert Kraus, t; Kurt Moll, bs; Christoph Grohmann, org; Bielefeld Musikverein Chorus; PH; Martin Stephani. Telefunken 6-42037.
++ARG 6-77 p19 -HFN 7-77 p109
+-Gr 7-77 p207 +RR 7-77 p84
+HF 6-77 p84
Te deum. cf Mass, no. 2, E minor.

BRUMEL, Antoine
Missa et ecce terrae motus: Gloria. cf HMV SLS 5049.
Songs: Du tout plongiet; Fors seulement, l'attente. cf HMV SLS 5049.
Vray dieu d'amours. cf HMV SLS 5049.

BRUNA
Variations on the Litany of the Virgin. cf Harmonia Mundi HM 759.

BRUNEAU, Alfred
L'Attaque du Moulin: Adieux. cf Rubini GV 38.

BRYANT, Allan
813 A bouncing little people planet. A rocket is a drum. Space guitars. Whirling take-off. Allan Bryant, original string instruments. CRI SD 366.
-ARG 9-77 p26 /MJ 10-77 p29
+FF 11-77 p15
A rocket is a drum. cf A bouncing little people planet.
Whirling take-off. cf A bouncing little people planet.

BUCHTEL
Polka dots. cf Nonesuch H 71341.

BUCK, Dudley
Concert variations on the Star-spangled banner, op. 23. cf Decca PFS 4416.
Concert variations on the Star-spangled banner, op. 23. cf Vista VPS 1038.
Festival overture on the American national air. cf PISTON: The incredible flutist.
Rock of ages. cf New World NW 220.

BUDD
New work, no. 5. cf Crystal S 361.

BULL, John
Les buffons. cf Saga 5447.
Dr. Bull's jewel. cf Argo ZRG 864.
Dr. Bull's my selfe. cf Argo ZRG 864.
Music from the Elizabethan court. cf Transatlantic XTRA 1169.
Revenant. cf BACH: Chorale prelude: O Gott, du frommer Gott, S 767.
Variations on the Dutch chorale "Laet ons met herten reijne". cf Philips 6500 926.

BULL, Ole
The chalet girl's Sunday. cf Decca DDS 507/1-2.

BULLOCK, Ernest
Fanfare for the coronation of Queen Elizabeth. cf CRD Britannia BR 1077.
Fanfare for the coronation of Queen Elizabeth II. cf Decca SPA 500.

BURIAN, Karel
American suite. cf Supraphon 111 1721/2.

BURROWES, John
Auld lang syne. cf Serenus SRS 12067.

BURTON, Eldin
814 Sonatina, flute and piano. COPLAND: Duo, flute and piano. PISTON:

Sonata, flute and piano. VAN VACTOR: Sonatina, flute and piano. Keith Bryan, flt; Karen Keys, pno. Orion ORS 76242.
+HF 5-77 p99 ++NR 10-76 p6

BURTON, Stephen

851 Songs of the Tulpehocken. Kenneth Riegel, t; Louisville Orchestra; Stephen Douglas Burton. Louisville LS 757.
+NR 12-77 p11

BUSATTI

Surrexit pastor bonus. cf HMV SQ ASD 3393.

BUSCH, Adolf

816 Sonata, violin and piano, no. 2, op. 56, A minor. Pina Carmarelli, vln; Rudolf Serkin, pno. Marlboro Recording MRS 9.
+ARG 12-77 p24 *MJ 4-77 p50

BUSCH, William

Songs: Come, o come, my life's delight; The echoing green; If thou wilt ease thine heart; The shepherd. cf Argo ZK 28/9.

BUSH, Alan

Voices of the prophets. cf Argo ZK 28/9.

BUSH, Geoffrey

Songs: Echo's lament for Narcissus; The wonder of wonders. cf Enigma VAR 1027.

BUSNOIS, Antoine

Fortune esperee. cf Argo ZK 24.

Songs: Fortuna desperata. cf HMV SLS 5049.

BUSONI, Ferrucio

Elegie, no. 4: Turandots Frauengemach. cf International Piano Library IPL 108.

Indian diary, Bk 1. cf International Piano Library IPL 104.

Short pieces for the cultivation of polyphonic playing (6). cf BARTOK: Etudes, op. 18.

Sonatina, no. 2. cf International Piano Library IPL 102.

BUSSOTTI, Sylvano

Pieces, piano, for David Tudor, no. 3. cf CP 3-5.

Ultima rara. cf DG 2530 561.

BUTTERWORTH, Arthur

817 Dales suite. HOLST: Moorside wuite. JAMES: Solitude. IRELAND: Downland suite. Besses o' th' Barn Band; Ifor Jones. Pye TB 3012.
+Gr 1-77 p1193

BUTTERWORTH, George

818 The banks of green willow. English idylls (2). A Shropshire lad. HOWELLS: Elegy, op. 15. Merry eye, op. 20b. Music for a prince. Herbert Downes, vla; Desmond Bradley, Gillian Eastwood, vln; Albert Cayzer, vla; Norman Jones, vlc; NPhO, LPO; Adrian Boult. Lyrita SRCS 69. (also HNH 4005)
+Gr 1-76 p1211 +RR 12-75 p50
++HFN 11-75 p153 ++St 10-77 p140
++NR 9-77 p1

The banks of green willow. cf BRITTEN: Variations on a theme by Frank Bridge, op. 10.

The banks of green willow. cf Angel S 37409.

The banks of green willow. cf Argo D26D4.

The banks of green willow. cf HMV (SQ) ASD 3338.

English idylls. cf The banks of green willow.

English idylls. cf BRITTEN: Variations on a theme by Frank Bridge, op. 10.

English idylls. cf Argo D26D4.

A Shropshire lad. cf The banks of green willow.
A Shropshire lad. cf BRITTEN: Variations on a theme by Frank Bridge, op. 10
A Shropshire lad: Rhapsody. cf Argo D26D4.
819 Songs: Bredon Hill: O fair enough are sky and plain; On the idle hill of summer, When the lad for longing sighs, With rue my heart is laden; A Shropshire lad. FINZI: Earth and air and rain, op. 15. Benjamin Luxon, bar; David Willison, pno. Argo ZRG 838.
++Gr 4-76 p1650 +NR 12-77 p10
+HFN 4-76 p103 ++RR 4-76 p68
+MM 5-77 p37 ++St 7-77 p113
+MT 8-76 p658

BUXTEHUDE, Dietrich
820 Cantatas: Alles, was ihr tut; Befiehl dem Engel, dass er komm; Mit Fried und Freud ich fahr dahin. STEFFANI: Stabat mater. Kurt Equiluz, Rudolf Resch, t; Nikolaus Simkowsky, Johannes Kunzel, bs; Greifswald Cathedral Choir, Vienna Boys Choir, Chorus Viennensis; Berlin Bach Orchestra, VCM; Hans Pflugbeil, Nikolaus Harnoncourt. CMS/Oryx 3C 303.
+St 3-77 p127
821 Cantatas: Hertlich lieb hab ich Dich, O Herr; Jesu, meine Freude; Wachet auf, ruft uns die Stimme. Herrad Wehrung, Gundula Bernat-Klein, s; Frauke Haasemann, con; Friedreich Melzer, c-t; Johannes Hoeffling, t; Wilhelm Pommerien, bs; Westphalian Choral Ensemble; South West German Chamber Orchstra; Wilhelm Ehmann. Nonesuch H 71332.
+ARG 3-77 p34 +NR 4-77 p7
+-HF 4-77 p97 +-St 3-77 p127
Canzona, C major. cf Organ works (Telefunken EX 6-35308).
Canzona, G major. cf Organ works (Telefunken EX 6-35309).
Canzona, G minor. cf Works, selections (Telefunken EK 6-35307).
Canzonettas, D minor, E minor, C major. cf Organ works (Telefunken EX 6-35308).
Canzonettas, G major, G minor. cf Organ works (Telefunken EK 6-35309).
Chaconne, E minor. cf Organ works (Titanic TI II).
822 Chorale preludes (6). Preludes and fugues, G major, G minor, E minor, C major. Toccatas, D minor, G major. Michel Chapuis, org. Telefunken AF 6-42001.
+HFN 10-76 p167 +-RR 10-76 p86
+NR 9-77 p14 ++SFC 6-26-77 p46
Chorale preludes: Auf meinen lieben Gott; Ach Gott und Herr; Ach Herr mich armen Sunder; Christ unser Herr zum Jordan kam; Durch Adams Fall ist ganz verderbt; Ein feste Burg ist unser Gott; Es ist das Heil uns kommen her; Jesus Christus, unser Heiland; Mensch, willt du leben seliglich; Te deum laudamus; Nimm von uns, Herr, du treuer Gott. cf Organ works (Telefunken EX 6-35308).
Chorale preludes: Erhalt uns Herr bein deinem Wort; Es spricht der unweisen Mund wohl; Gott der Vater wohn uns bei; Ich dank dir, lieber Herre; Ich dank dir schon durch deinen Sohn; Herr Jesu Christ, ich weiss gar wohl; Kommt her zu mir, spricht Gottes sohn; Nun bitten wir der Heiligen Geist; Nun lob mein Seel den Herren. cf Organ works (Telefunken EK 6-35309).
Chorale prelude: Es ist das Heil uns zu kommen. cf Organ works (Titanic TI II).

Chorale prelude: Wie schon leuchtet der Morgenstern. cf BACH: Choral prelude: O Gott, du frommer Gott, S 767.
Ciacona, C minor. cf Organ works (Telefunken EX 6-35308).
Ciacona, E major. cf Works, selections (Telefunken EK 6-35307).
Danket dem Herrn, denn er ist Freundlich. cf Works, selections (Telefunken EK 6-35307).
Fugues, C major, B major. cf Organ works (Telefunken EK 6-35309).
Gelobet seist du, Jesu Christ. cf Works, selections (Telefunken EK 6-35307).
Herr Christ, der einig Gottes Sohn. cf Works, selections (Telefunken EK 6-35307).
Ich ruf zu dir, Lobt Gott, ihr Christen. cf Organ works (Titanic TI 11).
In dulci jubilo. cf Works, selections (Telefunken EK 6-35307).
Komm heiliger Geist, Herre Gott. cf Organ works (Telefunken EK 6-35308).
Lobt Gott, ihr Christen allzugleich. cf Works, selections (Telefunken EK 6-35307).
Magnificat im 1 ton, 9 ton. cf Works, selections (Telefunken EK 6-35307).
Magnificat primi toni. cf Organ works (Telefunken EK 6-35309).
Nun freut euch, liebe Christen gmein. cf Works, selections (Telefunken EK 6-35307).
Nun komm der Heiden Heiland. cf Works, selections (Telefunken EK 6-35307).

823 Organ works: Chorale preludes: Auf meinen lieben Gott; Ach Gott und Herr; Ach Herr mich armen Sunder; Christ unser Herr zum Jordan kam; Durch Adams Fall ist ganz verderbt; Ein feste Burg ist unser Gott; Es ist das Heil uns kommen her; Jesus Christus unser Heiland; Mensch, willt du leben seliglich; Te deum laudamus; Nimm von uns, Herr, du treuer Gott. Ciacona, C minor. Canzona, C major. Canzonettas, D minor, E minor, C major. Preludes and fugues, A minor, B major, F major, F sharp minor, D major, G minor, F major. Komm heiliger Geist, Herre Gott. Vater unser im Himmelreich, variations. Michael Chapuis, org. Telefunken EX 6-35308 (2).
+Gr 5-77 p1713 -RR 4-77 p69
+HFN 5-77 p120

824 Organ works: Canzona, G major. Canzonettas, G major, G minor. Chorale preludes: Erhalt uns Herr bein deinem Wort; Es spricht der unweisen Mund wohl; Gott der Vater wohn uns bei; Ich dank dir, lieber Herre; Ich dank dir schon durch deinen Sohn; Herr Jesu Christ, ich weiss gar wohl; Kommt her zu mir, spricht Gottes sohn; Nun bitten wir der heiligen Geist; Nun lob mein Seel den Herren (2). Fugues, C major, B major. Preludes and fugues, D minor, A minor, G minor, C major, G major, E major. Magnificat primi toni. Toccatas, G major, F major. Michael Chapuis, org. Telefunken EK 6-35309 (2).
+-Gr 8-77 p323 +-RR 6-77 p80
+-HFN 7-77 p111

825 Organ works: Chorale preludes: Es ist das Heil uns zu kommen. Ich ruf zu dir, Lobt Gott, ihr Christen. Preludes and fugues, F major, E major, G minor. Chaconne, E minor. Toccata, F major. Mireille Lagace, org. Titanic TI 11.
++MU 4-77 p12 ++NR 11-76 p11

Passacaglia, D major. cf Works, selections (Telefunken EK 6-35307).
Passacaglia, D minor. cf Argo ZRG 806.

Praeludium und Fuge, G minor. cf Toccata 53623.
Praeludium und Fugen, C major, A major, E major, A minor, F minor. cf Works, selections (Telefunken EK 6-35307).
Prelude and fugue, A minor. cf Richardson RRS 3.
Preludes and fugues, G major, G minor, E minor, C major. cf Chorale preludes.
Preludes and fugues, A minor, B major, F major, F sharp minor, D minor, D major, G minor, F major. cf Organ works (Telefunken EX 6-35308).
Preludes and fugues, D minor, A minor, G minor, C major, G major, E major. cf Organ works (Telefunken EK 6-35309).
Preludes and fugues, F major, E major, G minor. cf Organ works (Titanic TI 11).
Prelude, fugue and chaconne, C major. cf Argo ZRG 806.
Puer natus in Bethlehem. cf Works, selections (Telefunken EK 6-35307).
Der Tag der ist so Freudenreich. cf Works, selections (Telefunken EK 6-35307).
Toccata, F major. cf Organ works (Titanic TI 11).
Toccatas, D minor, G major. cf Chorale preludes.
Toccatas, G major, F major. cf Organ works (Telefunken EK 6-35309).
Trauermusic. cf BACH: Chorale prelude: O Gott, du frommer Gott, S 767.
Vater unser im Himmelreich, variations. cf Organ works (Telefunken EX 6-35308).
Wie schon leuchtet der Morgenstern. cf Works, selections (Telefunken EK 6-35307).
Wir danken dir, Herr Jesu Christ. cf Works, selections (Telefunken EK 6-35307).

826 Works, selections: Canzona, G minor. Ciacona, E major. Danket dem Herrn, denn er ist Freundlich. Gelobet seist du, Jesu Christ. Herr Christ, der einig Gottes Sohn. In dulci jubilo. Lobt Gott, ihr Christen allzugleich. Magnificat im 1 ton, 9 ton. Nun freut euch, liebe Christen gmein. Nun komm der Heiden Heiland. Passacaglia, D major. Praeludium und Fugen, C major, A major, E major, A minor, F minor. Puer natus in Bethlehem. Der Tag der ist so Freudenreich. Wie schon leuchtet der Morgenstern. Wir danken dir, Herr Jesu Christ. Michael Chapuis, org. Telefunken EK 6-35307 (2).
+HFN 2-77 p113

BYFIELD, Jack
A Cornish pastiche. cf HMV SRS 5197.
A Cornish pastiche. cf Starline SRS 197.
Gabriel John. cf HMV SRS 5197.
Gabriel John. cf Starline SRS 197.
A pinch of salt. cf HMV SRS 5197.
A pinch of salt. cf Starline SRS 197.

BYRD, William
The barlye breake. cf Works, selections(Musical Heritage Society MHS 5460).
Browning. cf Works, selections (Musical Heritage Society MHS 5460).
Clarifica me, pater. cf RCA RL 2-5110.
Come woeful Orpheus. cf RCA RL 2-5110.
Content is rich. cf RCA RL 2-5110.
Earle of Oxford's march. cf Argo ZRG 823.
The Earle of Salisbury pavan. cf Amberlee ALM 602.

The Earl of Salisbury pavan. cf Decca PFS 4351.
The Earl of Salisbury, pavan and galliard. cf Saga 5447.
Emendemus in melius a 5. cf RCA RL 2-5110.
Fantasias (4). cf Works, selections (Musical Heritage Society MHS 5460).
Fantasia a 3. cf RCA RL 2-5110.
Fantasia a 4. cf RCA RL 2-5110.
Galliard (after Francis Tregian). cf Decca PFS 4351.
A gigg. cf Abbey LPB 765.
Haec dicit Dominus a 5. cf RCA RL 2-5110.
In nomine, a 4, a 5. cf Works, selections (Musical Heritage Society MHS 5460).
The leaves be green fantasy. cf BIS LP 57.
Libera me, Domine, et pone a 5. cf RCA RL 2-5110.
Lord Willobies welcome home. cf Works, selections (Musical Heritage Society MHS 5460).

827 Mass, 4 voices. Mass, 5 voices. Christ Church Cathedral Choir; Simon Preston. Argo ZRG 858 Tape (c) KZRC 858.

+-Gr 4-77 p1582 +-MT 10-77 p825
+-HF 11-77 p138 tape +NR 8-77 p8
+HFN 3-77 p101 +RR 3-77 p89
++HFN 5-77 p138 tape ++RR 5-77 p90 tape

828 Mass, 4 voices. Mass, 5 voices. St. Margaret's Singers; Richard Hickox. RCA RL 2-5070.

+Gr 6-77 p83 +-MT 10-77 p825
+HFN 8-77 p79 +-RR 6-77 p88

Mass, 5 voices. cf Mass, 4 voices (Argo ZRG 858).
Mass, 5 voices. cf Mass, 4 voices (RCA 2-5070).

829 Motets: Ave verum corpus; Domine, salva nos; Guadeamus omnes; Haec dies; Ne irascaris Domine; Vide Domino. TALLIS: The lamentations of Jeremiah. King's Singers. HMV (SQ) CSD 3779 Tape (c) TC CSD 3779.

+-Gr 6-77 p83 +-MT 11-77 p924
+HFN 6-77 p119 +-RR 7-77 p92
+HFN 10-77 p171 tape +-RR 10-77 p99 tape

Motets: Sacerdotes domini; Veni sancte spiritus; Senex puerum portabat. cf Vista VPS 1037.

830 My Lady Nevells booke. Christopher Hogwood, hpd. L'Oiseau-Lyre D29D4.

+Gr 1-77 p1160 +-MT 9-77 p735
+HFN 1-77 p103 +RR 2-77 p77

O Lord, make They servant Elizabeth. cf Guild GRSP 701.
Pavan a 5. cf Works, selections (Musical Heritage Society MHS 5460).
Praise our Lord, all ye gentiles. cf Abbey LPB 770.
Praise our Lord, all ye gentiles. cf RCA RL 2-5110.
Prelude and ground. cf Works, selections (Musical Heritage Society MHS 5460).
Songs: Hodie beata Virgo Maria; Senex puerum portabat. cf HMV ESD 7050.
Ut, Re. cf Argo ZRG 806.
Ut, Re mee fa sol la. cf Abbey LPB 752.
La volta. cf Saga 5425.
Wedded to will is witless. cf RCA RL 2-5110.
What pleasure have great princes. cf RCA RL 2-5110.
Will you walke the woods soe wylde. cf Works, selections (Musical Heritage Society MHS 5460).

Wolsey's wilde. cf Saga 5425.
831 Works, selections: The barley breake. Browning. Fantasias (4). In nomine a 4, a 5. Lord Willobies welcome home. Pavan a 5. Prelude and ground. The third pavan and galliard. Will you walke the woods soe wylde. Edward Smith hpd; New York Consort of Viols. Musical Heritage Society MHS 5460.
++St 7-77 p113

BYZANTINE HYMNS OF CHRISTMAS, EPIPHANY, EPITAPHIOS AND EASTER. cf Society for the Dissemination of National Greek Music SDNM 101/2, 107, 112.

CABALLERO
Chateau Margaux: Romanza de Angelita. cf London OS 26435.
Gigantes y cabezudos: Romanza de Pilar. cf London OS 26435.
El senor Joaquin: Balada y alborada. cf London OS 26435.

CABANILLES, Juan
Batalla imperial. cf London CS 7046.
Passacalles du 1er mode. cf Argo ZRG 806.

CABEZON, Antonio de
Diferencias sobre el Canto llano del caballero. cf London CS 7046.
Diferencias sobre "La dama le demanda". cf Nonesuch H 71326.
Diferencias sobre la Gallarda milanesa. cf London CS 7046.
Pavana con su glosa. cf London CS 7046.
Pavane and variations. cf RCA RL 2-5099.

CACCINI, Giulio
Songs: Amarilli mia bella; Belle rose porporine; Perfidissimo volto; Udite amante. cf DG 2533 305.

CAGE, John
832 Quartet, strings, 4 parts. LUTOSLAWSKI: Quartet, strings. LaSalle Quartet. DG 2530 735. (Reissue from DG 104988)
+Gr 2-77 p1296 +MT 8-77 p650
+HFN 2-77 p115 +RR 2-77 p67
833 Sonatas and interludes, prepared piano. John Tilbury, prepared piano. Decca HEAD 9.
+Gr 11-76 p838 ++MT 1-77 p43
+HFN 7-76 p85 +RR 7-76 p70
+MM 4-77 p40
834 27'10.554", percussion. DUCHAMP: The bride stripped bare by her bachelors, even, Erratum musical. Donald Knaack, perc. Finnadar SR 9017.
+MJ 10-77 p29 +St 11-77 p136
+NR 9-77 p15
Winter music. cf CP 3-5.

CAIX d'HERVELOIS, Louis de
Suite, A major. cf BIS LP 22.
Suite, op. 6, no. 3, G major. cf Saga 5425.

CAJKOVSKIJ
835 Quartet, strings, no. 4. JEZEK: Sonata, 2 violins. SCHULHOFF: Sonata, solo violin. Sukova Quartet. Panton 110 527.
+HFN 7-77 p111 +-RR 8-77 p72

CALABRO, Louis
836 Symphony, no. 3. Voyage. Bennington Community Chorus; Bennington College Motet Choir; Sage City Symphony Orchestra, Eastman-Rochester Symphony Orchestra; Louis Calabro, A. Clyde Roller. (Available from Sage City Symphony, P O Box 258, Shaftsbury, Vt)
+-ARG 3-77 p13
Voyage. cf Symphony, no. 3.

CALDARA, Antonio
Vaghe luci. cf HNH 4008.
CALESTANI, Vincenzo
Songs: Damigella tutta bella. cf DG 2533 305.
CALLEJA
Cancion triste. cf London CS 7015.
CALVI, Carlo
The Medici court. cf Saga 5420.
Suo corrente. cf Lyrichord LLST 7299.
CAMPIAN, Thomas
Never weather-beaten sail. cf Philips 6500 926.
Pieces, D major. cf d'ANGLETERRA: Carillo, G major.
Shall I come, sweet love to thee. cf Abbey LPB 712.
Songs: Beauty is but a painted hell; Come let us sound with melody; The cypress curtain of the night. cf Enigma VAR 1023.
Songs: Oft have I sighed; There is a garden in her face. cf Saga 5447.
CAMPO, Frank
Duet for equal trumpets. cf Crystal S 362.
CAMPRA, Andre
Les fetes venitiennes: Chanson du Papillon. cf Odyssey Y 33130.
CANN, Richard
Bonnylee. cf Odyssey Y 34139.
CANTELOUBE, Joseph
Chants d'Auvergne: Lo fiolaire. cf Seraphim 60274.

837 Songs of the Auvergne: L'antoueno; Pastourelle; L'aio de rotso; Ballero; Passo del prat; Malurous qu'o uno fenno; Brezairola. RACHMANINOFF: Vocalise, op. 34, no. 14 (arr. Dubensky). VILLA-LOBOS: Bachianas brasileiras, no. 5. Anna Moffo, s; American Symphony Orchestra; Leopold Stokowski. RCA LSB 4114 Tape (c) RK 42006.
/Gr 4-75 p1860 +-RR 4-75 p67
+-HFN 6-75 p97 +RR 6-77 p96 tape
+-HFN 5-77 p139 tape

CAPIROLA (16th century Italy)
Ricercare, nos. 1, 2, 10, 13. cf DG 2533 173.
CAPUA
O sol mio. cf HMV RLS 715.
CARA, Marchetto
Non e tempo. cf Nonesuch H 71326.
CARISSIMI, Giocomo

838 Historia di Jonas. CAVALLI: Missa pro defunctis. Louis Halsey Singers; Louis Halsey. L'Oiseau-Lyre SOL 347.
+-Gr 1-77 p1162 +MT 3-77 p214
++HFN 12-76 p140 +-RR 12-76 p90
+MM 8-77 p47

CARLSTEDT, Jan Axel

839 Quartet, strings, no. 3, op. 23. SHOSTAKOVITCH: Quartet, strings, no. 8, op. 110, C minor. Fresk Quartet. Caprice CAP 1052.
+-Gr 9-77 p500 +RR 4-77 p62
+-HFN 5-77 p120

840 Sinfonietta, 5 wind instruments. HOLMBOE: Notturno, op. 19. MORTENSEN: Quintet, winds, op. 4. SALMENHAARA: Quintet, winds. Goteborg Wind Quintet. BIS LP 24.
+Gr 3-77 p1457 ++RR 12-76 p72

841 Sonata, violin, op. 15. EKLUND: Quartet, strings, no. 3. KARKOFF: Japanese romances, op. 45 (5). LINDE: Songs (2). Berit Hall-

qvist, s; Emil Dekov, vln; Bo Linde, pno; Norrkoping Quartet. Swedish Society SLT 33199.
+RR 1-77 p69

CARLTON, Nicholas
Verse for 2 to play on one organ. cf Vista VPS 1039.

CARLTON, Richard
Calm was the air. cf DG 2533 347.

CAROSO, Fabrizio
Barriera. cf DG 2723 051.
Celeste Giglio. cf DG 2723 051.
Laura soave: Gagliarda, Saltarello (Balleto). cf DG 2530 561.

CAROUBEL, Pierre-Francisque
Courante (2). cf DG 2723 051.
Pavana de Spaigne. cf DG 2723 051.
Volte (2). cf DG 2723 051.

CARPENTER, John Alden
842 Adventures in a perambulator. MOORE: The pageant of P. T. Barnum. NELSON: Savannah river holiday. Eastman-Rochester Symphony Orchestra; Howard Hanson. Mercury SRI 75095.
++ARG 11-77 p38 +St 12-77 p140
+SFC 10-16-77 p47
Impromptu. cf Vox SVBX 5303
843 Sonata, violin and piano. FOOTE: Sonata, violin and piano, op. 20, G minor. Eugene Gratovich, vln; Regis Benoit, pno. Orion ORS 76243.
+-ARG 4-77 p21 /-NR 2-77 p6
+HF 6-77 p88 +NYT 7-3-77 pD11
+MJ 3-77 p46 ++SFC 1-24-77 p40

CARR, Benjamin
Variations on the Sicilian hymn. cf Advent ASP 4007.

CARTER
Canon for three. cf Crystal S 361.

CARTER, Elliot
844 Eight pieces, 4 timpani. Fantasy about Purcell's Fantasia upon one note. Quintet, brass. American Brass Quintet; Morris Lang, timpani. Odyssey Y 34137.
+HF 5-77 p81 +SR 11-13-76 p52
+-NR 1-77 p9
Elegy, string quartet. cf New England Conservatory NEC 115.
Fantasy about Purcell's Fantasia upon one note. cf Eight pieces, 4 timpani.
845 Quartets, strings, nos. 1 and 2. Composers Quartet. Nonesuch H 71249.
+ARG 7-71 p756 +NR 1-71 p6
+Gr 3-72 p1545 +NYT 2-7-71 pD27
++HF 2-71 p76 ++SFC 11-29-70 p34
++HF 4-71 p64 +St 2-71 p89
++HFN 3-72 p501 -St 7-76 p74
++MJ 3-71 p77 *St 5-77 p73
Quintet, brass. cf Eight pieces, 4 timpani.
Quintet, woodwinds. cf Vox SVBX 5307.
Songs: Defense of Corinth; Musicians wrestle everywhere. cf Vox SVBX 5353.

CARULLI, Fernando
846 Concerto, guitar, A major. HAYDN: Concerto, guitar, D major (arr. from Quartet, strings, op. 2, no. 2, E major). VIVALDI: Concerti, guitar, C major, D major. Milan Zelenka, gtr; Slovak

Chamber Orchestra; Bohdan Warchal. Royale ROY 2004.
+Gr 11-77 p853 +RR 12-77 p57
Serenade, op. 96. cf RCA ARL 1-0456.
Serenade, op. 109, no. 6, D major. cf BIS LP 60.

CARUSO
Dreams of long ago. cf HMV RLS 715.

CASALS
Song of the birds. cf Orion ORS 75181.

CASANOVES, Narcis
Paso, no. 7. cf Harmonia Mundi HM 759.

CASELLA, Alfredo
Contrasts, op. 31 (2). cf International Piano Library IPL 102.

CASERTA, Antonellus de
Amour m'a le cuer mis. cf HMV SLS 863.

CASSADO, Gaspar
Requiebros. cf Laurel Protone LP 13.

CASTBERG
Veste-blomme, enge-blomme. cf Rubini GV 43.

CASTELLO, dario
Sonata, mandora, theorbo and bass viol. cf Saga 5438.

CASTELNUOVO-TEDESCO, Mario
847 Concerto, guitar, no. 1, op. 99, D major. VILLA-LOBOS: Concerto, guitar. Narciso Yepes, gtr; LSO; Garcia Navarro. DG 2530 718 Tape (c) 3300 718.
/ARG 6-77 p21 +NR 6-77 p5
+-Gr 2-77 p1273 +-RR 2-77 p62
+Gr 5-77 p1738 tape +RR 8-77 p87 tape
+-HFN 2-77 p115 ++St 10-77 p147
+-HFN 3-77 p119 tape
Concerto, guitar, no. 1, op. 99, D major. cf ARNOLD: Serenade, guitar and strings.
La guarda cuydadosa. cf DG 2530 561.
Sonata, op. 77, D major. cf BRITTEN: Nocturnal, op. 70.
Sonatina, op. 205. cf BIS LP 30.
Songs, op. 207 (6). cf BRITTEN: Nocturnal, op. 70.
Tarantella. cf DG 2530 561.

CATALANI, Alfredo
Loreley: Non fui da um padre mai bendetta, Dove son, d'onde vengo ...Oforze recondite. cf Court Opera Classics CO 347.
Songs: Vieni, deh, vien. cf Columbia M 34501.
848 La Wally: Ebben, Ne andro lontana. DONIZETTI: Catarina Cornaro: Torna all'ospitetetto...Vieni o tu, che ognor io chiamo. Lucrezia Borgia: Com'e bello. Maria Stuarda: O nube, che lieve. Roberto Devereux: E Sara in questi orribili momenti...Vivi, ingrato, a lei d'accanto...Quel sangue versato. VERDI: La forza del destino: Pace, pace, mio Dio. Aida: O cieli azzuri. La traviata: Addio del passato. Il trovatore: Timor di me... D'amor sull'ali rosee. Leyla Gencer, s; RAI Torino Orchestra, Turin Symphony Orchestra; Arturo Basile, Gianandrea Gavazzeni. Cetra LPL 69001.
+-Gr 7-77 p221 +-HF 9-76 p105
La Wally: Ebben, Ne andro lontana. cf Court Opera Classics CO 347.
La Wally: Ebben, Ne andro lontana. cf Decca SXLR 6825.
La Wally: Ebben, Ne andro lontana. cf London OS 26497.
La Wally: Ebben, Ne andro lontana. cf Odyssey Y 33793.

CATELINET
Our American cousins suite. cf Rediffusion 15-56.

CATO, Diomedes
Fantasia. cf Hungaroton SLPX 11721.
Favorito. cf Hungaroton SLPX 11721.
Praeludium, Galliardas I, II. cf DG 2533 294.
Villanella. cf Hungaroton SLPX 11721.
CAURROY, Eustache de
Fantasia; Prince la France te veut. cf L'Oiseau-Lyre 12BB 203/6.
CAVALLI, Pietro Francesco
849 Arias: Aeneas farewell; Ah tristo scellarato; Campion di tua belta; In India vo tornar; Lament of Cassandra; Lament of Sofonisba; Mio core respira; O delle mie speranze. MONTEVERDI: Quel sguardo; Chiome d'oro; O sia tranquillo; Se vittorie si belle; Zefiro torno. Heather Harper, s; Gerald English, Hugues Cuenod, t; Bath Festival Ensemble; Raymond Leppard. Vanguard VRS 10124.
+NR 12-77 p7 ++SFC 11-13-77 p50
850 Messa concertata. Munich Vocal Soloists; Karl Heinz Klein, hpd; Franz Lehrndorfer, org; Bavarian State Orchestra Chamber Ensemble; Hans Ludwig Hirsch. Telefunken AW 6-41931.
/Gr 8-76 p324 +-MT 1-77 p46
+-HFN 8-76 p79 +-RR 7-76 p74
+-MM 8-77 p47
851 Missa pro defunctis. Jean Knibbs, Rosemary Hardy, s; Kevin Smith, Andrew Giles, c-t; Rogers Covey-Crump, Peter Hall, t; Stephen Varcoe, William Mason, bs; Louis Halsey Singers; Louis Halsey. L'Oiseau-Lyre SOL 347.
+Gr 1-77 p1162
Missa pro defunctis. cf CARISSIMI: Historia di Jonas.
CAVENDISH, Michael
Come, gentle swains. cf DG 2533 347.
Wand'ring in this place. cf L'Oiseau-Lyre 12BB 203/6.
CAVIE, C.T.
Variations on a theme by Lully. cf RCA PL 2-5046.
CENTRAL EUROPEAN LUTE MUSIC, 16th and 17th centuries. cf Hungaroton SLPX 11721.
CERNOHORSKY, Bohuslav
Laudetur Jesus Christ, excerpt. cf Supraphon 113 1323.
CERVETTI, Sergio
852 Aria suspendida. The bottom of the iceberg. CONSOLI: Sciui novi. Tre canzoni. Else Charlston, s; Elizabeth Szlek-Consoli, flt; Tsuyoshi Tsutsumi, vlc; Stuart Fox, gtr; Bryant Hayes, clt. CRI SD 359.
+-FF 11-77 p16 +-NR 10-77 p6
*MJ 10-77 p29
The bottom of the iceberg. cf Aria suspendida.
CESARINI
Songs: A la Barcillunisa; A la Vallelunghisa; Chiovu'Abballati; Cantu a Timuni; Firenze sogno; Muttetti di la palieu; Nota de li Lavannari. cf London SR 33221.
CHABRIER, Emanuel
Bourree fantasque. cf International Piano Library IPL 102.
Bourree fantasque. cf Vox SVBX 5483.
853 Espana. DEBUSSY: Prelude a l'apres-midi d'un faune. IBERT: Escales. RAVEL: Daphnis et Chloe: Suite, no. 2. Orchestra de Paris; Daniel Barenboim. CBS 76523 Tape (c) 40-76523. (also Columbia M 34500 Tape (c) MT 34500)

+Gr 10-76 p615 -NR 5-77 p5
+Gr 3-77 p1458 tape +-RR 11-76 p53
+HFN 10-76 p167 +-RR 5-77 p91 tape
+-HFN 1-77 p123 tape +-St 6-77 p148
+MJ 7-77 p70

854 Espana. GRANADOS: Danza espanola, op. 37: Andaluza. MOSZKOWSKI: Spanish dances, Bk 1, op. 12. RIMSKY-KORSAKOV: Capriccio espagnol, op. 34. LSO; Ataulfo Argenta. Decca ECS 797 Tape (c) KECC 797. (Reissues from LXT 5333, SXL 2020)
+Gr 6-77 p106 +RR 6-77 p62
+HFN 7-77 p125 +RR 10-77 p93 tape

855 Espana. PROKOFIEV: Symphony, no. 1, op. 25, D major.* RAVEL: Bolero. STRAVINSKY: Circus polka. Fireworks, op. 4.* NPhO; Rafael Fruhbeck de Burgos. HMV ESD 7019 Tape (c) TC ESD 7019. (*Reissues from ASD 2315, Columbia TWO 239)
-Gr 12-76 p1071 +HFN 3-77 p119 tape
+Gr 1-77 p1178 tape +-RR 12-76 p53
+-HFN 12-76 p152 +-RR 2-77 p98 tape

Espana. cf Works, selections (Angel (Q) S 37424).
Espana. cf Works, selections (Turnabout (Q) QTVS 34671).
Espana. cf Angel S 37250.
Gwendoline: Overture. cf Works, selections (Turnabout (Q) QTVS 34671).
Habanera. cf Works, selections (Angel (Q) S 37424).
Habanera. cf Works, selections (Turnabout (Q) QTVS 34671).
Impromptu. cf Vox SVBX 5483.
Marche joyeuse. cf Works, selections (Angel (Q) S 37424).
Marche joyeuse. cf Works, selections (Turnabout (Q) QTVS 34671).
Marche joyeuse. cf Angel S 37250.
Pieces, posth.: Ballabile. cf Vox SVBX 5483.
Le Roi malgre lui: Danse slave; Fete polonaise. cf Works, selections (Angel (Q) S 37424).
Le Roi malgre lui: Danse slave; Fete polonaise. cf Works, selections (Turnabout (Q) QTVS 34671).
Suite pastorale. cf Works, selections (Angel (Q) S 37424)
Suite pastorale. cf Works, selections (Turnabout (Q) QTVS 34671).

856 Works, selections: Espana. Habanera. Marche joyeuse. Le Roi malgre lui: Fete polonaise; Danse slave. Suite pastorale. Paris Opera Orchestra; Jean-Baptiste Mari. Angel (Q) S 37424 Tape (c) 4XS 37424. (also HMV SQ ESD 7046 Tape (c) TC ESD 7046)
+-Gr 10-77 p711 +RR 11-77 p50
+-Gr 12-77 p1145 tape +-RR 12-77 p94 tape
+HF 11-77 p110 +St 11-77 p135
+-HFN 10-77 p152

857 Works, selections: Le Roi malgre lui: Danse slave; Fete polonaise. Gwendoline: Overture. Habanera. Marche joyeuse. Suite pastorale. Luxembourg Radio Orchestra; Louis de Froment. Turnabout (Q) QTVS 34671.
-ARG 7-77 p18 +-St 9-77 p125
+NR 5-77 p3

CHADABE, Joel

858 Echoes. Flowers. Paul Zukofsky, vln. CP 2.
+MJ 10-77 p28 +-NR 11-77 p15

Flowers. cf Echoes.

CHADWICK, George Whitefield

Euterpe. cf Louisville LS 753/4.

CHAGRIN, Francis
Pieces (2). cf Orion ORS 77269.
CHAMBONNIERES, Jacques
Chaconne, G major. cf Argo ZRG 806.
859 Suites, A minor, D minor, C major, F major, G major. Lionel Party, hpd. Musical Heritage Society MHS 3557 Tape (c) MHC 5557.
+HF 12-77 p84
CHAMINADE, Cecile
860 Concertino, flute, op. 107. FAURE: Fantasie, flute (orch. Galway). IBERT: Concerto, flute. POULENC: Concerto, flute (orch. Berkeley). James Galway, flt; RPO; Charles Dutoit. RCA RL 2-5109.
+Gr 10-77 p648 +RR 10-77 p47
The flatterer. cf International Piano Library IPA 113.
Scarf dance. cf International Piano Library IPA 113.
CHAMPAGNE
Danse villageoise. cf Citadel CT 6011.
CHANLER, Theodore
Songs: The children; Once upon a time; The rose; Moo is a cow; These, my Ophelia; Thomas Logge. cf New World Records NW 243.
CHANTS FOR THE FEAST OF MARY. cf DG 2533 310.
CHAPI Y LORENTE, Ruperto
Las hijas del Zebedeo: Carceleras. cf London OS 26435.
La patria chica: Cancion de pastora. cf London OS 26435.
CHARPENTIER, Gustave
861 Louise. Ileana Cotrubas, Jane Berbie, s; Lyliane Guitton, ms; Placido Domingo, Michel Senechal, t; Gabriel Bacquier, bs-bar; Ambrosian Opera Chorus; NPhO; Georges Pretre. CBS 79302 (3). (also Columbia M3 34207)
+-ARG 2-77 p29 +-ON 2-5-77 p41
+-Gr 10-76 p643 +-OR 6/7-77 p777
+-HF 2-77 p92 ++RR 11-76 p34
+HFN 12-76 p140 ++SFC 11-14-76 p30
+-MJ 4-77 p33 +St 2-77 p116
+MT 5-77 p399 +ST 1-77 p777
++NR 1-77 p12 ++STL 10-10-76 p37
+OC 9-77 p54
CHARPENTIER, Marc-Antoine
862 L'ange a la trompette. DUPRE: Symphony, organ, no. 2, op. 26. Variations sur un Noel, op. 20. SAINT-SAENS: Fantaisie, op. 159, E flat major. VIERNE: Pieces de fantasie, op. 53: Feux follets. Naiades. Toccata, B flat major. Gillian Weir, org. Prelude PRS 2507.
+Gr 7-77 p202 +RR 7-77 p81
+HFN 10-77 p153
CHAUSSON, Ernst
863 Concerto, violin, piano and quartet, op. 21, D major. Mark Lubotsky, vln; Lyubov Yedlina, pno; Borodin Quartet. Melodiya 33 CM 02335/6.
+ARG 7-77 p41
864 Poeme, op. 25. RAVEL: Tzigane.* SAINT-SAENS: Introduction and rondo capriccioso, op. 28. Havanaise, op. 83. Itzhak Perlman, vln; Orchestre de Paris; Jean Martinon. Angel (Q) S 37118 Tape (c) 4XS 37118 (ct) 8XS 37118. (*Reissue from SLS 5016) (also HMV ASD 3125)
++Gr 1-76 p1191 ++NR 3-76 p13
+HF 12-75 p110 +NYT 7-24-77 pD11

++HF 2-76 p130 tape +RR 12-75 p61
+HFN 12-75 p152 +St 1-76 p110

865 Poeme, op. 25. FAURE: Berceuse, op. 16. SAINT-SAENS: Concerto, violin, no, 3, op. 61, B minor. Isaac Stern, vln; Orchestre de Paris; Daniel Barenboim. CBS 76530. Tape (c) 40-76530.
+Gr 4-77 p1542 +MT 11-77 p924
+HFN 4-77 p135 ++RR 4-77 p56
+HFN 5-77 p138 tape ++RR 7-77 p100 tape

866 Poeme, op. 25. SIBELIUS: Concerto, violin, op. 47, D minor. Miriam Fried, vln; Gisele Demoulin, pno; Belgian Radio and Television Orchestra; Rene Defossez. DG 2538 302 Tape (c) 3344 264.
+-Gr 6-74 p45 +-RR 10-77 p99 tape
+RR 6-74 p52

867 Poeme, op. 25. TCHAIKOVSKY: Concerto, violin, op. 35, D major. David Oistrakh, vln; USSR Symphony Orchestra, MPO; Kiril Kondrashin, Gennady Rozhdestvensky. HMV Melodiya SXLP 30220.
+Gr 12-77 p1079 +HFN 12-77 p167

Poeme, op. 25. Hungaroton SLPX 11825.
Quelques danses. cf Vox SVBX 5483.
Soir de fete, op. 32. cf Symphony, op. 20, B flat major (Pathe Marconi C 069-14086).

868 Songs: Poeme de l'amour et de la mer, op. 19. DEBUSSY: La damoiselle elui. Montserrat Caballe, s; Janet Coster, ms; Ambrosian Ladies Chorus; London Sinfonia; Wyn Morris. Rediffusion Symphonica SYM 6.
+-Gr 12-77 p1121

Songs: Le charme; Les papillons; Psyche. cf Orion ORS 77271.
Songs: Le colibri, op. 2, no. 7; Les papillons, op. 2, no. 3; Poeme de l'amour et la mer, op. 19; Les temps des lilas. cf HMV RLS 716.
Songs: Le temps des lilas. cf HMV RLS 719.

869 Symphony, op. 20, B flat major. Soir de fete, op. 32. Toulouse Orchestre du Capitole; Michel Plasson. Pathe Marconi C 069 14086.
++FF 11-77 p15

CHAVARRI, Lopez

El viejo castillo moro. cf RCA RL 2-5099.

CHAVEZ, Carlos

870 Concerto, piano. Eugene List, pno; VSOO; Carlos Chavez. Westminster WGS 8324. (Reissue)
+NR 10-76 p6 +St 1-77 p113

871 Toccata. HOVHANESS: Bacchanale. October mountain. SEREBRIER: Symphony. Tristan Fry Percussion Ensemble; John Eliot Gardiner. Gale GMFD 1-76-004.
+-HF 10-77 p112 ++HFN 9-77 p132

CHEDEVILLE, Nicolas

Musette. cf DG 2723 051.
Suite, no. 5, C major. cf Abbey LPB 765.

CHERUBINI, Luigi

Medea: Dei tuoi figli. cf HMV SLS 5057.

872 Quartets, strings, nos. 1-6. Melos Quartet. DG 2723 044 (3). (also DG 2710 018)
++Gr 8-76 p312 ++NR 1-77 p8
+-HF 1-77 p114 +RR 7-76 p67
+HFN 8-76 p83 +SR 11-13-76 p52
++MJ 12-76 p44 ++St 1-77 p84
+MT 9-76 p745

CHILCOT, Thomas
Suite, no. 1. cf Crescent BUR 1001.

CHILDS, Barney
873 Trio, clarinet, violoncello and piano. LENTZ: Songs of the sirens. NORGAARD: Spell. Montagnana Trio. ABC AB 67013 Tape (c) 5306-67013.
-NR 9-77 p5

CHOPIN, Frederic
874 Andante spianato and grande polonaise, op. 22, E flat major. Nocturne, op. 62, no. 1, B major. Polonaise-Fantaisie, op. 61, A flat major. Scherzo, no. 4, op. 54, E major. Emanuel Ax, pno. RCA ARL 1-1569.
++Gr 9-76 p443 +NR 9-76 p12
++HF 9-76 p90 +RR 9-76 p80
+HFN 1-77 p104 ++St 10-76 p121
+MJ 11-76 p45 +STL 10-10-76 p37

Andante spianato and grande polonaise, op. 22, E flat major. cf Piano works (DG 2530 826).
Andante spianato and grande polonaise, op. 22, E flat major. cf International Piano Library IPL 5001/2.
Ballades, nos. 1-4, opp. 23, 38, 47, 57. cf Piano works (DG 2740 163).
Ballade, no. 1, op. 23, G minor. cf Piano works (Connoisseur Society CS 2122).
Ballade, no. 1, op. 23, G minor. cf Piano works (HMV HQS 1375).
Ballade, no. 1, op. 23, G minor. cf Piano works (Reference Recordings RR 2.
Ballade, no. 1, op. 23, G minor. cf International Piano Library IPL 5001/2.
Ballade, no. 1, op. 23, G minor. cf Westminster WGM 8309.
Ballade, no. 3, op. 47, A flat major. cf Piano works (Decca DPA 563/4).
Ballade, no. 3, op. 47, A flat major. cf Piano works (DG 2548 215).
Ballade, no. 3, op. 47, A flat major. cf ALBENIZ: Espana, op. 165, no. 2: Tango.
Ballade, no. 3, op. 47, A flat major. cf International Piano Library IPA 114.
Ballade, no. 4, op. 52, F minor. cf Piano works (Eurodisc 89 836).
Ballade, no. 4, op. 52, F minor. cf BEETHOVEN: Sonata, piano, no. 8, op. 13, C minor.
Ballade, no. 4, op. 52, F minor. cf International Piano Archives IPA 5007/8.
Barcarolle, op. 60, F sharp major. cf Piano works (Decca DPA 563/4).
Barcarolle, op. 60, F sharp major. cf Piano works (Decca SXL 6801).
Barcarolle, op. 60, F sharp major. cf Piano works (Reference Recordings RR 2).
Barcarolle, op. 60, F sharp major. cf Piano works (Westminster WGS 8341).
Berceuse, op. 57, D flat major. cf Piano works (Decca DPA 563/4).
Berceuse, op. 57, D flat major. cf International Piano Library IPA 112.
Berceuse, op. 57, D flat major. cf International Piano Library IPL 5001/2.
Berceuse, op. 57, D flat major. cf RCA GL 4-2125.
Chants polonaise, op. 74, no. 1: The maiden's wish; no. 5: The warrior; no. 13: Spring. cf BELLINI: Songs (Cetra LPO 2003).
Chants polonaise, op. 74, no. 5. cf Decca SPA 473.

875 Concerto, piano, no. 1, op. 11, E minor. Mazurkas, op. 7, no. 1, B flat major; op. 67, no. 3, C major; op. 68, no. 2, A minor; op. posth., D major. Tamas Vasary, pno; BPhO; Jerzy Semkov. DG 2535 206 Tape (c) 3335 206. (Reissues)
+-Gr 1-77 p1141 +HFN 1-77 p119
+Gr 3-77 p1458 tape ++RR 12-76 p53
Concerto, piano, no. 1, op. 11, E minor. cf Piano works (Connoisseur Society CS 2029/30).
876 Concerto, piano, no. 2, op. 21, F minor. FALLA: Noches en los jardines de Espana. Alicia de Larrocha, pno; OSR; Sergiu Comissiona. London CS 6773 Tape (c) M 10252. (also Decca SXL 6528 Tape (c) KSXC 6528)
++ARG 12-72 p806 ++NR 4-72 p7
+-Ha 11-72 p127 +-RR 4-77 p91 tape
+-HF 7-72 p77 ++SFC 5-7-72 p46
+HFN 2-72 p306 +-St 6-72 p82
++HFN 7-75 p90 tape
877 Concerto, piano, no. 2, op. 21, F minor. SCHUMANN: Concerto, piano, op. 54, A minor. Eugene Istomin, pno; Columbia Symphony Orchestra, PO; Bruno Walter, Eugene Ormandy. Odyssey Y 34618.
++SFC 10-9-77 p40
878 Concerto, piano, no. 2, op. 21, F minor. LISZT: Concerto, piano, no. 2, G 125, A major. Frantisek Rauch, pno; Prague Symphony Orchestra; Vaclav Smetacek. Rediffusion Legend LGD 006 Tape (c) LGC 006.
-HFN 10-77 p167 +-RR 2-77 p96 tape
-RR 10-77 p48
Ecossaises, op. 72, no. 3, D flat major. cf Piano works (HMV HQS 1375).
879 Etudes (3). Preludes, nos. 1-26. Nikita Magaloff, pno. Philips 6580 118 Tape (c) 7317 184.
+HFN 1-77 p104 +HFN 11-77 p187 tape
880 Etudes, opp. 10 and 25. Wilhelm Backhaus, pno. Bruno Walter Society IGI 286.
+NR 4-77 p14
881 Etudes, opp. 10 and 25. Maurizio Pollini, pno. DG 2530 291 Tape (c) 3300 287.
+Gr 11-72 p928 +NYT 3-13-77 pD22
+HF 3-73 p82 ++RR 11-72 p80
+HF 7-76 p114 tape ++RR 1-74 p77 tape
++HFN 4-73 p792 tape ++SFC 3-4-73 p33
++NR 6-73 p12 ++SFC 7-15-73 p32
+NYT 4-1-73 pD28 ++St 8-73 p99
882 Etudes, opp. 10 and 25. Martha Goldstein, pno. Pandora Records 107.
+-CL 3-77 p8
883 Etudes, opp. 10 and 25. Mikita Magaloff, pno. Philips 6580 119.
+-Gr 7-77 p201 -RR 6-77 p80
+-HFN 6-77 p121
884 Etudes, opp. 10 and 25. Vladimir Ashkenazy, pno. Saga SAGA 5293. (Reissue)
+Gr 3-74 p1717 +RR 4-74 p64
+HFN 3-74 p104 ++St 1-77 p113
885 Etudes, opp. 10 and 25. Abby Simon, pno. Turnabout TV 34688.
+SFC 12-11-77 p61
Etudes, opp. 10 and 25. cf Piano works (DG 2740 163).
Etude, op. 10, no. 3, E major. cf Piano works (Decca DPA 563/4).

Etudes, op. 10, no. 3, E major. cf Piano works (Eurodisc 89 836).
Etudes, op. 10, no. 3, E major. cf Inner City IC 1006.
Etudes, op. 10, nos. 4-5, 7. cf International Piano Library IPL 101.
Etudes, op. 10, no. 4, C sharp minor. cf ALBENIZ: Espana, op. 165, no. 2: Tango.
Etudes, op. 10, no. 5, G flat major. cf Piano works (Decca DPA 563/4).
Etudes, op. 10, no. 5, G flat major (2 versions). cf International Piano Library IPL 104.
Etudes, op. 10, no. 5, G flat major. cf International Piano Library IPA 114.
Etudes, op. 10, no. 7, C major. cf Etudes, op. 25, no. 7, C sharp minor.
Etudes, op. 10, nos. 8, 12. cf Piano works (HMV HQS 1375).
Etudes, op. 10, no. 8, F major. cf Piano works (DG 2530 826).
Etudes, op. 10, no. 10, A flat major. cf Piano works (Connoisseur Society CS 2029/30).
Etudes, op. 10, no. 12, C minor. cf Piano works (Eurodisc 89 836).
Etudes, op. 10, no. 12, C minor. cf Decca PFS 4387.
Etudes, op. 25, nos. 1-2, 10, 11. cf International Piano Library IPL 101.
Etudes, op. 25, no. 1, A flat major. cf Piano works (Decca DPA 563/4).
Etudes, op. 25, no. 1, A flat major. cf International Piano Library IPA 114.
Etudes, op. 25, no. 2, F minor. cf Piano works (Decca DPA 563/4).
Etudes, op. 25, no. 5, E minor. cf International Piano Library IPL 104.

886 Etudes, op. 25, no. 7, C sharp minor. Etudes, op. 10, no. 7, C major. Nocturne, op. 15, no. 2, F sharp major. Nocturne, op. 15, no. 1, F major. Waltz, op. 42, A flat major. LISZT: Etude de concert, no. 2, G 144, F minor: La leggierezza. Hungarian rhapsody, no. 10, G 244. SCHUBERT: Impromptu, op. 142, no. 3, D 935, B flat major. Hark, hark the lark (Liszt). SCHUMANN: Fantasiestucke, op. 12, nos. 1-3: Des Abends; Aufschwang; Warum. Ignaz Jan Paderewski, pno. Pearl GEM 136. (Reissues from HMV 045560, 2-045511, 045554, 05620, 05616, 05718, 2-045505, 2-045506, 2-45500, 045549, 05714, DB 833)
+-Gr 4-77 p1577

Etudes, op. 25, no. 7, C sharp minor. cf ALBENIZ: Espana, op. 165, no. 2: Tango.
Etudes, op. 25, no. 9, G flat major. cf Piano works (Decca DPA 563/4).
Etudes, op. 25, no. 9, G flat major. cf International Piano Library IPL 5001/2).
Etudes, op. 25, no. 10, B minor. cf ALBENIZ: Espana, op. 165, no. 2: Tango.
Etudes, op. 25, no. 12, C minor. cf Piano works (Reference Recordings RR 2).
Fantasie-Impromptu, op. 66, C sharp minor. cf Piano works (Decca DPA 563/4).
Fantasie-Impromptu, op. 66, C sharp minor. cf Piano works (DG 2548 215).
Fantasie-Impromptu, op. 66, C sharp minor. cf Piano works (Eurodisc 89 836).
Fantasie-Impromptu, op. 66, C sharp minor. cf Piano works (HMV HQA 1375).

Impromptus, nos. 1-3, opp. 29, 36, 51. cf Piano works (DG 2548 215).
Impromptus, nos. 1-3, opp. 29, 36, 51. cf Piano works (DG 2740 163).
Impromptus, no. 1, op. 29, A flat major. cf Piano works (Decca DPA 563/4).
Introduction and variations on a German national air. cf Piano works (DG 2740 163).
Mazurka, op. 6, no. 2, C sharp minor. cf RCA GL 4-2125.
Mazurka, op. 7, no. 1, B flat major. cf Concerto, piano, no. 1, op. 11, E minor.
Mazurka, op. 7, no. 2, A minor. cf Piano works (Connoisseur Society CS 2122).
Mazurka, op. 17, no. 2, A minor. cf BRAHMS: Variations and fugue on a theme by Handel, op. 24, B flat major.
Mazurka, op. 17, no. 4, A minor. cf Decca PFS 4351.
Mazurka, op. 24, no. 1, G minor. cf Piano works (DG 2530 826).
Mazurka, op. 24, no. 2, C major. cf Piano works (DG 2530 826).
Mazurka, op. 24, no. 4, B flat minor. cf Piano works (DG 2530 826).
Mazurka, op. 24, no. 4, B flat minor. cf CBS 73589.
Mazurka, op. 30, no. 4, C sharp major. cf Piano works (Westminster WGS 8341).
Mazurka, op. 33, no. 2, D major. cf Piano works (Decca DPA 563/4).
Mazurkas, op. 41 (4). cf Piano works (DG 2548 215).
Mazurka, op. 50, no. 3, C sharp minor. cf Piano works (Connoisseur Society CS 2122).
Mazurkas, op. 63 (3). cf Piano works (Decca SXL 6801).
Mazurka, op. 63, no. 3, C sharp minor. cf BEETHOVEN: Sonata, piano, no. 14, op. 27, no. 2, C sharp minor.
Mazurkas, op. 67, nos. 2, 4. cf Piano works (Decca SXL 6801).
Mazurka, op. 67, no. 3, C major. cf Concerto, piano, no. 1, op. 11, E minor.
Mazurka, op. 67, no. 4, A minor. cf Piano works (Westminster WGS 8341).
Mazurka, op. 68, no. 2, A minor. cf Concerto, piano, no. 1, op. 11, E minor.
Mazurka, op. 68, no. 4. cf Piano works (Decca SXL 6801).
Mazurka, op. posth., D major. cf Concerto, piano, no. 1, op. 11, E minor.
Melancholy, op. 74, no. 12. cf Odyssey Y 33130.
My joys. cf International Piano Library IPL 103.
The nightingale's trill. cf Rubini GV 26.

887 Nocturnes, complete. Nikita Magaloff, pno. Philips 6780 024 (2).
+-Gr 3-77 p1429 +-RR 1-77 p70
+HFN 1-77 p104

888 Nocturnes, nos. 1-21. Tamas Vasary, pno. DG 2726 070 (2). (Reissues from SLPEM 136486, 136487)
/Gr 4-77 p1573 +-RR 3-77 p78
+HFN 8-77 p93

Nocturnes, opp. 9, 15, 27, 32, 37, 48, 55, 62, 72. cf Piano works (DG 2740 163).
Nocturnes, op. 1, no. 1, B flat minor. cf Piano works (HMV HQS 1375).
Nocturnes, op. 9, no. 2, E flat major. cf Piano works (Decca DPA 563/4).
Nocturnes, op. 9, no. 2, E flat major. cf Piano works (Eurodisc 89 836).

Nocturnes, op. 9, no. 2, E flat major. cf Piano works (Westminster WGS 8341).
Nocturnes, op. 9, no. 2, E flat major. cf Decca SPA 519.
Nocturnes, op. 9, no. 2, E flat major. cf International Piano Library IPL 5001/2.
Nocturnes, op. 9, no. 3, B major. cf International Piano Archives IPA 5007/8.
Nocturnes, op. 15, no. 1, F major. cf Etude, op. 25, no. 7, C sharp minor.
Nocturnes, op. 15, no. 1, F major. cf International Piano Library IPA 117.
Nocturnes, op. 15, no. 2, F sharp major. cf Etude, op. 25, no. 7, C sharp minor.
Nocturnes, op. 15, no. 2, F sharp major. cf Piano works (Connoisseur Society CS 2122).
Nocturnes, op. 15, no. 2, F sharp major. cf Piano works (Decca DPA 563/4).
Nocturnes, op. 15, no. 2, F sharp major. cf International Piano Library IPL 103.
Nocturnes, op. 15, no. 2, F sharp major. cf International Piano Library IPL 104.
Nocturnes, op. 15, no. 2, F sharp major. cf International Piano Library IPL 5001/2.
Nocturnes, op. 15, no. 2, F sharp major. cf RCA JRL 1-2315.
Nocturnes, op. 27, no. 2, D flat major. cf Piano works (Decca DPA 563/4).
Nocturnes, op. 27, no. 2, D flat major. cf BEETHOVEN: Sonata, piano, no. 14, op. 27, no. 2, C sharp minor.
Nocturnes, op. 32, no. 1, B major. cf International Piano Library IPL 109.
Nocturnes, op. 32, no. 2, A flat major. cf Piano works (Connoisseur Society CS 2122).
Nocturnes, op. 48, no. 1, C minor. cf Piano works (HMV HQS 1375).
Nocturnes, op. 48, no. 1, C minor. cf Piano works (Westminster WGS 8341).
Nocturnes, op. 48, no. 1, C minor. cf International Piano Library IPL 114.
Nocturnes, op. 55, no. 1, F minor. cf International Piano Library IPL 117.
Nocturnes, op. 55, no. 2, E flat major. cf Piano works (Connoisseur Society CS 2122).
Nocturnes, op. 62 (2). cf Piano works (Decca SXL 6801).
Nocturnes, op. 62, no. 1, B major. cf Andante spianato and grande polonaise, op. 22, E flat major.
Nocturnes, op. 62, no. 2, E major. cf Piano works (Westminster WGS 8341).
Nocturnes, op. posth., C sharp minor. cf Piano works (Reference Recordings RR 2).
Nocturnes, op. posth., C sharp minor. cf Discopaedia MB 1014.
Nouvelles etudes (3). cf Preludes, op. 28, nos. 1-24.

889 Piano works: Concerto, piano, no. 1, op. 11, E minor. Etude, op. 10, no. 10, A flat major. Polonaise, op. 44, F sharp minor. Scherzo, no. 4, op. 54, E major. Sonata, piano, no. 3, op. 58, B minor. Garrick Ohlsson, pno; Warsaw National Philharmonic Orchestra; Witold Rowicki. Connoisseur Society CS 2029/30 (3).

+Gr 3-75 p1677 ++St 12-71 p84
+NR 11-71 p10 +St 2-77 p80
+RR 2-75 p52

890 Piano works: Ballade, no. 1, op. 23, G minor. Mazurkas, op. 7, no. 2, A minor. Mazurkas, op. 50, no. 3, C sharp minor. Nocturnes, op. 15, no. 2, F sharp major. Nocturnes, op. 32, no. 2, A flat major. Nocturnes, op. 55, no. 2, E flat major. Preludes, nos. 1, 5, 7, 16, 17, 22. Ivan Moravec, pno. Connoisseur Society CS 2122.
+NR 6-77 p14 ++SFC 5-8-77 p46

891 Piano works: Ballade, no. 3, op. 47, A flat major. Barcarolle, op. 60, F sharp major. Berceuse, op. 57, D flat major. Etude, op. 10, no. 3, E major. Etude, op. 10, no. 5, G flat major. Etude, op. 25, no. 1, A flat major. Etude, op. 25, no. 2, F minor. Etude, op. 25, no. 9, G flat major. Fantasie-Impromptu, op. 66, C sharp minor. Impromptu, no. 1, op. 29, A flat major. Mazurka, op. 33, no. 2, D major. Nocturne, op. 9, no. 2, E flat major. Nocturne, op. 15, no. 2, F sharp major. Nocturne, op. 27, no. 2, D flat major. Polonaise, op. 40, no. 1, A major. Polonaise, op. 53, A flat major. Prelude, op. 28, no. 15, D flat major. Scherzo, no. 3, op. 39, C sharp minor. Sonata, piano, no. 2, op. 35, B flat minor. Waltz, op. 18, E flat major. Waltz, op. 64, no. 1, D flat major. Waltz, op. 64, no. 2, C sharp minor. Waltz, op. 70, no. 1, G flat major. Wilhelm Kempff, Freidrich Gulda, Peter Katin, Wilhelm Backhaus, Ilana Vered, Nikita Magaloff, Julius Katchen, pno. Decca DPA 563/4 (2).
*Gr 2-77 p1326 -RR 1-77 p70

892 Piano works: Barcarolle, op. 60, F sharp major. Mazurkas, op. 63 (3). Mazurkas, op. 67, nos. 2, 4. Mazurka, op. 68, no. 4. Nocturnes, op. 62 (2). Polonaise-Fantaisie, op. 61, A flat major. Waltzes, op. 64, nos. 1-3. Vladimir Ashkenazy, pno. Decca SXL 6801. (also London CS 7022)
+Gr 10-77 p661 +-RR 10-77 p75
+HFN 10-77 p153 ++SFC 12-11-77 p61

893 Piano works: Andante spianato and grande polonaise, op. 22, E flat major. Etude, op. 10, no. 8, F major. Mazurka, op. 24, no. 1, G minor. Mazurka, op. 24, no. 2, C major. Mazurka, op. 24, no. 4, B flat minor. Prelude, op. 28, no. 17, A flat major. Prelude, op. 28, no. 18, F minor. Scherzo, no. 4, op. 54, E major. Waltz, op. 34, no. 1, A flat major. Krystian Zimerman, pno. DG 2530 826.
+-Gr 2-77 p1307 +-MM 6-77 p48
+HF 10-77 p105 +NR 8-77 p12
+-HFN 2-77 p115 +-RR 2-77 p78
+-MJ 9-77 p34 +St 9-77 p125

894 Piano works: Ballade, no. 3, op. 47, A flat major. Fantasie-Impromptu, op. 66, C sharp minor. Impromptus, nos. 1-3, opp. 29, 36, 51. Mazurkas, op. 41 (4). Scherzo, no. 3, op. 39, C sharp minor. Stefan Askenase, pno. DG 2548 215 Tape (c) 3348 218. (Reissue from 2538 078)
-Gr 4-76 p1617 +HFN 4-76 p123
+Gr 4-77 p1603 tape +-RR 4-76 p62

895 Piano works: Ballades, nos. 1-4, opp. 23, 38, 47, 52. Etudes, opp. 10 and 25. Impromptus, nos. 1-3, opp. 29, 36 51. Nocturnes, opp. 9, 15, 27, 32, 37, 48, 55, 62, 72. Polonaise, op. 53, A flat major. Scherzi, nos. 1-4. Sonata, piano, no. 2, op. 35, B flat minor. Sonata, piano, no. 3, op. 58, B minor. Introduction and variations on a German national air. Waltzes, nos. 1-17. Tamas Vasary, pno. DG 2740 163 (6).
+HFN 5-77 p137 +-RR 5-77 p70

896 Piano works: Ballade, no. 4, op. 52, F minor. Etudes, op. 10, no. 3, E major; op. 10, no. 12, C minor. Fantasie-Impromptu, op. 66, C sharp minor. Nocturne, op. 9, no. 2, E flat major. Polonaise, op. 40, no. 1, A major. Preludes, op. 28, no. 6, B minor; no. 15, D flat major; no. 20, C minor. Robert Leonardy, pno. Eurodisc 89 836 XAK.
+HF 8-77 p78

897 Piano works: Ballade, no. 1, op. 23, G minor. Ecossaises, op. 72, no. 3, D flat major. Etudes, op. 10, nos. 8, 12. Fantasie-Impromptu, op. 66, C sharp minor. Nocturnes, op. 9, no. 1, B flat minor. Nocturnes, op. 48, no. 1, C minor. Scherzo, no. 2, op. 31, B flat minor. Waltzes, op. 34, nos. 1, 2. Daniel Adni, pno. HMV HQS 1375.
+-Gr 5-77 p1713 -RR 5-77 p70
+HFN 5-77 p120

898 Piano works: Ballade, no. 1, op. 23, G minor. Barcarolle, op. 60, F sharp major. Etude, op. 25, no. 12, C minor. Nocturne, op. posth., C sharp minor. Scherzo, no. 2, op. 31, B flat minor. Waltz, op. 64, no. 2, C sharp minor. Steven Gordon, pno. Reference Recordings RR 2.
+HF 10-77 p112

899 Piano works: Barcarolle, op. 60, F sharp major. Mazurka, op. 30, no. 4, C sharp major. Mazurka, op. 67, no. 4, A minor. Nocturne, op. 48, no. 1, C minor. Nocturne, op. 9, no. 2, E flat major. Nocturne, op. 62, no. 2, E major. Polonaise, op. 26, no. 2, E flat minor. Waltz, op. 34, no. 2, A minor. Waltz, op. 64, no. 1, D flat major. Yakov Flier, pno. Westminster WGS 8341.
+-ARG 9-77 p28 +-MJ 7-77 p70
+-CL 10-77 p8 +-NR 8-77 p12

900 Polonaises, complete. Garrick Ohlsson, pno. Angel SB 3794. (also HMV CSD 3732/3, HMV SLS 843)
+-Gr 5-73 p2064 +RR 5-73 p82
+HF 10-73 p101 ++St 7-73 p102
+HFN 6-73 p1174 +St 2-77 p80
++NR 6-73 p14

901 Polonaise, opp. 26, 40, 44, 53, 61, nos. 1-7. Maurizio Pollini, pno. DG 2530 659 Tape (c) 3300 569.
+-ARG 4-77 p19 ++MJ 3-77 p74
++Gr 12-76 p1021 ++NR 4-77 p14
+-HF 4-77 p97 +NYT 3-13-77 pD22
++HFN 12-76 p140 ++RR 12-76 p78
+HFN 3-77 p119 tape +-SFC 2-13-77 p33

Polonaise, op. 26, no. 2, E flat minor. cf Piano works (Westminster WGS 8341).
Polonaise, op. 26, no. 2, E flat minor. cf International Piano Library IPL 5007/8.
Polonaise, op. 40, no. 1, A major. cf Piano works (Decca DPA 563/4).
Polonaise, op. 40, no. 1, A major. cf Piano works (Eurodisc 89 836).
Polonaise, op. 40, no. 1, C minor. cf International Piano Library IPL 103.
Polonaise, op. 44, F sharp minor. cf Piano works (Connoisseur Society CS 2029/30).
Polonaise, op. 53, A flat major. cf Piano works (Decca DPA 563/4).
Polonaise, op. 53, A flat major. cf Piano works (DG 2740 163).

Polonaise, op. 53, A flat major. cf Decca SPA 519.
Polonaise, op. 53, A flat major. cf International Piano Library IPL 108.
Polonaise-Fantaisie, op. 61, A flat major. cf Andante spianato and grande polonaise, op. 22, E flat major.
Polonaise-Fantaisie, op. 61, A flat major. cf Piano works (Decca SXL 6801).
Preludes, nos. 1-26. cf Etudes.
Preludes, nos. 1, 5, 7, 16, 17, 22. cf Piano works (Connoisseur Society CS 2122).

902 Preludes, op. 28, nos. 1-24. Maurizio Pollini, pno. DG 2530 550 Tape (c) 3300 550.
+Gr 12-75 p1076 +NR 5-76 p13
++HF 11-76 p106 +NYT 3-13-77 pD22
+HFN 3-76 p113 tape ++RR 12-75 p80
++HF 4-77 p121 tape ++RR 3-76 p76
+MJ 10-76 p25 ++St 8-76 p92

903 Preludes, op. 28, nos. 1-24. Prelude, op. 45, C sharp minor. Prelude, op. posth., A flat major. Daniel Barenboim, pno. HMV ASD 3254.
-Gr 10-76 p627 -RR 10-76 p87
-HFN 1-77 p104

904 Preludes, op. 28, nos. 1-24. Prelude, op. 45, C sharp minor. Prelude, op. posth., A flat major. Nouvelles etudes (3). Nikita Magaloff, pno. Philips 6580 118.
-Gr 2-77 p1307 +-RR 11-76 p89

Preludes, op. 28, nos. 1, 7, 23. cf International Piano Library IPL 114.
Preludes, op. 28, no. 6, B minor. cf Piano works (Eurodisc 89 836).
Preludes, op. 28, no. 7, A major. cf International Piano Library IPL 104.
Preludes, op. 28, no. 15, D flat major. cf Piano works (Decca DPA 563/4).
Preludes, op. 28, no. 15, D flat major. cf Piano works (Eurodisc 89 836).
Preludes, op. 28, no. 17, A flat major. cf Piano works (DG 2530 836).
Preludes, op. 28, no. 18, F minor. cf Piano works (DG 2530 826).
Preludes, op. 28, no. 20, C minor. cf Piano works (Eurodisc 89 836).
Preludes, op. 28, no. 24, D minor. cf BEETHOVEN: Sonata, piano, no. 8, op. 13, C minor.
Preludes, op. 28, no. 24, D minor. cf CBS 73589.
Preludes, op. 45, C sharp minor. cf Preludes, op. 28, nos. 1-24 (HMV ASD 3254).
Preludes, op. 45, C sharp minor. cf Preludes, op. 28, nos. 1-24 (Philips 6580 118).
Preludes, op. posth., A flat major. cf Preludes, op. 28, nos. 1-24 (HMV ASD 3254).
Preludes, op. posth., A flat major. cf Preludes, op. 28, nos. 1-24 (Philips 6580 118).

905 Scherzi, nos. 1-4. Jeanne-Marie Darre, pno. Vanguard VCS 10122.
+MJ 5-77 p32

Scherzi, nos. 1-4. cf Piano works (DG 2740 163).
Scherzo, no. 1, op. 20, B minor (abbreviated). cf International Piano Library IPL 103.
Scherzo, no. 2, op. 31, B flat minor. cf Piano works (HMV HQS 1375).

Scherzo, no. 2, op. 31, B flat minor. cf Piano works (Reference Recordings RR 2).
Scherzo, no. 2, op. 31, B flat minor. cf International Piano Library IPL 117.
Scherzo, no. 3, op. 39, C sharp minor. cf Piano works (Decca DPA 563/4).
Scherzo, no. 3, op. 39, C sharp minor. cf Piano works (DG 2548 215).
Scherzo, no. 4, op. 54, E major. cf Andante spianato and grande polonaise, op. 22, E flat major.
Scherzo, no. 4, op. 54, E major. cf Piano works (Connoisseur Society CS 2029/30).
Scherzo, no. 4, op. 54, E major. cf Piano works (DG 2530 826).

906 Sonata, piano, no. 2, op. 35, B flat minor. MUSSORGSKY: Pictures at an exhibition. William Kapell, pno. Bruno Walter Society OP 83.
+-NR 8-77 p12

907 Sonata, piano, no. 2, op. 35, B flat minor. Sonata, piano, no. 3, op. 58, B minor. Tamas Vasary, pno. DG 2535 230. (Reissue from SLPEM 136450)
+Gr 12-77 p1106 +-RR 11-77 p85

Sonata, piano, no. 2, op. 35, B flat minor. cf Piano works (Decca DPA 563/4).
Sonata, piano, no. 2, op. 35, B flat minor. cf Piano works (DG 2740 163).
Sonata, piano, no. 2, op. 35, B flat minor. cf International Piano Library IPL 113.
Sonata, piano, no. 3, op. 58, B minor. cf Piano works (Connoisseur Society CS 2029/30).
Sonata, piano, no. 3, op. 58, B minor. cf Piano works (DG 2740 163).
Sonata, piano, no. 3, op. 58, B minor. cf Sonata, piano, no. 2, op. 35, B flat minor.
Sonata, violoncello, op. 65, G minor: Largo. cf Orion ORS 75181.
Songs: Dumka. cf BELLINI: Songs (Cetra LPO 2003).
Songs: The maiden's wish, op. 74, no. 1 (in French). cf Odyssey Y 33793.

908 Les sylphides (orch. Douglas). DELIBES: Coppelia: Suite. BPhO; Herbert von Karajan. DG 2535 189 Tape (c) 3335 189. (also DG 136257 Tape (ct) 86 257)
+-Gr 1-77 p1184 +HFN 1-77 p119
+Gr 1-77 p1183 tape -RR 12-76 p54

909 Les sylphides (Fistourlari). OFFENBACH (Rosenthal): Gaite parisienne. RPO; Anatole Fistoulari. Quintessence PMC 7029.
+St 11-77 p146

910 Variations on Mozart's "La ci darem la mano", op. 2, B flat major. PROKOFIEV: Concerto, piano, no. 1, op. 10, D flat major. TCHAIKOVSKY: Elegie. David Syme, pno; NPhO; Paul Freeman. Orion ORS 76221.
*MJ 9-77 p34 +NR 8-77 p12

911 Waltzes complete. Jeanne-Marie Darre, pno. Vanguard VCS 10115. (Reissue from VSD 71163)
-ARG 6-77 p21 +-MJ 5-77 p32

912 Waltzes (14). Gyorgy Cziffra, pno. Philips Tape (c) 7327 042.
+Gr 7-77 p225 tape +RR 7-77 p98 tape
+HFN 7-77 p127 tape

913 Waltzes, nos. 1-14. Stefan Askenase, pno. DG 2548 146 Tape (c) 3348 146. (Reissue from DG 136396)
+Gr 4-75 p1867 +HFN 10-75 p152
+Gr 2-77 p1325 tape +-RR 4-75 p50
+-Gr 5-77 p1738 tape +RR 9-76 p92 tape
Waltzes, nos. 1-17. cf Piano works (DG 2740 163).
914 Waltzes, nos. 1-19. Nikita Magaloff, pno. Philips 6580 173.
+-Gr 12-77 p1106 +-RR 11-77 p86
+HFN 11-77 p168
915 Waltz, op. 18, E flat major. Waltzes, op. 34, nos. 1-3. Waltz, op. 42, A flat major. Waltzes, op. 64, nos. 1-3. Waltzes, op. 69, nos. 1-2. Waltzes, op. 70, nos. 1-3. Waltzes, op. posth., E major, E minor, A minor. Peter Katin, pno. Decca SPA 486 Tape (c) KCSP 486. (Reissue from SDD 353)
+Gr 5-77 p1713 +HFN 6-77 p139 tape
+-Gr 5-77 p1738 tape +RR 4-77 p70
+HFN 4-77 p151
Waltz, op. 18, E flat major. cf Piano works (Decca DPA 563/4).
Waltz, op. 18, E flat major. cf International Piano Library IPL 5007/8.
Waltzes, op. 34, nos. 1, 2. cf Piano works (HMV HQS 1375).
Waltzes, op. 34, nos. 1-3. cf Waltz, op. 18, E flat major.
Waltz, op. 34, no. 1, A flat major. cf Piano works (DG 2530 826).
Waltz, op. 34, no. 2, A minor. cf Piano works (Westminster WGS 8341).
Waltz, op. 34, no. 2, A minor. cf International Piano Library IPL 109.
Waltz, op. 42, A flat major. cf Etude, op. 25, no. 7, C sharp minor.
Waltz, op. 42, A flat major. cf Waltz, op. 18, E flat major.
Waltz, op. 42, A flat major. cf International Piano Library IPL 5001/2.
Waltzes, op. 64, nos. 1-3. cf Piano works (Decca SXL 6801).
Waltzes, op. 64, nos. 1-3. cf Waltz, op. 18, E flat major.
Waltz, op. 64, no. 1, D flat major. cf Piano works (Decca DPA 563/4).
Waltz, op. 64, no. 1, D flat major. cf Piano works (Westminster WGS 8341).
Waltz, op. 64, no. 1, D flat major. cf International Piano Library IPL 117.
Waltz, op. 64, no. 1, D flat major. cf International Piano Library IPL 5007/8.
Waltz, op. 64, no. 1, D flat major. cf RCA JRL 1-2315.
Waltz, op. 64, no. 1, D flat major (arr. and orch. Gerhardt). cf RCA LRL 1-5094.
Waltz, op. 64, no. 1, D flat major: Messaggero amoroso. cf Decca SDD 507.
Waltz, op. 64, no. 2, C sharp minor. cf Piano works (Decca 563/4).
Waltz, op. 64, no. 2, C sharp minor. cf Piano works (Reference Recordings RR 2).
Waltz, op. 64, no. 2, C sharp minor. cf International Piano Library IPL 103.
Waltz, op. 64, no. 2, C sharp minor. cf International Piano Library IPL 109.
Waltz, op. 64, no. 2, C sharp minor. cf International Piano Library IPL 5001/2.
Waltz, op. 64, no. 3, A flat major. cf International Piano Library IPL 114.

Waltzes, op. 69, nos. 1-2. cf Waltz, op. 18, E flat major.
Waltzes, op. 70, nos. 1-3. cf Waltz, op. 18, E flat major.
Waltz, op. 70, no. 1, G flat major. cf Piano works (Decca 563/4).
Waltz, op. 70, no. 1, G flat major. cf International Piano Library IPL 114.
Waltz, op. 70, no. 1, G flat major. cf RCA GL 4-2125.
Waltz, op. posth., E minor. cf International Piano Library IPL 114.
Waltzes, op. posth., E major, E minor, A minor. cf Waltz, op. 18, E flat major.

CHOTEM
North country. cf Citadel CT 6011.

CHOU, Wen-Chung
Soliloquy of a Bhiksuni. cf Crystal S 361.

CHOUHAJIAN
Garineh: Horhor's aria. cf Golden Age GAR 1001.
CHRISTMAS CAROLS FROM WESTMINSTER CATHEDRAL. cf Enigma VAR 1016.
CHRISTMAS EVE AT THE CATHEDRAL OF ST. JOHN THE DIVINE. cf Vanguard SVD 71212.
CHRISTMAS MUSIC FROM KING'S. cf HMV ESD 7050.
CHRISTMAS MUSIC OF THE 15th AND 16th CENTURIES. cf ABC L 67002.
CHRISTOPHER PARKENING: Music of two centuries. cf Angel S 36053.

CHRISTOSKOV, Peter
916 Suite, no. 1. LINDE: Sonata, violin and piano, op. 10. GOLEMINOV: Suite. PERGAMENT: Chaconne. Emil Dekov, vln; Carin Gille-Rybrant, pno. BIS LP 9.
+-Gr 3-77 p1454

CILEA, Francesco
917 Adriana Lecouvreur. Renata Tebaldi, s; Mario del Monaco, t; Giuletta Simionato, ms; Orchestra. London OSA 13126 Tape (c) 5-13126.
/NYT 10-9-77 pD21
Adriana Lecouvreur: Arias. cf London OS 26499.
Adriana Lecouvreur: Poveri fiori. cf Angel S 37446.
Adriana Lecouvreur: Respiro appena...Io son l'umile ancella. cf HMV SLS 5057.
L'Arlesiana: E la solita storia. cf HMV RLS 715.

CIMAROSA, Domenico
918 Il matrimonio segreto. Julia Varady, Arleen Auger, ms; Julia Hamari, ms; Ryland Davies, t; Dietrich Fischer-Dieskau, bar; Alberto Rinaldi, bs; Richard Amner, hpd; ECO; Daniel Barenboim. DG 2709 069 (3).
+-ARG 12-77 p26 +NR 12-77 p9
+Gr 9-77 p470 +-RR 9-77 p36
+HF 11-77 p95 +-SFC 8-28-77 p46
+HFN 9-77 p140
Melodie. cf RCA FRL 1-3504.
Sonatas, guitar, C sharp minor, A major. cf RCA RK 1-1735.
CIMBALOM RECITAL. cf Hungaroton SLPX 11686.

CIRONE, Anthony
919 Triptych: Double concerto. HANNA: Sonic sauce. KEEZER: For 4 percussionists. Sonic Boom Percussion Ensemble. Crystal S 140.
-NR 9-77 p15

CITKOWITZ, Israel
Chamber music: Songs. cf New World Records NW 243.

CLARKE
Songs: The blind ploughman. cf Argo ZFB 95/6.

CLARKE, Herbert
Cousins. cf Nonesuch H 71341.
The maid of the mist, polka. cf Nonesuch H 71341.
Twilight dreams, waltz intermezzo. cf Nonesuch H 71341.

CLARKE, Jeremiah
Prince of Denmark's march. cf Argo SPA 507.
Prince of Denmark's march. cf Saga 5417.
Suite, D major. cf Pelca PSR 40571.
Trumpet voluntary. cf Decca PFS 4351.
Trumpet voluntary. cf Philips 6581 018.
Trumpet voluntary. cf RCA PL 2-5046.

CLARKE, M.
Tyrolean tubas. cf Virtuosi VR 7608.

CLEMENT, William-John
Alleluia, the Lord is king. cf Richardson RRS 3.

CLERAMBAULT, Louis Nicolas
920 Songs: Antienne de la Sainte Vierge; Gloria in excelsis; Motet de Saint Michel. Suite, organ, in the first mode and second mode. Mady Mesple, s; Gaston Litaize, org. Connoisseur Society (Q) CSQ 2126.
+MJ 5-77 p32 +SFC 12-18-77 p53
+NR 8-77 p11 ++St 7-77 p114

921 Suite du deuxieme ton. MARCHAND: Grand dialogue. SCHEIDT: Cantio sacra: Warum betrabst du dich, mein Herz. David McVey, org. Orion ORS 77264.
+-ARG 8-77 p35 +-NR 5-77 p15

CLUTSAM
Curly headed baby. cf Club 99-105.

COATES, Eric
The dambusters. cf HMV SQ ASD 3341.
Knightsbridge. cf DJM Records DJM 22062.
922 London every day. VAUGHAN WILLIAMS: Fantasia on "Greensleeves". Fantasia on a theme by Thomas Tallis. English folk song suite. Symphony Orchestra; Morton Gould. Quintessence PMC 7049.
+ARG 12-77 p43 ++SFC 12-18-77 p53
Oxford Street. cf DJM Records DJM 22062.
Springtime suite: Dance in the twilight. cf HMV SRS 5197.
Springtime suite: Dance in the twilight. cf Starline SRS 197.
The three Elizabeths: Elizabeth of Glamis. cf Decca DDS 507/1-2.
The three Elizabeths: Youth of Britain. cf Decca SB 715.
The three Elizabeths: Youth of Britain. cf Lismor LILP 5078.

COBBOLD, William
With wreaths of rose and laurel. cf DG 2533 347.

COCCIA
Per la patria: Bella Italia. cf Rubini GV 34.

COKE-JEPHCOTT, Norman
Bishop's promenade. cf St. John the Divine Cathedral Church.
Pieces, organ. cf St. John the Divine Cathedral Church.
Saint Anne with descant. cf St. John the Divine Cathedral Church.
Toccata on Saint Anne. cf St. John the Divine Cathedral Church.

COLERIDGE-TAYLOR, Samuel
Big lady moon. cf Prelude PRS 2505.
Hiawatha's wedding feast: Onaway, awake beloved. cf HMV HLP 7109.
923 Quintet, clarinet, F sharp minor. KREISLER: Quartet, strings, A minor. Georgina Dobree, clt; Amici String Quartet. Discourses ABM 23.
+Gr 4-77 p1568 ++RR 3-77 p72
+-HFN 5-77 p120

COMPERE, Louis
O bone Jesu, motet. cf HMV SLS 5049.
Virgo celesti. cf L'Oiseau-Lyre 12BB 203/6.
CONCERTO GROSSO FOR 7 VOICES. cf Supraphon 113 1323.
CONON DE BETHUNE
Ahi, amours. cf Argo D40D3.
CONSOLI, Marc-Antonio
Sciui novi. cf CERVETTI: Aria suspendida.
Tre canzoni. cf CERVETTI: Aria suspendida.
CONVERSE, Frederick Shepherd
Endymion's narrative, op. 10. cf Louisville LS 753/4.
Flivver ten million. cf Louisville LS 753/4.
COOK, John
Fanfare, organ. cf Vista VPS 1046.
COOLIDGE, Peggy Stuart
924 New England autumn. Rhapsody, harp and orchestra. Pioneer dances. Spirituals in sunshine and shadow. Westphalian Symphony Orchestra; Siegfried Landau. Turnabout QTV 34635.
+-MJ 3-77 p46 +St 1-77 p113
+NR 12-76 p5
Pioneer dances. cf New England autumn.
Rhapsody, harp and orchestra. cf New England autumn.
Spirituals in sunshine and shadow. cf New England autumn.
COOPER, Paul
925 Symphony, no. 4. JONES: Elegy, string orchestra. Let us now praise famous men. Houston Symphony Orchestra; Samuel Jones. CRI SD 347.
+MJ 7-76 p57 -RR 6-77 p56
+NR 7-76 p2 +SFC 6-6-76 p33
Variants, organ. cf ADLER: Xenia, a dialogue for organ and percussion.
COPE, David
Bright angel. cf ANTHEIL: Sonata, trumpet.
926 Navajo dedications. Folkways FTS 33869.
+MJ 3-77 p46
COPLAND, Aaron
927 Appalachian spring. Columbia Chamber Orchestra; Aaron Copland. Columbia M 32736 Tape (c) MT 32736 (ct) MA 32736 (Q) MQ 32736 Tape (ct) MAQ 32736.
+FF 11-77 p79 +NYT 4-21-74 pD26
+HF 9-74 p126 tape ++SFC 3-24-74 p27
++NR 5-74 p2 +SR 11-30-74 p40
928 Appalachian spring. Lincoln portrait. El salon Mexico. Melvyn Douglas, speaker, BSO; Serge Koussevitzky. RCA AVM 1-1739. (Reissue from RCA originals)
+ARSC Vol VIII, nos. 2-3 p85 +NYT 1-16-77 pD13
++SFC 8-8-76 p38
+-HF 11-76 p110
Appalachian spring. cf London CSA 2246.
Appalachian spring. cf BERNSTEIN: Candide: Overture.
Concerto, piano. cf BARBER: Concerto, piano, op. 38.
929 Dance symphony. PISTON: The incredible flutist suite. M.I.T. Symphony Orchestra; David Epstein. Turnabout QTVS 34670.
+ARG 12-77 p27 +St 10-77 p135
++NR 6-77 p2
930 Danzon cubano. Sonata, piano. Variations. Aaron Copland, pno; Orchestra; Leo Smit, Leonard Bernstein. New World Records NW

227.
+-MJ 3-77 p46
Danzon cubano. cf CBS 61780.
931 Duo. Vocalise (trans. Copland). DELLO JOIO: The developing flutist. MUCZYNSKI: Preludes, op. 18. PISTON: Sonata, flute and piano. Laila Padorr, flt; Anita Swearengin, pno. Laurel Protone LP 14.
+NR 4-77 p7 +SFC 12-25-77 p42
Duo, flute and piano. cf Sonata, violin and piano.
Duo, flute and piano. cf BURTON: Sonatina, flute and piano.
Episode. cf Vista VPS 1038.
Lincoln portrait. cf Appalachian spring.
Nonet, strings. cf Sonata, violin and piano.
932 Old American songs: At the river; The boatmen's dance; Ching-a-ring chaw; The dodger; The golden willow tree; I bought me a cat; The little horses; Long time ago; Simple gifts; Zion's walls. Poems of Emily Dickinson: The chariot; Dear March, come in; Going to heaven; Heart, we will forget him; I felt a funeral in my brain; I've heard an organ talk sometimes; Nature, the gentlest mother; Sleep is supposed to be; There came a wind like a bugle; When they come back; The world feels dusty; Why do they shut me out of heaven. Robert Tear, t; Philip Ledger, pno. Argo ZRG 862.
+Audio 9-77 p48 +HFN 3-77 p101
+Gr 3-77 p1437 +RR 3-77 p89
Piano blues (4). cf Supraphon 111 1721/2.
Pieces, treble choir (2). cf Vox SVBX 5353.
Poems of Emily Dickinson (12). cf Old American songs.
Poems of Emily Dickinson (12). cf BARBER: Hermit songs, op. 29.
Quiet city. cf Argo ZRG 845.
Rodeo: Hoe-down. cf CBS 61039.
Rodeo: Hoe-down. cf CBS 61780.
El salon Mexico. cf Appalachian spring. RCA 1-1739.
933 El salon Mexico. GERSHWIN: An American in Paris. Porgy and Bess (arr. Russell Bennett). LPO; John Pritchard. Classics for Pleasure CFP 40240 Tape (c) TC CFP 40240.
+Gr 2-77 p1325 -RR 3-76 p39
+HFN 3-76 p91 +RR 11-77 p117 tape
++HFN 12-76 p153 tape
934 Sonata, piano. SINDING: Sonata, piano, op. 91, B minor. Eva Knardahl, pno. BIS LP 52.
+HFN 11-77 p168 +-RR 11-77 p86
Sonata, piano. cf Danzon cubano.
935 Sonata, violin and piano. Duo, flute and piano. Nonet, strings. Isaac Stern, vln; Elaine Shaffer, flt; Aaron Copland, pno; Columbia String Ensemble; Aaron Copland. Columbia M 32737.
+AR 5-77 p19 +-SFC 3-24-74 p27
+-HF 5-74 p82 +-St 4-74 p108
+NYT 4-21-74 pD26
Song. cf New World Records NW 243.
Variations. cf Danzon cubano.
Variations. cf Vox SVBX 5303.
Vocalise. cf Duo.
CORBETTA, Francesco
Suite. cf Saga 5438.
CORDLE, Andrew
Interlude. cf Orion ORS 77269.

CORELLI, Arcangelo
Allemande. cf RCA FRL 1-3504.
936 Concerti grossi, op. 6, nos. 1-12. Slovak Chamber Orchestra; Bohdan Warchal. Opus 9111 0442/4 (3).
+-Gr 7-77 p173 +RR 3-77 p48
+-HFN 5-77 p121
937 Concerto grosso, op. 6, no. 8, G minor. LOCATELLI: Concerto grosso, op. 1, no. 8, F minor. MANFREDINI: Concerto grosso, op. 3, no. 12, C major. Slovak Chamber Orchestra; Bohdan Warchal. Opus 9111 0431.
+HFN 6-77 p121
Concerto grosso, op. 6, no. 8, G minor. cf ALBINONI: Adagio, G minor.
Concerto grosso, op. 6, no. 8, G minor. cf BACH: Cantata, no. 147, Jesu, joy of man's desiring.
Concerto grosso, op. 6, no. 8, G minor. cf Argo D69D3.
938 La folia (trans. Kreisler). NARDINI: Sonata, violin (trans. Flesch). TARTINI: Sonata, violin, G minor (trans. Kreisler). VITALI: Chaconne (trans. David). Ida Haendel, vln; Geoffrey Parsons, pno. HMV SQ ASD 3352.
++GR 8-77 p314 ++RR 9-77 p81
++HFN 8-77 p79
Sarabande and allegretto. cf HMV ASD 3346.
939 Sarabande, gigue and badinerie (arr. Arbos). MOZART: Serenade, no. 13, K 525, G major. PURCELL: Abdelazer: Suite. Slovak Chamber Orchestra; Bohdan Warchal. Opus 9111 0198.
+Gr 7-77 p173 +-RR 6-77 p61
-HFN 6-77 p121
Sarabande, gigue and badinerie. cf Supraphon 110 1890.
Sonata, op. 5, no. 8, D minor. cf Arion ARN 90416.
CORNYSHE, William
Songs: Ah, Robin. cf Enigma VAR 1020.
Songs: Ah, Robin. cf Saga 5444.
Songs: Ah, Robin. cf Saga 5447.
Songs: Ah, Robin; Adieu, mes amours; Blow thy horn, hunter. cf Argo ZK 24.
CORONATION CHORAL MUSIC 1953. cf Vista VPS 1053.
CORONATION MUSIC. cf Pye Nixa QS PCNHX 10.
CORRETTE, Michel
Menuets, nos. 1 and 2. cf DG 2723 051.
CORTECCIA, Francesco
940 St. John Passion. Arnold Foa, speaker; Schola Cantorum Francesco Cordini; Fosco Corti. DG 2533 301.
+-Gr 4-76 p1645 +MT 1-77 p43
++HFN 5-76 p99 +RR 5-76 p69
CORTES, Ramiro
Sonata, violin and piano. cf Orion ORS 76212.
COSENTINO
Misterios. cf Lyrichord LLST 7299.
COSTA
Serenata medioevale. cf Club 99-106.
COSTELEY, Guillaume
Helas, helas, que de mal. cf L'Oiseau-Lyre 12BB 203/6.
COTTENET
Chanson meditation. cf Discopaedia MB 1012.
COUCY, Le Chatelain de
Le noviaus tens. cf Argo D40D3.

COUPERIN, Armand Louis
941 Pieces de clavecin. COUPERIN, F.: Livres de clavecin, Bk IV, Ordre, no. 2. COUPERIN. L.: Suite, C major. Martin Pearlman, hpd. Titanic TI 9.
+-HF 12-77 p84
COUPERIN, Francois
L'Apotheose de Lully: Plainte des memes. cf Abbey LPB 765.
942 Concerto, trumpet and organ, no. 9: Ritratto dell'amore. La Steinkerque. MOURET: Suite, 3 trumpets and organ, D major. PURCELL, D.: Sonata, trumpet and organ, F major. VIVALDI: Sonata a 3, 2 trumpets and organ. Maurice Andre, Raymond Andre, Bernard Soustrot, tpt; Hedwig Bilgram, org. Erato STU 70760. (also Musical Heritage Society MHS 3311)
+MU 11-77 p17 +RR 10-73 p88
943 Concerts royaux. Nouveaux concerts. Thomas Brandis, vln; Heinz Holliger, ob; Aurele Nicolet, flt; Josef Ulsamer, Laurenzius Strehl, vla da gamba; Manfred Sax, bsn; Christianne Jaccottet, hpd, and others. DG 2723 046 (4). (also DG 2712 003)
++ARG 2-77 p31 ++NR 2-77 p8
+Gr 10-76 p616 ++RR 10-76 p48
++HF 4-77 p98 ++St 4-77 p115
+HFN 10-76 p167
Dialogue sur les grands jeux. cf Vista VPS 1033.
Les gouts reunis, no. 8, G major. cf BACH: Suite, orchestra, S 1068, D major.
Livres de clavecin, Bk II, Ordre no. 6: Les barricades mysterieuses. cf Angel S 36053.
Livres de clavecin, BK II, Order no. 6: Les moissonneurs; Les barricades mysterieuses; Le moucheron. cf Saga 5384.
Livres de clavecin, Bk III, Ordre no. 14: Le carillon de Cythere. cf International Piano Library IPL 112.
Livres de clavecin, Bk IV, Ordre, no. 23. cf COUPERIN, A.: Pieces de clavecin.
Nouveaux concerts. cf Concerts royaux.
Pieces en concert. cf BACH: Suite, orchestra, S 1068, D major.
944 Pieces en concert (5) (arr. Bazeliare). MENDELSSOHN: Trio, piano, no. 1, op. 49, D minor. SCHUMANN: Adagio and allegro, op. 70, A flat major. TRAD.: Song of the birds (arr. Casals). Pablo Casals, vlc; Alexander Schneider, vln; Mieczyslaw Horszowski, pno. CBS 61489. (Reissue from BRG 72035)
+-Gr 7-77 p198 +RR 6-77 p73
+-HFN 6-77 p137
Plein jeu. cf Vista VPS 1060.
Recit de tierce en taille. cf Vista VPS 1060.
La Steinkerque. cf Concerto, trumpet and organ, op. 9: Ritratto dell'amore.
La Steinquerque. cf Vox SVBX 5142.
La Sultane, D minor. cf Vox SVBX 5142.
COUPERIN, Louis
Chaconne, C major. cf Vista VPA 1030.
Suite, C major. cf COUPERIN, A.: Pieces de clavecin.
945 Suites, harpsichord, G minor, D major, A minor, F major. Alan Curtis, hpd. DG 2533 325 Tape (c) 3310 325.
+Gr 12-76 p1021 +NR 10-77 p14
++HF 12-77 p84 ++RR 12-76 p78
+HFN 1-77 p104
Sympnonies, violes (5). cf Vox SVBX 5142.

COWELL, Henry
Advertisement. cf Piano works (Finnadar SR 9016).
Advertisement. cf Vox SVBX 5303.
Aeolian harp. cf Vox SVBX 5303.
946 Air and scherzo. HUSA: Concerto, alto saxophone and concert band. IBERT: Concertino da camera. Robert Black, also sax; Patricia Black, pno. Brewster Records BR 1216.
++IN 3-77 p28
Amiable conversation. cf Piano works (Finnadar SR 9016).
Anger dance. cf Piano works (Finnadar SR 9016).
Antimony. cf Piano works (Finnadar SR 9016.
The banshee. cf Piano works (Finnadar SR 9016).
Dynamic motion. cf Piano works (Finnadar SR 9016).
Exultation. cf Vox SVBX 5303.
Fabric. cf Piano works (Finnadar SR 9016).
The harp of life. cf Piano works (Finnadar SR 9016).
The hero sun. cf Piano works (Finnadar SR 9016).
Hymn and fuguing tune, no. 10. cf Argo ZRG 845.
Invention. cf Vox SVBX 5303.
The lilt of the reel. cf Piano works (Finnadar SR 9016).
Luther's carol to his son. cf Vox SVBX 5353.
Maestoso. cf Piano works (Finnadar SR 9016).
947 Piano works: Advertisement. Amiable conversation. Anger dance. Antinomy. The banshee. Dynamic motion. Aeolin harp. Fabric. The harp of life. The hero sun. The lilt of the reel. Maestoso. Sinister resonance. Six ings. The tide of Manaunaun. Tiger. The trumpet of Angus Og. The voice of Lir. What's this. Doris Hays, pno. Finnadar SR 9016 Tape (c) CS 9016 (ct) TP 9016.
+CL 11-77 p10 ++NR 7-77 p11
+FF 9-77 p13 ++St 9-77 p125
+MJ 9-77 p34
Quartet euphometric. cf New England Conservatory NEC 115.
Sinister resonance. cf Piano works (Finnadar SR 9016).
Six ings. cf Piano works (Finnadar SR 9016).
948 Symphony, no. 16. ISOLFSSON: Introduction and passacaglia, F minor. LEIFS: Iceland overture, op. 9. Iceland Symphony Orchestra; William Strickland. CRI SD 179.
+-HFN 2-77 p115 +-RR 3-77 p49
The tide of Manaunaun. cf Piano works (Finnadar SR 9016).
Tiger. cf Piano works (Finnadar SR 9016).
The trumpet of Angus Og. cf Piano works (Finnadar SR 9016).
The voice of Lir. cf Piano works (Finnadar SR 9016).
What's this. cf Piano works (Finnadar SR 9016).
THE COZENS LUTE BOOK. cf L'Oiseau-Lyre DSLO 510.
CRAUS, Stephan
Chorea, Auff und nider. cf Hungaroton SLPX 11721.
Tantz, Hupff auff. cf Hungaroton SLPX 11721.
Die trunke pinter. cf Hungaroton SLPX 11721.
CRESPO, Gomez
Nortena, homenaje a Julian Aguirre. cf Saga 5412.
CRESTON
Celebration. cf MENC 76-11.
CRESTON, Paul
A rumor. cf Argo ZRG 845.
Toccata. cf Vista VPS 1046.

CROCE, Giovanni
Ave virgo. cf Argo ZRG 859.
CROFT, William
All people that on earth do dwell. cf CRD Britannia BR 1077.
CROIX (14th century)
S'amours eust point de poer. cf Turnabout TV 37086.
CROOKES
Way out west. cf Grosvenor GRS 1048.
CROSSE, Gordon
Concerto da camera. cf BIRTWISTLE: Tragoedia.
949 Purgatory. Glenville Hargreaves, bar; Peter Bodenham, t; Royal Northern School of Music Orchestra and Chorus; Michael Lankester. Argo ZRG 810.
+Gr 12-75 p1093 +MT 3-77 p214
+HFN 12-75 p152 +NR 2-76 p12
+MJ 7-76 p56 ++RR 11-75 p32
+MM 7-76 p37 +SR 4-17-76 p51
+MQ 10-77 p570
Some marches on a ground, op. 28. cf ADDISON: Concerto, trumpet, strings, and percussion.
CROUCH, Frederick
Kathleen Mavourneen. cf Philips 9500 218.
CRUMB, George
Night music. cf ERICKSON: Chamber concerto.
950 Night music I. ERICKSON: Chamber concerto. Louise Toth, s; Paul Parmelee, pno, celeste; David Burge, Thomas MacCluskey, perc; Hartt Chamber Players; Ralph Shapey. CRI SD 218.
+Gr 3-77 p1454 /RR 1-77 p79
951 Twelve fantasy pieces after the Zodiac (Makrokosmos II). Robert Miller, pno. Odyssey Y 34135.
++ARG 12-76 p25 ++NR 12-76 p13
+Ha 6-77 p88 ++St 2-77 p124
++HF 4-77 p99
CRUSE
Sonatas, trombones (3). cf ANTES: Chorales.
CRUSELL, Bernhard
952 Quartet, clarinet, no. 1, E flat major. Quartet, clarinet, no. 3, D major. Tapio Lotjonen, clt; Jorma Rahkonen, vln; Esa Kamu, vla; Esko Valsla, vlc. BIS LP 51.
+HFN 12-77 p168 +RR 11-77 p79
Quartet, clarinet, no. 3, D major. cf Quartet, clarinet, no. 1, E flat major.
Rondo, 2 clarinets and piano. cf BIS LP 62.
CSERMAK, Antal
953 Hungarian dances (6). The threatening dance or the love of the fatherland. ROZSAVOLGYI: Czardas. First Hungarian round dance. Hungarian Chamber Orchestra; Vilmos Tatrai. Hungaroton SLPX 11698.
+NR 9-76 p5 +SFC 1-16-77 p43
The threatening dance or the love of the fatherland. cf Hungarian dances.
CUI, Cesar
The little cloud. cf Rubini GV 26.
CUNDELL, Edric
Blackfriars, symphonic prelude. cf Virtuosi VR 7608.
CUNDICK
The west wind. cf Columbia M 34134.

CURTIS, E. de
Non m'ami piu. cf Club 99-105.
Songs: A canzone e Napule; Ti voglio tanto bene. cf London SR 33221.
Torna a Surriento. cf HMV RLS 715.

CUSTER, Arthur
954 Found objects, no. 7. RUSH: Hexahedron. THORNE: Sonata, piano. Dwight Peltzer, pno. Serenus SRS 12071.
+NR 12-77 p12

CUTTING, Francis
Galliard. cf RCA RL 2-5110.

CZIFFRA, Gyorgy
Fantaisie Roumaine (after the gypsy style). cf Connoisseur Society CS 2131.

DABROWSKI, Florian
955 Concerto, violin, 2 pianos and percussion. MEYER: Concerto, violin, op. 12. Jadwiga Kaliszewska, vln; Polish Philharmonic Orchestra, Poznan Percussion Ensemble; Renard Czajkowski. Muza SX 1054.
+NR 8-77 p6

DADMUN, J. W.
The babe of Bethlehem. cf New World Records NW 220.

DAGGERE
Downberry down. cf Argo ZK 24.

DALBY, Martin
Variations, violoncello and piano. cf BERKELEY: Duo, violoncello and piano.

DALLAPICCOLA, Luigi
956 Ciaconna, intermezzo e adagio. Parole di San Paolo. Studi (2). Tartiniana seconda. Sandro Materassi, vln; Pietro Scarpini, pno; Amedeo Baldovino, vlc; Magda Laszo, ms; Ensemble; Zoltan Pesko. CBS 16490.
+Te 6-77 p41
957 Cori di Michelangelo Buonarroti il Giovane, nos. 1 and 2. KODALY: Bilder aus der Matra-Gegend. LIDHOLM: Choruses (4). NAUMANN: Songs on Latin texts, mixed voices and some instruments, op. 24 (2). Uppsala Academic Chamber Choir, YMCA Chamber Choir; Dan-Olof Stenlund. Caprice CAP 1037.
+-HFN 12-77 p168 +-RR 11-77 p101
Parole di San Paolo. cf Ciaccona, intermezzo e adagio.
Studi (2). cf Ciaccona, intermezzo e adagio.
Tartiniana seconda. cf Ciaccona, intermezzo e adagio.

DALZA, Joanambrosio
Calata ala Spagnola. cf DG 2723 051.
Recercar; Suite ferrarese; Tastar de corde. cf L'Oiseau-Lyre 12BB 203/6.

DAMARE
Cleopatra. cf RCA PL 2-5046.

DAMASE, Jean-Michel
958 Concertino, harp and strings, op. 20. GLIERE: Concerto, harp, op. 74, E flat major. MOZART: Adagio, K 617, C minor. Olga Erdeli, hp; MRSO, Moscow Chamber Orchestra; Boris Khaikin, Gennady Rozhdestvensky. Westminster WG 8346. (Reissue)
+-ARG 9-77 p50 +-NR 9-77 p4
DAME NELLI MELBA, The London recordings 1904-1926. cf HMV RLS 719.

DA MOTTA, Jose Vianna
Cantiga de amor, op. 9, no. 1. cf International Piano Library IPL 108.

Chula (Danse Portugaise). cf International Piano Library IPL 108.
Valse caprichosa. cf International Piano Library IPL 108.
DA MOTTA, JOSE VIANNA, PIANO RECITAL. cf International Piano Library IPL 108.
DANCE MUSC FROM THE 15th-19th CENTURIES. cf DG 2723 051.

DANDRIEU, Jean
Armes, amours: O flour des flours. cf HMV SLS 863.
Deploration sur la mort de Machaut. cf CBS 76534.
The fifers. cf Kaibala 20B01.

DANIELSEN, Ragnar
Musical saunter. cf Amberlee ALM 602.

DANYEL, John
Dost thou withdraw thy grace. cf Abbey LPB 712.
Mistress Anne Grene her leaves be greene. cf L'Oiseau-Lyre 510.
Tyme cruell tyme. cf RCA RL 2-5110.

DANZI, Franz
959 Quintet, op. 56, no. 1, B flat major. REICHA: Quintet, op. 91, no. 5, A major. STAMITZ: Quartet, op. 8, no. 2, E flat major. Munich Residenz Quintet. Claves D 611.
+HFN 5-77 p121 -RR 4-77 p66
960 Quintet, op. 68, no. 2, F major. Quintet, op. 68, no. 3, D minor. Soni Ventorum Wind Quintet. Crystal S 251.
+IN 1-77 p20 +NR 6-76 p9
Quintet, op. 68, no. 3, D minor. cf Quintet, op. 68, no. 2, F major.
Quintet, op. 68, no. 3, D minor. cf BEETHOVEN: Adagio and allegro, music box.
Sonata, horn, op. 28, E flat major. cf BEETHOVEN: Sonata, horn, op. 17, F major.

DAQUIN, Louis-Claude
961 Noels (12). Gaston Litaize, org. Connoisseur Society CSQ 2125.
+MJ 5-77 p32 +St 10-77 p135
++NR 4-77 p14
962 Noels (12). MARCHAND: Basse de cromorne. Duo. Dialogue (2). Grand jeu. Fugue. Plein jeu. Quatuor. Tierce en taille. Arthur Wills, org. Saga 5433/4 (2).
+-ARG 12-77 p27 +-MT 7-77 p565
+Gr 12-76 p1021 +-RR 12-76 p83
+HF 10-77 p106 +-St 10-77 p135
963 Noels, nos. 1, 8-12. Michael Chapuis, org. Harmonia Mundi HM 531.
+-Gr 12-77 p1106
Noel no. VII en trio et en dialogue. cf Abbey LPB 752.
Noel Suisse. cf Argo ZRG 864.
Noel Suisse. cf Crescent ARS 109.
Noel Suisse. cf Vista VPS 1060.

DARGOMIZHSKY, Alexander
On our street. cf Rubini GV 26.
Russalka: Come unknown power. cf Club 99-105.

DARTER, Thomas
Sonatina. cf Golden Crest RE 7068.

DAVID, Felicien
La perle du Bresil: Charmant oiseau. cf Columbia 34294.
La perle du Bresil: Charmant oiseau. cf Rubini GV 57.

DAVIDOV
Night, love, moon. cf Club 99-96.
Night, love and moon. cf Rubini RS 301.

DAVIDOVSKY, Mario
Eletronic study, no. 2. cf AREL: Music for a sacred service: Prelude and postlude.
Synchronisms, no. 3. cf Delos DEL 25406.
Synchronisms, woodwind quintet and tape, no. 8. cf Vox SVBX 5307.
DAVIES
Solemn melody. cf Virtuosi VR 7608.
DAVIES, Walford
God be in my head. cf Vista VPS 1037.
R.A.F. march. cf CRD Britannia BR 1077.
R.A.F. march. cf HMV SQ ASD 3341.
DAVIS
Little drummer boy. cf RCA PRL 1-8020.
Songs: God will watch over you. cf Argo ZFB 95/6.
The West's awake. cf Philips 6599 227.
DAVIS, Sharon
Though men call us free. cf BROWN: Concertino, piano and band.
DAVY, Richard
The Bay of Biscay. cf HMV ESD 7002.
Songs: Ah mine heart. cf APPLEBY: Magnificat.
DEARNLEY, Christopher
Dominus regit me. cf Guild GRSP 7011.
Fanfare. cf Guild GRSP 7011.
DEBIASY
Dance. cf Saga 5421.
DEBUSSY, Claude
Ballades de Francois Villon (3). cf Works, selections (BIS LP 28).
Berceuse heroique. cf Works, selections (Angel S 37064).
964 Boite a joujoux. Printemps. French National Radio Orchestra; Jean Martinon. Angel S 37124 Tape (c) 4XS 37124.
+HF 10-77 p129 tape
Chansons de Bilitis (3). cf BIS LP 34.
Chansons de Charles d'Orleans: Dieu qu'il la fait bon regarder; Quant j'ai ouy le tambourin; Yver, vous n'estes qu'un villian. cf RCA RL 2-5112.
Children's corner suite. cf Works, selections (Angel S 37064).
Children's corner suite. cf Vox SVBX 5483.
965 Children's corner suite: Golliwog's cakewalk (arr. Choisnet). Sonata, violin and piano. Suite bergamasque: Clair de lune (arr. Relyans). RAVEL: Sonata, violin and piano. Tzigane. Andreas Kiss, vln; Katalin Lakatos, pno. Hungaroton SLPX 11796.
+-RR 2-77 p79
Children's corner suite: Golliwog's cakewalk. cf Grosvenor 1055.
Children's corner suite: Golliwog's cakewalk. cf Supraphon 111 1721/2.
Children's corner suite: Golliwog's cakewalk; Snow is dancing. cf BEETHOVEN: Sonata, piano, no. 14, op. 27, no. 2, C sharp minor.
La damoiselle elui. cf CHAUSSON: Songs (Reidffusion Symphonica SYM 6).
La damoiselle elui: Je voudrais qu'il fut deja pres de moi. cf Odyssey Y 33130.
Danse. cf Works, selections (Angel S 37064).
966 Danses sacree et profane. MOZART: Concerto, flute and harp, K 299, C major. Olga Erdeli, hp; Alexander Korneyev, flt; Moscow Chamber Orchestra; Rudolf Barshai. Westminster WGS 8334. (Reissue from Melodiya C 01131/2)

-ARG 4-77 p24 ++SFC 3-6-77 p34
+NR 5-77 p8

Danses sacree et profane. cf Works, selections (Angel S 37065).
En blanc et noir. cf Works, selections (BIS LP 28).
L'Enfant prodigue: Lia's recitative and aria. cf Odyssey Y 33130.

967 Estampes. Imgaes, Bks 1 and 2. L'Isle joyeuse. Masques. Jean-Philippe Collard, pno. Connoisseur Society (Q) CSQ 2136.
++NR 9-77 p12 +St 9-77 p128
++SFC 8-21-77 p46

Estampes: Soiree dans Grenade. cf CBS 73589.

968 Etudes. Paul Jacobs, pno. Nonesuch H 71322.
++Gr 9-76 p443 ++NR 8-76 p12
++HF 10-76 p104 +-RR 9-76 p80
+HFN 9-76 p119 +-SFC 10-31-76 p35
+MT 3-77 p214 +St 10-76 p121

Fantaisie, piano. cf Works, selections (Angel S 37065).
Images, Bks 1 and 2. cf Estampes.

969 Images pour orchestra: Iberia. Nocturnes, nos. 1-3. Prelude a l'apres midi d'un faune. Netherlands Radio Chorus; Netherlands Philharmonic Orchestra; Jean Fournet. Decca PFS 4317 Tape (c) KPFC 4317. (also London SPC 21104)
-Gr 10-75 p608 +-NR 7-77 p4
+-HF 8-77 p78 -RR 9-75 p35
+HFN 12-75 p153 -RR 1-77 p88 tape
+-MJ 7-77 p70

970 Images pour orchestra: Iberia. Nocturnes, nos. 1-3. Washington Oratorio Society; National Philharmonic Orchestra; Antal Dorati. Decca SXL 6742 Tape (c) KSXC 6742. (also London CS 6968 Tape (c) 5-6968)
+-ARG 4-77 p20 +-MJ 7-77 p70
+-Gr 11-76 p782 +-NR 7-77 p4
+-HF 8-77 p78 +-RR 11-76 p53
++HFN 10-76 p167 ++SFC 12-26-76 p34
+HFN 12-76 p155 tape

Images our orchestra: Iberia. cf RCA CRM 5-1900.
L'Isle joyeuse. cf Estampes.
Little shepherd. cf RCA Tape (c) RK 2-5030.
Khamma. cf Works, selections (Angel 37067/8).
Marche ecossaise. cf Works, selections (Angel 37067/8).

971 Le martyre de Saint-Sebastian (orch. Caplet). Printemps (orch. Busser). Orchestre de Paris; Daniel Barenboim. DG 2530 879.
+Gr 10-77 p627 +RR 9-77 p58
+HFN 11-77 p169

Masques. cf Estampes.

972 La mer. RAVEL: Daphnis and Chloe: Suite, no. 2. Pavane pour une infante defunte. CO; Georg Szell. CBS 61075 Tape (c) 40-61075.
+-HFN 12-75 p173 tape +-RR 5-77 p92 tape

973 La mer. Nocturnes. PhO; Carlo Maria Giulini. HMV Tape (c) TC EXE 185.
+HFN 4-76 p125 tape +-RR 2-77 p97 tape

974 La mer. Prelude a l'apres midi d'un faun. RAVEL: Bolero. CSO; Georg Solti. London CS 7033 Tape (c) 5-7033. (also Decca SXL 6813 Tape (c) KSXC 6813)
+-Gr 7-77 p235 tape +NR 9-77 p2
+MJ 7-77 p70 ++SFC 4-3-77 p43

975 La mer. Nocturnes. COA; Eduard van Beinum. Philips Fontana Tape (c) 7327 044.

+Gr 7-77 p225 tape +RR 7-77 p98
+HFN 7-77 p127 tape

976 La mer. Nocturnes. BSO; Pierre Monteux. Quintessence PMC 7027.
+ARG 12-77 p43 +St 11-77 p146
+HF 9-77 p108

977 La mer. Nocturnes. Prelude a l'apres-midi d'un faune. ORTF and Choir; Charles Munch. Turnabout TVS 34637 Tape Vox (c) CT 2119.
+-HF 8-77 p78

La mer. cf Preludes a l'apres-midi d'un faune.
La mer. cf Works, selections (Angel 37067/8).
La mer. cf RCA CRM 5-1900.
Nocturnes. cf La mer (HMV TC EXE 185).
Nocturnes. cf La mer (Philips 7327 044).
Nocturnes. cf La mer (Quintessence PMC 7027).
Nocturnes. cf La mer (Turnabout TVS 34637).
Nocturnes. cf Works, selections (Angel 37067/8).
Nocturnes, nos. 1-3. cf Images pour orchestra: Iberia (Decca 4317).
Nocturnes, nos. 1-3. cf Images pour orchestra: Iberia (Decca SXL 6742).

978 Nocturnes: Fetes. LUTOSLAWSKI: Variations on a theme by Paganini. POULENC: L'Embarquement pour Cythere. Sonata, 2 pianos. RAVEL: La valse. Monique Le Duc, Charles Engel, pno. Orion ORS 76238.
+NR 9-77 p12

979 Pelleas et Melisande. Elisabeth Soderstrom, s; Yvonne Minton, con; George Shirley, t; Denis Wicks, bar; Donald McIntyre, bs; ROHO and Chorus; Pierre Boulez. Columbia M3 30119 (3).
-ARG 3-71 p405 +NYT 10-9-77 pD24
/HF 2-71 p63 +ON 3-13-71 p32
+MJ 5-71 p66 ++SFC 12-20-70 p33
+NR 1-71 p8 ++SFC 6-13-71 p32

980 Pelleas et Melisande. Irene Joachim, Leila Ben-Sedira, s; Germaine Cernay, ms; Jacques Jansen, t; H. Etcheverry, bar; A. Marcon, Paul Cabanel, bs; Yvonne Gouverne Choeurs; Roger Desmormiere. EMI Odeon 2C 153 12513/5 (3).
+HF 4-75 p76 +NYT 10-9-77 pD21

981 Pelleas et Melisande. Erna Spoorenberg, s; Camille Maurane, bar; George London, bs-bar; OSR; Ernest Ansermet. London OSA 1379.
+NYT 10-9-77 pD21

Pelleas et Melisande: Mes longs cheveux. cf International Piano Library IPL 117.
Pelleas et Melisande: Voici ce qui'il ecrit; Tu ne sais pas pourquoi. cf HMV RLS 716.
Petite piece. cf L'Oiseau-Lyre DSLO 17.
Petite suite. cf Works, selections (Angel S 37064).
Il pleure dans mon coeur. cf ABC AB 67014.
La plus que lente. cf Works, selections (Angel S 37064).

982 Preludes, Bks 1 and 2. Pierre Huybregts, pno. Crystal S 161/2.
+-NR 9-77 p13

983 Preludes, Bk 1. Claude Helffer, pno. Harmonia Mundi HMU 951.
+-RR 11-77 p87

984 Preludes, Bk 1. Livia Rev, pno. Saga 5391 Tape (c) CA 5391.
+Gr 3-75 p1678 +-RR 4-75 p50
+Gr 2-77 p1322 tape +-RR 1-77 p88 tape
+HFN 12-76 p155 tape

Prelude, Bk 1, no. 8: The girl with the flaxen hair. cf Angel S 36053.

Prelude, Bk 1, no. 8: La fille aux cheveux de lin. cf Decca 473.
Prelude, Bk 1, no. 8: La fille aux cheveux de lin. cf London CS 7015.
Prelude, Bk 1, no. 8: La fille aux cheveux de lin. cf Orion ORS 75181.
Prelude, Bk 1, no. 12: Minstrels. cf HMV SQ ASD 3283.
985 Preludes, Bk 2. Theodore Paraskivesco, pno. Calliope CAL 1832.
+HFN 6-77 p121 +-RR 7-77 p80
986 Preludes, Bk 2. Livia Rev, pno. Saga 5442.
++Gr 12-77 p1106 +RR 11-77 p87
+HFN 12-77 p168
Preludes, Bk 2: Ondine. cf BEETHOVEN: Sonata, piano, no. 8, op. 13, C minor.
987 Prelude a l'apres-midi d'un faune. La mer. RAVEL: Bolero. CSO; Georg Solti. London CS 7033 Tape (c) 5-7033. (also Decca SXL 6813 Tape (c) KSXC 6813)
++ARG 10-77 p41 +HFN 6-77 p139 tape
+-Gr 3-77 p1395 +-RR 3-77 p57
++HF 8-77 p78 +-RR 7-77 p99 tape
+HFN 3-77 p101 +St 9-77 p140
Prelude a l'apres-midi u'un faune. cf Images pour orchestra: Iberia.
Prelude a l'apres-midi d'un faune. cf La mer (London CS 7033).
Prelude a l'apres-midi d'un faune. cf La mer (Turnabout TVS 34637).
Prelude a l'apres-midi d'un faune. cf Works, selections (Angel 36067/8).
Prelude a l'apres-midi d'un faune. cf BERLIOZ: Le Troyens: Royal hunt and storm.
Prelude a l'apres-midi d'un faune. cf CHABRIER: Espana.
Prelude a l'apres-midi d'un faune. cf Angel S 37409.
Prelude a l'apres-midi d'un faune. cf HMV (SQ) ASD 3338.
Prelude 1 l'apres-midi d'un faune. cf Seraphim SIB 6094.
Printemps. cf Boite a joujoux. cf Angel 37124.
Printemps. cf Le martyre de Saint-Sebastian.
Proses lyriques, no. 3: De fleurs. cf Odyssey Y 33130.
988 Quartet, strings, G minor. RAVEL: Quartet, strings, F major. Orford Quartet. Decca SDD 526.
+Gr 12-77 p1098 +-RR 11-77 p80
+HFN 11-77 p169
989 Quartet, strings, G minor. RAVEL: Quartet, strings, F major. Slovak Quartet. Opus 9111 0337.
+-Gr 5-77 p1707 +-RR 3-77 p69
+HFN 5-77 p121
990 Quartet, strings, op. 10, G minor. RAVEL: Quartet, strings, F major. Budapest Quartet. Columbia MS 6015.
+St 5-77 p73
Reverie. cf International Piano Library IPL 112.
Rhapsody, clarinet and orchestra. cf Works, selections (Angel S 37065).
Rhapsody, saxophone and orchestra. cf Works, selections (Angel S 37065).
Le Roi Lear: Fanfare; Le sommeil de Lear. cf Works, selections (Angel 37067/8).
Sonata, violin and piano. cf Children's corner suite: Golliwog's cakewalk.
Sonata, violin and piano. cf Works, selections (BIS LP 28).
Sonata, violoncello and piano. cf Works, selections (BIS LP 28).

991 Songs: Chansons de Bilitis: La flute de Pan; Le chevelure; Le tombeau des naiades. Fetes galantes: En sourdine; Fantoches; Clair de lune. Poemes de Baudelaire: Le balcon; Harmonie du soir; Le jet d'eau; Recueillement; La mort des amants. Mandoline. Romance. Voici que le printemps. Anna Moffo, s; Jean Casadesus, pno. RCS SD 6890.
+-Gr 11-75 p875 -MM 5-77 p36
+-HFN 10-75 p141 -RR 9-75 p68
Songs: Ballade des femmes de Paris; Chansons de Bilitis (3); Fetes galantes, I and II; Green. cf HMV RLS 716.
Songs: Green. cf Seraphim 60274.
Songs: Proses lyriques: Noel des enfants qui n'ont plus de maison. cf BIZET: Songs (Saga 5388).
Suite bergamasque: Clair de lune. cf Children's corner suite: Golliwog's cakewalk.
Suite bergamasque: Clair de lune. cf BEETHOVEN: Sonata, piano, no. 8, op. 13, C minor.
Suite bergamasque: Clair de lune. cf CBS 61039.
Suite bergamasque: Clair de lune. cf CBS 73589.
Suite bergamasque: Clair de lune. cf Decca PFS 4387.
Suite bergamasque: Clair de lune. cf International Piano Library IPL 113.
Suite bergamasque: Clair de lune. cf Prelude PRS 2512.
Suite bergamasque: Clair de lune. cf RCA JRL 1-2315.
Suite bergamasque: Clair de lune. cf RCA RK 2-5030.
Suite bergamasque: Clair de lune. cf Seraphim SIB 6094.
Syrinx, flute. cf Works, selections (BIS LP 28).
Syrinx, flute. cf Coronet LPS 3036.
Syrinx, flute. cf RCA RK 2-5030.

992 Works, selections: Berceuse heroique. Children's corner suite. Petite suite. Danse. La plus que lente. French National Radio Orchestra; Jean Martinon. Angel S 37064 Tape (c) 4XS 37064.
+HF 4-75 p82 ++SFC 12-8-74 p36
+HF 3-77 p123 tape ++St 4-75 p999

993 Works, selections: Danses sacree et profane. Fantaisie, piano. Rhapsody, clarinet and orchestra. Rhapsody, saxophone and orchestra. Marie-Claire Jamet, hp; Aldo Ciccolini, pno; Guy Dangain, clt; Jean-Marie Londeix, sax; ORTF; Jean Martinon. Angel S 37065 Tape (c) 4XS 37065.
+HF 4-75 p82 +SFC 4-4-75 p22
+HF 3-77 p123 tape +St 8-75 p95
+NR 4-75 p2

994 Works, selections: Khamma. Marche ecossaise. La mer. Nocturnes. Prelude a l'apres-midi d'un faune. Le Roi Lear: Fanfare; Le Sommeil de Lear. ORTF; ORTF Choir; Jean Martinon. Angel S 37067/8 (2) Tape (c) 4XS 37067/8.
+HF 10-75 p74 ++NR 8-75 p5
+HF 3-77 p123 tape ++SFC 8-3-75 p30
+HF 10-77 p129 tape

995 Works, selections: Ballades de Francois Villon (3). En blanc et noir. Sonata, violin and piano. Sonata, violoncello and piano. Syrinx, flute. Gunilla von Bahr, flt; Hans Palsson, Amalie Malling, pno; Arve Tellefsen, vln; Erik Saeden, bar; Frans Helmerson, vlc. BIS LP 28.
+RR 3-77 p78

DEFAYE

Danses (2). cf Boston Brass BB 1001.

Melancolie. cf RCA FRL 1-3504.
Sur un air de Bach. cf RCA FRL 1-3504.
Sur un air de Corelli. cf RCA FRL 1-3504.

DE GROOT, Cor
Cloches dans le matin. cf RCA GL 4-2125.

DELA
Dans tous les cantons, adagio. cf Citadel CT 6013.

DE LARA
Partir c'est mourir un peu. cf Rubini GV 57.

DE LA TORRE, Francisco
Alta. cf London CS 7046.

DEL BORGO
Canto, solo saxophone. cf Coronet LPS 3036.

DE LEEUW, Reinbert

996 Abschied. Humns and chorals. Radio Wind Ensemble; Rotterdam Philharmonic Orchestra; Edo de Waart, David Porcelijn. Donemus CV 7604.
+RR 8-77 p51

Hymns and chorals. cf Abschied.

DELIBES, Leo

997 Coppelia. OSR: Ernest Ansermet. Decca DPA 581/2 (2). Tape (c) KDPC 2-7045. (Reissue from SXL 2084/5) (also London STS 15371. Reissue from CSA 2201)
+Gr 5-77 p1749
-Gr 7-77 p225 tape
+HF 12-77 p87
+-HFN 5-77 p137
+HFN 7-77 p127 tape
+RR 5-77 p48
/-RR 9-77 p98 tape

998 Coppelia. Paris Opera Orchestra; Jean-Baptiste Mari. HMV SQ SLS 5091 (2) Tape (c) TC SLS 5091. (also Angel S 3843 Tape (c) 4XS 2-3843)
++Gr 9-77 p512
+-Gr 12-77 p1145 tape
+-HF 12-77 p87
+HFN 11-77 p169
+RR 11-77 p50

999 Coppelia: Ballet muisc. Sylvia: Ballet music. GOUNOD: Faust: Ballet music. ROSSINI: William Tell: Ballet music. NPhO; Charles Mackerras. Classics for Pleasure CFP 40229 Tape (c) TC CFP 40229.
+Gr 2-77 p1325
+HFN 10-75 p152
+HFN 12-76 p153 tape
+RR 10-75 p49

1000 Coppelia: Prelude et mazurka; Valse lente; Mazurka; Ballade de l'epi; Theme slave varie; Czardas; Valse de la poupee; Fete de la cloche; Galop final. Sylvia: Prelude; Les chasseresses; Intermezzo; Valse lente; Pas de Etheiopiens; March et cortege de Bacchus; Divertissement (a pizzicati); Apparition d'Endymion. OSR, NPhO; Richard Bonynge. Decca SXL 6776. (Reissues from SET 473/4, SXL 6635/6)
+-Gr 10-77 p711
+HFN 10-77 p165
+RR 10-77 p44

1001 Coppelia: Suite. Sylvia: Suite. LPO; Stanley Black. Decca PFS 4358 Tape (c) KPFC 4358. (also London 21147 Tape (c) 5-21147).
+Gr 8-76 p342
+HFN 8-76 p79
-NR 1-77 p4
+RR 8-76 p35

Coppelia: Suite. cf CHOPIN: Les sylphides.

1002 Coppelia: Suite, Act 1. Sylvia: Suite, Act 1. GOUNOD: Faust: Ballet music. ROSSINI: Guglielmo Tell: Ballet. NPhO; Charles Mackerras. Seraphim S 60284.
+NR 8-77 p6

Lakme: The bell song. cf Decca D65D3.
Lakme: The bell song. cf HMV HLM 7066.
Lakme: Dove l'Indiana bruna. cf VERDI: Rigoletto.
Lakme: Dove l'Indiana bruna. cf HMV SLS 5057.
Lakme: Pourquoi dans les grands bois; Bell song. cf Columbia 34294.
Sylvia: Ballet music. cf Coppelia: Ballet music.
Sylvia: Prelude; Les chasseresses; Intermezzo; Valse lente; Pas de Ehteiopiens; March et cortege de Bacchus; Divertissement (a pizzicati); Apparition d'Endymion. cf Coppelia: Prelude et mazurka; Valse lente; Mazurka; Ballade de l'epi; Theme slave varie; Czardas; Valse de la poupee; Fete de la cloche; Galop final.
Sylvia: Suite. cf Coppelia: Suite.
Sylvia: Suite, Act 1. cf Coppelia: Suite, Act 1.

DELISLE, Rouget

1003 Le Marseillaise (arr. Berlioz). MESSAGER: The two pigeons: Entrance of the gipsies; Entrance of Pepio and 2 pigeons pas de deux; Theme and variations; Entry; Hungarian dance; Finale. Isoline: Entr'acte...Pavane; Mazurka; Entrance of the first dancer...Seduction scene; Waltz. LALO: Scherzo. PIERNE: March of the little lead soldiers, op. 14, no. 6. Andrea Guiot, s; Claude Cales, t; Paris Opera Chorus; Orchestre de Paris; Jean-Pierre Jacquillat. HMV ESD 7048 Tape (c) TC ESD 7048. (Some reissues from Columbia TWO 264)
+Gr 10-77 p711 +HFN 11-77 p173
+Gr 12-77 p1145 tape +RR 11-77 p61

DELIUS, Frederick

Aquarelles. cf Works, selections (RCA RL 2-5079).
Brigg Fair. cf Works, selections (World Records SHB 32).

1004 Concerto, violin. Concerto, violin and violoncello. Yehudi Menuhin, vln; Paul Tortelier, vlc; RPO; Meredith Davies. Angel S 37262. (also HMV ASD 3343)
+FF 9-77 p14 +NYT 7-24-77 pD11
+-Gr 6-77 p47 +RR 6-77 p47
+HF 9-77 p100 +SFC 6-5-77 p45
+-HFN 6-77 p121 +SR 9-3-77 p42
+MT 10-77 p825 +-St 10-77 p136
+-NR 9-77 p3

Concerto, violin and violoncello. cf Concerto, violin.
Eventyr. cf Works, selections (World Records SHB 32).

1005 Fennimore and Gerda. Elisabeth Soderstrom, Kirsted Buhl-Moller, Bodil Kongsted, Ingeborg Junghans, s; Hedvig Rummel, ms; Robert Tear, Anthony Rolfe Johnson, Michael Hansen, t; Brian Rayner Cook, Peter Fog, Hans Christian Hansen, bar; Birger Brandt, Mogens Berg, bs; Danish State Radio Symphony Orchestra and Chorus; Meredith Davies. HMV SQ SLS 991 (2). (also Angel (Q) SX 3835)
+-Gr 12-76 p1045 +ON 2-12-77 p41
+HF 3-77 p98 +-RR 12-76 p42
++HFN 1-77 p105 +SR 5-28-77 p42
+-MT 5-77 p399 ++St 3-77 p136
+-NR 1-77 p9

Fennimore and Gerda: Intermezzo. cf Works, selections (RCA RL 2-5079).
Fennimore and Gerda: Intermezzo. cf Works, selections (World Records SHB 32).

Florida suite: La Calinda. cf Works, selections (World Records SHB 32).

Hassan: Intermezzo and serenade. cf Works, selections (HMV TC ASD 2477).

Hassan: Intermezzo and serenade. cf Works, selections (RCA RL 2-5079).

Hassan: Intermezzo, serenade, closing scene. cf Works, selections (World Records SHB 32).

In a summer garden. cf Works, selections (HMV TC ASD 2477).

In a summer garden. cf Works, selections (World Records SHB 32).

Irmelin: Prelude. cf Works, selections (RCA RL 2-5079).

Irmelin: Prelude. cf Works, selections (World Records SHB 32).

Koanga: La Calinda. cf Works, selections (HMV TC ASD 2477).

Koanga: La Calinda; Final scene. cf Works, selections (World Records SHB 32).

1006 A late lark. HERRMANN: The fantasticks. For the fallen. WARLOCK: Motets. Gillian Humphreys, s; Meriel Dickinson, con; John Amis, t; Michael Rippon, bs; Stephen Hicks, org; Thames Chamber Choir; National Philharmonic Orchestra; Bernard Herrmann, Louis Halsey. Unicorn RHS 340.

+Gr 9-76 p454 ++NR 10-77 p2
+-HF 12-76 p120 +-RR 8-76 p75
+-HFN 8-76 p79 ++St 11-76 p166

Late swallows. cf Works, selections (RCA RL 2-5079).

On hearing the first cuckoo in spring. cf Works, selections (HMV TC ASD 2477).

On hearing the first cuckoo in spring. cf Works, selections (RCA RL 2-5079).

On hearing the first cuckoo in spring. cf Works, selections (World Records SHB 32).

On hearing the first cuckoo in spring. cf Prelude PRS 2512.

On hearing the first cuckoo in spring. cf Pye Golden Hour GH 643.

Over the hills and far away. cf Works, selections (World Records SHB 32).

Quartet, strings: 3rd movement (Late swallows). cf Works, selections (HMV TC ASD 2477).

1007 Sonatas, violin (3). Sonata, violoncello. Alexander Kouguell, vlc; David Hancock, pno; Eleanor Hancock, pno. Classical Cassette Company CCC BP 57.

++HF 9-77 p119 tape

Sonata, violoncello. cf Sonatas, violin.

A song before sunrise. cf Works, selections (HMV TC ASD 2477).

A song before sunrise. cf Works, selections (RCA RL 2-5079).

1008 Songs: Midsummer song; On Craig Dhu; The splendour falls on castle walls; To be sung on a summer night on the water. ELGAR: Songs: As torrents in summer; Deep in my soul, op. 53, no. 2; The fountain, op. 71, no. 2; Love's tempest, op. 73, no. 1; My love dwelt in a northern land; Go, song of mine, op. 57; O wild west wind, op. 53, no. 3; Owls, op. 53, no. 4; The shower, op. 71, no. 1; There is sweet music, op. 53, no. 1. Ian Partridge, t; Louis Halsey Singers; Louis Halsey. Argo ZK 23. (Reissue from ZRG 607)

+Gr 10-77 p678 ++RR 10-77 p83
+HFN 10-77 p167

1009 Songs: Midsummer song; On Craig Dhu; The splendour falls on castle walls; To be sung of a summer night on the water. ELGAR: Songs: As torrents in summer; Deep in my soul, op. 53, no. 2; The

fountain, op. 71, no. 2; Go, song of mine, op. 57; Love's tempest, op. 73, no. 1; My love dwelt in a northern land; O wild west wind, op. 53, no. 3; Owls, op. 53, no. 4; The shower, op. 71, no. 1; There is sweet music, op. 53, no. 1. Louis Halsey Singers; Louis Halsey. Argo ZRG 607.
+St 4-77 p80

Songs: Appalachia; Le ciel est pardessus le toit (2); Cradle song; Evening voices; I Brasil; Irmelin Rose (2); Klein Venevil; Love's philosophy; La luna blanche; Mass of life: Prelude, part 2; The nightingale; Sea drift; So white, so soft; To the queen of the heart; The violet (2); Whither. cf Works, selections (World Records SHB 32).

Summer night on the river. cf Works, selections (HMV TC ASD 2477).

Summer night on the river. cf Works, selections (RCA RL 2-5079).

Summer night on the river. cf Works, selections (World Records SHB 32).

Summer night on the river. cf Prelude PRS 2512.

To daffodils. cf Argo ZK 28/9.

A village Romeo and Juliet: The walk to the paradise garden. cf Works, selections (World Records SHB 32).

1010 Works, selections: In a summer garden. Quartet, strings: 3rd movement (Late swallows). On hearing the first cuckoo in spring. A song before sunrise. Summer night on the river. Hassan: Intermezzo and serenade. Koanga: La Calinda. Halle Orchestra; John Barbirolli. HMV Tape (c) TC ASD 2477.
+Gr 8-77 p346 tape +-RR 1-77 p88 tape
+-HFN 4-76 p125 tape

1011 Works, selections: Aquarelles (2) (arr. Fenby). Fennimore and Gerda: Intermezzo (arr. Beecham). Hassan: Intermezzo and serenade (arr. Beecham). Irmelin: Prelude. Late swallows (arr. Fenby). On hearing the first cuckoo in spring. A song before sunrise. Summer night on the river. Bournemouth Sinfonietta; Norman Del Mar. RCA RL 2-5079 Tape (c) RK 2-5079.
+Audio 12-77 p48 ++HFN 8-77 p99
+Gr 6-77 p48 ++RR 6-77 p46
+Gr 8-77 p346 tape +RR 10-77 p96 tape
+HFN 8-77 p79

1012 Works, selections: Brigg Fair. Eventyr. Florida suite: La Calinda. Hassan: Intermezzo, serenade, closing scene. In a summer garden. On hearing the first cuckoo in spring. Over the hills and far away. Summer night on the river. Fennimore and Gerda: Intermezzo. Irmelin: Prelude. Koanga: La Calinda; Final scene. A village Romeo and Juliet: The walk to the paradise garden. Songs: Appalachia; Le ciel est pardessus le toit (2); Cradle song; Evening voices; I Brasil; Irmelin Rose (2); Klein Venevil; Love's philosophy; La luna blanche; Mass of life: Prelude, part 2; The nightingale; Sea drift; So white, so soft; To the queen of the heart; The violet (2); Whither. Dora Labbette, s; Heddle Nash, t; Gerald Moore, pno; LPO, RPO, Symphony Orchestra; Thomas Beecham. World Records SHB 32 (5). (Reissues)
++Gr 11-76 p781 +RR 1-77 p52
+HFN 1-77 p100

DELL'ACQUA

Villanelle. cf BIS LP 45.

Villanelle. cf Columbia 34294.

Villanelle. cf Court Opera Classics CO 342.

DELLO JOIO, Norman
The developing flutist. cf COPLAND: Duo.
A jubilant song. cf Columbia M 34134.
Meditations on Ecclesiastes. cf BARTOK: Concerto, orchestra.
DELMET
Envoi de fleurs. cf Club 99-107.
DEL STAIGER
Napoli. cf Grosvenor GRS 1048.
DEMESSIEUX, Jeanne
1013 Te deum, op. 11. Tryptyque, op. 7. DUPRE: Vision, op. 44. LANGLAIS: Fete. Graham Barber, org. Vista VPS 1032.
+Gr 2-77 p1308 +RR 2-77 p80
+RR 1-77 p71
Tryptyque, op. 7. cf Te deum, op. 11.
DE MONFRED, Avenir
In paradisum. cf Vista VPS 1046.
DE MONZA, Giovanni Battista
Ecclesiastical concerto (realized by J. P. Mathieu). cf Arion ARN 90416.
DENCKE
O, be glad, ye daughters of His people. cf Vox SVBX 5350.
DENISOV, Edison
Sonata, solo clarinet. cf BIS LP 62.
1014 Variations. SHOSTAKOVICH: Sonata, piano, no. 2, op. 61, B minor. SHCHEDRIN: Humoresque. SLONIMSKY: The bells. Lydia Majlingova, pno. Rediffusion Opus 9111 0342.
-Gr 7-77 p202 +-RR 11-77 p87
DENZA, Luigi
Songs: Culto; Occhi di fata. cf Rubini GV 34.
DE RILLE
The maryrs of the arena. cf BBC REC 267.
DESMARETS, Henri
Menuet. cf DG 2723 051.
Passepied. cf DG 2723 051.
DESPORTES
Pastorale joyeuse. cf BIS LP 60.
DESSAU, Paul
1015 Tierverse. EISLER: Legende von der Entstehung des Buches Taoteking; Elegien (3); Holderlin fragments; Zuchthaus-Kantate. LUTOSLAWSKI: Five songs on poems of Kazimiera Illakowicz. Roswitha Trexler, s.
+Te 6-77 p45
DETT, Robert Nathaniel
1016 In the bottoms. GRIFFES: Sonata, piano, F major. IVES: Three-page sonata. Clive Lythgoe, pno. Philips 9500 096.
+-ARG 11-76 p22 +NR 12-76 p13
+Gr 10-76 p627 +RR 10-76 p87
+-HF 12-76 p122 ++SFC 4-10-77 p30
+HFN 10-76 p167 +St 12-76 p144
+-MJ 12-76 p28
DEVIENNE, Francois
1017 Concerto, flute, no. 2, D major. IBERT: Concerto, flute. Peter-Lukas Graf, flt; ECO; Raymond Leppard. Claves P 501. (also HNH 4015)
+Gr 7-76 p168 ++NR 9-77 p3
+HFN 2-76 p95 +RR 6-76 p40
1018 Concerto, flute, no. 2, D major. GLUCK: Concerto, flute, G major.

GRETRY: Concerto, flute, C major. Michel Debost, flt; Toulouse Chamber Orchestra; Louis Auriacombe. Seraphim S 60287.
++NR 11-77 p9

1019 Symphonie concertante, G major. DIETER: Concerto concertant, D major. VIOTTI: Concerto, flute, A major. Jean-Pierre Rampal, Ransom Wilson, flt; I Solisti Veneti; Claudio Scimone. Musical Heritage Society MHS 3371.
+FF 9-77 p67

DIA, Comtesse de
A chantar m'er so qu'eu no volria. cf Telefunken 6-41126.

DIAMOND, David
1020 Quintet, clarinet, 2 violas and 2 violoncellos. HARRIS: Concerto, clarinet, piano and string quartet. Lawrence Sobol, clt; Peter Basquin, pno; Carol Webb, Ira Weller, vln; Louise Schulman, Linda Moss, vla; Timothy Eddy, Fred Sherry, vlc. Grenadilla GS 1007.
++FF 11-77 p27 ++St 11-77 p138
+NR 11-77 p10

Sonata, violoncello and piano. cf BARBER: Sonata, violoncello and piano, op. 6, C minor.

1021 Symphony, no. 4. MENNIN: Symphony, no. 7. NYP; Leonard Bernstein. New World Records NW 258.
+MJ 3-77 p46

DIEMENTE
Diary, Pt II. cf BACH: Prelude, bourree and gigue.

DIEREN, Bernard van
Songs: Dream pedlary; Take, o take those lips away. cf Argo ZK 28/9.

DIETER, Christian
Concerto concertant, D major. cf DEVIENNE: Symphonie concertante, G major.

DIEUPART, Charles
1022 Suite, flute, violin, bass and harpsichord, no. 3. DUVAL: Sonata, violin, viola da gamba and harpsichord. JACQUET DE LA GUERRE: Sonata, violin, bass and harpsichord, D minor. REBEL: Les elements. Pierre Sechet, flt; Frantisek Jaros, vln; Jean Lamy, bs viol; Antoine Geoffroy Dechaume, hpd; ORTF Lyric Orchestra; Andre Jouve. Inedits ORTF 995 039.
++St 2-77 p119

DIJON, Guiot de
Chanterei por mon corage. cf Argo D40D3.

DINICU, Dimitri
Hora staccato. cf Kelsey Records KEL 7601.
Hora staccato (arr. and orch. Gerhardt). cf RCA LRL 1-5094.
LA DIVINA: The art of Maria Callas. cf HMV SLS 5057.

DLUGORAJ, Adalbert
Carola polonesa. cf DG 2533 294.
Chorea polonica. cf Hungaroton SLPX 11721.
Fantasia. cf DG 2533 294.
Fantasia. cf Hungaroton SLPX 11721.
Finale (2). cf DG 2533 294.
Finale. cf Hungaroton SLPX 11721.
Kowaly. cf DG 2533 294.
Vilanella (2). cf DG 2533 294.
Vilanella polonica. cf Hungaroton SLPX 11721.

DODGSON, Stepehn
Concerto, guitar, no. 2. cf ARNOLD: Serenade, guitar and srings.
Sonata, brass. cf ADDISON: Divertimento, op. 9.
DOHNANYI, Ernst von
Pierrette's veil: Waltz. cf Mercury SRI 75098.
Rhapsody, op. 11, no. 3, C major. cf Decca SPA 473.
Ruralia Hungarica, op. 32. cf Discopaedia MB 1012.
1023 Sonata, violin and piano, op. 21, C sharp minor. SIBELIUS: Devotion, op. 77, no. 2. Berceuse, op. 79, no. 6. Sonatina, op. 80, E major. Souvenir, op. 79, no. 1. Diana Steiner, vln; David Berfield, pno. Orion ORS 76244.
+-ARG 4-77 p31 +NR 4-77 p7
+HF 6-77 p86 +St 7-77 p114
Suite, op. 19, F sharp minor. cf Variations on a nursery song, op. 25.
1024 Variations on a nursery song, op. 25. Suite, op. 19, F sharp minor. Bela Siki, pno; Seattle Symphony Orchestra; Milton Katims. Turnabout TVS 34623 Tape (c) KTVC 34623.
+Gr 2-77 p1274 +NR 8-76 p12
+HF 9-76 p90 +RR 2-77 p44
+-HFN 3-77 p101 +St 9-76 p119
+-HFN 4-77 p155 tape
DONATI, Baldassare
In te domine. cf HMV SQ ASD 3393.
DONAUDY
Songs: O del mio amato ben. cf Seraphim S 60280.
DONIZETTI, Gaetano
Anna Bolena: Deh non voler costringere. cf Orion ORS 77271.
Catarina Cornaro: Torna all'ospitetto...Vieni o tu, che ognor io chiamano. cf CATALANI: La Wally: Ebben, ne andro lontana.
La conocchia. cf L'Oiseau-Lyre SOL 345.
Don Pasquale: Arias. cf London OS 26499.
Don Pasquale: Che interminabile andrivieni. cf Seraphim S 60275.
1025 Don Pasquale: So anch'io la virtu magica. MOZART: Die Entfuhrung aus dem Serail, K 384: Ach, ich liebte. The magic flute, K 620: Ach, ich fuhl's. Le nozze di Figaro, K 492: Deh vieni, non tardar. PUCCINI: La boheme: Mi chiamano Mimi. La rondine: Chi il bel sogno di Doretta. Turandot: Tu, che di gel sei cinta. VERDI: La forza del destino: Pace, pace, mio Dio. Rigoletto: Caro nome. Ileana Cotrubas, s; NPhO; John Pritchard. Columbia M 34519. (also CBS 76521)
+Gr 6-77 p102 +NYT 4-24-77 pD27
/HF 7-77 p119 -OR 9/10-77 p31
+-HFN 6-77 p121 +-RR 6-77 p36
+NR 6-77 p8 ++St 7-77 p128
Don Sebastiano: Deserio in terra. cf RCA TVM 1-7201.
Il Duca d'Alba: Angelo casto e bel (completed by Matteo Salvi). cf Philips 9500 203.
Il Duca d'Alba: Angelo casto e bel. cf RCA TVM 1-7201.
1026 L'Elisir d'amore. Ileana Cotrubas, Lilian Watson, s; Placido Domingo, t; Geraint Evans, Ingvar Wixell, bar; ROHO and Chorus; John Pritchard. CBS 79210 (2) Tape (c) 40-79210. (also Columbia M3 34585)
+Gr 11-77 p885 +HFN 12-77 p187 tape
+-Gr 12-77 p1149 tape +RR 11-77 p36
+HFN 11-77 p169 +SFC 12-11-77 p61
1027 L'Elisir d'amore. Joan Sutherland, s; Luciano Pavarotti, t; ECO;

Richard Bonynge. London OSA 13101.
+NYT 10-9-77 pD21

1028 L'Elisir d'amore. Rosanna Carteri, s; Luigi Alva, t; Giuseppe Taddei, Rolando Panerai, bar; La Scala Orchestra; Tullio Serafin. Seraphim S 6001.
++NYT 10-9-77 pD21

L'Elisir d'amore: Arias. cf London OS 26499.

1029 L'Elisir d'amore: Prendi, per me sei libero. La figlia del reggimento: Convien partir. Lucrezia Borgia: Tranquillo ei pose ...Com'e bello. ROSSINI: La cenerentola: Nacqui all'affano. Guglielmo Tell: S'allontanano alfin...Selva opaca. Semiramide: Bel raggio lusinghier. Maria Callas, s; OSCCP; Nicola Rescigno. EMI Italy 3C 065 00592.
+-RR 8-77 p37

L'Elisir d'amore: Quanto e bella. cf RCA TVM 1-7203.

L'Elisir d'amore: Quanto e bella; Una furtiva lagrima. cf Decca SXL 6839.

L'Elisir d'amore: Udite, udite, o rustici. cf Decca ECS 811.

L'Elisir d'amore: Una furtiva lagrima. cf RCA CRM 1-1749.

1030 La favorita. Giulietta Simionato, s; Gianni Poggi, t; Ettore Bastianini, bar; Jerome Hines, bs; Maggio Musicale Fiorentino; Alberto Erede. Richmond 63510.
-NYT 10-9-77 pD21

La favorita: O mio Fernando. cf Club 99-103.

La favorita: O mio Fernando. cf Columbia/Melodiya M 33931.

La favorita: Quanto le soglie...Ah l'altro ardor, Fernando Fernando...Che fino al ciel. cf Club 99-106.

La favorita: Spirto gentil. cf Club 99-105.

1031 La fille du regiment. Joan Sutherland, s; Monica Sinclair, Edith Coates, con; Alan Jones, Luciano Pavarotti, t; Eric Garrett, bar; Jules Bruyere, Spiro Malas, bs; ROHO and Chorus; Richard Bonynge. Decca Tape (c) K23K22 (2).
+Gr 11-77 p900 tape
+HFN 12-77 p187 tape
+RR 12-77 p94 tape

La fille du regiment: Chacun le sait. cf Decca D65D3.

La fille du regiment: Chacun le sait, chacun le dit. cf HMV MLF 118.

La figlia del reggimento: Convien partir. cf L'Elisir d'amore: Prendi, per me sei libero.

La fille du regiment: Il faut partir. cf Columbia 34294.

La fille du regiment: March. cf Decca SB 713.

La fille du regiment: Pour me rapprocher de Marie. cf Decca SXL 6839.

1032 Gemma di Vergy. Montserrat Caballe, s; Natalya Chudy, ms; Luis Lima, t; Louis Quilico, bar; Paul Plishka, Mark Munkittrick, bs; Schola Cantorum; New York Opera Orchestra; Eve Queler. Columbia 34575 (3). (also CBS 79303)
+-ARG 5-77 p16
+-Gr 6-77 p95
+-HF 6-77 p86
+-HFN 7-77 p111
+-MJ 5-77 p68
-MT 11-77 p921
++NR 5-77 p11
+-NYT 4-17-77 pD17
+-ON 3-26-77 p32
+-RR 7-77 p38
-SFC 2-20-77 p39
+-St 7-77 p114

Linda di Chamounix: Linda, Linda...Da quel di che t'incontrai. cf BELLINI: La sonnambula: Perdona o mia diletta...Prendi l'anel ti dono.

Linda di Chamounix: O luce di quest'anima. cf Rubini GV 26.

1033 Lucia di Lammermoor. Maria Callas, s; Anna Maria Canali, ms; Giuseppe di Stefano, Vallano Natali, Gino Sarri, t; Tito Gobbi, bar; Rafaele Arie, bs; Maggio Musical, Florence, 1953, Orchestra and Chorus; Tullio Serafin. HMV SLS 5056 (2) Tape (c) TC SLS 5056. (Reissue from Columbia 33CX 1131/2)

+-Gr 3-77 p1444 +-HFN 4-77 p151
+-Gr 8-77 p346 tape +RR 3-77 p34

1034 Lucia di Lammermoor. Joan Sutherland, s; Huguette Tourangeau, ms; Luciano Pavarotti, Ryland Davies, t; Sherrill Milnes, bar; Nicolai Ghiaurov, bs; ROHO and Chorus; Richard Bonynge. London OSA 13103 (3) Tape (c) D 31210 (r) 90210. (also Decca SET 258/30 Tape (c) K2L22)

+ARG 10-72 p680 ++ON 11-72 p43
++Gr 5-72 p1928 +-Op 7-72 p636
+Gr 8-76 p341 tape +-RR 8-76 p82 tape
+Gr 8-77 p345 tape ++SFC 8-6-72 p31
+-HF 9-72 p78 +SR 8-12-72 p38
++HFN 5-72 p920 ++St 10-72 p81
+HFN 8-76 p94 tape +STL 6-11-72 p38
++NR 8-72 p12

1035 Lucia di Lammermoor. Montserrat Caballe, s; Ann Murray, ms; Jose Carreras, Claes Ahnsjo, Vicenzo Bello, t; Vicenzo Sardinero, bar; Samuel Ramey, bs; Ambrosian Singers; NPhO; Jesus Lopez-Cobos. Philips 6703 080 (3).

+Gr 9-77 p470 ++RR 9-77 p36
+HFN 9-77 p140 ++SFC 11-13-77 p50
+NYT 12-11-77 pD17

1036 Lucia di Lammermoor, excerpts. PUCCINI: Tosca, excerpts. ROSSINI: Elisabetta, Regina d'Inghilterra, excerpts. VERDI: Il Corsaro, exerpts. I due Foscari, excerpts. Jose Carreras, t; Various orchestras, choruses and conductors. Philips 6598 533.

+SFC 10-23-77 p45

Lucia di Lammermoor: Arias. cf London OS 26499.

Lucia di Lammermoor: Il dolce suono...Ardon gl'incensi. cf BELLINI: I puritani: Qui la voce...Vien diletto.

Lucia di Lammermoor: Mad scene. cf HMV RLS 719.

Lucia di Lammermoor: Oh giusto cielo; Ardon gl'incensi. cf HMV SLS 5057.

Lucia di Lammermoor: Per poco fra le tenebre. cf Seraphim S 60275.

Lucia di Lammermoor: Regnava nel silenzio. cf Decca D65D3.

Lucia di Lammermoor: Regnava nel silenzio. cf HMV SLS 5104.

Lucia di Lammermoor: Regnava nel silenzio...Quando rapita in estase; Mad scene. cf Columbia 34294.

Lucia di Lammermoor: Tombe degli ave...Fra poco a me ricovero. cf Club 99-105.

Lucia di Lammermoor: Tombe degli avi miei; Tu che a Dio spiegasti l'all. cf RCA RVM 1-7203.

Lucia di Lammermoor: Tu che a Dio spiegasti l'ali. cf Decca SXL 6839.

Lucia di Lammermoor: Tu che a Dio. cf Rubini GV 29.

Lucia paraphrase. cf Musical Heritage Society MHS 3611.

Lucrezia Borgia: Com'e bello. cf CATALANI: La Wally: Ebben, ne andro lontana.

Lucrezia Borgia: Tranquillo ei pose...Com'e bello. cf L'Elisir d'amore: Prendi, Per me sei libero.

Maria di Rohan: Alma soave e cara. cf Philips 9500 203.

Maria Stuarda: Ah, rimiro il bel sembiante. cf Decca SXL 6839.
Maria Stuarda: O nube, che lieve. cf CATALANI: La Wally: Ebben, ne andro lontana.
1037 Quartet, strings, D major (arr. for string orchestra). ROSSINI: Sonatas, strings, nos. 1-6. AMF: Neville Marriner. Argo ZK 26/7 (2). (Reissue from ZRG 603, 506)
+Gr 12-77 p1079 ++RR 12-77 p60
Roberto Devereux: E Sara in questi orribili momenti...Vivi, ingrato, a lei d'accanto...Quel sangue versato. cf CATALANI: La Wally: Ebben, ne andro lontana.
Songs: La corrispondenza amorosa; A mezzanotte; La mere et l'enfant. cf BELLINI: Songs (Cetra LPO 2003).
Songs: A mezzanotte; Me voglio fa na casa; La zingara. cf BELLINI: Songs (Westminster WG 1014).

DONJON, Johannes
Adagio nobile. cf Golden Crest RE 7064.
Le chant du vent. cf Golden Crest RE 7064.
Offertoire. cf Golden Crest RE 7064.
Pan. cf Golden Crest RE 7064.
Pipeaux. cf Golden Crest RE 7064.
Rossignolet. cf Golden Crest RE 7064.

DOPPLER, Albert Franz
Fantaisie pastorale hongroise, op. 26. cf RCA JRL 1-2315.
Fantaisie pastorale hongroise, op. 26. cf RCA LRL 1-5094.
Valse di bravura. cf Pearl SHE 533.

DOROW, Dorothy
Songs: Dream; Pastourelles, pastoureux. cf BIS LP 45.

DOSTAL, Nico
Clivia: Ich bin verliebt. cf HMV ESD 7043.
Die Ungarische Hochzeit: Spiel mir das Lied von Gluck und Treu. cf HMV ESD 7043.

DOWLAND, John
Can she excuse. cf RCA RL 2-5110.
Captain Digorie piper, his galliard. cf Saga 5425.
Captain Digorie piper's galliard. cf Works, selections (RCA ARL 1-1491).
Captain Digorie piper's galliard. cf CRD CRD 1019.
Dances (5). cf HMV SQ CSD 3781.
Earl of Essex, his galliard. cf CBS Tape (c) 40-72728.
A fancy (2). cf Works, selections (RCA ARL 1-1491).
A fancy. cf L'Oiseau-Lyre DSLO 510.
Fantasia. cf Saga 5438.
1038 Fantasias and dances for the lute. Joseph Bacon, lt. 1750 Arch S 1764.
+NR 11-77 p15 +SFC 11-6-77 p48
Farewell. cf Works, selections (RCA ARL 1-1491).
Fine knacks for ladies. cf Enigma VAR 1017.
1039 First booke of songes, 1597. Consort of Musicke; Anthony Rooley. L'Oiseau-Lyre DSLO 508/9 (2).
+Gr 11-76 p851 +-NR 8-77 p11
+HFN 11-76 p157 +RR 10-76 p20
+MJ 7-77 p70 ++SFC 11-6-77 p48
+-MM 8-77 p47 +St 6-77 p127
+MT 3-77 p215
Forlorn hope fancy. cf Works, selections (RCA ARL 1-1491).
Frogg galliard. cf L'Oiseau-Lyre DSLO 510.
Galliard to lachrimae. cf Works, selections (RCA ARL 1-1491).

The King of Denmark's galliard. cf CRD CRD 1019.
1040 Lachrimae 1604. Consort of Musicke; Anthony Rooley. L'Oiseau-Lyre DSLO 517.
++Gr 11-76 p851 +MT 3-77 p215
+HFN 10-76 p169 +RR 10-76 p20
+MJ 7-77 p70 +SFC 11-6-77 p48
+-MM 8-77 p47 ++St 9-77 p128
Lachrimae: Antiquae pavan. cf Philips 6500 926.
Lachrimae: Antiquae pavan. cf Saga 5425.
Lachrimae pavan. cf L'Oiseau-Lyre DSLO 510.
Melancholy galliard. cf Westminster WG 1012.
Mr. George Whitehead his almand. cf DG 2533 323.
Mr. Langton's galliard. cf Works, selections (RCA ARL 1-1491).
Mrs. Vaux's gigue. cf Westminster WG 1012.
Mrs. Winter's jump. cf DG 2723 051.
My Lord Chamberlain, his galliard. cf Works, selections (RCA ARL 1-1491).
My Lord Willoughby's welcome home. cf DG 2533 323.
My Lord Willoughby's welcome home. cf Works, selections (RCA ARL 1-1491).
Piper's pavan. cf Works, selections (RCA ARL 1-1491).
Queen Elizabeth, her galliard. cf CBS Tape (c) 40-72728.
Queen Elizabeth, her galliard. cf DG 2723 051.
Resolution. cf Works, selections (RCA ARL 1-1491).
Resolution. cf BIS LP 22.
1041 Second Booke of songs, 1600. Consort of Musicke; Anthony Rooley. L'Oiseau-Lyre DSLO 528/9 (2).
+Gr 9-77 p467 ++RR 8-77 p75
+HFN 9-77 p140
Semper Dowland, semper dolens. cf Westminster WG 1012.
Sir John Souche's galliard. cf Works, selections (RCA ARL 1-1491).
Songs: Can she excuse my wrongs; Sorrow, stay. cf Saga 5447.
Songs: Can she excuse my wrongs; Come heavy sleep; Come away, come sweet love; In darkness let me dwell; Say love if ever thou didst find; Sweet stay awhile. cf Enigma VAR 1023.
Songs: In darkness let me dwell; Lady if you so spite me. cf Abbey LPB 712.
1042 Works, selections: Captain Digorie piper's galliard. A fancy (2). Forlorn hope fancy. Farewell. Galliard to lachrimae. Mr. Langton's galliard. My Lord Chamberlain, his galliard. My Lord Willoughby's welcome home. Piper's pavan. Resolution. Sir John Souche's galliard. Julian Bream, lt. RCA ARL 1-1491.
+Gr 1-77 p1161 +NR 12-76 p15
+HF 7-77 p125 tape ++RR 2-77 p79
++HFN 4-77 p135 +SFC 11-6-77 p48
+MM 8-77 p47 ++St 8-77 p59
+MT 9-77 p735

DOWNEY, John
1043 Adagio lyrico. A dolphin. Octet, winds. What if. Daniel Nelson, t; John Downey, Anthony and Joseph Paratore, pno; University of Wisconsin Concert Choir; Instrumental Quartet, Wind Octet; Stanley DeRusha. Orion ORS 77267.
++NR 11-77 p9
A dolphin. cf Adagio lyrico.
Octet, winds. cf Adagio lyrico.
What if. cf Adagio, lyrico.

DRAGHI, Giovanni
Trio sonata, G minor. cf Telefunken AW 6-42129.
DRDLA, Franz
Souvenir. cf Angel (Q) S 37304.
Souvenir. cf Discopaedia MG 1012.
DRIGO, Riccardo
Les millions d'Arlequin: Serenade. cf Grosvenor GRS 1043.
Les millions d'Arlequin: Serenade (arr. and orch. Gamley. cf RCA LRL 1-5094.
Serenade. cf Angel (Q) S 37304.
DRUCKMAN, Jacob
Delizie contente che l'alme beate, woodwind quintet and tape. cf Vox SVBX 5307.
DRUZECKY, Jiri
1044 Parthia, no. 3, D sharp major. JIROVEC: Parthia, D sharp major. MASEK: Concerto, 2 harpsichord and wind octet, D major. VRANICKY: Marches in French style (3). Collegium Musicum, Prague; Frantisek Vajnar. Supraphon 111 1839.
++FF 9-77 p59 +HFN 9-77 p140
+Gr 8-77 p52 +NR 11-77 p10
1045 Partita, no. 4, E flat major. HUMMEL: Septet, op. 114, C major. Partita, E flat major. Bratislava Chamber Harmony; Justus Pavlik. Opus 9111 0409.
+-Gr 5-77 p1695 +RR 3-77 p71
+-HFN 5-77 p121
DUBOIS
Toccata. cf Argo ZRG 864.
DUBOIS, Pierre-Max
1046 Concerto, flute. IBERT: Concerto, flute. MARTIN: Ballade, flute, piano and string orchestra. Louise di Tullio, flt; ECO; Elgar Howarth. Crystal S 503.
++FF 9-77 p62 ++SFC 6-19-77 p46
+-HF 7-77 p119 ++St 7-77 p118
+NR 6-77 p5
DUCHAMP, Marcel
The bride stripped bare by her bachelors, even. cf CAGE: 27'10.554", percussion.
Erratum musical. cf CAGE: 27'10.554", percussion.
DUFAY, Guillaume
La belle se siet. cf HMV SLS 863.
Ce moys de may. cf HMV SLS 863.
Donnes l'assault. cf HMV SLS 863.
1047 Gloria ad modum tubae. Se la face ay pale, chanson. Se la face ay pale, keyboard (2 versions). Se la face ay pale, four-part instrumental version. Se la face ay pale, mass. Early Music Consort; David Munrow. HMV CSD 3751. (also Seraphim S 60267)
++Gr 5-74 p2049 +NYT 8-15-76 pD15
++HF 2-77 p110 +RR 7-74 p74
+NR 11-76 p10 ++St 11-76 p138
Gloria ad modern tubae. cf CBS 76534.
Helas mon dueil. cf HMV SLS 863.
Lamentatio Sanctae matris ecclesiae. cf HMV SLS 863.
Medieval dances from the dance book of Margaret of Austria. cf Missa sine nomine.
1048 Missa sine nomine. Medieval dances from the dance book of Margaret of Austria. Clemencic Consort; Rene Clemencic. Musical Heritage Society MHS 3496. (Reissue from Harmonia Mundi)
+MU 10-77 p8

Navre je suis. cf HMV SLS 863.
Par droit je puis. cf HMV SLS 863.
Pasce tuos. cf Argo ZRG 851.
Se la face ay pale, chanson. cf Gloria ad modum tubae.
Se la face ay pale, four-part instrumental version. cf Gloria ad modum tubae.
Se la face ay pale, keyboard (2 versions). cf Gloria ad modum tubae.
Se la face ay pale, mass. cf Gloria ad modum tubae.

1049 Songs: Alons ent, Resvelons vous; Ce jour de l'an; Les douleurs; En triumphant; Entre vous gentils amoureux; Dona i ardenti ray; Helas mon dueil; J'ay mis mon cuer; Je ne suy plus; Je sui povere de leesse; Malheureulx cueur; Ma tres douce, Tant que mon argent, Je vous pri; Puisque vous estez campieur; Mon bien m'amour; Par le regart; Resveillies vous. Musica Mundana; David Fallows. 1750 Arch Records 1751.
+ARG 9-77 p28 +St 1-76 p103

Songs: C'est bien raison de devoir essaucier; Je me complains piteusement; Invidia nimica; Malheureulx cuer que veux to faire; Par droit je puis bien complaindre. cf 1750 Arch S 1753.
Vergine bella. cf HMV SLS 863.
Vergine bella. cf Nonesuch H 71326.

DUKAS, Paul

Ariane et Barbe-bleue: Introduction, Act 3. cf Symphony, C major.

1050 La peri. ROUSSEL: Symphony, no. 3, op. 42, G minor. NYP; Pierre Boulez. Columbia (Q) M 34201 Tape (c) MT 34201. (also CBS 76519 Tape (c) 40-76519)
+-Gr 1-77 p1140 ++MT 8-77 p650
+-Gr 5-77 p1738 tape ++NR 1-77 p2
+-HF 6-77 p88 +RR 1-77 p53
+-HFN 1-77 p105 +RR 6-77 p96 tape
+-HFN 4-77 p155 tape ++SFC 12-19-76 p50
+MJ 4-77 p50 ++St 3-77 p128

1051 La peri: Fanfare. Polyeucte. The sorcerer's apprentice. CPhO; Antonio de Almeida. Supraphon 110 1560.
+Gr 7-76 p167 +NR 4-76 p3
+-HF 6-77 p88 +-RR 4-76 p40

La peri: Fanfare. cf Seraphim SIB 6094.
Polyeucte. cf La peri: Fanfare.
The sorcerer's apprentice. cf La peri: Fanfare.
The sorcerer's apprentice. cf DG 2584 004.
The sorcerer's apprentice. cf Philips 6747 327.

1052 Symphony, C major. Ariane et Barbe-bleue: Introduction, Act 3. ORTF; Jean Martinon. Connoisseur Society CS 2134.
+ARG 12-77 p28 ++NR 9-77 p2
/MJ 9-77 p35 +St 10-77 p36

Variations, interlude and finale on a theme by Rameau. cf Vox SVBX 5483.
Villanelle. cf Musical Heritage Society MHS 3547.

DUKE, John

Songs: Luke Havergal; Miniver Cheevy; Richard Cory. cf New World Records NW 243.

DUMITRESCU, Gheorghe

1053 Rascoala, excerpts. Valentin Loghin, bs; Silvia Voinea, s; Cornel Rusu, Dorin Teodorescu, soloists; Orchestra de Studio si Corul Radiotelevisioni; Carol Litvin. Electrecord STM 7CE 01275.
-FF 11-77 p18

DUNKLER
Dutch Grenadiers march. cf Decca SB 713.
DUNSTABLE, John
Songs: O rosa bella. cf BIS LP 3.
Songs: O rosa bella; Hastu mir. cf L'Oiseau-Lyre 12BB 203/6.
DUPARC, Henri
1054 Songs: Au pays ou se fait la guerre; Elegie; Extase; Chanson triste; L'invitation au voyage; Lamento; Le manoir de Rosamonde; Phydile; Serenade florentine; Soupir; Testament; La vague et la cloche; La vie anterieure. Danielle Galland, s; Bernard Kruysen, bar; Noel Lee, pno. Telefunken AS 6-42113.
+Gr 5-77 p1718 +RR 5-77 p80
Songs: Chanson triste. cf HMV RLS 719.
Songs: Chanson triste. cf Odyssey Y 33130.
Songs: Chanson triste; Extase; L'invitation au voyage; Phidyle. cf HMV RLS 716.
Songs: Extase. cf Decca PFS 4351.
Songs: L'invitation au voyage. cf Columbia 34294.
DUPHLY, Jacques
Allemande courante. cf Harpsichord works (ABC AB 67018).
Chaconne. cf Harpsichord works (ABC AB 67018).
La damanzy. cf Harpsichord works (ABC AB 67018).
La felix. cf Harpsichord works (ABC AB 67018).
La forqueray. cf Harpsichord works (ABC AB 67018).
Les graces. cf Harpsichord works (ABC AB 67018).
1055 Harpsichord works: Allemande courante. La damanzy. Chaconne. La felix. La forqueray. Les graces. La de belomere. Menuets. La pothouin. Gustav Leonhardt, hpd. ABC AB 67018.
++FF 11-77 p18 +SFC 9-11-77 p42
+NR 10-77 p14
La de belomere. cf Harpsichord works (ABC AB 67018).
Menuets. cf Harpsichord works (ABC AB 67018).
La pothouin. cf Harpsichord works (ABC AB 67018).
DUPONT
Les boeufs; Les sapins. cf Club 99-107.
DUPRE, Marcel
Chorale et fugue. cf Organ works (Advent 5014).
Cortege et litanie, op. 19. cf Organ works (Advent 5014).
Elevations, op. 32, no. 1. cf Organ works (Advent 5014).
Fileuse. cf Argo ZRG 864.
Fileuse. cf Vista VPS 1029.
Magnificat, op. 18, no. 10. cf RC Records RCR 101.
1056 Organ works: Cortege et litanie, op. 19. Chorale et fugue. Elevations, op. 32, no. 1. Pieces, op. 18, no. 3. Pieces, op. 27: Final. Prelude and fugue, G minor. Triptyque, op. 51: Musette. Michael Murray, org. Advent 5014.
+MU 3-77 p13
Pieces, op. 18, no. 3. cf Organ works (Advent 5014).
Pieces, op. 27: Final. cf Organ works (Advent 5014).
1057 Preludes and fugues, op. 36. MESSIAEN: Banquet celeste. Les Bergers. Marcel Dupre, org. Mercury SRI 75088.
+MJ 5-77 p32
Prelude and fugue, F minor. cf Calliope CAL 1922/4.
Prelude and fugue, G minor. cf Organ works (Advent 5014).
Prelude and fugue, op. 7, no. 1, B major. cf Vista VPS 1037.
Prelude and fugue, op. 7, no. 2, F minor. cf ALAIN: Variations sur un theme de Jannequin, op. 78.

Symphony, organ, no. 2, op. 26. cf CHARPENTIER: L'ange a la trompette.
Triptyque, op. 51: Musette. cf Organ works (Advent 5014).
Variations sur un Noel, op. 20. cf CHARPENTIER: L'ange a la trompette.
Visions, op. 44. cf DEMESSIEUX: Te deum, op. 11.

DURANTE, Francisco
Concerto, strings, no. 5, A major. cf Supraphon 110 1890.
Danza, danza fanciulla. cf HNH 4008.

DURKO, Zsolt
1058 Burial prayer. Chamber music. Erzsebet Tusa, Istvan Lantos, pno; Attila Fulop, t; Endre Uto, bs; Hungarian Radio and Television Chorus; Budapest Symphony Orchestra; Gyorgy Lehel. Hungaroton SLPX 11803.
+ARG 4-77 p20 +RR 12-76 p91
+NR 10-76 p7
Chamber music. cf Burial prayer.

DURUFLE, Maurice
Danses, op. 6: Danse lente. cf Requiem, op. 9.
1059 Requiem, op. 9. Danses, op. 6: Danse lente. Kiri Te Kanawa, s; Siegmund Nimsgern, bar; Ambrosian Singers; Desborough School Choir; NPhO; Andrew Davis. CBS 76633.
+Gr 10-77 p678 +-RR 10-77 p83
+HFN 11-77 p169
1060 Requiem, op. 9. Daniele Charpentier, ms; Pierre d'Hollander, bar; Gerard Letellier, org; Maitrise d'Enfants de la Resurrection; Manecanterie des Petits Chanteurs de Sainte Marie d'Antony; Chamber Orchestra; Paul Kuentz. RCA FRL 1-0153.
+Gr 4-77 p1582 +RR 7-77 p85
+HFN 8-77 p79
1061 Suite, op. 5: Toccata, B minor. FRANCK: Chorale, no. 2, B minor. LANGLAIS: Suite breve. George McPhee, org. Lismor LILP 5068.
+RR 8-77 p71
Suite, op. 5: Toccata, B minor. cf Vista VPS 1029.

DUSSEK, Johann Ladislaus (also DUSIK or DESSEK)
Andante, F major. cf Rediffusion Royale 2015.
La chasse. cf International Piano Library IPL 102.
My lodging is on the cold, cold ground. cf Serenus SRS 12067.
1062 Sonata, piano, op. 35, no. 1, B flat major. Sonata, piano, op. 77, F minor. Frederick Marvin, pno. Genesis GS 1068.
+ARG 5-77 p19 +-NR 10-77 p13
+-HF 9-77 p100 +-St 5-77 p106
1063 Sonata, piano, op. 35, no. 3, C minor. Sonata, piano, op. 61, F sharp minor. Frederick Marvin, pno. Genesis GS 1069.
+ARG 5-77 p19 +-NR 10-77 p13
+-HF 9-77 p100 +-St 5-77 p106
Sonata, piano, op. 61, F sharp minor. cf Sonata, piano, op. 35, no. 3, C minor.
Sonata, piano, op. 77, F minor. cf Sonata, piano, op. 35, no. 1, B flat major.

DUTILLEUX, Henri
1064 Sonatine, flute and piano. MULLER-ZURICH: Capriccio. QUANTZ: Sonata, flute, op. 1, no. 1, A minor. Sonata, flute, op. 1, no. 2, B flat major. Anne Diener Giles, flt; Allen Giles hpd and pno. Crystal S 312.
+ARG 11-76 p35 +IN 5-77 p28
++Audio 2-77 p80 ++NR 11-76 p7
+-HF 6-77 p102

DUVAL, Francois
Sonata, violin, viola da gamba and harpsichord. cf DIEUPART: Sonata, flute, violin, bass and harpsichord, no. 3.
DVORAK, Antonin
1065 The American flag, op. 102. American suite, op. 98, A major. Joseph Evans, t; Barry McDaniel, bar; St. Hedwigs Cathedral Choir, RIAS Chamber Choir; Berlin Radio Symphony Orchestra; Michael Tilson Thomas. Columbia M 34513. (also CBS 76510)
+Gr 2-77 p1308 +-RR 2-77 p84
+HF 7-77 p101 +SFC 4-3-77 p43
+HFN 2-77 p117 +SR 5-28-77 p42
+NR 7-77 p4 +-St 8-77 p111
American suite, op. 98, A major. cf The American flag, op. 102.
1066 Bagatelles, op. 47 (5). Quintet, piano and strings, op. 81, A major. Rudolf Firkusny, pno, harmonium; Juilliard Quartet. Columbia M 34515. (also CBS 76619)
+-Audio 11-77 p131 +NR 9-77 p8
+Gr 8-77 p314 +-RR 8-77 p66
+-HF 10-77 p106 +-SFC 9-4-77 p38
+HFN 8-77 p79 +St 9-77 p126
1067 Bagatelles, op. 47. Quintet, strings, no. 3, op. 97, E flat major. Vienna Philharmonia Quintet; Peter Planyavsky, harmonium. Decca SDD 487 Tape (c) KSDC 487.
+Gr 3-77 p1421 +-HFN 4-77 p155 tape
+Gr 6-77 p105 tape +MT 10-77 p826
+HFN 3-77 p101 +-RR 3-77 p70
Carnival overture, op. 92. cf Symphony, no. 9, op. 95, E minor.
1068 Concerto, piano, op. 33, G minor. Bruno Rigutto, pno; French National Radio Orchestra; Zdenek Macal. French Decca 7352.
+-Gr 7-77 p173 +RR 10-76 p49
1069 Concerto, piano, op. 33, G minor. Sviatoslav Richter, pno; Bavarian State Orchestra; Carlos Kleiber. HMV SQ ASD 3371. (also Angel S 37239)
+Gr 9-77 p424 ++RR 10-77 p44
+HF 12-77 p73 ++SFC 9-4-77 p38
+HFN 9-77 p141 +St 12-77 p140
+NR 10-77 p5
1070 Concerto, piano, op. 33, G minor. Sviatoslav Richter, pno; LSO; Kyril Kondrashin. Rococo 2118.
+-HF 12-77 p73 ++NR 8-76 p5
1071 Concerto, piano, op. 33, G minor. Rudolf Kirkusny, pno; St. Louis Symphony Orchestra; Walter Susskind. Vox Tape (c) CT 2145.
+HF 10-77 p129 tape
1072 Concerto, violin, op. 53, A minor. Romance, op. 11, F minor. Isaac Stern, vln; PO; Eugene Ormandy. CBS 61332 Tape (c) 40-61332. (Reissue from SBRG 72457)
+Gr 9-77 p424 +HFN 10-77 p169 tape
+HFN 9-77 p151 +-RR 9-77 p58
1073 Concerto, violoncello, op. 104, B minor. HAYDN: Concerto, violoncello, op. 101, D major. Emanuel Feuermann, vlc; Berlin State Opera Orchestra; Orchestra; Malcolm Sargent. Bruno Walter Society IGI 385.
+-NR 8-77 p6
1074 Converto, violoncello, op. 104, B minor. Gregor Piatigorsky, vlc; PO; Eugene Ormandy. Odyssey Y 34602. (Reissue from Columbia ML 4022)
+-ARG 6-77 p22

1075 Concerto, violoncello, op. 104, B minor. Silent woods, op. 68. Rondo, op. 94, G minor. Maurice Gendron, vlc; LPO; Bernard Haitink. Philips 6580 149 Tape (c) 7317 162. (Reissue from SAL 3675)
+Gr 4-77 p1547 +HFN 7-77 p127
++Gr 8-77 p346 tape +RR 4-77 p42
-HFN 4-77 p151 ++RR 10-77 p96
1076 Concerto, violoncello, op. 104, B minor. Lynn Harrell, vlc; LSO; James Levine. RCA ARL 1-1155 Tape (c) ARK 1-1155 (ct) ARS 1-1155 (Q) ARD 1-1155 Tape (c) ART 1-1155 (r) ERQ 1-1155. (also RK 11713)
+-Audio 4-76 p89 -HFN 7-76 p104 tape
++Gr 2-76 p1337 ++MJ 4-76 p31
-Gr 7-76 p230 tape ++NR 11-75 p7
++HF 1-76 p111 tape +-RR 2-76 p29
+-HF 2-76 p94 -RR 6-76 p86 tape
+-HF 2-76 p94 Quad +SR 11-1-75 p46
+HF 8-77 p96 Quad tape ++St 1-76 p103
+-HFN 2-76 p97
Cypresses. cf Quintet, strings, no. 3, op. 97, E flat major.
Cypresses. cf Works, selections (DG 2740 177).
1077 The golden spinning wheel, op. 109. The wood dove, op. 110. Bavarian Radio Symphony Orchestra; Rafael Kubelik. DG 2530 713.
+Gr 11-76 p782 +MT 2-77 p133
+HF 6-77 p88 +NR 5-77 p2
+HFN 11-76 p157 +RR 11-76 p63
+MJ 9-77 p35 ++St 7-77 p115
1078 The golden spinning wheel, op. 109. JANACEK: Taras bulba. Ostrava Janacek Philharmonic Orchestra; Otakar Trhlik. Supraphon 110 1889.
+-FF 11-77 p29 +NR 12-77 p2
+-Gr 9-77 p424 +RR 8-77 p56
Humoresque. cf CBS 61039.
Humoresque, op. 101, no. 7, G flat major. cf Amberlee ALM 602.
Humoresque, op. 101, no. 7, G flat major. cf Discopaedia MB 1015.
1079 In nature's realm overture, op. 91. My home overture, op. 62. SMETANA: Ma Vlast: Vltava; From Bohemia's woods and fields. CPhO; Karel Ancerl. Supraphon 110 1589. (Reissues from 50432, 50521/2)
/Gr 7-77 p226 +NR 9-75 p4
+-HFN 10-75 p141 +RR 9-75 p47
1080 Legends, op. 59. ECO; Rafael Kubelik. DG 2530 786.
+Gr 8-77 p296 +-RR 8-77 p52
+HFN 8-77 p79 ++SFC 11-20-77 p54
Legends, op. 59, nos. 4, 6-7. cf Symphonies, nos. 7-9.
My home overture, op. 62. cf In nature's realm overture, op. 91.
Nocturne, op. 40, B major. cf HMV SQ ESD 7001.
1081 The noonday witch, op. 108. Symphonic variations, op. 78. The water goblin, op. 107. Bavarian Radio Symphony Orchestra; Rafael Kubelik. DG 2530 712.
+-ARG 12-77 p28 +NR 12-77 p2
+Gr 11-76 p782 +RR 11-76 p63
+HFN 11-76 p157 ++SFC 9-4-77 p38
+-MT 2-77 p133 +St 12-77 p144
1082 Quartets, strings. Prague Quartet. DG 2740 177.
+HFN 12-77 p169
Quartets, strings, nos. 1-14. cf Works, selections (DG 2740 177)

Quartet, strings, no. 8, op. 80 (27), E major. cf Quartet, strings, no. 10, op. 51, E flat major.

1083 Quartet, strings, no. 10, op. 51, E flat major. Quartet, strings, no. 8, op. 80 (27), E major. Prague Quartet. DG 2530 719.
+Gr 11-76 p829 +-NR 9-77 p8
+HF 11-77 p110 +RR 10-76 p76
+HFN 10-76 p169 ++St 10-77 p88

1084 Quartet, strings, no. 11, op. 61, C major. Quartet, strings, no. 12, op. 96, F major. Talich Quartet. Calliope CAL 1617.
+HFN 6-77 p121 +-RR 6-77 p73

1085 Quartet, strings, no. 12, op. 96, F major. SMETANA: Quartet, strings, no. 1, E minor. Juilliard Quartet. Columbia MS 7144. (also CBS 61615. Reissue from 72719)
+-Gr 4-77 p1568 +-RR 3-77 p70
-HFN 3-77 p117 ++St 5-77 p73

1086 Quartet, strings, no. 12, op. 96, F major. Quartet, strings, no. 14, op. 105, A flat major. Prague Quartet. DG 2530 632.
+Gr 4-76 p1623 ++NR 1-77 p4
+-HF 12-76 p100 +RR 4-76 p61
++HFN 4-76 p103 ++St 2-77 p117

1087 Quartet, strings, no. 12, op. 96, F major. Quintet, strings, no. 3, op. 97, E flat major. Guarneri Quartet; Walter Trampler, vla. RCA ARL 1-1791 Tape (c) ARK 1-1791 (ct) ARS 1-1791.
+ARG 11-76 p20 +MJ 12-76 p44
+-HF 12-76 p100 ++NR 11-76 p6
+-HF 5-77 p101 tape +St 2-77 p117

1088 Quartet, strings, no. 12, op. 96, F major. HAYDN: Quartet, strings, op. 64, no. 5, D major. SCHUBERT: Quartet, strings, no. 12, D 703, C minor. Panocha Quartet. Supraphon 111 1683.
+Gr 7-77 p197 -NR 9-77 p8
-HFN 7-77 p112 +RR 6-77 p73

Quartet, strings, no. 12, op. 96, F major. cf Quartet, strings, no. 11, op. 61, C major.

Quartet, strings, no. 12, op. 96, F major. cf BORODIN: Quartet, strings, no. 2, D major.

Quartet, strings, no. 14, op. 105, A flat major. cf Quartet, strings, no. 12, op. 96, F major.

Quartettsatz (1873). cf Works, selections (DG 2740 177).

Quartettsatz, F major(1881): Fragment. cf Works, selections (DG 2740 177).

1089 Quintet, piano and strings, op. 81, A major. Emanuel Ax, pno; Cleveland Quartet. RCA ARL 1-2240 Tape (c) ARK 1-2240 (ct) ARS 1-2240.
+-HF 10-77 p146 +SFC 9-4-77 p38
+NR 9-77 p8 ++St 9-77 p126
+SR 9-3-77 p42

Quintet, piano and strings, op. 81, A major. cf Bagatelles, op. 47.

1090 Quintet, strings, no. 3, op. 97, E flat major. Cypresses. Dvorak Quartet; Josef Kodovsek, vla. Rediffusion Tape (c) LGC 016.
++RR 2-77 p96 tape

Quintet, strings, no. 3, op. 97, E flat major. cf Bagatelles, op. 47.

Quintet, strings, no. 3, op. 97, E flat major. cf Quartet, strings, no. 12, op. 96, F major.

Romance, op. 11, F minor. cf Concerto, violin, op. 53, A minor.

Rondo, op. 94, G minor. cf Concerto, violoncello, op. 104, B minor.

1091 Rusalka, op. 114. Alena Mikova, Milada Subrtova, s; Maria Ovcacikova, con; Ivo Zidek, t; Jiri Joran, bar; Eduard Haken, bs; Prague National Theatre Orchestra; Zdenek Chalabala. Supraphon ST 50440/3 (4). (also 87632)
++ARG 9-77 p30 +NR 11-76 p9
Rusalka: O silver moon. cf Virtuosi VR 7608.
Scherzo capriccioso. cf Symphonies, nos. 7-9.
Scherzo capriccioso, op. 66, D flat major. cf Symphony, no. 8, op. 88, G major.
1092 Serenade, strings, op. 22, E major. TCHAIKOVSKY: Serenade, strings op. 48, C major. AMF; Neville Marriner. Argo ZRG 848 Tape (c) KZRC 848. (Reissues from ZRG 670, 584)
++Gr 7-76 p175 +NR 3-77 p6
+Gr 6-77 p105 tape +-RR 5-76 p56
+HFN 6-76 p102 +RR 11-76 p110 tape
+HFN 7-76 p104 tape ++SFC 10-10-76 p32
+-MJ 4-77 p33 +St 7-77 p124
1093 Serenade, strings, op. 22, E major. PURCELL: Dido Aeneas: Dido's lament. VAUGHAN WILLIAMS: Fantasia on a theme by Thomas Tallis. LPO; Leopold Stokowski. Desmar 1011 Tape (c) E 1047.
+HF 4-77 p121 tape +SFC 12-18-77 p53
1094 Serenade, strings, op. 22, E major. TCHAIKOVSKY: Serenade, strings op. 48, C major. ECO; Raymond Leppard. Philips 9500 105 Tape (c) 7300 532.
+ARG 6-77 p46 +HFN 5-77 p138 tape
+Gr 3-77 p1395 +NR 7-77 p5
++Gr 6-77 p105 tape +-RR 3-77 p61
+-HF 8-77 p79 +RR 5-77 p95 tape
+HFN 3-77 p103 +St 7-77 p124
Silent woods, op. 68. cf Concerto, violoncello, op. 104, B minor.
1095 Slavonic dances, opp. 46 and 72. CPhO; Vaclav Neumann. Decca PFS 4396 Tape (c) KPFC 4396. (Reissue from Telefunken SAT 22523/5)
+-Gr 5-77 p1695 +-HFN 7-77 p127 tape
+-Gr 7-77 p225 tape +-RR 5-77 p53
+-HFN 5-77 p121
1096 Slavonic dances, op. 46, nos. 1-4. MOZART: Symphony, no. 38, K 504, D major. SMETANA: Ma Vlast: Vltava. BRSO, VSO; Lorin Maazel, Karel Ancerl. Philips Tape (c) 7327 041.
+HFN 7-77 p127 tape +RR 7-77 p98 tape
1097 Slavonic dances, op. 46, nos. 1-8. CPhO; Karel Sejna. Rediffusion Legend LGD 015 Tape (c) LGC 015. (Reissue from Supraphon SUAST 50105/6)
+Gr 7-77 p173 +RR 2-77 p96 tape
+HFN 8-77 p93 ++RR 7-77 p50
Slavonic dance, op. 46, no. 1. cf CBS 61780.
Slavonic dances, op. 46, nos. 2, 4, 6. cf Symphony, no. 8, op. 88, G major.
Slavonic dance, op. 46, no. 3, A flat major. cf BRAHMS: Concerto, violin and violoncello, op. 102, A minor.
1098 Slavonic dances, op. 72, nos. 1-8. CPhO; Karel Sejna. Rediffusion Legend LGD 039. (Reissue from Supraphon SUAST 50106)
+Gr 10-77 p627 ++RR 10-77 p44
+HFN 10-77 p165
Slavonic dance, op. 72, no. 2, E minor. cf BRAHMS: Concerto, violin and violoncello, op. 102, A minor.
Slavonic dance, op. 72, no. 2, E minor. cf Decca PFS 4351.

Songs, op. 55, no. 4. cf Discopaedia MB 1012.
Songs, op. 55: Songs my mother taught me. cf HMV RLS 723.

1099 Stabat mater, op. 58. Drahomira Tikalova, s; Marta Krasova, con; Beno Blachut, t; Karel Kalas, bs; Czech Singers Chorus; CPhO and Chorus; Vaclav Talich. Heritage HCN 8011/2 (2). (Reissue from LPM 24/6)
+Gr 7-77 p191 +RR 8-77 p76

Symphonic variations op. 78. cf The noonday witch, op. 108.

1100 Symphony, no. 6, op. 60, D major. CPhO; Karel Ancerl. Rediffusion Legend LGD 012 Tape (c) LGC 012. (Reissue from Supraphon SUAST 50746)
+Gr 10-77 p627 +HFN 10-77 p165
++HF 2-77 p96 tape ++RR 10-77 p44

1101 Symphony, no. 6, op. 60, C major. CPhO; Vaclav Neumann. Supraphon 110 1833. (Reissue from 110 1621/8)
+-Gr 12-76 p995 +-RR 9-76 p47
++NR 12-76 p6 +SFC 4-3-77 p43

1102 Symphonies, nos. 7-9. Legends, op. 58, nos. 4, 6-7. Scherzo capriccioso. Halle Orchestra; John Barbirolli. Pye GGCD 304/1-2 (2). (Reissues from CCL 30145, 30155, CSCL 70002)
+-Gr 2-77 p1274 +RR 1-77 p53
+-HFN 1-77 p119

1103 Symphony, no. 7, op. 70, D minor. LPO; Carlo Maria Giulini. Angel (Q) S 37270. (also HMV SQ ASD 3325 Tape (c) TC ASD 3325)
+Gr 4-77 p1542 +NR 10-77 p3
-Gr 8-77 p346 tape +-RR 4-77 p42
+HF 11-77 p110 +SFC 5-15-77 p50
+HFN 8-77 p99 tape +-St 9-77 p128
++MJ 9-77 p35

1104 Symphony, no. 7, op. 70, D minor. COA; Colin Davis. Philips 9500 132 Tape (c) 7300 535.
++ARG 5-77 p18 +-HFN 7-77 p127
+Gr 2-77 p1274 +MJ 4-77 p33
+Gr 7-77 p225 tape /-NR 3-77 p2
+Gr 8-77 p346 tape +RR 2-77 p44
++HF 4-77 p100 +SFC 1-23-77 p37
++HF 11-77 p138 tape ++St 9-77 p128
+HFN 2-77 p117

1105 Symphony, no. 7, op. 70, D minor. CPhO; Zdenek Kosler. Rediffusion Legend LGD 007 Tape (c) LGC 007. (Reissue from Supraphon SUAST 50647)
/Gr 8-77 p296 +-RR 2-77 p96 tape
+-HFN 8-77 p93 +-RR 7-77 p50

1106 Symphony, no. 7, op. 70, D minor. CPhO; Vaclav Neumann. Supraphon 110 1834.
+ARG 11-77 p14 +-RR 7-77 p50
+HFN 9-77 p151 ++SFC 11-20-77 p54
+NR 12-77 p2

1107 Symphony, no. 8, op. 88, G major. Columbia Symphony Orchestra; Bruno Walter. CBS 61274. (Reissue from SBRG 72097)
+Gr 12-77 p1079 +HFN 12-77 p47

1108 Symphony, no. 8, op. 88, G major. Scherzo capriccioso, op. 66, D flat major. LSO; Istvan Kertesz. Decca SXL 6044 Tape (c) KSXC 6044 (ct) ESXC 6044. (also London 6358)
+Gr 8-74 p400 tape +RR 6-74 p86 tape
+HFN 4-77 p155 tape

1109 Symphony, no. 8, op. 88, G major. The wood dove, op. 110. LAPO;

Zubin Mehta. Decca SXL 6750 Tape (c) KSXC 6750. (also London CS 6979 Tape (c) 5-6979)
+-Gr 2-77 p1279 +-NR 7-77 p5
+HF 12-77 p87 +-RR 2-77 p45
+HFN 2-77 p117 +-RR 5-77 p91 tape
+-HFN 4-77 p155 tape ++SFC 4-3-77 p43
+MJ 9-77 p35

1110 Symphony, no. 8, op. 88, G major. Slavonic dances, op. 46, nos. 2, 4, 6. COA; Bernard Haitink. Philips 6580 126 Tape (c) 7317 157. (Reissue from SAL 3451)
/Gr 11-76 p785 +-RR 10-76 p50
+HFN 11-76 p170 +-RR 10-77 p96 tape
+HFN 8-77 p99 tape

1111 Symphony, no, 8, op. 88, G major. Slovak Philharmonic Orchestra; Zdenek Kosler. Royale ROY 2003.
/Gr 8-77 p296

1112 Symphony, no. 9, op. 95, E minor. NPhO; Riccardo Muti. Angel (Q) S 37230 Tape (c) 4XS 37230 (ct) 8XS 37230. (also HMV ASD 3285 Tape (c) TC ASD 3285)
+-Gr 1-77 p1141 +RR 1-77 p53
+-HF 3-77 p98 +-RR 5-77 p91
+-HFN 1-77 p103 ++SFC 1-23-77 p37
+HFN 4-77 p155 tape ++St 5-77 p109
++NR 3-77 p6

1113 Symphony, no. 9, op. 95, E minor. SMETANA: Ma Vlast: The Moldau. BPhO; Herbert von Karajan. Angel S 37437 Tape (c) 4XS 37437.
++SFC 11-13-77 p50

1114 Symphony, no. 9, op. 95, E minor. Columbia Symphony Orchestra; Bruno Walter. CBS 61234. (Reissue from Philips SABL 152)
+Gr 12-77 p1079 +RR 12-77 p47

1115 Symphony, no. 9, op. 95, E minor. Orchestre de Paris; Georges Pretre. Connoisseur Society CS 2108 Tape Advent (c) E 1054.
-ARG 3-77 p14 +-MJ 1-77 p26

1116 Symphony, no. 9, op. 95, E minor. Carnival overture, op. 92. LAPO; Zubin Mehta. Decca SXL 6751 Tape (c) KSXC 6751. (also London 6980 Tape (c) 5-6980)
+-Gr 11-76 p785 +RR 11-76 p63
+HFN 11-76 p157 +-RR 1-77 p88 tape
+HFN 1-77 p123 tape ++SFC 10-10-76 p32
+MJ 2-77 p31 +St 5-77 p109
+-NR 1-77 p4

1117 Symphony, no. 9, E minor. NPhO; Vernon Handley. Enigma VAR 1018.
+Gr 1-77 p1141 +RR 1-77 p53
+HFN 1-77 p105

1118 Symphony, no. 9, op. 95, E minor. SMETANA: Ma Vlast: Vltava. BPhO; Herbert von Karajan. HMV SQ ASD 3407 Tape (c) TC ASD 3407. (also Angel S 35615)
+Gr 10-77 p627 ++HFN 11-77 p171
+-Gr 12-77 p1145 tape ++RR 10-77 p55

1119 Symphony, no. 9, op. 95, E minor. COA; Willem Mengelberg. Mack Records MACK 002. (Reissue from Telefunken originals)
+-ARSC vol 9, no. 1, 1977 p84

1120 Symphony, no. 9, op. 95, E minor. Slovak Philharmonic Orchestra; Zdenek Kosler. Opus 9110 0282.
+-Gr 8-77 p296 -RR 6-77 p47
+-HFN 6-77 p121

1121 Symphony, no. 9, op. 95, E minor. RPO; Jascha Horenstein. Quin-

tessence PMC 7001.
+ARG 12-77 p43 +St 11-77 p146
+-HF 9-77 p108

1122 Symphony, no. 9, op. 95, E minor. RPO; Jascha Horenstein. RCA GL 2-5060 Tape (c) GK 2-5060.
++Gr 7-77 p174 +RR 7-77 p50
+-HFN 7-77 p111 +-RR 10-77 p96 tape
+-HFN 8-77 p99 tape

1123 Symphony, no. 9, op. 95, E minor. CPhO; Karel Ancerl. Rediffusion Legend LGD 004 Tape (c) LGC 004.
+-Gr 8-77 p296 +RR 2-77 p96
-HFN 7-77 p125 +RR 6-77 p47

Symphony, no. 9, op. 95, E minor: Largo. cf Grosvenor GRS 1052.

1124 Trio, piano, op. 65, F minor. GLIERE: Duo, violin and violoncello, op. 39. HANDEL: Suite, harpsichord, no. 7: Passacaglia. STRAVINSKY: Suite italienne. Jascha Heifetz, vln; Gregor Piatigorsky, vlc; Leonard Pennario, pno. Columbia M 33447. (also CBS 76421)
+-ARG 2-77 p48 +-MJ 2-77 p31
++Audio 3-77 p92 +-MT 9-77 p735
+-Gr 3-77 p1422 +NR 4-77 p5
+-HF 6-77 p104 +RR 3-77 p70
+-HFN 3-77 p103 ++SFC 12-76 p34

Waltzes, op. 54 (2). cf Works, selections (DG 2740 177).
The water goblin, op. 107. cf The noonday witch, op. 108.
The wood dove, op. 110. cf The golden spinning wheel, op. 109.
The wood dove, op. 110. cf Symphony, no. 8, op. 88, G major.

1125 Works, selections: Cypresses. Quartets, strings, nos. 1-14. Quartettsatz (1874). Quartettsatz, F major (1881): Fragment. Waltzes, op. 54 (2). Prague Quartet. DG 2740 177 (12). (Reissues from 2530 719, 2530 632, 2530 480)
+Gr 12-77 p1105 ++RR 12-77 p59
+HFN 12-77 p169

DYSON, George
Be strong. cf Vista VPS 1053.

EAST, Michael
Hence stars. cf Argo ZK 25.
Hence stars too dim of light. cf DG 2533 347.
Peccavi. cf Crystal S 206.

EBERLIN, Johann
Toccata e fuga tertia. cf ABC ABCL 67008.
Toccata sexta. cf ABC ABCL 67008.

ECCARD, Johann
Ich steh an deiner Krippen hier. cf ABC L 67002.

ECCLES, Henry
Sonata, recorder and harpsichord, F major. cf Saga 5425.

EDWARDS
Take me home. cf BBC REC 267.

EDWARDS, George
Exchange misere. cf BRESNICK: B's garlands.

EDWARDS, Richard
Songs: In going to my naked bed; When griping griefs. cf BG HM 57/8.

EISLER, Hanns
Sonata, piano, no. 3. cf BRAXTON: P-JOS...4K-D (MIX).
Songs: Legende von der Entstehung des Buches Taoteking; Elegien (3); Holderlin fragments; Zuchthaus-Kantate. cf DESSAU: Tierverse.

EKLUND, Hans
Quartet, strings, no. 3. cf CARLSTEDT: Sonata, violin, op. 15.
ELGAR, Edward
Adieu. cf Piano works (Prelude PRS 2503).
Adieu. cf Works, selections (HMV SLS 5084).
Allegretto on a theme of five notes. cf Works, selections (Pearl 523).
The banner of St. George, op. 33: Epilogue. cf Works, selections (HMV RLS 713).
Bavarian dances, op. 27, nos. 1-3. cf Works, selections (HMV RLS 713).
Bavarian dances, op. 27, nos. 1-3. cf Works, selections (RCA LRL 1-5133).
Beau Brummel: Minuet (2). cf Works, selections (HMV RLS 713).
Bizarrerie, op. 13, no. 2. cf Works, selections (Pearl 523).
La capricieuse, op. 17. cf Works, selections (Pearl 523).
1126 Caractacus, op. 35. Sheila Armstrong, s; Robert Tear, t; Peter Glossop, bar; Brian Rayner Cook, bs-bar; Malcolm King, Richard Stuart, bs; Royal Liverpool Philharmonic Orchestra and Chorus; Charles Groves. HMV (SQ) SLS 998 (2) Tape (c) TC SLS 998.
+Gr 6-77 p84 +MT 11-77 p923
+HFN 7-77 p112 +RR 7-77 p85
+HFN 10-77 p169 tape ++RR 11-77 p119 tape
Caractacus, op. 35: Woodland interlude. cf Works, selections (RCA LRL 1-5133).
Caractacus, op. 35: Woodland interlude; Triumphal march. cf Works, selections (HMV RLS 713).
Carissima. cf Works, selections (CBS 76423).
Carissima. cf Works, selections (HMV RLS 713).
Carissima. cf Works, selections (HMV ESD 7009).
Chanson de matin, op. 15, no. 2. cf Works, selections (CBS 76423).
Chanson de matin, op. 15, no. 2. cf Works, selections (HMV RLS 713).
Chanson de matin, op. 15, no. 2. cf Works, selections (Pearl 523).
Chanson de matin, op. 15, no. 2. cf Works, selections (RCA LRL 1-5133).
Chanson de nuit, op. 15, no. 1. cf Works, selections (CBS 76423).
Chanson de nuit, op. 15, no. 1. cf Works, selections (HMV RLS 713)
Chanson de nuit, op. 15, no. 1. cf Works, selections (Pearl 523).
Chanson de nuit, op. 15, no. 1. cf Works, selections (RCA LRL 1-5133).
Chantant. cf Piano works (Prelude PRS 2503).
Civic fanfare. cf RCA RL 2-5081.
Civic fanfare and'God save the King'. cf Works, selections (HMV RLS 708).
1127 Cockaigne overture, op. 40. Enigma variations, op. 36. CSO, LPO; Georg Solti. Decca SXL 6795 Tape (c) KSXC 6795.
+Gr 11-76 p786 +MM 7-77 p36
+Gr 4-77 p1603 tape +RR 11-76 p65
+-HFN 12-76 p141 +-RR 5-77 p91 tape
+-HFN 1-77 p123 tape
Cockaigne overture, op. 40. cf Falstaff, op. 68.
Cockaigne overture, op. 40. cf Pomp and circumstance marches, op. 39, nos. 1-5.
Cockaigne overture, op. 40. cf Works, selections (CBS 79002).
Cockaigne overture, op. 40. cf Works, selections (HMV RLS 713).
Concert allegro, op. 41. cf Works, selections (HMV SLS 5084).

Concert allegro, op. 41. cf Piano works (Prelude PRS 2503).

1128 Concerto, violin, op. 61, B minor. Yehudi Menuhin, vln; NPhO; Adrian Boult. Angel S 36330. (also HMV ASD 2259)
+HF 4-71 p66 +ST 1-77 p779
+St 4-77 p79

1129 Concerto, violin, op. 61, B minor. Pinchas Zukerman, vln; LPO; Daniel Barenboim. CBS 76528 Tape (c) 40-76528. (also Columbia M 34517)
++Gr 11-76 p786 +NR 6-77 p6
-Gr 1-77 p1183 tape +-NYT 7-24-77 pD11
+HFN 11-76 p159 ++RR 11-76 p65
+HFN 1-77 p123 tape +-RR 4-77 p91 tape
+-MM 7-77 p36 +St 4-77 p79
+-MT 1-77 p43 +St 9-77 p132

1130 Concerto, violin, op. 61, B minor. Kyung-Wha Chung, vln; LPO; Georg Solti. Decca SXL 6842 Tape (c) KSXC 6842.
+Gr 10-77 p627 +HFN 12-77 p187 tape
+-Gr 12-77 p1142 ++RR 11-77 p55
+HFN 10-77 p153

1131 Concerto, violin, op. 61, B minor. BACH: Fantasia and fugue, S 537, C minor. Yehudi Menuhin, vln; LSO, Royal Albert Hall Orchestra; Edward Elgar. HMV HLM 7107.
+RR 2-77 p44

Concerto, violin, op. 61, B minor. cf Works, selections (HMV RLS 708).

Concerto, violin, op. 61, B minor. cf BACH: Fantasia and fugue, S 537, C minor.

1132 Concerto, violoncello, op. 85, E minor. Enigma variations, op. 36. Jacqueline du Pre, vlc; PO, LPO; Daniel Barenboim. CBS 76529 Tape (c) 40-76529. (also Columbia M 34530 Tape (c) MT 34530)
++Gr 11-76 p785 +-MT 1-77 p43
-Gr 1-77 p1183 tape +NYT 7-24-77 pD11
+HF 11-77 p112 +RR 11-76 p64
+-HFN 11-76 p159 +RR 5-77 p91 tape
+HFN 1-77 p123 tape +St 11-77 p82
+-MM 7-77 p36 +ST 8-77 p341

1133 Concerto, violoncello, op. 85, E minor. Introduction and allegro, op. 47. Serenade, strings, op. 20, E minor. Paul Tortelier, vlc; Rodney Friend, John Willison, vln; John Chambers, vla; Alexander Cameron, vlc; LPO; Adrian Boult. HMV ASD 2906 Tape (c) TC ASD 2906. (also Angel 37029)
+Gr 8-73 p330 ++RR 8-73 p41
++Gr 6-75 p106 tape ++RR 6-75 p91 tape
+-HF 8-75 p84 ++SFC 10-5-75 p38
+HFN 5-75 p142 tape ++St 6-75 p95
++NR 4-75 p5 ++St 4-77 p80

Concerto, violoncello, op. 85, E minor. cf Works, selections (HMV RLS 708).

Concerto, violoncello, op. 85, E minor. cf BRAHMS: Sonata, violoncello and piano, no. 2, op. 99, F major.

Concerto, violoncello, op. 85, E minor. cf HMV SLS 5068.

Contrasts, op. 10, no. 3. cf Works, selections (HMV RLS 713).

Contrasts, op. 10, no. 3. cf works, selections (HMV ESD 7009).

Contrasts, op. 10, no. 3. cf Works, selections (RCA LRL 1-5133).

1134 Coronation ode, op. 44. PARRY: I was glad when they said unto me. TRAD.: The national anthem (arr. Elgar). Felicity Lott, s; Alfreda Hodgson, con; Richard Morton, t; Stephen Roberts, bs; King's College Choir, Cambridge University Musical Society Chorus; Royal Military School of Music Band, NPhO; Philip Ledger. HMV (SQ) ASD 3345 Tape (c) TC ASD 3345.

+Audio 12-77 p48
+FF 11-77 p19
+Gr 5-77 p1718
-Gr 8-77 p346 tape
+HFN 6-77 p123
+HFN 8-77 p99 tape
++RR 5-77 p81
++RR 9-77 p98 tape
+ST 8-77 p341

1135 Coronation ode, op. 44. The spirit of England, op. 80. Teresa Cahill, s; Anne Collins, con; Anthony Rolfe-Johnson, t; Gwynne Howell, bs; Scottish National Orchestra and Chorus; Alexander Gibson. RCA RL 2-5074 (2).

+Audio 12-77 p48
+FF 11-77 p19
+Gr 5-77 p1718
+HFN 7-77 p112
+RR 5-77 p80
+ST 8-77 p341

Crown of India suite, op. 66. cf Imperial march, op. 32.
Crown of India suite, op. 66. cf Works, selections (CBS 79002).
Crown of India suite, op. 66. cf Works, selections (HMV RLS 713).
Dream children, op. 43. cf Works, selections (RCA LRL 1-5133).

1136 The dream of Gerontius, op. 38. Alfreda Hodgson, con; Robert Tear, t; Benjamin Luxon, bar; Scottish National Orchestra and Chorus; Alexander Gibson. CRD 1026/7 (2).

+ARG 9-77 p32
+Gr 3-77 p1438
+HFN 3-77 p95
+-MT 5-77 p399
+NR 11-77 p11
+RR 3-77 p90
++St 9-77 p132

The dream of Gerontius, op. 38: Kyrie eleison...All ye saints; Rescue him, O Lord...O Lord, into thy hands; Go in the name... Through the same; Praise to the Holies...Praise to the Holiest; And now the threshold...Most sure in all His ways; Jesu, by that shuddering dread...To that glorious home; Take me away... Praise to the Holiest. cf Works, selections (HMV RLS 713).
The dream of Gerontius, op. 38: So pray to me, my friends; O Jesu, help, Jesu by the shuddering dread; Take me away. cf Works, selections (HMV RLS 708).
Duett, trombone and double bass. cf Songs (Wealdon WS 152).

1137 Elegy strings, op. 58. Froissart overture, op. 19. Pomp and circumstance marches, op. 39. Sospiri, op. 70. PhO, NPhO; John Barbirolli. HMV ASD 2292 Tape (c) TC ASD 2292.

+-Gr 6-76 p102 tape
+-HFN 5-76 p117
++RR 4-77 p91 tape

Elegy, strings, op. 58. cf Enigma variations, op. 36.
Elegy, strings, op. 58. cf Works, selections (CBS 76423).
Elegy, strings, op. 58. cf Works, selections (HMV RLS 713).
Elegy, strings, op. 58. cf Argo D26D4.
Empire march. cf Pomp and circumstance marches, op. 39, nos. 1-5.

1138 Enigma variations, op. 36. Elegy, strings, op. 58. Serenade, strings, op. 20, E minor. CPhO, LSO; Leopold Stokowski, Ainslee Cox. Decca PFS 4338. (also London SPC 21136 Tape (c) SPC 5-21136)

+-Gr 8-75 p332
+HF 12-76 p100
+HF 4-77 p121
+-HFN 8-75 p75
+MJ 1-77 p26
+NR 12-76 p3
+-RR 8-75 p32
+-St 1-77 p114

1139 Enigma variations, op. 36. STRAUSS, R.: Don Juan, op. 20. COA; Bernard Haitink. Philips 6500 481 Tape (c) 7300 344.

+Gr 3-75 p1649 ++RR 4-75 p27
+Gr 6-75 p111 tape +RR 6-75 p91 tape
++HF 4-75 p84 -SFC 4-4-75 p22
++HFN 7-75 p90 tape ++St 5-75 p94
+MJ 2-75 p39 ++St 4-77 p79
++NR 2-75 p3

1140 Enigma variations, op. 36. Pomp and circumstance marches, op. 39, nos. 1-5. RPO; Norman Del Mar. DG 2535 217 Tape (c) 3335 217. (Reissues from Contour 2870 440)
+Gr 10-76 p595 ++HFN 3-77 p119
+-Gr 3-77 p1458 tape +-RR 11-76 p65
+HFN 11-76 p170 +-RR 5-77 p91 tape

1141 Enigma variations, op. 36. SCHOENBERG: Variations, op. 31. CSO; Georg Solti. London CS 6984 Tape (c) CS 5-6984.
+ARG 3-77 p16 -MJ 2-77 p31
+HF 4-77 p112 +-NR 2-77 p2

1142 Enigma variations, op. 36. BRAHMS: Variations on a theme by Haydn, op. 56a. LSO; Pierre Monteux. London STS 15188.
++NR 11-74 p2 ++St 4-77 p79
++St 1-75 p110

1143 Enigma variations, op. 36. Falstaff, op. 68. NPhO; Andrew Davis. Musical Heritage Society MHS 3628.
+-FF 9-77 p15

Enigma variations, op. 36. cf Cockaigne overture, op. 40 (Decca SXL 6795).
Enigma variations, op. 36. cf Concerto, violoncello, op. 85, E minor.
Enigma variations, op. 36. cf Works, selections (CBS 79002).
Enigma variations, op. 36. cf Works, selections (HMV RLS 708).
Enigma variations, op. 36, excerpts. cf HMV SLS 5073.

1144 Enigma variations, op. 36: Nimrod. Pomp and circumstance march, op. 39, no. 4, G major. TRAD.: The national anthem. LPO, CSO; Georg Solti. Decca F 13731. (Reissue from SXL 6795)
+Gr 8-77 p349

Enigma variations, op. 36: Nimrod. cf Decca PFS 4351.
Enigma variations, op. 36: Nimrod. cf Pye Golden Hour GH 643.

1145 Falstaff, op. 68. Cockaigne overture, op. 40. LPO; Daniel Barenboim. Columbia M 32599 (Q) MQ 32599 Tape (c) MAQ 32599. (also CBS 76284)
+-Gr 9-74 p495 ++RR 10-74 p45
+HF 8-74 p90 ++St 9-74 p124
++NR 7-74 p5 ++St 4-77 p79

Falstaff, op. 68. cf Enigma variations, op. 36.
Falstaff, op. 68. cf Works, selections (HMV RLS 708).
Falstaff, op. 68: Interludes. cf Works, selections (HMV RLS 713).
Falstaff, op. 68: Interludes (2). cf Works, selections (RCA LRL 1-5133).
Froissart overture, op. 19. cf Elegy, strings, op. 58.
Froissart overture, op. 19. cf Works, selections (HMV RLS 713).
Gavotte. cf Works, selections (Pearl 523).
Griffinesque. cf Piano works (Prelude PRS 2503).

1146 Imperial march, op. 32. Pomp and circumstance marches, op. 39, nos. 1-5. Crown of India suite, op. 66. LPO; Daniel Barenboim. Columbia M 32936 Tape (c) MT 32926 (ct) MA 32936 (Q) MQ 32936 Tape (ct) MAQ 32936. (also CBS 76248 Tape (c) 40-76248)

-Gr 9-74 p495 +-RR 10-74 p45
+-Gr 12-75 p1121 tape +-RR 6-76 p87 tape
-HF 4-75 p112 tape ++SFC 9-22-74 p22
+-HF 4-75 p112 Quad +St 4-77 p80
+-HFN 12-75 p173 tape

Imperial march, op. 32. cf Pomp and circumstance marches, op. 39, nos. 1-5.
Imperial march, op. 32. cf Works, selections (CBS 79002).
Imperial march, op. 32. cf BAX: Coronation march.
Imperial march, op. 32. cf Argo SPA 507.
Imperial march, op. 32. cf Decca SB 715.
In Smyrna. cf Piano works (Prelude PRS 2503).
In Smyrna. cf Songs (Wealdon WS 152).

1147 In the south overture, op. 50. Sea pictures, op. 37. Yvonne Minton, ms; LPO; Daniel Barenboim. CBS 76579 Tape (c) 40-76579.
+-Gr 8-77 p328 +RR 9-77 p87
+HFN 9-77 p141 +-RR 12-77 p1145 tape
+HFN 12-77 p187 tape +RR 12-77 p94 tape
+MT 12-77 p1014

1148 In the south overture, op. 50. VAUGHAN WILLIAMS: Fantasia on a theme by Thomas Tallis. The wasps: Overture. Bournemouth Symphony Orchestra; Constantin Silvestri. HMV ESD 7013 Tape (c) TC ESD 7013. (Reissue from ASD 2370)
+-Gr 10-76 p595 +-HFN 1-77 p123 tape
+Gr 1-77 p1178 tape +RR 10-76 p74
+HFN 11-76 p170 +RR 4-77 p91 tape

In the south overture, op. 50. cf Works, selections (HMV RLS 713).
Introduction and allegro, op. 47. cf Concerto, violoncello, op. 85, E minor.
Introduction and allegro, op. 47. cf Argo D26D4.
The kingdom, op. 51: Prelude. cf Works, selections (HMV RLS 708).
The land of hope and glory. cf Works, selections (HMV RLS 713).
The land of hope and glory. cf CRD Britannia BR 1077.
The light of life, op. 27: Meditation. cf Works, selections (HMV RLS 713).
May song. cf Piano works (Prelude PRS 2503).
May song. cf Works, selections (HMV RLS 713).
May song. cf Works, selections (HMV ESD 7009).
Mazurka, op. 10, no. 1. cf Works, selections (HMV RLS 713).
Mazurka, op. 10, no. 1. cf Works, selections (HMV ESD 7009).
Mina. cf Works, selections (HMV ESD 7009).
Minuet. cf Piano works (Prelude PRS 2503).
Minuet, op. 21. cf Works, selections (HMV RLS 713).
Minuet, op. 21. cf Works, selections (HMV ESD 7009).
Mot d'amour, op. 13, no. 1. cf Works, selections (Pearl 523).

1149 The music makers, op. 69. PARRY: Blest pair of sirens. Janet Baker, ms; LPO and Chorus; Adrian Boult. Vanguard VSD 71225.
+NR 12-77 p8 +SFC 12-11-77 p61

The music makers, op. 69: We are the music makers; A breath of our inspiration; For we are afar with the dawning. cf Works, selections (HMV RLS 708).
The national anthem. cf Works, selections (HMV RLS 713).
Nursery suite. cf Works, selections (HMV RLS 713).
O God, our help in ages past. cf Works, selections (HMV RLS 713).
Offertoire. cf Works, selections (Pearl 523).
Pastourelle, op. 4, no. 2. cf Works, selections (Pearl 523).
Piano improvisations (5). cf Works, selections (HMV RLS 713).

1150 Piano works: Adieu. Chantant. Concert allegro, op. 41. Griffinesque. In Smyrna. May song. Minuet. Serenade. Skizze. Sonatina. John McCabe, pno. Prelude PRS 2503.
+Gr 10-76 p627 +RR 10-76 p87
+-HFN 10-76 p169 +St 4-77 p80
+MT 12-76 p1003

Pomp and circumstance marches, op. 39. cf Elegy, strings, op. 58.

1151 Pomp and circumstance marches, op. 39, nos. 1-5. Cockaigne overture, op. 40. TRAD.: The national anthem. LPO; Georg Solti. Decca SXL 6848. (Reissues from F 13713, SXL 6795) (also London CS 7072)
++Gr 10-77 p628 +RR 10-77 p45
+HFN 10-77 p153 +SFC 10-30-77 p44

1152 Pomp and circumstance marches, op. 39, nos. 1-5. Empire march. Imperial march, op. 32. WALTON: Crown imperial. Orb and sceptre. LPO; Adrian Boult. HMV SQ ASD 3388 Tape (c) TC ASD 3388.
+Gr 10-77 p628 +HFN 11-77 p171
+Gr 12-77 p1145 tape +RR 11-77 p56

Pomp and circumstance marches, op. 39, nos. 1-5. cf Enigma variations, op. 36.

Pomp and circumstance marches, op. 39, nos. 1-5. cf Imperial march, op. 32.

Pomp and circumstance marches, op. 39, nos. 1-5. cf Works, selections (CBS 79002).

Pomp and circumstance marches, op. 39, nos. 1-4. cf Works, selections (HMV RLS 713).

Pomp and circumstance marches, op. 39, nos. 1, 2, 4. cf Works, selections (HMV RLS 713).

Pomp and circumstance marches, op. 39, nos. 1, 4. cf HMV SQ ASD 3341.

Pomp and circumstance march, op. 39, no. 1, D major. cf BAX: Coronation march.

Pomp and circumstance march, op. 39, no. 1, D major. cf BBC REH 290.

Pomp and circumstance march, op. 39, no. 1, D major. cf Philips 6747 327.

Pomp and circumstance march, op. 39, no. 4, G major. cf Enigma variations, op. 36: Nimrod.

Quartet, strings, op. 83, E minor. cf Works, selections (HMV SLS 5084).

Quintet, piano, op. 84, A minor. cf Works, selections (HMV SLS 5084).

Reminiscences, op. 1. cf Works, selections (Pearl 523).

Romance, op. 1. cf Works, selections (Pearl 523).

1153 Romance, op. 62. JACOB: Partita, bassoon. OSBORNE: Rhapsodie, bassoon. TANSMAN: Suite, bassoon and piano. Arthur Grossman, bsn; Silvia Kind, hpd; Randolph Hokanson, pno. Ravenna RAVE 761.
+HF 12-77 p115 +IN 2-77 p20

Romance, op. 62. cf Works, selections (CBS 76423).

Romance, op. 62. cf Works, selections (HMV ESD 7009).

Rosemary. cf Works, selections (CBS 76423).

Rosemary. cf Works, selections (HMV RLS 713).

Rosemary. cf Works, selections (HMV ESD 7009).

Salut d'amour, op. 12. cf Works, selections (CBS 76423).

Salut d'amour, op. 12. cf Works, selections (HMV RLS 713).

Salut d'amour, op. 12. cf Works, selections (Pearl 523).
Salut d'amour, op. 12. cf Works, selections (RCA LRL 1-5133).
Scherzo, rehearsal. cf Works, selections (HMV RLS 708).

1154 Sea pictures, op. 37. MAHLER: Ruckert songs (5). Janet Baker, ms; LSO, NPhO; John Barbirolli. Angel S 36796.
+ARG 5-72 p410 ++SFC 1-23-72 p30
+Audio 12-75 p104 ++St 3-72 p84
+NR 1-72 p12 ++St 4-77 p80
+-ON 3-4-72 p35

1155 Sea pictures, op. 37. NYSTROEM: Songs of the sea. Birgit Finnila, con; Geoffrey Parsons, pno. BIS LP 38.
++St 2-77 p117

Sea pictures, op. 37. cf In the south overture, op. 50.
Serenade. cf Piano works (Prelude PRS 2503).
Serenade. cf Works, selections (HMV SLS 5084).

1156 Serenade, strings, op. 20, E minor. HOLST: St. Paul's suite, op. 29, no. 2. IRELAND: Concertino pastorale. WARLOCK: Capriol suite. Bournemouth Sinfonietta; George Hurst. RCA RL 2-5071.
+Gr 9-77 p438 +-RR 9-77 p62
+HFN 12-77 p169

1157 Serenade, strings, op. 20, E minor. GRIEG: Holberg suite, op. 40. RESPIGHI: Ancient airs and dances: Suite, no. 3. Slovak Chamber Orchestra; Bohdan Warchal. Rediffusion Royale ROY 2002.
-Gr 7-77 p174 /-RR 6-77 p55
-HFN 6-77 p123

Serenade, strings, op. 20, E minor. cf Concerto, violoncello, op. 85, E minor.
Serenade, strings, op. 20, E minor. cf Enigma variations, op. 36.
Serenade, strings, op. 20, E minor. cf Works, selections (CBS 76423).
Serenade, strings, op. 20, E minor. cf Works, selections (HMV RLS 713).
Serenade, strings, op. 20, E minor. cf Argo D26D4.
Serenade lyrique. cf Works, selections (HMV RLS 713).
Serenade lyrique. cf Works, selections (HMV ESD 7009).
Serenade lyrique. cf Works, selections (RCA LRL 1-5133).
Serenade mauresque, op. 10, no. 2. cf Works, selections (HMV ESD 7009).
Severn suite, op. 87. cf Works, selections (HMV RLS 713).
Severn suite, op. 87. cf BALL: Sinfonietta, brass band.
Severn suite, op. 87. cf BLISS: Kenilworth.
Severn suite, op. 87. cf Decca SB 329.
Sevillana, op. 7. cf Works, selections (HMV ESD 7009).
The shower, op. 71, no. 1. cf RCA RL 2-5112.
Skizze. cf Piano works (Prelude PRS 2503).
Skizze. cf Songs (Wealdon WS 152).
Soliloquy, oboe (orch. Jacob). cf Works, selections (RCA LRL 1-5133).
Sonata, organ, no. 1, op. 28, G major. cf HMV HQS 1376.

1158 Sonata, violin, op. 82, E major. WALTON: Sonata, violin. Sidney Weiss, vln; Jeanne Weiss, pno. Unicorn RHS 341 Tape (c) ZCUN 341.
+Gr 7-76 p187 ++SFC 11-27-77 p66
+Gr 4-77 p1603 tape ++St 4-77 p80
+-HFN 7-76 p87 ++St 5-77 p109
+NR 11-77 p8

Sonata, violin op. 82, E minor. cf Works, selections (HMV SLS 5084).
Sonatina. cf Piano works (Prelude PRS 2503).
Sonatina. cf Works, selections (HMV SLS 5084).

1159 Songs: Caractacus: O'er-arch'd by leaves; Clapham town end; Grania and Diarmid: There are seven that pull the thread; Like to the damask rose; O salutaris hostia; Oh soft was the song, op. 59, no. 3; Pleading, op. 48, no. 1; Queen Mary's lute song; The river, op. 60, no. 2; Rondel, op. 16, no. 3; The shepherd's song, op. 16, no. 1; Spanish lady: Modest and fair; The starlight express, op. 78: The blue-eyed fairy; Still to be neat; The torch, op. 60, no. 1; Twilight, op. 59, no. 6. Mary Thomas, s; John Carol Case, bar; Daphne Ibbott, pno. Saga SAGA 5304. (Reissue from Alpha SPHA 3017)
-Gr 9-73 p518 +St 4-77 p80
+-St 12-76 p139

1160 Songs (vocal and choral): As torrents in summer; Ave verum, op. 2, no. 2; Doubt not thy father's care; Drake's Broughton; Fear not, O land; Fly, singing bird, op. 26, no. 2; I sing the birth; A poet's life; The prince of sleep; The rapid stream; Rondel; The shepherd's song, op. 16, no. 1; The snow, op. 26, no. 1; The woodland stream. Duett, trombone and double bass. In Smyrna. Skizze. Judith Robinson, s; Geoffrey Tomlinson, David Snelling, vln; Derek Roberts, Barry Collett, pno; John Skipp, trom; Gerald Bellamy, double bs; Uppingham Community College Choir; Barry Collett. Wealdon WS 152.
+-Gr 11-77 p873 +-RR 11-77 p102

Songs (choral): Angelus, op. 56, no. 1; Ave Maria, op. 2, no. 2; Ave maris stella, op. 2, no. 3; Ave verum, op. 2, no. 1; Give unto the Lord, op. 74; O hearken thou, op. 64. cf Suite, organ, op. 14: Introduction; Andante; Allegretto piacevole; Intermezzo; Poco lento.
Songs: As torrents in summer; Deep in my soul, op. 53, no. 2; The fountain, op. 71, no.2; Go, song of mine, op. 57; Love's tempest, op. 73, no. 1; My love dwelt in a northern land; O wild west wind, op. 53, no. 3; Owls, op. 53, no. 4; The shower, op. 71, no. 1; There is sweet music, op. 53, no. 1. cf DELIUS: Songs (Argo ARG 607).
Songs: As torrents in summer; Deep in my soul, op. 53, no. 2; The fountain, op. 71, no. 2; Love's tempest, op. 73, no. 1; My love dwelt in a northern land; Go, song of mine, op. 57; Love's tempest, op. 73, no. 1; O wild west wind, op. 53, no. 3; Owls, op. 53, no. 4; The shower, op. 71, no. 1; There is sweet music, op. 53, no. 1. cf DELIUS: Songs (Argo ZK 23).
Songs: Give unto the Lord, op. 67. cf Argo ZRG 871.
Sospiri, op. 70. cf Elegy, strings, op. 58 (HMV 2292).
Sospiri, op. 70. cf Works, selections (CBS 76423).
Sospiri, op. 70. cf Works, selections (Pearl 523).
Sospiri, op. 70. cf Argo D26D4.
The Spanish lady, op. 89: Suite. cf Argo D26D4.
The spirit of England, op. 80. cf Coronation ode, op. 44.

1161 The starlight express, op. 78. Valerie Masterson, s; Derek Hammond-Stroud, bar; LPO; Vernon Handley. HMV SQ SLS 5036 (2).
+FF 9-77 p16 +MT 7-76 p575
+Gr 5-76 p1757 +RR 4-76 p69
++HFN 4-76 p104 +-STL 5-9-76 p38

1162 Suite, organ, op. 14: Introduction; Andante; Allegretto piacevole; Intermezzo; Poco lento. Songs (choral): Angelus, op. 56, no. 1; Ave Maria, op. 2, no. 2; Ave maris stella, op. 2, no. 3; Ave verum, op. 2, no. 1; Give unto the Lord, op. 74; O hearken thou, op. 64. Harry Bramma, org; Worcester Cathedral Choir; Christopher Robinson. HMV CSD 3660
+St 4-77 p80

1163 Symphony, no. 1, op. 55, A flat major. LPO; Daniel Barenboim. Columbia M 32807 (Q) MQ 32807. (also CBS 76247)
+-Gr 9-74 p495 +-RR 10-74 p45
+HF 8-74 p90 +-SFC 7-21-74 p26
+NR 7-74 p5 ++St 9-74 p124
+-NYT 6-2-74 pD24 ++St 4-77 p79

1164 Symphony, no. 1, op. 55, A flat major. LPO; Adrian Boult. HMV ASD 3330 Tape (c) TC ASD 3330. (also Angel 37218)
+ARG 10-77 p27 +HFN 8-77 p99 tape
+Audio 9-77 p46 ++MT 8-77 p650
++Gr 4-77 p1547 ++RR 4-77 p45
-Gr 8-77 p346 tape ++RR 9-77 p98 tape
+-HFN 4-77 p135

Symphony, no. 1, op. 55, A flat major. cf Works, selections (HMV RLS 708).

1165 Symphony, no. 2, op. 63, E flat major. LPO; Daniel Barenboim. Columbia M 31997 (Q) MQ 31997 Tape (c) MT 31997. (also CBS 73094)
+-Gr 3-73 p1681 +SFC 10-14-73 p32
+HF 5-73 p79 +SR 3-73 p47
++HFN 3-73 p561 ++St 5-73 p112
+NR 4-73 p4 +St 4-77 p79
+-RR 3-73 p46

1166 Symphony, no. 2, op. 63, E flat major. LPO; Adrian Boult. HMV SQ ASD 3266 Tape (c) TC ASD 3266. (also Angel 37218)
++Audio 9-77 p46 ++NR 4-77 p2
+Gr 10-76 p595 ++RR 10-76 p50
++Gr 1-77 p1178 tape ++RR 4-77 p91 tape
+-HFN 10-76 p169 ++St 4-77 p79
+HFN 2-77 p135 tape ++St 5-77 p77

1167 Symphony, no. 2, op. 63, E flat major. Scottish National Orchestra; Alexander Gibson. RCS RL 2-5104.
+Gr 11-77 p837 +-RR 11-77 p56

Symphony, no. 2, op. 63, E flat major. cf Works, selections (HMV RLS 708).

Transcription of Bach's Fantasia and fugue, C minor. cf Works, selections(HMV RLS 708).

Virelai, op. 4, no. 3. cf Works, selections (Pearl 523).

Wand of youth suites, op. 1, nos. 1-2. cf Works, selections (HMV RLS 713).

Wand of youth suite, op. 1, no. 1, excerpts. cf Decca DDS 507/1-2.

1168 Works, selections: Carissima. Chanson de matin, op. 15, no. 2. Chanson de nuit, op. 15, no. 1. Elegy, strings, op. 58. Romance, op. 62. Rosemary. Salut d'amour, op. 12. Serenade, strings, op. 20, E minor. Sospiri, op. 70. Martin Gatt, bsn; ECO; Daniel Barenboim. CBS 76423 Tape (c) 40-76423. (also Columbia M 33584)
+-Audio 1-77 p85 +-NR 1-76 p4
+-Gr 11-75 p818 +-NYT 1-18-76 pD1
+-HFN 11-75 p153 +RR 11-75 p43

+-HFN 2-76 p116 tape +St 5-76 p113
+MT 7-76 p575 +St 4-77 p80

1169 Works, selections: Cockaigne overture, op. 40. Crown of India suite, op. 66. Enigma variations, op. 36. Imperial march, op. 32. Pomp and circumstance marches, op. 39, nos. 1-5. LPO; Daniel Barenboim. CBS 79002 (2). (Reissues from 76248, 76529, 76284)
+-Gr 8-77 p296 +-RR 8-77 p52
+-HFN 8-77 p95

1170 Works, selections: Concerto, violin, op. 61, B minor. Concerto, violoncello, op. 85, E minor. Civic fanfare and 'God save the King'. Dream of Gerontius, op. 38: So pray for me, my friends; O Jesu, help, Jesu by that shuddering dread; Take me away. Enigma variations, op. 36. Falstaff, op. 68. The kingdom, op. 51: Prelude. The music makers, op. 69: We are the music makers; O breath of our inspiration; For we are afar with the dawning. Scherzo, rehearsal. Symphony, no. 1, op. 55, A flat major. Symphony, no. 2, op. 63, E flat major. Transcription of Bach's Fantasia and fugue, C minor. Yehudi Menuhin, vln; Beatrice Harrison, vlc; Margaret Balfour, con; Tudor Davies, t; Horace Stevens, bs-bar; Herbert Brewer, org; Three Choirs Festival Chorus; Royal Albert Hall Orchestra, LSO, NSL, BBC Symphony Orchestra; Edward Elgar. HMV RLS 708 (5 discs and book "Images of Elgar"). (Reissues from 78s)
+-HFN 1-73 p109 +St 4-77 p79
+-RR 1-73 p40

1171 Works, selections: The banner of St. George, op. 33: Epilogue. Bavarian dances, op. 27, nos. 1-3. Beau Brummel: Minuets (2). Caractacus, op. 35: Woodland interlude; Triumphal march. Carissima. Chansons, op. 15, nos. 1-2. Cockaigne overture, op. 40 (2). Contrasts, op. 10, no. 3. Crown of India suite, op. 66. The dream of Gerontius, op. 38: Kyrie eleison...All ye saints; Rescue him, O Lord...O Lord, into thy hands; Go in the name...Through the same; Praise to the Holiest; Praise to the Holiest; And now the threshold...Most sure in all His ways; Jesu, by that shuddering dread...To that glorious home; Take me away...Praise to the Holiest. Elegy, strings, op. 58. Falstaff, op. 68: Interludes. Froissart overture, op. 19. In the south overture, op. 50. The land of hope and glory. The light of life, op. 29: Meditation. May song. Mazurka, op. 10, no. 1. Minuet, op. 21. The national anthem. Nursery suite. O God, our help in ages past. Rosemary. Piano improvisations (5). Pomp and circumstance marches, op. 39, nos. 1, 2, 4. Pomp and circumstance marches, op. 39, nos. 1-4. Salut d'amour, op. 12. Serenade, strings, op. 20, E minor. Serenade lyrique. Severn suite, op. 87. Wand of youth suites, op. 1, nos. 1-2. Film "Land of hope and glory" Elgar speaking and conducting. Various artists and orchestra; Edward Elgar. HMV RLS 713 (6).
++Gr 2-75 p1480 +St 4-77 p80
+RR 2-75 p34

1172 Works, selections: Adieu. Concert allegro, op. 41. Quartet, strings, op. 83, E minor. Quintet, piano, op. 84, A minor. Serenade. Sonata, violin, op. 82, E minor. Sonatina. Hugh Bean, vln; John Ogdon, pno; Music Group of London, Allegri Quartet. HMV SLS 5084 (2). (Reissues from HQS 1252, ASD 2501).
+Gr 6-77 p69 +RR 8-77 p66

1173 Works, selections: Carissima. Contrasts, op. 10, no. 3 (The gavotte, AD 1700 and 1900). May song. Mazurka, op. 10, no. 1. Minuet, op. 21. Mina. Romance, op. 62. Rosemary (That's for remembrance). Serenade lyrique. Serenade mauresque, op. 10, no. 2. Sevillana, op. 7. Michael Chapman, bsn; Northern Sinfonia Orchestra; Neville Marriner. HMV ESD 7009 Tape (c) TC ESD 7009. (Reissue from ASD 2638)
+Gr 11-76 p786 +RR 11-76 p64
+Gr 11-76 p887 tape +RR 1-77 p89 tape
+HFN 11-76 p175 +ST 1-77 p779

1174 Works, selections: Allegretto on a theme of five notes. Bizarrerie op. 13, no. 2. La capricieuse, op. 17. Chanson de matin, op. 15, no. 2. Chanson de nuit, op. 15, no. 1. Gavotte. Une idylle, op. 4, no. 1. Mot d'amour, op. 13, no. 1. Offertoire. Pastourelle, op. 4, no. 2. Reminiscences. Romance, op. 1. Salut d'amour, op. 12. Sospiri, op. 70. Virelai, op. 4, no. 3. John Georgiadis, vln; John Parry, pno. Pearl SHE 523.
++Gr 11-75 p848 +St 4-77 p80
+RR 12-75 p81

1175 Works, selections: Bavarian dances, op. 27, nos. 1-3. Caractacus, op. 35: Woodland interlude. Chanson de matin, op. 15, no. 2. Chanson de nuit, op. 15, no. 1. Contrasts, op. 10, no. 3. Dream children, op. 43. Falstaff, op. 68: Interludes (2). Serenade lyrique. Salut d'amor, op. 12. Soliloquy, oboe (orch. Jacob). Leon Goossens, ob; Bournemouth Sinfonietta; Norman Del Mar. RCA LRL 1-5133 Tape (c) 1-1746.
+Gr 11-76 p786 +HFN 3-77 p103
+Gr 1-77 p1185 tape +RR 11-76 p64

ELIZABETHAN AND JACOBEAN MADRIGALS. cf Enigma VAR 1017.
ELIZABETHAN SERENADE. cf Starline SRS 197.

ELWELL, Herbert
Blue symphony. cf BABIN: David and Goliath.

ELY, Robert
Trafalgar Square. cf DJM Records DJM 22062.

EMERSON, Keith
Concerto, piano, no. 1. cf Atlantic SD 2-7000.
Fanfare for the common man (arr. from Copland). cf Atlantic SD 2-7000.
Pirates. cf Atlantic SD 2-7000.
Tank. cf Atlantic SD 2-7000.

EMMANUEL, Maurice
1176 Symphony, no. 2, A major. POULENC: Concerto, 2 pianos, D minor. Marie-Jose Billard, Julien Azais, pno; ORTF Philharmonic Orchestra; ORTF National Orchestra; Jean Doussard, Maurice Suzan. Inedits ORTF 995 035.
+St 4-77 p119

EMMETT, Daniel
Dixie. cf Rubini GV 57.

ENCINA, Juan del
Ay triste que vengo. cf Nonesuch H 71326.
Congoxa mas. cf Enigma VAR 1024.
Si abra en este baldres. cf Enigma VAR 1024.
Todos los bienes. cf BIS LP 3.

ENESCO, Georges
1177 Cantabile e presto, flute and piano. POULENC: Sonata, flute and piano. PROKOFIEV: Sonata, flute and piano, op. 94, D major.

Albert Tipton, flt; Mary Norris, pno. Pandora 106.
+FF 11-77 p63

1178 Roumanian rhapsody, op. 11, no. 1, A minor. KACHATURIAN: Gayaneh: Ballet suite. LISZT: Hungarian rhapsody, no. 4, G 244, E flat major. PROKOFIEV: Love for three oranges: Suite. VPO; Constantin Silvestri. Classics for Pleasure CFP 40262.
++Gr 3-77 p1461 +RR 3-77 p48
+HFN 4-77 p151

Roumanian rhapsody, op. 11, no. 1, A minor. cf HMV ESD 7011.

1179 Roumanian rhapsodies, op. 11, nos. 1 and 2. LISZT: Hungarian rhapsodies, nos. 1-3, G 244. Orchestra; Leopold Stokowski. Quintessence PMC 7023.
-ARG 12-77 p43 +St 11-77 p146
+-HG 9-77 p108

1180 Songs, op. 15 (7). ROUSSEL: Songs: Adieu; A flower given to my daughter; Jazz dans la nuit; Light; Melodies, op. 20 (2); Odes anacreontiques, nos. 1, 5; Odelette; Poemes chinois, op. 12 (2); Poemes chinois, op. 35 (2). Yolanda Marcoulescou, s; Katja Phillabaum, pno. Orion ORS 75184. (also Saga 5416)
++Gr 5-76 p1788 +NYT 6-8-75 pD19
++HF 10-75 p78 +ON 7-77 p29
+-MT 8-76 p660 +RR 4-76 p70
+NR 7-75 p12 +-St 5-76 p114

1181 Symphony, no. 3, op. 21. Cluj-Napoca Symphony Orchestra and Chorus; Ion Baciu. Electrecord ST ECE 01234.
+FF 11-77 p20

ENGEL

Songs: Dear father; Kaddish of Reb Levi-Itzchok of Barditchev; Listen. cf Olympic OLY 105.

ENGLISH AND FRENCH PART-SONGS. cf RCA RL 2-5112.

THE ENGLISH MADRIGAL SCHOOL cf BG HM 57/8.

ENGLISH PIANO MUSIC. cf Saga 5445.

ENGLUND, Einar

1182 Epinikia. Symphony, no. 2. Helsingin Kaupunginorkesteri; Paavo Berglund, Pertti Pekkanen. Finnlevy SFX 34.
+RR 7-77 p55

Symphony, no. 2. cf Epinikia.

ERB, Donald

1183 Harold's trip to the sky. Sonneries. KOCH: Quartet, strings, no. 2. TIMM: The joiner and the diehard. Peggy Anne McMurray, ms; Marcia Ferritto, vla; Allyn Benedict, pno; Kent Williams, perc; Cleveland Orchestra String Quartet; Cleveland Orchestra Brass Section Percussion Ensemble; Donald Miller, Matthias Bamert. Crystal S 531.
+NR 4-77 p5

Sonneries. cf Harold's trip to the sky.

Trio for two. cf Finnadar SR 9015.

ERICKSON, Robert

1184 Chamber concerto. CRUMB: Night music. Hartt Chamber Players; Ralph Shapey. CRI 218.
+Gr 3-77 p1454 /RR1-77 p7

Chamber concerto. cf CRUMB: Night music I.

ERNST, Heinrich

Airs Hongrois varies, op. 22, A major. cf Discopaedia MB 1011.

1185 Concerto, violin, op. 23, F sharp minor. WIENIAWSKI: Concerto, violin, no. 1, op. 14, F sharp minor. Lukas David, vln; Prague Symphony Orchestra; Libor Hlavacek. Supraphon 110 1837.

+-FF 9-77 p18 +HFN 11-77 p171
+Gr 11-77 p837 ++RR 10-77 p45

l'ESCURIEL, Jehannot de
Amours, cent mille merciz. cf Turnabout TV 37086.

ETLER, Alvin
1186 Sonata, clarinet and piano. RACHMANINOFF (trans Pino): The harvest of sorrow, op. 4, no. 5. Oh stay my love, op. 4, no. 1. Vocalise, op. 34, no. 14. RIEPE: Studies on flight (3). SZALOWSKI: Sonatina. David Pino, clt; Frances Mitchum Webb, pno. Orion ORS 76256.
+NR 4-77 p5

EVANS
Pant y Fedwen. cf BBC REC 267.

EYBLER, Joseph
Polonaise. cf DG 2533 182.
Polonaise. cf DG 2723 051.

EYCK, Jacob van
Come again. cf BIS LP 48.
Fantasia en echo. cf BIS LP 48.

FAINI
Amore e Maggio. cf Club 99-106.
FAITH OF OUR FATHERS: Catholic hymns. cf Coimbra CCO 34.

FALLA, Manuel da
1187 El amor brujo. GRANADOS: Goyescas: Intermezzo. RAVEL: Miroirs: Alborada del gracioso. Pavane pour une infante defunte. Nati Mistral, ms; NPhO; Rafael Fruhbeck de Burgos. London STS 15358. (also Decca Tape (c) KSXC 6287)
+RR 4-77 p51 tape ++St 10-76 p121
El amor brujo: The magic circle. cf Decca DDS 507/1-2.
El amor brujo: Ritual fire dance. cf RCA CRL 1-2064.
El amor brujo (Love the magician): Ritual fire dance. cf ALBENIZ: Iberia: Triana; El corpus en Sevilla.
El amor brujo (Love the magician): Ritual fire dance. cf RCA GL 4-2125.
1188 El amor brujo: Suite. Fantasia baetica. Piezas espanolas: Aragonese, Cubana, Montanesca, Andaluza. El sombrero de tres picos: Three dances. Alicia de Larrocha, pno. Decca SXL 6683. (also London CS 6881 Tape (c) CS 5-6881)
++Gr 4-75 p1836 ++NYT 7-31-77 pD13
++HF 12-75 p92 ++RR 4-75 p50
+HF 9-77 p119 tape ++SFC 6-29-75 p26
+MM 4-76 p29 ++St 9-75 p112
Cancion del pescador and farruca. cf BIS LP 60.
1189 Concerto, harpsichord, flute, oboe, clarinet, violin and violoncello. The three-cornered hat. Jan DeGaetani, ms; Igor Kipnis, hpd; Paige Brook, flt; Harold Gomberg, ob; Stanley Drucker, clt; Eliot Chap, vln; Lorne Munroe, vlc; NYP; Pierre Boulez. Columbia M 33970 Tape (c) MT 33970 (Q) MQ 33970. (also CBS 76500 Tape (c) 40-76500)
+-Audio 10-76 p148 ++NR 7-76 p2
+Gr 8-76 p280 +RR 8-76 p36
+HF 8-76 p90 +RR 2-77 p97 tape
+HFN 8-76 p79 ++SFC 4-25-76 p30
+-HFN 12-76 p155 tape ++St 8-76 p94
+MT 1-77 p44
1190 Concerto, harpsichord, flute, oboe, clarinet, violin, and violoncello, D major. Fantasia baetica. Nights in the gardens of

Spain. Gyorgy Sandor, pno; Martin Galling, hpd; Robert Dohn, flt; Willy Schnell, ob; Hans Lemser, clt; Susanne Lautenbacher, vln; Thomas Blees, vlc; Luxembourg Radio Orchestra; Louis de Froment. Turnabout TV 34588 Tape (c) KTVC 34588.
+-Gr 2-77 p1279 /-RR 2-77 p45
+-HFN 5-77 p121 -RR 7-77 p99 tape
+HFN 4-77 p155 tape

Fantasia baetica. cf El amor brujo: Suite.
Fantasia baetica. cf Concerto, harpsichord, flute, oboe, clarinet, violin and violoncello, D major.
Fantasia baetica. cf Piano works (RCA TRL 1-7073).
Homenaje. cf Swedish Society SLT 33205.
Homenaje a Debussy. cf L'Oiseau-Lyre SOL 349.
Homenaje a Debussy. cf RCA RL 2-5099.
Homenaje a Paul Dukas. cf Piano works (RCA TRL 1-7073.

1191 Melodies (3). Spanish popular songs (7). GRANADOS: Tonadilla: La maja dolorosa. Tonadillas al estilo antiguo: Amor y odio; El majo discreto; El majo timido; El mirar de la maja; El tra la la y el punteado. TURINA: Poema en forma de canciones, op. 19. Jill Gomez, s; John Constable, pno. Saga 5409 Tape (c) CA 5409.
+Gr 2-76 p1373 ++RR 12-75 p88
+Gr 2-77 p1322 tape +-RR 1-77 p90 tape
+HFN 1-76 p107 +STL 2-8-76 p36
+HFN 10-76 p185 tape

Noches en los jardines de Espana (Nights in the gardens of Spain). cf Concerto, harpsichord, flute, oboe, clarinet, violin and violoncello, D major.
Noches en los jardines de Espana. cf CHOPIN: Concerto, piano, no. 2, op. 21, F minor.
Nocturno. cf Piano works (RCA TRL 1-7073).

1192 Piano works: Fantasia baetica. Homenaje a Paul Dukas. Nocturno. Piezas espanolas (4). Serenata andaluza. Vals capricho. Joaquin Achucarro, pno. RCA TRL 1-7073.
+Gr 12-76 p1021 +-MT 3-77 p215
+-HFN 3-77 p103 +RR 12-76 p83

Piezas espanolas (4). cf Piano works (RCA TRL 1-7073).
Piezas espanolas: Aragonese, Cubana, Montanesa, Andaluza. cf El amor brujo: Suite.
Serenata andaluza. cf Piano works (RCA TRL 1-7073).

1193 Spanish popular songs (7) (Canciones populares espanolas). GRANADOS: Tonadillas. Victoria de los Angeles, s; Gonzalo Soriano, pno. Angel S 37425.
++SFC 11-27-77 p66

Spanish popular songs (7). cf Melodies.
Spanish popular songs (7). cf Seraphim M 60291.
Canciones populares espanolas: Jota; Nana; El pano moruno. cf Orion ORS 77271.
Canciones populares espanolas, no. 6. cf Discopaedia MB 1012.

1194 The three-cornered hat (El sombrero de tres picos). Teresa Berganza, ms; BSO; Seiji Ozawa. DG 2530 823 Tape (c) 3300 823.
++FF 11-77 p21 +RR 6-77 p48
+-Gr 6-77 p48 +SFC 9-18-77 p42
+Gr 7-77 p225 tape +-St 12-77 p144
-HF 11-77 p112
+-HFN 6-77 p123

1195 The three-cornered hat. Victoria de los Angeles, s; PhO; Rafael Fruhbeck de Burgos. HMV SXLP 30187 Tape (c) TC EXE 188. (Reissue from ASD 608) (also Angel S 36235)
+Gr 2-76 p1338 ++HFN 5-76 p117 tape
+Gr 7-77 p225 tape +-RR 4-77 p91 tape
The three-cornered hat. cf Concerto, harpsichord, flute, oboe, clarinet, violin and violoncello.
El sombrero de tres picos: Dances (3). cf El amor brujo: Suite.
El sombrero de tres picos: Dances (3). cf Angel S 37409.
El sombrero de tres picos: Danse du corregidor; Danse du meunier. cf RCA RL 2-5099.
The three-cornered hat: Jota. cf BERLIOZ: Le Corsaire, op. 21.
The three-cornered hat: Jota. cf DG 2535 254.
The three-cornered hat: Suites, nos. 1 and 2. cf ALBENIZ: Iberia: Triana; El corpus en Sevilla.
The three-cornered hat: Suite, no. 2. cf HMV (SQ) ASD 3338.
Vals capricho. cf Piano works (RCA TRL 1-7073).
La vida breve: Spanish dance. cf ALBENIZ: Espana, op. 165, no. 2: Tango.

FANTINI, Girolamo
Sonata, 2 trumpets, B flat major. cf Philips 6500 926.

FARIS
Theme from Upstairs, Downstairs. cf Angel (Q) S 37304.

FARKAS, Ferenc
Passacaglia and postludium. cf Hungaroton SLPX 11808.
Walking in woods and meadows. cf Hungaroton SLPX 11823.

FARMER, John
Fair nymphs. cf Argo ZK 25.
Fair nymphs I heard one telling. cf DG 2533 347.

FARNABY, Giles
Giles Farnaby's dream. cf Argo ZRG 823.
His rest. cf Argo ZRG 823.
The new Sa-Hoo. cf Argo ZRG 823.
The old spagnoletta. cf Argo ZRG 823.
The old spagnoletta. cf National Trust NT 002.
Tell me, Daphne. cf Argo ZRG 823.
Tower hill. cf Saga 5425.
A toye. cf Argo ZRG 823.

FARNAM, Lynwood
Toccata on "O fillii et filiae". cf Orion ORS 76255.

FARNON, Robert
Portrait of a flirt. cf HMV SRS 5197.
Portrait of a flirt. cf Starline SRS 197.
Westminster waltz. cf DJM Records DJM 22062.

FARROW, Miles
Pieces. cf St. John the Divine Cathedral Church.

FASCH, Johann
1196 Concerto, guitar and strings, D minor. KREBS: Concerto, guitar and strings, G major. VIVALDI: Concerto, guitar and strings, D major. Konrad Ragossnig, gtr; South West German Chamber Orchestra; Paul Angerer. Turnabout TV 34547 Tape (c) KTVC 34547.
+Gr 8-76 p280 +RR 7-76 p65
/HFN 8-76 p76 +RR 3-77 p99 tape
Concerto, trumpet, D major. cf Argo D69D3.
1197 Concerto, trumpet, 2 oboes and strings, D major. Sinfonias, A major, G major. PACHELBEL: Canon, D major. Suites, B flat major, G major. Maurice Andre, tpt; Pierre Pierlot, Jacques

Chambon, ob; Jean-Francois Paillard Chamber Orchestra; Jean-Francois Paillard. RCA FRL 1-5468 Tape (c) FRK 1-5468 (ct) FRS 1-5468.
+-MJ 9-77 p35 +-NR 4-77 p15
Sinfonias, A major, G major. cf Concerto, trumpet, 2 oboes and strings, D major.

FAULKNER, Duncan
Intrada. cf Wealden WS 131.

FAURE, Gabriel
Apres un reve, op. 7, no. 1. cf Orion ORS 75181.
Berceuse, op. 16. cf CHAUSSON: Poeme, op. 25.

1198 La bonne chanson, op. 61. RAVEL: Chansons madecasses. POULENC: Le bal masque. Dietrich Fischer-Dieskau, bar; Wolfgang Sawallisch, pno; BPhO Soloists. BASF G 22765. (also HNH 4045)
+ARG 6-77 p52 ++St 8-77 p114
++SFC 12-25-77 p42

1199 Cantique de Jean Racine, op. 11. Requiem, op. 48. Benjamin Luxon, bar; Jonathon Bond, treble; Stephen Cleobury, org; St. John's College Chapel Choir; AMF; George Guest. Argo ZRG 841 Tape (c) 841.
-Gr 4-76 p1646 +MU 9-77 p14
+Gr 6-76 p102 tape *NR 1-77 p9
+HFN 5-76 p100 +-ON 2-5-77 p41
+HFN 7-76 p104 tape +-RR 4-76 p70
+MJ 1-77 p27 +RR 8-76 p84 tape
+MT 8-76 p660 /St 12-76 p139

1200 Chant funeraire, op. 117. KOECHLIN: Quelques chorals pour des fetes populaires. SCHMITT: Dionysiaques, op. 62, no. 1. Musique des Gardiens de la Paix; Desire Dondeyne. CRD CAL 1839. (also Musical Heritage Society MHS 3387)
+FF 9-77 p73 ++RR 11-75 p24
++Gr 10-75 p611

1201 Elegy, op. 24, C minor. LALO: Concerto, violoncello, D minor. SAINT-SAENS: Concerto, violoncello, no. 1, op. 33, A minor. Heinrich Schiff, vlc; NPhO; Charles Mackerras. DG 2530 793.
+ARG 11-77 p6 +NR 10-77 p5
++FF 11-77 p48 +NYT 7-24-77 pD11
+Gr 2-77 p1283 +RR 2-77 p58
+HF 11-77 p116 ++St 11-77 p135
+HFN 3-77 p103
Elegy, op. 24, C minor. cf ABC AB 67014.
Elegy, op. 24, C minor. cf Orion ORS 75181.
Fantaisie, flute. cf CHAMINADE: Concertino, flute, op. 107.

1202 Fantaisie, piano and orchestra, op. 111. RAVEL: Concerto, piano, for the left hand, D major. Concerto, piano, G major. Alicia de Larrocha, pno; LPO; Rafael Fruhbeck de Burgos, Lawrence Foster. Decca SXL 6680. (also London CS 6878 Tape (r) CSO 6876)
++Gr 11-74 p890 ++RR 10-74 p53
+HF 6-75 p102 ++SFC 4-27-75 p23
+HF 8-77 p96 tape ++St 7-75 p105
+NR 7-75 p6
Impromptu, no. 3, op. 34, A flat major. cf Vox SVBX 5483.
Impromptu, no. 5, op. 102, F sharp minor. cf Vox SVBX 5483.
Impromptu, no. 6, op. 86, D flat major. cf Works, selections (Decca ECS 805).

1203 Masques et bergamasques, op. 112. SAINT-SAENS: Carnival of the

animals. French Chamber Orchestra, Munich Philharmonic Orchestra; Albert Lizzio, Alfred Scholz. Pye GSGC 15020.
+HFN 1-77 p107 /RR 1-77 p58
Masques et bergamasques, op. 112. cf Works, selections (Decca ECS 805).
Masques et bergamasques, op. 112: Sicilienne. cf Prelude PRS 2512.
1204 Nell, op. 18, no. 1 (arr. Grainger). GERSHWIN: Love walked in; The man I love (arr. Grainger). GRAINGER: Country gardens. Irish tune from County Derry. Eastern intermezzo. Handel in the Strand. Knight and shepherd's daughter. Molly on the shore. To a Nordic princess. Over the hills and far away. Sailor's song. Shepherd's hey. Tribute to Foster: Lullaby. Walking tune. Daniel Adni, pno. HMV SQ HQS 1363.
+ARG 7-77 p40 +HFN 12-76 p143
+Gr 11-76 p843 +RR 11-76 p90
Nocturne, op. 74, C sharp minor. cf Vox SVBX 5483.
Papillon, op. 77. cf ABC AB 67014.
Pavane. cf Works, selections (Decca ECS 805).
1205 Pelleas et Melisande, op. 80. FRANCK: Symphony, D minor. NPhO; Andrew Davis. CBS 76526 Tape (c) 40-76526. (also Columbia M 34506 Tape (c) MT 34506)
+Gr 10-76 p596 +NR 5-77 p2
++HF 8-77 p82 +RR 11-76 p66
+HFN 10-76 p169 +-RR 1-77 p89 tape
+HFN 1-77 p123 tape +St 8-77 p114
+-MJ 7-77 p70
Pelleas et Melisande, op. 80: Incidental music. cf Works, selections (Decca ECS 805).
Penelope: Prelude. cf Works, selections (Decca ECS 805).
1206 Quartet, piano, op. 15, C minor. Quartet, strings, op. 121, E minor. Artur Rubinstein, pno; Guarneri Quartet. RCA ARL 1-0761.
+Gr 5-75 p1985 +RR 3-75 p38
+-HF 3-75 p78 ++SFC 11-24-74 p32
+MT 9-75 p798 +St 4-75 p100
++NR 1-75 p8 ++St 5-77 p73
Quartet, strings, op. 121, E minor. cf Quartet, piano, op. 15, C minor.
1207 Requiem, op. 48. Victoria de los Angeles, s; Dietrich Fischer-Dieskau, bar; Elisabeth Brasseur Chorale; Henriette Puig-Roget, org; OSCCP; Andre Cluytens. Classics for Pleasure CFP 40234 Tape (c) TC CFP 40234. (Reissue from HMV SAN 107)
+Gr 6-76 p71 +-RR 5-76 p69
+-HFN 3-76 p111 +-RR 12-77 p94 tape
1208 Requiem, op. 48. Alain Clement, treble; Philippe Huttenlocher, bar; Saint-Pierre-aux-Liens de Bulle Choir; Berne Symphony Orchestra; Philippe Corboz, org; Michel Corboz. Erato STU 70735.
+-Gr 6-73 p82 +RR 5-73 p94
+-Gr 3-77 p1438 +-RR 3-77 p91
-HFN 5-77 p123
1209 Requiem, op. 48. Pavane, op. 50. Elly Ameling, s; Bernard Kruysen, bar; Daniel Chorzempa, org; Netherlands Radio Chorus; Rotterdam Philharmonic Orchestra; Jean Fournet. Philips 6500 968 Tape (c) 7300 417.
+-Gr 3-76 p1493 +ON 2-5-77 p41
+HF 7-76 p77 +RR 2-76 p55

+HF 11-76 p153 ++St 8-76 p94
+NR 6-76 p12 +STL 2-8-76 p36

Requiem, op. 48. cf Cantique de Jean Racine, op. 11.
Requiem, op. 48: Pie Jesu. cf Argo ZK 11.
Les roses d'Ispahan, op. 39, no. 4. cf Columbia 34294.
Shylock, op. 57: Nocturne. cf Decca DDS 507/1-2.
Sicilienne, op. 78. cf ABC AB 67014.
Sicilienne, op. 78. cf HMV SQ ASD 3283.

1210 Sonata, violin, no. 1, op. 13, A major. Sonata, violin, no. 2, op. 108, E minor. Christiane Edinger, vln; Gerhard Puchelt, pno. Orion ORS 76258.
++NR 6-77 p7

Sonata, violin, no. 2, op. 108, E minor. cf Sonata, violin, no. 1, op. 13, A major.

1211 Songs, complete. Anne-Marie Rodde, Sonia Nigoghossian, s; Jacques Herbillon, bar; Theodore Paraskivesco, pno. Musical Heritage Society 3438/3443 (6).
+HF 8-77 p80 +-St 8-77 p120

1212 Songs: L'Absent; Accompangement; Apres un reve; Arpege; Au bord de l'eau; Au cimetiere; Aubade; L'Aurore; Automne; Barcarolle; Les berceaux; La bonne chanson; C'est la paix; Chanson; Chanson d'amour; La chanson d'Eve; Chanson du pecheur; Chanson de Shylock; Chant d'automne; Clair de lune; Dans le foret de Septembre; Dans les ruines d'une Abbaye; Le don silencieux; En priere; Le fee aux chansons; Fleur jetee; Le fleur qui va sur l'eau; L'horizon chimerique; Hymne; Ici-bas; Le jardin clos; Larmes; Lydia; Les matelots; Melisande's song; Mirages; Mai; Nocturne; Nell; Noel; Notre amour; Le papillon et la fleur; Le pays des reves; Pleurs d'or; Poeme d'un jour; Le plus doux chemin; Puisqu'Ici-bas toute ame; Les presents; Prison; La rancon; Le parfum imperissable; Le ramier; Reve d'amour; La rose; Les roses d'Ispahan; Le secret; Seule; Serenade Toscane; Serenade du bourgeois gentilhomme; Soir; Spleen; Tarantelle; Tristesse; Vocalise; Le voyageur. Elly Ameling, s; Gerard Souzay, bar; Dalton Baldwin, pno. Connoisseur Society CS 2127/8 (4). (also French EMI C 6512831/35)
+ARG 6-77 p5 +ON 7-77 p29
+HF 8-77 p80 +RR 2-77 p25
++MJ 5-77 p33 +SFC 2-27-77 p42
++NR 4-77 p11 ++St 8-77 p120

Songs: L'Absent, op. 5, no. 3; Apres un reve, op. 7, no. 1; Clair de lune, op. 46, no. 2; Dans les ruines d'une Abbaye, op. 2, no. 1; Ici-bas, op. 8, no. 3; Nell, op. 18, no. 1; Les roses d' Ispahan, op. 39, no. 4; Le secret, op. 23, no. 3; Soir, op. 83, no. 2. cf HMV RLS 716.

1213 Songs: Apres un reve, op. 7, no. 1; Au bord de l'eau, op. 8, no. 1; Chanson d'amour, op. 27, no. 1; Clair de lune, op. 46, no. 2; Lydia, op. 4, no. 2; Nell, op. 18, no. 1; Sylvie, op. 6, no. 3. HAHN: Songs: D'une prison; L'heure exquise; Mai; Le rossignol des lilas; Offrande; Paysage; Si mes vers avaient des ailes. MASSENET: Songs: Chant provencal; Elegie; Nuit d' espagne; Serenade d'automne; Stances; Un adieu; Vous aimerez demain. Martyn Hill, t; John Constable, pno. Saga 5419 Tape (c) CA 5419.
+Gr 12-76 p1042 +-MT 6-77 p487
+Gr 10-77 p706 tape +-RR 12-76 p93
+-HFN 1-77 p107 +RR 12-77 p94 tape
+HFN 11-77 p187 tape

1214 Songs: L'amour de moi (arr. Tiersot); L'horizon chimerique, op. 118; En venant de Lyon; Musette; Nocturne, op. 43, no. 2; Tambourin (arr. Tiersot); Poeme d'un jour. LULLY: Alceste: Air de Caron. POULENC: Songs: L'anguille; La belle jeunesse; C; Priez pour paix; Serenade. Martial Singher, bar; Alden Gilchrist, pno and hpd. 1750 Arch 1754.
+ARG 6-77 p23

Songs: Ave verum. cf Argo ZRG 871.
Songs: Le crucifix. cf Club 99-96.
Songs: Le crucifix. cf Rubini GV 26.
Songs: The crucifix. cf Rubin RS 301.

1215 Trio, piano, op. 120, D minor. SHOSTAKOVICH: Trio, piano, no. 2, op. 67, E minor. Hans Palsson, pno; Arve Tellefsen, vln; Frans Helmerson, vlc. BIS LP 26. (also HNH 4007)
+Gr 3-77 p1454 +-RR 9-76 p76
+HF 9-77 p108 +SFC 1-24-77 p40
+HFN 10-76 p169 ++St 5-77 p118
++NR 10-77 p6

1216 Works, selections: Impromptu, no. 6, op. 86, D flat major. Masques et bergamasques, op. 112. Pavane. Pelleas et Melisande, op. 80: Incidental music. Penelope: Prelude. Osian Ellis, hp; OSR, NSL; Ernest Ansermet, Raymond Agoult. Decca ECS 805. (Reissues from SXL 2303, L'Oiseau-Lyre SOL 308, RCA SF 5054)
+Gr 11-77 p837 +-RR 11-77 p56
+HFN 11-77 p185

FAYRFAX, Robert
I love, loved; Thatt was my woo. cf L'Oiseau-Lyre 12BB 203/6.

FELD, Jindrich

1217 Concerto, flute. Symphony, no. 1. Jean-Pierre Rampal, flt; CPhO; Antonio de Almeida, Vaclav Jiracek. Serenus SRS 12074.
+-St 12-77 p146

1218 Concerto, piano. Quintet, woodwinds, no. 2. Sonata, piano. Bozena Steinerova, pno; Prague Wind Quintet, CPhO; Otakar Trhlik. Serenus SRS 12075.
+NR 11-77 p10

Quintet. cf Crystal S 206.
Quintet, woodwinds, no. 2. cf Concerto, piano.
Quintet, woodwinds, no. 2. cf BARTA: Quintet, winds, no. 2.
Sonata, piano. cf Concerto, piano.
Symphony, no. 1. cf Concerto, flute.

FELDMAN, Morton

1219 For Frank O'Hara. Rothko chapel. Karen Phillips, vla; James Holland, perc; Gregg Smith Singers; Center of the Creative and Performing Arts, State University of New York at Buffalo, Members; Jan Williams, Gregg Smith. Odyssey Y 34138.
++ARG 12-76 p28 +NR 10-76 p7
+HF 12-76 p126 *ON 11-76 p98
+MJ 3-77 p46 +St 2-77 p124

Rothko chapel. cf For Frank O'Hara.

FELTON
Concerto, op. 2, no. 4, C major. cf Folkways FTS 32378.
Eighteenth century settings of the Star-Spangled Banner. cf Folkways FTS 32378.

FERANDIERE
Rondo. cf Lyrichord LLST 7299.

FERGUSON
The lark in the clear air. cf Philips 6599 227.

FERNANDEZ, Oscar
Reisado do pastoreio: Batuque. cf CBS 61780.
FERRABOSCO, Alfonso I
Pavan. cf RCA RL 2-5110.
Songs: Come my Celia; If all these cupids; It was no policy of court; So beauty on the waters stood; Yes were the loves. cf Enigma VAR 1023.
Spanish pavan. cf Saga 5438.
FERRARIS
Souvenir d'Ukraine. cf Rediffusion 15-57.
FETLER, Paul
1220 Pastorale suite. FOOTE: Trio, violin, violoncello and piano, op. 5, C minor. Macalester Trio. Golden Crest CRS 4153.
-ARG 5-77 p20
FIBICH, Zdenek
1221 Quintet, violin, clarinet, horn, violoncello and piano, op. 42, D major. Trio, piano, F minor. Karel Dlouhy, clt; Zdenek Tylsar, hn; Fibich Trio. Supraphon 111 1617.
+ARG 12-76 p29 +RR 8-76 p62
+Gr 9-76 p437 ++St 2-77 p117
+NR 11-76 p6
Trio, piano, F minor. cf Quintet, violin, clarinet, horn, violoncello and piano, op. 42, D major.
FIEDLER'S FAVORITE OVERTURES. cf Quintessence PMC 7013.
FIELD, John
Nocturne, no. 9, E minor. cf International Piano Library IPL 102.
1222 Nocturne, no. 11, E flat major. Nocturne, no. 12, G major. Pastorale, E major. Sonata, piano, no. 4, B major. MENDELSSOHN: Songs without words, op. 30, no. 6; op. 38, no. 6; op. 53, no. 4; op. 62, nos. 5, 6; op. 67, no. 5. Etudes, op. 104 (3). Richard Burnett, pno. Prelude PRS 2504.
+-Gr 12-76 p1027 +MT 9-77 p736
+HFN 12-76 p141 +-RR 12-76 p85
Nocturne, no. 12, G major. cf Nocturne, no. 11, E flat major.
Pastorale, E major. cf Nocturne, no. 11, E flat major.
Sonata, piano, no. 4, B major. cf Nocturne, no. 11, E flat major.
FILLMORE, Henry
Americans we. cf MENC 76-11.
Trombone family. cf Nonesuch H 71341.
FINCK, Hermann
Sauff aus und machs nit lang. cf Argo D40D3.
FINE, Irving
The choral New Yorker. cf Vox SVBX 5353.
Partita, wind quintet. cf Vox SVBX 5307.
FINZI, Gerald
1223 Concerto, clarinet, op. 31. Eclogue, op. 10. Grand fantasia and toccata, op. 38. John Denman, clt; Peter Katin, pno; NPhO; Vernon Handley. Lyrita HNH 4031. (also Lyrita SRCS 92)
+Gr 4-77 p1548 +RR 4-77 p45
+-HF 12-77 p88 ++St 11-77 p161
+-HFN 4-77 p136
Earth and air and rain, op. 15. cf BUTTERWORTH: Songs (Argo 838).
Eclogue, op. 10. cf Concerto, clarinet, op. 31.
God is gone up with a triumphant shout. cf Abbey LPB 783.
Grand fantasia and toccata, op. 38. cf Concerto, clarinet, op. 31.
White flowering days. cf RCA GL 2-5062.

FIOCCO, Joseph-Hector
Suite, no. 1, G major: Allegro. cf Discopaedia MB 1013.
FIORILLO, Federigo
Hanoverian air. cf Serenus SRS 12067.
FIRENZE, Lorenzo di
Da da a chi avareggia. cf Argo D40D3.
FISCHER
In tiefem Keller. cf Club 99-102.
FISCHER, Johann
Bouree. cf DG 2723 051.
Gigue. cf DG 2723 051.
Musikalisches Blumenbuschlein: Suite, no. 6, D major. cf Toccata 53623.
Preludes and fugues, B minor, D major, E flat major, C minor. cf ABC ABCL 67008.
Suite, no. 6, F major. cf Supraphon 111 1867.
FISCHER, William
I love to tell the story. cf New World NW 220.
FLAGG
Trumpet tune with fuguing section. cf Folkways FTS 32378.
FLAMENCO: Carabana gitana; Fandangos por verdiales; Farruca y rumba; Fiesta en Jerez; Garrotin; Jota; Lamento andaluz; Peterneras; Tanguillos Zorongo. cf GIULIANI: Concerto, guitar, op. 30, A major.
FLEGIER
Le cor. cf Club 99-107.
FLETCHER, Percy
Epic symphony. cf BALL: Sinfonietta, brass band.
Epic symphony: Elegy. cf Grosvenor GRS 1052.
The spirit of pageantry. cf Decca SB 715.
FLOSMAN, Oldrich
1224 Concerto, violin, no. 2. VALEK: Symphony, no. 10. Andre Gertler, Jiri Tomasek, vln; Josef Ruzicka, pno; Prague Radio Symphony Orchestra; Milos Konvalinka, Vladimir Valek. Supraphon 110 1750.
+ARG 9-77 p34 +NR 9-77 p3
+-Gr 9-77 p438 +RR 6-77 p70
+HFN 7-77 p113
Sonata, wind quintet and piano. cf BARTA: Quintet, winds, no. 2.
FLOTHUIS, Marius
1225 Songs: Hymnus; Per sonare ed ascoltare. HENKEMANS: Primavera; Tre aspetti d'amore. Erna Spoorenberg, s; Hubert Barwahser, flt; Netherlands Radio Chorus; COA, Netherlands Radio Chamber Orchestra; Bernard Haitink, Roelof Krol. Donemus CV 7603.
+-RR 2-77 p85
FLOTOW, Friedrich
Allesandro stradella: Overture. cf Supraphon 110 1637.
Martha: Ach, so fromm. cf RCA CRM 1-1749.
Martha: Ach, so fromm. cf Telefunken AJ 6-42232.
Martha: Cazone del porter. cf Rubini GV 39.
Martha: M'appari. cf RCA TVM 1-7201
Martha: M'appari. cf Seraphim S 60280.
FO (15th century Italy)
Tua voisi esser sempre mai. cf Nonesuch H 71326.
FOGLIANO, Giacomo
Tua volsi esser sempre mai. cf Enigma VAR 1024.

FONTANA, Giovanni
Sonata, violin. cf L'Oiseau-Lyre 12BB 203/6.
Sonata, prima. cf BIS LP 48.
FOOTE, Arthur
Characteristic pieces (3). cf Orion ORS 76247.
Francesca da Rimini. cf Louisville LS 753/4.
Sonata, violin and piano, op. 20, G minor. cf BEACH: Sonata, violin and piano, op. 34, A minor.
Sonata, violin and piano, op. 20, G minor. cf CARPENTER: Sonata, violin and piano.
Trio, violin, violoncello and piano, op. 5, C minor. cf FETLER: Pastoral suite.
FORD
The pill to purge melancholy. cf National Trust NT 002.
FORQUERAY, Antoine
1226 Pieces de clavecin. Gustav Leonhardt, hpd. ABC ABCL 67009.
++HF 3-77 p108 +NR 3-77 p14
+-MJ 2-77 p30 +St 5-77 p54
FORSTER, Georg
Vitrum nostrum gloriosum. cf L'Oiseau-Lyre 12BB 203/6.
FORSTER, Josef Bohuslav
1227 Quintet, winds, op. 95. KLUGHARDT: Quintet, winds, op. 79. Bohm Quintette. Orion ORS 76254.
+ARG 8-77 p21 +MJ 9-77 p35
+FF 11-77 p31 ++NR 9-77 p5
FOSCARINI, Giovanni
Il furioso. cf Saga 5420.
FOSS, Lukas
Behold I build an house. cf Vox SVBX 5353.
The cave of the winds. cf Vox SVBX 5307.
FOSTER, Stephen
Jeanie with the light brown hair. cf CBS 61039.
Jeanie with the light brown hair. cf Grosvenor GRS 1052.
Jeanie with the light brown hair. cf HMV RLS 715.
The old folks at home. cf Discopaedia MB 1012.
The old folks at home. cf HMV RLS 719.
Soiree polka. cf Songs (Nonesuch H 71333).
1228 Songs: Better times are coming; Come where my love lies dreaming; Come with thy sweet voice again; Hard times come again no more; The hour for thee and me; Katy Bell; Larry's good bye; Linger in blissful repose; My old Kentucky home; Summer longings; There are plenty of fish in the sea; The voice of bygone days; We are coming Father Abraham 300,000 more. Soiree polka. Village belle polka. Jan DeGaetani, ms; Leslie Guinn, bar; Camerata Chorus, Washington; Gilbert Kalish, pno, melodeon; Douglas Koeppe, flt, pic; Howard Bass, gtr; James Weaver, pno. Nonesuch H 71333.
+ARG 2-77 p50 +HFN 1-77 p107
+Gr 3-77 p1461 ++NR 8-77 p11
+Ha 3-77 p112 +ON 3-19-77 p40
+-HF 4-77 p100 +-RR 1-77 p80
1229 Songs: Ah, may the red rose live alway; Beautiful dreamer; Come where my love lies dreaming; De camptown races; I dream of Jeanie with the light brown hair; Massa's in de cold, cold ground; Oh Susanna; Old black Joe; Old folks at home; My old Kentucky home. Richard Crooks, t; The Balladeers; Frank La Forge, pno. RCA AVM 1-1738.

+NR 6-77 p13 +St 12-76 p149
*ON 11-76 p98

Village belle polka. cf Songs (Nonesuch H 71333).

FRANCAIX, Jean

1230 Concerto, piano and orchestra. Rhapsody, viola and chamber orchestra. Suite, violin and orchestra. Claude Paillard-Francaix, pno; Susanne Lautenbacher, vln; Ulrich Koch, vla; Luxembourg Radio Orchestra; Jean Francaix. Turnabout TV 34552.
/HF 4-77 p100 ++SFC 12-8-74 p36
++NR 1-75 p4

1231 Divertimento, flute and chamber orchestra. Suite, solo flute. Quintet, winds. Ransom Wilson, flt; Orpheus Chamber Ensemble, Musical Heritage Wind Quintet. Musical Heritage Society MHS 3286.
+HF 4-77 p100

Quintet, winds. cf Divertimento, flute and chamber orchestra.
Rhapsody, viola and chamber orchestra. cf Concerto, piano and orchestra.
Suite, solo flute. cf Divertimento, flute and chamber orchestra.
Suite, violin and orchestra. cf Concerto, piano and orchestra.
Suite Carmelite. cf Pelca PSR 40599.

FRANCHOS

Trumpet intrada. cf Argo ZRG 823.
FRANCIS CHAPELET: Organs of Spain. cf Harmonia Mundi HM 759.

FRANCISCUS (FRANCISQUE), Antoine

Phiton, Phiton. cf HMV SLS 863.

FRANCK, Cesar

1232 Le chasseur maudit. Nocturne. Psyche. Christa Ludwig, ms; Orchestre de Paris; Daniel Barenboim. DG 2530 771.
+Gr 12-76 p995 +NR 5-77 p13
+HFN 12-76 p143 +-RR 12-76 p55
/HF 4-77 p101

Chorale, B minor. cf RC Records RCT 101.
Chorale, no. 1, E major. cf Guild GRSP 7011.

1233 Chorale, no. 2, B minor. Piece heroique. LISZT: Variations on Bach's "Weinen, Klagen, Sorgen, Zagen", G 673. REGER: Pieces, op. 145, no. 2: Dankpsalm. SAINT-SAENS: Prelude and fugue, op. 99, no. 3, E flat major. Nicholas Danby, org. CBS 76514.
++Gr 9-76 p454 +-MM 7-77 p37
+HFN 9-76 p127 +-RR 2-77 p81

Chorale, no. 2, B minor. cf DURUFLE: Suite, op. 5: Toccata, B minor.
Chorale, no. 2, B minor. cf L'Oiseau-Lyre SOL 343.
Chorale, no. 2, B minor. cf Vista VPS 1030.
Chorale, no. 3, A minor. cf Odeon HQS 1356.
Chorale, no. 3, A minor. cf BRAHMS: Fugue, A flat minor.

1234 Les Djinns. Symphonic variations. d'INDY: Symphony on a French mountain air, op. 25. Aldo Ciccolini, pno; Orchestre de Paris, Orchestre de Liege; Paul Strauss, Serge Baudo. Angel (Q) S 37247.
+ARG 6-77 p21 +-NR 7-77 p13
+HF 8-77 p82 +SR 5-28-77 p42
+MJ 7-77 p68 +St 6-77 p127

1235 Les Djinns. Les eolides. Symphonic variations. Mark Westcott, pno; RPO; Paul Freeman. Musical Heritage Society MHS 3515 Tape (c) MHC 5515.
+-HF 8-77 p32 +St 6-77 p127
+MJ 7-77 p68

Les eolides. cf Les Djinns.
Nocturne. cf Le chasseur maudit.
Panis Angelicus. cf Argo ZK 11.
Panis Angelicus. cf Decca SPA 491.
Panis Angelicus. cf Decca SXL 6781.
Panis Angelicus. cf London OS 26437.
Panis Angelicus. cf Philips 9500 218.
Piece heroique. cf Chorale, no. 2, B minor.
Piece heroique. cf Abbey LPB 719.
Prelude, chorale and fugue. cf Vox SVBX 5483.

1236 Psyche. Belgian Radio Chorus; Orchestre de Liege; Paul Strauss. Connoisseur Society (Q) CSQ 2096. (also HMV SQ ASD 3164)
+-ARG 11-76 p21 +NR 9-76 p4
+Gr 3-76 p1464 +RR 2-76 p29
+-HF 6-77 p90 /SFC 1-2-77 p34
+-HFN 3-76 p92 ++St 12-76 p140

Psyche. cf Le chasseur maudit.
Redemption. cf Symphony, D minor.

1237 Sonata, flute and piano, A major. PROKOFIEV: Sonata, flute and piano, op. 94, D major. James Galway, flt; Martha Argerich, pno. RCA LRL 1-5095 Tape (c) RK 2-5029.
++Gr 11-75 p848 +-RR 10-75 p98
++HFN 12-75 p153 +-RR 7-77 p99 tape
+-HFN 5-77 p139 tape

1238 Sonata, violin and piano, A major. SZYMANOWSKI: Mythes, op. 30. Wanda Wilkomirska, vln; Antonio Barbosa, pno. Connoisseur Society CSQ 2050.
++MJ 2-77 p30

1239 Sonata, violin and piano, A major. PROKOFIEV: Sonata, violoncello and piano, op. 119, C major. Frans Helmerson, vlc; Hans Palsson, pno. HNH 4028. (also BIS LP 35)
+Gr 3-77 p1454 +RR 1-77 p73
+HFN 2-77 p119 ++St 5-77 p112

1240 Sonata, violin and piano, A major. PIERNE: Sonata, flute and piano, op. 36. Jean-Pierre Rampal, flt; Pierre Barbizet, pno. Odyssey Y 34615.
+NR 11-77 p9

1241 Sonata, violin and piano, A major. MILHAUD: Sonatas, viola and piano, nos. 1 and 2. Bernard Zaslav, vla; Naomi Zaslav, pno. Orion ORS 75186.
+ARG 6-77 p23 ++NR 7-76 p5

Sonata, violin and piano, A major. cf Hungaroton SLPX 11825.

1242 Symphonic variations. RACHMANINOFF: Concerto, piano, no. 3, op. 30, D minor. Walter Gieseking, pno; COA; Willem Mengelberg. Bruno Walter Society IGI 358.
+NR 12-77 p4

1243 Symphonic variations. GRIEG: Concerto, piano, op. 16, A minor. Gyorgy Cziffra, pno; Budapest Symphony Orchestra; Gyorgy Cziffra, Jr. Connoisseur Society CS 2090.
+HF 2-77 p99 +-St 12-76 p141
+-SFC 4-25-76 p30

1244 Symphonic variations. Symphony, D minor. Sylvia Kersenbaum, pno; Bournemouth Symphony Orchestra; Paavo Berglund. HMV ASD 3308 Tape (c) TC ASD 3308.
+Gr 2-77 p1279 +-HFN 5-77 p138 tape
-Gr 4-77 p1603 tape +RR 2-77 p46
+HFN 3-77 p103 +-RR 7-77 p99 tape

1245 Symphonic variations. Symphony, D minor. Pascal Roge, pno; CO; Lorin Maazel. London CS 7044 Tape (c) 5-7044. (also Decca SXL 6823 Tape (c) KSXC 6823)
+-Gr 4-77 p1548 +NR 10-77 p3
+-Gr 5-77 p1738 tape +RR 4-77 p46
-HF 8-77 p83 +-RR 7-77 p99 tape
+HFN 4-77 p136 ++SFC 5-15-77 p50
+HFN 5-77 p138 tape ++St 8-77 p114

1246 Symphonic variations. TCHAIKOVSKY: Concerto, piano, no. 1, op. 23, B flat minor. Earl Wild, pno; LPO; Anatole Fistoulari, Massimo Freccia. Quintessence PMC 7003.
+HF 9-77 p108 +St 11-77 p146

1247 Symphonic variations. Symphony, D minor. Ilse von Alpenheim, pno; RPO; Antal Dorati. Turnabout TVS 34663 Tape Vox (c) CT 2125.
/ARG 6-77 p24 +-NR 5-77 p2
+HF 8-77 p82 +-St 6-77 p127
+MJ 7-77 p70

Symphonic variations. cf Les Djinns (Angel 37247).
Symphonic variations. cf Les Djinns (Musical Heritage Society MHS 3515).
Symphonic variations. cf BACH: Concerto, harpsichord, no. 5, S 1056, F minor.

1248 Symphony, D minor. Redemption. Orchestre de Paris; Daniel Barenboim. DG 2530 707 Tape (c) 3300 707.
+Gr 10-76 p596 +NR 2-77 p3
+-HF 4-77 p101 -RR 9-76 p48
+HFN 10-76 p169 -SFC 11-28-76 p45
-MJ 4-77 p33 +-St 4-77 p119

1249 Symphony, D minor. BRSO; Lorin Maazel. DG 2535 156 Tape (c) 3335 156. (Reissue from SLPM 138693)
++Gr 4-76 p1598 +-RR 3-76 p39
+HFN 3-76 p109 -RR 2-77 p97 tape

1250 Symphony, D minor. WAGNER: Siegfried Idyll. VPO; Wilhelm Furtwangler, Hans Knappertsbusch. London R 23207. (Reissues from London LL 967, 1250)
/-HF 4-77 p101

1251 Symphony, D minor. LISZT: Les preludes, G 97. NSL; Adrian Boult. Quintessence PMC 7050.
+SFC 10-30-77 p44

1252 Symphony, D minor. RCA Victor Orchestra; Adrian Boult. RCA GL 2-5004 Tape (c) GK 2-5004. (Previously issued by Reader's Digest)
+Gr 10-76 p596 +RR 11-76 p66
+HFN 12-76 p143 +RR 12-76 p104 tape
+HFN 12-76 p153 tape +ST 1-77 p777

Symphony, D minor. cf Symphonic variations (HMV ASD 3308).
Symphony, D minor. cf Symphonic variations (London 7044).
Symphony, D minor. cf Symphonic variations (Turnabout TVS 34663).
Symphony, D minor. cf FAURE: Pelleas et Melisande, op. 80.

FRANCK, Melchior

1253 Fuhrwahr, Fuhrwahr. Hohenlied motets: Meine Schwester, liebe Braut; Ich such des Nachts; Fahet uns die Fuchse; O das ich dich, mein Bruder; Du bist aller Dinge schon. JOSQUIN DES PRES: Motets: In domino confido; Tribulatio et angustia; Jubilate Deo; O bone et dulcissime Jesu. The Canby Singers; Edward Tatnall Canby. Telarc 5024.

+Audio 8-77 p94 ++NR 9-77 p11

Hohenlied motets: Meine Schwester, liebe Braut; Ich sucht des Nachts; Fahet uns die Fuchse; O das ich dich, mein Bruder; Du bist aller Dinge schon. cf Fuhrwahr, Fuhrwahr.

FRANK PATTERSON: John McCormack favorites. cf Philips 9500 218.

FRANZ, Robert

O thank me not, op. 14, no. 1. cf HMV RLS 716.

Songs: Aus meinen grossen Schmerzen, op. 5, no. 1; Im Herbst, op. 17, no. 6. cf Club 99-108.

FRASER

A southern maid: Love's cigarette. cf Decca SDD 507.

FREDERIKSEN

Copenhagen march. cf Decca SB 713.

FREIRE

Ay, ay, ay. cf Philips 6392 023.

FRESCOBALDI, Girolamo

Air and variations. cf Classics for Pleasure CFP 40012.

Aria detto "La Frescobalda). cf RCA EK 1-1735.

Canzoni, D major, G major, C major. cf Arion ARN 90416.

Capriccio sopra un soggetto. cf Philips 6500 926.

Cento. cf Argo ZRG 806.

Partite sopra passacagli. cf Argo ZRG 806.

Toccata. cf L'Oiseau-Lyre 12BB 203/6.

1254 Toccatas, no. 1, G major; no. 9, F major. ROSSI: Toccata, no. 7, D major. SCARLATTI, D.: Sonata, harpsichord, G minor. SWEELINCK: Est-ce Mars. Fantasia ut re mi fa sol la. Toccata, A major. Ton Koopman, hpd, virginal. Telefunken AW 6-42157.

+Gr 11-77 p868 +HFN 11-77 p171
+HF 12-77 p112 +RR 10-77 p79

Toccata, no. 9, F major. cf Toccata, no. 1, G major.

FRICKER, Peter

Sonata, violoncello and piano, op. 28. cf BERKELEY: Duo, violoncello and piano.

1255 Symphony, no. 1, op. 9. Louisville Orchestra; Robert Whitney. RCA GL 2-5057.

+Gr 6-77 p48 +-RR 7-77 p55
+-HFN 8-77 p81 +ST 8-77 p343

FRIEDMANN

Slavonic rhapsody, no. 1. cf Decca SB 328.

FRIML, Rudolf

The vagabond king: Only a rose. cf HMV RLS 715.

FRITZ WUNDERLICH IN VIENNA. cf Polydor 2489 542.

FROBERGER, Johann

Capriccio, no. 8. cf ABC ABCL 67008.

Ricercare, no. 1. cf ABC ABCL 67008.

FROHLICH, Fredrich

Wem Gott will rechte Gunst erweisen. cf Telefunken DT 6-48085.

FRUMERIE, Gunnar de

1256 Songs: Det blir vackert dar du gar; Det kom ett brev; Karleckens visa; Lat mig ga vilse i ditt ijus; Saliga vantan. NYSTROEM: Bara hos dem; Vitt land; Onskan. RANGSTROM: Flickan under nymanen; Pan; Villemo Villemo; Notturno. STENHAMMAR: Det far att skepp; Flickan knyter i Johannenatten; Flickan Kom ifran sin alsklings mote; I skogen. Kerstin Meyer, con; Elisabeth Soderstrom, s; Jan Eyron, pno. Swedish Society SLT 33171.

+Gr 3-77 p1457 +RR 11-76 p102

FRY, William
Adieu. cf New World Records NW 257.
FUCIK, Julius
1257 Danubia, op. 229. Drachsel, op. 112. Elbthalgruss, op. 246. Die Erinnerung an Trient, op. 287. Fest und Treu, op. 177. Hortstein, op. 30. Kinizsi, op. 80. The Mississippi River, op. 160. Die Regimentskinder, op. 169. Salve imperator, op. 224. Sarajevo, op. 66. Schneidig vor, op. 79. Das Siegsschwert, op. 260. Unter der Admiralsflagge, op. 82. (arr. Urbanec) Czechoslovak Military Band; Rudolf Urbanec. Supraphon SQ 414 1714.
+FF 9-77 p19 +RR 8-77 p55
+Gr 9-77 p518
Donausagen, op. 333. cf Works, selections (Quintessence PMC 7038).
Drachsel, op. 112. cf Danubia, op. 229.
Elbthalgruss, op. 246. cf Danubia, op. 229.
Entry of the gladiators, op. 68. cf Works, selections (Quintessence PMC 7038).
Entry of the gladiators, op. 68. cf Grosvenor GRS 1055.
Die Erinnerung an Trient, op. 287. cf Danubia, op. 229.
Fearless and true march. cf Decca SB 713.
Fest und Treu, op. 177. cf Danubia, op. 229.
Florentine march, op. 214. cf Works, selections (Quintessence PMC 7038).
Herzogewina, op. 235. cf Works, selections (Quintessence PMC 7038).
Hortstein, op. 30. cf Danubia, op. 229.
Kinizsi, op. 80. cf Danubia, op. 229.
Marinarella, op. 215. cf Works, selections (Quintessence PMC 7038).
The Mississippi River, op. 160. cf Danubia, op. 229.
The old bear with a sore head. cf Works, selections (Quintessence PMC 7038).
Die Regimentskinder, op. 169. cf Danubia, op. 229.
Salve imperator, op. 224. cf Danubia, op. 229.
Sarajevo, op. 66. cf Danubia, op. 229.
Schneidig vor, op. 79. cf Danubia, op. 229.
Das Siegesschwert, op. 260. cf Danubia, op. 229.
Unter der Admiralsflagge, op. 82. cf Danubia, op. 229.
Winter storms, op. 184. cf Works, selections (Quintessence PMC 7038).
1258 Works, selections: Donausagen, op. 333. Entry of the gladiators, op. 68. Florentine march, op. 214. Herzogewina, op. 235. Marinarella, op. 215. The old bear with a sore head. Winter storms, op. 184. CPhO; Vaclav Neumann. Quintessence PMC 7038.
++ARG 12-77 p43 +St 11-77 p146
+FF 9-77 p20
FULDA
Dies est laeticiae. cf ABC L 67002.
FURTWANGLER, Wilhelm
1259 Symphony, no. 2. BPhO; Wilhelm Furtwangler. DG 2707 086 (2). (Reissue from LPM 18114/5)
+-HF 4-77 p101 +SFC 10-3-76 p33
++NR 10-76 p2 +St 3-77 p128
FUX, Johann
Sonata quinta. cf ABC ABCL 67008.
GABAYE, Pierre
Sonatine, flute and bassoon. cf Crystal S 351.

GABER, Harley
1260 The winds rise in the north. Linda Cummiskey, Malcolm Goldstein, Kathy Seplow, vln; Stephen Reynolds, vla; David Gibson, vlc. Titanic TI 16/7 (2).
-MJ 3-77 p46 -NR 6-77 p7
GABRIEL (15th century Spain)
De la dulce. cf Enigma VAR 1024.
GABRIELI, Andrea
La bataille de Marignan. cf ALBINONI: Adagio.
Conzona francese. cf L'Oiseau-Lyre 12BB 203/6.
Ricercar del secondo tono. cf Telefunken AW 6-42033.
Songs (choral): Jubilate Deo; Maria stabat; Egredimini et videte; O Rex gloriae; Te Deum patrem ingenitum. cf Argo ZRG 859.
GABRIELI, Domenico
Sonata, trumpet, D major. cf HMV ASD 3318.
GABRIELI, Giovanni
Canzon septimitoni. cf BIS LP 59.
Canzoni a 4, 5 and 6. cf Telefunken AW 6-42033.
Canzoni, nos. 5, 8. cf ALBINONI: Adagio.
Canzoni, nos. 5, 8. cf ALBINONI: Adagio, organ and strings, G minor.
1261 O magnum mysterium. JOSQUIN DES PRES: Ave Maria. PRAETORIUS: Courante. Songs (vocal and choral): Es ist ein Ros Entsprungen; In dulci jubilo; Joseph Lieber Joseph mein; Ein Kind geborn zu Bethlehem; Nun komm, der Heiden Heiland; Psallite ungenito Christo; Philov-Volte-Philov. ANON.: (Spanish villancicos) Dadme albricias; E la Don, Don, Verges Maria; Riu, riu, chiu. Waverly Consort; Michael Jaffee, dir. Columbia M 34554.
++St 12-77 p83
Sanctus Dominus Deus. cf L'Oiseau-Lyre 12BB 203/6.
Songs (choral): Cantate domine; Exultat iam Angelica; Deus, Deus meus; Beata es virgo. cf Argo ZRG 859.
GABURO, Kenneth
For Harry. cf AREL: Music for a sacred service: Prelude and postlude.
Lemon drops. cf AREL: Music for a sacred service: Prelude and postlude.
GADE
Festligt praeludium "Lover den Herre". cf Guild GRSP 7011.
GADE, J.
Jalousi (Tango Tzigane). cf RCA CRL 1-2064.
GALIANO, Marco da
1262 La Dafne. Mary Rawcliffe, Maurita Thornburgh, Sue Harmon, Robert White, Dale Terbeek; Musica Pacifica; Paul Vorwerk. ABC AB 67012/2 (2) Tape (c) 5306 67012. (Reissue from COMS 9004)
+ARG 12-77 p32
1263 La Dafne. Norma Lerer, Barbara Schlick, Ine Kollecker, s; Nigel Rogers, Ian Partridge, t; David Thomas, Berthold Possemeyer, bs; Hamburg Monteverdi Choir; Hamburg Camerata Accademica; Jurgen Jurgens. DG 2533 348.
+-Gr 9-77 p475 +-RR 8-77 p32
+HFN 8-77 p81
Songs: Valli profonde. cf DG 2533 305.
GALILEI, Michelangelo
For the Duke of Bavaria. cf Saga 5420.
Suite. cf L'Oiseau-Lyre SOL 349.

GALILEI, Vicenzo
Fuga a l'unisono. cf DG 2533 323.
GALLIARD, Johann
Sonata, bassoon, A minor. cf Orion ORS 77269.
GALLICULUS
Songs: In natali domini; Magnificat V toni. cf ABC L 67002.
GANNE
Marche Lorraine. cf Decca SB 713.
GARDNER, John
Theme and variations, op. 7. cf ADDISON: Divertimento, op. 9.
A GARLAND FOR DR. K. cf Universal Edition UE 15043.
A GARLAND FOR THE QUEEN. cf RCA GL 2-5062.
GARTNER
Aus Wien. cf Discopaedia MB 1012.
GATES
If. cf Decca SB 328.
Incantation and ritual. cf Coronet LPS 3036.
Oh, my luve's like a red, red rose. cf Columbia M 34134.
GAUCELM FAIDIT
Fortz chausa es. cf Argo D40D3.
GAULTIER, Denis
1264 La rhetorique des dieux: Suites, nos. 1-2, 12. Hopkinson Smith, lt. Telefunken AW 6-42122.
++Gr 12-77 p1117 ++RR 12-77 p61
GAY, John
1265 The beggar's opera (Pepusch arr. Austin). Argo Chamber Ensemble; Richard Austin. Argo DPA 591/2. (Reissue from RG 76/8)
+-Gr 12-77 p1134 +-RR 11-77 p37
+-HFN 11-77 p185
1266 The beggar's opera (arr. Austin). Elsie Morison, s; Monica Sinclair, Constance Shacklock, con; Alexander Young, t; John Cameron, bar; Ian Wallace, Owen Brannigan, bs; Pro Arte Chorus and Orchestra; Malcolm Sargent. HMV ESDW 704 (2) Tape (c) TC 2 ESDW 704. (Reissue from CLP 1052/3, CSD 1516/7) (also Seraphim S 6023)
+-Gr 6-77 p96 +HFN 10-77 p169 tape
+Gr 8-77 p346 tape +-RR 8-77 p32
+HFN 7-77 p125 +RR 11-77 p119 tape
GEEHL, Henry
For you alone (2). cf HMV RLS 715.
Romanza. cf Decca SB 329.
GEFORS, Hans
Songs about trusting (4). cf BIS LP 32.
GELBERT
Keep moving on. cf Olympic OLY 105.
GEMINIANI, Francesco
Concerto grosso, D minor (from Corelli's Sonata, violin, op. 5, no. 12). cf DG 2548 219.
1267 Concerti grossi, opp. 2, 3, 4. South West German Chamber Orchestra; Paul Angerer. Vox QSVBX 594 (3).
+-NR 10-77 p6
1268 Concerti grossi, op. 3, nos. 1-6. Jaap Schroder, vln; Academy of Ancient Music; Christopher Hogwood, hpd, cond. L'Oiseau-Lyre DSLO 526.
+Gr 10-77 p628 +RR 10-77 p46
++HFN 10-77 p155

1269 Sonatas, violoncello, op. 5, nos. 1-6. Anthony Pleeth, Richard Webb, vlc; Christopher Hogwood, hpd. L'Oiseau-Lyre DSLO 513.
+-ARG 3-77 p16 +HFN 6-76 p85
+-Gr 6-76 p61 ++RR 6-76 p71
+HF 5-77 p82

GENIN
Carnival in Venice. cf RCA JRL 1-2315.

GENZMER, Harald
Sonata, trumpet and organ. cf Musical Heritage Society MHS 3340.

GEORGE, Earl
Tuckets and sennets, trumpet and piano. cf ANTHEIL: Sonata, trumpet.

GERHARD, Roberto
1270 Astrological series: Libra; Gemini (Duo concertante); Leo. Nona Liddell, vln; John Constable, pno; London Sinfonietta; David Atherton. Decca HEAD 11.
+Gr 8-77 p296 ++RR 8-77 p66
++HFN 10-77 p155 ++Te 9-77 p42

GERMAN, Edward
As you like it: Suite. cf Works, selections (Rare Recorded Editions RRE 170/1).
Gipsy suite. cf Works, selections (Rare Recorded Editions RRE 170/1).
Henry VIII: incidental music. cf Works, selections (Rare Recorded Editions RRE 170/1).
Leeds suite: Valse gracieuse. cf Works, selections (Rare Recorded Editions RRE 170/1).
Merrie England: English rose. cf HMV HLP 7109.
Much ado about nothing, excerpts. cf Works, selections (Rare Recorded Editions RRE 170/1).
Nell Gwyn: Dances (3). cf Works, selections (Rare Recorded Editions RRE 170/1).
1271 Symphony, no. 2, A minor. JACOB: The barber of Seville goes to the devil. Hull Youth Orchestra; Geoffrey Heald-Smith. Gough & Davy DG 2001.
+Gr 4-77 p1458 +-RR 2-77 p46
1272 Welsh rhapsody. HARTY: With the wild geese. MacCUNN: The land of the mountain and the flood. SMYTH: The wreckers overture. Scottish National Orchestra; Alexander Gibson. HMV ASD 2400 Tape (c) TC ASD 2400.
+-HFN 8-76 p95 tape +-RR 1-77 p91 tape
1273 Works, selections: As you like it: Suite. Gipsy suite. Henry VIII: Incidental music. Much ado about nothing, excerpts. Leeds suite: Valse gracieuse. Nell Gwyn: Dances (3). Orchestra; Edward German. Rare Recorded Editions RRE 170/1 (2).
+HFN 12-77 p169

GERMAN MILITARY MARCHES. cf Olympic 6120.

GERNSHEIM, Friedrich
1274 Sonata, violoncello and piano, no. 1, op. 12, D minor. RUBINSTEIN: Sonata, violoncello and piano, no. 2, op. 39, G major. Gayle Smith, vlc; John Jensen, pno. Genesis GS 1060.
+-Gr 3-77 p1453 +NR 11-76 p14
+-HFN 2-77 p119 +SFC 7-11-76 p13

GERSHWIN, George
1275 An American in Paris. Rhapsody in blue. George Gershwin, pno; Columbia Jazz Band, NYP; Michael Tilson Thomas. Columbia M 34205 Tape (c) MT 34205 (ct) MA 34205. (also CBS 76509 Tape

(c) 40-76509)
+-Audio 3-77 p93
+Gr 4-77 p1603 tape
+Gr 2-77 p1279
+-HF 12-76 p100
+HFN 2-77 p111
+HFN 5-77 p138 tape
+MJ 2-77 p30
++NR 11-76 p13
+-RR 2-77 p47
+-RR 5-77 p91 tape
+-SFC 1-2-77 p34
+SR 11-13-76 p52

1276 An American in Paris. RUSSO: Street music, op. 65. San Francisco Symphony Orchestra; Corky Siegel, mouth organ, pno; Seiji Ozawa. DG 2530 788 Tape (c) 3300 788.
+-Gr 12-77 p1079
+-RR 12-77 p50

1277 An American in Paris. OFFENBACH: Gaite Parisienne, excerpts. Minneapolis Symphony Orchestra; Antal Dorati. Philips 6582 019. (Reissue from Mercury AMS 16135, 16005)
+-Gr 3-77 p1461
+HFN 3-77 p117
+-RR 3-77 p48

An American in Paris. cf Works, selections (Vox QSVBX 5132).
An American in Paris. cf BERNSTEIN: Candide: Overture.
An American in Paris. cf COPLAND: El salon Mexico.
An American in Paris. cf London CSA 2246.

1278 Blue Monday. A damsel in distress: The jolly tar and the maid; Sing of spring. Ming toy: In the Mandarin's orchid garden. The show is on: By Strauss. Let em eat cake: Order, orders... mine. Joyce Andrews, Catherine Aks, Rosalind Rees, s; Jeffrey Meyer, Thomas Bogdan, t; Patrick Mason, bar; Priscilla Magdam, alto; Walter Richardson, bs; Oresta Cybriwsky, pno; Gregg Smith Singers and Orchestra; Gregg Smith. Turnabout TVS 34638 Tape (c) Vox CT 210363.
+-ARG 3-77 p19
+-Audio 6-77 p128
+Gr 9-77 p476
+-HF 9-77 p92
+HFN 9-77 p141
+NR 6-77 p11
+-RR 9-77 p37
+-St 3-77 p128

1279 Concerto, piano, F major. Cuban overture. Stanislav Knor, pno; PSO; Vaclav Neumann. Everest 3405.
+-ARG 2-77 p34
+-NR 1-77 p5

Concerto, piano, F major. cf Works, selections (Vox QSVBX 5132).
Concerto, piano, F major. cf BERNSTEIN: West side story: Symphonic dances.
Cuban overture. cf Concerto, piano, F major.
Cuban overture. cf Works, selections (Vox QSVBX 5132).
A damsel in distress: The jolly tar and the maid; Sing of spring. cf Blue Monday.
Fascinatin' rhythm. cf Works, selections(CBS 30090).
A foggy day. cf Works, selections (CBS 30090).
Funny face: Overture. cf Overtures (CBS 76632).

1280 Gershwin's song book: Swanee; Somebody loves me; Who cares; I'll build a stairway to paradise; The man I love; Nobody but you; Do it again, 'S wonderful; O lady be good; Sweet and low-down; That certain feeling; Liza; I got rhythm. Preludes (3). Rhapsody in blue. Andre Watts, pno. Columbia M 34221 Tape (c) MT 34221 (ct) MA 34221. (also CBS 76508 Tape (c) 40-76508)
-ARG 11-76 p22
++Gr 9-76 p444
+-HF 12-76 p100
+HFN 9-76 p121
+HFN 11-76 p175 tape
+NR 10-76 p14
+RR 9-76 p81
+RR 1-77 p89 tape
-SFC 8-8-76 p38
+St 11-76 p140

Girl crazy, excerpts. cf Works, selections (CBS 30090).
Girl crazy: Overture. cf Overtures (CBS 76632).
Let 'em eat cake: Orders, orders...mine. cf Blue Monday.
Let 'em eat cake: Overture. cf Overtures (CBS 76632).
The little rhapsody in blue. cf Grosvenor GRS 1055.
Love walked in. cf Works, selections (CBS 30090).
Lullaby. cf Works, selections (Vox QSVBX 5132).
Madrigals (2). cf Vox SVBX 5353.
The man I love. cf Works, selections (CBS 30090).
Medley. cf MENC 76-11.
Ming Toy: In the Mandarin's orchid garden. cf Blue Monday.
Of thee I sing: Overture. cf Overtures (CBS 76632).
Oh, Kay: Overture. cf Overtures (CBS 76632).

1281 Overtures: Funny face. Girl crazy. Let 'em eat cake. Of thee I sing. Oh, Kay. Strike up the band. Buffalo Philharmonic Orchestra; Michael Tilson Thomas. CBS 76632.
++Gr 12-77 p1150

1282 Porgy and Bess. Willard White, Leona Mitchell, McHenry Boatwright, Florence Quivar, Barbara Hendricks, Barbara Conrad, Francois Clemmons; CO and Chorus and Children's Chorus; Lorin Maazel. London OSA 13116 (3) Tape (c) 5-13116 (ct) 8-13116. (also Decca SET 609/11 Tape (c) K3Q28)
++ARG 9-77 p10
++Gr 4-76 p1654
+Gr 7-76 p230 tape
+-HF 5-76 p77
+HF 8-76 p70 tape
++HFN 4-76 p105
++HFN 8-76 p94 tape
++MJ 7-76 p56
+MM 2-77 p32
++NR 6-76 p11
+-NYT 4-25-76 pD16
+ON 4-10-76 p32
+RR 4-76 p28
+-RR 8-76 p82 tape
++SFC 3-14-76 p27
+-SR 6-12-76 p48
++St 7-76 p73
+Te 9-76 p30

1283 Porgy and Bess. Cleo Laine; Ray Charles; Orchestra; Frank DeVol. RCA CPL 2-1831 (2).
+NR 2-77 p10
+ON 11-76 p98
+-SFC 1-2-77 p34

1284 Porgy and Bess. Clamma Dale, Wilma Shakesnider, Betty Lane, s; Carol Brice, ms; Larry Marshall, t; Donnie Ray Albert, Andrew Smith, Alexander Smalls, bar; Dick Hyman, pno; Children's Chorus; Houston Grand Opera Orchestra and Ensemble; John Demain. RCA ARL 3-2109 (3).
++ARG 9-77 p10
++Gr 9-77 p475
++HF 9-77 p92
+HFN 12-77 p169
+MJ 10-77 p27
++NR 7-77 p8
+NYT 7-10-77 pB10
++RR 9-77 p37
++SFC 6-12-77 p41
++SR 7-9-77 p49
++St 9-77 p130

1285 Porgy and Bess, excerpts (arr. DeVol). Cleo Laine, Ray Charles, vocals; Orchestra; Frank DeVol. Decca D31D2 (2).
-RR 2-77 p32

Porgy and Bess, excerpts (arr. Russell Bennett). cf COPLAND: El salon Mexico.
Porgy and Bess: Bess, you is my woman now. cf CBS 61039.
Porgy and Bess: Oh Bess, oh where's my Bess. cf Works, selections (CBS 30090).
Porgy and Bess: Six musical scenes. cf Rhapsody in blue.
Porgy and Bess: Suite. cf Works, selections (Vox QSVBX 5132).
Porgy and Bess: Summertime. cf HMV MLF 118.

Preludes. cf Supraphon 111 1721/2.
Preludes (3). cf Gershwin's song book: Thirteen songs.
Preludes (3). cf Vox SVBX 5303.
Promenade. cf Works, selections (CBS 30090).
Promenade. cf Works, selections (Vox QSVBX 5132).
1286 Rhapsody, no. 2. Rhapsody in blue. Variations on "I got rhythm". Teodor Moussov, pno; TVR Symphony Orchestra; Alexander Vladigerov. Monitor MCS 2153.
-ARG 9-77 p35 -RR 11-77 p22
+-MJ 9-77 p34
Rhapsody, no. 2. cf Works, selections (Vox QSVBX 5132).
1287 Rhapsody in blue. Porgy and Bess: Six musical scenes. Marden Abadi, pno. Orion ORS 77265.
+CL 4-77 p10 -NR 8-77 p14
Rhapsody in blue. cf An American in Paris (Columbia M 34205).
Rhapsody in blue. cf Gershwin's song book.
Rhapsody in blue. cf Rhapsody, no. 2.
Rhapsody in blue. cf Works, selections (Vox QSVBX 5132).
'S wonderfu. cf Works, selections (CBS 30090).
The show is on: By Strauss. cf Blue Monday.
Someone to watch over me. cf Works, selections (CBS 30090).
Songs: Love walked in; The man I love. cf FAURE: Nell, op. 18, no. 1.
Strike up the band: Overture. cf Overtures (CBS 76632).
Strike up the band. cf Works, selections (CBS 30090).
Strike up the band (arr. Green). cf RCA CRL 1-2064.
Variations on "I got rhythm". cf Rhapsody, no. 2.
Variations on "I got rhythm". cf Works, selections (Vox QSVBX 5132).
Wintergreen for president. cf Works, selections (CBS 30090).
1288 Works, selections: Fascinatin' rhythm. A foggy day. Girl crazy, excerpts. Love walked in. The man I love. Porgy and Bess: Oh Bess, oh where's my Bess. Promenade. 'S wonderful. Someone to watch over me. Strike up the band. Wintergreen for president. Orchestra; Andre Kostelanetz. CBS 30090 Tape (c) 40-30090.
+-HFN 6-77 p139 tape +-RR 3-77 p48
1289 Works, selections: An American in Paris. Concerto, piano, F major. Cuban overture. Lullaby. Porgy and Bess: Suite. Promenade. Rhapsody, no. 2. Rhapsody in blue. Variations on "I got rhythm". Jeffrey Siegel, pno; St. Louis Symphony Orchestra; Leonard Slatkin. Vox (Q) QSVB 5132 Tape (c) CT 2122. (also Turnabout TV 37080/2)
+-Gr 7-75 p176 +-NR 12-74 p2
/HF 4-75 p85 +RR 6-75 p36
+HF 11-77 p138 tape +SFC 10-27-74 p3
+-HFN 6-75 p88 ++St 2-75 p110

GERSTER, Robert
Bird in the spirit. cf Crystal S 351.

GERVAISE, Claude
Branle de Bourgogne. cf DG 2723 051.
Branle de Champagne. cf DG 2723 051.
French Renaissance dances. cf HMV ASD 3318.

GESUALDO, Carlo
Canzona francese; Mille volte il dir moro. cf L'Oiseau-Lyre 12BB 203/6.
Gagliarda del Principi di Venosa. cf DG 2723 051.

GHISELIN, Johannes
Songs: Ghy syt die werste boven al (Verbonnet). cf HMV SLS 5049.
GIANNEO, Luis
Suite Argentine. cf BEETHOVEN: Sonatas, mandolin.
GIANNINI
Preludium and allegro. cf MENC 76-11.
GIBBONS, Edward
Songs: Love live fair Oriana; Round about her charret. cf DG 2533 347.
GIBBONS, Orlando
Alman (2). cf Works, selections (L'Oiseau-Lyre DSLO 515).
Alman, "The King's jewel". cf Works, selections (L'Oiseau-Lyre DSLO 515).
Coranto. cf Works, selections (L'Oiseau-Lyre DSLO 515).
The Earl of Salisbury pavan. cf Works, selections (L'Oiseau-Lyre DSLO 515).
A fancy. cf Works, selections (L'Oiseau-Lyre DSLO 515).
Fantasias (2). cf Works, selections (L'Oiseau-Lyre DSLO 515).
Fantasia of 4 parts. cf Works, selections (L'Oiseau-Lyre DSLO 515).
Fantasia of 4 parts. cf Vista VPA 1047.
Galliards (3). cf Works, selections (L'Oiseau-Lyre DSLO 515).
Galliard. cf DG 2723 051.
Ground, A major. cf Works, selections (L'Oiseau-Lyre DSLO 515).
Hosanna to the Son of David. cf HMV ESD 7050.
In nomine. cf Argo ZRG 823.
Italian ground. cf Works, selections (L'Oiseau-Lyre DSLO 515).
Lincoln's Inn masque. cf Works, selections (L'Oiseau-Lyre DSLO 515).
Now each flowery bank. cf L'Oiseau-Lyre 12BB 203/6.
O clap your hands together. cf Pye Nixa QS PCNHX 10.
Pavans (2). cf Works, selections (L'Oiseau-Lyre DSLO 515).
Prelude. cf Works, selections (L'Oiseau-Lyre DSLO 515).
The Queens command (2). cf Works, selections (L'Oiseau-Lyre DSLO 515).
Royal pavane. cf Argo ZRG 823.
The silver swan. cf Enigma VAR 1017.

1290 Songs: First song of Moses. Glorious and powerful God. Psalm, no. 145, verses, 1-14, Second preces; Second service: Voluntary, no. 1; Te Deum; Voluntary, nos. 1 and 2; Jubilate; This is the record of John; See, the word is incarnate. King's College Chapel Choir; Simon Preston, org; Jacobean Consort of Viols; David Willcocks. Argo ZK 8 Tape (c) KZKC 8. (Reissue from ZRG 5151)

+Gr 4-77 p1582 +HFN 5-77 p138 tape
+Gr 5-77 p1738 tape +RR 4-77 p81
+HFN 4-77 p151 +RR 7-77 p99

1291 Songs (church music): Hosanna to the son of David; Hymns and songs of the church, nos. 1, 3-5, 9, 13, 18, 20, 22, 24, 31, 47, 69; I am the resurrection; Lord we beseech Thee; O clap your hands; O Lord in Thy wrath; Praise the Lord, O my soul; See, see, the word is incarnate. The Clerkes of Oxenford; David Wulstan. Calliope CAL 1611.

++Gr 11-76 p851 +RR 2-77 p85

1292 Songs (Madrigals and motets, Set I, 5 parts): Ah, dear heart; Dainty fine bird; Fair is the rose; Fair ladies that to love; Farewell all joys; How art thou thralled; I feign not friendship; I see ambition never pleased; I tremble not at noise of war; I weigh not fortunes frown; Lais now old; Mongst thousand good; Nay let me weep; Ne'er let the sun; Now each flowery bank of May; O that the learned poets; The silver swan; Trust not too much fair youth; What is our life; Yet if that age. Consort of Musicke; Anthony Rooley. L'Oiseau-Lyre DSLO 512.

+-Gr 1-76 p1222 +MQ 1-77 p146
+-HF 1-77 p137 +MT 4-76 p321
++HFN 2-76 p97 ++NR 6-76 p13
+MJ 11-76 p45 +-RR 1-76 p56
+-MM 5-76 p34 ++St 11-76 p142

Verse. cf Works, selections (L'Oiseau-Lyre DSLO 515).

1293 Works, selections: Alman (2). Alman, "The King's jewel". Coranto. A fancy. Fantasias (2). Fantasia of 4 parts. Galliard (3). Ground, A major. Italian ground. Lincoln's Inn masque. The Earl of Salisbury pavan. Pavan (2). Prelude. The Queens command (2). Verse. Christopher Hogwood, hpd, org, spinet. L'Oiseau-Lyre DSLO 515.

++Gr 2-76 p1359 +MT 4-76 p321
+HFN 2-76 p97 ++NR 7-76 p13
+MJ 11-76 p45 +RR 1-76 p48
++MM 5-76 p34 +St 11-76 p142
+MQ 1-77 p146

GIBBS, Cecil Armstong

Fancy dress: Dusk. cf Citadel CT 6013.

Songs: The fields are full; A song of shadows. cf Enigma VAR 1027.

GIGOUT, Eugene

Scherzo. cf Calliope CAL 1922/4.

GILBERT, David

Poems VI and VII. cf BEHRENS: The feast of life.

GILLES DE PUSIEUX

Ida capillorum. cf DG 2710 019.

Rachel plorat filios. cf DG 2710 019.

GILMOUR, Howard

Slumber song. cf Club 99-108.

GIMENEZ

El barbero de Sevilla: Me llaman la primorosa. cf London OS 26435.

GINASTERA, Alberto

1294 Concerto, piano, no. 2, op. 37. Quintet, piano and strings. A. Black, A. Edelberg, vln; Jacob Glick, vla; Seymour Barab, vlc; Hilde Somer, pno. Orion ORS 76241.

++ARG 6-77 p25 ++NR 11-76 p5
++CL 3-77 p8 +St 1-77 p114

Estancia: Danza final. cf DG 2584 004.

Quintet, piano and strings. cf Concerto, piano, no. 2, op. 37.

GIORDANI, Tommaso

Songs: Caro mio ben. cf HNH 4008.

GIORDANO, Umberto

Adriana Lecouvreur; Io sono l'umile ancella; Poveri fiori. cf Club 99-100.

1295 Andrea Chenier. Renata Scotto, s; Jean Kraft, Gwendolyn Killebrew, Maria Ewing, ms; Placido Domingo, Piero de Palma, Michel Senechal, t; Sherrill Milnes, Allan Monk, bar; Enzo Dara, bs; John Alldis Choir; National Philharmonic Orchestra; James Levine. RCA ARL 3-2046 (3).

++ARG 10-77 p29 +NR 6-77 p10
+FF 11-77 p22 +NYT 4-17-77 pD17
++Gr 8-77 p338 +RR 8-77 p32
+-HF 8-77 p71 ++SFC 5-8-77 p46
++HFN 10-77 p147 +St 6-77 p128
+MJ 7-77 p68

Andrea Chenier: Arias. cf Free Lance FLPS 675.
Andrea Chenier: Come un bel di di maggio. cf HMV RLS 715.
Andrea Chenier: Come un bel di. cf HMV ASD 3302.
Andrea Chenier: La mamma morta. cf Decca SXLR 6825.
Andrea Chenier: La mamma morta. cf London OS 26497.
Andrea Chenier: La mamma morta. cf Odyssey Y 33793.
Andrea Chenier: La mamma morta. cf Seraphim IB 6105.
Andrea Chenier: Un di all'azzuro spazio. cf RCA TVM 1-7201.
Andrea Chenier: Vecchia madlon. cf Club 99-106.
Andrea Chenier: Vicino a te...La morte nostra. cf PUCCINI: Madama Butterfly.
Fedora: Amor ti vieta. cf HMV RLS 715.
Fedora: Arias. cf London OS 26499.
Fedora: Morte di Fedora. cf Club 99-100.
Siberia: Nel suo amore; Non odi la il martis. cf Club 99-100.
Siberia: Nel suo amore rianimata la coscienza. cf Court Opera Classics CO 347.

GIULIANI, Mauro

1296 Concerto, guitar, op. 30, Amajor. RODRIGO: Concierto madrigal, 2 guitars and orchestra. Pepe and Angel Romero, gtr; AMF; Neville Marriner. Philips 6500 918 Tape (c) 7300 369.
+Gr 1-76 p1197 ++NR 2-76 p4
+Gr 5-77 p1743 tape +-RR 12-75 p48
+-HFN 1-76 p107 ++SFC 2-8-76 p26
+MM 8-76 p35 +St 1-77 p71

1297 Concerto, guitar, op. 30, A major. RODRIGO: Concerto madrigal. Flamenco: Carabana gitana; Fandangos por verdiales; Farruca y rumba; Fiesta en Jerez; Garrotin; Jota; Lamento andaluz; Peterneras; Tanguillos Zorongo. Pepe Romero, gtr. Philips Tape (c) 7328 013.
+RR 7-77 p98 tape

1298 Divertimento. MANINNO: Mini quintet, op. 74. A little music for 3 friends. MARGOLA: Partita, flute and oboe. MONTANARI: Invections, wind quartet (5). Halle Wind Quintet; Michael Davis, vln; John Adams, vla; Ian Rudge, vlc. Amberlee Euroson ALF 701.
+HFN 4-77 p138 +RR 3-77 p71

Grand overture, op. 61. cf Angel S 36093.
Grande ouverture, op. 61. cf DG 2530 561.

1299 Introduction, theme with variations and polonaise, op. 65. RODRIGO: Fantasia para un gentilhombre. Pepe Romero, gtr; AMF; Neville Marriner. Philips 9500 042 Tape (c) 7300 442.
++ARG 3-77 p33 +HFN 2-77 p135 tape
+-Gr 11-76 p791 +NR 8-77 p16
+Gr 5-77 p1743 tape +RR 11-76 p77
+-HFN 11-76 p159

1300 Le Rossiniane, op. 121 and op. 119. SOR: Sonata, guitar, op. 25, C major. Julian Bream, gtr. RCA ARL 1-0711 Tape (c) ARK 1-0711 (ct) ARS 1-0711.
++Gr 3-75 p1678 +RR 8-77 p86 tape
+-HFN 5-77 p139 tape ++SFC 3-2-75 p24

++MJ 3-75 p25 ++St 4-75 p106
+NR 3-75 p14 +St 1-77 p73
++RR 3-75 p47

1301 Sonata, flute and guitar, op. 85, A major. LOEILLET: Sonata, flute op. 1. VISEE: Suite for guitar. Jean-Pierre Rampal, flt; Rene Bartoli, gtr. Everest SDBR 3408.
++NR 11-77 p9

Sonata, flute and guitar, op. 85, A major. cf BIS LP 30.

Sonata, guitar, op. 15: 1st movement. cf CBS 72526.

1302 Sonata, violin and guitar, op. 25. PAGANINI: Cantabile, op. 17, D major. Centone di sonata, op. 64, no. 1. Sonata, violin and guitar, op. 3, no. 6, E minor. Sonata, volion and guitar. A major. Itzhak Perlman, vln; John Williams, gtr. CBS 76525 Tape (c) 40-76525. (also Columbia M 34508 Tape (c) MT 34508)
+-Gr 11-76 p829 ++NR 9-77 p7
+Gr 4-77 p1603 tape ++RR 11-76 p91
+-HFN 11-76 p159 ++RR 2-77 p97 tape
+-HFN 1-77 p123 tape ++St 10-77 p142
++MJ 9-77 p35

Variations on a theme by Handel, op. 107. cf CBS 40-72728.

Variazioni concertante, op. 130. cf RCA ARL 1-0456.

GIUSTINI, Lodovico

Sonata, no. 8, A major. cf BACH, C.P.E.: Sonata, harpsichord, A minor: Poco adagio.

GLASS, Philip

1303 Music in twelve parts, pts 1 and 2. Philip Glass, electronic org; Jon Gibson, sax, flt; Dickie Landry, sax, flt; Richard Peck, sax; Joan LaBarbara, voice, electronic org; Michael Riesman, electronic org. Virgin CA 2010.
+FF 9-77 p22

1304 Strung out. SCELSI: Anahit. XENAKIS: Mikka. Mikka "s". Paul Zukofsky, vln. CF 6.
+MJ 10-77 p28 -NR 11-77 p15

GLAZUNOV, Alexander

Concerto waltz, op. 41, E flat major. cf Sonata, piano, no. 1, op. 74, B flat minor.

1305 Concerto, piano, no. 1, op. 92, F minor. YARDUMIAN: Passacaglia, recitative and fugue. John Ogdon, pno; Bournemouth Symphony Orchestra; Paavo Berglund. HMV SQ ASD 3367.
+Gr 8-77 p305 -RR 8-77 p55
+HFN 9-77 p143

1306 Concerto, piano, no. 2, op. 100, B major. Concerto, violin, op. 82, A minor. Meditation, violin. Ruggiero Ricci, vln; Michael Ponti, pno; PH, Westphalian Symphony Orchestra; Reinhard Peters, Siegfried Landau. Turnabout QTS S 34621.
+Audio 1-77 p84 ++SFC 7-25-76 p29
+-NR 5-76 p6

Concerto, saxophone. cf Citadel CT 6012.

Concerto, saxophone, op. 109, E flat major. cf Symphony, no. 6, op. 58, C minor.

1307 Concerto, violin, op. 82, A minor. MENDELSSOHN: Concerto, violin, op. 64, E minor. Konstanty Kulka, vln; Warsaw Philharmonic Orchestra; Jerzy Katlewicz. Telefunken AG 6-42078.
-Gr 12-77 p1080 ++NR 12-77 p4
+HFN 9-77 p143 +-RR 10-77 p57

Concerto, violin, op. 82, A minor. cf Concerto, piano, no. 2, op. 100, B major.

Concerto, violin, op. 82, A minor. cf BRUCH: Concerto, violin, no. 1, op. 26, G minor.
Cortege solennel, op. 91. cf Symphony, no. 4, op. 48, E flat minor.
Etude, op. 31, no. 1, C major. cf Pye PCNH 9.
From the Middle Ages, op. 79, E major. cf Symphony, no. 5, op. 55, B flat major.
Meditation, violin. cf Concerto, piano, no. 2, op. 100, B major.
Poeme lyrique, op. 12. cf Symphony, no. 4, op. 48, E flat minor.
1308 Quartet, strings, no, 3, op. 26b, G major. Quartet, strings, no. 5, op. 70, D major. Dartington Quartet. Pearl SHE 536.
+ARG 10-77 p56 +RR 12-76 p72
/Gr 2-77 p1296
Quartet, strings, no. 5, op. 70, D major. cf Quartet, strings, no. 3, op 26b, G major.
Raymonda, op. 57: Entr'acte. cf HMV SXLP 30256.
Raymonda, op. 57: Entr'acte. cf HMV SXLP 30259.
1309 Sonata, piano, no. 1, op. 74, B flat minor. LISZT: Sonata, piano, G 178, B minor. Anton Kuerti, pno. Aquitane XM 90414.
+ARG 12-77 p35
1310 Sonata, piano, no. 1, op. 74, B flat minor. Theme and variations, op. 72. Richard Tetley-Kardos, pno. Orion ORS 76233.
+-NR 12-77 p13 +-St 11-77 p136
1311 Sonata, piano, no. 1, op. 74, B flat minor. Sonata, piano, no. 2, op. 75, E minor. Concert waltz, op. 41, E flat major. Leslie Howard, pno. Pearl SHE 538.
+Gr 5-77 p1714 +-RR 5-77 p71
+HFN 5-77 p123
1312 Sonata, piano, no. 2, op. 75, E minor. MEDTNER: Sonata, piano, op. 22, G minor. Emil Gilels, pno. Odyssey Y 34611.
+-NR 12-77 p13 ++St 11-77 p136
Sonata, piano, no. 2, op. 75, E minor. cf Sonata, piano, no. 1, op. 74, B flat minor.
1313 Song of the troubadour, op. 71. SHOSTAKOVICH: Concerto, violoncello, no. 2, op. 126, G major. Mstislav Rostropovich, vlc; BSO; Seiji Ozawa. DG 2530 653 Tape (c) 3300 653.
+Gr 11-76 p791 +RR 9-76 p66
+HF 2-77 p102 ++SFC 12-5-76 p58
+-HFN 10-76 p171 ++St 3-77 p142
+MJ 4-77 p33 +STL 9-19-76 p36
++NR 1-77 p5
1314 Symphony, no. 4, op. 48, E flat minor. Poeme lyrique, op. 12. Cortege solennel, op. 50. MRSO; Nathan Rakhlin, Gennady Rozhdestvensky. HMV ASD 3238.
+-Gr 1-77 p1141 -MT 6-77 p485
+-HFN 1-77 p107 +-RR 1-77 p54
1315 Symphony, no. 5, op. 55, B flat major. From the Middle Ages, op. 79, E major. Moscow Philharmonic Orchestra; Vladimir Fedoseyev. Columbia M 34522.
+FF 11-77 p24 /SFC 7-17-77 p42
+NR 8-77 p2
1316 Symphony, no. 5, op. 55, B flat major. Moscow Radio and TV Symphony Orchestra; Vladimir Fedoseyev. Melodiya C10 06469-70.
++FF 9-77 p23
1317 Symphony, no. 6, op. 58, C minor. Concerto, saxophone, op. 109, E flat major. Lev Mikhailov, sax; MRSO; Alexander Korneyev, Vladimir Fedoseyev. HMV Melodiya ASD 3383.
+-Gr 9-77 p429 +-RR 9-77 p59
+HFN 10-77 p155

Theme and variations, op. 72. cf Sonata, piano, no. 1, op. 74, B flat minor.

GLIERE, Reinhold

The bronze horseman, op. 89: Suite, no. 2. cf Symphony, no. 3, op. 42, B minor.

Concerto, harp, op. 74, E flat major. cf DAMASE: Concertino, harp and strings, op. 20.

Concerto, harp, op. 74, E flat major: 3rd movement. cf Musical Heritage Society MHS 3611.

Duo, violin and violoncello, op. 39. cf DVORAK: Trio, piano, op. 65, F minor.

Red Army march. cf HMV CSD 3782.

1318 The red poppy, op. 70: Ballet suite. RIMSKY-KORSAKOV: The legend of Sadko, op. 5. SHOSTAKOVICH: Age of gold, op. 22. Seattle Symphony Orchestra; Milton Katims. Turnabout TV 34644.
+Audio 7-77 p103 +NR 6-77 p2
+MJ 5-77 p32 +SFC 4-10-77 p30

The red poppy, op. 70: Sailors dance. cf BORODIN: In the Steppes of Central Asia.

The red poppy, op. 70: Sailors dance. cf CBS 61780.

The red poppy, op. 70: Sailors dance. cf CBS 61781.

The red poppy, op. 70: Sailors dance. cf Seraphim S 60277.

Rondo, op. 43, no. 2. cf HMV Melodiya SXLP 30256.

Rondo, op. 43, no. 2. cf HMV SXLP 30259.

1319 Symphony, no. 3, op. 42, B minor. The bronze horseman, op. 89: Suite, no. 2. MRSO, Bolshoi Theatre Orchestra; Nathan Rakhlin, Algis Zuraitis. HMV SLS 5062 (2).
+Gr 8-76 p280 +MT 6-77 p485
+-HFN 8-76 p80 +RR 8-76 p36

GLINKA, Mikhail

Barcarolle, G major. cf Piano works (Musical Heritage Society MHS 1973).

Mazurkas, C minor, A minor. cf Piano works (Musical Heritage Society MHS 1973).

Nocturne, F minor. cf Piano works (Musical Heritage Society MHS 1793).

1320 Piano works: Barcarolle, G major. Mazurkas, C minor, A minor. Nocturne, F minor. Waltz, G major. Trio pathetique, D minor. Variations on Alabiev's song "The nightingale". Thomas Hrynkiv, pno; Esther Lamneck, clt; Michael McCraw, bsn; New American Trio. Musical Heritage Society MHS 1973 Tape (c) C 1973.
++HF 9-75 p85 ++St 7-76 p109
+HF 11-77 p138 tape

1321 Russlan and Ludmilla: Overture. MUSSORGSKY: A night on the bare mountain. TCHAIKOVSKY: Overture, the year 1812, op. 49. WAGNER: Lohengrin: Prelude, Act 3. LPO, Welsh Guards Band; Charles Mackerras. Classics for Pleasure CFP 101 Tape (c) TC CFP 101.
+-Gr 4-71 p1637 +RR 11-77 p118 tape
+-HFN 1-71 p121

Russlan and Ludmila: Overture. cf BORODIN: In the Steppes of Central Asia (DG 2543 247).

Russlan and Ludmilla: Overture. cf BORODIN: In the Steppes of Central Asia (Legend LGD 008).

Russlan and Ludmilla: Overture. cf Angel S 37409.

Russlan and Ludmilla: Overture. cf Decca SXL 6782.

Russlan and Ludmilla: Overture. cf HMV (SQ) ASD 3338.

1322 Sextet, piano and strings, E flat major. TANEYEV: Trio, 2 violins and viola, op. 21, D major. Tamara Fidler, pno; Vladimir Ovcharek, Grigori Lutsky, vln; Vissarion Solovyov, vla; Iosif Levinzon, vlc; Sergei Akopov, bs. Melodiya 33D 029887/8.
+ARG 8-77 p42

1323 Songs: Barcarolle; Doubt; How sweet it is to be with you; I remember the wonderful moment; The lark; Night in Venice; No sooner did I know you; To her. RACHMANINOFF: Songs: Music, op. 34, no. 8; Night is mournful, op. 26, no.12; O cease thy singing, maiden fair, op. 4, no. 4; Spring waters, op. 14, no. 11; Vocalise, op. 34, no. 14. Galina Vishnevskaya, s; Mstislav Rostropovich, pno. DG 2530 725.
+-ARG 9-77 p45
+-Gr 3-77 p1443
+HF 7-77 p104
+HFN 2-77 p119
+MT 9-77 p737
+ON 8-77 p56
+RR 2-77 p88
+St 8-77 p130

Songs: Do not tempt me needlessly. cf Club 99-96.
Songs: Do not tempt me needlessly. cf Rubini RS 301.

1324 Trio pathetique, D minor. SCHUMANN: Marchenerzahlungen, op. 132. WEBER: Duo concertant, op. 48, E flat major. The Music Party. L'Oiseau-Lyre DSLO 524.
-Gr 7-76 p187
+HFN 6-76 p85
+MJ 2-77 p30
+-MT 10-76 p832
+NR 12-76 p8
+-RR 6-76 p65
+STL 9-19-76 p36

Trio pathetique, D minor. cf Piano works (Musical Heritage Society MHS 1973).
Trio pathetique, D minor. cf BRUCH: Pieces, piano, violoncello and clarinet, op. 83.
Variations on Alabiev's song "The nightingale". cf Piano works (Musical Heritage Society MHS 1973).
Waltz, G major. cf Piano works (Musical Heritage Society MHS 1973).

GLOGAUER LIEDERBUCH

Zwe Lider. cf BIS LP 3.

GLOVER

Songs: Rose of Tralee. cf Argo ZFB 95/6.
Songs: The rose of Tralee. cf Philips 9500 218.

GLUCK, Christoph

Alceste: Divinites du Styx. cf Arias (Philips 9500 023).
Alceste: Divinites du Styx. cf HMV SLS 5104.
Alceste: Divinities du Styx. cf Orion ORS 77271.
Alceste: Divinities du Styx. cf Philips 6767 001.
Alceste: Overture. cf SCHUBERT: Symphony, no. 8, D 759, B minor.

1325 Alceste: Pantomime, Act 1; Divinites du Styx. Don Juan: Largetto ...Allegro non troppo. Paride ed Elena: O del mio dolce ardor; Chaconne. Orfeo ed Euridice: Dance of the blessed spirits; Che puro ciel; Che faro senza Euridice. Teresa Berganza, ms; Claude Monteux, flt; LSO, ROHO, AMF, Geraint Jones Orchestra, Stuttgart Chamber Orchestra; Pierre Monteux, Alexander Gibson, Neville Marriner, Geraint Jones, Karl Munchinger. Decca ECS 806. (Reissues from SXL 6612, 2251, 6639, 2265, LXT 5273/6).
+Gr 12-77 p1134
+HFN 11-77 p185
+RR 11-77 p38

1326 Arias: Alceste: Divinites du Styx. Armide: Le perfide Renaud. Iphigenie en Aulide: Vous essayez en vain...Par la crainte;

Adieu conservez votre ame. Iphigenie en Tauride: Non cet affreux devoir. Orfeo ed Euridice: Che puro ciel; Che faro senza Euridice. Paride ed Elena: Spaigge amate; Oh, del mio dolce ardor; Le belle immagini; Dit te scordarmi. La rencontre imprevue: Bel inconnu; Je cherche a vous fair. Janet Baker, ms; ECO; Raymond Leppard. Philips 9500 023 Tape (c) 7300 440.

+-ARG 3-77 p17
+Gr 10-76 p644
+-Gr 3-77 p1485 tape
+HF 1-77 p116
+HFN 10-76 p171
++HFN 2-77 p135 tape
+-MJ 2-77 p30
+-MT 4-77 p307
+NR 1-77 p13
-ON 1-12-77 p44
+RR 12-76 p43
++RR 3-77 p98 tape
++SFC 11-14-76 p30
+St 2-77 p129

Armide: Le perfide Renaud. cf Arias (Philips 9500 023).
Armide: La perfide Renaud. cf Philips 6767 001.
Armide: Pius j'observe ces lieux. cf Rubini GV 38.
Armide: Sicilienne. cf Seraphim SIB 6094.
Concerto, flute, G major. cf DEVIENNE: Concerto, flute, no. 2, D major.
Don Juan: Allegretto. cf DG 2533 182.
Don Juan: Allegretto. cf DG 2723 051.
Don Juan: Larghetto...Allegro non troppo. cf Alceste: Pantomime, Act 1; Divinites du Styx.
Gavotte. cf International Piano Library IPL 101.
Gavotte. cf International Piano Library IPL 103.
Iphiginia: Gavotte. cf International Piano Library IPL 114.

1327 Iphigenie en Aulide (ed. Wagner) (sung in German). Anna Moffo, Arleen Auger, s; Trudeliese Schmidt, ms; Ludovic Spiess, t; Dietrich Fischer-Dieskau, Bernd Weikl, bar; Thomas Stewart, bs; Bavarian Radio Chorus; Munich Radio Orchestra; Kurt Eichhorn. Eurodisc 86271 X$ (2). (also RCA ARL 2-1104)

+-Gr 12-76 p1046
+HF 4-73 p74
+-HFN 3-77 p96
+MJ 3-76 p24
+NR 2-76 p11
+ON 2-21-76 p32
+-RR 1-77 p34
+-SFC 1-18-76 p38
+-SR 3-6-76 p41
++St 5-73 p86
++St 5-76 p114

Iphigenie en Aulide: Lento. cf Seraphim SIB 6094.
Iphigenie en Aulide: Vous essayez en vain...Par la crainte; Adieu conservez votre ame. cf Arias (Philips 9500 023).
Iphigenie en Aulide: Vous essayez en vain...Par la crainte; Adieu conservez dans votre ame. cf Philips 6767 001.
Iphigenie en Tauride: Non cet affreux devoir. cf Arias (Philips 9500 023).
Iphigenie en Tauride: Non cet affreux devoir. cf Philips 6767 001.
Melodie. cf Enigma VAR 1025.
Orphee ed Euridice, excerpt. cf RCA RK 2-5030.
Orfeo ed Euridice: Ballet. cf DG 2533 182.
Orfeo ed Euridice: Ballet. cf DG 2723 051.
Orfeo ed Euridice: Ballet music. cf BACH: Works, selections (DG 2530 647).
Orfeo ed Euridice: Che puro ciel; Che faro senza Euridice. cf Arias (Philips 9500 023).
Orfeo ed Euridice: Che puro ciel; Cara sposa...Che faro senza Euridice. cf Philips 6580 174.
Orfeo ed Euridice: Che puro ciel; Che faro senza Euridice. cf Philips 6767 001.

Orphee ed Eurydice: Dance of the blessed spirits. cf Discopaedia MB 1014.
Orfeo ed Euridice: Dance of the blessed spirits. cf HMV SQ ASD 3375.
Orfeo ed Euridice: Dance of the blessed spirits. cf RCA LRL 1-5094.
Orfeo ed Euridice: Dance of the blessed spirits; Che puro ciel; Che faro senza Euridice. cf Alceste: Pantomime, Act 1; Divinites du Styx.

1328 Orfeo ed Eurydice: Melodie (arr. Kreisler). KREISLER: Andantino, in the style of Padre Martini. Liebesfreud. Liebesleid. Praeludium and allegro, in the style of Pugnani. MENDELSSOHN (Kreisler): Songs without words, op. 62, no. 1: May breezes. TARTINI (Kreisler): Sonata, violin, G minor (The devil's trill). Variations on a theme by Corelli. Beverly Somach, vln; Fritz Jahoda, pno. Musical Heritage Society MHS 3612.
+St 12-77 p148

Orfeo ed Euridice: Melodie. cf International Piano Library IPL 114.
Orfeo ed Euridice: Minuet. cf RCA JRL 1-2315.
Paride ed Elena: O del mio dolce ardor; Chaconne. cf Alceste: Pantomime, Act 1; Divinites du Styx.
Paride ed Elena: Spiagge amate; Oh, del mio dolce ardor; Le belle immagini; Dit te scordarme. cf Arias (Philips 9500 023).
Paride ed Elena: Spiagge amate; Oh, del mio dolce amor; Le belle immagini; Dit te scordarmi. cf Philips 6767 001.
La rencontre imprevue: Bel inconnu; Je cherche a vous faire. cf Arias (Philips 9500 023).
La rencontre imprevue: Bel inconnu; Je cherche a vous faire. cf Philips 6767 001.

1329 Sinfonia, G major. HAYDN: Divertimento, E flat major. PURCELL: Suite, strings, E minor. VIVALDI: Sinfonia, no. 3, G major. Sofia Chamber Orchestra; Vassil Kasandjiev. Denon PCM OX 7044.
+HF 10-77 p112

GNATTALI, Radames
Sonata, guitar and violoncello. cf ALMEIDA: Jass-tuno at the mission.

GODARD
Embarquez-vous. cf Rubini GV 39.

GODARD, Benjamin
Adagio pathetique. cf Discopaedia MOB 1020.
Chanson d'Estelle. cf HMV RLS 716.
Jocelyn: Berceuse. cf Philips 9500 218.
Jocelyn: Berceuse. cf Seraphim IB 6105.
Pieces, op. 116: Waltz. cf RCA LRL 1-5094.
La vivandiere: Viens avec nous petit. cf Rubini GV 57.

GODFREY, Daniel
Progression. cf Quartet, strings.

1330 Quartet, strings. Progression. Rowe String Quartet. Orion ORS 77262.
+ARG 6-77 p26 ++NR 4-77 p5

GODOWSKY, Leopold
The gardens of Buitenzorg. cf International Piano Library IPL 102.

GOEHR, Alexander
Choruses (2). cf BENNETT: Calendar.

GOETZ, Hermann
Leichte Stucke, violin and violoncello, op. 2. cf Quartet, op. 6, E major.

1331 Quartet, op. 6, E major. Quintet, op. 16, C minor. Trio, op. 1, G minor. Leichte Stucke, violin and violoncello, op. 2. Gerald Robbins, pno; Glenn Dicterow, vln; Terry King, vlc; Dennis Trembly, double bs; Alan de Vertich, vla. Genesis GS 1037/8.
+-Audio 7-76 p73 ++NR 4-76 p7
+-Gr 3-77 p1453 +RR 1-77 p66
+-HFN 2-77 p119
Quintet, op. 16, C minor. cf Quartet, op. 6, E major.
Trio, op. 1, G minor. cf Quartet, op. 6, E major.

GOETZE
Still as the night. cf HMV RLS 716.

GOLDENBERG
Kojak, theme. cf EMI TWOX 1058.

GOLDMAN, Richard
1332 Sonata, violin and piano. Sonatina, 2 clarinets. PERSICHETTI: Parable. WYLIE: Psychogram. Theodore Cole, Thomas Falcone, clt; Berl Senofsky, vln; Ellen Mack, pno; Arthur Weisberg, bsn. CRI SD 353.
+-MJ 3-77 p46 +-RR 6-77 p81
+-NR 7-76 p5
Sonatina, 2 clarinets. cf Sonata, violin and piano.

GOLEMINOV, Marin
Suite. cf CHRISTOSKOV: Suite, no. 1.

GOLUB
Tanchum. cf Olympic OLY 105.

GOMBERT, Nicolas
Caeciliam cantate. cf L'Oiseau-Lyre 12BB 203/6.

GOMES, Antonio
1333 Salvator Rosa: Di sposo di padre le gioie serene. HALEVY: La Juive: Si la rigueur. MEYERBEER: Les Huguenots: Seigneur rampart et suel soutien, Piff paff. Robert le Diable: Nonnes qui reposez. VERDI: Don Carlo: Ella giammai m'amo. Ernàni: Che mai vegg'io...Infelice e tu credevi. Nabucco: Vieni o Levita... Tu sul labbro dei veggenti. Simon Boccanegra: Il lacerato spirito. Cesare Siepi, bs; Santa Cecilia Orchestra; Alberto Erede. London R 23218.
+NR 5-77 p10
Salvator Rosa: Mia piccirella. cf Odyssey Y 33793.

GOODMAN, Joseph
Jadis III. cf Crystal S 351.

GOOSSEN, Frederic
Clausulae. cf BASART: Fantasy.
Temple music. cf BASART: Fantasy.

GOOSSENS, Eugene
1334 Chamber music, op. 51: 6 songs. Concerto, oboe, op. 45.* Divertissement. Islamite dance. Old Chinese folk-song, op. 4. Searching for lambs, op. 49. When thou art dead, op. 43. Meriel Dickinson, ms; Leon Goossens, ob; David Lloyd, Peter Dickinson, pno; National Philharmonic Orchestra, PhO; Gaspar Chiarelli, Walter Susskind. Unicorn RHS 348. (*Reissue from Columbia DX 1578/9).
+Gr 9-77 p429 +RR 7-77 p25
1335 Concerto, oboe, op. 45. VAUGHAN WILLIAMS: Variants on "Dives and Lazarus" (5). WILLIAMSON: Sinfonietta. Guy Henderson, ob; Sydney Symphony Orchestra, Melbourne Symphony Orchestra; Robert Pikler, Carl Pini, Yuval Zaliouk. RCA GL 4-0542.
+Gr 7-77 p174 +RR 7-77 p62
+-HFN 8-77 p81

Concerto, oboe, op. 45. cf Chamber music, op. 51: 6 songs.
Divertissement. cf Chamber music, op. 51: 6 songs.
Islamite dance. cf Chamber music, op. 51: 6 songs.
Kaleidoscope, op. 18. cf Saga 5445.
Old Chinese folk-song, op. 4. cf Chamber music, op. 51: 6 songs.
Searching for lambs, op. 49. cf Chamber music, op. 51: 6 songs.
When thou art dead, op. 43. cf Chamber music, op. 51: 6 songs.

GORING
Ma voisine. cf Rubini GV 57.

GOSS, John
If we believe that Jesus died. cf Argo ZK 3.
Praise, my soul, the King of heaven. cf Guild GRSP 701.
Psalms, nos. 127 and 128. cf Guild GRS 7008.

GOSSEC, Francois
Tambourin. cf BIS LP 60.

GOTOVAC, Jakov
1336 Ero s onoga svijeta (Ero the joker). Drago Bernardic, Marijana Radev, Branka Oblak-Stilinovic, Josep Gostic, Vladimir Ruzdjak, Bozena Gatolin, Franjo Vuckovic; Zagreb National Theatre Orchestra and Chorus; Jakov Gotovac. Jugoton LPYV 94/6.
++ARG 5-77 p55

GOTTSCHALK, Louis
La bananier, op. 5. cf Piano works (Musical Heritage Society MHS 3430).
Concert paraphrase on national airs, op. 48: The Union. cf Piano works (Vanguard VSD 71218).
L'Etincelle, op. 21. cf Piano works (Vanguard VSD 71218).
La gallina, op. 53. cf Piano works (Musical Heritage Society MHS 3430).
La gallina, op. 53. cf Piano works (Vanguard VSD 71218).
Grand tarantelle. cf GOULD: Latin American symphonette.
La jota aragonesa, op. 14. cf Piano works (Musical Heritage Society MHS 3430).
La jota aragonesa, op. 14. cf Piano works (Vanguard VSD 71218).
Marche de nuit, op. 17. cf Piano works (Musical Heritage Society MHS 3430).
Marche de nuit, op. 17. cf Piano works (Vanguard VSD 71218).
Night in the tropics. cf GOULD: Latin American symphonette.
Ojos criollas, op. 37. cf Piano works (Musical Heritage Society MHS 3430).
Orfa, op. 71. cf Piano works (Vanguard VSD 71218).
Ouverture de Guillaume Tell (Rossini). cf Piano works (Musical Heritage Society MHS 3430).
1337 Piano works: La bananier, op. 5. La gallina, op. 53. La jota aragonesa, op. 14. Marche de nuit, op. 17. Ojos criollas, op. 37. Ouverture de Guillaume Tell (Rossini). Reponds-moi, op. 50. Radieuse, op. 72. Ses yeux, op. 66. David and Deborah Apter, pno. Musical Heritage Society MHS 3430.
+ARG 7-77 p22
1338 Piano works: Concert paraphrase on national airs, op. 48: The union. L'Etincelle, op. 21. La gallina, op. 53. La jota aragonesa, op. 14. Marche de nuit, op. 17. Orfa, op. 71. Printemps d'amour, op. 40. Radieuse, op. 72. Reponds-moi, op. 50. Ses yeux, op. 66. Souvenirs d'Andalousie, op. 22. Tremolo, op. 58. Eugene List, Joseph Werner, Cary Lewis, pno. Vanguard VSD 71218.

++ARG 7-77 p22 +MJ 3-77 p46
++Audio 5-77 p99 +NR 5-77 p14
++HF 3-77 p100 +St 10-76 p124

Printemps d'amour, op. 40. cf Piano works (Vanguard VSD 71218).
Radieuse, op. 72. cf Piano works (Musical Heritage Society MHS 3430).
Radieuse, op. 72. cf Piano works (Vanguard VSD 71218).
Reponds-moi, op. 50. cf Piano works (Musical Heritage Society MHS 3430).
Reponds-moi, op. 50. cf Piano works (Vanguard VSD 71218).
Romance. cf New World Records NW 257.
Ses yeux, op. 66. cf Piano works (Musical Heritage Society MHS 3430).
Ses yeux, op. 66. cf Piano works (Vanguard VSD 71218).
Souvenirs d'Andalousie, op. 22. cf Piano works (Vanguard VSD 71218).
Tremolo, op. 58. cf Piano works (Vanguard VSD 71218).

GOULD
Songs: The curfew. cf Argo ZFB 95/6.
Fourth of July. cf MENC 76-11.

GOULD, Morton
1339 Latin American symphoneete. GOTTSCHALK: Night in the tropics. Grand tarantelle (orch. Kay). Reid Nibley, pno; Utah Symphony Orchestra; Maurice Abravanel. Vanguard S 275 Tape (c) BCS 0275.
+HF 11-77 p138 tape

Symphonette, no. 2. cf BAKER: Le chat qui peche.

GOUNOD, Charles
Ave Maria (Bach). cf Decca SXL 6781.
1340 Faust. BERLIOZ: The damnation of Faust, op. 24, excerpts. Mireille Berthon, s; Marthe Coiffier, ms; Jeanne Montfort, alto; Cesar Vezzani, t; Louis Musy, Michel Cozette, bar; Marcel Journet, Louis Morturier, bs; Paris Opera Orchestra and Chorus; St. Gervais Chorus; Pasdeloup Concerts Orchestra; Henri Busser, Piero Coppola. Club 99 OP 1000 (3). (Reissues from French EMI originals)
+HF 7-76 p78 +ON 1-1-77 p48
1341 Faust (sung in German). Emmy Destinn, s; Maria Gotze, ms; Ida von Scheele-Muller, alto; Karl Jorn, t; Desider Zador, Arthur Neuendahn, bar; Paul Knupfer, bs; Orchestra and Chorus; Bruno Seidler-Winkler. Discophilia KS 4/6 (3). (Reissued from DG originals, 1908)
+-HF 7-76 p78 +ON 1-1-77 p48
+NR 2-76 p11
1342 Faust. Montserrat Caballe, Anita Terzian, s; Jocelyne Taillon, ms; Giacomo Aragall, t; Philippe Huttenlocher, Jean Brun, bar; Paul Plishka, bs; Rhine Opera Chorus; Strasbourg Philharmonic Orchestra; Alain Lombard. Erato STU 71031/4 (4). (also RCA FRL 4-2493)
+-FF 11-77 p24 +NR 10-77 p8
-Gr 9-77 p476 +-RR 9-77 p38
+-HF 12-77 p90 +-SFC 8-28-77 p46
+-MJ 11-77 p28 +St 12-77 p150
1343 Faust. MASSENET: Le Cid. Bolshoi Theatre Orchestra; Boris Khaikin, Gennady Orzhdestvensky. Westminster/Melodiya WGS 8329.
+-NR 5-77 p5 +ON 2-5-77 p41

Faust: All hail thou dwelling pure and lowly. cf HMV HLP 7109.
1344 Faust: Ballet music. OFFENBACH: Gaite parisienne (arr. Rosenthal).

ROHO; Georg Solti. Decca JB 12 Tape (c) KJBC 12. (Reissue from SXL 2280)
+Gr 9-77 p512 +-RR 9-77 p66
+Gr 11-77 p899 tape +-RR 12-77 p97 tape
++HFN 9-77 p151

1345 Faust: Ballet music. OFFENBACH: Gaite parisienne, excerpts (orch. Rosenthal). WALDTEUFEL: The skaters, op. 183: Waltz. PhO; Herbert von Karajan. HMV SXLP 30224 Tape (c) TC SXLP 30224. (Reissues from Columbia SAX 2274, 2404)
++Gr 1-77 p1184 +RR 1-77 p56
+HFN 3-77 p119 tape

1346 Faust: Ballet music. MASSENET: Le Cid: Overture; Ballet music. Bolshoi Theatre Orchestra; Boris Khaikin, Gennady Rozhdestvensky. Westminster WGS 8329.
+-NR 5-77 p5 +ON 2-5-77 p41

Faust: Ballet music. cf DELIBES: Coppelia: Ballet music.
Faust: Ballet music. cf DELIBES: Coppelia: Suite, Act 1.
Faust: Ballet music. cf CBS 30091.

1347 Faust: The hour is late. PUCCINI: La boheme: Your tiny hand is frozen; They call me Mimi; Lovely maid in the moonlight. Madama Butterfly: Ah, love me a little. Tosca: Mario, Mario, Mario. VERDI: Aida: I see thee again, my sweet Aida. Joan Hammond, s; Charles Craig, t; RPO; Vilem Tausky. HMV ESD 7033 Tape (c) TC ESD 7033. (Reissue from ASD 384)
+Gr 4-77 p1599 +HFN 8-77 p99 tape
+-HFN 5-77 p137 +RR 4-77 p36

Faust: Je voudrais savoir...It etait un Roi de Thule. cf Seraphim IB 6105.
Faust: The jewel song. cf Decca D65D3.
Faust: The jewel song; Final trio. cf HMV RLS 719.
Faust: Salut, demeure chaste et pure. cf HMV RLS 715.
Faust: Salut, demeure. cf HMV ASD 3302.
Faust: Salut, demeure chaste et pure. cf RCA CRM 1-1749.
Faust: Salve dimora. cf RCA TVM 1-7203.
Faust: Salve dimora. cf Rubini GV 29.
Faust: Serenade; Le veau d'or. cf Rubini GV 39.
Faust: Soldiers chorus. cf BBC REC 267.
Romeo et Juliette: Ah leve toi, soleil (2). cf HMV RLS 715.
Romeo et Juliette: Allons jeunes gens. cf Rubini GV 39.
Romeo et Juliette: Depuis hier je cherche en vain. cf CBS 76522.
Romeo et Juliette: Invocation, Dieu qui fis l'homme a ton image. cf Club 99-107.
Romeo et Juliette: Je veux vivre dans ce reve. cf Court Opera Classics CO 342.
Romeo et Juliette: Madrigal; Juliette est vivan. cf Rubini GV 38.
Romeo et Juliette: Mab, la reine des mensonges. cf Philips 6580 174.
Romeo et Juliette: Waltz song. cf HMV RLS 719.
Romeo et Juliet: Waltz song. cf HMV HLM 7066.
Songs: Barcarolle; Serenade. cf Rubini GV 57.
Songs: Primavera. cf Club 99-107.
Songs: Quando canti "Quand tu chantes". cf Club 99-106.

GOWERS, Patrick

1348 Chamber concerto, guitar. Rhapsody, guitar, electric guitars and electric organ. John Williams, gtr; John Scott, sax, flt; Patrick Halling, vln; Stephen Shingles, vla; Denis Vigay, vlc; Herbie Flowers, bs guitar; Patrick Gowers, org; Tristan Fry,

drum; Godfrey Salmon. CBS 61790 Tape (c) 40-61790. (Reissues from 72979, 73350)
+-Gr 10-77 p628 +-RR 8-77 p67
+HFN 9-77 p151 +-RR 11-77 p119
+HFN 10-77 p169 tape
Rhapsody, guitar, electric guitars and electric organ. cf Chamber concerto, guitar.

GRAINGER, Percy
Bold William Taylor. cf Argo ZK 28/9.
Country gardens. cf FAURE: Nell, op. 18, no. 1.
Country gardens. cf Decca SPA 519.
Handel in the Strand. cf FAURE: Nell, op. 18, no. 1.
Irish tune from County Derry. cf FAURE: Nell, op. 18, no. 1.
Knight and shepherd's daughter. cf FAURE: Nell, op. 18, no. 1.
Molly on the shore. cf FAURE: Nell, op. 18, no. 1.
Over the hills and far away. cf FAURE: Nell, op. 18, no. 1.
Sailor's song. cf FAURE: Nell, op. 18, no. 1.
Shepherd's hey. cf FAURE: Nell, op. 18, no. 1.
1349 Songs (folksong arrangements): The British waterside; Died for love; The lost lady found; The pretty maid milkin' her cow; Shallow Brown; Six Dukes went afishing; The sprig of thyme; Willow willow. HOLST: Songs (folksong arrangements): Folksongs of England; Lovely Joan; The maid of Islington; O who is that that rapts at my window; The seeds of love; The unquiet grave. VAUGHAN WILLIAMS: Songs (folksong arrangements): Died for love; Abroad I as I was walking; O who is that; Our ship she lies in harbour; The willow tree. Robin Doveton, t; Victoria Hartnung, pno. Premier PMS 1502
+MT 8-77 p651 +-RR 2-77 p84
To a Nordic princess. cf FAURE: Nell, op. 18, no. 1.
Tribute to Foster: Lullaby. cf FAURE: Nell, op. 18, no. 1.
Walking tune. cf FAURE: Nell, op. 18, no. 1.

GRANADOS, Enrique
A la Cubana, op. 36. cf Works, selections (CRD CRD 1035).
Allegro do concierto, C major. cf Piano works (Vox SVBX 5484).
Aparicion. cf Piano works (Vox SVBX 5484).
Aparicion. cf Works, selections (CRD CRD 1035).
Barcarola. cf Piano works (Vox SVBX 5484).
Capricho espanol, op. 39. cf Piano works (Vox SVBX 5484).
Cartas de amor, op. 44: Valses intimos. cf Works, selections (CRD CRD 1035).
Cuentos para la juventud. cf Piano works (Vox SVBX 5484).
1350 Danzas espanolas (Spanish dances), op. 37 (12). Gonzalo Soriano, pno. Connoissuer Society CS 2105.
+MJ 1-77 p38 +SR 9-18-76 p50
++SFC 12-26-76 p34 ++St 10-76 p122
Danse espagnole, op. 37. cf ALBENIZ: Espana, op. 165, no. 2: Tango.
Danzas espanolas, op. 37: Andaluza. cf ALBENIZ: Iberia: Triana; El corpus en Sevilla.
Danzas espanolas, op. 37: Andaluza. cf CHABRIER: Espana.
Spanish dances, op. 37: Andaluza. cf RCA GL 4-2125.
Danzas espanolas, op. 37, nos. 6 and 11. cf RCA ARL 1-0456.
Spanish dances, nos. 7 and 10. cf International Piano Library IPL 109.
Danza caracteristica. cf Works, selections (CRD CRD 1035).
Danza lenta. cf Piano works (Vox SVBX 5484).
Escenas poeticas, 2nd series. cf Works, selections (CRD CRD 1035)

Estudio. cf Piano works (Vox SVBX 5484).
Estudios expresivos (6). cf Piano works (Vox SVBX 5484).
1351 Goyescas. Francisco Aybar, pno. Connoisseur Society CS 2091.
-HFN 10-77 p106 ++SFC 5-30-76 p24
+MJ 10-76 p52
1352 Goyescas. Alicia de Larrocha, pno. London CS 7009 Tape (c) CS 5-7009 (ct) CS 8-7009. (also Decca SXL 6785)
++Gr 12-77 p1107 +NYT 7-31-77 pD13
++HF 10-77 p106 +St 10-77 p138
++HFN 12-77 p171
Goyescas. cf Piano works (Vox SVBX 5484).
Goyescas: Intermezzo. cf FALLA: El amor brujo.
Goyescas: El Pelele. cf International Piano Library IPL 109.
Impromptus (2). cf Piano works (Vox SVBX 5484).
1353 Piano works: Allegro de concierto, C major. Aparicion. Barcarola. Capricho espanol, op. 29. Danza lenta. Estudio. Cuentos para juventud. Estudios expresivos (6). Impromptus (2). Goyescas. Pieces on popular Spanish songs (6). Marylene Dosse, pno. Vox SVBX 5484 (3).
+HF 10-77 p106 +-NYT 7-31-77 pD13
+-NR 10-77 p12
Pieces on popular Spanish songs (6). cf Piano works (Vox SVBX 5484).
Quintet, piano, G minor. cf Works, selections (CRD CRD 1035).
Tonadillas. cf FALLA: Spanish popular songs.
Tonadillas. cf Seraphim M 60291.
Tonadillas: La maja dolorosa. cf FALLA: Melodies.
Tonadillas al estilo antiguo: Amor y odio; El majo discreto; El majo timido; El mirar de la maja; El tra-la-la y el punteado. cf FALLA: Melodies.
Tonadillas al estilo antiguo: La maja de Goya. cf Angel S 36094.
1354 Works, selections: A la Cubana, op. 36. Aparicion. Cartas de amor, op. 44: Valses intimos. Danza carateristica. Excenas poeticas, 2nd series. Quintet, piano, G minor. Thomas Rajna, pno; Alberni Quartet. CRD CRD 1035.
+Gr 8-77 p314 ++RR 8-77 p67
+HFN 8-77 p83
GRAND OPERA CHORUSES. cf Decca SXL 6826.
GRANDES HEURES LITURGIQUES. cf Delos FY 001.

GRANDI, Alessandro
1355 Motets: Ave Regina; Dixit Dominus; Exaudi Deus; Jesu mi dulcissime; O quam tu pulchra es; O vos omnes; Ploraba die ac nocte; Vulnerasti cor meum. Paul Esswood, c-t; Edgar Fleet, Nigel Rogers, t; Trinity Boys Choir; Accademia Monteverdiana; Denis Stevens. Nonesuch H 71329.
+ARG 12-76 p30 +ON 3-12-77 p40
+HFN 9-76 p171 +-RR 2-77 p86
+MT 1-77 p44 +St 12-76 p140
+NR 9-76 p8
Songs: O vos omnes; O beate benedicte. cf HMV SQ ASD 3393.

GRANTHAM
When the crimson sun has set. cf HMV SQ CSD 3784.

GRAUPNER, Johann
1356 Concerto, trumpet, D major. HAYDN, M.: Concerto, trumpet, D major. QUERFURTH: Concerto, trumpet, E flat major. RICHTER: Concerto, trumpet, D major. Don Smithers, tpt; Concerto Amsterdam; Jaap Schroder. BASF MPS 21778.
+-NR 4-77 p4

GRAYSON, Richard
Promenade. cf Orion ORS 77263.
GREAT CONTINENTAL MARCHES. cf Decca SB 713.
GREAT SOPRANOS OF THE CENTURY. cf Seraphim 60274.
GREAT TENORS OF TODAY. cf HMV ASD 3302.
GREATEST MUSIC IN THE WORLD. cf Philips 6747 327.
GREEN
Sunset. cf Grosvenor GRS 1043.
GREENE, Maurice
Lord, let me know mine end. cf Guild GRS 7008.
1357 Songs (anthems): Arise, shine, O Zion; I will sing of the pow'r, O God; Like as the hart; Lord, let me know mine end; Magnificat and nunc dimittis, C major; My God, my God, look upon me; O clap your hands together all ye people. St. Alban's Abbey Choir; Peter Hurford. Argo ZRG 832.
+-Gr 6-76 p71 +MT 1-77 p45
+HFN 5-76 p100 +RR 5-76 p69
Voluntaries, G major, C minor. cf Cambridge CRS 2540.
Voluntary, D major. cf Nonesuch H 71279.
GREGORIAN ANTHOLOGY. cf London OS 26493.
GREGORIAN CHANTS. cf Coimbra CCO 73.
GREGORIAN CHANTS. cf DG 2533 310.
GREGORIAN CHANTS. cf Everest SDBR 3402.
GREGORIAN CHANTS. cf Philips 6580 105.
GREGORIAN CHANTS. cf Westminster WG 1008.
GREGORIAN CHANTS (Sunday compline). cf Coimbra CCO 37.
GREGORIAN CHANTS: Easter liturgy and Christmas cycle. cf Seraphim 60269.
GREGORIAN CHANTS: Midnight mass for Christmas eve; Mass for Christmas day. cf Peters PLE 013.
GREGORIAN CHANTS: Noel Provencal. cf Arion ARN 34348.
GREGORIAN CHANTS FOR PALM SUNDAY. cf DG 2533 320.
GRESSEL, Joel
Points in time. cf Odyssey Y 34139.
GRETRY, Andre
1358 Cephale et Procris, suite de ballet. Danses Villageoises. Overtures: L'Epreuve Villageoise, Les mariages Samnites, Richard Coeur de Lion. Orchestre de Liege; Paul Strauss. Seraphim S 60268.
++NR 6-76 p2 +St 12-76 p141
+SR 2-19-77 p42
Concerto, flute, C major. cf DEVIENNE: Concerto, flute, no. 2, D major.
Danses Villageoises. cf Cephale et Procris, suite de ballet.
L'Epreuve Villageoise overture. cf Cephale et Procris, suite de ballet.
Les mariages Samnites overture. cf Cephale et Procris, suite de ballet.
Richard Coeur de Lion overture. cf Cephale et Procris, suite de ballet.
Zemire et Azor: La fauvette. cf Columbia 34294.
GRIEG, Edvard
Ballade, op. 24, G minor. cf BEETHOVEN: Sonata, piano, no. 26, op. 81a, E flat major.
1359 Concerto, piano, op. 16, A minor. SCHUMANN: Concerto, piano, op. 54, A minor. John Ogdon, pno; NPhO; Paavo Berglund. Vanguard VCS 10112.

+-ARG 5-77 p21 +-NR 4-77 p4
+HF 6-77 p90 +-St 5-77 p112

Concerto, piano, op. 16, A minor. cf FRANCK: Symphonic variations.
Concerto, piano, op. 16, A minor. cf EMI Italiana C 153 52425/31.
Concerto, piano, op. 16, A minor. cf HMV SLS 5068.
Concerto, piano, op. 16, A minor. cf HMV SLS 5094.
Elegiac melodies, op. 34 (2). cf Works, selections (Vox QSVBS 5140).

1360 Holberg suite, op. 40. Norwegian melodies, op. 63: Cowkeeper's tune and country dance. Peer Gynt, op. 46: Morning mood; The death of Aase; Anitra's dance; In the hall of the mountain king; Solveig's song. Sigurd Jorsalfar, op. 56: Incidental music. LSO, Stuttgart Chamber Orchestra, London Proms Orchestra; Oivin Fjeldstad, Karl Munchinger, Charles Mackerras. Decca SPA 421 Tape (c) KCSP 421. (Reissues)
+Gr 1-77 p1184 ++RR 12-76 p55
+HFN 1-77 p119

Holberg suite, op. 40. cf Works, selections (Vox QSVBS 5140).
Holberg suite, op. 40. cf ELGAR: Serenade, strings, op. 20, E minor.
In autumn, op. 11. cf Works, selections (Vox QSVBS 5140).
Lyric pieces, op. 43, no. 1: Butterfly. cf Decca SPA 519.
Lyric pieces, op. 43, no. 6: To the spring. cf Discopaedia MB 1012.

1361 Lyric pieces, op. 54: Suite. Peer Gynt, opp. 46 and 55: Suites, nos. 1 and 2. PSO; Vaclav Neumann. Everest SDBR 3401.
-ARG 3-77 p18 -NR 2-77 p3

1362 Lyric pieces, op. 54: Suite. Norwegian dances, op. 35 (4). Peer Gynt, opp. 46 and 55: Overture; Dance of the mountain king's daughter; Norwegian bridal procession. Sigurd Jorsalfar, op. 56: Hommage march. Halle Orchestra; John Barbirolli. HMV SXLP 30254 Tape (c) TC SXLP 30254. (Reissues from ASD 2773, Columbia TWO 269)
+Gr 9-77 p512 +RR 11-77 p57
+Gr 11-77 p899 tape

Lyric pieces, op. 54: Suite. cf Works, selections (Vox QSVBX 5140).
Lyric pieces, op. 68, no. 4: Evening in the mountains; no. 5: At the cradle. cf Works, selections (Vox QSVBS 5140).
Norwegian dance. cf Grosvenor GRS 1048.

1363 Norwegian dances, op. 35 (4) (orch. Sitt). Peer Gynt, op. 46: Suite, no. 1; op. 55: Suite no. 2. ECO; Raymond Leppard. Philips 9500 106 Tape (c) 7300 513.
+ARG 3-77 p18 ++RR 11-76 p66
+Gr 11-76 p791 ++RR 3-77 p99 tape
++HFN 11-76 p160 ++SFC 12-5-76 p58
+HFN 2-71 p135 tape +St 6-77 p132
++NR 2-77 p3

Norwegian dances, op. 35. cf Lyric pieces, op. 54: Suite.
Norwegian dances, op. 35, no. 2. cf CBS 61780.
Norwegian dances, op. 35, no. 2. cf International Piano Library IPL 109.
Norwegian dances, op. 35, no. 2. cf Seraphim S 60277.

1364 Norwegian dances, op. 35, nos. 2 and 3. Peer Gynt, opp. 46 and 55: Suites, nos. 1 and 2. Symphonic dance, op. 64, no. 3. Oslo Philharmonic Orchestra; Odd Gruner-Hegge, Morton Gould. Quintessence PMC 7016.
+St 11-77 p146

Norwegian melodies, op. 63: Cowkeeper's tune and country dance. cf Holberg suite, op. 40.

Norwegian melodies, op. 63: In popular folk style; Cowkeeper's tune and country dance. cf HMV SQ ESD 7001.

Old Norwegian melody with variations, op. 51. cf Works, selections (Vox QSVBS 5140).

Peer Gynt, op. 46, excerpts. cf HMV SLS 5073.

Peer Gynt, op. 46: In the hall of the mountain king. cf DG 2548 148.

Peer Gynt, op. 46: Morning mood; The death of Aase; Anitra's dance; In the hall of the mountain king; Solveig's song. cf Holbert suite, op. 40.

1365 Peer Gynt, opp. 46 and 55: Morning; The death of Aase; Anitra's dance; In the hall of the mountain king; Ingrid's lament; Solveig's song. TCHAIKOVSKY: The nutcracker suite, op. 71. VPO; Herbert von Karajan. Decca JB 16 Tape (c) KJBC 16. (Reissue from SXL 2308)

+Gr 9-77 p512 | +-RR 9-77 p68
++Gr 11-77 p899 tape | +-RR 12-77 p99 tape
+HFN 9-77 p151

Peer Gynt, op. 46: Suite, no. 1. cf Works, selections (Vox QSVBS 5140).

Peer Gynt, op. 46: Suite, no. 1: Morning. cf Pye Golden Hour GH 643.

Peer Gynt, op. 46: Der Winger mag scheiden. cf Court Opera Classics CO 342.

Peer Gynt, opp. 46 and 55: Overture; Dance of the mountain king's daughter; Norwegian bridal procession. cf Lyric pieces, op. 54: Suite.

1366 Peer Gynt, op. 46: Suite, no. 1; op. 55: Suite, no. 2. Songs: Fra Monte Pincio, op. 39, no. 1; Jeg Elsker Dig, op. 5, no. 3; Princessen; En Svane, op. 25, no. 2; Verdens Gang, op. 48, no. 3. Elisabeth Soderstrom, s; NPhO; Andrew Davis. Columbia M 34531 Tape (c) 34531. (also CBS 76527 Tape (c) 40-76527)

+Gr 2-77 p1280 | +HFN 4-77 p155 tape
+-Gr 5-77 p1738 tape | +-RR 5-77 p91 tape
+-HF 11-77 p114 | +-St 11-77 p138
+-HFN 2-77 p119

1367 Peer Gynt, op. 46: Suite, no. 1; op. 55: Suite, no. 2. SCHUBERT: Rosamund, op. 26, D 797: Overture; Entr'acte, no. 3, B flat major; Ballet music, no. 2, G major. Annette de la Bije, s; COA; Jean Fournet, Georg Szell. Philips Fontana 6530 054. (Reissues from SABE 2029, ABL 3238)

/Gr 10-77 p711 | +-RR 10-77 p46
+-HFN 10-77 p165

Peer Gynt, opp. 46 and 55: Suites, nos. 1 and 2. cf Lyric pieces, op. 54: Suite.

Peer Gynt, op. 46: Suite, no. 1; op. 55: Suite, no. 2. cf Norwegian dances, op. 35.

Peer Gynt, opp. 46 and 55: Suites, nos. 1 and 2. cf Norwegian dances, op. 35, nos. 2 and 3.

Peer Gynt, op. 55: Suite, no. 2. cf Works, selections (Vox QSVBS 5140).

Scenes from peasant life, op. 19, no. 2: Bridal procession passing by. cf Works, selections (Vox QSVBS 5140).

Sigurd Jorsalfar, op. 56. cf Works, selections (Vox QSVBS 5140)

Sigurd Jorsalfar, op. 56: Hommage march. cf Lyric pieces, op. 54: Suite.

Sigurd Jorsalfar, op. 56: Homage march. cf DG 2548 148.
Sigurd Jorsalfar, op. 56: Incidental music. cf Holberg suite, op. 40.
Sonata, piano, op. 7, E minor: Finale. cf International Piano Library IPL 117.
1368 Sonatas, violin and piano, nos. 1-3. Henri Temianka, vln; James Fields, pno. Orion ORS 75193.
+-HF 5-76 p84 ++NR 4-76 p8
+IN 4-76 p16 +St 7-76 p109
+MJ 9-77 p34
Sonata, violin and piano, no. 2, op. 13, G minor. cf Clear TLC 2586.
Sonata, violin and piano, no. 2, op. 13, G minor. cf RCA CRM 6-2264.
1369 Sonata, violoncello and piano, op. 36, A minor. RHEINBERGER: Sonata, violoncello, op. 92, Cmajor. Ludwig Hoelscher, vlc; Kurt Rapf, pno. BASF 212 2397-3.
+-ARG 8-77 p42 +-RR 11-75 p78
+-HFN 11-75 p154
1370 Sonata, violoncello and piano, op. 36, A minor. SCHUMANN: Fantasiestucke, op. 73. Gilbert Reese, vlc; Ralph Linsley, pno. Crystal S 134.
+MJ 9-77 p34
1371 Sonata, violoncello and piano, op. 36, A minor. SCHUBERT: Sonata, arpeggione and piano, D 821, A minor. Paul Tortelier, vlc; Robert Weisz, pno. HMV HQS 1398. (Reissue from World Records CM 26)
+Gr 11-77 p854 +-RR 11-77 p95
+HFN 12-77 p185
1372 Songs: Den Aergjerrige, op. 26, no. 3; Det foorste mode, op. 21, no. 1; En drom, op. 48, no. 6; Eros, op. 70, no. 1; Fra Monte Pincio, op. 39, no. 1; Der gynger baad paa boige, op. 69, no. 1; Jeg elsker dig, op. 5, no. 3; Jeg giver mit digt til Vaaren, op. 21, no. 3; Liden bojt deroppe, op. 39, no. 3; Liden Kirsten, op. 60, no. 1; Med en primula veris, op. 26, no. 4; Med en vandlilje, op. 25, no. 4; Millom Rosor, op. 39, no. 4; Hytten, op. 18, vol. 11, no. 3. Kirsten Flagstad, s; Edwin McArthur, pno. London R 23220. (Reissue from LL 1547)
+ARG 6-77 p26 ++NR 9-77 p11
Songs: Fra Monte Pincio, op. 39, no. 1; Jeg Elsker Dig, op. 5, no. 3; Princessen; En Svane, op. 25, no. 2; Verdens Gang, op. 48, no. 3. cf Peer Gynt, op. 46: Suite, no. 1; op. 55: Suite, no. 2.
Songs: From the fatherland; I love you; The first primrose; The dairy maid; The poet's heart; The swan. cf Kaibala 40D03.
Songs: Mens jeg venter; En Svane. cf Rubini GV 43.
Symphonic dances, op. 64. cf Works, selections (Vos SQVBS 5140).
Symphonic dances, op. 64, no. 3. cf Norwegian dances, op. 35, nos. 2 and 3.
1373 Works, selections: Scenes from peasant life, op. 19, no. 2: Bridal procession passing by. Elegiac melodies, op. 34 (2). Holberg suite, op. 40. In autumn (concert overture), op. 11. Lyric pieces, op. 54: Suite. Lyric pieces, op. 68, no. 4: Evening in the mountains; no. 5: At the cradle. Old Norwegian melody with variations, op. 51. Peer Gynt, op. 46: Suite, no. 1; op. 55: Suite, no. 2. Sigurd Jorsalfar, op. 56. Symphonic dances, op. 64. Utah Symphony Orchestra; Maurice Abravanel. Vox QSVBS 5140 (3).

+Audio 1-77 p85 ++NR 9-76 p3
+HF 11-76 p118 ++SFC 8-15-76 p38
+MJ 11-76 p45

GRIFFES, Charles

Roman sketches, op. 7, nos. 1 and 3. cf Vox SVBX 5303.

1374 Sonata, piano. Tone pictures, op. 5: The lake at evening; The night winds; The vale of dreams. RAVEL: Tombeau de Couperin. Susan Starr, pno. Orion ORS 77270.
++CL 9-77 p11 ++NR 7-77 p11
*MT 12-77 p1016 ++St 12-77 p150

Sonata, piano, F major. cf DETT: In the bottoms.

1375 Songs: German songs (4); Impressions (4); Song of the dagger. The pleasure dome of Kubla-Khan, op. 8. Poems of Fiona MacLeod, op. 11 (3). Tone-pictures, op. 5 (3). Phyllis Bryn-Julson, s; Olivia Stapp, ms; Sherrill Milnes, bar; New World Ensemble, BSO; Seiji Ozawa. New World Records NW 273.
+-FF 11-77 p25 +ON 6-76 p52
+MJ 11-76 p44

Tone pictures, op. 5: The lake at evening; The night winds; The vale of dreams. cf Sonata, piano.

1376 The white peacock. MacDOWELL: Sonata, piano, no. 1, op. 45, G minor. MEINEKE: Madam de Neuville's favourite waltz, with variations. Barry Snyder, pno. Golden Crest RE 7063.
+-CL 9-77 p10 +-MJ 1-77 p38
+-HF 6-77 p93

GRIGNY, Nicolas de

Hymns: Verbum supernum; Veni creator. cf Mass.

1377 Mass, organ, Pt 2. Veni creator spiritus. Verbum supernum. John Fesperman, org. Orion ORS 76253.
+NR 2-77 p13

Recit de tierce en taille. cf RC Records RCR 101.

Tierce en taille. cf Argo ZRG 864.

Veni creator. cf BACH: Chorale prelude: O Gott du frommer Gott, S 767.

GRIMACE, Magister

A l'arme, a l'arme. cf HMV SLS 863.

A l'arme, a l'arme. cf 1750 Arch S 1753.

GRIPPE

1378 Musique Douze. JOHNSON: Disappearances. MELBY: Stevens songs. PERERA: Alternate routes. Phyllis Bryn-Julson, s; tape computed at Digital Computing Lab, Univ. of Illinois, converted at Godfrey Winham Lab, Princeton University; realized at Bregman Electronic Music Studio, Dartmouth College; realized at Electronmusikstudion EMS, Stockholm; realized at Fylkingen's Studio, Stockholm. CRI SD 364.
-MJ 3-77 p46 +NR 5-77 p15

GRISON, Jules

1379 Toccata, F major. JONGEN: Sonata, organ, op. 94. WIDOR: Symphony, organ, no. 5, op. 42, no. 1, F minor. Jane Parker-Smith, org. HMV SQ HQS 1406.
+-Gr 11-77 p868 +RR 12-77 p73

GROBE, Charles

United States grand waltz, op. 43. cf New World Records NW 257.

GRODSKI (Grodsky)

The seagull's cry. cf Club 99-96.

The seagull's cry. cf Rubini RS 301.

GRUBER, Franz
Stille Nacht. cf HMV ESD 7024.
Stille Nacht. cf RCA PRL 1-8020.
GUAMI, Gioseffo
La brillantina. cf L'Oiseau-Lyre 12BB 203/6.
Canzona a 8. cf HMV SQ ASD 3393.
GUANIERI, Carmargo
Brazilian dance. cf CBS 61780.
A GUIDE TO GREGORIAN CHANT. cf Vanguard VSD 71217.
GUILAIN, Jean
1380 Suites, organ, nos. 1-4. Jean Jacquenod, org. Harmonia Mundi HM 1206.
+Gr 12-77 p1108
GUILLOU
Toccata. cf Marc MC 5355.
GUILMONT, Felix Alexandre
March on a theme by Handel. cf Argo SPA 507.
Morceau symphonique, op. 88. cf Boston Brass BB 1001.
1381 Sonata, organ, no. 1, op. 42, D minor: Finale. WIDOR: Symphony, organ, no. 6, op. 42, no. 2, G major: Allegro. Andre Isoir, org. Calliope CAL 1922
+Gr 3-77 p1454
Sonata, organ, no. 1, op. 42, D minor: Finale. cf Calliope CAL 1922/4.
Sonata, organ, no. 3, op. 56, C minor. cf Vista VPS 1034.
Sonata, organ, no. 5, op, 80, C minor. cf Vista VPA 1042.
GUIMARAES
Sounds of bells. cf London CS 7015.
GUIRAUT DE BORNELH
Leu chansonet a vil. cf Telefunken 6-41126.
GUITAR MUSIC OF THE TWENTIETH CENTURY. cf DG 2530 802.
GULIELMUS, M.
Bassa danza a 2. cf DG 2723 051.
Falla con misuras. cf Nonesuch H 71326.
GUMBERT
Cheerfulness. cf Nonesuch H 71341.
GUNGL, Joseph
Amoretten Tanze, op. 161. cf Classics for Pleasure CFP 40213.
Casino dances. cf Angel (Q) S 37304.
GURNEY, Ivor
Songs: Nine of the clock; Ploughman singing; Under the greenwood tree. cf Enigma VAR 1027.
HADLEY, Patrick
I sing of a maiden. cf HMV CSD 3774.
HAGEN, Joachim
Sonata, lute, B major. cf Telefunken AW 6-42155.
HAHN, Gunnar
Suite in folk style, solo flute. cf BIS LP 60.
HAHN, Reynaldo
Songs: Dernier voeu; L'heure exquise. cf Rubini GV 57.
Songs: D'une prison; L'heure exquise; Mai; Le rossignol des lilas; Offrande; Paysage; Si mes vers avaient des ailes. cf FAURE: Songs (Saga 5419).
Songs: En sourdine; L'heure exquise; L'offrande; Si mes vers avaient des ailes; Mozart: Etre adore; L'adieu. cf HMV RLS 716.
Songs: Si mes vers avaient des ailes. cf HMV RLS 719.
Songs: Si mes vers avaient des ailes. cf Odyssey Y 33130.

HAINES
Sonata, harp. cf Orion ORS 75207.
HALEVY, Jacques
Charles VI; Guerre aux tyrans. cf Club 99-107.
La Juive: Dieu que ma voix tremblante. cf Rubini GV 38.
La Juive: Il va venir. cf BASF 22-22645-3.
La Juive: Rachel quand du Seigneur. cf RCA CRM 1-1749.
La Juive: Rachel quand du Seigneur. cf RCA TVM 1-7201.
La Juive: Si la rigueur. cf GOMES: Salvator Rosa: Di sposo di padre le gioie serene.
HALFFTER, Cristobal
Oda para felicitar a un amigo. cf Universal Edition UE 15043.
HALFFTER, Ernesto
1382 Concerto, guitar. BACARISSE: Concertino, guitar and orchestra, op. 72, A minor. Narciso Yepes, gtr; Spanish Radio and Television Symphony Orchestra; Odon Alonso. DG 2530 326.
+-Audio 2-76 p95 +-RR 7-73 p42
+-Gr 6-73 p187 +SR 6-14-75 p46
-NR 6-75 p8 +St 1-77 p72
HALLNAS, Hilding
Songs (3). cf BIS LP 34.
HAMBRAEUS, Bengt
Shogaku. cf BACK: ...for Eliza.
HAMILTON, Thomas
1383 Pieces for Kohn. ARP synthesizers. Somnath Records KH 120.
+-ARG 12-77 p32
HANDEL, Georg Friedrich
Acis and Galatea: O ruddier than the cherry. cf Club 99-102.
Agrippina: Overture. cf Concerto a due cori, no. 1, B flat major.
Agrippina: Overture. cf Overtures (DG 2535 242).
1384 Ah, che troppo inegali. Look down harmonious Saint (Praise of harmony). Nel dolce dell'oblio. Joseph: Overture. Silete venti, motet. Elly Ameling, Halina Lukomska, s; Theo Altmeyer, t; Collegium Aureum; Instrumentalists and various conductors. BASF 39 21687/2 (2).
+-ARG 12-77 p33
Air with variations, B flat major. cf BRAHMS: Variations and fugue on a theme by Handel, op. 24, B flat major.
Alcina: Dream music. cf HMV RLS 717.
Alcine: Entree, Act 3; Il ballo. cf Works, selections (HMV ESD 7031).
Alcina: Overture. cf Overtures (DG 2535 242).
Alcina: Tiranna, gelosia. cf Decca D65D3.
1385 Alessandro: Lusinghe piu care. Atalanta: Care selve. Ode for Queen Anne's birthday: Eternal source of light divine. Samson: Let the bright seraphim. A choice sett of aires: Overture; Allegro; Bourree; Aire; March. SCARLATTI, A.: Su le sponde del Tebro. Judith Blegen, s; Kenneth Cooper, hpd; Columbia Chamber Ensemble; Gerard Schwarz, tpt and cond. Columbia M 34518. (also CBS 76636)
+Gr 10-77 p685 +NR 8-77 p11
+HF 8-77 p88 +RR 10-77 p85
+HFN 10-77 p155 +St 10-77 p145
+MU 9-77 p14
Alexander's feast: Revenge, Timotheus cries. cf Arias (HMV 1367).
Allegro. cf RCA FRL 1-3504.
Allegro, F major. cf Works, selections (ABC ABCL 67005/3).

Andante, B minor. cf Works, selections (ABC ABCL 67005/3).
Arianna: Overture. cf Concerto a due cori, no. 1, B flat major.

1386 Arias: Alexander's feast: Revenge, Timotheus cries. Ezio: Se un bell'ardire. Giulio Cesare: Dall'andoso periglio...Aure, deh per pieta. Judas Maccabaeus: I feel the deity within...Arm, arm ye brave. Samson: Honour and arms. Semele: Where'er you walk. Serse: Frondi tenere...Ombra mai fu. Tolomeo: Che piu si tarda...Stille amare. Marco Bakker, bar; ECO; Kenneth Montgomery. HMV SQ HQS 1367.
+-Gr 4-77 p1587 +RR 4-77 p36
+-HFN 4-77 p136

Ariodante: Dopo notte. cf Philips 6767 001.
Ariodante: Dream music. cf Works, selections (HMV SQ ESD 7031).
Ariodante: Overture. cf Works, selections (London CSA 2247).
Arioso. cf Discopaedia MOB 1020.
Arminio: Overture. cf Works, selections (London CSA 2247).
Atalanta: Care selve. cf Alessandro: Lusinghe piu care.
Atalanta: Care selve. cf Philips 6767 001.

1387 Belshazzar. Felicity Palmer, s; Maureen Lehane, con; Paul Esswood, c-t; Robert Tear, Thomas Sunnegaardh, t; Peter van der Bilt, Steffan Sandlund, bs; Stockholm Chamber Chorus; VCM; Nikolaus Harnoncourt. Telefunken 6-35326 (4).
+Gr 2-77 p1311 +RR 1-77 p80
++HFN 1-77 p109 ++SFC 6-26-77 p46
+-MT 11-77 p923 ++St 9-77 p134
+NR 9-77 p9

Belshazzar: Overture. cf Overtures (DG 2535 242).
Belshazzar: Overture. cf Works, selections (London CSA 2247).

1388 Berenice: Overture. Concerti, oboe, nos. 1-3. Concerto, oboe, no. 2, F major (variant). Solomon: Arrival of the Queen of Sheba. Roger Lord, ob; AMF; Neville Marriner. Argo ZK 2 Tape (c) KZKC 2. (Reissue from ZRG 5442)
+Gr 12-76 p95 +RR 1-77 p54
+HFN 1-77 p119 ++RR 5-77 p92
+HFN 2-77 p135 tape

Berenice: Overture. cf Works, selections (HMV SQ ESD 7031).
Berenice: Overture. cf Works, selections (London CSA 2247).
Bourree. cf RCA FRL 1-3504.
A choice sett of aires: Overture; Allegro; Bourree; Aire; March. cf Alessandro: Lusinghe piu care.

1389 Concerto, no. 1, F major. Concerto, no. 3, D major. Concerto a due cori, no. 2, F major. Royal fireworks music. LSO; Charles Mackerras. HMV SQ ASD 3395. (also Angel S 37404 Tape (c) 4XS 37404)
+Gr 11-77 p838 ++RR 12-77 p47
++HFN 12-77 p171

Concerto, no. 3, D major. cf Concerto, no. 1, F major.
Concerti, oboe, nos. 1-3. cf Berenice: Overture.
Concerto, oboe, no. 3, G minor. cf Argo D69D3.

1390 Concerti, organ, nos. 1-6. Lionel Rogg, org; Toulouse Chamber Orchestra; Georges Armand. Connoisseur Society CSQ 2115/6 (4). (also Pathe Marconi)
+ARG 5-77 p21 +NR 5-77 p8
++HF 1-77 p116 +SFC 9-11-77 p42
+-MJ 5-77 p32

1391 Concerti, organ, opp. 4 and 7. Herbert Tachezi, org; VCM; Nikolaus Harnoncourt. Telefunken 6-35282 (3) Tape (c) MH 14-35282.

+Audio 4-77 p41
+-Gr 7-77 p174
+Gr 10-77 p698 tape
+HF 3-76 p88
+-HFN 9-77 p143
++NR 3-76 p4
+PRO 7/8-76 p18
+RR 5-77 p53
++RR 9-77 p97 tape
++SFC 1-25-76 p30

1392 Concerti, organ, op. 4, nos. 1-6; op. 7, nos. 13-18; nos. 19-20. Sonata, organ and strings (Il trionofo del tempo e del disinganno). George Malcom, org and hpd; AMF; Neville Marriner. Argo D3D4 (4).
+ARG 8-77 p14
+Gr 9-76 p415
+HF 6-77 p90
+-HFN 9-76 p121
++MJ 9-77 p35
++NR 5-77 p6
+RR 9-76 p53
+-SFC 12-4-77 p56

1393 Concerti, organ, op. 4, nos. 1-6; op. 7, nos. 13-18, 19-20. Daniel Chorzempa, org; Concerto Amsterdam; Jaap Schroder. Philips 6709 009 (5).
+ARG 3-77 p23
+Audio 4-77 p91
+Gr 9-76 p415
+-HF 6-77 p90
+HFN 9-76 p121
+MJ 2-77 p30
++NR 3-77 p7
+RR 10-76 p50
+SFC 12-12-76 p55
+SR 2-19-77 p42

Concerto, organ, op. 7, no. 4, D minor. cf Argo D69D3.
Concerto, trumpet, B flat major. cf Philips 6500 926.
Concerto, trumpet, D minor. cf ALBINONI: Concerto, trumpet, op. 7, no. 3, B flat major.

1394 Concerti a due cori, nos. 1-3 (double concerti). ECO; Raymond Leppard. Philips 6580 212. (Reissue from SAL 3707)
+Gr 8-77 p305
+HFN 7-77 p125
++RR 7-77 p56

1395 Concerto a due cori, no. 1, B flat major. Concerto a due cori, no. 3, F major. Agrippina: Overture. Arianna: Overture. AMF: Neville Marriner. HMV SQ ASD 3182. (also Angel S 37176)
+-Gr 5-76 p1757
++HFN 4-76 p107
++HF 5-77 p82
+NR 3-77 p3
++RR 4-76 p41
++SFC 12-12-76 p55

Concerto a due cori, no. 2, F major. cf Concerto, no. 1, F major.
Concerto a due cori, no. 3, F major. cf Concerto a due cori, no. 1, B flat major.
Concerto grosso. cf RCA FRL 1-3504.

1396 Concerti grossi (2). Royal fireworks music. Orchestra; Arthur Davison. Classics for Pleasure (c) TC CFP 105.
+Gr 2-77 p1325 tape
+-HFN 12-76 p153 tape

Concerto grosso, op. 3, no. 1. cf Argo D69D3.

1397 Concerti grossi, op. 6 (12). La Grande Ecurie et la Chambre du Roy; Jean-Claude Malgoire. CBS 79306 (3).
+-Gr 10-77 p635
+HFN 11-77 p171
+RR 11-77 p58

1398 Concerti grossi, op. 6, nos. 1-2, 4, 6, 8, 11. BPhO; Herbert von Karajan. DG 2726 068 (2). (Reissues from 139042)
+-Gr 5-77 p1696
+HFN 8-77 p93
+-RR 3-77 p49

1399 Coronation anthems: The king shall rejoice. Let thy hand be strenthened. My heart is inditing. Zadok the priest. Hudders, field Choral Society; Northern Sinfonia Orchestra; Keith Rhodes, org; John Pritchard. Enigma VAR 1030.
+-Gr 8-77 p331
=HFN 7-77 p113
+-RR 7-77 p86

Dank sei dir, Herr. cf Golden Age GAR 1001.
Deidamia: Overture. cf Overtures (DG 2535 242).
Deidamia: Overture. cf Works, selections (London CSA 2247).
Dettingen Te Deum, no. 17. cf Discopaedia MB 1013.
1400 Dixit dominus domino meo. Helen Donath, Trudy Koeleman, s; Aafje Heynis, con; Gerard van Dolder, t; David Hollestelle, bs; NCRV Vocal Ensemble; Amsterdam Chamber Orchestra; Marinus Voorberg. Philips 6580 135. (Reissue from 6500 044)
+Gr 3-77 p1438 +RR 2-77 p87
+HFN 3-77 p118
Duo, F major. cf Abbey LPB 765.
Esther: Overture. cf Works, selections (London CSA 2247).
Ezio: Se un bell'ardire. cf Arias (HMV SQ HQS 1367).
Fantasia, C major. cf Crescent BUR 1001.
Faramondo: Overture. cf Works, selections (HMV SQ ESD 7031).
Faramondo: Overture. cf Works, selections (London CSA 2247).
Giulio Cesare (Julius Caesar): Dall'andoso periglio...Aure, deh, per pieta. cf Arias (HMV SQ HQS 1367).
Julius Caesar: Overture. cf Works, selections (London CSA 2247).
Giulio Cesare: V'adoro pupillo. cf Decca D65D3.
Hercules: Where shall I fly. cf Philips 6767 001.
1401 Israel in Egypt. Elizabeth Gale, Lilian Watson, s; James Bowman, alto; Ian Partridge, t; Tom McDonnell, Alan Watt, bs; Christ Church Cathedral Choir; ECO; Simon Preston. Argo ZRG 817/8.
+-ARG 11-76 p24 +MT 3-77 p216
+-Gr 4-76 p1646 +-NR 4-77 p11
+-HFN 4-76 p104 +RR 4-76 p70
+-MJ 1-77 p27 +-St 12-76 p144
Israel in Egypt: But as for his people; Sing ye to the Lord. cf Works, selections (CBS 61139).
Israel in Egypt: The Lord is a man of war. cf Club 99-102.
Jeptha: Deeper and deeper still...Waft her angels. cf HMV HLP 7109.
Jeptha: Overture. cf Overtures (DG 2535 242).
Jeptha: Sinfonia. cf Works, selections (London CSA 2247).
Joseph: Overture. cf Ah, che troppo inegali.
Joshua: O had I Jubal's lyre. cf Philips 6767 001.
Joy to the world. cf RCA PRL 1-8020.
1402 Judas Maccabaeus. Felicity Palmer, s; Janet Baker, ms; Paul Esswood, c-t; Ryland Davies, t; John Shirley-Quirk, bar; Christopher Keyte, bs; Wandsworth School Choir; ECO; Charles Mackerras. DG 2723 050 Tape (c) 3376 011. (also DG 2710 021 Tape (c) 3376 021)
++Gr 9-77 p467 ++HFN 11-77 p163
++Gr 11-77 p900 tape ++RR 9-77 p88
1403 Judas Maccabaeus, highlights. Heather Harper, s; Helen Watts, alto; Alexander Young, t; John Shirley-Quirk, bs; Amor Artis Chorale, Wandsworth School Boys Choir; ECO; Johannes Somary. Vanguard VCS 10126. (Reissue from VCS 10105/7)
+ARG 2-77 p34
Judas Maccabaeus: Arm, arm ye brave. cf Club 99-102.
Judas Maccabaeus: I feel the deity within...Arm, arm ye brave. cf Arias (HMV SQ HQS 1367).
Judas Maccabaeus: Overture. cf Works, selections (London CSA 2247).
Judas Maccabaeus: See, the conquering hero comes; Sing unto God. cf Works, selections (CBS 61139).

Judas Maccabaeus: Sound an alarm. cf BBC REC 267.
The king shall rejoice. cf Coronation anthems (Enigma VAR 1030).
The king shall rejoice. cf Pye NIXA SQ PCNHX 10.
Let they hand be strengthened. cf Coronation anthems (Enigma VAR 1030).
Look down harmonious saint (Praise of harmony). cf Ah, che troppo inegali.
Love in Bath: Suite. cf HMV SLS 5073.
Lucrezia: Solo cantata. cf Philips 6767 001.

1404 Messiah. Elly Ameling, s; Anna Reynolds, con; Philip Langridge, t; Gwynne Howell, bs; AMF; Neville Marriner. Argo D18D3 (3) Tape (c) K18K32.
+ARG 3-77 p19
+Audio 6-77 p133
+Gr 11-76 p852
+-Gr 1-77 p1183 tape
+-HF 4-77 p103
++HFN 11-76 p160
++HFN 1-77 p123 tape
+MM 2-77 p33
+-MT 3-77 p216
+-RR 11-76 p95
+RR 1-77 p90 tape
++St 4-77 p119

1405 Messiah. Felicity Palmer, s; Helen Watts, con; Ryland Davies, t; John Shirley-Quirk, bs; Philip Jones, tpt; Leslie Pearson, org; ECO and Chorus; Raymond Leppard. Erato STLU 70921/3 (3). (also RCA CRL 3-1426, also Musical Heritage Society MHS 3273/5)
+ARG 12-76 p6
+Gr 2-76 p1364
+HF 9-76 p91
+HFN 2-76 p99
+MJ 10-76 p24
+-MM 2-77 p33
++MT 7-76 p575
++MU 10-76 p16
+NR 2-76 p12
++RR 2-76 p57
++SFC 4-18-76 p23

1406 Messiah. Rae Woodland, s; Norma Procter, con; Paul Esswood, c-t; David Johnston, t; Stephen Roberts, bar; London Choral Society Members; English Symphony Orchestra; John Tobin. GHF Records QS GHF 1-4 (4).
+-Gr 1-77 p1162
+MT 3-77 p216
+RR 1-77 p81

1407 Messiah. Rae Woodland, s; Norma Procter, con; Paul Esswood, c-t; David Johnston, t; Stephen Roberts, bar; London Choral Society; English Symphony Orchestra; John Tobin. Lyntone Recordings LYN 3574/81.
+MM 2-77 p33

1408 Messiah. Soloists; Handel and Haydn Society of Boston; Thomas Dunn. Sine Qua Non Superba 2015/3.
+SFC 12-11-77 p60

1409 Messiah, excerpts. Soloists; Pro Musica Antiqua Chamber Orchestra and Chorus; Randolph Jones. Everest SDBR 3398.
-NR 4-77 p11

1410 Messiah: Amen; And the glory of the Lord; Behold the Lamb of God; Come unto Him, confort ye; Ev'ry valley; For unto us a child is born; Hallelujah; He shall feed his flock; He was despised; I know that my Redeemer liveth; Then shall the eyes; Worthy is the lamb. Elsie Morison, s; Marjorie Thomas, con; Richard Lewis, t; Huddersfield Choral Society; Royal Liverpoor Philharmonic Orchestra; Malcolm Sargent. Classics for Pleasure CFP 40020 Tape (c) TC CFP 40020. (Reissue, 1959 and SAX 2365)
-Gr 2-77 p1325 tape
+HFN 2-73 p347
+HFN 12-76 p153 tape
+RR 2-73 p82

1411 Messiah: And the glory of the Lord; And he shall purify; For unto us a child is born; Glory to God; His yoke is easy; Behold the lamb of God; Surely, he hath borne our griefs; And with his stripes; All we like sheep; He trusted in God; Lift up your heads; Let all the angels of God; The Lord gave the word; Let us break their bonds; Hallelujah; Since my man came death; Worthy is the lamb; Amen. King's College Chapel Choir; AMF: David Willcocks. HMV CSD 3778 Tape (c) TC CSD 3778. (Reissue from SLS 845)

+-Gr 12-76 p1028 +HFN 3-77 p119 tape
+-HFN 12-76 p151 +RR 12-76 p92

1412 Messiah: Comfort ye; Every valley; And the glory of the Lord; For unto us a child is born; And suddenly there was with the angel; Glory to God; Then shall the eyes of the blind be opened; He shall feed his flock; He was despised; How beautiful are the feet; Why do the nations so furiously rage together; Let us break their bonds asunder; Hallelujah; I know that my redeemer liveth; Behold I tell you a mystery; The trumpet shall sound; Worthy is the lamb; Amen. Jennifer Vyvyan, s; Norma Procter, con; George Maran, t; Owen Brannigan bs; LPO and Chorus; George Malcolm, hpd; Ralph Downes, org; Adrian Boult. Decca ECS 791. (Reissue from LXT 2921/4)

+-Gr 5-77 p1723 +RR 3-77 p92
+-HFN 3-77 p118

1413 Messiah: Confort ye, my people...Every valley shall be exalted; Thus saith the Lord...But who may abide; For unto us a child is born; Then shall the eyes...He shall feed his flock; He was despised; Thy rebuke hath broken his heart...Behold and see; He was cut off; But thou didst not leave; How beautiful are the feet; Why do the nations; Let us break their bonds; Hallelujah; I know that my redeemer liveth; Behold I tell you a mystery...The trumpet shall sound. Dora Labbette, s; Muriel Brunskill, con; Hubert Eisdell, t; Harold Williams, bar; BBC Chorus; Orchestra; Thomas Beecham. HMV HLM 7053. (Reissue from Columbia L 2018/35)

+-Gr 4-77 p1581 +RR 4-77 p81
+HFN 4-77 p136

1414 Messiah: Ev'ry valley; And the glory of the Lord; Behold, a virgin; O thou that tellest; For unto us a child is born; And suddenly there was with the angel; Glory to God; Rejoice greatly; Behold the lamb of God; He was despised; All we like sheep; Hallelujah; I know that my redeemer liveth; The trumpet shall sound; Worthy is the lamb; Amen. Elly Ameling, s; Anna Reynolds, con; Philip Langridge, t; Gwynne Howell, bs; AMF; Neville Marriner. Argo ZRG 879. (Reissue from D183D3)

+-Gr 11-77 p873 +-RR 11-77 p103
+HFN 11-77 p185

Messiah: For unto us a child is born; Hallelujah. cf Works, selections (CBS 61139).
Messiah: Hallelujah chorus. cf Philips 6747 327.
Messiah: Hallelujah chorus. cf CBS SQ 79200.
Messiah: He shall feed his flock. cf Orion ORS 77271.
Messiah: Overture; Pastoral symphony. cf Works, selections (HMV SQ ESD 7031).
Minuet, D major. cf Angel S 36053.
Minuet, E minor. cf Works, selections (ABC ABCL 67005/3).
Minuet, G minor. cf BACH: Works, selections (DG 2530 647).
My heart is inditing. cf Coronation anthems (Enigma VAR 1030).

Nel dolce dell'oblio. cf Ah, che troppo inegali.
Ode for Queen Anne's birthday: Eternal source of light divine. cf Alessandro: Lusinghe piu care.
1415 Overtures: Alcina. Agrippina. Belshazzar. Deidamia. Jeptha. Radamisto. Rinaldo. Rodelinda. Susanna. LPO; Karl Richter. DG 2535 242 Tape (c) 3335 242. (Reissue from 2530 342)
+-Gr 12-77 p1080 +HFN 11-77 p183
+Gr 12-77 p1146 tape +RR 11-77 p58
1416 Overture, D minor (arr. Elgar). Royal firewords music: Suite. Samson: Overture (arr. Sargent). Water music: Suite. RPO; Malcolm Sargent. Seraphim S 60276.
+NR 8-77 p2
Partenope: Combattono il mio core. cf Philips 6580 174.
Passacaglia. cf RCA CRM 6-2264.
1417 Il pensieroso: Sweet bird. PUCCINI: La boheme: Entrate...C'e Rodolfo; Donde lieta usci; Addio di Mimi; Addio dolce svergliare alla mattina,; Gavotta...Minuetto sono andati; Io Musetta... Oh come e bello e morbido. THOMAS: Hamlet: Mad scene. VERDI: Otello: Piangea cantando; Ave Maria. Rigoletto: Caro nome. La traviata: Dita alla giovine. Lord Stanley's address and Melba's farewell speech. Mellie Melba, s. Bruno Walter Society IP 84.
+NR 10-77 p8
Il pensieroso: Sweet bird. cf HMV RLS 719.
Radamisto: overture. cf Overtures (DG 2535 242).
Radamisto: Overture. cf Works, selections (London CSA 2247).
Rigaudon, bourree and march. cf Kaibala 20B01.
1418 Rinaldo. Ileana Cotrubas, Jeanette Scovotti, s; Carolyn Watkinson, con; Paul Esswood, Charles Brett, c-t; Ulrik Cold, bs; La Grande Ecurie et la Chambre du Roy; Jean-Claude Malgoire. CBS 79308 (3).
+Gr 11-77 p885 +RR 11-77 p22
++HFN 11-77 p163
Rinaldo: Overture. cf Overtures (DG 2535 242).
Rinaldo: Overture. cf Works, selections (London CSA 2247).
Rodelinda: Dove sei. cf Decca SPA 450.
Rodelinda: Overture. cf Overtures (DG 2535 242).
Rodelinda: Pompe vane di morte...Dove sei. cf Philips 6767 001.
1419 Royal fireworks music. Water music, excerpts. Leslie Pearson, hpd; ECO; Raymond Leppard. Philips 6580 147. (Reissues from 6500 369, 6500 047)
+Gr 5-77 p1696 +RR 4-77 p46
+-HFN 4-77 p151
1420 Royal fireworks music. Water music: Suite. Minnesota Orchestra; Stanislaw Skrowaczewski. Turnabout QTV 34632 Tape (c) KTVC 34632. (also Vox tape (c) CT 2108)
+Gr 10-77 p635 +NR 10-76 p3
+-HFN 9-77 p143 +RR 8-77 p55
+MJ 1-77 p26 +RR 10-77 p96 tape
Royal fireworks music. cf Concerto, no. 1, F major.
Royal fireworks music. cf BACH, J.C.: Concerto, harp, op. 1, no. 6, D major.
Royal fireworks music: Suite. cf Overture, D minor.
Sampson: Awake the trumpet's lofty sound; Let their celestial concerts unite. cf Works, selections (CBS 61139).
Sampson: Honour and arms. cf Arias (HMV SQ HQS 1367).
Sampson: Honour and arms. cf Club 99-102.

Samson: Let the bright seraphim. cf Alessandro: Lusinghe piu care.
Samson: let the bright seraphim. cf Decca D65D3.
Samson: Overture. cf Overture, D minor.
Sarabande, D major. cf Philips 6833 159.
Sarabande and variations. cf Angel S 36053.
Saul: Dead march. cf Works, selections (HMV SQ ESD 7031).
Saul: Dead march. cf Argo SPA 507.
Saul: Welcome welcome mighty king; David, his ten thousands slew; How excellent thy name. cf Works, selections (CBS 61139).
Scipione: March. cf Works, selections (HMV SQ ESD 7031).
Scipione: Overture. cf Works, selections (London CSA 2247).

1421 Semele. Sheila Armstrong, s; Helen Watts, alto; Felicity Palmer, s; Mark Deller, Robert Tear, Edgar Fleet, t; Justino Diaz, bs; Amor Artis Chorale; ECO; Johannes Somary. Vanguard VCS 10127/9 (3) (Q) 30022/4. (Reissue from VSD 71180/2)
+ARG 3-77 p21 +SFC 9-11-77 p42
+NR 4-77 p11

Semele: Sinfonia. cf Works, selections (London CSA 2247).
Semele: Where'er you walk. cf Arias (HMV SQ HQS 1367).
Semele: Where'er you walk. cf HMV HLP 7109.
Semele: Where'er you walk; O sleep, why dost thou leave me. cf Philips 9500 218.
Serse (Xerxes): Frondi tenere...Ombra mai fu. cf Arias (HMV 1367).
Xerxes: Grove so beautiful and stately...Shadows so sweet. cf HMV HLP 7109.
Xerxes: Holy art thou. cf Works, selections (CBS 61139).
Xerxes: Largo. cf HMV SQ ASD 3375.
Serse: Ombra mai fu. cf Argo ZK 11.
Serse: Ombra mai fu. cf Philips 6767 001.
Serse: Ombra mai fu. cf RCA CRM 1-1749.
Silete venti, motet. cf Ah, che troppo inegali.
Sing unto God, anthem. cf BACH: Cantata, no. 131, Aus der Tiefe rufe ich, Herr (HMV CSD 3741).
Sing unto God, anthem. cf BACH: Cantata, no. 131, Aus der Tiefe rufe ich, Herr (Nonesuch H 71294).
Solomon: Arrival of the Queen of Sheba. cf Berenice: Overture.
Solomon: Arrival of the Queen of Sheba. cf Works, selections (HMV SQ ESD 7031).
Solomon: Overture. cf Works, selections (London CSA 2247).
Sonata, flute, F major. cf BOISMORTIER: Sonata, trumpet, G minor.
Sonatas, flute, op. 1, nos. 1-2, 4-5, 7, 11. cf Works, selections (ABC ABCL 67005/3).
Sonata, flute, op. 1, no. 9, B minor. cf Saga 5425.
Sonata, flute, op. 1, no. 9, D minor: Movements. cf Works, selections (ABC ABCL 67005/3).
Sonata, flute, op. 1, no. 11, F major. cf Cambridge CRS 2826.

1422 Sonatas, flute and harpsichord. Jiri Valek, flt; Josef Hala, hpd; Frantisek Slama, vlc. Supraphon 111 1891/2.
+RR 10-77 p77

1423 Sonata, oboe, op. 1, no. 6, G minor. Sonata, oboe, op. 1, no. 8, C minor. Trio sonata, no. 2, D minor. Trio sonata, no. 3, E flat major. Ronald Roseman, Virginia Brewer, ob; Timothy Eddy, vlc; Donald MacCourt,bsn; Edward Brewer, hpd. Nonesuch H 71339.
+FF 11-77 p26 ++SFC 9-11-77 p42
+HF 11-77 p114 +St 12-77 p150

Sonata, oboe, op. 1, no. 8, C minor. cf Sonata, oboe, op. 1, no. 6, G minor.
Sonata, oboe, op. 1, no. 8, C minor. cf Works, selections (ABC ABCL 67005/3).
Sonata, organ and strings (Il trionfo del tempo e del disinganno). cf Concerti, organ, op. 4, nos. 1-6.
Sonatas, recorder, B minor, D minor, B flat major. cf Works, selections (ABC ABCL 67005/3).
Sonata, violin, op. 1, no. 5, G major: Minuet. cf Discopaedia MOB 1020.
Sonata, violin, op. 1, no. 15, E major. cf RCA CRM 6-2264.
Sosarme: Overture. cf Works, selections (London CSA 2247).
Suite, harpsichord, no. 3, D minor. cf Toccata 53623.
Suite, harpsichord, no. 5, E major (The harmonious blacksmith). cf Saga 5384.
Suite, harpsichord, no. 5, E major: Air and variations. cf International Piano Library IPL 112.
Suite, harpsichord, no. 7: Passacaglia. cf DVORAK: Trio, piano, op. 65, F minor.
Susanna: Overture. cf Overtures (DG 2535 242).
Teseo: Overture. cf Works, selections (London CSA 2247).
Tolomeo: Che piu si tarda...Stille amare. cf Arias (HMV SQ 1367).
Tolomeo: Silent worship. cf L'Oiseau-Lyre DSLO 20.
Trio sonata, no. 2, D minor. cf Sonata, oboe, op. 1, no. 6, G minor.
Trio sonata, no. 3, E flat major. cf Sonata, oboe, op. 1, no. 6, G minor.

1424 Water music. Virtuosi of England; Arthur Davison. Classics for Pleasure CFP 40092 Tape (c) TC CFP 40092.
/Gr 2-75 p1491 +-RR 1-75 p28
+HFN 6-76 p105 tape /RR 11-77 p118 tape

1425 Water music. COA; Eduard van Beinum. Philips Tape (c) 7327 027.
+Gr 7-77 p225 tape ++RR 7-77 p98
+HFN 7-77 p127 tape

Water music, excerpts. cf Royal fireworks music.
Water music: Aria. cf RCA FRL 1-3504.
Water music: Lento; Hornpipe; Air. cf Works, selections (HMV SQ ESD 7031).
Water music: Minuet. cf Supraphon 113 1323.
Water music: Suite. cf Overture, D minor.
Water music: suite. cf Royal fireworks music (Turnabout QTV 34632).

1426 Works, selections: Allegro, F major. Andante, B minor. Minuet, E minor. Sonatas, flute, op. 1, nos. 1-2, 4-5, 7, 11. Sonata, flute, op. 1, no. 9, D minor: Movements. Sonata, oboe, op. 1, no. 8, C minor. Sonatas, recorder, B minor, D minor, B flat major. Bruce Haynes, ob; Bob van Asperen, hpd, org; Anner Bylsma, vlc; Hans Jurg Lange, bsn; Frans Bruggen, rec, cond. ABC ABCL 67005/3 (3).
++ARG 2-77 p35 +NR 4-77 p6
+-FF 11-77 p27 +PRO 5-77 p25
++HF 3-77 p108 +SFC 12-12-76 p55
+MJ 2-77 p30 ++St 5-77 p54

1427 Works, selections: Israel in Egypt: But as for his people; Sing ye to the Lord. Judas Maccabaeus: See, the conquering hero comes; Sing unto God. Messiah: For unto us a child is born; Hallelujah. Sampson: Awake the trumpet's lofty sound; Let

their celestial concerts unite. Saul: Welcome, welcome mighty king; David, his ten thousands slew; How excellent thy name. Xerxes: Holy art thou. Zadok the priest. Mormon Tabernacle Choir; PO; Eugene Ormandy. CBS 71139.
+-Gr 7-77 p226 +-RR 5-77 p82
+-HFN 6-77 p137

1428 Works, selections: Alcina: Entree, Act 3; Il ballo. Ariodante: Dream music. Berenice: Overture. Faramondo: Overture. Messiah: Overture; Pastoral symphony. Saul: Dead march. Scipione: March. Solomon: Arrival of the Queen of Sheba. Water music: Lento; Hornpipe; Air. Bournemouth Sinfonietta; Kenneth Montgomery. HMV SQ ESD 7031 Tape (c) TC ESD 7031.
+-Gr 4-77 p1547 +-HFN 8-77 p99 tape
+-HFN 5-77 p123 +RR 5-77 p53

1429 Works, selections (Overtures and sinfonias): Ariodante. Arminio. Belshazzar. Berenice. Esther. Faramondo. Jeptha. Judas Maccabaeus. Julius Caesar. Deidamia. Radamisto. Rinaldo. Scipione. Semele. Sosarme. Solomon. Teseo. ECO; Richard Bonynge. London CSA 2247 (2). (Reissues from LCS 6586, 6711)
+NR 4-77 p2 +-SFC 12-12-76 p55

Zadok the priest. cf Coronation anthems (Enigma VAR 1030).
Zadok the priest. cf Works, selections (CBS 61139).
Zadok the priest. cf CRD Britannia BR 1077.
Zadok the priest. cf Decca SPA 500.
Zadok the priest. cf Pye Nixa SQ PCNHX 10.
Zadok the priest. cf Vista VPS 1053.

HANFF, Johann
Chorale preludes: Gott vom Himmel sieh darein; Auf meinem lieben Gott; Ein feste Burg ist unser Gott; Erbarm dich mein lieber Gott; Helft mir Gottes Gute preisen; War Gott nicht mit uns diese Zeit. cf BACH: Chorale preludes (Pelca PSR 40583).

HANMER
Stephen Foster fantasy. cf Rediffusion 15-56.

HANNA, Stephen
Sonic sauce. cf CIRONE: Triptych: Double concerto.

HANNEBERG
Triplets of the finest, concert polka. cf Nonesuch H 71341.

HANSON, Howard
Psalm, no. 150. cf Columbia M 34134.

1430 Symphony, no. 2, op. 30. National Philharmonic Orchestra; Charles Gerhardt. RCA GL 2-5021.
+ST 1-77 p779

d'HARDELOT
Because. cf HMV RLS 715.
Because. cf HMV RLS 716.
Because. cf Rubini GV 43.
Three green bonnets. cf HMV RLS 719.

HARRIS, Donald
Ludis II. cf Delos DEL 25406.

HARRIS, Roy
Concerto, clarinet, piano and string quartet. cf DIAMOND: Quintet, clarinet, 2 violas and 2 violoncellos.
Sonata, piano, op. 1. cf Vox SVBX 5303.

1431 Symphony, no. 3. IVES: Three places in New England. PO; Eugene Ormandy. RCA ARL 1-1682 Tape (c) ARK 1-1682 (ct) ARS 1-1682.
+MJ 11-76 p44 ++SFC 7-3-77 p34
++NR 9-76 p2

1432 Symphony, no. 5. MARTINU: Symphony, no. 5. Louisville Orchestra; Robert Whitney. RCA GL 2-5058.
+-Gr 7-77 p179 +-RR 7-77 p56
+-HFN 7-77 p113

HARRIS, William
Let my prayer come up. cf Vista VPS 1053.

HARRISON
Songs: Give me a ticket to heaven. cf Argo ZFB 95/6.

HARRISON, Lou
Concerto, organ and percussion. cf ADLER: Xenia, a dialogue for organ and percussion.
1433 Elegaic symphony. HUGHES: Cadences. Oakland Symphony Youth Orchestra; Denis de Coteau, Robert Hughes. 1750 Arch S 1772.
+-SFC 11-27-77 p66

HARTLEY, Fred
My love she's but a lassie yet. cf HMV SRS 5197.
Rouge et noire. cf HMV SRS 5197.
Rouge et noire. cf Starline SRS 197.

HARTLEY, Walter
Orpheus. cf Crystal S 206.

HARTSOUGH
Gwahoddiaa. cf BBC REC 267.

HARTY, Hamilton
With the wild geese. cf GERMAN: Welsh rhapsody.

HARWOOD, Basil
Paean, op. 15, no. 2. cf Vista VPS 1042.

HASLAM David
1434 Juanita the Spanish lobster. PROKOFIEV: Peter and the wolf, op. 67. Johnny Morris, narrator; Northern Sinfonia Orchestra; David Haslam. CRD CRD 1032.
+-Gr 3-77 p1402 ++RR 2-77 p56
+HFN 2-77 p119

HASPROIS, Jehan
Ma douce amour. cf HMV SLS 863.

HASSE, Johan
Concerto, mandolin, G major. cf BEETHOVEN: Sonatas, mandolin.

HASSLER, Hans
Cantata domino. cf Richardson RRS 3.
Canzon. cf DG 2533 323.

HAUBENSTOCK-RAMATI, Roman
1435 Quartet, strings, no. 1. URBANNER: Quartet, strings, no. 3. WEBERN: Bagatelles, string quartet, op. 9 (6). Movements, string quartet, op. 5 (5). Quartet, strings, op. 28. Alban Berg Quartet. Telefunken 6-41994.
++Gr 9-76 p437 +NR 2-77 p7
+-HF 3-77 p115 +RR 8-76 p65
++HFN 9-76 p121 ++St 12-76 p149
Rounds. cf Universal Edition UE 15043.

HAUSSMANN, Valentin
Catkanei. cf DG 2723 051.
Galliard. cf DG 2723 051.
Paduan. cf DG 2723 051.
Tantz. cf DG 2723 051.

HAVELKA, Sviatopluk
1436 Rose of wounds. Symphonic image "Ernesto Che Guevara". JAROCH: Symphony concertante, no. 3. Miroslav Frydlewicz, t; Antonin Novak, vln; Czech Radio Orchestra; Jiri Kout, Alois Klima.

Panton 110 480.
+-RR 1-77 p81

HAYDN, Josef

1437 Andante and variations, F minor. MOZART: Fantasia, K 397, D minor. Sonata, piano, no. 9, K 311, D major. Sonata, piano, no. 10, K 330, C major. Alicia de Larrocha, pno. London CS 7008 Tape (c) CS 5-7008. (also Decca SXL 6784)
+Gr 12-76 p1021 ++NYT 8-8-76 pD13
++HF 5-77 p101 tape +-RR 12-76 p84
+HFN 12-76 p144 ++St 8-76 p96
++NR 9-76 p12

Arianna a Naxos: Berenice che fai. cf Philips 6767 001.

Capriccio, G major. cf Sonatas, piano, nos. 2-3, 5, 7, 11, 13, 16, 19, 34, 40, 44, 49, 51, 58.

Capriccio, C major. cf Sonatas, piano, nos. 54-62.

Concerto, guitar, D major (arr. from Quartet, strings, op. 2, no. 2, E major). cf CARULLI: Concerto, guitar, A major.

1438 Concerto, harpsichord, op. 21, D major. MOZART: Concerto, piano, no. 8, K 246, C major. Ana-Maria Vera, pno; Rotterdam Philharmonic Orchestra; Edo de Waart. Philips 6833 199.
+ARG 6-77 p35 +SFC 2-13-77 p33
+-Audio 11-77 p127 +SR 5-28-77 p42
+MJ 4-77 p33 +St 6-77 p134
+NR 5-77 p8

1439 Concerto, horn, no. 1, D major. Concerto, organ, no. 1, C major. Concerto, trumpet, E flat major. Alan Stringer, tpt; Barry Tuckwell, hn; Simon Preston, org; AMF; Neville Marriner. Argo ZK 6. (Reissues from ZRG 543, 5498, 631)
+Gr 1-77 p1141 +RR 1-77 p55
+HFN 1-77 p119

1440 Concerto, horn, no. 1, D major. Concerto, oboe, C major. Concerto, trumpet, E flat major. Gerard Schwarz, tpt; Martin Smith, hn; Ronald Roseman, ob; Philharmonia Virtuosi, New York; Richard Kapp. Vox Tape (c) CT 2147.
+HF 10-77 p129 tape

1441 Concerto, oboe, C major (attrib.). MOZART: Concerto, oboe, K 314, C major. Ingo Goritzki, ob; South West German Chamber Orchestra; Paul Angerer. Claves D 606.
+-Gr 9-77 p504 +RR 4-77 p54
+HFN 5-77 p129

Concerto, oboe, C major. cf Concerto, horn, no. 1, D major.

Concerto, organ, no. 1, C major. cf Concerto, horn, no. 1, D major.

1442 Concerti, piano, C major, G major, D major, F major. Valentina Kamenikova, pno; Virtuosi Pragenses; Libor Hlavacek. Supraphon 110 1861/2 (2).
+ARG 11-77 p17 +-HFN 9-77 p143
+-FF 9-77 p25 +NR 12-77 p5
+Gr 9-77 p429 +-RR 7-77 p56

1443 Concerti, piano, nos. 2-4, 9, 11, D major, G major, F major, G major, D major, C major. Ilse von Alpenheim, pno; Bamberg Symphony Orchestra; Antal Dorati. Turnabout TV 37090/2 (3).
+-Gr 11-77 p838 +RR 11-77 p59
+-HFN 11-77 p172

1444 Concerto, trumpet, E flat major. L'Incontro improvviso overture. Sinfonia concertante, B flat major. Maurice Andre, tpt; Otto Winter, ob; Helman Jung, bsn; Walter Forchert, vln; Hans Haub-

lein, vlc; Bamberg Symphony Orchestra; Theodor Guschlbauer. Erato STU 70652.
+Gr 9-76 p421 +RR 9-76 p55
+-HFN 1-77 p119

1445 Concerto, trumpet, E flat major. HUMMEL: Concerto, trumpet, E flat major. NERUDA: Concerto, trumpet, E flat major. William Lang, tpt; Northern Sinfonia Orchestra; Christopher Seaman. Unicorn RHS 337.
+Gr 4-76 p1598 ++RR 4-76 p41
+HFN 4-76 p107 +St 10-76 p122
+NR 11-77 p7

Concerto, trumpet, E flat major. cf Concerto, horn, no. 1, D major (Argo ZK 6).

Concerto, trumpet, E flat major. cf Concerto, horn, no. 1, D major (Vox CT 2147).

Concerto, trumpet, E flat major. cf HMV SLS 5068.

1446 Concerti, violin, B flat major, F major. Petr Adamec, pno; Bohuslav Matousek, vln; PCO; Libor Hlavacek. Supraphon 110 1767.
-NR 4-77 p4

1447 Concerto, violin, no. 1, C major. Sinfonia concertante, op. 84, B flat major. Pinchas Zukerman, vln; Ronald Leonard, vlc; Barbara Winters, ob; David Breidenthal, bsn; LAPO; Pinchas Zukerman. DG 2530 907.
+Gr 12-77 p1080 +RR 12-77 p48
+-HFN 12-77 p171

Concerto, violin, no. 1, C major. cf BACH: Concerto, violin and strings, S 1041, A minor.

1448 Concerti, violoncello, C major, D major. Maurice Gendron, vlc. Philips Tape (c) 7317 115.
+-Gr 5-77 p1738 tape

1449 Concerti, violoncello, C major, D major. Samuel Mayes, vlc; University of Michigan Chamber Orchestra; Paul Makanowitzky. University of Michigan Records SM 0005.
+ARG 12-77 p34

1450 Concerto, violoncello, C major. Concerto, violoncello, op. 101, D major. AMF; Mstislav Rostropovich, vlc and cond. Angel S 37193. (also HMV SQ ASD 3255)
++Gr 9-76 p421 +NR 11-76 p5
+HF 12-76 p102 +RR 9-76 p54
+HFN 9-76 p123 +-St 1-77 p115
/MJ 4-77 p33

Concerto, violoncello, op. 101, D major. cf Concerto, violoncello, C major.

Concerto, violoncello, op. 101, D major. cf DVORAK: Concerto, violoncello, op. 104, B minor.

Divertimento, B flat major. cf Richardson RRS 3.

Divertimento, E flat major. cf GLUCK: Sinfonia, G major.

Divertimento, no. 15, F major. cf Telefunken EK 6-35334.

1451 Divertimenti, baryton, viola and violoncello. Jorg Eggebrecht, baryton; Deinhart Goritzki, vla; Willi Schmid, vlc. Claves D 609.
+HFN 5-77 p123 +RR 4-77 p66

Fantasia, C major. cf Sonatas, piano, nos. 6, 10, 18, 33.

1452 Fantasia, C major (Capriccio). Sonatas, piano, nos. 40-42, 47-52. Variations, Hob XVII:5-6. Dezso Ranki, pno. Hungaroton SLPX 11625/7 (3).

++ARG 12-76 p31 +RR 1-77 p70
++HF 10-77 p108

1453 La fedelta premiata. Ileana Cotrubas, Kari Lovaas, s; Frederica von Stade, ms; Lucia Valentini, con; Tonny Landy, Luigi Alva, t; Alan Titus, Maurizio Mazzieri, bar; OSR Chorus; Lausanne Chamber Orchestra; Antal Dorati. Philips 6707 028 (4). (also 9500 072/5)
+Gr 9-76 p466 +OC 12-77 p49
+HF 6-76 p68 +-ON 9-76 p70
+HFN 9-76 p122 +RR 9-76 p19
+MJ 10-76 p24 +SR 9-18-76 p50
++MT 4-77 p304 ++St 8-76 p94
++NR 5-76 p11

Feldpartita, B flat major. cf Saga 5417.
Glorious things. cf St. John the Divine Cathedral Church.
L'Incontro improvviso overture. cf Concerto, trumpet, E flat major.

1454 L'Infedelta delusa. Magda Kalmar, Julia Paszthy, s; Istvan Rozsos, Attila Fulop, t; Jozsef Gregor, bs; Ferenc Liszt Chamber Orchestra; Frigyes Sandor. Hungaroton SLPX 11832/4 (3).
+-ARG 10-77 p31 +HFN 5-77 p125
+-Gr 4-77 p1593 +RR 4-77 p34
+-HF 8-77 p83

Insanae et vanae curae. cf Abbey LPB 776.
Kleine Tanze fur die Jugend: Minuets (7). cf Piano works (Decca SHDN 112/5).

1455 Mass, no. 5, B flat major. MOZART: Mass, no. 7, K 167, C major. Elly Ameling, s; Peter Planyavsky, org; Vienna State Opera Chorus; VPO; Karl Munchinger. Decca SXL 6747. (also London 26443)
+-Gr 3-76 p1494 +-ON 1-22-77 p33
+-HF 11-76 p121 +-RR 3-76 p69
+HFN 3-76 p95 ++St 3-77 p138
++NR 3-77 p11

Minuets (2). cf DG 2533 182.
Minuets (2). cf DG 2723 051.

1456 Minuets (24). PH; Antal Dorati. Decca HDNW 90/1 (2). (also London STS 15359/60)
+Gr 10-76 p601 +RR 9-76 p54
+HFN 9-76 p122 ++SFC 5-13-77 p34
++NR 8-77 p2 ++St 5-77 p81
+NYT 6-5-77 pD19

1457 Nocturni (8). The Music Party; Alan Hacker. L'Oiseau-Lyre DSLO 521/2 (2).
+Gr 3-77 p1421 ++NR 8-77 p7
+HFN 4-77 p137 +RR 3-77 p49
++MJ 11-77 p30 ++St 10-77 p135
+MT 7-77 p565

1458 Orlando Paladino. Arleen Auger, Elly Ameling, s; Gwendolyn Killebrew, ms; George Shirley, Claes Ahnsjo, t; Benjamin Luxon, Domenico Trimarchi, bar; Maurizio Mazzieri, bs; Lausanne Chamber Orchestra; Antal Dorati. Philips 6707 029 (4).
+ARG 11-77 p16 ++MJ 11-77 p28
+Gr 9-77 p481 +NR 12-77 p9
++HF 12-77 p107 +RR 9-77 p39
+HFN 9-77 p144 ++SFC 8-28-77 p46

1459 Piano works: Sonatas, nos. 8, 12, 14-15, 29, 37, 61-62; B flat major (attrib. Schwanenberg); no. 28, D major (completed McCabe); E flat major (anon., formerly attrib. Haydn). Kleine Tanze fur die Jugend: Minuets (7). The seven last words of Christ, op. 15. Variations, E flat major, D major. John McCabe, pno. Decca SHDN 112/5 (4).
+Gr 10-77 p661 +RR 10-77 p77

Pieces, flute clock (7). cf BEETHOVEN: Adagio and allegro, music box.

1460 Quartets, strings, op. 0, E flat major; op. 1, nos. 1-4, 6; op. 2, nos. 1-2, 4, 6. Aeolian Quartet. London STS 15328/32 (5).
+HF 10-77 p108 ++NR 8-77 p7

Quartets, strings, op. 1, nos. 1-4, 6. cf Quartet, strings, op. 0, E flat major.

Quartets, strings, op. 2, nos. 1-2, 4, 6. cf Quartet, strings, op. 0, E flat major.

1461 Quartets strings, op. 3. The seven last words of Christ, op. 51. Peter Pears, narrator; Aeolian Quartet. Argo HDNV 82/4 (3).
+Gr 9-77 p444 +RR 9-77 p75
+HFN 9-77 p143

1462 Quartets, strings, op. 20, no. 2, C major. Quartets, strings, op. 20, no. 4, D major. Esterhazy Quartet. ABC L 67011.
++IN 12-77 p28 ++PRO 5-77 p25
+-MJ 2-77 p30 -SFC 12-12-76 p55
++NR 3-77 p8 +St 5-77 p54

Quartets, strings, op. 20, no. 4, D major. cf Quartets, strings, op. 20, no. 2, C major.

1463 Quartets, strings, op. 50, nos. 1-6. Tokyo Quartet. DG 2709 060 (3). (also DG 2740 135)
+-HF 2-77 p98 ++NR 2-77 p7
+MJ 4-77 p50 ++St 2-77 p82

1464 Quartets, strings, op. 64, nos. 1-6. Medici Quartet. HMV SLS 5077 (3).
+Gr 4-77 p1569 +RR 4-77 p66
+HFN 4-77 p137

1465 Quartets, strings, op. 64, no. 5, D major. Quartets, strings, op. 76, no. 2, D minor. Cleveland Quartet. RCA ARL 1-1409.
+HF 8-76 p91 ++SFC 7-18-76 p31
+MJ 10-76 p24 ++St 5-77 p72
+NR 6-76 p10

Quartets, strings, op. 64, no. 5, D major. cf DVORAK: Quartets, strings, no. 12, op. 96, F major.

1466 Quartets, strings, opp. 71 and 74. Aeolian Quartet. London STS 15325/7 (3).
/HF 11-76 p112 ++NR 1-77 p7
+-MJ 1-77 p26 ++SFC 9-12-76 p31

Quartets, strings, op. 76, no. 2, D minor. cf Quartet, strings, op. 64, no. 5, D major.

1467 Quartets, strings, op. 76, no. 3, C major. Quartets, strings, op. 76, no. 4, B flat major. Aeolian Quartet. Argo ZK 16 Tape (c) KZKC 16. (Reissue from HDNP 57/60)
+Gr 7-77 p197 +RR 10-77 p96 tape
+HFN 7-77 p125 ++RR 12-77 p95 tape
++RR 7-77 p66

1468 Quartets, strings, op. 76, no. 3, C major. Quartets, strings, op. 76, no. 4, B flat major. Quartetto Italiano. Philips 9500 157.

+Audio 12-77 p48 ++HFN 5-77 p125
+Gr 5-77 p1707 ++RR 9-77 p76

Quartets, strings, op. 76, no. 4, B flat major. cf Quartets, strings, op. 76, no. 3, C major (Argo 16).

Quartets, strings, op. 76, no. 4, B flat major. cf Quartets, strings, op. 76, no. 3, C major (Philips 9500 157).

1469 Quartets, strings, op. 77, no. 1, G major. Quartets, strings, op. 77, no. 2. Tatrai Quartet. Hungaroton SLPX 11776.
*NR 1-77 p7 ++St 4-77 p87

Quartets, strings, op. 77, no. 2. cf Quartets, strings, op. 77, no. 1, G major.

1470 Il ritorno di Tobia. Klara Takacs, alto; Veronica Kincses, Magda Kalmar, s; Attila Fulop, t; Zsolt Bende, bar; Budapest Madrigal Choir; HSO; Ferenc Szekeres. Hungaroton SLPX 11660/3 (4).
/Gr 4-75 p1849 +RR 3-75 p62
+HF 6-75 p81 +SFC 5-22-77 p42
++NYT 8-25-74 pD20

St. Antoni chorale. cf Rediffusion Royale 2015.

1471 Die Schopfung (The creation) (sung in German). Mimi Coertse, s; Julius Patzak, t; Vienna Singverein; Vienna Volksoper Orchestra; Jascha Horenstein. Classical Cassette Club (c) CCC 20.
+-HF 1-77 p151 tape

1472 Die Schopfung. Lucia Popp, Helena Dose, s; Werner Hollweg, t; Kurt Moll, Benjamin Luxon, bs; Brighton Festival Chorus; RPO; Antal Dorati. Decca D50D2 (2).
+HFN 12-77 p171 +RR 12-77 p77

1473 The seasons (Die Jahreszeiten). Gundula Janowitz, s; Peter Schreier, t; Martti Talvela, bs; Vienna Singverein; VSO; Karl Bohm. DG 2709 026 Tape (c) 3371 028.
++Gr 11-77 p900 tape ++RR 10-77 p91 tape

1474 Die Sieben Letzten Worte Unseres Erlosers am Kreuze, op. 51 (The seven last words of Christ). Virginia Babikian, Ina Dressel, s; Eunice Alberts, alto; John van Kesteren, t; Otto Wiener, bs; VSOO; Vienna Academy Chorus; Hermann Scherchen. Westminster WGS 8342. (Reissue from WST 17006)
+ARG 10-77 p30 +NR 9-77 p9

The seven last words of Christ, op. 51. cf Piano works (Decca SHDN 112/5).

The seven last words of Christ, op. 51. cf Quartets, strings, op. 3.

Sinfonia concertante, B flat major. cf Concerto, trumpet, E flat major.

Sinfonia concertante, op. 84, B flat major. cf Concerto, violin, no. 1, C major.

Sonata, piano, B flat major (attrib. Schwanenberg). cf Piano works (Decca SHDN 112/5).

Sonata, piano, E flat major (anon., formerly attrib. Haydn). cf Piano works (Decca SHDN 112/5).

1475 Sonatas, piano, nos. 1-19. Variations, A major. Variations, D major. Zsuzsa Pertis, Janos Sebestyen, hpd. Hungaroton SLPX 11614/7 (4).
+HF 1-77 p106 +RR 6-76 p72
++NR 5-76 p14 ++SFC 4-17-77 p42
+NYT 5-15-77 pD17 +-St 11-76 p143

1476 Sonatas, piano, nos. 1-26, 29-34. Rudolf Buchbinder, pno. Telefunken FK 6-35088 (6).

+ARG 9-77 p36
+-Gr 5-77 p1995
+MJ 7-77 p70
+NR 7-77 p13
+-NYT 5-15-77 pD17
+-RR 3-75 p48

1477 Sonatas, piano, nos. 1-34. Rudolf Buchbinder, pno. Telefunken SHD 25123 (6).
+SFC 4-17-77 p42

Sonata, piano, no. 1, C major: Tempo di menuetto. cf HMV RLS 723.

1478 Sonatas, piano, nos. 2-3, 5, 7, 11, 13, 16, 19, 34, 40, 44, 49, 51, 58. Capriccio, G major. John McCabe, pno. Decca 4HDN 109/11 (3).
+-Gr 4-77 p1573
+RR 4-77 p72

1479 Sonatas, piano, no. 6, 10, 18, 33, 38-39, 47, 50, 52, 60. Fantasia, C major. Variations, F minor. John McCabe, pno. Decca 1 HSN 100/2 (3). (also London STS 15343/5 (3))
+-Gr 10-75 p657
+HF 1-77 p106
+HFN 10-75 p142
-MJ 1-77 p26
+MT 11-76 p914
+NR 12-76 p12
+NYT 5-15-77 pD17
+-RR 10-75 p76
+SFC 4-17-77 p42
+-St 11-76 p143
+STL 6-6-76 p37

Sonatas, piano, nos. 8, 12, 14-15, 29, 37, 61-62. cf Piano works (Decca SHDN 112/5).

1480 Sonatas, piano, nos. 8, 14-15, 28-29, 37, 62; sonatas, B flat major, E flat major. John McCabe, pno. Decca 5HDN 112/5 (4).
+HFN 10-77 p155

Sonata, piano, no. 12, A major. cf BEETHOVEN: Sonata, piano, no. 22, op. 54, F major.

1481 Sonatas, piano, nos. 19, 37, 44. Variations, F minor. Gilbert Kalish, pno. Nonesuch H 71328.
-ARG 11-76 p26
+-Gr 6-77 p70
+-HF 1-77 p106
+HFN 4-77 p137
+-NR 9-76 p12
+NYT 5-15-77 pD17
+-RR 4-77 p72
+SFC 4-17-77 p42
+St 3-77 p129

1482 Sonatas, piano, nos. 23, 35, 46, 51. Vasso Devetzi, pno. Monitor MCS 2147.
+-MJ 10-76 p52
+NR 9-76 p12
+-St 3-77 p129

Sonata, piano, no. 28, D major (completed McCabe). cf Piano works (Decca SHDN 112/5).

1483 Sonatas, piano, nos. 32, 34, 46, 51. Gilbert Kalish, pno. Nonesuch H 71318.
+HF 1-77 p106
+-HFN 9-76 p123
++NR 3-76 p12
+NYT 5-15-77 pD17
+-RR 10-76 p88
++St 4-76 p79

Sonatas, piano, nos. 40-42, 47-52. cf Fantasia, C major.

1484 Sonata, piano, no. 46, A flat major. Sonata, piano, no. 48, C major. MOZART: Adagio, K 540, B minor. Andante, K 616, F major. Eine kleine gigue, K 574, G major. Minuet, K 355, D major. Rondo, piano, K 511, A minor. Renee Sandor, pno. Hungaroton LXP 11638.
+-NR 9-75 p11
+-RR 8-75 p57
++SFC 2-13-77 p33
++St 2-76 p112

1485 Sonatas, piano, nos. 48-50, 53. Christoph Eschenbach, pno. DG 2530 736 Tape (c) 3300 736.
-Gr 10-76 p628
+HFN 10-76 p171
+-RR 10-76 p88
+-RR 1-77 p90 tape

Sonata, piano, no. 48, C major. cf Sonata, piano, no. 46, A flat major.

1486 Songs (canzonettas): Content; Despair; Fidelity; The mermaid; A pastoral song; Piercing eyes; Pleasing pain; The sailor's song; She never told me her love; Recollections; Sympathy; The wanderer. Songs: The spirit's song; O tuneful voice. James Griffett, t; Bryan Vickers, fortepiano. Pearl SHE 540.
+Gr 6-77 p84 +RR 5-77 p82
+-HFN 6-77 p123

1487 Symphony, no. 22, E flat major. Symphony, no. 23, G major. Symphony, no. 24, D major. Prague Chamber Orchestra; Bernhard Klee. DG 2530 885 Tape (c) 3300 885.
+Gr 10-77 p635 +RR 9-77 p59
++Gr 11-77 p899 tape +-RR 12-77 p95 tape
+HFN 11-77 p172

Symphony, no. 23, G major. cf Symphony, no. 22, E flat major.
Symphony, no. 24, D major. cf Symphony, no. 22, E flat major.

1488 Symphony, no. 43, E flat major. Symphony, no. 59, A major. AMF; Neville Marriner. Philips 9500 159 Tape (c) 7300 524.
+Gr 1-77 p1142 ++RR 1-77 p55
+HFN 1-77 p109 +RR 12-77 p95 tape

1489 Symphony, no. 44, E minor. Symphony, no. 49, F minor. ECO; Daniel Barenboim. DG 2530 708 Tape (c) 3300 708.
+ARG 7-77 p22 +HFN 3-77 p119 tape
+Gr 2-77 p1280 /RR 2-77 p47
+HF 10-77 p110 +-RR 4-77 p92 tape
+HFN 2-77 p121

Symphony, no. 49, F minor. cf Symphony, no. 44, E minor.
Symphony, no. 59, A major. cf Symphony, no. 43, E flat major.

1490 Symphonies, nos. 82-87. ECO; Daniel Barenboim. HMV SLS 5065 (3). (also EMI/Capitol SLS 5065)
++Gr 10-76 p601 +HFN 9-76 p122
+HF 10-77 p110 +-RR 9-76 p55

1491 Symphony, no. 86, D major. Symphony, no. 92, G major. LSO, OSCCP; Bruno Walter. Bruno Walter Society BWS 999.
-NR 1-77 p4

1492 Symphony, no. 88, G major. Symphony, no. 96, D major. LSO; Andre Previn. Angel SQ 37274. (also HMV ASD 3328 Tape (c) TC ASD 3328)
+ARG 7-77 p22 +-HFN 4-77 p136
+Audio 9-77 p46 ++NR 7-77 p5
+Gr 4-77 p1549 -RR 4-77 p46
-Gr 6-77 p105 tape /SFC 7-17-77 p42

1493 Symphony, no. 88, G major. Symphony, no. 99, E flat major. COA; Colin Davis. Philips 9500 138 Tape (c) 7300 534.
+-Audio 9-77 p46 +HFN 7-77 p127 tape
+Gr 4-77 p1549 ++NR 10-77 p3
+Gr 9-77 p508 tape +-RR 4-77 p46
+HFN 4-77 p137 +RR 9-77 p99 tape

1494 Symphony, no. 90, C major. Symphony, no. 91, E flat major. Esterhazy Orchestra; David Blum. Vanguard C 10044 Tape (c) D 10044.
+HF 11-77 p138 tape

Symphony, no. 91, E flat major. cf Symphony, no. 90, C major.
Symphony, no. 92, G major. cf Symphony, no. 86, D major.

1495 Symphony, no. 94, G major. Symphony, no. 103, E flat major. LPO; Raymond Leppard. Classics for Pleasure CFP 40269.
+Gr 12-77 p1080 +-RR 10-77 p47
+HFN 10-77 p155

1496 Symphony, no. 94, G major. Symphony, no. 101, D major. PH; Antal Dorati. Decca SPA 494 Tape (c) KCSP 494. (Reissue from HDNJ 41/6)
+Gr 8-77 p305 /-RR 7-77 p57
+HFN 7-77 p125
1497 Symphony, no. 96, D major. Symphony, no. 99, E flat major. COA; Bernard Haitink. Philips 6580 151. (Reissue from SAL 3721)
+Gr 7-77 p179 +-RR 6-77 p55
+HFN 6-77 p137
Symphony, no. 96, D major. cf Symphony, no. 88, G major.
Symphony, no. 96, D major. cf BARTOK: Concerto, orchestra.
1498 Symphonies, nos. 99-104. PH; Antal Dorati. Decca SDD 503/5 (3). (Reissues from HDNJ 41/6)
+Gr 5-77 p1696 +-RR 4-77 p47
+HFN 5-77 p137
1499 Symphony, no. 99, E flat major. Symphony, no. 101, D major. NYP; Leonard Bernstein. CBS 76580.
+-HFN 8-77 p83 +-RR 7-77 p57
1500 Symphony, no. 99, E flat major. Symphony, no. 100, G major. NYP; Leonard Bernstein. Columbia M 34126 Tape (c) MT 34126.
-Gr 8-77 p305 +MJ 11-76 p44
+HF 2-77 p98 +NR 7-76 p3
Symphony, no. 99, E flat major. cf Symphony, no. 88, G major.
Symphony, no. 99, E flat major. cf Symphony, no. 96, D major.
1501 Symphony, no. 100, G major. Symphony, no. 103, E flat major. AMF; Neville Marriner. Philips 9500 255 Tape (c) 7300 543.
++Gr 10-77 p635 +HFN 12-77 p187 tape
+-Gr 12-77 p1145 tape +-RR 10-77 p47
+HFN 10-77 p155 +-RR 12-77 p95 tape
Symphony, no. 100, G major. cf Symphony, no. 99, E flat major.
Symphony, no. 101, D major. cf Symphony, no. 94, G major.
Symphony, no. 101, D major. cf Symphony, no. 99, E flat major.
Symphony, no. 101, D major. cf BACH, J.C.: Concerto, harp, op. 1, no. 6, D major.
Symphony, no. 103, E flat major. cf Symphony, no. 94, G major.
Symphony, no. 103, E flat major. cf Symphony, no. 100, G major.
1502 Symphony, no. 104, D major. SCHUBERT: Symphony, no. 8, D 759, B minor. BPhO; Herbert von Karajan. HMV SQ ASD 3203 Tape (c) TC ASD 3203. (also Angel S 37058 Tape (c) 4XS 37058)
++Gr 9-76 p416 +-HFN 11-76 p175 tape
++HF 10-77 p114 +-RR 9-76 p60
+HFN 11-76 p160 +-RR 1-77 p91 tape
1503 Trios, baryton, nos. 37, 48, 70-71, 85, 96-97, 109, 113, 117, 121. Esterhazy Baryton Trio. HMV SQ SLS 5095 (2).
+Gr 9-77 p451 +RR 10-77 p66
+HFN 11-77 p172
1504 Trio, piano, no. 7, D major. Trio, piano, no. 9, A major. Trio, piano, no. 12, E minor. Beaux Arts Trio. Philips 9500 326.
+Gr 8-77 p314 +-RR 7-77 p66
+HFN 7-77 p113
Trio, piano, no. 9, A major. cf Trio, piano, no. 7, D major.
Trio, piano, no. 12, E minor. cf Trio, piano, no. 7, D major.
1505 Trios, piano, nos. 13, 16, 17, C minor, D major, F major. Beaux Arts Trio. Philips 9500 035.
++ARG 5-77 p22 ++NR 3-77 p8
+Gr 2-77 p1301 +RR 2-77 p73
+HFN 3-77 p105 ++St 7-77 p115
-MJ 7-77 p70

1506 Trio, piano, no. 14, A flat major. Trio, piano, no. 15, G major. Beaux Arts Trio. Philips 9500 034.
+Gr 11-76 p830 ++NR 1-77 p7
+HF 1-77 p118 +-RR 11-76 p86
+HFN 11-76 p161 ++SFC 4-17-77 p42
+-MJ 2-77 p31

Trio, piano, no. 15, G major. cf Trio, piano, no. 14, A flat major.

1507 Trios, piano, nos. 19, 27, 29, G minor and C major. Amade Trio. Titanic TI 12.
++St 7-77 p115

1508 Trio, piano, no. 25, G major. Trio, piano, no. 26, F sharp minor. Trio, piano, no. 27, C major. Beaux Arts Trio. Philips 6500 023.
++St 5-77 p71

Trio, piano, no. 26, F sharp minor. cf Trio, piano, no. 25, G major.
Trio, piano, no. 27, C major. cf Trio, piano, no. 25, G major.
Variations, Hob XVII:5-6. cf Fantasia, C major.
Variations, A major. cf Sonatas, piano, nos. 1-19.
Variations, C major, F minor. cf Sonatas, piano, nos. 54-62.
Variations, D major. cf Sonatas, piano, nos. 1-19.
Variations, E flat major, D major. cf Piano works (Decca SHDN 112/5).
Variations, F minor. cf MOZART: Fantasia, K 397, D major.
Variations, F minor. cf Sonatas, piano, nos. 6, 10, 18, 33 (Decca 100/2).
Variations, F minor. cf Sonatas, piano, nos. 19, 37, 44.

1509 La vera costanza. Jessye Norman, Helen Donath, Kari Lovaas, s; Claes Ahnsjo, Anthony Rolfe Johnson, t; Domenico Trimarchi, bar; Wladimiro Ganzarolli, bs; Lausanne Chamber Orchestra; Antal Dorati. Philips 6703 077 (3).
++ARG 8-77 p14 ++NR 8-77 p10
+-Gr 6-77 p96 +NYT 4-17-77 pD17
+HF 8-77 p84 +RR 6-77 p36
+HFN 6-77 p125 ++SFC 5-22-77 p42
+MJ 7-77 p68 ++St 8-77 p115

HAYDN, Michael

Concerto, trumpet, D major. cf GRAUPNER: Concerto, trumpet, D major.

1510 Die Hochzeit auf der Alm. Pietas in Hostem. MOZART: Concerto, harpsichord, no. 1, K 107, D major. Galithmathias Musicum, K 32. Isolde Ahlgrimm, hpd; Camerata Academica; Bernhard Paumgartner. Amadeo AVRS 19030. (Reissues)
+HFN 11-77 p172 +-RR 11-77 p60

Pietas in Hostem. cf Die Hochzeit auf der Alm.
Songs: Anima nostra; In dulci jubilo. cf RCA PRL 1-8020.
Symphony, G major (with introduction by Mozart, K 444). cf BACH, J.C.: Sinfonia, op. 6, no. 6, C minor.

HAYES

Concertino, saxophone. cf Citadel CT 6012.

HAYNE VON GHIZEGHEM

Songs: De tous biens plaine; A la audienche. cf HMV SLS 5049.

HEAD, Michael

A piper. cf BIS LP 37.

HEATH

Air and rondo. cf Rediffusion 15-56.

High spirits. cf Rediffusion 15-56.

HECKEL, Wolf (Wolfgang)

Mille regretz; Nach willen dein. cf L'Oiseau-Lyre 12BB 203/6.

Ein ungarischer Tantz, Proportz auff den ungarischen Tantz. cf Hungaroton SLPX 11721.

HEGAR

Waltzes, nos. 2 and 4. cf Discopaedia MOB 1020.

HEIDEN, Bernhard

Sonata, horn and piano. cf Musical Heritage Society MHS 3547.

1511 Variations, solo tuba and 9 horns. WILDER: Suite, horn, tuba and piano, no. 2. Harvey Phillips, tuba; John Barrows, hn; Milton Kaye, pno; The Valhalla Horn Choir; Bernhard Heiden. Golden Crest CRSO 4147.

+ARG 5-77 p23

HEINICHEN, Johann

Concerto, 4 recorders and strings. cf BOISMORTIER: Concerto, 5 recorders without bass, D minor.

HEINRICH, Anthony

The Elssler dances: The laurel waltz. cf New World Records NW 257.

HEINTZ, Wolff

Da truncken sie. cf BIS LP 3.

HEISS, John

1512 Inventions, contours and colors. HODKINSON: Megalith trilogy. UNG: Mohori. Barbara Martin, s; William Albright, org; Speculum Musicae, Contemporary Chamber Ensemble; Richard Fitz, Arthur Weisberg. CRI SD 363.

+ARG 3-77 p41 +NR 10-77 p16
-MJ 3-77 p46

HEKKING, Gerard

Villageoise. cf HMV SQ ASD 3283.

HELLERMANN, William

On the edge of a node. cf BRESNICK: B's garland.

HELPS, Robert

The running sun. cf New World Records NW 243.

HELY-HUTCHINSON, Victor

1513 Carol symphony. VAUGHAN WILLIAMS: Fantasia on Christmas carols. WARLOCK: Adam lay y bounden. Bethlehem down. TRAD. (arr. Vaughan Williams): And all in the morning. Wassail song. Pro Arte Orchestra; Guildford Cathedral Choir; Barry Rose. HMV ESD 7021 Tape (c) TC ESD 7021.

+HFN 1-77 p119 +RR 12-76 p97
+HFN 4-77 p155 tape +ST 1-77 p781

HENKEMANS, Hans

Songs: Primavera; Tre aspetti d'amore. cf FLOTHUIS: Songs: Hymnus; Per sonare ed ascoltare.

HENRY VIII, King

Songs: En vray amoure; O my heart. cf Saga 5444.

Songs: Pastime with good company; Taunder naken. cf Argo ZK 24.

Songs: Taunder naken. cf National Trust NT 002.

HENSCHEL, George

Spring. cf HMV RLS 719.

HENZE, Hans Werner

1514 In memoriam: Die Weisse Rose. Kammermusik. Philip Langridge, t; Timothy Walker, gtr; London Sinfonietta; Hans Werner Henze. L'Oiseau-Lyre DSLO 5.

++Audio 5-77 p100 +NR 2-76 p3
+Gr 11-75 p875 +-NYT 12-21-75 pD18

+HF 3-76 p89 +RR 9-75 p69
+HFN 12-75 p155 +STL 4-11-76 p36
++MT 9-76 p746

Kammermusik. cf In memoriam: Die Weisse Rose.
Ragtimes and habaneras. cf BIRTWISTLE: Grimethorpe aria.
1515 Tristan. Homero Francesch, pno; Cologne Radio Symphony Orchestra; Hans Werner Henze. DG 2530 834.
+Gr 8-77 p305 +MT 12-77 p1014
+HFN 10-77 p156 +RR 9-77 p62

HERBECK, Johann
Songs: Angels we have heard on high; Kommet ihr Hirten; Pueri concinite. cf RCA PRL 1-8020.

HERBERT, Victor
L'Encore. cf Angel (Q) S 37304.
The fortune teller, excerpts. cf Angel (Q) S 37304.
March of the toys. cf EMI TWOX 1058.

HERBST
God was in Jesus. cf Vox SVBX 5350.

HERMAN
Milk and honey, Shalom. cf HMV MLF 118.

HERMAN, Johan
Lobt Gott ihr Christen. cf ABC L 67002.

HERMANSON, Ake
Appell I-IV, op. 10. cf ROSENBERG: Concerto, orchestra, no. 3.
In nuce, op. 7. cf ROSENBERG: Concerto, orchestra, no. 3.
Invoco, op. 4. cf ROSENBERG: Concerto, orchestra, no. 3.
Sound of a flute. cf BIS LP 32.
Winter flute. cf BIS LP 32.

HERMANSSON, Christen
Shadow play. cf BIS LP 59.

HEROLD, Louis Joseph
Le pre aux clercs: Le rendezvous de noble compagnie. cf Club 99-107.
1516 Zampa: Overture. LALO: Le Roi d'Ys: Overture. MASSANET: Phedre: Overture. THOMAS: Mignon: Overture. Raymond: Overture. MRSO; Gennady Rozhdestvensky. Westminster WGS 8331.
+NR 5-77 p5 +-SFC 8-21-77 p46

Zampa: Overture. cf HMV ESD 7010.
Zampa: Overture. cf Quintessence PMC 7013.
Zampa: Perche tremar. cf Rubini GV 34.

HERRMANN, Bernard
The fantasticks. cf DELIUS: A late lark.
For the fallen. cf DELIUS: A late lark.
1517 Symphony. National Philharmonic Orchestra; Bernard Herrmann. Unicorn RHS 331.
+Gr 2-76 p1345 +NR 10-77 p2
++HF 2-76 p97 ++St 7-76 p110

HERTEL, Johann
Concerto, trumpet, E flat major. cf ALBINONI: Concerto, trumpet, op. 7, no. 3, B flat major.

HEUBERGER, Richard
Der Opernball (The opera ball): In chambre separee. cf Angel (Q) S 37304.
The opera ball: Midnight bells. cf HMV ASD 3346.
Der Opernball: Overture. cf Classics for Pleasure CFP 40236.

HEUSSENSTAMM, George
Tubafour, op. 30. cf Crystal S 221.

HEWITT-JONES, Tony
Fanfare. cf Vista VPS 1047.
HICKS, J. W.
The last hymn. cf New World Records NW 220.
HIDDEN MELODIES: Cheek to cheek (In the style of Mozart); Love walked in (Delius); Wouldn't it be loverly (Brahms); O what a beautiful morning (Grieg); The Vicar of Bray (Chopin); Crib, the melodies revealed. cf Decca SPA 519.
HIDDEN MELODIES: I've got you under my skin (In the style of Mendelssohn); Three blind mice (Bach); Waltzing Matilda (Scarlatti); I saw three ships come sailing by (Schumann); When Johnny comes marching home (Schubert); For he's a jolly good fellow (Chopin); The Lambeth walk (Rachmaninoff); The Londonderry air (Brahms); Three blind mice (Debussy). cf Decca SPA 473.
HIGHLIGHTS FROM THE LAST NIGHT OF THE PROMS, 1974. cf BBC REH 290.
HILTON, John
Fair Oriana, beauty's queen. cf DG 2533 347.
Fantasies, nos. 1-3. cf Telefunken AW 6-42033.
HINDEMITH, Paul
1518 Canonic variations, 2 violins. Duets, violin and clarinet (2). Kleine Klaviermusik, op. 45, no. 4. Sonata, bassoon and piano. Paul Shure, Bonnie Douglas, vln; Hugo Raimondi, clt; Don Christlieb, bsn; Dolores Stevens, pno. GSC 1.
+-HF 11-73 p108 ++NR 9-73 p8
+MQ 10-73 p652 +NYT 8-14-77 pD13
1519 Concert music, wind band, op. 14. Geschwindmarsch. Symphony, band, B flat major. University of Michigan School of Music Wind Ensemble and Symphony Band; H. Robert Reynolds. University of Michigan SM 0003.
+ARG 6-77 p27 +St 7-77 p132
+IN 11-77 p24
1520 Duet, viola and violoncello. Echo, flute and piano. Sonata, flute and piano. Sonata, solo violoncello, op. 25, no. 3. Louise di Tullio, flt; Lincoln Mayorga, pno; Kurt Reher, vlc; Sven Reher, vla. GSC 3.
+-HF 11-73 p108 ++NR 9-73 p8
+MQ 10-73 p652 +NYT 8-14-77 pD13
Duets, violin and clarinet (2). cf Canonic variations, 2 violins.
Echo, flute and piano. cf Duet, viola and violoncello.
1521 The four temperaments. STRAVINSKY: Capriccio. Kjara Havlikova, pno; Bratislava Radio Orchestra; Otakar Trhlik. Aurora AUR 5052.
+-Gr 5-77 p1696 +-HFN 6-77 p125
Geschwindmarsch. cf Concert music, wind band, op. 41.
1522 Kammermusik (7). Concerto Amsterdam Soloists. Telefunken SLT 43110/2 (3).
++FF 11-77 p77
Kammermusik, no. 5, op. 36, no. 4. cf Der Schwanendreher.
1523 Kleine Kammermusik, op. 24, no. 2. IBERT: Pieces breves (3). JANACEK: Mladi (Youth). LIGETI: Pieces, wind quintet (10). Vienna Wind Soloists; Horst Hajek, clt. Decca SDD 523.
+Gr 7-77 p197 +MT 12-77 p1015
+HFN 7-77 p113 ++RR 7-77 p67
Kleine Klaviermusik, op. 45, no. 4. cf Canonic variations, 2 violins.
1524 Das Marienleben, op. 27. Peggy Bonini, s; Ingolf Dahl, pno. GSC 7.
+NYT 8-14-77 pD13

Morgenmusik. cf BIS LP 59.
Pieces, bassoon and violoncello (4). cf Gasparo GS 103.
Ploner Musiktag: Trio. cf BIS LP 57.
1525 Quartet, strings, no. 2, op. 16, C major. Quartet, strings, no. 3, op. 22. Kreuzberg String Quartet. Telefunken AW 6-42077.
+Gr 8-77 p317 +RR 7-77 p66
+HFN 7-77 p115
Quartet, strings, no. 3, op. 22. cf Quartet, strings, no. 2, op. 16, C major.
Rondo, 3 guitars. cf Trio for soprano and 2 alto recorders.
1526 Der Schwanendreher. Kammermusik, no. 5, op. 36, no. 4. Igor Boguslavsky, vla; MRSO Chamber Group, Gennady Rozhdestvensky. Westminster WGS 8330.
/ARG 5-77 p20 ++NR 3-77 p7
Der Schwanendreher. cf BLOCH: Suite hebraique.
1527 Die Serenaden, op. 35. Sonata, trumpet and piano, B flat major. Marni Nixon, s; Gordon Pope, ob; Sven Reher, vla; Kurt Reher, vlc; Thomas Stevens, tpt; Delores Stevens, pno. GSC 2.
+-HF 11-73 p108 ++NR 9-73 p8
+MQ 10-73 p652 +NYT 8-14-77 pD13
Sonata, bassoon and piano. cf Canonic variations, 2 violins.
1528 Sonatas, bass instruments and piano. PO Members; Glenn Gould, pno. Columbia X 1398 (2).
+MJ 10-77 p29
Sonata, harp. cf Trio for soprano and 2 alto recorders.
1529 Sonata, horn and piano. Sonata, solo viola, op. 25, no. 1. William Kosinski, English hn; Sven Reher, vla; Lincoln Mayorga, pno. GSC 4.
+NR 8-75 p7 +NYT 8-14-77 pD13
Sonata, horn and piano. cf Works, selections (Columbia M2 33971).
Sonata, horn and piano, E flat major. cf Works, selections (Columbia M2 33971).
1530 Sonatas, organ (3). George Baker, org. Delos FY 026.
++HF 9-77 p100 ++NR 10-76 p15
++MU 10-76 p16 ++St 10-77 p136
1531 Sonatas, organ (3). George Markey, org. Psallite 106 100870.
+MJ 5-77 p32
1532 Sonatas, organ, nos. 1-3. Ivan Sokol, org. Opus 9111 0195.
+-Gr 5-77 p1714 +-RR 3-77 p79
Sonata, string bass and piano. cf Three pieces, 5 instruments.
Sonata, trombone and piano. cf Works, selections (Columbia M2 33971).
Sonata, trumpet. cf Crystal S 361.
Sonata, trumpet and piano, B flat major. cf Die Serenaden, op. 35.
Sonata, trumpet and piano, B flat major. cf Works, selections (Columbia M2 33971).
Sonata, bass tuba and piano. cf Works, selections (Columbia M2 33971).
Sonata, solo viola, op. 25, no. 1. cf Sonata, horn and piano.
Sonata, violin, no. 3, E major. cf Discopaedia MB 1016.
Sonata, violin, op. 31, no. 2. cf Discopaedia MB 1016.
Sonata, violin and piano, E major. cf BARTOK: Hungarian folk songs.
Sonata, solo violoncello, op. 25, no. 3. cf Duet, viola and violoncello.
Suite "1922", op. 26. cf Three pieces, 5 instrument.
Suite "1922", op. 26. cf Supraphon 111 1721/2.
Symphony, band, B flat major. cf Concert music, wind band, op. 41.

1533 Three pieces, 5 instruments. Suite, "1922", op. 26. Sonata, string bass and piano. Gary Karr, bs; Harmon Lewis, Zita Carno, pno; Julian Spear, clt; Malcolm McNab, tpt; Israel Baker, vln; Buell Neidlinger, bs. GSC S 6.
+NR 11-77 p9 +NYT 8-14-77 pD13
Trio, viola, heckelphone and piano. cf Trio for soprano and 2 alto recorders.
1534 Trio for soprano and 2 alto recorders (Ploner Musiktag). Rondo, 3 guitars. Sonata, harp. Trio, viola, heckelphone and piano, op. 47. Myra Kestenbaum, vla; John Ellis, heckelphone; Delores Stevens, pno; Laurindo Almeida, gtr; Robin Howell, rec; Gail Laughton, hp. GSC 5.
++NR 6-75 p10 +NYT 8-14-77 pD13
1535 Works, selections: Sonata, horn and piano. Sonata, alto horn and piano, E flat major. Sonata, trombone and piano. Sonata, trumpet and piano, B flat major. Sonata, bass tuba and piano. Glenn Gould, pno; Philadelphia Brass Ensemble Members. Columbia M2 33971 (2).
++HF 5-77 p83 +SFC 6-6-76 p33
+MQ 1-77 p144 ++St 10-76 p123
+-NR 7-76 p5
HINE, William
A flute piece. cf Vista VPS 1060.
HODDINOTT, Alun
1536 Concertino, viola and small orchestra, op. 14. Dives and Lazarus, op. 39. Night music, op. 48. Sinfonietta, no. 1, op. 56. Felicity Palmer, s; Thomas Allen, bar; Csabo Erdelyi, vla; Welsh National Opera Chorus; NPhO; David Atherton. Argo ZRG 824.
+Gr 7-76 p175 +MT 6-77 p487
+-HFN 7-76 p89 +RR 7-76 p75
1537 Concerto, piano, no. 3, op. 44. Landscapes, op. 86. Sinfonietta, no. 2, op. 67. Roger Woodward, pno; NPhO; Hans-Hubert Schonzeler. RCA RL 2-5082 Tape (c) RK 2-5082.
+-Gr 6-77 p55 +-MT 12-77 p1015
+-HFN 8-77 p83 +-RR 6-77 p55
Dives and Lazarus, op. 39. cf Concertino, viola and small orchestra, op. 14.
Landscapes, op. 86. cf Concerto, piano, no. 3, op. 44.
Night music, op. 48. cf Concertino, viola and small orchestra, op. 14.
Sinfonietta, no. 1, op. 56. cf Concertino, viola and small orchestra, op. 14.
Sinfonietta, no. 2, op. 67. cf Concerto, piano, no. 3, op. 44.
HODKINSON, Sydney
Megalith trilogy. cf HEISS: Inventions, contours and colors.
HOFFMAN, Richard
Dixiana. cf New World Records NW 257.
In memoriam L.M.G. cf New World Records NW 257.
HOFFMEISTER, Franz
1538 Concerto, flute, G major. TELEMANN: Suite, A minor. Ingrid Dingfelder, flt; ECO; Lawrence Leonard. Enigma VAR 1026.
+Gr 6-77 p55 +-RR 5-77 p60
1539 Concerto, flute, G major. MOZART: Rondo, K Anh 184, D major (trans. Hoffmeister). STAMITZ: Concerto, flute, op. 29, G minor. Jean-Pierre Rampal, flt; Saar Chamber Orchestra, VSO, Vienna Baroque Ensemble; Karl Ristenpart, Theodore Guschlbauer.

RCA ARL 1-2091 Tape (c) ARK 1-2091 (ct) ARS 1-2091.
-MJ 9-77 p35 +SFC 3-6-77 p34
++NR 5-77 p7

1540 Concerto, viola, D major. PAGANINI: Sonata, viola, op. 35, C minor. STAMITZ: Concerto, viola, D major. Atar Arad, vla; PH; Reinhard Peters. Telefunken AW 6-42007.
++Gr 11-76 p792 +NR 1-77 p6

Duo, violin and viola, G major. cf MOZART: Duo, violin and viola, K 423, G major.

1541 Duo concertante, F major. LOEILLET: Trio sonata, E minor. MOZART: Trio, clarinet, K 498, E flat major. James Carter Chamber Ensemble. Reference Recordings RR 4.
+HF 10-77 p112

Parthia, E flat major. cf Telefunken EK 6-35334.

HOFHEIMER, Paul

Nach willen dein. cf L'Oiseau-Lyre 12BB 203/6.

HOFMANN, Josef

Chromaticon, piano and orchestra. cf International Piano Library IPL 5001/2.

Kaleideskop, op. 40. cf International Piano Library IPL 5007/8.

Mignonettes: Nocturne. cf International Piano Library IPL 103.

Penguine. cf International Piano Library IPL 5007/8.

HOFSTETTER, Romanus

Quartet, strings, F major: Serenade. cf Decca SPA 491.

HOLBORNE, William

Consorts, 5 recorders. cf Abbey LPB 765.

The funerals: Pavan. cf DG 2723 051.

Heigh ho holiday: Coranto. cf DG 2723 051.

The image of melancholly. cf National Trust NT 002.

Noel's galliard. cf DG 2723 051

The widowes myte. cf RCA RL 2-5110.

HOLCOMBE

Three aires. cf Crescent BUR 1001.

HOLEWA, Hans

Concertino, no. 3. cf BIS LP 32

HOLLER, Karl

Choral variations on "Jesu meine Freude", op. 22, no. 2. cf Musical Heritage Society MHS 3340.

1542 Ciacona, op. 54. KARG-ELERT: Passacaglia and fugue on B-A-C-H, op. 150. REGER: Pieces, op. 145, no. 2: Dankpsalm. Andrew Armstrong, org. Vista VPS 1028.
+-Gr 7-76 p199 +MU 7-77 p14
+HFN 2-77 p125 +-RR 10-76 p91

HOLLIGER, Heinz

Elis. cf BERIO: Rounds.

HOLLINS, Alfred

Concert overture, C minor. cf Vista VPS 1035.

Trumpet minuet. cf Vista VPS 1042.

HOLLIS, W.

John Blundeville's last farewell. cf L'Oiseau-Lyre DSLO 510.

HOLMBOE, Vagn

Notturno, op. 19. cf CARLSTEDT: Sinfonietta, 5 wind instruments.

1543 Symphony, no. 10, op. 105. NYSTROEM: Symphony, no. 1. Gothenburg Symphony Orchestra; Sixten Ehrling. Caprice CAP 1116.
+RR 11-77 p60

HOLMES, Augusta

Petite pieces (3). cf BOHM: Fantasy on a theme by Schubert.

HOLMES, John
Thus bonny-boots. cf Argo ZK 25.
Thus bonny-boots the birthday celebrated. cf DG 2533 347.
HOLST, Gustav
Beni Mora suite, op. 29, no. 1, E minor. cf Works, selections (Pearl GEM 126).
Fantasia on the Dargason. cf Grosvenor GRS 1052.
Marching song. cf Suite, no. 1, op. 28, no. 1, E flat major.
Moorside suite. cf BLISS: Kenilworth.
Moorside suite. cf BUTTERWORTH: Dales suite.
The perfect fool, op. 39: Ballet music. cf BBC REH 290.
1544 The planets, op. 32. Vienna State Opera Chorus; VPO; Herbert von Karajan. Decca JB 30. (Reissue from SXL 2305)
+Gr 12-77 p1087 +-RR 12-77 p48
1545 The planets op. 32 (trans. Gleeson). Eu Polyphonic Synthesizer; Patrick Gleeson. Mercury SRI 80000.
+NR 11-76 p14 -St 3-77 p129
1546 The planets, op. 32. PO; Eugene Ormandy. RCA Tape (c) RK 1-1797.
+-HFN 6-77 p139 tape
1547 The planets, op. 32 (arr. Tomita). Isao Tomita, electronic synthesizer. RCA ARL 1-1919.
++HFN 6-77 p125 +-St 3-77 p130
1548 The planets, op. 32. Mendelssohn Club Chorus, Women's Voices; PO; Eugene Ormandy. RCA CRL 1-1921 Tape (c) CRK 1-1921 (ct) CRS 1-1921. (also RCA RL 1-1797 Tape (c) RK 1-1797)
++Gr 4-77 p1549 ++NR 11-76 p2
+-HF 12-76 p104 +RR 4-77 p47
+-HFN 6-77 p125 +-RR 8-77 p86 tape
+MJ 1-77 p26 +-St 3-77 p129
1549 The planets, op. 32. St. Louis Symphony Orchestra; Walter Susskind. Vox Turnabout (Q) QTVS 34598 Tape (c) KTVC 34598.
-Gr 4-77 p1549 ++NR 12-75 p3
+-HF 4-76 p111 +-RR 2-77 p47
+-HFN 12-76 p155 tape +-RR 2-77 p97 tape
+HFN 2-77 p121 ++SFC 11-16-75 p32
The planets, op. 32: Jupiter. cf BEETHOVEN: Egmont, op. 84: Overture.
The planets, op. 32: Jupiter. cf Philips 6747 327.
The planets, op. 32: Jupiter, the bringer of jollity. cf HMV ESD 7011.
St. Paul's suite, op. 29, no. 2. cf Works, selections (Pearl GEM 126).
St. Paul's suite, op. 29, no. 2. cf ELGAR: Serenade, strings, op. 20, E minor.
Songs (folksong arrangements): Folksongs of England; Lovely Joan; The maid of Islington; O who is that that rapts at my window; The seeds of love; The unquiet grave. cf GRAINGER: Songs (folksong arrangements) (Premier PMS 1502).
Songs: A little music; The floral bandit; The thought. cf Enigma VAR 1027.
Songs, voice and violin, op. 35 (4). cf Works, selections (Pearl GEM 126).
Songs without words, op. 22, nos. 1, 2. cf Works, selections (Pearl GEM 126).
Songs without words, op. 22b: Marching song. cf HMV SQ ASD 3341.
1550 Suite, no. 1, op. 28, no. 1, E flat major. Suite, no. 2, op. 28, no. 2, F major. Marching song (arr. Leidzen). VAUGHAN WILLIAMS:

Sea songs. Ground Self Defence Force Central Band, Tokyo; Yoshio Tamamashi, Akira Arimasa. RCA GL 4-5043.
+-Gr 6-77 p64 +RR 6-77 p56
+HFN 8-77 p83

Suite, no. 1, op. 28, no. 1, E flat major: March. cf RCA PL 2-5046.

Suite, no. 2, op. 28, no. 2, F major. cf Suite, no. 1, op. 28, no. 1, E flat major.

1551 Works, selections: Beni Mora suite, op. 29, no. 1, E minor. St. Paul's suite, op. 29, no. 2. Songs, voice and violin, op. 35 (4). Songs without words, op. 22, nos. 1, 2. Dora Labette, s; W. H. Reed, vln; LSO; String Orchestra; Gustav Holst. Pearl GEM 126. (Reissues)
+-ARSC vol 9, no. 1, 1977, p98 +Gr 3-75 p1650
-RR 3-75 p27

HOLYOKE

Masconic processional march. cf Folkways FTS 32378.

HONEGGER, Arthur

Choral. cf Abbey LPB 719.

Fugue. cf Abbey LPB 719.

1552 Jeanne d'Arc au Bucher. Nelly Borgeaud, Michel Favory, narrators; CPhO and Chorus; Serge Baudo. Supraphon 412 1651/2 (2).
+HFN 12-77 p172

Pastorale d'ete. cf Prelude PRS 2512.

1553 Poemes (4). SATIE: La diva de l'empire, Dapheneo; Le chapelier; Je te veux; La statue de bronze; Tendrement. SCHMITT: Chants, op. 98 (3); Poemes de Ronsard, op. 100 (4). Yolanda Marcoulescou, s; Katja Phillabaum, pno. Orion ORS 76240.
+ARG 4-77 p30 ++NYT 6-5-77 pD19
+HF 4-77 p116 +ON 7-77 p29
+NR 4-77 p12

Romance. cf Golden Crest RE 7064.

HOPKINS, Antony

1554 Riding to Canonbie. WILLIAMSON: The brilliant and the dark. April Cantelo, Salle Le Sage, s; Alfreda Hodgson, Norma Procter, con; Janet Canetty-Clarke, Simon Campion, pno; Avalon Singers, Lewellyn Singers; ECO; Antony Hopkins. Readers Digest RDS 9351/2 (2).
+-HFN 6-77 p127

HOPKINS, James

Diferencias sobre una tema original. cf BLOCH: Nocturnes.

HOPKINS, Jerome

The wind demon, op. 11. cf New World Records NW 257.

HOPKINSON, Francis

Songs: A toast; Beneath a weeping willow's shade; Come fair Rosina; Enraptur'd I gaze; My days have been so wond'rous free; My gen'rous heart disdains; My love is gone to sea; O'er the hills far away; See, down Maria's blushing cheek; The traveler benighted and lost. cf Vox SVBX 5350.

HORN, Charles

Cherry ripe. cf HMV ESD 7002.

Cherry ripe. cf Prelude PRS 2505.

HOTTETERRE, Jean

Bourree. cf DG 2723 051.

Suite "Le festin". cf Abbey LPB 765.

HOVE, Joachim van den

Galliarde. cf DG 2533 302.

Lieto godea. cf DG 2533 323.

HOVHANESS, Alan
1555 And God created great whales. Fantasy on Japanese woodprints, op. 211. Floating world, op. 209. Meditation on Orpheus, op. 155. The Ruaiyat of Omar Khayyam. Sunrise. Orchestra; Andre Kostelanetz. Columbia M 34537 Tape (c) MT 34537.
+-FF 11-77 p28 +NR 11-77 p6
Bacchanale. cf CHAVEZ: Toccata.
Fantasy on Japanese woodprints, op. 211. cf And God created great whales.
1556 Firdausi, op. 252. HUSA: Evocations de Slovaquie. Gloria Agostini, hp; Neal Boyar, perc; Louise Schulman, vla; Timothy Eddy, vlc; Lawrence Sobol, clt. Grenadilla GS 1008.
+NR 2-77 p7
Floating world, op. 209. cf And God created great whales.
1557 Magnificat, op. 157. Audrey Nossaman, s; Elizabeth Johnson, con; Thomas East, t; Richard Dales, bar; Louisville University Chorus; Louisville Orchestra; Robert Whitney. Poseidon 1018.
+NR 2-76 p8 +-SFC 7-31-77 p40
Meditation on Orpheus, op. 155. cf And God created great whales.
October mountain. cf CHAVEZ: Toccata.
The Rubaiyat of Omar Khayyam. cf And God created great whales.
Sonata, trumpet and organ: 1st movement. cf Crystal S 362.
Sunrise. cf And God created great whales.
HOWARD
I wonder who's kissing her now. cf Rubini GV 43.
HOWARTH, Elgar
Fireworks, brass band. cf BIRTWISTLE: Grimethorpe aria.
HOWELL, T.
Robin Adair. cf Serenus SRS 12067.
HOWELLS, Herbert
Behold, O god our defender. cf Vista VPS 1053.
Elegy, op. 15. cf BUTTERWORTH: The banks of green willow.
Exultate Deo. cf Vista VPA 1037.
A hymn for St. Cecilia. cf Vista VPS 1037.
Inheritance. cf RCA GL 2-5062.
Master Tallis testament. cf Vista VPS 1060.
Merry eye, op. 20b. cf BUTTERWORTH: The banks of green willow.
Music for a prince. cf BUTTERWORTH: The banks of green willow.
1558 Pieces organ: Preludio; Saraband for the morning of Easter; Master Tallis testament; Fugue, chorale and epilogue; Saraband; Paean. Siciliano for a high ceremony. Michael Nicholas, org. Vista VPS 1031.
+Gr 1-77 p1161 +-RR 1-77 p71
+MU 8-77 p10
1559 Psalm prelude. PACHELBEL: Wie schon leuchtet der Morgenstern, chorale prelude. REGER: Fantasia on Straf mich in deinem Zorn. WALTHER: Concerto, organ (arr.). John Cooper Green, org. Wealden WS 164.
+Gr 8-77 p327
Psalm prelude, op. 32, no. 1. cf Vista VPS 1047.
Siciliano for a high ceremony. cf Pieces, organ.
A spotless rose. cf HMV CSD 3774.
HOWET, Gregorio
Fantasie. cf DG 2533 302.
HOYOUL, Baudion
Gelobet seist du Jesu Christ. cf ABC L 67002.

HUBAY, Jeno
Hejre Kati, op. 32. cf Kelsey Records KEL 7601.
Hungarian poems (6). cf Kelsey Records KEL 7601.
Poeme Hongroise, nos. 1 and 2 (Szigeti). cf Discopaedia MOB 1019.
HUE, Georges
Soir paien. cf HMV RLS 719.
HUGGENS
Chorale and rock-out. cf Grosvenor GRS 1052.
HUGGLER, John
Quintet, no. 1. cf Crystal S 204.
HUGHES, Herbert
The stuttering lover. cf Philips 6599 227.
HUGHES, Robert
Cadences. cf HARRISON: Elegaic symphony.
HUHN
Songs: Invictus. cf Argo ZFB 95/6.
HUME
BB AND CF march. cf Grosvenor GRS 1043.
HUME, Tobias
Musick and mirth. cf L'Oiseau-Lyre 12BB 203/6.
Songs: Touch me lightly; Tickle me quickly. cf RCA RL 2-5110.
HUMFREY, Pelham
A hymn to God the Father. cf Abbey LPB 712.
Songs (anthems): O give thanks; Hear, O heav'ns; By the waters of Babylon. cf BLOW: Sing unto the Lord O ye saints, anthem.
HUMMEL, Johann
Concerto, trumpet, E flat major. cf HAYDN: Concerto, trumpet, E flat major.
Partita, E flat major. cf DRUZECKY: Partita, no. 4, E flat major.
1560 Septet, op. 74, D minor. Neckar Septet. Oryx 1810.
-NR 9-77 p7
Septet, op. 114, C major. cf DRUZECKY: Partita, no. 4, E flat major.
1561 Sonata, piano, op. 81, F sharp minor. Sonata, piano, op. 106, D major. Malcolm Binns, fortepiano. L'Oiseau-Lyre DSLO 530.
+Gr 8-77 p323 +RR 7-77 p82
+-HFN 8-77 p83
Sonata, piano, op. 106, D major. cf Sonata, piano, op. 81, F sharp minor.
Theme and variations, op. 102. cf Decca SB 329.
HUMPERDINCK, Engelbert
Hansel and Gretel, excerpts. cf HMV SLS 5073.
Hansel und Gretel: Ein Mannlein steht im Walde. cf Rubini RS 301.
Hansel and Gretel: Overture. cf Classics for Pleasure CFP 40263.
HUNT, Thomas
Hark, did ye ever hear. cf Argo ZK 25.
Hark, did ye ever hear. cf DG 2533 347.
HURE, Jean
Communion sur un Noel. cf Calliope CAL 1922/4.
HUS, John
Communion hymn. cf ANTES: Chorales.
HUSA, Karol
Concerto, alto saxophone and concert band. cf COWELL: Air and scherzo.
Divertimento, brass quintet. cf University of Iowa Press unnumbered.
Evocations de Slovaquie. cf HOVHANESS: Firdausi, op. 252.
Preludes, flute, clarinet and bassoon. cf Vox SVBX 5307.

HUZELLA, Elek
Miser catulle. cf Hungaroton SLPX 11811.
IANNACONNE, Anthony
1562 Bicinia, flute and alto saxophone. Partita, piano. Rituals, violin and piano. Sonatine, trumpet and tuba. Rodney Hill, flt; Max Plank, alto sax; Alfio Pignotti, vln; Dady Mehta, Joseph Gurt, pno; Carter Eggers, tpt; John Smith, tuba. Coronet LPS 3038.
+Audio 12-77 p133 +MJ 3-77 p46
+IN 3-77 p32 +NR 10-76 p7
Partita, piano. cf Bicinia, flute and alto saxophone.
Rituals, violin and piano. cf Bicinia, flute and alto saxophone.
Sonatine, trumpet and tuba. cf Bicinia, flute and alto saxophone.
IBERT, Jacques
Concertino da camera. cf BACH: Prelude, bourree and gigue.
Concertino da camera. cf COWELL: Air and scherzo.
Concerto, flute. cf CHAMINADE: Concertino, flute, op. 107.
Concerto, flute. cf DEVIENNE: concerto, flute, no. 2, D major.
Concerto, flute. cf DUBOIS: Concerto, flute.
Entr'acte. cf BIS LP 30.
Escales. cf CHABRIER: Espana.
Escales: 2nd movement. cf Prelude PRS 2512.
Pieces (3). cf Abbey LPB 719.
Pieces breves (3). cf HINDEMITH: Kleine Kammermusik, op. 24, no. 2.
ICHIYANGANGI, Toshi
Piano media. cf CP 3-5.
IMBRIE, Andrew
1563 Symphony, no. 3. SCHUMAN: Credendum (Article of faith).* LSO, PO; Harold Farberman, Eugene Ormandy. CRI SD 308. (*Reissue from Columbia ML 5185)
+Gr 3-77 p1454 ++RR 4-75 p27
+HF 11-73 p109 +RR 2-77 p48
+HFN 5-75 p128 ++St 8-74 p114
+NR 11-73 p2 ++Te 3-75 p48
d'INDIA, Sigismondo
Isti sunt duae olivae. cf HMV SQ ASD 3393.
Songs: Cruda amarilli; Intenerite voi, lagrime mie. cf DG 2533 305.
d'INDY, Vincent
Chant des bruyeres. cf Vox SVBX 5483.
1564 Symphony on a French mountain air, op. 25. RAVEL: Concerto, piano, for the left hand, D major. Gabriella Torma, pno; Budapest Philharmonic Orchestra; Tamas Pal. Hungaroton SLPX 11789.
+-FF 9-77 p27
Symphony on a French mountain air, op. 25. cf FRANCK: Les Djinns.
1565 Trio, op. 29, B flat major. RAMEAU: Pieces de clavecin en concert. Montagnana Trio. Delos DEL 25431.
+NR 12-77 p6
INGALLS, Jeremiah
Northfield. cf Vox SVBX 5350.
1566 Songs: Christain song; Crostic; Delay; Election hymn; Falmouth; Farewell humn; Lamentation; Love divine; New Jerusalem; Northfield; Tranquility. MOORS: By the offence of one-anthem; Cavendish; Charlotte; Dorset; Fairfax; I will praise thee, anthem; Moretown; Orwell; Mount Holly; Plainfield; Pittsford; Shirley. University of Vermont Choral Union; James Chapman.

Philo 1038.
++St 4-77 p85
INSTRUMENTS OF THE MIDDLE AGES AND RENAISSANCE. cf HMV SLS 988.
INSTRUMENTS OF THE MIDDLE AGES AND RENAISSANCE. cf Vanguard VSD 71219/20.

IPPOLITOV-IVANOV, Mikhail
Bless the Lord, O my soul. cf St. John the Divine Cathedral Church.
Caucasian sketches, op. 10: Procession of the Sardar. cf CBS 61781.
Caucasian sketches, op. 10: Procession of the Sardar. cf RCA CRL 1-2064.
Caucasian sketches, op. 10: Procession of the Sardar. cf RCA CRL 3-2026.
Jubilee march. cf HMV CSD 3782.
Treachery: Aria of Erekle. cf Club 99-105.

IRELAND, John
Amberley wild brooks. cf Piano works (Lyrita SRCS 88).
April. cf Piano works (Lyrita SRCS 88).
Aubade. cf Saga 5445.
Bergomask. cf Piano works (Lyrita SRCS 88).
Comedy overture. cf BLISS: Kenilworth.
Concertino pastorale. cf ELGAR: Serenade, strings, op. 20, E minor.
Concerto, piano, E flat major. cf HMV SLS 5080.
The darkened valley. cf Piano works (Lyrita SRCS 88).
Downland suite. cf BUTTERWORTH: Dales suite.
Equinox. cf Piano works (Lyrita SRCS 88).
For remembrance. cf Piano works (Lyrita SRCS 88).
The hills. cf RCA GL 2-5062.
On a birthday morning. cf Piano works (Lyrita SRCS 88).

1567 Piano works: April. Amberley wild brooks. Bergomask. The darkened valley. Equinox. For remembrance. On a birthday morning. Soliloquy. Sonata, piano. Summer evening. Eric Parkin, pno. Lyrita SRCS 88.
+Gr 4-77 p1573 +RR 4-77 p72
+HFN 4-77 p137

Sea fever. cf L'Oiseau-Lyre DSLO 20.
Soliloquy. cf Piano works (Lyrita SRCS 88).
Sonata, piano. cf Piano works (Lyrita SRCS 88).
Songs: Friendship in misfortune; The land of lost content; Love and friendship; The one hope; The trellis. cf Argo ZK 28/9.
Summer evening. cf Piano works (Lyrita SRCS 88).
IRISH SONGS. cf Philips 6599 227.

ISAAC, A.
Ne piu bella di queste; Palle, palle; Quis dabit pacem. cf L'Oiseau-Lyre 12BB 203/6.

ISAAC, Heinrich
E qui la dira. cf Enigma VAR 1024.
Heliogierons nous. cf Argo D40D3.
Innsbruck, ich muss dich lassen. cf Argo D40D3.
Innsbruck, ich muss dich lassen. cf Nonesuch H 71326.
La la ho ho. cf L'Oiseau-Lyre 12BB 203/6.
La mi la sol. cf Argo ZK 24.
Maudit seyt. cf Argo D40D3.
La Mora. cf Argo D40D3.
Songs: Donna di dentro di tua casa; Missa la bassa danz: Agnus Dei; A la battaglia. cf HMV SLS 5049.

ISELE, David Clark
1568 Mass. Vespers. Notre Dame Chapel Choir; Sue Seid-Martin. University of Notre Dame, unnumbered.
+-MU 7-77 p13
Vespers. cf Mass.
ISHII
Aphorismen II fur einen Pianisten. cf CP 3-5.
ISOLFSSON, Pall
Introduction and passacaglia, F minor. cf COWELL: Symphony, no. 16.
ISRAEL, Brian
Dance variations. cf Golden Crest RE 7068.
ITALIAN OPERA CHORUSES. cf Seraphim S 60275.
ITALIAN RENAISSANCE LUTE MUSIC. cf DG 2533 173.
IVANOVICI, Iosif
Waves of the Danube. cf Classics for Pleasure CFP 40213.
IVES, Charles
Adeste fidelis in an organ prelude. cf Vista VPS 1038.
The anti-abolitionist riots. cf Vox SVBX 5303.
The bells of Yale. cf Piano works (Vox SVBX 5482).
Central Park in the dark. cf Symphony, no. 4.
Election songs (2). cf Vox SVBX 5353.
Emerson transcriptions (4). cf Piano works (Vox SVBX 5482).
Hallowe'en. cf Works, selections (vox SVBX 564).
1569 Holidays. Temple University Concert Choir; PO; Eugene Ormandy. RCA ARL L-1249 Tape (c) ARK 1-1249 (ct) ARS 1-1249 (Q) ARD 1-1249.
-HF 6-76 p70
+-HF 1-77 p151 tape
++NR 2-76 p3
++SFC 2-29-76 p25
+St 6-76 p104
1570 Holidays: Decoration day. Symphony, no. 2. Variations on "America" (orch. William Schuman). LAPO; Zubin Mehta. Decca SXL 6753.
++Gr 7-76 p176
++HFN 7-76 p89
-MT 3-77 p216
+RR 7-76 p52
Holidays: Decoration day. cf London CSA 2246.
In re con moto et al. cf Works, selections (Vox SVBX 564).
The innate (Adagio cantabile), piano and string quartet. cf Works, selections (Vox SVBX 564).
Largo, violin and piano. cf Works, selections (Vox SVBX 564).
Largo, violin, clarinet and piano. cf Works, selections (Vox SVBX 564).
Largo, violin, clarinet and piano. cf Delos DEL 25406.
Largo risoluto I and II. cf Works, selections (Vox SVBX 564).
Marches, G major and D major. cf Piano works (Vox SVBX 5482).
1571 Piano works: The bells of Yale (arr. Deutsch). Emerson transcriptions (4). Marches, G major and D major. The seen and unseen. Sonatas, piano, nos. 1 and 2. Three-page sonata. Variations on "America" (arr. Deutsch). Waltz-rondo. Nina Deutsch, pno. Vox SVBX 5482 (3).
+-ARG 8-77 p16
-HF 9-77 p101
+-MJ 5-77 p32
/NR 5-77 p13
+SFC 4-10-77 p30
Quarter-tone pieces, 2 pianos. cf Works, selections (Vox SVBX 564)
Quartet, strings, no. 2. cf BARBER: Quartet, strings, op. 11, B minor.
Scherzo, string quartet. cf BARBER: Quartet, strings, op. 11, B minor.

The seen and unseen. cf Piano works (Vox SVBX 5482).
Some southpaw pitching. cf Vox SVBX 5303.
Sonatas, piano, nos. 1 and 2. cf Piano works (Vox SVBX 5482)

1572 Sonata, piano, no. 2. Gilbert Kalish, pno; John Graham, vla; Samuel Baron, flt. Nonesuch H 71337.
++HF 9-77 p100 ++St 8-77 p124
++NR 7-77 p11

Sonata, piano, no. 2: The Alcotts. cf Vox SVBX 5303.
Sonatas, violin and piano, nos. 1-4. cf Works, selections (Vox SVBX 564).

1573 Songs: Abide with me; Ann Street; At the river; Autumn; The children's hour; A Christmas carol; Disclosure; Elegie; A farewell to land; Feldeinsamkeit; Ich grolle nicht; In Flanders fields; Swimmers; Tom sails away; Two little flowers; Weil auf mir; West London; Where the eagle; The white gulls. Dietrich Fischer-Dieskau, bar; Michael Ponti, pno. DG 2530 696.
+-ARG 2-77 p36 +ON 3-19-77 p40
+Gr 7-76 p205 ++RR 7-76 p76
+HF 4-77 p103 +SFC 7-3-77 p34
++HFN 8-76 p81 +St 3-77 p131
+-MT 3-77 p216 +STL 8-8-76 p29
++NR 1-77 p13

1574 Songs: Ann Street; At the river; The cage; A Christmas carol; The circus band; A farewell to land; From "Paracelsus; The Housatonic at Stockbridge; The Indians; The innate; Like a sick eagle; In the mornin'; Majority; Memories (A--very pleasant; B--rather sad); Serenity; The things our fathers loved; Thoreau. Jan Degaetani, ms; Gilbert Kalish, pno. Nonesuch H 71325.
+Gr 11-76 p852 ++NYT 7-4-76 pD1
++HF 8-76 p84 +ON 6-76 p52
+HFN 11-76 p161 +RR 11-76 p96
++MT 3-77 p216 +SFC 8-8-76 p38
+NR 7-76 p9 ++St 9-76 p86

Symphony, no. 2. cf Holidays: Decoration day.
Symphony, no. 2. cf London CSA 2246.
Symphony, no. 3. cf Argo ZRG 845.

1575 Symphony, no. 4. Central Park in the dark. Jerome Rosen, pno; Tanglewood Festival Chorus; BSO; Seiji Ozawa. DG 2530 787.
-ARG 10-77 p32 ++NR 7-77 p2
+-Gr 2-77 p1280 +RR 2-77 p48
+-HFN 3-77 p105 +SFC 7-3-77 p34
+MJ 9-77 p35

1576 Symphony, no. 4. LPO; Jose Serebrier. RCA ARL L-0589 (Q) ARD 1-0589 Tape (Q) ERQ 1-0589.
+Gr 10-74 p688 +MT 10-76 p831
+HF 10-74 p81 +NR 11-74 p2
+HF 2-75 p110 Quad ++NYT 10-20-74 pD26
+HF 2-77 p118 tape +-RR 10-74 p47
+MJ 12-74 p45 ++SFC 10-6-74 p26

Three-page sonata. cf Piano works (Vox SVBX 5482).
Three-page sonata. cf DETT: In the bottoms.
Three places in New England. cf HARRIS: Symphony, no. 3.
Trio, violin, violoncello and piano. cf Works, selections (Vox SVBX 564).
The unanswered question. cf Westminster WGS 8338.
Variations on "Ameria" (orch. William Schuman). cf Holiday: Decoration day.

Variations on "America". cf Piano works (Vox SVBX 5482).
Variations on "America". cf Delos FY 025.
Variations on "America". cf London CSA 2246.
Variations on "America". cf Orion ORS 76247.
Variations on "America". cf Vista VPS 1038.
Waltz-rondo. cf Piano works (Vox SVBX 5482).

1577 Works, selections: Hallowe'en. In re con moto et al. The innate (adagio cantabile), piano and string quartet. Largo, violin, clarinet and piano. Largo, violin and piano. Largo risoluto I and II. Quarter-tone pieces, 2 pianos (3). Sonatas, violin and piano, nos. 1-4. Trio, violin, violoncello and piano. Millard Taylor, John Celentano, vln; Francis Tursi, vla; Alan Harris, vlc; Stanley Hasty, clt; Artur Balsam, Frank Glazer, pno. Vox SVBX 564 (3).
+-HF 11-76 p114 +SFC 4-10-77 p30
+MJ 11-76 p44 +St 12-76 p144
+NR 8-76 p6

JACKSON, Nicholas
Carillon. cf Wealden WS 159.
Songs: Allelui, laudate pueri; Evening hymn; Impromptu. cf BAIRSTOW: Pange lingua.

JACOB, Gordon
The barber of Seville goes to the devil. cf GERMAN: Symphony, no. 2, A minor.
A consort of recorders. cf Abbey LPB 765.

1578 Divertimento, harmonica and string quartet. MOODY: Quintet, harmonica. Tommy Reilly, harmonica; Hindar Quartet. Argo ZDA 206.
+Audio 8-77 p75 +NR 2-76 p5
+Gr 6-75 p61 +RR 8-75 p52

Music for a festival: Interludes for trumpets and trombones, no. 1, Intrada; no. 2, Round of seven parts; no. 3, Interlude; no. 4, Saraband; no. 5, Madrigal. cf RCA RL 2-5081.
Music for the festival: Overture and march. cf Lismor LILP 5078.
The national anthem. cf Vista VPS 1053.
Partita, bassoon. cf ELGAR: Romance, op. 62.
Partita, bassoon. cf Gasparo GS 103.

1579 Pieces (5). MOODY: Little suite. TAUSKY: Concertino. VAUGHAN WILLIAMS: Romance. Tommy Reilly, harmonica; AMF; Neville Marriner. Argo ZRG 856 Tape (c) KZRC 856.
+Gr 2-77 p1296 +RR 2-77 p50
++HFN 3-77 p113 ++SFC 8-14-77 p50
++HFN 5-77 p139 tape +St 11-77 p162
+NR 12-77 p14

JACOBSON
Chanson de Marie Antoinette. cf Columbia 34294.

JACQUET DE LA GUERRE, Elisabeth
Sonata, violin, bass and harpsichord, D minor. cf DIEUPART: Suite, flute, violin, bass and harpsichord, no. 3.

JAMES, Ifor
Solitude. cf BUTTERWORTH: Dales suite.

JANACEK, Leos
1580 Amarus. SUK: Under the apple tree, op. 20. Eva Gubauerova, s; Bohuslava Jelinkova, con; Jiri Zahradnicek, t; Rene Tucek, bar; Czech Philharmonic Chorus; Ostrava Janacek Philharmonic Orchestra; Otakar Trhlik. Supraphon 112 1678.

+ARG 11-77 p8 +HFN 9-77 p144
-FF 11-77 p29 +NR 10-77 p9
+Gr 8-77 p331 +RR 6-77 p88

1581 In the mists. Sonata, piano (Z Ulice). SZYMANOWSKI: Masques, op. 34. Jan Latham Koenig, pno. Prelude PMS 1503.
+Gr 7-77 p201 +RR 9-77 p82
+HFN 8-77 p83

1582 Katya Kabanova. Elisabeth Soderstrom, Jitka Pavlova, s; Libuse Marova, Gertrude Jahn, ms; Petr Dvorsky, Zdenek Svehla, Vladimir Krejcik, t; Nadezda Kniplova, con; Jiri Soucek, bar; Dalibor Jedlicka, bs; Vienna State Opera Chorus; VPO; Charles Mackerras. Decca D51D2 (2). (also London 12109 Tape (c) 5-12109)
++Gr 10-77 p691 ++NYT 9-11-77 pD30
++HFN 10-77 p156 ++St 12-77 p151

Mladi. cf HINDEMITH: Kleine Kammermusik, op. 24, no. 2.

Sonata, piano (Z Ulice). cf In the mists.

1583 Songs (choral works): Hradcany songs; Kaspar Rucky; Rikadia; Wolf tracks. Various soloists; CPhO and Chorus; Josef Veselka. Supraphon 112 1486.
++ARG 9-77 p37 +-HFN 8-76 p81
+Gr 8-76 p327 +NR 10-76 p7
+HF 12-76 p104 ++RR 8-76 p75

1584 Suite. STRAUSS, R.: Capriccio: Introduction for sextet. SUK: Serenade. Los Angels Chamber Orchestra. Argo ZRG 792.
+MJ 4-77 p33

Taras Bulba. cf DVORAK: The golden spinning wheel, op. 109.

JANNEQUIN, Clement

La bataille de Marignan. cf ALBINONI: Adagio, organ and strings, G minor.

Les cris de Paris. cf L'Oiseau-Lyre 12BB 203/6.

1585 Songs (Chansons nouvelles): L'amour, la mort et la vie; Baisez moy tost; Une belle jeune espousee; L'espoir confus; Il estoit une filette; Guillot ung jour; J'atens le temps; Jehanneton fut l'aultre jour; M'amye a eu de Dieu; O fortune n'estois tu pas contente; Ou mettra l'on ung baiser; Plus ne suys; Las qu'on congneust; Secouez moy; Si come il chiaro sole; Sy celle la qui oncques; Ung gay bergier; Ung jour Robin; Va rossignol. Ensemble Polyphonique de France; Charles Ravier. Telefunken AW 6-42120.
+-Gr 12-77 p1122 +-RR 11-77 p104
+-HFN 11-77 p172

JANSEN, Sigurd

Fragments. cf Amberlee ALM 602.

JAROCH, Jiri

Symphony concertante, no. 3. cf HAVELKA: Rose of wounds.

JARRETT, Keith

Crystal moment. cf Works, selections (ECM/Polydor 1033/4).

Fughata. cf Works, selections (ECM/Polydor 1033/4).

In the cave, in the light. cf Works, selections (ECM/Polydor 1033/4).

Metamorphosis. cf Works, selections (ECM/Polydor 1033/4).

Pagan hymn. cf Works, selections (ECM/Polydor 1033/4).

Quartet, strings. cf Works, selections (ECM/Polydor 1033/4).

Quintet, brass. cf Works, selections (ECM/Polydor 1033/4).

Short piece. cf Works, selections (ECM/Polydor 1033/4).

1586 Works, selections: Crystal moment. Fughata. In the cave, in the light. Metamorphosis. Pagan hymn. Quartet, strings. Quintet, brass. Short piece. Willi Freivogel, flt; Ralph Towner, gtr; Keith Jarrett, pno; Fritz Sonnleitner Quartet, American Brass Quintet, Sudfunk Symphony Orchestra, Stuttgart; Keith Jarrett. ECM/Polydor 1033/4 (2) Tape (c) CF 1/2-1033 (ct) 8F 1/2-1033.
+-ARG 10-77 p34

JAZZ INSPIRED PIANO COMPOSITIONS. cf Supraphon 111 1721/2.

JELIC, Vinco

1587 Motets: Justum deduxit; Oculi tui Deus. LUKACIC: Motets: Ex ore infantum. Accademia Monteverdiana; Denis Stevens. Musical Heritage Society MHS 3497.
+MU 10-77 p8

JENEY, Zoltan

1588 Alef: Homage a Schonberg. Round. Soliloquium, flute, no. 1. SARY: Catocoustics, 2 pianos. Immaginario, no. 1. Incanto. Istvan Matuz, flt; Margit Bognar, hp; Zsuzsa Pertis, hpd; Nora Schmidt, Istvan Nagy, Zoltan Benko, pno; Gesualdo Vocal Quintet, Hungarian Radio and Television Orchestra, Gyor Philharmonic Orchestra; Peter Eotvos, Janos Sandor. Hungaroton SLPX 11589.
++NR 10-75 p13 +Te 6-77 p44
++RR 9-75 p59

Round. cf Alef: Homage a Schonberg.

Soliloquium, flute, no. 1. cf Alef: Homage a Schonberg.

JENKINS, Cyril

Life divine. cf Grosvenor GRS 1043.

JENKINS, John

Newark siege. cf National Trust NT 002.

JENSEN, Adolf

Erotikon, op. 44: Eros. cf International Piano Library IPL 102.

JEZEK, Jaroslav

Sonata, 2 violins. cf CAJKOVSKIJ: Quartet, strings, no. 4.

JIROVEC (Gyrowetz), Adalbert

1589 Parthia, B flat major. KOZELUH: Pastorale. KRAMAR-KROMMER: Harmonie, op. 71, E flat major. STAMITZ: Parthia, E flat major. Collegium Musicum Pragense; Frantisek Vajnar. Panton 110 522.
+HFN 12-77 p172

Parthia, D sharp major. cf DRUZECKY: Parthia, no. 3, D sharp major.

JOACHIM, Joseph

Romance. cf Discopaedia MOB 1019.

JOHNSON

Disappearances. cf GRIPPE: Musique Douze.

Song of the heart. cf Orion ORS 77271.

JOHNSON, E.

Come blessed bird. cf DG 2533 347.

JOHNSON, Robert

The satyres masque. cf RCA RL 2-5110.

Songs: Defiled is my name; Benedicam Domino. cf BG HM 57/8.

Treble to a ground. cf DG 2533 323.

Where the bee sucks. cf Saga 5447.

JOLIVET, Andre

Arioso barocco. cf Musical Heritage Society MHS 3340.

JONES
Four movements, 5 brass. cf Golden Crest CRS 4148.
Ironside, theme. cf EMI TWOX 1058.
Mirimachi ballad: The Jones boys. cf Citadel CT 6011.
JONES, Daniel
1590 Quartet, strings, no. 9. Sonata, 3 timpani. Trio, strings. Gabrieli Quartet; Tristan Fry, timpani. Argo ZRG 772.
/Gr 8-76 p317 +-MT 6-77 p487
++HFN 6-76 p83 +RR 6-76 p64
Sonata, 3 timpani. cf Quartet, strings, no. 9.
Trio, strings. cf Quartet, strings, no. 9.
JONES, Jeffrey
Piece mouvante. cf BRESNICK: B's garland.
JONES, Robert
Fair Oriana, seeming to wink. cf DG 2533 347.
Songs: Love is a babel; Love is a pretty frenzy; Now what is love. cf Enigma VAR 1023.
Songs: To sigh and to bee sad; Wither runneth my sweet hart. cf RCA RL 2-5110.
JONES, Samuel
Elegy, string orchestra. cf COOPER: Symphony, no. 4.
Let us now praise famous men. cf COOPER: Symphony, no. 4.
JONGEN, Joseph
Legende, op. 89, no. 1. cf Decca DDS 507/1-2.
Sonata, organ, op. 94. cf GRISON: Toccata, F major.
Toccata. cf L'Oiseau-Lyre SOL 343.
JOPLIN, Scott
The cascades. cf Works, selections (RCA ARL 1-2243).
The cascades. cf Golden Crest CRS 4148.
The chrysanthemum. cf Works, selections (RCA ARL 1-2243).
The easy winners. cf Works, selections (RCA ARL 1-2243).
Elite syncopations. cf Works, selections (RCA ARL 1-2243).
The entertainer. cf Works, selections (RCA ARL 1-2243).
Maple leaf rag. cf Works, selections (RCA ARL 1-2243).
Original rags. cf Works, selections (RCA ARL 1-2243).
Paragon rag. cf Works, selections (RCA ARL 1-2243).
Pine apple rag. cf Works, selections (RCA ARL 1-2243).
Scott Joplin's new rag. cf Works, selections (RCA ARL 1-2243).
Sugar cane. cf Works, selections (RCA ARL 1-2243).
1591 Treemonisha. Carmen Balthrop, s; Betty Allen, ms; Curtis Rayam, t; Willard White, bs; Orchestra and Chorus; Gunther Schuller. DG 2707 083 (2) Tape (c) 3370 012.
+-Audio 6-77 p128 ++NR 4-76 p9
+Gr 7-76 p206 +ON 6-76 p52
+-Gr 12-76 p1066 tape /RR 7-76 p28
+HF 7-76 p72 /SFC 5-16-76 p28
+HF 11-76 p153 tape +St 5-76 p116
+HFN 7-76 p90 +-STL 8-8-76 p29
+-MJ 7-76 p56 +Te 3-77 p17
The weeping willow. cf Works, selections (RCA ARL 1-2243).
1592 Works, selections: The cascades. The chrysanthemum. The easy winners. Elite syncopations. The entertainer. Maple leaf rag. Original rags. Paragon rag. Pine apple rag. Scott Joplin's new rag. Sugar cane. The weeping willow. James Levine, pno. RCA ARL 1-2243 Tape (c) ARK 1-2243 (ct) ARS 1-2243.
+NR 10-77 p14 +St 10-77 p140

JORDAN, Sverre
Baeken. cf Rubini GV 43.
JOSEPHSON
Songs: Serenade; Sjung, sjung du underbara sang; Tro ej galdjen. cf Swedish SLT 33209.
JOSQUIN DES PRES (also Des Pres, Depres)
Ave Maria. cf GABRIELI: O magnum mysterium.
La Bernardina. cf HMV SLS 5049.
Mille regretz. cf L'Oiseau-Lyre 12BB 203/6.
Motets: Benedicta es caelorum Regina; De profundis; Inviolata, integra et casta es, Maria. cf HMV SLS 5049.
Motets: In domino confido; Tribulatio et angustia; Jubilate Deo; O bone et dulcissime Jesu. cf FRANCK: Fuhrwahr, Fuhrwahr.
Songs: Allegez moy, doulce plaisant brunette; Adieu mes amours; El grillo e buon cantore; Guillaume se va; Scaramella va alla guerra. cf HMV SLS 5049.
La Spagna. cf HMV SLS 5049.
Tu pauperum refugium. cf Pelca PSR 40607.
Vive le roy. cf HMV SLS 5049.
JOUBERT, John
1593 Hymns to St. Oswald, op. 74. Pro pace motets: Libera plebem, op. 19; O tristia secla priora, op. 32; Solus ad victimam, op. 29. Louis Halsey Singers; Martin Neary, org; Louis Halsey. Pearl SHE 534.
+Gr 5-77 p1723 +RR 6-77 p89
Pro pace motets: Libera plebem, op. 19; O tristia secla priora, op. 32; Solus ad victimam, op. 29. cf Hymns to St. Oswald, op. 74.
THE JOY OF CHRISTMAS. cf HMV SQ CSD 3784.
JUDENKUNIG, Hans
Ellend bringt peyn. cf DG 2533 302.
Hoff dantz. cf DG 2533 302.
JULIAN
Akasha. cf Finnadar SR 9015.
KABALEVSKY, Dmitri
1594 Colas Breugnon, opp. 24/90. Valentina Kayevchenko, Albina Chitikova, s; Nina Isakova, ms; Anatol Mishchevsky, Nikolai Gutorovich, t; Yevgeny Maximenko, bar; Leonid Boldin, bs-bar; Georgy Dudarev, bs; Stanislavsky Nemirovich-Danchenko Musical Theatre Orchestra and Chorus; Georgi Zhemchuzhin. Columbia/Melodiya M3 33588 (3).
+-HF 4-77 p104 +-NYT 4-17-77 pD17
+MJ 4-77 pD33 +OC 6-77 p44
+NR 2-77 p11 +ON 1-22-77 p33
The comedians: Galop. cf Seraphim S 60277.
KAGEL, Mauricio
Pandoras box. cf Orion ORS 77263.
KALABIS, Victor
Little chamber music, wind quintet, op. 27. cf BARTA: Quintet, winds, no. 2.
Symphony, no. 4, op. 34. cf BOHAC: Suita drammatica.
KALLINIKOV, Wassilij (Vassili)
Songs: I will love thee. cf Argo ZRG 871.
1595 Symphony, no. 1, G minor. State Academic Orchestra; Yevgeny Svetlanov. Columbia/Melodiya M 34523.
++ARG 9-77 p38 ++St 12-77 p152
++NR 8-77 p2

KALMAN, Emmerich
Die Czardasfurstin: Heut Nacht hab ich getraumt von dir. cf HMV RLS 715.
Grafin Maritza: Lustige Zigeunerweisen; Hore ich Zigeunergeigen. cf HMV ESD 7043.
Grafin Maritza: Tassilo's song; Maritza and Tassilo duet; Ausschnitt. cf Das Veilchen vom Montmartre: Heut nacht hab ich getraumt von dir.
Das Veilchen vom Montmartre: Du Veilchen vom Martmartre. cf HMV RLS 715.
1596 Das Veilchen vom Montmartre: Heut nacht hab ich getraumt von dir. Grafin Maritza: Tassilo's song, Maritza and Tassilo duet, Ausschnitt. LEHAR: Friederike: Poet's song. Das Land des Lachelns: Su Tshongs's songs. STRAUSS, O.: Rund und die Liebe: Hans songs. ZELLER: Der Vogelhandler: Adam's song. Jozsef Simandy, t; Choral and Orchestral accompaniments. Qualiton SXLP 16581.
-NR 3-76 p10 +St 2-77 p132
KALMAR, Laszlo
Canons. cf Works, selections (Hungaroton SLPX 11744).
Cycles. cf Works, selections (Hungaroton SLPX 11744).
Nocturno, no. 1. cf Works, selections (Hungaroton LSPX 11744).
Scotto voce. cf Works, selections (Hungaroton SLPX 11744).
Song. cf Works, selections (Hungaroton SLPX 11744).
Trio. cf Works, selections (Hungaroton SLPX 11744).
1597 Works, selections: Canons (4). Cycles. Nocturno, no. 1. Scotto voce. Song. Trio. Various artists. Hungaroton SLPX 11744.
/HFN 2-77 p121 /RR 2-77 p73
KALNINS, Janis
The long night. cf APKALNS: Burial songs.
KANCELI, Gija
1598 Symphony, no. 4. PETROV: Poem, 4 trumpets, 2 klaviers, organ, strings and percussion (To commemorate the siege of Leningrad). Czech Radio Symphony Orchestra; Milos Konvalink. Panton 110 582.
+RR 9-77 p66
KANTOR, Joseph
Playthings of the wind. cf BEACH: Then said Isaiah.
KAPSBERGER, Giovanni Girolamo
Canzona, no. 2. cf Telefunken AW 6-42155.
Toccata. cf Saga 5438.
Toccata, no. 7. cf Telefunken AW 6-42155.
KARAS
Cafe Mozart waltz. cf Rediffusion 15-57.
Harry Lime theme. cf Rediffusion 15-57.
KARG-ELERT, Siegfried
Chorale improvisation, op. 65: Nun danket alle Gotte; March triomphale. cf Decca PFS 4416.
In dulci jubilo, op. 75, no. 2. cf TR 1006.
Jerusalem, du hochgebaute Stadt, op. 65. cf Vista VPS 1033.
March triomphale "Now thank we all our God". cf Argo SPA 507.
Passacaglia and fugue on B-A-C-H, op. 150. cf HOLLER: Ciacona, op. 54.
KARJINSKY
Equisse. cf HMV SQ ASD 3283.
KARKOFF, Maurice
Japanese romances, op. 45. cf CARLSTEDT: Sonata, violin, op. 15.

KARLOWICZ, Mieczyslaw
1599 Symphony, op. 7, E minor. Bydgoszcz Pomeranian Philharmonic Orchestra; Bohdan Wodiszko. Muza SXL 1072.
+ARG 5-77 p55

KASTALSKY, Alexander
1600 Motets: I see thy bridal chamber adorned; The Lord is God and hath appeared unto us; It is very meet to bless thee; Thou art immortal. RACHMANINOFF: Vesper mass, op. 37. Meriel Dickinson, con; Wynford Evans, t; Bruckner-Mahler Choir; Wyn Morris. Philips 6747 246 (2).
+-Gr 11-77 p873 +NR 1-77 p9
+HFN 10-77 p157 +RR 11-77 p108
+MJ 1-77 p27 +St 1-77 p119

KATTNIGG, Rudolf
Balkanliebe: Leise erklingen Glocken. cf HMV ESD 7043.

KAUFMAN, Jeffrey
1601 Reflections, clarinet and piano. REIF: Duo for three. VOLLINGER: More than conquerors. Barbara Martin, s; Bruce Fifer, bar; Lawrence Sobol, clt; Peter Basquin, pno; Richard Locker, vlc. Grenadilla 1009.
+NR 12-77 p10

KAY, Ulysses
Suite, organ, no. 1. cf Orion ORS 76255.

KEEL, Frederick
Trade winds. cf L'Oiseau-Lyre DSLO 20.

KEEZER, Ronald
For 4 percussionists. cf CIRONE: Triptych: Double concerto.

KELER BELA
1602 Rakoczi overture, op. 76. WALDTEUFEL: Dans tes yeux. Fleurs et baisers. Gouttes de rosee. Vision. Concerti Allegri of London; Grahma Nash. Rare Recorded Editions SRRE 172.
+Gr 12-77 p1155 +-HFN 12-77 p172

KELLNER, David
Fantasy, A minor. cf BACH: Prelude, fugue and allegro, S 998, E flat major.
Fantasy, C major. cf BACH: Prelude, fugue and allegro, S 998, E flat major.

KELLNER, Johann
Was Gott tut. cf Vista VPS 1042.

KELLY
Songs: Last week I took a wife; The mischievous bee. cf Vox SVBX 5350.

KELLY, Bryan
Divertimento. cf Grosvenor GRS 1048.

KENINS, Talivaldis
1603 Concerto, violin. Symphony, no. 4. Steven Staryk, vln; CBC Vancouver Chamber Orchestra; John Avison. C.B.C. SM 293.
/-FF 11-77 p30
Symphony, no. 4. cf Concerto, violin.

KENNEDY, Russell
Land of hearts desire. cf HMV RLS 716.

KENNY
Serenade for a gondolier. cf Rediffusion 15-57.

KERLL, Johann
Canzona, G minor. cf ABC ABCL 67008.
Toccata con durezza e ligature. cf ABC ABCL 67008.

KETELBY, Albert
Cockney Suite: Appy Ampstead. cf DJM Records DJM 22062.
With honour crowned. cf Decca SB 715.
KHACHATURIAN, Aram
Dance, G minor. cf Works, selections (Genesis GS 1062).
Fughetta. cf Works, selections (Genesis GS 1062).
1604 Gayaneh, ballet. National Philharmonic Orchestra; Loris Tjeknavorian. RCA RL 2-5035 (2).
+Gr 4-77 p1549 +RR 4-77 p47
+HFN 6-77 p127
Gayaneh: Ballet suite. cf ENESCO: Roumanian rhapsody, op. 11, no. 1, A minor.
1605 Gayaneh: Lezghinka; Lullaby; Storm; Sabre dance; Mountaineers; Invention. Spartacus: Variation of Aegina; Adagio of Spartacus and Phrygia; Entrance of Harmodius and Adagio of Aegina and Harmodius; Dance of the Gaditanae; The rebels approach. LSO; Aram Khachaturian. HMV SQ ASD 3347 Tape (c) TC ASD 3347. (also Angel S 37411 Tape (c) 4XS 37411)
+-Gr 5-77 p1696 +HFN 8-77 p99 tape
-Gr 8-77 p346 tape +NR 8-77 p2
+HF 10-77 p110 +-RR 6-77 p57
+HFN 6-77 p127
Gayaneh: Sabre dance. cf DG 2535 254.
Gayaneh: Sabre dance. cf DG 2584 004.
Gayaneh: Sabre dance. cf RCA CRL 1-2064.
Gayaneh: Sabre dance. cf Seraphim S 60277.
1606 Gayaneh: Suite. MASSENET: El Cid: Ballet music. Netherlands Radio Orchestra, LSO; Stanley Black. London SPC 21133.
+NR 1-77 p4 +SFC 10-24-76 p35
1607 Gayaneh: Suites, nos. 1-3. National Philharmonic Orchestra; Loris Tjeknavorian. RCA CRL 2-2263 (2).
+ARG 8-77 p19 +MJ 7-77 p70
+HF 10-77 p110 +NR 6-77 p2
Nocturne. cf Hungaroton SLPX 11825.
Poem. cf Works, selections (Genesis GS 1062).
Sonata, piano. cf Works, selections (Genesis GS 1062).
Sonatina. cf Works, selections (Genesis GS 1062).
Spartacus: Variation of Aegina; Adagio of Spartacus and Phrygia; Entrance of Harmodius and Adagio of Aegina and Harmodius; Dance of the Gaditanae; The rebels approach. cf Gayaneh: Lezghinka; Lullaby; Storm; Sabre dance; Mountaineers; Invention.
To the heroes. cf HMV CSD 3782.
Toccata. cf Works, selections (Genesis GS 1062).
Valse caprice. cf Works, selections (Genesis GS 1062).
1608 Works, selections: Dance, G minor. Fughetta. Poem. Sonata, piano. Sonatina. Toccata. Valse caprice. David Dubal, pno. Genesis GS 1062.
+-HF 11-77 p116 +St 6-77 p136
+NR 7-77 p13
KHRENNIKOV
Much ado about nothing: Nightingale and the rose. cf Club 99-105.
KICKHAM
She lived beside the Anner. cf Philips 6599 227.
KILPINEN, Yrjo
Songs of death, op. 62. cf BARTOK: Songs, op. 16.
Songs of love, opp. 60 and 61. cf BARTOK: Songs, op. 16.

KING
Garland entree. cf MENC 76-11.
KIRBYE, George
With angel's face. cf DG 2533 347.
KIRKPATRICK
Away in a manger. cf HMV CSD 3774.
KJERULF, Halfdan
Berceuse, op. 12, no. 5. cf Piano works (Polydor NKF 30004).
Caprice, op. 12, no. 4. cf Piano works (Polydor NKF 30004).
Folk dances (3). cf Piano works (Polydor NKF 30004).
Folk songs (2). cf Piano works (Polydor NKF 30004).
Impromptu, op. 12, no. 6. cf Piano works (Polydor NKF 30004).
Idylle, op. 4, no. 2. cf Piano works (Polydor NKF 30004).
1609 Piano works: Berceuse, op. 12, no. 5. Caprice, op. 12, no. 4. Folk dances (3). Folk songs (2). Impromptu, op. 12, no. 6. Idylle, op. 4, no. 2. Sketches, op. 28, nos. 3, 5-6. Wiegenlied, op. 4, no. 3. Jan Henrik Kayser, pno. Polydor NKF 30004.
+HFN 9-77 p144 +-RR 7-77 p82
1610 Romances. Olav Eriksen, bar; Einar Steen-Nakleberg, pno. Polydor Norway NKF 30003.
+HFN 3-77 p109 +RR 1-77 p81
Sketches, op. 28, nos. 3, 5-6. cf Piano works (Polydor NKF 30004).
Wiegenlied, op. 4, no. 3. cf Piano works (Polydor NKF 30004).
KJORLING
Songs: Aftonstamning. cf RCA LSC 9884.
KLAMI, Uuno
1611 Cheremissian fantasy. Kalevala suite. Arto Noras, vlc; Helsinki Philharmonic Orchestra; Jorma Panula. Finnlevy SFX 4.
+-RR 8-77 p56
Kalevala suite. cf Cheremissian fantasy.
KLEMPERER, Otto
1612 Merry waltz. STRAUSS, J. II: Die Fledermaus, op. 363: Overture. Kaiser Walzer, op. 437. Wiener Blut, op. 354. WEILL: Kleine Dreigroschenmusik. PhO; Otto Klemperer. HMV SXLP 30226 Tape (c) SXLP 30226. (Reissue from Columbia SAX 2460)
+-Gr 1-77 p1184 +HFN 9-77 p155 tape
+HFN 3-77 p117 +-RR 2-77 p60
+HFN 5-77 p138 tape
KLUGHARDT, August
Quintet, winds, op. 79. cf FORSTER: Quintet, winds, op. 95.
KNIGHT
Songs: Rocked in the cradle of the deep. cf Argo ZFB 95/6.
KNIPPER, Lev
Cossack patrol. cf Grosvenor GRS 1055.
KOCH, Erland von
Canto e danza. cf BIS LP 30.
1613 Oxberg variations. STENHAMMAR: Serenade, op. 31, F major. Stockholm Philharmonic Orchestra; Rafael Kubelik, Stig Westerberg. Swedish Society SLT 33227.
+Gr 3-77 p1457 +RR 10-76 p66
KOCH, Frederich
Quartet, strings, no. 2. cf ERB: Harold's trip to the sky.
KOCSAR, Miklaus
Sextet, brass. cf Hungaroton SLPX 11811.
KODALY, Zoltan
Bilder aus der Matra-Gegend. cf DALLAPICCOLA: Cori di Michelangelo Buonarroti il Giovane, nos. 1 and 2.

1614 Concerto, orchestra. Dances of Galanta. Dances of Marosszek. Theatre overture. PH; Antal Dorati. Decca SXL 6712. (Reissue from SXLM 6665/7) (also London CS 6862)
+-Gr 12-75 p1032 ++NR 7-77 p2
+HFN 12-75 p171 +-RR 12-75 p50
Concerto, orchestra. cf Works, selections (Hungaroton HLX 90053/5)
Dances of Galanta. cf Concerto, orchestra.
Dances of Marosszek. cf Concerto, orchestra.
1615 Duo, op. 7. Sonata, solo violoncello, op. 8. Paul Olefsky, vlc; Leonard Posner, vln. Musical Heritage Society Tape (r) C 3047.
+HF 11-77 p138 tape
Duo, op. 7. cf BEETHOVEN: Trio, strings, op. 8, D major.
Die Engel und die Hirten. cf RCA PRL 1-8020.
1616 Hary Janos: Suite. PROKOFIEV: Lieutenant Kije, op. 60: Suite. Netherlands Radio Orchestra; Antal Dorati. Decca PFS 4355 Tape (c) KPFC 4355. (also London 21146 Tape (c) 5-21146)
+-ARG 3-77 p31 -MM 11-76 p43
+Gr 5-76 p1758 +NR 3-77 p3
+-HF 4-77 p105 +-RR 4-76 p47
+HFN 5-76 p101 +-RR 6-76 p88 tape
+HFN 5-76 p117 tape
1617 Hary Janos: Suite. PROKOFIEV: Lieutenant Kije, op. 60: Suite. PO; Eugene Ormandy. RCA ARL 1-1325 Tape (c) ARS 1-1325 (ct) ARK 1-1325.
+-HF 1-77 p151 tape ++NR 3-76 p1
+-HF 4-77 p105 ++SFC 2-8-76 p29
+MJ 4-76 p31
Missa brevis. cf Works, selections (Hungaroton HLX 90053/5).
Praeludium. cf Hungaroton SLPX 11808.
Psalmus Hungaricus, op. 13. cf Works, selections (Hungaroton HLX 90053/5).
St. Gregory procession. cf Hungaroton SLPX 11823.
Sonata, solo violoncello, op. 8. cf Duo, op. 7.
Sonata, solo violoncello, op. 8. cf BACH: Suite, solo violoncello, no. 3, S 1009, C major.
Summer evening. cf Works, selections (Hungaroton HLX 90053/5).
Te Deum (Buda Castle). cf Works, selections (Hungaroton HLX 90053/5).
Theatre overture. cf Concerto, orchestra.
1618 Works, selections: Concerto, orchestra. Missa brevis. Psalmus Hungaricus, op. 13. Summer evening. Te Deum (Buda Castle). Iren Szecsody, Maria Gyurkovics, Edit Gancs, Timea Cser, s; Magda Tiszay, con; Endre Rosler, Tibor Udvardy, t; Andras Farago, Gyorgy Littasy, bs; Sandor Margittay, org; Budapest Choir; HSO, Budapest Philharmonic Orchestra; Zoltan Kodaly. Hungaroton HLX 90053/5 (3).
+ARG 9-77 p51

KOECHLIN, Charles
Quelques chorals pour des fetes populaires. cf FAURE: Chant funeraire, op. 117.
Songs: Si tu le veux. cf Odyssey Y 33130.

KOETSIER, Jan
Partita, English horn and organ. cf TR 1006.

KOHLER, Ernesto
Fantasia on a theme by Chopin. cf Pearl SHE 533.

KOHLERS, Siegfried
Sonata, horn and piano. cf Musical Heritage Society MHS 3547.

KOLB, Barbara

1619 Looking for Claudio. Spring river flowers moon night. MAYS: Concerto, alto saxophone and chamber ensemble. RHODES: Divertimento, small orchestra. David Starobin, gtr, mand; Gordon Gottlieb, perc; Alexandria Ivanoff, s; Patrick Mason, bar; Robert Phillips, Franco Renzulli, pno; John Sampen, sax; Brooklyn College Ensemble, Wichita State Faculty Chamber Ensemble, St. Paul Chamber Orchestra; Barbara Kolb, Walter Mays, Dennis Russel. CRI SD 361.
+ARG 3-77 p41 +-NR 2-77 p6
+MJ 3-77 p46 +-RR 8-77 p67

Spring river flowers moon night. cf Looking for Claudio.

KOLOVRATEK

Parthia pastoralis, F major. cf Supraphon 113 1323.

KONDO

Air I, amplified piano with trumpet. cf CP 3-5.

KORLING (Kjorling)

Songs: Aftonstamning. cf RCA LSC 9884.

Vita rosor. cf HMV RLS 715.

KORNGOLD, Wolfgang

1620 Don Quixote. Der Schneemann. Sonata, piano, no. 1, D minor. Antonin Kubalek, pno. Citadel CT 6009.
+NR 8-77 p15

Die Kathrin: Letter song; Prayer. cf Songs and arias (Entr'acte ERS 6502).

Love letter, op. 9. cf Songs and arias (Entr'acte ERS 6502).

1621 Much ado about nothing, op. 11. WEILL: Quodlibet, op. 9. Westphalian Symphony Orchestra; Siegfried Landau. Candide CE 31091.
-MJ 5-77 p32 +-NR 8-76 p4
+-MQ 7-77 p446

Much ado about nothing, op. 11: The page's song. cf Songs and arias (Entr'acte ERS 6502).

The private lives of Elizabeth and Essex: Come live with me. cf Songs and arias (Entr'acte ERS 6502).

1622 Quartet, strings, no. 1, op. 16, A major. Quartet, strings, no. 3, op. 34, D major. Chilingirian Quartet. RCA RL 2-5097.
+Gr 10-77 p653 +RR 9-77 p76
+HFN 12-77 p172

Quartet, strings, no. 3, op. 34, D major. cf Quartet, strings, no. 1, op. 16, A major.

1623 Quintet, piano and strings, op. 15, E major. Sonata, piano, no. 3, op. 25, C major. Endre Granat, Sheldon Sanov, vln; Milton Thomas, vla; Douglas David, vlc; Harold Gray, pno. Genesis GS 1063.
+-Gr 3-77 p1454 ++NR 10-75 p4
+HFN 2-77 p121 +-RR 2-77 p73

Der Ring des Polykrates: Diary song. cf Songs and arias (Entr'acte ERS 6502).

Der Schneemann. cf Don Quixote.

The sea hawk: Old Spanish song. cf Songs and arias (Entr'acte ERS 6502).

The silent serenade: Song of bliss; Without you. cf Songs and arias (Entr'acte ERS 6502).

Sonata, piano, no. 1, D minor. cf Don Quixote.

Sonata, piano, no. 3, op. 25, C major. cf Quintet, piano and strings, op. 15, E major.

1624 Songs and arias: Die Kathrin: Letter song; Prayer. Love letter, op. 9. Much ado about nothing, op. 11: The page's song. The private lives of Elizabeth and Essex: Come live with me. Der Ring des Polykrates: Diary song. The sea hawk: Old Spanish song. The silent serenade: Song of bliss; Without you. Die tote Stadt, op. 12: Lute song. Polly Jo Baker, s; George Calusdian, pno. Entr'acte ERS 6502.
+Gr 12-76 p1033 +NR 7-76 p9
+HFN 1-77 p109 +RR 11-76 p39
Die tote Stadt, op. 12: Der erste der Lieb mich gelehrt. cf Seraphim IB 6105.
Die tote Stadt, op. 12: Lute song. cf Songs and arias (Entr'acte ERS 6502).
Die tote Stadt, op. 12: Marietta's Lied. cf Advent S 5023.
1625 Violanta: Prelude and carnival. WAGNER: The flying Dutchman: Overture. Siegfried Idyll. Tannhauser: Venusberg music. Beecham Choral Society; RPO; Jascha Horenstein. Quintessence PMC 7047.
+ARG 12-77 p43

KOTTER, Hans
Hochersperger Spanieler. cf Argo D40D3.

KOZELUH, Leopold
Pastorale. cf JIROVEC: Parthia, B flat major.

KRAFT, William
1626 The imagistes. ROSENMAN: Chamber music, no. 2. The New Muse, Los Angeles Percussion Ensemble; Sue Harmon, s; Ellen Geer, Michael Kermoyan, readers; William Kraft. Delos DEL 25432.
++NR 12-77 p6
In memoriam Igor Stravinsky. cf Orion ORS 76212.

KRAL, Johann
Hoch Hapsburg. cf Decca SB 713.

KRAMAR-KROMMER, Frantisek
Harmonie, op. 71, E flat major. cf JIROVEC: Parthia, B flat major.
Serenade, C minor. cf Telefunken EK 6-35334.

KREBS, Johann
Chorale preludes: Allein Gott in der Hoh sei Ehr; Von Gott will ich nicht lassen; Jesu, meine Freude. cf Crescent ARS 109.
Concerto, guitar and strings, G major. cf FASCH: Concerto, guitar and strings, D minor.
Jesu meine Freude. cf ABC ABCL 67008.
Jesus meine Zuversicht. cf ABC ABCL 67008.
Liebster Jesu wir sind hier. cf Pelca PSR 40571.
Prelude and fugue, F sharp major. cf Pelca PSR 40571.
Von Gott will ich nicht lassen. cf ABC ABCL 67008.
Wachet auf, ruft uns die Stimme. cf Pelca PSR 40571.
Wachet auf, ruft uns die Stimme. cf Nonesuch H 71279.

KREIN, Jascha
Gipsy carnival. cf Kelsey Records KEL 7601.

KREISLER, Fritz
Andantino, in the style of Padre Martini. cf GLUCK: Orfeo ed Eurydice: Melodie.
Caprice Viennois. cf Works, selections (Decca PFS 4423).
Caprice Viennois. cf Works, selections (RCA ARL 1-2365).
Caprice Viennois. cf ALBENIZ: Espana, op. 165, no. 2: Tango.
La chasse. cf Works, selections (RCA ARL 1-2365).
Concerto, violin, C major. cf ALBENIZ: Espana, op. 165, no. 2: Tango.

Concerto in the style of Vivaldi. cf Works, selections (Decca PFS 4423).
Danse espagnole. cf Works, selections (Decca PFS 4423).
La gitana. cf Works, selections (RCA ARL 1-2365).
La gitana. cf Discopaedia MOB 1018.
1627 Liebesfreud. Liebesleid (arr. Rachmaninoff). RACHMANINOFF: Etudes tableaux, op. 39 (9). Ruth Laredo, pno. Columbia M 34532.
+CL 11-77 p10 +HF 11-77 p122
Liebesfreud. cf Works, selections (Decca PFS 4423).
Liebesfreud. cf Works, selections (RCA ARL 1-2365).
Liebesfreud. cf ALBENIZ: Espana, op. 165, no. 2: Tango.
Liebesfreud. cf GLUCK: Orfeo ed Eurydice: Melodie.
Liebesfreud. cf Angel S 37219.
Liebesfreud. cf Orion ORS 76247.
Liebesfreud. cf RCA JRL 1-2315.
Liebesleid. cf Liebesfreud.
Liebesleid. cf Works, selections (Decca PFS 4423).
Liebesleid. cf Works, selections (RCA ARL 1-2365).
Liebesleid. cf ALBENIZ: Espana, op. 165, no. 2: Tango.
Liebesleid. cf GLUCK: Orfeo ed Eurydice: Melodie.
Liebesleid. cf Angel S 37219.
Liebesleid. cf CBS 61039.
Liebesleid. cf Discopaedia MB 1012.
Liebesleid. cf RCA JRL 1-2315.
Minuet. cf Works, selections (RCA ARL 1-2365).
The old refrain. cf Works, selections (RCA ARL 1-2365).
Polichinelle. cf Discopaedia MB 1012.
Praeludium and allegro. cf Works, selections (RCA ARL 1-2365).
Praeludium and allegro, in the style of Pugnani. cf GLUCK: Orfeo ed Eurydice: Melodie.
Quartet, strings, A minor. cf COLERIDGE-TAYLOR: Quintet, clarinet, F sharp minor.
Recitative and scherzo. cf Works, selections (RCA ARL 1-2365).
Recitative and scherzo-caprice, solo violin, op. 6. cf ALBENIZ: Espana, op. 165, no. 2: Tango.
Rondino on a theme by Beethoven. cf Discopaedia MB 1014.
Schon Rosmarin. cf Works, selections (Decca PFS 4423).
Schon Rosmarin. cf Works, selections (RCA ARL 1-2365).
Schon Rosmarin. cf ALBENIZ: Espana, op. 165, no. 2: Tango.
Schon Rosmarin. cf Discopaedia MB 1012.
Sicilienne and rigaudon. cf Works, selections (RCA ARL 1-2365).
Syncopation. cf Works, selections (Decca PFS 4423).
Syncopation. cf ALBENIZ: Espana, op. 165, no. 2: Tango.
Tambourin Chinois, op. 3. cf Works, selections (Decca PFS 4423).
Tambourin Chinois, op. 3. cf Works, selections (RCA ARL 1-2365).
Tambourin Chinois, op. 3. cf ALBENIZ: Espana, op. 165, no. 2: Tango.
Tambourin Chinois, op. 3. cf Discopaedia MOB 1020.
Viennois, op. 2. cf Discopaedia MB 1012.
1628 Works, selections: Caprice Viennois, op. 3. Concerto in the style of Vivaldi. Danse espagnole. Liebesfreud. Liebesleid. Syncopation. Schon Rosmarin. Tambourin Chinois. Erich Gruenberg, vln; NPhO; Elgar Howarth. Decca PFS 4423.
+-HFN 12-77 p173
1629 Works, selections: Caprice Viennois. La chasse. La gitana. Liebesfreud. Liebesleid. Minuet. The old refrain. Praelud-

ium and allegro. Recitative and scherzo. Schon Rosmarin. Sicilienne and rigaudon. Tambourin Chinois, op. 3. Eugene Fodor, vln; Stephen Swedish, pno. RCA ARL 1-2365.
+NR 11-77 p15 +-St 12-77 p148

KRENEK, Ernst

1630 Aulokithara, op. 213a. Echoes from Austria, op. 166. Sacred pieces, op. 210 (3). Wechselrahmen, op. 189. James Ostryniec, ob; Karen Lindquist, hp; Beverly Ogdon, s; Ernst Krenek, pno; COD Vocal Ensemble; John Norman. Orion ORS 76246.
+-NR 10-76 p6 +Te 9-77 p38

Echoes from Austria, op. 166. cf Aulokithara, op. 213a.

Kleine Blasmusik, op. 70a. cf MAXWELL DAVIES: Saint Michael sonata.

Merry marches, op. 44. cf MAXWELL DAVIES: Saint Michael sonata.

1631 O lacrymosa (3 songs). Sante Fe timetable, chorus. Tape and double, 2 pianos. Toccata, accordion. Genevieve Weide, s; Young McMahan, accordion; John Dare, Patricia Marcus, William Tracy, pno; California State University Northridge Chamber Singers; John Alexander. Orion ORS 75204.
+-HF 6-76 p80 +NR 2-76 p8
++MJ 7-76 p57 +Te 9-77 p38

1632 Reisebuch aus den osterreichischen Alpen, op. 62. Waldemar Kmentt, t; Richard Elsinger, pno. Preiser SPR 3269.
+ARG 5-77 p24

Sacred pieces, op. 210. cf Aulokithara, op. 213a.

Santa Fe timetable, chorus. cf O lacrymosa.

Tape and double, 2 pianos. cf O lacrymosa.

Toccata, accordion. cf O lacrymosa.

Wechselrahmen, op. 189. cf Aulokithara, op. 213a.

KREUTZENHOFF

Frisch und frolich wollen wir leben. cf Argo D40D3.

KREUTZER, Conradin

1633 Das Nachtlager in Granada. Ladislav Illavsky, bar; Anita Ammersfeld; Wolfgang Fassler, Kurt Ruzicka, Gerd Fussi, Wolfgang Kandutsch; Arnold Schoenburg Chor; Akademische Orchesterverein; Karl Etti. Preiser SPR 3271/2 (2).
+ARG 5-77 p24

Das Nachtlager von Granada: Overture. cf Supraphon 110 1637

KRIEGER, Arthur

Short piece. cf Odyssey Y 34139.

KRUGER

Suite (sonata). cf BOHM: Fantasy on a theme by Schubert.

KUBIK, Gail

Fanfare for the century. cf BIS LP 59.

KUCERA, Vaclav

Diario. cf DG 2530 802.

KUGELMANN

Dies est laeticiae. cf ABC L 67002.

KUHN

Ausfahrt. cf Telefunken DT 6-48085.

KUISMA, Rainer

1634 Concertpiece for percussion. MILHAUD: Saudades do Brasil. NILSSON: Fragments. SALLINEN: Symphony, no. 2. Rainer Kuisma, perc; Lucia Negro, pno; Norrkoping Symphony Orchestra; Okko Kamu. Caprice CAP 1073.
+-HFN 6-77 p127 +-RR 4-77 p48

KUNNEKE
Die Lockende Flamme: Lind ist die Nacht. cf HMV ESD 7043.
KURTAG, Gyorgy
Duos, violin and cimbalom, op. 4. cf Hungaroton SLPX 11686.
In memory of a winter twilight, op. 8. cf Hungaroton SLPX 11686.
Splinters, op. 6c. cf Hungaroton SLPX 11686.
KYNASTON, Trent
Dawn and jubilation. cf BACH: Prelude, bourree and gigue.
LABRIOMA
Sailor's song. cf Club 99-105.
LAKE, Greg
Songs: C'est la vie; Closer to believing; Hallowed be thy name; Lend your love to me tonight; Nobody loves you like I do. cf Atlantic SD 2-7000.
LALO, Edouard
1635 Concerto, violoncello, D minor. SCHUMANN: Concerto, violoncello, op. 129, A minor. Csaba Onczay, vlc; Hungarian Radio and Television Orchestra; Antal Jancsovics. Hungaroton SLPX 11705.
+-NR 3-76 p4 ++SFC 1-2-77 p34
+-RR 3-76 p48
Concerto, violoncello, D minor. cf FAURE: Elegy, op. 24, C minor.
1636 Rapsodie norvegienne, op. 21. Le Roi d'Ys: Overture. Symphony, G minor. Monte Carlo Opera Orchestra; Antonio de Almeida. Philips 6500 927.
*Audio 8-76 p78 ++MT 9-76 p747
+Gr 4-76 p1603 +NR 6-76 p2
+-HF 4-77 p106 +-RR 3-76 p45
+-HFN 3-76 p95 +SFC 5-23-76 p36
Le Roi d'Ys: Aubade. cf HMV RLS 719.
Le Roi d'Ys: Overture. cf Rapsodie norvegienne, op. 21.
Le Roi d'Ys: Overture. cf HEROLD: Zampa: Overture.
Le Roi d'Ys: Overture. cf Virtuosi VR 7608.
Scherzo. cf DELISLE: La Marseillaise.
1637 Symphonie espagnole, op. 21. TCHAIKOVSKY: Serenade melançolique, op. 26. Miklos Szenthelyi, vln; HSO; Ervin Lukacs. Hungaroton SLPX 11826.
+-RR 2-77 p49
Symphony, G minor. cf Rapsodie norvegienne, op. 21.
LAMB
Songs: The volunteer organist. cf Argo ZFB 95/6.
LAMBE
Salve regina. cf Coimbra CCO 44.
LANDINI, Francesco
La bionda treccia. cf Argo D40D3.
Cara mie donna. cf Argo D40D3.
Con bracchi assai. cf Argo D40D3.
De dimni tu. cf Argo D40D3.
Donna'l tuo partimento. cf Argo D40D3.
Ecco la primavera. cf Argo D40D3.
Ecco la primavera. cf BIS LP 75.
Giunta vaga bilta. cf Argo D40D3.
Questa fanciulla amor. cf Argo D40D3.
Songs: Se la nimica mie; Adiu adiu. cf 1750 Arch S 1753.
LANG, C. S.
Tuba tune, op. 15, D major. cf Vista VPS 1034.
LANG, Istvan
Cassazione. cf Hungaroton SLPX 11811.

1638 Concerto bucolico. In memoriam N. N. Three sentences from Romeo and Juliet. Magda Kalmar, s; Gabor Lehotka, org; HRT Chorus; Budapest Symphony Orchestra, Liszt Ferenc Chamber Orchestra; Ferenc Sapszon, Frigyes Sandor, Janos Sandor, Geza Oberfrank. Hungaroton SLPX 11784.
+ARG 7-77 p23 +RR 5-77 p82
++NR 11-77 p5
Improvisation, cimbalom. cf Hungaroton SLPX 11686.
In memoriam N. N. cf Concerto bucolico.
Three sentences from Romeo and Juliet. cf Concerto bucolic.

LANGFORD, Gordon
North country fantasie. cf Decca SB 329.
The seventies set. cf Decca SB 328.

LANGLAIS, Jean
Dialogue sur les mixtures. cf Argo ZRG 864.
Fete. cf DEMESSIEUX: Te deum, op. 11.
Suite breve. cf DURUFLE: Suite, op. 5: Toccata, B minor.
Suite breve: Jeux; Cantilene; Plainte; Dialogue sur les mixtures. cf ALAIN: Organ works (Cathedral CRMS 858).
Suite francaise. cf Wealden WS 159.
Triptyque. cf Vista VPS 1029.

LANNER, Joseph
Bruder Halt, Dampf-Waltzer and Galopp, op. 94. cf Works, selections (HMV SQ ESD 7045).
Favorit-polka, op. 201. cf Saga 5421.
Hofballtanze, op. 161. cf Works, selections (HMV SQ ESD 7045).
Marienwalzer, op. 143. cf Works, selections (HMV SQ ESD 7045).
Neujahrsgalopp, op. 61, no. 2. cf Works, selections (HMV ESD 7045).
Pesther-Walzer, op. 93. cf Works, selections (HMV SQ ESD 7045).
Die Schonbrunner, op. 200. cf Classics for Pleasure CFP 40213.
Die Schonbrunner, op. 200. cf Mercury SRI 75098.
Tarantel-Galopp, op. 125. cf Works, selections (HMV SQ ESD 7045).
Ungarischer Galopp. cf DG 2723 051.
Die Werber, op. 103. cf Works, selections (HMV SQ ESD 7045).
1639 Works, selections: Bruder Halt, Dampf-Waltzer and Galopp, op. 94. Hofballtanze, op. 16. Marienwalzer, op. 143. Neujahrsgalopp, op. 61, no. 2. Pesther-Walzer, op. 93. Tarantel-Galopp, op. 125. Die Werber, op. 103. Johann Strauss Orchestra; Willi Boskovsky. HMV SQ ESD 7045 Tape (c) TC ESD 7045.
++Gr 11-77 p910 ++RR 12-77 p48
+-Gr 12-77 p1145 tape -RR 12-77 p96 tape

LANSKY, Paul
Mild und Leise. cf Odyssey Y 34139.

LAPARRA, Raoul
L'Illustre Fregona: Melancolique tombe le soir. cf HMV RLS 715.

LAPPI, Pietro
Le negrona. cf HMV SQ ASD 3393.

LARSSON, Lars-Erik
1640 Concertino, op. 45, no. 9. ROMAN: Symphony, no. 20, E minor. SALLINEN: Kamarimusikki, no. 1, op. 38. TELEMANN: Concerto, 2 violas and strings, G major. Stockholm Chamber Ensemble. HNH 4011. (also BIS LP 46)
++FF 11-77 p50 +-HFN 2-77 p131
+Gr 3-77 p1457 +NR 9-77 p3
+HF 9-77 p108 +-RR 1-77 p58
Concertino, bassoon, op. 45, no. 4. cf BEETHOVEN: Trio, bassoon, flute and piano.

Concertino, flute, op. 45, no. 1. cf BEETHOVEN: Trio, bassoon, flute and piano.
Concertino, piano, op. 45, no. 12. cf BEETHOVEN: Sonata, piano, no. 22, op. 54, F major.
The disguised god: Lyric suite; Prelude. cf Decca DDS 507/102.

LASERNA, Blas
Tonadillas. cf HMV RLS 723.

LASSUS, Roland de
Fantasias. cf Kaibala 20B01.
Madrigal dell'eterna. cf Argo ZRG 823.
Morescas: Cathalina, apra finestra; Matona mia cara. cf L'Oiseau-Lyre 12BB 203/6.
Ricercar a 2. cf Telefunken AW 6-42033.

1641 Songs: Alma redemptoris mater; Omnes de Saba venient; Psalmus poenitentalis V: Domine exaudi orationem meam, Tui sunt coeli; Salve regina. Christ Church Cathedral Choir; Simon Preston. Argo ZRG 795.
+Gr 9-76 p454 +-NR 7-77 p8
+HFN 7-76 p89 ++RR 7-76 p76
+MT 1-77 p45, 2-77 p133

LATOUR
Variations on "God save the king". cf Serenus SRS 12067.

LAURENCINI (17th century England)
Fantasia. cf L'Oiseau-Lyre DSLO 510.

LAURO, Antonio
1642 Danza negra. Suite venezolano: Valse. Valse, no. 3. PONCE: Campo. SAINZ DE LA MAZA: Homenaje a la guitarra. VILLA-LOBOS: Choro typico. Preludes, nos. 1-5. Julian Byzantine, gtr. Classics for Pleasure CFP 40209 Tape (c) TC CFP 40209.
+Gr 10-75 p658 +HFN 12-76 p153 tape
+Gr 2-77 p1325 tape +RR 9-75 p62
+HFN 11-75 p155 tape
Suite venezolano: Valse. cf Danza negra.
Suite venezolano: Vals. cf Saga 5412.
Valse, no. 3. cf Danza negra.
Valse criollo. cf Classics for Pleasure CFP 40012.
Vals venezolano. cf L'Oiseau-Lyre SOL 349.
Venezuelan waltzes (4). cf BARRIOS: Las abejas.

LAVIGNE, Philippe de
Sonata, recorder, C major. cf Cambridge CRS 2826.

LAW, Andrew
Bunker Hill. cf Vox SVBX 5350.

LAWES, Henry
Sweet stay awhile. cf National Trust NT 002.

LAWES, William
Gather ye rosebuds. cf National Trust NT 002.

LAZAROF, Henri
1643 Adieu. Chamber concerto, 12 soloists, no. 3. Cadence VI, tuba and tape. Duo 1973. Roger Bobo, tuba; Gabor Rejto, vlc; Gary Gray, clt; Alice Rejto, Irma Vallecillo, pno; Ensemble; Henri Lazarof. Avant AV 1019.
+ARG 12-77 p35
Cadence VI, tuba and tape. cf Adieu.
Chamber concerto, 12 soloists, no. 3. cf Adieu.
1644 Concerto, flute. Spectrum. James Galway, flt; Thomas Stevens, tpt; Utah Symphony Orchestra, NPhO; Henri Lazarof. CRI SD 373.
+ARG 11-77 p18 ++NR 9-77 p3
++FF 11-77 p32

Duo 1973. cf Adieu.
Spectrum. cf Concerto, flute.

LAZZARO
Chitarra romana. cf London SR 33221.

LEAF, Robert
Let the whole creation cry. cf Columbia M 34134.

LEAR
Shylock: Polka. cf Grosvenor GRS 1043.

LEBEGUE, Nicolas
1645 Elevation, G major. Magnificat mode IV. Offertory, F major. Suite in the sixth mode. Symphony, F major. NIVERS: Suite, Bk 3, in the second mode. Xavier Darasse, org. Musical Heritage Society MHS 3317.
+-MU 11-77 p17
Magnificat mode IV. cf Elevation, G major.
Offertory, F major. cf Elevation, G major.
Suite on the sixth mode. cf Elevation, G major.
Symphony, F major. cf Elevation, G major.

LECLAIR, Jean-Marie
Sonata, violin, op. 9, no. 3, D major. cf Discopaedia MB 1013.

LECLERC, Michel
Par monts et par vaux. cf University of Iowa Press unnumbered.

LECOCQ, Charles
1646 La fille de Madame Angot. Mady Mesple, Christiane Stutzmann, s; Denise Benoit, ms; Charles Burles, Jacques Loreau, t; Bernard Sinclair, Michel Roux, bar; Gerard Chapuis, bs; Paris Opera Chorus; Opera-Comique Orchestra; Jean Doussard. Connoisseur Society CS 2-2135 (2). (also CRD Pathe C 161 12500/1)
+ARG 10-77 p36 +-MJ 9-77 p34
++Gr 1-77 p1187 +-NR 8-77 p10
+HF 9-77 p104 ++SFC 6-19-77 p46

LEEMANS
March of the Belgian parachutists. cf Decca SB 713.

LEES, Benjamin
1647 Symphony, no. 3. TURINA: Danzas gitanas, op. 55. Louisville Orchestra. Louisville LS 752.
+MJ 3-77 p46

LEGRANT, Guillaume
Entre vous, noviaux maries. cf Nonesuch H 71326.

LEGRENZI, Giovanni
Sonata, strings, no. 6, E minor. cf Supraphon 110 1890.

LEHAR, Franz
1648 The Count of Luxembourg, excerpts. Soloists, Choir and Orchestra. Telefunken AF 6-23067 Tape (c) CH 4-23067.
+-HFN 12-77 p187 tape -RR 11-77 p38
Friederike: Poet's song. cf KALMAN: Das Veilchen vom Montmartre: Heut nacht hab ich getraumt von dir.
Giuditta: Meine Lippen sie kussen so heiss. cf HMV ESD 7043.
Gold und Silber, op. 79. cf Classics for Pleasure CFP 40213.
Das Land des Lachelns: Su Tshong's songs. cf KALMAN: Das Veilchen vom Montmartre: Heut nacht hab ich getraumt von dir.
1649 Die lustige Witwe (The merry widow). Hilde Gueden, Emmy Loose, s; Edith Winkler, ms; Per Grunden, Waldemar Kmentt, Kurt Equiluz, t; Karl Donch, Hans Duhan, Peter Preses, bar; Marjan Rus, Ljubomir Pantscheff, bs; VSOO and Chorus; Robert Stolz. Decca DPA 573/4. (Reissue from SXL 2022/3)
+-Gr 3-77 p1462 +-RR 2-77 p32
+HFN 3-77 p118

1650 Die lustige Witwe, excerpts. Adelaide Singers; Adelaide Symphony Orchestra; John Lanchbery. Angel S 37092 Tape (c) 4XS 37092.
+HF 11-76 p97 +MJ 12-76 p28
+HF 6-77 p103 tape

1651 Die lustige Witwe, excerpts. Soloists; VSOO and Chorus; Robert Stolz. Telefunken AF 6-22992 Tape (c) CH 4-22992.
+-HFN 12-77 p187 tape -RR 11-77 p38

Die lustige Witwe: Es lebt eine Vilja. cf HMV ESD 7043.

1652 The merry widow: Introduction...A highly respectable wife; Entry of widow; I'm off to Chez Maxim; All's one to all men where there's gold; Finale; Introduction...Dance...Vilia; Jogging in a one-horse gig; You're back where you first began; Red as the rose in Maytime; Finale; The cake-walk; Can can; Love unspoken; Finale; You're back where you first began. Catherina Wilson, Patricia Hay, s; Jonny Blanc, David Hillman, David Fieldsend, t; William McCue, Gordon Sandison, bar; Scottish Opera Chorus; Scottish National Symphony Orchestra; Alexander Gibson. Classics for Pleasure CFP 40276.
+-Gr 9-77 p517 +-RR 9-77 p40
+-HFN 8-77 p85

1653 The merry widow: Pontevedro in Paree; A highly respectable wife; Gentlemen, no more; I'm still a Pontevedrian; I'm off to Chez Maxim; Ladies choice; Young lovers all awake; Come away to the ball; Heigh-ho; Vilia; Jogging in a one-horse gig; You're back where you first began; Red as the rose in Maytime; Look where a leafy bower lies; Quite a la mode Paree; Oh how splendid; There once were two royal children; Eh, voila les belles grisettes; Love unspoken. June Bronhill, s; Anne Howard, ms; David Hughes, t; Jeremy Brett, bar; John McCarthy Singers; Orchestra; Vilem Tausky. EMI Note NTS 103. (Reissue from Columbia TWO 234)
+Gr 1-77 p1187

1654 Die lustige Witwe: Verehrteste Damen und Herren; So kommen Sie; Bitte, meine Herren; Oh, Vaterland; Damenwahl; Es lebt eine Vilja Heia, Madel, aufgeschaut; Wie die Weiber man behandelt; Wie eine Rosenknospe; Den Herrschaften hab ich was zu erzahlen; Ja wir sind es, die Grisetten; Lippen schweigen; Ja das Studium der Weiber ist schwer. Elizabeth Harwood, Teresa Stratas, s; Rene Kollo, Werner Hollweg, Donald Grobe, Werner Krenn, t; Zoltan Keleman, bar; German Opera Chorus; BPhO; Herbert von Karajan. DG 2530 729. (Reissue from 2707 070)
+-Gr 11-76 p893 +RR 11-76 p39
+HFN 1-77 p121

The merry widow: Vilia. cf HMV MLV 118.
The merry widow: Waltz. cf Mercury SRI 75098.

LEHMANN, Liza (Elizabetta)
The cuckoo. cf Decca SDD 507.

LEICESTER
Parliament Square. cf DJM Records DJM 22062.

LEIFS, Jon
Iceland overture, op. 9. cf COWELL: Symphony, no. 16.

1655 Saga symphony, no. 26. Iceland Symphony Orchestra; Jussi Jalas. ITM 2.
+RR 6-77 p58

LEIGHTON, Kenneth
Dialogues on the Scottish psalm-tune "Martyrs", op. 73. cf Vista VPS 1039.

Festival fanfare. cf Vista VPS 1034.
God is ascended. cf Abbey LPB 776.
Lully, lulla, thou little tiny child. cf HMV CSD 3774.

LE JEUNE, Claude
Fiere cruelle. cf L'Oiseau-Lyre 12BB 203/6.
1656 Missa ad placitum. TITELOUZE: Quatre versets sur "Veni creator". Michel Chapuis, org; Deller Consort; Alfred Deller.
+St 2-77 p118

LENNON-MCCARTNEY
I want to hold your hand. cf RCA CRL 1-2064.
Yesterday. cf ALMEIDA: Jazz-tuno at the mission.

LENTZ, Daniel
Songs of the sirens. cf CHILDS: Trio, clarinet, violoncello and piano.

LEONCAVALLO, Ruggiero
Au clair de la lune. cf Club 99-105.
I Pagliacci. cf MASCAGNI: Cavalleria rusticana (DG 3371 011).
I Pagliacci: Arias. cf Free Lance FLPS 675.
I Pagliacci: Ballatella. cf Odyssey Y 33793.
I Pagliacci: Prologue. cf Seraphim S 60280.
I Pagliacci: Quel fiamma avea nel guardo. cf HMV SLS 5104.
I Pagliacci: Recitar...Vesti la giubba. cf HMV RLS 715.
I Pagliacci: Recitar...Vesti la giubba. cf RCA TVM 1-7203.
I Pagliacci: Recitar...Vesti la giubba; No Pagliaccio non son. cf RCA TVM 1-7201.
1657 I Pagliacci: Si puo; Signore, Signori; Un tal gioco; Don din don din; Qual flamma avea nel guardo...Stridono lassu; Recitar... Vesti la giubba; Non, Pagliaccio non son. MASCAGNI: Cavalleria rusticana: Gli aranci olezzano; Voi lo sapete, o mamma; Oh, Il Signore vi manda; Intermezzo; Mamma, quel vino e generoso. Joan Carlyle, s; Fiorenza Cossotto, ms; Carlo Bergonzi, Ugo Benelli, t; Giuseppe Taddei, Rolando Panerai, Gianciacomo Guelfi, bar; Marie Gracis Allegri, con; La Scala Orchestra and Chorus; Herbert von Karajan. DG 2535 199. (Reissue from SLPM 139205/7)
++Gr 12-76 p1046 +RR 1-77 p35
+HFN 1-77 p121
I Pagliacci: Vesti la giubba. cf Decca ECS 811.
I Pagliacci: Vesti la giubba. cf RCA CRM 1-1749.
I Pagliacci: Vesti la giubba; No Pagliaccio non son. cf Rubini GV 43.
Songs: Aprile. cf Decca SDD 507.
Songs: Serenade francaise; Serenade napolitaine. cf Columbia M 34501.
Zaza: Dir che ci sono al mondo. cf Club 99-100.

LEONI, Franco
1658 L'Oracolo. The prayer of the sword. Joan Sutherland, s; Huguette Tourangeau, ms; Ryland Davies, t; Tito Gobbi, bar; Richard Van Allan, Clifford Grant, bs; John Alldis Choir; National Symphony Orchestra; Richard Bonynge. London OSA 12107 (2) Tape (c) 5-12107. (also Decca D34D2 Tape (c) K34K22)
+ARG 9-77 p38 +NR 8-77 p8
+FF 11-77 p32 +-NYT 6-5-77 pD19
+Gr 7-77 p212 +-ON 9-77 p69
+Gr 9-77 p508 tape ++RR 7-77 p38
+HF 8-77 p87 +-RR 11-77 p120 tape
+HFN 7-77 p115 +SFC 5-8-77 p46
+HFN 10-77 p169 tape ++St 8-77 p115

The prayer of the sword. cf L'Oracolo.

LEONINUS (12th century)

Alleluia pascha nostrum. cf DG 2710 019.

Guade Maria virgo. cf DG 2710 019.

Locus iste. cf DG 2710 019.

Viderunt omnes. cf DG 2710 019.

LEROUX, Xavier

Astarte: Adieu d'Hercule. cf Rubini GV 38.

Le ROY, Adrien

Basse dance. cf ATTAINGNANT: Basse dance.

Branle de Bourgogne. cf DG 2723 051.

Branle gay. cf ATTAINGNANT: Basse dance.

Destre amoureux. cf ATTAINGNANT: Basse dance.

Haulberroys. cf ATTAINGNANT: Basse dance.

Passemeze. cf ATTAINGNANT: Basse dance.

LESCUREL, Jehannot de

A vous douce debonaire. cf HMV SLS 863.

LESUR, Daniel

1659 Andrea del Sarto, excerpts. Andree Esposito, Daniele Perreirs, s; Alain Vanzo, t; Gabriel Bacquier, bar; Jacques Mars, bs; Henri Gui, bar; ORTF; ORTF Choirs; Manuel Rosenthal. Inedits 995 037.
+HF 7-74 p90 +St 2-77 p116
+RR 1-74 p30

LEVITZKI, Mischa

Valse, A major. cf International Piano Library IPL 114.

Valse de concert. cf International Piano Library IPL 114.

LEWIN, Gordon

The poacher. cf Amberlee ALM 602.

LEWIS

Monophony VII. cf Crystal S 361.

LIADOV, Anatol

Prelude and pastorale. cf HMV SXLP 30256.

Prelude and pastorale. cf HMV SXLP 30259.

LIAPUNOV, Sergey

Rhapsody on Ukrainian themes, op. 28. cf BALAKIREV: Concerto, piano, no. 2, E flat major.

LIDDELL

Abide with me. cf Prelude PRS 2505.

LIDHOLM, Ingvar

Choruses (4). cf DALLAPICCOLA: Cori di Michelangelo Buonarroti il Giovane, nos. 1 and 2.

Invention, clarinet and bass clarinet. cf BIS LP 62.

1660 Nausicaa alone. PETTERSSON: Concerto, string orchestra, no. 1. Elisabeth Soderstrom, s; Swedish Radio Symphony Orchestra and Chorus; Stig Westerberg. Caprice CAP 1110.
++St 2-77 p118

Songs (6). cf BIS LP 34.

LIEBERSON, Peter

1661 Concerto, 4 groups of instruments. Fantasy, piano. LUNDBORG: Music forever, no. 2, excerpts. Passacaglia. Ursula Oppens, pno; Speculum Musicae, Light Fantastic Players. CRI SD 350.
-MJ 3-77 p46 +NR 7-76 p4

Fantasy, piano. cf Concerto, 4 groups of instruments.

LIEURANCE

By the waters of Minnetonka. cf HMV RLS 719.

LIGETI, Gyorgy

Concerto. flute and oboe. cf Works, selections (BIS LP 53).

1662 Concerto, flute, oboe and orchestra. Concerto, 13 instrumentalists. Melodien, orchestra. London Sinfonietta; Aurele Nicolet, flt; Heinz Holliger, ob; David Atherton. Decca HEAD 12.
++Gr 8-76 p286 ++MT 8-77 p651
+Gr 9-77 p499 ++RR 7-76 p54
+-HFN 7-76 p89 +-Te 12-76 p32
+MM 4-77 p40

Concerto, 13 instrumentalsists. cf Concerto, flute, oboe and orchestra.

Continuum. cf Works, selections (BIS LP 53).

1663 Etude, no. 1. Lux aeterna. Quartet, strings, no. 2. Volumina. LaSalle Quartet; North German Radio Chorus; Gerd Zacher, org; Helmut Franz. DG 2530 392. (Reissues from 2720 025, 104988/93)
++Gr 1-77 p1160 +RR 10-76 p81
++HFN 11-76 p161 +-Te 12-76 p32

Lux aeterna. cf Etude, no. 1.

Melodien, orchestra. cf Concerto, flute, oboe and orchestra.

Nocturnes. cf Works, selections (BIS LP 53).

Pieces, wind quintet (1). cf HINDEMITH: Kleine Kammermusik, op. 24, no. 2.

Quartet, strings, no. 1 (Metamorphoses). cf Works, selections (BIS LP 53).

Quartet, strings, no. 2. cf Etude, no. 1.

San Franciso polyphony. cf Works, selections (BIS LP 53).

Volumina. cf Etude, no. 1.

Volumina. cf BACK: ...for Eliza.

1664 Works, selections: Concerto, flute and oboe. Continuum. Nocturnes. Quartet, strings, no. 1 (Metamorphoses). San Francisco Polyphony. Gunilla von Bahr, flt; Torleif Lannerholm, ob; Voces Intimae; Eva Nordwall, hpd; Swedish Radio Symphony Orchestra; Elgar Howarth. BIS LP 53.
+Gr 9-77 p499 +RR 4-77 p48

LILY PONS: Coloratura assoluta. cf Columbia 34294.

LINCKE, Paul

Father Rhine. cf Decca SB 713.

Folies Bergere. cf Angel (Q) S 37304.

Frau Luna: Schlosser die im Monde liegen. cf HMV ESD 7043.

The glow worm. cf Angel (Q) S 37304.

LINDBERG, Nils

1665 Noah's ark. Alice Babs, Jan Malmsjo; Orchestra and Chorus; Nils Lindberg. Swedish Discofil SLT 33216.
+HFN 9-77 p144 +RR 9-77 p88

LINDBLAD

Songs: Aftonen; Am Aarensee; Manntro; En ung flickas morgonbetraktelse. cf Swedish SLT 33209.

LINDE, Bo

Sonata, violin and piano, op. 10. cf CHRISTOSKOV: Suite, no. 1.

Songs (2). cf CARLSTEDT: Sonata, violin, op. 15.

LINDE, Hans-Martin

Amarilli mia bella. cf BIS LP 48.

LINDENFELD, Harris

Combinations I: The last gold of perished stars. cf Golden Crest RE 7068.

LINLEY, Thomas

Allegretto. cf Crescent BUR 1001.

LISLEY, John

Fair Cytherea presents her doves. cf DG 2533 347.

LISZT, Franz
(G refers to Grove's number, 5th ed.)

1666 Ad nos, ad salutarem undam, G 259. Adagio, G 263, D flat major. Ave Maria (Von Arcadelt), G 659, F major. Prelude and fugue on the name B-A-C-H, G 260. Lionel Rogg, org. Connoisseur CSQ 2100.
+MJ 5-77 p32 +NR 8-76 p11

1667 Ad nos, ad salutarem undam, G 259 (arr. Busoni). REUBKE: Sonata, piano, B flat minor. Hamish Milne, pno. L'Oiseau-Lyre DSLO 21.
+Gr 6-77 p70 +MT 12-77 p1015
+HFN 7-77 p115 +-RR 6-77 p84
+MM 11-77 p46

Adagio, G 263, D flat major. cf Ad nos, ad salutarem undam, G 259.
Adagio, G 263, D flat major. cf Organ works (Argo ZRG 784).
Angelus, G 378: Priere aux anges gardiens. cf Organ works (Argo ZRG 784).

1668 Annees de pelerinage, G 160, G 161, G 163. Lazar Berman, pno. DG 2709 076 (3).
+-Gr 12-77 p1107 ++RR 12-77 p69
+HFN 12-77 p173

1669 Annees de pelerinage, 1st year, G 160. Gyorgy Cziffra, pno. Connoisseur Society (Q) CSQ 2141.
+MJ 10-77 p27 +SR 9-3-77 p42
+NR 9-77 p12 ++St 11-77 p141

Annees de pelerinage, 1st year, G 160: No. 7, Eglogue. cf International Piano Library IPL 108.

1670 Annees de pelerinage, 2nd year, G 161. Gyorgy Cziffra, pno. Connoisseur Society (Q) CSQ 2142.
+MJ 10-77 p27 +SR 9-3-77 p42
+NR 9-77 p12 ++St 11-77 p141

1671 Annees de pelerinage, 2nd year, G 161. Milan Klicnik, pno. Supraphon 111 1766.
-FF 9-77 p28 +-NR 10-77 p14
+-HFN 9-77 p145 -RR 8-77 p71

1672 Annees de pelerinage, 2nd year, G 161: Dante sonata. Annees de pelerinage, 3rd year, G 163: Les jeux d'eau a la Villa d'Este. Etudes d'execution transcendente, G 139 (12). Russell Sherman, pno. Vanguard SRV 354/5 (2). (Reissue from Advent cassette E 1010)
+-HF 6-77 p92 +NR 4-77 p13

1673 Annees de pelerinage, 2nd year, G 161: Sonnets of Petrarch. Leyla Gencer, s. Cetra LPS 69008.
+-Gr 7-77 p221

1674 Annees de pelerinage, 2nd year, G 161: Sonetti del Petrarca (3).* SCHUMANN: Symphonic etudes, op. 13. Alexis Weissenberg, pno. Connoisseur Society CS 2109. (*Reissue from Angel S 36383)
+-HF 6-77 p99 ++NR 1-77 p14
+-MJ 12-76 p28 +St 3-77 p142

Annees de pelerinage, 2nd year, G 161: Sonnetto del Petrarca, no. 123. cf Piano works (Argo ZK 9).

1675 Annees de pelerinage, 2nd year, G 161: Sposalizio, Il pensieroso; Canzonetta del Salvator Rosa; Sonetti del Petrarca, nos. 47, 104, 123; Dante sonata. David Bean, pno. Westmisnter WGS 8339.
+CL 4-77 p10 +St 3-77 p127
+NR 4-77 p13

1676 Annees de pelerinage, 3rd year, G 163. Gyorgy Cziffra, pno. Connoisseur Society (Q) CSQ 2143.

+MJ 10-77 p27 +SR 9-3-77 p42
+NR 9-77 p12 ++St 11-77 p141

Annees de pelerinage, 3rd year, G 163: Les jeux d'eau a la Villa d'Este. cf Annees de pelerinage, 2nd year, G 161: Dante sonata.

Annees de pelerinage, 3rd year, G 163: Sunt lacrymae rerum. cf Piano works (International Piano Library IPL 111).

Ave Maria (Von Arcadelt), G 659, F major. cf Ad nos, ad salutarem undam, G 259.

Ballade, no. 2, G 171, B minor. cf Piano works (International Piano Library IPL 111).

1677 Concerto, piano, no. 1, G 124, E flat major. Concerto, piano, no. 2, G 125, A major. Garrick Ohlsson, pno; NPhO; Moshe Atzmon. Angel S 37145 (Q) SQ 37145 Tape (c) 4XS 37145. (also HMV (Q) SQ ASD 3159)
++Gr 4-76 p1603 ++RR 2-76 p31
+HF 4-76 p111 ++St 3-76 p116 Quad
++HF 5-77 p101 tape +St 2-77 p80
+-HFN 2-76 p101 +STL 3-7-76 p37
+-NR 2-76 p13

1678 Concerto, piano, no. 1, G 124, E flat major. Concerto, piano, no. 2, G 125, A major. Lazar Berman, pno; VSO; Carlo Maria Giulini. DG 2530 770 Tape (c) 3300 770.
++Gr 11-76 p792 +-NR 2-77 p4
+Gr 5-77 p1743 tape ++RR 11-76 p68
++HF 2-77 p95 +SFC 1-30-77 p36
++HFN 12-76 p144 ++St 3-77 p90
+MJ 5-77 p32

1679 Concerto, piano, no. 1, G 124, E flat major. TCHAIKOVSKY: Concerto, piano, no. 1, op. 23, B flat minor. Horacio Gutierrez, pno; LSO; Andre Previn. HMV SQ ASD 3262. (also Angel S 37177 Tape (c) 4XS 37177)
+-Gr 9-76 p421 ++NR 2-77 p5
+HF 2-77 p95 ++RR 9-76 p67
++HFN 9-76 p130 ++SFC 11-21-76 p35
+MJ 1-77 p26 +St 4-77 p133
-MM 4-77 p41 ++STL 9-19-76 p36

1680 Concerto, piano, no. 1, G 124, E flat major. Totentanz, G 126. Gyula Kiss, pno; Hungarian State Orchestra; Janos Ferencsik. Hungaroton SLPX 11792.
+ARG 12-76 p35 +NR 1-77 p6
+HF 2-77 p95

1681 Concerto, piano, no. 1, G 124, E flat major. Concerto, piano, no. 2, G 125, A major. Michele Campanella, pno; LPO; Hubert Soudant. Pye Nixa PCNHX 7.
+Gr 1-77 p1142 ++RR 12-76 p61
+-HFN 1-77 p109

1682 Concerto, piano, no. 1, G 124, E flat major. Concerto, piano, no. 2, G 125, A major. Emil von Sauer, pno; OSCCP; Felix Weingartner. Turnabout THS 65098.
+MJ 10-77 p27 +SFC 9-4-77 p38
+NR 12-77 p4 +St 11-77 p141

Concerto, piano, no. 1, G 124, E flat major. cf BEETHOVEN: Concerto, piano, no. 3, op. 37, C minor.

Concerto, piano, no. 1, G 124, E flat major. cf EMI Italiana C 153 52425/31.

Concerto, piano, no. 1, G 124, E flat major. cf HMV SLS 5068.

Concerto, piano, no. 2, G 125, A major. cf Concerto, piano, no. 1, G 124, E flat major (Angel 37145).
Concerto, piano, no. 2, G 125, A major. cf Concerto, piano, no. 1, G 124, E flat major (DG 2530 770).
Concerto, piano, no. 2, G 125, A major. cf Concerto, piano, no. 1, G 124, E flat major (Pye Nixa 7).
Concerto, piano, no. 2, G 125, A major. cf Concerto, piano, no. 1, G 124, E flat major (Turnabout THS 65098).
Concerto, piano, no. 2, G 125, A major. cf CHOPIN: Concerto, piano, no. 2, op. 21, F minor.
Consolations, G 172. cf Piano works (Argo ZK 9).
Consolation, no. 3, G 172, D flat major. cf Discopaedia MB 1014.
Consolation, no. 3, G 172, D flat major. cf RCA GL 4-2125.

1683 Duo (sonata), violin and piano, G 127, C sharp minor. Elegie, no. 2, G 131. Elegie, no. 1, G 130. Epithalam, G 129. Grand duo concertant, G 128. Romance oubliee, G 132. Endre Granat, vln; Francoise Regnat, pno. Orion ORS 76210.
+Audio 5-77 p100 +St 5-77 p122
+NR 11-76 p7

Elegie, no. 1, G 130. cf Duo (sonata), violin and piano, G 127, C sharp minor.
Elegie, no. 2, G 131. cf Duo (sonata), violin and piano, G 127, C sharp minor.
En reve, G 207. cf Piano works (International Piano Library IPL 111).
Episodes from Lenau's "Faust". cf Les preludes, G 97.
Epithalam, G 129. cf Duo (sonata), violin and piano, G 127, C sharp minor.

1684 Etudes de concert, G 144. Magyar Dallok, no. 12, G 242, E minor. Polonaise, no. 1, G 223, C minor. Marguerite Wolff, pno. Liszt Society DF 1.
-RR 12-77 p70

Etude de concert, no. 2, G 144, F minor: La leggierezza. cf CHOPIN: Etude, op. 25, no. 7, C sharp minor.
Etude de concert, no. 3, G 144, D flat major. cf Piano works (Argo ZK 9).

1685 Etude de concert, no. 2, G 145: Gnomenreigen. Harmonies poetiques et religieuses, G 173: Funerailles. Liebestraum, no. 3, G 541, A flat major. Faust waltz, G 407. Sonata, piano, G 178, B minor. Simon Barere, pno. Turnabout THS 65001.
++FF 11-77 p77

1686 Etudes d'execution transcendente, G 139. Hungarian rhapsody, no. 3, G 244, B flat major. Rhapsodie espagnole, G 254. Lazar Berman, pno. Columbia/Melodiya M2 33928 (2). (also HMV/Melodiya SLS 5040)
+-Gr 2-76 p1359 +MT 10-77 p825
+-HF 5-76 p75 +NR 4-76 p12
+-HFN 1-76 p110 +RR 1-76 p50
++MJ 10-76 p25 +-SFC 2-8-76 p26

Etudes d'execution transcendente, G 139 (12). cf Annees de pelerinage, 2nd year, G 161: Dante sonata.

1687 Etudes d'execution transcendente, nos. 5, 12, G 139: Feux follets; Chasse neige. Mephisto waltz, no. 1, G 514. Sonata, piano, G 178, B minor. Janina Fialkowska, pno. RCA FRL 1-0142.
+Gr 4-77 p1574 -RR 4-77 p74
+HFN 6-77 p129 +St 12-77 p152
+NR 10-77 p14

1688 Etudes d'execution transcendente d'apres Paganini, no. 3, G 140, A flat minor: La campanella. Faust waltz, G 407. Hungarian fantasia, G 123. Totentanz, G 126. Gyorgy Cziffra, pno; Orchestre de Paris; Gyorgy Cziffra, Jr. Connoisseur Society CS 2092.

+HF 2-77 p99 +St 12-76 p144
+MJ 5-77 p32

Etudes, d'execution transcendante d'apres Paganini, no. 3, G 140, A flat minor: La campanella. cf International Piano Library IPL 114.

Etudes d'execution transcendente d'apres Paganini, no. 5, G 140: La chasse. cf Westminster WGM 8309.

Faust waltz, G 407. cf Etudes de concerto, no. 2, G 145: Gnomenreigen.

Faust waltz, G 407. cf Etudes d'execution transcendente d'apres Paganini, no. 3, G 140, A flat minor: La campanella.

Grand duo concertant, G 128. cf Duo (sonata), violin and piano, G 127, C sharp minor.

1689 Harmonies poetiques et religieuses, G 173: Benediction de Dieu dans la solitude; Funerailles. Liebestraume, G 541. Mephisto waltz, no. 1, G 514. Garrich Ohlsson, pno. HMV HQS 1361. (also Angel S 37125)

+Gr 9-76 p444 ++St 3-76 p116
+HF 5-76 p86 +St 2-77 p80
+-NR 2-76 p13 +STL 8-8-76 p29
+-RR 9-76 p82

1690 Harmonies poetiques et religieuses, G 173: Funerailles. Hungarian rhapsody, no. 9, G 244, E flat major. SCHUBERT (Liszt): Ave Maria, D 839. Der Erlkonig, op. 1, D 328. Die junge Nonne, D 828. Die schone Mullerin, op. 25, no. 2, D 795: Wohin. Die Winterreise, op. 89, nos. 19 and 24, D 911. Lazar Berman, pno. Everest 3407. (Reissues from Melodiya D 08677/8, D 016151/2)

+-ARG 2-77 p42 +-NR 4-77 p13

Harmonies poetiques et religieuses, G 173: Funerailles. cf Etudes de concert, no. 2, G 145: Gnomenreigern.

Hungarian battle march, G 119. cf Paraphrases (transcriptions).

Hungarian fantasia, G 123. cf Etudes d'execution transcendente d'apres Paganini, no. 3, G 140, A flat minor: La campanella.

Hungarian fantasia, G 123. cf HMV SLS 5094.

Hungarian rhapsodies, nos. 1-3, G 244. cf ENESCO: Roumanian rhapsodies, op. 11, nos. 1 and 2.

1691 Hungarian rhapsodies, nos. 1-15, G 244. Gyorgy Cziffra, pno. HMV SLS 5089 (2).

-Gr 9-77 p456 -RR 9-77 p82
+-HFN 9-77 p145

Hungarian rhapsodies, nos. 1, 4, 6, G 359. cf Paraphrases (transcriptions).

1692 Hungarian rhapsody, no. 1, F minor (Doppler). SMETANA: Ma Vlast: Vltava. STRAUSS, J. II: Emperor waltz, op. 437. TCHAIKOVSKY: Capriccio italien, op. 45. BPhO, Bamberg Symphony Orchestra, Berlin Radio Orchestra; Ferenc Fricsay, Ferdinand Leitner, Richard Kraus. DG 2548 249.

+HFN 1-77 p121 +RR 1-77 p63

1693 Hungarian rhapsody, no. 2, G 244, C sharp minor. Totentanz, G 126. MENDELSSOHN: Capriccio brillante, op. 22, B minor. Rondo brillante, op. 29, E falt major. Peter Katin, pno; LPO;

Jean Martinon. Decca ECS 784. (Reissue from LXT 2932, LW 5134)
+-Gr 4-77 p1549 -RR 3-77 p50
+-HFN 3-77 p118
Hungarian rhapsody, no. 2, G 244, C sharp minor. cf Decca PFS 4387.
Hungarian rhapsody, no. 2, G 244, C sharp minor. cf International Piano Library 103.
Hungarian rhapsody, no. 2, G 244, C sharp minor. cf International Piano Library 117.
Hungarian rhapsody, no. 3, G 244, B flat major. cf Etudes d'execution transcendente, G 139.

1694 Hungarian rhapsodies, nos. 4 and 5, G 359. Tasso, lamento e trionfo, G 96. BPhO; Herbert von Karajan. DG 2530 698.
++Gr 8-76 p286 +-RR 8-76 p41
+HFN 8-76 p83 +-NR 5-77 p5
+-MJ 2-77 p31 +SFC 12-26-76 p34

Hungarian rhapsody, no. 4, G 244, E flat major. cf ENESCO: Roumanian rhapsody, op. 11, no. 1, A minor.
Hungarian rhapsody, no. 9, G 244, E flat major. cf Harmonies poetiques et religieuses, no. 7, G 173: Funerailles.
Hungarian rhapsody, no. 9, G 244, E flat major. cf Westminster WGM 8309.
Hungarian rhapsody, no. 10, G 244. cf CHOPIN: Etude, op. 25, no. 7, C sharp minor.
Hungarian rhapsody, no. 13, G 244, A minor. cf International Piano Library IPL 109.
Hungarian rhapsody, no. 13, G 244, A minor (abbreviated). cf International Piano Library IPL 104.
Hungarian rhapsody, no. 15, G 244. cf Paraphrases (transcriptions).
Hungarian rhapsody, no. 15, G 244. cf International Piano Library IPL 117.
Hungarian rhapsody, no. 15, G 244: Rakoczy march. cf Piano works (Argo ZK 9).
Ich liebe dich, G 315. cf Piano works (Argo ZK 9).
Kirchliche Festouverture, G 675. cf Organ works (Argo ZRG 784).
Legendes, G 175. cf Piano works (International Piano Library IPL 111).
Liebestraum, G 541. cf Harmonies poetiques et religieuses, G 173: Benediction de Dieu dans la solitude; Funerailles.
Liebestraum, no. 3, G 541, A flat major. cf Etudes de concert, no. 2, G 145: Gnomenreigen.
Liebestraum, no. 3, G 541, A flat major. cf Piano works (Argo ZK 9).
Liebestraum, no. 3, G 541, A flat major. cf Decca PFS 4387.
Liebestraum, no. 3, G 541, A flat major. cf RCA GL 4-2125.
Magyar Dallok, no. 12, G 242, E minor. cf Etudes de concert, G 144.
Mazeppa, G 135. cf TCHAIKOVSKY: Symphony, no. 6, op. 74, B minor.

1695 Mephisto waltz. Tasso, lamento e trionfo, op. 2, G 96. Von der Wiege bis zum Grabe (From the cradle to the grave), G 107. CSO; Georg Solti. London 6925 Tape (c) CS 5-6925.
++HF 10-77 p129 tape

Mephisto waltz. cf HMV RLS 717.
Mephisto waltz, no. 1, G 514. cf Etudes d'execution transcendente, nos. 5 and 12, G 139: Feux follets; Chasse neige.

Mephisto waltz, no. 1, G 514. cf Harmonies poetiques et religieuses, G 173: Benediction de Dieu dans la solitude; Funerailles.

1696 Missa choralis, G 10. Via crucis, G 53. BBC Northern Singers; Gordon Thorne. Saga 5432. (Reissue from SIX 5079, SIC 5105)
+-ARG 12-77 p36 +RR 6-77 p89
+Gr 7-77 p207 /SFC 8-14-77 p50

1697 Organ works: Adagio, G 263, D flat major. Angelus, G 378: Priere aux anges gardiens. Kirchliche Festouverture, G 675. Trauerode, no. 2, G 268, E minor. Variationen on Bach's "Weinen, Klagen, Sorgen, Zagen", G 673. Peter Planyavsky, org. Argo ZRG 784.
+Gr 9-76 p445 +NR 5-77 p15
+HFN 8-76 p83 ++RR 2-77 p81
+MM 7-77 p37

1698 Parapharases (transcriptions): Hungarian battle march, G 119. Hungarian rhapsody, no. 15, G 244. Hungarian rhapsodies, nos. 1, 4, 6, G 359. PH; Willi Boskovsky. HMV SQ ESD 7039 Tape (c) TC ESD 7039. (also Angel S 37277)
/Gr 8-77 p349 +HFN 9-77 p144
+-Gr 10-77 p698 tape +-HFN 10-77 p169 tape
+HF 12-77 p92 +RR 9-77 p62

1699 Paraphrases (Transcriptions): Bellini: Norma, G 394. Donizetti: Lucia di Lammermoor, G 397. Wagner: Der fliegende Hollander: The spinning chorus, G 440; Senta's ballad, G 441. Lohengrin: Bridal procession to the minister, G 445, no. 2. Tristan und Isolde: Isoldens Liebestod, G 447. David Wilde, pno. Saga 5437.
+Gr 5-77 p1714 +RR 5-77 p72
+HFN 5-77 p125

1700 Paraphrases and transcriptions: Auber: La muette de Portici: Grand tarantelle bravoure. Mendelssohn: A midsummer night's dream: Wedding march. Tchaikovsky: Eugene Onegin: Polonaise. Verdi: Rigoletto: Paraphrase. Wagner: Tannhauser: Overture. Gyorgy Cziffra, pno. Connoisseur Society CS 2130. (Reissues from French Pathe Marconi FALP 520)
++MJ 5-77 p32 +-St 9-77 p146
+-NR 6-77 p13

Paraphrases (transcriptions): Saint-Saens: Danse macabre, G 555. Verdi: Rigoletto, G 434: Paraphrase de concert. Wagner: Tristan und Isolde, G 447: Isoldens Liebestod. cf Piano works (Argo ZK 9).

1701 Piano works: Annees de pelerinage, 2nd year, G 161: Sonnetto del Petrarca, no. 123. Etudes de concert, no. 3, G 144, D flat major. Ich liebe dich, G 315. Consolations, G 172. Hungarian rhapsodies, no. 15, G 244: Rakoczy march. Liebestraum, no. 3, G 541, A flat major. Valse oubliee, no. 1, G 215, F sharp minor. Transcriptions: Saint-Saens: Danse macabre, G 555. Verdi: Rigoletto, G 434: Paraphrase de concert. Wagner: Tristan und Isolde, G 447: Isoldens Liebestod. Rhondda Gillespie, pno. Argo ZK 9.
+Gr 4-77 p1574 -RR 4-77 p74
/HFN 4-77 p138

1702 Piano works: Annees de pelerinage, 3rd year, G 163: Sunt lacrymae rerum. Ballade, no. 2, G 171, B minor. En reve, G 207. Legendes, G 175. Schwanengesang, G 560: Abschied. Ervin Nyiregyhazi, pno. International Piano Library IPL 111.
+FF 11-77 p33 +NYT 6-5-77 pD19
+NR 12-77 p11

Polonaise, no. 1, G 223, C minor. cf Etudes de concert, G 144.
1703 Les preludes, G 97. Episodes from Lenau's "Faust" (2). Westphalian Symphony Orchestra; Siegfried Landau. Turnabout QTVS 34597.
+Audio 1-77 p85
Les preludes, G 97. cf FRANCK: Symphony, D minor.
Les preludes, G 97. cf HMV RLS 717.
Les preludes, no. 3, G 97. cf BEETHOVEN: Egmont, op. 84: Overture.
Prelude and fugue on the name B-C-C-H, G 260. cf Ad nos, ad salutarem undam, G 259.
Prelude and fugue on the name B-A-C-H, G 260. cf Decca SDD 499.
Prelude and fugue on the name B-A-C-H, G 260. cf Odeon HQS 1356.
Prelude and fugue on the name B-A-C-H, G 260. cf Vista VPS 1030.
Rhapsodie espagnole, G 254. cf Etudes d'execution transcendente, G 139.
Romance oubliee, G 132. cf Duo (sonata), violin and piano, G 127, G sharp minor.
Schwanengesang, G 560: Abschied. cf Piano works (International Piano Library IPL 111).
1704 Sonata, piano, G 178, B minor. SCHUMANN: Fantasia, op. 17, C major. Alexis Weissenberg, pno. Connoisseur Society CS 2137. (Reissues from Angel S 36383, S 36616)
+NR 9-77 p12 +-SFC 5-29-77 p42
Sonata, piano, G 178, B minor. cf Etudes de concerto, no. 2, G 145: Gnomenreigen.
Sonata, piano, G 178, B minor. cf Etudes d'execution transcendente, nos. 5 and 12, G 139: Feux follets; Chasse neige.
Sonata, piano, G 178, B minor. cf BEETHOVEN: Sonata, piano, no. 23, op. 57, F minor.
Sonata, piano, G 178, B minor. cf GLAZUNOV: Sonata, piano, no. 1, op. 74, B flat minor.
1705 A symphony on Dante's "Divina Commedia", G 109. Bolshoi Theatre Orchestra and Chorus; Boris Khaikin. HMV SXLP 30234.
+-Gr 2-77 p1283 +RR 2-77 p49
+HFN 2-77 p121
Tasso, lamento e trionfo, G 96. cf Hungarian rhapsodies, nos. 4 and 5, G 359.
Tasso, lamento e trionfo, op. 2, G 96. cf Mephisto waltz.
Totentanz, G 126. cf Concerto, piano, no. 1, G 124, E flat major.
Totentanz, G 126. cf Etudes d'execution transcendente d'apres Paganini, no. 3, G 140, A flat minor: La campanella.
Totentanz, G 126. cf Hungarian rhapsody, no. 2, G 244, C sharp minor.
Totentanz, G 126. cf International Piano Library IPL 108.
Trauerode, no. 2, G 268, E minor. cf Organ works (Argo ZRG 784).
Valse oubliee, no. 1, G 215, F sharp minor. cf Piano works (Argo ZK 9).
Variationen on Bach's "Weinen, Klagen, Sorgen, Zagen", G 673. cf Organ works (Argo ZRG 784).
Variations on Bach's "Weinen, Klagen, Sorgen, Zagen", G 673. cf BRAHMS: Fugue, A flat minor.
Variations on Bach's "Weinen, Klagen, Sorgen, Zagen", G 673. cf FRANCK: Chorale, no. 2, B minor.
Via crucis, G 53. cf Missa choralis, G 10.
Von der Wiege bis zum Grabe (From the cradle to the grave), G 107. cf Mephisto waltz.
Waldesrauschen, G 145. cf International Piano Library IPL 103.

LITAIZE, Gaston
Toccata sue le veni creator. cf Vista VPS 1029.
LOBEL, Michael
1706 Trip-tych. Electronic tape and sounds. Serenus SRS 12076.
+NR 12-77 p14
LOCATELLI, Pietro
Concerto grosso, op. 1, no. 8, F minor. cf CORELLI: Concerto grosso, op. 6, no. 8, G minor.
1707 Concerto grosso, op. 8, no. 1, F minor. MANFREDINI: Concerto grosso, op. 3, no. 12, C major. TORELLI: Concerto grosso, op. 8, no. 6, G minor. Slovak Chamber Orchestra; Bohdan Warchal. Opus 9111 0431.
+Gr 6-77 p55 +RR 6-77 p58
LOCKE, Matthew
Music for His Majesty's sackbuts and cornetts. cf The tempest.
Music for His Majesty's sackbuts and cornetts. cf BIS LP 59.
Songs (anthems): The king shall rejoice; When the son of man shall come in his glory. cf BLOW: Sing unto the Lord, O ye saints, anthem.
Suites, nos. 1 and 2. cf Telefunken AW 6-42129.
1708 The tempest. Music for His Majesty's sackbuts and cornetts. Judith Nelson, Emma Kirkby, Prudence Lloyd, s; Martyn Hill, Rogers Covey-Crump, Richard Morton, Alan Byers, t; David Thomas, Geoffrey Shaw, bs; John York Skinner, Charles Brett, c-t; Academy of Ancient Music, Michael Laird Cornett and Sackbut Ensemble; Christopher Hogwood, hpd. L'Oiseau-Lyre Florilegium DSLO 507.
+Gr 7-77 p212 +RR 7-77 p68
+HFN 7-77 p115
LOEFFLER, Charles
1709 Rhapsodies, oboe, viola and piano (2). MOZART: Quartet, oboe, K 370, F major. John Mack, ob; Daniel Majeske, vln; Abraham Skernick, vla; S. Geber, vlc; E. Podis, pno. Advent S 5017.
+NR 6-76 p9 ++St 2-77 p118
LOEILLET, Jean-Baptiste
Corente. cf DG 2723 051.
Gigue. cf DG 2723 051.
Saraband. cf DG 2723 051.
Sonata, flute, op. 1. cf GIULIANI: Sonata, flute and guitar, op. 85, A major.
Sonata, recorder, G major. cf Cambridge CRS 2826.
Sonata, recorder and harpsichord, G major. cf Saga 5425.
Sonata, trumpet, C major. cf HMV ASD 3318.
Suite, G minor. cf Saga 5384.
Trio sonata, E minor. cf HOFFMEISTER: Duo concertante, F major.
LOEWE, F.
My fair lady: Selections. cf RCA CRL 1-2064.
LOHR
Songs: When Jack and I were children. cf Argo ZFB 95/6.
LONDON, Edwin
Bjorne Enstabile's Christmas music. cf Advance FGR 18.
1710 Choral music on old testament text: Day of desolation; Better is; Sacred hair; Dream thing on biblical episodes. University of Connecticut Concert Choir, Smith College Chamber Singers, University of Illinois Concert Choir; John Poellein, Iva Dee Hiatt, Edwin London, Harold Decker. Ubres CS 302.
*NR 10-77 p9

LOPEZ DE BELASCO
Versos de quatro tono. cf Harmonia Mundi HM 759.
LORD STANLEY'S ADDRESS AND MELBA'S FAREWELL SPEECH. cf HANDEL: Il pensieroso: Sweet bird.
LORTZING, Gustav
Undine: Vater, Mutter. cf Telefunken AJ 6-42232.
Zar und Zimmermann: Lebe wohl, mein flandrisch Madchen. cf Telefunken AJ 6-42232.
LOTTI, Antonio
Pur dicesti. cf HMV RLS 719.
LOVER, Samuel
The low-back'd car. cf Philips 6599 227.
LOWRY
Shall we know each other there. cf New World Records NW 220.
LUBECK, Vicenz
Prelude and fugue, F major. cf Pelca PSR 40571.
LUKACIC, Ivan
Motet: Ex ore infantum. cf JELIC: Motets (Musical Heritage Society MHS 3497).
LUKE, Ray
1711 Concerto, bassoon. WELCHER: Concerto da camera, bassoon. Leonard Sharrow, bsn; Crystal Chamber Orchestra; Ernest Gold. Crystal S 852.
++NR 11-77 p10
LULLY, Jean
1712 Alceste. Felicity Palmer, Anne-Marie Rodde, Sonia Nigoghossian, s; Bruce Brewer, John Elwes, t; Max von Egmond, Pierre-Yves Le Maigat, bs; Raphael Passaquet Vocal Ensemble; Le Grande Ecurie et la Chambre du Roy; Jean-Claude Malgoire. CBS 79301 (3). (also Columbia M3 34580)
+-Gr 1-76 p1235 /NYT 7-10-77 pD13
+-HF 10-77 p95 ++RR 11-75 p32
+HFN 12-75 p156 +-SR 9-3-77 p42
+-MJ 11-77 p28 +St 12-77 p142
+-MT 2-76 p142 +STL 1-11-76 p36
1713 Alceste: Prologe, Overture; Rondeau; Act 1, conclusion; La guerra; Air d'Alceste; Duo Alceste and Admete; Scene funebre "Alceste est morte"; Choeur funebre; Scene de Charon; La fete infernale; Choeur des suivants de Pluton; Les demons. Felicity Palmer, Renee Auphan, Sonia Nighogossian, Anne-Marie Rodde, s; Bruce Brewer, John Elwes, Francois Loup, t; Pierre-Yves La Maigat, bar; Max van Egmond, Marc Vento, bs; Maitrise National d'Enfants, Raphael Passaquet Vocal Ensemble; La Grande Ecurie et la Chambre du Roy; Jean-Claude Malgoire. CBS 76551. (Reissue from 79301).
+-Gr 5-77 p1724 +-RR 5-77 p34
+-HFN 5-77 p137
Amadis de Gaule: Amour que veux-tu. cf Rubini GV 57.
Ballets du Roy. cf Discopaedia MB 1015.
Gigas. cf Westminster WGM 8309.
La grotte de Versailles: Overture. cf Telefunken AW 6-42155.
Une noce de village: Derniere entree. cf DG 2723 051.
Les sourdines d'Armide. cf Telefunken AW 6-42155.
Suite, C major. cf Supraphon 111 1867.
LUNA
El nino Judio: De Espana vengo. cf London OS 26435.
LUNDBORG, Erik
Music forever, no. 2, excerpts. cf LIEBERSON: Concerto, 4 groups of instruments.

Passacaglia. cf LIEBERSON: Concerto, 4 groups of instruments.
LUPO, Thomas
Dance. cf RCA RL 2-5110.
LUTE AND GUITAR RECITAL. cf Westminster WG 1012.
LUTE MUSIC OF THE DUTCH RENAISSANCE. cf DG 2533 302.
LUTOSLAWSKI, Witold
Five songs on poems of Kazimiera Illakowicz. cf DESSAU: Tierverse.
1714 Quartet, strings. PETROVICS: Quartet, strings. New Budapest Quartet. Hungaroton SLPX 11847.
+FF 9-77 p31
Quartet, strings. cf CAGE: Quartet, strings, 4 parts.
Variations on a theme by Paganini. cf DEBUSSY: Nocturnes: Fetes.
LUYTENS, Elizabeth
Bagatelles, op. 49 (5). cf BENNETT: Telegram.
Intermezzi (5). cf BENNETT: Telegram.
Piano e forte, op. 43. cf BENNETT: Telegram.
Plenum I, op. 87. cf BENNETT: Telegram.
Plenum IV, op. 100. cf Vista VPS 1039.
LVOV, Alexis
To thy heavenly banquet (Communian anthem). cf Vanguard SVD 71212.
LYNE, Peter
Epigrams (3). cf BIS LP 57.
LYNN, Robert
Vino. cf Orion ORS 76212.
LYON
Friendship. cf Vox SVBX 5350.
LYRA
Wanderschaft. cf Telefunken DT 6-48085.
McBETH
Brass. cf Crystal S 206.
McCABE, John
1715 Fantasy on a theme by Liszt. Studies, nos. 1-4. Variations. John McCabe, pno. RCA RL 2-5076.
+Gr 6-77 p75 +MT 11-77 p923
+HFN 8-77 p85 +RR 6-77 p82
Notturni ed alba, soprano and orchestra. cf Symphony, no. 2.
Partita, solo violoncello. cf BERKELEY: Duo, violoncello and piano.
Studies, nos. 1-4. cf Fantasy on a theme by Liszt.
1716 Symphony, no. 2. Notturni ed alba, soprano and orchestra. Jill Gomez, s; Birmingham City Symphony Orchestra; Louis Fremaux. HMV ASD 2904.
+-Gr 11-73 p928 -MJ 10-77 p29
Variations. cf Fantasy on a theme by Liszt.
McCALL, J. P.
Songs: Kelly, the boy from Killane; Boolavogue. cf Philips 6599 227.
MacCUNN, Hamish
The land of the mountain and the flood. cf GERMAN: Welsh rhapsody.
The land of the mountain and the flood: Overture. cf Grosvenor GRS 1043.
MacDOWELL, Edward
1717 Fantasy pieces, op. 17 (2). Woodland sketches, op. 51. NEVIN: Un giorno in Venezia. Songs. Water scenes, op. 13. Paulina Drake, pno. Genesis GS 1067.
+ARG 5-77 p26 +St 6-77 p136
+-HF 6-77 p93

Sonata, piano, no. 1, op. 45, G minor. cf GRIFFES: The white peacock.

1718 Sonata, piano, no. 2, op. 50, G minor. Woodland sketches, op. 51. Clive Lythgoe, pno. Philips 9500 095.
++Gr 11-76 p838 +NR 7-76 p12
+-HF 11-76 p118 +NYT 7-4-76 pD1
+HFN 11-76 p161 ++RR 11-76 p91
+-MJ 12-76 p28 +SFC 12-26-76 p34
+MT 5-77 p401 +St 10-76 p126

Sonata, piano, no. 4, op. 59. cf BARBER: Excursions.

Woodland sketches, op. 51. cf Fantasy pieces, op. 17.

Woodland sketches, op. 51. cf Sonata, piano, no. 2, op. 50, G minor.

Woodland sketches, op. 51, no. 1: To a wild rose. cf Amberlee ALM 602.

Woodland sketches, op. 51, no. 1: To a wild rose. cf International Piano Library IPL 102.

Woodland sketches, op. 51, no. 1: To a wild rose; no. 6: To a water lily. cf Decca SPA 519.

Woodland sketches, op. 51, nos. 3, 7-10. cf Vox SVBX 5303.

MacMILLAN, Ernest

A Saint Malo. cf Citadel CT 6011.

McPHEE, Colin

1719 Concerto, piano with wind octet. WILSON: Concert piece, violin and piano. Music for solo flute. Harvey Sollberger, flt; Rolf Schulte, vln; Ursula Oppens, Grant Johannesen, pno; Wind Octet; Carlos Surinach. CRI SD 315.
+-MJ 7-74 p48 +SFC 7-17-77 p42
+NR 2-74 p4

MacPHERSON, Charles

A little organ book: Andante, G major. cf Guild GRS 7008.

MACHAUT, Guillaume

Amours me fait desirer. cf HMV SLS 863.

Christe qui lux es. cf DG 2710 019.

Dame se vous m'estes. cf HMV SLS 863.

De bon espoir: Puis que la douce. cf HMV SLS 863.

De toutes flours. cf HMV SLS 863.

Douce dame jolie. cf HMV SLS 863.

Hareu, hareu: Helas, ou sera pris confors. cf HMV SLS 863.

Hoquetus David. cf DG 2710 019.

Lasse comment oublieray. cf DG 2710 019.

Ma fin est mon commencement. cf HMV SLS 863.

Mes esperis se combat. cf HMV SLS 863.

Moult suis de bonne heure nee. cf CBS 76534.

Phyton, le merveilleus serpent. cf HMV SLS 863.

Quant j'ay l'espart. cf HMV SLS 863.

Quant je sui mis. cf HMV SLS 863.

Quant Theseus: Ne quier veoir. cf HMV SLS 863.

Qui es promesses. cf DG 2710 019.

Se je souspir. cf HMV SLS 863.

Trop plus est belle: Biaute paree; Je ne sui. cf HMV SLS 863.

MADERNA, Bruno

Aulodia per Lothar. cf CBS 61453.

1720 Il giardino religioso. SCHULLER: Tre invenzione. Contours. Instrumental ensemble, Contemporary Chamber Ensemble; Gunther Schuller, Arthur Weisberg. Odyssey Y 34141.
+-NR 1-77 p9 ++St 2-77 p124

Y despues. cf DG 2530 802.

MADETOJA, Leevi

Comedy overture. cf Symphony, no. 3.

1721 Pohjalaisia. Maija Lokka, Maiju Kuusoja, Eero Erkkila, Hannu Heikkila, Jorma Hymninen, Kalevi Koskinen, Kauko Vayrynen, Raita Karpo; Finnish National Opera Orchestra and Chorus; Jorma Panula. Finnlevy SFX 22/24.
+RR 9-77 p40

1722 Symphony, no. 3. Comedy overture. Helsinki Philharmonic Orchestra; Jorma Panula. Finnlevy SFX 20.
/FF 11-77 p35 +-RR 7-76 p54

MADRIGUERA, Paquita

Humorada. cf Lyrichord LLST 7299.

MAHLER, Gustav

1923 Kindertotenlieder. Songs of the wayfarer. Dietrich Fischer-Dieskau, bar; PhO, BPhO; Wilhelm Furtwangler, Rudolf Kempe. Seraphim 60272.
+NR 10-77 p11 +St 5-77 p114

1724 Des Knaben Wunderhorn. Jessye Norman, s; John Shirley-Quirk, bar; COA; Bernard Haitink. Philips 9500 316 Tape (c) 7300 572.
++ARG 11-77 p21 ++HFN 11-77 p172
+Gr 11-77 p874 +NR 12-77 p10
+HF 12-77 p94 +RR 11-77 p106

Des Knaben Wunderhorn: Das irdische Leben; Rheinlegendchen. cf Prelude PRS 2505.

1725 Das Lied von der Erde. James King, Dietrich Fischer-Dieskau, bar; VPO; Leonard Bernstein. Decca JB 13. (Reissue)
+HFN 12-77 p185 +-RR 12-77 p79

1726 Das Lied von der Erde. Ruckert Lieder (5). Christa Ludwig, ms; Rene Kollo, t; BPhO; Herbert von Karajan. DG 2707 082 (2) Tape (c) 3581 015. (also Tape (c) 3370 007)
++Gr 12-75 p1089 +NR 2-76 p13
+-Gr 6-77 p105 tape +-NYT 4-4-76 pREC 1
+-HF 4-76 p112 +RR 12-75 p88
+-HFN 1-76 p111 +-RR 10-77 p91 tape
++MJ 3-76 p25 ++SFC 1-4-76 p27
+MT 5-77 p401 ++St 5-76 p118

1727 Das Lied von der Erde. Yvone Minton, ms; Rene Kollo, t; CSO; Georg Solti. London OS 26292. (also Decca SET 555 Tape (c) KCET 555)
++Gr 11-72 p939 -NYT 3-11-73 pD30
+-HF 4-73 p84 ++RR 11-72 p104
++HFN 12-76 p155 tape +-RR 2-77 p97 tape
+-MJ 4-73 p8 ++SFC 1-21-73 p41
+NR 6-73 p11 -SR 2-73 p54

1728 Das Lied von der Erde. Janet Baker, ms; James King, t; COA; Bernard Haitink. Philips 6500 831 Tape (c) 7300 362.
+ARG 4-77 p21 +MJ 1-77 p26
+-Gr 10-76 p634 +MT 5-77 p401
++Gr 6-77 p105 tape ++NR 1-77 p13
++HF 1-77 p118 +-RR 10-76 p97
+HFN 10-76 p173

Lieder eines fahrenden Gesellen. cf Symphony, no. 5, C sharp minor.

Lieder eines fahrenden Gesellen. cf BRUCKNER: Symphony, no. 7, E major.

Lieder eines fahrenden Gesellen (Songs of the wayfarer). cf Kindertotenlieder.

Ruckert Lieder (3). cf Symphony, no. 4, G major.
Ruckert songs. cf ELGAR: Sea pictures, op. 37.
Ruckert Lieder (5). cf Das Lied von der Erde.

1729 Symphonies, nos. 1-9. Symphony, no. 10, F sharp minor: Adagio. Sheila Armstrong, Reri Grist, Erna Spoorenberg, Gwyneth Jones, Gwenyth Annear, s; Janet Baker, Martha Lipton, ms; Anna Reynolds, Norma Proctor, con; John Mitchinson, t; Vladimir Ruzdjak, bar; Donald McIntyre, bs; Edinburg Festival Chorus, Schola Cantorum Women's Chorus, Church of the Transfiguration Boys Choir, Leeds Festival Chorus, Orpington Junior Singers, Highgate Schools Boys Choir, Finchley Children's Music Group; NYP, LSO; Leonard Bernstein. Columbia GMS 765 (15). (also CBS GM 15. Reissues from 72649, 78249, SBRG 72065/6, 77507, 72182/3, 77215, 72427/8, 72491/2, 72691/1, 76475)
+-Gr 1-77 p1142 +-HFN 1-77 p119
+HF 4-77 p66 +RR 1-77 p55

1730 Symphony, no. 1, D major. LPO; Gaetano Delogu. Classics for Pleasure CFP 40264.
+Gr 11-77 p838 +-RR 8-77 p57
+-HFN 8-77 p85

1731 Symphony, no. 1, D major. RPO; Carlos Paita. Decca PFS 4402.
-HFN 10-77 p157

1732 Symphony, no. 1, D major. Dresden Staatskapelle Orchestra; Otmar Suitner. DG Tape (c) 3348 123.
+-Gr 2-77 p1325 -RR 9-76 p92 tape

1733 Symphony, no. 1, D major. Israel Philharmonic Orchestra; Zubin Mehta. London CS 7004 Tape (c) 5-7004. (also Decca SXL 6779 Tape (c) KSXC 6779)
-ARG 7-77 p26 +NR 12-76 p6
+-Gr 9-76 p422 ++RR 9-76 p56
-Gr 11-76 p887 tape +-RR 1-77 p90 tape
/HF 12-76 p98 -SFC 11-14-76 p30
+-HFN 9-76 p125 ++St 4-77 p122
+-HFN 11-76 p175 tape

1734 Symphony, no. 1, D major. LSO; James Levine. RCA ARL 1-0894 Tape (c) ARK 1-0894 (ct) ARS 1-0894.
+Gr 5-75 p1967 ++RR 5-75 p33
-HF 6-75 p98 +-RR 5-77 p92 tape
++HFN 5-75 p134 ++SFC 5-11-75 p23
+-HFN 5-77 p139 tape +-SR 5-3-75 p33
-NR 5-75 p5 +St 5-75 p96
+NYT 4-4-76 pREC 1 +-ST 10-76 p78

1735 Symphony, no. 2, C minor. Marilyn Horne, ms; Carol Neblett, s; CSO and Chorus; Claudio Abbado. DG 2707 094 (2) Tape (c) 3370 015.
-FF 9-77 p32 +NR 11-77 p4
++Gr 6-77 p55 ++RR 6-77 p60
+Gr 6-77 p105 tape +SFC 9-18-77 p42
+-HF 11-77 p118 ++St 11-77 p142
+HFN 6-77 p129

1736 Symphony, no. 2, C minor. Edith Mathis, s; Norma Procter, con; Bavarian Radio Orchestra and Chorus; Rafael Kubelik. DG 2726 026 (2). (Reissue from 139332/3)
+-Gr 3-77 p1396 +-RR 3-77 p50
+HFN 8-77 p93

1737 Symphony, no. 3, D minor. Marilyn Horne, ms; Glen Ellyn Children's Chorus; CSO and Chorus; James Levine. RCA ARL 2-1757 (2) Tape

(c) CRK 2-1757 (ct) CRS 2-1757.
+ARG 5-77 p25 ++MJ 1-77 p26
-FF 9-77 p33 +-NR 1-77 p5
++Gr 3-77 p1396 +RR 2-77 p49
+Gr 6-77 p105 tape +RR 6-77 p96 tape
++HF 3-77 p101 ++SFC 11-14-76 p30
++HF 5-77 p101 tape ++St 3-77 p131
+HFN 5-77 p125

1738 Symphony, no. 3, D minor. Norma Procter, con; Wandsworth Boys School Choir; Ambrosian Singers; LSO; Jascha Horenstein. Unicorn Tape (c) ZCUN 302/3. (also Nonesuch/Advent 73023 Tape (c) E 1009)
+ARG 5-71 p577 +NYT 4-18-71 pD26
+Gr 7-72 p247 tape +-RR 6-77 p96 tape
+Gr 12-70 p994 +SFC 4-25-71 p32
+HF 12-74 p146 tape ++St 9-71 p88
+HF 7-71 p80 +St 10-76 p78
++NR 7-71 p3

Symphony, no. 3, D minor: 1st movement. cf Pandora PAN 2001.

1739 Symphony, no. 4, G major. Ruckert Lieder (3). SCHUBERT: Symphony, no. 8, D 759, B minor. Elisabeth Schwarzkopf, s; VPO; Bruno Walter. Bruno Walter Society BWS 705 (2).
+-NR 3-77 p2

1740 Symphony, no. 4, G major. Margaret Ritchie, s; COA; Eduard van Beinum. London R 23211.
+NYT 1-16-77 pD13

1741 Symphony, no. 4, G major. Judith Blegen, s; CSO; James Levine. RCA ARL 1-0895 Tape (c) ARK 1-0895 (ct) ARS 1-0895. (also RK 11733 Tape (c) RK 1173)
++Gr 10-75 p612 +NYT 4-4-76 pREC 1
++Gr 11-76 p887 tape ++RR 11-75 p43
++HF 6-75 p98 +-RR 1-77 p90 tape
-HFN 11-75 p155 ++St 8-75 p100
+-HFN 11-76 p175 tape -St 10-76 p79
+NR 5-75 p5 +ST 2-76 p739

1742 Symphony, no. 4, G major. Netania Davrath, s; Utah Symphony Orchestra; Maurice Abravanel. Vanguard C 10042 Tape (c) GRT 8193 10042E (r) D 10042.
+HF 5-76 p114 tape +-HF 11-77 p138 tape

1743 Symphony, no. 4, G major. Galina Pisarenko, s; MRSO; Kiril Kondrashin. Westminster WGS 8328.
++ARG 4-77 p22 +SFC 10-23-77 p45
+NR 5-77 p6

1744 Symphony, no. 5, C sharp minor. Symphony, no. 10, F sharp minor: Adagio. LAPO; Zubin Mehta. Decca SXL 6806/7 (2). (also London CSA 2248)
+-Gr 11-77 p838 -RR 11-77 p61
+HFN 11-77 p173 ++SFC 10-17-77 p47

1745 Symphony, no. 5, C sharp minor. Lieder eines fahrenden Gesellen. Dietrich Fischer-Dieskau, bar; Bavarian Radio Orchestra; Rafael Kubelik. DG 2726 064 (2). (Reissue from 2720 033)
+-Gr 3-77 p1396 +RR 3-77 p50

1746 Symphony, no. 5, C sharp minor. PETTERSSON: Barefoot songs (8). Erik Saeden, bs-bar; Stockholm Philharmonic Orchestra; Antal Dorati. HNH 4003/4 (2).
+NR 11-77 p5 +St 10-77 p140

1747 Symphony, no. 9, D major. CSO; Carlo Maria Giulini. DG 2707 097 (2) Tape (c) 3370 018.
++ARG 7-77 p24
+Gr 4-77 p1550
+Ha 9-77 p108
+HF 7-77 p101
++HFN 4-77 p138
++MJ 9-77 p35
++NR 7-77 p2
+NYT 12-4-77 pD18
+-RR 4-77 p53
++SFC 5-15-77 p50
++St 8-77 p116
Symphony, no. 10, F sharp minor: Adagio. cf Symphony, no. 5, C sharp minor.
MAINERIO, Giorgio
Dances (2). cf HMV SQ ASD 3393.
Primo libro de balli. cf Telefunken AW 6-42033.
Schiarazula marazula. cf DG 2723 051.
Tedesca: Saltarello. cf DG 2723 051.
Ungaresca: Saltarello. cf DG 2723 051.
MALATS, Joaquin
Serenata. cf International Piano Library IPL 109.
Spanish serenade. cf Classics for Pleasure CFP 40012.
MALOTTE, Albert
The Lord's prayer. cf L'Oiseau-Lyre DSLO 20.
THE MAN WITH THE GOLDEN FLUTE, James Galway. cf RCA RK 2-5030.
MANDEL
The shadow of your smile. cf EMI TWOX 1058.
MANEN, Juan
Concerto da camera. cf Clear TLC 2586.
MANERI, Raffaele
Salve regina. cf L'Oiseau-Lyre SOL 343.
MANFREDINI, Francesco
Concerto grosso, op. 3, no. 12, C major. cf CORELLI: Concerto grosso, op. 6, no. 8, G minor.
Concerto grosso, op. 3, no. 12, C major. cf LOCATELLI: Concerto grosso, op. 8, no. 1, F minor.
MANFREDINI, Vicenzo
Concerto grosso, op. 3, no. 9, D major. cf Supraphon 110 1890.
MANINNO
A little music for 3 friends. cf GIULIANI: Divertimento.
Mini quintet, op. 74. cf GIULIANI: Divertimento.
MANZIARLY, Marcelle de
Dialogue. cf Laurel Protone LP 13.
MARAIS, Marin
La gamme. cf Vox SVBX 5142.
Les matelots: Air. cf Telefunken AW 6-42155.
Pieces de violes, Bk 2: Suite, B minor. cf Vox SVBX 5142.
1748 Suites, recorder, F major, E minor. Hotteterre Quartet. Telefunken AW 6-42035.
+Gr 11-77 p859
+HFN 11-77 p173
+RR 11-77 p80
1749 Suite, viola da gamba and harpsichord, E minor. SAINTE-COLOMBE: Concerts a deux violes esgales (3). Catharina Meints, James Caldwell, vla da gamba; James Weaver, hpd. Cambridge CRS 2201.
+ARG 4-77 p23
++NR 3-75 p6
+NR 10-76 p7
+SFC 11-14-76 p30
MARCABRU (12th century)
Pax in nomine Domini. cf Argo D40D3.
MARCELLO
Adieu a Venise. cf RCA FRL 1-3504.

Final. cf RCA FRL 1-3504.
Sonata, A minor. cf Arion ARN 90416.

MARCELLO, Alessandro

1750 Concerto, oboe and strings. SAMMARTINI: Concerto, soprano recorder and strings. SCARLATTI. A.: Concerto, alto recorder and 2 violins. STRADELLA: Sonata, trumpet and strings. Hermann Sauter, tpt; Gunther Holler, rec; Helmut Hucke, ob; South West German Chamber Orchestra; Paul Angerer. Turnabout (Q) QTV 34573.
+-Audio 1-77 p84 ++NR 4-75 p5

MARCHAND, Louis

Basse de cromorne. cf DAQUIN: Noels.
Basse de trompette. cf Vista VPS 1030.
Dialogue (2). cf DAQUIN: Noels.
Duo. cf DAQUIN: Noels.
Fugue. cf DAQUIN: Noels.
Grand dialogue. cf CLERAMBAULT: Suite du deuxieme ton.
Grand jeu. cf DAQUIN: Noels.
Plein jeu. cf DAQUIN: Noels.
Quatuor. cf DAQUIN: Noels.
Tierce en taille. cf DAQUIN: Noels.

MARENZIO, Luca

O voi che sospirate; Occhi lucenti. cf L'Oiseau-Lyre 12BB 203/6.

MARGOLA

Partita, flute and oboe. cf GIULIANI: Divertimento.

MAROS, Rudolf

Bagatelles. cf Hungaroton SLPX 11808

1751 Consort. Gemma. Monumentum. Musica da camera per II. Notices. Trio. Budapest Symphony Orchestra, Budapest Chamber Ensemble, Hungarian Wind Quintet; Eszler Perenyi, vln; Zoltan Toth, vla; Eva Maros, hp; Gyorgy Lehel, Andras Mihaly. Hungaroton SLPX 11775.
+ARG 9-77 p40 +NR 11-77 p6

Gemma. cf Consort.
Momentum. cf Consort.
Musica da camera per II. cf Consort.
Notices. cf Consort.
Trio. cf Consort.

MARQUINA, Pascual

Espani cani. cf Decca SB 713.

MARSHALL

I hear you calling me. cf Philips 9500 218.

MARSHALL, Philip

Reveille pavan. cf Vista VPS 1037.

MARSON, George

The nymphs and shepherds. cf Argo ZK 25.
The nymphs and shepherds danced. cf DG 2533 347.

MARTEAU, Henri

Cakewalk. cf Discopaedia MOB 1020.
Valse fantastic. cf Discopaedia MOB 1020.

MARTIN, Frank

Ballade, flute, piano and string orchestra. cf DUBOIS: Concerto, flute.
Christmas songs (3). cf BIS LP 37.
Passacaille. cf Abbey LPB 719.

1752 Preludes, piano (8). Trio on Irish folksongs. Trio, violin, viola and violoncello. Werner Genuit, pno; New String Trio. BASF 2521638-1.
/FF 9-77 p34

1753 Quintet, piano. MARTINU: Quintet, piano, no. 1. Zurich Piano Quintet. Jecklin 159.
+FF 9-77 p34
Sonata da chiese, viola d'amore and strings. cf BLOCH: Suite hebraique.
Trio, violin, viola and violoncello. cf Preludes, piano (8).
Trio on Irish folksongs. cf Preludes, piano.

MARTINI, Giovanni
Aria variata, thema and 4 variationen. cf Pelca PSR 40599.

MARTINI IL TEDESCO, Johann
Songs: Plaisir d'amour. cf Philips 6747 327.
Songs: Plaisir d'amour. cf Philips 9500 218.

MARTINO, Donald
1754 A set for clarinet. REYNOLDS: Caprices (4). TALMA: Duologues (3). WEBSTER: Pieces, solo clarinet (5). Michael Webster, clt; Beveridge Webster, pno. CRI SD 374.
+FF 11-77 p59 +NR 11-77 p8
++HF 12-77 p108

MARTINU, Bohuslav
1755 Concerto, piano, no. 1. SHOSTAKOVICH: Preludes and fugues, op. 87 (2). Dagmar Baloghova, pno; Czech Radio Orchestra; Josef Hrncir. Panton 110 533.
++FF 11-77 p35
Preludes (8). cf Supraphon 111 1721/2.
Quintet, piano, no. 1. cf MARTIN: Quintet, piano.
1756 Sinfonietta la jolla. Toccata e due canzoni. PCO; Zdenek Hnat, pno. Supraphon 110 1619.
+ARG 12-76 p36 +NR 11-76 p2
++Gr 11-76 p797 +RR 9-76 p57
+HF 10-77 p110
Sonata, flute. cf BIS LP 50.
Symphony, no. 5. cf HARRIS: Symphony, no. 5.
Toccata a due canzoni. cf Sinfonietta la jolla.
1757 Trio, flute, violoncello and piano. ROREM: Book of the hours. Ingrid Dingfelder, flt; Martine Geliot, hp; Jerome Carrington, vlc; Anita Gordon, pno. CRI SD 362.
+-ARG 7-77 p30 +NR 10-77 p6
+MJ 10-77 p29 ++St 11-77 p142

MASCAGNI, Pietro
L'Amico Fritz: Ed anche...Oh amore. cf HMV ASD 3302.
L'Amico Fritz: Don pochi. cf Angel S 37446.
1758 Cavalleria rusticana. LEONCAVALLO: I Pagliacci. Maria Gracia Allegri, Joan Carlyle, s; Fiorenza Cossotto, Adriane Martino, ms; Carlo Bergonzi, Ugo Benelli, t; Giangiacomo Guelfi, Giuseppe Taddei, Rolando Panerai, bar; La Scala Orchestra and Chorus; Herbert von Karajan. DG 2709 020 Tape (c) 3371 011.
+Gr 11-74 p974 +RR 8-74 p84 tape
+NYT 10-9-77 pD21
1759 Cavalleria rusticana. Renata Tebaldi, s; Ettore Bastianini, bar; Jussi Bjorling, bs; Florence May Festival Orchestra and Chorus; Alberto Erede. London OSA 12101 (2).
++FF 11-77 p78
Cavalleria rusticana: Addio alla madre. cf RCA TVM 1-7201.
1760 Cavalleria rusticana: Duet of Santuzza and Turiddu, Turridu's aria. PUCCINI: La boheme: Rodolfo's aria. Girl of the golden west: Johnson's aria. Tosca: Cavaradossi's aria, Acts 1, 3. VERDI: La traviata: Alfred's aria. Rigoletto: Duke's aria.

Il trovatore: Duet of Azucena and Manrico. Janis Zabers, t; Orchestral accompaniments. Kaibala 40D02.
+-NR 1-77 p11
Cavalleria rusticana: Gli aranci olezzano; Voi lo sapete, o mamma; Oh, Il Signore vi manda; Intermezzo; Mamma, quel vino e generoso. cf LEONCAVALLO: I Pagliacci: Si puo; Signore, Signori.
Cavalleria rusticana: Ineggiamo, Il Signor non e morto. cf HMV 5104.
Cavalleria rusticana: Intermezzo. cf RCA CRL 1-2064.
Cavalleria rusticana: Intermezzo. cf RCA PL 2-5046.
Cavalleria rusticana: Intermezzo. cf Rediffusion 15-57.
Cavalleria rusticana: Mamma, quel vino e generoso. cf Decca ECS 811.
Cavalleria rusticana: O Lola, bianca come fior di spino; Mamma que vino e generoso. cf HMV RLS 715.
Cavalleria rusticana: Regina coeli...Inneggiamo. cf Seraphim S 60275.
Cavalleria rusticana: Siciliana; Addio alla madre. cf Rubini GV 43.
Cavalleria rusticana: Siciliana; Brindisi. cf Rubini GV 29.
Cavalleria rusticana: Voi lo sapete. cf BASF 22-22645-3.
Cavalleria rusticana: Voi lo sapete. cf Club 99-100.
Cavalleria rusticana: Voi lo sapete. cf Club 99-109.
Cavalleria rusticana: Voi lo sapete. cf Decca SXLR 6825.
Cavalleria rusticana: Voi lo sapete. cf HMV SLS 5057.
Cavalleria rusticana: Voi lo sapete. cf London OS 26497.
Cavalleria rusticana: Voi lo sapete. cf Rubini GV 57.
Iris: Apri la tua finestra. cf RCA TVM 1-7203.
Songs: Ave Maria. cf Argo ZFB 95/6.
Songs: M'ama, non m'ama; La luna; Serenata. cf Columbia M 34501.

MASCHERONI, Eduardo
Eternamente. cf Odyssey Y 33793.
Lorenza: Susanna al bagno. cf Club 99-100.

MASEK, Vaclav
Concerto, 2 harpsichords and wind octet, D major. cf DRUZECKY: Parthia, no. 3, D sharp major.

MASON, Daniel Gregory
Prelude and fugue, op. 20. cf BEACH: Concerto, pianoforte, op. 45.

MASON, John
Songs: Quales sumus; Vae nobis miserere. cf APPLEBY: Magnificat.

MASON, William
A pastoral novellette. cf New World Records NW 257.
Silver spring, op. 6. cf New World Records NW 257.
THE MASS AT DOWNHAM MARKET. cf Coimbra CC 1.

MASSE, Victor
Galathee: Grand air. cf Club 99-107
La mule de Pedro: Ma mule qui chaque semaine. cf Club 99-107.

MASSENET, Jules
1761 Ariane: Lamento d'Ariane. Le Cid: Ballet music. MEYERBEER: Les patineurs: Ballet suite (arr. Lambert). NPhO; Richard Bonynge. Decca SXL 6812. (also London CS 7032 Tape CS 5-7032)
+ARG 3-77 p81
+Audio 6-77 p134
+Gr 11-76 p797
+HF 4-77 p106
+HF 9-77 p119 tape
+HFN 11-76 p161
++NR 3-77 p3
+RR 11-76 p69
++SFC 4-10-77 p30
Cendrillon: Enfin, je suis ici. cf CBS 76522.

1762 Cendrillon: Marche des princesses. Scenes dramatiques. Scenes alsaciennes. National Philharmonic Orchestra; Richard Bonynge. Decca SXL 6827. (also London CS 7048)
+Gr 12-77 p1087 +RR 12-77 p48
++HFN 12-77 p173 +SFC 9-4-77 p38
+NR 9-77 p2

1763 Le Cid. Grace Bumbry, con; Eleanor Bergquist; Placido Domingo, t; Paul Plishka, bs; Arnold Voketaitis; Bryne Camp Chorale; New York Opera Orchestra; Eve Queler. Columbia M3 34211 (3). (also CBS 79300)
+ARG 2-77 p30 +OC 12-77 p49
+Gr 3-77 p1449 +ON 12-11-76 p48
+-HF 3-77 p102 +RR 3-77 p34
+-HFN 5-77 p125 ++SFC 11-14-76 p30
++NR 1-77 p10 +St 3-77 p132

1764 Le Cid. MEYERBEER: Les patineurs. National Philharmonic Orchestra; Richard Bonynge. Decca SXL 6812 (c) KSXC 6812. (also London CS 7032)
+HFN 1-77 p123 tape +MJ 2-77 p30

Le Cid. cf GOUNOD: Faust.

Le Cid: Air de Don Diegue, Il a fait noblement. cf Club 99-107.

1765 Le Cid: Ballet music. Scenes pittoresques. La vierge: The last sleep of the virgin. Birmingham City Symphony Orchestra; Elisabeth Robinson, cor anglais; Anthony Moroney, flt; Hilary Robinson, vlc; Louis Fremaux. HMV SQ ESD 7040 Tape (c) TC ESD 7040. (Reissue from Columbia TWO 350)
++Gr 9-77 p512 +HFN 10-77 p169 tape
++Gr 10-77 p698 tape +RR 9-77 p63
++HFN 10-77 p165

Le Cid: Ballet music. cf Ariane: Lamento d'Ariane.

Le Cid: Ballet music. cf KHACHATURIAN: Gayaneh: Suite.

Le Cid: Navarraise. cf Seraphim S 60277.

Le Cid: Overture, Ballet music. cf GOUNOD: Faust: Ballet music.

Le Cid: Pleurez mes yeux. cf HMV RLS 719.

Le Cid: Pleurez mes yeux. cf HMV SLS 5057.

Don Cesar de Bazan: Sevillana (2). cf HMV RLS 719.

Elegie. cf HMV RLS 716.

1766 Esclarmonde. Joan Sutherland, s; Huguette Tourangeau, ms; Giacomo Aragall, Ryland Davies, Ian Caley, Graham Clark, t; Louis Quilico, bar; Robert Lloyd, Clifford Grant, bs; Finchley Children's Music Group, John Alldis Choir; NPhO; Richard Bonynge. Decca SET 612/4 (3). (also London 13118 Tape (c) 5-13118)
+-ARG 3-77 p24 +NR 2-77 p10
++Gr 11-76 p867 +ON 12-11-76 p48
+-HF 1-77 p120 +OR 6/7-77 p31
+HFN 12-76 p144 +RR 11-76 p40
-MJ 2-77 p30 ++SFC 11-28-76 p45
+MT 6-77 p487 +-St 1-77 p82

Le mage: Ah, parais. cf Rubini GV 38.

Manon: Adieu, notre petite table. cf HMV SLS 5104.

Manon: Ah dispar vision. cf Rubini GV 29.

Manon: Ancor son io tutt attonita; Addio nostro piccolo desco. cf Club 99-100.

Manon: Les grands mots que voila...Epouse quelque brave fille. cf Philips 6580 174.

Manon: Il le faut...Adieu notre petite table. cf Seraphim IB 6105.

Manon: Le reve. cf Club 99-105.

Phedre: Overture. cf HEROLD: Zampa: Overture.
Roi de Lahore: Viaggia o bella. cf Club 99-106.
Sappho: Pendant un an je fus ta femme. cf Rubini GV 57.
Scenes alsaciennes. cf Cendrillon: Marche des princesses.
Scenes dramatiques. cf Cendrillon: Marche des princesses.
Scenes pittoresques. cf Le Cid: Ballet music.

1767 Songs: L'ame des fleurs; Les amoureuses sont des folles; Ce que disent les cloches; Elle s'en est allee; L'eventail; Je t'aime; La melodie des baisers; Nuit d'Espagne; On dit; Passionnement; Le petit Jesus; Pensee d'automne; Pitchounette; Printemps dernier; Roses d'Octobre; Le sais-tu; Serenade d'automne; Souhait; Souvenance; Les yeux clos. Huguette Tourangeau, ms; Richard Bonynge, pno; Reginald Kilbey, vlc. Decca SXL 6765.
+-Gr 7-77 p208 +RR 7-77 p87
+HFN 7-77 p115

Songs: Chant provencal; Elegie; Nuit d'Espagne; Serenade d'automne; Stances; Un adieu; Vous aimerez demain. cf FAURE: Songs (Saga 5419).

Songs: Crepuscule; Premiere danse; Le poete et le fantome; Ouvre tes yeux bleus; Le saistu; Serenade. cf BIZET: Songs (Musical Heritage Society MHS 3433).

Songs: Serenade du passant. cf Club 99-107.

1768 Thais. Beverly Sills, Norma Burrowes, Anne-Marie Connors, s; Ann Murray, Patricia Kern, ms; Nicolai Gedda, t; Sherrill Milnes, bar; Richard Van Allan, bs; John Alldis Choir; NPhO; Lorin Maazel. Angel SCLX 3832 (3) Tape (c) 4XS 3832. (also HMV SLS 993)
+-Gr 1-77 p1173 +OC 3-77 p53
+-HF 1-77 p122 +ON 12-11-76 p48
+HFN 5-77 p127 +RR 1-77 p35
+NR 1-77 p12 +St 12-76 p142
-NYT 10-9-77 pD21

1769 Thais. Anna Moffo, s; Jose Carreras, t; Gabriel Bacquier, bs-bar; Justino Diaz, bs; Ambrosian Opera Chorus; NPhO; Julius Rudel. RCA ARD 3-0842 (3) (Q) ARD 3-0842.
+-Gr 5-75 p2013 -NYT 10-9-77 pD21
-HF 5-75 p65 +-OC 9-75 p48
+HFN 5-75 p134 +-ON 3-8-75 p34
-MT 12-75 p1071 +SR 3-22-75 p36
+-NR 3-75 p7 -St 6-75 p98
-NYT 2-16-75 pD19

1770 Thais, abridged. Renee Doria, s; Michel Senechal, t; Robert Massard, bar; Orchestra; Jesus Etcheverry. Westminster 8203.
-NYT 10-9-77 pD21

Thais: Ah je suis seule...Dis-moi que je suis belle. cf BASF 22-22645-3.
Thais: Meditation. cf Discopaedia MB 1012.
Thais: Meditation. cf Rediffusion 15-57.
Thais: Meditation. cf Rediffusion Royale 2015.
Thais: Voila donc la terrible cite. cf Philips 6580 174.
La vierge: The last sleep of the virgin. cf Le Cid: Ballet music.
Werther: Air des larmes. cf Rubini RS 301.
Werther: Air des larmes; Tears. cf Club 99-96.
Werther: Des cris joyeux. cf HMV SLS 5057.
Werther: Letter scene. cf Seraphim IB 6105.
Werther: Pourquoi me reveiller. cf Club 99-105.
Werther: Pourquoi me reveiller. cf Rubini GV 29.

Werther: Va, laisse les couler mes larmes. cf CBS 76522.
MASTERS OF THE BOW, Fritz Kreisler. cf Discopaedia MB 1012.
MASTERS OF THE BOW, Georg Kulenkampff. cf Discopaedia MB 1015.
MASTERS OF THE BOW, Henri Marteau and pupils, Goran Olsson-Follinger, Florizel von Reuter, Ferenc Aranyi. cf Discopaedia MOB 1020.
MASTERS OF THE BOW, Joseph Joachim, Marie Soldat-Roger, Deszo Szigeti. cf Discopaedia MOB 1019.
MASTERS OF THE BOW, Nathan Milstein. cf Discopaedia MB 1014.
MASTERS OF THE BOW, Ossy Renardy. cf Discopaedia MB 1011.
MASTERS OF THE BOW, Paul Kochanski. cf Discopaedia MOB 1018.
MASTERS OF THE BOW, Ruggiero Ricci. cf Discopaedia MB 1016.
MASTERS OF THE BOW, Yehudi Menuhin. cf Discopaedia MB 1013.

MATHIAS, William

1771 Divertimento, flute, oboe and piano, op. 24. Quartet, strings, op. 38. Quintet, winds, op. 22. William Bennett, flt; Anthony Camden, ob; Levon Chilingirian, vln; Martin Jones, Clifford Benson, pno; Gabrieli Quartet, Nash Ensemble. Argo ZRG 771.
+Gr 11-76 p830 +MT 6-77 p487
++HFN 6-76 p83 +RR 6-76 p65

Quartet, strings, op. 38. cf Divertimento, flute, oboe and piano, op. 24.
Quintet, winds, op. 22. cf Divertimento, flute, oboe and piano, op. 24.

1772 This worlde's joie. Janet Price, s; Kenneth Bowen, t; Michael Rippon, bar; Bach Choir, St. George's Chapel Choir; NPhO; David Willcocks. HMV ASD 3301.
++Gr 12-76 p1033 +RR 1-77 p82
+HFN 1-77 p111

MATSUDAIRA, Yori-Aki

Allotropy for pianist. cf CP 3-5.

MATSUDAIRA, Yoritsune

Somaksah. cf CBS 61453.

MATTEO

Bella donna. cf Club 99-105.

MAURER, Ludwig

Pieces, brass quintet (4). cf Argo ZRG 851.

MAXWELL DAVIES, Peter

1773 Dark angels. WERNICK: Songs of remembrance. Jan DeGaetani, ms; Oscar Ghiglia, gtr; Philip West, shawm, English hn, ob. Nonesuch H 71342.
+-FF 11-77 p36 +NR 12-77 p11
++HF 12-77 p86

1774 Hymn to St. Magnus. Psalm, no. 124. Renaissance Scottish dances. Mary Thomas, s; The Fires of London; Peter Maxwell Davies. L'Oiseau-Lyre DSLO 12.
+Gr 1-77 p1167 ++MT 3-77 p214
++HFN 12-76 p141 +RR 12-76 p90
++MM 6-77 p49 ++Te 6-77 p42

Leopardi fragments. cf BENNETT: Calendar.
Psalm, no. 124. cf Hymn to St. Magnus.
Renaissance Scottish dances. cf Hymn to St. Magnus.

1775 Saint Michael sonata. KRENEK: Kleine Blasmusik, op. 70a. Merry marches, op. 44 (3). Louisville Orchestra; Jorge Mester. Louisville LS 756.
*MJ 10-77 p29 +SFC 11-27-77 p66
+NR 10-77 p2 +St 7-77 p132

1776 Songs for a mad king (8). John d'Armand, reciter; University of Massachusetts Group for New Music; Charles Fussell. Opus One 26.
+-ARG 12-77 p27

MEYER, William
Octagon, piano and orchestra. cf BARBER: Symphony, no. 1, op. 9, in one movement.

MAYERL, Billy
Marigold, op. 78. cf Decca SPA 519.

MAYS, Walter
Concerto, alto saxophone and chamber ensemble. cf KOLB: Looking for Claudio.

1777 Invocations to the Svara mandala. WERNICK: A prayer for Jerusalem. Jan DeGaetani, ms; Glen Steele, perc; Wichita State University Percussion Orchestra; J. C. Combs. CRI SD 344.
-MJ 3-77 p46 +NR 11-76 p16
+-NR 8-76 p14

MECHEM
Make a joyful noise unto the Lord (Psalm 110). cf Columbia M 34134.
MEDIEVAL GERMAN PLAINCHANT AND POLYPHONY. cf Nonesuch H 71312.
MEDIEVAL PARIS, music of the city. cf Turnabout TV 37086.

MEDINS, Janis
Caress. cf Kaibala 40D03.
Suite concertante. cf BRUCH: Kol Nidrei, op. 47.

MEDTNER, Nikolai
Concerto, piano, no. 1, op. 33, C minor. cf BALAKIREV: Concerto, piano, no. 1, op. 1, F sharp minor.
Dythyramb, op. 10, no. 2. cf Piano works (CRD CRD 1038/9).
Elegy, op. 59a, no. 2. cf Piano works (CRD CRD 1038/9).
Forgotten melodies, op. 39, nos. 1, 3. cf Piano works (CRD CRD 1038/9).
Hymns in praise of toil, op. 49 (3). cf Piano works (CRD CRD 1038/9).

1778 Piano works: Dythyramb, op. 10, no. 2. Elegy, op. 59a, no. 2. Forgotten melodies, op. 39, nos. 1, 3. Hymns in praise of toil, op. 49 (3). Skazki (Fairy tales), op. 9, no. 3; op. 14, nos. 1, 2; op. 26, no. 2; op. 35, no. 4; D minor (1915). Sonata, piano, op. 25, no. 2, E minor. Sonata triad, op. 11. Hamish Milne, pno. CRD CRD 1038/9.
+HFN 12-77 p173 +RR 11-77 p90

Skazki (Fairy tales), op. 9, no. 3; op. 14, nos. 1, 2; op. 26, no. 2; op. 35, no. 4; D minor (1915). cf Piano works (CRD CRD 1038/9).

1779 Skazki, op. 26 (4). Sonate ballade, op. 27, F sharp major. Sonata orageuse, op. 53, no. 2, F minor. Malcolm Binns, pno. Pearl SHE 535.
+Gr 5-77 p1714 +RR 5-77 p74

Sonata, piano, op. 22, G minor. cf BEETHOVEN: Sonata, piano, no. 3, op. 2, no. 3, C major.
Sonata, piano, op. 22, G minor. cf GLAZUNOV: Sonata, piano, no. 2, op. 75, E minor.
Sonata, piano, op. 25, no. 2, E minor. cf Piano works (CRD CRD 1038/9).
Sonata orageuse, op. 53, no. 2, F minor. cf Skazki, op. 26.
Sonata triad, op. 11. cf Piano works (CRD CRD 1038/9).
Sonate ballade, op. 27, F sharp major. cf Skazki, op. 26.

MEHUL, Etienne
Joseph: O toi le digne appui. cf Club 99-107.
MEINEKE, Christopher
Madam de Neuville's favourite waltz, with variations. cf GRIFFES: The white peacock.
MELARTIN, Erkki Gustav
1780 Der Traurige Garten, op. 52. PALMGREN: Contrasts: Les adieux; Arlequin. May night, op. 27, no. 4. Nocturnal scenes, op. 72: The stars twinkle; Song of the night; Dawn. Preludes, op. 17, nos. 12, 14, 24. Ralf Gothoni, pno. Finnlevy SFX 6.
+-RR 8-77 p71
MELBY, John
Stevens songs. cf GRIPPE: Musique Douze.
MELLII, Pietro
For the Emperor Matthias. cf Saga 5420.
MELLNAS, Arne
Fragments for family flute. cf BIS LP 50.
MELNOTTE, Claude
Angels visits. cf New World Records NW 220.
MENDELSSOHN, Felix
Athalia, op. 74: March of the priests. cf Works, selections (Philips 6833 204).
Calm sea and prosperous voyage, op. 27. cf Overtures (DG 2530 782).
Calm sea and prosperous voyage, op. 27. cf Symphony, no. 3, op. 56, A minor.
Capriccio brillante, op. 22, B minor. cf LISZT: Hungarian rhapsody, no. 2, G 244, C sharp minor.
Chant populaire. cf HMV SQ ASD 3283.
Characteristic pieces, op. 7, no. 4, A major. cf International Piano Library IPL 112.
1781 Concerto, violin, op. 64, E minor. MOZART: Concerto, violin, no. 3, K 216, G major. Leonid Kogan, vln; OSCCP; Constantin Silvestri. Connoisseur Society CS 2111.
+MJ 1-77 p27 +SFC 12-76 p34
++NR 12-76 p7 +St 1-77 p115
1782 Concerto, violin, op. 64, E minor. TCHAIKOVSKY: Concerto, violin, op. 35, D major. Ruggiero Ricci, vln; Netherlands Radio Orchestra; Jean Fournet. Decca PFS 4345 Tape (c) KPFC 4345. (also London SPC 21116 Tape (c) 5-21116)
+-Gr 1-76 p1198 +-RR 12-75 p68
+-HFN 1-76 p111 -RR 4-77 p92
+NR 12-76 p7 +St 1-77 p115
1783 Concerto, violin, op. 64, E minor. TCHAIKOVSKY: Concerto, violin, op. 35, D major. Nathan Milstein, vln; NYP, CSO; Bruno Walter, Frederick Stock. Odyssey Y 34064. (Reissue from ML 4001, ML 4053)
+ARG 7-77 p27 +Audio 9-77 p101
1784 Concerto, violin, op. 64, E minor. Concerto, violin, D minor. Salvatore Accardo, vln; LPO; Charles Dutoit. Philips 9500 154 Tape (c) 7300 522.
+Gr 3-77 p1396 ++NR 6-77 p6
++HFN 3-77 p107 +RR 3-77 p55
++HFN 5-77 p138 tape +RR 6-77 p97 tape
+MJ 4-77 p33 ++SFC 5-13-77 p34
1785 Concerto, violin, op. 64, E minor. TCHAIKOVSKY: Concerto, violin, op. 35, D major. Henryk Szeryng, vln; COA; Bernard Haitink. Philips 9500 321.

+-Gr 10-77 p636 ++RR 10-77 p60
+HFN 10-77 p157

Concerto, violin, op. 64, E minor. cf Symphony, no. 4, op. 90, A major.
Concerto, violin, op. 64, E minor. cf Works, selections (Decca DPA 557/8).
Concerto, violin, op. 64, E minor. cf BACH: Concerto, violin and strings, S 1041, A minor.
Concerto, violin, op. 64, E minor. cf BEETHOVEN: Concerto, violin, op. 61, D major.
Concerto, violin, op. 64, E minor. cf BRAHMS: Concerto, violin, op. 77, D major.
Concerto, violin, op. 64, E minor. cf BRUCH: Concerto, violin, op. 26, G minor (HMV 2926).
Concerto, violin, op. 64, E minor. cf BRUCH: Concerto, violin, no. 1, op. 26, G minor (HMV SXLP 30245).
Concerto, violin, op. 64, E minor. cf GLAZUNOV: Concerto, violin, op. 82, A minor.
Concerto, violin, op. 64, E minor. cf Clear TLC 60.
Concerto, violin, op. 64, E minor. cf HMV SLS 5068.
Concerto, violin, op. 64, E minor: 2nd movement. cf Works, selections (Philips 6833 204).
Concerto, violin, D minor. cf Concerto, violin, op. 64, E minor.
Concerto, violin, op. posth., D minor. cf Concerto, violin and piano D minor.

1786 Concerto, violin and piano, D minor. Concerto, violin, op. posth. D minor. Susanne Lautenbacher, vln; Marylene Dosse, pno; Wurttemberg Chamber Orchestra; Jorg Faerber. Turnabout TV 34662 Tape (c) KTVC 34662.

+Gr 8-77 p306 +RR 8-77 p57
+-HFN 8-77 p87

Elijah: Ye people rend your hearts. cf HMV HLP 7109.

1787 Die erste Walpurgnisnacht, op. 60. Symphony, no. 1, op. 11, C minor. Marie Luise Gilles, ms; Horst Laubenthal, t; Wolfgang Schone, bar; Tadao Yoshie, bs; Frankfurt Singakadamie; Frankfurt Opera and Museum Orchestra; Christoph von Dohnanyi. Turnabout TV 34651.

++NR 5-77 p2

1788 Etudes, op. 104 (3). Fantasy, op. 28, F sharp minor. Rondo capriccioso, op. 14, E major. Songs without words: Elegy, Restlessness. Variations serieuses, op. 54, D minor. Constance Keene, pno. Laurel Protone LP 12.

+Audio 3-77 p91 ++NR 7-76 p12
+HF 7-77 p102 ++SFC 10-31-76 p35
-MJ 1-77 p38

Etudes (studies), op. 104 (3). cf Preludes and fugues, op. 35.
Etudes (studies), op. 104 (3). cf FIELD: Nocturne, no. 11, E flat major.
The fair Melusine, op. 32. cf Overtures (DG 2530 782).
The fair Melusine, op. 32. cf BERLIOZ: Le Corsaire, op. 21.
The fair Melusine, op. 32. cf BRAHMS: Tragic overture, op. 81.

1789 Fantasies, op. 16 (3). Rondo capriccioso, op. 14, E major. Sonata, piano, op. 106, B flat major. Variations serieuses, op. 54, D minor. Ilse von Alpenheim, pno. Philips 9500 162.

+ARG 6-77 p28 +-MJ 7-77 p70
+Gr 4-77 p1574 +-NR 12-77 p12
+-HF 7-77 p102 +-RR 4-77 p75
++HFN 4-77 p138

Fantasy, op. 28, F sharp minor. cf Etudes, op. 104.
Hebrides overture (Fingals cave), op. 26. cf Overtures (DG 2530 782).
Hebrides overture, op. 26. cf Symphony, no. 3, op. 56, A minor (Classics for Pleasure CFP 40270).
Hebrides overture, op. 26. cf Symphony, no. 3, op. 56, A minor (Decca SPA 503).
Hebrides overture, op. 26. cf Symphony, no. 4, op. 90, A major.
Hebrides overture, op. 26. cf Works, selections (Decca DPA 557/8).
Hebrides overture, op. 26. cf Works, selections (Philips 6833 204).
Die Lorely, op. 98: Nel verde maggia. cf RCA TVM 1-7203.
A midsummer night's dream, op. 21: Overture. cf Overtures (DG 2530 782).
A midsummer night's dream, op. 21: Overture. cf BERLIOZ: Le Corsaire, op. 21.
A midsummer night's dream, op. 21: Scherzo. cf RACHMANINOFF: Concerto, piano, no. 2, op. 18, C minor.

1790 A midsummer night's dream, opp. 21/61: Incidental music. Jennifer Vyvyan, Marion Lowe, s; ROHO Women's Chorus; LSO; Peter Maag. Decca SPA 451 Tape (c) KCSP 451. (Reissue from SXL 2060)
+Gr 9-76 p422 +RR 8-76 p41
+HFN 8-76 p93 +-RR 4-77 p92 tape

1791 A midsummer night's dream, opp. 21/61: Incidental music. Lilian Watson, s; Delia Wallis, ms; Finchley Children's Music Group; LSO; Andre Previn. HMV SQ ASD 3377 Tape (c) TC ASD 3337. (also Angel S 37268)
++Gr 9-77 p430 +RR 10-77 p48
+Gr 11-77 p899 tape ++SFC 9-18-77 p42
+HFN 10-77 p157

A midsummer night's dream, opp. 21/61: Incidental music. cf Symphony, no. 4, op. 90, A major.
A midsummer night's dream, opp. 21/61: Incidental music. cf RCA CRM 5-1900.
A midsummer night's dream, opp. 21/61: Overture. cf HMV SLS 5073.
A midsummer night's dream, opp. 21/61: Overture, Scherzo, Nocturne, Wedding march. cf Symphony, no. 4, op. 90, A major (Decca 4359).
A midsummer night's dream, opp. 21/61: Overture, Scherzo, Nocturne, Wedding march. cf Symphony, no. 4, op. 90, A major (Philips 9500 078).
A midsummer night's dream, opp. 21/61: Overture, Scherzo, Nocturne, Wedding march. cf Works, selections (Decca DPA 557/8).
A midsummer night's dream, opp. 21/61: Wedding march. cf Works, selections (Philips 6833 204).
A midsummer night's dream, op. 61: Scherzo. cf Angel S 37219.
A midsummer night's dream, op. 61: Wedding march; Fairies march. cf DG 2548 148.

1792 Octet, strings, op. 20, E flat major. Janacek Quartet, Smetana Quartet. Vanguard/Supraphon SU 4. (Reissue from Westminster, 1959)
+Audio 6-76 p97 ++SFC 8-3-75 p30
+-HF 5-75 p74 +St 5-75 p102
++NR 3-75 p5 ++St 5-77 p74

Octet, strings, op. 20, E flat major: Scherzo. cf Works, selections (Philips 6833 204).
On wings of song. cf Philips 6392 023.

1793 Overtures: Calm sea and prosperous voyage, op. 27. Hebrides, op. 26. The fair Melusine, op. 32. A midsummer night's dream, op. 21. Ruy Blas, op. 95. LSO; Gabriel Chmura. DG 2530 782 Tape (c) 3300 782.
+-Gr 11-77 p843 +HFN 12-77 p173
+Gr 12-77 p1146 tape +RR 12-77 p49

1794 Preludes and fugues, op. 35 (6). Etudes, op. 104 (3). Daniel Adni, pno. HMV SQ HQS 1394.
+Gr 8-77 p324 -RR 9-77 p83
+HFN 9-77 p145

1795 Quartet, strings, op. 44, no. 1, D major. SCHUMANN: Quartet, strings, op. 41, no. 1, A minor. Budapest Quartet. Odyssey 34603. (Recorded 1959, 1961)
+-Audio 6-77 p129 +NR 5-77 p9
+-HF 5-77 p83 +-St 7-77 p118

Rondo brillante, op. 29, E flat major. cf LISZT: Hungarian rhapsody, no. 2, G 244, C sharp minor.

Rondo capriccioso, op. 14, E major. cf Etudes, op. 104.

Rondo capriccioso, op. 14, E major. cf Fantasies, op. 16.

Ruy Blas overture, op. 95. cf Overtures (DG 2530 782).

Ruy Blas overture, op. 95. cf Classics for Pleasure CFP 40263.

Ruy Blas overture, op. 95. cf HMV ESD 7010.

1796 St. Paul (Paulus). Helen Donath, Hanna Schwartz, s; Werner Hollweg, t; Dietrich Fischer-Dieskau, bar; Dusseldorf Musikverein and Boys Choir; Dusseldorf Symphony Orchestra; Rafael Fruhbeck de Burgos. Angel SC 3842 (3).
+-NYT 10-2-77 pD19 +SFC 12-4-77 p56

St. Paul: How lovely are the messengers. cf Guild GRS 7008.

Scherzo, op. 16, no. 2, E minor. cf International Piano Library IPL 101.

Son and stranger: Heimkehr aus der Fremde. cf Club 99-102.

Sonata, clarinet and piano. cf BIS LP 62.

Sonata, organ, no. 2. cf BIS LP 27.

Sonata, piano, op. 106, B flat major. cf Fantasies, op. 16.

1797 Sonata, violoncello and piano, no. 1, op. 45, B flat major. Sonata, violoncello and piano, no. 2, op. 58, D major. Songs without words, op. 109, D major. Variations concertantes, op. 17. Friedrich-Jurgen Sellheim, vlc; Eckart Sellheim, pno. CBS 76547.
+Gr 11-76 p830 +MT 3-77 p217
+-HFN 11-76 p163 ++RR 12-76 p84

1798 Sonata, violoncello and piano, no. 2, op. 58, D major. SCHUBERT: Sonata, arpeggione and piano, D 821, A minor. Lynn Harrell, vlc; James Levine, pno. RCA ARL 1-1568 Tape (c) ARK 1-1568 (ct) ARS 1-1568.
+Gr 9-76 p438 +MJ 11-76 p45
+-HF 3-77 p104 ++NR 9-76 p7
+HF 7-77 p125 tape +RR 9-76 p83
+-HFN 9-76 p125

Sonata, violoncello and piano, no. 2, op. 58, D major. cf Sonata, violoncello and piano, no. 1, op. 45, B flat major.

Songs: Above all praise and majesty. cf Guild GRS 7008.

Songs: Abschied vom Walde; Fruhlingsgruss; Die Nachtigall, op. 59, no. 4. cf Telefunken DT 6-48085.

Songs: Adeste fideles; Greensleeves; Hark, the herald angels sing. cf RCA PRL 1-8020.

Songs: Auf Flugeln des Gesanges, op. 34, no. 2. cf Bruno Walter Society BWS 729.

Songs: Auf Flugeln des Gesanges, op. 34, no. 2. cf Club 99-108.
Songs: Auf Flugeln des Gesanges, op. 34, no. 2; Fruhlingslied, op. 34, no. 3; Der Mond, op. 86, no. 5. cf Swedish SLT 33209.
Songs: Hark the herald angels sing. cf Works, selections (Philips 6833 204).
Songs: Hark the herald angels sing. cf HMV CSD 3774.
Songs: Hark the herald angels sing. cf HMV SQ CSD 3784.
Songs: Hear my prayer. cf Works, selections (Decca DPA 557/8).
Songs: O for the wings of a dove. cf HMV RLS 719.
Songs without words: Elegy, Restlessness. cf Etudes, op. 104.
Songs without words: Spring song. cf Works, selections (Philips 6833 204).
Songs without words: Spring song; The bees wedding. cf Works, selections (Decca DPA 557/8).
Songs without words, op. 16, nos. 2, 6. cf International Piano Library IPL 101.
Songs without words, op. 19, no. 3. cf International Piano Library IPL 101.
Songs without words, op. 30, no. 6. cf FIELD: Nocturne, no. 11, E flat major.
Songs without words, op. 38, no. 6. cf FIELD: Nocturne, no. 11, E flat major.
Songs without words, op. 38, no. 6. cf Westminster WGM 8309.
Songs without words, op. 53, no. 4. cf FIELD: Nocturne, no. 11, E flat major.
Songs without words, op. 62, no. 1. cf HMV ASD 3346.
Songs without words, op. 62, no. 1. cf GLUCK: Orfeo ed Eurydice: Melodie.
Songs without words, op. 62, nos. 5, 6. cf FIELD: Nocturne, no. 11, E flat major.
Songs without words, op. 67, no. 4: Spinning song. cf Citadel CT 6013.
Songs without words, op. 67, no. 4: Spinning song. cf International Piano Library IPL 5001/2.
Songs without words, op. 67, no. 4: Spinning song. cf RCA GL 4-2125.
Songs without words, op. 67, nos. 4, 6. cf International Piano Library IPL 101.
Songs without words, op. 67, no. 5. cf FIELD: Nocturne, no. 11, E flat major.
Songs without words, op. 109, D major. cf Sonata, violoncello and piano, no. 1, op. 45, B flat major.
Songs without words, op. 109, D major. cf HMV RLS 723.

1799 Symphony, no. 1, op. 11, C minor. Symphony, no. 5, op. 107, D minor. VPO; Christoph von Dohnanyi. London CS 7038. (also Decca SXL 6818 Tape (c) KSXC 6818)

+Gr 2-77 p1283
+HFN 2-77 p123
++HFN 4-77 p155 tape
*MJ 9-77 p35
+MT 6-77 p488
+NR 10-77 p3
+-NYT 12-4-77 pD18
+RR 2-77 p50
+-St 9-77 p136

Symphony, no. 1, op. 11, C minor. cf Die erste Walpurgnisnacht, op. 60.

1800 Symphony, no. 3, op. 56, A minor. Calm sea and prosperous voyage, op. 27. NPhO; Riccardo Muti. Angel (Q) S 37168. (also HMV SQ ASD 3184 Tape (c) TC ASD 3184)

+Gr 6-76 p49
-MJ 10-76 p25

-Gr 10-76 p658 tape +-NR 6-76 p5
+-HF 3-77 p98 +-RR 5-76 p45
+HFN 6-76 p91 +St 10-76 p127

1801 Symphony, no. 3, op. 56, A minor. Hebrides overture, op. 26. Scottish National Orchestra; Alexander Gibson. Classics for Pleasure CFP 40270.
+Gr 10-77 p636 +RR 10-77 p48
/-HFN 10-77 p157

1802 Symphony, no. 3, op. 56, A minor. Hebrides overture, op. 26. LSO; Peter Maag. Decca SPA 503. (Reissue from SXL 2246)
+Gr 10-77 p636 +RR 10-77 p48
+HFN 10-77 p165

1803 Symphony, no. 4, op. 90, A major. Hebrides overture, op. 26. WEBER: Oberon: Overture. CO; Georg Szell. CBS 61019 Tape (c) 40-61019. (Reissue from Columbia SAX 2524)
+-Gr 3-77 p1396 +-RR 3-77 p53
+-HFN 3-77 p117 +RR 4-77 p92 tape
+-HFN 4-77 p155 tape

1804 Symphony, no. 4, op. 90, A major. A midsummer night's dream, opp. 21/61: Incidental music. LPO; James Lockhart. Classics for Pleasure CFP 40224.
+-Gr 4-77 p1550 +-HFN 3-77 p107

1805 Symphony, no. 4, op. 90, A major. A midsummer night's dream, opp. 21/61: Overture, Scherzo, Nocturne, Wedding march. RPO; Hans Vonk. Decca PFS 4359. (also London SPC 21145)
+-Gr 6-76 p50 +-NR 10-77 p3
+-HFN 5-76 p101 +-RR 5-76 p46

1806 Symphony, no. 4, op. 90, A major. Concerto, violin, op. 64, E minor. Michele Auclair, vln; VSO; Wolfgang Sawallisch. Fontana 6530 006.
+-Gr 2-77 p1326

1807 Symphony, no. 4, op. 90, A major. SCHUMANN: Symphony, no. 4, op. 120, D minor. NPhO; Riccardo Muti. HMV SQ ASD 3365 Tape (c) TC ASD 3365. (also Angel S 37412)
+Gr 7-77 p179 +NR 12-77 p2
+-Gr 10-77 p698 tape +NYT 12-4-77 pD18
+HFN 10-77 p157 +-RR 10-77 p48

1808 Symphony, no. 4, op. 90, A major. A midsummer night's dream, opp. 21/61: Overture, Scherzo, Nocturne, Wedding march. BSO; Colin Davis. Philips 9500 068 Tape (c) 7300 480.
+-Gr 6-77 p105 tape /NR 1-77 p2
-HF 4-77 p106 +SFC 12-3-77 p37
+MJ 1-77 p26 ++St 6-77 p133

Symphony, no. 4, op. 90, A major. cf Works, selections (Decca 557/8).

Symphony, no. 4, op. 90, A major: 1st movement. cf Works, selections (Philips 6833 204).

1809 Symphony, no. 5, op. 107, D minor. SCHUBERT: Symphony, no. 5, D 485, B flat major. NBC Symphony Orchestra; Arturo Toscanini. RCA AT 123. (Reissues from HMV ALP 1267; Recorded 1953)
++FF 11-77 p79 +HFN 3-74 p109
+-Gr 4-74 p1914 +RR 3-74 p40

Symphony, no. 5, op. 107, D minor. cf Symphony, no. 1, op. 11, C minor.

1810 Symphony, strings, no. 9, C major. Symphony, strings, no. 10, B minor. Symphony, strings, no. 12, G minor. AMF; Neville Marriner. Argo ZK 7 Tape (c) KZKC 7. (Reissue from ZRG 5467)
+Gr 5-77 p1697 ++RR 4-77 p53

++HFN 5-77 p137 +RR 6-77 p97 tape
+HFN 5-77 p139 tape

Symphony, strings, no. 10, B minor. cf Symphony, strings, no. 9, C major.

Symphony, strings, no. 12, G minor. cf Symphony, strings, no. 9, C major.

1811 Trio, piano, no. 1, op. 49, D minor. Trio, piano, no. 2, op. 66, C minor. Beaux Arts Trio. Philips 6580 211. (Reissue from SAL 3646)
+Gr 8-77 p317 +NR 12-77 p5
+HFN 8-77 p93 +RR 7-77 p70

Trio, piano, no. 1, op. 49, D minor. cf COUPERIN: Pieces en concert.

Trio, piano, no. 1, op. 49, D minor. cf HMV RLS 723.

Trio, piano, no. 2, op. 66, C minor. cf Trio, piano, no. 1, op. 49, D minor.

Variations concertantes, op. 17. cf Sonata, violoncello and piano, no. 1, op. 45, B flat major.

Variations serieuses, op. 54, D minor. cf Etudes, op. 104.

Variations serieuses, op. 54, D minor. cf Fantasies, op. 16.

1812 Works, selections: Concerto, violin, op. 64, E minor. Hebrides overture, op. 26. A midsummer night's dream, opp. 21/61: Overture, Scherzo, Nocturne, Wedding march. Songs without words: Spring song; The bees wedding. Songs: Hear my prayer. Symphony, no. 4, op. 90, A major. Alastair Roberts, treble; Wilhelm Backhaus, pno; Ruggiero Ricci, vln; St. John's College Choir; LSO, OSR; Peter Maag, Pierino Gamba, Ernest Ansermet. Decca DPA 557/8 (2) Tape (c) KDPC 557/8.
++Gr 2-77 p1326 tape +-RR 1-77 p56
/HFN 12-76 p155 tape

1813 Works, selections: Concerto, violin, op. 64, E minor: 2nd movement. Hebrides overture, op. 26. Athalia, op. 74: March of the priests. A midsummer night's dream, opp. 21/61: Wedding march. Songs: Hark the herald angels sing. Octet, strings, op. 20, E flat major: Scherzo. Songs without words: Spring song. Symphony, no. 4, op. 90, A major: 1st movement. Werner Haas, pno; Arthur Grumiaux, vln; COA, Hague Philharmonic Orchestra, BBC Symphony Orchestra, I Musici; Winchester Cathedral Choir; Bernard Haitink, Georg Szell, Willem van Otterloo, Colin Davis. Philips 6833 204.
++Gr 2-77 p1331 +-RR 2-77 p50
+HFN 4-77 p153

MENNIN, Peter

Symphony, no. 7. cf DIAMOND: Symphony, no. 4.

MERCADANTE, Giuseppe

Le sette ultime parole di Nostro Signore sulla croce: Qual'ciglia candido (Parola quinta). cf Decca SXL 6781.

Le sette ultime parole di Nostro Signore sulla croce: Qual'ciglia candido (Parola quinta). cf London OS 26437.

MERCANDANTE, Saverio

Il Giuramento: Bella adorata incognita; Compiuta e omai. cf Philips 9500 203.

MERIKANTO, Aare

1814 Concert piece, violoncello. Concerto, piano, no. 2. Partita, orchestra. Veikko Hoyla, vlc; Eero Heinenen, pno; Finnish National Opera Orchestra; Ulf Soderblom. HMV 9C 063 36024.
+-FF 11-77 p37

Concerto, piano, no. 2. cf Concert piece, violoncello.
Partita, orchestra. cf Concert piece, violoncello.

MERKEL, Gustav
Sonata, organ, op. 30, D minor. cf Vista VPS 1039.

MERUCO (14th century)
De home vray. cf HMV SLS 863.

MERULA, Tarquinio
Un cromatico ovvero capriccio. cf ABC ABCL 67008.

MERULO, Claudio
Canzona francese. cf L'Oiseau-Lyre 12BB 203/6.

MESSAGER, Andre
Isoline: Entr'acte...Pavane; Mazurka; Entrance of the first dancer...Seduction scene; Waltz. cf DELISLE: La Marseillaise.
Monsieur Beaucaire: Philomel; I do not know; Lightly, lightly; What are the names. cf HMV RLS 716.

1815 Monsieur Beaucaire: Prologue; Red rose; A little more; Come with welcome; I do not know; Who is this; English maids; Lightly lightly; That's a woman's way; Philomel; Say no more; We are not speaking now; Under the moon; Finale. Martha Angelici, Liliane Berton, s; Rene Lenoty, t; Michel Dens, Gilbert Moryn, bar; Raymond Saint-Paul Chorus; Lamoureux Orchestra; Jules Gressier. CRD Pathe C051 12086.
+Gr 1-77 p1187

The two pigeons: Entrance of the gipsies; Entrance of Pepio and 2 pigeons pas de deux; Theme and variations; Entry; Hungarian dance; Finale. cf DELISLE: La Marseillaise.

MESSIAEN, Oliver

1816 Apparition de l'eglise eternelle. L'Ascension. Le banquet celeste. Charles Krigbaum, org. Lyrichord LLST 7297.
++MU 7-77 p16 ++NR 8-76 p11

1817 L'Ascension. Hymne. Les offrandes oubliees. ORTF; Marius Constant. Erato STU 70673.
+-Gr 4-77 p1550 +MT 9-77 p736
+-HFN 4-77 p138

L'Ascension. cf Apparition de l'eglise eternelle.
L'Ascension: Alleluias sereins; Transports de joie. cf Decca SDD 499.
Le banquet celeste. cf Apparition de l'eglise eternelle.
Le banquet celeste. cf DUPRE: Preludes and fugues, op. 36.
Le banquet celeste. cf Wealden WS 131.
Les Bergers. cf DUPRE: Preludes and fugues, op. 36.
Canteyodiaya. cf BERIO: Rounds.
Etudes de rythme (4). cf BARTOK: Etudes, op. 18.
Hymne. cf L'Ascension.
Mode de valeurs et d'intensities pour piano. cf CP 3-5.
La nativite du Seigneur: Dieu parmi nous. cf Vista VPS 1033.

1818 O sacrum convivium. POULENC: Exultate Deo. Litanies a la vierge noire de Rocamadour. Salve regina. VIERNE: Mass, C sharp minor. Worcester Cathedral Choir; Harry Bramma, Paul Trepte, org; Donald Hunt. Abbey LPB 780.
+Gr 7-77 p211 +MT 10-77 p838
+HFN 8-77 p81 +RR 8-77 p76

Les offrandes oubliees. cf L'Ascension.
Poemes pour mi. cf BIS LP 37.

1819 Quartet for the end of time. Peter Serkin, pno; Ida Kavafian, vln; Fred Sherry, vlc; Richard Stoltzman, clt. RCA ARL 1-1567.

++Gr 7-77 p197
+Ha 6-77 p88
+HF 1-77 p124
+HFN 10-77 p157
+-MJ 11-76 p45
+MT 12-77 p1015
+NR 9-76 p6
+RR 7-77 p70
++SFC 8-29-76 p29
++St 10-76 p128
++St 5-77 p73

1820 La rousserolle effarvatte. Vingt regards sur l'enfant Jesus. Peter Serkin, pno. RCA ARL 3-0759 (3).
+-Te 3-77 p31

1821 Vingt regards sur l'enfant Jesus. Michel Beroff, pno. Connoisseur Society CS 2-2133 (2). (Reissue from French EMI CO 65-1067-78)
++ARG 8-77 p21
++MJ 5-77 p32
+NR 8-77 p14
++SFC 2-13-77 p33
++St 8-77 p116

Vingt regards sur l'enfant Jesus. cf La rousserolle effarvatte.
METROPOLITAN OPERA 25th ANNIVERSARY. cf Advent S 5023.

MEYER
Appalachian echoes. cf Orion ORS 75207.

MEYER, Krzysztof
Concerto, violin, op. 12. cf DABROWSKI: Concerto, violin, 2 pianos and percussion.

MEYERBEER, Giacomo
L'Africaine: Adamastor, roi de vagues profondes. cf Philips 6580 174.
L'Africana: Di qui si vede il mar...Quai celesti concenti. cf Court Opera Classics CO 347.
L'Africaine: O paradiso. cf RCA CRM 1-1749.
L'Africana: O paradiso. cf Seraphim S 60280.
L'Africaine: Pays merveilleux...O paradis. cf Rubini GV 38.
Dinorah: Ombre legere. cf Columbia 34294.
Dinorah: Ombra leggiera. cf VERDI: Rigoletto.
Dinorah: Ombra leggiera. cf HMV SLS 5057.
Dinorah: Ombra leggiera. cf Seraphim 60274.
L'Etoile du nord: O jours heureux. cf Rubini GV 39.
Les Huguenots: Choral de Luther; Piff, paff. cf Club 99-107.
Les Huguenots: Nobles seigneurs, salut. cf CBS 76522.
Les Huguenots: O beau pays. cf Decca D65D3.
Les Huguenots: O beau pays; Beaute divine, enchanteresse. cf Rubini GV 26.
Les Huguenots: Seigneur rampart et suel soutien, Piff paff. cf GOMES: Salvator Rosa: Di sposo di padre le gioie serene.
Les Huguenots: Tu l'as dit; Beaute divine enchanteresse. cf Rubini GV 38.
Les Huguenots: Vaga donna, No no no giammai. cf Club 99-106.
Les patineurs. cf MASSENET: Le Cid.
Les patineurs: Ballet suite. cf MASSENET: Ariane: Lamento d' Ariane.

1822 Le prophete. Renata Scotto, s; Marilyn Horne, ms; James McCracken, Jean Dupouy, t; Christian du Plessis, bar; Jules Bastin, Jerome Hines, bs; Ambrosian Opera Chorus; RPO; Henry Lewis. Columbia M4 34340 (4). (also CBS 79400)
+-ARG 6-77 p29
+-Gr 5-77 p1729
+-HF 3-77 p87
+HFN 5-77 p127
++MJ 7-77 p68
+NR 4-77 p9
+ON 1-29-77 p44
+RR 5-77 p35
+-St 5-77 p114

Robert le diable, excerpt. cf Grosvenor GRS 1055.
Robert le diable: Invocation. cf Rubini GV 39.
Robert le diable: Nonnes qui reposez. cf GOMES: Salvator Rosa: Di sposo di padre le gioie serene.
Robert le diable: Nonnes qui reposez. cf Club 99-107.

MEYER-HELMUND, Erik
In the morning I bring you violets. cf Club 99-96.
Violets. cf Rubini RS 301.

MIASKOVSKY, Nikolai
Army march. cf HMV CSD 3782.

MICHALSKY, Donal
Concerto in re. cf Pandora PAN 2001.

MICHNA Z OSTRADOVIC, Adam
Christmas music. cf BRIXI: Mass.

MIDDENDORF
Stand up for America. cf MENC 76-11.

MIDDLETON, John
Fantasie. cf Vista VPS 1047.

MIEG, Peter
Les charmes de Lostorf. cf Works, selections (Claves P 610).
Hermance (Les etudes de Czerny). cf Works, selections (Claves P 610).
Les jouissances de Mauensee. cf Works, selections (Claves P 610).
Lettres a Goldoni. cf Works, selections (Claves P 610).
Morceau elegant. cf Works, selections (Claves P 610).
Les plaisirs de Rued. cf Works, selections (Claves P 610).
Sur les rives du lac Leman. cf Works, selections (Claves P 610).

1823 Works, selections: Les charmes de Lostorf. Hermance (Les etudes de Czerny). Les jouissances de Mauensee. Lettres a Goldoni. Morceau elegant. Les plaisirs de Rued. Sur les rives du lac Leman. Peter Lukas Graf, Anne Utagawa, Dominique Hunziger, flt; Ursula Holliger, hp; Alexander van Wijnkoop, Eva Zurbrugg, Thomas Furi, vln; Walter Grimmer, vlc; Ernst Gerber, hpd; Dinorah Varsi, Urs Voegelin, pno. Claves P 610.
+HFN 6-77 p129 +RR 6-77 p74

MIHALY, Andras
Little tower music. cf Hungaroton SLPX 11811.

MILAN, Luis
Fantasias, nos. 10-12, 16. cf Pavanas, nos. 1-6.
Pavana. cf National Trust NT 002.
Pavanas, 1 and 2. cf DG 2723 051.

1824 Pavanas, nos. 1-6. Fantasias, nos. 10-12, 16. MUDARRA: Diferencias sobre El Conde claros. Fantasia que contrahaza la harpa en la manera de Ludovico. Gallarda. O guardame las vacas. Pavana de Alexandre. NARVAEZ: Baxa de contrapunto. Diferencias sobre "Guardame las vacas". Fantasia. Mille regretz. Konrad Ragossnig, lt. DG 2533 183.
++Gr 7-75 p223 ++RR 7-75 p48
++HFN 7-75 p87 +St 8-77 p59
++NR 3-76 p13

Songs: Toda mi vida os ame. cf National Trust NT 002.

MILANO, Francesco da
Fantasia. cf DG 2533 173.

MILHAUD, Darius
1825 L'Automne. La bal martiniquias. Paris. Le printemps. Scaramouche. Christian Ivaldi, Noel Lee, Michel Beroff, Jean-Philippe Collard, pno. Connoisseur Society CS 2101.

+MJ 5-77 p32 ++SFC 8-29-76 p29
+NR 8-76 p12 ++St 11-76 p148

Le bal martiniquais. cf L'Automne.
Le carnaval d'Aix: Le Capitaine Cartuccia. cf DG 2548 148.
Catalogue de fleurs. cf BIS LP 34.

1826 Chansons de Ronsard, op. 223 (4). Symphony, no. 6, op. 343. Paula Seibel, s; Louisville Orchestra; Jorge Mester. Louisville LS 744. (also RCA GL 2-5020)
+-Gr 11-76 p797 +RR 10-76 p61
++HF 8-75 p90 ++SFC 5-25-75 p17
+-HFN 12-76 p145 /ST 1-77 p777

Chansons de Ronsard, op. 223. cf Columbia 34294.

1827 La cheminee du Roi Rene, op. 205. SAINT-SAENS: Septet, op. 65, E flat major. Leningrad State Philharmonic Woodwind Quintet, Taneyev Quartet. Westminster WG 8348. (Reissue from Melodiya, 1963)
+ARG 10-77 p37 ++SFC 6-19-77 p46
++NR 9-77 p6

1828 Four seasons: Concertino d'hiver, op. 327. MOZART: Concerto, flute, no. 2, K 314, D major. VILLA-LOBOS: Dance of seven notes. Lev Pechersky, bsn; Victor Venglovsky, trom; Dmitri Bida, flt; Leningrad Chamber Orchestra; Lazar Gozman. Westminster WGS 8336.
+-ARG 5-77 p51 +-HF 12-77 p113

Four season: Concertino d'hiver, op. 327. cf BASSETT: Suite, solo trombone.
Hymne de glorification. cf Vox SVBX 5483.
Paris. cf L'Automne.
Le printemps. cf L'Automne.
Romances. cf Vox SVBX 5483.
Saudades do Brasil. cf BARRAUD: Symphonie concertante, trumpet and orchestra.
Saudades do Brasil. cf KUISMA: Concert piece for percussion.
Scaramouche. cf L'Automne.
Sonatas, viola and piano, nos. 1 and 2. cf FRANCK: Sonata, violin and piano, A major.
Symphony, no. 1. cf Prelude PRS 2512.
Symphony, no. 6, op. 343. cf Chansons de Ronsard, op. 223.
MILITARY BAND OF THE QUEEN'S REGIMENT. cf Olympic 6122.

MILLOCKER, Karl
Der Bettelstudent: Ich setz den Fall; Ich hab kein Gold. cf HMV RLS 715.
Drei Paar Schuhe: I und mei Bua. cf Decca SDD 507.

MILNER
Songs: The hunter; In Cheyder. cf Olympic OLY 105.

MILTON, John
Fair Orian, in the morn. cf DG 2533 347.

MIMAROGLU, Ilhan
1829 Face the windmills, turn left. Prelude, op. 14. Realized at the Columbia-Princeton Electronic Music Center. Finnadar SR 9012.
+Audio 9-77 p100 ++NR 5-77 p16
+MJ 9-77 p35

Prelude, op. 14. cf Face the windmills, turn left.

MINGUS
Goodbye, porkpie hat. cf Finnadar SR 9015.

MINKUS, Ludwig
Don Quixote: Pas de deux. cf ADAM: Giselle: Pas de deux; Grand pas de deux and finale.

MIYAGI, Michio
Haru no umi (arr. and orch. Gerhardt). cf RCA LRL 1-5094.
MIZUNO
Tone for piano. cf CP 3-5.
MODENA (16th century Spain)
Recercare a 4. cf L'Oiseau-Lyre 12BB 203/6.
MOERAN, Ernest
Rhapsody, F sharp major. cf BRIDGE: Phantasm rhapsody.
Songs: The merry month of May. cf Argo ZK 28/9.
1830 Symphony, G minor. NPhO; Adrian Boult. Lyrita SRCS 70. (also HNH 4014)
++Gr 7-75 p187 ++NR 9-77 p1
+HF 9-77 p108 +RR 7-75 p28
++HFN 7-75 p84
MOLINARO, Simone
Ballo detto "Il Conte Orlando". cf DG 2533 173.
Ballo detto "Il Conte Orlando". cf DG 2533 323.
Ballo detto "Il Conte Orlando": Saltarello. cf DG 2723 051.
Fantasias, nos. 1, 9-10. cf DG 2533 173.
Saltarello. cf DG 2533 323.
Saltarello. cf DG 2723 051.
Saltarello (2). cf DG 2533 173.
MOLINS, Pierre de
Amis tout dous. cf HMV SLS 863.
MOLIQUE, Wilhelm Bernard
1831 Concerto, flute, op. 69, B minor. ROMBERG: Concerto, flute, op. 30, B minor. John Wion, flt; Orchestra; Arthur Bloom. Musical Heritage Society MHS 3551.
-St 12-77 p153
Concertino, oboe, G minor. cf BELLINI: Concerto, oboe, E flat major.
MOLLEDA
Variations on a theme. cf RCA ARL 1-1323.
MOLLER, John C.
Sinfonia. cf Orion ORS 76255.
MOLLOY
Love's old sweet song. cf Prelude PRS 2505.
MOLTER, Johann
1832 Concerti, clarinet (4). Georgina Dobree, clt; Carlos Villa Ensemble. Chantry ABM 22. (Reissue from EMI HQS 1119)
+-HFN 5-77 p127 +RR 3-77 p55
MONASTERIO
Sierra morena. cf Discopaedia MB 1013.
MONDELLO
Siciliana. cf Orion ORS 75207.
MONOD, Jacques-Louis
1833 Cantus contra cantum I. SHIFRIN: Quartet, strings, no. 4.* WEBER: Quartet, strings, no. 2, op. 35.* Merja Sargon, s; New Music Quartet, Fine Arts Quartet; Chamber Orchestra; Jacques-Louis Monod. CRI SD 358. (*Reissues)
+ARG 8-77 p35 ++NR 11-76 p8
+-MJ 3-77 p46
MONROE, Ervin
Sketches, solo flute. cf Golden Crest RE 7064.
MONTANARI, Nunzio
Invections, wind quartet (5). cf GIULIANI: Divertimento.

MONTEMEZZI, Italo

1834 L'Amore dei tre re. Anna Moffo, Alison MacGregor, Elaine Tomkinson, s; Michael Sanderson, boy soprano; Elizabeth Bainbridge, ms; Placido Domingo, Ryland Davies, Alan Byers, t; Pablo Elvira, bar; Cesare Siepi, bs; Ambrosian Opera Chorus; LSO; Nello Santi. RCA ARL 2-1945 (2) Tape (c) ARS 2-1945 (ct) ARK 2-1945.

+ARG 11-77 p24
+-FF 9-77 p36
+Gr 10-77 p692
+HF 8-77 p67
++HF 11-77 p138 tape
+NR 8-77 p9
+-NYT 7-10-77 pD13
+ON 9-77 p69
++RR 10-77 p30
+-SFC 8-28-77 p46
+-St 10-77 p87

MONTEVERDI, Claudio

Il combattimento di Tancredi e Clorinda: Paraphrase on the madrigal by Monteverdi, op. 28. cf L'Oiseau-Lyre DSLO 17.

La favola d'Orfeo: Tusei morta mia vita. cf Philips 6580 174.

L'Incoronazione di Poppea: Disprezzata Regina; Tu che dagli avi miei...Maestade, che prega; Addio Roma. cf Songs (Telefunken 6-41956).

Lamento d'Olimpia. cf L'Oiseau-Lyre 12BB 203/6.

1835 Madrigals: Ardo e scoprir; Bel pastor; Di far sempre gioire; Ecco mormorar l'onde; Eccomi pronta ai baci; Hor ch'el ciel e la terra; Io mi son giovinetta; La mia turca; Lamento della ninfa; Lasciatemi morire; Maladetto sia l'aspetto; Non e di gentil core; Ohime ch'io cado; Perche t'em fuggi; Presso un fiume tranquillo; Questi vaghi concenti; Sfogava con le stelle; Si ch'io vorrei morire; Si dolce e il tormento; Tirsi e Clori; Vorrei baciarti. Jennifer Smith, Wally Stampfli, s; Nicole Rossier, ms; Hanna Schaer, con; Oliver Dufour, John Elwes, t; Philippe Huttenlocher, bar; Michel Brodard, bs; Catherine Eisenhofer, hp; Jurg Hubscher, lt; Marcal Cervera, vla da gamba; Christiane Jaccotet, hpd; Philippe Corboz, org; Lausanne Vocal Ensemble and Chamber Orchestra; Michel Corboz. Erato STU 70848/9 (2).

+-Gr 4-77 p1587
+-HFN 4-77 p141
+-RR 4-77 p82

1836 Madrigals: Altri canti d'amor; Bel pastor; Ecco mormorar l'onde; Eccomi pronta ai baci; La fiera vista; Gira il nemico insidioso; Hor che il ciel e la terra; Io son pur vezzosetta; Lamento della ninfa; La mia turca; O come sei gentile; Ohime ch'io cado; Perche fuggi; Presso un fiume tranquillo; Qual si puo dir; Si ch' io vorrei morire; Soave liberate; Su, su, su pastorelli vezzosi; Tempro la cetra. Lausanne Chamber Orchestra and Chorus; Michel Corboz. RCA CRL 1-1973 (2).

+ARG 3-77 p26
+-MU 3-77 p14
+NR 2-77 p9

Magnificat. cf Vespro della beata vergine.

Mass, In illo tempore. cf Vespro della beata vergine.

L'Orfeo: Mira, deh mira, Orfeo...In un fiorito prato. cf Songs (Telefunken 6-41956).

1837 Songs: Lamento d'Arianna; Lettera amorosa; Partenza amorosa; A quest'olmo; Questi vaghi. Hertha Topper, alto; Vocal and Instrumental Soloists. Musical Heritage Society MHS 3457.

-St 6-77 p134

1838 Songs: Arias, canzonettas and recitatives. Con che soavita; Lamento d'Arianna; Lettera amorosa. L'Orfeo: Mira, deh mira, Orfeo

...In un fiorito prato. L'Incoronazione di Poppea: Disprezzata Regina; Tu che dagli avi miei...Maestade, che prega; Addio Roma. Cathy Berberian, s; VCM; Nikolaus Harnoncourt. Telefunken 6-41956.

+-HFN 6-76 p91 ++NR 3-76 p12
+-MJ 2-77 p30 +-St 4-77 p122

Songs: Laudate dominum terzo. cf Argo ZRG 859.

Songs (Madrigals): Quel sguardo; Chiome d'oro; O sia tranquillo; Se vittorie si belle; Zefiro torno. cf CAVALLI: Arias (Vanguard VRS 10124).

1839 Vespro della beata vergine. Mass, In illo tempore. Magnificat. Paul Esswood, Kevin Smith, c-t; Ian Partridge, John Elwes, t; David Thomas, Christopher Keyte, bs; Edward Tarr, Ralph Bryant, Richard Cook, cor; Fritz Brodersen, Harald Strutz, Walfried Kohlert, trom; Sebastian Kelber, Klaus Holsten, flt and Renaissance rec; Eduard Melkus, Spiros Rantos, Thomas Weaver, vln; Lilo Gabriel, David Becker, vla; Klaus Storck, Eugene Eicher, vlc; Laurenzius Strehl, vla da gamba and violone; Dieter Kirsch, lt; Hubert Gumz, Gerd Kaufmann, org; Regensburg Domspatzen; Hanns-Martin Schneidt. DG 2723 043 (3) Tape (c) 3376 010.

+Gr 11-75 p876 +MT 6-76 p495
++Gr 11-77 p900 tape +RR 11-75 p86
++HFN 11-75 p155 +STL 11-2-75 p38

1840 Vespro della beata vergine. Elly Ameling, Norma Burrowes, s; Charles Brett, c-t; Anthony Rolfe Johnson, Robert Tear, Martyn Hill, t; Peter Knapp, John Noble, bs; King's College Chapel Choir; Early Music Consort; Philip Ledger. HMV SQ SLS 5064 (2) Tape (c) TC SLS 5064. (also Angel S 3837)

+-Gr 11-76 p857 +NR 7-77 p8
++HF 8-77 p87 +RR 1-77 p90 tape
+-HFN 9-76 p125 +-RR 9-76 p85
+HFN 12-76 p153 tape +-St 8-77 p116
-MT 1-77 p46 +STL 9-19-76 p36

MONTI

Czardas. cf Angel (Q) S 37304.

Czardas. cf Kelsey Records KEL 7601.

MONTSALVATGE, Bassols Xavier

1841 Concerto breve. SURINACH: Concerto, piano. Alicia de Larrocha, pno; RPO; Rafael Fruhbeck de Burgos. London CS 6990 Tape (c) 5-6990. (also Decca SXL 6757 Tape (c) KSXC 6757)

+Gr 9-77 p431 ++RR 9-77 p63
++HFN 10-77 p173 ++RR 12-77 p96 tape
+MJ 10-77 p29 +SR 7-9-77 p48
++NR 8-77 p6 ++St 9-77 p136

MOODY, James

Little suite. cf JACOB: Pieces.

Quintet, harmonica. cf JACOB: Divertimento, harmonica and string quartet.

MOORE

The last rose of summer. cf Columbia 34294.

MOORE, Douglas

1842 The ballad of Baby Doe. Beverly Sills, s; Walter Cassel, bar; Frances Bible, ms; Lynn Taussig, Helen Baisley, Grant Williams, Chester Ludgin, Beatrice Krebs, Jack DeLon, Joshua Hecht, soloists; New York City Opera Orchestra and Chorus; Emerson Buckley. DG 2709 061 (3). (Reissue from MGM 3GC 1, Heliodor H 25035/3) (also 2584 009/11)

*ARG 11-76 p34 +-MT 5-77 p402
+-Gr 7-76 p206 +NR 10-76 p11
+HFN 7-76 p91 +ON 7-76 p43
*MJ 11-76 p44 +RR 7-76 p29

The pageant of P. T. Barnum. cf CARPENTER: Adventures in a perambulator.

MOORE, Thomas

Songs: Believe me, if all those endearing young charms; The minstrel boy; The young May moon. cf Philips 6599 227.

MOORS, Hezekiah

Songs: By the offence of one-anthem; Cavendish; Charlotte; Dorset; Fairfax; I will praise thee, anthem; Moretown; Orwell; Mount Holly; Plainfield; Pittsford; Shirley. cf INGALLS: Songs (Philo 1038).

MORALES, Cristobal de

1843 Andreas Christi famulus. Emendemus in melius. Jubilate Deo omnis terra. Lamentabatur Jacob. Magnificat secundi toni. Pastores dicite quidnam vidistis. Pro Cantione Antiqua; Early Music Consort; Bruno Turner. DG 2533 321.

+-ARG 8-77 p22 ++RR 12-76 p93
+-Gr 1-77 p1167 ++SFC 6-26-77 p46
+HFN 1-77 p111 ++St 8-77 p118
+NR 8-77 p11

Emendemus in melius. cf Andreas Christi famulus.
Jubilate Deo omnis terra. cf Andreas Christi famulus.
Lamentabatur Jacob. cf Andreas Christi famulus.
Magnificat secundi toni. cf Andreas Christi famulus.
Pastores dicite quidnam vidistis. cf Andreas Christi famulus.

MORET

Le Nelumbo. cf Odyssey Y 33130.
Silver heels. cf Angel (Q) S 37304.

MORGAN, Justin

Songs: Amanda; Despair; Montgomery. cf Vox SVBX 5350.

MORLEY, Thomas

Alman. cf RCA RL 2-5110.
La caccia a 2. cf Philips 6500 926.
A deception. cf Hungaroton SLPX 11823.
Fancy. cf BIS LP 22.
The frog galliard. cf Abbey LPB 765.
It was a lover and his lass. cf HMV ESD 7002.
It was a lover and his lass. cf Saga 5447.
Lamento. cf BIS LP 22.
Now is the gentle season. cf BIS LP 75.
La sampogna. cf Philips 6500 926.
See mine own sweet jewel. cf RCA RL 2-5110.
Songs: April is my mistress' face; Daemon and Phyllis; Fire, fire; I love, alas; Leave, alas, this tormenting; My bonny lass; Now is the month of Maying; O grief, even on the bud; Those dainty daffadillies; Though Philomela lost her love. cf Enigma VAR 1017.
Songs: Arise, awake; Hard by a crystal fountain. cf Argo ZK 25.
Songs: Arise, awake, awake, awake; Hard by a crystal fountain. cf DG 2533 347.
Songs: He who comes here; Sweet nymph. cf BG HM 57/8.
MORMAN TABERNACLE CHOIR: A jubilant song. cf Columbia M 34134.

MORRILL, Dexter

Studies, trumpet and computer. cf Golden Crest RE 7068.

MORTENSEN, Finn
Quintet, winds, op. 4. cf CARLSTEDT: Sinfonietta, 5 wind instruments.
MOSCHELES, Ignaz
Concertante, flute and oboe, F major. cf BELLINI: Concerto, oboe, E flat major.
German dances. cf DG 2723 051.
1844 Grande sonate symphonique, op. 112. PIXIS: Concerto, violin, piano and strings. Mary Louise Boehm, Pauline Boehm, pno; Kees Kooper, vln; Westphalian Symphony Orchestra; Siegfried Landau. Turnabout TV 34590.
+Gr 9-76 p445 +NR 12-75 p7
-HFN 8-76 p83 +RR 8-76 p49
+MT 8-77 p651
MOSS, Katie
Songs: The floral dance. cf Argo ZFB 95/6.
Songs: The floral dance. cf L'Oiseau-Lyre DSLO 20.
MOSZKOWSKI, Moritz
Caprice espagnole, op. 37. cf International Piano Library IPL 5001/2.
La jongleuse. cf International Piano Library IPL 114.
Spanish dances, Bk 1, op. 12. cf CHABRIER: Espana.
Valse, op. 34. cf International Piano Library IPL 102.
MOTTU, Alexander
Prelude et chorale. cf Abbey LPB 719.
MOURET, Jean Joseph
Fanfares (Rondeau). cf Golden Crest CRS 4148.
Rondeau. cf University of Iowa Press Unnumbered.
Suite, 3 trumpets and organ, D major. cf COUPERIN: Concerto, trumpet and organ, no. 9: Ritratto dell'amore.
MOUTON, Charles
La, la, la l'oysillon du bois. cf L'Oiseau-Lyre 12BB 203/6.
Nesciens Mater virgo virum. cf HMV SLS 5049.
MOZART, Leopold
Cassation, orchestra and toys, G major: Toy symphony. cf HMV SQ ASD 3375.
MOZART, Wolfgang Amadeus
Adagio, C major. cf Discopaedia MOB 1020.
1845 Adagio, K 261, E major. Concerto, violin, no. 1, K 207, B flat major. Rondos, violin, K 269, B flat major; K 373, C major. Wolfgang Schneiderhan, vln; BPhO; Wolfgang Schneiderhan. DG 2535 205. (Reissue from SLPM 139350/2)
-Gr 12-76 p996 +RR 1-77 p56
+HFN 12-76 p151
Adagio, K 261, E major. cf Works, selections (Angel SD 3789).
Adagio, K 261, E minor. cf Works, selections (Philips 7699 048).
1846 Adagio, K 540, B minor. Sonata, piano, no. 11, K 331, A major. Sonata, piano, no. 13, K 333, B flat major. Alfred Brendel, pno. Philips 9500 025 Tape (c) 7300 474.
++Gr 6-76 p70 -MJ 3-77 p74
+-HF 4-77 p106 +NR 4-77 p14
++HFN 6-76 p93 +-RR 6-76 p73
Adagio, K 540, B minor. cf HAYDN: Piano works (Hungaroton 11638).
Adagio, K 617, C minor. cf DAMASE: Concertino, harp and strings, op. 20.
Adagio and allegro, organ. cf Works, selections (Philips 6747 384).

Allegro and andante, K 533, F major. cf Sonata, piano, no. 11, K 331, A major.
Alleluja. cf Works, selections (Philips 6747 384).

1847 Andante, K 315, C major. Concerto, flute, no. 1, K 313, G major. Concerto, flute, no. 2, K 314, D major. Eugenia Zukerman, flt; ECO; Pinchas Zukerman. Columbia M 34520 Tape (c) MT 34520. (also CBS 76594)
+Gr 5-77 p1697 +RR 5-77 p54
+HFN 5-77 p129 +-St 11-77 p144
+-NR 6-77 p6

1848 Andante, K 315, C major. Concerto, flute, no. 1, K 313, G major. Concerto, flute (oboe), no. 2, K 314, D major. Michel Debost, flt; Maurice Bourgue, ob; Orchestre de Paris; Daniel Barenboim. HMV ASD 3320. (also Angel S 37269)
-Gr 3-77 p1401 +-RR 3-77 p55
+HFN 3-77 p107 -SFC 4-17-77 p42
+MJ 9-77 p35

1849 Andante, K 315, C major. Concerto, flute, no. 1, K 313, G major. Concerto, flute, no. 2, K 314, D major. James Galway, flt; Festival Strings; Rudolf Baumgartner. RCA LRL 1-5109 Tape (c) RK 11732. (also RCA ARL 1-2159 (c) ARK 1-2159 (ct) ARS 1-2159)
+-ARG 7-77 p27 ++RR 3-76 p45
+-Gr 3-76 p1467 ++RR 1-77 p90 tape
+-HFN 4-76 p109 ++SFC 4-17-77 p42
+-HFN 11-76 p175 tape +SR 7-9-77 p48
-MJ 9-77 p35 ++St 8-77 p118
++NR 5-77 p7

1850 Andante, K 315, C major. Concerto, flute, no. 1, K 313, G major. Concerto, flute, no. 2, K 314, D major. Jean-Pierre Rampal, flt; VPO; Theodor Guschlbauer. RCA FRL 1-5330 Tape (c) FRK 1-5330 (ct) FRS 1-5330.
++ARG 3-77 p28 +MJ 2-77 p30
+HF 4-77 p123 tape ++NR 1-77 p6

1851 Andante, K 315, C major. Concerto, flute, no. 1, K 313, G major. Concerto, flute, no. 2, K 314, D major. Julius Baker, flt; I Solisti di Zagreb, VSOO; Antonio Janigro, Felix Prohaska. Vanguard SRV 364.
++NR 11-77 p7

Andante, K 315, C major. cf RCA RK 2-5030.
Andante, K 616, F major. cf BEETHOVEN: Adagio and allegro, music box.
Andante, K 616, F major. cf HAYDN: Sonata, piano, no. 46, A flat major.
Andante, organ. cf Works, selections (Philips 6747 384).

1852 Andantino, violoncello and piano, K 374, B minor. Quartet, flute, K 285, D major. Quartet, oboe, K 370, F major. Quartet, piano, K 478, G minor. Maurice Bourgue, ob; Michel Debost, flt; Hepzibah Menuhin, pno; Yehudi Menuhin, vln; Luigi Alberto Biachi, vla; Maurice Gendron, vlc. HMV ASD 3329.
+-Gr 3-77 p1421 +RR 4-77 p66
+-HFN 4-77 p141

Apollo et Hyacinthus, K 38: Overture. cf Overtures (Eurodisc SQ 27257).
Apollo et Hyacinthus, K 38: Overture. cf Overtures (Turnabout TV 34628).

1853 Arias: Ah lo previdi, K 272. Bella mia fiamma, K 528. Ch'io mi scordi di te, K 505. Idomeneo, Re di Creta, K 366: D'Oreste, d'ajace. Sylvia Sass, s; Andras Schiff, pno; Hungarian State Opera Orchestra; Ervin Lukacs. Hungaroton SLPX 11812.
++FF 9-77 p39 +OC 6-77 p45
+-HF 4-77 p89 +-OR 6/7-77 p35
+-HFN 2-77 p123 ++RR 2-77 p87
+NR 3-77 p10 +SFC 2-27-77 p42
+-NYT 4-24-77 pD27 +St 5-77 p120

1854 Arias: Cosi fan tutte, K 588: Un aura amorosa; Ah lo veggio; In qual fiero. Don Giovanni, K 527: Il mio tesoro; Dalla sua pace. Die Entfuhrung aus dem Serail, K 384: Hier soll ich dich; Wenn der Freude; Constanze; Ich baue ganz. Idomeneo, Re di Creta, K 366: Fuor del mar. Die Zauberflote, K 620: Dies Bildnis ist bezaubernd schon; Wie stark ist nicht dein Zauberton. Stuart Burrows, t; LSO, LPO; John Pritchard. L' Oiseau-Lyre DSLO 13.
+ARG 9-77 p40 +ON 6-77 p44
+-Gr 10-76 p649 +-RR 10-76 p28
+HFN 11-76 p164 ++SFC 2-27-77 p42
+NR 5-77 p10 ++St 8-77 p124

1855 Arias: La clemenza di Tito, K 621: Parto, parto. Don Giovanni, K 527: In quali accessi...Mi tradi quell'alma ingrata; Crudele, Ah no, mio bene...Non mi dir. Die Entfuhrung aus dem Serail, K 384: Martern aller Arten. Idomeneo, Re di Creta, K 366: Parto, e l'unico oggetto. Le nozze di Figaro, K 492: Giunse alfin il momento...Deh vieni non tardar; E Susanna no vien... Dove sono; Voi che sapete. Il Re pastore, K 208: L'amero saro costante. Margaret Price, s; ECO; James Lockhart. RCA SER 5675. (also RCA AGL 1-1532)
+Gr 10-73 p727 +-OR 6/7-77 p35
+-HF 4-77 p89 +-RR 11-73 p28
+MJ 12-76 p28 ++SFC 9-12-76 p31
+NR 9-76 p9 ++St 11-76 p89
++ON 9-76 p70 +ST 2-76 p741
+Op 12-73 p1104

Ascanio in Alba, K 111: Overture. cf Overtures (Eurodisc SQ 27257).

Ascanio in Alba, K 111: Overture. cf Overtures (Turnabout TV 34628).

Ave Maria. cf Works, selections (Philips 6747 384).

Ave verum corpus, K 618. cf Works, selections (Philips 6747 384).

Ave verum corpus, K 618. cf Works, selections (Philips 7699 052).

1856 Bastien und Bastienne, K 50. Edith Mathis, s; Claes Ahnsjo, t; Walter Berry, bs; Leopold Hager, hpd; Salzburg Mozarteum Orchestra; Leopold Hager. BASF G 22772.
+ARG 6-77 p33 +St 6-77 p134

1857 Bastien und Bastienne, K 50. Die Entfuhrung aus dem Serail, K 384. Zaide, K 344. Die Zauberflote, K 620. Soloists; VSO, Berlin State Orchestra, Bavarian State Orchestra and Chorus, RIAS Kammerchor and Symphony Orchestra; John Pritchard, Bernhard Klee, Eugen Jochum, Ferenc Fricsay. Philips 6747 387 (8).
+-Gr 10-77 p686 +RR 9-77 p28
+-HFN 9-77 p153

Bastien und Bastienne, op. 50: Overture. cf Overtures (Eurodisc SQ 27257).

Bastien und Bastienne, K 50: Overture. cf Overtures (Turnabout TV 34628).
Betulia liberata, K 74c: Overture. cf Overtures (Eurodics 27257).
Cassations (2). cf Works, selections (Philips 6747 378).
Cassation (serenade) K 62a, D major (with March, K 62, D major). cf BACH, J. C.: Sinfonia, op. 6, no. 6, C minor.
Cassation, K 63, G major: Adagio. cf Works, selections (ABC ABCL 67010/2).

1858 La clemenza di Tito, K 621. Lucia Popp, s; Janet Baker, Yvonne Minton, Frederica von Stade, ms; Stuart Burrows, t; Robert Lloyd, bs; ROHO and Chorus; Colin Davis. Philips 6703 079 (3).
+Gr 11-77 p886 +RR 11-77 p38
+HFN 11-77 p175

1859 La clemenza di Tito, K 621: Non piu di fiori; Parto, parto. Don Giovanni, K 527: Vedrai, carino. Le nozze di Figaro, K 492: Non so piu; Voi che sapete. ROSSINI: Il barbiere di Siviglia: Una voce poco fa. La cenerentola: Nacqui all'affanno. Otello: Assisa a pie d'un salice. Frederica von Stade, ms; Rotterdam Philharmonic Orchestra; Edo de Waart. Philips 9500 098 Tape (c) 7300 511.
+Gr 2-77 p1317 +-NR 3-77 p10
+-HF 3-77 p118 +-ON 1-29-77 p44
+HFN 2-77 p123 +RR 2-77 p37
+HFN 7-77 p127 tape +RR 9-77 p99 tape
+-MJ 2-77 p30 ++St 5-77 p121

La clemenza di Tito, K 621: Overture. cf Overtures (Eurodisc 27257).
La clemenza di Tito, K 621: Overture. cf Overtures (Turnabout TV 34628).
La clemenza di Tito, K 621: Parto, parto. cf Arias (RCA SER 5675).
La clemenza di Tito, K 621: Parto, parto, ma tu ben mio; Deh per questo istante solo. cf Philips 6767 001.

1860 Concerto, bassoon, K 191, B flat major. Concerto, clarinet, K 622, A major. March, K 249, D major. Thamos, King of Egypt, K 345: Entr'acte, no. 2. Jack Brymer, clt; Gwydion Brooke, bsn; RPO; Thomas Beecham. HMV SXLP 30246 Tape (c) TC SXLP 30246. (Reissues from ASD 344, 423, 259)
++Gr 9-77 p429 +-HFN 10-77 p167
+-Gr 11-77 p899 tape +RR 9-77 p64

Concerto, bassoon, K 191, B flat minor. cf Works, selections (Philips 7699 048).

1861 Concerto, clarinet, K 622, A major. Concerto, flute and harp, K 299, C major. Alfred Prinz, clt; Werner Tripp, flt; Hubert Jellinek, hp; VPO; Karl Munchinger. Decca SPA 495 Tape (c) KCSP 495. (Reissue from SXL 6054)
+Gr 7-77 p179 ++RR 7-77 p57
+HFN 7-77 p125

Concerto, clarinet, K 622, A major. cf Concerto, bassoon, K 191, B flat major.
Concerto, clarinet, K 622, A major. cf Works, selections (Philips 7699 048).
Concerto, flute. cf Works, selections (Philips 7699 048).

1862 Concerto, flute, no. 1, K 313, G major. Concerto, flute, no. 2, K 314, C major. Michel Debost, flt; Maurice Bourgue, ob; Orchestre de Paris; Daniel Barenboim. Angel S 37269.
+NR 5-77 p7

Concerto, flute, no. 1, K 313, G major. cf Andante, K 315, C major (Columbia M 34520).
Concerto, flute, no. 1, K 313, G major. cf Andante, K 315, C major (HMV ASD 3320).
Concerto, flute, no. 1, K 313, G major. cf Andante, K 315, C major (RCA 1-5109).
Concerto, flute, no. 1, K 313, G major. cf Andante, K 315, C major (RCA 1-5330).
Concerto, flute, no. 1, K 313, G major. cf Andante, K 315, C major (Vanguard SRV 364).
Concerto, flute, no. 2, K 314, D major. cf Andante, flute, K 315, C major (Columbia M 34520).
Concerto, flute (oboe), no. 2, K 314, D major. cf Andante, K 315, C major (HMV ASD 3320).
Concerto, flute, no. 2, K 314, D major. cf Andante, K 315, C major (RCA 1-5109).
Concerto, flute, no. 2, K 314, D major. cf Andante, K 315, C major (RCA 1-5330).
Concerto, flute, no. 2, K 314, D major. cf Andante, K 315, C major (Vanguard SRV 364).
Concerto, flute, no. 2, K 314, D major. cf Concerto, flute, no. 1, K 313, G major.
Concerto, flute, no. 2, K 314, D major. cf MILHAUD: Four seasons: Concertino d'hiver, op. 327.

1863 Concerto, flute and harp, K 299, C major. Sinfonia concertante, K 297b, E flat major. Wolfgang Schulz, flt; Nicanor Zabaleta, hp; Walter Lehmayer, ob; Peter Schmidl, clt; Gunther Hogner, hn; Fritz Faltl, bsn; VPO; Karl Bohm. DG 2530 715.
++ARG 10-77 p38
+NR 9-77 p4
+-Gr 11-76 p798
+RR 10-76 p61
+HFN 10-76 p175

Concerto, flute and harp, K 299, C major. cf Concerto, clarinet, K 622, A major.
Concerto, flute and harp, K 299, C major. cf Works, selections (Philips 7699 048).
Concerto, flute and harp, K 299, C major. cf DEBUSSY: Danses sacree et profane.
Concerto, harpsichord, no. 1, K 107, D major. cf HAYDN, M.: Die Hochzeit auf der Alm.

1864 Concerti horn and strings, nos. 1-4, K 412, K 417, K 447, K 495. James Brown, hn; Virtuosi of England; Arthur Davison. Classics for Pleasure CFP 148 Tape (c) TC CFP 148.
+Gr 11-76 p794
+HFN 12-76 p153 tape
+-Gr 2-77 p1325 tape

1865 Concerti, horn and strings, nos. 1-4, K 412, K 417, K 447, K 495. Erich Penzel, hn; VSO; Bernard Baumgartner. Fontana 6530 059. (Reissue from Philips SGL 5838)
+-Gr 11-77 p843
+RR 10-77 p57
+-HFN 10-77 p167

Concerti, horn and strings, nos. 1-4, K 412, K 417, K 447, K 495. cf Works, selections (Philips 7699 048).
Concerto, horn and strings, no. 4, K 495, E flat major. cf Works, selections (CBS 30088).
Concerto, horn and strings, no. 4, K 495, E flat major. cf HMV SLS 5068.
Concerto, horn and strings, no. 4, K 495, E flat major. cf HMV SLS 5073.

Concerto, oboe (flute), K 314, C major. cf Works, selections (Philips 7699 048).
Concerto, oboe, K 314, C major. cf HAYDN: Concerto, oboe, C major.
1866 Concerti, piano, nos. 1-4. Daniel Barenboim, pno; ECO; Daniel Barenboim. HMV ASD 3218 Tape (c) TC ASD 3218. (Reissue from SLS 5031)
+-Gr 1-77 p1147 +-HFN 4-77 p155 tape
1867 Concerti, piano, nos. 6, 12, 20, 22. Karl Engel, pno; Salzburg Mozarteum Orchestra; Leopold Hager. Telefunken AW 6-42040/1.
+-Gr 7-77 p179 +-RR 5-77 p55
Concerto, piano, no. 8, K 246, C major. cf HAYDN: Concerto, harpsichord, op. 21, D major.
1868 Concerti, piano, nos. 9, 17, 21, 23, 25, 27. Rudolf Serkin, pno; Columbia Symphony Orchestra, Marlboro Festival Orchestra; Georg Szell, Alexander Schneider. Odyssey Y3 34642 (3). (Reissues from Columbia ML 5013, 5169, 5209, 5297)
+-ARG 7-77 p28
1869 Concerto, piano, no. 9, K 271, E major. Concerto, piano, no. 21, K 467, C major. Murray Perahia, pno; ECO; Murray Perahia. CBS 76584 Tape (c) 40-76584.
+-Gr 4-77 p1555 +HFN 6-77 p139 tape
+-Gr 7-77 p225 tape +RR 4-77 p54
+-HFN 4-77 p138 +-RR 8-77 p86 tape
Concerto, piano, no. 9, K 271, E flat major. cf EMI Italiana C 153 52425/31.
1870 Concerto, piano, no. 13, K 415, C major. Concerto, piano, no. 22, K 482, E flat major. Sonata, piano, no. 12, K 332, F major. POULENC: Concerto champetre, harpsichord and orchestra. Wanda Landowska, pno and hpd; Symphony Orchestra; Artur Rodzinski, Leopold Stokowski. International Piano Library IPL 106/7.
+ARG 2-77 p17 ++NR 8-73 p4
+-Gr 12-72 p764 +-NR 2-77 p5
+-Gr 3-77 p1453 +NYT 2-20-77 pD13
+MJ 2-77 p30 +RR 3-77 p26
1871 Concerto, piano, no. 13, K 415, C major. Concerto, piano, no. 15, K 450, B flat major. Karl Engel, pno; Salzburg Mozarteum Orchestra; Leopold Hager. Telefunken AW 6-42043.
+HFN 11-77 p175 +-RR 11-77 p61
1872 Concerto, piano, no. 14, K 449, E flat major. Concerto, piano, no. 24, K 491, C minor. Murray Perahia, pno; ECO; Murray Perahia. CBS 76481 Tape (c) 40-76481. (also Columbia M 34219 Tape (c) MT 34219)
++ARG 4-77 p25 +-MJ 4-77 p33
+Gr 5-76 p1761 +-NR 2-77 p4
+Gr 9-76 p497 tape +RR 5-76 p46
+HFN 6-76 p93 ++St 4-77 p124
+-HFN 10-76 p185 tape
1873 Concerto, piano, no. 14, K 449, E flat major. Concerto, piano, no. 23, K 488, A major. Ivan Moravec, pno; Czech Chamber Orchestra; Josef Vlach. Supraphon 110 1768.
+-ARG 12-77 p38 +HFN 9-77 p145
-FF 12-77 p38 +NR 11-77 p7
+-Gr 9-77 p431 +St 12-77 p153
1874 Concerto, piano, no. 14, K 449, E flat major. Concerto, piano, no. 25, K 503, C major. Karl Engel, pno; Salzburg Mozarteum Orchestra; Leopold Hager. Telefunken 6-41925 Tape (c) 4-41925.
+-HF 7-77 p125 tape +-RR 2-76 p32
++HFN 2-76 p105

1875 Concerto, piano, no. 15, K 450, B flat major. Concerto, piano, no. 17, K 453, G major. Ingrid Haebler, pno; Vienna Concert Society Chamber Orchestra, Bamberg Symphony Orchestra; Heinrich Hollreiser. Saga 5436. (Reissues from Vox PL 8300, PL 9390)
+Gr 1-77 p1147 +-RR 6-77 p60

Concerto, piano, no. 15, K 450, B flat major. cf Concerto, piano, no. 13, K 415, C major.

Concerto, piano, no. 17, K 453, G major. cf Concerto, piano, no. 15, K 450, B flat major.

1876 Concerto, piano, no. 18, K 456, B flat major. Symphony, no. 37, K 444, G major. Beveridge Webster, pno; Earl's Court Orchestra; Baird Hastings. Educo 4032.
-Audio 11-77 p129

1877 Concerto, piano, no. 18, K 456, B flat major. Concerto, piano, no. 24, K 491, C minor. Karl Engel, pno; Salzburg Mozarteum Orchestra; Leopold Hager. Telefunken AW 6-41926.
-Gr 3-76 p1468 ++NR 11-77 p7
++HFN 2-76 p105 +-RR 2-76 p32

1878 Concerto, piano, no. 19, K 459, F major. Concerto, piano, no. 23, K 488, A major. Maurizio Pollini, pno; VPO; Karl Bohm. DG 2530 716 Tape (c) 3300 716.
+Gr 1-77 p1147 ++NR 2-77 p4
++Gr 7-77 p225 tape +NYT 3-13-77 pD22
-HFN 1-77 p111 ++RR 1-77 p56
+-HFN 4-77 p155 tape ++SFC 2-13-77 p33
++MJ 3-77 p74 +St 5-77 p115

1879 Concerto, piano, no. 19, K 459, F major. Concerto, piano, no. 27, K 595, B flat major. Geza Anda, pno; Salzburg Mozarteum Orchestra; Geza Anda. DG Tape (c) 3335 244.
++Gr 12-77 p1146

1880 Concerto, piano, no. 19, K 459, F major. Concerto, piano, no. 23, K 488, A major. Alfred Brendel, pno; AMF; Neville Marriner. Philips 6500 283 Tape (c) 7300 227 (r) L 45283.
+Gr 11-72 p915 +MJ 1-74 p10
+Gr 7-77 p225 tape ++NR 3-73 p5
+-HF 3-73 p86 ++RR 11-72 p55
++HF 7-74 p128 tape ++St 4-73 p118
-MJ 12-73 p9

1881 Concerti, piano, nos. 20-27. Geza Anda, pno; Salzburg Mozarteum Orchestra; Geza Anda. DG 2740 138 (4). (no. 24 reissue from 138916; no. 27 from 139477)
+Gr 10-76 p602 +NR 1-77 p6
++HFN 10-76 p181 +-RR 3-77 p55

1882 Concerto, piano, no. 20, K 466, D minor. Concerto, 2 pianos, K 365, E flat major. Andre Previn, Radu Lupu, pno; LSO; Andre Previn. Angel (Q) SQ 37291. (also HMV SQ ASD 3337 Tape (c) TC ASD 3337)
+-Gr 9-77 p430 +NR 9-77 p4
+-Gr 11-77 p899 tape +-RR 9-77 p64
+HFN 9-77 p147 ++St 11-77 p144

1883 Concerto, piano, no. 20, K 466, D minor. Concerto, piano, no. 23, K 488, A major. Alan Schiller, piano; LPO; Charles Mackerras. Classics for Pleasure CFP 40249.
+-Gr 1-77 p1147 +-RR 12-76 p63
+-HFN 1-77 p111

1884 Concerto, piano, no. 20, K 466, D minor. SZYMANOWSKI: Masques, op. 34, no. 1: Sheherazade. Peter Toperczer, pno; Slovak Philharmonic Orchestra; Ladislav Slovak. Rediffusion Royale ROY 2005.
+-Gr 10-77 p637 +-RR 9-77 p64
Concerto, piano, no. 20, K 466, D minor. cf Decca D62D4.
Concerto, piano, no. 20, K 466, D minor. cf EMI Italiana C 153 52425/31.
1885 Concerto, piano, no. 21, K 467, C major. Quartet, oboe. Serenade, no. 13, K 525, G major. Moura Lymphany, pno; Ian Wilson, ob; Virtuosi of England, Gabrieli Quartet; Arthur Davison. Classics for Pleasure CFP 40009 Tape (c) TC CFP 40009.
+-RR 11-77 p118 tape
1886 Concerto, piano, no. 21, K 467, C major. Serenade, no. 6, K 239, D major. Serenade, no. 13, K 525, G major. Annerose Schmidt, pno; Dresden State Orchestra; Otmar Suitner. Philips 6580 112.
+-Gr 5-77 p1697 +-HFN 8-77 p93
+HFN 4-77 p138 +-RR 6-77 p60
Concerto, piano, no. 21, K 467, C major. cf Concerto, piano, no. 9, K 271, E major.
Concerto, piano, no. 21, K 467, C major. cf HMV SLS 5068.
Concerto, piano, no. 21, K 467, C major: Andante. cf Decca SPA 491.
1887 Concerto, piano, no. 22, K 482, E flat major. Rondo, piano, K 382, D major. Rondo, piano, K 386, A major. Alfred Brendel, pno; AMF; Neville Marriner. Philips 9500 145 Tape (c) 7300 521.
++Gr 10-77 p636 ++HFN 11-77 p187 tape
+Gr 11-77 p899 tape +RR 11-77 p62
++HFN 10-77 p159
1888 Concerto, piano, no. 22, K 482, E flat major. Concerto, piano, no. 25, K 503, C major. Edwin Fischer, pno; Orchestra; PhO; Josef Krips, John Barbirolli. Turnabout THS 65094.
+MJ 10-77 p27 +-NR 11-77 p7
Concerto, piano, no. 22, K 482, E flat major. cf Concerto, piano, no. 13, K 415, C major.
1889 Concerto, piano, no. 23, K 488, A major. Concerto, piano, no. 24, K 491, C minor. Wilhelm Kempff, pno; Bamberg Symphony Orchestra; Ferdinand Leitner. DG 2535 204. (Reissue from SLPM 138 645)
+Gr 12-76 p996 +-RR 2-77 p55
+HFN 12-76 p151
Concerto, piano, no. 23, K 488, A major. cf Concerto, piano, no. 14, K 449, E flat major.
Concerto, piano, no. 23, K 488, A major. cf Concerto, piano, no. 19, K 459 F major (DG 2530 716).
Concerto, piano, no. 23, K 488, A major. cf Concerto, piano, no. 19, K 459, F major (Philips 6500 283).
Concerto, piano, no. 23, K 488, A major. cf Concerto, piano, no. 20, K 466, D minor.
1890 Concerti, piano, nos. 24 - 27. Geza Anda, pno; Salzburg Mozarteum Camerata Academica. DG 2726 060 (2). (Reissues from SLPM 139 196, 139384, 139113, 139447)
+Gr 3-77 p1401 +-RR 3-77 p55
1891 Concerto, piano, no. 24, K 491, C minor. Concerto, 3 pianos, K 242, F major. Ingrid Haebler, Ludwig Hoffman Sas Bunge, pno; LSO; Colin Davis, Alceo Galliera. Philips 6580 144 Tape (c) 7317 158. (Reissues from SAL 3642, AXS 12000/1-12)

+-Gr 12-76 p1001 +-HFN 10-77 p169 tape
+HFN 11-76 p171

Concerto, piano, no. 24, K 491, C minor. cf Concerto, piano, no. 14, K 449, E flat major.

Concerto, piano, no. 24, K 491, C minor. cf Concerto, piano, no. 18, K 456, B flat major.

Concerto, piano, no. 24, K 491, C minor. cf Concerto, piano, no. 23, K 488, A major.

Concerto, piano, no. 24, K 491, C minor. cf EMI Italiana C 153 52425/31.

Concerto, piano, no. 24, K 491, C minor: 2nd movement. cf Philips 6580 114.

1892 Concerto, piano, no. 25, K 503, C major. Concerto, piano, no. 27, K 595, B flat major. Friedrich Gulda, pno; VPO; Claudio Abbado. DG 2530 642 Tape (c) 3300 642.
+-Gr 7-76 p181 +NR 12-77 p5
-Gr 10-76 p658 tape +-RR 6-76 p45
+-HFN 7-76 p91

Concerto, piano, no. 25, K 503, C major. cf Concerto, piano, no. 14, K 449, E flat major.

Concerto, piano, no. 25, K 503, C major. cf Concerto, piano, no. 22, K 482, E flat major.

Concerto, piano, no. 25, K 503, C major. cf EMI Italiana C 153 52425/31.

Concerto, piano, no. 27, K 595, B flat major. cf Concerto, piano, no. 19, K 459, F major

Concerto, piano, no. 27, K 595, B flat major. cf Concerto, piano, no. 25, K 503, C major.

1893 Concerti, 2 pianos, nos. 10, 12, 18, 20. Robert Casadesus, Gaby Casadesus, pno; Columbia Symphony Orchestra; Georg Szell. Odyssey Y2 34641 (2). (Reissues from Columbia ML 5151, 5276)
++ARG 7-77 p28

Concerto, 2 pianos, K 365, E flat major. cf Concerto, piano, no. 20, K 466, D minor.

Concerto, 3 pianos, K 242, F major. cf Concerto, piano, no. 24, K 491, C minor.

Concerti, violin, nos. 1-5. cf Works, selections (Angel SD 3789).

Concerto, violin, no. 1, K 207, B flat major. cf Adagio, K 261, E major.

Concerto, violin, no. 1, K 207, B flat major. cf Works, selections (ABC ABCL 67010/2).

1894 Concerto, violin, no. 2, K 211, D major. Concerto, violin, no. 4, K 218, D major. Hermann Krebbers, vln; Netherlands Chamber Orchestra; David Zinman. Philips 6580 120 Tape (c) 7317 148.
++Gr 6-76 p51 +RR 6-76 p46
/HFN 8-76 p85 /RR 10-77 p97 tape
+HFN 8-77 p99 tape

Concerto, violin, no. 2, K 211, D major. cf Works, selections (ABC ABCL 67010/2).

Concerti, violin, nos. 3-5. cf Works, selections (Philips 7699 048).

1895 Concerto, violin, no. 3, K 216, G major. Sinfonia concertante, K 364, E flat major. Isaac Stern, vln; CO, LSO; Walter Trampler, vla; Isaac Stern, Georg Szell. CBS 61810 Tape (c) 40-61810. (Reissue from 72662)
+-Gr 9-77 p431 +-RR 9-77 p66
+-HFN 9-77 p151 +-RR 12-77 p96 tape
/HFN 10-77 p169 tape

1896 Concerto, violin, no. 3, K 216, G major. Concerto, violin, no. 5, K 219, A major. Jose Luis Garcia, vln; ECO; Jose Luis Garcia. HNH 4030.
++SFC 10-9-77 p40 ++St 12-77 p154
Concerto, violin, no. 3, K 216, G major. cf MENDELSSOHN: Concerto, violin, op. 64, E minor.
Concerto, violin, no. 3, K 216, G major: Adagio. cf Discopaedia MB 1013.
Concerto, violin, no. 4, K 218, D major. cf Concerto, violin, no. 2, K 211, D major.
1897 Concerto, violin, no. 5, K 219, A major. SCHUMANN: Concerto, violin, D minor. Georg Kulenkampff, vln; BPhO, Berlin State Opera Orchestra; Hans Schmidt-Isserstedt, Artur Rother. Telefunken AJ 6-42216.
+HFN 12-77 p175 +RR 12-77 p51
Concerto, violin, no. 5, K 219, A major. cf Concerto, violin, no. 3, K 216, G major.
Concerto, violin, no. 5, K 219, A major. cf BACH: Concerto, violin and strings, S 1042, E major.
Concerto, violin, no. 5, K 219, A major. cf RCA CRM 6-2264.
Concerto, violin, no. 5, K 219, A major: 1st movement. cf Discopaedia MOB 1019.
Concerto, violin, no. 7, K 271a, D major. cf Concerto, violin, K Anh 294a, D major.
1898 Concerto, violin, K Anh 294a, D major. Concerto, violin, no. 7, K 271a, D major. Yehudi Menuhin, vln; Menuhin Festival Orchestra; Yehudi Menuhin. HMV ASD 3198. (also Angel S 37167)
+Gr 6-76 p51 +NR 10-76 p5
+-HFN 6-76 p93 +-RR 6-76 p46
+-MJ 1-77 p27
1899 Concerti, winds, complete. Jack Brymer, clt; Michael Chapman, bsn; Claude Monteux, flt; Neil Black, ob; Osian Ellis, hp; Alan Civil, hn; AMF; Neville Marriner. Philips 6747 377 (4).
+Gr 9-77 p490 +RR 9-77 p28
++HFN 9-77 p153
1900 Concertone, 2 violins , K 190 (K 166b), C major. Sinfonia concertante, K 364, E flat major. Soloists; Jean-Francois Paillard Orchestra; Jean-Francois Paillard. Denon PCM OX 7022ND.
+-HF 10-77 p112
Concertone 2 violins, K 190 (K 166b), C major. cf Works, selections (Angel SD 3789).
1901 Contradanses (5). Divertimenti, C major, G minor, D major, F major, B flat major, E flat major. AMF; Neville Marriner. Philips 6833 222. (Reissue from 6500 367)
+RR 9-77 p66
Contradanses, F major (3), G major, C major. cf Works, selections (Philips 6500 367).
Contradanses, K 535. cf BEETHOVEN: German dances, WoO 8.
Contradanses, K 609 (5). cf BEETHOVEN: German dances, WoO 8.
Contradances, K 609 (5). cf DG 2533 182.
Contradanses, K 609 (5). cf DG 2723 051.
1902 Cosi fan tutte, K 588. Elisabeth Schwarzkopf, Hanny Steffek, s; Christa Ludwig, ms; Alfredo Kraus, t; Guiseppe Taddei, bar; Walter Berry, bs; PhO and Chorus; Karl Bohm. Angel S 3631. (also HMV SLS 5028 Tape (c) TC SLS 5028).
+-Gr 10-75 p679 +MT 1-76 p41
+-Gr 5-77 p1743 tape +RR 11-75 p34

+-HFN 11-75 p159 +St 4-75 p68
+HFN 5-77 p138 tape

1903 Cosi fan tutte, K 588. Pilar Lorengar, Jane Berbie, s; Teresa Berganza, ms; Ryland Davies, t; Tom Krause, bar; Gabriel Bacquier, bar; ROHO Chorus; LPO; Georg Solti. Decca D56D4 (4). (Reissue from SET 575/8)
+Gr 8-77 p339 +-RR 8-77 p35
+HFN 8-77 p95

1904 Cosi fan tutte, K 588. Don Giovanni, K 527. Le nozze di Figaro, K 492. Mirella Freni, Jessye Norman, Lillian Watson, Martina Arroyo, Montserrat Caballe, Ileana Cotrubas, s; Maria Casula, Janet Baker, ms; Yvonne Minton, Kiri Te Kanawa, con; Robert Tear, David Lennox, Stuart Burrows, Nicolai Gedda, t; Wladimiro Ganzarolli, Ingvar Wixell, Paul Hudson, bar; Clifford Grant, Luigi Roni, Richard van Allan, bs; BBC Symphony Orcestra and Chorus, ROHO and Chorus; Colin Davis. (Includes "Mozart the Musician") Philips 6747 280. (Reissues)
+-Gr 4-77 p1567 +-RR 3-77 p35
+HFN 4-77 p151

Cosi fan tutte, K 588: Un aura amorosa; Ah lo veggio; In qual fiero. cf Arias (L'Oiseau-Lyre DSLO 13).

Cosi fan tutte, K 588: Overture. cf Overtures (DG 2535 229).

Divertimenti (6). cf Works, selections (Philips 6747 378).

Divertimenti, C major, G minor, D major, F major, B flat major, E flat major. cf Contradanses (Philips 6833 222).

Divertimenti, C major, G minor, D major, F major, B flat major, E flat major. cf Works, selections (Philips 6500 367).

Divertimenti, K 136-138. cf Quartets, strings, nos. 1-13.

1905 Divertimento, no. 1, K 136, D major. Quintet, clarinet, K 581, A major. Thea King, clt; Aeolian Quartet. Saga Tape (c) CA 5291.
+-Gr 2-77 p1322 tape +RR 11-76 p109 tape
+-HFN 10-76 p185 tape

1906 Divertimenti, no. 3, K 166, E flat major; no. 4, K 186, B flat major; K Anh 226, E flat major; K Anh 227, B flat major. Vienna Philharmonic Wind Ensemble. DG 2530 703 Tape (c) 3300 703.
++Gr 7-76 p181 +-NR 3-77 p8
+Gr 9-76 p497 tape ++RR 7-76 p61
++HFN 8-76 p85 ++RR 11-76 p109 tape

Divertimento, no. 3, K 166, E flat major. cf Westminster WGS 8338.

Divertimento, no. 5, K 187, C major. cf Works, selections (Supraphon 111 1671/2).

1907 Divertimento, string trio, K 563, E flat major. Isaac stern, vln; Pinchas Zukerman, vla; Leonard Rose, vlc. Columbia M 33266. (also CBS 76381)
++Gr 8-75 p342 +RR 8-75 p52
+HF 8-75 p90 +-SFC 5-18-75 p23
+HFN 8-75 p79 +St 8-75 p101
+-NR 6-75 p8 ++St 5-77 p71
+NYT 8-8-76 pD13

Divertimento, string trio, K 563, E flat major. cf RCA CRM 6-2264.

Divertimento, K Anh 229, B flat major. cf Works, selections (Supraphon 111 1671/2).

1908 Don Giovanni, K 527. Ljuba Welitsch, Elisabeth Schwarzkopf, Irmgard Seefried, s; Tito Gobbi, Alfred Poell, bar; Anton Dermota, t; Erich Kunz, Josef Greindl, bs; VPO; Wilhelm Furtwangler. Bruno Walter Society RR 407 (3).
+-NR 7-77 p9

1909 Don Giovanni, K 527. Birgit Nilsson, Leontyne Price, Eugenia Ratti, s; Cesare Valletti, t; Heinz Blankenburg, bar; Cesare Siepi, Fernando Corena, Arnold van Mill, bs; Vienna State Opera Chorus; VPO; Erich Leinsdorf. Decca D10D4 (4). (Reissue from RCA SER 4528/31)
-Gr 12-76 p1046
+-HFN 1-77 p121
+RR 12-76 p44

1910 Don Giovanni, K 527. Joan Sutherland, Elisabeth Schwarzkopf, Graziella Sciutti, s; Luigi Alva, t; Eberhard Wachter, Giuseppe Taddei, Piero Cappuccilli, bar; Gottlob Frick, bs; PhO and Chorus; Carlo Maria Giulini. HMV SLS 5083 (3) Tape (c) TC SLS 5083. (Reissue from Columbia SAX 2369/72) (also Angel S 3605)
++Gr 7-77 p216
+HFN 8-77 p95
+NYT 10-9-77 pD21
+RR 7-77 p39

1911 Don Giovanni, K 527. Martina Arroyo, Kiri Te Kanawa, Mirella Freni, s; Stuart Burrows, t; Ingvar Wixell, Wladimiro Ganzarolli, bar; Luigi Roni, Richard Van Allan, bs; ROHO and Chorus; Colin Davis. Philips 6707 022 (4).
+Gr 11-73 p979
+HF 7-74 p77
+HFN 11-73 p2320
+MJ 7-74 p50
+NR 6-74 p8
+NYT 4-7-74 pD28
+NYT 10-9-77 pD21
+ON 4-13-74 p28
-Op 1-74 p52
+-RR 11-73 p28
++SFC 5-5-74 p28
+St 8-74 p117

1912 Don Giovanni, K 527. Soloists; Glyndebourne Festival Orchestra; Fritz Busch. Turnabout THS 65084/6.
+NYT 10-9-77 pD21

Don Giovanni, K 527. cf Cosi fan tutte, K 588.

Don Giovanni, K 527: Deh vieni alla finestra; Fin ch'han dal vino calda la testa. cf Philips 6580 174.

Don Giovanni, K 527: In quali accessi...Mi tradi quell'alma ingrata; Crudele, Ah no, mio bene...Non mi dir. cf Arias (RCA SER 5675).

Don Giovanni, K 527: Madamina. cf Works, selections (CBS 30088).

Don Giovanni, K 527: Madamina. cf Decca ECS 811.

Don Giovanni, K 527: Madamina; Finch'han dal vino. cf Golden Age GAR 1001.

Don Giovanni, K 527: Mine be her burden; Speak for me to my lady. cf HMV HLP 7109.

Don Giovanni, K 527: Il mio tesoro. cf Philips 9500 218.

Don Giovanni, K 527: Il mio tesoro; Dalla sua pace. cf Arias (L'Oiseau-Lyre DSLO 13).

Don Giovanni, K 527: Nur ihrem Frieden; Folget der Heissgeliebten. cf Telefunken AJ 6-42232.

Don Giovanni, K 527: Or sai chi l'onore. cf Decca D65D3.

Don Giovanni, K 527: Overture. cf Overtures (DG 2535 229).

Don Giovanni, K 527: Overture. cf Overtures (Eurodisc SQ 27257).

Don Giovanni, K 527: Overture. cf Pye PCNHX 6.

Don Giovanni, K 527: Vedrai, carino. cf La clemenza di Tito, K 621: Non piu di fiori; Parto, parto.

Duettino concertante (arr. Busoni). cf International Piano Library IPL 108.

Duos, 2 horns, K 487 (12). cf Works, selections (Supraphon 111 1671/2).

1913 Duo, violin and viola, K 423, G major. Duo, violin and viola, K 424, B flat major. HOFFMEISTER: Duo, violin and viola, G major. Arthur Grumiaux, vln; Arrigo Pelliccia, vla. Philips 839747.

+HF 7-71 p81 ++SFC 10-24-71 p34
+NR 6-71 p6 ++St 6-71 p92
+SR 6-26-71 p46 ++St 5-77 p71

Duo, violin and viola, K 424, B flat major. cf Duo, violin and viola, K 423, G major.

Duo, violin and viola, K 424, B flat major. cf RCA CRM 6-2264.

1914 Die Entfuhrung aus dem Serail, K 384. Lois Marshall, Ilse Hollweg, s; Leopold Simoneau, Gerhard Unger, t; Gottlob Frick, bs; Beecham Choral Society; RPO; Thomas Beecham. EMI 2C 16701541/2.

+RR 9-77 p40

Die Entfuhrung aus dem Serail, K 384. cf Bastien und Bastienne, K 50.

Die Entfuhrung aus dem Serail, K 384: Ach, ich liebte. cf DONIZETTI: Don Pasquale: So anch'io la virtu magica.

Die Entfuhrung aus dem Serail, K 384: Hier soll ich dich denn sehen; Konstanze...O wie angstlich; Im Mohrenland. cf Telefunken AJ 6-42232.

Die Entfuhrung aus dem Serail, K 384: Hier soll ich dich; Wenn der Freude; Constanze; Ich baue ganz. cf Arias (L'Oiseau-Lyre DSLO 13).

Die Entfuhrung aus dem Seraglio, K 384: I was heedless in my rapture. cf HMV HLM 7066.

Die Entfuhrung aus dem Serail, K 384: Martern aller Arten. cf Arias (RCA SER 5675).

Die Entfuhrung aus dem Serail, K 384: Overture. cf Overtures (DG 2535 229).

Die Entfuhrung aus dem Serail, K 384: Overture. cf Overtures (Eurodisc SQ 27257).

Exsultate jubilate, K 165. cf BACH: Cantata, no. 202, Weichet nur betrubte Schatten.

Exsultate jubilate, K 165. cf Works, selections (CBS 30088).

Exsultate jubilate, K 165. cf Works, selections (Philips 6747 384).

Exsultate jubilate, K 165. cf Works, selections (Philips 7699 052).

Fantasia, K 396, C minor. cf BACH: Fantasia, S 906, C minor.

Fantasia, K 396, C minor. cf BEETHOVEN: Sonata, piano, no. 22, op. 54, F major.

1915 Fantasia, K 397, D major. Sonata, piano, no. 9, K 311, D major. Sonata, piano, no. 10, K 330, C major. HAYDN: Variations, F minor. Alicia de Larrocha, pno. London CS 7008 Tape (c) 5-7008.

+HF 2-77 p100

Fantasia, K 397, D minor. cf BACH, C.P.E.: Fantasia, Wq 254, C minor.

Fantasia, K 397, D minor. cf HAYDN: Andante and variations, F minor.

1916 Fantasia, K 475, C minor. Sonata, piano, no. 14, K 457, C minor. Sonata, piano, no. 15, K 545, C major. Maria Jaoa Pires, pno. Denon PCM OX 7057ND.

+HF 10-77 p112

1917 Fantasia, K 475, C minor. Sonata, piano, no. 8, K 310, A minor.

Sonata, piano, no. 14, K 457, C minor. Daniel Barenboim, pno. Westminster WG 1010.
+-Gr 3-77 p1429 +-RR 6-77 p82
Fantasia, K 475, C minor. cf BEETHOVEN: Concerto, piano, no. 1, op. 15, C major.
Fantasia, K 594, F minor. cf BACH: Chorale preludes, S 645-50.
Fantasia, K 608, F minor. cf BACH: Chorale preludes, S 645-50.
Fantasia, organ. cf Works, selections (Philips 6747 384).
Fantasia, organ, K 608, F minor. cf BEETHOVEN: Adagio and allegro, music box.
Fantasia, organ, K 608, F minor. cf Wealden WS 131.
1918 Fantasia and fugue, K 394, C major. Sonata, piano, no. 13, K 333, B flat major. Variations on "Ein Weib ist das herrlichste Ding", K 613. David Ward, pno. Saga 5435.
-Gr 2-77 p1307 +-RR 5-77 p75
+-HFN 1-77 p111
La finta giardiniera, K 199 and 207a: Overture. cf Overtures (Eurodisc SQ 27257).
La finta semplice, K 46a: Overture. cf Overtures (Eurodisc SQ 27257).
1919 Fugue, 2 pianos, K 426, C minor. Sonata, 2 pianos, K 448, D major. Sonata, piano, 4 hands, K 521, C major. Alfons Kontarsky, Aloys Kontarsky, pno. CMS/Oryx 3C 322.
+NR 9-77 p13 +-St 3-77 p138
Galithmathias Musicum, K 32. cf HAYDN, M.: Die Hochzeit auf der Alm.
German dances, K 509 (6). cf BEETHOVEN: German dances, WoO 8.
German dances, no. 3, K 605. cf CBS 61780.
Idomeneo, Re di Creta, K 366: D'Oreste, d'ajace. cf Arias (Hungaroton SLPX 11812).
Idomeneo, Re di Creta, K 366: Fuor del mar. cf Arias (L'Oiseau-Lyre DSLO 13).
Idomeneo, Re di Creta, K 366: Overture. cf Overtures (Turnabout TV 34628).
Idomeneo, Re di Creta, K 366: Overture and ballet music, K 367. cf Overtures(Eurodisc SQ 27257).
Idomeneo, Re di Creta, K 366: Parto, el l'unico oggetto. cf Arias (RCA SER 5675).
Eine kleine Gigue, K 574, G major. cf HAYDN: Sonata, piano, no. 46, A flat major.
Kyrie. cf Works, selections (Philips 6747 384).
Kyrie, K 341, D minor. cf Works, selections (Philips 7699 052)
Landler, K 606 (6). cf DG 2533 182.
Landler, K 606 (6). cf DG 2723 051.
Lucia Silla, K 135: Overture. cf Overtures (Eurodisc SQ 27257).
Lucia Silla, K 135: Overture. cf Overtures (Turnabout TV 34628).
Lullaby. cf HMV HLM 7066.
Lullaby. cf Kaibala 40D03.
Marches (6). cf Works, selections (Philips 6747 378).
March, K 215, D major. cf Serenade, no. 5, K 213a (K 204), D major.
March, K 237, D major. cf DG 2548 148.
March, K 249, D major. cf Concerto, bassoon, K 191, B flat major.
March, K 249, D major. cf Serenade, no. 7, K 250, D major.
Marches, K 335 (K 320a) (2). cf Serenade, no. 9, K 320, D major.
March, K 408, no. 2, D major. cf BEETHOVEN: German dances, WoO 8.
March, K 408, no. 3, C major. cf BEETHOVEN: German dances, WoO 8.

1920 Mass, no. 4, K 139, C minor. Gundula Janowitz, s; Frederica von Stade, ms; Wieslaw Ochman, t; Kurt Moll, bs; Rudolf Scholz, org; Vienna State Opera Chorus; VPO; Claudio Abbado. DG 2530 777 Tape (c) 3300 777.
+ARG 9-77 p20 +HFN 3-77 p107
+Gr 3-77 p1438 ++NR 9-77 p9
+HF 9-77 p102 +RR 3-77 p93
Mass, no. 4, K 139, C minor. cf Works, selections (Philips 6747 384).
Mass, no. 6, K 192, F major. cf Works, selections (Philips 6747 384).
Mass, no. 7, K 167, C major. cf HAYDN: Mass, no. 5, B flat major.
Mass, no. 8, K 220, C minor. cf Works, selections (Philips 7699 052).
Mass, no. 10, K 229, C major. cf Works, selections (Philips 6747 384).
Mass, K 257, C major. cf Works, selections (Philips 6747 384).
Mass, K 257, C major. cf Works, selections (Philips 7699 052).
Mass, no. 13, K 259, C major. cf Works, selections (Philips 6747 384).
Mass, no. 14, K 317, C minor. cf Works, selections (Philips 7699 052).
1921 Mass, no. 16, K 317, C major. Mass, no. 18, K 427, C minor. Lamoureux Orchestra and Chorus; Igor Markevitch. DG 2535 148 Tape (c) 3355 148.
+Gr 7-77 p225 tape
1922 Mass, no. 16, K 317, C major. Vesperae solennes de confessore, K 339, C major. Edda Moser, s; Julia Hamari, ms; Nicolai Gedda, t; Dietrich Fischer-Dieskau, bar; Bavarian Radio Orchestra and Chorus; Eugen Jochum. HMV SQ ASD 3373 Tape (c) TC ASD 3373. (also Angel S 37283)
+Gr 10-77 p685 +-RR 12-77 p79
-Gr 12-77 p1146 tape +-SFC 12-11-77 p61
Mass, no. 16, K 317, C major. cf Works, selections (Philips 6747 384).
Mass, no. 16, K 317, C major. cf BRUCKNER: Te deum.
Mass, no. 18, K 427, C minor. cf Mass, no. 16, K 317, C major.
Mass, no. 18, K 427, C minor. cf Works, selections (Philips 6747 384).
Mass, no. 18, K 427, C minor. cf Works, selections (Philips 7699 052).
1923 Mass, no. 19, K 626, D minor (ed. Beyer). Ileana Cotrubas, s; Helen Watts, con; Robert Tear, t; John Shirley Quirk, bs; AMF and Chorus; Neville Marriner. Argo ZRG 876.
+Gr 12-77 p1122 +RR 12-77 p79
+HFN 12-77 p175
1924 Mass, no. 19, K 626, D minor. Elly Ameling, s; Marilyn Horne, ms; Ugo Benelli, t; Tugomir Franc, bs; Vienna State Opera Chorus; VPO; Istvan Kertesz. Decca SPA 476 Tape (c) KCSP 476. (Reissue from SET 302)
+-Gr 6-77 p84 +HFN 2-77 p135 tape
+Gr 7-77 p225 tape +-RR 2-77 p97 tape
+HFN 4-77 p153 /-RR 4-77 p82
1925 Mass, no. 19, K 626, D minor. Anna Tomowa-Sintow, s; Agnes Baltsa, con; Werner Krenn, t; Jose van Dam, bs; Wienna Singverein; BPhO; Herbert von Karajan. DG 2530 705 Tape (c) 3300 705.

+ARG 3-77 p27
+Gr 12-76 p1034
+Gr 7-77 p25 tape
+HFN 11-76 p164
+-MJ 3-77 p74
++NR 3-77 p11
+RR 11-76 p97
-ON 3-5-77 p33
++SFC 7-31-77 p40

1926 Mass, no. 19, K 626, D minor. Elly Ameling, s; Barbara Scherler, alto; Louis Devos, t; Roger Soyer, bs; Gulbenkian Foundation Symphony Orchestra and Chorus; Michel Corboz. RCA AGL 1-1533. (also Erato STU 70943)

+-Gr 6-76 p72
+MJ 10-76 p24
+-MU 10-76 p16
+-NR 8-76 p10
-ON 3-5-77 p33
-RR 6-76 p79
-St 11-76 p148

1927 Mass, no. 19, K 626, D minor. Carole Bogard, s; Ann Murray, ms; Richard Lewis, t; Michael Rippon, bs; Amor Artis Chorale; ECO: Johannes Sommary. Vanguard VSD 71211.

+-MJ 3-77 p74
++NR 10-76 p9
++St 11-76 p148

Mass, no. 19, K 626, D minor. cf Works, selections (Philips 7699 052).

Mass, no. 19, K 626, D minor. cf Works, selections (Philips 6747 384).

Minuet, K 355, D major. cf HAYDN: Sonata, piano, no. 46, A flat major.

Minuets, K 463 (2). cf BEETHOVEN: German dances, WoO 8.

Mitridate, Re di Ponto: Overture. cf Overtures (Eurodisc SQ 27257).

Mitridate, Re di Ponto: Overture. cf Overtures (Turnabout TV 34628).

Nocturne, K 286. cf Works, selections (Philips 6747 378).

1928 Le nozze di Figaro, K 492. Sesto Bruscantini, bar; Orchestra; Fernando Previtali. Cetra LPS 3219. (Reissue from LPC 1219)

-Gr 7-77 p222

1929 Le nozze di Figaro, K 492. Lisa della Casa, Hilde Gueden, Suzanne Danco, s; Alfred Poell, bar; Fernando Corena, Cesare Siepi, bs; Vienna State Opera Chorus; VPO; Erich Kleiber. Decca Tape (c) K79K32 (2).

++Gr 12-77 p1146 tape
++RR 12-77 p96 tape

1930 Le nozze di Figaro, K 492. Gundula Janowitz, Edith Mathis, Barbara Bogel, s; Tatiana Troyanas, Patricia Johnson, ms; Erwin Wohlfahrt, Martin Vantin, t; Dietrich Fischer-Dieskau, Hermann Prey, bar; Peter Lagger, Klaus Hirte, bs; German Opera Orchestra and Chorus; Karl Bohm. DG 2740 139 (4). (Reissue from SLPM 139276/9)

+Gr 12-77 p1134
+HFN 12-77 p185
+RR 12-77 p32

1931 Le nozze di Figaro, K 492. Heather Harper, Judith Blegen, Elizabeth Gale, Elizabeth Ritchie, Patricia O'Neill, s; Teresa Berganza, Birgit Finnila, ms; John Fryatt, John Robertson, t; Dietrich Fischer-Dieskau, Geraint Evans, Malcolm Donnelly, bar; William McCue, bs; John Alldis Choir; ECO; Daniel Barenboim. HMV SQ SLS 995 (4).

+Gr 7-77 p215
+HFN 8-77 p75
+-RR 8-77 p35

Le nozze di Figaro, K 492. cf Cosi fan tutte, K 588.

Le nozze di Figaro, K 492: Deh vieni, non tardar. cf DONIZETTI: Don Pasquale: So anch'io la virtu magica.

Le nozze di Figaro, K 492: Deh viene, non tardar. cf Seraphim IB 6105.

Le nozze di Figaro, K 492: Dove sono. cf Advent S 5023.
Le nozze di Figaro, K 492: E Susanna non viene...Dove sono. cf Angel S 37446.
Le nozze di Figaro, K 492: Giunse alfin il momento...Deh vieni non tardar; E Susanna no viene...Dove sono; Voi che sapete. cf Arias (RCA SER 5675).
Le nozze di Figaro, K 492: Hai gia vinta la causa...Vedro mentr'io sospiro. cf Philips 6580 174.
Le nozze di Figaro (The marriage of Figaro), K 492: Non piu andrai. cf Decca SPA 450.
Le nozze di Figaro, K 492: Non piu andrai. cf Decca ECS 811.
Le nozze di Figaro, K 492: Non so piu; Voi, che sapete. cf La clemenza di Tito, K 621: Non piu di fiori; Parto, parto.
Le nozze di Figaro, K 492: Overture. cf Overtures (DG 2535 229).
Le nozze di Figaro, K 492: Overture. cf Overtures (Eurodisc SQ 27257).
Le nozze di Figaro (The marriage of Figaro), K 492: Overture. cf Symphony, no. 41, K 551, C major.
Le nozze di Figaro (The marriage of Figaro), K 492: Overture. cf Classics for Pleasure CFP 40236.
Le nozze di Figaro, K 492: Porgi amor. cf Club 99-109.
Le nozze di Figaro, K 492: Porgi amor. cf HMV RLS 719.
Le nozze di Figaro (The marriage of Figaro), K 492: Se vuol ballare. cf Works, selections (CBS 30088).
Le nozze di Figaro, K 492: Voi che sapete. cf Abbey LPB 778.
Le nozze di Figaro, K 492: Voi che sapete. cf Rubini GV 57.

1932 Overtures: Cosi fan tutte, K 588. Don Givanni, K 527. Die Entfuhrung aus dem Serail, K 384. Le nozze di Figaro, K 492. Der Schauspieldirektor, K 486. Symphony, no. 32, K 318, G major (Overture in the Italian style). Die Zauberflote, K 620. Dresden Staatskapelle Orchestra, BPhO, German Opera Orchestra, Prague National Theatre Orchestra; Karl Bohm. DG 2535 229 Tape (c) 3335 229. (Reissues from 2740 112, 2709 017, 2709 059, 2711 007, 2711 006, SLPM 138112)
++Gr 8-77 p306 +-HFN 7-77 p125
+-Gr 9-77 p509 +RR 7-77 p58

1933 Overtures: Apollo et Hyacinthus, K 38. Ascanio in Alba, K 111. Bastien und Bastienne, K 50. Betulia liberata, K 74c. La clemenza di Tito, K 621. Don Giovanni, K 527. Die Entfuhrung aus dem Serail, K 384. La finta giardiniera, K 199 & 207a. La finta semplice, K 46a. Idomeneo, Re di Creta, K 366: Overture and ballet music, K 367. Lo sposo deluso, K 430. Lucio Silla, K 135. Mitridate, Re di Ponto. Le nozze di Figaro, K 492. Il Re pastore, K 208. Der Schauspieldirektor, K 486. Il sogno di Scipione, K 126. Die Zauberflote, K 620. Basel Symphony Orchestra; Moshe Atzmon. Eurodisc SQ 27257 XDR (3).
+ARG 5-77 p56

1934 Overtures: Apollo et Hyacinthus, K 38. Ascanio in Alba, K 111. Bastien und Bastienne, K 50. La clemenza di Tito, K 621. Idomeneo, Re di Creta, K 366. Lucio Silla, K 135. Mitridate, Re di Ponto. Il Re pastore, K 208. Der Schauspieldirektor, K 486. Il sogno di Scipione, K 126. Wurttemberg Chamber Orchestra; Jorg Faerber. Turnabout TV 34628 Tape (c) KTVC 34628.
+-Gr 9-77 p430 +-RR 8-77 p57
+HFN 9-77 p145 +-RR 10-77 p97 tape

Les petites riens, K Anh 10. cf Symphony, no. 31, K 297, D major.
Prelude on "Ave verum", K 580a. cf Works, selections (CBS 30088).

1935 Quartet, flute, K 285, D major. PERGOLESI (attrib): Concertino, no. 6, B flat major. TELEMANN: Concerto, alto recorder, viola da gamba and strings, A minor. Aston Magna; Albert Fuller. Cambridge CRS 2827.
/HF 7-77 p114 +NR 3-77 p9
++IN 9-77 p28
Quartet, flute, K 285, D major. cf Andantino, violoncello and piano, K 374, B minor.
Quartet, oboe. cf Concerto, piano, no. 21, K 467, C major.
1936 Quartet, oboe, K 370, F major. Quintet, clarinet, K 581, A major. Gervase de Peyer, clt; Lothar Koch, ob; Amadeus Quartet. DG 2530 720 Tape (c) 3300 720.
+Gr 2-77 p1301 +HFN 4-77 p155 tape
+Gr 7-77 p225 tape +MT 7-77 p566
++HF 11-77 p118 +RR 2-77 p73
+HFN 2-77 p123 +RR 4-77 p92 tape
Quartet, oboe, K 370, F major. cf Andantino, violoncello and piano, K 374, B minor.
Quartet, oboe, K 370, F major. cf Works, selections (Supraphon 111 1671/2).
Quartet, oboe, K 370, F major. cf LOEFFLER: Rhapsodies, oboe, viola and piano.
Quartet, piano, K 478, G minor. cf Andantino, violoncello and piano, K 374, B minor.
1937 Quartets, strings, nos. 1-13. Divertimenti, K 136-138. Amadeus Quartet. DG 2740 165 (4).
+-Gr 9-77 p451 +-RR 10-77 p67
+HFN 11-77 p175
1938 Quartet, strings, no. 14, K 387, G major. Quartet, strings, no. 15, K 421, D minor. Melos Quartet. DG 2530 898.
+Gr 11-77 p860 +RR 10-77 p67
++HFN 11-77 p175
1939 Quartet, strings, no. 14, K 387, G major. Quartet, strings, no. 15, K 421, D minor. Alban Berg Quartet. Telefunken AW 6-42039.
++Gr 10-77 p654 ++RR 9-77 p76
+HFN 9-77 p147
Quartet, strings, no. 15, K 421, D minor. cf Quartet, strings, no. 14, K 387, G major (DG 2530 898).
Quartet, strings, no. 15, K 421, D minor. cf Quartet, strings, no. 14, K 387, G major (Telefunken AW 6-42039).
1940 Quartet, strings, no. 16, K 428, E flat major. Quartet, strings, no. 17, K 458, B flat major. Melos Quartet. DG 2530 800 Tape (c) 3300 800.
+Gr 5-77 p1707 +RR 5-77 p62
+HFN 5-77 p129
Quartet, strings, no. 17, K 458, B flat major. cf Quartet, strings, no. 16, K 428, E flat major.
1941 Quartet, strings, no. 19, K 465, C major. Quartet, strings, no. 22, K 589, B flat major. Tokyo Quartet. DG 2530 468.
+Gr 3-75 p1674 ++SFC 12-15-74 p33
+HF 3-75 p85 ++SR 2-22-75 p47
++NR 1-75 p8 ++St 4-75 p103
+RR 3-75 p38 ++St 5-77 p72
1942 Quartets, strings, nos. 20-23. Juilliard Quartet. Columbia MG 33976 (2). (also CBS 79204)

+Gr 10-76 p621 +NYT 8-8-76 pD13
+HF 2-77 p99 +RR 10-76 p81
++HFN 10-76 p175 ++SFC 7-18-76 p31
+NR 7-76 p6 +-St 9-76 p122

1943 Quartet, strings, no. 21, K 575, D major. Quartet, strings, no. 23, K 590. F major. Kuchl Quartet. Decca SDD 509 Tape (c) KSDC 509.
+Gr 11-76 p830 +-HFN 12-76 p155
+Gr 7-77 p225 tape +-HFN 2-77 p135 tape
+-HFN 10-76 p175 +RR 11-76 p86

1944 Quartet, strings, no. 22, K 589, B flat major. Quartet, strings, no. 23, K 590, F major. Alban Berg Quartet. Telefunken AW 6-42042.
+-Gr 5-77 p1707 ++RR 5-77 p65

Quartet, strings, no. 22, K 589, B flat major. cf Quartet, strings, no. 19, K 465, C major.

Quartet, strings, no. 23, K 590, F major. cf Quartet, strings, no. 21, K 575, D major.

Quartet, strings, no. 23, K 590, F major. cf Quartet, strings, no. 22, K 589, B flat major.

1945 Quintet, clarinet, K 581, A major. WEBER: Quintet, clarinet, op. 34, B flat major. Gervase de Peyer, clt; Melos Ensemble. HMV HQS 1395 Tape (c) TC HQS 1395. (Reissues from ASD 605, 2374)
++Gr 9-77 p452 +RR 8-77 p68
+MT 12-77 p1015 +RR 10-77 p96 tape

Quintet, clarinet, K 581, A major. cf Divertimento, no. 1, K 136, D major.

Quintet, clarinet, K 581, A major. cf Quartet, oboe, K 370, F major.

Quintet, horn, K 407, E flat major. cf Works, selections (Supraphon 111 1671/2).

1946 Quintet, piano and winds, K 452, E flat major. Trio, clarinet, K 498, E flat major. Vienna Octet Members; Walter Panhöffer, pno. Decca ECS 796. (Reissue from LXT 5293)
+-Gr 8-77 p317 +RR 6-77 p74
+HFN 7-77 p125

Quintet, strings, K 46, B flat major: Adagio. cf Decca DDS 507/1-2.

1947 Quintet, strings, no. 2, K 406, C minor. Quintet, strings, no. 5, K 593, D major. Aeolian Quartet; Kenneth Essex, vla. Argo ZK 12.
+Gr 5-77 p1707 +-RR 5-77 p65
++HFN 5-77 p129

Quintet, strings, no. 2, K 406, C minor. cf Quintet, strings, no. 4, K 516, G minor.

1948 Quintet, strings, no. 3, K 515, C major. Kenneth Essex, vla; Aeolian Quartet. Argo ZK 17.
+HFN 10-77 p159 +-RR 10-77 p67

1949 Quintet, strings, no. 4, K 516, G minor. Quintet, strings, no. 2, K 406, C minor. Grumiaux Quintet. Philips 6500 620.
++HF 6-76 p90 +-NYT 8-8-76 pD13
+MJ 10-77 p24 +-St 5-77 p74
++NR 3-76 p5

Quintet, strings, no. 5, K 593, D major. cf Quintet, strings, no. 2, K 406, C minor.

1950 Il Re pastore, K 208. Reri Grist, Lucia Popp, Arlene Saunders, s; Nicola Monti, Luigi Alva, t; Denis Vaughan, hpd; Naples Orchestra; Denis Vaughan. RCA PVL 2-9086 (2). (Reissue from SER 5567/8)

+Gr 12-76 p1046 +RR 1-77 p36
+HFN 4-77 p153

Il Re pastore, K 208: L'amero saro costante. cf Arias (RCA SER 5675).
Il Re pastore, K 208: L'amero saro costante. cf BIS LP 45.
Il Re pastore, K 208: L'amero saro costante. cf HMV RLS 719.
Il Re pastore, K 208: Overture. cf Overtures (Eurodisc SQ 27257).
Il Re pastore, K 208: Overture. cf Overtures (Turnabout TV 34628).
Rondo, flute, K Anh 184, D major. cf HOFFMEISTER: Concerto, flute, G major.
Rondo, horn, K 371, E minor. cf Works, selections (Philips 7699 048).
Rondo, piano, K 382, D major. cf Concerto, piano, no. 22, K 482, E flat major.
Rondo, piano, K 382, D major. cf BEETHOVEN: Concerto, piano, no. 3, op. 37, C minor.
Rondo, piano, K 386, A major. cf Concerto, piano, no. 22, K 482, E flat major.
Rondo, piano, K 386, A major. cf BEETHOVEN: Concerto, piano, no. 3, op. 37, C minor.
Rondo, piano, K 494, F major. cf Sonata, piano, no. 11, K 331, A major.
Rondo, piano, K 511, A minor. cf HAYDN: Sonata, piano, no. 46, A flat major.
Rondo, violin, G major. cf Finnlevy SFLP 8569.
Rondo, violin, K 269, B flat major. cf Adagio, K 261, E major.
Rondo, violin, K 269 (K 261a), B flat major. cf Works, selections (Angel SD 3789).
Rondo, violin, K 373, C major. cf Adagio, K 261, E major.
Rondo, violin, K 373, C major. cf Works, selections (Angel SD 3789).

1951 Der Schauspieldirektor, K 486. Peter Ustinov, narrator; Mady Mesple, s; Edda Moser, ms; Nicolai Gedda, t; Klaus Hirte, bar; Bavarian State Orchestra; Eberhard Schoener. Angel S 37405.
++SFC 5-22-77 p42

1952 Der Schauspieldirektor, K 486. Lo sposo deluso, K 430. Ruth Welting, Felicity Palmer, Ileana Cotrubas, s; Anthony Rolfe Johnson, Robert Tear, t; Clifford Grant, bar; LSO; Colin Davis. Philips 9500 011 Tape (c) 7300 472.
+-ARG 11-76 p29 +NR 12-76 p11
+-Gr 2-77 p1312 +OC 9-77 p54
+-HF 2-77 p106 +-ON 12-4-76 p60
+HF 4-77 p121 tape +-RR 1-77 p36
+HFN 1-77 p111 +SFC 9-12-76 p31
+-MJ 12-76 p28 +-St 1-77 p118

Der Schauspieldirektor, K 486: Overture. cf Overtures (DG 2535 229).
Der Schauspieldirektor, K 486: Overture. cf Overtures (Eurodisc SQ 27257).
Der Schauspieldirektor, K 486: Overture. cf Overtures (Turnabout TV 34628).
Serenades (6). cf Works, selections (Philips 6747 378).
Serenade, E flat major. cf Telefunken EK 6-35334.

1953 Serenade, no. 1, K 100 (K 62a), D major. Serenade, no. 2, K 101, F major. Serenade, no. 6, K 239, D major. Die Wiener Solisten; Wilfried Boettcher. Amadeo AVRS 19048.
+HFN 11-77 p175 +RR 11-77 p62

Serenade, no. 2, K 101, F major. cf Serenade, no. 1, K 100 (K62a), D major.

1954 Serenade, no. 3, K 167a (185), D major. Serenade, no. 13, K 525, G major. Vienna Mozart Ensemble; Willi Boskovsky. Decca JB 19. (Reissue from SXL 6420)
+Gr 10-77 p636
+HFN 9-77 p151
-RR 9-77 p65

Serenade, no. 3, K 185, D major: Andante and allegro. cf Works selections (ABC ABCL 67010/2).

Serenade, no. 4, K 203, D major: Andante and menuetto. cf Works, selections (ABC ABCL 67010/2).

1955 Serenade, no. 5, K 213a (K204), D major. March, K 215, D major. Uto Ughi, vln; Dresden State Orchestra; Edo de Waart. Philips 6500 967.
+-ARG 3-77 p29
++Gr 3-76 p1467
+HF 7-77 p102
+HFN 3-76 p101
+MJ 2-77 p30
+NR 3-77 p6
+-RR 3-76 p46

Serenade, no. 5, K 213a (K 204), D major. cf Works, selections (Philips 7699 049).

Serenade, no. 5, K 213a (K 204), D major: Andante moderato, Menuetto. cf Works, selections (ABC ABCL 67010/2).

1956 Serenade, no. 6, K 239, D major. Serenade, no. 13, K 525, G major. Symphony, no. 21, K 134, A major. New London Soloists Ensemble; Ronald Thomas. CRD 1040.
+HFN 10-77 p159

Serenade, no. 6, K 239, D major. cf Concerto, piano, no. 21, K 467, C major.

Serenade, no. 6, K 239, D major. cf Serenade, no. 1, K 100 (K 62a), D major.

Serenade, no. 6, K 239, D major. cf Works, selections (Philips 7699 049).

Serenade, no. 6, K 239, D major. cf Symphony, no. 29, K 201, A major.

1957 Serenade, no. 7, K 250, D major. Yehudi Menuhin, vln; Bath Festival Orchestra. Classics for Pleasure CFP 40275. (Reissue from HMV ASD 627).
+Gr 8-77 p306
+HFN 6-77 p137
+RR 6-77 p60

1958 Serenade, no. 7, K 250, D major. March, K 249, D major. Uto Ughi, vln; Dresden Staatskapelle; Edo de Waart. Philips 6500 966.
+Gr 6-76 p50
+HF 11-76 p122
+HFN 6-76 p91
+MJ 2-77 p30
++NR 11-76 p2
+NYT 8-8-76 pD13
+-RR 6-76 p45
++SFC 8-15-76 p38
+STL 6-6-76 p37

Serenade, no. 7, K 250, D major. cf Works, selections (Philips 7699 049).

Serenade, no. 7, K 250, D major: Rondo. cf HMV ASD 3346.

Serenade, no. 8, K 286, D major. cf Works, selections (Philips 7699 049).

1959 Serenade, no. 9, K 320, D major. Marches, K 335 (K 320a) (2). LPO; Hans-Hubert Schonzeler. Classics for Pleasure CFP 40258.
+-Gr 9-77 p430
+HFN 8-77 p87
+RR 8-77 p58

Serenade, no. 9, K 320, D major. cf Works, selections (Philips 7699 049).
1960 Serenade, no. 10, K 361, B flat major. Marlboro Music Festival Soloists; Marcel Moyse. Marlboro Recording Society MRS 11.
+HF 10-77 p114
Serenade, no. 10, K 361, B flat major. cf Works, selections (Philips 7699 049).
Serenade, no. 12, K 388, C minor. cf Works, selections (Philips 7699 049).
1961 Serenade, no. 13, K 525, G major. SAINT-SAENS: The carnival of the animals.* Alfons and Aloys Kontarsky, pno; VPO; Karl Bohm. DG 2530 731. (*Reissue from 2530 588)
+Gr 12-76 p1071 +-RR 12-86 p64
+HFN 1-77 p119
1962 Serenade, no. 13, K 525, G major. Sinfonia concertante, K 364, E flat major. Igor Oistrakh, vla; Moscow Symphony Orchestra Solo Ensemble; Igor Oistrakh. Westminster WGS 8343.
+-NR 7-77 p5
Serenade, no. 13, K 525, G major. cf Concerto, piano, no. 21, K 467, C major (Classics for Pleasure CFP 40009).
Serenade, no. 13, K 525, G major. cf Concerto, piano, no. 21, K 467, C major (Philips 6580 112).
Serenade, no. 13, K 525, G major. cf Serenade, no. 3, K 167a (K 185), D major.
Serenade, no. 13, K 525, G major. cf Works, selections (Philips 7699 049).
Serenade, no. 13, K 525, G major. cf Serenade, no. 6, K 239, D major.
Serenade, no. 13, K 525, G major. cf Symphony, no. 29, K 201, A major.
Serenade, no. 13, K 525, G major. cf Symphony, no. 41, K 551, G major.
Serenade, no. 13, K 525, G major. cf ALBINONI: Adagio, G minor.
Serenade, no. 13, K 525, G major. cf BACH: Concerto, 2 violins and strings, S 1043, D minor.
Serenade, no. 13, K 525, G major. cf CORELLI: Sarabande, gigue and badinerie.
Serenade, no. 13, K 525, G major. cf HMV SQ ASD 3375.
Serenade, no. 13, K 525, G major. cf HMV SLS 5073.
Sinfonia concertante, K 297b, E flat major. cf Concerto, flute and harp, K 299, C major.
Sinfonia concertante, K 364, E flat major. cf Concerto, violin, no. 3, K 216, G major.
Sinfonia concertante, K 364, E flat major. cf Concertone, 2 violins, K 190 (K 166b), C major.
Sinfonia concertante, K 364, E flat major. cf Serenade, no. 13, K 525, G major.
Sinfonia concertante, K 364, E flat major. cf Works, selections (Angel SD 3789).
Il sogno di Scipione, K 126: Overture. cf Overtures (Eurodisc SQ 27257).
Il sogno di Scipione, K 126: Overture. cf Overtures (Turnabout TV 34628).
Sonata, bassoon, K 292, B flat major. cf Gasparo GS 103.
Sonata, flute and harpsichord, no. 4, K 13, F major. cf BEETHOVEN: Sonata, flute, B flat major.

Sonatas, organ, nos. 1-17. cf Works, selections (Philips 6747 384).
Sonata, piano, no. 8, K 310, A minor. cf Fantasia, K 475, C minor.
Sonata, piano, no. 9, K 311, D major. cf Fantasia, K 397, D major.
Sonata, piano, no. 9, K 311, D major. cf HAYDN: Andante and variations, F minor.
Sonata, piano, no. 10, K 330, C major. cf Fantasia, K 397, D major.
Sonata, piano, no. 10, K 330, C major. cf HAYDN: Andante and variations, F minor.

1963 Sonata, piano, no. 11, K 331, A major. Allegro and andante, K 533, F major. Rondo, piano, K 494, F major. Ingrid Haebler, pno. Philips 6580 153. (Reissue from AXS 6001)

++Gr 1-77 p1161 +HFN 1-77 p121

Sonata, piano, no. 11, K 331, A major. cf Adagio, K 540, B minor.
Sonata, piano, no. 11, K 331, A major: Rondo alla turca. cf Amberlee ALM 602.
Sonata, piano, no. 11, K 331, A minor: Rondo alla turca. cf Decca SPA 519.
Sonata, piano, no. 11, K 331, A major: Rondo alla turca. cf Philips 6747 327.
Sonata, piano, no. 13, K 333, B flat major. cf Adagio, K 540, B minor (Philips 9500 025).
Sonata, piano, no. 13, K 333, B flat major. cf Fantasia and fugue, K 394, C major.
Sonata, piano, no. 14, K 457, C minor. cf Fantasia, K 475, C minor (Denon PCM OX 7057).
Sonata, piano, no. 14, K 457, C minor. cf Fantasia, K 475, C minor (Westminster WG 1010).
Sonata, piano, no. 15, K 545, C major. cf Fantasia, K 475, C minor.
Sonata, piano, no. 17, K 576, D major. cf BACH: Fantasia, S 906, C minor.
Sonata, piano, 4 hands, K 521, C major. cf Fugue, 2 pianos, K 426, C minor.
Sonata, 2 pianos, K 448, D major. cf Fugue, 2 pianos, K 426, C minor.
Sonatas, violin and piano, K 12 and K 13, A major, F major. cf BRAHMS: Serenade, no. 2, op. 16, A major.

1964 Sonata, violin and piano, no. 17, K 296, C major. Sonata, violin and piano, no. 18, K 301, G major. Sonata, violin and piano, 19, K 302, E flat major. Sonata, violin and piano, no. 21, K 304, E minor. Sonata, violin and piano, no. 22, K 305, A major. Sonata, violin and piano, no. 26, K 378, B flat major. Sonata, violin and piano, no. 28, K 380, E flat major. Szymon Goldberg, vln; Radu Lupu, pno. London CSA 2244 (2). (also Decca 13BB 207/12)

+Gr 11-75 p853 +NR 5-77 p8
+HF 6-77 p94 ++SFC 5-22-77 p42
++HFN 10-75 p145 ++MJ 7-77 p70
+MT 1-76 p42

Sonata, violin and piano, no. 17, K 296, C major. cf RCA CRM 6-2264.

1965 Sonatas, violin and piano, nos. 18, 21-22, 28. Szymon Goldberg, vln; Radu Lupu, pno. Decca SDD 515/6 (2). (Reissue from 13BB 207/13)

+-Gr 10-77 p654 +RR 10-77 p78
++HFN 10-77 p167

Sonata, violin and piano, no. 18, K 301, G major. cf Sonata, violin and piano, no. 17, K 296, C major.

Sonata, violin and piano, no. 19, K 302, E flat major. cf Sonata, violin and piano, no. 17, K 296, C major.

1966 Sonatas, violin and piano, nos. 20, 23-24, 32-33. Szymon Goldberg, vln; Radu Lupu, pno. Decca SDD 513/4 (2). (Reissue from 13BB 207/12)

+-Gr 5-77 p1708 +RR 5-77 p75
++HFN 5-77 p137

1967 Sonata, violin and piano, no. 20, K 303, C major. Sonata, violin and piano, no. 23, K 306, D major. Sonata, violin and piano, no. 24, K 376, F major. Sonata, violin and piano, no. 32, K 454, B flat major. Sonata, violin and piano, no. 33, K 481, E flat major. Radu Lupu, pno; Szymon Goldberg, vln. London CSA 2243 (2). (also Decca 13BB 207/12)

++ARG 12-76 p37 +MT 1-76 p42
+Gr 11-75 p853 ++NR 1-77 p8
+HF 6-77 p94 +RR 11-75 p80
++HFN 10-75 p145 ++SFC 5-22-77 p42
+MJ 1-77 p27 ++St 11-76 p89

Sonata, violin and piano, no. 21, K 304, E minor. cf Sonata, violin and piano, no. 17, K 296, C major.

Sonata, violin and piano, no. 21, K 304, E minor. cf Sonata, violin and piano, no. 24, K 376, F major.

Sonata, violin and piano, no. 22, K 305, A major. cf Sonata, violin and piano, no. 17, K 296, C major.

Sonata, violin and piano, no. 23, K 306, D major. cf Sonata, violin and piano, no. 20, K 303, C major.

1968 Sonata, violin and piano, no. 24, K 376, F major. Sonata, violin and piano, no. 21, K 304, E minor. Sonata, violin and piano, no. 28, K 380, E flat major. Sonya Monosoff, vln; Malcolm Bilson, fortepiano. Pleiades P 104.

+-Gr 9-77 p504 +NR 12-77 p15
+-MJ 9-77 p35 +St 11-77 p144

Sonata, violin and piano, no. 24, K 376, F major. cf Sonata, violin and piano, no. 20, K 303, C major.

1969 Sonata, violin and piano, no. 25, K 377, F major. Sonata, violin and piano, no. 27, K 379, G major. Sonata, violin and piano, K 526, A major. Sonata, violin and piano, K 547, F major. Szymon Goldberg, vln; Radu Lupu, pno. London CSA 2245 (2).

++MJ 10-77 p27 ++SFC 10-9-77 p40
++NR 11-77 p11

Sonata, violin and piano, no. 26, K 378, B flat major. cf Sonata, violin and piano, no. 17, K 296, C major.

Sonata, violin and piano, no. 26, K 378, B flat major. cf RCA CRM 6-2264.

Sonata, violin and piano, no. 27, K 379, G major. cf Sonata, violin and piano, no. 25, K 377, F major.

Sonata, violin and piano, no. 28, K 380, E flat major. cf Sonata, violin and piano, no. 17, K 296, C major.

Sonata, violin and piano, no. 28, K 380, E flat major. cf Sonata, violin and piano, no. 24, K 376, F major.

Sonata, violin and piano, no. 32, K 454, B flat major. cf Sonata, violin and piano, no. 20, K 303, C major.

Sonata, violin and piano, no. 32, K 454, B flat major. cf BEETHOVEN: Sonata, violin and piano, no. 5, op. 24, F major.

Sonata, violin and piano, no. 32, K 454, B flat major. cf BEETHOVEN: Sonata, violin and piano, no. 9, op. 47, A major.

Sonata, violin and piano, no. 32, K 454, B flat major. cf RCA CRM 6-2264.

Sonata, violin and piano, no. 33, K 481, E flat major. cf Sonata, violin and piano, no. 20, K 303, C major.

Sonata, violin and piano, K 526, A major. cf Sonata, violin and piano, no. 25, K 377, F major.

Sonata, violin and piano, K 547, F major. cf Sonata, violin and piano, no. 25, K 377, F major.

Sonata, violoncello and bassoon, K 292. cf BOCCHERINI: Quintet, guitar.

1970 Songs: Adoramus te, K 109; Ave verum corpus, K 618; De profundis, K 93; Ergo interest an quis, K 73a; God is our refuge, K 20; Justum deduxit Dominus, K 93d; Kyrie, K 73k; Regina coeli, K 74d. Jill Gomez, s; St. Bartholomew the Great Choir, St. Bartholomew's Hospital Choral Society and Orchestra; Andrew Morris, Robert Anderson. Abbey LPB 773.
+RR 5-77 p83

1971 Songs: Abendempfidung, K 523; An Chloe, K 524; Die betrogene Welt, K 474; Der Fruhling; Ich wurd auf meinem Pfad, K 340; Das Kinderspiel; Die kleine Friedrichs Geburtstag, K 529; Komm, liebe Zither, komm, K 351; Lied der Freiheit, K 506; Das Lied der Trennung, K 510; Sehnsucht nach dem Fruhling, S 596; Sei du mein Trost, K 391; Das Traumbild, K 530; Das Veilchen, K 476; Wie unglicklich bin ich, K 125; Die Zufriedenheit, K 473. Peter Schreier, t; Jorg Demus, pno. Eurodisc 27822.
+FF 11-77 p38

1972 Songs: Abendempfidung, K 523; Als Luise die Briefe, K 520; An Chloe, K 524; Dans un bois solitaire, K 308; Ich wurd auf meinem Pfad, K 340; Das Kinderspiel, K 598; Manner suchen stets zu naschen, K 433; Oiseaux si tous les ans, K 307; Ridente la calma, K 152; Sehnsucht nach dem Fruhling, K 596; Trennungslied, K 519; Das Veilchen, K 476; Die Verschweigung, K 518; Der Zauberer, K 472; Die Zufriedenheit, K 473. Patricia Corbett, s; Galen Lurwick, pno. Orion ORS 77268.
/FF 11-77 p39 +-NR 10-77 p11

1973 Songs: Abendempfidung, K 523; Als Luise die Briefe, K 520; An die Freundschaft, K 125h; An die Hoffnung, K 390; An die Einsamkeit; Dans un bois solitaire, K 308; Gesellen Reise, K 468; Die Grossmutige Gelassenheit, K 125d; Eine kleine Deutsche Kantata, K 619; Die kleine Spinnerin; Das Lied der Trennung, K 510; Oiseau, si tous le ans, K 307; Ridente la calma, K 152; Sehnsucht nach dem Fruhling, K 596; Das Veilchen, K 476; Der Zauberer, K 472; Die Zufriedenheit, K 473. Jill Gomez, s; John Constable, pno. Saga 5441.
+ARG 9-77 p48 +RR 6-77 p89
+Gr 6-77 p89 +SFC 6-19-77 p46
+-HFN 6-77 p131

Songs: Abendempfidung, K 523; Das Veilchen, K 476. cf Philips 6767 001.

Songs: Sehnsucht nach dem Fruhling, K 596. cf Telefunken DT 6-48085.

Songs: Sehnsucht nach dem Fruhling, K 596; Das Veilchen, K 476; Warnung, K 433. cf Bruno Walter Society BWS 729.

Lo sposo deluso, K 430. cf Overtures (Eurodisc SQ 27257).
Lo sposo deluso, K 430. cf Die Schauspieldirektor, K 486.

1974 Symphonies, nos. 21-41. (Includes book "Mozart the Man") COA; Josef Krips. Philips 6747 130 (8)
+-Gr 4-77 p1567 +-RR 3-77 p18
+-HFN 4-77 p151

1975 Symphony, no. 21, K 134, A major. Symphony, no. 36, K 425, C major. COA: Josef Krips. Philips 6500 525.
+ARG 10-77 p39 +SFC 12-25-77 p42
+NR 11-77 p4

Symphony, no. 21, K 134, A major. cf Serenade, no. 6, K 239, D major.

1976 Symphony, no. 22, K 162, C major. Symphony, no. 27, K 199, G major. Symphony, no. 29, K 201, A major. COA; Josef Krips. Philips 6500 528.
++ARG 6-77 p34 +SFC 3-27-77 p41
+MJ 9-77 p35 ++St 7-77 p89
++NR 5-77 p1

1977 Symphony, no. 23, K 181, D major. Symphony, no. 28, K 200, C major. Symphony, no. 30, K 202, D major. COA; Josef Krips. Philips 6500 527.
++ARG 9-77 p41 +NR 8-77 p5
+MJ 9-77 p35 +SFC 12-25-77 p42

1978 Symphony, no. 24, K 182, B flat major. Symphony, no. 25, K 183, G minor. Symphony, no. 26, K 184, E flat major. COA; Josef Krips. Philips 6500 529.
++ARG 6-77 p34 ++NR 5-77 p1
+-Audio 9-77 p101 +SFC 3-27-77 p41
+MJ 9-77 p35 ++St 7-77 p89

1979 Symphonies, nos. 25-41. PhO; Otto Klemperer. HMV SLS 5048 (6).
+-HFN 4-77 p151

1980 Symphony, no. 25, K 183, G minor. Symphony, no. 29, K 201, A major. NPhO; Riccardo Muti. HMV (SQ) ASD 3326 Tape (c) TC ASD 3326. (also Angel S 37257)
+Gr 5-77 p1697 +MJ 9-77 p35
+HFN 5-77 p127 ++NR 7-77 p5
+HFN 8-77 p99 tape +RR 9-77 p65

1981 Symphony, no. 25, K 183, G minor. Symphony, no. 38, K 504, D major. LSO; Georg Solti. London R 23238. (Reissue from LL 1034)
+-ARSC vol 9, no. 1, 1977 p79

Symphony, no. 25, K 183, G minor. cf Symphony, no. 24, K 182, B flat major.
Symphony, no. 26, K 184, E flat major. cf Symphony, no. 24, K 182, B flat major.
Symphony, no. 27, K 199, G major. cf Symphony, no. 22, K 162, C major.

1982 Symphony, no. 28, K 200, C major. Symphony, no. 29, K 201, A major. Vienna Chamber Orchestra; Philippe Entremont. CBS 76581.
+Gr 6-77 p56 /RR 6-77 p61
+HFN 6-77 p129

Symphony, no. 28, K 200, C major. cf Symphony, no. 23, K 181, D major.

1983 Symphony, no. 29, K 201, A major. Serenade, no. 6, K 239, D major. Serenade, no. 13, K 525, G major. New London Soloists Ensemble; Ronald Thomas. CRD CRD 1040.
-Gr 9-77 p431 +-RR 9-77 p64
+HFN 10-77 p159

Symphony, no. 29, K 201, A major. cf Symphony, no. 22, K 162, C major.

Symphony, no. 29, K 201, A major. cf Symphony, no. 25, K 183, G minor.

Symphony, no. 29, K 201, A major. cf Symphony, no. 28, K 200, C major.

Symphony, no. 30, K 202, D major. cf Symphony, no. 23, K 181, D major.

1984 Symphony, no. 31, K 297, D major. Symphony, no. 36, K 425, C major. Les petites riens, K Anh 10. Bavarian Radio Symphony Orchestra; Ferdinand Leitner. DG Tape (c) 3348 220.
+Gr 2-77 p1325 tape +RR 9-76 p92 tape

1985 Symphony, no. 31, K 297, D major. Symphony, no. 36, K 425, C major. Slovak Philharmonic Orchestra; Ludovic Rajter. Opus 9110 0258. (Reissue from Supraphon SUAST 50873)
+-Gr 7-77 p180 +-RR 3-77 p56
/-HFN 5-77 p127

1986 Symphony, no. 31, K 297, D major. Symphony, no. 38, K 504, D major. COA; Josef Krips. Philips 6500 466.
++ARG 9-77 p41 +NR 8-77 p5
+MJ 9-77 p35

1987 Symphony, no. 32, K 318, G major. Symphony, no. 33, K 319, B flat major. Symphony, no. 34, K 338, C major. COA: Josef Krips. Philips 6500 526.
+ARG 10-77 p39 +NR 11-77 p4

Symphony, no. 32, K 318, G major (Overture in the Italian style). cf Overtures (Turnabout TV 34628).

Symphony, no. 33, K 319, B flat major. cf Symphony, no. 32, K 318, G major.

1988 Symphony, no. 34, K 338, C major. Symphony, no. 39, K 543, E flat major. Israel Philharmonic Orchestra; Zubin Mehta. Decca SXL 6833 Tape (c) KSXC 6833. (also London 7055 Tape (c) 5-7055)
+-Audio 9-77 p46 +-NR 10-77 p3
+-Gr 3-77 p1401 +-RR 3-77 p56
+HFN 3-77 p107 -RR 5-77 p92 tape
+HFN 4-77 p155 tape +SFC 11-20-77 p54

Symphony, no. 34, K 338, C major. cf Symphony, no. 32, K 318, G major.

1989 Symphony, no. 35, K 385, D major. RUGGLES: Men and mountains. TELEMANN: Concerto, D major. VIVALDI: Concerto, piccolo, C major. Soloists; New Hampshire Music Festival Orchestra; Thomas Nee. Hammar SD 150.
+-Audio 11-77 p132

Symphony, no. 36, K 425, C major. cf Symphony, no. 21, K 134, A major.

Symphony, no. 36, K 425, C major. cf Symphony, no. 31, K 297, D major (DG 3348 220).

Symphony, no. 36, K 425, C major. cf Symphony, no. 31, K 297, D major (Opus 9110 0258).

Symphony, no. 37, K 444, G major. cf Concerto, piano, no. 18, K 456, B flat major.

1990 Symphony, no. 38, K 504, D major. SCHUBERT: Symphony, no. 8, D 759, B minor. ECO; Benjamin Britten. London CS 6741.
+NR 10-77 p3 +SFC 12-25-77 p42

Symphony, no. 38, K 504, D major. cf Symphony, no. 25, K 183, G minor.

Symphony, no. 38, K 504, D major. cf Symphony, no. 31, K 297, D major.

Symphony, no. 38, K 504, D major. cf DVORAK: Slavonic dances, op. 46, nos. 1-4.

1991 Symphony, no. 39, K 543, E flat major. Symphony, no. 40, K 550, G minor. COA; Josef Krips. Philips 6500 430 Tape (c) 7300 271.
++ARG 6-77 p34
+-Gr 5-73 p2051
+-HFN 5-73 p986
+MJ 9-77 p35
++NR 5-77 p1
-RR 5-73 p57
-RR 1-75 p71 tape
+St 7-77 p89

Symphony, no. 39, K 543, E flat major. cf Symphony, no. 34, K 338, C major.

1992 Symphony, no. 40, K 550, G minor. Symphony, no. 41, K 551, C major. LPO; Charles Mackerras. Classics for Pleasure CFP 40253 Tape (c) TC CFP 40253.
+-Gr 11-76 p798
*Gr 2-77 p1329
+-HFN 10-76 p173
+HFN 12-76 p153
+-RR 10-76 p62

1993 Symphony, no. 40, K 550, G minor. Symphony, no. 41, K 551, C major. NPhO; Carlo Maria Giulini. Decca JB 8 Tape (c) KJCB 8. (Reissue from SXL 6225)
+-Gr 9-77 p431
+-Gr 11-77 p899 tape
+HFN 9-77 p151
+-HFN 12-77 p187 tape
+-RR 9-77 p65
+-RR 12-77 p96 tape

1994 Symphony, no. 40, K 550, G minor. Symphony, no. 41, K 551, C major. VPO; Karl Bohm. DG 2530 780 Tape (c) 3300 780.
+Gr 6-77 p56
++HFN 6-77 p129
+RR 6-77 p61

1995 Symphony, no. 40, K 550, G minor. Symphony, no. 41, K 551, C major. London Sinfonia; Anthony Collins. Fanfare SIT 60036.
+Gr 2-77 p1329

1996 Symphony, no. 40, K 550, G minor. Symphony, no. 41, K 551, C major. LSO; Antal Dorati, Hans Schmidt-Isserstedt. Fontana 6531 006 Tape (c) 7328 006.
+Gr 2-77 p1329
+-HFN 7-77 p127
+-RR 7-77 p98 tape

1997 Symphony, no. 40, K 550, G minor. Symphony, no. 41, K 551, C major. Mozarteum Orchestra; Leopold Hager. Turnabout QTVS 34563.
-Audio 1-77 p86

Symphony, no. 40, K 550, G minor. cf Symphony, no. 39, K 543, E flat major.

Symphony, no. 40, K 550, G minor. cf BEETHOVEN: Egmont, op. 84: Overture.

Symphony, no. 40, K 550, G minor. cf BEETHOVEN: Symphony, no. 5, op. 67, C minor.

Symphony, no. 40, K 550, G minor: 1st movement. cf Works, selections (CBS 30088).

Symphony, no. 40, K 550, G minor: 1st movement. cf Philips 6580 114.

1998 Symphony, no. 41, K 551, C major. The marriage of Figaro, K 492: Overture. Serenade, no. 13, K 525, G major. RPO; Hans Vonk. Decca PFS 4425. (also London CS 7048)
+-Gr 12-77 p1087
+-HFN 12-77 p175
+RR 12-77 p49

Symphony, no. 41, K 551, C major. cf Symphony, no. 40, K 550, G minor (Classics for Pleasure CFP 40253).
Symphony, no. 41, K 551, C major. cf Symphony, no. 40, K 550, G minor (Decca JB 8).
Symphony, no. 41, K 551, C major. cf Symphony, no. 40, K 550, G minor (DG 2530 780).
Symphony, no. 41, K 551, C major. cf Symphony, no. 40, K 550, G minor (Fanfare SIT 60036).
Symphony, no. 41, K 551, C major. cf Symphony, no. 40, K 550, G minor (Fontana 6531 006).
Symphony, no. 41, K 551, C major. cf Symphony, no. 40, K 550, G minor (Turnabout QTVS 34563).
Thamos, King of Egypt, K 345: Entr'acte, no. 2. cf Concerto, bassoon, K 191, B flat major.

1999 Thamos, Konig in Agypten, K 345: Incidental music. Charlotte Lehmann, s; Rose Scheible, con; Oly Pfaff, t; Bruce Abel, bs; Heilbronner Vocal Ensemble; Wurttemberg Chamber Orchestra; Jorg Faerber. Turnabout QTV 34679.
+NR 12-77 p2

Trio, clarinet, K 498, E flat major. cf Quintet, piano and winds, K 452, E flat major.
Trio, clarinet, K 498, E flat major. cf HOFFMEISTER: Duo concertante, F major.
Variations on a minuet by Duport, K 573. cf BACH: Fantasia, S 906, C minor.
Variations on "Ein Weib ist das herrlichste Ding", K 613. cf Fantasia and fugue, K 394, C major.
Das Veilchen (La Violetta), K 476. cf Seraphim S 60280.
Vesperae solennes de confessore, K 339, C major. cf Mass, no. 16, K 317, C major.
Vesperae solennes de confessore, K 339, C major. cf Works, selections (Philips 7699 052).

2000 Works, selections: Cassation, K 63, G major: Adagio. Concerto, violin, no. 1, K 207, B flat major. Concerto, violin, no. 2, K 211, D major. Serenade, no. 3, K 185, D major: Andante and allegro. Serenade, no. 4, K 203, D major: Andante and menuetto. Serenade, no. 5, K 213a (K 204), D major: Andante moderato, Menuetto. Jaap Schroder, baroque violin; Amsterdam Mozart Ensemble; Frans Bruggen. ABC ABCL 67010/2 (2).
+-HF 3-77 p108 +PRO 5-77 p25
+IN 10-77 p24 -SFC 12-12-77 p28
++NR 2-77 p4 +St 5-77 p54

2001 Works, selections: Adagio, K 261, E major. Concerti, violin, nos. 1-5. Concertone, 2 violins, K 190 (K 166b), C major. Rondo, violin, K 373, C major. Rondo, violin, K 269 (K 261a), B flat major. Sinfonia concertante, K 364, E flat major. David Oistrakh, Igor Oistrakh, vln; David Oistrakh, vla; BPhO; David Oistrakh. Angel SD 3789. (also HMV SLS 828 Tape (c) TC SLS 828; ASD 2839/42)
+Gr 10-72 p703 +RR 10-72 p66
+HF 2-73 p90 +RR 10-77 p97 tape
+HFN 10-72 p1898 ++St 2-73 p124
+HFN 10-77 p169 tape +STL 12-10-72 p35
+-NR 12-72 p7

2002 Works, selections: Concerto, horn and strings, no. 4, K 495, E flat major.* Exsultate jubilate, K 165. Don Giovanni, K 527: Madamina. The marriage of Figaro, K 492: Se vuol ballere.

Prelude on "Ave verum, K 580a.* Symphony, no. 40, K 550, G minor: 1st movement. Judith Raskin, s; Ezio Pinza, bs; E. Power Biggs, org; Mason Jones, hn; Marlboro Festival Orchestra, CO, Metropolitan Opera Orchestra, PO; Pablo Casals, Georg Szell, Bruno Walter, Eugene Ormandy. CBS 30088 Tape (c) 40-30088. (*Reissues from SBRG 72477, 61095)

+Gr 12-76 p1071 +-HFN 4-77 p155 tape
+-HFN 4-77 p153 +RR 12-76 p62

2003 Works, selections: The eight-year-old Mozart in Chelsea: Contradanses, F major (3), G major, C major. Divertimenti, C major, G minor, D major, F major, B flat major, E flat major. AMF; Neville Marriner. Philips 6500 367.

+Gr 10-72 p718 +MJ 2-73 p28
+Gr 9-77 p490 +-NR 3-73 p3
-HF 1-73 p92 +RR 10-72 p63
+HFN 10-72 p1913 +ST 1-73 p110

2004 Works, selections: Cassations (2). Divertimenti (6). Marches (6). Nocturne, K 286. Serenades (6). Uto Ughi, vln; Dresden Philharmonic Orchestra, Dresden State Orchestra, Berlin Philharmonic Octet; Gunther Herbig, Edo de Waart. Philips 6747 378 (10).

+-Gr 9-77 p490 +-RR 9-77 p28
+HFN 9-77 p153

2005 Works, selections: Adagio and allegro, organ. Alleluja. Ave Maria. Ave verum corpus, K 618. Andante, organ. Exsultate jubilate, K 165. Fantasia, organ. Kyrie. Masses, no. 4, K 139, C minor; no. 6, K 192, F major; no. 13, K 259, C major; no. 16, K 317, C major; no. 10, K 220, C major; K 257, C major; no. 18, K 427, C minor; no. 19, K 626, D minor. Sonatas, organ, nos. 1-17. Elly Ameling, s; Daniel Chorzempa, org; Leipzig Radio Symphony Orchestra and Chorus, Vienna Boys Choir; Dom Orchestra, ECO, John Alldis Choir, LSO and Chorus, BBC Symphony Orchestra, German Bach Soloists; Herbert Kegel, Ferdinand Grossman, Raymond Leppard, Colin Davis, Helmut Winschermann. Philips 6747 384 (10).

+-Gr 9-77 p490 +-RR 9-77 p28
+HFN 9-77 p153

2006 Works, selections: Adagio, K 261, E minor. Concerto, bassoon, K 191, B flat minor. Concerto, clarinet, K 622, A major. Concerto, flute. Concerto, flute and harp, K 299, C major. Concerti, horn, nos. 1-4. Concerto, oboe (flute), K 314, C major. Concerti, violin, nos. 3-5. Rondo, horn, K 371, E minor. Jack Brymer, clt; Michael Chapman, bsn; Alan Civil, hn; Neil Black, ob; Osian Ellis, hp; Claude Monteux, flt; Henrik Szeryng, vln; PhO; Neville Marriner, Alexander Gibson. Philips Tape (c) 7699 048 (3).

+RR 12-77 p90 tape

2007 Works, selections: Serenade, no. 5, K 213a (K 204), D major. Serenade, no. 6, K 239, D major. Serenade, no. 8, K 286, D major. Serenade, no. 7, K 250, D major. Serenade, no. 9, K 320, D major. Serenade, no. 10, K 361, B flat major. Serenade, no. 12, K 388, C minor. Serenade, no. 13, K 525, G major. Dresden State Orchestra, Berlin Philharmonic Octet, NWE; Edo de Waart. Philips Tape (c) 7699 049.

+RR 12-77 p90 tape

2008 Works, selections: Ave verum corpus, K 618. Exsultate jubilate, K 165. Kyrie, K 341, D minor. Mass, no. 8, K 220, C minor.

Mass, K 257, C major. Mass, no. 14, K 317, C minor. Mass, no. 18, K 427, C minor. Mass, no. 19, K 626, D minor. Vesperae solennes de confessore, K 339, C major. Vienna Sangerknaben und Dom Orchestra; LSO and Chorus, BBC Symphony Orchestra; Ferdinand Grossmann, Colin Davis. Philips Tape (c) 7699 052 (3).
+RR 12-77 p90 tape

2009 Works, selections: Divertimento, K Anh 229, B flat major. Divertimento, no. 5, K 187, C major. Duos, 2 horns, K 487 (12). Quartet, oboe, K 370, F major. Quintet, horn, K 407, E flat major. Czech Philharmonic Wind Ensemble, CPhO Members. Supraphon 111 1671/2 (2).
++ARG 3-77 p26 +NR 11-75 p8

2010 Zaide, K 344. Edith Mathis, s; Peter Schreier, Armin Ude, Werner Hollweg, t; Ingvar Wixell, bar; Reiner Suss, bs; Berlin State Opera Orchestra; Bernhard Klee. Philips 6700 097 (2).
+ARG 11-76 p31 +ON 12-4-76 p60
+-HF 2-77 p100 +RR 10-76 p28
+HFN 10-76 p173 ++SFC 9-12-76 p31
+-MJ 12-76 p28 ++St 12-76 p145
+MT 3-77 p217 +ST 1-77 p777
+OC 9-77 p54

Zaide, K 344. cf Bastien und Bastienne, K 50.
Die Zauberflote (The magic flute), K 620. cf Bastien und Bastienne, K 50.
Die Zauberflote, K 620: Ach, ich fuhl's. cf Club 99-109.
The magic flute, K 620: Ach, ich fuhl's. cf DONIZETTI: Don Pasquale: So anch'io la virtu magica.
Die Zauberflote, K 620: Dies Bildnis. cf Telefunken AJ 6-42232.
Die Zauberflote, K 621: Dies Bildnis ist bezaubernd schon. cf Decca ECS 812.
Die Zauberflote, K 620: Dies Bildnis ist bezaubernd schon; Wie stark ist nicht dein Zauberton. cf Arias (L'Oiseau-Lyre DSLO 13).
Die Zauberflote, K 620: O zittre nicht. cf Decca D65D3.
Die Zauberflote, K 620: Overture. cf Overtures (DG 2535 229).
Die Zauberflote, K 620: Overture. cf Overtures (Eurodisc SQ 27257).
The magic flute, K 620: Overture. cf Decca SB 328.
The magic flute, K 620: Overture. cf Transatlantic XTRA 1169.
The magic flute, K 620: Overture and airs. cf Telefunken EK 6-35334.

MUCZYNSKI, Robert
Preludes, op. 18. cf COPLAND: Duo.

2011 Sonata, violoncello and piano, op. 25. RACHMANINOFF: Sonata, violoncello and piano, op. 19, G minor. Gordon Epperson, vlc; Robert Muczynski, pno. Coronet S 3000.
+MJ 3-77 p46 +NR 5-73 p14

MUDARRA, Alonso de
Diferencias sobre El Conde claros. cf MILAN: Pavanas, nos. 1-6.
Diferencias sobre El Conde claros. cf CBS 72526.
Dulces exuviae. cf L'Oiseau-Lyre 12BB 203/6.
Fantasia. cf Angel S 36093.
Fantasia. cf CBS 72526.
Fantasia que contrahaza la harpa en la manera de Ludovico. cf MILAN: Pavanas, nos. 1-6.
Fantasia que contrahaza la harpa en la manera de Ludovico. cf L'Oiseau-Lyre SOL 349.

Gallarda. cf Harmonia Mundi HM 759.
Gallarda. cf MILAN: Pavanas, nos. 1-6.
Gallarda. cf Angel S 36093.
O guardame las vacas. cf MILAN: Pavanas, nos. 1-6.
Pavana de Alexandre. cf MILAN: Pavanas, nos. 1-6.
Romanesca Guarda me las vacas. cf DG 2723 051.
Tiento and fantasia. cf RCA RL 2-5099.

MUDD, Thomas
Let thy merciful ears, O Lord. cf Abbey LPB 779.
Let thy merciful ears, O Lord. cf Vista VPS 1037.

MUFFAT, Georg
Fugue, G minor. cf ABC ABCL 67008.
2012 Componimenti musicali. Susanne Shapiro, hpd. Musical Heritage Society Tape (c) MHC 5548/9.
+HF 10-77 p129 tape

MULET, Henri
Tu es Petrus. cf Argo ZRG 864.

MULLER, Johann
2013 Quintets, nos. 1-3. Richards Quintet. Crystal S 252.
+NR 6-77 p7

MULLER-ZURICH, Paul
Capriccio. cf DUTILLEUX: Sonatine, flute and piano.

MUNDY, John
Robin. cf National Trust NT 002.
Were I a king. cf Enigma VAR 1017.

MUNDY, William
Lightly she whipped. cf Argo ZK 25.
Lightly she whipped o'er the dales. cf DG 2533 347.
O Lord the maker of all things. cf Abbey LPB 779.

MURADIAN
Drunken with love. cf Golden Age GAR 1001.

MURPHY
Connemara cradle song. cf Philips 6392 023.

MURRAY, Alan
Songs: I'll walk beside you. cf Argo ZFB 95/6.
Songs: I'll walk beside you. cf L'Oiseau-Lyre DSLO 20.

MURTULA, Giovanni
Tarantella. cf DG 2530 561.

MUSET, Colin
Quant je voi. cf BIS LP 3.
Quant je voy yver. cf Nonesuch H 71326.
Quant je voi yver retorner. cf Turnabout TV 37086.

MUSGRAVE, Thea
Concerto, clarinet. cf BANKS: Concerto, horn.
2014 Concerto, horn. Concerto, orchestra. Barry Tuckwell, hn; Keith Pearson, clt; Scottish National Orchestra; Thea Musgrave, Alexander Gibson. Decca HEAD 8. (also London HEAD 8)
+Gr 6-75 p46
+-HFN 6-75 p89
++MJ 3-76 p24
+MQ 10-77 p570
++MT 10-75 p886
++NR 2-76 p6
++RR 5-75 p36
+SR 1-24-76 p53
++STL 6-8-76 p36
++Te 3-76 p31
Concerto, orchestra. cf Concerto, horn.
Music, horn and piano. cf Musical Heritage Society MHS 3547.

MUSHEL, George
Toccata. cf Vista VPS 1060.

MUSIC FOR A GREAT CATHEDRAL. cf Guild GRS 7008.
MUSIC FOR LUTE AND GAMBA. cf BIS LP 22.
MUSIC FOR RECORDER AND HARPSICHORD. cf Saga 5425.
MUSIC FOR THE COLONIAL BAND. cf Folkways FTS 32378.
MUSIC FOR THE VYNE. cf National Trust NT 002.
MUSIC FOR TWO GUITARS. cf Saga 5412.
MUSIC FROM THE TIME OF THE POPES AT AVIGNON. cf CBS 76534.
MUSIC OF SPAIN, ZARZUELA ARIAS. cf London OS 26435.
MUSIC OF THE FRENCH BAROQUE. cf Vox SVBX 5142.
MUSIC OF THE GOTHIC ERA. cf DG 2710 019.
MUSIC OF THE MINSTRELS. cf Telefunken 6-41928.
MUSIC OF VENICE. cf DG 2548 219.
MUSIC TO ENTERTAIN ELIZABETH I. cf Argo ZK 25.
MUSIC TO ENTERTAIN HENRY VIII. cf Argo ZK 24.
MUSICKE OF SUNDRE KINDES: Renaissance secular music, 1480-1620. cf L'Oiseau-Lyre 12BB 203/6.

MUSSORGSKY, Modest

2015 Boris Godunov. Boris Christoff, bs; OSCCP; Andre Cluytens. Angel SD 3633.
+NYT 10-9-77 pD24

2016 Boris Godunov. George London, bs-bar; Bolshoi Theatre Orchestra; Alexander Melik-Pashayev. Columbia D4S 696.
+NYT 10-9-77 pD24

2017 Boris Godunov (ed. Rimsky-Korsakov). Nadia Dobrianova, s; Alexandrina Miltcheva-Noneva, Reni Penkova, Neli Bojkova, Boika Kosseva, s; Dimiter Damianov, Lyubomir Bodurov, Kiril Diulgherov, Verter Vratchovsky, Georgi Tomov, Dimiter Dimitrov, t; Sabin Markov, Peter Bakardjiev, bar; Nikolai Ghiuselev, Assen Tchavdarov, Boyan Katsarski, Peter Petrov, bs; Bodra Smyana Children's Chorus; Sofia Opera Orchestra and Chorus; Assen Naidenov. Harmonia Mundi HMU 4-144 (4).
-HF 5-77 p83

2018 Boris Godunov (original version). Halina Lukomska, s; Bozena Kinasz, Wiera Baniewicz, Stefania Toczyska, ms; Nicolai Gedda, Bohdan Paprocki, Paulos Raptis, t; Andrzej Hiolski, Jan Goralski, Wlodzimierz Zalewski, bar; Kazimierz Sergiel, bs-bar; Leonard Mroz, Martti Talvela, Aage Haugland, bs; Cracow Philharmonic Boys Chorus; Polish Radio Orchestra and Chorus; Jerzy Semkov. HMV SQ SLS 1000 (4) Tape (c) TC SLS 1000. (also Angel SX 3844)
++FF 11-77 p39
+-Gr 9-77 p482
+Gr 12-77 p1149 tape
++HFN 10-77 p159
+NR 11-77 p12
++NYT 10-9-77 pD24
+RR 9-77 p41
+-SFC 10-17-77 p47

2019 Boris Godunov. Galina Vishnevskaya, s; Ludovico Spiess, Aleksei Maslennikov, t; Nicolai Ghiaurov, Martti Talvela, bs; VPO; Herbert von Karajan. London OSA 1439.
+NYT 10-9-77 pD24

2020 Boris Godunov (ed. Rimsky-Korsakov). Eva Kruglikova, s; Maria Maksakova, Bronislava Zlatogorova, Yevgenia Verbitskaya, ms; Georgy Nelepp, Nikander Khanayev, Ivan Kozlovsky, V. Yakushenko, t; I. Bogdanov, bar; Mark Reizen, Maxim Mikhailov, V. Lubenchov, Sergei Krasovsky, Ivan Sipaev, bs; Bolshoi Theatre Orchestra and Chorus; Nicolai Golovanov. Recital Records RR 440 (3). (Reissue from Russian originals)
+-HF 5-77 p83

2021 Boris Godunov (ed. Rimsky-Korsakov). Ludmila Lebedeva, s; Eugenia

Zareska, Lydia Romanova, ms; Nicolai Gedda, Andre Bielecki, Wassili Pasternak, Gustav Ustinov, Raymond Bonte, t; Kim Borg, bs-bar; Boris Christoff, Stanislav Pieczora, Eugene Bousquet, bs; Choeurs Russes de Paris; French National Radio Orchestra; Issay Dobrowen. Seraphim ID 6101 (4). (Reissue from RCA LHMV 6400, Captiol GDR 7164). (also HMV SLS 5072 Reissue from ALP 1044/7)

+ARG 3-77 p30 +-HFN 4-77 p153
+-GR 2-77 p1318 +NYT 10-9-77 pD24
+-HF 5-77 p83 +-RR 2-77 p32

Boris Godunov: Introduction and polonaise. cf Works, selections (HMV 3101).
Boris Godunov: Vali suda. cf Decca SXL 6826.
Capriccio "In the Crimea". cf Pye PCNH 9.
Fair at Sorochinsk: Gopak. cf Amberlee ALM 602.
Fair at Sorochinsk: Gopak. cf Angel S 37219.
Fair at Sorochinsk: Introduction; Gopak. cf Works, selections (HMV 3101).
Gopak. cf Impromptu passione.
Gopak (arr. Rachmaninoff). cf Impromptu passione.
Hebrew song. cf Golden Crest RE 7064.

2022 Impromptu passione. Pictures at an exhibition. Sonata, 2 pianos. Gopak. Gopak (arr. Rachmaninoff). John Browning, pno. Delos DEL 25430.

-NR 12-77 p13

Intermezzo, B minor. cf Works, selections (HMV 3101).

2023 Khovanschina. Sofia Opera Orchestra; Athanas Margaritov. Classical Cassette Company CCC CP 48/49.

+-HF 9-77 p119 tape.

Khovanschina: Dawn on the Moscow River; Dance of the Persian slaves; Galitzin's journey. cf Works, selections (HMV 3101).
Khovanschina: Marfa's prophecy. cf Columbia/Melodiya M 33931.
Khovanschina: Prelude. cf Pictures at an exhibition.
Khovanschina: Prelude, Act 1; Dance of the Persian maidens; Entracte, Act 4. cf BORODIN: In the Steppes of Central Asia.
Mlada: Triumphal march. cf Works, selections (HMV 3101).

2024 A night on the bare mountain (arr. Rimsky-Korsakov). RESPIGHI: The pines of Rome. RIMSKY-KORSAKOV: Capriccio espagnol, op. 34. BPhO; Lorin Maazel. DG 2548 267 Tape (c) 3348 267. (Reissue from SLPM 138033)

+Gr 7-77 p180 ++RR 10-77 p98
++RR 6-77 p62

2025 A night on the bare mountain. Pictures at an exhibition. St. Louis Symphony Orchestra; Leonard Slatkin. Turnabout TV 34633 Tape (c) CT 2109, KTVC 34633.

+Gr 8-77 p396 +-HFN 10-77 p169 tape
+Gr 9-77 p511 tape +MJ 11-76 p45
+HF 4-77 p109 +NR 8-76 p2
+-HFN 8-77 p87 +-RR 8-77 p58

A night on the bare mountain. cf Works, selections (HMV 3101).
A night on the bare mountain. cf BALAKIREV: King Lear: Overture.
A night on the bare mountain. cf BORODIN: In the Steppes of Central Asia (Philips 6530 022).
A night on the bare mountain. cf BORODIN: In the Steppes of Central Asia (Quintessence PMC 7026).
A night on the bare mountain. cf BORODIN: Prince Igor: Polovtsian dances (DG 2536 379).

A night on the bare mountain. cf BORODIN: Prince Igor: Polovtsian dances (HMV 7006).
A night on the bare mountain. cf GLINKA: Russlan and Ludmilla: Overture.
A night on the bare mountain. cf DG 2584 004.
2026 Pictures at an exhibition. Scherzo, B flat major (arr. Liapunov). Turkish march (arr. Chernov). Michel Beroff, pno. Angel S 37223.
++ARG 3-77 p30 +NR 11-76 p13
+HF 12-76 p108
2027 Pictures at an exhibition (orch. Ravel). RAVEL: Pavane pour une infante defunte. Tokyo Metropolitan Symphony Orchestra; Louis Fremaux. Denon PCM OX 7072ND.
+-HF 10-77 p112
2028 Pictures at an exhibition (orch. Ravel). PROKOFIEV: Symphony, no. 1, op. 25, D major. CSO; Carlo Maria Giulini. DG 2530 783 Tape (c) 3300 783.
+-ARG 11-77 p37 +MJ 7-77 p70
+Gr 4-77 p1555 ++NR 6-77 p2
++HF 7-77 p101 ++RR 4-77 p55
++HFN 4-77 p141 +-SFC 5-15-77 p50
2029 Pictures at an exhibition. Gina Bachauer, pno. Pictures at an exhibition (orch. Ravel). PhO; Lorin Maazel. HMV SXLP 30233 Tape (c) TC SXLP 30233. (Reissues from DLP 1154, Columbia SAX 2484)
+-Gr 3-77 p1401 +HFN 5-77 p138 tape
+HFN 4-77 p151 +-RR 4-77 p54
2030 Pictures at an exhibition. PROKOFIEV: Visions fugitives, op. 22. Tedd Joselson, pno. RCA ARL 1-2158 Tape (c) ARK 1-2158 (ct) ARS 1-2158.
+-ARG 6-77 p40 +-NR 8-77 p15
+-HF 7-77 p102 +-St 7-77 p118
+-MJ 7-77 p70
2031 Pictures at an exhibition (arr. Ravel). Khovanschina: Prelude. NPhO; Charles Mackerras. Vanguard VSD 71188 (Q) VSQ 30032. (also VCS 10116)
+Gr 12-74 p1143 +-RR 12-74 p36
+HF 11-74 p117 +St 9-74 p127
+NR 3-77 p3
Pictures at an exhbition. cf A night on the bare mountain (Turnabout TV 34633).
Pictures at an exhibition. cf Impromptu passione.
Pictures at an exhibition. cf CHOPIN: Sonata, piano, no. 2, op. 35, B flat minor.
Pictures at an exhibition: Hut on fowls' legs, Great gate at Kiev. cf Seraphim SIB 6094.
Scherzo, B flat major (arr. Liapunov). cf Pictures at an exhibition.
Scherzo, B flat major. cf Works, selections (HMV 3101).
Sonata, 2 pianos. cf Impromptu passione.
2032 Works, selections: Boris Godunov: Introduction and polonaise (arr. Rimsky-Korsakov). Fair at Sorochinsk: Introduction, Gopak. Intermezzo, B minor (arr. Rimsky-Korsakov). Khovanschina: Dawn on the Moscow River; Dance of the Persian slaves; Galitzin's journey (arr. Rimsky-Korsakov). Mlada: Triumphal march (arr. Rimsky-Korsakov). A night on the bare mountain (arr. Rimsky-Korsakov). Scherzo, B flat major. USSR Symphony Orch-

estra; Yevgeny Svetlanov. HMV ASD 3101. (also Angel SR 40273).
+-Gr 8-75 p326 +RR 8-75 p40
+-HFN 8-75 p80 +-SFC 10-10-76 p32
+MJ 2-77 p31 +St 5-77 p115
+NR 12-76 p4

MYSLIVECEK, Josef
2033 Quintets, strings, nos. 1-4, 6. Czech Chamber Soloists; Miroslav Matyas. Supraphon 110 1880.
+Gr 7-77 p197 +NR 10-77 p7
+HFN 8-77 p147 +-RR 6-77 p61

NAPRAVNIK, Eduard
Dubrovsky: French duet; Masha's air. cf Club 99-96.
Dubrovsky: Never to see her. cf Rubini RS 301.
Dubrovsky: Vladimir's recitative and romance. cf RACHMANINOFF: Francesca da Rimini, op. 25.
Harold: Cradle song. cf Club 99-96.
Harold: Lullaby. cf Rubini RS 301.

NARDINI, Pietro
Sonata, violin. cf CORELLI: La folia.

NARES, James
Introduction and fugue, F major. cf BAIRSTOW: Pange lingua.
The souls of the righteous. cf BAIRSTOW: Pange lingua.
The souls of the righteous. cf Argo ZK 3.

NARVAEZ, Luis de
Baxa de contrapunto. cf MILAN: Pavanas, nos. 1-6.
Diferencias (variations) on "Guardame las vacas". cf Angel S 36093.
Diferencias sobre "Guardame las vacas". cf MILAN: Pavanas, nos. 1-6.
Fantasia. cf MILAN: Pavanas, nos. 1-6.
Fantasia. cf L'Oiseau-Lyre 12BB 203/6.
Mille regretz. cf MILAN: Pavanas, nos. 1-6.
Mille regretz. cf L'Oiseau-Lyre 12BB 203/6.

NAUMANN, Siegfried
Songs on Latin texts, mixed voices and some instruments, op. 24. cf DALLAPICCOLA: Cori si Michelangelo Buonarroti il Giovane, nos. 1 and 2.

NAYLOR, Bernard
O Lord, almighty god. cf Abbey LPB 770.

NEALE
Good King Wenceslas. cf HMV SQ CSD 3784.

NEBRA, Manuel Blasco de
Sonata, no. 5. cf RCA RL 2-5099.

NEGRI, Marc Antonio
Il bianco fiore. cf DG 2533 173.
Lo spagnoletto. cf DG 2533 173.

NEIDHART VON REUENTHAL
Winder, diniu meil; Meie, din liehter schin; Meienzit. cf BIS LP 75.

NELHYBEL, Vaclav
Scherzo concertante. cf Musical Heritage Society MHS 3547.
Trittico. cf MENC 76-11.

NELSON, Ron
Savannah river holiday. cf CARPENTER: Adventures in a perambulator.

NEPOMUK, David Johann
2034 Chaconne and fugue. Choralwerk II: Macht hoch die Tur, die Tor Macht weit und Komm, heiliger Geist, Herr Gott. Choralwerk

VIII: Es sungen drei Engel ein sussen Gesang. Prelude and fugue, G major. Graham Barber, org. Vista VPS 1048.
+HFN 6-77 p131
Choralwerk II: Macht hoch die Tur, die Tor Macht weit und Komm, heiliger Geist, Herr Got. cf Chaconne and fugue.
Choralwerk VIII: Es sungen drei Engel ein sussen Gesang. cf Chaconne and fugue.
Prelude and fugue, C major. cf Chaconne and fugue.

NERUDA, Jan Krtitel
Concerto, trumpet, E flat major. cf HAYDN: Concerto, trumpet, E flat major.

NEUSIDLER, Hans
The Burgher of Nuremberg. cf Saga 5420.
Ein guter Venezianer Tantz. cf Hungaroton SLPX 11721.
Hie folget ein welscher Tantz Wascha Mesa, Der hupff auf. cf Hungaroton SLPX 11721.
Der Judentanz. cf DG 2533 302.
Der Judentanz. cf DG 2723 051.
Der Juden Tantz, Der hupff auf zur Juden Tantz. cf Hungaroton SLPX 11721.
Der polnisch Tantz, Der hupff auf. cf Hungaroton SLPX 11721.
Preamble. cf DG 2533 302.
Welscher Tantz Wascha mesa. cf DG 2533 302.
Welscher Tanz Wascha mesa: Hupfauf. cf DG 2723 051.

NEVILLE
Shrewsbury fair. cf EMI TWOX 1058.

NEVIN, Ethelbert
Un giorno in Venezia. cf MacDOWELL: Fantasy pieces, op. 17.
The rosary. cf Prelude PRS 2505.
Songs. cf MacDOWELL: Fantasy pieces, op. 17.
Water scenes, op. 13. cf MacDOWELL: Fantasy pieces, op. 17.
NEW CHORAL MUSIC: The ineluctable modality. cf Advance FGR 18.
NEW MUSIC FOR CONTRABASS. cf Finnadar SR 9015.
NEW SWEDISH MUSIC. cf BIS LP 32.

NEWELL
Ryo-nen. cf Advance FGR 18.

NEWMAN (16th century)
Pavan. cf ABC ABCL 67008.
A pavyon. cf Abbey LPB 765.

NIBELLE
Carillon orleanais. cf Calliope CAL 1922/4.

NICHOLS
We've only just begun. cf Grosvenor GRS 1055.

NICHOLSON, Richard
No more, good herdsman, of thy song. cf National Trust NT 002.
Sing, shepherds all. cf DG 2533 347.

NICOLAI, Otto
2035 Die lustige Weiber von Windsor (The merry wives of Windsor). Edith Mathis, Helen Donath, s; Hanna Schwarz, ms; Peter Schreier, Karl-Ernst Mercker, Claude Dormoy, t; Bernd Weikl, bar; Kurt Moll, Siegfried Vogel, bs; German Opera Orchestra and Chorus; Bernhard Klee. DG 2709 065 (3) Tape (c) 3371 026. (also DG 2740 159)
+-ARG 9-77 p44 +NR 8-77 p9
+Gr 4-77 p1593 ++NYT 7-10-77 pD13
+-HF 9-77 p102 +OC 9-77 p53
+HFN 4-77 p131 +RR 4-77 p35

+-MJ 9-77 p34 ++SFC 6-5-77 p45
++MT 9-77 p737 ++St 8-77 p79

Die lustige Weiber von Windsor: Horch, die Lerche. cf Telefunken AJ 6-42232.
The merry wives of Windsor: Overture. cf Classics for Pleasure CFP 40263.
The merry wives of Windsor: Overture. cf Quintessence PMC 7013.
Die lustige Weiber von Windsor: Overture. cf Supraphon 110 1637.

NIELSEN, Carl

At a young artist's bier. cf Works, selections (Seraphim SIC 6097).
Bohemian Danish folk melody. cf Works, selections (Seraphim SIC 6097).
Chaconne, op. 32. cf Piano works (Decca SDD 476).

2036 Concerto, clarinet. Concerto, flute. Gilbert Jesperson, flt; Ib Erikson, clt; Danish State Radio Orchestra; Thomas Jensen, Mogens Woldike. Decca ECS 800. (Reissue from LXT 2979)
+Gr 9-77 p432 +RR 7-77 p58
+-HFN 7-77 p125

2037 Concerto, clarinet, op. 57. Concerto, flute. Concerto, violin, op. 33. Symphonic rhapsody. Kjell-Inge Stevensson, clt; Frantz Lemsser, flt; Arve Tellefsen, vln; Danish State Radio Orchestra; Herbert Blomstedt. Seraphim (Q) SIB 6106 (2). (Reissues from SIXC 6097/8)
+NR 10-77 p5 +St 11-77 p81

Concerto, flute. cf Concerto, clarinet.
Concerto, flute. cf Concerto, clarinet, op. 57.
Concerto, violin, op. 33. cf Concerto, clarinet, op. 57.
Dance of the lady's maids. cf Piano works (Decca SDD 476).
Festival prelude. cf Piano works (Decca SDD 475).
Festpraeludium. cf Guild GRSP 7011.
Helios overture, op. 17. cf Works, selections (Seraphim SIC 6097).
Humoresque bagatelles, op. 11. cf Piano works (Decca SDD 476).

2038 Hymnus amoris, op. 12. Sleep, op. 18. Kirsten Schultz, Bodil Gobel, s; Tonny Landy, t; Bent Norup, bs-bar; Megens Schmidt Johansen, Hans Christian Andersen, bs; Copenhagen Boys Choir; Danish State Radio Orchestra and Chorus; Mogens Woldike. HMV SQ ASD 3358.
+-Gr 9-77 p468 +RR 9-77 p89
+HFN 10-77 p159

Little suite, op. 1, A minor. cf HMV SQ ESD 7001.
Pan and syrinx, op. 49. cf Symphonies, nos. 4-6.
Piano music for young and old, op. 53. cf Piano works (Decca SDD 475).

2039 Piano works: Festival prelude. Piano music for young and old, op. 53. Pieces, piano, op. 3 (5). Pieces, piano, op. 59 (3). Symphonic suite, op. 8. John McCabe, pno. Decca SDD 475.
++Gr 1-76 p1218 +MT 6-77 p488
+HFN 11-75 p161 +-RR 11-75 p81
+-MM 12-76 p43

2040 Piano works: Chaconne, op. 32. Dance of the lady's maids. Humoresque bagatelles, op. 11. Suite, op. 45. Theme and variations, op. 40. John McCabe, pno. Decca SDD 476.
++Gr 1-76 p1218 +MT 6-77 p488
+HFN 11-75 p161 +-RR 11-75 p81
+-MM 12-76 p43

Pieces, piano, op. 3. cf Piano works (Decca SDD 475).
Pieces, piano, op. 59. cf Piano works (Decca SDD 475).

Rhapsodie overture. cf Symphonies, nos. 4-6.
Saga-Drom, op. 39. cf Symphonies, nos. 4-6.
2041 Saul and David. Elisabeth Soderstrom, Sylvia Fisher, Bodil Gobel, s; Willy Hartmann, Alexander Young, t; Boris Christoff, Michael Langdon, Kim Borg, bs; John Alldis Choir; Danish State Radio Orchestra and Chorus; Jascha Horenstein. Unicorn RHS 343/5 (3).
+-ARSC vol 9, no. 1, 1977 p101
+Gr 7-76 p211
++HF 11-76 p96
+HFN 7-76 p93
+NR 9-77 p11
+-ON 10-77 p70
+RR 8-76 p22
+SFC 8-29-76 p28
++St 11-76 p150
+STL 8-8-76 p29
Sleep, op. 18. cf Hymnus amoris, op. 12.
Suite, op. 45. cf Piano works (Decca SDD 476).
Symphonic rhapsody. cf Concerto, clarinet, op. 57.
Symphonic suite, op. 8. cf Piano works (Decca SDD 475).
Symphony, no. 1, op. 7, G minor. cf Works, selections (Seraphim SIC 6097).
Symphony, no. 2, op. 16. cf Works, selections (Seraphim SIC 6097).
Symphony, no. 3, op. 27. cf Works, selections (Seraphim SIC 6097).
2042 Symphonies, nos. 4-6. Saga-Drom, op. 39. Pan and syrinx, op. 49. Rhapsodie overture. Danish State Radio Symphony Orchestra; Herbert Blomstedt. Seraphim SIC 6098 (3).
+-ARG 5-77 p27
+HF 2-77 p100
+NR 6-77 p3
++SFC 2-13-77 p33
2043 Symphony, no. 6. LSO; Ole Schmidt. Unicorn RHS 329. (Reissue from RHS 324/30)
+Gr 9-77 p432
+HFN 9-77 p151
++RR 6-77 p61
Theme and variations, op. 40. cf Piano works (Decca SDD 476).
2044 Works, selections: At a young artist's bier (Andante lamentoso). Bohemian Danish folk melody. Helios overture, op. 17. Symphony, no. 1, op. 7, G minor. Symphony, no. 2, op. 16. Symphony, no. 3, op. 27. Kirsten Schultz, s; Peter Rasmussen, bar; Danish State Radio Orchestra; Herbert Blomstedt. Seraphim (Q) SIC 6097 (3).
+-ARG 5-77 p27
+HF 2-77 p100
+NR 3-77 p4
++SFC 12-5-76 p58
+-St 3-77 p138
NIGEL ROGERS, Canti amorosi. cf DG 2533 305.

NIGHTINGALE, James
Entente. cf Orion ORS 77263.

NILSSON, Bo
Fragments. cf KUISMA: Concertpiece for percussion.
2045 Quantitaten. PAULSON: Modi, op. 108b. SCHUMANN: Kinderscenen, op. 15. STRAVINSKY: Serenade, A major. Hans Palsson, pno. Caprice RIKS LP 79. (also CAP 1079)
+RR 10-75 p80
+RR 6-76 p75
+St 6-77 p136

NIN, Joaquin
Granadina. cf HMV SQ ASD 3283.
Spanish suite. cf Laurel Protone LP 13.

NIVERS, Guillaume
Suite, Bk 3, in the second mode. cf LEBEGUE: Elevation, G major.

NOBLE
Scherzino. cf Marc NC 5355.

NODA
Improvisation, alto saxophone, no. 1. cf Coronet LPS 3036.
NOLA, Gian Domenico da
Madonna nui sapima. cf Enigma VAR 1024.
NONO, Luigi
2046 Como una ola de fuerza y luz. Y entonces comprendio. Slavka Taskova, Mary Lindsay, Liliana Poli, Gabriela Ravazzi, s; Kadigia Bove, Miriam Acevedo, Elena Vicini, speakers; Maurizio Pollini, pno; Bavarian Radio Symphony Orchestra; RAI Rome Chamber Choir; Claudio Abbado, Nino Antonellini; Magnetic tape, Luigi Nono, sound direction. DG 2530 436.
+-Gr 7-74 p253 +-NR 8-74 p13
+HF 9-74 p98 +NYT 3-13-77 pD22
+MJ 10-75 p39 +RR 7-74 p77
Y entonces comprendio. cf Como una ola de fuerza y luz.
NORCOME, Daniel
With angel's face. cf DG 2533 347.
NORDHEIM, Arne
Dinosaurus. cf Orion ORS 77263.
NORGAARD, Per
Arcana. cf Orion ORS 77263.
Spell. cf CHILDS: Trio, clarinet, violoncello and piano.
NOUGES, Jean
Quo Vadis: Amica l'ora. cf Rubini GV 34.
NOVACEK, Ottokar
Caprices, op. 5, no. 4. cf Discopaedia MB 1013.
Perpetuum mobile. cf CBS 73589.
NUCIUS
Freut euch ihr Auserwahlte. cf ABC L 67002.
NYSTROEM, Gosta
The merchant of Venice: Nocturne. cf Decca DDS 507/102.
Songs (3). cf BIS LP 34.
Songs: Bara hos dem; Vitt land; Onskan. cf FRUMERIE: Songs (Swedish Society SLT 33171).
Songs of the sea. cf ELGAR: Sea pictures, op. 37.
Symphony, no. 1. cf HOLMBOE: Symphony, no. 10, op. 105.
OBRECHT, Jacob
Haec deum caeli laudemus nunc Dominum. cf HMV SLS 5049.
Ic draghe de mutze clutze; Mijn morken gaf; Pater noster. cf L'Oiseau-Lyre 12BB 203/6.
La stangetta. cf Enigma VAR 1024.
Tsaat een meskin. cf Enigma VAR 1024.
OCHSENKUHN, Sebastian
Innsbruck, ich muss dich lassen. cf DG 2533 302.
OCKEGHEM, Johannes
Missa au travail suis. cf Missa ma maistresse.
2047 Missa ma maistresse. Missa au travail suis. Motets: Alma redemptoris; Au travail suis; Ave Maria; Ma maistresse. Pomerium Musices; Alexander Blachly. Nonesuch H 71336.
+NR 8-77 p11 ++St 9-77 p138
Motets: Alma redemptoris; Au travail suis; Ave Maria; Ma maistresse. cf Missa ma maistresse.
Motets: Intemerata Dei mater. cf HMV SLS 5049.
Songs: Prenez sui moi; Ma bouche rit. cf HMV SLS 5049.
OFFENBACH, Jacques
Barbe bleue: Overture. cf Overtures (HMV ESD 7034).
2048 La belle Helene. Janine Linda, s; Loly Valdarnini, ms; Andre

Dran, Jean Mollien, t; Roger Giraud, Jacques Linsolas, bar; Paris Philharmonic Orchestra and Chorus; Rene Leibowitz. Everest S 458/2 (2). (Reissue from Nixa PLP 206/1-2)
+Gr 1-77 p1187

La belle Helene: Au mont Ida. cf HMV RLS 715.
La belle Helene: Overture. cf Overtures (HMV ESD 7034).
Les contes d'Hoffmann (The tales of Hoffmann): Doll's song. cf HMV HLM 7066.
Les contes d'Hoffmann: Elle a fui la tourterelle. cf Seraphim IB 6105.
Les contes d'Hoffmann: Elle a fui, la tourterelle; Les oiseaux dans la charmille. cf Decca D65D3.
Les contes d'Hoffmann: Les oiseaux dans la charmille. cf Columbia 34294.
Les contes d'Hoffmann: Les oiseaux dans la charmille. cf Court Opera Classics CO 342.

2049 La fille du Tambour Major: Legende du fruit defendu; Par une chaleur si forte; Chanson de l'ane; De grace ayez pitie de moi; Couplets du tailleur; Petit Francais, brave Francais; C'etait une princesse; Examiner ma figure; Quatuor du billet du logement; Couplets de Claudine; Valse and ensemble; Tenez, j'aurai la franchise; Chanson de la fille du Tambour Major; Tarantella; Duo de la confession; Finale. Liliane Chatel, Nadine Sautereau, s; Suzanne Lafaye, ms; Thierry Peyron, Remy Corazza, t; Michel Dens, Dominique Tirmont, Jacques Provost, bar; Rene Duclos Chorus; OSCCP; Felix Novolone. CRD Pathe C051 12192.
+Gr 1-77 p1187

2050 La fille du Tambour Major: Overture. Gaite parisienne (Rosenthal). Monte Carlo Opera Orchestra; Manuel Rosenthal. Angel (Q) S 37209 Tape (c) 4XS 37209 (ct) 8XS 37209. (also HMV ASD 3311 Tape (c) TC ASD 3311)
+-Gr 2-77 p1329 +-NR 5-77 p3
+-Gr 7-77 p225 tape +RR 2-77 p55
+HF 11-77 p120 +SFC 5-13-77 p34
+HFN 2-77 p125 +St 6-77 p134
+HFN 5-77 p138 tape

Gaite parisienne (Rosenthal). cf La fille du Tambour Major: Overture.
Gaite parisienne. cf CHOPIN: Les sylphides.
Gaite parisienne. cf GOUNOD: Faust: Ballet music.

2051 Gaite parisienne, excerpts (arr. Rosenthal). STRAUSS, J. II: Le beau Danube, op. 314, ballet (arr. Desormiere). Berlin Radio Symphony Orchestra; Paul Strauss. DG 2548 248 Tape (c) 3348 248. (Reissues from SLPE 13301, SLPEM 136026)
+Gr 1-77 p1184 +HFN 1-77 p119
+Gr 7-77 p225 tape +RR 1-77 p56

Gaite parisienne, excerpts. cf GERSHWIN: An American in Paris.
Gaite parisienne, excerpts. cf GOUNOD: Faust: Ballet music.

2052 Gaite parisienne: Ballet (arr. Rosenthal). STRAUSS, J. II: Graduation ball, op. 97, excerpts (arr. Dorati). PhO; Charles Mackerras. Classics for Pleasure CFP 40268. (Reissue from HMV CSD 1533)
+Gr 10-77 p711 +RR 10-77 p57
+HFN 10-77 p165

Genevieve de Brabant: Duo des gendarmes. cf Club 99-102.

2053 La grande duchesse de Gerolstein. Regine Crespin, Mady Mesple, s; Alain Vanzo, Charles Burles, Tibert Raffali, t; Robert Massard, Claude Meloni, bar; Francois Loup, bs; Toulouse Capitole Theatre Orchestra and Chorus; Michel Plasson. Columbia M2 34576 (2). (also CBS 79207)

+-FF 9-77 p40 +NYT 4-17-77 pD17
+Gr 4-77 p1604 +RR 4-77 p36
+-HF 9-77 p104 ++SFC 6-19-77 p46
+HFN 4-77 p131 +St 11-77 p145
+NR 8-77 p9

2054 La grande duchesse de Gerolstein, abridged. Gisele Prevet, s; Eugenia Zareska, ms; Andre Dran, Georges Lacour, Gabriel Bonton, Jean Mollien, t; Georges Thery, bar; John Riley, bs; Choeur Lyrique de Paris; Pasdeloup Concerts Orchestra; Rene Leibowitz. Saga 5446. (Reissue from Opera Society OPS 118/9).

+Gr 7-77 p229 -RR 7-77 p39
+-HFN 7-77 p117

La grande duchesse de Gerolstein: Dites lui. cf CBS 76522.
La grande duchesse de Gerolstein: Overture. cf Overtures (HMV ESD 7034).
Orpheus in the underworld: Overture. cf Overtures (HMV ESD 7034).
Orpheus in the underworld: Overture. cf Quintessence PMC 7013.

2055 Overtures: Barbe bleue. La belle Helene. La grande duchesse de Gerolstein. Orpheus in the underworld. La vie parisienne. Birmingham Symphony Orchestra; Louis Fremaux. HMV ESD 7034 Tape (c) TC ESD 7034. (Reissue from Columbia TWO 388)

+Gr 4-77 p1604 +-HFN 8-77 p99 tape
+-Gr 7-77 p225 tape +RR 4-77 p55
/HFN 4-77 p151 +RR 8-77 p86 tape

2056 La perichole. Regine Crespin, s; Alain Vanzo, t; Jacques Trigeau, Gerard Friedmann, bar; Jules Bastin, bs; Rhine Opera Chorus; Strasbourg Philharmonic Orchestra; Alain Lombard. Musical Heritage Society MHS 3518/9 (2) Tape (c) MHC 5518/9. (also Erato STU 70994/5, RCA FRL 2-5994)

+-HF 9-77 p104 +SFC 12-11-77 p61
+OC 6-77 p44

La perichole: Ah, que diner je viens de faire. cf CBS 76522.
La perichole: O mon cher amant. cf Rubini GV 57.

2057 La vie parisienne. Regine Crespin, Mady Mesple, Eliane Lublin, Christiane Chateau, s; Michel Senechal, t; Michel Trempont, Luis Masson, Jean-Christoph Benoit, Michel Jarry, bar; Toulouse Symphony Orchestra and Choruses; Michel Plasson. Angel (Q) SBLX 3839 (2). (also HMV SLS 5076)

+Gr 5-77 p1749 +-ON 4-16-77 p37
++HF 6-77 p96 ++RR 5-77 p35
+HFN 5-77 p131 ++SFC 5-13-77 p34
+NR 6-77 p9 +St 6-77 p134
+NYT 4-17-77 pD17

La vie parisienne: Overture. cf Overtures (HMV ESD 7034).

OHANA, Maurice

2058 Tres graficos para guitarra y orquesta. RUIZ PIPO: Tablas para guitarra y orquesta. Narciso Yepes, gtr; LSO; Rafael Fruhbeck de Burgos. DG 2530 585.

+Gr 3-76 p1468 ++NR 4-76 p14
+HFN 5-76 p103 ++RR 3-76 p47
+MJ 7-76 p56 ++St 1-77 p72

OLAGUE, Bartolomeo
Xacara. cf London CS 7046.
OLD SWEDISH ORGANS. cf BIS LP 27.
OLDFIELD, Mike
2059 Collaborations (with David Bedford). Hergest ridge. Ommadawn. Tubular bells. Mike Oldfield, David Beford and others. Virgin VBOX 1-4 (4).
+-Te 3-77 p27
Hergest ridge. cf Collaborations.
Ommadawn. cf Collaborations.
Tubular bells. cf Collaborations.
OLSSON-FOLLINGER, Goran
Mejlikan. cf Discopaedia MOB 1020.
Polska, A major. cf Discopaedia MOB 1020.
ORFF, Carl
2060 Carmina burana. Norma Burrowes, s; Louis Devos, t; John Shirley-Quirk, bar; Brighton Festival Chorus, Southend Boys Choir; RPO; Antal Dorati. Decca PFS 4368 Tape (c) KPFC 4368. (also London 21153 Tape (c) 5-21153)
+-ARG 2-77 p37 +HFN 1-77 p123 tape
+-Gr 12-76 p1034 +-NR 4-77 p8
+-HF 4-77 p109 +-RR 12-76 p94
+-HFN 1-77 p123 +-RR 3-77 p99 tape
2061 Carmina burana. Celestina Casapietra, s; Horst Hiestermann, t; Karl-Heinz Stryczek, bar; Dresden Boys Choir; Leipzig Radio Symphony Orchestra and Chorus; Herbert Kegel. Philips 9500 040 Tape (c) 7300 444 (ct) 7750 096.
+-ARG 2-77 p37 ++MJ 1-77 p27
/Gr 10-76 p639 +NR 4-77 p8
++HF 12-76 p110 ++RR 10-76 p98
++HFN 10-76 p161 +St 3-77 p139
2062 Lamenti. Lucia Popp, s; Hanna Schwarz, con; Rose Wagemann, ms; Hermann Prey, bar; Karl Ridderbusch, bs; Carl Orff, speaker; Bavarian Radio Chorus; Munich Radio Orchestra; Kurt Eichhorn. BASF 44-22458/9 (2).
+-FF 11-77 p42 ++NR 3-77 p11
2063 Songs (choral works): Concento di voce: 3, Sermio; Der gute Mensch; Nanie und Dithyrambe; Veni creator spiritus; Vom Fruhjahr, Oeltank, und vom Fliegen. Czech Philharmonic Chorus; Instrumental Ensemble; Vaclav Smetacek. Supraphon 112 1137.
+-Gr 9-76 p459 +NR 10-76 p7
+-HF 12-76 p110 ++RR 8-76 p76
+-HFN 8-76 p85 +St 3-77 p139
ORGAN MUSIC FROM KING'S. cf Odeon HQS 1356.
ORGAN MUSIC FROM THE CITY OF LONDON. cf Vista VPS 1047.
ORGAN RECITAL, Earl Barr. cf TR 1006.
ORNSTEIN, Leo
2064 A la chinoise op. 39. Arabesques, op. 42. Morning in the woods. Sonata, no. 4. Martha Anne Verbit, pno. Genesis GS 1066.
+ARG 10-77 p39 +NR 7-77 p13
A la chinoise, op. 39. cf Piano works (Orion ORS 75194).
Arabesques, op. 42. cf A la chinoise, op. 39.
Chant of the Hindou priests. cf Piano works (Orion ORS 75194).
Melancholy landscape. cf Piano works (Orion ORS 75194).
Morning in the woods. cf A la chinoise, op. 39.
Nocturne and dance of the fates. cf Louisville LS 753/4.
2065 Piano works: A la chinoise, op. 39. Chant of the Hindou priests.

Melancholy landscape. Ten poems. Three moods. Wild men's dance. Michael Sellers, pno. Orion ORS 71594.
+Audio 12-76 p88 +-NR 11-76 p13

Quintet, piano and strings, op. 92. cf Three moods.

Sonata, no. 4. cf A la chinoise, op. 39.

Ten poems. cf Piano works (Orion ORS 75194).

2066 Three moods. Quintet, piano and strings, op. 92. William Westney, pno; Daniel Stepner, Michael Strauss, vln; Peter John Sacco, vla; Thomas Mansbacher, vlc. CRI SD 339.
*Audio 4-77 p92 +NYT 4-11-76 pD23
+HF 5-76 p88 +SR 9-18-76 p50
++MJ 3-76 p24 ++St 5-76 p119
++NR 1-76 p7

Three moods. cf Piano works (Orion ORS 75194).

Wild man's dance. cf Piano works (Orion ORS 75194).

ORTIZ, Diego

Dulce memoire; Recercada. cf L'Oiseau-Lyre 12BB 203/6.

Quinta pars. cf BIS LP 22.

Recercadas primera y segunda. cf BIS LP 22.

Recercadas primera y segunda. cf Nonesuch H 71326.

OSBORNE, L.

Lullaby for Penelope. cf HMV SRS 5197.

Lullaby for Penelope. cf Starline SRS 197.

My northern hills. cf HMV SRS 5197.

My northern hills. cf Starline SRS 197.

OSBORNE, Willson

Rhapsodie, bassoon. cf ELGAR: Romance, op. 62.

OXENFORD

The ash grove. cf HMV ESD 7002.

PACHELBEL, Johann

Alle Menschen mussen sterben. cf ABC ABCL 67008.

Canon. cf ALBINONI: Adagio, G minor.

Canon, D major. cf ALBINONI: Adagio, organ and strings, G minor.

Canon, D major. cf FASCH: Concerto, trumpet, 2 oboes and strings, D major.

Canon, D major. cf DG 2548 219.

Chaconne, A minor. cf Argo ZRG 806.

Magnificat: Fugues, nos. 4-5, 10, 13. cf ABC ABCL 67008.

Suites, B flat major, G major. cf FASCH: Concerto, trumpet, 2 oboes and strings, D major.

Toccata and fugue, B flat major. cf ABC ABCL 67008.

Was Gott tut, das ist wohlgetan. cf BACH: Chorale preludes (Pelca PSR 40583).

Wie schon leuchtet der Morgenstern, chorale prelude. cf HOWELLS: Psalm prelude.

PACOLINI (16th century Italy)

Padoana commun; Passamezzo commun. cf L'Oiseau-Lyre 12BB 203/6.

Passamezzo della battaglia. cf DG 2533 323.

Saltarello della traditoria. cf DG 2533 323.

Saltarello milanese. cf DG 2533 323.

PADEREWSKI, Ignace Jan

Legende, op. 16, no. 1, A flat major. cf International Piano Library IPL 102.

2067 Symphony, B minor. Bydgoszcz Pomeranian Philharmonic Orchestra; Bohdan Wodiszko. Muza SXL 0968.
+ARG 5-77 p55

PAGANINI, Niccolo

Cantabile, op. 17, D major. cf GIULIANI: Sonata, violin and guitar, op. 25.

Cantabile, op. 17, D major. cf Enigma VAR 1025.

2068 Caprices, op. 1, nos. 1-24. Ruggiero Ricci, vln. Decca ECS 803. (Reissue from SXL 2194)
+Gr 7-77 p202 ++RR 6-77 p83
+HFN 7-77 p125

2069 Caprices, op. 1, nos. 1-24. Itzhak Perlman, vln. HMV ASD 3384 Tape (c) TC ASD 3384. (Reissue from SLS 832) (also Angel S 36860)
++Gr 10-77 p662 ++RR 11-77 p93
+Gr 11-77 p900 tape

2070 Caprices, op. 1, nos. 1-24. Michael Rabin, vln. Seraphim SIB 6096 (2). (Reissue from Capitol SPBR 8477)
++HF 4-77 p109 +MJ 11-76 p45

Caprices, op. 1, nos. 1-24. cf Discopaedia MB 1011.

Caprices, op. 1, nos. 5, 13. cf Enigma VAR 1025.

Caprice, op. 1, no. 24, A minor. cf CBS 40-72728.

Centone di sonate, op. 64, no. 1. cf GIULIANI: Sonata, violin and guitar, op. 25.

2071 Centone di sonate, op. 64, no. 2, D major; no. 5, E major. Sonata, violin and guitar, A major. Sonatas, violin and guitar, op. 2 (6). Gyorgy Terebesi, vln; Sonja Prunnbauer, gtr. Telefunken 6-41936.
+Gr 3-77 p1421 ++NR 3-76 p12
+HF 7-76 p83 +SR 3-6-76 p41

2072 Concerto, violin, no. 1, op. 6, D major.* Le Streghe, variations on a theme by Sussmayr, op. 8. Salvatore Accardo, vln; LPO; Charles Dutoit. DG 2530 714. (*Reissue from 2740 121)
+Gr 11-76 p803 +NR 6-77 p6
+HFN 1-77 p119 +RR 11-76 p70
++MJ 9-77 p34

2073 Concerto, violin, no. 1, op. 6, D major. Concerto, violin, no. 2, op. 7, B minor. Shmuel Ashkenasi, vln; VSO; Heribert Esser. DG 2535 207 Tape (c) 3335 207. (Reissue from SLPM 139424)
+Gr 12-76 p1001 +RR 12-76 p63
+HFN 12-76 p151 +-RR 2-77 p98 tape

2074 Concerto, violin, no. 1, op. 6, D major. Boris Belkin, vln; Israel Philharmonic Orchestra; Zubin Mehta. London CS 7019. (also Decca SXL 6798 Tape (c) KSXC 6798)
+Gr 2-77 p1284 +-RR 2-77 p55
+HFN 2-77 p125 +-RR 6-77 p97 tape
+HFN 4-77 p155 tape -SFC 8-28-77 p46
-MJ 9-77 p34 ++St 9-77 p138
+-NR 11-77 p6

2075 Concerto, violin, no. 1, op. 6, D major. Concerto, violin, no. 4, D minor. Henryk Szeryng, vln; LSO; Alexander Gibson. Philips 9500 069 Tape (c) 7300 477.
+Gr 10-76 p602 +MJ 1-77 p27
++HF 1-77 p151 tape +NR 1-77 p6
++HFN 10-76 p175 ++RR 10-76 p64

2076 Concerto, violin, no. 2, op. 7, B minor. Introduction and variations on "Non piu mesta" from Rossini's "La cenerentola", op. 12. Sonata, violin, A major. Salvatore Accardo, vln; LPO; Charles Dutoit. DG 2530 900 Tape (c) 3300 900. (Reissue from 2740 121)

++Gr 10-77 p637 +HFN 11-77 p177
+Gr 12-77 p1146 ++RR 9-77 p66

Concerto, violin, no. 2, op. 7, B minor. cf Concerto, violin, no. 1, op. 6, D major.

Concerto, violin, no. 2, op. 7, B minor. cf International Piano Library IPL 117.

Concerto, violin, no. 2, op. 7, B minor: Ronde a la clochette. cf Discopaedia MB 1014.

2077 Concerto, violin, no. 3, E major.* Sonata, viola, op. 35, C minor. Salvatore Accardo, vln; Dino Asciolla, vla; LPO; Charles Dutoit. DG 2530 629 Tape (c) 3300 629. (*Reissue from 2470 121)

++Gr 5-76 p1765 +NR 7-76 p4
+HF 1-77 p151 tape +RR 5-76 p47
+HFN 4-76 p109

Concerto, violin, no. 4, D minor. cf Concerto, violin, no. 1, op. 6, D major.

2078 Concerto, violin, no. 6, op. posth., E minor. Salvatore Accardo, vln; LPO; Charles Dutoit. DG 2530 467 Tape (c) 3300 412.

+HF 1-77 p151 tape

Fantasia on "Dal tuo stellato soglio", C minor. cf Sonatas, violin and guitar, opp. 2 and 3.

Introduction and variations on "Di tanti palpiti", op. 13. cf Enigma VAR 1025.

Introduction and variations on "Non piu mesta" from Rossini's "La cenerentola", op. 12. cf Concerto, violin, no. 2, op. 7, B minor.

2079 Maestosa sonata sentimentale. I Palpiti, op. 13. Perpetuela. Sonata con variazioni. Sonata Napoleone, op. 31. Salvatore Accardo, vln; LPO; Charles Dutoit. DG 2536 376 Tape (c) 3336 376.

+Gr 12-77 p1087 +HFN 12-77 p176
+Gr 12-77 p1146 tape ++RR 12-77 p71

Moses fantasy on the G string. cf Discopaedia MOB 1020.

Moto perpetuo, op. 11, no. 2 (arr. and orch. Gerhardt). cf RCA LRL 1-5094.

Moto perpetuo, op. 11, no. 2. cf RCA RK 2-5030.

I Palpiti. cf Maestosa sonata sentimentale.

Perpetuela. cf Maestosa sonata sentimentale.

2080 Sonata, viola, op. 35, C minor. ROLLA: Concerto, violin, A major. ROSSINI: Duet, violoncello and double bass. Susanne Lautenbacher, vln; Ulrich Koch, vla; Georges Mallach, vlc; Jean Poppe, double-bs; Wurttemberg Chamber Orchestra, Luxembourg Radio Orchestra; Jorg Faerber, Pierre Cao. Turanbout QTV 34606. (also Decca TVS 34606)

+Gr 3-77 p1401 +NR 5-76 p6
+HFN 3-77 p109 +RR 3-77 p58
+MT 6-77 p489

Sonata, viola, op. 35, C minor. cf Concerto, violin, no. 3, E major.

Sonata, viola, op. 35, C minor. cf HOFFMEISTER: Concerto, viola, D major.

Sonata, violin, A major. cf Concerto, violin, no. 2, op. 7, B minor.

Sonata, violin, op. 25, C major. cf DG 2530 561.

Sonata, violin, op. posth., A major. cf Discopaedia MB 1011.

Sonata, violin and guitar, A major. cf Centone di sonate, op. 64, no. 2, D major; no. 5, E major.
Sonata, violin and guitar, A major. cf GIULIANI: Sonata, violin and guitar, op. 25.

2081 Sonatas, violin and guitar, opp. 2 and 3. Fantasia on "Dal tuo stellato soglio", C minor. Tarantella, A minor. Gyorgy Terebesi, vln; Sonja Prunnbauer, gtr. Telefunken 6-41995.
+-Gr 2-77 p1301 +-NR 7-76 p6
+-HFN 2-77 p125 +-RR 1-77 p72

Sonatas, violin and guitar, op. 2 (6). cf Centone di sonate, op. 64, no. 2, D major; no. 5, E major.

2082 Sonatas, violin and guitar, op. 3, nos. 1-6. VIOTTI: Duets, 2 violins, op. 29, nos. 2 and 3. WIENIAWSKI: Etudes caprices (3). Desmond Bradley, Donald Weekes, vln; Hermann Leeb, gtr. RCA GL 2-5095.
+Gr 9-77 p452 +RR 10-77 p78
+-HFN 12-77 p176

Sonata, violin and guitar, op. 3, no. 6, E minor. cf GIULIANI: Sonata, violin and guitar, op. 25.
Sonata con varizioni. cf Maestosa sonata sentimentale.
Sonata Napoleone. cf Maestosa sonata sentimentale.
Le Streghe, variations on a theme by Sussmayr, op. 8. cf Concerto, violin, no. 1, op. 6, D major.
Tarantella, A minor. cf Sonatas, violin and guitar, opp. 2 and 3.
Variations on etude, no. 6. cf Musical Heritage Society MHS 3611.

PAGGI (19th century Italy)
Rimembranze Napolitane. cf Pearl SHE 533.

PAINE, John Knowles
Prelude, op. 19, no. 1, D flat major. cf Orion ORS 76255.
Variations on the Austrian hymn. cf Advent ASP 4007.

PAISIBLE, James
Sonata, 4 recorders. cf HMV SQ CSD 3781.

PAIX, Jacob
Schiarazula Marazula. cf DG 2723 051.
Ungaresca: Saltarello. cf DG 2723 051.

PALESTRINA, Giovanni de
Dum complerentur dies Pentecostes. cf Abbey LPB 783.
Ricercar a 4. cf Telefunken AW 6-42033.
Sanctus and benedictus (Soriano). cf Vanguard SVD 71212.

2083 Songs (choral works): Hodie beata virgo. Litaniae de beata Virgine Maria a 8 vocum. Magnificat a 8 voci (primi toni). Senex puerum portabat. Stabat mater. King's College Choir; David Willcocks. Argo ZK 4 Tape (c) KZKS 4. (Reissue from ZRG 5398)
+-Gr 12-76 p1034 +-HFN 4-77 p155 tape
+HFN 1-77 p121 +RR 1-77 p82

2084 Songs (choral works): Missa aeterna Christi munera. Oratio Jeremiae prophetae. Motets: Sicut cervus desiderat; Super flumina Babylonis; O bone Jesu. Pro Cantione Antiqua; Bruno Turner. DG 2533 322 Tape (c) 3310 322.
+-Gr 12-76 p1034 +NR 8-77 p8
+HFN 1-77 p113 +RR 12-76 p93
+-MT 2-77 p134 ++St 11-77 p145

A PALM COURT CONCERT. cf Angel (Q) S 37304.

PALMER, Greg
Food for your soul (South). cf Atlantic SD 2-7000.
L. A. nights. cf Atlantic SD 2-7000.

New Orleans. cf Atlantic SD 2-7000.
Tank. cf Atlantic SD 2-7000.

PALMER, Robert

2085 Sonata, trumpet, B flat major. ROLLIN: Reflections on ruin by the sea. SHAPERO: Sonata, trumpet and piano, C major. Marice Stith, tpt; Stuart Raleigh, pno. Redwood ES 4.
+-NR 7-77 p15

PALMGREN, Selim

Contrasts: Les adieux; Arlequin. cf MELARTIN: Der Traurige Garten, op. 52.
May night, op. 27, no. 4. cf MELARTIN: Der Traurige Garten, op. 52.
Nocturnal scenes, op. 72: The stars twinkle; Song of the night; Dawn. cf MELARTIN: Der Traurige Garten, op. 52.
Preludes, op. 17, nos. 12, 14, 24. cf MELARTIN: Der Traurige Garten, op. 52.

PAMER, Michael

Waltz, E major. cf DG 2723 051

PARADIS, Maria Theresia von

Sicilienne. cf Enigma VAR 1025.

PARADISI, Pietro

Sonata, harpsichord, no. 6, A major. cf Saga 5384.

PARKER, Charlie

Au privave. cf Crystal S 221.

PARKER, Horatio

Fugue, op. 26, no. 3, C minor. cf Orion ORS 76255.

PARKES

Big Ben. cf DJM Records DJM 22062.

PARMA, Nicola

Aria del Gran Duca. cf DG 2533 173.
Ballo del serenissimo Duca di Parma. cf DG 2533 173.
La Cesarina. cf DG 2533 173.
Corenta. cf DG 2533 173.
Gagliarda Manfredina. cf DG 2533 173.
La Mutia. cf DG 2533 173.
La ne mente per la gola. cf DG 2533 173.

PARRY, Charles

Blest pair of sirens. cf ELGAR: The music makers, op. 69.

PARRY, Hubert

Chorale prelude on "St. Thomas". cf Vista VPS 1051.
Elegy, A flat major. cf Vista VPS 1035.
I was glad. cf Decca SPA 500.
I was glad. cf Guild GRSP 701.
I was glad. cf Pye Nixa SQ PCNHX 10.
I was glad. cf Vista VPS 1053.
I was glad when they said unto me. cf ELGAR: Coronation ode, op. 44.
Jerusalem, op. 208. cf BBC REH 290.
Jerusalem, op. 208. cf CRD Britannia BR 1077.
Old 100th, chorale fantasia. cf Vista VPS 1035.

2086 Songs (Partsongs): How sweet the answer; If I had but two little wings; Music when soft voices die; O love they wrong thee much; Since thou O fondest; There rolls the deep; What voice of gladness; Ye thrilled me once. STANFORD: Partsongs, op. 110: Heraclitus; op. 119: The blue bird; Farewell my joy; The inkbottle; Chillingham; My heart is thine; The swallow; The train; The witch; op. 127: The haven. Richard Hickox Singers;

Richard Hickox. Prelude PRS 2506.
+Gr 7-77 p208 +RR 7-77 p91
+-HFN 8-77 p91
Songs of farewell: My soul there is a country. cf Abbey LPB 783.
Songs of farewell: My soul there is a country. cf Guild GRS 7008.
Toccata and fugue (The wanderer). cf Vista VPS 1034.

PARTCH, Harry

2087 And on the seventh day petals fell in Petaluma. Gate 5 Ensemble; Harry Partch. CRI SD 213.
+RR 1-77 p72

2088 The bewitched: Scene 10 and epilogue. Castor and Pollux. The letter. Cloud-chamber music. Windsong. Gate 5 Ensemble of Sausalito; University of Illinois Musical Ensemble; John Garvey. CRI SD 193.
+Gr 3-77 p1457 +-RR 10-76 p88
Castor and Pollux. cf The bewitched: Scene 10 and epilogue.
Cloud-chamber music. cf The bewitched: Scene 10 and epilogue.
The letter. cf The bewitched: Scene 10 and epilogue.
Windsong. cf The bewitched: Scene 10 and epilogue.

PASCHA, Edmund

2089 Christmas carols. Christmas mass, F major. Harmonia pastoralis. Prosae pastoralis. Stefan Nosal, t; Josef Raninec, bar; Ondrej Malachovsky, bs; Prague Madrigal Singers; Instrumentalists; Miroslav Venhoda. Opus 9112 0208. (also Supraphon 112 0821)
+ARG 12-77 p56
Christmas mass, F major. cf Christmas carols.
Harmonia pastoralis. cf Christmas carols.
Little Slovak suite: Andante cantabile, con brio. cf Supraphon 113 1323.
Prosae pastoralis. cf Christmas carols.
Tota pulchra. cf Supraphon 113 1323.

PASQUINI, Bernardo

Canzone francese. cf ABC ABCL 67008.
Ricercare. cf ABC ABCL 67008.

PASSEREAU (16th century France)

Il est bel et bon. cf Argo ZRG 823.

PATACHICH, Ivan

Ritmi dispari. cf Hungaroton SLPX 11811.

PATTERSON, Paul

Fluorescences. cf Works, selections (HMV CSD 3780).
Gloria. cf Works, selections (HMV CSD 3780).
Kyrie. cf Works, selections (HMV CSD 3780).
Trilogy, organ. cf Works, selections (HMV CSD 3780).
Visions. cf Works, selections (HMV CSD 3780).

2090 Works, selections: Fuorescences. Gloria. Kyrie. Trilogy, organ. Visions. Arthur Wills, org; London Chorale; Roy Wales. HMV CSD 3780.
+Gr 6-77 p89 +RR 6-77 p90
+HFN 7-77 p117

PAULSON, Gustaf

Modi, op. 108b. cf NILSSON: Quantitaten.

PAUMANN, Conrad

Ellend du hast. cf BIS LP 3.

PAYNE

London medley. cf DJM Records DJM 22062.

PAYNE, Anthony
2091 Paean. Phoenix mass. The world's winter. Jane Manning, s; Susan Bradshaw, pno. BBC Singers; Philip Jones Brass Ensemble, Nash Ensemble; John Poole, Lionel Friend. BBC REH 297.
+Gr 12-77 p1122
Phoenix mass. cf Paean.
The world's winter. cf Paean.
PECK
Questions from the electric chairman. cf Advance FGR 18.
PECK, Russell
2092 Automobile. PENN: Fantasy. Preludes, marimba (4). WILSON: Echoes. Diane Ragains, s; Peter Middleton, flt; Karyl Louwenaar, hpd; Leigh Howard Stevens, marimba; Phillip Rehfeldt, clt; David Johnson, perc; Aventino Calvetti, bs viol. CRI SD 367.
+ARG 9-77 p50 *MJ 10-77 p29
+FF 11-77 p58
PEELE, George
Songs: Descend ye sacred daughters; Long may she come. cf Argo ZK 25.
PEERSON, Martin
Fall of leafe. cf BIS LP 75.
PEETERS, Flor
2093 Concerto, organ and piano, op. 74. STANFORD: Fantasia and toccata, op. 57, D minor. Short preludes and postludes, op. 101, Set II. STANLEY: Voluntary, E minor (arr. Campbell). Ronald Perrin, org. Vista VPS 1040.
+-Gr 3-77 p1444 +MU 7-77 p14
+-HF 7-77 p95 /-RR 3-77 p83
Koncertstuck, op. 52a. cf Vista VPS 1033.
PENDERECKI, Krzysztof
Capriccio per Siegfried Palm. cf BACH: Suite, solo violoncello, no. 1, S 1007, G major.
Miniatures, clarinet and piano. cf BIS LP 62.
St. Luke passion: Miserere. cf Pelca PSR 40607.
PENN, William
Fantasy. cf PECK: Automobile.
Preludes, marimba. cf PECK: Automobile.
2094 Ultra mensuram. ROSS: Concerto, trombone and orchestra. Prelude, fugue and big apple, bass trombone and tape. SCHWANTNER: Modus caelestis. Western Michigan University Wind Ensemble; Per Brevig, trom; Bergen Symphony Orchestra; Gregg Shearer, flt; New England Conservatory Repertory Orchestra; Karsten Andersen, Richard Pittman. CRI SD 340.
+HF 12-75 p110 +SFC 10-30-77 p44
+-NR 9-75 p8
PEPPER
Songs: Over the rolling sea. cf Argo ZFB 95/6.
PERAZA, Francisco de
Tiento de medio registro alto de primero tono. cf Harmonia Mundi HM 759.
PERERA, Ronald
Alternate routes. cf GRIPPE: Musique Douze.
PERGAMENT, Moses
Chaconne. cf CHRISTOSKOV: Suite, no. 1.
2095 The Jewish song. Birgit Nordin, s; Sven-Olof Eliasson, t; Stockholm Philharmonic Orchestra and Chorus; James De Priest.

Caprice CAP 2003 (2).
+ARG 7-77 p29 *RR 4-77 p84
+HFN 6-77 p131
Little suite, 2 recorders. cf BIS LP 57.
Piece, 4 flutes. cf BIS LP 50.
Sonata, flute. cf BIS LP 50.
Sonatina. cf BIS LP 32.
Who plays in the night. cf BIS LP 37.
PERGOLESI, Giovanni
Concertino, no. 6, B flat major. cf MOZART: Quartet, flute, K 285, D major.
Sinfonia, F major. cf BASSETT: Suite, solo trombone.
Songs: Tre giorni son che nina. cf HNH 4008.
PERI, Jacopo
2096 Euridice. Nerina Santini, s; Rodolfo Farolfi, t; Gastone Sarti, bar; Franco Ghitti, t; Frederico Davis, bs; Milano Coro Polifonico; I Solisti di Milano; Angelo Ephrikian. Telefunken 6-35014. (also Musical Heritage Society OR 344/5, Telefunken SAWT 9603/4)
+-ARG 7-77 p29 +ON 4-9-77 p37
+-MJ 7-77 p68 +-NR 8-77 p9
Songs: O durezza di ferro; Tre le donne; Bellissima regina. cf DG 2533 305.
PERLE, George
Inventions. cf Orion ORS 77269.
PERLEA, Jonel
2097 Don Quichotte. Variations symphoniqes sur un theme propre. RTV Symphony Orchetra, Banatul Philharmonic Orchestra; Emanuel Elenescu, Nicolae Boboc. Electrecord ECE 01236.
++FF 11-77 p43
Variations symphoniques sur un theme propre. cf Don Quichotte.
PEROTIN LE GRAND
Alleluja. cf Turnabout TV 37086.
Sederunt principes. cf DG 2710 019.
Viderunt omnes. cf DG 2710 019.
PERSICHETTI, Vincent
Drop, drop slow tears. cf Delos FY 025.
O cool is the valley. cf MENC 76-11.
Parable. cf GOLDMAN: Sonata, violin and piano.
Parable, op. 123. cf Coronet LPS 3036.
2098 Quartets, nos. 1-4. New Art String Quartet. Arizona State University unnumbered.
++ARG 5-77 p6 ++St 4-77 p124
+SR 9-18-76 p50
Serenade, no. 6, op. 44. cf BASSETT: Suite, solo trombone.
PERT, Morris
2099 Chromosphere. Japanese verses (4). Luminos. Georgina Dobree, basset hn; Morris Pert, pno; Veronica Hayward, s; Suntreader. Discourses ABM 21.
+-Gr 8-76 p317 +-RR 10-76 p90
+-HFN 5-76 p105 +Te 3-77 p36
*MT 9-76 p747
Japanese verses. cf Chromosphere.
Luminos. cf Chromosphere.
PERUSIO, Matheus de
Andray soulet. cf HMV SLS 863.
Andray soulet. cf 1750 Arch S 1753.

Le greygnour bien. cf HMV SLS 863.

PESSARD, Emile

Bolero. cf Golden Crest RE 7064.

PETER, J. F.

Songs: I will freely sacrifice to thee; I will make an everlasting covenant. cf Vox SVBX 5350.

PETER, S.

Songs: Look ye, how my servants shall be feasting; O, there's a sight that rends my heart. cf Vox SVBX 5350.

PETERSON-BERGER, Olaf

Songs: Bland skogens hoga furustammar; Nar jag for mig sjalv i morka skogen gar. cf RCA LSC 9884.

PETRARQUE

Non al so amante. cf CBS 76534.

PETROV

Don't believe me child. cf Club 99-105.

PETROV, Andrei

Poem, 4 trumpets, 2 klaviers, organ strings and percussion (To commemorate the siege of Leningrad). cf KANCELI: Symphony, no. 4.

2100 Songs of our days (To the memory of those who died during the blockade of Leningrad). Leningrad Philharmonic Orchestra; Arvid Yansons. Columbia M 34526.
+NR 11-77 p5 +-St 12-77 p154

PETROVICS, Emil

Cassazione. cf Hungaroton SLPX 11811.

Nocturne. cf Hungaroton SLPX 11686.

Quartet, strings. cf LUTOSLAWSKI: Quartet, strings.

PETRUS DE CRUCE

Aucun ont trouve. cf DG 2710 019.

PETTERSSON, Gustaf Allan

Barefoot songs. cf MAHLER: Symphony, no. 5, C sharp minor.

Concerto, string orchestra, no. 1. cf LIDHOLM: Nausicaa alone.

2101 Symphony, no. 2. Swedish Radio Symphony Orchestra; Stig Westerberg. Swedish Society SLT 33219.
++ARG 8-77 p23 +RR 9-76 p58
+Gr 3-75 p1658 ++St 2-77 p119

2102 Symphony, no. 6. Norrkoping Symphony Orchestra; Okko Kamu. CBS 76553.
+-ARG 8-77 p23 *MT 10-77 p826
+-Gr 2-77 p1284 +RR 2-77 p56
+HFN 2-77 p127 +Te 3-77 p38

2103 Symphony, no. 7. Stockholm Philharmonic Orchestra; Antal Dorati. Decca SXL 6538. (also London CS 6740)
+ARG 8-77 p23 ++NR 3-73 p2
+Gr 5-72 p1894 ++SR 3-73 p47
+HF 5-73 p88 ++St 5-73 p118
+HFN 6-72 p1116 +St 6-75 p51
++MJ 5-73 p35

Symphony, no. 10. cf BLOMDAHL: Symphony, no. 2.

2104 Vox humana. Marianne Mellnas, s; Margot Rodin, con; Sven-Erik Alexandersson, t; Erland Hagegard, bar; Swedish Radio Orchestra and Chorus; Stig Westerberg. BIS LP 55.
+RR 7-77 p88

PEZEL, Johann

Pieces (3). cf Golden Crest CRS 4148.

Sonata, trumpet and bassoon, C major. cf Nonesuch H 71279.

Suite, C major. cf Saga 5417.

PFITZNER, Hans

Symphony, op. 46, C major. cf BEETHOVEN: Symphony, no. 4, op. 60, B flat major.

Symphony, op. 46, C major. cf BEETHOVEN: Symphony, no. 9, op. 125, D minor.

PHALESE, Pierre

Dances. cf BIS LP 3.

Passamezzo: Saltarello. cf DG 2723 051.

Passamezzo d'Italye: Reprise, Gaillarde. cf DG 2723 051.

PHYLE

President's march. cf Richardson RRS 3.

PIANO MUSIC IN AMERICA, 1900-1945. cf Vox SVBX 5303.

PICCININI, Alessandro

Canzona. cf DG 2533 323.

Corrente. cf Telefunken AW 6-42155.

Toccata. cf Saga 5438.

Toccata, no. 10. cf Telefunken AW 6-42155.

PICK

March and troop. cf Saga 5417.

Suite, B flat major. cf Saga 5417.

PICON

Marche des bouffons. cf Grosvenor GRS 1052.

PIEFKE, G.

Kutschke polka. cf Angel (Q) S 37304.

PIERNE, Gabriel

2105 Cydalise et le chevre-pied: Suites, nos. 1 and 2. Ramuntcho: Overture on Basque themes. Paris Opera Orchestra; Jean-Baptiste Mari. Angel (Q) S 37281.

+HF 9-77 p108 +SR 9-3-77 p42
+NR 10-77 p2 ++St 9-77 p138

March of the little lead soldiers, op. 14, no. 6. cf DELISLE: La Marseillaise.

Pieces, op. 29 (3). cf Vista VPS 1060.

Prelude, G minor. cf Calliope CAL 1922/4.

Ramuntcho: Overture on Basque themes. cf Cydalise et le chevre-pied: Suites, nos. 1 and 2.

Serenade. cf Discopaedia MOB 1018.

Sonata, flute and piano, op. 36. cf FRANCK: Sonata, violin and piano, A major.

PIERO

Con dolce brama. cf Argo D40D3.

PILLIN, Boris

2106 Serenade. Sonata, violoncello and piano. Tune, C minor. Sharon Davis, pno; Westwood Wind Quintet; Barry Silverman, perc; Douglas Davis, vlc. WIM WIMR 11.

+NR 2-77 p6

Sonata, violoncello and piano. cf Serenade.

Tune, C minor. cf Serenade.

PILSS

Scherzo. cf Golden Crest CRS 4148.

PISTON, Walter

2107 The incredible flutist. BUCK: Festival overture on the American national air. Louisville Orchestra; Jorge Mester. Louisville LS 755.

+-ARG 6-77 p36 +SFC 12-10-76 p50
++MJ 2-77 p31

The incredible flutist. cf COPLAND: Dance symphony.
Passacaglia. cf Vox SVBX 5303.
Psalm and prayer of David. cf Vox SVBX 5353.
Sonata, flute and piano. cf BURTON: Sonatina, flute and piano.
Sonata, flute and piano. cf COPLAND: Duo.
Trio, piano. cf BLOCH: Nocturnes.

PIXIS, Johann
Concerto, violin, piano and strings. cf MOSCHELES: Grande sonate symphonique, op. 112.

PIZZETTI
Canti ad una Giovane Fidanzata: Affettuoso. cf Discopaedia MB 1014.
Songs: I pastori. cf Columbia M 34501.

PLANQUETTE, Robert
2108 Les cloches de Corneville. Mady Mesple, Christiane Stutzmann, Annie Tallard, Arta Verlen, s; Charles Burles, Jean Giraudeau, Jean Bussard, t; Bernard Sinclair, Jean-Christoph Benoit, bar; Charles Roeder, bs; Paris Opera Chorus; Opera Comique Orchestra; Jean Doussard. Connoisseur Society CS 2-2107 (2).
+ARG 11-76 p35 +-ON 1-29-77 p44
+-HF 9-76 p95 +SFC 10-24-76 p35
*MJ 12-76 p28 +-St 11-76 p152

THE PLAY OF DANIEL. cf Calliope CAL 1848.
THE PLEASURE OF THE ROYAL COURTS. cf Nonesuch H 71326.

PLEKHCHEYEV
The crown of roses. cf HMV SQ CSD 3784.

PLESKOW, Raoul
Pentimento. cf BESTOR: Sonata, piano.

PLIHAL
Fugue, F major. cf Supraphon 113 1323.

PLOG, Anthony
Fanfare, 2 trumpets. cf Crystal S 362.

POGLIETTI, Alessandro
Balletto. cf DG 2723 051.

POLAK, Jakub
Praeludium. cf DG 2533 294.

POLDINI, Eduard
Marionnettes: Poupee valsante (Dancing doll). cf Amberlee ALM 602.
Marionnettes: Poupee valsante. cf Discopaedia MB 1015.
Marionnettes: Dancing doll. cf HMV ASD 3346.

POLLONOIS
Courante. cf Hungaroton SLPX 11721.

PONCE, Manuel
Campo. cf LAURO: Danza negra.
Canciones populares mexicanas (3). cf Saga 5412.
Folie d'Espagne: Thema and variations with fugue. cf BARRIOS: Las abejas.
Preludes (3). cf Laurel Protone LP 13.
Prelude, E major. cf RCA ARL 1-0864.
Star of love. cf HMV HLM 7066.
Valse. cf Saga 5412.

PONCHIELLI, Amilcare
Il figliuol prodigo: Tenda natal. cf Philips 9500 203.
La gioconda: Arias. cf Free Lance FLPS 675.
La gioconda: Cielo e mar. cf RCA CRM 1-1749.
La gioconda: Cielo e mar. cf RCA TVM 1-7201.
La gioconda: Cielo e mar. cf Rubini GV 29.

La gioconda: Dance of the hours. cf Angel S 37250.
La gioconda: Dance of the hours. cf CBS 30091.
La gioconda: E un anatema...L'amo come il fulgor del creato, Cosi mantieni il patto. cf Court Opera Classics CO 347.
La gioconda: Suicidio. cf BASF 22-22645-3.
La gioconda: Suicidio. cf Club 99-100.
La gioconda: Suicidio. cf Decca SXLR 6825.
La gioconda: Suicidio. cf HMV SLS 5057.
La gioconda: Suicidio. cf London OS 26497.
La gioconda: Voce di donna. cf Club 99-103.
La gioconda: Voce di donna. cf Club 99-106.

POPPER, David
Dance of the elves, op. 39. cf HMV SQ ASD 3283.
POPULAR MUSIC FROM THE TIME OF HENRY VIII. cf Saga 5444.

PORPORA, Nicola
2109 Concerto, violoncello, G major. SAMMARTINI: Concerto, viola pomposa, C major. VIVALDI: Concerto, viola d'amore, A major. Thomas Blees, vlc; Ulrich Koch, vla d'amore, vla pomposa; South West German Orchestra; Paul Angerer. Turnabout TV 34574.
+-Gr 4-76 p1607
+-HFN 8-76 p85
++NR 6-75 p6
+RR 3-76 p51
++SFC 6-26-77 p46

PORTA
Corda deo dabimus. cf HMV SQ ASD 3393.

PORTER, Walter
Thus sang Orpheus. cf L'Oiseau-Lyre 12BB 203/6.

POULENC, Francis
Le bal masque. cf FAURE: La bonne chanson, op. 61.
Concert champetre, harpsichord and orchestra. cf Concerto, organ, strings and timpani, G minor.
Concert champetre, harpsichord and orchestra. cf MOZART: Concerto, piano, no. 13, K 415, C major.
Concerto, flute. cf CHAMINADE: Concertino, flute, op. 107.
2110 Concerto, organ, strings and timpani, G minor. Concert champetre, harpsichord and orchestra. Marie-Claire Alain, org; Robert Veyron-Lacroix, hpd; ORTF; Jean Martinon. Musical Heritage Society MHS 1595. (also Erato STU 70637)
+Gr 7-76 p181
+HF 10-73 p109
+HFN 4-77 p141
++RR 6-76 p51
++St 2-74 p118
2111 Concerto, piano. Gloria, G major. Norma Burrowes, s; Christina Ortiz, pno; Birmingham City Symphony Orchestra and Chorus; Louis Fremaux. HMV SQ ASD 3299 Tape (c) TC ASD 3299. (also Angel (Q) S 37246)
++ARG 5-77 p30
++Gr 12-76 p1041
+-HF 4-77 p110
+HFN 1-77 p113
+-HFN 4-77 p155 tape
++NR 8-77 p8
++RR 1-77 p82
+SFC 2-13-77 p33
++St 5-77 p116
2112 Concerto, 2 pianos. Sonata, 2 pianos. Yarbrough and Cowan, pno; NPhO; Paul Freeman. Musical Heritage Society MHS 3576.
+FF 9-77 p42
Concerto, 2 pianos, D minor. cf EMMANUEL: Symphony, no. 2, A major.
L'Embarquement pour Cythere. cf DEBUSSY: Nocturnes: Fetes.
Exultate Deo. cf MESSIAEN: O sacrum convivium.

2113 Gloria, G major. STRAVINSKY: Symphony of psalms. Judith Blegen, s; Westminster Choir, English Bach Festival Chorus; LSO; Leonard Bernstein. Columbia M 34551 Tape (c) MT 34551.
+-SFC 12-11-77 p61
Gloria, G major. cf Concerto, piano.
Litanies a la vierge noire de Rocamadour. cf MESSIAEN: O sacrum convivium.
2114 Melancolie. Pieces: Pastorale; Toccata; Hymne. Suite francaise. ROUSSEL: Pieces, op. 49 (3). Sonatine, op. 16. Andre Previn, pno. CBS 61782. (Reissue)
++Gr 7-77 p202 +-RR 7-77 p83
+HFN 8-77 p95
Pastourelle. cf Angel S 36053.
Pastourelle. cf RCA GL 4-2125.
Pieces: Pastorale; Toccata; Hymne. cf Melancolie.
Salve regina. cf MESSIAEN: O sacrum convivium.
Sarabande. cf DG 2530 802.
Sextet, piano. cf BEETHOVEN: Quintet, piano, op. 16, E flat major.
Un soir de neige. cf RCA RL 2-5112.
2115 Sonata, clarinet and bassoon. Trio, clarinet, bassoon and piano. VILLA-LOBOS: Duo, oboe and bassoon. Fantaisie concertante. Soni Ventorum Wind Quintet Members. Musical Heritage Society Tape (c) 3187.
+HF 11-77 p138 tape
Sonata, flute and piano. cf ENESCO: Cantabile e presto, flute and piano.
Sonata, 2 pianos. cf Concerto, 2 pianos.
Sonata, 2 pianos. cf DEBUSSY: Nocturnes: Fetes.
Songs: L'Anguille; La belle jeunesse; C; Priez pour paiz; Serenade. cf FAURE: Songs (1750 Arch 1954).
2116 The story of Babar the little elephant. SAINT-SAENS: The carnival of the animals. Eleanor Bron, narrator; Susan Bradshaw, Richard Rodney Bennett, pno; Vesuvius Ensemble. EMI BRNA 502 Tape (c) TC BRNA 502.
+-Gr 9-77 p432 +RR 9-77 p67
+-HFN 9-77 p147
2117 The story of Babar the little elephant (orch. Francaix). SAINT-SAENS: The carnival of the animals. Peter Ustinov, narrator; Aldo Ciccolini, Alexis Weissenberg, pno; OSCCP; Georges Pretre. HMV ESD 7020 Tape (c) TC ESD 7020. (Reissues from ASD 2286, 2316)
+Gr 11-76 p888 +HFN 9-77 p155 tape
+HFN 1-77 p121 +RR 11-76 p70
+HFN 5-77 p77 p139 tape +RR 5-77 p93 tape
Suite francaise. cf Melancolie.
Trio, clarinet, bassoon and piano. cf Sonata, clarinet and bassoon.

POUSSEUR, Henri
Echoes II de votre Faust. cf Universal Edition UE 15043.

POWELL
The Zelanski medley. cf Advance FGR 18.

POWELL, John
2118 Sonata Teutonica, op. 24. Roy Hamlin Johnson, pno. CRI SD 368.
+ARG 8-77 p24 ++NR 7-77 p12
++HF 12-77 p94 ++St 10-77 p142
+-MJ 10-77 p29

PRAEGER, Heinrich
Introduction, theme and variations, op. 21, A minor. cf BIS LP 60.

PRAETORIUS, MIchael
Courante. cf GABRIELI: O magnum mysterium.
Galliarde de la guerre. cf DG 2723 051.
Galliarde de Monsieur Wustron. cf DG 2723 051.
Peasant dances. cf CRD CRD 1019.
Reprinse. cf DG 2723 051.
Songs: Deck the halls; Es ist ein Ros entsprungen; Gloria, Gott in der Hoh. cf RCA PRL 1-8020.
Songs (vocal and choral): Es ist ein Ros Entsprungen; In dulci jubilo; Joseph Lieber Joseph mein; Ein Kind geborn zu Bethlehem; Nun komm, der Heiden Heiland; Psallite ungenito Christo; Philov-Volte-Philov. cf GABRIELI: O magnum mysterium.
Terpsichore: Ballet. cf CBS 72526.
Terpsichore: Fire dance; Stepping dance; Windmills; Village dance; Sailor's dance; Fisherman's dance; Festive march. cf CRD CRD 1019.
Winter. cf BIS LP 75.
PRATT, Charles
Put my little shoes away. cf New World Records NW 220.
PREMRU, Raymond
Tissington variations. cf BIS LP 59.
PRENTZL (17th century Germany)
Sonata, trumpet and bassoon, C major. cf Nonesuch H 71279.
PRESTON, Thomas
Beatus Laurentius. cf APPLEBY: Magnificat.
PREVIN, Andre
Outings for brass (4). cf Argo ZRG 851.
PRIGOZIN, Lucian
2119 Sonata, violin and piano, no. 2. SHOSTAKOVICH: Preludes, op. 34 (24). Jiri Tomasek, vln; Josef Ruzicka, Dagmar Baloghova, pno. Panton 110 488.
+FF 11-77 p35
PRIOLI
Canzona prima a 12. cf HMV SQ ASD 3393.
A PROCESSION OF VOLUNTARIES. cf Cambridge CRS 2540.
PROCH
Theme and variations. cf Columbia 34294.
PROKOFIEV, Serge
2120 Alexander Nevsky, op. 78. Lili Chookasian, con; Westminster Symphony Choir; NYP; Thomas Schippers. CBS 61769 Tape (c) 40-61769. (Reissue from SBRG 72081) (also Odyssey Y 31014)
+-Gr 1-77 p1167 +-HFN 4-77 p155 tape
+-Gr 9-77 p511 tape +-RR 12-76 p94
+-HFN 1-77 p119
2121 Alexander Nevsky, op. 78. Betty Allen, ms; Mendelssohn Club Chorus; PO; Eugene Ormandy. RCA ARL 1-1151 Tape (c) ARK 1-1151 (ct) ARS 1-1151 (Q) ARD 1-1151.
+Audio 3-76 p64 ++MT 2-77 p134
+Gr 4-76 p1649 +-NR 11-75 p9
+HF 2-76 p102 +-ON 1-24-76 p51
+-HF 2-76 p102 Quad +-RR 4-76 p73
++HF 2-77 p118 tape ++SFC 11-9-75 p22
+-HFN 4-76 p111 ++St 3-76 p120
+-MM 11-76 p42
2122 Autumnal, op. 8. Concerto, piano, no. 3, op. 26, C major. Symphony, no. 1, op. 25, D major. Vladimir Ashkenazy, pno; LSO; Andre Previn, Vladimir Ashkenazy. Decca SXL 6768 Tape (c) KSXC 6768. (Reissues from 15BB 218)

+-Gr 11-76 p803 +-RR 12-76 p63
+HFN 12-76 p151 +-RR 4-77 p92
+MT 6-77 p488

2123 Cinderella, op. 87. MRSO; Gennady Rozhdestvensky. HMV SXDW 3026 (2). (Reissue from ASD 2429/30) (also Melodiya/Angel S4102)
+Gr 3-77 p1402 +RR 10-76 p65

2124 Cinderella, op. 87: Pavan; Gavotte; Summer fairy; Winter fairy; Orientalia; Passepied; Capriccio; Bourree; Adagio; Quarrel; Waltz (Cinderella leaves the ball); Pas-e-chat; Amoroso. Romeo and Juliet, op. 64: National dance; Scena; Menuet; Juliet; Masks; Montecchi and Capuletti; Father Lorenzo; Mercutio; Dance of young girls with lilies; Romeo and Juliet before separation. Cristina Ortiz, pno. HMV SQ HQS 1393.
/Gr 8-77 p324 ++RR 8-77 p72
+HFN 7-77 p117

2125 Concerto, piano, no. 1, op. 10, D flat major. Sonata, piano, no. 8, op. 84, B flat major. Toccata, op. 11. Lazar Berman, pno; HSO; Andras Korodi. Hungaroton SHLX 90048.
+-Gr 2-77 p1287 +-RR 2-77 p56
+-HFN 2-77 p127

2126 Concerto, piano, no. 1, op. 10, D flat major. RACHMANINOFF: Concerto, piano, no. 1, op. 1, F sharp minor. Sviatoslav Richter, pno; MSO, MRSO; Kiril Kondrashin, Kurt Sanderling. Odyssey Y 34610.
+-NR 9-77 p4

Concerto, piano, no. 1, op. 10, D flat major. cf CHOPIN: Variations on Mozart's "La ci darem la mano", op. 2, B flat major.

2127 Concerto, piano, no. 2, op. 16, G minor. Concerto, piano, no. 5, op. 55, G major.* Jorge Bolet, Alfred Brendel, pno; Cincinnati Symphony Orchestra, VSOO; Thor Johnson, Jonathan Sternberg. Turnabout TV 34543. (*Reissue from PLP 527)
-Gr 1-76 p1203 +-RR 11-75 p47
+HFN 11-75 p173 +SFC 1-30-77 p36

2128 Concerto, piano, no. 3, op. 26, C major. Concerto, piano, no. 4, op. 53, B flat major. Vladimir Ashkenazy, pno; LSO; Andre Previn. London CS 6964. (Reissue)
+NR 9-77 p4

2129 Concerto, piano, no. 3, op. 26, C major. RACHMANINOFF: Rhapsody on a theme by Paganini, op. 43. Marian Lepsansky, pno; Slovak Philharmonic Orchestra; Ladislav Slovak. Rediffusion Royale ROY 2007.
+-Gr 10-77 p638 +RR 8-77 p58

Concerto, piano, no. 3, op. 26, C major. cf Autumnal, op. 8.

2130 Concerto, piano, no. 4, op. 53, B flat major. Concerto, piano, no. 5, op. 55, G major. Vladimir Ashkenazy, pno; LSO; Andre Previn. Decca SXL 6769 Tape (c) KSX 6769. (Reissue from 15BB 218)
+Gr 5-77 p1697 ++RR 5-77 p92 tape
+HFN 5-77 p139 tape

Concerto, piano, no. 4, op. 53, B flat major. cf Concerto, piano, no. 3, op. 26, C major.

Concerto, piano, no. 5, op. 55, G major. cf Concerto, piano, no. 2, op. 16, G minor.

Concerto, piano, no. 5, op. 55, G major. cf Concerto, piano, no. 4, op. 53, B flat major.

2131 Concerto, violin, no. 1, op. 19, D major. Concerto, violin, no. 2, op. 63, G minor. Nathan Milstein, vln; PhO, NPhO; Carlo Maria

Giulini, Rafael Fruhbeck de Burgos. HMV SXLP 30235 Tape (c) SXLP 30235. (Reissue from Columbia SAX 5275)
+Gr 3-77 p1402 +-HFN 4-77 p151
+Gr 9-77 p511 tape +-RR 3-77 p57

2132 Concerto, violin, no. 1, op. 19, D major. Concerto, violin, no. 2, op. 63, G minor. Kyung-Wha Chung, vln; LSO; Andre Previn. London CS 6997 Tape (c) CS 5-6997. (also Decca SXL 6773 Tape (c) KSXC 6773)
++Gr 3-77 p1402 ++NYT 7-24-77 pD11
+Gr 9-77 p511 tape ++RR 3-77 p57
+-HF 11-77 p120 ++RR 7-77 p99 tape
+-HFN 3-77 p111 ++RR 10-77 p98 tape
+NR 9-77 p4

2133 Concerto, violin, no. 1, op. 19, D major. Concerto, violin, no. 2, op. 63, G minor. Stoika Milanova, vln; Bulgarian Radio and Television Symphony Orchestra; Vassil Stefanov. Monitor HS 90101.
+HF 10-74 p110 ++NR 9-74 p5
+MJ 1-77 p27 ++St 10-74 p135

Concerto, violin, no. 2, op. 63, G minor. cf Concerto, violin, no. 1, op. 19, D major (Connoisseur Society CS 6997).

Concerto, violin, no. 2, op. 63, G minor. cf Concerto, violin, no. 1, op. 19, D major (HMV SXLP 30235).

Concerto, violin, no. 2, op. 63, G minor. cf Concerto, violin, no. 1, op. 19, D major (Monitor HS 90101).

2134 The gambler, op. 24. Nina Poliakova, s; Tamara Antipova, Anna Matushina, ms; Vladimir Makhov, Andrei Sokolov, t; Gennady Troitsky, bs-bar; All-Union Radio Orchestra and Chorus; Gennady Rozhdestvensky. Columbia/Melodiya M3 34579 (3).
+ARG 11-77 p26 +-NYT 7-10-77 pD13
+MJ 11-77 p28 +ON 8-77 p56
+-NR 7-77 p10 +St 11-77 p148

Lieutenant Kije, op. 60: Kije's wedding; Troika. cf CBS 61781.

Lieutenant Kije, op. 60: Suite. cf KODALY: Hary Janos: Suite (Decca 4355).

Lieutenant Kije, op. 60: Suite. cf KODALY: Hary Janos: Suite (RCA 1-1325).

The love for three oranges, op. 33: March. cf CBS 61781.

The love for three oranges, op. 33: March. cf RCA CRL 3-2026.

The love for three oranges, op. 33: March. cf Westminster WGM 8309.

The love for three oranges: Suite. cf ENESCO: Roumanian rhapsody, op. 11, no. 1, A minor.

March, op. 99, B flat major. cf DG 2548 148.

March, op. 99, B flat major. cf HMV CSD 3782.

2135 Peter and the wolf, op. 67. SAINT-SAENS: Carnival of the animals. Angela Rippon, reader; Anthony Goldstone, Ian Brown, pno; RPO; Owain Arwel Hughes. Enigma VAR 1047.
+-Gr 12-77 p1087 +-RR 12-77 p49
+-HFN 12-77 p176

Peter and the wolf, op. 67. cf BRITTEN: The young person's guide to the orchestra, op. 34.

Peter and the wolf, op. 67. cf HASLAM: Juanita the Spanish lobster.

Pieces, piano, op. 12: Rigaudon. cf Pye PCNH 9.

2136 Romeo and Juliet, op. 64, complete ballet score. CO; Lorin Maazel. London CSA 2312 Tape (4) R 480275 (c) D 10275, CSA 5-2312. (also Decca SXL 6620/22 Tape (c) K20K32)

++Gr 9-73 p489 ++NR 12-73 p2
++HF 12-73 p94 ++RR 10-73 p77
+HF 6-77 p103 +RR 4-77 p90 tape
++HF 12-74 p146 tape +SR 12-4-73 p33
++HFN 5-77 p138 tape ++SFC 9-16-73 p33
+MJ 2-74 p12 ++St 11-73 p128

Romeo and Juliet, op. 64, excerpts. cf HMV ESD 7011.
Romeo and Juliet, op. 64: Montagues and Capulets; Juliet, the little girl; Dance of the maids with the lilies; Romeo and Juliet's grave. cf Symphony, no. 5, op. 100, B flat major.
Romeo and Juliet, op. 64: National dance; Scena; Menuet; Juliet; Masks; Montecchi and Capuletti; Father Lorenzo; Mercutio; Dance of young girls with lilies; Romeo and Juliet before separation. cf Cinderella, op. 87: Pavan (HMV HQS 1393).
Scythian suite, op. 20: The enemy god dances with the black spirit. cf Atalntic SD 2-7000.
Sonata, flute and piano, op. 94, D major. cf ENESCO: Cantabile e presto, flute and piano.
Sonata, flute and piano, op. 94, D major. cf FRANCK: Sonata, flute and piano, A major.

2137 Sonata, piano, no. 2, op. 14, D minor. Sonata, piano, no. 8, op. 84, B flat major. Tedd Joselson, pno. RCA ARL 1-1570.
+-Gr 9-76 p446 +-NR 8-77 p14
+-HF 12-76 p110 +-RR 9-76 p83
+HFN 9-76 p127 +St 11-76 p155
+-MJ 1-77 p38

Sonata, piano, no. 2, op. 14, D minor. cf BARTOK: Suite, op. 14.

2138 Sonata, piano, no. 7, op. 83, B flat major. STRAVINSKY: Petrouchka: Three movements. Maurizio Pollini, pno. DG 2530 225.
+NYT 3-13-77 pD22

2139 Sonata, piano, no. 7, op. 83, B flat major. WILDER: Suite, piano. Barry Snyder, pno. Golden Crest RE 7058.
+CL 9-77 p10 +MJ 1-77 p38

Sonata, piano, no. 8, op. 84, B flat major. cf Concerto, piano, no. 1, op. 10, D flat major.
Sonata, piano, no. 8, op. 84, B flat major. cf Sonata, piano, no. 2, op. 14, D minor.
Sonata, violoncello and piano, op. 119, C major. cf FRANCK: Sonata, violin and piano, A major.

2140 Symphony, no. 1, op. 25, D major. Symphony, no. 7, op. 131, C sharp minor. LSO; Walter Weller. Decca SXL 6702. (also London CS 6897)
++Gr 5-75 p1968 +NR 7-77 p4
+HF 6-77 p97 +-RR 5-75 p38
+HFN 5-75 p137 ++SFC 5-15-77 p50
+-MJ 7-77 p70

Symphony, no. 1, op. 25, D major. cf Autumnal, op. 8.
Symphony, no. 1, op. 25, D major. cf CHABRIER: Espana.
Symphony, no. 1, op. 25, D major. cf MUSSORGSKY: Pictures at an exhibition.
Symphony, no. 1, op. 25, D major. cf RCA CRL 3-2026.
Symphony, no. 1, op. 25, D major: Gavotte. cf Amberlee ALM 602.

2141 Symphony, no. 5, op. 100, B flat major. LSO; Walter Weller. Decca SXL 6887 Tape (c) KSXC 6787.
+-Audio 9-77 p46 +-HFN 5-77 p138 tape
+-Gr 3-77 p1402 +RR 5-77 p92 tape
+-HFN 3-77 p111 +RR 3-77 p56

2142 Symphony, no. 5, op. 100, B flat major. PO; Eugene Ormandy. RCA ARL 1-1869 Tape (c) ARK 1-1869 (ct) ARS 1-1869.
++SFC 11-13-77 p50
2143 Symphony, no. 5, op. 100, B flat major. Romeo and Juliet, op. 64: Montagues and Capulets; Juliet the little girl; Dance of the maids with the lilies; Romeo and Juliet's grave. BSO; Serge Koussevitzky. RCA VL 1-2021.
++Gr 7-77 p180 +NYT 1-16-77 pD13
+HFN 9-77 p147 +RR 7-77 p58
+MJ 10-77 p27
Symphony, no. 7, op. 131, C sharp minor. cf Symphony, no. 1, op. 25, D major.
Toccata, op. 11. cf Concerto, piano, no. 1, op. 10, D flat major.
Visions fugitives, op. 22. cf MUSSORGSKY: Pictures at an exhibition.

PRYOR, Arthur
Blue bells of Scotland. cf Nonesuch H 71341.
Blue bells of Scotland. cf Pandora PAN 2001.
Exposition echoes, polka. cf Nonesuch H 71341.
Thought of love, valse de concert. cf Nonesuch H 71341.

PUCCINI, Domenico Vincenzo Maria
2144 Concerto, piano, B flat major. VIOTTI: Concerto, piano, G minor. Eugene List, pno; Vienna Tonkunstler Orchestra; Zlatko Topolski. Oryx Musical Heritage Society MHS 709 Tape (c) MHS 7BCC 0709.
+-Gr 11-73 p935 +HF 8-77 p96 tape

PUCCINI Giocomo
2145 La boheme. Rosanna Carteri, s; Franco Tagliavini, t; Giuseppe Taddei, bar; Cesare Siepi, bs; Orchestra and conductor. Cetra LPS 3237. (Reissue from OLPC 1237)
+-Gr 7-77 p221
2146 La boheme. Renata Tebaldi, Gianna d'Angelo, s; Carlo Bergonzi, Piero de Palma, t; Ettore Bastianini, Attilio d'Orazi, bar; Renato Cesari, Cesare Siepi, Fernando Corena, Giorgio Onesti, bs; Rome, Santa Cecilia Orchestra and Chorus; Tullio Serafin. Decca D5D2 (2) Tape (c) K5K22. (Reissue from SXL 2170/1)
+Gr 8-76 p332 +HFN 12-77 p187 tape
/Gr 11-77 p900 tape +RR 8-76 p23
+HFN 8-76 p85 +RR 12-77 p98 tape
2147 La boheme. Victoria de los Angeles, Lucine Amara, s; Jussi Bjorling, William Nahr, t; Robert Merrill, John Reardon, Thomas Powell, George del Monte, bar; Giorgio Tozzi, Fernando Corena, bs; Columbus Boys Choir; RCA Victor Chorus and Orchestra; Thomas Beecham. HMV SLS 896 (2) Tape (c) TC SLS 896. (Reissue from ALP 1409/10) (also Seraphim SIB 6099)
++Gr 11-74 p963 +RR 11-74 p28
+Gr 10-75 p721 tape +RR 1-76 p67 tape
+NYT 10-9-77 pD21 +St 7-75 p110
2148 La boheme. Renata Tebaldi, s; Carlo Bergonzi, bar; Santa Cecilia Orchestra; Tullio Serafin. London OSA 1208.
+NYT 10-9-77 pD21
2149 La boheme. Kingsway Symphony Orchestra; Salvador Camarata. London SPC 21159.
+NR 8-77 p6 +SFC 8-28-77 p46
La boheme: Addio di Mimi; On m'appelle Mimi; Si mia chiamano Mimi; Entrate...C'e Rodolfo; Donde lieta usci; Addio dolce svegliare alla mattina; Gavotta...Minuetto; Sono andati; Io Musetta...Oh come e bello e morbido. cf HMV RLS 719.

2150 La boheme: Al quartiera Latin qui attende Momus; Che guardi... Viva Parpignol; Quando m'en vo; Mimi...Donde lieta usci; In un coupe; O Dio Mimi. Victoria de los Angeles, Lucine Amara, s; Jussi Bjorling, William Nahr, t; Robert Merrill, John Reardon, bar; Giorgio Tozzi, Fernando Corena, bs; Columbus Boys Choir; RCA Victor Orchestra and Chorus; Thomas Beecham. HVV ESD 7023 Tape (c) TC ESD 7023. (Reissue from ALP 1409/10, SLS 896)
+Gr 9-77 p487 +HFN 10-77 p171 tape
+HFN 9-77 p153 +RR 8-77 p36

La boheme: Arias. cf London OS 26499.
La boheme: Che gelida manina. cf Decca SPA 491.
La boheme: Che gelida manina. cf RCA CRM 1-1749.
La boheme: Che gelida manina. cf Rubini GV 29.
La boheme: Che gelida manina. cf Rubini GV 43.
La boheme: Che gelida manina. cf Seraphim S 60280.
La boheme: Che gelida manina; O soave fanciulla. cf HMV RLS 715.
La boheme: Che gelida manina; Testa adorata. cf RCA TVM 1-7201.
La boheme: Donde lieta usci. cf HMV SLS 5057.
La boheme: D'un pas leger. cf Seraphim M 60291.
La boheme: Entrate...C'e Rodolfo, Donde lieta usci, Addio di Mimi, Addio dolce svegliare alla mattina, Govotta...Minuetto sono andati, Io Musetta...Oh come e bello e morbido. cf HANDEL: Il pensieroso: Sweet bird.
La boheme: Mi chiamano Mimi. cf DONIZETTI: Don Pasquale: So anch' io la virtu magica.
La boheme: Mi chiamano Mimi; Donde lieta. cf Club 99-100.

2151 La boheme: Non sono in vena...Che gelida manina...Si, mi chiamano Mimi...Momus, momus...O soave fanciulla; Finale, La commedia stupenda...Quando me'n vo; Finale, Addio che vai...Donde lieta usci...Addio dolce svegliare; In un coupe...O Mimi, tu piu non torni; Vecchia zimarra; Finale, Sono andati. Renata Tebaldi, Gianna d'Angelo, s; Carlo Bergonzi, t; Ettore Bastianini, bar; Cesare Siepi, Fernando Corena, Renato Cesari, bs; Santa Cecilia Orchestra and Chorus; Tullio Serafin. Decca JB 11 Tape (c) KJBC 11. (Reissue from SXL 2170/1)
+Gr 9-77 p487 +HFN 10-77 p171 tape
+Gr 10-77 p706 tape +RR 9-77 p42
+HFN 9-77 p153

2152 La boheme: O soave fanciulla; O Mimi, tu piu non torni. Madama Butterfly: Bimba dagli occhi pieni di malia; Una nave da guerra. Tosca: Mari Mario. VERDI: La forza del destino: Solenne in quest'ora. Renata Tebaldi, s; Giuseppe di Stefano, Mario del Monaco, Carlo Bergonizi, t; Fiorenza Cossotto, ms; Leonard Warren, Ettore Bastianini, bar; Various orchestra; Fernando Previtali, Francesco Molinari-Pradelli, Tullio Serafin. Decca SPA 496.
+HFN 11-77 p185 +-RR 11-77 p44

La boheme: Rodolfo's aria. cf MASCAGNI: Cavalleria rusticana: Duet of Santuzza and Turiddu, Turridu's aria.
La boheme: Si, mi chiamano Mimi. cf Decca ECS 811.
La boheme: Si, mi chiamano Mimi. cf HMV SLS 5104.
La boheme: Si, mi chiamano Mimi. cf Odyssey Y 33793.
La boheme: Si, mi chiamano Mimi. cf Seraphim IB 6105.
La boheme: Your tiny hand is frozen. cf HMV HLP 7109.
La boheme: Your tiny hand is frozen; They call me Mimi; Lovely maid in the moonlight. cf GOUNOD: Faust: The hour is late.

2153 Edgar. Renata Scotto, s; Gwendolyn Killebrew, ms; Carlo Bergonzi, t; Vicente Sardinero, bar; Schola Cantorum; New York Opera Society Orchestra; Eve Queler. Columbia M2 34584 (2).
+NYT 12-11-77 pD17 +SFC 11-13-77 p50

2154 La fanciulla del West (The girl of the golden West). Birgit Nilsson, s; Joao Gibin, Renato Ercolani, t; Andrea Mongelli, Enzo Sordello, bar; Antonio Cassinelli, Nicola Zaccaria, Jake Wallace, Carlo Forti, bs; La Scala Orchestra and Chorus; Lovro von Matacic. Seraphim SCL 6074 (3). (Reissue from Angel SCL 3593) (also HMV SLS 5079 Tape (c) TC SLS 5079. Reissue from Columbia SAX 2286/8)
+Gr 6-77 p101 +NR 1-74 p11
+-Gr 9-77 p511 tape +-RR 6-77 p37
+HF 2-74 p99 ++St 2-74 p119
+HFN 6-77 p137

La fanciulla del West: Ch'ella mi creda libero. cf HMV RLS 715.
Girl of the golden West: Johnson's aria. cf MASCAGNI: Cavalleria rusticana: Duet of Santuzza and Turridu, Turridu's aria.

2155 Gianni Schicchi. Ileana Cotrubas, Scilly Fortunato, s; Anna Di Stasio, Stefania Malagu, ms; Placido Domingo, Florindo Andreoli, t; Tito Gobbi, Carlo Del Bosco, Nicola Troisi, bar; Alfredo Marriotti, Giancarlo Luccardi, Leo Pudis, Guido Mazzini, bs; LSO; Lorin Maazel. CBS 76563. (also Columbia M 34534)
++Gr 5-77 p1729 +RR 5-77 p36
++HF 12-77 p95 +SFC 9-18-77 p42
+HFN 5-77 p133 +St 12-77 p94
+NR 10-77 p9

2156 Gianni Schicchi. Giuseppe Taddei, bar; other soloists, orchestra and conductor. Cetra LPS 39. (Reissue from LPC 50028)
+-Gr 7-77 p221

2157 Gianni Schicchi. Suor Angelica. Il tabarro. Margaret Mas, Sylvia Bertona, Victoria de los Angeles, Lidia Marimpietri, Santa Chissari, Anna Marcangeli, Giuliana Raymondi, s; Miriam Pirazzini, Fedora Barbieri, Mina Doro, Corinna Vozza, Teresa Cantarini, Maria Huder, Anna Maria Canali, ms; Giacinto Prandelli, Piero de Palma, Carlo del Monte, Adelio Zagonara, Claudio Cornoldi, Renato Ercolani, t; Tito Gobbi, Fernando Valentini, bar; Plinio Clabassi, Paolo Montarsolo, Alfredo Mariotti, Saturno Meletti, bs; Rome Opera Orchestra and Chorus; Vicenzo Bellezza, Tullio Serafin, Gabriele Santini. HMV SLS 5066 (3) Tape (c) TC SSL 5066. (Reissues from ALP 1355, 1577, ASD 295)
+Gr 10-76 p650 +-HFN 2-77 p135 tape
+HFN 12-76 p151 +-RR 12-76 p44

Gianni Schicchi: O mio babbino caro. cf Angel S 37446.
Gianni Schicchi: O mio babbino caro. cf HMV SLS 5104.

2158 Madama Butterfly. Renata Scotto, s; Carlo Bergonzi, t; Rolando Panerai, bar; Rome Opera Orchestra and Chorus; John Barbirolli. Angel SCL 3702 (3).
+HF 4-71 p67 +NYT 10-9-77 pD21

2159 Madama Butterfly. Clara Petrella, s; Franco Tagliavini, t; Giuseppe Taddei, bar; Orchestra. Cetra LPS 3248. (Reissue from OLPC 1248)
+Gr 7-77 p221

2160 Madama Butterfly. Arias: GIORDANO: Andrea Chenier: Vicino a te... La morte nostra. PUCCINI: Madama Butterfly: Ancora un passo, Bimba dagli occhi, E questo...Che tua madre, Un bel di. Manon Lescaut: Tu, tu, Amore...Tentratrice. VERDI: Otello: Gia la

notte, Ave Maria. Margaret Sheridan, s; Ida Mannerini, ms; Lionel Cecil, Nello Palai, t; Vittorio Weinberg, bar; A. Gelli, bs; La Scala Orchestra; Carlo Sabajno. Club 99-1001.
+-HF 5-77 p87 +-NR 5-76 p10

2161 Madama Butterfly. Montserrat Caballe, s; Silvana Mazzieri, ms; Bernabe Marti, Piero de Palma, t; Franco Bordoni, Dalmacio Gonzalez, bar; Juan Pons, bs; Gran Teatro Liceo Coro; Barcelona Sinfonica Orquesta; Armando Gatto. London OSA 13121 (3) Tape (c) OSA 5-13121.
+HF 12-77 p90 +-ON 9-77 p69
-NYT 7-10-77 pD13 +-St 10-77 p142

Madama Butterfly: Ah, love me a little. cf GOUNOD: Faust: The hour is late.

Madama Butterfly: Ancora un passo, Bimba dagli occhi, E questo... Che tua madre, Un bel di. cf Madama Butterfly (Club 99-1001).

Madama Butterfly: Bimba dagli occhi pieni di malia; Una nave da guerra. cf La boheme: O soave fanciulla; O Mimi, tu piu non torni.

Madama Butterfly: Butterfly's entrance. cf Seraphim IB 6105.

Madama Butterfly: Con onor muore...Tu, tu, piccolo iddio. cf HMV SLS 5057.

2162 Madama Butterfly: Dovunque al mondo...A quanto cielo; Viene la sera; Un gel di vedremo; Ah, m'ha scordata; E questo...Che tua madre; Una nava da guerra...Scuoti quella fronda; Humming chorus; Io so che alle sua pene...Addio fiorito asil; Con onor muore...Tu, tu, piccolo iddio. Mirella Freni, s; Christa Ludwig, Elke Schary, ms; Luciano Pavarotti, Michel Senechal, t; Robert Kerns, Giorgio Stendoro, Siegfried Rudolf Frese, bar; Marius Rintzler, Hans Helm, bs; VSOO Chorus; VPO; Herbert von Karajan. Decca SET 605. (Reissue from SET 584/6)
+-Gr 11-76 p868 +HFN 2-77 p135 tape
++HFN 11-76 p173 +-RR 11-76 p42

Madama Butterfly: Humming chorus. cf Decca SXL 6826.

Madama Butterfly: Love duet. cf Decca SPA 450.

2163 Madama Butterfly: Opening scene...Love duet; One fine day; Telescope duet...Flower duet; Trio and Pinkerton's farewell; Death scene. Maria Collier, s; Ann Robson, ms; Charles Craig, t; Gwyn Griffiths, bar; Sadler's Wells Orchestra; Bryan Balkwill. HMV ESD 7030 Tape (c) TC ESD 7030. (Reissue from CSD 1290)
+-Gr 4-77 p1594 +HFN 8-77 p99 tape
+HFN 4-77 p153 +-RR 4-77 p39

2164 Madama Butterfly: Un bel di vedremo. Manon Lescaut: In quelle trine morbide; Sola, perduta, abbandonata. Tosca: Vissi d'arte. Turandot: In questa reggia. VERDI: Aida: Ritorna vincitor. Macbeth: Sleepwalking scene. I Lombardi: O madre, dal cielo... Se vano, se vano...No, no, giusta causa. Sylvia Sass, s; LSO; Lamberto Gardelli. London OS 26524 Tape (c) OS 5-26524. (also Decca SXL 6841 Tape (c) KSXC 6841)
+Gr 9-77 p488 +-ON 6-77 p44
+Gr 12-77 p1149 tape +RR 9-77 p42
+-HF 7-77 p121 +RR 11-77 p121 tape
+HFN 9-77 p149 +SFC 7-31-77 p40
+-NR 11-77 p13 -SR 7-9-77 p48
+-NYT 4-24-77 pD27 +-St 7-77 p128

Madama Butterfly: Un bel di vedremo. cf HMV SLS 5104.

Madama Butterfly: Un bel di. cf Seraphim 60274.

Madama Butterfly: Un bel di. cf Advent S 5023.

Madama Butterfly: Un bel di. cf Angel S 37446.
Madama Butterfly: Un bel di. cf Decca ECS 811.
Manon Lescaut: Arias. cf Free Lance FLPS 675.
Manon Lescaut: Arias. cf London OS 26499.
Manon Lescaut: Donna non vidi mai. cf HMV RLS 715.
Manon Lescaut: Donna non vidi mai. cf HMV ASD 3302.
Manon Lescaut: Donna non vidi mai. cf RCA TVM 1-7201.
Manon Lescaut: In quelle trine morbide. cf Seraphim IB 6105.
Manon Lescaut: In quelle trine morbide; Sola, perduta, abbandonata. cf Madama Butterfly: Un bel di vedremo.
Manon Lescaut: Intermezzo. cf Rediffusion 15-57.
Manon Lescaut: Sola perduta, abbandonata. cf HMV SLS 5057.
Manon Lescaut: Sola, perduta abbandonata. cf HMV SLS 5104.
Manon Lescaut: Tu, tu, amore...Tentatrice. cf Madama Butterfly.

2165 Mass, A major. Kari Lovaas, s; Werner Hollweg, t; Barry McDaniel, bar; West German Radio Chorus; Frankfurt Radio Symphony Orchestra; Eliahu Inbal. Philips 9500 009.
+Gr 2-77 p1311
+HF 3-77 p106
+HFN 2-77 p127
+MJ 3-77 p74
+NR 2-77 p9
+ON 2-26-77 p33
+RR 2-77 p88
+SFC 2-27-77 p42
+St 8-77 p119

La rondine: Chi il bel sogno di Doretta. cf DONIZETTI: Don Pasquale: So anch'io la virtu magica.
La rondine: Chi il bel sogno il Doretta. cf Angel S 37446.
Songs: E l'uccelino ninna nanna. cf Club 99-106.
Songs: Sole e amore; Menti all'avviso. cf Columbia M 34501.

2166 Suor Angelica. Renata Scotto, Ileana Cotrubas, s; Marilyn Horne, Patricia Payne, Gillian Knight, ms; Ambrosian Opera Chorus; NPhO; Lorin Maazel. Columbia M 34505. (also CBS 76570)
+Gr 4-77 p1594
+HF 7-77 p104
+HFN 4-77 p143
+NR 5-77 p10
+ON 3-19 p40
+OR 9/10-77 p29
+RR 4-77 p26
+SFC 8-28-77 p46
+-St 7-77 p119

Suor Angelica. cf Gianni Schicchi.
Suor Angelica: Intermezzo. cf Decca DDS 507/1-2.
Suor Angelica: Senza mama. cf Advent S 5023.

2167 Il tabarro. Renata Scotto, Yvonne Kenney, s; Gillian Knight, ms; Placido Domingo, Michel Senechal, Peter Jeffes, t; Ingvar Wixell, bar; Denis Wicks, bs; Ambrosian Opera Chorus; NPhO; Lorin Maazel. CBS 76641.
+-Gr 12-77 p1134
-HFN 12-77 p176
+RR 12-77 p32

Il tabarro. cf Gianni Schicchi.

2168 Tosca. Maria Callas, s; Giuseppe di Stefano, t; Tito Gobbi, bar; La Scala Opera Orchestra and Chorus; Victor de Sabata. Angel 3508 Tape (c) Ampex C 3655.
++HF 4-71 p65
+HF 12-71 p136 tape
++NYT 10-9-77 pD21

2169 Tosca. Franco Tagliavini, t; Giangiacomo Guelfi, bar; Other soloists and Orchestra. Cetra LPS 3261. (Reissue from OLPC 1261)
+Gr 7-77 p221

2170 Tosca. Leontyne Price, s; Giuseppe di Stefano, Piero de Palma, t; Giuseppe Taddei, bar; Fernando Corena, Leonardo Monreale, Alfredo Mariotti, bs; Vienna State Opera Chorus; VPO; Herbert von

Karajan. Decca 5BB 123/4 Tape (c) K59K22. (Reissue from RCA SER 5507/8) (also London 1284 Tape (c) M 33170)
+Gr 1-73 p1366 ++Op 7-73 p624
+Gr 10-77 p706 tape ++RR 1-73 p36
+HFN 1-73 p127 +RR 12-77 p98 tape
+HFN 10-77 p169 tape

2171 Tosca. Galina Vishnevskaya, s; Franco Bonisolli, Mario Guggia, t; Matteo Manugeurra, bar; Guido Mazzini, Antonio Zerbini, Domenico Medici, Giocomo Bertasi, bs; French National Orchestra and Chorus; Mstislav Rostropovich. DG 2707 087 (2) Tape (c) 3370 008.
+Gr 12-76 p1051 +-ON 1-1-77 p48
-HF 12-76 p112 -RR 1-77 p39
+-HFN 1-77 p113 -SFC 12-26-76 p34
+MJ 12-76 p28 +-St 1-77 p119
+-NR 12-76 p10

2172 Tosca. Montserrat Caballe, Ann Murray, s; Piero de Palma, Jose Carreras, t; Ingvar Wixell, Domenico Trimarchi, bar; Samuel Ramey, William Elvin, bs; ROHO and Chorus; Colin Davis. Philips 7600 108 (2) Tape (c) 7699 034.
+-ARG 6-77 p42 +NR 6-77 p11
+-Gr 5-77 p1730 +ON 4-9-77 p37
-HF 6-77 p97 +-RR 5-77 p39
+HF 7-77 p125 tape +RR 8-77 p86 tape
++HFN 8-77 p99 tape +-SFC 3-20-77 p34
+HFN 5-77 p115 +-St 6-77 p138
+MJ 5-77 p32

Tosca, excerpts. cf DONIZETTI: Lucia di Lammermoor, excerpts.
Tosca: Arias. cf Free Lance FLPS 675.
Tosca: Arias. cf London OS 26499.
Tosca: Amaro sol per te. cf Seraphim IB 6105.
Tosca: Amaro sol per te m'era il morire. cf Court Opera Classics CO 347.
Tosca: Cavaradossi's aria, Acts 1, 3. cf MASCAGNI: Cavalleria rusticana: Duet of Santuzza and Turiddu, Turridu's aria.
Tosca: E lucevan le stelle. cf HMV ASD 3302.
Tosca: E lucevan le stelle; Recondita armonia. cf HMV RLS 715.
Tosca: Mario Mario. cf La boheme: O soave fanciulla; O Mimi, tu piu non torni.
Tosca: Mario, Mario, Mario. cf GOUNOD: Faust: The hour is late.
Tosca: Recondita armonia. cf Decca SPA 450.
Tosca: Recondita armonia: E lucevan le stelle. cf RCA CRM 1-1749.
Tosca: Recondita armonia; E lucevan le stelle. cf RCA TVM 1-7201.
Tosca: Recondita armonia; E lucevan le stelle. cf Rubini GV 29.
Tosca: Recondita armonia; E lucevan le stelle. cf Rubini GV 43.

2173 Tosca: Recondita armonia; Non la sospiri la nostra casetta; Quale occhia al mondo; Vissi d'arte; E lucevan le stelle; O dolci mani; Trionfa di nova speme. Vladimir Atlantov, t; Tamara Milashkina, s; Bolshoi Theatre Orchestra; Mark Ermler. Columbia M 34516.
+NR 5-77 p10 -SFC 3-20-77 p34
+-ON 6-77 p44 -St 8-77 p119
+-SR 5-28-77 p42

Tosca: Recondita armonia; O dolci mani. cf Seraphim S 60280.
Tosca: Vissi d'arte. cf Madama Butterfly: Un bel di vedremo.
Tosca: Vissi d'arte. cf Advent S 5023.
Tosca: Vissi d'arte. cf Angel S 37446.

Tosca: Vissi d'arte. cf Bruno Walter Society BWS 729.
Tosca: Vissi d'arte. cf Club 99-96.
Tosca: Vissi d'arte. cf Decca D65D3.
Tosca: Vissi d'arte. cf Decca SDD 507.
Tosca: Vissi d'arte. cf HMV RLS 719.
Tosca: Vissi d'arte. cf Decca ECS 811.
Tosca: Vissi d'arte. cf HMV SLS 5104.
Tosca: Vissi d'arte. cf Rubini RS 301.
Tosca: Vissi d'arte; Quanto...Gia mi docon venal. cf Club 99-100.
Turandot: Arias. cf Free Lance FLPS 675.
Turandot: Del primo pianto. cf Seraphim IB 6105.
Turandot: Gira le cote. cf Seraphim S 60275.
Turandot: In questa reggia. cf Madama Butterfly: Un bel di vedremo.
Turandot: In questa reggia. cf Decca SXLR 6825.
Turandot: In questa reggia. cf HMV SLS 5057.
Turandot: In questa reggia. cf HMV SLS 5104.
Turandot: In questa reggia. cf London OS 26497.
Turandot: Nessun dorma. cf HMV RLS 715.
Turandot: Nessun dorma. cf HMV ASD 3302.
Turandot: Non piangere, Liu. cf Decca SXL 6839.
Turandot: Signore, ascolta. cf Angel S 37446.
Turandot: Tu, che di gel sei cinta. cf DONIZETTI: Don Pasquale: So anch'io la virtu magica.

PUJOL VILARRUBI, Emilio

Guajira. cf Enigma VAR 1015.
Tango. cf Enigma VAR 1015.

PULITI, Gabriello

Concerti, nos, 4 and 5 (realized by J. P. Mathieu). cf Arion ARN 90416.

PURCELL, Daniel

Sonata, trumpet and organ, F major. cf COUPERIN: Concerto, trumpet and organ, no. 9: Ritratto dell'amore.

PURCELL, Henry

2174 Abdelazer. The history of Dioclesian.* The old bachelor. The virtuous wife.* Laurence Boulay, hpd; Rouen Chamber Orchestra; Albert Beauchamp. Philips 6581 020. (*Reissue from 6797 001)
+-Gr 6-77 p56 +-MT 9-77 p737
+HFN 5-77 p137 +-RR 5-77 p55

Abdelazer: Incidental music. cf Works, selections (L'Oiseau-Lyre DSLO 504).
Abdelezer: Suite. cf CORELLI: Sarabande, gigue and badinerie.
Allegro and air. cf Crystal S 221.
Chaconne, F major. cf Abbey LPB 765.
Chaconne "Two in one upon a ground". cf Telefunken AW 6-42129.

2175 Chacony, G minor. Fantasia, 3 parts to a ground, D major. Pavans, nos. 1-5. Sonata, violin, bass viol and organ. Songs: Elegy on the death of John Playford; Elegy on the death of Thomas Farmer; Elegy on the death of Matthew Locke. Martyn Hill, t; Christopher Keyte, bs; Academy of Ancient Music; Christopher Hogwood. L'Oiseau-Lyre DSLO 514.
+Gr 10-77 p678 +RR 10-77 p84
+HFN 10-77 p161

Chacony, G minor. cf ALBINONI: Adagio, organ and strings, G minor.

2176 Chapel songs. Tavern songs. Catherine Mackintosh, Monica Huggett, vln; Jane Ryan, vla da gamba; Robert Elliott, chamber org, hpd; Deller Consort; Alfred Deller. Harmonia Mundi HM 242.
+RR 11-77 p106

2177 Dido and Aeneas. Tatiana Troyanos, Felicity Palmer, Elizabeth Gale, Linn Maxwell, sl Alfreda Hodgson, Patricia Kern, ms; Richard Stilwell, Philip Langridge, t; ECO and Chorus; Raymond Leppard. Erato STU 71091.
+-Gr 10-77 p692 +RR 10-77 p31
2178 Dido and Aeneas. Josephine Veasey, Helen Donath, Delia Wallis, Gillian Knight, s; Elizabeth Bainbridge, s; Frank Patterson, t; John Shirley-Quirk, Thomas Allen, bar; John Alldis Choir; AMF; Colin Davis. Philips 7400 131 Tape (c) 7300 073 (r) L 5131.
+AR 8-72 p97 ++MJ 5-77 p33
+ARG 8-72 p610 +NR 3-72 p10
+Gr 9-71 p495 +NYT 4-16-72 pD26
++HF 4-72 p98 +ON 1-6-73 p30
+HFN 9-71 p1643 ++Op 9-71 p810
+HF 8-72 p108 tape
Dido and Aeneas: Dido'a lament. cf DVORAK: Serenade, strings, op. 22, E major.
Distrsssed innocence: Incidental music. cf Works, selections (L'Oiseau-Lyre DSLO 504).
Distressed innocence: Rondeau; Air; Minuet. cf National Trust NT 002.
Fairy Queen. cf BACH: Suite, lute, S 997, C minor.
Fantasia, 3 parts to a ground, D major. cf Chacony, G minor.
2179 Funeral music for Queen Mary. Songs (anthems): Blessed are they that fear the Lord; Hear my prayer; My beloved spake; Rejoice in the Lord alway; Remember not Lord our offences. King's College Chapel Choir; Philips Jones Brass Ensemble, AMF; Philip Ledger. HMV ASD 3316 Tape (c) TC ASD 3316. (also Angel S 73282)
+ARG 11-77 p27 +NR 9-77 p8
+-Gr 3-77 p1443 +-RR 3-77 p94
+HFN 3-77 p111 +RR 7-77 p100 tape
+HFN 5-77 p138 tape +-St 12-77 p154
+-MT 10-77 p826
Funeral music for Queen Mary. cf Songs: Come ye sons of art.
The Gordian knot untied: Incidental music. cf Works, selections (L'Oiseau-Lyre DSLO 504).
2180 Harpsichord works, complete. Janos Sebestyen, hpd. Vox SVBX 5481 (3).
++SFC 9-11-77 p42
The history of Dioclesian. cf Abdelazer.
The history of Dioclesian: Chaconne. cf HMV SQ CSD 3781.
I was glad. cf Decca SPA 500.
2181 The Indian Queen. Timon of Athens: Masque. Honor Sheppard, Jean Knibbs, s; Mark Deller, Alfred Deller, c-t; Paul Elliott, Malcolm Knowles, t; Maurice Bevan, bar; King's Music, Deller Consort; Alfred Deller. Harmonia Mundi HM 243 (2).
+Gr 12-77 p1137 +-RR 8-77 p37
+-HFN 8-77 p87
The Indian Queen: Trumpet overture. cf Philips 6580 114.
King Arthur: Fairest isle. cf HMV RLS 716.
King Arthur: Hornpipe. cf Seraphim SIB 6094.
The married beau: Incidental music. cf Works, selections (L'Oiseau-Lyre DSLO 504).
The old bachelor. cf Abdelazer.
Pavans, nos. 1-5. cf Chacony, G minor.
The Queen's dolour. cf HMV ASD 3318.

Sonata, violin, bass viol and organ. cf Chacony, G minor.

2182 Songs: Come ye sons of art (Birthday ode for Queen Mary). Funeral music for Queen Mary. Felicity Lott, s; Charles Brett, John Williams, c-t; Thomas Allen, bs; Monteverdi Orchestra and Choir, Equale Brass Ensemble; John Eliot Gardiner. Erato STU 70911.
+-Gr 8-77 p331 +RR 8-77 p77
+HFN 11-77 p179

2183 Songs (choral works) In guilty night; Saul and the witch of Endor; Man that is born of woman; Te deum and jubilate Deo, D major. Robert Elliott, org; Deller Consort, Stour Music Festival Chorus; Orchestra; Alfred Deller. Harmonia Mundi HMU 207. (Reissue from RCA LSB 4038)
+-Gr 12-77 p1122 +-RR 7-77 p88

2184 Songs (Birthday odes for Mary II): Come, ye sons of art; Love's goddess sure. Norma Burrows, s; James Bowman, Charles Brett, c-t; Robert Lloyd, bs; Christopher Hogwood, hpd and org; Oliver Brooks, vla da gamba; Early Music Consort; David Munrow. HMV ASD 3166. (also Angel (Q) S 37251)
+-Gr 4-76 p1650 +MT 7-76 p578
++HF 8-77 p88 +NR 8-77 p11
+HFN 4-76 p111 +RR 4-76 p74
+MJ 9-77 p35 +St 8-77 p122
+-MM 8-76 p34

Songs (Anthems): Blessed are they that fear the Lord; Hear my prayer; My beloved spake; Rejoice in the Lord alway; Remember not Lord our offences. cf Funeral music for Queen Mary.

Songs: Elegy on the death of John Playford; Elegy on the death of Thomas Farmer; Elegy on the death of Matthew Locke. cf Chacony, G minor.

Songs: Evening hymn; The fatal hour; Man is for the woman made; The Queen's epicedium; Music for a while; There's nothing so fatal; Twas within a furlong of Edinburgh town. cf Abbey LPB 712.

Songs: Lucinda is bewitching fair; See where repenting Celia lies. cf Works, selections (L'Oiseau-Lyre DSLO 504)

Suite, strings, E minor. cf GLUCK: Sinfonia, G major.

Suite of dances. cf Philips 6581 018.

Symphonies, nos. 1 and 2. cf Telefunken AW 6-42199.

Tavern songs. cf Chapel songs.

Timon of Athens: Masque. cf The Indian Queen.

Trumpet tune. cf Argo SPA 507.

Trumpet tune (The cebell), D major. cf Philips 6500 926.

Trumpet tune and air. cf Philips 6500 926.

Trumpet tune and air. cf Philips 6581 018.

Trumpet tune and air. cf St. John the Divine Cathedral Church.

The virtuous wife. cf Abdelazer.

Voluntary, G major. cf Vista VPS 1047.

Voluntary, organ, D minor. cf Nonesuch H 71279.

2185 Works, selections: Abdelazer: Incidental music. Distressed innocence: Incidental music. The Gordian knot untied: Incidental music. The married beau: Incidental music. Songs: Lucinda is bewitching fair; See where repenting Celia lies. Joy Roberts, s; Academy of Ancient Music; Christopher Hogwood. L'Oiseau-Lyre DSLO 504.
+-Gr 6-76 p51 ++NR 1-77 p8
+-HF 3-77 p106 +RR 6-76 p51
+HFN 6-76 p95 +SFC 12-12-76 p55
+MJ 1-77 p26

Yorkshire feast song. cf Philips 6580 114.
PYKINI (14th century)
Plasanche or tost. cf HMV SLS 863.
QUANTZ, Johann
Sonata, flute, op. 1, no. 1, A minor. cf DUTILLEUX: Sonatine, flute and piano.
Sonata, flute, op. 1, no. 2, B flat major. cf DUTILLEUX: Sonatine, flute and piano.
QUARANTA
O ma charmante. cf Rubini GV 34.
QUERFURTH, Franz
Concerto, trumpet, E flat major. cf GRAUPNER: Concerto, D major.
QUILTER, Roger
Songs: Go lovely rose; O mistress mine. cf Enigma VAR 1027.
Songs: Now sleeps the crimson petal, op. 3, no. 2. cf Argo ZFG 95/6.
Songs: Now sleeps the crimson petal, op. 3, no. 2. cf HMV RLS 716.
RACHMANINOFF, Sergei
Aleko: Intermezzo and women's dance. cf Symphony, no. 3, op. 44, A minor.
Alfred de Musset, op. 21, no. 6, fragment. cf Angel S 37219.
2186 The bells, op. 35. Vocalise, op. 34, no. 14. Sheila Armstrong, s; Robert Tear, t; John Shirley-Quirk, bar; LSO and Chorus; Andre Previn. HMV SQ ASD 3284 Tape (c) TC ASD 3284. (also Angel (Q) S 37169)
++ARG 3-77 p31 -NR 2-77 p8
+-Gr 12-76 p1041 -ON 8-77 p56
+HF 1-77 p127 ++RR 11-76 p100
++HFN 12-76 p145 +St 3-77 p140
++HFN 4-77 p155 tape
Caprice bohemian, op. 12. cf BALAKIREV: Symphony, no. 1, C major.
2187 Concerti, piano, nos. 1-4. Rhapsody on a theme by Paganini, op. 43. Vladimir Ashkenazy, pno; LSO; Andre Previn. London CSA 2311 (3) Tape (c) CSA 5-2311. (also Decca SXL 6654/6 Tape (c) KSXC 6654, K43K33)
++Gr 1-73 p1364 ++HFN 5-77 p138 tape
++Gr 9-72 p492 +MJ 1-74 p40
++Gr 9-74 p595 tape +-NR 11-72 p7
+Gr 5-77 p1743 tape ++RR 9-72 p68
/HF 12-72 p66 +-RR 4-77 p90 tape
+-HF 6-77 p103 ++SFC 8-20-72 p28
++HFN 9-72 p56 ++St 11-72 p119
2188 Concerti, piano, nos. 1-4. Rhapsody on a theme by Paganini, op. 43. Rafael Orozco, pno; RPO; Edo de Waart. Philips 6747 397 (3). (Reissue from 6500 540)
+-Gr 12-77 p1088 +-RR 11-77 p62
+HFN 11-77 p179
2189 Concerto, piano, no. 1, op. 1, F sharp minor. Rhapsody on a theme by Paganini, op. 43. Daniel Wayneberg, Malcolm Binns, pno; PhO, LPO; Christoph von Dohnanyi, Alexander Gibson. Classics for Pleasure CFP 40267.
+-HFN 11-77 p179 +-RR 11-77 p67
2190 Concerto, piano, no. 1, op. 1, F sharp minor. Concerto, piano, no. 2, op. 18, C minor. Tamas Vasary, pno; LSO; Yuri Ahronovitch. DG 2530 717 Tape (c) 3300 717.
+-Gr 11-76 p803 +-NR 6-77 p5
+HF 6-77 p98 +-RR 11-76 p76
+-HFN 11-76 p165 ++St 6-77 p139
+MJ 5-77 p32

Concerto, piano, no. 1, op. 1, F sharp minor. cf PROKOFIEV: Concerto, piano, no. 1, op. 10, D flat major.
Concerto, piano, no. 1, op. 1, F sharp minor: Finale. cf Works, selections (CBS 30089).
2191 Concerto, piano, no. 2, op. 18, C minor. Prelude, op. 23, no. 5, G minor. BIZET (Rachmaninoff): L'Arlesienne: Suite, no. 1: Minuet. MENDELSSOHN (Rachmaninoff): A midsummer night's dream, op. 21: Scherzo. Gyorgy Cziffra, pno; NPhO; Gyorgy Cziffra, Jr.
+-HF 2-77 p99 +SFC 7-11-76 p13
-MJ 5-77 p32
2192 Concerto, piano, no. 2, op. 18, C minor. Rhapsody on a theme by Paganini, op. 43. Earl Wild, pno; RPO; Jascha Horenstein. Quintessence PMC 7006.
+HF 9-77 p108 +St 11-77 p146
2193 Concerto, piano, no. 2, op. 18, C minor. Rhapsody on a theme by Paganini, op. 43. Abbey Simon, pno; St. Louis Symphony Orchestra; Leonard Slatkin. Turnabout QTV 34658 Tape (c) CT 2148.
+NR 9-77 p4 ++SFC 5-29-77 p42
Concerto, piano, no. 2, op. 18, C minor. cf Concerto, piano, no. 1, op. 1, F sharp minor.
Concerto, piano, no. 2, op. 18, C minor. cf Works, selections (Decca DPA 565/6).
Concerto, piano, no. 2, op. 18, C minor. cf Decca D62D4.
Concerto, piano, no. 2, op. 18, C minor. cf HMV SLS 5068.
2194 Concerto, piano, no. 3, op. 30, D minor. Lazar Berman, pno; LSO; Claudio Abbado. CBS 76597 Tape (c) 40-76597. (also Columbia M 34540)
++FF 11-77 p43 +HFN 8-77 p99 tape
+Gr 6-77 p63 +NR 12-77 p5
+Gr 8-77 p346 tape ++RR 6-77 p24
+-HFN 7-77 p117 +-SFC 10-17-77 p47
2195 Concerto, piano, no. 3, op. 30, D minor. Alicia de Larrocha, pno; LSO; Andre Previn. Decca SXL 6746. (also London CS 6977 Tape (c) 5-6977)
+-ARG 11-76 p37 +HFN 2-76 p111
+-Gr 3-76 p1471 ++NR 12-76 p6
+-HF 3-77 p106 +-RR 2-76 p35
2196 Concerto, piano, no. 3, op. 30, D minor. Tamas Vasary, pno; LSO; Yuri Ahronovitch. DG 2530 859 Tape (c) 3300 859.
+Gr 6-77 p56 +-RR 6-77 p62
-HFN 6-77 p131 ++SFC 8-21-77 p46
+NR 10-77 p5 +-St 11-77 p148
2197 Concerto, piano, no. 3, op. 30, D minor. Joseph Alfidi, pno; Belgian Radio Orchestra; Rene Defossez. DG 2548 262 Tape (c) 3348 262.
-Gr 9-77 p432 -RR 7-77 p58
-HFN 7-77 p117
2198 Concerto, piano, no. 3, op. 30, D minor. Andrei Gavrilov, pno; USSR Symphony Orchestra; Alexander Lazarev. HMV ESD 7032 Tape (c) TC ESD 7032.
+Gr 4-77 p1555 +-HFN 4-77 p143
+-Gr 8-77 p349 tape +-RR 4-77 p55
2199 Concerto, piano, no. 3, op. 30, D minor. Earl Wild, pno; RPO; Jascha Horenstein. Quintessence PMC 7030.
+ARG 12-77 p43 +SFC 10-30-77 p44
2200 Concerto, piano, no. 3, op. 30, D minor. Vladimir Ashkenazy, pno; PO; Eugene Ormandy. RCA ARL 1-1324 Tape (c) ARK 1-1324 (ct)

ARS 1-1324 (Q) ARD 1-1324.
++Gr 5-76 p1765 +MJ 4-76 p30
++HF 5-76 p90 ++NR 4-76 p4
++HF 8-76 p70 tape ++RR 5-76 p48
+HFN 7-76 p94 ++SFC 4-25-76 p30
+-HFN 5-77 p139 tape ++St 8-76 p98

Concerto, piano, no. 3, op. 30, D minor. cf FRANCK: Symphonic variations.

2201 Concerto, piano, no. 4, op. 40, G minor. Rhapsody on a theme by Paganini, op. 43. Tamas Vasary, pno; LSO; Yuri Ahronovitch. DG 2530 905 Tape (c) 3300 905.
+Gr 12-77 p1088 -RR 12-77 p50
+HFN 12-77 p177 ++SFC 10-30-77 p44

2202 Concerto, piano, no. 4, op. 40, G minor. Rhapsody on a theme by Paganini, op. 43. Vladimir Ashkenazy, pno; LSO; Andre Previn. London CS 6776 Tape (c) 5-6776 (ct) 8-6776.
+HF 6-77 p103 +-NR 6-75 p7

Concerto, piano, no. 4, op. 40, G minor: Largo. cf Works, selections (CBS 30089).

Daisies, op. 38, no. 3. cf Angel S 37219.

Elegie, op. 3, no. 1. cf RCA GL 4-2125.

Etudes tableux, op. 39. cf KREISLER: Liebesfreud.

2203 Francesca da Rimini, op. 25. Aria recital: BORODIN: Prince Igor: Vladimir's recitative and cavatina. NAPRAVNIK: Dubrovsky: Vladimir's recitative and romance. TCHAIKOVSKY: Pique Dame: Arioso, Act 1; Forgive me heavenly creature; What is life; It is a game. Makvala Kasrashvili, s; Vladimir Atlantov, Alexander Laptev, t; Mikhail Maslov, bar; Yevgeny Nesterenko, bs; Bolshoi Theatre Orchestra and Chorus; Mark Ermler. Columbia/Melodiya MS 34577 (2).
+ARG 10-77 p40 +ON 8-77 p56
*MJ 11-77 p28 +SFC 5-29-77 p42
+NR 7-77 p10 +St 11-77 p148
+NYT 7-10-77 pD13

The harvest of sorrow, op. 4, no. 5. cf ETLER: Sonata, clarinet and piano.

Humoresque, op. 10, no. 5 (revised version). cf Angel S 37219.

Hymn of the cherubim. cf Abbey LPB 770.

2204 The isle of the dead, op. 29. Symphonic dances, op. 45. LSO; Andre Previn. HMV SQ ASD 3259 Tape (c) TC ASD 3259. (also Angel (Q) S 37158)
+ARG 3-77 p32 +MJ 1-77 p26
+Gr 9-76 p423 ++NR 1-77 p3
++HF 1-77 p124 +RR 9-76 p59
+-HFN 9-76 p127 ++SFC 10-10-76 p32
+HFN 11-76 p175 tape +-St 3-77 p140

Lilacs, op. 21, no. 5. cf Angel S 37219.

Melodie, op. 3, no. 3, E major (revised version). cf Angel S 37219.

2205 Moments musicaux, op. 16. Sonata, piano, no. 2, op. 36, B flat minor. Claudette Sorel, pno. Musical Heritage Society MHS 3338.
+CL 4-77 p10 +MJ 3-77 p74

2206 Nocturnes (4). Preludes (7). Claudette Sorel, pno. Musical Heritage Society.
+CL 4-77 p10

Oh stay my love, op. 4, no. 1. cf ETLER: Sonata, clarinet and piano.

Pieces, piano, op. 10, no. 3: Barcarolle. cf Pye PCNH 9.

Polichinelle, op. 3, no. 4, F sharp minor. cf Decca SPA 519.
Polka de V. R. cf Pye PCNH 9.
Preludes (7). cf Nocturnes.

2207 Prelude, op. 3, no. 2, C sharp minor. Preludes, op. 23, nos. 1-10. Preludes, op. 32, nos. 1-13. Vladimir Ashkenazy, pno. London CSA 2241 Tape (c) 5-2241 (r) CSAO 2241. (also Decca 5BB 221/2)
+Gr 2-76 p1360 ++NR 11-76 p12
+HF 8-77 p96 tape ++RR 2-76 p52
+HFN 4-76 p111 ++St 10-76 p129
++MJ 12-76 p28

Prelude, op. 3, no. 2, C sharp minor. cf Works, selections (CBS 30089).
Prelude, op. 3, no. 2, C sharp minor. cf Works, selections (Decca DPA 565/6).
Prelude, op. 3, no. 2, C sharp minor. cf Decca SPA 473.
Prelude, op. 3, no. 2, C sharp minor. cf Decca PFS 4351.
Prelude, op. 3, no. 2, C sharp minor. cf Decca PFS 4387.
Prelude, op. 3, no. 2, C sharp minor. cf International Piano Library IPL 103.
Preludes, op. 23, nos. 1-10. cf Prelude, op. 3, no. 2, C sharp minor.
Prelude, op. 23, no. 3, D minor. cf Works, selections (CBS 30089).
Prelude, op. 23, no. 5, G minor. cf Concerto, piano, no. 2, op. 18, C minor.
Prelude, op. 23, no. 5, G minor. cf International Piano Library IPL 103.
Prelude, op. 23, no. 5, G minor. cf International Piano Library IPL 117.
Prelude, op. 23, no. 5, G minor. cf International Piano Library IPL 5001/2.
Prelude, op. 23, no. 5, G minor. cf Philips 6747 327.
Prelude, op. 23, no. 6, E flat major. cf Works, selections (CBS 30089).
Preludes, op. 32, nos. 1-13. cf Prelude, op. 3, no. 2, C sharp minor.
Prelude, op. posth., D minor. cf Angel S 37219.
Rhapsody on a theme by Paganini, op. 43. cf Concerti, piano, nos. 1-4 (Decca SXL 6654/6).
Rhapsody on a theme by Paganini, op. 43. cf Concerti, piano, nos. 1-4 (Philips 6747 397).
Rhapsody on a theme by Paganini, op. 43. cf Concerto, piano, no. 1, op. 1, F sharp minor.
Rhapsody on a theme by Paganini, op. 43. cf Concerto, piano, no. 2, C minor (Quintessence PMC 7006).
Rhapsody on a theme by Paganini, op. 43. cf Concerto, piano, no. 2, op. 18, C minor (Turnabout QTV 34658).
Rhapsody on a theme by Paganini, op. 43. cf Concerto, piano, no. 4, op. 40, G minor (DG 2530 905).
Rhapsody on a theme by Paganini, op. 43. cf Concerto, piano, no. 4, op. 40, G minor (London 6776).
Rhapsody on a theme by Paganini, op. 43. cf Works, selections (Decca DPA 565/6).
Rhapsody on a theme by Paganini, op. 43. cf PROKOFIEV: Concerto, piano, no. 3, op. 26, C major.
The rock, op. 7. cf Symphonies, nos. 1-3.
The rock, op. 7. cf Symphony, no. 3, op. 44, A minor.

Serenade, op. 3, no. 5. cf HMV SXLP 30256.
Serenade, op. 3, no. 5. cf HMV SXLP 30259.

2208 Sonata, piano, no. 2, op. 36, B flat minor (original version). Variations on a theme by Corelli, op. 42. Jean-Philippe Collard, pno. Connoissuer Society CS 2082 (Q) CSQ 2082. (also HMV HQS 1366)

+-Gr 2-77 p1307
++HF 12-75 p102
+HFN 2-77 p127
++MJ 4-76 p30
++MT 10-77 p827
++NR 11-75 p13
+RR 2-77 p82
++SFC 9-28-75 p30
++St 2-76 p107 Quad

Sonata, piano, no. 2, op. 36, B flat major. Moments musicaux, op. 16.
Sonata, violoncello and piano, op. 19, G minor. cf MUCZYNSKI: Sonata, violoncello and piano, op. 25.
Sonata, violoncello and piano, op. 19, G minor: Andante. cf CBS SQ 79200.

2209 Songs: A-oo, op. 38, no. 6; Daisies, op. 38, no. 3; Dissonance, op. 34, no.13; Dreams, op. 38, no. 5; The harvest of sorrow, op. 4, no. 5; How fair this spot, op. 21, no. 7; In my garden at night, op. 38, no. 1; The morn of life, op. 34, no. 10; The muse, op. 34, no. 1; Oh, never sing to me again, op. 4, no. 4; The pied piper, op. 38, no. 4; The poet, op. 34, no. 9; The storm, op. 34, no. 3; To her, op. 38, no. 2; Vocalise, op. 34, no. 14; What wealth of rapture, op. 34, no. 12. Elisabeth Soderstrom, s; Vladimir Ashkenazy, pno. Decca 6718. (also London OS 26428)

+ARG 2-77 p38
++Gr 7-75 p228
+-HF 9-76 p95
++HFN 7-75 p87
+MJ 12-76 p28
+NR 11-76 p11
++RR 7-75 p59
+SR 11-13-76 p52
+St 10-76 p86

2210 Songs: Arion, op. 34, no. 5; Believe it not, op. 14, no. 7; Day to night comparing went the wind her way, op. 34, no. 4; A dream, op. 8, no. 5; Fate, op. 21, no. 1; I wait for thee, op. 14, no. 1; In the silent night, op. 4, no. 3; The little island, op. 14, no. 2; Midsummer nights, op. 14, no. 5; Music, op. 34, no. 8; The raising of Lazarus, op. 34, no. 6; So dread a fate I'll ne'er believe, op. 34, no. 7; So many hours, so many fancies, op. 4, no. 6; The soldier's wife, op. 8, no. 4; Spring waters, op. 14, no. 11; The world would see thee smile, op. 14, no. 6. Elisabeth Soderstrom, s; Vladimir Ashkenazy, pno. Decca SXL 6772. (also London OS 26453)

+Gr 12-76 p1041
+-HF 7-77 p104
++HFN 1-77 p115
++MT 9-77 p737
+NR 7-77 p11
+ON 8-77 p56
++RR 12-76 p94
++SFC 9-4-77 p38
++St 7-77 p119

2211 Songs: The answer, op. 21, no. 4; Before my window, op. 26, no. 10; Before the image, op. 21, no. 10; By the grave, op. 21, no. 2; The fountains, op. 26, no. 11; The lilacs, op. 21, no. 5; Loneliness, op. 21, no. 6; Melody, op. 21, no. 9; Night is mournful, op. 26, no. 12; No prophet, op. 21, no. 11; On the death of a linnet, op. 21, no. 8; Powder'd paint (arr. Rachmaninoff/trans. Ashkenazy); The ring, op. 26, no. 14; Sorrow in springtime, op. 21, no. 12; To the children, op. 26, no. 7; Twilight, op. 21, no. 3. Elisabeth Soderstrom, s; Vladimir

Ashkenazy, pno. Decca SXL 6832.
+Gr 12-77 p1122 ++RR 12-77 p80
+HFN 12-77 p177

Songs: Here beauty dwells; Oh cease thy singing maiden fair, op. 4, no. 4; Vocalise, op. 34, no. 14. cf Columbia 34294.

Songs: Hymn of the cherubim. cf Argo ARG 871.

Songs: In the silent night, op. 4, no. 3; Lilacs, op. 21, no. 5. cf HMV RLS 715.

Songs: Music, op. 34, no. 8; Night is mournful, op. 26, no. 12; O cease thy singing, maiden fair, op. 4, no. 4; Spring waters, op. 14, no. 11; Vocalise, op. 34, no. 14. cf GLINKA: Songs (DG 2530 725).

Symphonic dances, op. 45. cf The isle of the dead, op. 29.

Symphonic dances, op. 45, no. 1. cf Works, selections (CBS 30089).

2212 Symphonies, nos. 1-3. The rock, op. 7. LPO, OSR; Walter Weller. Decca D9D3 Tape (c) K9K33.
++Gr 12-76 p1066 -RR 9-76 p59
+Gr 9-76 p477 +-RR 2-77 p98 tape
+HFN 10-76 p181

2213 Symphony, no. 2, op. 27, E minor. Halle Orchestra; James Loughran. Classics for Pleasure CFP 40065 Tape (c) 40-40065. (also Sine Qua Non Tape (c) SQN 2008)
++Gr 9-74 p511 +-HFN 12-76 p153 tape
++Gr 2-77 p1325 tape +RR 9-74 p58
+-HF 9-77 p119 tape +-RR 11-77 p118 tape

2214 Symphony, no. 2, op. 27, E minor. PO; Eugene Ormandy. RCA ARL 1-1150 Tape (c) ARK 1-1150 (ct) ARS 1-1150.
+Gr 4-76 p1607 +MT 8-76 p661
-HF 12-75 p103 ++NR 11-75 p3
/HF 3-76 p116 tape +RR 4-76 p52
+-HFN 5-77 p139 tape +-St 3-76 p120

Symphony, no. 2, op. 27, E major. cf Works, selections (Decca DPA 565/6).

Symphony, no. 2, op. 27, E minor: Adagio. cf Works, selections (CBS 30089).

2215 Symphony, no. 3, op. 44, A minor. Aleko: Intermezzo and women's dance. LSO; Andre Previn. Angel S 37260 Tape (c) 4XS 37260. (also HMV SQ ASD 3369 Tape (c) TC ASD 3369)
+Gr 8-77 p306 +HFN 10-77 p171 tape
+Gr 10-77 p698 tape ++NR 8-77 p2
+HF 10-77 p114 +RR 8-77 p58
+HFN 8-77 p89

2216 Symphony, no. 3, op. 44, A minor. Vocalise, op. 34, no. 14. National Philharmonic Orchestra; Leopold Stokowski. Desmar DSM 1007 Tape (c) E 1046.
++Gr 4-76 p1608 ++NR 2-76 p2
+-HF 4-76 p116 ++RR 5-76 p48
+HF 4-77 p121 tape ++St 5-76 p120
+MJ 3-76 p25

2217 Symphony, no. 3, op. 44, A minor. The rock, op. 7. Rotterdam Philharmonic Orchestra; Edo de Waart. Philips 9500 302.
++SFC 12-25-77 p42

Variations on a theme by Corelli, op. 42. cf Sonata, piano, no. 2, op. 36, B flat minor.

2218 Vespers, op. 37. KASTALSKY: Motets (4). Meriel Dickinson, con; Wynford Evans, t; Bruckner-Mahler Choir; Wyn Morris. Philips 6747 246 (2).

+-Gr 11-77 p873 +NR 1-77 p9
+HFN 10-77 p157 +RR 11-77 p108
+MJ 1-77 p27 +St 1-77 p119

2219 Vespers, op. 37 (sung in English). Los Angeles Orthodox Concert Choir; Alexander Ruggieri. Turnabout TVS 34641.
+-ARG 6-77 p43 +NR 8-77 p8
++MJ 5-77 p33

Vocalise, op. 34, no. 14. cf The bells, op. 35.
Vocalise, op. 34, no. 14. cf Symphony, no. 3, op. 44, A minor.
Vocalise, op. 34, no. 14. cf Works, selections (Decca DPA 565/6).
Vocalise, op. 34, no. 14. cf CANTELOUBE: Songs of the Auvergne.
Vocalise, op. 34, no. 14. cf ETLER: Sonata, clarinet and piano.
Vocalise, op. 34, no. 14. cf ABC AB 67014.
Vocalise, op. 34, no. 14. cf Clear TLC 2586.
Vocalise, op. 34, no. 14. cf Discopaedia MB 1016.
Vocalise, op. 34, no. 14. cf Discopaedia MOB 1018.
Vocalise, op. 34, no. 14. cf HMV SQ ASD 3283.

2220 Works, selections: Concerto, piano, no. 1, op. 1, F sharp minor: Finale. Concerto, piano, no. 4, op. 40, G minor: Largo. Preludes, op. 3, no. 2, C sharp minor; op. 23, no. 3, D minor; op. 23, no. 6, E flat major. Symphonic dance, op. 45, no. 1. Symphony, no. 2, op. 27, E minor: Adagio. Philippe Entremont, pno; PO; Eugene Ormandy. CBS 30089 Tape (c) 40-30089.
+Gr 2-77 p1326 +-HFN 2-77 p135 tape
+HFN 4-77 p153 +-RR 12-76 p64

2221 Works, selections: Concerto, piano, no. 2, op. 18, C minor. Prelude, op. 3, no. 2, C sharp minor. Rhapsody on a theme by Paganini, op. 43. Symphony, no. 2, op. 27, E major. Vocalise, op. 34, no. 14. Julius Katchen, Ilana Vered, Vladimir Ashkenazy, pno; Elisabeth Soderstrom, s; LSO, LPO; Georg Solti, Hans Vonk, Adrian Boult. Decca DPA 565/6 Tape (c) KDPC 565/6.
+Gr 2-77 p1326 +HFN 1-77 p123 tape
+-HFN 12-76 p152 +-RR 12-76 p63

RADECK
Fredericus Rex. cf Decca SB 713.

RAFF, Joachim
Cavatina. cf Discopaedia MOB 1018.
La fileuse, op. 157, no. 2. cf Decca SPA 519.
Rigaudon, op. 204, no. 3. cf International Piano Library IPL 102.

RAINIER, Priaulx
Cycle for declamation. cf Argo ZK 28/9.

RAISON, Andre
Trio en passacaille. cf Argo ZRG 806.

RAMEAU, Jean
Gavotte with 6 doubles. cf Kaibala 20B01.

2222 Harpsichord works, complete. George Malcolm, hpd. Argo ZK 32/33 (2). (Reissues)
+HFN 12-77 p185 +-RR 12-77 p71

2223 Nouvelles suites de pieces de clavecin. Pieces de clavecin. Pieces de clavecin en concert, excerpts. Premier livre. Scott Ross, hpd. Telefunken FK 6-35346 (4).
+-Gr 5-77 p1717 +RR 5-77 p75

Pieces de clavecin. cf Nouvelles suites de pieces de clavecin.
Pieces de clavecin en concert. cf d'INDY: Trio, op. 29, B flat major.
Pieces de clavecin en concert, excerpts. cf Nouvelles suites de pieces de clavecin.

Pieces de clavecin en concert: Sonata, no. 4, G major. cf Vox SVBX 5142.
Premier livre. cf Nouvelles suites de pieces de clavecin.
Le Reppel des oiseaux. cf Westminster WGM 8309.
2224 Suites, D major/minor; G major/minor. Trevor Pinnock, hpd. CRD CRD 1030.
+Gr 5-77 p1717 ++RR 5-77 p76
++NR 10-77 p14
Zoroastre: Dances (7). cf DG 2723 051.
RANDS, Bernard
Monotone. cf Universal Edition UE 15043.
RANGSTROM, Ture
Songs: Flickan under nymanen; Pan; Villemo Villemo; Notturno. cf FRUMERIE: Songs (Swedish Society SLT 33171).
RARIG, John
Dance episode. cf Citadel CT 6012.
RASBACH, Otto
Songs: Trees. cf Argo ZFB 95/6.
Songs: Trees. cf L'Oiseau-Lyre DSLO 20.
RASI, Francesco
Indarno Febo. cf DG 2533 305.
RATHBURN, Eldon
2225 The metamorphic ten. RODBY: Concerto for 29. London Sinfonietta, Instrumental Ensemble; Elgar Howarth, Eldon Rathburn. Crystal S 504.
+NR 9-77 p6
RAUH
Ach Elslein. cf BIS LP 22.
RAUTAVAARA, Einojohani
Sonata, clarinet and piano, op. 53. cf BIS LP 62.
Sonata, flute and guitar. cf BIS LP 30.
Varietude, solo violin, op. 82. cf Finnlevy SFLP 8569.
RAVEL, Maurice
A la maniere de Borodine. cf Piano works (Decca SXL 6700).
A la maniere de Chabrier. cf Piano works (Decca SXL 6700).
2226 Bolero. Miroirs: Alborada del gracioso. Pavane pour une infante defunte. La valse. Halle Orchestra; James Loughran. Classics for Pleasure CFP 40036 Tape (c) TC CFP 40036.
+-Gr 2-77 p1325 tape +HFN 12-76 p153 tape
2227 Bolero. Daphnis and Chloe: Suite, no. 2. Ma mere l'oye. La valse. LAPO; Zubin Mehta. Decca SXL 6488 Tape (c) KXSC 16488.
++HFN 7-75 p91 tape +RR 5-77 p92 tape
2228 Bolero. Daphnis et Chloe: Suite, no. 2. La valse. Strasbourg Philharmonic Orchestra; Alain Lombard. Erato STU 70930.
+Gr 4-77 p1556 +-RR 3-77 p57
++HFN 4-77 p143
2229 Bolero. Pavane pour une infante defunte. Le tombeau de Couperin. La valse. COA; Bernard Haitink. Philips 9500 314.
+-Gr 11-77 p843 ++RR 11-77 p67
+HFN 11-77 p179
2230 Bolero. Daphnis and Chloe: Suite, no. 2. Miroirs: Alborada del gracioso. Pavane for a dead princess. CSO, Symphony Orchestra; Jean Martinon, Morton Gould. Quintessence PMC 7017.
+ARG 12-77 p43 ++SFC 10-30-77 p44
+-FF 11-77 p44
Bolero. cf Works, selections (Decca DPA 561/2).
Bolero. cf Works, selections (Vox SVBX 5133).

Bolero. cf CHABRIER: Espana.
Bolero. cf DEBUSSY: La mer.
Bolero. cf DEBUSSY: Prelude a l'apres-midi d'un faune.
Bolero. cf Bruno Walter Society RR 443.
Bolero. cf DG 2535 254.
Chansons madecasses. cf BOCCHERINI: Quintet, guitar.
Chansons madecasses. cf FAURE: La bonne chanson, op. 61.
Concerto, piano, G major. cf FAURE: Fantaisie, piano and orchestra, op. 111.
Concerto, piano, for the left hand, D major. cf FAURE: Fantaisie, piano and orchestra, op. 111.
Concerto, piano, for the left hand, D major. cf d'INDY: Symphony on a French mountain air, op. 25.

2231 Daphnis et Chloe. Camerata Singers; NYP; Pierre Boulez. Columbia M 33523 Tape (c) MT 33523 (Q) MQ 33523. (also CBS 76425 Tape (c) 40-76425)
+Audio 6-76 p96 +-MJ 12-75 p41
+-Gr 12-75 p1045 ++NYT 1-18-76 pD1
+HF 12-75 p104 ++ON 3-6-76 p42
++HF 6-77 p103 tape ++RR 12-75 p58
+HFN 12-75 p163 ++St 11-75 p132
+HFN 2-76 p116

2232 Daphne et Chloe. ROHO Chorus; LSO; Pierre Monteux. Decca Tape (c) KSDC 170.
+RR 5-77 p92 tape

2233 Daphnis et Chloe: Suites, nos. 1 and 2. Ma mere l'oye. St. Olaf Choir; Minnesota Orchestra; Kenneth Jennings, Stanislaw Skrowaczewski. Turnabout TV 34603 Tape (c) KTVC 34603.
+Gr 9-77 p432 ++RR 9-77 p67
+HFN 9-77 p149 ++RR 10-77 p98 tape
+-HFN 10-77 p171

Daphnis et Chloe: Suites, nos. 1 and 2. cf Works, selections (Vox SVBX 5133).

2234 Daphnis et Chloe: Suite, no. 2. Miroirs: Alborada del gracioso. Pavane pour une infante defunte. Rapsodie espagnole. PhO, NPhO; Carlo Maria Giulini. HMV SXLP 30198 Tape (c) TC EXE 181. (Reissues from Columbia SAX 2476, 5265)
+Gr 4-76 p1608 +RR 2-76 p36
+HFN 3-76 p109 +RR 5-77 p92 tape
-HFN 5-76 p119 tape

Daphnis and Chloe: Suite, no. 2. cf Bolero (Decca 6488).
Daphnis et Chloe: Suite, no. 2. cf Bolero (Erato STU 70930).
Daphnis and Chloe: Suite, no. 2. cf Bolero (Quintessance PMC 7017).
Daphnis et Chloe: Suite, no. 2. cf Works, selections (Decca DPA 561/2).
Daphnis et Chloe: Suite, no. 2. cf CHABRIER: Espana.
Daphnis and Chloe: Suite, no. 2. cf DEBUSSY: La mer.
Empress of the pagodas. cf Angel S 36053.
Fanfare for "L'Eventail de Jeanne." cf Works, selections (Vox SVBX 5133).

2235 Gaspard de la nuit. Serenade grotesque. STRAVINSKY: Les cinq doigts. Petrouchka: Scenes (3). Valse pour les enfants. Idil Biret, pno. Finnadar SR 9013 Tape (c) CS 9013F.
+FF 11-77 p45 +MJ 5-77 p32
+-HF 5-77 p88 +NR 5-77 p15
++HF 4-77 p121 tape ++St 6-77 p136

Gaspard de la nuit. cf Piano works (Decca SXL 6700).

2236 Gaspard de la nuit: Le gibet. Miroirs: Oiseaux tristes; La vallee des cloches. Pavane pour une infante defunte. Le tombeau de Couperin: Toccata. Maurice Ravel, pno. Everest 3403 Tape (c) 3403. (Reissue from Duo Art piano rolls)
-ARG 2-77 p39 +NR 5-77 p15
+-HF 10-77 p129 tape

Habanera. cf Enigma VAR 1025.

Habanera. cf RCA JRL 1-2315.

Introduction and allegro. cf Works, selections (Decca DPA 561/2).

2237 Jeux d'eau. Ma mere lyoye. Miroirs. Pascal Roge, Denise Francoise Roge, pno. Decca SXL 6715. (also London CS 6936)
+Gr 11-75 p864 +NR 2-76 p14
+-HF 4-76 p120 +RR 11-75 p82
+HFN 11-75 p163 +RR 1-77 p73
+-MJ 4-76 p30 /St 7-76 p112

2238 Ma mere l'oye. Menuet antique. Le tombeau de Couperin. BSO; Seiji Ozawa. DG 2530 752.
++ARG 9-77 p45 ++SFC 8-21-77 p46
+NR 9-77 p2

Ma mere l'oye. cf Bolero.

Ma mere l'oye. cf Daphnis et Chloe: Suites, nos. 1 and 2.

Ma mere l'oye. cf Jeux d'eau.

Ma mere l'oye. cf Works, selections (Decca DPA 561/2).

Ma mere l'oye. cf Works, selections (Vox SVBX 5133).

2239 Menuet antique. Miroirs: Alborada del gracioso. Rapsodie espagnole. Valses nobles et sentimentales. COA: Bernard Haitink. Philips 9500 347.
++Gr 10-77 p637 ++RR 11-77 p67
+HFN 11-77 p179 ++SFC 9-18-77 p42
++NR 12-77 p3

Menuet antique. cf Ma mere l'oye.

Menuet antique. cf Piano works (Decca SXL 6700).

Menuet antique. cf Works, selections (Vox SVBX 5133).

Menuet sur le nom de Haydn. cf Piano works (Deccs SXL 6700).

2240 Miroirs. Pavane pour une infante defunte. Serenade grotesque. Valses nobles et sentimentales. Therese Dussaut, pno. Pandora PAN 108.
-FF 11-77 p45

Miroirs. cf Jeux d'eau.

Miroirs: Alborada del gracioso. cf Bolero (Classics for Pleasure CFP 40036)

Miroirs: Alborada del gracioso. cf Bolero (Quintessence PMC 7017).

Miroirs: Alborada del gracioso. cf Daphnis et Chloe: Suite, no. 2.

Miroirs: Alborada del gracioso. cf Menuet antique.

Miroirs: Alborada del gracioso. cf Works, selections (Decca DPA 561/2).

Miroirs: Alborada del gracioso. cf FALLA: El amor brujo.

Miroirs: Alborada del gracioso. cf Seraphim S 60277.

2241 Miroirs: Alborada del gracioso; Une barque sur l'ocean. Pavane pour une infante defunte. Valses nobles et sentimentales. BSO; Seiji Ozawa. DG 2530 753. (Reissue from 2711 015)
+-ARG 8-77 p25 +NR 7-77 p4

Miroirs: Alborada del gracioso; Une barque sur l'ocean. cf Works, selections (Vox SVBX 5133).

Miroirs: Oiseaux tristes; La vallee des cloches. cf Gaspard de la nuit: Le gibet.

Pavane of the sleeping beauty. cf Angel S 36053.
Pavane pour une infante defunte. cf Bolero (Classics for Pleasure CFP 40036).
Pavane pour une infante defunte. cf Bolero (Philips 9500 314).
Pavane pour une infante defunte (Pavane for a dead princess). cf Bolero (Quintessence PMC 7017).
Pavane pour une infante defunte. cf Daphnis et Chloe: Suite, no. 2.
Pavane pour une infante defunte. cf Gaspard de la nuit: Le gibet.
Pavane pour une infante defúnte. cf Miroirs.
Pavane pour une infante defunte. cf Miroirs: Alborada del gracioso; Une barque sur l'ocean.
Pavane pour une infante defunte. cf Piano works (Decca SXL 6700).
Pavane pour une infante defunte. cf Works, selections (Decca DPA 561/2).
Pavane pour une infante defunte. cf Works, selections(Vox SVBX 5133).
Pavane pour une infante defunte. cf DEBUSSY: La mer.
Pavane pour une infante defunte. cf FALLA: El amor brujo.
Pavane pour une infante defunte. cf MUSSORGSKY: Pictures at an exhibition.
Pavane pour une infante defunte. cf HMV SLS 5073.
Pavane pour une infante defunte. cf Prelude PRS 2512.
Pavane pour une infante defunte. cf RCA RK 1-1735.

2242 Piano works: A la maniere de Borodine. A la maniere de Chabrier. Gaspard de la nuit. Menuet antique. Menuet sur le nom de Haydn. Pavane pour une infante defunte. Prelude. Pascal Roge, pno. Decca SXL 6700. (also London CS 6873)
+Audio 8-75 p80 ++RR 3-75 p51
+Gr 3-75 p1682 +RR 1-77 p73

Piece en forme de habanera. cf Orion ORS 75181.
Prelude. cf Piano works (Decca SXL 6700).
Quartet, strings, F major. cf DEBUSSY: Quartet, strings, G minor (Decca SDD 526).
Quartet, strings, F major. cf DEBUSSY: Quartet, strings, G minor (Opus 9111 0337).
Quartet, strings, F major. cf DEBUSSY: Quartet, strings, op. 10, G minor.
Rapsodie espagnole. cf Daphnis et Chloe: Suite, no. 2.
Rapsodie espagnole. cf Menuet antique.
Rapsodie espagnole. cf Works, selections (Decca DPA 561/2).
Rapsodie espagnole. cf Works, selections (Vox SVBX 5133).
Serenade grotesque. cf Gaspard de la nuit.
Serenade grotesque. cf Miroirs.
Serenade grotesque. cf Vox SVBX 5483.
Sheherazade. cf BERLIOZ: Les nuits d'ete, op. 7.
Sonata, violin and piano. cf DEBUSSY: Children's corner suite: Golliwog's cakewalk.

2243 Sonatine. Le tombeau de Couperin. Valses nobles et sentimentales. Pascal Roge, pno. Decca SXL 6674. (also London CS 6873)
+Gr 11-74 p922 +RR 1-77 p73
+-HF 4-75 p90 ++SFC 5-4-75 p35
+NR 7-75 p15 +-SR 2-8-75 p37
+RR 11-74 p83

2244 Songs: Chansons madecasses; Histoires naturelles; Melodies populaires grecques (6); Poemes de Stephane Mallarme (3). Felicity Palmer, s; Judith Pearse, flt; Christopher van Kempen, vlc;

Clifford Benson, John Constable, pno; Nash Ensemble; Simon Rattle. Argo ZRG 834.
+-Gr 3-76 p1494 +-MM 5-77 p36
+-HFN 2-76 p111 +RR 2-76 p62

2245 Songs: Don Quichotte a Dulcinee; Epigrammes de Clement Marto (2); Histoires naturelles; Melodies hebraiques (2); Reves; Ronsard a son ame; Sainte; Sur l'herbe. Jacques Herbillon, bar; Theodore Paraskivesco, pno. Calliope CAL 1856.
+HFN 6-77 p133 +-RR 6-77 p91

Songs: D'Anne jouant delespinette; D'Anne que me jecta de la neige; Sheherazade. cf HMV RLS 716.

Songs: L'Enfant et les sortileges; Toi, le coeur de la rose. cf Odyssey Y 33130.

Songs: Melodies populaires grecques. cf BIS LP 34.

Songs: Nicolette; Trois beaux oiseaux de Paradis; Ronde. cf RCA RL 2-5112.

2246 Le tombeau de Couperin. STRAVINSKY: Petrouchka: 3 movements. Alexis Weissenberg, pno. Connoisseur Society CS 2114.
-HF 5-77 p88 +-NR 7-77 p12
+MJ 5-77 p32 ++St 1-77 p130

Le tombeau de Couperin. cf Bolero.

Le tombeau de Couperin. cf Ma mere l'oye.

Le tombeau de Couperin. cf Sonatine.

Le tombeau de Couperin. cf Works, selections (Vox SVBX 5133).

Le tombeau de Couperin. cf GRIFFES: Sonata, piano.

Le tombeau de Couperin: Toccata. cf Gaspard de la nuit: Le gibet.

2247 Trio, violin, violoncello and piano, A minor (1914). SHOSTAKOVICH: Trio, piano, no. 2, op. 67, E minor. Rostislav Dubinsky, vln; Valentin Berlinsky, vlc; Lubov Yedlina, pno. Westminster WGS 8332.
+-ARG 5-77 p36 +NR 4-77 p6

Trio, violin, violoncello and piano, A minor. cf ARENSKY: Trio, op. 32, D minor.

Tzigane. cf BRAHMS: Sonata, violin and piano, no. 3, op. 108, D minor.

Tzigane. cf CHAUSSON: Poeme, op. 25

Tzigane. cf DEBUSSY: Children's corner suite: Golliwog's cakewalk.

La valse. cf Bolero (Classics for Pleasure CFP 40036).

La valse. cf Bolero(Decca 6488).

La valse. cf Bolero (Erato STU 70930).

La valse. cf Bolero (Philips 9500 314).

La valse. cf Works, selections (Decca DPA 561/2).

La valse. cf Works, selections (Vox SVBX 5133).

La valse. cf DEBUSSY: Nocturnes: Fetes.

Valses nobles et sentimentales. cf Menuet antique.

Valses nobles et sentimentales. cf Miroirs.

Valses nobles et sentimentales. cf Miroirs: Alborada del gracioso; Une barque sur l'ocean.

Valses nobles et sentimentales. cf Sonatine.

Valses nobles et sentimentales. cf Works, selections (Vox SVBX 5133).

2248 Works, selections: Bolero. Daphnis et Chloe: Suite, no. 2. Introduction and allegro. Ma mere l'oye. Miroirs: Alborada del gracioso. Pavane pour une infante defunte. Rapsodie espagnole. La valse. Melos Ensemble, OSR; Ossian Ellis, hp; Ernest Ansermet. Decca DPA 561/2.
+-HFN 12-76 p152 +-RR 1-77 p57

2249 Works, selections: Bolero. Daphnis et Chloe: Suites, nos. 1 and 2. Fanfare for "L'Eventail de Jeanne." Ma mere l'oye. Menuet antique. Miroirs: Alborada del gracioso; Une barque sur l' ocean. Pavane pour une infante defunte. Rapsodie espagnole. Le tombeau de Couperin. La valse. Valses nobles et sentimentales. Minnesota Orchestra; Stanislaw Skrowaczewski. Vox SVBX 5133 (4) (Q) QSVBX 5133.
++Audio 1-77 p83 ++NR 6-75 p1
++HF 7-75 p66 +St 8-75 p102 Quad

RAVENSCROFT, Thomas
Now Robin lend to me thy bow. cf Enigma VAR 1020.
Remember, O thou man. cf HMV CSD 3774.

RAVINA, Jean Henri
Etude de style. cf International Piano Library IPL 102.

RAWSTHORNE, Alan
Canzonet. cf RCA GL 2-5062.
Concerto, piano, no. 1. cf HMV SLS 5080.
Concerto, piano, no. 2. cf HMV SLS 5080.
Symphonic studies. cf Symphony, no. 1.
2250 Symphony, no. 1. Symphonic studies. LPO; John Pritchard. Lyrita SRCS 90. (also HNH 4044)
++Gr 4-77 p1555 +HFN 4-77 p143
+FF 11-77 p47 +RR 4-77 p55

RAXACH, Enrique
The looking glass. cf BACK: ...for Eliza.

RAYE-PRINCE
The boogie-woogie bugle boy of Company B. cf RCA CRL 1-2064.

READ
Songs: Down steers the bass; Russia. cf Vox SVBX 5350.

REBEL, Jean-Ferry
Les elements. cf DIEUPART: Suite, flute, violin, bass and harpsichord, no. 3.
RECORDER MUSIC OF THE ITALIAN RENAISSANCE. cf Telefunken AW 6-42033.

REDDLE
Gypsy fantasy. cf Citadel CT 6012.

REDFORD, John
Rejoice in the Lord. cf Vista VPS 1053.

REED
Testament of an American. cf Menc 76-11.

REEVE
I am a friar of orders gray. cf Club 99-102.

REGER, Max
Aus meinem Tagebuche, op. 82: Adagio and vivace. cf International Piano Library IPL 102.
Fantasia and fugue on the chorale "Hallelujah, Gott zu loben bleibe meine Seelen Freund", op. 52, no. 3. cf Abbey LPB 752.
Fantasia on Straf mich in deinem Zorn. cf HOWELLS: Psalm prelude.
2251 Ein feste Burg, op. 27. Pieces, op. 59, nos. 5, 6. Pieces, op. 145. Wie schon leuchtet, op. 40, no. 1. George Markey, org. Psallite 101 280770.
+MJ 5-77 p32
Ein feste Burg, op. 67, no. 6. cf Vista VPS 1035.
Introduction and passacaglia, D minor. cf Vista VPS 1035.
Phantasie uber den choral "Straf mich nicht in deinem Zorn", op. 40, no. 2. cf Pelca PSR 40599.
Pieces, op. 59, nos. 2, 5. cf BIS LP 27.

Pieces, op. 59, nos. 5, 6. cf Ein feste Burg, op. 27.
Pieces, op. 59, no. 9: Benedictus. cf Decca SDD 499.
Pieces, op. 59, no. 9: Benedictus. cf Wealden WS 131.
Pieces, op. 80: Toccata and fugue. cf Wealden WS 159.
Pieces, op. 145. cf Ein feste Burg, op. 27.
Pieces, op. 145, no. 2: Dankpsalm. cf FRANCK: Chorale, no. 2, B minor.
Pieces, op. 145, no. 2: Dankpsalm. cf HOLLER: Ciacona, op. 54.
Sonata, clarinet and piano, no. 3, op. 107, B flat major. cf BACH: Canons on the First 8 bass notes of the "Goldberg variations" aria, S 1087.
Toccata and fugue, op. 59. cf BRAHMS: Fugue, A flat minor.
Vater unser. cf Pelca PSR 40607.
Wie schon leuchtet, op. 40, no. 1. cf Ein feste Burg, op. 27.

REICH, Steve

2252 Drumming. Music for mallet instruments, voices and organ. Six pianos. Steve Reich and musicians. DG 2740 106 (3).
++FF 9-77 p90 +-St 6-75 p110
+Gr 1-75 p1353

Music for mallet instruments, voices and organ. cf Drumming.
Six pianos. cf Drumming.

REICHA (REJCHA), Antonin

Quintet, op. 91, no. 5, A major. cf DANZI: Quintet, op. 56, no. 1, B flat major.

REIF, Paul

Duo for three. cf KAUFMAN: Reflections, clarinet and piano.

2253 Vignettes, 4 singers (8). THOMPSON: The peaceable kingdom. Gregg Smith Singers, Soloists; Pepperdine University A Cappella Choir; Gregg Smith, Lawrence McCommas. Orion ORS 76228.
+NR 10-77 p9

REINAGLE, Alexander

2254 Sonatas, piano, nos. 1-3. Jack Winerock, pno. Musical Heritage Society MHS 3359.
+-HF 3-77 p107

REINECKE, Carl

Ballade, op. 288. cf BOHM: Fantasy on a theme by Schubert.

2255 Sonata, flute and piano, op. 167, E minor. SCHUBERT: Introduction and variations, op. 160, D 802, E minor. SCHUMANN: Romances, op. 94 (3). Jean-Pierre Rampal, flt; Robert Veyron-Lacroix, pno. RCA ARL 1-2092 Tape (c) ARK 1-2092 (ct) ARS 1-2092.
++ARG 8-77 p37 +NR 5-77 p7
+-MJ 9-77 p35 ++SFC 5-8-77 p46

Sonata, flute and piano, op. 167, E minor. cf Pearl SHE 533.

REIZENSTEIN, Franz

Prelude and fugue, no. 8, op. 32, D major. cf Prelude, no. 11, op. 32, B major.

2256 Prelude, no. 11, op. 32, B major. Prelude and fugue, no. 8, op. 32, D major. Sonata, violin, op. 20, G sharp major. Sonata, solo violin, op. 46. Erich Gruenberg, vln; David Wilde, pno. L'Oiseau-Lyre SOL 348.
+Gr 2-77 p1301 +RR 2-77 p82
++HFN 2-77 p127

Sonata, violin, op. 20, G sharp major. cf Prelude, no. 11, op. 32, B major.
Sonata, solo violin, op. 46. cf Prelude, no. 11, op. 32, B major.

RENIE

Concerto, harp, in one movement. cf Musical Heritage Society MHS 3611.

RENWICK
Dance. cf Crystal S 206.
RESPIGHI, Ottorino
2257 Ancient airs and dances. Los Angeles Chamber Orchestra; Neville Marriner. HMV ASD 3188 Tape (c) TC ASD 3188. (also Angel S 37301 Tape (c) 4XS 37301)
+HFN 10-77 p171 tape
2258 Ancient airs and dances: Suites, nos. 1-3. PH; Antal Dorati. Philips 6892 010 Tape (c) 7321 022. (Reissue from Mercury AMS 16028)
+Gr 4-76 p1608 +HFN 7-77 p127 tape
+-HFN 3-76 p109 +RR 6-77 p97 tape
Ancient airs and dances: Suite, no. 2: Bergamasca. cf Philips 6747 327.
Ancient airs and dances: Suite, no. 3. cf ELGER: Serenade, strings, op. 20, E minor.
2259 Belfagor: Overture. Fountains of Rome. The pines of Rome. LSO; Lamberto Gardelli. HMV SQ ASD 3372. (also Angel S 37402 Tape (c) 4XS 37402)
+Gr 11-77 p844 +NR 12-77 p2
Feste Romane (Roman festivals). cf The pines of Rome.
Feste Romane. cf RCA CRM 5-1900.
Roman festivals. cf Fountains of Rome.
2260 Fountains of Rome. The pines of Rome. Roman festivals. PO; Eugene Ormandy. RCA Tape (c) ARS 1407 (ct) ARK 1-1407.
++HF 1-77 p151 tape +MJ 1-77 p26
Fountains of Rome. cf Belfagor: Overture.
2261 Gli uccelli (The birds). Trittico botticelliano (Three Botticelli pictures). AMF; Neville Marriner. Angel (Q) S 37252. (also HMV ASD 3327 Tape (c) TC ASD 3327)
+Gr 3-77 p1409 ++HFN 10-77 p171 tape
+Gr 10-77 p698 tape -NR 10-77 p2
++HF 7-77 p106 +-RR 3-77 p58
++HFN 3-77 p111 ++St 9-77 p140
2262 The pines of Rome. Feste Romane (Roman festivals). CO; Lorin Maazel. Decca SXL 6822. (also London CS 7043)
+Gr 11-77 p844 +RR 11-77 p67
+HFN 11-77 p179 +SFC 12-11-77 p61
2263 The pines of Rome. STRAUSS, R.: Don Juan, op. 20. Salome, op. 54: Dance. RPO; Rudolf Kempe, Antal Dorati. Quintessence PMC 7005.
+ARG 12-77 p43 +St 11-77 p146
+HF 9-77 p108
The pines of Rome. cf Belfagor: Overture.
The pines of Rome. cf Fountains of Rome.
The pines of Rome. cf MUSSORGSKY: A night on the bare mountain.
Re Enzo: Stornellactrice. cf Club 99-96.
Songs: Au milieu du jardin; Povero core; Razzolan; Soupir. cf Columbia M 34501.
Songs: Pioggia. cf Orion ORS 77271.
Stornellactrice, 1st and 2nd edition. cf Rubini RS 301.
Trittico botticelliano (Three Botticelli pictures). cf Gli uccelli.
REUBKE, Julius
Sonata, piano, B flat minor. cf LISZT: Ad nos, ad salutarem undam, G 259.
Sonata on the 94th psalm. cf BRIDGE: Prelude and minuet.

Sonata on the 94th psalm. cf Vista VPS 1046.

REUSNER, Esiais

Suite. cf DG 2723 051.

Suite, no. 2, C minor: Paduana. cf CBS 72526.

REVEULTAS, Silvestre

Caminos. cf Works, selections (RCA ARL 1-2320).

Itinerarios. cf Works, selections (RCA ARL 1-2320).

Janitzio. cf Works, selections (RCA ARL 1-2320).

Redes. cf Works, selections (RCA ARL 1-2320).

Sensemaya. cf Works, selections (RCA ARL 1-2320).

2264 Works, selections: Caminos. Itinerarios. Janitzio. Redes. Sensemaya. NPhO; Eduarda Mata. RCA ARL 1-2320.
++ARG 11-77 p29 ++SFC 9-4-77 p38
++NR 10-77 p4 +St 11-77 p150

REYER, Ernest (also Rey, Louis)

Sigurd: Chant du barde: Au nom de roi. cf Club 99-107.

Sigurd: Esprit, gardiens; Duet de la fontaine. cf Rubini GV 38.

Le statue: Grand recitative. cf Rubini GV 38.

REYNOLDS, Verne

Caprices (4). cf MARTINO: A set for clarinet.

Music, 5 trumpets. cr Crystal S 362.

REZNICEK, Emil

Donna Diana: Overture. cf HMV ESD 7010.

Donna Diana: Overture. cf Supraphon 110 1637.

RHEINBERGER, Josef

2265 Sonata, organ, no. 1, op. 27, C minor. Sonata, organ, no. 20, op. 196, F major. Conrad Eden, Timothy Farrell, org. Vista VPS 1011.
+-Gr 5-75 p1996 +MU 7-77 p15
+HFN 5-75 p138 +-RR 4-75 p53
+-MT 7-76 p578

2266 Sonata, organ, no. 2, op. 63, A flat major. Sonata, organ, no. 9, op. 142, B flat minor. Conrad Eden, Robert Munns, org. Vista VPS 1013.
/Gr 8-75 p348 +MU 7-77 p15
+-HFN 8-75 p82 +-RR 9-75 p61
+-MT 7-76 p578

2267 Sonata, organ, no. 3, op. 88, G major. Sonata, organ, no. 19, op. 192, G minor. Conrad Eden, Timothy Farrell, org. Vista VPS 1015.
+-Gr 3-76 p1487 +MU 7-77 p15
+HFN 4-76 p113 +RR 3-76 p63
+-MT 7-76 p578

2268 Sonatas, organ, nos. 4-5, 7-8, 10, 13-16, 18. Robert Munns, Roger Fisher, Timothy Farrell, Conrad Eden, org. Vista VPS 1016/20 (5).
+Gr 4-76 p1639 +MU 7-77 p15
+HFN 4-76 p113 +-RR 6-76 p74
+-MT 7-76 p578

2269 Sonata, organ, no. 6, op. 119, E flat minor. Sonata, organ, no. 11, op. 148, D minor. Robert Munns, Roger Fisher, org. Vista VPS 1012.
+Gr 6-75 p69 +MU 7-77 p15
++HFN 6-75 p91 +RR 7-75 p53
++MT 7-76 p578

Sonata, organ, no. 9, op. 142, B flat minor. cf Sonata, organ, no. 2, op. 63, A flat major.

Sonata, organ, no. 11, op. 148, D minor. cf Sonata, organ, no. 6, op. 119, E flat minor.

2270 Sonata, organ, no. 12, op. 154, D flat major. Sonata, organ, no. 17, op. 181, B major. Roger Fisher, Timothy Farrell, org. Vista VPS 1014.

+Gr 10-75 p658 +MU 7-77 p15
+HFN 12-75 p163 +RR 12-75 p84
++MT 7-76 p578

Sonata, organ, no. 17, op. 181, B major. cf Sonata, organ, no. 12, op. 154, D flat major.

Sonata, organ, no. 19, op. 192, G minor. cf Sonata, organ, no. 3, op. 88, G major.

Sonata, organ, no. 20, op. 196, A major. cf Sonata, organ, no. 1, op. 27, C minor.

Sonata, violoncello, op. 92, C major. cf GRIEG: Sonata, violoncello and piano, op. 36, A minor.

RHENE-BATON (Baton, Rene)

Heures d'ete. cf HMV RLS 716.

RHODES, Philip

Divertimento, small orchestra. cf KOLB: Looking for Claudio.

RIBAYAZ, Lucas Ruiz de

Hachas. cf RCA RL 2-5099.

RICHAFORT, Jean

De mon triste deplaisir. cf Argo ZK 24.

RICHARD I, King of England

Je nus hons pris. cf Argo D40D3.

Je nus hons pris. cf Enigma VAR 1020.

RICHARDSON, Clive

Beachcomber. cf HMV SQ CSD 3781.

RICHTER, Franz (Frantisek)

Concerto, flute and strings, D major. cf BENDA: Concerto, flute and strings, E minor.

Concerto, trumpet, D major. cf GRAUPNER: Concerto, trumpet, D major.

Sonata, harpsichord, flute and violoncello. cf BACH, J. C. F.: Sonata, fortepiano, flute and violoncello, D major.

RIDOUT, Alan

Fall fair. cf Citadel CT 6011.

RIEGGER, Wallingford

New and old: The twelve tones, The tritone, Seven times seven, Tone clusters, Twelve upside down, Fourths and fourths. cf Vox SVBX 5303.

Who can revoke. cf Vox SVBX 5353.

RIEPE, Russell

Studies on flight (3). cf ETLER: Sonata, clarinet and piano.

RIES, Ferdinand

La capriccioso. cf Discopaedia MB 1013.

RIETI, Vittorio

2271 Concerto, harpsichord and orchestra. Partita, harpsichord, flute, oboe and string quartet. Sylvia Marlow, hpd; Samuel Baron, flt; Ronald Roseman, ob; Charles Libove, Anahid Ajemian, vln; Harry Zaratzian, vla; Charles McCracken, vlc; Chamber Orchestra; Samuel Baron. CRI SD 312. (Reissue from Decca)

+-RR 2-77 p57 ++St 3-74 p118
++SFC 2-17-74 p28

2272 Conundrum, ballet suite. Sestetto pro Gemini. Second Avenue waltzes. Elda Beretta, Maria Madini-Moretti, pno; Harkness

Ballet Orchestra, Gemini Ensemble; Jorge Mester. Serenus SRS 12073.
+St 12-77 p155
Incisioni. cf Crystal S 204.
Partita, harpsichord, flute, oboe and string quartet. cf Concerto, harpsichord and orchestra.
Second Avenue waltzes. cf Conundrum, ballet suite.
Sestetto pro Gemini. cf Conundrum, ballet suite.

RIETZ, Julius
Concert overture, op. 7. cf BRUCH: Symphony, no. 2, op. 36, F minor.
Konzertstuck, oboe, op. 33, F minor. cf BELLINI: Concerto, oboe, E flat major.

RIMMER
Rule Britannia. cf Grosvenor GRS 1052.

RIMSKY-KORSAKOV, Nikolai
Capriccio espagnol, op. 34. cf BORODIN: Prince Igor: Polovtsian dances (DG 2536 379).
Capriccio espagnol, op. 34. cf BORODIN: Prince Igor: Polovtsian dances (HMV 7006).
Capriccio espagnol, op. 34. cf CHABRIER: Espana.
Capriccio espagnol, op. 34. cf MUSSORGSKY: A night on the bare mountain.
Capriccio espagnol, op. 34. cf Seraphim S 60277.
Le coq d'or (The golden cockerel): Hymne au soleil. cf Columbia 34294.
The golden cockerel: Suite. cf BORODIN: In the Steppes of Central Asia.
Le coq d'or: Wedding march (Bridal procession). cf RCA CRL 3-2026.
Fantasy on Russian themes, op. 33. cf ARENSKY: Concerto, violin op. 54, A minor.
Kashchei, the deathless: The night descends. cf Columbia/Melodiya M 33931.
The legend of Sadko, op. 5. cf BALAKIREV: King Lear: Overture.
The legend of Sadko, op. 5. cf GLIERE: The red poppy, op. 70: Ballet suite.
The legend of Sadko, op. 5: Song of India. cf Columbia 34294.
The legend of Sadko, op. 5: Chanson Indoue. cf RCA TVM 1-7203.
The legend of Sadko, op. 5: Chanson hindoue. cf HMV RLS 719.
Maid of Pskov: Overture. cf CBS 73589.

2273 May night. Olga Pastushenko, Tamara Antipova, s; Lyudmilla Sapegina, Nina Derbina, ms; Anna Matyushina, Lyuziya Rashkovets, con; Konstantin Lisovsky, Yuri Yelnikov, t; Ivan Budrin, bs-bar; Alexei Krivchenya, Gennady Troitsky, bs; MRSO and Chorus; Vladimir Fedoseyev. DG 2709 063 (3).

+ARG 6-77 p44	+-NR 7-77 p10
++FF 11-77 p47	+OR 9/10-77 p28
+-Gr 11-76 p873	+-RR 10-76 p30
+-HF 9-77 p105	+SFC 5-8-77 p46
+-HFN 11-76 p165	+SR 7-9-77 p48
+MJ 7-77 p68	++St 7-77 p90

May night: Oveture. cf BORODIN: In the Steppes of Central Asia.
Quintet, piano, B flat major. cf BERWALD: Quartet, piano, E flat major.
The rose and the nightingale. cf Columbia 34294.
Russian Easter festival overture, op. 36. cf BORODIN: In the Steppes of Central Asia (Philips 6530 022).

Russian Easter festival overture, op. 36. cf BORODIN: In the Steppes of Central Asia (Quintessence PMC 7026).
Russian Easter festival overture, op. 36. cf BORODIN: Prince Igor: Polovtsian dances (DG 2536 379).
Russian Easter festival overture, op. 36. cf BORODIN: Prince Igor: Polovtsian dances (HMV 7006).

2274 Scheherazade, op. 35. PhO; Lovro von Matacic. Classics for Pleasure SIT 60042 Tape (c) TC SIT 60042.
+Gr 2-77 p1329 +HRN 12-76 p153 tape

2275 Scheherazade, op. 35. RPO; Thomas Beecham. HMV SXLP 30253 Tape (c) TC SXLP 30253. (Reissue from ASD 251)
++Gr 10-77 p635 +-HFN 10-77 p165
++Gr 11-77 p899 ++RR 10-77 p58

2276 Scheherazade, op. 35. Minneapolis Symphony Orchestra; Antal Dorati. Philips 6547 028.
-Gr 2-77 p1329

2277 Scheherazade, op. 35. Erich Gruenberg, vln; RPO; Leopold Stokowski. RCA ARL 1-1182 Tape (c) ARK 1-1182 (ct) ARS 1-1182.
+ARG 12-76 p39 +HFN 8-77 p99 tape
++Gr 4-77 p1556 +MJ 12-76 p28
/Gr 8-77 p349 tape +-NR 10-76 p3
+-HF 12-76 p104 +RR 7-77 p59
++HFN 7-77 p119 +St 2-77 p120

2278 Scheherazade, op. 35. St. Louis Symphony Orchestra; Jerzy Semkov. Turnabout (Q) QTVS 34667 Tape (c) CT 2136.
+-NR 6-77 p2 /St 8-77 p122
++SFC 5-15-77 p50

Scheherazade, op. 35. cf BORODIN: Prince Igor: Polovtsian dances (Decca SDD 496).
Scheherazade, op. 35, excerpt. cf Decca SPA 491.
Scheherazade, op. 35: Festival of Bagdad; The sea; The shipwreck. cf BEETHOVEN: Egmont, op. 84: Overture.
Skazka, op. 29. cf BALAKIREV: Russia.

2279 The snow maiden. Valentina Sokolik, Lidiya Zakharenko, s; Irina Arkhipova, Nina Derbina, Anna Matushina, ms; Alexander Arkhipov, Yuri Yelnikov, Anton Grigoriev, Vladimir Ermakov, t; Alexander Moksyakov, bar; Alexander Vedernikov, Ivan Budrin, Vladimir Matorin, Vladimir Makhov, bs; MRSO and Chorus; Vladimir Fedoseyev. HMV Melodiya SLS 5102 (4).
++Gr 11-77 p893

2280 The snow maiden: Aria of spring (Prologue and Act 4); Song of Lel (Acts 1-3). The Tsar's bride: Scene and duet of Lyubasha and Gryanznoi; Intermezzo and scene of Lyubasha and Bomelius. Irina Arkhipova, ms; Yevgeni Nechipailo, bar; Georgi Schulpin, t; Bolshoi Theatre Orchestra and Chorus; Aleksandr Melik-Pashayev. Westminster WGS 8333. (Reissue from Melodiya C 0799-80)
+ARG 4-77 p28 +OR 6/7-77 p35
+-NR 4-77 p10

The snow maiden: Dance of the tumblers. cf CBS 61781.
The tale of the Tsar Sultan: Flight of the bumblebee. cf Angel S 37219.
The tale of the Tsar Sultan: Flight of the bumblebee. cf CBS 61039.
The tale of the Tsar Sultan: Flight of the bumblebee. cf CBS 73589.
The tale of the Tsar Sultan: Flight of the bumblebee. cf Citadel CT 6013.
The tale of the Tsar Sultan: Flight of the bumblebee. cf Connoisseur Society CS 2131.

The tale of the Tsar Sultan: Flight of the bumblebee. cf Discopaedia MB 1014.
The tale of the Tsar Sultan: Flight of the bumblebee. cf Enigma VAR 1025.
The tale of the Tsar Sultan: Flight of the bumblebee. cf HMV RLS 723.
The tale of the Tsar Sultan: Flight of the bumblebee. cf HMV SQ ASD 3283.
The tale of the Tsar Sultan: Flight of the bumblebee (arr. and orch. Gerhardt). cf RCA LRL 1-5094.
The Tsar's bride: Lyubasha's aria. cf Columbia/Melodiya M 33931.
The Tsar's bride: Scene and duet of Lyubasha and Gryanznoi; Intermezzo and scene of Lyubasha and Bomelius. cf The snow Maiden: Aria of spring (Prologue and Act 4); Song of Lel.

RITTER, Christian
Sonatina. cf BIS LP 27.

ROBB, John Donald
Dialogue. cf BEHRENS: The feast of life.

ROBERTS
Prelude and trumpetings. cf Delos FY 025.

ROBERTS, Myron
Pastorale and aviary. cf RC Records RCR 101.

ROBINSON, McNeil
Improvisation on a submitted theme. cf L'Oiseau-Lyre SOL 343.

ROBINSON, Thomas
Lessons for the lute. cf Saga 5420.
The Queenes good night. cf BIS LP 22.
Spanish pavan. cf L'Oiseau-Lyre DSLO 510.
A toy. cf DG 2533 323.

ROCHBERG, George
2281 Duo concertante. Quartet, strings, no. 1. Ricordanza, soliloquy for piano and violoncello. Mark Sokol, vln; Norman Fischer, vlc; George Rochberg, pno; Concord Quartet. CRI SD 337.
+-Gr 3-77 p1454 ++NR 1-76 p7
+HF 5-76 p90 +-RR 3-77 p72
++MQ 10-76 p337 +St 7-76 p111
Quartet, strings, no. 1. cf Duo concertante.
Ricordanza, soliloquy for piano and violoncello. cf Duo concertante.
2282 Songs in praise of Krishna. Neva Pilgrim, s; George Rochberg, pno. CRI SD 360.
+-HFN 12-77 p177 +NR 4-77 p12
+MJ 3-77 p46 ++St 10-77 p144

RODBY, John
Concerto for 29. cf RATHBURN: The metamorphic ten.

RODGERS, Richard
Slaughter on tenth avenue. cf EMI TWOX 1058.
Victory at sea. cf EMI TWOX 1058.

RODRIGO, Joaquin
2283 Concierto de Aranjuez, guitar and orchestra. Fantasia para un gentilhombre. Angel Romero, gtr; LSO; Andre Previn. HMV SQ ASD 3415. (also Angel S 37440 Tape (c) 4XS 37440)
+Gr 12-77 p1089
Concierto de Aranjuez, guitar and orchestra. cf BACARISSE: Concertino, guitar and orchestra, op. 72, A minor.
Concierto de Aranjuez, guitar and orchestra. cf BERKELEY: Concerto, guitar (RCA 1-1181).

Concierto de Aranjuez, guitar and orchestra. cf Classics for Pleasure CFP 40012.
Concierto de Aranjuez, guitar and orchestra. cf HMV SLS 5068.
Concierto de Aranjuez, guitar and orchestra: Adagio. cf Philips 6747 327.
Concierto madrigal. cf GIULIANI: Concerto, guitar, op. 30, A major (Philips 6500 918).
Concierto madrigal. cf GIULIANI: Concerto, guitar, op. 30, A major (Philips 7328 013).
En los trigales. cf RCA RK 1-1735.
Fandango. cf Angel S 36094.
Fandango. cf Enigma VAR 1015.
Fantasia para un gentilhombre. cf Concierto de Aranjuez, guitar and orchestra.
Fantasia para un gentilhombre. cf GIULIANI: Introduction, theme with variations and polonaise, op. 65.
Pequenas sevillanas. cf Enigma VAR 1015.
Ya se van los pastores. cf Enigma VAR 1015.

RODRIGUEZ, Xavier Robert
Variations, violin and piano. cf Orion ORS 76212.

ROGALSKI
Roumanian dances (2). cf Westminster WGS 8338.

ROLLA, Alessandro
Concerto, violin, A major. cf PAGANINI: Sonata, viola, op. 35, C minor.

ROLLIN, Robert
Reflections on ruin by the sea. cf PALMER: Sonata, trumpet, B flat major.

ROMAN, Johann Helmich
2284 Drottningholmsmusiken. Symphony, no. 16, D major. Symphony, no. 20, E minor. Drottingholm Chamber Orchestra; Stig Westerberg. Swedish Society SLT 33140.
+RR 8-76 p52 +St 3-77 p140
Symphony, no. 16, D major. cf Drottningholmsmusiken.
Symphony, no. 20, E minor. cf Drottningholmsmusiken.
Symphony, no. 20, E minor. cf LARSSON: Concertino, op. 45, no. 9.

ROMANI, Equitus
Exercitium. cf L'Oiseau-Lyre DSLO 510.
ROMANTIC MUSIC FOR HARP ENSEMBLE AND SOLO HARP. cf Musical Heritage Society MHS 3611.

ROMBERG, Bernhard
Concerto, flute, op. 30, B minor. cf MOLIQUE: Concerto, flute, op. 69, B minor.

RONALD, Landon
Songs: Away on the hill; Down in the forest; Sounds of earth. cf HMV RLS 719.

RONCALLI, Lodovico
Suite, G major. cf DG 2530 561.

RONTANI, Raffaello
Nerinda bella. cf L'Oiseau-Lyre 12BB 203/6.

ROPARTZ, Joseph Guy
Piece, E flat minor. cf Boston Brass BB 1001.
Prelude funebre. cf Calliope CAL 1922/4.

RORE, Cipriano de
A la dolc'ombra. cf Telefunken AW 6-42033.
Contrapunto sopra "Non mi toglia il ben mio". cf Saga 5438.
De la belle contrade. cf L'Oiseau-Lyre 12BB 203/6.

Pero piu ferm. cf Telefunken AW 6-42033.

ROREM, Ned

Book of hours. cf MARTINU: Trio, flute, violoncello and piano.

ROSA

Quand on aime. cf Rubini GV 57.

ROSELL, Lars-Erik

Poem in the dark. cf BIS LP 32.

ROSENBERG, Hilding

2285 Concerto, orchestra, no. 3. HERMANSON: Appell I-IV, op. 10. In nuce, op. 7. Invoco, op. 4. Mircea Saulesco, vln; Bjorn Sjorgren, vla; Ake Olofsson, vlc; Swedish Radio Symphony Orchestra, South West German Symphony Orchestra; Stig Westerburg, Bruno Maderna. Swedish Society SLT 33215.
+-RR 7-77 p59

Quartet, strings, no. 12. cf BENTZON: Quartet, strings, no. 8, op. 228.

ROSENMAN, Leonard

Chamber music, no. 2. cf KRAFT: The imagistes.

ROSS, Walter

Concerto, trombone and orchestra. cf PENN: Ultra mensuram.

Fancy dances, 3 bass tubas. cf Crystal S 221.

Prelude, fugue and big apple, bass trombone and tape. cf PENN: Ultra mensuram.

ROSSETER, Philip

And would you see my msitress face. cf Saga 5425.

What then is love. cf Enigma VAR 1023.

What then is love but mourning. cf Abbey LPB 712.

ROSSI, Michel Angelo

Toccata, no. 7, D major. cf FRESCOBALDI: Toccatas, no. 1, G major; no. 9, F major.

ROSSI, Salomone

Gagliarda "Norsina" a 5. cf Telefunken AW 6-42033.

Sinfonia grave a 5. cf Telefunken AW 6-42033.

ROSSINI, Gioacchino

L'Assedio di Corinto (The siege of Corinth): Cielo che diverro... Si ferite, il chieggo, il merto...Dah soggiorno degli estinti... No non piu spero, oh Dio. cf BELLINI: I Capuleti ed i Montecchi: Eccomi in lieta vesta...Oh quante volte.

Le siege de Corinthe: Ballet music. cf Guillaume Tell: Ballet music.

L'Assedio di Corinto: Overture. cf Overtures (DG 2530 559).

The siege of Corinth: Overture. cf Overtures (Everest SDBR 3396).

2286 Il barbiere di Siviglia. Giulietta Simionato, ms; Giuseppe Taddei, bar; Luigi Infantino. Cetra LPS 3211. (Reissue from OLPC 1211)
+-Gr 7-77 p222

2287 Il barbiere di Siviglia. Giulietta Simionato, Rina Cavallari, ms; Alvinio Misciano, Giuseppe Zampieri, t; Ettore Bastianini, Arturo la Porta, bar; Fernando Corena, Cesare Siepi, bs; Maggio Musical Fiorentino Orchestra and Chorus; Alberto Erede. Decca D38D3 (3). (Reissue from LXT 5283/5)
+-Gr 9-77 p487 +-RR 6-77 p37
+-HFN 7-77 p119

Il barbiere di Siviglia (The barber of Seville): Largo al factotum. cf Decca SPA 491.

The barber of Seville: Overture. cf Overtures (Decca PFS 4386).

Il barbiere di Siviglia: Overture. cf Overtures (DG 2530 559).

Il barbiere di Siviglia: Overture. cf Overtures (Everest SBDR 3396).

Il barbiere di Siviglia: Overture and storm music. cf Overtures (DG 2548 171).
The barber of Seville: There is a voice within my heart. cf HMV HLM 7066.
The barber of Seville: Una voce poco fa. cf BELLINI: I puritani: Qui la voce...Vien diletto.
Il barbiere di Siviglia: Una voce poco fa. cf MOZART: La clemenza di Tito, K 621: Non piu di fiori; Parto, parto.
Il barbiere di Siviglia: Una voce poco fa. cf VERDI: Rigoletto.
Il barbiere di Siviglia: Una voce poca fa. cf Court Opera Classics CO 342.
Il barbiere di Siviglia: Una voce poco fa. cf Decca SPA 450.
Il barbiere di Siviglia: Una voce poca fa. cf HMV SLS 5057.
Il barbiere di Siviglia: Una voce poco fa. cf HMV SLS 5104.
Il barbiere di Siviglia: Una voce poco fa; Contro un cor che accende amore...Cara immagine ridente. cf BELLINI: I Capuleti ed i Montecchi: Eccomi in lieta vesta...Oh quante volte.

2288 La boutique fantasque: Ballet suite. Suite Rossiniana. (arr. Respighi) RPO; Anatole Fistoulari. Decca PFS 4407.
+-Gr 12-77 p1089 +RR 11-77 p68
++HFN 10-77 p161

Cambiale di matrimonio: Overture. cf BERLIOZ: Le Corsaire, op. 21.

2289 La cenerentola. Giulietta Simionato, ms; Orchestra. Cetra LPS 3208. (Reissue from OLPC 1208)
+Gr 7-77 p222

La cenerentola: Nacqui all'affano. cf DONIZETTI: L'Elisir d'amore: Prendi, Per me sei libero.
La cenerentola: Nacqui all'affano. cf MOZART: La clemenza di Tito, K 621: Non piu di fiori; Parto, parto.
La cenerentola: Nacqui all'affano...No no tergete il ciglio. cf Seraphim M 60291.
La cenerentola: Overture. cf Overtures (DG 2530 559).
La cenerentola: Overture. cf Overtures (Everest SDBR 3396).
Cinderella: Overture. cf Overtures (Decca PFS 4386).
La danza. cf Connoisseur Society CS 2131.
Duet, violoncello and double bass. cf PAGANINI: Sonata, viola, op. 35, C minor.

2290 Elisabetta Regina d'Inghilterra. Montserrat Caballe, Valerie Masterson, s; Rosanne Creffield, ms; Jose Carreras, Ugo Benelli, Neil Jenkins, t; Ambrosian Singers; LSO; Gianfranco Masini. Philips 6703 067 (3).
+ARG 2-77 p39 +-ON 1-15-77 p33
+Gr 9-76 p467 +OR 6/7-77 p31
++HF 2-77 p106 +RR 9-76 p22
+HFN 9-76 p115 +-SFC 11-21-76 p35
+MJ 2-77 p30 ++St 3-77 p140
+NR 2-77 p11

Elisabetta, Regina d'Inghilterra, excerpts. cf DONIZETTI: Lucia di Lammermoor, excerpts.
La gazza ladra (The thieving magpie): Il mio piano e preparato. cf Decca ECS 811.
La gazza ladra: Overture. cf Overtures (Classics for Pleasure CFP 40077).
The thieving magpie: Overture. cf Overtures (Decca PFS 4386).
La gazza ladra: Overture. cf Overtures (DG 2530 559).
La gazza ladra: Overture. cf Overtures (DG 2548 171).
La gazza ladra: Overture. cf Overtures (Everest SDBR 3396).

La gazza ladra: Overture. cf Overtures (Seraphim S 60282).
La gazza ladra: Overture. cf BERLIOZ: Le Corsaire, op. 21.
The thieving magpie: Overture. cf Decca SXL 6782.

2291 Guillaume Tell: Ballet Music. Moise: Ballet music. Otello: Ballet music. Le siege de Corinthe: Ballet music. Monte Carlo Opera Orchestra; Antonio de Almeida. Philips 6780 027 (2).
-Gr 12-76 p1001 /NR 3-77 p6
+-HF 3-77 p110 -ON 1-15-77 p33
+HFN 11-76 p167 ++RR 11-76 p77
+MJ 2-77 p30 +-SFC 11-21-76 p35
-MT 5-77 p402

Guglielmo Tell: Ballet. cf DELIBES: Coppelia: Suite, Act 1.
Guillaume Tell (William Tell): Ballet music. cf DELIBES: Coppelia: Ballet music.
William Tell: Ballet music. cf Pye Golden Hour GH 643.
Guillaume Tell: Overture. cf Overtures (Classics for Pleasure CFP 40077).
William Tell: Overture. cf Overtures (Decca PFS 4386).
Guillaume Tell: Overture. cf Overtures (DG 2548 171).
Guglielmo Tell: Overture. cf Overtures (Everest SDBR 3396).
Guglielmo Tell: Overture. cf Overtures (Seraphim S 60282).
William Tell: Overture. cf Connoisseur Society CS 2131.
William Tell: Overture. cf HMV SLS 5073.
Guglielmo Tell: Overture. cf Orion ORS 76247.
William Tell: Overture. cf Pye PCNHX 6.
William Tell: Overture. cf Quintessence PMC 7013.
Guglielmo Tell: S'allontanano alfin...Selva opaca. cf DONIZETTI: L'Elisir d'amore: Prende, Per me sei libero.
L'Inganno felice: Overture. cf Overtures (Everest SDBR 3396).
Introduction and variations, clarinet and orchestra. cf Rediffusion 15-57.
L'Invito. cf L'Oiseau-Lyre SOL 345.
L'Italiana in Algeri (The Italian girl in Algiers): Overture. cf Overtures (Classics for Pleasure CFP 40077).
The Italian girl in Algiers: Overture. cf Overtures (Decca PFS 4386).
L'Italiana in Algeri: Overture. cf Overtures (DG 2530 559).
L'Italiana in Algeri: Overture. cf Overtures (Everest SDBR 3396).
L'Italiana in Algeri: Overture. cf Overtures (Seraphim S 60282).
Moise: Ballet music. cf Guillaume Tell: Ballet music.
Mose in Egitto: Dal duo stellato soglio. cf Seraphim S 60275.
Otello: Assisa a pie d'un salice. cf MOZART: La clemenza di Tito, K 621: Non piu di fiori; Parto, parto.
Otello: Ballet music. cf Guillaume Tell: Ballet music.

2292 Overtures: La gazza ladra. Guillaume Tell. L'Italiana in Algeri. Semiramide. Il Signor Bruschino. RPO; Colin Davis. Classics for Pleasure CFP 40077 Tape (c) TC CFP 40077. (Reissues)
-Gr 2-77 p1325 tape +RR 5-74 p39
+HFN 12-76 p153 tape

2293 Overtures: The barber of Seville. Cinderella. Italian girl in Algiers. Semiramide. The thieving magpie. William Tell. RPO; Carlos Paita. Decca PFS 4386. (also London SPC 21164)
++Gr 12-76 p1071 ++NR 10-77 p4
+HFN 2-77 p129 +RR 1-77 p58

2294 Overtures: L'Assedio di Corinto. Il barbiere di Siviglia. La cenerentola. La gazza ladra. L'Italiana in Algeri. Il Signor Bruschino. LSO; Claudio Abbado. DG 2530 559 Tape (c) 3300 497.

+Gr 1-76 p1244 +-ON 3-12-77 p40
+HFN 1-76 p117 +RR 1-76 p37
+HFN 2-76 p116 tape +-RR 9-76 p95 tape
++MJ 5-76 p28 ++SFC 4-4-76 p34
-NR 5-76 p1 +St 7-76 p111

2295 Overtures: Il barbiere di Siviglia: Overture and storm music. La gazza ladra. Guillaume Tell. La scala de seta. Semiramide. Rome Opera Orchestra; Tullio Serafin. DG 2548 171 Tape (c) 3348 171. (Reissue from 136395)
++Gr 11-75 p840 +HFN 10-75 p152
+Gr 2-77 p1325 tape

2296 Overtures: Il barbiere di Siviglia. La cenerentola. Guglielmo Tell. La gazza ladra. L'Inganno felice. L'Italia in Algeri. La scala di seta. Semiramide. Siege of Corinth. Il Signor Bruschino. Tancredi. Santa Cecilia Orchestra; Fernando Previtali. Everest SDBR 3396.
-NR 3-77 p6

2297 Overtures: La gazza ladra. Guglielmo Tell. L'Italiana in Algeri. Il Signor Bruschino. Semiramide. RPO; Colin Davis. Seraphim S 60282.
+-NR 10-77 p4

La scala di seta: Overture. cf Overtures (DG 2548 171).
La scala de seta: Overture. cf Overtures (Everest SDBR 3396).
Semiramide: Bel raggio lusinghier. cf DONIZETTI: L'Elisir d'amore: Prendi, Per me sei libero.
Semiramide: Bel raggio. cf Decca D65D3.
Semiramide: Overture. cf Overtures (Classics for Pleasure CFP 40077).
Semiramide: Overture. cf Overtures (Decca PFS 4386).
Semiramide: Overture. cf Overtures (DG 2548 171).
Semiramide: Overture. cf Overtures (Everest SDBR 3396).
Semiramide: Overture. cf Overtures (Seraphim S 60282).
Il Signor Bruschino: Overture. cf Overtures (Classics for Pleasure CFP 40077).
Il Signor Bruschino: Overture. cf Overtures (DG 2530 559).
Il Signor Bruschino: Overture. cf Overtures (Everest SDBR 3396).
Il Signor Bruschino: Overture. cf Overtures (Seraphim S 60282).
Il Signor Bruschino: Overture. cf Philips 6747 327.
Sonata, strings, nos. 1-6. cf DONIZETTI: Quartet, strings, D major.
Sonata, 2 violins, violoncello and double bass, no. 3, C major. cf DG 2548 219.
Songs: La danza; L'orgia; La promessa. cf BELLINI: Songs (Cetra LPO 2003).
Songs: La fioraia fiorentina; Mi lagnero tacendo; L'invito; La promesa. cf BELLINI: Songs (Westminster WG 1014).

2298 Stabat mater. Sung-Sook Lee, s; Florence Quivar, ms; Kenneth Riegel, t; Paul Plishka, bs; Cincinnati May Festival Chorus; CnSO; Thomas Schippers. Turnabout (Q) QTVS 34634.
+MJ 3-77 p74 ++SFC 6-27-76 p29
/NR 10-76 p9 ++St 12-76 p146
+-NYT 11-27-77 pD15

Stabat mater: Cumus animam. cf Decca SXL 6839.
Stabat mater: Inflammatus. cf Club 99-108.
Stabat mater: Pro peccatis. cf Rubini GV 39.
Suite Rossiniana. cf La boutique fantasque: Ballet suite.

2299 Tancredi. Hannah Francis, s; Patricia Price, Elisabeth Stokes, con; Keith Lewis, Peter Jeffes, t; Tom McDonnell, bar; London

Voices; Centre d'Action Musicale de l'Ouest; John Perras. Arion ARN 338 010 (3).
+-Gr 9-77 p487 -RR 9-77 p42
Tancredi: Overture. cf Overtures (Everest SDBR 3396).
2300 Il Turco in Italia. Maria Callas, s; Jolanda Gardino, ms; Nicolai Gedda, Piero de Palma, t; Mariano Stabile, bar; Nicola Rossi-Lemeni, Franco Calabrese, bs; La Scala Orchestra and Chorus; Gianandrea Gavazzeni. Seraphim IB 6095 (2). (Reissue from Angel S 3535)
++ARG 3-77 p34 +ON 1-15-77 p33
+HF 5-77 p89
Il Turco in Italia: Non si da follia maggiore. cf HMV SLS 5104.

ROTA, Nino
2301 Il cappello di Paglia di Firenze (The Italian straw hat). Daniela Mazzuccato Meneghini, Edith Martelli, s; Viorica Cortez, ms; Ugo Benelli, Mario Carlin, Angelo Mercuriali, Pier Francesco Poli, Sergio Tedesco, t; Mario Basiola, Giorgio Zancanaro, bar; Alfred Mariotti, Enrico Campi, bs; Rome Symphony Orchestra and Chorus; Nino Rota. RCA TRL 2-1153 (2).
+-Gr 3-77 p1449 +-RR 8-77 p37
++HFN 10-77 p161

ROTOLI
Fior che langue. cf Club 99-96.
Fior che langue. cf Rubini RS 301.
Songs: La gondola nera; Mia sposa sara la mia bandiera. cf Rubini GV 34.

ROUSE, Christopher
Subjectives VIII. cf Golden Crest RE 7068.

ROUSSEL, Albert
2302 Bacchus et Ariane, op. 43: Suite, no. 2. Psalm, no. 80, op. 37. John Mitchinson, t; Stephane Caillat Chorus; Orchestre de Paris; Serge Baudo. Connoisseur Society CS 2124.
+ARG 5-77 p31 +SFC 6-12-77 p41
+NR 8-77 p8 ++St 7-77 p119
Pieces, piano, op. 49. cf Vox SVBX 5483.
Pieces, piano, op. 49 (3). cf POULENC: Melancolie.
Poemes de Ronsard (2). cf BIS LP 45.
Psalm, no. 80, op. 37. cf Bacchus et Ariane, op. 43: Suite, no. 2.
Segovia, op. 29. cf RCA RK 1-1735.
Sonatine, op. 16. cf POULENC: Melancolie.
Songs: Adieu; A flower given to my daughter; Jazz dans la nuit; Light; Melodies, op. 20; Poemes chinois, op. 12; Poemes chinois, op. 35; Odes anacreontiques, nos. 1, 5; Odelette. cf ENESCO: Songs, op. 15.
Suite, op. 14: Bourree. cf Vox SVBX 5483.
Symphony, no. 3, op. 42, G minor. cf DUKAS: La peri.

ROUTH, Francis John
Sonatina, op. 9. cf RC Records RCR 101.
ROYAL MUSIC FROM ST. PAULS. cf Guild GRSP 701.

ROYLLART
Rex Karole. cf DG 2710 019.

ROZSAVOLGYI, Mark
Czardas. cf CSERMAK: Hungarian dances.
First Hungarian round dance. cf CSERMAK: Hungarian dances.

RUBBRA, Edmund
Concerto, piano, no. 2, op. 85, G major. cf HMV SLS 5080.
Fanfare for Europe, op. 142. cf RCA RL 2-5081.

Improvisation, op. 89. cf BRITTEN: Concerto, violin, op. 15, D minor.
Improvisations on virginal pieces by Giles Farnaby, op. 50. cf Symphony, no. 10, op. 145.
Meditazioni sopra "Coeurs desoles". cf HMV SQ CSD 3781.
Salutation, op. 82. cf RCA GL 2-5062.

2303 Symphony, no. 7, op. 88, C major. VAUGHAN WILLIAMS: Fantasia on a theme by Thomas Tallis. LPO; Adrian Boult. Musical Heritage Society MHS 1397.
+FF 9-77 p43 ++St 2-73 p126

2304 Symphony, no. 10, op. 145. Improvisations on virginal pieces by Giles Farnaby, op. 50. A tribute, op. 56. Bournemouth Sinfonietta; Hans-Hubert Schonzler. RCA RL 2-5027. (also RCA LRL 1-5137)
+Gr 2-77 p1284 +MT 9-77 p737
+-HFN 4-77 p143 +RR 2-77 p57

A tribute, op. 56. cf Symphony, no. 10, op. 145.
Variations on "The shining river", op. 101. cf BALL: Sinfonietta, brass band.

RUBENS

The Maggie Teyte encore song "Ladies and Gentlemen". cf Decca SDD 507.

RUBINSTEIN, Anton

Concerto, piano, no. 4, op. 70, D minor. cf International Piano Library IPL 5001/2.
The demon: Sweet friends; The night is warm and quiet. cf Rubini GV 26.
Etude, op. 23, no. 2. cf Pye PCNH 9.
Melody, F major. cf Inner City IC 1006.
Melody, F major. cf Rediffusion 15-56.
Melody, op. 3, no. 1, F major. cf International Piano LIbrary IPL 103.
Melody, op. 3, no. 1, F major. cf International Piano Library IPL 113.
Prelude and fugue, op. 53, no. 2. cf International Piano Library IPL 102.

2305 Sonatas, violin (4). Robert Murray, vln; Daniel Graham, pno. Musical Heritage Society MHS 3385/6 (2).
+-Audio 12-77 p133

Sonata, violoncello and piano, no. 2, op. 39, G major. cf GERNSHEIM: Sonata, violoncello and piano, no. 1, op. 12, D minor.
Songs: Night. cf Club 99-96.
Songs: Night. cf Rubini RS 301.

RUDHYAR, Dane

2306 Tetragram, nos. 1-3. WEIGL: Night fantasies. Dwight Peltzer, pno. Serenus SRS 12072.
+NR 11-77 p13

2307 Tetragrams, nos. 4 and 5. Transmutation. Marcia Mikulak, pno. CRI SD 372.
+NR 11-77 p13

Transmutation. cf Tetragrams, nos. 4 and 5.

RUE, Pierre de la

Missa Ave santissima Maria: Sanctus. cf HMV SLS 5049.
Pour ung jamais. cf L'Oiseau-Lyre 12BB 203/6.

RUGGLES, Carl

Evocations. cf Vox SVBX 5303.
Men and mountains. cf MOZART: Symphony, no. 35, K 385, D major.

RUIZ-PIPO, Antonio
Cancion y danza. cf L'Oiseau-Lyre SOL 349.
Cancion y danza. cf RCA RL 2-5099.
Cancion y danza, no. 1. cf Swedish Society SLT 33205.
Endecha. cf RCA RL 2-5099.
Estancias. cf DG 2530 802.
Tablas para guitarra y orquesta. cf OHANA: Tres graficos para guitarra y orquesta.
RUSH, Oren
Hexahedron. cf CUSTER: Found objects, no. 7.
2308 A little traveling music. Oh Susanna. Soft music, hard music. Dwight Peltzer, pno. Serenus SRS 12070.
+NR 12-77 p12
Oh Susanna. cf A little traveling music.
Soft music, hard music. cf A little traveling music.
RUSSEL, Armand
Suite concertante. cf BACH: Sonata, flute, E flat major.
RUSSELL
By appointment: White roses. cf HMV RLS 716.
RUSSIAN LITURGICAL CHANTS. cf Philips 6504 135.
RUSSIAN MARCHES. cf HMV CSD 3782.
RUSSO, William
Street music, op. 65. cf GERSHWIN: An American in Paris.
RYBA, Jan Yakub
2309 Czech Christmas mass. Pastorella. Jaroslava Vymazalova, Helena Tattermuschova, s; Marie Mrazova, con; Beno Blachut, t; Zdenek Kroupa, bs; Czech Philharmonic Chorus; PSO; Vaclav Smetacek. Supraphon 50768.
+-ARG 12-77 p55 ++RR 12-75 p91
Pastorella. cf Czech Christmas mass.
RYTERBAND, Roman
2310 Sonata, piano, no. 1, D major. Suite polonaise. Vladimir Pleshakov, pno. Orion ORS 76222.
+NR 8-77 p15
Suite polonaise. cf Sonata, piano, no. 1, D major.
RZEWSKI, Frederic
Variations on "No place to go but around". cf BRAXTON: P-JOS.. 4K-D (MIX).
SAEGUSA
Baire's theorem. cf CP 3-5.
SAENGER
Concert miniatures, op. 130, no. 2. cf Discopaedia MB 1013.
SAEVERUD, Harald
Ballade of revolt, op. 22, no. 5. cf BARTOK: Suite, op. 14.
2311 Peer Gynt, op. 28: Suites, nos. 1 and 2. Oslo Philharmonic Orchestra; Militiades Caridis. Phonogram 6507 006.
+RR 7-77 p59
Rondo amoroso, op. 14, no. 7. cf BARTOK: Suite, op. 14.
SAGRERAS, Julio
El colibri. cf London CS 7015.
SAINT-MARTIN, Leonce de
Toccata de la liberation. cf Vista VPS 1029.
SAINT-SAENS, Camille
Africa, op. 89. cf Concerti, piano, nos. 1-5.
Africa, op. 89: Improvised cadenza. cf International Piano Library IPL 117.
Album, op. 72. cf Piano works (Vox 5476/7).

Allegro appassionato, op. 43. cf Works, selections (Vox 5134).
Allegro appassionato, op. 70. cf Piano works (Vox 5476/7).
Ascanio: Ballet music, Adagio and variations. cf RCA LRL 1-5094.
Bacchanale. cf Works, selections (Decca ECS 808).
Bagatelles, op. 3. cf Piano works (Vox 5476/7).
Caprice, violin, op. 52. cf Works, selections (HMV SQ SLS 5103).
Caprice, violin, op. 122. cf Works, selections (Vox 5134).
Caprice andalous, op. 122. cf Works, selections (HMV SQ SLS 5103).
Caprice arabe, op. 96. cf Piano works (Vox 5476/7).
Caprice heroique, op. 106. cf Piano works (Vox 5476/7).

2312 Caprice on Danish and Russian airs, op. 79. Sonata, bassoon and piano, op. 168. Sonata, clarinet and piano, op. 167. Sonata, oboe and piano, op. 166. Paul Freed, pno; John Miller, bsn; Minneapolis Chamber Ensemble. Musical Heritage Society MHS 3324 Tape (r) BCC 3324.
+HF 8-77 p96 tape

Caprice sur des aires de ballet d'Alceste. cf Piano works (Vox 5476/7).
The carnival of the animals. cf FAURE: Masques et bergamasques, op. 112.
The carnival of the animals. cf MOZART: Serenade, no. 13, K 525, G major.
The carnival of the animals. cf POULENC: The story of Babar the little elephant (EMI BRNA 502).
The carnival of the animals. cf POULENC: The story of Babar the little elephant (HMV 7020).
The carnival of the animals. cf PROKOFIEV: Peter and the wolf, op. 67.
Cavatine, op. 144. cf Boston Brass BB 1001.
Les cloches du soir, op. 85. cf Piano works (Vox 5476/7).

2313 Concerti, piano, nos. 1-5. Africa, op. 89. Rapsodie d'Auvergne, op. 73. Wedding cake, op. 76. Gabriel Tacchino, pno; Luxembourg Radio Orchestra; Louis de Froment. Vox SVBX 5143 (3).
+HF 12-77 p96
+-NR 6-77 p6
+NYT 6-5-77 pD19
+SFC 6-19-77 p46

2314 Concerto, piano, no. 1, op. 17, D major. Concerto, piano, no. 5, op. 103, F major. Philippe Entremont, pno; Toulouse Orchestre du Capitole; Michel Plasson. CBS 76532 Tape (c) 40-76532. (also Columbia M 34512 Tape (c) MT 34512)
-Gr 1-77 p1148
+-HF 12-77 p96
+HFN 1-77 p115
+-HFN 5-77 p138 tape
+HFN 9-77 p155 tape
+-NR 6-77 p6
+RR 1-77 p58

2315 Concerto, piano, no. 2, op. 22, G minor. Concerto, piano, no. 5, op. 103, F major. Gabriel Tacchino, pno; Luxembourg Radio Orchestra; Louis de Froment. Candide (Q) QCE 31080.
++Audio 1-77 p85
++HF 7-75 p82
++NR 1-75 p5
+SFC 12-22-74 p20

Concerto, piano, no. 3, op. 29, E flat major: Allegro. cf Piano works (Vox 5476/7).
Concerto, piano, no. 5, op. 103, F major. cf Concerto, piano, no. 1, op. 17, D major.
Concerto, piano, no. 5, op. 103, F major. cf Concerto, piano, no. 2, op. 22, G minor.
Concerti, violin (3). cf Works, selections (Vox 5134).
Concerti, violin, nos. 1-4. cf Works, selections (HMV SLS 5103).
Concerto, violin, no. 1, op. 20, A major. cf Discopaedia MB 1011.

2316 Concerto, violin, no. 3, op. 61, B minor. VIEUXTEMPS: Concerto, violin, no. 5, op. 37, A minor. Kyung-Wha Chung, vln, LSO; Lawrence Foster. Decca SXL 6759 Tape (c) KSXC 6759. (also London 6992 Tape (c) 5-6992)

+-ARG 2-77 p41
+Gr 9-76 p424
+Ha 9-77 p108
+HF 4-77 p111
++HFN 9-76 p129
+-HFN 11-76 p175
+-MJ 4-77 p33
++NR 4-77 p4
+RR 9-76 p68
+-RR 3-77 p99 tape
+St 4-77 p125

Concerto, violin, no. 3, op. 61, B minor. cf CHAUSSON: Poeme, op. 25.
Concerti, violoncello (2). cf Works, selections (Vox 5134).
Concerto, violoncello, no. 1, op. 33, A minor. cf BRAHMS: Concerto, violin and violoncello, op. 102, A minor.
Concerto, violoncello, no. 1, op. 33, A minor. cf FAURE: Elegy, op. 24, C minor.
Danse macabre, op. 40. cf Works, selections (Decca ECS 808).
Danse macabre, op. 40. cf Works, selections (Vox 5144).
Danse macabre, op. 40. cf DG 2584 004.
Le deluge, op. 45: Prelude. cf Works, selections HMV SLS 5103).
Duettino, op. 11. cf Piano works (Vox 5476/7).

2317 Etudes, opp. 52 and 111. Annie d'Arco, pno. Calliope CAL 1858.

+-RR 3-77 p84

Etudes, op. 52. cf Piano works (Vox 5476/7).
Etudes, op. 111. cf Piano works (Vox 5476/7).
Etudes, op. 135. cf Piano works (Vox 5476/7).
Etude, op. 135: Bourree. cf Vox SVBX 5483.
Fantaisie, op. 159, E flat major. cf CHARPENTIER: L'ange a la trompette.
Feuillet d'album, op. 81. cf Piano works (Vox 5476/7).
Feuillet d'album, op. 169. cf Piano works (Vox 5476/7).
Fugues, op. 161. cf Piano works (Vox 5476/7).
Gavotte, op. 23. cf Piano works (Vox 5476/7).
Havanaise, op. 83. cf Works, selections (Decca ECS 808).
Havanaise, op. 83. cf Works, selections (HMV SLS 5103).
Havanaise, op. 83. cf Works, selections (Vox 5134).
Havanaise, op. 83. cf CHAUSSON: Poeme, op. 25.
Introduction and rondo capriccioso, op. 28. cf Works, selections (Decca ECS 808).
Introduction and rondo capriccioso, op. 28. cf Works, selections (HMV SLS 5103).
Introduction and rondo capriccioso, op. 28. cf Works, selections (Vox 5134).
Introduction and rondo capriccioso, op. 28. cf CHAUSSON: Poeme, op. 25.
Le jeunesse d'Hercule, op. 50. cf Works, selections (Vox 5144).
Marche heroique, op. 34. cf Works, selections (Vox 5144).
Marche interalliee, op. 155. cf Piano works (Vox 5476/7).
Mazurkas, opp. 21, 24, 66. cf Piano works (Vox 5476/7).
Menuet et valse, op. 56. cf Piano works (Vox 5476/7).
Morceau de concert, op. 62. cf Works, selections (Vox 5134).
La muse et le poete, op. 132. cf Works, selections (HMV SLS 5103).
Une nuit a Lisbonne, op. 63. cf Piano works (Vox 5476/7).
Pas redouble, op. 86. cf Piano works (Vox 5476/7).
Phaeton, op. 39. cf Works, selections (Vox 5144).

2318 Piano works: Album, op. 72. Concerto, piano, no. 3, op. 29, E flat major: Allegro. Allegro appassionato, op. 70. Bagatelles, op. 3 (6). Caprice arabe, op. 96. Caprice heroique, op. 106. Caprice sur des airs de ballet d'Alceste. Les cloches du soir, op. 85. Duettino, op. 11. Etudes, op. 52 (6). Etudes, op. 111 (6). Etudes, op. 135 (6). Feuillet d'album, op. 81. Feuillet d'album, op. 169. Fugues, op. 161 (6). Gavotte, op. 23. Marche interalliee, op. 155. Mazurkas, opp. 21, 24, 66 (3). Menuet et valse, op. 56. Une nuit a Lisbonne, op. 63. Pas redouble, op. 86. Polonaise, op. 77. Scherzo, op. 87. Souvenir d'Ismailia, op. 100. Souvenir d'Italie, op. 80. Suite, op. 90. Theme varie, op. 97. Valse canariote, op. 88. Valse gaie, op. 139. Valse langoureuse, op. 120. Valse mignonne, op. 104. Valse nonchalante, op. 110. Variations on a theme by Beethoven, op. 35. Marylene Dosse, Annie Petit, pno. Vox SVBX 5476/7 (6).
+-HF 9-75 p79 +SFC 5-15-77 p50
++NR 6-75 p14

Polonaise, op. 77. cf Piano works (Vox 5476/7).

Prelude and fugue, op. 99, no. 3, E flat major. cf FRANCK: Chorale, no. 2, B minor.

Rapsodie d'Auvergne, op. 73. cf Concerti, piano, nos. 1-5.

2319 Requiem, op. 54. Danielle Galland, s; Jeannine Collard, con; Francis Bardot, t; Jacques Villisech, bs; Micheline Lagache, org; Contrepoint Choral Ensemble; ORTF Orchestre Lyrique; Jean-Gabriel Gaussens. RCA AGL 1-1968. (Reissue from SB 6864)
/ARG 2-77 p41 *MU 2-77 p8
+-HF 4-77 p111 +NR 2-77 p9
+MJ 3-77 p74 ++SFC 1-23-77 p37
+ON 2-5-77 p41

Romance, op. 37, D flat major. cf Works, selections (HMV SLS 5103).

Romance, op. 48, C major. cf Works, selections (HMV SLS 5103).

Romance, op. 48, C major. cf Works, selections (Vox 5134).

Romance, op. 67. cf BEETHOVEN: Sonata, horn, op. 17, F major.

Le rouet d'Omphale, op. 31. cf Works, selections (Decca ECS 808).

Le rouet d'Omphale, op. 31. cf Works, selections (Vox 5144).

Samson et Dalila: Aprile foriero, S'apre per te il mio cor. cf Club 99-103.

Samson et Dalila: Aprile foriero, S'apre per te il mio cor. cf Club 99-106.

Samson et Dalila: Arretez mes freres. cf HMV ASD 3302.

Samson et Dalila: Bacchanale. cf CBS 30091.

Samson et Dalila: Gloire a Dagon. cf Club 99-107.

Samson et Dalila: Mon coeur s'ouvre a ta voix. cf Works, selections (Decca ECS 808).

Samson et Dalila: Mon coeur s'ouvre a ta voix; Amour, viens aider. cf Columbia/Melodiya M 33931.

Samson et Dalila: Printemps que commence. cf HMV SLS 5104.

Scherzo, op. 87. cf Piano works (Vox 5476/7).

Septet, op. 65, E flat major. cf MILHAUD: La cheminee du Roi Rene, op. 205.

Sonata, bassoon and piano, op. 168. cf Caprice on Danish and Russian airs, op. 79.

Sonata, clarinet and piano, op. 167. cf Caprice on Danish and Russian airs, op. 79.

Sonata, oboe and piano, op. 166. cf Caprice on Danish and Russian airs, op. 79.

2320 Sonata, violoncello and piano, no. 1, op. 32, C minor. Sonata, violoncello and piano, no. 2, op. 123, D major. Andre Navarra, vlc; Annie d'Arco, pno. Calliope CAL 1818.
+-RR 3-77 p84

Sonata, violoncello and piano, no. 2, op. 123, D major. cf Sonata, violoncello and piano, no. 1, op. 32, C minor.

Songs: L'Attente; Aimons-nous; Pourquoi rester seulette. cf BIZET: Songs (Musical Heritage Society MHS 3433).

Songs: Calme des nuits; Les fleurs et les arbres. cf RCA RL 2-5112.

Souvenir d'Ismailia, op. 100. cf Piano works (Vox 5476/7).

Souvenir d'Italie, op. 80. cf Piano works (Vox 5476/7).

Suite, op. 90. cf Piano works (Vox 5476/7).

Suite algerienne, op. 60: Marche militarie francaise. cf Rediffusion Royale 2015.

The swan. cf Virtuosi VR 7608.

2321 Symphony, no. 1, op. 2, E flat major. Symphony, no. 2, op. 55, A minor. Frankfurt National Radio Orchestra; Eliahu Inbal. Philips 9500 079.
+FF 11-77 p48 +MT 9-77 p738
++Gr 10-76 p607 +-RR 11-76 p77
++HFN 10-76 p175

Symphony, no. 1, op. 2, E flat major. cf Works, selections (Vox 5144).

Symphony, no. 2, op. 55, A minor. cf Symphony, no. 1, op. 2, E flat major (Philips 9500 079).

Symphony, no. 2, op. 55, A minor. cf Works, selections (Vox 5144).

2322 Symphony, no. 3, op. 78, C minor. Gaston Litaize, org; CSO; Daniel Barenboim. DG 2530 619 Tape (c) 3300 619.
++Gr 4-76 p1608 +NR 8-76 p3
++Gr 7-76 p230 tape +-RR 4-76 p53
+HF 10-76 p118 +-RR 4-77 p94 tape
+HF 2-77 p118 tape +SFC 5-23-76 p36
+HFN 4-76 p113 ++St 10-76 p131
+HFN 6-76 p105 tape ++STL 4-11-76 p36

2323 Symphony, no. 3, op. 78, C minor. Christopher Robinson, org; Birmingham City Symphony Orchestra; Louis Fremaux. HMV SQ ESD 7038. (Reissue from Columbia TWO 404)
++Gr 10-77 p638 +RR 11-77 p69
++HFN 10-77 p167

2324 Symphony, no. 3, op. 78, C minor. Anita Priest, org; LAPO; Zubin Mehta. London C 6680. (also Decca SXL 6482 Tape (c) KSXC 6482)
+Gr 1-71 p1160 +RR 4-77 p94 tape
+HF 9-71 p104 ++SFC 10-3-71 p37
+HFN 11-76 p175 tape +St 12-71 p94
+NR 9-71 p65

2325 Symphony, no. 3, op. 78, C minor. Wedding cake, op. 76. Daniel Chorzempa, pno and org; Rotterdam Philharmonic Orchestra; Edo de Waart. Philips 9500 306.
+SFC 11-13-77 p50

Symphony, no. 3, op. 78, C minor. cf Works, selections (Vox 5144).

Theme varie, op. 97. cf Piano works (Vox 5476/7).

Le timbre d'Argent: Le bonheur est une chose legere. cf Columbia 34294.

Valse canariote, op. 88. cf Piano works (Vox 5476/7).

Valse gaie, op. 139. cf Piano works (Vox 5476/7).

Valse langoureuse, op. 120. cf Piano works (Vox 5476/7).

Valse mignonne, op. 104. cf Piano works (Vox 5476/7).
Valse nonchalante, op. 110. cf Piano works (Vox 5476/7).
Variations on a theme by Beethoven, op. 35. cf Piano works (Vox 5476/7).
Wedding cake, op. 76. cf Concerti, piano, nos. 1-5.
Wedding cake, op. 76. cf Symphony, no. 3, op. 78, C minor.

2326 Works, selections: Bacchanale. Danse macabre, op. 40. Havanaise, op. 83. Introduction and rondo capriccioso, op. 28. Le rouet d'Omphale, op. 31. Samson et Dalila: Mon couer s'ouvre a ta voix. OSCCP, LSO, ROHO; Jean Martinon, Pierino Gamba, Edward Downes, Anatole Fistoulari. Decca ECS 808.
+HFN 11-77 p185 +RR 11-77 p68

2327 Works, selections: Caprice, violin, op. 52 (orch. Ysaye). Caprice andalous, op. 122. Concerti, violin, nos. 1-4. Le deluge, op. 45: Prelude. Havanaise, op. 83. Introduction and rondo capriccioso, op. 28. La muse et le poete, op. 132. Romance, op. 48, C major. Romance, op. 37, D flat major. Ulf Hoelscher, vln; Ralph Kirshbaum, vlc; NPhO; Pierre Dervaux. HMV SQ SLS 5103 (3).
+Gr 12-77 p1089 +RR 12-77 p51
+-HFN 12-77 p179

2328 Works, selections: Allegro appassionato, op. 43. Caprice, violin, op. 122. Concerti, violoncello (2). Concerti, violin (3). Havanaise, op. 83. Introduction and rondo capriccioso, op. 28. Morceau de concert, op. 62. Romance, op. 48, C major. Ruggiero Ricci, vln; Laszlo Varga, vlc; Luxembourg Radio Orchestra, Westphalian Symphony Orchestra, PH; Pierre Cao, Reinhard Peters, Siegfried Landau. Vox (Q) SVBX 5134 (3).
+-HF 1-76 p94 +NYT 6-5-77 pD19
++NR 10-75 p3 ++SFC 9-28-75 p30

2329 Works, selections: Danse macabre, op. 40. Le jeunesse d'Hercule, op. 50. Marche heroique, op. 34. Phaeton, op. 39. Le rouet d'Omphale, op. 31. Symphony, no. 1, op. 2, E flat major. Symphony, no. 2, op. 55, A minor. Symphony, no. 3, op. 78, C minor. Luxembourg Radio Orchestra; Louis de Froment. Vox SVBX 5144 (3).
+-NR 9-77 p3 +NYT 6-5-77 pD19

SAINTE-COLOMBE, Sieur de
Concerts, 2 bass violes (2). cf Vox SVBX 5142.
Concerts a duex violes esgales (3). cf MARAIS: Suite, viola da gamba and harpsichord, E minor.

2330 Concerts a deux violes esgales, nos. 27, 41, 44, 48, 54. Wieland Kuijken, Jordi Savall, vla da gamba. Telefunken AW 6-42123.
+RR 12-77 p72

SAINZ DE LA MAZA, Eduardo
Campanas del alba. cf London CS 7015.
Homenaje a la guitarra. cf LAURO: Danza negra.

SALIERI, Antonio
Menuetto. cf DG 2723 051.
Minuet. cf DG 2533 182.

SALLINEN, Aulis
Chaconne. cf Marc MC 5355.

2331 Chamber music II, alto flute and strings, op. 41. Elegy for Sebastian Knight, op. 10. Quartet, strings, no. 3, op. 19. Quarttro per quattro, op. 12. Gunilla von Bahr, flt; Ari Angervo, vln; Veikko Hoyla, vlc; Eva Nordwall, hpd; Frans Helmerson, vlc; Stockholm Chamber Ensemble, Voces Intimae Quartet; Okko Kamu. BIS LP 64.

++FF 11-77 p49 +RR 2-77 p73
+Gr 9-77 p500 +St 10-77 p144

2332 Chorali. Symphony, no. 1. Symphony, no. 3. Finnish Radio Symphony Orchestra, Helsinki Philharmonic Orchestra; Okko Kamu, Paavo Berglund. BIS LP 41.

++FF 9-77 p44 ++HF 2-77 p102
+Gr 3-77 p1457 +RR 1-77 p62

Elegy for Sebastian Knight, op. 10. cf Chamber music II, alto flute and strings, op. 41.
Kamarimusikki, no. 1, op. 38. cf LARSSON: Concertino, op. 45, no. 9.
Quartet, strings, no. 3, op. 19. cf Chamber music II, alto flute and strings, op. 41.
Quattro per quattro, op. 12. cf Chamber music II, alto flute and strings, op. 41.
Symphony, no. 1. cf Chorali.
Symphony, no. 2. cf KUISMA: Concertpiece for percussion.
Symphony, no. 3. cf Chorali.

SALMENHAARA, Erkki
Quintet, winds. cf CARLSTEDT: Sinfonietta, 5 wind instruments.

SALOME, T.
Grand choeur. cf Vista VPS 1033.

SALZEDO, Carlos
Piece concertante, op. 27. cf Boston Brass BB 1001.

SAMAZEUIHL, Gustave
Chant d'Espagne. cf Discopaedia MB 1013.
Serenade. cf RCA ARL 1-1323.

SAMMARTINI, Giuseppe
Concerto, soprano recorder and strings. cf MARCELLO, A.: Concerto, oboe and strings.
Concerto, viola pomposa, C major. cf PORPORA: Concerto, violoncello, G major.

SAMPSON
High command. cf Decca SB 329.

SAN SEBASTIAN
Preludios vascos: Dolor. cf RCA ARL 1-1323.

SANDERS, John
Festival te deum. cf Abbey LPB 783.

SANDERSON
Songs: Friend o' mine. cf Argo ZFB 95/6.

SANDRIN, Pierre
Doulce memoire. cf L'Oiseau-Lyre 12BB 203/6.

SANDSTROM, Sven-David
Close to. cf BIS LP 32.

SANTOS
O lonely moon. cf Philips 6392 023.

SANZ, Gaspar
Canarios. cf DG 2723 051.
Canarios. cf Enigma VAR 1015.
Espanoletas. cf DG 2723 051.
Espanoletas. cf Enigma VAR 1015.
Gallarda y villano. cf DG 2723 051.
Passacalle de la Cavalleria de Napoles. cf DG 2723 051.
Suite espanola. cf Angel S 36093.

SARACINI, Claudio
Songs: Da te parto; Deh, come invan chiedete; Giovinetta vezzosetta; Io moro; Quest'amore, quest'arsura. cf DG 2533 305.

SARASATE, Pablo

Carmen fantasy. cf Discopaedia MOB 1020.

Danzas espanolas: Romanza andaluza. cf Discopaedia MB 1014.

Danzas espanolas, no. 6. cf Discopaedia MB 1013.

Danzas espanolas (Spanish dances), opp. 21, 23, 23, 26 (8). cf Navarra, op. 33.

Danzas espanolas, op. 21, no. 1: Malaguena. cf Discopaedia MOB 1018.

Danzas espanolas, op. 21, no. 2: Habanera. cf Discopaedia MB 1016.

Danzas espanolas, op. 21, no. 2: Habanera. cf Discopaedia MOB 1020.

Danzas espanolas, op. 23, no. 2: Zapateado. cf Enigma VAR 1025.

Introduction and tarantelle, op. 43. cf Discopaedia MB 1016.

2333 Navarra, op. 33. Spanish dances, opp. 21, 22, 23, 26 (8). Alfredo Campoli, Belinda Bunt, vln; Daphne Ibbott, pno. L'Oiseau-Lyre DSLO 22.

+Gr 8-77 p324 ++RR 9-77 p84
+HFN 8-77 p89

Zigeunerweisen, op. 20, no. 1. cf Discopaedia MB 1016.

Zigeunerweisen, op. 20, no. 1. cf Kelsey Records KEL 7601.

SARY, Laszlo

Catocoustics, 2 pianos. cf JENEY: Alef: Homage a Schonberg.

Immaginario, no. 1. cf JENEY: Alef: Homage a Schonberg.

Incanto. cf JENEY: Alef: Homage a Schonberg.

Sounds for cimbalom, excerpts. cf Hungaroton LSPX 11686.

SATIE, Erik

Croquis et agaceries d'un gros bon homme en bois. cf Vox SVBX 5483.

Gnossiennes, nos. 1, 4-5. cf Piano works (Saga 5387).

Gnossienne, no. 2. cf Works, selections (Unicorn RHS 338).

Gymnopedies, nos. 1-3. cf Piano works (Saga 5387).

Gymnopedies, nos. 1-3. cf Angel S 36053.

Gymnopedies, nos. 1 and 3. cf RCA GL 4-2125.

Gymnopedie, no. 1. cf Works, selections (Unicorn RHS 338).

Gymnopedie, no. 1. cf Decca PFS 4387.

Jack in the box. cf Supraphon 111 1721/2.

Nocturne, no. 1. cf Piano works (Saga 5387).

Parade: Rag-time. cf Piano works (Saga 5387).

Passacaille. cf Piano works (Saga 5387).

2334 Piano works: Gnossiennes, nos. 1, 4-5. Gymnopedies, nos. 1-3. Nocturne, no. 1. Parade: Rag-time (arr. Ourdine). Passacaille. Pieces: Desespoir agreable; Effronterie; Poesie; Prelude canin; Profondeur; Songe creux. Sarabandes, nos. 1, 3. Sonatine bureaucratique. Sports et divertissements. Veritable preludes flasques. Vieux sequins et vieilles cuirasses. John McCabe, pno. Saga 5387 Tape (c) CA 5387.

+-Gr 12-74 p1182 +RR 12-74 p63
+-Gr 2-77 p1322 tape +RR 1-77 p91 tape
+-HFN 10-76 p185 tape

Le piccadilly. cf Works, selections (Unicorn RHS 338).

Pieces: Desespoir agreable; Effronterie; Poesie; Prelude canin; Profondeur; Songe creux. cf Piano works (Saga 5387).

Pieces froides: Airs a faire fuir. cf Works, selections (Unicorn RHS 338).

Poudre d'or. cf Works, selections (Unicorn RHS 338).

Poudre d'or. cf Vox SVBX 5483.

Preludes veritables frasques pour un Chien. cf Vox SVBX 5483.
Sarabandes, nos. 1, 3. cf Piano works (Saga 5387).
Sonatine bureaucratique. cf Piano works (Saga 5387).
Songs: Chanson medievale; La diva de l'empire; Elegie; Hymne; Salut drapeau; Genevieve de Brabant: Air de Genevieve; Je te veux; Petit air; Les anges, Les fleurs, Sylvie; Tendrement. cf Works, selections (Unicorn RHS 338).
Songs: La diva de l'empire; Dapheneo; Le chapelier; Je te veux; La statue de bronze; Tendrement. cf HONEGGER: Poemes.
Sports et divertissements. cf Piano works (Saga 5387).
Veritable preludes flasques. cf Piano works (Saga 5387).
Vexations. cf Works, selections (Unicorn RHS 338).
Vieux sequins et vieilles cuirasses. cf Piano works (Saga 5387).

2335 Works, selections: Pieces froides: Airs a faire fuir. Gnossienne, no. 2. Gymnopedie, no. 1. Poudre d'or. Le piccadilly. Vexations. Songs: Chanson; Chanson medievale; La diva de l'empire; Elegie; Hymne: Salut drapeau; Genevieve de Brabant: Air de Genevieve; Je te veux; Petit air. Songs: Les anges, Les fleurs, Sylvie; Tendrement. Peter Dickinson, pno; Merial Dickinson, ms. Unicorn RHS 338
+Gr 11-76 p858 ++RR 11-76 p101
+-HFN 11-76 p167 +St 2-77 p120

SATOH
Calligraphy for piano. cf CP 3-5.

SAVIO, Isaias
Batucada. cf L'Oiseau-Lyre SOL 349.
Seroes. cf L'Oiseau-Lyre SOL 349.

SCARLATTI
Preambulo and allegro vivo. cf Angel S 36053.

SCARLATTI, Alessandro
Concerto, recorder and strings, A minor. cf BOISMORTIER: Concerto, 5 recorders without bass, D minor.
Concerto, alto recorder and 2 violins. cf MARCELLO, A.: Concerto, oboe and strings.

2336 La Griselda, excerpts. Carole Bogard, Judith Nelson, s; Kari Windingstad, ms; Daniel Collins, c-t; Riccardo Cascio, t; University of California, Berkeley, Orchestra Members; Lawrence Moe. Cambridge CRS 2903.
+-ARG 5-77 p32 +-ON 3-26-77 p32
+-HF 6-77 p72 +St 6-77 p143
+NR 6-77 p11

2337 Madrigals: Arsi un tempo; Cor mio, deh non languire; Intenerite voi, lagrime mie; Mori, mi dici; O morte, agl'altri fosca; O selce, o tigre, o ninfa; Or che da te, mio bene; Sdegno la fiamma estinse. Monteverdi Choir, Hamburg; Jurgen Jurgens. DG 2533 300.
+Gr 8-76 p328 ++NR 7-76 p7
++HF 8-76 p93 +RR 7-76 p78
+HFN 8-76 p86 +St 11-76 p155
+MT 3-77 p217

Pastorale and capriccio. cf International Piano Library IPL 103.
Songs: Gia il sole dal gange; O cessate di piagarmi; Su venite a consiglio; Toglietemi la vita ancor; Le violette. cf HNH 4008.

2338 Stabat mater. Mirella Freni, s; Teresa Berganza, ms; Kuentz Chamber Orchestra; Charles Mackerras. DG 2533 324 Tape (c) 3310 324.

+-ARG 3-77 p34 +NR 4-77 p12
+-Gr 1-77 p1168 +ON 3-26-77 p32
+-HF 5-77 p90 +RR 2-77 p89
++HFN 12-76 p147 +-St 4-77 p125

Su le sponde del Tebro. cf HANDEL: Alessandro: Lusinghe piu care.

2339 Il trionfo dell'onore. Orchestra; Carlo Maria Giulini. Cetra LPC 1223.

-Gr 7-77 p221

SCARLATTI, Domenico

Sonatas, L 104, C major; L 352, C minor; L 325, E minor; L 413, D minor; L 366, D minor; L 93, A minor; L 424, D minor; L 465. cf Kaibala 20B01.

Sonata, L 250 (K 190), B flat major. cf International Piano Library IPL 109.

Sonatas, guitar (2). cf RCA ARL 1-0864.

Sonata, guitar, A minor. cf London CS 7015.

Sonata, guitar, A major. cf London CS 7015.

Sonata, guitar, E minor. cf RCA RK 1-1735.

Sonatas, guitar, L 83, A major; L 352, E minor; L 423, A minor; L 483, D major. cf Angel S 36093.

2340 Sonatas, harpsichord (6). VILLA-LOBOS: Preludes (5). John Williams, gtr. Columbia M 34198 Tape (c) MT 34198. (also CBS 73545 Tape (c) 40-73545)

+ARG 12-76 p39 +MU 3-77 p14
+Gr 7-76 p194 ++NR 8-77 p16
+Gr 4-77 p1603 tape ++RR 7-76 p73
+HFN 10-76 p183 +RR 2-77 p99 tape
++HFN 10-76 p185 tape +-St 1-77 p126

2341 Sonatas, harpsichord, E major, A major, B minor, D major, E minor, F major, C major, F major, F minor, A major, G major, D major, Longo/Kirpatrick nos. 21/162; 483/322, 449/27, 365/96, 275/394, 384/17, 2/420, 116/518, 475/519, 238/208, 286/427/ 164/491. Andras Schiff, pno. Hungaroton SLPX 11806.

+ARG 9-77 p46 +-FF 9-77 p45

2342 Sonatas, harpsichord, K 9, 14, 20, 72, 126, 132, 133, 159, 184, 198, 264, 376-377, 424-425, 430, 474, 543. Luciano Sgrizzi, hpd. Erato STU 72001.

+Gr 3-77 p1429 +RR 3-77 p84
+HFN 4-77 p143

2343 Sonatas, harpsichord, K 12 (L 489), G minor; K 25 (L 481), F sharp minor; K 45 (L 255), D major; K 188 (L 122), D major; K 187 (L 285), F minor; K 197 (L 147), B minor; K 201 (L 129), C major; K 213 (L 108), D minor; K 409 (L 150), B minor; K 481 (L 187), F minor; K 517 (L 266), D minor; K 545 (L 500), B flat major. Elaine Comparone, hpd. Musical Heritage Society MHS 3330 Tape (c) MHC 5330.

+-CL 11-77 p12 +HF 11-77 p124

2344 Sonatas, harpsichord, K 28, K 208, K 209, K 490, K 491, K 492, K 52, K 123, K 133, K 544, K 545. Valda Aveling, hpd. HMV HQS 1365.

+Gr 2-77 p1308 ++RR 1-77 p74
++HFN 1-77 p115

2345 Sonatas, harpsichord, K 206, K 212, K 222, K 364, K 365, K 370, K 371, K 481, K 501, K 502, K 513, K 524, K 525, K 532. Colin Tilney, hpd. Argo ZK 5 Tape (c) KZKC 5.

+Gr 12-76 p1022 +RR 12-76 p86
+HFN 1-77 p115 ++RR 2-77 p98 tape
+-HFN 2-77 p135 tape

2346 Sonatas, harpsichord, Kk 430, D major; Kk 394, E minor; Kk 395, E major; Kk 259, G major; Kk 260, G major; Kk 308, C major; Kk 309, C major; Kk 429, A major; Kk 460, C major; Kk 461, C major; Kk 9, D minor; Kk 402, E minor; Kk 403, E major; Kk 446, F major; Kk 206, E major. Blandine Verlet, hpd. Philips 6581 028.
+Gr 4-77 p1574 +RR 4-77 p76
+-HFN 4-77 p143
2347 Sonatas, harpsichord, L 17/K 450; L 120/K 541; L 459/K 270; L 155/K 271; L 441/K 314; L 235/K 315; L 399/K 228; L 199/K 229; L 67/K 294; L 270/K 295; L 491/K 456; L 292/K 457. Gilbert Rowland, hpd. Keyboard KGR 1003.
+HFN 1-77 p115 +RR 1-77 p74
2348 Sonatas, harpsichord, K 447/L 294, F sharp minor; K 448/L 485, F sharp minor; K 232/L 62, E minor; K 233/L 467, E minor; K 396/L 110, D minor; K 397/L 208, D major; K 225/L 351, C major; K 226/L 112, C minor; K 300/L 92, A major; K 546/L 312, G minor; K 547/L 28, G major. Gilbert Rowland, hpd. Keyboard KGR 1004.
++HFN 7-77 p119 +RR 7-77 p83
Sonata, harpsichord, G minor. cf FRESCOBALDI: Toccatas, no. 1, G major.
Sonata, harpsichord, no. 6, A major. cf Saga 5384.
Sonata, keyboard, L 345, A major. cf International Piano Library IPL 112.

SCELSI, Giacinto
Anahit. cf GLASS: Strung out.

SCHAFFER, Boguslaw
Free form, no. 2: Evocazioni. cf Finnadar SR 9015.
Project. cf Finnadar SR 9015.

SCHARWENKA, Xaver
2349 Dances, op. 40 (2). Polish dances, op. 3. Theme and variations, op. 48. Evelinde Trenkner, pno. Orion ORS 76230.
+NR 5-77 p14
Polish dances, op. 3. cf Dances, op. 40.
Theme and variations, op. 48. cf Dances, op. 40.

SCHEIDT, Samuel
Cantio sacra: Warum betrabst du dich, mein Herz. cf CLERAMBAULT: Suite du deuxieme ton.
Cantus, 5 voices on "O Nachbar Roland". cf BIS LP 57.
Canzon a 5 and imitationem bergamas. cf BIS LP 59.
Suite, C major. cf Abbey LPB 765.
Suite, C major. cf Supraphon 111 1867.
2350 Tabulatura nova, excerpts. Aniko Horvath, hpd. Hungaroton SLPX 11848.
+FF 9-77 p46

SCHENKER, Friedrich
2351 In memoriam Martin Luther King. Leipzig Radio Symphony Orchestra; Herbert Kegel. Nova 885 106.
+-FF 11-77 p51

SCHIEBE
Trio, F major. cf Pelca PSR 40571.

SCHMELZER, Johann
2352 Sacro-profanus concentus musicus. VCM; Nikolaus Harnoncourt. Telefunken 6-42100. (Reissue from SAWT 9563/4)
+HF 12-77 p99 ++St 11-77 p150

SCHMID, Bernard
Englischer Tanz. cf DG 2723 051.
Tanz Du hast mich wollen nemmen. cf DG 2723 051.

SCHMIDT

Cassation. cf Supraphon 113 1323.

SCHMIDT, Franz

2353 Das Buch mit sieben Siegeln (The book of the seven seals). Hanny Steffek, s; Hertha Topper, ms; Julius Patzak, Erich Majkut, t; Otto Wiener, bar; Frederick Guthrie, bs; Franz Illenberger, org; Graz Cathedral Choir; Munich Philharmonic Orchestra; Anton Lippe. Musical Heritage Society MHS 3350/1 (2). (Reissue from Amadeo 5004/5)

+-FF 9-77 p48 +-HF 11-76 p122

2354 Das Buch mit sieben Siegeln. Anton Dermota, Robert Holl, Margarite Kyriaki, Hertha Topper, Thomas Moser, Artur Korn; Rudolf Scholz, org; Grazer Concertchor; Tonkunstler Orchestra; Alois Hochstrasser. Preiser SQPR 3263/4 (2).

+FF 9-77 p48

2355 Quintet, piano, G major. Variations on a theme by Josef Labor. Alfred Prinz, clt; Vienna Philharmonia Quintet. Decca SDD 491. (also London STS 15401)

+FF 9-77 p48 +NR 11-77 p8
+-Gr 5-76 p1781 +RR 5-76 p62
+HFN 5-76 p107 +St 11-77 p150
+MM 2-77 p35

Variations on a Hussar's song. cf ARRIAGA Y BALZOLA: Symphony, D major.

Variations on a theme by Josef Labor. cf Quintet, piano, G major.

SCHMIDT, William

Vendor's call. cf BROWN: Concertino, piano and band.

SCHMITT, Florent

Dionysiaques, op. 62, no. 1. cf FAURE: Chant funeraire, op. 117.

Songs: Chants, op. 98 (3); Poemes de Ronsard, op. 100 (4). cf HONEGGER: Poemes.

SCHOECK, Othmar

Concerto, horn, op. 65, D minor. cf Westminster WGS 8338.

SCHOENBERG, Arnold

2356 Das buch der hangenden Garten, op. 15 (The book of hanging gardens). SCHUBERT: Songs: An mein Herz, D 860; Blondel zu Marien, D 626; Ganymed, D 544; Heidenroslein, D 257; Der Musensohn, D 764; Nur wer die Sehnsucht kennt, D 877/4; Schafers Klagelied, D 121; Sprache der Liebe, D 412; Rastlose Liebe, D 138. Jan DeGaetani, ms; Gilbert Kalish, pno. Nonesuch H 71320.

+Gr 9-76 p459 +-MT 4-77 p306
++HF 3-76 p75 +-NR 2-76 p12
+-HFN 9-76 p129 +-ON 2-7-76 p34
-MQ 7-76 p456 ++St 10-76 p132

2357 Chamber symphony, no. 1, op. 9, E major. Verklarte Nacht, op. 4. London Sinfonietta; David Atherton. Decca SDD 519. (Reissue from SXLK 6660/4)

+Gr 5-77 p1698 ++RR 5-77 p55
++HFN 5-77 p137

2358 Chamber symphony, no. 9, op. 9, E major (arrangement for orchestra, op. 9b). Chamber symphony, no. 2, op. 38, E flat major. Frankfurt Radio Symphony Orchestra; Eliahu Inbal. Philips 6500 923.

+-Gr 12-75 p1045 ++NR 7-76 p2
+HF 4-77 p112 +RR 11-75 p55
+HFN 12-75 p163 /-St 4-77 p128
+MJ 9-77 p35

Chamber symphony, no. 1, op. 9, E major. cf Pierrot Lunaire, op. 21.
Chamber symphony, no. 1, op. 9, E major. cf Works, selections (Decca SXLK 6660/4).
Chamber symphony, no. 2, op. 38, E flat major. cf Chamber symphony, no. 1, op. 9, E major (arr. for orchestra, op. 9b).
Die eiserne Brigade. cf Works, selections (Decca SXLK 6660/4).
Fantasia, violin and piano, op. 47. cf Works, selections (Decca SXLK 6660/4).

2359 Gurrelieder. Jeannette Vreeland, s; Rose Bampton, ms; Paul Althouse, Robert Betts, t; Abrasha Rabofsky, bs; Benjamin de Loache, speaker; Princeton Glee Club, Fortnightly Club, Mendelssohn Club of Philadelphia; PO; Leopold Stokowski. RCA AVM 2-2017 (2). (Reissue from RCA M 127, LCT 6012, HMV DB 1769/82).
+ARG 8-77 p5 +-MJ 4-77 p33
+Gr 4-77 p1587 +-RR 7-77 p89
+-HF 5-77 p91 ++SFC 1-30-77 p36
+-HFN 9-77 p149

Herzgewachse, op. 20. cf Works, selections (Decca SXLK 6660/4).
Lied der Waldtaube. cf Works, selections (Decca SXLK 6660/4).
Little pieces, piano, op. 19 (6). cf Piano works (DG 2530 531).

2360 Moses und Aron. Felicity Palmer, Jane Manning, s; Gillian Knight, ms; Richard Cassilly, John Winfield, t; John Noble, Roland Hermann, bar; Richard Angas, Michael Rippon, bs; Gunter Reich, speaker; BBC Singers, Orpheus Boys Choir; BBC Symphony Orchestra; Pierre Boulez. CBS 79201 (2). (also Columbia M2 33594)
+Gr 11-75 p900 +ON 4-3-76 p56
++HF 6-76 p92 +-RR 11-75 p35
++HFN 11-75 p167 +SFC 8-29-76 p28
+-MT 3-77 p217 ++St 10-76 p131
+NR 4-76 p10

Nachtwandler. cf Works, selections (Decca SXLK 6660/4).
Der neue Klassizimus, op. 28, no. 3. cf Works, selections (Decca SXLK 6660/4).
Ode to Napoleon, op. 41. cf Works, selections (Decca SXLK 6660/4).

2361 Piano works: Pieces, op. 11 (3). Pieces, op. 23 (5). Pieces, opp. 33a and 33b. Little pieces, op. 19 (6). Suite, op. 25. Maurizio Pollini, pno. DG 2530 531.
++Gr 5-75 p1999 +NR 12-75 p13
+HF 11-75 p116 +NYT 3-13-77 pD22
++HFN 6-75 p92 ++RR 6-75 p76
+MT 7-76 p578 ++St 3-76 p121

Pieces, chamber orchestra (3). cf Songs (Decca SDD 520).
Pieces, chamber orchestra (3). cf Works, selections (Decca SXLK 6660/4).
Pieces, orchestra, op. 16 (5). cf BLOCH: Concerto grosso, no. 1.
Pieces, piano, op. 11 (3). cf Piano works (DG 2530 531).
Pieces, piano, op. 23 (5). cf Piano works (DG 2530 531).
Pieces, piano, opp. 33a and 33b. cf Piano works (DG 2530 531).

2362 Pierrot Lunaire, op. 21. Erika Stiedry-Wagner, speaker; Rudolf Kolisch, vln and vla; Stefan Auber, vlc; Leonard Posella, flt and pic; Kalman Bloch, clt; Edward Steuermann, pno; Arnold Schoenberg. Odyssey Y 33791. (Reissue from Columbia ML 4471).
+ARSC Vol VIII, no. 2-3 p83 +-HF 9-76 p96
+NYT 5-16-76 pD19
+-Audio 11-77 p126

2363 Pierrot Lunaire, op. 21. Chamber symphony, no. 1, op. 9, E major. Mary Thomas, s; The Fires of London; Peter Maxwell Davies. Unicorn RHS 319.
+Gr 9-74 p535 +RR 9-74 p34
+HF 5-75 p77 ++SFC 6-8-75 p23
++NR 4-75 p7 +-St 4-75 p105
+ON 4-2-77 p48

Pierrot Lunaire, op. 21. cf Songs (Decca SDD 520).
Pierrot Lunaire, op. 21. cf Works, selections (Decca SXLK 6660/4)
Quartet, strings, D major. cf Quartets, strings, nos. 1-4.

2364 Quartet, strings, nos. 1-4. Quartet, strings, D major. Juilliard Quartet. Columbia M3 34581 (3).
++NR 12-77 p6 ++SFC 10-2-77 p44
+NYT 10-23-77 pD15

2365 Quintet, wind instruments, op. 26. Vienna Wind Soloists. DG 2530 825.
+Gr 8-77 p317 +RR 8-77 p68
+HFN 8-77 p89

Quintet, wind instruments, op. 26. cf Works, selections (Decca SXLK 6660/4).
Rondo. cf Works, selections (Decca SXLK 6660/4).

2366 Serenade, op. 24. Kenneth Bell, bs; Light Fantastic Players; Daniel Shulman. Nonesuch H 71331.
+-Gr 1-77 p1160 ++NR 2-77 p7
+-HF 4-77 p111 +ON 4-2-77 p48
+-HFN 2-77 p129 ++RR 1-77 p67
+MT 9-77 p738 +St 4-77 p128

Serenade, op. 24. cf Works, selections (Decca SXLK 6660/4).

2367 Songs: Herzgewachse, op. 20; Nachtwandler. Pieces, chamber orchestra (3). Pierrot Lunaire, op. 21. Ein Stelldichein. Mary Thomas, June Barton, s; London Sinfonietta; David Atherton. Decca SDD 520. (Reissue from SXLK 6660/4)
+Gr 5-77 p1723 +RR 5-77 p84
+HFN 6-77 p137

Ein Stelldichein. cf Songs (Decca SDD 520).
Ein Stelldichein. cf Works, selections (Decca SXLK 6660/4).
Suite, op. 25. cf Piano works (DG 2530 531).
Suite, op. 29. cf Works, selections (Decca SXLK 6660/4).
Variations, op. 31. cf ELGAR: Enigma variations, op. 36.
Verklarte Nacht, op. 4. cf Chamber symphony, no. 1, op. 9, E major.
Verklarte Nacht, op. 4. cf Works, selections (Decca SXLK 6660/4).
Verklarte Nacht, op. 4. cf BERG: Lyric suite.
Weihnachtsmusik. cf Works, selections (Decca SXLK 6660/4).

2368 Works, selections: Chamber symphony, no. 1, op. 9, E major. Die eiserne Brigade. Fantasia, violin and piano, op. 47. Herzgewachse, op. 20. Lied der Waldtaube. Nachtwandler. Der neue Klassizimus, op. 28, no. 3. Ode to Napoleon, op. 41. Pieces, chamber orchestra (3). Pierrot Lunaire, op. 21. Quintet, wind instruments, op. 26. Rondo. Serenade, op. 24. Ein Stelldichein. Suite, op. 29. Verklarte Nacht, op. 4. Weihnachtsmusik. Der wunsch des Liebhabers, op. 27, no. 4. Nona Liddell, vln; John Constable, pno; Mary Thomas, June Barton, s; Anna Reynolds, ms; John Shirley Quirk, bar; Gerald English, speaker; London Sinfonietta and Chorus; David Atherton. Decca SXL 6660/4 (5). (also London SXLK 6660/4)

+Gr 9-74 p535 +RR 9-74 p34
+-HF 7-76 p89 ++SFC 12-14-75 p39
++MJ 2-76 p32 ++St 4-77 p128
+-MQ 7-76 p456 ++Te 9-74 p34
+NR 2-76 p4

Der wunsch des Liebhabers, op. 27, no. 4. cf Works, selections (Decca SXLK 6660/4).

SCHOLEFIELD

St. Clement, hymn tune. cf Grosvenor GRS 1043.

SCHRADER

I de lyse naetter. HMV RLS 715.

SCHUBERT, Franz

Alfonso und Estrella, D 732: Konnt ich ewig hier verweilen. cf BEETHOVEN: Ah perfido, op. 65.

Alfonso und Estrella, D 732: Konnt ich ewig hier verwielen. cf Philips 6767 001.

Alfonso und Estrella, D 732: Von Fels und Wald umrungen; Wer bist du, holdes Wesen; Freundlich bist du mir erschienen; Konnt ich ewig hier verweilen; Lass dir als Erinnerungszeiche. cf Works, selections (Philips 9500 170).

Ave Maria, D 839. cf LISZT: Harmonies poetiques et religieuses, no. 7, G 173: Funerailles.

Ave Maria, D 839. cf CBS 61039.

Ave Maria, D 839. cf Decca SXL 6781.

Ave Maria, D 839. cf Discopaedia MB 1015.

Ave Maria, D 839. cf Philips 9500 218.

The bee, op. 13, no. 9. cf HMV SQ ASD 3283.

Die Burgschaft, D 435: Die Mutter sucht ihr liebes Kind; Welche Nacht hab ich erlebt; Horch die Seufzer uns'rer Mutter. cf Works, selections (Philips 9500 170).

Claudine von Villa Bella, D 239: Hin und wieder fliegen die Pfeile; Liebe schwarmt auf allen Wegen. cf Works, selections (Philips 9500 170).

Deutsche Tanze (German dances), D 336 (4). cf Sonata, piano, no. 17, op. 53, K 850, D major

Ecossaises, D 529, nos. 1-3, 5; D 783, nos. 1 and 2. cf Saga 5421.

Der Erlkonig, op. 1, D 328. cf LISZT: Harmonies poetiques et religieuses, no. 7, G 173: Funerailles.

2369 Fantasia, op. 15, D 760, C major. Sonata, piano, no. 16, op. 42, D 845, A minor. Maurizio Pollini, pno. DG 2530 473 Tape (c) 3300 504.

++Gr 1-75 p1372 +-NYT 3-13-77 pD22
++HF 5-75 p86 -RR 12-74 p64
++HF 10-76 p147 tape ++RR 3-76 p77 tape
++NR 4-75 p12 ++SFC 3-2-75 p24

Die Freunde von Salamanka, D 326: Gelagert unter'm hellen Dach. cf Works, selections (Philips 9500 170).

2370 Grand duo, D 812, C major. Variations on an original theme, D 603, B flat major. Bracha Eden, Alexander Tamir, pno. Decca SXL 6794.

+Gr 6-77 p75 +MT 11-77 p924
+HFN 6-77 p133 +-RR 6-77 p84

Hark hark the lark. cf CHOPIN: Etude, op. 25, no. 7, C sharp minor.

2371 Impromptus, op. 90, D 988 and op. 142, D 935. Augustin Anievas, pno. HMV SQ SQS 1397.

+-Gr 10-77 p662 +-RR 11-77 p95
+-HFN 10-77 p161

2372 Impromptus, op. 90, D 899 and op. 142, D 935. Alfred Brendel, pno. Philips 9500 357. (Reissue from 6747 175)
+Gr 12-77 p1108 +RR 11-77 p95
+HFN 11-77 p185

2373 Impromptus, op. 90, D 899, nos. 1-4. Impromptus, op. 142, D 935, nos. 1-4. Alfred Brendel, pno. Turnabout TV 34141 Tape (c) KTV 34141. (also Vox tape CT 2130)
+HF 10-77 p129 tape ++RR 6-74 p87 tape
/HFN 2-74 p349 tape

Impromptu, op. 90, no. 4, K 899, A flat major. cf Philips 6747 327.

Impromptus, op. 142, D 935, nos. 1-4. cf Impromputus, op. 90, D 899, nos. 1-4.

Impromptu, op. 142, no. 3, D 935, B flat major. cf BEETHOVEN: Concerto, piano, no. 4, op. 58, G major.

Impromptu, op. 142, no. 3, D 935, B flat major. cf CHOPIN: Etude, op. 25, no. 7, C sharp minor.

Introduction and variations, op. 160, D 802, E minor. cf BEETHOVEN: Sonata, flute and piano, B flat major (Coronet LPS 3037).

Introduction and variations, op. 160, D 802, E minor. cf BEETHOVEN: Sonata, flute and piano, B flat major (Enigma VAR 1029).

Introduction and variations, op. 160, D 802, E minor. cf REINECKE: Sonata, flute and piano, op. 167, E minor.

Die junge Nonne, D 828. cf LISZT: Harmonies poetiques et religieuses, no. 7, G 173: Funerailles.

Landler, D 354 (4). cf DG 2723 051.

Lazarus, D 689: So schlummert auf Rosen. cf BEETHOVEN: Ah perfido, op. 65.

Lazarus, D 689: So schlummert auf Rosen. cf Philips 6767 001.

Marche militaire, no. 1, D 733, D major. cf Rediffusion Royale 2015.

2374 Mass, no. 5, D 678, A flat major. Wendy Eathorne, s; Bernadette Greevy, con; Wynford Evans, t; Christopher Keyte, bs; St. John's College Chapel Choir; AMF; George Guest. Argo ZRG 869.
+Gr 11-77 p874 +RR 10-77 p85
+HFN 10-77 p161

2375 Mass, no. 5, D 678, A flat major. Marlee Sabo, s; Jan DeGaetani, ms; Paul Sperry, t; Leslie Guinn, bar; Carleton College Choir, Chamber Singers, Festival Chorus; Saint Paul Chamber Orchestra; Dennis Russell Davies. Nonesuch H 71135.
+NR 9-77 p9 ++St 8-77 p80
+-ON 6-77 p44

2376 Mass, no. 6, D 905, E flat major. St. Paul Chamber Orchestra; Dennis Russell Davies. Nonesuch H 71335.
+NYT 11-27-77 pD15

Minuets, D 89. cf Symphony, no. 5, D 485, B flat major.

Minuets, D 89 (5). cf DG 2723 051.

2377 Moments musicaux, op. 94, D 780. Sonata, piano, no. 19, op. posth., D 958, C minor. Ingrid Haebler, pno. Philips 6580 128. (Reissue from SAL 3647, 6741 002)
+Gr 10-76 p628 +-RR 1-77 p75
+HFN 11-76 p171

Moment musical, op. 94, no. 3, D 780, F minor. cf ALBENIZ: Espana, op. 165, no. 2: Tango.

Moment musical, op. 94, no. 3, D 780, F minor. cf Decca PFS 4351.

Moment musical, op. 94, no. 3, D 780, F minor. cf Decca PFS 4387.

Moment musical, op. 94, no. 3, D 780, F minor. cf International Piano Library IPL 112.

Moment musical, op. 94, no. 3, D 780, F minor. cf International Piano Library IPL 5007/8.

Moment musical, op. 94, no. 3, D 780, F minor. cf Rediffusion Royale 2015.

2378 Nocturne, op. 148, D 897, E flat major. Quintet, piano, op. 114, D 667, A major. Jorg Demus, pno; Franzjosef Maier, vln; Heinz-Otto Graf, vla; Rudolf Mandalka, vlc; Paul Breuer, double bs. BASF KHB 20314.
++St 5-77 p74

2379 Nocturne, op. 148, D 897, E flat major. Sonata, piano, violin and violoncello, D 28, B flat major. Trio, piano, no. 1, op. 99, D 898, B flat major. Trio, piano, no. 2, op. 100, D 929, E flat major. Vienna Trio. Telefunken 6-35055 (2).
+MJ 11-77 p30 +St 9-77 p142
+NR 9-77 p6

2380 Octet, op. 166, D 803, F major. New Vienna Octet. Decca SDD 508.
+Gr 12-77 p1105 ++RR 12-77 p60
+HFN 12-77 p179

2381 Octet, op. 166, D 803, F major. Vienna Octet. London 6051.
++St 5-77 p74

2382 Quartet, strings, no. 8, D 112, B flat major. Quartet, strings, no. 10, D 87, E flat major. Melos Quartet. DG 2530 899. (Reissue from 2740 123)
+Gr 12-77 p1105 ++RR 11-77 p82
+HFN 11-77 p179

2383 Quartets, strings, nos. 9, 12-15. Amadeus Quartet. DG 2733 008 (3). (Reissues from SLPM 139194, 138048, 139103)
+Gr 4-77 p1569 ++RR 4-77 p67
+-HFN 4-77 p151

Quartet, strings, no. 10, D 87, E flat major. cf Quartet, strings, no. 8, D 112, B flat major.

2384 Quartet, strings, no. 12, D 703, C minor. Quartet, strings, no. 15, D 887, G major. Gabrieli Quartet. Decca SDD 512 Tape (c) KSDC 512.
+Gr 5-77 p1708 +HFN 6-77 p139 tape
+HFN 5-77 p133 +RR 5-77 p66

2385 Quartet, strings, no. 12, D 703, C minor. Quintet, piano, op. 114, D 667, A major. Emil Gilels, pno; Norbert Brainin, vln; Peter Schidlof, vla; Martin Lovett, vlc; Rainer Zepperitz, double bs; Amadeus Quartet. DG 2530 646.
+Gr 8-76 p317 +-NR 2-77 p6
+-HF 6-77 p98 +-RR 6-76 p65
+HFN 6-76 p97 +-St 2-77 p120

2386 Quartet, strings, no. 12, op. posth., D 703, C minor. Quintet, strings, op. 163, D 956, C major. Weller Quartet. London STS 15300.
++FF 9-77 p88

2387 Quartet, strings, no. 12, D 703, C minor. Quartet, strings, no. 13, op. 29, D 804, A minor. Guarneri Quartet. RCA LSC 3285.
++St 5-77 p72

Quartet, strings, no. 12, K 703, C minor. cf BRAHMS: Sextet, strings, no. 1, op. 18, B flat major.

Quartet, strings, no. 12, D 703, C minor. cf DVORAK: Quartet, strings, no. 12, op. 96, F major.

Quartet, strings, no. 13, op. 29, D 804, A minor. cf Quartet, strings, no. 12, D 703, C minor.

2388 Quartet, strings, no. 14, D 810, D minor. WOLF: Italian serenade. Guarneri Quartet. RCA ARL 1-1994 Tape (c) ARK 1-1994 (ct) ARS 1-1994.
+-ARG 5-77 p36 +NR 2-77 p6
++HF 5-77 p101 tape ++St 5-77 p116
+-MJ 4-77 p50

2389 Quartet, strings, no. 14, D 810, D minor. Trio, strings, no. 1, D 471, B flat major. Vladimir Ovcharek, Grigori Lutsky, vln; Vissarion Solovyov, vla; Iosif Levinzon, vlc. Westminster WGS 8349.
+NR 12-77 p5

Quartet, strings, no. 15, D 887, G major. cf Quartet, strings, no. 12, D 703, C minor.

2390 Quintet, piano, op. 114, D 667, A major. Moura Lympany, pno; LSO. Classics for Pleasure CFP 40085 Tape (c) TC CFP 40085.
+Gr 9-74 p536 +-HFN 12-76 p153 tape
+Gr 2-77 p1325 tape -RR 9-74 p69

2391 Quintet, piano, op. 114, D 667, A major. Beaux Arts Trio; Samuel Rhodes, vla; Georg Hortnagel, double bs. Philips 9500 071 Tape (c) 3300 481.
+Gr 8-76 p317 +MJ 12-76 p44
++HF 6-77 p98 ++NR 12-76 p8
+HFN 7-76 p95 ++RR 7-76 p68
++HF 5-77 p101 tape ++St 2-77 p120

2392 Quintet, piano, op. 114, D 667, A major. Peter Serkin, pno; Ida Kavafian, vla; Fred Sherry, vlc; Joseph Silverstein, vln; Buell Neidlinger, double bs. RCA ARL 1-1882 Tape (c) ARK 1-1882 (ct) ARS 1-1882.
+HF 6-77 p98 +NR 5-77 p9
-MJ 4-77 p50 +St 5-77 p116

Quintet, piano, op. 114, D 667, A major. cf Nocturne, op. 148, D 897, E flat major.

Quintet, piano, op. 114, D 667, A major. cf Quartet, strings, no. 12, D 703, C minor.

2393 Quintet, strings, op. 163, D 956, C major. Alberni Quartet; Thomas Igloi, vlc. CRD CRD 1018 Tape (c) 4018.
+Gr 11-75 p854 ++RR 11-75 p65
+Gr 10-77 p705 tape +RR 12-77 p98 tape
+HFN 12-75 p164 ++SFC 9-25-77 p50
+-HFN 11-77 p187 tape ++St 3-76 p80
+NR 10-77 p7

2394 Quintet, strings, op. 163, D 956, C major. Trio, strings, no. 1, K 471, B flat maor. Richard Harand, vlc; Vienna Philharmonic Quartet Members. London STS 15386.
+NR 10-77 p7

Quintet, strings, op. 163, D 956, C major. cf Quartet, no. 12, op. posth., D 703, C minor.

Rosamunde, op. 26, D 797: Entr'acte, B flat major. cf HMV SQ ASD 3375.

2395 Rosamunde, op. 26, D 797: Incidental music. Rohangiz Yachmi, con; Vienna State Opera Chorus; VPO; Karl Mumchinger. London OS 26444 Tape (c) OS 5-26444. (also Decca SXL 6748 Tape (c) KSXC 6748)
+ARG 12-76 p39 +NR 9-77 p9
+Gr 3-76 p1471 +ON 6-77 p44
+Gr 9-76 p497 tape +RR 2-76 p36
+HFN 2-76 p111 +-RR 4-76 p80 tape

+HFN 5-76 p117 tape +SFC 9-25-77 p50
+MJ 2-77 p31 +-St 6-77 p114

Rosamunde, op. 26, D 797: Overture. cf Symphony, no. 8, D 759, B minor.

Rosamunde, op. 26, D 797: Overture. cf Classics for Pleasure CFP 40236.

Rosamunde, op. 26, D 797: Overture. cf Pye PCNHX 6.

Rosamunde, op. 26, D 797: Overture; Entr'acte, no. 3, B flat major; Ballet music, no. 2, G major. cf GRIEG: Peer Gynt, op. 46: Suite, no. 1; op. 55: Suite, no. 2.

Rosamunde, op. 26, D 797: Der Vollmond strahlt. cf BEETHOVEN: Ah perfido, op. 65.

Rosamunde, op. 26, D 797: Der Vollmond strahlt. cf Philips 6767 001.

Scherzi, D 593 (2). cf Sonata, piano, no. 18, op. 78, D 894, G major.

2396 Die schone Mullerin, op. 25, D 795. Winterreise, op. 89, D 911. Schwanengesang, D 957. Dietrich Fischer-Dieskau, bar; Gerald Moore, pno. DG 2720 059 (4) Tape (c) 3371 029.

+-Gr 1-73 p1353 +RR 12-72 p92
++Gr 11-77 p900 tape +RR 10-77 p91 tape
+NR 2-73 p13 +St 12-72 p134
+NYT 12-17-73 pD32 +STL 12-10-72 p35

2397 Die schone Mullerin, op. 25, D 795. Schwanengesang, D 957. Winterreise, op. 89, D 911. Hermann Prey, bar; Leonard Hokanson, Wolfgang Sawallisch, Gerald Moore, pno. Philips 6767 300 (4).

-Gr 12-77 p1128 +-RR 11-77 p108
+HFN 12-77 p179

Die schone Mullerin, op. 25, D 795: Impatience. cf Kaibala 40D03.

Die schone Mullerin, op. 25, D 795: Morgengruss. cf International Piano Library IPL 113.

Die schone Mullerin, op. 25, no. 2, D 795: Wohin. cf LISZT: Harmonies poetiques et religieuses, no. 7, G 173: Funerailles.

Die schone Mullerin, op. 25, D 795: Wohin. cf Angel S 37219.

Die schone Mullerin, op. 25, D 795: Wohin (arr. Liszt). cf International Piano Library IPL 108.

Schwanengesang, D 957. cf Die schone Mullerin, op. 25, D 795 (DG 2720 059).

Schwanengesang, D 957. cf Die schone Mullerin, op. 25, D 795 (Philips 6767 300).

2398 Schwangengesang, D 957: Liebesbotschaft. Songs: An die Laute, D 905; An die Musik, D 547; An Silvia, D 891; Auflosung, D 807; Bei dir Alleine, D 866; Dass sie hier gewesen, D 775; Der Einsame, D 800; Fischerweise, D 881; Die Forelle, D 550; Ganymed, D 544; Der Schiffer, D 536; Standchen, D 889; Die Sterne, D 939; Uber Wildemann, D 884; Der Wanderer an den Mond, D 870; Wanderers Nachtlied, D 768. Ian Partridge, t; Jennifer Partridge, pno. Enigma VAR 1019.

+-HFN 1-77 p115

Serenade. cf Discopaedia MOB 1020.

Serenade. cf Inner City IC 1006.

Serenade. cf Kaibala 40D03.

Sonata, arpeggione and piano, D 821, A minor. cf GRIEG: Sonata, violoncello and piano, op. 36, A minor.

Sonata, arpeggione and piano, D 821, A minor. cf MENDELSSOHN: Sonata, violoncello and piano, no. 2, op. 58, D major.

2399 Sonata, piano, no. 2, D 279, C major. Sonata, piano, no. 21, op. posth., D 960, B flat major. Wilhelm Kempff, pno. DG 2535 240 Tape (c) 3335 240.
+-Gr 12-77 p1146 +HFN 11-77 p185

Sonata, piano, D 537, A minor. cf BEETHOVEN: Sonata, piano, no. 12, op. 26, A flat major.

2400 Sonata, piano, no. 5, D 557, A flat major. Sonata, piano, no. 20, op. posth., D 959, A major. Radu Lupu, pno. Decca SXL 6771 Tape (c) KSXC 6771.
+Gr 8-77 p324 ++HFN 10-77 p171 tape
+Gr 10-77 p705 tape +-RR 9-77 p84
++HFN 9-77 p149

2401 Sonata, piano, no. 6, D 566, E minor. Sonata, piano, no. 8, D 575, B major. Sonata, piano, no. 13, D 664, A major. Sonata, piano, no. 15, D 840, C major. Sonata, piano, no. 16, D 845, A minor. Sonata, piano, no. 18, D 894, G major. Walter Klien, pno. Turnabout TV 37096/8 (3).
+-Gr 12-77 p1108 +HFN 12-77 p179

Sonata, piano, no. 8, D 575, B major. cf Sonata, piano, no. 6, D 566, E minor.

Sonata, piano, no. 13, D 664, A major. cf Sonata, piano, no. 6, D 566, E minor.

2402 Sonata, piano, no. 14, op. 143, D 784, A minor. Sonata, piano, no. 21, op. posth., D 960, B flat major. Ingrid Haebler, pno. Philips 6580 133. (Reissue from SAL 3756)
+-Gr 11-76 p838 +-RR 3-77 p85
+HFN 11-76 p171

Sonata, piano, no. 15, D 840, C major. cf Sonata, piano, no. 6, D 566, E minor.

Sonata, piano, no. 16, op. 42, D 845, A minor. cf Fantasia, op. 15, D 760, C major.

Sonata, piano, no. 16, D 845, A minor. cf Sonata, piano, no. 6, D 566, E minor.

2403 Sonata, piano, no. 17, op. 53, K 850, D major. German dances, D 336 (4). Vladimir Ashkenazy, pno. London CS 6961. (also Decca SXL 6739 Tape (c) KSXC 6739)
+Audio 9-77 p48 ++HFN 5-77 p138 tape
+Gr 4-77 p1577 +NR 2-77 p14
+Gr 5-77 p1743 tape +RR 5-77 p93
++HF 12-76 p116 +RR 4-77 p77
+HFN 4-77 p145

2404 Sonata, piano, no. 18, op. 78, D 894, G major. Scherzi, D 593 (2). Radu Lupu, pno. London CS 6966. (also Decca SXL 6741)
+Gr 5-76 p1787 +NR 2-77 p14
+-HF 12-76 p116 +RR 5-76 p66
+HFN 5-76 p107

2405 Sonata, piano, no. 18, op. 78, D 894, G major. Christian Zacharias, pno. Seraphim S 60285.
+FF 9-77 p52 +SR 7-9-77 p48
+MJ 9-77 p34 ++St 9-77 p140
+NR 9-77 p13

Sonata, piano, no. 18, D 894, G major. cf Sonata, piano, no. 6, K 566, E minor.

Sonata, piano, no. 18, op. 78, D 894, G major: Minuetto. cf International Piano Library IPL 108.

Sonata, piano, no. 19, op. posth., D 958, C minor. cf Moments musicaux, op. 94, D 780.

2406 Sonata, piano, no. 20, op. posth., D 959, A major. Rudolf Serkin, pno. CBS 61645. (Reissue from SBRG 72432)
++Gr 12-76 p1022 ++RR 1-77 p75
+HFN 1-77 p121

Sonata, piano, no. 20, op. posth., D 959, A major. cf Sonata, piano, no. 5, D 557, A flat major.

Sonata, piano, no. 21, op. posth., D 960, B flat major. cf Sonata, piano, no. 2, D 279, C major.

Sonata, piano, no. 21, op. posth., D 960, B flat major. cf Sonata, piano, no. 14, op. 143, D 784, A minor.

Sonata (trio), piano, violin and violoncello, D 28, B flat major. cf Nocturne, op. 148, D 897, E flat major.

Sonata, violin and piano, op. 162, D 574, A major: Finale. cf Discopaedia MOB 1020.

2407 Sonatina, violin and piano, no. 1, op. 137, K 384, D major. Sonatina, violin and piano, no. 2, op. 137, D 385, A minor. Sonatina, violin and piano, no. 3, op. 137, K 408, G minor. Ilse Mathieu, vln; Jorg Ewald Dahler, pno. Claves D 608.
+HFN 4-77 p145 +RR 6-77 p84

Sonatina, violin and piano, no. 2, op. 137, D 385, A minor. cf Sonatina, violin and piano, no. 1, op. 137, K 384, D major.

Sonatina, violin and piano, no. 3, op. 137, K 408, G minor. cf Sonatina, violin and piano, no. 1, op. 137, K 384, D major.

Sonatina, violin and piano, no. 3, op. 137, K 408, G minor: Andante; Allegro giusto. cf Supraphon 113 1323.

2408 Songs: Ach deinem feuchten Schwingen, D 717; Ave Maria, D 839; Du bist die Ruh, D 776; Die Forelle, D 550; Jager ruhe von der Jagd, D 838; Im Freien, K 880; Raste Krieger, D 837; Schwestergruss, D 762; Was debeutet die Bewegung, D 720. Gundula Janowitz, s; Irwin Gage, pno. DG 2530 858.
+Gr 11-77 p874 +-RR 10-77 p86

2409 Songs: An die Laute, D 905; An die Musik, D 547; An Sylvia, D 891; Auflosung, D 807; Bei dir Allein, D 866; Das sie hier gewesen, D 775; Der Einsame, D 800; Fischerweise, D 881; Die Forelle, D 550; Ganymed, D 544; Liebesbotschaft, D 957; Der Schiffer, D 536; Standchen, D 957; Die Sterne, D 939; Uber Wildemann, D 844; Der Wanderer an den Mond, D 870; Wanderers Nachtlied, D 768. Ian Partridge, t; Jennifer Partridge, pno. Enigma VAR 1019.
+-Gr 1-77 p1168 +-RR 3-77 p94
+St 12-77 p157

2410 Songs: An Schwager Kronos, D 369 (trans. Brahms); Erlkonig, D 328 (trans. Berlioz); Erlkonig, D 328 (trans. Liszt); Geheimes, D 719 (trans. Brahms); Gruppe aus dem Tartarus, D 583 (trans. Reger); Harpers songs from Wilhelm Meisters Lehrjahre: Wer sich der Einsamkeit ergibt, D 478, Wer nie sein Brot mit Tranen ass, D 480, An die Turen will ich schleichen, D 479 (trans. Reger); Im Abendrot, D 799 (trans. Reger); Memnon, D 541 (trans. Brahms); Standchen, D 957 (trans. Offenbach); Prometheus, D 674 (trans. Reger). Hermann Prey, bar; Munich Philharmonic Orchestra; Gary Bertini. RCA RL 3-0453.
+Gr 11-77 p879 +RR 12-77 p81

Songs: Die abgebluhte Linde, D 514; Heimliches Lieben, D 922; Minnelied, D 429; Der Musensohn, D 764. cf BRAHMS: Songs (Saga 5277).

Songs: An die Laute, K 905; An die Musik, D 547; An Silvia, D 891; Auflosung, D 807; Bei dir Allein, D 866; Dass sie hier gewesen,

D 775; Der Einsame, D 800; Fischerweise, D 881; Die Forelle, D 550; Ganymed, D 544; Der Schiffer, D 536; Standchen, D 889; Die Sterne, D 939; Uber Wildemann, D 884; Der Wanderer an den Mond, D 870; Wanderers Nachtlied, D 768. cf Schwanengesang, D 957: Liebesbotschaft.

Songs: An die Musik, D 547; Der Tod und das Madchen, D 531. cf Seraphim IB 6105.

Songs: An mein Herz; Blondel zu Marien, D 626; Ganymed, D 544; Heidenroslein,; Der Musensohn, D 764; Nur wer die Sehnsucht kennt, D 877/4; Schafers Klagelied, D 121; Sprache der Liebe, D 410; Rastlose Liebe, D 138. cf SCHOENBERG: Das Buch der hangenden Garten, op. 15.

Songs: Auf dem Strome, op. 119, D 943. cf AMON: Quartet, horn, op. 20, no. 1.

Songs: Ave Maria. cf Seraphim 60274.

Songs: Ave Maria, D 839; Mille cherubini in coro (arr. Melichar). cf London OS 26437.

Songs: Du bist die Ruh, D 776; Erlkonig, D 328; Gretchen am Spinnrade, D 118. cf Club 99-108.

Songs: Du bist die Ruh, D 776; Die Forelle, D 550; Gretchen am Spinnrade, D 118; Heidenroslein, D 257. cf L'Oiseau-Lyre SOL 345.

Songs: Der Erlkonig, D 328. cf Decca SDD 507.

Songs: Horch, horch, die Lerch; Der Musensohn, D 764; Lachen und Weinen, D 777; The shepherd on the rock, D 965. cf Abbey LPB 778.

Songs: Salve regina. cf Argo ZK 11.

Songs: Standchen. cf Bruno Walter Society BWS 729.

Songs: Standchen. cf Philips 6392 023.

Songs: Zogernd Leise, D 920. cf BEETHOVEN: Ah perfido, op. 65.

Songs: Zogernd Leise, D 920. cf Philips 6767 001

2411 Symphony, no. 5, K 485, B flat major. Symphony, no. 8, D 759, B minor. Georges Enesco Symphony Orchestra; John Barbirolli. Everest SDBR 3411.

-NR 12-77 p3

2412 Symphony, no. 5, D 485, B flat major. Symphony, no. 8, D 759, B minor. COA; Bernard Haitink. Philips 9500 099 Tape (c) 7300 512.

++ARG 5-77 p34
+-Audio 6-77 p134
+-HF 10-77 p114
+-MJ 4-77 p33
++NR 5-77 p4

2413 Symphony, no. 5, D 485, B flat major. Minuets, D 89. Moscow Chamber Orchestra; Rudolf Barshai. Westminster WGS 8335.

+-ARG 5-77 p35
+HF 10-77 p114
/NR 5-77 p4

Symphony, no. 5, D 485, B flat major. cf MENDELSSOHN: Symphony, no. 5, op. 107, D minor.

2414 Symphony, no. 8, D 759, B minor. Rosamunde, op. 26, D 797: Overture. GLUCK: Alceste: Overture. BPhO; Welhelm Furtwangler. DG 2535 804.

++Gr 5-77 p1690
+HFN 6-77 p137
++RR 5-77 p56

Symphony, no. 8, D 759, B minor. cf Symphony, no. 5, D 485, B flat major (Everest SDBR 3411).

Symphony, no. 8, D 759, B minor. cf Symphony, no. 5, D 485, B flat major (Philips 9500 099).

Symphony, no. 8, D 759, B minor. cf BEETHOVEN: Symphony, no. 5, op. 67, C minor.

Symphony, no. 8, D 759, B minor. cf BRAHMS: Tragic overture, op. 81.
Symphony, no. 8, D 759, B minor. cf HAYDN: Symphony, no. 104, D major.
Symphony, no. 8, D 759, B minor. cf MAHLER: Symphony, no. 4, G major.
Symphony, no. 8, D 759, B minor. cf MOZART: Symphony, no. 38, K 504, D major.

2415 Symphony, no. 9, D 944, C major. Cologne Radio Symphony Orchestra; Erich Kleiber. Amadeo AVRS 19015. (Reissue from Decca CL 237)
-HFN 12-77 p179 -RR 11-77 p69

2416 Symphony, no. 9, C major. LSO; Bruno Walter. Bruno Walter Society BWS 727.
+-NR 1-77 p4

2417 Symphony, no. 9, D 944, C major. LSO; Josef Krips. Decca SPA 467 Tape (c) KCSP 467. (Reissue from SXL 2045)
+Gr 4-77 p1556 +-HFN 5-77 p138
+-Gr 5-77 p1743 tape +-RR 4-77 p56
+-HFN 4-77 p151 +-RR 5-77 p93 tape

2418 Symphony, no. 9, D 944, C major. Israel Philharmonic Orchestra; Zubin Mehta. Decca SXL 6729 Tape (c) KSCX 6729.
+Gr 7-77 p180 +HFN 10-77 p171 tape
++Gr 10-77 p698 tape -RR 7-77 p60
+HFN 7-77 p119 -RR 12-77 p98

2419 Symphony, no. 9, D 944, C major. BPhO; Wilhelm Furtwangler. DG 2535 808. (Reissued from DG 18015/6)
++Gr 5-77 p1690 +RR 5-77 p56
+HFN 6-77 p137

2420 Symphony, no. 9, D 944, C major. COA; Bernard Haitink. Philips 9500 097 Tape (c) 7300 510.
++ARG 8-77 p26 +-HFN 7-77 p127 tape
+-Audio 6-77 p134 ++MJ 9-77 p35
++Gr 2-77 p1289 +-NR 8-77 p3
+HF 10-77 p114 +-RR 2-77 p58
++HF 11-77 p138 tape +-RR 6-77 p98 tape
+-HFN 2-77 p129 ++SFC 9-25-77 p50

Symphony, no. 9, D 944, C major. cf RCA CRM 5-1900.

2421 Trio, piano, no. 1, op. 99, D 898, B flat major. Trio, piano, no. 2, op. 100, D 929, E flat major. Henryk Szeryng, vln; Pierre Fournier, vlc; Artur Rubinstein, pno. RCA ARL 2-0731 (2).
+Gr 7-75 p205 +NR 5-75 p8
++HF 6-75 p90 +-RR 8-75 p51
++HFN 8-75 p83 ++St 7-75 p106
++MJ 9-75 p51 ++St 5-77 p71

Trio, piano, no. 1, op. 99, D 898, B flat major. cf Nocturne, op. 148, D 897, E flat major.
Trio, piano, no. 2, op. 100, D 929, E flat major. cf Nocturne, op. 148, D 897, E flat major.
Trio, piano, no. 2, op. 100, D 929, E flat major. cf Trio, piano, no. 1, op. 99, D 898, B flat major.
Trio, strings, no. 1, D 471, B flat major. cf Quartet, strings, no. 14, D 810, D minor.
Trio, strings, no. 1, D 471, B flat major. cf Quintet, strings, op. 163, D 956, C major.
Variations on an original theme, D 603, B flat major. cf Grand duo, D 812, C major.

Die Verschworenen, D 787: Ich schleiche band und still herum. cf Works, selections (Philips 9500 170).
Der Vierjahrige Posten, D 190: Overture. cf Works, selections (Philips 9500 170).
Wiener Tanze. cf BERTE: Lilac time.
Winterreise, op. 89, D 911. cf Die schone Mullerin, op. 25, D 795 (DG 2720 059).
Winterreise, op. 89, D 911. cf Die schone Mullerin, op. 25, D 795 (Philips 6767 300).
Winterreise, op. 89, nos. 19 and 24, D 911. cf LISZT: Harmonies poetiques et religieuses, no. 7, G 173: Funerailles.
Winterreise, op. 89, D 911: Gute Nacht. cf International Piano Library IPL 113.
Winterreise, op. 89, D 911: The post; The organ grinder. cf Kaibala 40D03.

2422 Works, selections: Alfonso und Estrella, D 732: Von Fels und Wald umrungen; Wer bist du, holdes Wesen; Freundlich bist du mir erschienen; Konnt ich ewig hier verweilen; Lass dir als Erinnerungszeichen. Die Burgschaft, D 435: Die Mutter sucht ihr liebes Kind; Welche Nacht hab ich erlebt; Horch die Seufzer uns'rer Mutter. Claudine von Villa Bella, D 239: Hin und wieder fliegen die Pfeile; Liebe schwarmt auf allen Wegen. Die Freunde von Salamanka, D 326: Gelagert unter'm hellen Dach. Die Verschworenen, D 787: Ich schleiche band und still herum. Der Vierjahrige Posten, D 190: Overture. Die Zwillingsbruder, D 647: Der Vater mag wohl immer Kind mich nennen. Elly Ameling, s; Claes Ahnsjo, t; Rotterdam Philharmonic Orchestra; Edo de Waart. Philips 9500 170.

++ARG 6-77 p45
+Gr 2-77 p1317
+HFN 2-77 p129
+NR 6-77 p12
+ON 6-77 p44
++RR 2-77 p37
+St 6-77 p143

2423 Die Zwillingsbruder. Helen Donath, s; Nicolai Gedda, t; Dietrich Fischer-Dieskau, bar; Kurt Moll, Hans-Joachim Gallus, bs; Bavarian Radio Orchestra and Chorus; Wolfgang Sawallisch. EMI Electrola 065 28833. (also HMV ASD 3300)

-Gr 12-76 p1051
+HFN 12-76 p133
++MT 3-77 p217
++RR 12-76 p45
++St 9-76 p124

Die Zwillingsbruder, D 647: Der Vater mag wohl immer Kind mich nennen. cf Works, selections (Philips 9500 170).

SCHULHOFF, Erwin
Esquisses de jazz. cf Supraphon 111 1721/2.
Rag music. cf Supraphon 111 1721/2.
Sonata, solo violin. cf CAJKOVSKIJ: Quartet, strings, no. 4.

SCHULLER, Gunther
Contours. cf Maderna: Il giardino religioso.
Moods for tuba quartet. cf Crystal S 221.
Quartet, strings, no. 1. cf New England Conservatory NEC 115.
Quintet, woodwinds. cf Vox SVBX 5307.
Tre invenzione. cf MADERNA: Il giardino religioso.

SCHULZ, Johann
Ihr Kinderlein kommet. cf RCA PRL 1-8020.
Der Mond ist aufgegangen. cf Telefunken DT 6-48085.

SCHUMAN, William
Chester overture. cf Menc 76-11.
Credendum (Article of faith). cf IMBRIE: Symphony, no. 3.

New England triptych. cf BENNETT, R. R.: The fun and faith of William Billings, American.
Prelude. cf Vox SVBX 5353.
2424 Symphony, no. 8. SUDERBURG: Concerto "Within the memory of time". Bela Siki, pno; Seattle Symphony Orchestra, NYP; Milton Katims, Leonard Bernstein. Odyssey Y 34140. (Reissue from Columbia MS 6512)
++ARG 11-76 p39 +NR 12-76 p7
+-HF 12-76 p126 ++SFC 7-3-77 p34
++MJ 3-77 p46
Three-score set. cf Vox SVBX 5303.
Variations on "America" (after Charles Ives). cf London CSA 2246.
SCHUMANN, Robert
Abendlied, op. 107, no. 6 (Soldat). cf Discopaedia MOB 1019.
Adagio and allegro, op. 70, A flat major. cf AMON: Quartet, horn, op. 20, no. 1.
Adagio and allegro, op. 70, A flat major. cf BEETHOVEN: Sonata, horn, op. 17, F major.
Adagio and allegro, op. 70, A flat major. cf COUPERIN: Pieces en concert.
2425 Album fur die Jugend, op. 68. Kinderscenen, op. 15. Alexis Weissenberg, pno. Connoisseur Society CS 2-2110 (2).
-HF 7-77 p106 ++SFC 10-31-76 p35
++NR 1-77 p14 +St 11-77 p150
Album fur die Jugend, op. 68. cf Piano works (Vox SVBX 5470).
Album fur die Jugend, op. 68, nos. 1-43. cf Piano works (Turnabout TV 37093/5).
Albumblatter, op. 124. cf Piano works (Telefunken FK 6-35287).
Am springbrunne. cf International Piano Library IPL 101.
2426 Arabeske, op. 18, C major. Fantasiestucke, op. 12. Kinderscenen, op. 15. Homero Francesch, pno. DG 2530 644.
+Gr 3-77 p1430 +-RR 3-77 p85
+HFN 3-77 p111
Arabeske, op. 18, C major. cf Piano works (Turnabout TV 37093/5).
2427 Blumenstuck, op. 19, D major. Kinderscenen, op. 15. Papillons, op. 2. Romances, op. 28 (3). Claudio Arrau, pno. Philips 6500 395.
+-HF 11-77 p124 +-NYT 9-25-77 pD19
+NR 11-77 p14
Blumenstuck, op. 19, D major. cf Piano works (Turnabout TV 37093/5).
Bunte Blatter, op. 99. cf Piano works (Vox SVBX 5470).
Bunte Blatter, op. 99, nos. 1-14. cf Piano works (Telefunken FK 6-35287).
Canon on "To Alexis", op. posth. cf Piano works (Turnabout TV 37093/5).
Canons, op. 56, no. 5, B minor. cf Argo SPA 507.
Canons, op. 56, no. 5, B minor. cf Vista VPS 1035.
2428 Carnaval, op. 9. Waldscenen, op. 82. Sequeira Costa, pno. Supraphon 111 2026.
+-Gr 9-77 p459 /NR 9-77 p13
+-HFN 7-77 p119 +-RR 6-77 p85
Carnaval, op. 9. cf Piano works (Turnabout TV 37093/5).
Carnaval, op. 9. cf BEETHOVEN: Sonata, piano, no. 26, op. 81a, E flat major.
Clavierstucke, op. 32, no. 3: Romanze. cf International Piano Library IPL 101.
2429 Concerto, piano, op. 54, A minor. Symphony, no. 1, op. 38, B flat

major. Friedrich Gulda, pno; VPO, LSO; Volkmar Andreae, Josef Krips. Decca SPA 493. (Reissues from LXT 5280, 5347, SXL 2223)
+Gr 11-77 p844 +-RR 10-77 p58
+HFN 10-77 p167
Concerto, piano, op. 54, A minor. cf CHOPIN: Concerto, piano, no. 2, op. 21, F minor.
Concerto, piano, op. 54, A minor. cf GRIEG: Concerto, piano, op. 16, A minor.
Concerto, piano, op. 54, A minor. cf EMI Italiana C 153 52425/31.
Concerto, piano, op. 54, A minor. cf HMV SLS 5094.
2430 Concerto, violin, D minor. Concerto, violoncello, op. 129, A minor. Susanne Lautenbacher, vln; Laszlo Varga, vlc; Luxembourg Radio Orchestra, Westphalian Symphony Orchestra; Pierre Cao, Siegfried Landau. Turnabout TV 34631 Tape (c) KTVC 34631.
-Gr 8-77 p306 +-RR 8-77 p62
+-HFN 8-77 p89 /-RR 12-77 p98 tape
-HFN 10-77 p171 tape ++SFC 10-9-77 p40
Concerto, violin, D minor. cf MOZART: Concerto, violin, no. 5, K 219, A major.
2431 Concerto, violoncello, op. 129, A minor. Pieces in folk style, op. 102 (5). Pablo Casals, vlc; Leopold Mannes, pno; Prades Festival Orchestra; Eugene Ormandy. CBS 71298. (Reissue from Philips ABR 4035, 61579)
+Gr 7-77 p187 +RR 6-77 p67
+-HFN 6-77 p137
Concerto, violoncello, op. 129, A minor. cf Concerto, violin, D minor.
Concerto, violoncello, op. 129, A minor. cf BLOCH: Schelomo.
Concerto, violoncello, op. 129, A minor. cf LALO: Concerto, violoncello, D minor.
Der contrabandiste, op. 74. cf International Piano Library IPL 117.
Der contrabandiste, op. 74 (arr. Liszt). cf Westminster WGM 8309.
Les deux grenadiers. cf Rubini GV 39.
Dicterliebe, op. 48. cf CBS SQ 79200.
Dichterliebe, op. 48: Ich grolle nicht. cf BEETHOVEN: An die ferne Geliebte, op. 98.
Etudes, op. 3. cf Piano works (Vox SVBX 5470).
Etudes on caprices by Paganini, op. 10 (6). cf Piano works (Telefunken FK 6-35287)
Etudes on caprices by Paganini, op. 10. cf Piano works (Vox SVBX 5470).
2432 Fantasia, op. 17, C major. Sonata, piano, no. 1, op. 11, F sharp minor. Maurizio Pollini, pno. DG 2530 379.
++Gr 5-74 p2046 +-NYT 3-13-77 pD22
+HF 2-74 p102 +RR 4-74 p70
+-MJ 3-74 p10 ++SFC 12-23-73 p18
++NR 3-74 p11 ++St 3-74 p120
Fantasia, op. 17, C major. cf LISZT: Sonata, piano, G 178, B minor.
Fantasiestucke, op. 12. cf Arabeske, op. 18, C major.
Fantasiestucke, op. 12. cf BEETHOVEN: Sonata, piano, no. 18, op. 31, no. 3, E flat major.
Fantasiestucke, op. 12, nos. 1-3: Des Abends; Aufschwang; Warum. cf CHOPIN: Etude, op. 25, no. 7, C sharp minor.
Fantasiestucke, op. 12, no. 7: Traumeswirren. cf Westminster WGM 8309.
Fantasiestucke, op. 12, no. 9. cf Piano works (Vox SVBX 5470).

Fantasiestucke, op. 73. cf GRIEG: Sonata, violoncello and piano, op. 36, A minor.
Fantasiestucke, op. 73. cf L'Oiseau-Lyre DSLO 17.
Faschingsschwank aus Wien, op. 26. cf Piano works (Turnabout TV 37093/5).

2433 Frauenliebe und Leben, op. 42. Liederkreis, op. 39. Janet Baker, ms; Daniel Barenboim, pno. Angel S 37222. (also HMV ASD 3217)
+Gr 7-76 p205 ++RR 7-76 p78
+-HF 11-76 p124 +St 5-77 p117
++HFN 8-76 p86 +-STL 9-19-76 p36
++NR 11-76 p11

2434 Frauenliebe und Leben, op. 42. Liederkreis, op. 39. Mildred Miller, ms; John Wustman, pno. Musical Heritage Society MHS 3556 Tape (c) MHC 5556.
+HF 10-77 p116

2435 Frauenliebe und Leben, op. 42. Liederkreis, op. 39. Jessye Norman, s; Irwin Gage. pno. Philips 9500 110.
-ARG 7-77 p31 +HFN 6-77 p133
+Audio 12-77 p50 +NR 6-77 p12
+-Gr 6-77 p89 +-RR 6-77 p91
-HF 10-77 p116 +-St 9-77 p142

Frauenliebe und Leben, op. 42. cf BRAHMS: Songs (Gale GMFD 7-86-006).
Frauenliebe und Leben, op. 42. cf BRAHMS: Songs (Saga 5277).
Fugues (4). cf Piano works (Vox SVBX 5470).
Fugues, op. 72. cf Piano works (Telefunken FK 6-35287).
Humoreske, op. 20, B flat major. cf Piano works (Telefunken FK 6-35287).
Humoreske, op. 20, B flat major. cf Piano works (Vox SVBX 5470).

2436 Impromptus on a theme by Clara Wieck, op. 5. Sonata, piano, no. 3, op. 14, F minor. Jean-Philippe Collard, pno. Connoisseur Society CS 2081.
+Audio 12-76 p86 +NR 11-75 p14
+HF 12-75 p106 +St 1-77 p126
+-MJ 4-76 p30

Kinderscenen, op. 15. cf Album fur die Jugend, op. 68.
Kinderscenen, op. 15. cf Arabeske, op. 18, C major.
Kinderscenen, op. 15. cf Blumenstuck, op. 19, D major.
Kinderscenen, op. 15. cf NILSSON: Quantitaten.
Kinderscenen, op. 15, no. 7: Traumerei. cf Decca PFS 4387.
Kinderscenen, op. 15, no. 7: Traumerei. cf HMV RLS 723.
Klavierstucke, op. 32 (4). cf Piano works (Telefunken FK 6-35287).
Klavierstucke, op. 85: Abendlied. cf Discopaedia MB 1014.
Klavierstucke, op. 85: Abendlied. cf Discopaedia MB 1015.

2437 Kreisleriana, op. 16. Variations on a theme by Clara Wieck, op. 14. Vladimir Horowitz, pno. Columbia MS 7264 Tape (c) 1611 0214. (also CBS 72841)
++FF 9-77 p91 ++HF 4-71 p67
++Gr 12-70 p1010 ++HF 9-71 p132

Kreisleriana, op. 16. cf Piano works (Vox SVBX 5470).
Kreisleriana, op. 16. cf International Piano Library IPL 5007/8.
Liederkreis, op. 24. cf Songs (DG 2530 543).
Liederkreis, op. 24: Schone Wiege meiner Leiden. cf BEETHOVEN: An die ferne Geliebte, op. 98.
Liederkreis, op. 39. cf Frauenliebe und Leben, op. 42 (Angel 37222).
Liederkreis, op. 39. cf Frauenliebe und Leben, op. 42 (Musical Heritage Society MHS 3556).

Liederkreis, op. 39. cf Frauenliebe und Leben, op. 42 (Philips 9500 110).
Liederkreis, op. 39. cf Songs (DG 2740 167).
Liederkreis, op. 39: In der Fremde. cf Bruno Walter Society BWS 729.
Manfred overture, op. 115. cf Symphonies, nos. 1-4.
Manfred overture, op. 115. cf BRAHMS: Tragic overture, op. 81.
Marchenerzahlungen, op. 132. cf GLINKA: Trio pathetique, D minor.
Myrthen, op. 25. cf Songs (DG 2530 543).
Myrthen, op. 25: Widmung, Freisinn, Der Nussbaum, Sitz ich allein, Setze mir nich du Grobian, Die Lotosblume, Talismane, Hochlanders Abschied, Mein Herz ist schwer, Ratsel, Leis rudern hier, Wenn durch die Piazzetta, Hauptmanns Weib, Was will die einsame Trane, Niemand, Du bist wie eine Blume, Ich sende einen Gruss wie Duft der Rosen, Zum Schluss. cf Songs (DG 2740 167).
Nachtstucke, op. 23. cf Piano works (Telefunken FK 6-35287).
Nachtstucke, op. 23. cf Piano works (Vox SVBX 5470).
Overture to Goethe's "Hermann und Dorothea", B minor. cf Symphony, no. 1, op. 38, B flat major.

2438 Papillons, op. 2. Symphonic etudes, op. 13. Murray Perahia, pno. Columbia M 34539 Tape (c) MT 34539.
+NR 12-77 p12

Papillons, op. 2. cf Blumenstuck, op. 19, D major.
Papillons, op. 2. cf Piano works (Vox SVBX 5470).

2439 Piano works: Albumblatter, op. 124. Bunte Blatter, op. 99, nos. 1-14. Etudes on caprices by Paganini, op. 10 (6). Fugues, op. 72 (4). Humoreske, op. 20, B flat major. Klavierstucke, op. 32 (4). Nachtstucke, op. 23. Toccata, op. 7, C major. Variationen uber eigenes Thema. Waldscenen, op. 82. Karl Engel, pno. Telefunken FK 6-35287 (4).
+-Gr 5-76 p1787 -MJ 3-77 p47
-HF 9-77 p110 +NR 4-77 p13

2440 Piano works: Arabeske, op. 18, C major. Album fur die Jugend, op. 68, nos. 1-43. Blumenstuck, op. 19, D major. Carnaval, op. 9. Canon on "To Alexis", op. posth. Faschingsschwank aus Wien, op. 26. Symphonic studies, op. 13. Supplement to Symphonic studies, op. posth. Toccata, op. 7, C major. Peter Frankl, pno. Turnabout TV 37093/5 (3).
+-Gr 11-77 p867 +-RR 11-77 p95
+HFN 11-77 p181

2441 Piano works: Album fur die Jugend, op. 68. Bunte Blatter, op. 99. Etudes, op. 3. Fantasiestucke, op. 12, no. 9. Fugues (4). Humoreske op. 20, B flat major. Kreisleriana, op. 16. Nachtstucke, op. 23. Papillons, op. 2. Pieces, op. 32 (4). Etudes on caprices by Paganini, op. 10. Peter Frankl, pno. Vox SVBX 5470 (3).
+HF 4-77 p112 +NR 4-77 p13
+MJ 12-76 p29 +SFC 10-3-76 p33

Pieces, piano, op. 32 (4). cf Piano works (Vox SVBX 5470).
Pieces in folk style, op. 102 (5). cf Concerto, violoncello, op. 129, A minor.

2442 Quartet, piano, op. 47, E flat major. Quintet, piano, op. 44, E flat major. Thomas Rajna, pno; Alberni Quartet. CRD 1024 Tape (c) CRD 4024.
+Gr 8-76 p317 +-HFN 11-77 p187
+Gr 10-77 p705 tape +-RR 7-76 p68

+-HF 3-77 p111 ++St 11-76 p158
+-HFN 7-76 p95

2443 Quartet, piano, op. 47, E flat major. Quintet, piano, op. 44, E flat major. Beaux Arts Trio; Samuel Rhodes, vla; Dolf Bettelheim, vln. Philips 9500 065.
++ARG 3-77 p35 ++NR 2-77 p6
+-Gr 5-76 p1781 +RR 5-76 p62
+-HF 3-77 p110 ++St 8-77 p122
/HFN 5-76 p109 ++STL 8-8-76 p29
+-MJ 2-77 p31

Quartet, piano, op. 47, E flat major. cf BRAHMS: Quartet, piano and strings, no. 3, op. 60, C minor.

Quartets, strings, op. 41, nos. 1-3. cf BRAHMS: Quartets, strings, complete.

2444 Quartet, strings, op. 41, no. 1, A minor. Quartet, strings, op. 41, no. 2, F major. Alberni Quartet. CRD CRD 1033.
++Gr 3-77 p1422 +NR 10-77 p7
+HFN 6-77 p133 +-RR 3-77 p73

Quartet, strings, op. 41, no. 1, A minor. cf BRAHMS: Quartet, strings, no. 3, op. 67, B flat major.

Quartet, strings, op. 41, no. 1, A minor. cf MENDELSSOHN: Quartets, strings, op. 44, no. 1, D major.

Quartet, strings, op. 41, no. 2, F major. cf Quartet, strings, op. 41, no. 1, A minor.

Quintet, piano, op. 44, E flat major. cf Quartet, piano, op. 47, E flat major (CRD 1024).

Quintet, piano, op. 44, E flat major. cf Quartet, piano, op. 47, E flat major (Philips 9500 065).

Quintet, piano, op. 44, E flat major. cf BRAHMS: Trio, horn, op. 40, E flat major.

2445 Requiem, op. 148. Requiem fur Mignon, op. 98b. Eva Andor, Katalin Szokefalvy-Nagy, s; Zsuzsa Barlay, Livia Budai, con; Gyorgy Korondy, t; Jozsef Gregor, bs; Budapest Chorus; HSO; Miklos Forrai. Hungaroton SLPX 11809.
-FF 9-77 p52 /NYT 6-5-77 pD19
+-Gr 4-77 p1588 +-RR 2-77 p89
+-HFN 5-77 p133

Requiem fur Mignon, op. 98b. cf Requiem, op. 148.

Romances, op. 28. cf Blumenstuck, op. 19, D major.

Romance, op. 28, no. 1, B flat minor. cf International Piano Library IPL 112.

Romance, op. 28, no. 2, F sharp major. cf International Piano Library IPL 101.

Romances, op. 94. cf REINECKE: Sonata, flute and piano, op. 167, E minor.

2446 Sonata, piano, no. 1, op. 11, F sharp minor. Sonata, piano, no. 2, op. 22, G minor. Lazar Berman, pno. Columbia/Melodiya M 34528. (also HMV ASD 3322)
+FF 11-77 p51 +-NR 9-77 p13
+Gr 3-77 p1429 ++RR 4-77 p77
+-HF 10-77 p118 +St 11-77 p151
+HFN 4-77 p145

2447 Sonata, piano, no. 1, op. 11, F sharp minor. Sonata, piano, no. 2, op. 22, G minor. Murray Perahia, pno. Columbia M 34539.
+NYT 9-25-77 pD19

Sonata, piano, no. 1, op. 11, F sharp minor. cf Fantasia, op. 17, C major.

Sonata, piano, no. 2, op. 22, G minor. cf Sonata, piano, no. 1, op. 11, F sharp minor (Columbia 34528).

Sonata, piano, no. 2, op. 22, G minor. cf Sonata, piano, no. 1, op. 11, F sharp minor (Columbia M 34539).

2448 Sonata, piano, no. 3, op. 14, F minor. SCRIABIN: Sonata, piano, no. 5, op. 53, F sharp major. Vladimir Horowitz, pno. RCA ARL 1-1766 Tape (c) ARK 1-1766 (ct) ARS 1-1766.

+Gr 12-76 p1022
++HF 1-77 p124
++HF 4-77 p121 tape
+MJ 1-77 p27
+HFN 4-77 p145
+MT 12-77 p1016
+NR 4-77 p13
++RR 12-76 p87
++SFC 10-10-76 p32
++St 1-77 p126

Sonata, piano, no. 3, op. 14, F minor. cf Impromptus on a theme by Clara Wieck, op. 5.

Sonata, violin, no. 1, op. 105, A minor. cf BRAHMS: Sonata, violin and piano, no. 2, op. 100, A major.

2449 Songs: Liederkreis, op. 24. Myrthen, op. 25. Dietrich Fischer-Dieskau, bar; Christoph Eschenbach, pno. DG 2530 543.

+Gr 8-75 p355
+HF 12-75 p106
++HFN 3-76 p103
+NR 12-75 p12
*ON 5-76 p48
+-RR 1-76 p61
++SFC 11-16-75 p32
+SR 9-3-77 p42
++St 3-76 p121

2450 Songs: Der frohe Wandersmann, op. 77, no. 1; Gedichte, op. 30 (3); Gedichte, op. 36 (6); Gedichte aus Ruckerts Liebesfruhling, op. 37: Der Himmel hat eine Trane geweint, Ich hab in mich gesogen, Flugel Flugel um zu fliegen, Rose Meer und Sonne; Gesange, op. 31 (3); Lieder, op. 40 (5); Lieder und Gesange, op. 27 (5); Liederkreis, op. 39; Myrthen, op. 25: Widmung, Freisinn, Der Nussbaum, Sitz ich allein, Setze mir nich du Grobian, Die Lotosblume, Talismane, Hochlanders Abschied, Mein Herz ist schwer, Ratsel, Leis rudern hier, Wenn durch die Piazzetta, Hauptmanns Weib, Was will die einsame Trane, Niemand, Du bist wie eine Blume, Ich sende einen Gruss wie Duft der Rosen, Zum Schluss; Romanzen und Balladen, op. 45: Der Schatzgraber, Fruhlingsfahrt. Dietrich Fischer-Dieskau, bar; Christoph Eschenbach, pno. DG 2740 167 (3). (Reissues from 2530 543)

+-Gr 11-77 p879
+HFN 11-77 p181
+RR 11-77 p109

2451 Songs: Der contrabandiste, op. 74. Gedichte and requiem, op. 90 (6). Der Handschuh, op. 87. Heitere Gesange, op. 125 (5). Husarenlieder, op. 117. Lieder, op. 89 (6). Lieder und Gesange, opp. 77, 96, 127. Soldatenlied. Waldlieder, op. 119. Bernard Kruysen, bar; Noel Lee, pno. Telefunken EK 6-48097 (2).

+-Gr 9-77 p468
++HFN 8-77 p91
+RR 8-77 p78

Songs: An den Sonnenschein, op. 38, no. 4; Dicterliebe, op. 48: Im wunderschonen Monat Mai, Die Rose die Lilie; Roselein, op. 89, no. 6. cf Swedish SLT 33209.

Songs: Auftrage, op. 77, no. 5; Volksliedchen, op. 51, no. 2. cf Bruno Walter Society BWS 729.

Songs: Die beiden Grenadiere, op. 49, no. 1; Belsazar, op. 57; Die Lotosblume, op. 25, no. 7; Widmung, op. 25, no. 1. cf BEETHOVEN: An die ferne Geliebte, op. 98.

Songs: Fruhlingsgrass; Gute Nacht. cf Telefunken DT 6-48085.
Songs: Die Lotosblume, op. 25, no. 7; Der Nussbaum, op. 25, no. 3. cf Seraphim IB 6105.
Songs: Die Lotosblume, op. 25, no. 7; Widmung, op. 25, no. 1. cf Club 99-108.
Supplement to Symphonic studies, op. posth. cf Piano works (Turnabout TV 37093/5).
Symphonic etudes, op. 13. cf Papillons, op. 2.
Symphonic etudes, op. 13. cf Piano works (Turnabout TV 37093/5).
Symphonic etudes, op. 13. cf ALBENIZ: Espana, op. 165, no. 2: Tango.
Symphonic etudes, op. 13. cf LISZT: Annees de pelerinage, 2nd year, G 161: Sonetti del Petrarca (3).

2452 Symphonies, nos. 1-4. Manfred overture, op. 115. St. Louis Symphony; Jerzy Semkow. Vox SVBX 5146 (3).
+NYT 11-27-77 pD15 +SFC 12-11-77 p61

2453 Symphony, no. 1, op. 38, B flat major. Symphony, no. 4, op. 120, D minor. VPO; Zubin Mehta. Decca SXL 6819 Tape (c) KSXC 6819. (also London 7039 Tape (c) 5-7039)
+Gr 6-77 p63 -NYT 12-4-77 pD18
+Gr 9-77 p511 tape +-RR 6-77 p67
+-HFN 6-77 p133 +-RR 9-77 p99 tape
+HFN 8-77 p99 tape

2454 Symphony, no. 1, op. 38, B flat major. Symphony, no. 4, op. 120, D minor. CSO; Daniel Barenboim. DG 2530 660 Tape (c) 3300 660.
+-Gr 12-77 p1145 +-RR 12-77 p51
+-HFN 12-77 p179

2455 Symphony, no. 1, op. 38, B flat major. Overture to Goethe's "Hermann und Dorothea", B minor. Leipzig Gewandhaus Orchestra; Kurt Masur. Musical Heritage Society MHS 3595.
+NYT 12-4-77 pD18

Symphony, no. 1, op. 38, B flat major. cf Concerto, piano, op. 54, A minor.
Symphony, no. 4, op. 120, D minor. cf Symphony, no. 1, op. 38, B flat major (Decca SXL 6819).
Symphony, no. 4, op. 120, D minor. cf Symphony, no. 1, op. 38, B flat major (DG 3530 660).
Symphony, no. 4, op. 120, D minor. cf MENDELSSOHN: Symphony, no. 4, op. 90, A major.
Toccata, op. 7, C major. cf Piano works (Telefunken FK 6-35287).
Toccata, op. 7, C major. cf Piano works (Turnabout TV 37093/5).
Toccata, op. 7, C minor. cf Westminster WGM 8309.
Trio, piano, no. 1, op. 63, D minor. cf HMV RLS 723.
Variationen uber eigenes Thema. cf Piano works (Telefunken FK 6-35287).
Variations on a theme by Clara Wieck, op. 14. cf Kreisleriana, op. 16.
Waldscenen, op. 82. cf Carnaval, op. 9.
Waldscenen, op. 82. cf Piano works (Telefunken FK 6-35287).

SCHUTZ, Heinrich

Das ist je gewisslich wahr. cf Pelca PSR 40607.

2456 Songs (Italian madrigals): Alma afflitta; Cosi morir debb'io; d'Orrida selce alpina; Di marmo siete voi; Dunque addio, care selve; Feriteve, viperette mordaci; Fiamma ch'allaccia; Fuggio o mio core; Giunto e pur; Io moro; Mi saluta costei; O dolcezze amarissime; O primavera; Quella damma son io; Ride la primavera; Selve beate; Sospir che del bel petto; Tornate a cari

baci; Vasto mar. Monteverdi Choir; Jurgen Jurgens. DG 2708 033 (2).
+-Gr 11-76 p858 +-RR 2-77 p89
+HFN 10-76 p175

SCHWANTNER, Joseph
Autumn canticles. cf BLOCH: Nocturnes.
Consortium I. cf Delos DEL 25406.
In aeternum. cf Delos DEL 25406.
Modus caelestis. cf PENN: Ultra mensuram.

SCOTT
Annie Laurie. cf HMV RLS 719.
Goodnight. cf HMV RLS 719.

SCOTT, Cyril
2457 Concerto, piano, no. 2. Early one morning. John Ogdon, pno; LPO; Bernard Herrmann. Lyrita SRCS 82.
+Gr 4-77 p1556 ++RR 4-77 p59
+HFN 4-77 p145
Early one morning. cf Concerto, piano, no. 2.
Water wagtail, op. 71, no. 3. cf Saga 5445.

SCOTTO (1500 c. Italy)
O fallace speranza. cf Enigma VAR 1024.

SCOTT-WOOD
Shy serenade. cf Rediffusion 15-57.

SCRIABIN, Alexander
Etude, op. 12, no. 12. cf Pye PCNH 9.
Poeme de l'extase, op. 54. cf RCA CRL 3-2026.
2458 Sonata, piano, no. 1, op. 6, F minor. Sonata, piano, no. 3, op. 23, F sharp minor. Lazar Berman, pno. HMV ASD 3396.
+Gr 12-77 p1108 +RR 12-77 p72
+HFN 12-77 p180
2459 Sonatas, piano, nos. 3-5, 9. Vladimir Ashkenazy, pno. Decca SXL 6705 Tape (c) KSXC 6705. (also London CS 6920)
+Gr 11-75 p864 ++NR 11-76 p12
+HFN 12-75 p164 ++RR 11-75 p84
+HFN 8-76 p95 tape ++St 11-76 p158
+MJ 12-76 p28 +-St 1-77 p128
Sonata, piano, no. 3, op. 23, F sharp minor. cf Sonata, piano, no. 1, op. 6, F minor.
Sonata, piano, no. 5, op. 53, F sharp major. cf SCHUMANN: Sonata, piano, no. 3, op. 14, F minor.
2460 Universe (Nemtin). Aleksei Lyubimov; pno; Irina Orlova, org; RSFSR Yurlov Chorus; MPO; Kiril Kondrashin. Angel/Melodiya SR 40260. (also HMV Melodiya ASD 3201)
+Gr 5-76 p1795 +-RR 7-76 p79
+-HF 6-76 p94 *SFC 3-14-76 p27
+HFN 7-76 p95 +-St 10-76 p132
-MJ 5-76 p28 +STL 5-9-76 p38
+MT 8-76 p662 +-Te 3-77 p29
*NR 5-76 p8

SEARLE, Humphrey
Aubade, op. 28. cf BANKS: Concerto, horn.

SEEGER
Chant. cf Vox SVBX 5353.

SEGERSTAM, Leif
2461 Quartets, strings, nos. 4 and 5. Segerstam Quartet. Swedish Society Discofil SLT 33222.
++ARG 7-77 p33

2462 Quartet, strings, no. 6. SEGERSTAM/WERNER: Rituals in La. Segerstam Quartet; Lasse Werner, Leif Segerstam pno. BIS LP 20.
++ARG 7-77 p33 -Gr 3-77 p1457
2463 Quartet, strings, no. 7. Three moments of parting. Segerstam Quartet; Hannele Segerstam, vln; Ralf Gothoni, pno. BIS LP 39.
++ARG 7-77 p33 +HFN 12-77 p180
Rituals in La. cf SEGERSTAM: Quartet, strings, no. 6.
Three moments of parting. cf Quartet, strings, no. 7.

SEGNI, Julio
Fantasien. cf Telefunken AW 6-42033.

SEGOVIA, Andres
Dos anecdotas. cf Lyrichord LLST 7299.

SEIBER, Matyas
Concertino, clarinet and strings. cf ADDISON: Concerto, trumpet, strings and percussion.

SELBY, William
Anthem for Christmas day. cf Folkways FTS 32378.
A lesson. cf Folkways FTS 32378.
Songs: O be joyful to the Lord; Ode for the New Year. cf Vox SVBX 5350.
Voluntary, A major. cf Orion ORS 76255.

SEMEGEN, Daria
Electronic composition, no. 1. cf Odyssey Y 34139.

SENAILLE
Allegro spiritoso. cf Kaibala 20B01.

SENFL, Ludwig
Ach Elslein, liebe Elselein. cf Argo D40D3.
Entlaubet ist der Walde. cf Argo D40D3.
Das glaut zu Speyer. cf Argo D40D3.
Gottes Namen fahren wir. cf Argo D40D3.
Ich stund an einem Morgen. cf Argo D40D3.
Ich weiss nit, was er ihr verhiess. cf Argo D40D3.
Laub, Gras u Blut. cf BIS LP 75.
Lieder (5). cf BIS LP 3.
Meniger stellt nach Geld. cf Argo D40D3.
Mit Lust tritt ich an diesen Tanz. cf Argo D40D3.
Non wollt ihr horen neue Mar. cf Nonesuch H 71326.
Quis dabit oculis nostris. cf Argo D40D3.
Was wird es doch. cf Argo D40D3.

SEREBRIER, Jose
Symphony. cf CHAVEZ: Toccata.

SERMISY, Claudine de
Content desir; La, je j'y plaine. cf L'Oiseau-Lyre 12BB 203/6.

SERRANO Y RUIZ, Emilio
La cancion del olvido: Cancion de Marinella. cf Discopaedia 1013.
El carro del sol: Cancion veneciana. cf London OS 26435.

SESSIONS, Roger
Chorale, no. 1, G major. cf Vista VPS 1038.
2464 Concerto, violin. Paul Zukofsky, vln; ORTF; Gunther Schuller. CRI SD 220.
+Gr 3-77 p1454 +RR 12-76 p64
From my diary. cf Vox SVBX 5303.
On the beach at Fontana. cf New World Records NW 243.
Pieces, violoncello (6). cf BACH: Suite, solo violoncello, no. 1, S 1007, G major.
Sonata, violin. cf BLACKWOOD: Sonata, violin, no. 2.
Turn O libertad. cf Vox SVBX 5353.

SEVENTY-FIVE YEARS OF CATHEDRAL MUSIC: Saint John the Divine, New York. cf St. John the Divine Cathedral Church.

SEVERAC, Deodat de
Pippermint-get. cf Vox SVBX 5483.
Sous les Lauriers roses. cf Vox SVBX 5483.

SHAPERO, Harald
Sonata, trumpet and piano, C major. cf PALMER: Sonata, trumpet, B flat major.

SHAPEY, Ralph
2465 Praise. Paul Geiger, bs-bar; University of Chicago Contemporary Chamber Players; Paul Shapey. CRI SD 355.
-Gr 9-77 p500
+HFN 12-77 p180
+MJ 3-77 p46
+NR 10-76 p7
+-RR 8-77 p78
+St 2-77 p124

SHCHEDRIN, Rodion
Basso ostinato, clarinet and piano. cf BIS LP 62.
Humoresque. cf DENISOV: Variations.
In the style of Albeniz. cf Hungaroton SLPX 11825.

SHEPHERD
O happy dames. cf BG HM 57/8.

SHEPHERD, Arthur
Sonata, violin and piano. cf BABIN: David and Goliath.

SHEPHERD, John
Verbum caro factum est. cf Abbey LPB 776.

SHIELDS, Alice
Wildcat songs. cf BEHRENS: The feast of life.

SHIFRIN, Seymour
Quartet, strings, no. 4. cf MONOD: Cantus contra cantum I.

SHINOHARA, Makoto
Fragmente. cf BIS LP 48.

SHOSTAKOVICH, Dmitri
Age of gold, op. 22. cf GLIERE: The red poppy, op. 70: Ballet suite.
Age of gold, op. 22. cf CBS 61781.
Concertino, 2 pianos, op. 94. cf Works, selections (HMV RLS 721).
Concerto, piano, no. 2, op. 101, F major. cf Concerto, piano and trumpet, no. 1, op. 35, C minor.
Concerto, piano, no. 2, op. 101, F major. cf Works, selections (HMV RLS 721).
2466 Concerto, piano and trumpet, no. 1, op. 35, C minor. Concerto, piano, no. 2, op. 102, F major. Fantastic dances, op. 5 (3). Cristina Ortiz, pno; Rodney Senior, tpt; Bournemouth Symphony Orchestra; Paavo Berglund. HMV ASD 3081 Tape (c) TC ASD 3081. (also Angel S 37109)
+Gr 6-75 p55
+Gr 9-77 p511 tape
+-HF 9-75 p90
++HFN 6-75 p93
+-HFN 10-77 p171 tape
+NR 7-75 p6
+-RR 6-75 p53
+-RR 10-77 p98 tape
+St 11-75 p136

Concerto, piano and trumpet, no. 1, op. 35, C minor. cf BACH: Concerto, harpsichord, no. 5, S 1056, F minor.
Concerto, violoncello, no. 2, op. 126, G major. cf GLAZUNOV: Song of the troubadour, op. 71.
Fantastic dances, op. 5 (3). cf Concerto, piano and trumpet, no. 1, op. 35, C minor.
Fantastic dances, op. 5 (3). cf Hungaroton SLPX 11825.
Festive march. cf HMV CSD 3782.

From Jewish folk poetry, op. 79. cf Works, selections (HMV RLS 721).

2467 The gadfly, op. 97a. USSR Cinema Orchestra; Emin Khachaturian. HMV ASD 3309.
+Gr 2-77 p1289 +-MT 8-77 p651
+HFN 2-77 p131 +RR 2-77 p58

2468 Hamlet, op. 32. New Babylon, op. 18. MPO; Gennady Rozhdestvensky. HMV Melodiya ASD 3381.
+Gr 12-77 p1090 +RR 12-77 p54
+HFN 12-77 p180

2469 The new Babylon, op. 18. Moscow Philharmonic Orchestra, Soloists; Gennady Rozhdestvensky. Columbia/Melodiya M 34502. (also Columbia/Melodiya X698)
++Audio 7-77 p103 ++NR 3-77 p4
+MJ 7-77 p70 +St 5-77 p117
++HF 11-77 p126

New Babylon, op. 18. cf Hamlet, op. 32.

2470 The nose, op. 15. Soloists; Moscow Musical Chamber Theatre Chamber Orchestra; Gennady Rozhdestvensky. Soviet Melodiya 333 10-07007/10 (2). (also HMV Melodiya SLS 5088, Melodiya/Eurodisc 89502)
++ARG 4-77 p11 +HFN 9-77 p149
+Gr 9-77 p487 +-RR 9-77 p43
+-HF 12-77 p69

2471 Preludes, op. 34 (34). Sonata, piano, no. 2, op. 61, B minor. Inger Wikstrom, pno. RCA GL 2-5003.
+ARG 5-77 p56 +HFN 12-76 p147
+-Gr 11-76 p843 +RR 10-76 p91

Preludes, op. 34 (24). cf PRIGOZIN: Sonata, violin and piano, no. 2.

Preludes, op. 34, nos. 10, 15-16, 24. cf Works, selections (HMV RLS 721).

Prelude, op. 34, no. 14, E flat minor. cf CBS 73589.

Prelude and fugue, A flat major. cf HMV SXLP 30256.

Preludes and fugues, op. 87 (2). cf MARTINU: Concerto, piano, no. 1.

Preludes and fugues, op. 87, nos. 2-3, 5-7, 16, 20, 23. cf Works, selections (HMV RLS 721).

Prelude and fugue, op. 87, no. 17, A flat major. cf HMV SXLP 30259.

2472 Quartet, strings, no. 4, op. 83, D major. Quartet, strings, no. 12, op. 133, D flat major. Fitzwilliam Quartet. L'Oiseau-Lyre DSLO 23.
++Gr 11-77 p860 +RR 11-77 p82
+HFN 12-77 p180

2473 Quartet, strings, no. 7, op. 108, F sharp minor. Quartet, strings, no. 13, op. 138, B flat minor. Quartet, strings, no. 14, op. 142, F sharp major. Fitzwilliam Quartet. L'Oiseau-Lyre DSLO 9.
+Audio 6-76 p96 ++NR 2-76 p6
+Gr 12-75 p1065 +NYT 8-7-77 pD13
++HF 5-76 p94 +RR 11-75 p65
+HFN 12-75 p164 ++St 8-76 p68
+MJ 12-76 p44 +STL 4-11-76 p36
++MT 12-76 p1005 ++Te 9-76 p26

2474 Quartet, strings, no. 8, op. 110, C minor. Quartet, strings, no. 15, op. 144, E flat minor. Fitzwilliam Quartet. L'Oiseau-Lyre DSLO 11.

++ARG 12-76 p40 +NYT 8-7-77 pD13
++Gr 4-76 p1624 ++RR 4-76 p61
++HF 2-77 p102 +SR 11-13-76 p52
++HFN 4-76 p113 +St 5-77 p117
+MJ 7-77 p70 +STL 4-11-76 p36
++MT 9-76 p748 +Te 9-76 p26

Quartet, strings, no. 8, op. 110, C minor. cf BORODIN: Quartet, strings, no. 2, D major.

Quartet, strings, no. 8, op. 110, C minor. cf CARLSTEDT: Quartet, strings, no. 3, op. 23.

Quartet, strings, no. 12, op. 133, D flat major. cf Quartet, strings, no. 4, op. 83, D major.

2475 Quartet, strings, no. 13, op. 138, B flat minor. Romances on poems by Pushkin, op. 46. Ze ctyr Monologu, op. 91. TCHAIKOVSKY: Partita, violoncello and chamber ensemble. Suk Quartet; Jaromir Vavruska, singer; Jiri Pokorny, npo; Jaroslav Chovanec, vlc; Chamber Ensemble; Eduard Fischer. Panton 110 420.
+-RR 1-77 p67

Quartet, strings, no. 13, op. 138, B flat minor. cf Quartet, strings, no. 7, op. 108, F sharp minor.

2476 Quartet, strings, no. 14, op. 142, F sharp major. Quartet, strings, no. 15, op. 15, op. 144, E flat major. Taneyev Quartet. Columbia/Melodiya M 34527.
+NR 10-77 p6 +St 11-77 p151
+NYT 8-7-77 pD13

2477 Quartet, strings, no. 14, op. 142, F sharp major. Quartet, strings, no. 15, op. 144, E flat minor. Beethoven Quartet. HMV SQS 1362.
+Gr 11-76 p830 +RR 9-76 p77
+HFN 9-76 p129 +Te 3-77 p36

Quartet, strings, no. 14, op. 142, F sharp major. cf Quartet, strings, no. 7, op. 108, F sharp minor.

Quartet, strings, no. 15, op. 144, E flat minor. cf Quartet, strings, no. 8, op. 110.

Quartet, strings, no. 15, op. 144, E flat minor. cf Quartet, strings, no. 14, op. 142, F sharp major (HMV 1362).

Quartet, strings, no. 15, op. 144, E flat major. cf Quartet, strings, no. 14, op. 142, F sharp major (Columbia/Melodiya M 34527).

Quintet, piano, op. 57, G minor. cf Works, selections (HMV RLS 721).

Romances on poems by Pushkin, op. 46. cf Quartet, strings, no. 13, op. 138, B flat minor.

2478 Romances on verses of English poets, op. 140 (6). Suite of 6 songs to poems by Marina Tsvetaeva, op. 143a. Suite on verses by Michelangelo, op. 145a. Irina Bogacheva, ms; Yevgeny Nestorenko, bs; MRSO, Moscow Chamber Orchestra; Maxim Shotakovich, Rudolf Barshai. HMV Melodiya SLS 5078 (2).
++Gr 5-77 p1723 +RR 5-77 p84
+HFN 5-77 p133

Romances on words of Alexander Block, op. 127 (7). cf BIS LP 37.

Sonata, piano, no. 2, op. 61, B minor. cf Preludes, op. 34.

Sonata, piano, no. 2, op. 61, B minor. cf DENISOV: Variations.

2479 Sonata, viola and piano, op. 147. Milan Telecky, vla; Lydia Mailingova, pno. Aurora AUR 5051.
+Gr 5-77 p1708 +-RR 6-77 p85
+HFN 6-77 p134

2480 Sonata, viola and piano, op. 147. Sonata, violin and piano, op. 134.* Fyodor Druzhinin, vla; David Oistrakh, vln; Michael Muntyan, Sviatoslav Richter, pno. HMV Melodiya HQS 1369. (*Reissue from ASD 2718)
+Gr 5-77 p1708 ++RR 6-77 p85
+HFN 6-77 p134

Sonata, violin and piano, op. 134. cf Sonata, viola and piano, op. 147.

Sonata, violoncello, op. 40, D minor. cf Works, selections (HMV RLS 721).

Suite of 6 songs to poems by Marina Tsvetaeva, op. 143a. cf Romances on verses of English poets, op. 140.

Suite on verses by Michelangelo, op. 145a. cf Romances on verses of English poets, op. 140.

Suite on verses by Michelangelo, op. 145: Istina, Razluka, Gnyev, Smyert, Bessmyertiye. cf Symphony, no. 1, op. 10, F minor.

2481 Symphony, no. 1, op. 10, F minor. Suite on verses by Michelangelo, op. 145: Istina, Razluka, Gnyev, Smyert, Bessmyertiye. Richard Novak, bs; Cyril Klimes, pno; CPhO; Jiri Kout. Panton 110 604.
+-Gr 12-77 p1090 +-HFN 12-77 p180

2482 Symphony, no. 5, op. 47, D minor. Bournemouth Symphony Orchestra; Paavo Berglund. Angel S 37279. (also HMV SLS 5044).
+Ha 6-77 p88 +-NR 4-77 p1
+-HF 4-77 p113

2483 Symphony, no. 5, op. 47, D minor. Bournemouth Symphony Orchestra; Paavo Berglund. HMV SQ ESD 7029 Tape (c) TC ESD 7029. (Reissue from SLS 5044)
+Gr 2-77 p2189 ++HFN 5-77 p138 tape
++HFN 4-77 p151 +-RR 2-77 p59

2484 Symphony, no. 5, op. 47. PO; Eugene Ormandy. RCA ARL 1-1149 Tape (c) ARK 1-1149 (ct) ARS 1-1149 (Q) ARD 1-1149 Tape (ct) ART 1-1149 (r) ERQ 1-1149 (c) RK 11712.
++Audio 6-76 p96 +-HFN 8-76 p87
+-Gr 8-76 p298 +MJ 11-75 p20
+HF 1-76 p95 ++NR 11-75 p2
+-HF 1-76 p95 Quad -RR 1-77 p91 tape
++HF 4-76 p148 tape +RR 8-76 p52
+HF 8-77 p96 Quad tape ++SFC 9-14-75 p28

2485 Symphony, no. 5, op. 47, D minor, excerpts. Symphony, no. 15, op. 141, A major, excerpts. LPO; Joseph Eger. Charisma CAS 1128.
-HFN 11-77 p181 -RR 10-77 p58

2486 Symphony, no. 7, op. 60, C major. CPhO; Karel Ancerl. Everest SDBR 3404.
+-NR 2-77 p2

2487 Symphony, no. 7, op. 60, C major. CPhO; Karel Ancerl. Rediffusion HCN 8003. (Reissue from Parliament PLP 127)
+-Gr 7-77 p187 ++RR 7-77 p60
+HFN 8-77 p93

2488 Symphony, no. 7, op. 60, C major. Symphony, no. 9, op. 70, E flat major. CPhO; Vaclav Neumann. Supraphon 110 1771/2 (2).
-Gr 8-76 p298 +NR 11-76 p3
+-HF 3-77 p111 -RR 6-76 p53

Symphony, no. 9, op. 70, E flat major. cf Symphony, no. 7, op. 60, C major.

2489 Symphony, no. 10, op. 93, E minor. Bournemouth Symphony Orchestra; Paavo Berglund. Angel S 37280. (also HMV SQ ESD 7049 Tape (c) TC ESD 7049. Reissue from SLS 5044)
+Gr 10-77 p638 +NR 7-77 p1
+Gr 12-77 p1146 tape +RR 11-77 p69
+-HF 6-77 p99 ++SFC 4-10-77 p30
+MJ 7-77 p70

2490 Symphony, no. 10, op. 93, E minor. LPO; Bernard Haitink. Decca SXL 6838. (also London CS 7061)
+Gr 10-77 p638 +RR 10-77 p59
+-HFN 10-77 p163

2491 Symphony, no. 14, op. 135. Galina Vishnevskaya, s; Mark Reshetin, bs; MPO; Mstislav Rostropovich. HMV Melodiya ASD 3090. (also Columbia/Melodiya M 34507)
++ARG 8-77 p27 +-MJ 7-77 p70
+Gr 12-75 p1046 ++NR 7-77 p1
+Ha 6-77 p88 +NYT 8-7-77 pD13
++HF 6-77 p71 ++RR 1-76 p38
++HFN 1-76 p119 +St 7-77 p120

Symphony, no. 15, op. 141, A major, excerpts. cf Symphony, no. 5, op. 47, D minor, excerpts.

Trio, piano, no. 2, op. 67, E minor. cf FAURE: Trio, piano, op. 120, D minor.

Trio, piano, no. 2, op. 67, E minor. cf RAVEL: Trio, violin, violoncello and piano, A minor.

2492 Works, selections: Concerto, piano, no. 2, op. 101, F major. Concertino, 2 pianos, op. 94. From Jewish folk poetry, op. 79. Quintet, piano, op. 57, G minor.* Sonata, violoncello, op. 40, D minor.** Preludes and fugues, op. 87, nos. 2-3, 5-7, 16, 20, 23.*** Preludes, piano, op. 34, nos. 10, 15-16, 24. Dmitri Shostakovich, Maxim Shotakovich, pno; Leonid Kogan, vln; Mstislav Rostropovich, vlc; Nina Dorliak, s; Zara Dolukhanova, ms; Alexei Maslennikov, t; MRSO, Beethoven Quartet; Alexander Gauk. HMV RLS 721 (3). (*Reissue from Parlophone PMA 1040, ** PMA 1043, *** nos. 6, 7, 20 from PMC 1056)
+Gr 10-76 p607 +-RR 10-76 p66
+-HFN 11-76 p167 +ST 1-77 p775

Ze ctyr Monologu, op. 91. cf Quartet, strings, no. 13, op. 138, B flat minor.

SHUKUR, Salman

2493 Festival in Baghdad. Improvisations by Shukur on a theme of Hajji "Adb ul-Ghaffar" from a takya in Tikrit: Prelude in Rast; Taqsim in Rast; Sama'i in Rast. The mountain fairy. Romance. Salman Shukur, oud. Decca HEAD 16.
+Gr 9-77 p459 +RR 8-77 p69

Improvisations by Shukur on a theme of Hajji "Adb ul-Ghaffar" from a takya in Tikrit: Prelude in Rast; Taqsim in Rast; Sam'i in Rast. cf Festival in Baghdad.

The mountain fairy. cf Festival in Baghdad.

Romance. cf Festival in Baghdad.

SIBELIUS, Jean

2494 Andante festivo. King Christian II, op. 27: Suite. Swanwhite suite, op. 54. HSO; Jussi Jalas. London CS 7005. (also Decca SDD 506)
/ARG 3-77 p35 -NR 4-77 p2
+-Gr 7-77 p187 +RR 6-77 p67
+-HFN 6-77 p135 +SFC 4-10-77 p30
+MJ 9-77 p34 /St 4-77 p132

Andante festivo. cf Works, selections (HMV SQ ASD 3287).
Aspen, op. 75, no. 3. cf Piano works (RCA GL 4-2229).
2495 The bard, op. 64. Karelia, op. 10. King Christian II, op. 27: Suite. Scenes historiques, op. 25: Festivo. Scottish National Orchestra; Alexander Gibson. Classics for Pleasure CFP 40273. (Reissue from HMV SQS 1070)
+Gr 12-77 p1090 +RR 12-77 p55
+HFN 12-77 p185
The bard, op. 64. cf Symphony, no. 4, op. 63, A minor.
Berceuse, op. 79, no. 6. cf DOHNANYI: Sonata, violin and piano, op. 21, C sharp minor.
Berceuse, op. 79, no. 6. cf Finnlevy SFLP 8569.
Canzonetta, op. 62a. cf Works, selections (Decca SDD 489).
Canzonetta, op. 62a. cf Works, selections (HMV SQ ASD 3287).
Capriccio, op. 24, no. 3. cf Piano works (RCA GL 4-2229).
2496 Concerto, violin, op. 47, D minor. Humoreske, op. 89, no. 3, E flat major. Serenade, no. 1, op. 69a, D major. Serenade, no. 2, op. 69b, G minor. Ida Haendel, vln; Bournemouth Symphony Orchestra; Paavo Berglund. HMV SQ ASD 3199 Tape (c) TC ASD 3199.
+Gr 6-76 p52 +HFN 8-76 p95
++Gr 9-76 p497 tape ++RR 5-76 p51
+HFN 7-76 p97 ++RR 9-77 p99 tape
Concerto, violin, op. 47, D minor. cf Works, selections (CBS 61804/9).
Concerto, violin, op. 47, D minor. cf Works, selections (HMV Odeon ASD 3100).
Concerto, violin, op. 47, D minor. cf BEETHOVEN: Romances, nos. 1 and 2, opp. 40, 50.
Concerto, violin, op. 47, D minor. cf CHAUSSON: Poeme, op. 25.
Dance intermezzo, op. 45, no. 2. cf Works, selections (HMV SQ ASD 3287).
Devotion, op. 77, no. 2. cf DOHNANYI: Sonata, violin and piano, op. 21, C sharp minor.
The Dryad, op. 45, no. 1. cf Works, selections (HMV SQ ASD 3287).
Etude, op. 76, no. 2. cf Piano works (RCA GL 4-2229).
2497 Finlandia, op. 26. Legends, op. 22: The swan of Tuonela. En Saga, op. 9. Tapiola, op. 112. BPhO; Herbert von Karajan. Angel S 37408 Tape (c) 4XS 37408.
+SFC 12-25-77 p42
2498 Finlandia, op. 26. Karelia suite, op. 11. Legends, op. 22: The swan of Tuonela. En saga, op. 9. VPO; Malcolm Sargent. Classics for Pleasure CFP 40247 Tape (c) TC CFP 40247.
+-Gr 2-77 p1325 tape -HFN 12-76 p153 tape
+HFN 6-76 p102 +-RR 5-76 p50
2499 Finlandia, op. 26. Karelia suite, op. 11. Legends, op. 22: The swan of Tuonela. Kuolema, op. 44: Valse triste. NPhO; Kazimierz Kord. Decca PFS 4378 Tape (c) KPFC 4378.
+-Gr 5-77 p1698 +-RR 4-77 p59
+HFN 4-77 p145 +RR 10-77 p98 tape
++HFN 6-77 p139 tape
2500 Finlandia, op. 26. Legends, op. 22: The swan of Tuonela. En saga, op. 9. Tapiola, op. 112. BPhO; Herbert von Karajan. HMV SQ ASD 3374. (also Angel S 37415)
++Gr 12-77 p1090 ++RR 12-77 p54
+HFN 12-77 p181

2501 Finlandia, op. 26. Kuolema, opp. 44 and 62. Scenes historiques, opp. 25 and 66. HSO; Jussi Jalas. London CS 6956. (also Decca SDD 489)
+Gr 11-77 p844 ++SFC 8-29-76 p29
+NR 12-76 p3 +-St 4-77 p132
-RR 10-77 p59
2502 Finlandia, op. 26. Kuolema, op. 44: Valse triste. Legends, op. 22: Lemminkainen's return; The swan of Tuonela. Pohjola's daughter, op. 49. Orchestra; Morton Gould. Quintessence PMC 7022.
+HF 9-77 p108 +St 11-77 p146
Finlandia, op. 26. cf Symphonies, nos. 1-7.
Finlandia, op. 26. cf Symphony, no. 1, op. 39, E minor.
Finlandia, op. 26. cf Works, selections (Decca SDD 489).
Finlandia, op. 26. cf Seraphim SIB 6094.
Granen, op. 75, no. 5. cf Piano works (RCA GL 4-2229).
Humoreske, op. 89, no. 3, E flat major. cf Concerto, violin, op. 47, D minor.
Humoreske, op. 89, no. 3, E flat major. cf Works, selections (HMV Odeon ASD 3100).
In memoriam, op. 59. cf Legends, op. 22.
Intrada, op. 111a. cf Guild GRSP 7011.
Kalevala, op. 41: Kyllikki. cf Sonatinas, op. 67, nos. 1-3.
2503 Kalevala, op. 41: Kyllikki pieces (3). Sonatinas, op. 67, nos. 1-3. Sonata, piano, op. 12, F major. David Rubinstein, pno. Musical Heritage Society 1218 Tape (r) BCC 1218.
++HF 8-77 p96 tape
Karelia overture, op. 10. cf The bard, op. 64.
2504 Karelia suite, op. 11. Legends, op. 22. Helsinki Radio Symphony Orchestra; Okko Kamu. DG 2530 656 Tape (c) 3300 656.
+ARG 5-77 p37 +MJ 9-77 p34
++Gr 10-76 p608 +NR 4-77 p2
+HF 7-77 p106 ++RR 11-76 p78
+HF 10-76 p175 +RR 6-77 p97 tape
+HFN 5-77 p139 tape ++SFC 4-10-77 p30
+-HFN 9-77 p155 tape ++St 4-77 p132
Karelia suite, op. 11. cf Finlandia, op. 26 (Classics for Pleasure CFP 40247).
Karelia suite, op. 11. cf Finlandia, op. 26 (Decca PFS 4378).
Karelia suite, op. 11. cf HMV SLS 5073.
King Christian II, op. 27: Suite. cf Andante festivo.
King Christian II, op. 27: Suite. cf The bard, op. 64.
Kuolema, opp. 44 and 62. cf Finlandia, op. 26.
Kuolema, op. 44: Scene with cranes; Valse triste. cf Works, selections (Decca SDD 489).
Kuolema, op. 44: Valse triste. cf Finlandia, op. 26 (Decca KPFC 4378).
Kuolema, op. 44: Valse triste. cf Finlandia, op. 26 (Quintessence PMC 7022).
Kuolema, op. 44: Valse triste. cf Works, selections (CBS 61804/9).
Kuolema, op. 44: Valse triste. cf BEETHOVEN: Symphony, no. 3, op. 55, E flat major.
2505 Legends, op. 22. Royal Liverpool Philharmonic Orchestra; Charles Groves. Angel S 37106. (also HMV ASD 3092 Tape (c) TC ASD 3092)
+Gr 9-75 p472 +RR 8-75 p43
+HF 11-75 p126 +RR 2-77 p98 tape

+HFN 9-75 p104 +SFC 8-31-75 p20
++NR 8-75 p6

Legends, op. 22. cf Karelia suite, op. 11.

Legends, op. 22: Lemminkainen's return; The swan of Tuonela. cf Finlandia, op. 26.

2506 Legends, op. 22: Lemminkainen and the maidens of the island; The swan of Tuonela; Lemminkainen in Tuonela; Lemminkainen's return. In memoriam, op. 59. HSO; Jussi Jalas. Decca SDD 488 Tape (c) KSDC 488. (Reissue from DPA 531/2) (also London CS 6955)

++Audio 5-77 p102 +-MJ 9-77 p34
+-Gr 9-76 p431 +-NR 9-76 p5
-HF 11-76 p126 +-RR 8-76 p53
+HFN 10-76 p185 tape +-St 4-77 p132
+-HFN 9-76 p129

Legends, op. 22: The swan of Tuonela. cf Finlandia, op. 26 (Angel S 37408).

Legends, op. 22: The swan of Tuonela. cf Finlandia, op. 26 (Classics for Pleasure CFP 40247).

Legends, op. 22: The swan of Tuonela. cf Finlandia, op. 26 (Decca KPFC 4378).

Legends, op. 22: The swan of Tuonela. cf Finlandia, op. 26 (HMV SQ ASD 3374).

Legends, op. 22: The swan of Tuonela. cf Symphonies, nos. 1-7.

Legends, op. 22: The swan of Tuonela. cf Symphony, no. 1, op. 39, E minor.

Legends, op. 22: Swan of Tuonela. cf Seraphim SIB 6094.

Luonnotar, op. 70. cf Symphony, no. 4, op. 63, A minor.

Luonnotar, op. 70. cf Works, selections (CBS 61804/9).

Luonnotar, op. 70. cf Works, selections (HMV Odeon ASD 3100).

2507 The oceanides, op. 73. Pelleas and Melisande, op. 46: Incidental music. Tapiola, op. 112. RPO; Thomas Beecham. HMV SXLP 30197 Tape (c) TC EXE 180. (Reissues from ASD 468, 518)

++Gr 2-76 p1349 +RR 2-76 p41
+-HFN 2-76 p115 +RR 2-77 p98
+-HFN 5-76 p117 tape

The oceanides, op. 73. cf Symphony, no. 4, op. 63, A minor.

Pan and echo, op. 53a. cf Works, selections (HMV SQ ASD 3287).

Pelleas and Melisande, op. 46: Incidental music. cf The oceanides, op. 73.

Pensee melodique, op. 40, no. 6. cf Piano works (RCA GL 4-2229).

2508 Piano works: Aspen, op. 75, no. 3. Capriccio, op. 24, no. 3. Etude, op. 76, no. 2. Granen, op. 75, no. 5. Pensee melodique, op. 40, no. 6. Piece enfantine, op. 76, no. 8. Romance, op. 24, no. 9, D flat major. Rondino, op. 68, no. 1, G sharp minor. Scene romantique, op. 101, no. 5. Sonatinas, op. 67, nos. 1-3. Ervin Laszlo, pno. RCA GL 4-2229. (Reissue from VICS 1538)

/-Gr 12-77 p1117 -HFN 12-77 p181

Piece enfantine, op. 76, no. 8. cf Piano works (RCA GL 4-2229).

Pohjola's daughter, op. 49. cf Finlandia, op. 26.

Pohjola's daughter, op. 49. cf Works, selections (CBS 61804/9).

Pohjola's daughter, op. 49. cf Works selections (HMV Odeon ASD 3100).

2509 Rakastava, op. 14. Scenes historiques, op. 25: All'overtura; Scene; Festivo. Scenes historiques, op. 66: The chase; Lovesong; At the drawbridge. Valse lyrique, op. 96, no. 1. Scottish National Orchestra; Alexander Gibson. RCA RL 2-5051 Tape (c) RK 2-5051.

+Gr 4-77 p1561
+HFN 6-77 p134
+MT 9-77 p738
++RR 6-77 p68
+RR 8-77 p87 tape

Romance, op. 24, no. 9, D flat major. cf Piano works (RCA GL 4-2229).
Romance, op. 42. C major. cf Works, selections (HMV SQ ASD 3287).
Rondino, op. 68, no. 1, G sharp minor. cf Piano works (RCA GL 4-2229).
En saga, op. 9. cf Finlandia, op. 26 (Angel S 37408).
En saga, op. 9. cf Finlandia, op. 26 (Classics for Pleasure CFP 40247).
En saga, op. 9. cf Finlandia, op. 26 (HMV SQ ASD 3374).
En saga, op. 9. cf Symphony, no. 5, op. 82, E flat major (Classics for Pleasure 40218).
En saga, op. 9. cf BEETHOVEN: Symphony, no. 3, op. 55, E flat major.
En saga, op. 9. cf BEETHOVEN: Symphony, no. 5, op. 67, C minor.
Scenes historiques, opp. 25 and 66. cf Finlandia, op. 26.
Scenes historiques, opp. 25 and 65: Suites, nos. 1 and 2. cf Works, selections (Decca SDD 489).
Scenes historiques, op. 25. cf Symphony, no. 1, op. 39, E minor.
Scenes historiques, op. 25: All'overtura; Scene; Festivo. cf Rakastava, op. 14.
Scenes historiques, op. 25: All'overtura; Scene; Festivo. cf Symphony, no. 1, op. 39, E minor.
Scenes historiques, op. 25: Festivo. cf The bard, op. 64.
Scenes historiques, op. 66: The chase; Love-song; At the drawbridge. cf Rakastava, op. 14.
Scene romantique, op. 101, no. 5. cf Piano works (RCA GL 4-2229).
Serenade, no. 1, op. 69a, D major. cf Concerto, violin, op. 47, D minor.
Serenade, no. 2, op. 69b, G minor. cf Concerto, violin, op. 47, D minor.
Serenade, no. 2, op. 69b, G minor. cf Works, selections (HMV Odeon ASD 3100).
Sonata, piano, op. 12, F major. cf Kalevala, op. 41: Kyllikki pieces (3).

2510 Sonatinas, op. 67, nos. 1-3. Kalevala, op. 41: Kyllikki. Glenn Gould, pno. Columbia M 34555.
+SFC 12-25-77 p42

Sonatinas, op. 67, nos. 1-3. cf Kalevala, op. 41: Kyllikki pieces (3).
Sonatinas, op. 67, nos. 1-3. cf Piano works (RCA GL 4-2229).
Sonatina, op. 80, E major. cf DOHNANYI: Sonata, violin and piano, op. 21, C sharp minor.

2511 Songs: And I questioned them no further, op. 17, no. 1; Arioso, op. 3; Autumn night, op. 38, no. 1; Black roses, op. 36, no. 1; Come away death, op. 60, no. 1; The diamond on the March snow, op. 36, no. 6; The first kiss, op. 37, no. 1; But my bird is long in homing, op. 36, no. 2; On a balcony by the sea, op. 38, no. 2; Whisper O reed, op. 36, no. 4; Spring is flying, op. 13, no. 4; To evening, op. 17, no. 6; The tryst, op. 37, no. 5; Was it a dream, op. 37, no. 4. Kirsten Flagstad, s; LSO; Oivin Fjeldstad. Decca ECS 794 Tape (c) KECC 794. (Reissue from LXT 5444) (also London SR 33216)
++Gr 4-77 p1588
+HFN 3-77 p118
++RR 3-77 p95
+RR 5-77 p93

+HFN 5-77 p138 tape +SFC 6-13-76 p30
+HFN 9-77 p155

2512 Songs: Black roses, op. 36, no. 1; The diamond on the March snow, op. 36, no. 6; Driftwood, op. 17, no. 7; The first kiss, op. 37, no. ; From an anxious breast; The hunter boy; Hymn to Thais; The Jewish girl's song; Little lasse; The maiden's message; On a balcony by the sea, op. 38, no. 2; Paddle paddle little mallard; Romeo, op. 61, no. 4; Sigh reeds sigh; To evening, op. 17, no. 6. Aulikki Rautawara, s; Gerald Moore, Jussi Jalas, pno; Orchestra; Jussi Jalas. Finnlevy SFLP 8570.
+-RR 7-77 p90

Songs: Be still my soul. cf L'Oiseau-Lyre DSLO 20.
Songs: Demanten pa marssnon; Sav sav susa. cf RCA LSC 9884.
Souvenir, op. 79, no. 1. cf DOHNANYI: Sonata, violin and piano, op. 21, C sharp minor.
Spring song, op. 16. cf Works, selections (HMV SQ ASD 3287).
Suite champetre, op. 98b. cf Works, selections (HMV SQ ASD 3287).
Suite mignonne, op. 98. cf Works, selections (HMV SQ ASD 3287).
Swanwhite suite, op. 54. cf Andante festivo.

2513 Symphonies, nos. 1-7. VPO; Lorin Maazel. Decca D7D4 Tape (c) KE 9. (Reissue)
+-Gr 9-76 p477 +RR 4-76 p81
+Gr 9-77 p443 +-RR 9-76 p66
+-HFN 10-76 p181

2514 Symphonies, nos. 1-7. Finlandia, op. 26. Legends, op. 22: The swan of Tuonela. Tapiola, op. 112. BSO; Colin Davis. Philips 6709 011 (5).
+Gr 9-77 p443 +-RR 9-77 p67
++HFN 9-77 p151

Symphonies, nos. 1-7. cf Works, selections (CBS 61804/9).

2515 Symphony, no. 1, op. 39, E minor. Legends, op. 22: The swan of Tuonela. National Philharmonic Orchestra; Leopold Stokowski. Columbia M 34538 Tape (c) MT 34548.
+SFC 12-25-77 p42

2516 Symphony, no. 1, op. 39, E minor. Scenes historiques, op. 25. Bournemouth Symphony Orchestra; Paavo Berglund. HMV SQ ASD 3216. (also Seraphim S 60289)
+-ARG 10-77 p55 +-HF 11-77 p126
+-ARG 11-77 p31 +HFN 1-77 p117
+-Gr 12-76 p1002 +RR 1-77 p62

2517 Symphony, no. 1, op. 39, E minor. Finlandia, op. 26. BSO; Colin Davis. Philips 9500 140 Tape (c) 7300 517.
+-Audio 11-77 p126 ++HFN 5-77 p138 tape
+Gr 3-77 p1409 +NR 3-77 p4
+Gr 8-77 p349 tape +RR 3-77 p61
++HF 6-77 p100 +RR 7-77 p100
+HF 9-77 p119 tape +SFC 1-23-77 p37
+HFN 3-77 p111 ++St 7-77 p124

2518 Symphony, no. 1, op. 39, E minor. Scenes historiques, op. 25. Bournemouth Symphony Orchestra; Paavo Berglund. Seraphim S 60289.
+NR 11-77 p6

2519 Symphony, no. 2, op. 43, D major. Pittsburgh Symphony Orchestra; Andre Previn. HMV SQ ASD 3414.
+-Gr 11-77 p844 +-RR 12-77 p55
+-HFN 12-77 p180

2520 Symphony, no. 2, op. 43, D major. BSO; Colin Davis. Philips 9500 141 Tape (c) 7300 518.
++ARG 10-77 p41
+Gr 4-77 p1556
+Gr 8-77 p349 tape
+HF 9-77 p119 tape
++HF 12-77 p98 tape
++HFN 4-77 p145
+HFN 8-77 p99 tape
+MJ 9-77 p35
++NR 7-77 p2
+RR 4-77 p59
++RR 9-77 p99
+St 11-77 p156

2521 Symphony, no. 2, op. 43, D major. RPO; John Barbirolli. Quintessence PMC 7008.
+HF 9-77 p108
+St 11-77 p146

2522 Symphony, no. 2, op. 43, D major. RPO; John Barbirolli. RCA GL 2-5011. (Previously issued by Reader's Digest)
++Gr 10-76 p608
+Gr 2-77 p1325 tape
+-HFN 12-76 p147
+-RR 11-76 p83
+RR 12-76 p104
+ST 1-77 p779

2523 Symphony, no. 3, op. 52, C major. Symphony, no. 6, op. 104, D minor. BSO; Colin Davis. Philips 9500 142 Tape (c) 7300 519.
+Gr 8-77 p309
+-Gr 10-77 p705 tape
++HFN 9-77 p151
++HFN 11-77 p187 tape
+RR 9-77 p68
+RR 12-77 p98 tape

2524 Symphony, no. 4, op. 63, A minor. The bard, op. 64. Bournemouth Symphony Orchestra; Paavo Berglund. HMV (SQ) ASD 3340 Tape (c) TC ASD 3340.
+Gr 6-77 p63
-Gr 8-77 p349 tape
+-HFN 7-77 p119
+HFN 8-77 p99 tape
+RR 6-77 p68

2525 Symphony, no. 4, op. 63, A minor. Tapiola, op. 112. BSO; Colin Davis. Philips 9500 143 Tape (c) 7300 520.
+Gr 8-77 p310
+Gr 10-77 p705 tape
++HFN 9-77 p150
++HFN 11-77 p187 tape
+-RR 9-77 p68
+RR 12-77 p98 tape

2526 Symphony, no. 4, op. 63, A minor. Symphony, no. 5, op. 82, E flat major. RPO; Loris Tjeknavorian. RCA LRL 1-5134 Tape (c) 1-1747.
+Gr 11-76 p804
-Gr 8-77 p349 tape
+HFN 3-77 p111
+RR 11-76 p16

2527 Symphony, no. 4, op. 63, A minor. Luonnotar, op. 70. The oceanides, op. 73. Helmi Liukkonen, s; Finnish National Symphony Orchestra, BBC Symphony Orchestra; Georg Schneevoigt, Adrian Boult. World Records SH 237.
+Gr 3-77 p1409
+-HFN 2-77 p131
+-RR 1-77 p63

2528 Symphony, no. 5, op. 82, E flat major. En saga, op. 9. Scottish National Orchestra; Alexander Gibson. Classics for Pleasure CFP 40218 Tape (c) TC CFP 40218.
+Gr 9-75 p466
+-Gr 2-77 p1325 tape
/HFN 8-75 p83
+-HFN 12-76 p153
+RR 9-75 p47

Symphony, no. 5, op. 82, E flat major. cf Symphony, no. 4, op. 63, A minor.

Symphony, no. 6, op. 104, D minor. cf Symphony, no. 3, op. 52, C major.

Symphony, no. 6, op. 104, D minor. cf Works, selections (HMV Odeon ASD 3100).

Tapiola, op. 112. cf Finlandia, op. 26 (Angel S 37408).

Tapiola, op. 112. cf Finlandia, op. 26 (HMV SQ ASD 3374).

Tapiola, op. 112. cf The oceanides, op. 73.
Tapiola, op. 112. cf Symphonies, nos. 1-7.
Tapiola, op. 112. cf Symphony, no. 4, op. 63, A minor.

2529 The tempest, op. 109: Prelude; Suites, nos. 1 and 2. Royal Liverpool Philharmonic Orchestra; Charles Groves. HMV Tape (c) TC ASD 2961.
++RR 2-77 p98 tape

Valse lyrique, op. 96, no. 1. cf Rakastava, op. 14.
Valse romantique, op. 62b. cf Works, selections (Decca SDD 489).
Valse romantique, op. 62b. cf Works, selections (HMV SQ ASD 3287).

2530 Works, selections: Symphonies, nos. 1-7. Concerto, violin, op. 47, D minor. Kuolema, op. 44: Valse triste. Luonnotar, op. 70. Pohjola's daughter, op. 49. Phyllis Curtin, s; Zino Francescatti, vln; NYP; Leonard Bernstein. CBS 61804/9 (2) Tape (c) 40-61804/9. (Reissues from 30029, 72732, 72733, 72736, 73162, 72686, 73086, 72356, 72351)
+-Gr 8-77 p309
+Gr 10-77 p705 tape
+-HFN 10-77 p171 tape
+-HFN 8-77 p93
+-RR 8-77 p63
+-RR 10-77 p98 tape

2531 Works, selections: Canzonetta, op. 62a. Finlandia, op. 26. Kuolema, op. 44: Scene with the cranes; Valse triste. Scenes historiques, opp. 25 and 65: Suites, nos. 1 and 2. Valse romantique, op. 62b. HSO; Jussi Jalas. Decca SDD 489.
+-HFN 10-77 p163

2532 Works, selections: Concerto, violin, op. 47, D minor. Humoreske, op. 89, no. 3, E flat major. Luonnotar, op. 70. Pohjola's daughter, op. 49. Serenade, no. 2, op. 69b, G minor. Symphony, no. 6, op. 104, D minor. Taru Valjakka, s; Ida Haendel, vln; Bournemouth Symphony Orchestra; Paavo Berglund. HMV Odeon ASD 3100.
+ARG 4-77 p11

2533 Works, selections: Andante festivo. Canzonetta, op. 62a. Dance intermezzo, op. 45, no. 2. The Dryad, op. 45, no. 1. Pan and echo, op. 53a. Romance, op. 42, C major. Spring song, op. 16. Suite champetre, op. 98b. Suite mignonne, op. 98. Valse romantique, op. 62b. Royal Liverpool Philharmonic Orchestra; Charles Groves. HMV SQ ASD 3287.
+Gr 11-76 p811
+-HFN 12-76 p148
+-MT 12-77 p1016
+RR 11-76 p78

SIEBERT

Hawaiian hoe-down. cf Rediffusion 15-56.

SIEGMEISTER, Elie

American harp. cf Orion ORS 75207.

2534 Concerto, clarinet. Concerto, flute. Peter Lloyd, flt; Jack Brymer, clt; LSO, Members; Elie Siegmeister. Turnabout TVS 34640.
+ARG 7-77 p34
+MJ 10-77 p29
+NR 6-77 p5

Concerto, flute. cf Concerto, clarinet.

2535 Fantasy and soliloquy, solo violoncello. On this ground, solo piano. Sonata, violin and piano, no. 4. Robert Sylvester, vlc; Nancy Mandel, vln and pno; Alan Mandel, pno. Orion ORS 7284.
+HF 12-72 p118
+MJ 12-73 p33
+MJ 3-77 p46
+NR 10-72 p6

On this ground, solo piano. cf Fantasy and soliloquy, solo violoncello.

Sonata, violin and piano, no. 4. cf Fantasy and soliloquy, solo violoncello.

2536 Songs: Elegies for Garcia Lorca; Evil; Five Cummings songs; For my daughters; Johnny Appleseed; Lazy afternoon; Nancy Hanks; The strange funeral in Braddock; Two songs of the city. Elizabeth Kirkpatrick, s; Herbert Beattie, bs-bar; Alan Mandel, pno. Orion ORS 76220.
+MJ 3-77 p46 +-NR 1-77 p13

SILCHER, Friedrich
Die Lorelei. cf Club 99-108.

SIMON, Anton
Quartet, in form of a sonata. cf Argo ZRG 851.

SIMONS
Marta, rambling rose of the wildwood. cf Philips 6392 023.

SIMONS, Gardell
Atlantic zephyrs. cf Pandora PAN 2001.

SIMPSON
On the track. cf EMI TWOX 1058.

SIMPSON, Christopher
Alman. cf DG 2723 051.

SIMPSON, Robert
Canzona for brass. cf RCA RL 2-5081.

SINDING, Christian
Pieces, piano. cf Symphony no. 1, op. 21, D major.
Rustle of spring, op. 32, no. 3. cf Decca SPA 473.
Rustle of spring, op. 32, no. 3. cf International Piano Library IPL 113.
Sonata, piano, op. 91, B minor. cf COPLAND: Sonata, piano.
Suite, op. 10, A minor. cf RCA CRM 6-2264.

2537 Symphony, no. 1, op. 21, D major. Pieces, piano. Kjell Baekkelund, Robert Levin, pno; Oslo Philharmonic Orchestra; Oivin Fjeldstad. Polydor Norway NKF 30011, NKF 30014.
+HFN 3-77 p109 +-RR 1-77 p75
+RR 1-77 p63

SINGER, Andre

2538 Parables to Kafka's America. Serial pieces, piano (3). Sonata, 2 pianos. Kenneth Wentworth, Jean Wentworth, pno; Kenneth Wentworth, narrator. Grenadilla 1011.
++NR 11-77 p13

Serial pieces, piano (3). cf Parables to Kafka's America.
Sonata, 2 pianos. cf Parables to Kafka's America.

SINGER, Lawrence
Musica a 2. cf CBS 61453.

SINOPOLI, Giuseppe
Numquid. cf CBS 61453.

SJOBERG, C. L.
Songs: I drommen du ar mig nara; Tonerna. cf HMV RLS 715.
Songs: Tonerna. cf RCA LSC 9884.

SJOGREN, Emil
Evit dig til hjerlet trykke. cf Rubini GV 43.

SKALKOTTAS, Nikos

2539 Octet. Quartet, strings, no. 3. Variations on a Greek folk tune (8). Robert Masters, vln; Derek Simpson, vlc; Marcel Gazelle, pno; Dartington Quartet. Argo ZRG 753. (Reissue from HMV ASD 2289).
+Gr 11-74 p921 ++RR 11-74 p73
++MQ 1-76 p139 ++SFC 7-17-77 p42
++NR 2-75 p6 ++St 6-75 p107

Quartet, strings, no. 3. cf Octet.
Variations on a Greek folk tune. cf Octet.

SLATER, Gordon
For life, with all it yields. cf Vista VPS 1037.
SLAVIC ORTHODOX LITURGY PROGRAMS. cf Classical Cassette Company CP 52.

SLAVICKY
Fresco. cf Marc MC 5355.

SLONIMSKY, Sergei
The bells. cf DENISOV: Variations.

SMART, Henry
Postlude, D major. cf Vista VPS 1035.

SMETANA, Bedrich
The bartered bride: Dance of the comedians. cf HMV SLS 5073.
The bartered bride: Overture. cf Pye Golden Hour GH 643.
Characteristic pieces (6). cf Memories of Bohemia, opp. 12 and 13.

2540 Czech dances, Sets I and II. Memories of Bohemia, opp. 12 and 13. Polkas, opp. 7 and 8. Jan Novotny, pno. Supraphon 111 1901/2.
++FF 9-77 p54 +NR 10-77 p13
++Gr 8-77 p324 +RR 8-77 p72
+HFN 8-77 p89

2541 Hubic (The kiss). Ludmila Cervinkova, Stefa Petrova, s; Marta Krasova, con; Beno Blachut, Karel Hruska, t; Premsyl Koci, bar; Karel Kalas, bs; Prague Theatre Orchestra and Chorus; Zdenek Chalabala. Rediffusion Heritage HCNL 8006/7. (Reissue from Supraphon LPV 142/4)
+Gr 10-77 p692 +-RR 8-77 p38
+HFN 10-77 p167

2542 Ma Vlast. VPO; Rafael Kubelik. Decca DPA 575/6 (2). (Reissue from SXL 2064/5)
+Gr 3-77 p1410 -RR 2-77 p59
-HFN 3-77 p117

2543 Ma Vlast. BSO; Rafael Kubelik. DG 2720 032 (2) Tape (c) 3581 008, 3300 895 (r) 47054.
+-Gr 12-77 p899 tape ++NR 11-71 p1
+-HF 12-71 p108 tape +SFC 12-31-72 p22 tape
+HFN 12-71 p2316 ++St 12-71 p104
+HFN 6-72 p1123 tape

2544 Ma Vlast. CPhO; Vaclav Talich. Rediffusion Heritage HCN 8001/2 (2). (Reissue from LPV 247/8)
+Gr 7-77 p191 ++RR 7-77 p61

Ma Vlast: Andantino. cf Discopaedia MB 1012.
Ma Vlast: The Moldau. cf DVORAK: Symphony, no. 9, op. 95, E minor.
Ma Vlast: The Moldau. cf Musical Heritage Society MHS 3611.
Ma Vlast: Vltava. cf DVORAK: Slavonic dances, op. 46, nos. 1-4.
Ma Vlast: Vltava. cf DVORAK: Symphony, no. 9, op. 95, E minor.
Ma Vlast: Vltava. cf LISZT: Hungarian rhapsody, no. 1, F minor.
Ma Vlast: Vltava. cf HMV SLS 5073.
Ma Vlast: Vltava; From Bohemia's woods and fields. cf DVORAK: In nature's realm overture, op. 91.

2545 Memories of Bohemia, opp. 12 and 13. Characteristic pieces (6). Antonin Kubalek, pno. Citadel CT 6010.
+NR 10-77 p13 +St 2-77 p122

Memories of Bohemia, opp. 12 and 13. cf Czech dances, Sets I and II.

Polkas, opp. 7 and 8. cf Czech dances, Sets I and II.
2546 Quartet, strings, no. 1, E minor. Quartet, strings, no. 2, D minor. Smetana Quartet. Supraphon 50448.
++RR 1-77 p68
Quartet, strings, no. 1, E minor. cf DVORAK: Quartet, strings, no. 12, op. 96, F major.
Quartet, strings, no. 2, D minor. cf Quartet, strings, no. 1, E minor.
SMIT, Leo
At the corner of the sky. cf BLANK: Songs (2).
Songs of wonder. cf BLANK: Songs (2).
SMITH
The cascades, polka brilliant. cf Nonesuch H 71341.
SMITH, John Christoph
Lesson, op. 2, no. 7: Fugue. cf Crescent BUR 1001.
SMITH, Julia
2547 Daisy, highlights. Scott Miller, Linda Smalley, Elizabeth Volkman, Larry Gerber, Janet Ariosto, David Rae Smith, Deborah Osborne; Charlotte Opera Association Orchestra and Chorus; Charles Rosekrans. Orion ORS 76248.
+ARG 4-77 p29 +NR 3-77 p10
SMITH, William
Fancies for clarinet alone. cf BABBITT: Phenomena, soprano and piano.
Straws. cf Crystal S 351.
SMOKER, Paul
Brass in spirit. cf University of Iowa Press unnumbered.
SMYTH, Ethel
The wreckers overture. cf GERMAN: Welsh rhapsody.
SMYTHE, Thomas
Galliards (2). cf L'Oiseau-Lyre DSLO 510.
SODERMAN, John
Songs: Kung Heimer och Aslog. cf RCA LSC 9884.
SODERO
Crisantemi. cf Odyssey Y 33793.
SOJO, Vicente
Pieces from Venezuela (5). cf BARRIOS: Las abejas.
SOKOL, Thomas
Sonatina. cf Golden Crest RE 7068.
SOLAGE (14th century France)
Fumeux, fume. cf HMV SLS 863.
Helas, je voy mon cuer. cf HMV SLS 863.
SOLER, Antonio
Fandango. cf London CS 7046.
Sonatas (2). cf RCA RL 2-5099.
Sonatas, B minor, D minor. cf London CS 7046.
2548 Sonatas, harpsichord (10). Fernando Valenti, hpd. Desmar 1001 Tape (c) E 1050.
+-Gr 2-76 p1363 ++NR 3-76 p11
+-HF 4-77 p121 tape +RR 1-76 p52
+-MJ 12-75 p38 +St 1-76 p106
SOLLBERGER, Harvey
2549 Riding the wind I. WYNER: Intermedio. Susan Davenny Wyner, s; Patricia Spencer, flt; String Orchestra, Da Capo Chamber Players; Yehudi Wyner, Harvey Sollberger. CRI SD 352.
+ARG 12-76 p44 -MJ 3-77 p46
+-HF 7-77 p112 +NR 11-76 p8

SONDHEIM, Stephen
A little night music: Send in the clowns. cf HMV MLF 118.
A little night music: Send in the clowns. cf Grosvenor GRS 1048.
SOR, Fernando
Los adioses. cf Lyrichord LLST 7299.
Andante, C minor. cf RCA ARL 1-0864.
Fantasia, no. 2: Largo. cf Westminster WG 1012.
Fantasia elegiaca, op. 59. cf Enigma VAR 1015.
Introduction and variations on "Malbrough a'en va-t-en guerre". cf RCA ARL 1-1323.
Minuets, C major, A major, C major. cf RCA ARL 1-0864.
Minuet, op. 11, no. 6. cf Swedish Society SLT 33205.
Minuet, op. 22. cf Swedish Society SLT 33205.
Sicilienne, D minor. cf RCA ARL 1-1323.
Sonata, guitar, op. 2: Rondo allegretto. cf Westminster WG 1012.
Sonata, guitar, op. 25, C major. cf GIULIANI: Le Rossiniane, op. 121 and op. 119.
Variations on a theme by Mozart, op. 9. cf Angel S 36093.
Variations on a theme by Mozart, op. 9. cf CBS Tape (c) 40-72728.
Variations on a theme by Mozart, op. 9. cf L'Oiseau-Lyre SOL 349.
SOTHCOTT, John
Fanfare. cf CRD CRD 1019.
SOUSA, John Philip
Anchor and star. cf Works, selections (Philips SON 036).
The Atlantic City pageant. cf Works, selections (Philips SON 036).
The beau ideal. cf Works, selections (Philips 9500 151).
The bells of Chicago. cf Works, selections (Philips 9500 151).
The bride elect. cf Works, selections (Antilles AN 7015).
The bride elect. cf Works, selections (Philips 9500 151).
The bride elect: Waltzes. cf Orion ORS 76247.
2550 El Capitan (arr. Rogers). King cotton (arr. Rogers). Semper fidelis (arr. Rogers). Stars and stripes forever (ballet, arr. Kay). National Philharmonic Orchestra; Henry Lewis. London SPC 21161. (also Decca PFS 4382)
-Gr 7-77 p229 +NR 7-77 p4
-HF 11-77 p120 +St 10-77 p146
A century of progress. cf Works, selections (Philips SON 036).
The charlatan. cf Works, selections (Antilles AN 7015).
Coquette. cf Works, selections (Antilles AN 7015).
The crusader. cf Works, selections (Philips SON 036).
The crusader. cf Works, selections (Philips 9500 151).
The diplomat. cf Works, selections (Philips 9500 151).
The directorate. cf Works, selections (Philips 9500 151).
Esprit de corps. cf Works, selections (Philips SON 036).
Fairest of the fair. cf Lismor LILP 5078.
George Washington bicentennial. cf Works, selections (Philips SON 036).
The gladiator. cf Works, selections (Philips 9500 151).
Guide right. cf Works, selections (Philips SON 036).
Guide right. cf Works, selections (Philips 9500 151).
King cotton. cf El Capitan.
Liberty bell. cf Transatlantic XTRA 1169.
Liberty bell. cf Rediffusion Royale 2015.
The loyal legion. cf Works, selections (Philips SON 036).
Manhattan Beach. cf Grosvenor GRS 1052.
Mother Hubbard. cf Works, selections (Antilles AN 7015).
National fencibles. cf Works, selections (Philips SON 036).

National fencibles. cf Works, selections (Philips 9500 151).
Nymphalin. cf Works, selections (Antilles AN 7015).
The Occidental. cf Works, selections (Philips 9500 151).
On parade. cf Works, selections (Philips 9500 151).
The pathfinder of Panama. cf Works, selections (Philips SON 036).
The red man. cf Works, selections (Antilles AN 7015).
La reine de la mer. cf Works, selections (Antilles AN 7015).
The Royal Welsh fusiliers. cf Works, selections (Philips SON 036).
The Salvation Army. cf Works, selections (Philips SON 036).
Semper fidelis. cf El Capitan.
Semper fidelis. cf Works, selections (Philips 9500 151).
Stars and stripes forever. cf El Capitan.
Stars and stripes forever. cf Works, selections (Philips 9500 151).
Stars and stripes forever. cf Menc 76-11.
Stars and stripes forever. cf RCA CRL 1-2064.
The summer girl. cf Works, selections (Antilles AN 7015).
The triumph of time. cf Works, selections (Antilles AN 7015).
The wolverine. cf Works, selections (Philips SON 036).

2551 Works, selections: The bride elect. Coquette. The charlatan. Mother Hubbard. Nymphalin. The red man. The summer girl. La reine de la mer. The triumph of time. Antonin Kubalek, pno. Antilles AN 7015.
+NR 9-76 p13 +St 3-77 p143

2552 Works, selections: Anchor and star. The Atlantic City pageant. A century of progress. The crusader. Esprit de corps. George Washington bicentennial. Guide right. The loyal legion. National fencibles. The pathfinder of Panama. The Royal Welsh fusilers. The Salvation Army. The wolverine. Scots Guards Band; James Howe. Philips SON 036 Tape (c) SOC 036. (Reissue from Fontana LPS 16253)
+Gr 5-77 p1749

2553 Works, selections: The beau ideal. The bells of Chicago. The bride elect. The crusader. The diplomat. The directorate. The gladiator. Guide right. The national fencibles. The Occidental. On parade. Semper fidelis. Stars and stripes forever. Eastman Wind Ensemble; Donald Hunsberger. Philips 9500 151.
+ARG 2-77 p50 +NR 3-77 p15
+MJ 3-77 p46 ++St 3-77 p143

SOUSTER, Tim

2554 Afghan amplitudes. Arcane artefact. Music from afar. Spectral. Surfit. Tim Souster, synthesisers, keyboards; Tony Greenwood, perc; Peter Britton, vibraphone, marimbaphone. Transatlantic TRAG 343.
+-HFN 12-77 p181 +-Te 9-77 p43
+-RR 9-77 p84

Arcane artefact. cf Afghan amplitudes.
Music from afar. cf Afghan amplitudes.
Spectral. cf Afghan amplitudes.
Surfit. cf Afghan amplitudes.

SOUTHERS, Leroy

Spheres. cf Crystal S 362.

SOWERBY, Leo

Prelude on "The King's majesty". cf Orion ORS 76255.

2555 Symphony, G major. David Mulbury, org; Second Presbyterian Church, Indianapolis. Lyrichord LLST 7306.
++MU 7-77 p16 ++NR 3-77 p14

SPALDING, Albert
Etchings, op. 5. cf Clear TLC 2586.
Sonata, solo violin, E minor. cf Clear TLC 2586.
SPANISH MUSIC FOR HARP. cf RCA RL 2-5099.
SPEAKS, Oley
On the road to Mandalay. cf L'Oiseau-Lyre DSLO 20.
SPEER, Daniel
Sonatas, brass (3). cf Crystal S 206.
Sonatas, brass, G major/C major. cf Crystal S 204.
SPENDIARIAN
Oh, rose. cf Golden Age GAR 1001.
SPINACINO, Francesco
Ricercare. cf DG 2533 173.
SPOHR, Louis (Ludwig)
Concerto, violin, no. 8, op. 47, A minor. cf Discopaedia MB 1015.
Concerto, violin, no. 9, op. 55, D minor: Adagio (Soldat). cf Discopaedia MOB 1019.
Duet, no. 1, D minor. cf Discopaedia MB 1013.
2556 Nonet, op. 31, F major. Octet, op. 32, E minor. Quintet, op. 52, C minor. Septet, op. 147, A minor. Danzi Quintet, Members; Jaap Schroder, vln; Anner Bylsma, vlc; Werner Genuit, pno. BASF 22132-6 (2).
+ARG 8-77 p42
Octet, op. 32, E minor. cf Nonet, op. 31, F major.
Quintet, op. 52, C minor. cf Nonet, op. 31, F major.
Septet, op. 147, A minor. cf Nonet, op. 31, F major.
Songs: Zwiegesang. cf Abbey LPB 778.
SPONTINI, Gasparo
La vestale: Tu che invoco. cf HMV SLS 5057.
STAINER, John
The crucifixion: God so loved the world. cf Decca SPA 491.
Lord Jesus, think on me. cf Guild GRS 7008.
STAMITZ
Concerto, violin, B flat major: Adagio. cf Discopaedia MB 1014.
STAMITZ, Johann
2557 Concerto, clarinet, B flat major. Symphony, op. 4, no. 2, D major. Symphony, G major. Symphony, op. 3, no. 2, D major. Alan Hacker, clt; Academy of Ancient Music; Christopher Hogwood. L'Oiseau-Lyre DSLO 505.

++Gr 2-76 p1349	+NR 6-76 p5
+HF 2-77 p103	+RR 1-76 p38
+HFN 5-76 p109	+SFC 3-6-77 p34
+MJ 10-76 p24	+St 6-76 p110

Symphony, G major. cf Concerto, clarinet, B flat major.
Symphony, op. 3, no. 2, D major. cf Concerto, clarinet, B flat major.
Symphony, op. 4, no. 2, D major. cf Concerto, clarinet, B flat major.
STAMITZ, Karl
Concerto, flute, op. 29, G minor. cf HOFFMEISTER: Concerto, flute, G major.
Concerto, viola, D major. cf HOFFMEISTER: Concerto, viola, D major.
Parthia, E flat major. cf JIROVEC: Parthia, B flat major.
Quartet, op. 8, no. 2, E flat major. cf DANZI: Quintet, op. 56, no. 1, B flat major.
STANFORD, Charles
Biblical songs, op. 113, nos. 1-6. cf Argo ZK 11.

The blue bird, op. 119, no. 3. cf RCA RL 2-5112.
The critic, op. 144: Masque. cf Works, selections (Pearl GEM 123).
Evening service, C major. cf Vista VPS 1037.
Fantasia and toccata, op. 57. D minor. cf PEETERS: Concerto, organ and piano, op. 74.
Gloria in excelsis. cf Pye Nixa QS PCNHX 10.
Gloria in excelsis. cf Vista VPS 1053.
Irish rhapsody, no. 1, op. 78. cf Works, selections (Pearl GEM 123)
Magnificat, G major. cf Abbey LPB 779.
Magnificat and nunc dimittis, B flat major. cf Abbey LPB 776.
Postlude, D minor. cf Vista VPS 1047.
Postlude, D minor. cf Vista VPS 1051.
Shamus O'Brien, op. 61: Overture. cf Works, selections (Pearl GEM 123).
Short preludes and postludes, op. 101, Set II. cf PEETERS: Concerto, organ and piano, op. 74.

2558 Songs (Motets): English motets, op. 135: Ye holy angels bright; Eternal Father; Glorious and powerful God. Latin motets, op. 51: Justorum animae; Coelis ascendit; Beati quorum via. WOOD: Expectans expectavi; God omnipotent reigneth; Glory and honour and laud; Hail, gladdening light; Tis the day of resurrection; O thou the central orb. Magdalen College Choir; Ian Crabbe, org; Bernard Rose. Argo ZRG 852.
+Gr 11-76 p858 +MT 1-77 p45
+-HFN 10-76 p177 +RR 11-76 p102

2559 Songs (Biblical songs and hymns): Biblical songs, op. 113: I came forth from the mouth of the most high; I will lift up mine eyes unto the hills; If the Lord himself had not been on our side; Out of the deep have I called unto thee O Lord; There shall come forth a rod out of the stem of Jesse; When the Lord turned again the captivity of Sion. Hymns: In thee is gladness; Let us with gladsome mind; Praise to the Lord, the almighty; Pray that Jerusalem may have peace and felicity; Purest and highest; O for a closer walk with God. Maurice Bevan, bar; Barry Rose Singers; Barry Rose, John Dexter, org. Guild GRS 7009.
+Gr 7-77 p208 +RR 7-77 p91
+HFN 6-77 p135

Songs (Partsongs), op. 110: Heraclitus, op. 119: The blue bird; Farewell my joy; The inkbottle; Chillingham; My heart is thine; The swallow; The train; The witch; op. 127: The haven. cf PARRY: Songs (Prelude PRS 2506).
Songs: Songs of the fleet, op. 117: no. 1, Sailing at dawn; no. 2, The song of the sou'wester; no. 3, The middle watch; no. 4, The little admiral; no. 5, Farewell. cf Works, selections (Pearl GEM 123).
Songs: The monkey's carol. cf Abbey LPB 778.
Songs of the sea, op. 91, no. 4: Drake's drum. cf L'Oiseau-Lyre DSLO 20.
Suite of ancient dances, op. 58, no. 1, Morris dance; no. 2, Sarabande. cf Works, selections (Pearl GEM 123).

2560 Works, selections: The critic, op. 144: Masque. Irish rhapsody, no. 1, op. 78. Shamus O'Brien, op. 61: Overture. Suite of ancient dances, op. 58, no. 1, Morris dance; no. 2, Sarabande. Songs: Songs of the fleet, op. 117: no. 1, Sailing at dawn; no. 2, The song of the sou'wester; no. 3, The middle watch; no. 4, The little admiral; no. 5, Farewell. Harold Williams, bar;

Symphony Orchestra, LSO; Charles Stanford. Pearl GEM 123.
(Reissues from HMV D 191/3; Columbia 939/51)
+-ARSC Vol 9, no. 1 1977, p98 +Gr 9-74 p518
+RR 8-74 p41

STANLEY, John
Concerto, D major. cf Folkways FTS 32378.
Sonata, flute, G major. cf BACH, J.C.F.: Sonata, fortepiano, flute and violoncello, D major.
Suite of trumpet voluntaries, D major. cf Nonesuch H 71279.
Trumpet voluntary, D major. cf Philips 6500 926.
Voluntaries, D major, A minor. cf Cambridge CRS 2540.
Voluntary, D major. cf Pelca PSR 40599.
Voluntary, E minor. cf PEETERS: Concerto, organ and piano, op. 74.
Voluntary, op. 5, no. 1, C major. cf Vista VPS 1047.

STARER, Robert
Prelude. cf Orion ORS 75207.

STAROKADOMSKY, Mikhail
Victory march. cf HMV CSD 3782.

STARZER, Josef
Diane et Endimione: Dances. cf DG 2723 051.
Gli Orazi e gli Zuriazi: Dances. cf DG 2723 051.
Roger et Bradamante: Dances. cf DG 2723 051.

STECK
Birdcage walk. cf DJM Records DJM 22062.

STEFFANI, Agostino
Stabat mater. cf BUXTEHUDE: Cantatas (CMS/Oryx 3C 303).

STEFFE, William
Glory, glory. cf Transatlantic XTRA 1169.

STELZMULLER, Vinzenz
Dance. cf Saga 5421.

STENHAMMAR, Wilhelm
Serenade, op. 31, F major. cf KOCH: Oxberg variations.
Songs: Det far att skepp; Flickan knyter i Johannenatten; Flickan Kom ifran sin alsklings mote; I skogen. cf FRUMERIE: Songs (Swedish Society SLT 33171).
Songs: Sverige. cf RCA LSC 9884.

STEPHENS
Copa cabana. cf Grosvenor GRS 1055.

STEVENS
Policewoman, theme. cf EMI TWOX 1058.
Sigh no more, ladies. cf HMV ESD 7002.

STEVENS, Halsey
2561 Sonata, horn and piano. TUFTS: Sonata, horn and piano. VERRALL: Sonata, horn and piano. Christopher Leuba, hn; Kevin Aanerud, pno. Crystal S 372.
+NR 9-77 p5

STEVENS, John
Music, 4 tubas. cf Crystal S 221.

STEVENSON
Behold I bring you glad tidings. cf Vox SVBX 5350.

STEWART, Richard
Prelude, organ and tape. cf Abbey LPB 752.

STOBAEUS, Johann
Alia Chorea polonica. cf Hungaroton SLPX 11721.

STOCKHAUSEN, Karlheinz
2562 Bird of passage. Ceylon. Harald Boje, electronium; Peter Eotvos, camel bells, triangles, synthesizer; Aloys Kontarsky, pno;

Joachim Krist, tam tam; Tim Souster, sound projections; Markus Stockhausen, tpt, electric tpt and flugel horn; John Miller, tpt; Karlheinz Stockhausen, chromatic rin, lotus flute, Indian bells, bird whistle, voice, Kandy drum. Crysalis CHR 1110.
+-Gr 9-76 p438 +St 2-77 p122
+-MM 3-77 p44 +Te 3-77 p32
+-RR 8-76 p64

Ceylon. cf Bird of passage.

Fur Dr. K. cf Universal Edition UE 15043.

Klavierstuck XI. cf CP 3-5.

2563 Mikrophonie I and II. Aloys Kontarsky, Fred Alings, tam tam; Johannes Fritsch, Harald Boje, microphones; Hugh Davies, Jaap Spek, Karlheinz Stockhausen, electronics; WDR Choir, Cologne Studio Choir; Alfons Kontarsky, org; Herbert Schernus. DG 2530 583. (Reissue from CBS SBRG 72647)
+Gr 1-77 p1178 +RR 9-76 p77
++HFN 11-76 p167 +Te 3-77 p32
+-MM 3-77 p44

2564 Momente (Europa version 1972). Gloria Davy, s; West German Radio Chours; Musique Vivante Ensemble Instrumentalists; Karlheinz Stockhausen. DG 2709 055 (3).
+-MM 3-77 p44 ++RR 10-76 p67

2565 Prozession (1971 version). Harald Boje, electronium; Christoph Caskel, tam tam; Joachim Krist, microphone; Peter Eotvos, electrochord with synthesizer; Aloys Kontarsky pno; Karlheinz Stockhausen, electronics. DG 2530 582.
+Gr 1-77 p1178 +-RR 9-76 p77
+HFN 11-76 p167 +-Te 3-77 p32
+-MM 3-77 p44

2566 Punkte. Songs (choral): Atmen gibt das Leben; Chore fur Doris; Choral. Irmgard Jacobeit, Susanne Denman, s; Ulf Kenklies, t; North German Radio Chorus; North German Radio Orchestra and Chorus; Karlheinz Stockhausen. DG 2530 641.
+Gr 9-77 p469 +RR 9-77 p89
+HFN 10-77 p163

Songs (choral): Atmen gibt das Leben; Chore fur Doris; Choral. cf Punkte.

2567 Trans (2). South West German Radio Orchestra, Saarbrucken Radio Orchestra; Ernest Bour, Hans Zender. DG 2530 726.
+-Gr 3-77 p1410 +RR 2-77 p59
+-HFN 2-77 p131

STOELZEL, Gottfried

Concerto, trumpet, D major. cf DG 2530 792.

Concerto, trumpet, D major. cf Philips 6581 018.

STOKOWSKI (STOJOWSKI), Sigismund

Caprice oriental. cf International Piano Library IPL 5007/8 (2).

STOLTZER, Thomas

2568 German psalms: Erzurne Dich nicht; Hilf, Herr, die Heiligen haben abgenommen; Herr, wie lang; Herr, neige deine Ohren; Missa Duplex per totum annum. Songs: Accessit ad Pedes; O admirable commercium; In Gottes Namen fahren wir; Konig, Ein Herr ob Alle Reich; Ich Klag den Tag; De Sancto Martino. Octo tonorum melodiae: Fantasias (4). Munich Capella Antiqua; Konrad Ruhland. ABC ABCL 67003/2 (2).
++ARG 5-77 p38 +NR 4-77 p7
++HF 3-77 p108 +ON 3-12-77 p40
+-MJ 2-77 p30 ++St 5-77 p54

STOLZ, Robert
Two hearts swing in three-quarter time. cf Philips 6392 023.
STORACE, Bernardo
Ballo della battaglia aus. cf ABC ABCL 67008.
STORACE, Stephen
Sonata, harpsichord, D major: 1st movement. cf Crescent BUR 1001.
STRADELLA, Alessandro
Sonata, trumpet and strings. cf MARCELLO, A.: Concerto, oboe and strings.
Songs: Pieta Signore. cf Decca SXL 6781.
Songs: Pieta Signore. cf London OS 26437.
STRAUSS, Eduard
Bahn frei, op. 45. cf Decca JB 28.
Unter der Enns, op. 12. cf Saga 5421.
STRAUSS, Johann I
Austrian jubilation sounds waltz, op. 179. cf Works, selections (Musical Heritage Society MHS 3396).
Artists ball dances, op. 84: Waltz. cf Works, selections (Musical Heritage Society MHS 3396).
Ball-Raketen waltz, op. 96. cf Works, selections (Musical Heritage Society MHS 3396).
Chinese galop, op. 20. cf STRAUSS, J. II: Works, selections (DG 2584 008).
Hungarian galop (Frischka), op. 36. cf Works, selections (Musical Heritage Society MHS 3396).
Jubilation quadrille, op. 130. cf Works, selections (Musical Heritage Society MHS 3396).
Ketternbrucke Walzer, op. 4. cf Lorelei Rheinklange, op. 154.
Ladies souvenir polka, op. 236. cf Works, selections (Musical Heritage Society MHS 3396).
2569 Lorelei Rheinklange, op. 154. Ketternbrucke Walzer, op. 4. Radetzky march, op. 228. Sperl galopp, op. 42. STRAUSS, J. II: Accelerationen, op. 234. Champagne polka, op. 211. Pesther Csardas, op. 23. Tritsch-Tratsch, op. 214. Unter Donner und Blitz, op. 324. Wiener Blut, op. 354. STRAUSS, J. II/Josef: Piccicato polka. London Concert Orchestra; John Georgiadis. Polydor 2460 266 Tape (c) 3170 288.
+Gr 2-77 p1330
Lorelei Rheinklange, op. 154. cf Saga 5421.
Radetzky march, op. 228. cf Lorelei Rheinklange, op. 154.
Radetzky march, op. 228. cf Decca JB 28.
Radetzky march, op. 228. cf DG 2548 148.
Sperl galopp, op. 42. cf Lorelei Rheinklange, op. 154.
Tauberin waltz, op. 1. cf Works, selections (Musical Heritage Society MHS 3396).
2570 Works, selections: Austrian jubilation sounds waltz, op. 179. Artists ball dances, op. 84: Waltz. Ball-Raketen waltz, op. 96. Hungarian galop (Frischka), op. 36. Jubilation quadrille, op. 130. Ladies souvenir polka, op. 236. Tauberin waltz, op. 1. Classic Vienna Strauss-Lanner Orchestra; Kurt Rapf. Musical Heritage Society MHS 3396.
+St 5-77 p118
STRAUSS, Johann II
Accelerationen, op. 234. cf STRAUSS J. I: Lorelei Rheinklange, op. 154.
An der schonen blauen Donau (The beautiful blue Danube), op. 314. cf Works, selections (HMV ESD 7025).

The beautiful blue Danube, op. 314. cf Works, selections (Quintessence PMC 7051).
An der schonen blauen Donau, op. 314. cf Works, selections (RCA GL 2-5019).
Le beau Danube, op. 314. cf Die Fledermaus, op. 363: Overture.
2571 The beautiful blue Danube, op. 314. Artists life, op. 316. Emperor waltz, op. 437. Wine, women and song, op. 333. LPO; Theodor Guschlabuer. Classics for Pleasure Tape (c) TC CFP 165.
+RR 11-77 p119 tape
The beautiful blue Danube, op. 314. cf Connoisseur Society CS 2131.
The beautiful blue Danube, op. 314. cf Seraphim SIB 6094.
Le beau Danube, op. 314, ballet. cf OFFENBACH: Gaite parisienne, excerpts.
Annen polka, op. 117. cf Works, selections (Classics for Pleasure CFP 40048).
Annen polka, op. 117. cf Works, selections (RCA ARL 1-2266).
Auf der Jagd, op. 373. cf Works, selections (RCA ARL 1-2266).
Cagliostrowalzer, op. 370. cf Works, selections (Everest SDBR 3406).
Casanova: Sul mare lucica. cf HMV ESD 7043.
Champagne polka, op. 211. cf Works, selections (Classics for Pleasure CFP 40048).
Champagne polka, op. 211. cf STRAUSS, J. I: Lorelei Rheinklange, op. 154.
Demolirer, op. 269. cf Decca JB 28.
Du und Du, op. 367. cf Works, selections (Everest SDBR 3406).
Du und Du, op. 367. cf Decca JB 28.
Eljen a Magyar, op. 332. cf Works, selections (Classics for Pleasure CFP 40048).
2572 Die Fledermaus. Julia Varady, Lucia Popp, Evi List, s; Rene Kollo, Perry Gruber, t; Hermann Prey, Benno Kusche, Bernd Weikl, Nikolai Lugovoi, bar; Ivan Rebroff, Franz Muxeneder, bs; Bavarian State Opera Chorus; Bavarian State Orchestra; Carlos Kleiber. DG 2707 088 (2) Tape (c) 3370 009.
+ARG 3-77 p36 /NR 3-77 p9
+Gr 10-76 p650 -ON 12-4-76 p60
+-HF 2-77 p103 +RR 10-76 p30
++HFN 10-76 p159 ++SFC 11-7-76 p33
+-MJ 5-77 p32 +-St 3-77 p145
2573 Die Fledermaus, op. 363, excerpts. Gundula Janowitz, Renate Holm, s; Sylvia Lukan, ms; Wolfgang Windgassen, Waldemar Kmentt, t; Eberhard Wachter, Heinz Holecek, Erich Kunz, Erich Kuchar, bar; Vienna State Opera Chorus; VPO; Karl Bohm. Decca SET 600 Tape (c) KCET 600. (Reissue from SET 540/1)
+Gr 7-77 p1230 +-RR 5-77 p95 tape
+HFN 7-77 p125 +RR 6-77 p38
2574 Die Fledermaus, op. 363, excerpts. Soloists; Graunke Symphony Orchestra; Peter Falk. Telefunken AF 6-22995 Tape (c) CH 6-22995.
+HFN 12-77 p187 -RR 11-77 p43
Die Fledermaus, op. 363, excerpt. cf Decca JB 28.
Die Fledermaus, op. 363: Czardas; Mein Herr Marquis. cf HMV MLF 118.
Die Fledermaus, op. 363: Klange der Heimat. cf HMV ESD 7043.
Die Fledermaus, op. 363: Mein Herr was dachten Sie von mir. cf Seraphim IB 6105.

2575 Die Fledermaus, op. 363: Overture. Geschichten aus dem Wienerwald, op. 325. Rosen aus dem Suden, op. 388. Der Zigeunerbaron, op. 420: Overture. NYP; Leonard Bernstein. Columbia M 34125. (also CBS 61779 Tape (c) 40-61779)
+-Gr 4-77 p1604
+-HF 12-76 p118
-HFN 4-77 p147
+-HFN 6-77 p139 tape
+-MJ 12-76 p28
+NR 7-76 p3
/RR 4-77 p60

2576 Die Fledermaus, op. 363: Overture. Le beau Danube, op. 314 (arr. Desormier). STRAUSS, J. II/J. I: Bal de Vienne (arr. Gamley). National Philharmonic Orchestra; Richard Bonynge. Decca SXL 6701 Tape (c) KSXC 6701. (also London CS 6896 Tape (c) CS 5-6896)
+Gr 12-75 p1098
+HF 11-76 p97
+HFN 12-75 p165
+HFN 2-76 p116 tape
++NR 1-77 p4
+RR 5-76 p78
+RR 12-75 p62

Die Fledermaus, op. 363: Overture. cf Works, selections (Classics for Pleasure CFP 40048).

Die Fledermaus, op. 363: Overture. cf Works, selections (RCA ARL 1-2266).

Die Fledermaus, op. 363: Overture. cf KLEMPERER: Merry waltz.

Die Fledermaus, op. 363: Overture. cf Classics for Pleasure CFP 40236.

2577 Die Fledermaus, op. 363: Overture; Taubchen, das Entflattert ist; Komm mit mir Souper; Trinke liebchen, trinke schnell; Mein Herr, was dachten sie von mir; Mein schones, grosses Vogelhaus; Ich lade gern mir Gaste ein; Mein Herr Marquis; Die Klange der Heimat; Im feuerstrom der Reben; Bruderlein; Bruderlein und Schwesterlein; Ich stehe voll Zagen; O Feldermaus; Champagner hat's verschuldet; Die Majestat wird anerkannt. Anneliese Rothenberger, Renate Holm, s; Nicolai Gedda, Adolf Dallapozza, t; Brigitte Fassbaender, con; Dietrich Fischer-Dieskau, bar; Walter Berry, bs-bar; Vienna State Opera Chorus; VSO; Willi Boskovsky. HMV ASD 2891 Tape (c) TC ASD 2891 (Q) Q4ASD 2891.
+-Gr 11-74 p963
+Gr 11-74 p970 Quad
+-Gr 11-76 p893
+HFN 1-77 p121
+HFN 5-77 p138 tape
+HFN 9-77 p155 tape
+RR 7-74 p31
+RR 11-76 p42

Die Fledermaus, op. 363: Reminiscences. cf Connoisseur Society CS 2131.

Freut euch des Lebens, op. 340. cf Works, selections (Everest SDBR 3406).

Fruhlingsstimmen (Voices of spring), op. 410. cf Court Opera Classics CO 342).

Voices of spring, op. 410. cf HMV RLS 717.

Voices of spring, op. 410. cf HMV HLM 7066.

Voices of spring, op. 410. cf Mercury SRI 75098.

Geschichten aus dem Wienerwald (Tales from the Vienna Woods), op. 325. cf Die Fledermaus, op. 363.

Geschichten aus dem Wienerwald, op. 325. cf Works, selections (HMV ESD 7025).

Tales from the Vienna Woods, op. 325. cf Works, selections (Quintessence PMC 7051).

Geschichten aus dem Wienerwald, op. 325. cf Works, selections (RCA GL 2-5019).

Graduation ball, op. 97, excerpts. cf OFFENBACH: Gaite parisienne: Ballet.

Im Krapfenwald, op. 336. cf Works, selections (DG 2584 008).
Kaiser Walzer (Emperor waltz), op. 437. cf The beautiful blue Danube, op. 314.
Emperor waltz, op. 437. cf Works, selections (DG 2584 008).
Kaiser Walzer, op. 437. cf Works, selections (HMV ESD 7025).
Emperor waltz, op. 437. cf Works, selections (Quintessence PMC 7051).
Emperor waltz, op. 437. cf Works, selections (RCA ARL 1-2266).
Kaiser Walzer, op. 437. cf Works, selections (RCA GL 2-5019).
Kaiser Walzer, op. 437. cf KLEMPERER: Merry waltz.
Emperor waltz, op. 437. cf LISZT: Hungarian rhapsody, no. 1, F minor.
Emperor waltz, op. 437. cf Angel S 37409.
Emperor waltz, op. 437. cf HMV (SQ) ASD 3338.
Kunstlerleben (Artists life), op. 316. cf The beautiful blue Danube, op. 314.
Artists life, op. 316. cf Works, selections (RCA ARL L-2266).
Artists life, op. 316. cf Mercury SRI 75098.
Kusswalzer, op. 400. cf Works, selections (Everest SDBR 3406).
Lagunenwalzer, op. 411. cf Works, selections (Everest SDBR 3406).
Leichtes Blut, op. 319. cf Works, selections (Classics for Pleasure CFP 40048).
Man lebt nur einmal, op. 167. cf Works, selections (Everest SDBR 3406).
Man lebt nur einmal, op. 167. cf Saga 5421.
Marchen aus dem Orient, op. 444. cf Works, selections (DG 2584 008).
Myrthenblutenwalzer. cf Works, selections (Everest SDBR 3406).

2578 Eine Nacht in Venedig: Introduction...Wenn vom Lido...Ihr Veneziāner, hort, Seht, o seht...Frutti di mare; S'ist wahr, ich bin nicht allzu klug; Evviva, Caramello, Caramello, Annina; Alle maskiert; Komm in die Gondel; Messer Delacqua...Zur serenade; Was mir der Zufall gab; Ach was ist das...Ja, beim Tanz; Solch ein Wirtshaus lob ich mir...Noch sah Ciboletta ich nicht; Ninana, ninana; Kommt, o komt, ihr holden Tauben; Ach wie so herrlich zu schau'n. Elisabeth Ebert, Rosmarie Ronisch, s; Harald Neukirch, Martin Ritzmann, t; Reiner Suss, Siegfried Vogel, bar; Leipzig Radio Chorus; Dresden Philharmonic Orchestra; Heinz Rogner. Fontana 6530 047.
+-Gr 11-77 p910 +-RR 11-77 p43
+-HFN 12-77 p181

2579 Eine Nacht in Venedig: Wenn vom Lido sacht; Seht, o seht; Eviva Caramello; Duet; Alle maskiert; Hier ward es still; Komm in die Gondel; Entr'acte; Venedigs Frauen herzufuhren; Was mir der Zufall gab; So sind wir endlich denn allein; Solch ein Wirthaus lob ich mir; Ninana, hier will ich singen; Lasset die Ander'n nur tanzen; Karneval ruft uns zum Ball; Ach, wie so herrlich zu schau'n; Ein Herzog, reich und machtig; Die Tauben von San Marco. Esther Rethy, s; Ruthilde Bosch, Maria Schober, ms; Karl Friedrich, t; Alfred Jerger, Kurt Preger, bar; Bregenz Festival Chorus; VSO; Anton Paulik. Saga 5423.
+-Gr 1-77 p1188 +-MT 3-77 p218
+HFN 2-77 p131 -RR 1-77 p40

Neue Wien (New Vienna), op. 342. cf Works, selections (Everest SDBR 3406).
New Vienna, op. 342. cf Works, selections (DG 2584 008).
O schoner Mai, op. 365. cf Works, selections (Everest SDBR 3406).

Perpetuum Mobile, op. 257. cf Works, selections (Classics for Pleasure CFP 40048).
Perpetuum Mobile, op. 257. cf Works, selections (DG 2584 008).
Persian march, op. 289. cf Works, selections (Classics for Pleasure CFP 40048).
Pesther Csardas, op. 23. cf STRAUSS, J. I: Lorelei Rheinklange, op. 154.
Rosen aus dem Suden (Roses from the south), op. 388. cf Die Fledermaus, op. 363: Overture.
Roses from the south, op. 388. cf Works, selections (RCA ARL 1-2266).
Rosen aus dem Suden, op. 388. cf Works, selections (HMV ESD 7025).
Roses from the south, op. 388. cf Mercury SRI 75098.
Seid umschlungen Millionen, op. 443. cf Works, selections (Everest SDBR 3406).
Sinngedichte, op. 1. cf Saga 5421.
Spanischer Marsch, op. 433. cf Decca JB 28.
Ein Tausend und eine Nacht (A thousand and one nights), op. 346. cf Works, selections (Everest SDBR 3406).
Tritsch-Tratsch, op. 214. cf Works, selections (Classics for Pleasure CFP 40048).
Tritsch-Tratsch, op. 214. cf Works, selections (DG 2584 008).
Tritsch-Tratsch, op. 214. cf STRAUSS, J. I: Lorelei Rheinklange, op. 154.
Tritsch-Tratsch, op. 214. cf Connoisseur Society CS 2131.
Unter Donner und Blitz (Thunder and lightning), op. 324. cf Works, selections (Classics for Pleasure CFP 40048).
Thunder and lightning, op. 324. cf Works, selections (RCA ARL 1-2266).
Unter Donner und Blitz, op. 324. cf STRAUSS, J. I: Lorelei Rheinklange, op. 154.
Vienna bonbons, op. 307. cf Works, selections (Everest SDBR 3406).
Waltzes (5). cf HMV SLS 5073.
Wein, Weib und Gesang (Wine, women and song), op. 333. cf The beautiful blue Danube, op. 314.
Wein, Weib und Gesang, op. 333. cf Works, selections (HMV ESD 7025).
Wine, women and song, op. 333. cf Works, selections (Quintessence PMC 7051).
Wein, Weib und Gesang, op. 333. cf Works, selections (RCA GL 2-5019).
Wein, Weib und Gesang, op. 333. cf HMV RLS 717.
Wine, women and song, op. 333. cf RCA CRL 1-2064.

2580 Wiener Blut, op. 354. Anneliese Rothenberger, Renate Holm, Gabriele Fuchs, s; Nicolai Gedda, Heinz Zednik, t; Klaus Hirte, bar; Cologne Opera Chorus; Wiener Schrammeln; PH; Willi Boskovsky. Angel (Q) SBLX 3831 (2). (also HMV SLS 5074)

+Gr 3-77 p1462 +NR 6-77 p9
+-HF 7-77 p108 +RR 3-77 p36
+HFN 3-77 p113 ++St 6-77 p144

Wiener Blut (Vienna blood), op. 354. cf Works, selections (HMV ASD 7025).
Vienna blood, op. 354. cf Works, selections (Quintessence PMC 7051).
Wiener Blut, op. 354. cf Works, selections (RCA GL 2-5019).
Wiener Blut, op. 354. cf KLEMPERER: Merry waltz.
Wiener Blut, op. 354. cf STRAUSS, J. I: Lorelei Rheinklange, op. 154.

Wo die Zitronen bluhn, op. 364. cf Works, selections (Everest 3406).

2581 Works, selections: Annen polka, op. 117. Champagne polka, op. 211. Eljen a Magyar, op. 332. Die Fledermaus, op. 363: Overture. Leichtes Blut, op. 319. Persian march, op. 289. Perpetuum Mobile, op. 257. Tritsch-Tratsch, op. 214. Unter Donner und Blitz, op. 324. Der Zigeunerbaron, op. 420: Overture. LPO; Theodor Guschlbauer. Classics for Pleasure CFP 40048 Tape (c) TC CFP 40048.
+Gr 2-77 p1325 tape ++RR 8-73 p54
++HFN 12-76 p153 tape

2582 Works, selections: Emperor waltz, op. 437. Im Krapfenwald, op. 336. New Vienna, op. 342. Perpetuum Mobile, op. 257. Marchen aus dem Orient, op. 444. Tritsch-Tratsch, op. 214. Der Zigeunerbaron, op. 420: March, Act 3. STRAUSS, J. I: Chinese galop, op. 20. BPO; Arthur Fiedler. DG 2584 008. (also DG 2535 231)
+FF 11-77 p52 +HFN 2-77 p131
+Gr 2-77 p1330 +RR 2-77 p60
+HF 11-77 p128

2583 Works, selections: Cagliostrowalzer, op. 370. Du und Du, op. 367. Freut euch des Lebens, op. 340. Kusswalzer, op. 400. Lagunenwalzer, op. 411. Man lebt nur einmal, op. 167. Myrthenblutenwalzer. Neue Wien, op. 342. O schoner Mai, op. 365. Seid umschlungen Millionen, op. 443. A thousand and one nights, op. 346. Vienna bonbons, op. 307. Wo die Zitronen bluhn, op. 364. VSO; Josef Krips. Everest SDBR 3406.
-ARG 2-77 p41 +NR 9-77 p2

2584 Works, selections: An der schonen blauen Donau, op. 314. Geschichten aus dem Wiernerwald, op. 325. Kaiser Walzer, op. 437. Rosen aus dem Suden, op. 388. Wein, Weib und Gesang, op. 333. Wiener Blut, op. 354. Johann Strauss Orchestra; Willi Boskovsky. HMV ESD 7025 Tape (c) TC ESD 7025. (Reissues from TWO 368, 389, SLS 5017)
+Gr 12-76 p1072 +HFN 5-77 p155 tape
+HFN 3-77 p117 +RR 12-76 p65
+HFN 5-77 p138 tape +RR 8-77 p87 tape

2585 Works, selections: The beautiful blue Danube, op. 314. Emperor waltz, op. 437. Tales from the Vienna Woods, op. 325. Vienna blood, op. 354. Wine, women and song, op. 333. VSOO; Jascha Horenstein. Quintessence PMC 7051.
+ARG 12-77 p43 +St 11-77 p146
-SFC 10-30-77 p44

2586 Works, selections: Annen polka, op. 117. Auf der Jagd, op. 373. Artists life, op. 316. Emperor waltz, op. 437. Die Fledermaus, op. 363: Overture. Roses from the south, op. 388. Thunder and lightning, op. 324. STRAUSS, Josef: Feuerfest, op. 269. PO; Eugene Ormandy. RCA ARL 1-2266.
+MJ 9-77 p35 +NR 8-77 p6

2587 Works, selections: An der schonen blauen Donau, op. 314. Geschichten aus dem Wienerwald, op. 325. Kaiser Walzer, op. 437. Wein, Weib und Gesang, op. 333. Wiener Blut, op. 354. VSO; Jascha Horenstein. RCA GL 2-5019 Tape (c) GK 2-5019.
+-Gr 10-76 p661 +-HFN 12-76 p153
+Gr 2-77 p1325 tape /RR 10-76 p68
+HFN 12-76 p148 +RR 12-76 p104 tape

Der Zigeunerbaron, op. 420: March, Act 3. cf Works, selections (DG 2584 008).

Der Zigeunerbaron, op. 420: Overture. cf Die Fledermaus, op. 363: Overture.
Der Zigeunerbaron, op. 420: Overture. cf Works, selections (Classics for Pleasure CFP 40048).
Der Zigeunerbaron, op. 420: Wer uns getraut. cf HMV RLS 715.

STRAUSS, Johann II/Johann I
Bal Vienne. cf STRAUSS, J. II: Die Fledermaus, op. 363: Overture.

STRAUSS, Johann II/Josef
Pizzicato polka. cf STRAUSS, J. I: Lorelei Rheinklange, op. 154.
Pizzicato polka. cf Decca JB 28.

STRAUSS, Johann II/Josef/Eduard
Schutzenquadrille. cf Decca JB 28.

STRAUSS, Josef
Brennende Liebe, op. 129. cf Decca JB 28.
Dorfschwalben aus Oesterreich (Village swallows from Austria), op. 164. cf Works, selections (Decca SXL 6817).
Dorfschwalben aus Oesterreich, op. 164. cf Court Opera Classics CO 342.
Village swallows from Austria, op. 164. cf Mercury SRI 75098.
Feuerfest, op. 269. cf Works, selections (Decca SXL 6817).
Feuerfest, op. 269. cf STRAUSS, J. II: Works, selections (RCA ARL 1-2266).
Galoppin polka, op. 237. cf Saga 5421.
Heiterer Mut, op. 281. cf Works, selections (Decca SXL 6817).
Im Fluge, op. 230. cf Works, selections (Decca SXL 6817).
Jokey, op. 278. cf Works, selections (Decca SXL 6817).
Mein Lebenslauf ist Lieb und Lust, op. 263. cf Works, selections (Decca SXL 6817).
Ohne Sorgen, op. 271. cf Works, selections (Decca SXL 6817).
Plappermaulchen, op. 245. cf Works, selections (Decca SXL 6817).
Transaktionen, op. 184. cf Works, selections (Decca SXL 6817).
Transaktionen, op. 184. cf Decca JB 28.

2588 Works, selections: Delirien, op. 212. Dorfschwalben aus Oesterreich, op. 164. Feuerfest, op. 269. Heiterer Mut, op. 281. Jokey, op. 278. Im Fluge, op. 230. Mein Lebenslauf ist Lieb und Lust, op. 263. Plappermaulchen, op. 245. Ohne Sorgen, op. 271. Transaktionen, op. 184. VPO; Willi Boskovsky. Decca SXL 6817 Tape (c) KSXC 6817.
++Gr 12-76 p1075 +HFN 2-77 p135 tape
+HFN 12-76 p148 +RR 12-76 p65

STRAUSS, Oscar
The chocolate soldier: My hero. cf HMV MLF 118.
Rund und die Liebe: Hans songs. cf KALMAN: Das Veilchen vom Montmartre: Heut nacht hab ich getraumpt von dir.

STRAUSS, Richard
2589 An Alpine symphony, op. 64. LAPO; Zubin Mehta. London CS 6981 Tape (c) CS 5-6981. (also Decca SXL 6752 Tape (c) KSXC 6752)
+-Gr 4-76 p1611 +-MT 12-76 p1006
+-Gr 6-76 p102 tape +NR 8-76 p8
+HF 11-76 p126 +RR 5-76 p51
+HFN 4-76 p115 ++SFC 6-20-76 p26
++HFN 7-76 p104 tape +St 11-76 p158
++MJ 1-77 p26

2590 Also sprach Zarathustra, op. 30. BSO; William Steinberg. DG 2535 209. (Reissue from 2530 160)
+Gr 7-77 p187 -RR 6-77 p69
+HFN 7-77 p125

2591 Also sprach Zarathustra, op. 30. Salome, op. 54: Dance of the seven veils. Till Eulenspiegels lustige Streiche, op. 28. Dresden Staatskapelle; Rudolf Kempe. HMV ESD 7026 Tape (c) TC ESD 7026. (Reissues from SLS 861, 894)
+-Gr 4-77 p1561 ++RR 2-77 p60
+HFN 5-77 p139 tape ++RR 9-77 p99 tape
2592 Also sprach Zarathustra, op. 30. Till Eulenspiegels lustige Streiche, op. 28. CSO; Georg Solti. London CS 6978 Tape (c) 5-6978 (ct) 6-6978.
++FF 9-77 p89
2593 Also sprach Zarathustra, op. 30. Till Eulenspiegels lustige Streiche, op. 28. VPO; Clemens Krauss. London R 23208.
+-NR 12-76 p6 +NYT 1-16-77 pD13
2594 Also sprach Zarathustra. Dresden State Orchestra; Rudolf Kempe. Seraphim (Q) S 60283.
++NR 6-77 p3 ++St 10-77 p146
Also sprach Zarathustra, op. 30: Opening. cf HMV MLF 118.
Arabella, op. 79: Mein Elemer. cf Seraphim IB 6105.
2595 Ariadne auf Naxos, op. 60: Es gibt ein Reich. Capriccio, op. 85: Wo ist mein Bruder. Songs: Einerlei, op. 69, no. 3; Befreit, op. 39, no. 4; Ich wollt ein Strausslein binden, op. 68, no. 2; Schlechtes Wetter, op. 69, no. 5; Vier letze Lieder, op. posth. Lisa Della Casa, s; Franz Bierbach, bs; Karl Hudez, pno; VPO; Henrich Hollreiser, Karl Bohm. Decca ECM 778. (Reissues from LXT 5017, 5258, LW 5056)
+-Gr 3-76 p1507 +-ON 2-12-77 p41
+-HFN 3-76 p112 +RR 2-76 p22
Ariadne auf Naxos: Es gibt ein Reich. cf Decca ECS 812.
2596 Aus italien, op. 16, G major. Dresden Staatskapelle Orchestra; Rudolf Kempe. HMV ASD 3319 Tape (c) TC ASD 3319. (Reissue from SLS 894)
+Gr 2-77 p1289 ++RR 6-77 p69
+HFN 5-77 p139 tape
2597 Aus italien, op. 16. VPO; Clemens Krauss. London R 23210.
+NYT 1-16-77 pD13
Le bourgeois gentilhomme, op. 60. cf Works, selections (DG 2740 160).
2598 Burleske, D minor. Concerto, violin, op. 8, D minor. Malcolm Frager, pno; Ulf Hoelscher, vln; Dresden State Orchestra; Rudolf Kempe. Angel S 37267. (also HMV SQ ASD 3399 Tape (c) TC ASD 3399. Reissue from SLS 5067)
+Gr 11-77 p853 +SFC 8-14-77 p50
+-Gr 12-77 p1142 +SR 9-3-77 p42
+HF 10-77 p118 ++St 11-77 p156
++RR 11-77 p69
Burleske, D minor. cf Works, selections (HMV SQ SLS 5067).
Capriccio: Introduction for sextet. cf JANACEK: Suite.
Capriccio, op. 85: Wo ist mein Bruder. cf Ariadne auf Naxos, op. 60: Es gibt ein Reich.
Concertino, clarinet, bassoon, harp and string orchestra. cf Works, selections (HMV SQ SLS 5067).
Concerto, horn, no. 1, op. 11, E flat major. cf Works, selections (HMV SQ SLS 5067).
Concerto, horn, no. 2, op. 11, E flat major. cf Works, selections (HMV SQ SLS 5067).
Concerto, oboe, D major. cf Works, selections (HMV SQ SLS 5067).
Concerto, oboe, D major. cf BACH: Concerto, oboe d'amore, A major.

Concerto, violin, op. 8, D minor. cf Burleske, D minor.
Concerto, violin, op. 8, D minor. cf Works, selections (HMV SQ SLS 5067).

2599 Don Juan, op. 20. Der Rosenkavalier, op. 59: First and second waltz sequence. Till Eulenspiegels lustige Streiche, op. 28. COA; Eugen Jochum. Philips 6580 129 Tape (c) 7317 151. (Reissue from SABL 201)
+-Gr 8-76 p299 +-RR 7-76 p63
+HFN 2-77 p135 tape

2600 Don Juan, op. 20. Der Rosenkavalier, op. 59: Suite. Till Eulenspiegels lustige Streiche, op. 28. PO; Eugene Ormandy. RCA ARL 1-1408 Tape (c) ARK 1-1408 (ct) ARS 1-1408.
+-HF 2-77 p104 /SFC 6-6-76 p33
+-NR 7-76 p2

2601 Don Juan, op. 20. Macbeth, op. 23. Dresden State Orchestra; Rudolf Kempe. Seraphim S 60288.
++HF 11-77 p128 ++St 12-77 p155
+NR 10-77 p4

2602 Don Juan, op. 20. Der Rosenkavalier, op. 59: 1st waltz sequence. Salome, op. 54: Dance of the seven veils. Till Eulenspiegels lustige Streiche, op. 28. Cincinnati Symphony Orchestra; Thomas Schippers. Turanbout QTV 34666 Tape (c) CT 2138.
+NR 7-77 p2 +SFC 6-5-77 p45
+NYT 11-27-77 pD15

Don Juan, op. 20. cf BEETHOVEN: Egmont, op. 84: Overture.
Don Juan, op. 20. cf Works, selections (DG 2740 160).
Don Juan, op. 20. cf BEETHOVEN: Symphony, no. 9, op. 125, D minor.
Don Juan, op. 20. cf ELGAR: Enigma variations, op. 36.
Don Juan, op. 20. cf RESPIGHI: The pines of Rome.

2603 Don Quixote, op. 35. Samuel Mayes, vlc; Joseph de Pasquale, vla; PO; Eugene Ormandy. RCA ARL 1-2287 Tape (c) ARK 1-2287 (ct) ARS 1-2287.
-HF 10-77 p118 +-NR 10-77 p4

Don Quixote, op. 35. cf Works, selections (DG 2740 160).

2604 Elektra, op. 58. Jean Madeira, Inge Borkh, Lisa Della Casa, s; Max Lorenz, t; Kurt Bohme, bar; Salzburg Festival Orchestra; Dimitri Mitropoulos. Bruno Walter Socity SID 731 (2).
+NR 5-77 p12

Elektra, op. 58: Allein weh ganz allein. cf BASF 22-22645-3.

2605 Die Frau ohne Schatten, op. 65. Leonie Rysanek, Christel Goltz, s; Hans Hopf, t; Paul Schoeffler, bs; VPO; Karl Bohm. Richmond 64503.
+NYT 10-9-77 pD21

2606 Ein Heldenleben, op. 40. Gerhart Hetzel, vln; VPO; Karl Bohm. DG 2530 781 Tape (c) 3300 781.
+-Gr 6-77 p64 +RR 6-77 p69
++HFN 6-77 p135

2607 Ein Heldenleben, op. 40. Scipione Guidi, vln; NYP; Willem Mengelberg. RCA AVM 1-2019. (Reissue from Victor 78s)
++ARG 8-77 p5 +NR 4-77 p2
++HF 4-77 p90 +NYT 1-16-77 pD13
+MJ 4-77 p33 +SR 2-19-77 p42

2608 Ein Heldenleben, op. 40. VPO; Clemens Krauss. Richmond R 23209. (Reissue)
+-ARG 12-76 p41 +NYT 1-16-77 pD13

Ein Heldenleben, op. 40. cf Works, selections (DG 2740 160).
Intermezzo, op. 72: Symphonic interlude. cf Works, selections (DG 2740 160).

Japanische Festmusik, op. 84. cf Works, selections (DG 2740 160).
Macbeth, op. 23. cf Don Juan, op. 20.
Panathenaenzug, op. 74. cf Works, selections (HMV SLS 5067).
Parergon to Symphonia domestica, op. 73. cf Works, selections (HMF SQ SLS 5067).

2609 Quartet, piano, op. 13, C minor. Los Angeles String Trio; Irma Vallecillo, pno. Desmar DSM 1002 Tape (c) E 1049.
-Gr 2-76 p1356
-HF 5-76 p96
+-HF 5-77 p101 tape
++MJ 12-75 p38
++NR 12-75 p9
+RR 2-76 p46
+SR 3-6-76 p41
+St 1-76 p106

2610 Der Rosenkavalier, op. 59. Elisabeth Schwarzkopf, Ljuba Welitsch, s; Christa Ludwig, ms; Teresa Stich-Randall, con; PhO; Herbert von Karajan. Angel S 3563 (4). (also HMV SLS 810 Tape (c) TC SLS 810)
+HFN 8-76 p94 tape
+-NYT 10-9-77 pD21
+Op 12-71 p1088
+RR 11-76 p110 tape
+St 4-75 p70

2611 Der Rosenkavalier, op. 59. Regine Crespin, Helen Donath, Emmy Loose, s; Yvonne Minton, Anne Howells, ms; Murray Dickie, Luciano Pavarotti, t; Otto Wiener, bar; Manfred Jungwirth, Herbert Lachner, bs; VPO; Vienna Staatsoper Chorus; Georg Solti. London OSA 1435 Tape (c) 131165 (r) 90165. (also Decca Tape (c) K3N23)
+Gr 8-76 p341 tape
+HF 4-71 p68
++HFN 8-76 p94 tape
++NYT 10-9-77 pD21
+-RR 8-76 p82 tape
++St 6-72 p109 tape

2612 Der Rosenkavalier, op. 59. Evelyn Lear, Ruth Welting, Nelly Morpurgo, Renee van Haarlem, s; Frederica von Stade, Sophia van Sante, ms; Jose Carreras, James Atherton, Wouter Goedhardt, Matthijs Coppens, Adriaan van Limpt, t; Derek Hammond Stroud, Henk Smit, bar; Jules Bastin, bs; Netherlands Opera Chorus; Rotterdam Philharmonic Orchestra; Edo de Waart. Philips 6707 030 (4) Tape (c) 7699 045.
+-ARG 10-77 p43
++FF 9-77 p56
+Gr 8-77 p339
+HF 9-77 p110
-HF 11-77 p138 tape
+HFN 7-77 p103
+HFN 9-77 p155 tape
+-MJ 9-77 p34
+MT 11-77 p924
+NR 8-77 p10
+-NYT 7-10-77 pD13
+ON 8-77 p56
+-OR 9/10-77 p29
+RR 7-77 p40
+-RR 10-77 p99 tape
+SFC 6-5-77 p45
+St 10-77 p146

2613 Der Rosenkavalier, op. 59. Maria Reining, Hilde Gueden, Sena Jurinac, s; Ludwig Weber, bs; VPO; Vienna State Opera Chorus; Erich Kleiber. Richmond 64001. (Reissue from London A 4404)
++HF 4-71 p65
++NYT 10-9-77 pD21

2614 Der Rosenkavalier, op. 59, abridged. Lotte Lehmann, s; Orchestra. Seraphim 6041.
+NYT 10-9-77 pD21

Der Rosenkavalier, op. 59: Da geht er hin, Die Zeit die ist ein sonderbar Ding. cf BASF 22-22645-3.
Der Rosenkavalier, op. 59: First waltz sequence. cf Don Juan, op. 20.
Der Rosenkavalier, op. 59: First and second waltz sequence. cf Don Juan, op. 20.
Der Rosenkavalier, op. 59: Kann mich auch an ein Madel erinnern,

Oh sei Er got Quinquin...Die Zeit die ist ein sonderbar Ding. cf Seraphim IB 6105.
Der Rosenkavalier, op. 59: Suite. cf Don Juan, op. 20.
Der Rosenkavalier, op. 59: Waltzes. cf Works, selections (DG 2740 160).
Salome, op. 54: Ah du wolltest mich deinen Mund. cf BASF 22-22645-3.
Salome, op. 54: Dance. cf RESPIGHI: The pines of Rome.
Salome, op. 54: Dance of the seven veils. cf Also sprach Zarathustra, op. 30.
Salome, op. 54: Dance of the seven veils. cf Don Juan, op. 20.
Salome, op. 54: Dance of the seven veils. cf DG 2535 254.
Salome, op. 54: Jochanaan, ich bin verliebt. cf Club 99-109.
Salome, op. 54: Salome's dance. cf Works, selections (DG 2740 160).
Sonata, violoncello and piano, op. 6. cf BRUCH: Kol Nidrei, op. 47.

2615 Songs: Als mir dein Lied erklang, op. 68, no. 4; Befreit, op. 39, no. 4; Einerlei, op. 69, no. 3; Freundliche Vision, op. 48, no. 1; Heimkehr, op. 15, no. 5; Ich wollt ein Strausslein binden, op. 68, no. 2; Meinem Kinde, op. 37, no. 3; Die Nacht, op. 10, no. 3; Sausle, liebe Myrte, op. 68, no. 3; Schlagende Herzen, op. 29, no. 2; Schlechtes Wetter, op. 69, no. 5; Der Stern, op. 69, no. 1; Wie sollten wir geheim sie halten, op. 19, no. 4. Hilde Gueden, s; Friedrich Gulda, pno. London R 23212.
+NR 11-76 p11 +-ON 2-12-77 p41

2616 Songs: Amor, op. 68, no. 5; Einkehr, op. 47, no. 4; Heimkehr, op. 15, no. 5; Ich schwebe, op. 48, no. 2; Ich wollt ein Strausslein binden, op. 68, no. 2; Sausle, liebe Myrte, op. 68, no. 3; Schlagende Herzen, op. 29, no. 2; Der Stern, op. 69, no. 1. WOLF: Songs: Eichendorff Lieder: Waldmadchen; Verschwiegene Liebe. Goethe Lieder: Die Bekehrte; Epiphanias; Die Sprode. Morike Lieder: Schlafendes Jesuskind; Zum neuen Jahre. Spanish songbook: Ach des Knaben Augen; Die ihr schwebet; Nun wandre, Maria. Judith Blegen, s; Martin Katz, pno. RCA ARL 1-1571.
+ARG 3-77 p46 +-NR 2-77 p12
+Gr 9-76 p466 +ON 12-4-76 p60
+-HF 1-77 p134 +-RR 10-76 p99
+HFN 10-76 p177 +SFC 2-27-77 p42
++MJ 3-77 p74 +-SR 2-19-77 p42
++St 11-76 p164

Songs: Einerlei, op. 69, no. 3; Befreit, op. 39, no. 4; Ich wollt ein Strausslein binden, op. 68, no. 2; Schlechtes Wetter, op. 69, no. 5; Vier letze Lieder, op. posth. cf Ariadne auf Naxos, op. 60: Es gibt ein Reich.
Songs: Morgen, op. 27, no. 4. cf Seraphim IB 6105.
Songs: Liebeshymnus, op. 32, no. 3; Muttertandelei, op. 43, no. 2; Das Rosenband, op. 36, no. 1; Ruhe, meine Seele, op. 27, no. 1. cf BRAHMS: Alto rhapsody, op. 53.
Songs: Standchen, op. 17, no. 2. cf Club 99-108.
Songs: Standchen, op. 17, no. 2; Traum durch die Dammerung, op. 29, no. 1; Zueignung, op. 10, no. 1. cf Bruno Walter Society BWS 729.
Till Eulenspiegels lustige Streiche, op. 28. cf Also sprach Zarathustra, op. 30 (HMV ESD 7026).
Till Eulenspiegels lustige Streiche, op. 28. cf Also sprach Zarathustra, op. 30 (London CS 6978).
Till Eulenspiegels lustige Streiche, op. 28. cf Also sprach Zarathustra, op. 30 (London R 23208).

Till Eulenspiegels lustige Streiche, op. 28. cf Don Juan, op. 20 (Philips 6580 129).
Till Eulenspiegels lustige Streiche, op. 28. cf Don Juan, op. 20 (RCA ARL L-1408).
Till Eulenspiegels lustige Streiche, op. 28. cf Don Juan, op. 20 (Turnabout QTV 34666).
Till Eulenspiegels lustige Streiche, op. 28. cf Works, selections (DG 2740 160).
Tod und Verklarung, op. 24. cf Works, selections (DG 2740 160).
Tod und Verklarung, op. 24. cf RCA CRM 5-1900.

2617 Works, selections: Don Juan, op. 20. Le bourgeois gentilhomme, op. 60. Don Quixote, op. 35. Ein Heldenleben, op. 40. Intermezzo, op. 72: Symphonic interlude. Japanische Festmusik, op. 84. Der Rosenkavalier, op. 59: Waltzes. Salome, op. 54: Salome's dance. Till Eulenspiegels lustige Streiche, op. 28. Tod und Verklarung, op. 24. Berlin State Opera Orchestra, BPhO; Richard Strauss. DG 2740 160. (Reissues from Polydor LY 6087/91, 69849/51, CA 8126/7, 69867, 95392/6, 69854, 66887/8, CA 8017, 67599/600, 67756/60, DG 30538)
+Gr 1-77 p1148 +MT 2-77 p134
+HFN 2-77 p131 +RR 1-77 p24

2618 Works, selections: Burleske, D minor. Concertino, clarinet, bassoon, harp and string orchestra. Concerto, horn, no. 1, op. 11, E flat major. Concerto, horn, no. 2, op. 11, E flat major. Concerto, oboe, D major. Concerto, violin, op. 8, D minor. Panathenaenzug, op. 74. Parergon to Symphonia domestica, op. 73. Peter Damm, hn; Ulf Hoelscher, vln; Manfred Clement, ob; Manfred Weise, clt; Wolfgang Liebscher, bsn; Peter Rosel, Malcolm Frager, pno; Dresden Staatskapelle; Rudolf Kempe. HMV SQ SLS 5067 (4) Tape (c) TC SLS 5067.
+Gr 10-76 p611 +HFN 2-77 p135 tape
+Gr 9-77 p511 tape +MT 12-76 p1006
+HFN 11-76 p168 +RR 10-76 p67

STRAVINSKY, Igor

2619 Apollon musagette. The fairy's kiss. Orpheus. Pulcinella. Irene Jordan, s; George Shirley, t; Donald Gramm, bs; Columbia Symphony Orchestra, CSO; Igor Stravinsky. CBS 77376 (3). (Reissues from SBRG 72452, 72355, 72407)
+-Gr 12-77 p1097 +RR 12-77 p56
+-HFN 12-77 p185

Apollon musagete: Variation. cf Works, selections (Crystal S 302).

2620 Ave Maria. Mass. Les noces. Pater noster. Zora Mojsilovic, s; Aleksandra Ivanovic, ms; Dusan Cvejic, t; Lazar Ivkov, bs; Belgrade Radio and TV Orchestra; Borivoje Simic. Everest SDBR 3399.
+NR 1-77 p9

2621 Le baiser de la fee (The fairy's kiss). Duo concertante. Pulcinella: Suite italienne. Itzhak Perlman, vln; Bruno Canino, pno. Angel S 37115. (also HMV ASD 3219)
++Gr 9-76 p438 +NYT 7-24-77 pD11
++HF 5-76 p101 ++RR 9-76 p84
+HFN 9-76 p130 ++SFC 7-25-76 p29
+NR 3-76 p6 +St 1-77 p128

Le baiser de la fee (The fairy's kiss). cf Apollon musagette.
The fairy's kiss. cf Works, selections (Crystal S 302).

2622 Canticum sacrum. Symphony of psalms. Christ Church Cathedral Choir, Oxford; Philip Jones Ensemble; Simon Preston. Argo ARG 799.

++Audio 6-76 p97
+Gr 10-75 p675
+-HF 10-76 p120
+-HFN 10-75 p150
+-MT 8-76 p663
++NR 5-76 p7
+RR 10-75 p90
+SFC 6-12-77 p41
+-St 9-76 p125
+ST 2-76 p739
+-Te 3-76 p28

Capriccio. cf HINEMITH: The four temperaments.
Les cinq doigts. cf RAVEL: Gaspard de la nuit.
Circus polka. cf CHABRIER: Espana.

2623 Concerto, 16 wind instruments, E flat major. Octet, wind instruments. The soldier's tale: Suite. Nash Ensemble; Elgar Howarth. Classics for Pleasure CFP 40098. (also Sine Qua Non SQN 2011)

+Gr 2-75 p1510
+HF 9-77 p119
++HFN 5-75 p140
++RR 3-75 p35
+-ST 2-76 p739

Double canon. cf New England Conservatory NEC 115.
Duo concertante. cf Le baiser de la fee.
Etudes, op. 7 (4). cf BARTOK: Etudes, op. 18.
Fanfare for a new theater. cf Crystal S 361.
Fireworks, op. 4. cf CHABRIER: Espana.

2624 L'Histoire du soldat (The soldier's tale). Glenda Jackson, narrator; Rudolph Nureyev, the soldier; Michael MacLiammoir, the devil; Ensemble; Gennady Zalkowich. Argo ZNF 15 Tape (c) KZNC 15.

+Audio 12-77 p48
+-Gr 7-77 p198
-HFN 7-77 p121
+-HFN 10-77 p171 tape
+-RR 7-77 p61

2625 L'Histoire du soldat (English version by Michael Flanders and Kitty Black). Sir John Gielgud, narrator; Tom Courtenay, soldier; Ron Moody, the devil; Boston Symphony Chamber Players. DG 2530 609 Tape (c) 3300 609.

+-Gr 3-76 p1471
+-HF 8-76 p96
+HF 1-77 p151 tape
++HFN 3-76 p105
++HFN 3-76 p113 tape
++NR 7-76 p2
+RR 3-76 p49
+-RR 4-76 p82
++SFC 5-16-76 p28
+-SR 11-13-76 p52
+St 11-76 p159

2626 L'Histoire du soldat. Pastorale. Pieces, solo clarinet (3). Septet. Suite italienne. Tashi Ensemble. RCA ARL 1-2449.

++SFC 11-27-77 p66

2627 L'Histoire du soldat. Madeleine Milhaud, narrator; Jean Pierre Aumont, soldier; Martial Singher, devil; Instrumental Ensemble; Leopold Stokowski. Vanguard 71165, VCS 10121 Tape (c) 5183-10121, 71166 (ct) 8183-10121.

++Audio 11-77 p126
++FF 11-77 p53
++SFC 6-5-77 p45

The soldier's tale: Suite. cf Concerto, 16 wind instruments, E flat major.
Jeu de cartes. cf The firebird: Suite.
The king of the stars. cf The rite of spring.
Mass. cf Ave Maria.
Mavra: Chanson Russe. cf Works, selections (Crystal S 302).
Les noces. cf Ave Maria.
Octet, wind instruments. cf Concerto, 16 wind instruments, E flat major.

2628 L'Oiseau de feu (The firebird)(1910 version). RPO; Antal Dorati.

Enigma VAR 1022 Tape (c) TC VAR 1022.
+Gr 4-77 p1562 +RR 4-77 p60
+HFN 4-77 p147 +RR 8-77 p85 tape

2029 The firebird, excerpts. Petrouchka, excerpts. The rite of spring, excerpts. LPO; Joseph Eger. Charisma CAS 1129.
-HFN 11-77 p181 -RR 10-77 p60

L'Oiseau de feu: Berceuse, Scherzo. cf Works, selections (Crystal S 302).

The firebird: Danse infernale. cf DG 2584 004.

2630 The firebird: Suite. Jeu de cartes. LSO; Claudio Abbado. DG 2530 537 Tape (c) 3300 483.
+Gr 8-75 p331 ++NYT 1-18-76 pD1
+HF 2-77 p104 +-RR 4-76 p82 tape
+HFN 8-75 p84 +RR 8-75 p44
+HFN 10-75 p155 tape ++SFC 12-14-75 p40
/NR 1-76 p4

The firebird: Suite. cf BERLIOZ: Les Troyens: Royal hunt and storm.

L'Oiseau de feu: Suite. cf RCA CRL 3-2026.

Orpheus. cf Apollon musagette.

Pastorale. cf L'Histoire du soldat.

Pater noster. cf Ave Maria.

2631 Petrouchka. Tamas Vasary, pno; LSO; Charles Dutoit. DG 2530 711 Tape (c) 3300 711.
-ARG 11-77 p32 ++RR 5-77 p56
-FF 11-77 p53 ++SFC 10-30-77 p44
++Gr 5-77 p1698 -St 11-77 p156
++Gr 6-77 p105 tape
++HFN 5-77 p135
/NR 10-77 p4

2632 Petrouchka. Warsaw National Philharmonic Orchestra; Witold Rowicki. Muza SX 1368.
-NR 7-77 p4

2633 Petrouchka (1974 version). The rite of spring. Minneapolis Symphony Orchestra; Antal Dorati. Philips 6582 021. (Reissue from Mercury AMS 16065, 16056)
+-Gr 8-77 p310 +RR 7-77 p61
+-HFN 7-77 p121

2634 Petrouchka (1911 version). LSO; Charles Mackerras. Vanguard VSD 71177 Tape (c) ZCVSM 71177 (Q) VSQ 30021 Tape (r) VSS 23. (also VCS 10113)
+-Gr 2-74 p1566 ++NR 2-77 p2
+-HF 8-73 p104 ++RR 11-73 p53
+HF 5-76 p114 Quad tape +-RR 2-75 p76
+-HFN 12-73 p2617 ++St 6-73 p123

Petrouchka, excerpts. cf The firebird, excerpts.

Petrouchka, excerpts. cf Pye PCNH 9.

Petrouchka: Danse Russe. cf Works, selections (Crystal S 302).

Petrouchka: Russian dance. cf Hungaroton SLPX 11825.

Petrouchka: Scenes (3). cf RAVEL: Gaspard de la nuit.

2635 Petrouchka: 3 movements. TCHAIKOVSKY: Concerto, piano, no. 1, op. 23, B flat minor. Ilana Vered, pno; LSO; Kazimierz Kord. London SPC 21148 Tape (c) 5-21148 (ct) 8-21148. (also Decca PFS 4362 Tape (c) KPFC 4362)
+Gr 4-76 p1611 ++NR 8-76 p5
-HF 11-76 p130 +-RR 4-76 p54
+HF 10-76 p147 tape +-RR 12-76 p108 tape

+-HFN 4-76 p117 ++SFC 3-7-76 p27
+HFN 6-76 p105 tape ++St 9-76 p128
+-MM 4-77 p41

Petrouchka: 3 movements. cf PROKOFIEV: Sonata, piano, no. 7, op. 83, B flat major.
Petrouchka: 3 movements. cf RAVEL: Le tombeau de Couperin.
Pieces, solo clarinet (3). cf L'Histoire du soldat.
Pieces, solo clarinet (3). cf BIS LP 62.
Pulcinella. cf Apollon musagette.
Pulcinella: Suite, excerpts. cf BARTOK: Hungarian folk songs.
Pulcinella: Suite italienne. cf Le baiser de la fee.
Rag-time. cf Hungaroton SLPX 11686.

2636 Le sacre du printemps (The rite of spring). OSCCP; Pierre Monteux. Decca ECS 750. (Reissue from RCA RB 16007) (also London STS 15318)
+-Gr 9-74 p518 ++MJ 5-76 p28
+HF 2-77 p104 +-RR 11-74 p61

2637 The rite of spring. VPO; Lorin Maazel. Decca SXL 6735. (also Longon 6954 Tape (c) 5-6954)
+-ARG 12-76 p42 +MJ 2-77 p31
-Gr 8-76 p299 +NR 12-76 p3
-HF 2-77 p104 +-RR 6-76 p56
+HFN 6-76 p97 +-St 2-77 p123

2638 The rite of spring. LSO; Claudio Abbado. DG 2530 635.
+ARG 12-76 p42 +NR 11-76 p3
+Gr 5-76 p1766 +-RR 5-76 p52
+HF 2-77 p104 ++St 2-77 p123
+MJ 1-77 p26

2639 The rite of spring. The kind of the stars. BSO; New England Conservatory Male Chorus; Michael Tilson Thomas. DG 2535 222 Tape (c) 3335 222. (Reissue from 2530 252)
+Gr 9-77 p437 +HFN 10-77 p171 tape
++HFN 10-77 p167

2640 The rite of spring. National Youth Orchestra; Simon Rattle. Enigma MID 5001.
+-HFN 11-77 p181 +RR 11-77 p72

2641 Le sacre du printemps. Budapest Philharmonic Orchestra; Ken-Ichiro Kobayashi. Hungaroton SLPX 11841.
-HFN 2-77 p133 -RR 2-77 p61
-NR 3-77 p4

2642 The rite of spring. CSO; Georg Solti. London CS 6885 Tape (c) CS 5-6885. (also Decca SXL 6691 Tape (c) KSXC 6691)
++Audio 9-75 p70 +-NYT 12-15-74 pD21
+Gr 11-74 p909 ++RR 11-74 p60
+Gr 2-75 p1562 tape +RR 4-75 p78 tape
+HF 2-75 p102 ++SFC 12-8-74 p36
++HF 1-77 p151 tape ++St 3-75 p106
+NR 2-75 p4

The rite of spring. cf Petrouchka.
The rite of spring, excerpts. cf The firebird, excerpts.
The rite of spring: Danse sacrale. cf DG 2535 254.
Septet. cf L'Histoire du soldat.
Serenade, A major. cf NILSSON: Quantitaten.
Songs: Anthem; Ave Maria; Pater noster. cf BEACH: Then said Isaiah.
Songs: Russian maiden's song. cf Enigma VAR 1025.
Songs: La rosee sainte. cf HMV RLS 716.
Suite italienne. cf L'Histoire du soldat.

Suite italienne. cf DVORAK: Trio, piano, op. 65, F minor.
Suite on themes by Pergolesi. cf Works, selections (Crystal S 302).
Symphony of psalms. cf Canticum sacrum.
Symphony of psalms. cf POULENC: Gloria, G major.
Valse pour les enfants. cf RAVEL: Gaspard de la nuit.

2643 Works, selections: Apollon musagete: Variation. The fairy's kiss. Mavra: Chanson Russe. Petrouckha: Danse Russe. L'Oiseau de feu: Berceuse, Scherzo. Suite on themes by Pergolesi. Eudice Shapiro, vln; Ralph Berkowitz, pno. Crystal S 302.
++Audio 11-76 p108 ++SFC 5-2-76 p38
+IN 9-76 p20 +St 1-77 p128
++NR 6-76 p7

STUART
Soldiers of the Queen. cf CRD Britannia BR 1077.

SUBOTNIK, Morton
2644 Until spring. Created on the electric music box. Odyssey Y 34158.
+Audio 5-77 p98 +-NR 2-77 p15
+HF 12-76 p125

SUCHON, Eugen
2645 Krutnava (The whirlpool). Soloists; Slovak Philharmonic Chorus, Bratislava Radio Choir; Tibor Freso. Opus 9112 0246/8.
+RR 3-77 p36

SUDERBURG, Robert
Concerto "Within the memory of time". cf SCHUMAN: Symphony, no. 8.

SUK, Josef
Pieces, op. 17: Burleska. cf Discopaedia MB 1014.
Serenade. cf JANACEK: Suite.
Under the apple tree, op. 20. cf JANACEK: Amarus.

SULLIVAN, Arthur
2646 The gondoliers (with dialogue). Marmion. Julia Goss, Barbara Lilley, Glynis Prendergast, Anne Egglestone, s; Jane Metcalfe, Caroline Baker, ms; Beti Lloyd-Jones, con; Geoffrey Shovelton, Meston Reid, Barry Clark, t; John Reed, Michael Rayner, James Conroy-Ward, bar; Kenneth Sandford, Michael Buchan, bs; D'Oyly Carte Opera Chorus; RPO; Royston Nash. Decca SKL 5277/8 (2) Tape (c) K73K22.
-Gr 7-77 p230 +HFN 10-77 p169 tape
+HFN 8-77 p91 +-RR 7-77 p43
+-RR 12-77 p99 tape

2647 The gondoliers (without dialogue). Edna Graham, Elsie Morison, s; Monica Sinclair, Helen Watts, Margorie Thomas, con; Alexander Young, Richard Lewis, t; Geraint Evans, John Cameron, bar; James Milligan bs-bar; Glyndebourne Festival Chorus; Pro Arte Orchestra; Malcolm Sargent. HMV SXCW 3027 (2) Tape (c) SXDW 3027. (Reissue from ASD 265/6)
+Gr 9-76 p478 +HFN 12-76 p155
+-Gr 9-77 p511 tape +-RR 9-76 p34
+HFN 11-76 p173

The gondoliers: For the merriest fellow; Buon giorno, signorine; We're called gondolieri; In enterprise of martial kind; There lived a king. cf Works, selections (Classics for Pleasure CFP 40260).
The gondoliers: In enterprise of martial kind; Let all your doubts take wing. cf Works, selections (Decca SKL 5254).

The gondoliers: Take a pair of sparkling eyes. cf HMV HLP 7109.
The gondoliers: Take a pair of sparkling eyes; I am a courtier, grave and serious...Gavotte; Dance a cachucha. cf Works, selections (Classics for Pleasure CFP 40238).

2648 The grand Duke. Barbara Lilley, Julia Goss, Anne Eggleston, Glynis Prendergast, s; Jane Metcalfe, Patricia Leonard, ms; Lyndsie Holland, Beti Lloyd-Jones, con; Meston Reid, t; John Reed, Kenneth Sandford, Michael Rayner, John Ayldon, James Conroy-Ward, bar; Jon Ellison, bs; D'Oyly Carte Opera Company Chorus; RPO; Royston Nash. Decca SKL 5239/40 (2) Tape (c) K17K22. (also London OSA 12106 Tape (c) OSA 5-12106)

+-ARG 11-77 p15 ++NR 8-77 p12
+FF 9-77 p21 +-NYT 7-10-77 pD13
+Gr 12-76 p1071 +-ON 7-77 p29
+-HF 11-77 p138 tape +RR 12-76 p45
+HFN 12-76 p149 +SFC 6-5-77 p45
++HFN 12-76 p153 ++St 8-77 p112

The grand Duke: When you find you're a broken down critter. cf Works, selections (Decca SKL 5254).
Henry VIII: Graceful dance. cf Decca SB 715.
H.M.S. Pinafore: Opening chorus...Little Buttercup's song; Captain's song; Finale. cf Works, selections (Classics for Pleasure CFP 40238).
H.M.S. Pinafore: Overture. cf Grosvenor GRS 1048.
Iolanthe: Entrance and march of peers; If we're weak enough to tarry; Final chorus. cf Works, selections (Classics for Pleasure CFP 40238).
Iolanthe: The law is the true enforcement; When I went to the bar as a very young man; When all night long a chap remains; If you go in. cf Works, selections (Decca SKL 5254).
Iolanthe: Overture. cf Transatlantic XTRA 1169.
Iolanthe: Tripping hither tripping thither; When all night long a chap remains; Strephon's a member of the Parliament; When Britain really ruled the waves. cf Works, selections (Classics for Pleasure CFP 40260).
Marmion. cf The gondoliers.

2649 The Mikado. Valerie Masterson, Pauline Wales, s; Peggy Ann Jones, ms; Lyndsie Holland, con; Colin Wright, t; Michael Rayner, John Reed, bar; John Ayldon, Kenneth Sandford, John Broad, bs; D'Oyly Carte Opera Chorus; RPO; Royston Nash. Decca SKL 5158/9 (2) Tape (c) KSKC 5158/9 (ct) ESKC 5158/9. (also London OSA 12103 Tape (c) OSA 5-12103)

+-Audio 2-76 p95 +MJ 10-75 p45
+-Gr 1-74 p1409 +NR 5-75 p12
+Gr 4-74 p1919 tape ++RR 4-74 p93 tape
+HF 3-75 p74 ++SFC 2-23-75 p23
+-HF 7-77 p125 tape +St 7-75 p99
+HFN 2-74 p345

2650 The Mikado. Barbara Troxell, s; Martyn Green, bar; James Pease, bs; North German Radio Orchestra; Richard Korn. Everest SDBR 3412.

+-NR 12-77 p10

2651 The Mikado. Elsie Griffin, Beatrice Elburn, s; Aileen Davies, ms; Bertha Lewis, con; Derek Oldham, t; Henry Lytton, Leo Sheffield, George Baker, bar; Darrell Fancourt, T. Penry Hughes, bs; D'Oyly Carte Opera Chorus; Light Opera Orchestra; Mr. Norris. Pearl GEM 137/8 (2). (Reissue from HMV D 1172/82)

+-Gr 7-77 p229 +-RR 5-77 p40

The Mikado: Behold the Lord High Executioner; Taken from the county jail; Three little maids; There is a beauty in the bellow of the blast. cf Works, selections (Classics for Pleasure CFP 40260).

The Mikado: Opening chorus...A wand'ring minstrel I; The sun whose rays; Final chorus. cf Works, selections (Classics for Pleasure CFP 40238).

2652 Patience. Elsie Morison, Elizabeth Harwood, Heather Harper, s; Marjorie Thomas, Monica Sinclair, con; Alexander Young, t; John Shaw, George Baker, John Cameron, bar; Trevor Anthony, bs; Glyndebourne Festival Chorus; Pro Arte Orchestra; Malcolm Sargent. HMV SXDW 3031 (2) Tape (c) TC SXDW 3031. (Reissue from ASD 484/
+Gr 8-77 p350 /RR 8-77 p38
+-Gr 9-77 p511 tape

Patience: When I first put the uniform on; If you're anxious for to shine; Love is a plaintive song; So go to him and say to him. cf Works, selections (Classics for Pleasure CFP 40260).

2653 Pineapple poll (arr. Mackerras). RPO; Charles Mackerras. HMV ESD 7028 Tape (c) TC ESD 7028. (Reissue from CSD 1399)
++Gr 2-77 p1329 +RR 2-77 p61
+HFN 3-77 p117 ++RR 5-77 p95 tape
+HFN 5-77 p138 tape

2654 Pirates of Penzance. Soloists; D'Oyly Carte Orchestra; Isidore Godfrey. Decca Tape (c) K61K22.
+Gr 9-77 p511 tape

Pirates of Penzance: Major General's song; Oh, is there not one maiden breast...Poor wand'ring one; Sergeant of police's song. cf Works, selections (Classics for Pleasure CFP 40238).

Pirates of Penzance: Oh better far to live and die; When the foeman bares his steel. cf Works, selections (Classics for Pleasure CFP 40260).

2655 Princess Ida. Soloists; D'Oyly Carte Orchestra; Malcolm Sargent. Decca Tape (c) K66K22.
+Gr 9-77 p511 tape +RR 12-77 p99 tape
+HFN 9-77 p155 tape

2656 Ruddigore. Elsie Morison, Elizabeth Harwood, s; Pamela Bowden, Monica Sinclair, con; Richard Lewis, t; George Baker, bar; Owen Brannigan, Harold Blackburn, Joseph Rouleau, bs; Glyndebourne Festival Chorus; Pro Arte Orchestra; Malcolm Sargent. HMV SXDW 3029 (2) Tape (c) TC SXDW 3029. (Reissue from ASD 563/4)
+Gr 4-77 p1604 +HFN 9-77 p155 tape
+Gr 9-77 p511 tape +-RR 2-77 p38
+HFN 5-77 p139 tape

Songs: The lost chord. cf L'Oiseau-Lyre DSLO 20.

Songs: The lost chord. cf Prelude PRS 2505.

Trial by jury: When first my old old love I knew; When I good friends was called to the bar. cf Works, selections (Classics for Pleasure CFP 40260).

Trial by jury: When I good friends was called to the bar. cf Works selections (Decca SKL 5254).

Twilight, op. 12. cf Pearl SHE 533.

2657 Utopia limited. D'Oyly Carte Orchestra, RPO; Roystan Nash. London SA 12105 (2) Tape (c) 5-12105.
+-HF 7-77 p125 tape +ON 7-77 p29
+NR 10-76 p13 +SFC 11-7-76 p33

Utopia limited: Let all your doubts take wing. cf Works, selections (Decca SKL 5254).

2658 Works, selections: The gondolirs: Take a pair of sparkling eyes; I am a courtier, grave and serious...Gavotte; Dance a cachucha. Iolanthe: Entrance and march of peers; If we're weak enough to tarry; Final chorus. H.M.S. Pinafore: Opening chorus...Little Buttercup's song; Captain's song; Finale. The Mikado: Opening chorus...A wand'ring minstrel I; The sun whose rays; Final chorus. Pirates of Penzance: Sergeant of police's song. The yeomen of the guard: When maiden loves; I have a song to sing. Soloists; Glyndebourne Festival Chorus; Pro Arte Orchestra; Malcolm Sargent. Classics for Pleasure CFP 40238 Tape (c) CFP 40238.
+Gr 2-77 p1325 tape +HFN 12-76 p153 tape
+HFN 3-76 p111 +RR 3-76 p29
+HFN 5-76 p115

2659 Works, selections: The gondoliers: For the merriest fellow; Buon giorno, signorine; We're called gondolieri; In enterprise of martial kind; There lived a king. Iolanthe: Tripping hither tripping thither; When all night long a chap ramains; Strephon's a member of the Parliament; When Britain really ruled the waves. The Mikado: Behold the Lord High Executioner; Taken from the county jail; Three little maids; There is a beauty in the blast. Patience: When I first put the uniform on; If you're anxious for to shine; Love is a plaintive song; So go to him and say to him. Pirates of Penzance: Oh better far to live and die; When the foeman bares his steel. Trial by jury: When first my old old love I knew; When I good friends was called to the bar. Glyndebourne Festival Chorus; Pro Arte Orchestra; Malcolm Sargent. Classics for Pleasure CFP 40260.
/RR 6-77 p38

2660 Works, selections: The gondoliers: In enterprise of martial kind; Let all your doubts take wing. The grand Duke: When you find you're a broken down critter. Iolanthe: The law is the true enforcement; When I went to the bar as a very young man; When all night long a chap remains; If you go in. Trial by jury: When I good friends was called to the bar. Utopia limited: Let all your doubts take wing. John Reed, bar; D'Oyly Carte Opera Chorus; Various orchestras and conductors. Decca SKL 5254.
+RR 2-77 p38

2661 The yeomen of the guard. Elizabeth Harwood, s; Elsie Maynard; Orchestra; Malcolm Sargent. Decca Tape (c) K60K22.
+Gr 9-77 p511 tape +RR 12-77 p99 tape
+HFN 9-77 p155 tape

2662 The yeomen of the guard. Elsie Morison, s; Marjorie Thomas, Monica Sinclair, con; Richard Lewis, t; Geraint Evans, John Carol Case, bar; Owen Brannigan, bs; Glyndebourne Festival Chorus; Pro Arte Orchestra; Malcolm Sargent. HMV SXDW 3033 (2). (Reissue)
+Gr 12-77 p1155 +-RR 11-77 p43
+HFN 12-77 p185

The yeomen of the guard: When maiden loves; I have a song to sing. cf Works, selections (Classics for Pleasure CFP 40238).

SULYOK, Imre

Te deum. cf Hungaroton SLPX 11808.

SUMSION, Herbert

Magnificat, G major. cf Abbey LPB 783.

Nunc dimittis. cf Abbey LPB 783.

SUPPE, Franz von
The beautiful Galathea. cf Overtures (Philips 6531 012).
2663 The beautiful Galathea, excerpts. Soloists; Graunke Symphony Orchestra; Peter Falk. Telefunken AF 6-23065.
-RR 11-77 p44
Boccaccio. cf Overtures (Philips 6531 012).
Light cavalry: Overture. cf Overtures (Philips 6531 012).
Light cavalry: Overture. cf HMV ESD 7010.
Light cavalry: Overture. cf Quintessence PMC 7013.
Morning, noon and night. cf Overtures (Philips 6531 012).
2664 Overtures: Beautiful Galathea. Boccaccio. Light cavalry. Morning, noon and night. Pique Dame. Detroit Symphony Orchestra; Paul Paray. Philips 6531 012 (2).
+Gr 2-77 p1329
Pique Dame. cf Overtures (Philips 6531 012).
Poet and peasant: Overture. cf Classics for Pleasure CFP 40236.
Poet and peasant: Overture. cf Decca SPA 491.
SURINACH, Carlos
Concerto, piano. cf MONTSALVATGE: Concerto breve.
SUSATO, Tielman
La bataille. cf Argo ZRG 823.
Bergeret sans roch. cf Argo ZRG 823.
Branle quatre branles. cf Argo ZRG 823.
Dances. cf BIS LP 3.
Mon amy, ronde. cf Argo ZRG 823.
La Mourisque. cf Argo ZRG 823.
Ronde. cf Argo ZRG 823.
Ronde. cf DG 2723 051.
SVENDSEN, Johan
2665 Concerto, violin, op. 6. Concerto, violoncello, op. 7. Norwegian rhapsodies, nos. 1-4. Symphony, no. 1, op. 4, D major. Arve Tellefsen, vln; Hege Waldeland, vlc; Oslo Philharmonic Orchestra, Bergen Symphony Orchestra; Miltiades Caridis, Karsten Andersen. Polydor Norway NKF 20001/2, 30006.
+Gr 3-77 p1457 +RR 1-77 p64
+HFN 3-77 p109
Concerto, violoncello, op. 7. cf Concerto, violin, op. 6.
2666 Festival polonaise, op. 12. Norwegian artists carnival, op. 16. Romeo and Juliet, op. 18. Zorahayda, op. 11. Bergen Symphony Orchestra; Karsten Andersen. Polydor NKF 30016.
+HFN 8-77 p89 +RR 8-77 p63
Norwegian artists carnival, op. 16. cf Festival polonaise, op. 12.
Norwegian rhapsodies, nos. 1-4. cf Concerto, violin, op. 6.
Romance, op. 26, G major. cf Discopaedia MB 1014.
Romeo and Juliet, op. 18. cf Festival polonaise, op. 12.
2667 Symphony, no. 1, op. 4, D major. Oslo Philharmonic Orchestra; Miltiades Caridis. Norsk Kulturrads NFK 30001.
/FF 11-77 p54
Symphony, no. 1, op. 4, D major. cf Concerto, violin, op. 6.
2668 Symphony, no. 2, op. 15, B flat major. Oslo Philharmonic Orchestra; Oivin Fjeldstad. Norsk Kulturrads NFK 30009.
/FF 11-77 p54
Zorahayda, op. 11. cf Festival polonaise, op. 12.
SVETLANOV, Yevgeny
Aria, string orchestra. cf HMV SXLP 30256.
Aria, string orchestra. cf HMV SXLP 30259.
Dawn in the field. cf Works, selections (Melodiya 33C 10-06077-82)

Pictures of Spain. cf Works, selections (Melodiya 33C 10-06077-82).
Poem, violin and orchestra. cf Works, selections (Melodiya 33C 10-06077-82).
The red Guelder rose. cf Works, selections (Meldoiya 33C 10-06077-82).
Romantic ballad. cf Works, selections (Melodiya 33C 10-06077-82).
Songs: Three Russian songs. cf Works, selections (Melodiya 33C 01-06077-82).
Symphonic meditations. cf Works, selections (Melodiya 33C 10-06077-82).
Symphony, op. 13, B minor. cf Works, selections (Melodiya 33C 10-06077-82).

2669 Works, selections: Dawn in the field. Pictures of Spain. Poem, violin and orchestra. The red Guelder rose. Romantic ballad. Symphonic meditations. Symphony, op. 13, B minor. Songs: Three Russian songs. Raisa Bobrineva, s; Alexandra Strelchenko, alto; Eduard Grach, vln; USSR Academic Symphony Orchestra; Yevgeny Svetlanov. Melodiya 33C 10-06077-82 (3).
+ARG 7-77 p40

SVIRIDOV, Georgy

2670 Songs (choral): Concert in memory of Alexander Yurlov; Incidental music to Tolstoi's "Tsar Fedor Ivanovich": Choruses; Miniatures; Spring cantata. Yurlov Choir; MRSO; Yuri Ukhov. Columbia M 34525.
+NR 10-77 p9

SWAN

China. cf Vox SVBX 5350.

SWEELINCK, Jan

Balletto del Granduca. cf Vista VPS 1042.
Est-ce Mars. cf FRESCOBALDI: Toccatas, no. 1, G major.
Fantasia ut re mi fa sol la. cf FRESCOBALDI: Toccatas, no. 1, G major.
Hodie Christus natus est. cf HMV ESD 7050.
Hodie Christus natus est. cf Vanguard SVD 71212.
Mein junges Leben hat ein End. cf Argo ZRG 864.
Psalms, nos. 5, 23. cf DG 2533 302.
Toccata, A major. cf FRESCOBALDI: Toccatas, no. 1, G major.
Unter der Linden grune. cf Wealden WS 131.

SWIFT, Richard

Quartet, strings, no. 4. cf New England Conservatory NEC 115.

SZABO, Ferenc

2671 Ludas Matyi suite. Lyric suite. HSO; Gyula Nemeth. Hungaroton SLPX 11780.
+ARG 6-77 p46 +RR 2-77 p61
*NR 3-77 p2

Lyric suite. cf Ludas Matyi suite.

SZALOWSKI, Antoni

Sonatina. cf ETLER: Sonata, clarinet and piano.

SKOKOLAY, Sandor

Lament and cultic dance. cf Hungaroton SLPX 11686.

SZONYI, Erzsebet

Concerto, organ. cf Hungaroton SLPX 11808.

SZULC, Jozsef

Clair de lune, op. 81, no. 1. cf HMV RLS 716.
Clair de lune, op. 81, no. 1. cf HMV RLS 719.

SZYMANOWSKI, Karol
La berceuse d'Aitacho Enia, op. 52. cf Works, selections (Telarc S 5025).
Chant de Roxanne, op. 46. cf Works, selections (Telarc S 5025).
Harnasie, op. 55: Dance. cf Works, selections (Telarc S 5025).
Masques, op. 34. cf JANACEK: In the mists.
Masques, op. 34, no. 1: Sheherazade. cf MOZART: Concerto, piano, no. 20, K 466, D minor.
2672 Mazurkas, opp. 50 and 62. Carol Rosenberger, pno. Delos DEL 25417.
+NR 12-77 p13 +St 9-77 p142
Mythes, op. 30. cf Works, selections (Telarc S 5025).
Mythes, op. 30. cf FRANCK: Sonata, violin and piano, A major.
Notturno e tarantella, op. 28. cf Works, selections (Telarc S 5025).
Notturno e tarantella, op. 28, nos. 1 and 2. cf Finnlevy SFLP 8569.
Romance, op. 23. cf Works, selections (Telarc S 5025).
2673 Works, selections: La berceuse d'Aitacho Enia, op. 52. Chant de Roxanne, op. 46. Harnasie, op. 55: Dance. Mythes, op. 30. Notturno e tarantella, op. 28. Romance, op. 23. Hanna Lachert, vln. Telarc S 5025.
++NR 11-77 p15

TAGLIAFERRO
Mandolinata. cf Club 99-105.

TAKAHASHI, Yugi
Meander. cf CP 3-5.

TAKEMITSU, Toru
Garden rain. cf BIRTWISTLE: Grimethorpe aria.
Piano distance. cf CP 3-5.
Uninterrupted rests. cf CP 3-5.

TALLIS, Thomas
If ye love me. cf Abbey LPB 779.
The lamentations of Jeremiah. cf BYRD: Motets (HMV CSD 3779).
2674 Motets: Derelinquat impius; In ieiunio et fletu; In manus tuas; Ecce tempus idoneum; O nata lux; Salvator mundi; Sancte Deus; Spem in alium; Te lucis (2); Veni redemptor; Videte miraculum. The lamentations of Jeremiah the Prophet. Organ lesson. Andrew Davis, John Langdon, org; King's College Choir, Cambridge University Musical Society Chorus; David Willcocks. Argo ZK 30/1 (2). (Reissue from ZRG 5436, 5479)
+RR 12-77 p82
2675 Motets: Ecce tempus idoneum; Gaude gloriosa; Hear the voice and the prayer; If ye love me; Lamentations I; Loquebantur variis linguis; O nata lux de lumine; Spem in alium. Clerkes of Oxenford; David Wulstan. Seraphim S 60256.
++HF 11-76 p130 +NR 11-76 p10
+MJ 4-77 p12 ++St 1-77 p130
O nata lux. cf Abbey LPB 776.
Songs: Like as the doleful dove. cf BG HM 57/8.

TALMA, Louise
Duologues (3). cf MARTINO: A set for clarinet.
Let's touch the sky. cf Vox SVBX 5353.

TANENBAUM, Elias
2676 Rituals and reactions. TOWER: Breakfast rhythms I and II. Hexachords. Patricia Spencer, flt; Elizabeth Reel, s; DaCapo Chamber Players, Manhattan School of Music Instrumental Ensemble and Chorus; Daniel Paget. CRI S 354.

+Audio 11-77 p128 +NR 10-77 p6
+MJ 10-77 p29

TANEYEV, Sergi
Trio, 2 violins and viola, op. 21, D major. cf GLINKA: Sextet, piano and strings, E flat major.

TANSMAN, Alexandre
Suite, bassoon and piano. cf ELGAR: Romance, op. 62.
A TAPESTRY OF MUSIC FOR CHRISTOPHER COLUMBUS AND HIS CREW. cf Enigma VAR 1024.

TARREGA, Francisco
Adelita. cf Angel S 36094.
La alborada. cf L'Oiseau-Lyre 349.
Capricho arabe. cf Enigma VAR 1015.
Estudio brillante. cf Angel S 36094.
Estudio brillante. cf Swedish Society SLT 33205.
Gran jota. cf Enigma VAR 1015.
Lagrima. cf Lyrichord LLST 7299.
Lagrima. cf L'Oiseau-Lyre SOL 349.
Maria. cf Angel S 36094.
Maria. cf Enigma VAR 1015.
Marieta. cf Angel S 36094.
Mazurka. cf Angel S 36094.
Preludes, nos. 2, 5. cf Angel S 36094.
Recuerdos de la Alhambra. cf L'Oiseau-Lyre SOL 349.
Recuerdos de la Alhambra. cf Philips 6833 159.
Recuerdos de la Alhambra. cf Swedish Society SLT 33205.
Sueno. cf Enigma VAR 1015.

TARTINI, Giuseppe
Concerto, violoncello, A major. cf Bruno Walter Society IGI 323.
Concerto, violoncello, D major: Grave ed espressivo. cf HMV RLS 723.
Sonata, violin, G minor. cf HMV ASD 3346.
Sonata, violin, G minor. cf CORELLI: La folia.
Sonata, violin, G minor. cf GLUCK: Orfeo ed Eurydice: Melodie.
Sonata, violin, G minor. cf Clear TLC 2586.
Variations on a theme by Corelli. cf GLUCK: Orfeo ed Eurydice: Melodie.

TAUSKY, Vilem
Concertino. cf JACOB: Pieces.

TAVERNER, John
2677 Canciones espanolas. Requiem for Father Malachy. James Bowman, Kevin Smith, c-t; King's Singers; Nash Ensemble; John Tavener. RCA LRL 1-5104.
+Gr 10-76 p604 +-MT 2-77 p135
+HFN 11-76 p168 +RR 9-76 p86
++MM 6-77 p49
Magnificat III. cf Coimbra CCO 44.
Requiem for Father Malachy. cf Canciones espanolas.

TAYLOR, Master
Pavan and galliard. cf ABC ABCL 67008.

TAYLOR, Raynor
Variations on Adeste fideles. cf Advent ASP 4007.

TCHAIKOVSKY, Peter
2678 Capriccio italien, op. 45. Eugene Onegin, op. 24: Waltz. Romeo and Juliet: Overture. RPO; Hans Vonk. Decca PFS 4388 Tape (c) KPFC 4388.

+-Gr 3-77 p1410 -RR 3-77 p61
+-HFN 3-77 p113 +-RR 4-77 p94 tape
+-HFN 5-77 p138 tape

2679 Capriccio italien, op. 45. Overture, the year 1812, op. 49. Romeo and Juliet: Fantasy overture. COA; Igor Markevitch, Bernard Haitink. Philips 6530 009 (2).
+-Gr 2-77 p1329

Capriccio italien, op. 45. cf BORODIN: In the Steppes of Central Asia.

Capriccio italien, op. 45. cf LISZT: Hungarian rhapsody, no. 1, F minor.

Capriccio italien, op. 45. cf RCA CRL 3-2026.

Chanson triste, op. 40, no. 2. cf Piano works (Seraphim S 60250).

Chants sans paroles (Songs without words). cf Discopaedia MOB 1018.

Chants sans paroles, op. 2: Souvenirs de Hapsal. cf Discopaedia MB 1012.

Songs without words, op. 2, no. 3, F major. cf Piano works (Seraphim S 60250).

Chants sans paroles, op. 40, no. 6, A minor. cf Decca PFS 4351.

2680 Concerto, piano, no. 1, op. 23, B flat minor. Peter Katin, pno; LPO; John Pritchard. Classics for Pleasure CFP 115 Tape (c) TC CFP 115.
+Gr 2-77 p1325 tape +-HFN 12-76 p153 tape

2681 Concerto, piano, no. 1, op. 23, B flat minor. Earl Wild, pno; RPO; Anatole Fistoulari. RCA GL 2-5013 Tape (c) GK 2-5013. (Previously issued by Reader's Digest)
+Gr 10-76 p612 +-HFN 12-76 p153 tape
+Gr 2-77 p1325 +-RR 10-76 p68
+HFN 12-76 p149 +RR 12-76 p104

Concerto, piano, no. 1, op. 23, B flat minor. cf BRAHMS: Concerto, piano, no. 1, op. 15, D minor.

Concerto, piano, no. 1, op. 23, B flat minor. cf FRANCK: Symphonic variations.

Concerto, piano, no. 1, op. 23, B flat minor. cf LISZT: Concerto, piano, no. 1, G 124, E flat major.

Concerto, piano, no. 1, op. 23, B flat minor. cf STRAVINSKY: Petrouchka: 3 movements.

Concerto, piano, no. 1, op. 23, B flat minor. cf Decca D62D4.

Concerto, piano, no. 1, op. 23, B flat minor. cf HMV SLS 5068.

Concerto, piano, no. 1, op. 23, B flat minor. cf HMV SLS 5094.

Concerto, piano, no. 1, op. 23, B flat minor: Theme. cf Inner City IC 1006.

2682 Concerto, piano, no. 2, op. 44, G major. Sylvia Kersenbaum, pno; ORTF; Jean Martinon. Connoisseur Society CS 2076 (Q) CSQ 2076.
+Audio 11-77 p132 +MJ 4-76 p30
+-HF 1-76 p97 ++SFC 11-9-75 p22
++NR 12-75 p2 +St 10-75 p118

2683 Concerto, violin, op. 35, D major. Valse scherzo, op. 34. Boris Belkin, vln; NPhO; Vladimir Ashkenazy. Decca SXL 6854.
+Gr 10-77 p636 +RR 10-77 p60
++HFN 10-77 p163

2684 Concerto, violin, op. 35, D major. Valse scherzo, op. 34. Souvenir d'un lieu cher, op. 42: Meditation, Scherzo (arr. Glazunov). Nathan Milstein, vln; Pittsburgh Symphony Orchestra; William Steinberg, Robert Irving. HMV SXLP 30225. (Reissue from Capitol SP 8512, Columbia SAX 2563)

+Gr 1-77 p1153 +RR 1-77 p65
+-HFN 1-77 p119

2685 Concerto, violin, op. 35, D major. Serenade melancolique, op. 26. Arthur Grumiaux, vln; NPhO; Jan Krenz. Philips 9500 086 Tape (c) 7300 490.
-ARG 4-77 p27 ++NR 3-77 p7
++Gr 6-76 p52 ++RR 6-76 p58
++HFN 6-76 p101 +St 4-77 p133
+MJ 4-77 p33

2686 Concerto, violin, op. 35, D major. Serenade melancolique, op. 26. Valse scherzo, op. 34. Salvatore Accardo, vln; BBC Symphony Orchestra; Colin Davis. Philips 9500 146 Tape (c) 7300 514.
+-Gr 1-77 p1153 ++RR 1-77 p65
+HFN 1-77 p117 ++SFC 8-28-77 p46
+HFN 5-77 p138 tape

Concerto, violin, op. 35, D major. cf CHAUSSON: Poeme, op. 25.
Concerto, violin, op. 35, D major. cf MENDELSSOHN: Concerto, violin, op. 64, E minor (Decca 4345).
Concerto, violin, op. 35, D major. cf MENDELSSOHN: Concerto, violin, op. 64, E minor (Odyssey 34064).
Concerto, violin, op. 35, D major. cf MENDELSSOHN: Concerto, violin, op. 64, E minor (Philips 9500 321).
Concerto, violin, op. 35, D major. cf Clear TLC 2586.
Cradle song, op. 16, no. 1. cf Angel S 37219.

2687 Danse hongroise. Drdla. Gabriel-Marie. Melodie. Ring-Hager. Serenade-Badine. Souvenir. WIEDOEFT: Dans l'Orient. Saxarella. Saxophobia. Valse Erica. Valse Marilyn. Valse Mazanetta. Valse Llewellyn. Valse Vanite. Ted Hegvik, sax; Ferde Malenke, pno. Golden Crest CRS 4155.
+ARG 3-77 p81 +IN 5-77 p26

2688 Dmitri the imposter. Suite, no. 1, op. 43, D minor. MRSO; Arvid Yansons, Yevgeny Akulov. HMV SXLP 30244.
+-Gr 9-77 p437 +RR 9-77 p68
+-HFN 10-77 p167

Drdla. cf Danse hongroise.
Dumka, op. 59. cf Piano works (Seraphim S 60250).
Elegie. cf CHOPIN: Variations on Mozart's "La ci darem la mano", op. 2, B flat major.
Elegie, G major. cf HMV SQ ESD 7001.
L'Espiegle, op. 72, no. 12. cf Pye PCNH 9.

2689 Eugene Onegin, op. 24. Teresa Kubiak, s; Anna Reynolds, Julia Hamari, Enid Hartle, ms; Stuart Burrows, Michel Senechal, t; Bernd Weikl, bar; William Mason, Nicolai Ghiaurov, Richard Van Allan, bs; John Alldis Choir; ROHO; Georg Solti. Decca SET 596/8 (3). (also London OSA 13112)
+-Gr 6-75 p84 +ON 12-6-75 p52
+-HF 2-76 p107 +RR 5-75 p18
++HFN 6-75 p96 ++SFC 11-2-75 p28
+MJ 3-76 p25 +SR 3-6-76 p38
++NR 2-76 p10 ++St 1-76 p75
/NYT 10-9-77 pD21

2690 Eugene Onegin, op. 24. Mstislav and Galina Rostropovich. Melodiya/Angel SRCL 4115 (3). (also HMV SLS 951/3)
+ARG 8-71 p822 +ON 12-5-70 p34
+NR 1-71 p9 ++SFC 10-4-70 p32
+-NYT 10-9-77 pD27 +STL 2-28-71 p27

Eugene Onegin, op. 24: Letter scene, excerpt. cf Odyssey Y 33793.
Eugene Onegin, op. 24: Polonaise. cf BERLIOZ: Le Corsaire, op. 21.
2691 Eugene Onegin, op. 24: Scene and Lensky's arioso; Tatiana's letter scene; Onegin's aria; Entr'acte and waltz; Lensky's aria; Polonaise; Gremin's aria; Scene and Onegin's arioso; Final scene. Teresa Kubiak, s; Anna Reynolds, Julia Hamari, Enid Hartle, ms; Stuart Burrows, Michel Senechal, t; Bernd Weikl, bar; Nicolai Ghiaurov, William Mason, Richard Van Allan, bs; John Alldis Choir; ROHO; Georg Solti. Decca SET 599. (Reissue from SET 596/8)
+Gr 2-77 p1318 +RR 2-77 p38
+HFN 3-77 p118
Eugene Onegin, op. 24: Waltz. cf Capriccio italien, op. 45.
Eugene Onegin, op. 24: Waltz. cf Philips 6747 327.
Eugene Onegin, op. 24: Waltz, Polonaise. cf CBS 61781.
2692 Fatum, op. 77. Overture, F major. The storm, op. 76. The Voyevode, op. 78. MSO; Veronika Dudarova. Angel/Melodiya SR 40271.
*ARG 3-77 p51 ++NR 1-77 p2
2693 Francesca da Rimini, op. 32. Romeo and Juliet. PO; Eugene Ormandy. RCA ARL 1-2490 Tape (c) ARK 1-2490 (ct) ARS 1-2490.
++SFC 10-30-77 p44
2694 Francesca da Rimini, op. 32. Hamlet overture, op. 67. Utah Symphony Orchestra; Maurice Abravanel. Turnabout QTV 34601.
+Audio 1-77 p83 ++NR 2-76 p4
Francesca da Rimini, op. 32. cf Serenade, strings, op. 48, C major.
Francesca da Rimini, op. 32. cf Works, selections (Vox 5129/31).
Francesca da Rimini, op. 32. cf BIZET: Symphony, C major.
Gabriel-Marie. cf Danse hongroise.
Hamlet overture, op. 67a. cf Francesca da Rimini, op. 32 (Turnabout QTV 34601).
Hamlet overture, op. 67a. cf Works, selections (Vox 5129/31).
Humoresque, op. 10. cf CBS 73589.
Humoresque, op. 10, no. 2. cf Piano works (Seraphim S 60250).
Humoresque, op. 10, no. 2. cf HMV SXLP 30256.
Humoresque, op. 10, no. 2. cf HMV SXLP 30259.
Impromptu, op. 72, no. 1. cf Piano works (Seraphim S 60250).
The maid of Orleans: Joan's aria. cf Columbia/Melodiya M 33931.
2695 Manfred symphony, op. 58. LSO; Yuri Ahronovich. DG 2530 878 Tape (c) 3300 878.
+-Gr 10-77 p648 +-HFN 11-77 p181
+Gr 12-77 p1145 tape -RR 10-77 p59
2696 Manfred symphony, op. 58. LSO; Andre Previn. HMV ASD 3018. (also Angel S 37018)
+Gr 12-74 p1150 +NR 8-75 p2
+Gr 10-77 p698 tape +-NYT 1-18-76 pD1
++HF 11-75 p126 +-RR 12-74 p40
+-HFN 10-77 p171 tape +-St 1-76 p110
Manfred symphony, op. 58. cf Symphonies, nos. 1-6.
Manfred symphony, op. 58. cf Works, selections (Vox 5129/31).
Marche slav, op. 31. cf Symphony, no. 5, op. 64, E minor. (Philips 7321 023).
Marche slav, op. 31. cf Symphony, no. 5, op. 64, E minor (Quintessence PMC 7002).
Marche slav, op. 31. cf HMV ESD 7011.
Marche slav, op. 31. cf RCA CRL 3-2026.
2697 Mazeppa. V. Davydova, Alexei Ivanov, Ivan Petrov, N. Pokrovskaya, G. Bolshakov; Bolshoi Theatre Orchestra and Chorus; Vasili

Nebolsin. Melodiya D 014757/62.
+-ARG 10-77 p56
Mazeppa: Gopak. cf Philips 6747 327.
Melodie. cf Danse hongroise.
Melodie. cf Discopaedia MOB 1018.
The months (The seasons), op. 37: January, December, November, June, August, March. cf Piano works (Seraphim S 60250).
The months, op. 37: Barcarolle. cf International Piano Library IPL 113.
The seasons, op. 37: Barcarolle; Autumn song. cf HMV SXLP 30256.
The seasons, op. 37: Barcarolle; Autumn song. cf HMV SXLP 30259.
The months, op. 37: November. cf International Piano Library IPL 114.
Nocturne, op. 19, no. 4. cf Piano works (Seraphim S 60250).
Nocturne, op. 19, no. 4. cf ABC AB 67014.
None but the lonely heart, op. 6, no. 6. cf CBS 61039.
None but the lonely heart, op. 6, no. 6. cf Decca SPA 491.

2698 The nutcracker, op. 71. LSO; Ambrosian Singers; Andre Previn. Angel SB 3788 (2) Tape (c) 4X2S 3788. (also HMV SLS 834 Tape (c) TC SLS 834)
+Audio 6-77 p133
++Gr 1-73 p1333
++HF 3-73 p96
+HFN 1-73 p123
+HFN 3-76 p113 tape
+-NR 4-73 p2
+RR 1-73 p52
+RR 5-76 p78 tape
++SFC 11-19-72 p31

2699 The nutcracker, op. 71. Bolshoi Theatre Orchestra; Gennady Rozhdestvensky. Columbia/Melodiya M2 33116 (2). (also HMV Melodiya SXDW 3028 Tape (c) SXDW 3028. Reissue from Artia).
+Gr 11-76 p812
+Gr 3-77 p1458 tape
+-HF 3-75 p72
+-HFN 11-76 p170
+HFN 9-77 p155
+HFN 5-77 p138 tape
+-NR 2-75 p2
-RR 11-76 p84
+-St 3-75 p109

2700 The nutcracker, op. 71. OSR; Ernest Ansermet. Decca DPA 569/70 (2). (Reissue from SXL 2092/3) (also London STS 15433/4)
+-Gr 2-77 p1326
+HFN 3-77 p117
+RR 2-77 p62
++SFC 12-11-77 p61

2701 The nutcracker, op. 71. OSR; Ernest Ansermet. Decca Tape (c) KDPC 7043 (2).
+-HFN 5-77 p138 tape
+HFN 9-77 p155 tape
+RR 6-77 p97 tape

2702 The nutcracker, op. 71. COA; St. Bavo Cathedral Boys Choir; Antal Dorati. Philips 6747 257 (2) Tape (c) 7505 076. (also 6747 364)
+Audio 6-77 p133
+Gr 1-77 p1148
++HF 1-77 p128
++HFN 5-77 p138 tape
++HFN 1-77 p117
+MJ 1-77 p26
++NR 12-76 p5
++RR 1-77 p64
++SFC 10-24-76 p35
+St 2-77 p123

2703 The nutcracker, op. 71, excerpts. Sinfonia of London; John Hollingsworth. Classics for Pleasure CFP 40272.
/RR 12-77 p56
The nutcracker, op. 71: Ballet suites. cf Amberlee ALM 602.
The nutcracker (Casse-Noisette), op. 71: March. DG 2548 148.

2704 The nutcracker, op. 71: Miniature overture; March; Dance of the sugar-plum fairy; Russian dance; Arabian dance; Chinese dance; Dance of the reed-pipes; Waltz of the flowers; Scene and waltz

of the snowflakes; Spanish dance; Pas de deux; Coda; Waltz finale and apotheosis. Slavonic march, op. 31.* Minneapolis Symphony Orchestra; Antal Dorati. Philips 6582 018. (*Reissue from Mercury AMS 16059)
+Gr 6-77 p106 +-RR 6-77 p69
+HFN 6-77 p137

The nutcracker, op. 71: Pas de deux, Act 2. cf ADAM: Giselle: Pas de deux; Grand pas de deux and finale.

2705 The nutcracker, op. 71: Suite. Overture, the year 1812, op. 49. Kraft and Alexander, ARP synthesizers. Decca PFS 4395. (also London 21168)
++HFN 9-77 p150 +RR 9-77 p70
-NR 7-77 p5

2706 The nutcracker, op. 71: Suite. Swan Lake, op. 20, excerpts. NSL; Adrian Boult. Quintessence PMC 7010.
+St 11-77 p146

The nutcracker, op. 71: Suite. cf GRIEG: Peer Gynt, opp. 46 and 55: Morning; The death of Aase.

The nutcracker, op. 71: Trepak. cf CBS 61780.

The nutcracker, op. 71: Waltz of the flowers. cf RCA CRL 1-2064.

Overture, F major. cf Fatum, op. 77.

Overture, the year 1812, op. 49. cf Capriccio italien, op. 45.

Overture, the year 1812, op. 49. cf The nutcracker, op. 71: Suite.

Overture, the year 1812, op. 49. cf Works, selections (Vox 5129/31

Overture, the year 1812, op. 49. cf BORODIN: In the Steppes of Central Asia.

Overture, the year 1812, op. 49. cf GLINKA: Russlan and Ludmilla: Overture.

Overture, the year 1812, op. 49. cf Bruno Walter Society RR 443.

Partita, violoncello and chamber ensemble. cf SHOSTAKOVICH: Quartet, strings, no. 13, op. 138, B flat minor.

Pezzo capriccioso, op. 62. cf ABC AB 67014.

2707 Piano works: Chanson triste, op. 40, no. 2. Dumka, op. 59. Humoresque, op. 10, no. 2. Impromptu, op. 72, no. 1. Nocturne, op. 19, no. 4. Russian danse, op. 40, no. 10. Scherzo, op. 40, no. 11. The seasons, op. 37: January, December, November, June, August, March. Songs without words, op. 2, no. 3, F major. Valse, op. 40, no. 8. Danielle Laval, pno. Seraphim S 60250.
+NR 12-77 p13 +St 12-77 p156

2708 Pique dame (Queen of spades), op. 68. Galina Visnevskaya, Christine Mitlehner, Lucia Popp, Hanna Schwarz, s; Regina Resnik, con; Ewa Dobrowska, ms; Peter Gougaloff, Fausto Tenzi, Heinz Kruse, t; Bernd Weikl, Dan Iordachescu, bar; Dimiter Petkov, Rudolf Alexander Sutey, bs; Tchaikovsky Choir, French Radio Childrens Chorus; French National Orchestra; Mstislav Rostropovich. DG 2740 176 (4).
+-Gr 12-77 p1137 +-RR 12-77 p35
+-HFN 12-77 p181

Pique dame, op. 68: Air de Lisa. cf Club 99-96.

Pique dame, op. 68: Arioso, Act 1; Forgive me, heavenly creature; What is life; It is a game. cf RACHMANINOFF: Francesca da Rimini, op. 25.

The queen of spades, op. 68: Ich muss am Fenster lehnen; Es geht auf Mitternacht. cf Decca ECS 812.

The queen of spades, op. 68: Twill soon be midnight. cf Rubini RS 301.

Un poco di Chopin, op. 72, no. 11. cf Pye PCNH 9.
2709 Quartets, strings, nos. 1-3. Gabrieli Quartet. Decca SDD 524/5 (2). (also London STS 15424/5)
+Gr 10-77 p654 +-RR 10-77 p68
+-HFN 11-77 p181
Ring-Hager. cf Danse hongroise.
Romance. cf Pye PCNH 9.
Romeo and Juliet. cf Francesca da Rimini, op. 32.
Romeo and Juliet. cf Works, selections (Vox QSVBX 5129/31).
Romeo and Juliet, excerpts. cf HMV SLS 5073.
Romeo and Juliet: Fantasy overture. cf Capriccio italien, op. 45.
Romeo and Juliet: Fantasy overture. cf Symphony, no. 2, op. 17, C minor.
Romeo and Juliet: Fantasy overture. cf BERLIOZ: Romeo and Juliet, op. 17: Love scene.
Romeo and Juliet: Fantasy overture. cf BERLIOZ: Romeo and Juliet, op. 17: Romeo's reverie and fete of the Capulets; Love scene; Queen Mab scene.
Romeo and Juliet: Overture. cf Capriccio italien, op. 45.
Romeo and Juliet: Overture. cf RCA CRL 3-2026.
Russian danse, op. 40, no. 10. cf Piano works (Seraphim S 60250).
Scherzo, op. 40, no. 11. cf Piano works (Seraphim S 60250).
Scherzo, op. 42, no. 2. cf HMV SXLP 30256.
Scherzo, op. 42, no. 2. cf HMV SXLP 30259.
Scherzo humoristique, op. 19, no. 2. cf HMV SXLP 30256.
Scherzo humoristique, op. 19, no. 2. cf HMV SXLP 30259.
2710 Serenade, strings, op. 48, C major. Francesca da Rimini, op. 32. LSO; Leopold Stokowski. Philips 6500 921 Tape (c) 7300 364.
+Gr 5-76 p1766 /MJ 11-76 p44
-HF 11-76 p132 +-NR 6-76 p4
+HF 4-77 p121 tape +RR 5-76 p58
+-HFN 4-76 p117 +RR 11-76 p110 tape
Serenade, strings, op. 48, C major. cf ARENSKY: Variations on a theme by Tchaikovsky, op. 35a.
Serenade, strings, op. 48, C major. cf DVORAK: Serenade, strings, op. 22, E major (Argo 848).
Serenade, strings, op. 48, C major. cf DVORAK: Serenade, strings, op. 22, E major (Philips 9500 105).
Serenade-Badine. cf Danse hongroise.
Serenade melancolique, op. 26. cf Concerto, violin, op. 35, D major (Philips 9500 086).
Serenade melancolique, op. 26. cf Concerto, violin, op. 35, D major (Philips 9500 146).
Serenade melancolique, op. 26. cf LALO: Symphonie espagnole, op. 21.
Serenade melancolique, op. 26. cf Finnlevy SFLP 8569.
2711 Sextet (Souvenir de Florence), op. 70, D minor. VERDI: Quartet, strings, E minor. Netherlands Chamber Orchestra; David Zinman. Philips 9500 104.
+Gr 6-77 p64 +RR 6-77 p69
+HFN 6-77 p135 +SFC 9-4-77 p38
+NR 10-77 p7
Slavonic march, op. 31. cf The nutcracker, op. 71: Miniature overture; March (Philips 6582 018).
Slavonic march, op. 31. cf DG 2548 148.
2712 Sleeping beauty, op. 66: Ballet suites. Swan Lake, op. 20: Ballet suites. VSO; Karel Ancerl. Philips 6530 033.
+-Gr 2-77 p1329

Sleeping beauty, op. 66: Blue bird pas de deux. cf ADAM: Giselle: Pas de deux; Grand pas de deux and finale.

2713 Sleeping beauty, op. 66: Prologue, introduction; March; Pas de six (Adagio and variations, nos. 4-6); Valse, Scene; Pas d' action (Adagio); Farandole; Panorama; Polacca; Pas de quatre (The silver fairy; The diamond fairy); Pas de caractere (Puss-in-boots and the white cat); Pas de quatre (Cinderella and Prince Fortune; The blue bird and Prince Florine); Pas de caractere (Red Riding-hood and the wolf); Pas de deux; Finale (Apotheosis). LSO; Andre Previn. HMV ASD 3370 Tape (c) TC ASD 3370. (Reissue from SLS 5001)

+-Gr 8-77 p349 +RR 9-77 p68
+-HFN 10-77 p167 +-RR 11-77 p121 tape
+HFN 10-77 p171 tape

Sleeping beauty, op. 66: Waltz. cf Angel S 37250.
Sleeping beauty, op. 66: Waltz. cf Argo ZRG 851.
Sleeping beauty, op. 66: Waltz. cf CBS 61781.
Sleeping beauty, op. 66: Waltz. cf RCA PL 2-5046.
Sleeping beauty, op. 66: Waltz, Act 1. cf Seraphim S 60277.

2714 Songs (choral): Blessed is he that smiles; Evening; The golden cloud had slept; Hymn to St. Cyril and St. Methodius; Jurisprudence students song; Legend, op. 54, no. 5; The merry voice grew silent; Morning prayer, op. 39, no. 1; The nightingale; 'Tis not the cuckoo in the damp woods; To sleep; Without time or reason. Vadim Korshunov, t; USSR Russian Chorus; Aleksander Sveshnikov. HMV Melodiya ASD 3165.

+Gr 3-76 p1499 +MT 1-77 p47
+HFN 3-76 p107 +RR 2-76 p63

Songs: At the ball, op. 38, no. 3. cf Rubini RS 301.
Souvenir. cf Danse hongroise.

2715 Souvenir de Florence (Sextet), op. 70. VERDI: Quartet, strings, E minor. Netherlands Chamber Ensemble; David Zinman. Philips 9500 104.

+Gr 6-77 p64 +RR 6-77 p69
+HFN 6-77 p135 +SFC 9-4-77 p38
+NR 10-77 p7

Souvenir d'un lieu cher, op. 42: Meditation. cf BRAHMS: Sonata, violin and piano, no. 3, op. 108, D minor.
Souvenir d'un lieu cher, op. 42: Meditation, Scherzo. cf Concerto, violin, op. 35, D major.
The storm, op. 76. cf Fatum, op. 77.

2716 Suites, orchestra, nos. 1-4. NPhO; Antal Dorati. Mercury SRI 77008 (3).

++MJ 1-77 p26

Suite, no. 1, op. 43, D minor. cf Dmitri the imposter.
Suite, no. 4, op. 61, G major: Prayer. cf Decca DDS 507/1-2.

2717 Swan Lake, op. 20. National Philharmonic Orchestra; Richard Bonynge. Decca D37D3 (3) Tape (c) K37K33.

+-Gr 9-77 p517 +RR 9-77 p69
++HFN 9-77 p150

2718 Swan Lake, op. 20. LSO; Ida Haendel, vln; Douglas Cummings, vlc; Andre Previn. HMV SQ SLS 5070 (3) Tape (c) TC SLS 5070. (also Angel SX 3834 Tape (c) 4X3S 3834)

+-Gr 12-76 p1009 +HFN 5-77 p138 tape
+-Gr 3-77 p1458 tape +HFN 9-77 p155 tape
+-Gr 5-77 p1743 tape +MJ 7-77 p70
+HF 3-77 p112 +NR 4-77 p3

+HF 6-77 p103 tape +RR 12-76 p66
++HFN 12-76 p149 +-St 8-77 p122

2719 Swan Lake, op. 20. NBC Symphony Orchestra; Leopold Stokowski. Quintessence PMC 7007.
+HF 9-77 p108 +St 11-77 p146

2720 Swan Lake, op. 20, excerpts. Netherlands Radio Orchestra; Anatole Fistoulari. Decca PFS 4375 Tape (c) KPFC 4375. (Reissue from 10BB 168/70)
+Gr 3-77 p1461 +HFN 9-77 p155 tape
+HFN 3-77 p117 -RR 3-77 p62
+-HFN 5-77 p138 tape +-RR 6-77 p97 tape

Swan Lake, op. 20, excerpts. cf The nutcracker, op. 71: Suite.

Swan Lake, op. 20: Ballet suites. cf Sleeping beauty, op. 66: Ballet suites.

Swan Lake, op. 20: Neapolitan dance. cf Discopaedia MB 1015.

Swan Lake, op. 20: Pas de deux, Act 3. cf ADAM: Giselle: Pas de deux; Grand pas de deux and finale.

Swan Lake, op. 20: Waltz, Act 1. cf HMV SLS 5073.

2721 Symphonies, nos. 1-6. Manfred symphony, op. 58. LPO; Mstislav Rostropovich. Angel (Q) SGE 3847 (7). (also HMV SQ SLS 5099)
+ARG 12-77 p14 +NR 11-77 p5
+Gr 10-77 p647 ++RR 11-77 p72
+-HF 12-77 p100 +-SFC 9-4-77 p38
++HFN 12-77 p163 +-St 11-77 p140
+MJ 11-77 p25

Symphonies, nos. 1-6. cf Works, selections (Vox 5129/31).

2722 Symphonies, nos. 1-3. COA; Antal Dorati. Mercury SRI 77009 (3).
+MJ 1-77 p26

2723 Symphony, no. 1, op. 13, G minor. NPhO; Riccardo Muti. HMV SQ ASD 3213 Tape (c) TC ASD 3213. (also Angel S 37114)
+ARG 2-77 p43 /HFN 12-76 p155 tape
++Audio 5-77 p101 ++NR 1-77 p2
+Gr 7-76 p181 -RR 7-76 p64
+-HF 3-77 p98 +STL 6-6-76 p37
+-HFN 7-76 p98

2724 Symphony, no. 2, op. 17, C minor. Romeo and Juliet: Fantasy overture. VPO; Lorin Maazel. Decca JB 21. (Reissues from SXL 6162, 6206)
+-Gr 12-77 p1097 +RR 12-77 p56
+-HFN 12-77 p185

2725 Symphonies, nos. 4-6. PO; Eugene Ormandy. RCA CRL 3-1835 (3).
+MJ 1-77 p26

2726 Symphony, no. 4, op. 36, F minor. VPO; Claudio Abbado. DG 2530 651 Tape (c) 3300 651.
++GR 11-76 p812 +-RR 11-76 p84
+HF 10-77 p122 +SFC 3-20-77 p34
++HFN 11-76 p168 +-St 7-77 p126
+NR 4-77 p1

2727 Symphony, no. 4, op. 36, F minor. LSO; Georg Szell. London CS 6987. (also Decca SPA 206)
+-HF 10-77 p122 +St 7-77 p126
++NR 8-77 p4

2728 Symphony, no. 4, op. 36, F minor. LSO; Antal Dorati. Philips 6582 022. (Reissue from Mercury AMS 16118)
+-Gr 5-77 p1698 +-RR 5-77 p60
+HFN 5-77 p137

2729 Symphony, no. 4, op. 36, F minor. LPO; Hubert Soudant. Pye PCNHX 8.
-Gr 2-77 p1290 -RR 2-77 p62
-HFN 2-77 p133
Symphony, no. 4, op. 36, F minor: Scherzo. cf Seraphim SIB 6094.
2730 Symphony, no. 5, op. 64, E minor. VPO; Lorin Maazel. Decca JB 24. (Reissue from SXL 6085)
-Gr 12-77 p1097 +RR 12-77 p57
+HFN 12-77 p185
2731 Symphony, no. 5, op. 64, E minor. NPhO; Leopold Stokowski. Decca SDD 493. (Reissue from PFS 4129)
+-Gr 1-77 p1153 +-RR 12-76 p66
+HFN 12-76 p151
2732 Symphony, no. 5, op. 64, E minor. BPhO; Herbert von Karajan. DG 2530 699 Tape (c) 3300 699.
+Gr 8-76 p300 ++MJ 1-77 p26
++Gr 10-76 p658 tape -NR 1-77 p3
++HF 1-77 p130 +-SFC 10-24-76 p35
+HFN 10-76 p177 ++St 2-77 p129
2733 Symphony, no. 5, op. 64, E minor. BSO; Seiji Ozawa. DG 2530 888.
++HFN 12-77 p181 ++RR 12-77 p56
2734 Symphony, no. 5, op. 64, E minor. BPhO; Rudolf Kempe. HMV SXLP 30216. (Reissue from ASD 379)
+Gr 11-76 p812 +RR 1-77 p65
2735 Symphony, no. 5, op. 64, E minor. CSO; Georg Solti. London CS 6983 Tape (c) CS 5-6983. (also Decca SXL 6754 Tape (c) KSXC 6754)
-Gr 4-76 p1612 +NR 12-76 p6
+HF 10-76 p120 +-RR 4-76 p54
+-HFN 4-76 p117 +-RR 8-76 p85 tape
+HFN 7-76 p104 tape -SFC 5-30-76 p24
+MJ 1-77 p26 +-St 9-76 p128
2736 Symphony, no. 5, op. 64, E minor. Marche slav, op. 31. LSO, Minneapolis Symphony Orchestra; Antal Dorati. Philips Tape (c) 7321 023.
+-HFN 7-77 p127 tape +-RR 8-77 p87 tape
2737 Symphony, no. 5, op. 64, E minor. Marche slav, op. 31. NPhO, LPO; Adrian Boult. Quintessence PMC 7002.
+-ARG 12-77 p43 +St 11-77 p146
+HF 9-77 p108
2738 Symphony, no. 6, op. 74, B minor. LISZT: Mazeppa, G 138. RPO, BPhO; Oskar Fried. Bruno Walter Society BWS 734.
+NR 5-77 p5
2739 Symphony, no. 6, op. 74, B minor. CSO; Georg Solti. Decca SXL 6814 Tape (c) KSXC 6814. (also London CS 7034)
+Gr 9-77 p437 +-RR 9-77 p68
+-Gr 12-77 p1145 tape -RR 12-77 p99
++HFN 9-77 p150 ++SFC 11-20-77 p54
2740 Symphony, no. 6, op. 74, B minor. BPhO; Herbert von Karajan. DG 2530 774 Tape (c) 3300 774.
+Gr 6-77 p64 +-NR 11-77 p5
+Gr 6-77 p105 tape +RR 6-77 p70
+-HF 12-77 p102 +SFC 10-16-77 p47
+HFN 6-77 p135
2741 Symphony, no. 6, op. 74, B minor. BPhO; Wilhelm Furtwangler. DG 2535 165.
+-Gr 5-76 p1772 +STL 1-9-77 p35
+RR 5-76 p22

2742 Symphony, no. 6, op. 74, B minor. CPhO; Vaclav Talich. Heritage HCN 8013. (Reissue from Parliament PLP 113)
+Gr 7-77 p171 ++RR 8-77 p63
2743 Symphony, no. 6, op. 74, B minor. LSO; Antal Dorati. Philips 6582 014.
+-Gr 2-77 p1290 +-RR 1-77 p65
2744 Symphony, no. 6, op. 74, B minor. LPO; Hubert Soudant. Pye Nixa PCNHX 12 Tape (c) ZCPNH 12.
+-Gr 12-77 p1145 tape /-RR 12-77 p57
/-HFN 12-77 p181
2745 Symphony, no. 6, op. 74, B minor. LSO; Loris Tjeknavorian. RCA LRL 1-5129.
+Gr 11-76 p812 +RR 11-76 p16
++HFN 3-77 p113
2746 Symphony, no. 6, op. 74, B minor. LSO; Jascha Horenstein. Vanguard VSC 10114.
++ARG 2-77 p42 ++NR 1-77 p2
-HF 4-77 p113 +-St 5-77 p119
-MJ 2-77 p31
Symphony, no. 6, op. 74, B minor. cf RCA CRM 5-1900.
Trepak, op. 72, no. 18. cf International Piano Library IPL 117.
2747 Trio, piano, op. 50, A minor. Grigori Feigin, vln; Valentin Feigin, vlc; Igor Zhukov, pno. HMV Melodiya HQS 1381.
+Gr 7-77 p198 -RR 7-77 p72
+-HFN 7-77 p121
Trio, piano, op. 50, A minor: Pezzo elegiaco. cf CBS SQ 79200.
Valse, op. 40, no. 2. cf Piano works (Seraphim S 60250).
Valse scherzo, op. 34. cf Concerto, violin, op. 35, D major (Decca SXL 6854).
Valse scherzo, op. 34. cf Concerto, violin, op. 35, D major (HMV SXLP 30225).
Valse scherzo, op. 34. cf Concerto, violin, op. 35, D major (Philips 9500 146).
Valse scherzo, op. 34. cf BRAHMS: Sonata, violin and piano, no. 3, op. 108, D minor.
Valse sentimentale, op. 51, no. 6. cf HMV SXLP 30256.
Valse sentimentale, op. 51, no. 6. cf HMV SXLP 30259.
The Voyevode, op. 78. cf Fatum, op. 77.
2748 Works, selections: Symphonies, nos. 1-6. Hamlet, op. 67a. Overture, the year 1812, op. 49. Manfred symphony, op. 58. Francesca da Rimini, op. 32. Romeo and Juliet. Utah Symphony Orchestra; Maurice Abravanel. Vox QSVBX 5129/31 (30.
+Audio 1-77 p83 +NR 1-75 p2
+-HF 7-75 p63

TCHEREPNIN, Alexander
2749 Concerto, piano, no. 2. Symphony, no. 2. Alexander Tcherepnin, pno; Louisville Orchestra; Robert Whitney. RCA GL 2-5059.
+-Gr 9-77 p437 +RR 9-77 p70
+HFN 12-77 p183
Symphony, no. 2. cf Concerto, piano, no. 2.

TELEMANN, Georg Philipp
L'amour. cf BOISMORTIER: Sonata, trumpet, G minor.
L'armement. cf BOISMORTIER: Sonata, trumpet, G minor.
Concerto, D major. cf MOZART: Symphony, no. 35, K 385, D major.
2750 Concerto, recorder and horn, F major. Sonatas, recorder, C major, F minor. Trio sonata, recorder and violin, D minor. Trio sonata, recorder and oboe, E minor. Denmark Concentus Musicus. Nonesuch H 71065.

+-HFN 7-77 p123 +RR 6-77 p74

Concerto, recorder and strings, F major. cf BOISMORTIER: Concerto, 5 recorders without bass, D minor.

Concerto, alto recorder, viola da gamba and strings, A minor. cf MOZART: Quartet, flute, K 285, D major.

Concerto 2 recorders and strings, B minor. cf BOISMORTIER: Concerto, 5 recorders without bass, D minor.

Concerto, trumpet, D major. cf ALBINONI: Concerto, trumpet, op. 7, no. 3, B flat major.

Concerto, trumpet, D major. cf Argo D69D3.

Concerto, trumpet and strings, D major. cf DG 2530 792.

Concerto, trumpet and 2 oboes, D major. cf Works, selections (RCA FRL 1-8081).

Concerto, viola, G major. cf Argo D69D3.

Concerto, 2 violas and strings, G major. cf LARSSON: Concertino. op. 45, no. 9.

Divertissement, 2 trumpets and strings, D major. cf Philips 6581 018.

Don Quichotte. cf BACH, J. C.: Sinfonias, op. 18, nos. 2, 4, 6.

La douceur. cf BOISMORTIER: Sonata, trumpet, G minor.

L'esperance. cf BOISMORTIER: Sonata, trumpet, G minor.

2751 Fantasias, flute (12). Jean-Pierre Rampal, flt. Odyssey Y 33200.

++Audio 10-76 p150 +SFC 1-2-77 p34
++NR 1-76 p15 ++St 8-76 p99

La gaillardise. cf BOISMORTIER: Sonata, trumpet, G minor.

La generosite. cf BOISMORTIER: Sonata, trumpet, G minor.

La Grace. cf BOISMORTIER: Sonata, trumpet, G minor.

2752 Hamburger Ebb und Fluth, C major. Overture des nations anciens et modernes, G major. Overture, C major. AMF; Neville Marriner. Argo ZRG 837 Tape (c) KZRC 837.

+Gr 4-77 p1562 ++RR 4-77 p60
+HF 10-77 p122 ++RR 8-77 p87
+HFN 4-77 p147 +St 10-77 p147
++NR 8-77 p5

La Majeste. cf BOISMORTIER: Sonata, trumpet, G minor.

2753 Musique de table. Frans Bruggen, rec; Concerto Amsterdam. Telefunken 6-35298. (aso 2648 006/8)

++NR 1-77 p8 ++SFC 8-15-76 p38

Overture, C major. cf Hamburger Ebb und Fluth, C major.

Overture des nations anciens et modernes, G major. cf Hamburger Ebb und Fluth, C major.

Partita, recorder, no. 2, E minor. cf Works, selections (RCA FRL 1-8081).

2754 Pimpinone (incorporating Tessarini's Violin concerto, op. 1, no. 7, B flat major; Albinoni's Oboe concerto, op. 9, no. 8, B flat major; Vivaldi's Violin concerto, op. 7, no. 2, C major). Uta Spreckelsen, s; Siegmund Nimsgern, bar; Florilegium Musicum Ensemble; Hans Ludwig Hirsch. Telefunken 6-35285ER (2).

++Audio 2-77 p80 +RR 4-76 p30
+-HF 7-76 p94 ++SFC 1-18-76 p38
+HFN 5-76 p111 ++St 6-76 p114
+NR 2-76 p11

La rejoussance. cf BOISMORTIER: Sonata, trumpet, G minor.

Sonata, bassoon, F minor. cf Orion ORS 77269.

2755 Sonatas, 2 flutes, op. 2, nos. 1-6. Michael Debost, James Galway, flt. HMV SQS 1368.

++Gr 4-77 p1569 ++RR 4-77 p78
++HFN 4-77 p147

Sonatas, flute and harpsichord, G major, C major. cf Works, selections (RCA FRL 1-8081).
Sonata, flute, recorder and harpsichord, A major. cf Works, selections (RCA FRL 1-8081).
Sonatas, recorder, C major, F minor. cf Concerto, recorder and horn, F major.
Sonata, recorder, E minor. cf BIS LP 48.
Sonata, recorder, F major. cf Cambridge CRS 2826.
Sonata, recorder and harpsichord, F major. cf Works, selections (RCA FRL 1-8081).
Sonata, viola da gamba and harpsichord, A minor. cf BACH: Sonatas, viola da gamba and harpsichord, nos. 1-3, S 1027-1029.
Suite, A minor. cf HOFFMEISTER: Concerto, flute, G major.
Suite, D major. cf Pelca PSR 40571.
Suite, F major. cf Supraphon 111 1867.
Trio sonata, recorder and oboe, E minor. cf Concerto, recorder and horn, F major.
Trio sonata, recorder and violin, D minor. cf Concerto, recorder and horn, F major.
La Vaillance. cf BOISMORTIER: Sonata, trumpet, G minor.

2756 Works, selections: Concerto, trumpet and 2 oboes, D major. Partita, recorder, no. 2, E minor. Sonatas, flute and harpsichord, G major, C major. Sonata, recorder and harpsichord, F major. Sonata, flute, recorder and harpsichord, A major. Maurice Andre, tpt; Jean-Pierre Rampal, flt; Pierre Pierlot, Jacques Chambon, ob; Mario Duschenes, rec; Paul Hongne, bsn; Robert Veyron-Lacroix, hpd. RCA FRL 1-8081 Tape (c) FRK 1-8081 (ct) FRS 1-8081.

+ARG 7-77 p34 ++NR 9-77 p7
+HF 10-77 p129 tape ++SR 7-9-77 p48
++MJ 9-77 p35

TERANO, da
Rosetta. cf Argo D40D3.

TERZI, Giovanni
Ballo tedesco e francese. cf DG 2533 173.
Canzona francese. cf L'Oiseau-Lyre 12BB 203/6.
Tre parti di gagliarde. cf DG 2533 173.

THALBEN-BALL, George
Tune, E major. cf Vista VPS 1046.

THIBAUT DE CHAMPAGNE
Au tens plain de felonnie. cf Argo D40D3.

THOMAS, Ambroise
La Caid: Air du Tambour-major. cf Rubini GV 39.
Hamlet: A vos jeux...Partagez-vous me fleurs...Et maintenant ecoutez ma chanson. cf HMV SLS 5057.
Hamlet: Mad scene. cf HANDEL: Il pensieroso: Sweet bird.
Hamlet: Mad scene. cf HMV RLS 719.
Hamlet: O vin, dissipe ma tristesse. cf Philips 6580 174.
Mignon: Ah, non credevi tu. cf Seraphim S 60280.
Mignon: Behold, Titania. cf HMV HLM 7066.
Mignon: Berceuse. cf Club 99-107.
Mignon: Connais-tu le pays. cf CBS 76522.
Mignon: Connais-tu le pays. cf Seraphim M 60291.
Mignon: Connais-tu le pays; Elle est la pres de lui...Elle est aimee. cf Seraphim IB 6105.

Mignon: Non consoci il bel suol. cf Club 99-100.
Mignon: Overture. cf HEROLD: Zampa: Overture.
Mignon: Overture. cf HMV ESD 7010.
Mignon: Overture. cf Quintessence PMC 7013.
Raymond: Overture. cf HEROLD: Zampa: Overture.
Songs: Songe d'une nuit d'ete; Couplets de Falstaff, alors que tout s'apprete. cf Club 99-107.

THOMAS, Andrew

2757 The death of Yukio Mishima. Dirge in the woods. WIDDOES: From a time of snow. Jeanne Ommerle, s; Notes from Underground; Peter Leonard. Opus One 28.
++St 2-77 p124

Dirge in the woods. cf The death of Yukio Mishima.

2758 Pricksong. WOOD: Sonata, violin and piano. Alyssa Hess, hp; Leonard Velberg, vln; George Robert, pno. Opus One 30.
+ARG 2-77 p51

THOMPSON

March: The Nybbs. cf RCA PL 2-5046.

THOMPSON, R.

Alleluia. cf Vox SVBX 5353.

THOMPSON, Randall

Glory to God in the highest. cf Columbia M 34134.
The peaceable kingdom. cf REIF: Vignettes, 4 singers.

THOMSON, Virgil

2759 Autumn (Concertino, harp, strings and percussion). The plow that broke the plains. The river. Ann Mason Stockton, hp; Los Angeles Chamber Orchestra; Neville Marriner. Angel (Q) S 37300. (also HMV ASD 3294)
+Gr 1-77 p1153 ++NR 5-76 p4
-HF 9-76 p98 ++RR 12-76 p67
+HFN 12-76 p132 +SFC 6-13-76 p30
++MJ 7-76 p57 ++St 7-76 p83 Quad

2760 The mother of us all. Mignon Dunn, Aviva Orvath, Ashley Putnam, s; Batyah Godfrey, Linn Maxwell, con; James Atherton, t; Joseph McKee, Gene Ives, bar; Philip Booth, bs; Santa Fe Opera Orchestra; Raymond Leppard. New World Records NW 288/9 (2).
+ARG 8-77 p30 +OC 12-77 p49
+HF 7-77 p92 +-ON 3-19-77 p40
+MJ 7-77 p68 +OR 6/7-77 p32
++NR 6-77 p11 ++St 6-77 p89

The plow that broke the plains. cf Autumn.
The river. cf Autumn.
Sonata, piano, no. 3. cf Vox SVBX 5303.
Southern hymns. cf Vox SVBX 5353.
Ten etudes: Parallel chords, Ragtime bass. cf Vox SVBX 5303.
Variations on Shall we gather at the river. cf Advent ASP 4007.

THORNE, Francis

Sonata, piano. cf CUSTER: Found objects, no. 7.

THRANE

Fjallvisan. cf Swedish SLT 33209.

THUILLE, Ludwig

Lobetanz: An allen Zweigen. cf Club 99-109.

TIBURTINO, Giuliano

Ricercare a 3. cf L'Oiseau-Lyre 12BB 203/6.

TIERNEY

Irene: Alice blue gown. cf Angel (Q) S 37304.

TIMM, Kenneth
The joiner and the diehard. cf ERB: Harold's trip to the sky.
TINCTORIS, Johannes de
Missa 3 vocum: Kyrie. cf HMV SLS 5049.
TIPPETT, Michael
2761 A child of our time. The midsummer marriage: Ritual dances. Elsie Morison, s; Pamela Bowden, con; Richard Lewis, t; Richard Staden, bs; Royal Liverpool Philharmonic Orchestra and Chorus; ROHO; John Pritchard. Decca DPA 571/2 (2). (Reissues from Pye CCL 30114/5)
+Gr 3-77 p1443 +-RR 4-77 p84
++HFN 4-77 p153
Concerto, piano. cf HMV SLS 5080.
Concerto, 2 string orchestras. cf Argo D26D4.
Dance, clarion air. cf RCA GL 2-5062.
Fanfare, brass, no. 1. cf RCA RL 2-5081.
Fantasia concertante on a theme by Corelli. cf Argo D26D4.
Little music, string orchestra. cf Argo D26D4.
A midsummer marriage: Ritual dances. cf A child of our time.
2762 Quartets, strings, nos. 1-3. Lindsay Quartet. L'Oiseau-Lyre DSLO 10.
++Gr 12-75 p1065 ++NR 7-76 p6
++HF 4-77 p113 ++RR 12-75 p74
+HFN 12-75 p167 +St 10-76 p136
+MJ 12-76 p44 ++STL 1-4-76 p36
+MM 8-76 p31 +Te 3-76 p31
+MT 11-76 p915
Songs for Ariel. cf Argo ZK 28/9.
2763 Suite for the birthday of Prince Charles. Symphony, no. 1. LSO; Colin Davis. Philips 9500 107.
++ARG 11-76 p40 +-NR 12-76 p5
+Audio 5-77 p99 +RR 10-76 p74
+Gr 10-76 p612 *SFC 12-19-76 p50
++HF 1-77 p130 ++St 12-76 p148
++HFN 10-76 p179 +Te 6-77 p43
++MJ 12-76 p44
TITELOUZE, Jean
Quatre versets sur "Veni creator". cf LE JEUNE: Missa ad placitum.
TOCH, Ernest
Impromptu, op. 90. cf ABC AB 76014.
TOMASEK, Vaclav Jan
2764 Dithyrambs, piano, op. 65 (3). Goethe songs: Am Flusse; Erster Verlust, op. 56; Erlkonig; Das Geheimnis; Der Konig in Thule; Der Fischer, op. 59; Rastlose Liebe; Die Sorge; Die Spinnerin, op. 55; Das Veilchen, op. 57; Wanderers Nachtlied, op. 58. Libuse Marova, s; Jindrich Jindrak, bar; Dagmar Simonkova, Alfred Holecek, pno. Panton 110 516.
+-Gr 5-77 p1724 +RR 12-76 p96
+HFN 12-76 p149
2765 Eclogues, op. 35, nos. 5 and 6; op. 39, nos. 1 and 3; op. 47, nos. 2, 3, 6; op. 51, no. 5; op. 63, no. 1; op. 66, no. 6; op. 83, no. 6. Pavel Stepan, pno. Supraphon 111 1488.
+Gr 9-77 p459 +NR 10-77 p13
+HFN 7-77 p123 +RR 6-77 p86
Eclogues, op. 39, nos. 1 and 3. cf Eclogues, op. 35, nos. 5 and 6.
Eclogues, op. 47, nos. 2, 3, 6. cf Eclogues, op. 35, nos. 5 and 6.
Eclogues, op. 51, no. 5. cf Eclogues, op. 35, nos. 5 and 6.

Eclogues, op. 63, no. 1. cf Eclogues, op. 35, nos. 5 and 6.
Eclogues, op. 66, no. 6. cf Eclogues, op. 35, nos. 5 and 6.
Eclogues, op. 83, no. 6. cf Eclogues, op. 35, nos. 5 and 6.
Goethe songs: Am Flusse; Erster Verlust, op. 56; Erlkonig; Der Fischer, op. 59; Das Geheimnis; Der Konig in Thule; Rastlose Liebe; Die Sorge; Die Spinnerin, op. 55; Das Veilchen, op. 57; Wanderers Nachtlied, op. 58. cf Dithyrambs, piano, op. 65.

TOMASI, Henri
Holy week at Cuzco. cf Musical Heritage Society MHS 3340.
Le muletier des Andes. cf London CS 7015.

TOMKINS, Thomas
Fancy for 2 to play. cf Vista VPS 1039.
The fauns and satyrs. cf Argo ZK 25.
The fauns and satyrs tripping. cf DG 2533 347.
The King shall rejoice. cf Guild GRSP 701.
Worster braules. cf Argo ZRG 864.

TOMLINSON, Ernest
Overture on famous English airs. cf RCA PL 2-5046.

TORELLI, Giuseppe
Concerto, trumpet and strings, D major. cf DG 2530 792.
Concerto grosso, op. 8, no. 6, G minor. cf LOCATELLI: Concerto grosso, op. 8, no. 1, F minor.
Sonata a 5, no. 7, D major. cf Philips 6580 114.

TORRE, Francisco de la
Adormas te, Senor; La Spagna. cf L'Oiseau-Lyre 12BB 203/6.
Danza alta a 3. cf DG 2723 051.
Pascua d'espiritu santo. cf Enigma VAR 1024.

TORRENS
Songs: How pansies grow. cf Decca SDD 507.

TORROBA, Federico
Aires de la Mancha. cf CBS 72526.
Burgalesa. cf Enigma VAR 1015.
Burgalesa. cf Swedish Society SLT 33205.
Madronos. cf Angel S 36094.
Nocturno. cf Philips 6833 159.
Nocturne. cf Swedish Society SLT 33205.
Sonatine, A major. cf Westminster WG 1012.

TORTELIER, Paul
Miniatures, 2 violoncellos. cf HMV SQ ASD 3283.
Valse, no. 1. cf HMV SQ ASD 3283.

TOSELLI, Enrico
Serenade. cf Philips 6392 023.
Serenata, op. 6, no. 1. cf HMV RLS 715.

TOSTI, Francesco
Dopo. cf Club 99-100.
Penso. cf Club 99-96.
Penso. cf Rubini RS 301.
Songs: Goodbye; Mattinata; La serenata. cf HMV RLS 719.
Songs: Mattinata; Ideale. cf HMV RLS 715.
Songs: Parted. cf Argo ZFB 95/6.
Songs: Serenata; Malia. cf Columbia M 34501.

TOURNEMIRE, Charles
Cantilene improvisee. cf Wealden WS 159.
Chorale sur le victimae Paschali. cf Vista VPS 1029.
Messe de l'assomption: Paraphrase carillon. cf Calliope CAL 1922/4.

2766 L'Orgue mystique, op. 56, no. 25 (Pentecost mass). Gerald Farrell, org. Liturgical Press.
++MU 4-77 p11

Te deum improvisee. cf Wealden WS 159.
TOURS
Songs: Mother o' mine. cf Argo ZFB 95/7.
TOWER, Joan
Breakfast rhythms I and II. cf TANENBAUM: Rituals and reactions.
Hexachords. cf TANENBAUM: Rituals and reactions.
TRAVIS, Roy
Concerto, piano. Songs and epilogues. Symphonic allegro. Harold Enns, bs-bar; Irma Vallecillo, pno; RPO, Utah Symphony Orchestra; Jan Popper. Orion ORS 76219.
+NR 2-77 p15
Songs and epilogues. cf concerto, piano.
Symphonic allegro. cf Concerto, piano.
TREDINNICK, Noel
Brief encounters. cf Vista VPS 1046.
TREVALSA
Songs: My treasure. cf Decca SDD 507.
THE TRIUMPHS OF ORIANA. cf DG 2533 347.
TROMBONCINO, Bartolomeo
Ave Maria; Hor ch'el ciel e la terra; Ostinato vo seguire. cf L'Oiseau-Lyre 12BB 203/6.
Io son l'ocello. cf Enigma VAR 1024.
TROUBADOUR SONGS AND INSTRUMENTAL WORKS. cf Telefunken 6-41126.
THE TRUMPET IN CONTEMPORARY CHAMBER SETTINGS. cf Crystal S 362.
TUDOR CHURCH MUSIC. cf Coimbra CCO 44.
TUFTS, Paul
Sonata, horn and piano. cf STEVENS: Sonata, horn and piano.
TURCO, Giovanni del
Songs: Occhi belli. cf DG 2533 305.
TURETZKY, Bertram
Gamelan music. cf Finnadar SR 9015.
TURINA, Joaquin
Danzas fantasticas. cf ALBENIZ: Iberia.
Danzas gitanas, op. 55. cf LEES: Symphony, no. 3.
Fandanguillo. cf Angel S 36094.
Fandanguillo, op. 36. cf Westminster WG 1012.
Garrotin. cf Angel S 36094.
Homage a Tarrega. cf Swedish Society SLT 33205.
La oracion del torero, op. 34. cf Seraphim SIB 6094.
Poema en forma de canciones, op. 19. cf FALLA: Melodies.
Rafaga. cf Angel S 36094.
Sacro-monte, op. 55, no. 5. cf RCA RL 2-5099.
Soleares. cf Angel S 36094.
TURNER, John
How far is it to Bethlehem. cf Abbey LPB 776.
TWENTIETH CENTURY MUSIC FOR TRUMPET AND ORGAN. cf Musical Heritage Society MHS 3340.
TYE, Christopher
In nomine. cf L'Oiseau-Lyre 12BB 203/6.
TYSON
One little cloud. cf Orion ORS 77271.
UNG, Chinary
Mohori. cf HEISS: Inventions, contours and colors.
URBANNER, Erich
2767 Concerto, double bass and chamber orchestra. VANHAL: Concerto, double bass and chamber orchestra. Ludwig Streicher, double bs; Innsbruck Chamber Orchestra; Erich Urbanner, Othmar Costa. Telefunken AW 6-42045.

-Gr 7-77 p188 +RR 5-77 p61
Quartet, strings, no. 3. cf HAUBENSTOCK-RAMATI: Quartet, strings, no. 1.
USSACHEVSKY, Vladimir
Linear contrasts. cf AREL: Music for a sacred service: Prelude and postlude.
Metamorphosis. cf AREL: Music for a sacred service: Prelude and postlude.
VACCAI, Nicola
Giullietta e Romeo: Tu dormi, sveglati. cf Club 99-106.
VAILLANT, Jean
Par maintes foys. cf 1750 Arch S 1753.
Tres doulz amis: Ma dame; Cent mille fois. cf HMV SLS 863.
VALDAMBRINI, Francesco
Dioe. cf CBS 61453.
VALEK, Jiri
2768 Symphony, no. 8. Symphony, no. 9. Jarmila Vrchtova-Patova, s; Jiri Tomasek, vln; Hubert Simacek, vla; Vaclav Bernasek, vlc; Ostrava Janacek Philharmonic Orchestra, Prague Chamber Orchestra; Otakar Trhlik, Eduard Fischer. Supraphon 110 1569.
/ARG 10-77 p45 +NR 10-77 p2
+-Gr 9-77 p438 +RR 6-77 p70
+HFN 8-77 p91
Symphony, no. 9. cf Symphony, no. 8.
Symphony, no. 10. cf FLOSMAN: Concerto, violin, no. 2.
VALEN, Fartein
2769 Concerto, violin, op. 37. Epithalamion, op. 19. Trio, violin, violoncello and piano, op. 5. Arve Tellefsen, Stig Nilsson, vln; Hege Waldeland, vlc; Eva Knardahl, pno; Bergen Symphony Orchestra; Karsten Andersen. Phonogram 6507 039.
+RR 10-77 p65
Epithalamion, op. 19. cf Concerto, violin, op. 37.
Trio, violin, violoncello and piano, op. 5. cf Concerto, violin, op. 37.
VALENTI, Antonio
Gavotte. cf HMV RLS 723.
VALLET, Nicolas
Galliarde. cf DG 2533 302.
Prelude. cf DG 2533 302.
Slaep, soete, slaep. cf DG 2533 302.
Suite. cf Saga 5438.
VALVERDE, Joaquin
Clavelitos. cf Seraphim 60274.
VAN VACTOR, David
Sonatina, flute and piano. cf BURTON: Sonatina, flute and piano.
VANHAL, Jan
Concerto, double bass and chamber orchestra. cf URBANNER: Concerto, double bass and chamber orchestra.
Parthia, B flat major. cf Telefunken EK 6-35334.
Sonata, B flat major: Adagio cantabile; Rondo. cf Supraphon 113 1323.
Sonata, clarinet, B flat major. cf BERNSTEIN: Sonata, clarinet and piano.
VACQUIERAS, Raimbault de
Kalenda meia. cf Telefunken 6-41126.
VASQUEZ, Juan
Lindos ojos aveys, senora. cf L'Oiseau-Lyre 12BB 203/6.

VASSEUR
Songs: La cruche cassee, chanson espagnol. cf Decca SDD 507.
VAUGHAN WILLIAMS, Ralph
Benedicite. cf Songs (HMV SLS 5082).
Coastal command: Dawn patrol. cf HMV SQ ASD 3341.
Concerto, oboe, A minor. cf BACH: Concerto, oboe d'amore, A major.
2770 Concerto grosso. Fantasia on a theme by Thomas Tallis. Partita, 2 string orchestras. LPO; Adrian Boult. HMV SQ ASD 3286 Tape (c) TC ASD 3286. (also Angel (Q) S 37211)
+Gr 11-76 p812 ++NR 2-77 p2
+HF 3-77 p108 ++RR 11-76 p85
+HFN 12-76 p149 ++RR 3-77 p99 tape
+HFN 5-77 p139 tape ++SFC 12-26-76 p34
+HFN 9-77 p155 tape ++St 3-77 p145
Dona nobis pacem. cf Songs (HMV SLS 5082).
2771 English folk song suite. Fantasia on the "Old 104th" psalm tune. The wasps. Peter Katin, pno; LSO, LPO and Chorus; Adrian Boult. Angel S 37276.
++NR 6-77 p2 +SFC 12-18-77 p53
English folk song suite. cf COATES: London every day.
Fantasia on a theme by Thomas Tallis. cf Concerto grosso.
Fantasia on a theme by Thomas Tallis. cf COATES: London every day.
Fantasia on a theme by Thomas Tallis. cf DVORAK: Serenade, strings, op. 22, E major.
Fantasia on a theme by Thomas Tallis. cf ELGAR: In the south overture, op. 59.
Fantasia on a theme by Thomas Tallis. cf RUBBRA: Symphony, no. 7, op. 88, C major.
Fantasia on a theme by Thomas Tallis. cf Argo D26D4.
Fantasia on Christmas carols. cf Songs (HMV SLS 5082).
Fantasia on Christmas carols. cf HELY-HUTCHINSON: Carol symphony.
Fantasia on "Greensleeves". cf COATES: London every day.
Fantasia on "Greensleeves". cf Argo D26D4.
Fantasia on "Greensleeves". cf HMV ESD 7011.
Fantasia on "Greensleeves". cf Pye Golden Hour GH 643.
Fantasia on the "Old 104th" psalm tune. cf English folk song suite.
Fantasia (Quasi variazione) on the "Old 104th psalm tune". cf Songs (HMV SLS 5082).
Flos campi suite. cf Songs (HMV SLS 5082).
Greensleeves. cf HMV RLS 716.
Greensleeves. cf HMV SQ CSD 3781.
Hodie, A Christmas cantata. cf Songs (HMV SLS 5082).
The lark ascending. cf Argo D26D4.
Magnificat. cf Songs (HMV SLS 5082).
Mass, G minor. cf Songs (HMV SLS 5082).
Mass, G minor: Credo. cf Guild GRSP 701.
Mystical songs (5). cf Songs (HMV SLS 5082).
O clap your hands. cf Songs (HMV SLS 5082).
2772 Old King Cole. The wasps: Overture. On Wenlock edge. Songs of travel: The roadside fire. Gervase Elwes, t; Frederick Kiddle, pno; London String Quartet, Aeolian Orchestra; Ralph Vaughan Williams. Pearl GEM 127. (Reissue)
+-ARSC vol 9, no. 1, 1977 p98
On Wenlock edge. cf Old King Cole.
An Oxford elegy. cf Songs (HMV SLS 5082).
Partita, 2 string orchestras. cf Concerto grosso.

2773 Phantasy quintet. Sonata, violin and piano, A minor. Studies in English folksong (6). Hugh Bean, Frances Mason, vln; David Parkhouse, pno; Christopher Wellington, Ian Jewel, vla; Eileen Croxford, vlc. HMV SQS 1327.
+ARG 9-77 p52 +RR 8-74 p45
+Gr 8-74 p370

Preludes on Welsh hymn tunes (3). cf HMV HQS 1376.
Preludes on Welsh hymn tunes: Rhosymedre. cf Odeon HQS 1356.
Preludes on Welsh hymn tunes: Rhosymedre. cf Vista VPS 1047.
Romance. cf JACOB: Pieces.
Sancta civitas. cf Songs (HMV SLS 5082).
Sea songs. cf HOLST: Suite, no. 1, op. 28, no. 1, E flat major.
Serenade to music. cf Songs (HMV SLS 5082).
Serenade to music. cf BBC REH 290.
Silence and music. cf RCA GL 2-5062.
Sonata, violin and piano, A minor. cf Phantasy quintet.

2774 Songs (choral): Benedicite. Dona nobis pacem. Fantasia on Christmas carols. Fantasia (Quasi variazione) on the "Old 104th psalm tune". Flos campi suite. Hodie, A Christmas cantata. Magnificat. Mass, G minor. Mystical songs (5). O clap your hands. An Oxford elegy. Sancta civitas. Serenade to music. Toward the unknown region. Tudor portraits (5). Variants on "Dives and Lazarus" (5). Sheila Armstrong, Heather Harper, Norma Burrowes, Susan Longfield, Marie Hayward, s; Janet Baker, ms; Elizabeth Bainbridge, Helen Watts, Alfreda Hodgson, Gloria Jennings, Shirley Minty, Meriel Dickinson, con; Nigel Perrin, Ian Partridge, Bernard Dickerson, Wynford Evans, Kenneth Bowen, Richard Lewis, t; John Shirley-Quirk, John Carol Case, bar; David van Asch, Richard Angas, John Noble, Christopher Keyte, bs; Gavin Williams, Philip Ledger, org; Cecil Aronowitz, vla; King's College Chapel Choir, Guildford Cathedral Choir, Ambrosian Singers, Westminster Abbey Choir; ECO, LPO and Chorus, LSO, NPhO, Nova of London, Jacques Orchestra; David Willcocks, Adrian Boult, Barry Rose, Meredith Davies. HMV SLS 5082 (7). (Reissues from ASD 2458, 2962, 2422, 2489, CSD 3580, ASD 2699, 2538, 2581, 2487, Columbia SCX 3570)
+Gr 6-77 p89 ++RR 7-77 p93
+HFN 8-77 p95

Songs: All people that on earth do dwell; O taste and see. cf Guild GRSP 701.
Songs: The cloud-capp'd towers; Full fathom five; Over hill over dale. cf RCA RL 2-5112.
Songs (Folksong arrangements): Died for love; Abroad as I was walking; O who is that; Our ship she lies in harbour; The willow tree. cf GRAINGER: Songs (Premier PMS 1502).
Songs: For all the saints. cf Abbey LPB 776.
Songs: Linden Lea. cf Abbey LPB 778.
Songs: Linden Lea. cf L'Oiseau-Lyre DSLO 20.
Songs: O taste and see. cf Decca SPA 500.
Songs: O taste and see; The old hundredth. cf Pye Nixa QS PCNHX 10.
Songs (anthems): The old hundredth; O taste and see. cf Vista VPS 1053.
Songs of travel: The roadside fire. cf Old King Cole.
Songs of travel: The vagabond. cf L'Oiseau-Lyre DSLO 20.
Studies in English folksong (6). cf Phantasy quintet.
Studies in English folksong (6). cf BERNSTEIN: Sonata, clarinet and piano.

Suite for pipes. cf HMV SQ CSD 3781.
Te deum, G major. cf Abbey LPB 779.
Toward the unknown region. cf Songs (HMV SLS 5082).
Tudor portraits (5). cf Songs (HMV SLS 5082).
Variants on "Dives and Lazarus" (5). cf Songs (HMV SLS 5082).
Variants on "Dives and Lazarus" (5). cf GOOSENS: Concerto, oboe, op. 45.
Variants on "Dives and Lazarus" (5). cf Argo D26D4.
The wasps. cf English folk song suite.
The wasps: Overture. cf Old King Cole.
The wasps: Overture. cf ELGAR: In the south overture, op. 50.
Wither's rocking carol. cf HMV ESD 7024.

VAUTOR, Thomas
Mother I will have a husband. cf BG HM 57/8.
Sweet Suffolk owl. cf Enigma VAR 1017.

VAZZANA, Anthony
Incontri. cf Orion ORS 76212.

VECCHI, Orazio
Saltarello. cf Argo ZRG 823.

VEJVANOVSKY, Pavel
Sonata, no. 14 a 6 Campanarum. cf Supraphon 111 1867.

VELLONES
Rhapsody for saxophone. cf Citadel CT 6012.

VENZANO
O che assorta. cf Rubini GV 26.

VERDELOT, Philippe
Ave sanctissima Maria. cf HMV SLS 5049.
Madonna, qual certezza. cf L'Oideau-Lyre 12BB 203/6.

VERDI, Giuseppe

2775 Aida. Montserrat Caballe, Esther Casas, s; Fiorenza Cossotto, ms; Placido Domingo, Nicola Martinucci, t; Piero Cappuccilli, bar; Nicolai Ghiaurov, Luigi Roni, bs; NPhO; ROHO Chorus; Riccardo Muti. Angel SCLX 3815 (3) Tape (c) 4X3S 3815. (also HMV SLS 977 Tape (c) TC SLS 977)
+Gr 2-75 p1548 +-ON 1-18-75 p32
+Gr 10-75 p721 tape +-RR 2-75 p23
+-HF 2-75 p102 +RR 9-75 p79 tape
+HF 4-77 p121 tape ++St 1-75 p118
+HFN 9-75 p110 tape +-STL 2-9-75 p37

2776 Aida. Franco Corelli, t; Orchestra and conductor. Cetra LPS 3262. (Reissue from OLPC 1262)
+Gr 7-77 p221

2777 Aida. Renata Tebaldi, s; Giuletta Simionato, ms; Carlo Bergonzi, t; Cornell MacNeil, bar; Vienna Singverein; VPO; Herbert von Karajan. Decca Tape (c) K2A20. (also London Tape (c) OSA 5-1313)
+Gr 8-76 p341 tape ++HFN 8-76 p94 tape
+HF 3-77 p123 tape +RR 8-76 p82 tape

2778 Aida. Leontyne Price, Mietta Singhele, s; Rita Gorr, ms; Jon Vickers, Franco Ricciardi, t; Robert Merrill, bar; Giorgio Tozzi, Plinio Clabassi, bs; Rome Opera House Orchestra and Chorus; Georg Solti. Decca Tape (c) K64K32 (2).
+Gr 11-77 p900 tape +HFN 12-77 p187 tape

2779 Aida. Leontyne Price, s; Grace Bumbry, ms; Placido Domingo, Bruce Brewer, t; Sherrill Milnes, bar; Ruggero Raimondi, Hans Sotin, bs; John Alldis Choir; LSO; Erich Leinsdorf. RCA Tape (c) RK 40005 (2).
-RR 4-77 p94 tape

2780 Aida, excerpts. Montserrat Caballe, Esther Casas, s; Fiorenza Cossotto, ms; Placido Domingo, Nicola Martinucci, t; Piero Cappuccilli, bar; Nicolai Ghiaurov, Luigi Roni, bs; NPhO; ROHO Chorus; Riccardo Muti. HMV ASD 3292 Tape (c) TC ASD 3292.
+HFN 5-77 p139 tape +HFN 9-77 p155 tape
Aida: Arias. cf Free Lance FLPS 675.
Aida: Ballet music. cf Angel S 37250.
Aida: Ballet music. cf CBS 30091.
Aida: Celeste Aida. cf Decca SPA 450.
Aida: Celeste Aida. cf RCA CRM 1-1749.
Aida: Celeste Aida. cf Seraphim S 60280.
Aida: Celeste Aida. cf Works, selections (Decca SPA 447).
Aida: Ciel mio padre, A te grave cagion m'adduce, Aida, Pur ti riveggo. cf Court Opera Classics CO 347.
Aida: La fatal pietra...O terra, addio. cf BELLINI: La sonnambula: Perdona o mia diletta...Prendi l'anel ti dono.
Aida: Fu la sorte dell'armi. cf Decca ECS 811.
Aida: Gloria all'egitto. cf Seraphim S 60275.
Aida: Grand march. cf Decca SB 328.
Aida: I see thee again, my sweet Aida. cf GOUNOD: Faust: The hour is late.
Aida: O cieli azzuri. cf CATALANI: La Wally: Ebben, ne andro lontana.
Aida: O patria mia. cf HMV SLS 5104.
Aida: O patria mia. cf Seraphim 60274.
Aida: Prelude. cf Overtures (DG 2707 090).
Aida: Ritorna vincitor. cf PUCCINI: Madama Butterfly: Un bel di vedremo.
Aida: Ritorna vincitor. cf Advent S 5023.
Aida: Ritorna vincitor. cf BASF 22-22645-3.
Aida: Ritorna vincitor. cf Club 99-100.
Aida: Ritorna vincitor; O patria mia. cf Club 99-109.
Aida: Se quel guerrier io fossi...Celeste Aida. cf Arias (Philips 6580 150).
Aida: Se quel guerrier io fossi...Celeste Aida. cf Works, selections (Philips 6833 223).
Aida: Se quel guerrier io fossi...Celeste Aida. cf HMV RLS 715.
Aida: Se quel guerrier...Celeste Aida. cf HMV ASD 3302.
Aida: What if tis I am chosen...Heavenly Aida. cf HMV HLP 7109.
Alzira: Da Gusman, su fragil barca. cf Works, selections (RCA AGL 1-1283).
2781 Arias: Aida: Se quel guerrier io fossi...Celeste Aida. Un ballo in maschera: Di tu se fedele. Ernani: Merce, diletti amici... Come rugiada. La forza del destino: La vita e inferno...Oh tu che in seno agli angeli. Luisa Miller; Oh, fede negar potessi ...Quando le sere al placido. Macbeth: O figli...Ah, la paterna mano. Otello: Dio, mio potevi scagliar; Niun mi tema. Rigoletto: Questa o quella; La donna e mobile. La traviata: Lunge de lei...De miei bollenti spiriti. Il trovatore: Il presagio funesto...Ah si, ben mio...Di quella pira. Carlo Bergonzi, t; NPhO; Nello Santi. Philips 6580 150 Tape (c) 7317 160. (Reissue from 6747 193)
+Gr 1-77 p1177 +HFN 10-77 p171 tape
+Gr 10-77 p706 tape +RR 1-77 p40
+HFN 1-77 p118
2782 Arias: Attila: Tregua e cogl unni-dagl immortali vertici. Un ballo in maschera: Alzati, la tuo figlio, Eri tu macchiavi quell

anima. Don Carlo: Son io mio Carlo, Per me giunto. Falstaff: E sogno, o realta. La forza del destino: Morir, tremenda cosa; Urna fatale del mio destino. Otello: Vanne, la tua meta gia vedo; Credo in un Dio credel. Rigoletto: Cortigiani vil razza danata. Il trovatore: Tutto e deserto; Il balen del suo sorriso. Ingvar Wixell, bar; Dresden State Opera Orchstra; Silvio Varviso. Philips 6580 171.
+-ARG 9-77 p40 +NR 8-77 p8
+-Gr 2-77 p1317 +-RR 4-77 p39
+HFN 3-77 p113 +SFC 5-8-77 p46

Aroldo: Ah, dagli scanni eterei. cf Works, selections (RCA AGL 1-1283).
Aroldo: Overture. cf Overtures (DG 2707 090).
Attila: Oh nel fuggente nuvolo. cf Works, selections (RCA AGL 1-1283).
Attila: Oh nel fuggente nuvolo. cf HMV SLS 5057.
Attila: Overture. cf Overtures (DG 2707 090).
Attila: Te sol quest'anima. cf RCA TVM 1-7203.
Attila: Te sol quest'anima. cf Seraphim S 60280.
Attila: Tregua e cogl unni-dagl immortali vertici. cf Arias (Philips 6580 171).
Ave Maria. cf L'Oiseau-Lyre SOL 345.

2783 Un ballo in maschera. Ferruccio Tagliavini, t; Giuseppe Valdengo, bar; Orchestra. Cetra LPS 3250. (Reissue from LPC 1250)
+-Gr 7-77 p221

Un ballo in maschera: Alzati la tuo figlio, Eri tu che macchiavi quell anima. cf Arias (Philips 6580 171).
Un ballo in maschera: Di tu se fedele. cf Arias (Philips 6580 150).
Un ballo in maschera: Ecco l'orrido campo. cf HMV SLS 5057.
Un ballo in maschera: Ma dall arido. cf Club 99-109.
Un ballo in maschera: Ma dall arido stelo divulsa; Morro, ma prima in grazia. cf Decca ECS 811.
Un ballo in maschera: Ma se m'e forza perderti. cf Philips 9500 203.
Un ballo in maschera: Morro, ma prima in grazia. cf Advent S 5023.
Un ballo in maschera: Morro, ma prima in grazia. cf HMV SLS 5104.
Un ballo in maschera: Morro, ma prima in grazia. cf Seraphim 60274.

2784 Un ballo in maschera: Posa in pace; Sire...Che leggo, il bando ad una donna; Ecco l'ordo campo; Alzati la tuo figlio...Eri tu; Ah, perche qui fuggite. Christina Deutekom, Patricia Hay, s; John Robertson, Charles Craig, t; Jan Derksen, bar; William McCue, Pieter van den Berg, bs; Scottish Opera Chorus; Scottish National Orchstra; Alexander Gibson. Classics for Pleasure CFP 40252.
+-Gr 1-77 p1174 +RR 1-77 p40
+HFN 12-76 p149

Un ballo in maschera: Prelude. cf Overtures (DG 2707 090).
Un ballo in maschera: Re dell abisso. cf Club 99-106.
Un ballo in maschera: Le rivedra nell'estas; Di tu se fedele. cf Decca SXL 6839.
La battaglia di Legnano (The battle of Legnano): Overture. cf Overtures (DG 2707 090).
The battle of Legnano: Overture. cf Overtures (HMV SQ ASD 3366).
Il Corsaro, excerpt. cf DONIZETTI: Lucia di Lammermoor, excerpts.

Il Corsaro: Non so le tetre immagini. cf Works, selections (RCA AGL 1-1283).
Il Corsaro: Overture. cf Overtures (DG 2707 090).
Don Carlo: Arias. cf London OS 26499.
Don Carlo: Duet, Act 2. cf Works, selections (Decca SPA 447).
Don Carlo: Ella giammai m'amo. cf GOMES: Salvator Rosa: Di sposo di padre le gioie serene.
Don Carlo: Monologue of Philip. cf Rubini GV 39.
Don Carlo: O don fatale. cf BASF 22-22645-3.
Don Carlo: O don fatale. cf Columbia/Melodiya M 33931.
Don Carlo: Per me giunto; Morro, ma lieta in corre. cf Rubini GV 34.
Don Carlo: Son io mio Carlo, Per me giunto. cf Arias (Philips 6580 171).
Don Carlo: Tu che le vanita. cf HMV SLS 5057.
I due Foscari, excerpts. cf DONIZETTI: Lucia di Lammermoor, excerpts.
I due Foscari: Tu al au sguardo onnimpossente. cf Works, selections (RCA AGL 1-1283).
Ernani: Che mai vegg'io...Infelice e tu credevi. cf GOMES: Salvator Rosa: Di sposo di padre le gioie serene.
Ernani: Merce, diletti amici...Come rugiada. cf Arias (Philips 6580 150).
Ernani: Prelude. cf Overtures (DG 2707 090).
Ernani: Surta e la notte. cf HMV SLS 5057.
Falstaff: O sogno, o realta. cf Arias (Philips 6580 171).

2785 La forza del destino. Leontyne Price, s; Fiorenza Cossotto, Gillian Knight, ms; Michel Senechal, Placido Domingo, t; Sherrill Milnes, Gabriel Bacquier, Malcolm King, bar; Bonaldo Giaiotti, Kurt Moll, bs; John Alldis Choir; LSO; James Levine. RCA ARL 4-1864 (4).

+ARG 6-77 p47
+Gr 8-77 p340
+-HF 7-77 p71
+HFN 10-77 p163
+MJ 7-77 p68
+-NR 4-77 p8
+-ON 3-12-77 p40
+RR 8-77 p41
++SFC 2-20-77 p39
++St 4-77 p84

La forza del destino: Arias. cf Free Lance FLPS 675.
La forza del destino: Al suon del tamburo, Rataplan. cf Club 99-106.
La forza del destino: Morir, tremenda cosa; Urna fatale del mio destino. cf Arias (Philips 6580 171).
La forza del destino: O tu che in seno. cf HMV ASD 3302.
La forza del destino: O tu che in seno agli angeli. cf Philips 9500 203.
La forza del destino: Overture. cf Works, selections (Decca SPA 447).
La forza del destino: Overture. cf Overtures (DG 2707 090).
La forza del destino (The force of destiny): Overture. cf Overtures (HMV ASD 3366).
La forza del destino: Overture. cf Works, selections (Philips 6833 223).
The force of destiny: Overture. cf Decca SXL 6782.
La forza del destino: Pace, pace, mio Dio. cf CATALANI: La Wally: Ebben, ne andro lontana.
La forza del destino: Pace, pace, mio Dio. cf DONIZETTI: Don Pasquale: So anch'io la virtu magica.
La forza del destino: Pace, pace, mio Dio. cf Advent S 5023.

La forza del destino: Pace, pace, mio Dio. cf HMV SLS 5104.
La forza del destino: Il santo nome di Dio. cf Simon Boccanegra.
La forza del destino: Solenne in quest'ora. cf PUCCINI: La boheme: O soave fanciulla; O mimi, tu piu non torni.
La forza del destino: Solenne in quest'ora. cf RCA TVM 1-7203.
La forza del destino: La vita e inferno...Oh tu che inseno agli angeli. cf Arias (Philips 6580 150).

2786 Un giorno di regno. Lina Pagliughi, s; Juan Oncina, t; Cristiano Dalamangas, Sesto Bruscantini; Orchestra; Alfredo Simonetto. Cetra LPS 3225. (Reissue from LPC 1225)
+-Gr 7-77 p221

Un giorno di regno: Grave a core innamorato. cf Works, selections (RCA AGL 1-1283).
Un giorno di regno: Overture. cf Overtures (DG 2707 090).
Giovanna d'Arco (Joan of Arc): Overture. cf Overtures (DG 2707 090).
Joan of Arc: Overture. cf Overtures (HMV ASD 3366).
Jerusalem: O mes amis. cf Philips 9500 203.
I Lombardi: La mia letizia infondere. cf Philips 9500 203.
I Lombardi: Non fu sogna. cf Works, selections (RCA AGL 1-1283).
I Lombardi: O madre, dal cielo...Se vano, se vano...No, no, giusta causa. cf PUCCINI: Madama Butterfly: Un bel di vedremo.
I Lombardi: O signore, dal tetto natio. cf Works, selections (Philips 6833 223).
I Lombardi: Qual volutta trascorrere. cf Seraphim S 60280.

2787 Luisa Miller. Montserrat Caballe, Annette Celine, s; Anna Reynolds, ms; Luciano Pavarotti, t; Sherrill Milnes, bar; Richard Van Allan, Bonaldo Giaiotti, bs; London Opera Chorus; National Philharmonic Orchestra; Peter Maag. London OSA 13114 (3) Tape (c) 5-13114 (r) OSAO 13114. (also Decca SET 606/8 Tape (c) K26K25)

+Gr 5-76 p1790
+Gr 8-76 p341 tape
++HF 9-76 p98
+HF 8-77 p96 tape
+HFN 5-76 p111
+HFN 8-76 p94 tape
+MT 11-76 p916
+ON 10-76 p72
+NR 11-76 p8
+-RR 5-76 p27
+RR 8-76 p82
+SFC 6-27-76 p29
+St 10-76 p136
+STL 6-6-76 p37

Luisa Miller: Arias. cf London OS 26499.
Luisa Miller: Oh, fede negar potessi...Quando le sere al placido. cf Arias (Philips 6580 150).
Luisa Miller: Overture. cf Overtures (DG 2707 090).
Luisa Miller: Overture. cf Overtures (HMV ASD 3366).
Luisa Miller: Overture. cf Works, selections (Philips 6833 223).
Luisa Miller: Quando le sere al placido. cf Works, selections (Decca SPA 447).
Luisa Miller: Quando le sere al placido. cf Philips 9500 203.

2788 Macbeth. Shirley Verrett, Stefania Malagu, ms; Placido Domingo, Antonio Savastano, t; Piero Cappuccilli, bar; Nicolai Ghiaurov, Carlo Zardo, Giovanni Foiani, Alfredo Mariotti, Sergio Fontana, bs; La Scala Orchestra and Chorus; Claudio Abbado. DG 2709 062 (3) Tape (c) 3371 022.

++ARG 5-77 p40
++Gr 10-76 p653
+Gr 12-76 p1066 tape
+-HF 1-77 p131
+HFN 10-76 p179
++MJ 12-76 p28
+MT 8-77 p653
++NR 12-76 p9
+ON 3-5-77 p33
+OR 6/7-77 p32
++RR 10-76 p32
++SFC 11-14-76 p30

2789 Macbeth. Fiorenza Cossotto, s; Maria Borgato, ms; Jose Carreras, Giuliano Bernardi, Leslie Fyson, t; Sherrill Milnes, John Noble, Neilson Taylor, bar; Ruggero Raimondi, Carlo Del Bosco, bs; Ambrosian Opera Chorus; NPhO; Riccardo Muti. HMV SQ SLS 992 (3) Tape (c) TC SLS 992. (also Angel SX 3833 Tape (c) 4X3S 3833)

+ARG 5-77 p40 · +NR 6-77 p10
++Gr 12-76 p1052 · +ON 3-5-77 p33
+-Gr 11-77 p909 tape · +OR 6/7-77 p32
+-HF 5-77 p92 · +RR 1-77 p41
+HFN 1-77 p117 · +SFC 2-20-77 p39
+MT 8-77 p653

Macbeth: Ballet music. cf Works, selections (Philips 6833 223).
Macbeth: La luce langue. cf BASF 22-22645-3.
Macbeth: La luce langue. cf HMV SLS 5104.
Macbeth: Nel di della vittoria. cf HMV SLS 5057.
Macbeth: Nel di della vittoria...Vieni t'affretta...Or tutti sorgete. cf Decca SXLR 6825.
Macbeth: O figli...Ah, la paterna mano. cf Arias (Philips 6580 150).
Macbeth: O figli miel...La paterna mano. cf Decca SXL 6839.
Macbeth: Patria opressa. cf Seraphim S 60275.
Macbeth: Prelude. cf Overtures (DG 2707 090).
Macbeth: Sleepwalking scene. cf PUCCINI: Madama Butterfly: Un bel di vedremo.
Macbeth: Viene t'affretta...Or tutti sorgete. cf London OS 26497.
I masnadieri: Prelude. cf Overtures (DG 2707 090).

2790 Nabucco. Caterina Mancini, s; Mario Binci, t; Paolo Silveri, bar; Orchestra. Cetra LPS 3216. (Reissue from LPC 1216)

+-Gr 7-77 p221

Nabucco, excerpts. cf Works, selections (Decca SPA 447).
Nabucco: Ben io t'invenni. cf HMV SLS 5057.
Nabucco: Chorus of the Hebrew salves. cf Virtuosi VR 7608.
Nabucco: Gli arredi festivi giu cadano infranti; Va, pensiero. cf Decca SXL 6826.
Nabucco: Overture. cf Overtures (DG 2707 090).
Nabucco: Overture. cf Overtures (HMV ASD 3366).
Nabucco: Overture. cf HMV ESD 7010.
Nabucco: Sperate, O figli...D'Egitto la sui lidi; Oh chi piange... Del futuro nel bujo. cf Simon Boccanegra.
Nabucco: Va pensiero. cf Works, selections (Philips 6833 223).
Nabucco: Va pensiero. cf Seraphim S 60275.
Nabucco: Va pensiero...O chi piange. cf Decca SPA 450.
Nabucco: Vieni o Levita...Tu sul labbro dei veggenti. cf GOMES: Salvator Rosa: Di sposo di padre le gioie serene.
Oberto, Conte de San Bonifacio: Overture. cf Overtures (DG 2707 090).

2791 Otello. Mirella Freni, s; Stefania Malagu, ms; Jon Vickers, Aldo Bottion, Michel Senechal, t; Peter Glossop, Hans Helm, bar; Jose van Dam, Mario Machi, bs; BPhO; Berlin Deutsche Oper Chorus; Herbert von Karajan. Angel SCLX 3809 (3). (also HMV SLS 975 Tape (c) TC SLS 975)

+Gr 10-74 p755 · +-NYT 10-9-77 pD21
+-Gr 1-75 p1402 tape · +-ON 12-28-74 p48
+-HF 1-75 p90 · +-RR 10-74 p28
+NR 12-74 p11 · +RR 3-75 p73 tape
-NYT 11-10-74 pD1 · +-St 12-74 p132

2792 Otello. Renata Tebaldi, s; Mario del Monaco, t; Aldo Protti, Tom Krause, bar; Fernando Corena, bs; Vienna State Opera Chorus; VPO; Herbert von Karajan. Decca D55D3 Tape (c) K2A21. (also London OSA 5-1324) (Reissue fom SET 209/11)
+Gr 8-76 p341 tape
+Gr 8-77 p345
+HF 3-77 p123
+- HFN 8-76 p94 tape
+HFN 9-77 p153
+-NYT 10-9-77 pD21
+RR 8-76 p82 tape
+-RR 8-77 p42

2793 Otello. Elisabeth Rethberg, s; Thelma Votipka, ms; Giovanni Martinelli, Alessio de Paolis, t; Lawrence Tibbett, George Cehanovsky, bar; Nicola Moscona, bs; Metropolitan Opera Orchestra; Ettore Panizza. MET 4. (From Metropolitan Opera)
+NYT 6-12-77 pD17
+ON 7-77 p29

2794 Otello. Leonie Rysanek, s; Miriam Pirazzini, ms; John Vickers, Florindo Andreolli, Mario Carlin, t; Tito Gobbi, Robert Kerns, bar; Ferruccio Mazzoli, Franco Calabrese, bs; Rome Opera House Orchestra and Chorus; Tullio Serafin. RCA Tape (c) RK 40001 (2).
+-RR 4-77 p94 tape

Otello: Ave Maria. cf Angel S 37446.
Otello: Dio, mio potevi scagliar; Niun mi tema. cf Arias (Philips 6580 150).
Otello: Fuoco di gioia. cf Seraphim S 60275.
Otello: Gia la notte, Ave Maria. cf PUCCINI: Madama Butterfly.
Otello: Gia nella notte densa. cf BELLINI: La sonnambula: Perdona o mia diletta...Prendi l'anel ti dono.
Otello: Niun mi tema. cf HMV ASD 3302.
Otello: Piangea cantando; Ave Maria. cf HANDEL: Il pensieroso: Sweet bird.
Otello: Piangea cantando; Ave Maria piena di grazia. cf HMV RLS 719.
Otello: Vanne, la tua met gia vedo; Credo in un Dio crudel. cf Arias (Philips 6580 171).
Otello: Una vela una vela. cf Decca SXL 6826.
Otello: Willow song. cf Works, selections (Decca SPA 447).
Otello: Willow song; Ave Maria. cf Decca D65D3.

2795 Overtures: Aroldo. Atilla. La battaglia di Legnano. Il Corsaro. La forza del destino. Giovanna d'Arco. Un giorno di regno. Luisa Miller. Nabucco. Oberto, Conte de San Bonifacio. I vespri siciliani. Preludes: Aida. Un ballo in maschera. Ernani. Macbeth. I masnadieri. Rigoletto. La traviata. BPhO; Herbert von Karajan. DG 2707 090 (2) Tape (c) 3370 090.
+ARG 2-77 p44
+FF 11-77 p55
+Gr 8-76 p305
+HFN 9-76 p131
+MJ 4-77 p33
+NR 4-77 p3
+ON 3-12-77 p40
+RR 9-76 p68
+SFC 12-26-76 p34
+St 9-77 p125
++STL 8-8-76 p29

2796 Overtures: Battle of Legnano. The force of destiny. Joan of Arc. Luisa Miller. Nabucco. Sicilian vespers. NPhO; Riccardo Muti. HMV SQ ASD 3366 Tape (c) TC ASD 3366. (also Angel S 37407)
+Gr 10-77 p648
+GR 11-77 p909 tape
+HFN 10-77 p163
+RR 10-77 p65

Pezzi sacri: Te deum. cf BRUCKNER: Te deum.
Quartet, strings, E minor. cf TCHAIKOVSKY: Sextet, op. 70, D minor.
Quartet, strings, E minor. cf TCHAIKOVSKY: Souvenir de Florence, op. 70.

2797 Requiem. Leontyne Price, s; Janet Baker, ms; Veriano Luchetti, t; Jose van Dam, bs; CSO and Chorus; Georg Solti. RCA RL 2-2476 (2) Tape (c) ARK 2-2476 (ct) ARS 2-2476.
++Gr 12-77 p1127 +-RR 12-77 p82
+-NR 12-77 p7 -SFC 10-23-77 p45

Requiem: Ingemisco. cf Decca SXL 6839.

2798 Rigoletto. Maria Callas, s; Giuseppe di Stefano, t; Tito Gobbi, bar; Orchestra. Angel 3537.
+NYT 10-9-77 pD21

2799 Rigoletto. Lina Pagliughi, s; Franco Tagliavini, t; Giuseppe Taddei, bar; Orchestra. Cetra LPS 3247. (Reissue from LPC 1247)
+Gr 7-77 p221

2800 Rigoletto. Renata Scotto, s; Carlo Bergonzi, t; Dietrich Fischer-Dieskau, bar; La Scala Orchestra and Chorus; Rafael Kubelik. DG 2709 014 Tape (c) 3371 001.
+HF 4-71 p68 +RR 4-74 p90
+NYT 10-9-77 pD21

2801 Rigoletto. Operatic arias: DELIBES: Lakme: Dov e l'Indiana bruna. MEYERBEER: Dinorah: Ombra leggiera. ROSSINI: Il barbiere di Siviglia: Una voce poco fa. VERDI: I vespri siciliani: Merce, dilette amiche. Maria Callas, Elvira Galassi, s; Adriana Lazzarini, Luisa Mandelli, ms; Giuseppe di Stefano, Renato Ercolani, t; Tito Gobbi, William Dickie, bar; Nicola Zaccaria, Plinio Clabassi, Carlo Forti, bs; La Scala Opera Orchestra and Chorus; PhO; Tullio Serafin. HMV SLS 5018 (3) Tape (c) TC SLS 5018. (Reissues from 33CXS 1324, 1325/6, 1231)
+-Gr 3-76 p1507 +-RR 2-76 p23
+-HFN 3-76 p107 +-RR 10-77 p99 tape
+HFN 10-77 p169 tape

2802 Rigoletto. Joan Sutherland, s; Huguette Tourangeau, ms; Luciano Pavarotti, t; Sherrill Milnes, bar; Martti Talvela, Clifford Grant, bs; Ambrosian Opera Chorus; LSO; Richard Bonynge. London OSA 13105 Tape (r) R 490225. (also Decca SET 542/4 Tape (c) K2A3)
++Gr 5-73 p2090 +NYT 10-9-77 pD21
+Gr 5-75 p2031 tape +-ON 6-73 p34
+-HF 6-73 p108 +-Op 7-73 p620
+HFN 5-73 p990 +-RR 5-73 p36
++HFN 6-75 p109 tape +-RR 5-75 p76 tape
+-LJ 2-75 p38 tape ++SFC 4-1-73 p30
+-RR 5-73 p36 ++SR 4-73 p72
+NR 5-73 p9 ++St 6-73 p81
/NYT 3-4-73 pD28

Rigoletto: Arias. cf London OS 26499.
Rigoletto: Ah veglia o donna; Si vendetta. cf Rubini GV 26.
Rigoletto: Caro nome. cf BELLINI: I puritani: Qui la voce...Vien diletto.
Rigoletto: Caro nome. cf DONIZETTI: Don Pasquale: So anch'io virtu magica.
Rigoletto: Caro nome. cf HANDEL: Il pensieroso: Sweet bird.
Rigoletto: Caro nome. cf Columbia 34294.
Rigoletto: Caro nome. cf Decca D65D3.
Rigoletto: Caro nome. cf Decca SPA 450.
Rigoletto: Caro nome. cf HMV MLF 118.
Rigoletto: Caro nome. cf HMV SLS 5104.
Rigoletto: Caro nome; Quartet. cf HMV RLS 719.

Rigoletto: Cortigiani vil razza danata. cf Arias (Philips 6580 171).
Rigoletto: La donna e mobile. cf Works, selections (Decca SPA 447).
Rigoletto: La donna e mobile. cf Works, selections (Philips 6833 223).
Rigoletto: La donna e mobile. cf Rubini GV 43.
Rigoletto: Duke's aria. cf MASCAGNI: Cavalleria rusticana: Duet of Santuzza and Turiddu, Turridu's aria.
Rigoletto: E il sol dell'anima. cf Court Opera Classics CO 342.
Rigoletto: Ella mi fu rapita...Parmi veder le lagrime. cf Rubini GV 29.
Rigoletto: Gualtier Malde...Caro nome. cf HMV SLS 5057.
Rigoletto: Gualtier Malde...Dearest name. cf HMV HLM 7066.
Rigoletto: Prelude. cf Overtures (DG 2707 090).
Rigoletto: Questa o quella; La donna e mobile. cf Arias (Philips 6580 150).
Rigoletto: Questa o quella; La donna e mobile. cf HMV RLS 715.
Rigoletto: Questa o quella; La donna e mobile. cf RCA CRM 1-1749.
Rigoletto: Questa o quella; Ella mi fu rapita...Parmi veder le lagrime. cf Decca SXL 6839.

2803 Simon Boccanegra. Mirella Freni, s; Jose Carreras, t; Piero Cappuccilli, Jose van Dam, bar; Nicolai Ghiaurov, bs; La Scala Orchestra and Chorus; Claudio Abbado. DG 2709 017 (3) Tape (c) 3371 032.
+Gr 11-77 p893 +NYT 12-11-77 pD17
+Gr 12-77 p1149 tape ++RR 11-77 p44
++HFN 12-77 p183

2804 Simon Boccanegra. Arias: BELLINI: Norma: Ite sul colle. VERDI: La forza del destino: Il santo nome di Dio. Nabucco: Sperate, O figli...D'Egitto la sui lidi; Oh chi piange...Del futuro nel bujo. Victoria de los Angeles, Silvia Bertona, s; Giuseppe Campora, Paolo Caroli, t; Tito Gobbi, Walter Monachesi, Paolo Dari, bar; Boris Christoff, bs; Rome Opera Orchestra and Chorus; Gabriele Santini, Vittorio Gui. HMV SLS 5090 (3) Tape (c) TC SLS 5090. (Reissues from ALPS 1634, ALP 1635/6, 1585)
+-Gr 8-77 p345 +-HFN 10-77 p169 tape
+HFN 10-77 p167 +-RR 9-77 p43

Simon Boccanegra: Come in quest ora bruna. cf BASF 22-22645-3.
Simon Boccanegra: Il lacerato spirito. cf GOMES: Salvator Rosa: Di sposo di padre le gioie serene.
Songs: Ave Maria; Pater noster. cf Argo ZRG 871.
Songs: Chu chin chow: Cobblers song; The mighty deep, The windmill, My old shako; Flotsam and jetsam; What's the matter with Rachmaninoff; Schubert's toy shop, Polonaise in the mall, Simon the bootlegger, Song of the air, Only a few of us left, The Alsatian and the Pekingese. cf Club 99-102.
Songs: Perduta ho la pace; Ad una stella; Stornello; Lo spazzacamino. cf BELLINI: Songs (Westminster WG 1014).
Tosca: Vissi d'arte. cf HMV SLS 5073.

2805 La traviata. Joan Sutherland, s; Carlo Bergonzi, Piero de Palma, t; Robert Merrill, bar; Maggio Musicale Orchestra and Chorus; John Pritchard. Decca Tape (c) K19K32.
+HFN 12-76 p153 tape +-RR 2-77 p99 tape

2806 La traviata. Ileana Cotrubas, Helena Jungwirth, s; Stefania Malagu, ms; Placido Domingo, Walter Gullino, t; Sherrill Milnes, Bruno Grella, bar; Alfredo Giacometti, Giovanni Fioani, bs;

Bavarian State Opera Orchestra and Chorus; Carlos Kleiber. DG 2707 103 (3).
+Gr 11-77 p894

2807 La traviata. Victoria de los Angeles, Santa Chissari, Silvia Bertona, s; Carlo del Monte, Sergio Tedesco, Renato Ercolani, t; Mario Sereni, Vico Polotto, bar; Silvio Maionica, Bonaldo Giaiotti, bs; Rome Opera Orchestra and Chorus; Tullio Serafin. HMV 5097 (3). (Reissue from ASD 359/61)
+Gr 11-77 p894 +-RR 12-77 p36
+HFN 12-77 p185

2808 La traviata. Pilar Lorengar, s; Dietrich Fischer-Dieskau, bar; Berlin German Opera Orchestra; Lorin Maazel. London OSA 1279.
+NYT 10-9-77 pD21

2809 La traviata. Anna Moffo, s; Richard Tucker, t; Robert Merrill, bar; Rome Opera Orchestra; Fernando Previtali. RCA LSC 6154 Tape (c) ARK 3-6154.
+NYT 10-9-77 pD21

2810 La traviata, highlights. Maria Cebotari, s; Helge Roswange, t; Heinrich Schlusnus, bar; Berlin State Opera Orchestra and Chorus; Hans Steinkopf. BASF 10 21498-2.
+NR 4-77 p9

La traviata, excerpts. cf Philips 6747 327.

La traviata: Addio del passato. cf CATALANI: La Wally: Ebben, ne andro lontana.

La traviata: Ah, fors'e lui. cf Decca SDD 507.

La traviata: Ah, fors'e lui...Follie. cf Court Opera Classics CO 342.

La traviata: Ah, fors'e lui...Sempre libera. cf BELLINI: I puritani: Qui la voce...Vien diletto.

La traviata: Ah, fors'e lui...Sempre libera. cf Columbia 34294.

La traviata: Ah fors'e lui...Sempre libera. cf HMV MLF 118.

La traviata: Ah, fors'e lui...Sempre libera. cf Seraphim 60274.

La traviata: Ah, fors'e lui; Sempre libera; Dite alla giovine. cf HMV RLS 719.

La traviata: Alfred's aria. cf MASCAGNI: Cavalleria rusticana: Duet of Santuzza and Turiddu, Turridu's aria.

La traviata: De miei bollenti spiriti. cf RCA TVM 1-7203.

La traviata: Di provenza il mar. cf Golden Age GAR 1001.

La traviata: Dita alla giovine. cf HANDEL: Il pensieroso: Sweet bird.

La traviata: E strano...Ah fors'e lui. cf Decca D65D3.

La traviata: E strano...Ah, fors'e lui...Follie Follie. cf Angel S 37446.

La traviata: E strano...Ah, fors'e lui...Follie, Follie...Sempre libera. cf BELLINI: I Capuleti ed i Montecchi: Eccomi in lieta vesta...Oh quante volte.

La traviata: Libiamo ne'lieti calici; Un di felice. cf Rubini GV 29.

La traviata: Libiamo ne lieti calici; Un di felice; Signora... Che t'accadde...Parigi, O cara. cf BELLINI: La sonnambula: Perdona o mia dilletta...Prendi l'anel ti dono.

La traviata: Lunge de lei...De miei bollenti spiriti. cf Arias (Philips 6580 150).

La traviata: Prelude. cf Overtures (DG 2707 090).

La traviata: Prelude, Act 1. cf Decca SPA 450.

La traviata: Prelude, Act 1. cf Decca SPA 491.

La traviata: Prelude, Act 1; Di provenza il mar. cf Works, selections (Decca SPA 447).
La traviata: Prelude, Act 1; E strano...Ah, fors'e lui...Sempre libera. cf Works, selections (Philips 6833 223).
La traviata: Tis wondrous...Ah was it he...What folly. cf HMV HLM 7066.
La traviata: Un di felice. cf Rubini GV 26.

2811 Il trovatore. Raina Kabaivanska, s; Viorica Cortez, ms; Franco Bonisolli, t; Gisela Pohl, con; Giancarlo Luccardi, bs; Giorgio Zancanaro, Johannes Bier; Deutsche Staatsoper Orchestra and Chorus; Bruno Bartoletti. Ariola-Eurodisc 28 169 XFR (3).
+ARG 9-77 p46

2812 Il trovatore. Joan Sutherland, Norma Burrowes, s; Marilyn Horne, ms; Luciano Pavarotti, Graham Clark, Wynford Evans, t; Ingvar Wixell, bar; Nicolai Ghiaurov, Peter Knapp, bs; London Opera Chorus; National Philharmonic Orchestra; Richard Bonynge. Decca D82D3 (3). (also London 13124 Tape (c) 5-13124)
+-Gr 10-77 p697 +-RR 10-77 p31
+-HFN 10-77 p165 +SFC 9-18-77 p42
+NYT 10-9-77 pD21 +St 12-77 p152

2813 Il trovatore. Zinka Milanov, s; Fedora Barbieri, Margaret Roggero, ms; Jussi Bjorling, Nathaniel Sprinzena, t; Leonard Warren, George Cehanovsky, bar; Nicola Moscona, bs; Robert Shaw Chorale; RCA Orchestra; Renato Cellini. RCA AVM 2-0699 (2). (Reissue from LM 6008)
+-HF 4-75 p92 ++NYT 10-9-77 pD21
+NR 5-75 p10

2814 Il trovatore. Leontyne Price, s; Fiorenza Cossotto, Elizabeth Bainbridge, ms; Placido Domingo, Ryland Davies, t; Sherrill Milnes, Neilson Taylor, bar; Bonaldo Giaiotti, bs; Ambrosian Opera Chorus; NPhO; Zubin Mehta. RCA Tape (c) RK 40002 (2).
+RR 4-77 p94 tape

Il trovatore: Arias. cf Free Lance FLPS 675.
Il trovatore: Arias. cf London OS 26499.
Il trovatore: Ah si ben mio, coll'essere io tuo; Di quella pira. cf HMV RLS 715.
Il trovatore: Ai nostri monti. cf Club 99-106.
Il trovatore: Anvil chorus; Stride la vampa. cf Works, selections (Decca SPA 447).
Il trovatore: Che piu t'arresti...Tacea la notte...Di tale amor. cf Decca SXLR 6825.
Il trovatore: D'amor sull ali rosee. cf BASF 22-22645-3.
Il trovatore: D'amor sull ali rosee. cf Club 99-109.
Il trovatore: Di quella pira. cf RCA CRM 1-1749.
Il trovatore: Di quella pira. cf Seraphim S 60280.
Il trovatore: Duet of Azucena and Manrico. cf MASCAGNI: Cavalleria rusticana: Duet of Santuzza and Turiddu, Turridu's aria.

2815 Il trovatore: Fled was that golden vision; Anvil chorus; Fierce flames are raging; All the stars that shine above us; Finale; As from the dread pyre; Miserere; See all the bitter tears I shed; Home to our mountains; Ah, a light is glimm'ring; Rather than live. Elizabeth Fretwell, Rita Hunter, s; Patricia Johnson, ms; Charles Craig, t; Peter Glossop, Donald McIntyre, bar; Sadler's Wells Opera Orchestra and Chorus; Michael Moores. HMV ESD 7027 Tape (c) TC ESD 7027. (Reissue from CSD 1440)
+Gr 1-77 p1174 +HFN 9-77 p155 tape
+HFN 3-77 p118 +RR 1-77 p42
+HFN 5-77 p138 tape +RR 7-77 p100 tape

Il trovatore: Mira, d'acerbe; Vivra, contende. cf Rubini GV 34.
Il trovatore: Il presagio funesto...Ah si, ben mio...Di quella pira. cf Arias (Philips 6580 150).
Il trovatore: Stride la vampa. cf Club 99-103.
Il trovatore: Tacea la notte: D'amor sull'ali rosee. cf Odyssey Y 33793.
Il trovatore: Tacea la notte placida...Di tale amor. cf HMV SLS 5104.
Il trovatore: Tacea la notte...Di tale amor. cf London OS 26497.
Il trovatore: Timor di me...D'amor sull'ali rosee. cf CATALANI: La Wally: Ebben, ne andro lontana.
Il trovatore: Timor di me...D'amor sull'ali rosee. cf HMV SLS 5057.
Il trovatore: Tutto e deserto; Il balen del suo sorriso. cf Arias (Philips 6580 171).
Il trovatore: Vedi, le fosche notturne. cf Seraphim S 60275.
I vespri siciliani (Sicilian vespers): Bolero. cf Works, selections (Decca SPA 447).
I vespri siciliani: Merce, dilette amiche. cf Rigoletto.
I vespri siciliani: Merce, dilette amiche; Arrigo, Ah parli a un core. cf HMV SLS 5057.
I vespri siciliani: O tu palermo. cf Club 99-102.
I vespri siciliani: Overture. cf Overtures (DG 2707 090).
Sicilian vespers: Overture. cf Overtures (HMV ASD 3366).

2816 Works, selections: Aida: Celeste Aida. Don Carlo: Duet, Act 2. La forza del destino: Overture. Luisa Miller: Quando le sere al placido. Nabucco, excerpts. Otello: Willow song. Rigoletto: La donna e mobile. La traviata: Prelude, Act 1; Di provenza il mar. Il trovatore: Anvil chorus; Stride la vampa. I vespri siciliani: Bolero. Joan Sutherland, Marilyn Horne, s; Luciano Pavarotti, t; Maria Chiara; Various orchestras and conductors. Decca SPA 447 Tape (c) KCSP 447.
+-HFN 4-77 p153 +-RR 4-77 p39
+-HFN 5-77 p138 tape ++RR 7-77 p100 tape

2817 Works, selections: Aida: Se quel guerrier io fossi...Celeste Aida. La forza del destino: Overture. I Lombardi: O signore, dal tetto natio. Luisa Miller: Overture. Macbeth: Ballet music. Nabucco: Va, pensiero. Rigoletto: La donna e mobile. La traviata: Prelude, Act 1; E strano...Ah, fors'e lui...Sempre libera. Cristina Deutekom, s; Carlo Bergonzi, t; Various orchestras and choruses; Horst Kegel, Lamberto Gardelli, Antonio de Almeida, Nello Santi, Igor Markevitch, Carlo Franci. Philips 6833 223.
+HFN 11-77 p185 -RR 11-77 p44

2818 Works, selections: Alzira: Da Gusman, su fragil barca. Aroldo: Ah, dagli scanni eterei. Attila: Oh nel fuggente nuvolo. Il Corsaro: Non so le tetre immagini. I due Foscari: Tu al au sguardo onnimpossente. Un giorno di regno: Grave a core innamorato. I Lombardi: Non fu sogna. Montserrat Caballe, s; RCA Italiano Opera Orchestra and Chorus; Anton Guadagno. RCA AGL 1-1283.
++FF 9-77 p87

VERRALL, John
Sonata, horn and piano. cf STEVENS: Sonata, horn and piano.

VERSCHRAEGEN, Gabriel
Partita octavi toni super Veni Creator. cf Vista VPS 1060.

VICTORIA, Tomas de
O magnum mysterium. cf Vanguard SVD 71212.
Songs: O magnum mysterium; Senex puerum portabat. cf HMV ESD 7050.

VIDAL, Peire
Baron de mon dan covit. cf Telefunken 6-41126
VIENNESE DANCE MUSIC FROM THE CLASSICAL PERIOD. cf DG 2533 182.
VIENNESE OVERTURES. cf Classics for Pleasure CFP 40236.
VIERNE, Louis
Berceuse. cf Odeon HQS 1356.
Canzona. cf St. John the Divine Cathedral Church.
Carillon de Westminster. cf Decca PFS 4416.
Carillon de Westminster. cf Wealden WS 159.
2819 Impromptu. Symphony, no. 3, op. 28, F minor. Toccata. TOURNEMIRE: Messe de l'assomption: Paraphrase carillon. Andre Isoir, org. Calliope CAL 1923.
+Gr 3-77 p1454
Impromptu. cf Argo ZRG 864.
Impromptu. cf Calliope CAL 1922/4.
Mass, C sharp minor. cf MESSIAEN: O sacrum convivum.
Naiades. cf CHARPENTIER: L'Ange a la trompette.
2820 Pieces de faintaisie (24). Gaston Litaize, org. Connoisseur Society CS 2-2119 (2).
+MJ 5-77 p32 +NR 4-77 p14
Pieces de fantaisie, op. 53: Feux follets. cf CHARPENTIER: L'ange a la trompette.
Pieces in style libre, op. 31: Cortege; Berceuse; Divertissement; Le carillon de Longpont. cf Symphony, organ, no. 3, op. 28, F minor.
Pieces in style libre, op. 31, no. 5: Prelude. cf Vista VPS 1034.
Symphony, no. 1, op. 14, D minor: Finale. cf Argo SPA 507.
2821 Symphony, no. 3, op. 28, F minor. Pieces en style libre, op. 31: Cortege; Berceuse; Divertissement; Le carillon de Longpont. Arthur Wills, org. Saga 5456.
+-Gr 11-77 p867 +-RR 11-77 p97
Symphony, no. 3, op. 28, F minor. cf Impromptu.
Symphony, no. 3, op. 28, F minor. cf Calliope CAL 1922/4.
Symphony, no. 3, op. 28, F minor. cf Vista VPS 1051.
Toccata. cf Impromptu.
Toccata. cf Calliope CAL 1922/4.
Toccata, B flat major. cf CHARPENTIER: L'ange a la trompette.
VIEUXTEMPS, Henri
Concerto, violin, no. 5, op. 37, A minor. cf SAINT-SAENS: Concerto, violin, no. 3, op. 61, B minor.
VILLA-LOBOS, Heitor
2822 Bachianas brasileiras, no. 4: Prelude. Prole do bebe, Bk I. Rudepoema. The three Marias. Nelson Freire, pno. Telefunken SAT 22547.
+Gr 9-74 p553 ++RR 8-74 p55
+MJ 7-77 p70 ++St 9-75 p115
++NR 6-75 p13
Bachianas brasileiras, no. 5. cf CANTELOUBE: Songs of the Auvergne.
Carnaval das criancas. cf Piano works (DG 2530 634).
Choro typico. cf LAURO: Danza negra.
Choros, no. 1, E major. cf Saga 5412.
Cirandinha, no. 1: Therezinha de Jesus. cf Saga 5412.
Cirandinha, no. 10: A canoa virou. cf Saga 5412.
Concerto, guitar. cf CASTELNUOVO-TEDESCO: Concerto, guitar, no. 1, op. 99, D major.
2823 Concerto, guitar and small orchestra. Mystic sextet. Preludes (5). Turibio Santos, gtr; Maxence Larrieu, flt; Lucien Debray,

ob; Henri-Rene Pollin, sax; Francois-Joel Thiollier, celesta; Lily Laskine, hp; Jean-Francois Paillard Chamber Orchestra; Jean-Francois Paillard. Musical Heritage Society MHS 3397.
+St 4-77 p134
Dance of seven notes. cf MILHAUD: Four seasons: Concertino d'hiver, op. 327.
Duo, oboe and bassoon. cf POULENC: Sonata, clarinet and bassoon.
2824 Etudes, guitar. Preludes (5). Suite popular brasileira. Eric Hill, gtr. Saga 5453.
++RR 11-77 p97
Etude, guitar, no. 1. cf Philips 6833 159.
Fantaisie concertante. cf POULENC: Sonata, clarinet and bassoon.
Festa no sertao. cf BARTOK: Suite, op. 14.
A fiandeira. cf Piano works (DG 2530 634).
Impressoes seresteiras. cf BARTOK: Suite, no. 14.
A lendo do caboclo. cf Piano works (DG 2530 634).
Mystic sextet. cf Concerto, guitar and small orchestra.
New York skyline (1957 version). cf Piano works (DG 2530 634).
2825 Piano works: A fiandeira. A lenda do caboclo. Carnaval das criancas. New York skyline (1957 version). Rudepoema. Saudades das selvas brasileiras. Suite floral, op. 97 (1949 revision). Roberto Szidon, Richard Metzler, pno. DG 2530 634 Tape (c) 3300 634.
++ARG 6-77 p49 +MJ 5-77 p32
+-Gr 10-76 p628 ++NR 8-77 p15
+-HF 5-77 p94 ++RR 9-76 p84
+HFN 10-76 p179 +RR 1-77 p92 tape
Preludes (5). cf Concerto, guitar and small orchestra.
Preludes (5). cf Etudes, guitar.
Preludes (5). cf SCARLATTI, D.: Sonatas, harpsichord.
Preludes, nos. 1-5. cf LAURO: Danza negra.
Preludes, no. 1, E minor. cf L'Oiseau-Lyre SOL 349.
Preludes, no. 1, E minor. cf Westminster WG 1012.
Preludes, no. 2, E major. cf CBS 72526.
Preludio, no. 3. cf Lyrichord LLST 7299.
Preludes, no. 4, E minor. cf CBS 72526.
Preludes, no. 5, D major. cf L'Oiseau-Lyre SOL 349.
A prole do Bebe, Bk I. cf Bachianas brasileiras, no. 4: Prelude.
Rudepoema. cf Bachianas brasileiras, no. 4: Prelude.
Rudepoema. cf Piano works (DG 2530 634).
Saudades das selvas brasileiras. cf Piano works (DG 2530 634).
Song of the black swan. cf Laurel-Protone LP 13.
Suite floral, op. 97. cf Piano works (DG 2530 634).
Suite popular brasileira. cf Etudes, guitar.
The three Marias. cf Bachianas brasileiras, no. 4: Prelude.

VILLASENOR
Paisaje. cf Marc MC 5355.

VINTER, Gilbert
Entertainments: Elegy. cf Grosvenor GRS 1055.
Portuguese party. cf Decca SB 328.
Salute to youth. cf Decca SB 329.

VIOTTI, Giovanni
Concerto, flute, A major. cf DEVIENNE: Symphonie concertante, G major.
Concerto, piano, G minor. cf PUCCINI, D.: Concerto, piano, B flat major.
Duets, 2 violins, op. 29, nos. 2 and 3. cf PAGANINI: Sonatas, violin and guitar, op. 3, nos. 1-6.

VIRGILIO
Jana: Morte di Jane. cf Club 99-100.
VISEE, Robert de
Giga. cf Angel S 36053.
Pascalle. cf Lyrichord LLST 7299.
Suite, D minor. cf d'ANGLETERRA: Carillon, G major.
Suite, G major. cf BACH: Suite, lute, S 997, C minor.
Suite for guitar. cf GIULIANI: Sonata, flute and guitar, op. 85, A major.
VISETTI
Diva, the Adelina Patti waltz. cf Decca SDD 507.
VITALI, Giovanni
Chaconne. cf CORELLI: La folia.
VITRY, Philippe de
Cum statua. cf DG 2710 019.
Impudenter. cf DG 2710 019.
VIVALDI, Antonio
2826 Beatus vir. Gloria. Mary Burgess, Jocelyne Chamonin, s; Carolyn Watkinson, con; Yves Poucel, baroque ob; Daniele Salzer, org; Raphael Passaquet Vocal Ensemble; Le Grande Ecurie et la Chambre du Roy; Jean-Claude Malgoire. CBS 76596 Tape (c) 40-76596.
-Gr 10-77 p685 +HFN 11-77 p187 tape
-Gr 11-77 p900 tape +RR 10-77 p86
+HFN 10-77 p165 +RR 12-77 p99 tape
2827 Beatus vir. Credo. Gloria. Lauda Jerusalem. Melinda Lugosi, s; Katalin Szokefalvi-Nagy, Maria Zempleni, s; Klara Takacs, con; Janos Rolla, vln; Budapest Madrigal Choir; Liszt Academy Chamber Orchestra; Ferenc Szekeres. Hungaroton SLPX 11695.
++HF 3-77 p114 +NR 10-76 p9
2828 Beatus vir. Invicti bellate. Kyrie. Melinda Lugosi, s; Klara Takacs, ms; Zsolt Bende, bar; Budapest Madrigal Choir; Liszt Academy Chamber Orchestra; Ferenc Szekeres. Hungaroton SLPX 11830.
/NR 7-77 p8 +-RR 5-77 p85
2829 Il cimento dell'armonia e dell'invenzione, op. 8. I Solisti Veneti; Claudio Scimone. Erato STU 70680 (3).
+Gr 4-77 p1562 +-MT 11-77 p924
+Gr 12-72 p1157 -RR 11-72 p75
++HFN 4-77 p149
2830 Il cimento dell'armonia e dell'invenzione, op. 8. Felix Ayo, vln; I Musici. Philips 6747 311 (3).
+ARG 8-77 p32 +NR 7-77 p7
2831 Concerti, op. 3, nos. 6, 8-10. Henryk Szeryng, Gerard Poulet, Claire Bernard, Maurice Hasson, vln; ECO; Henryk Szeryng. Philips 9500 158.
+-Gr 2-77 p1295 ++RR 2-77 p65
+HFN 2-77 p133
2832 Concerti, op. 4. I Solisti Veneti; Claudio Scimone. Erato STU 70955 (2).
+-Gr 3-77 p1410 +-MT 11-77 p924
+HFN 4-77 p149 +RR 3-77 p63
2833 Concerto, bassoon, A minor. Concerto, flute, C minor. Concerto, oboe, F major. Concerto, 2 oboes, bassoon, 2 horns and violin, F major. Neil Black, Celia Nicklin, ob; Martin Gatt, bsn; Timothy Brown, Robin Davis, hn; Iona Brown, vln; William Bennett, flt; AMF; Neville Marriner. Argo ZRG 839.

+Gr 11-77 p853 +RR 11-77 p74
+HFN 11-77 p183

2834 Concerto, bassoon and strings, E minor. Concerto, recorder and strings, C minor (ed. Giegling). Concerto, flute and strings, D major (ed. Negri). Concerto, oboe, A minor (ed. Negri). Leo Driehuys, ob; Marco Constantini, bsn; Severino Gazzelloni, flt; I Musici. Philips 6580 152.
+Gr 2-77 p1290 +RR 11-76 p85
+HFN 11-76 p171

Concerto, bassoon, flute, oboe and violin, C major. cf Works, selections (Odyssey Y 34614).

Concerto, bassoon, flute and violin, D minor. cf Works, selections (Odyssey Y 34614).

Concerto, bassoon, flute, oboe and violin, G minor. cf Works, selections (Odyssey Y 34614).

Concerto, flute, C minor. cf Concerto, bassoon, A minor.

2835 Concerti, flute, op. 10, nos. 1-6. Claude Veilhan, rec; La Grande Ecurie et la Chambre du Roy; Jean-Claude Malgoire. CBS 76595.
+-Gr 5-77 p1703 +-MT 9-77 p738
++HFN 5-77 p135 +RR 5-77 p61

2836 Concerti, flute, op. 10, nos. 1-6. Stephen Preston, rec; Academy of Ancient Music; Christopher Hogwood. L'Oiseau-Lyre DSLO 519.
+Gr 4-77 p1562 +NR 8-77 p6
+HFN 4-77 p149 +RR 3-77 p63
+-MT 9-77 p738 +St 12-77 p156

2837 Concerto, flute, P 79, C major. Concerto, viola d'amore, lute and strings, P 266, D minor. Concerto, 2 violins, harpsichord and strings, P 222, A major. Concerto, violoncello and strings, E minor. Pierre Fournier, vlc; Walter Prystawski, Herbert Hover, vln; Hans-Martin Linde, rec; Monique Frasca-Colombier, vla d'amore; Narciso Yepes, gtr; Lucerne Festival Strings, Emil Seiler Chamber Orchestra, Paul Kuentz Chamber Orchestra; Rudolf Baumgartner, Wolfgang Hofmann, Paul Kuentz. DG 2535 200. (Reissues from SLPM 138986, 138947, 2530 211, SAPM 198318)
+-Gr 1-77 p1153

2838 Concerto, 4 flutes, 4 violins, 2 organs and strings, P 226, A major. Concerto, oboe, violin, organ, harpsichord and strings, P 36, C major. Concerto, violin, organ, harpsichord and strings, P 274, F major. Concerto, violin, organ, harpsichord and strings, P 311, D minor. Monique Frasca-Colombier, vln; Andre Isoir, positive organ; Michel Giboureau, ob; Paul Kuentz Chamber Orchestra; Paul Kuentz. DG 2530 652 Tape (c) 3300 652.
++ARG 4-77 p27 +NR 4-77 p3
+-Gr 8-76 p305 +RR 7-76 p65
+-Gr 10-76 p658 tape +RR 12-76 p110 tape
+HFN 8-76 p91 +St 5-77 p120

Concert, flute and bassoon, op. 10, no. 2, G minor. cf RCA RK 2-5030.

Concerto, flute and strings, D major. cf Concerto, bassoon and strings, A minor.

Concerto, flute and violin, D major. cf Works, selections (Odyssey Y 34614).

Concerti, guitar, C major, D major. cf CARULLI: Concerto, guitar, A major.

Concerto, guitar and strings, D major. cf FASCH: Concerto, guitar and strings, D minor.

Concerto, 2 mandolins: Adagio. cf L'Oiseau-Lyre SOL 349.

2839 Concerti, oboe, P 42, A minor; P 41, C major; P 50, C major; op. 11, no. 6, G minor. Heinz Holliger, ob; I Musici. Philips 9500 044 Tape (c) 7300 443.
++Gr 6-76 p57 ++RR 5-76 p59
++HFN 5-76 p113 ++St 5-77 p119
++MJ 4-77 p50 +STL 7-4-76 p36
++NR 4-77 p4

Concerto, oboe, A minor. cf Concerto, bassoon and strings, E minor.

Concerto, oboe, F major. cf Concerto, bassoon, A minor.

Concerto, oboe, violin, organ, harpsichord and strings, P 36, C major. cf Concerto, 4 flutes, 4 violins, 2 organs and strings, P 226, A major.

Concerto, 2 oboes, bassoon, 2 horns and violin, F major. cf Concerto, bassoon, A minor.

2840 Concerto, orchestra, G minor. Nisi dominus. Stabat mater. James Bowman, c-t; Academy of Ancient Music; Christopher Hogwood. L'Oiseau-Lyre DSLO 506.
+Gr 2-77 p1311 +-MT 4-77 p308
+-HF 10-77 p124 -NR 9-77 p11

Concerto, piccolo and strings, C major. cf ALBINONI: Adagio.

Concerto, piccolo and strings, C major. cf ALBINONI: Adagio, organ and strings, G minor.

Concerto, piccolo and strings, C major. cf MOZART: Symphony, no. 35, K 385, D major.

Concerto, recorder and strings, C minor. cf Concerto, bassoon and strings, E minor.

Concerto, trumpet, D major. cf Philips 6581 018.

Concerto, 2 trumpets, P 75. cf Argo D69D3.

Concerto, 2 trumpets and strings, C major. cf DG 2530 792.

2841 Concerti, viola d'amore, P 166, D major; P 37, A minor; P 287, D minor; P 233, A major; P 288, D minor; P 289, D minor; P 286, F major; P 266, D minor. Orlando Cristoforetti, lt; Nane Calabrese, vla d'amore; I Solisti Veneti; Claudio Scimone. Erato STU 70826/7 (2). (Reissue)
+-Gr 2-75 p1503 ++RR 2-75 p45
+Gr 3-77 p1415 +RR 3-77 p62
+HFN 5-77 p137 ++STL 2-9-75 p37
+-MT 11-77 p924

Concerto, viola d'amore, A major. cf PORPORA: Concerto, violoncello, G major.

Concerto, viola d'amore, lute and strings, P 266, D minor. cf Concerto, flute, P 79, C major.

Concerto, violin, op. 4, no. 1, B flat major. cf Argo D69D3.

2842 Concerti, violin, op. 1, nos. 1-12 (La cetra). I Solisti Veneti; Claudio Scimone. Erato STU 70897 (2).
++Gr 4-77 p1562 +-MT 11-77 p924
+HFN 4-77 p147

2843 Concerti, violin (oboe) and strings, op. 7 (12). Heinz Holliger, ob; Salvatore Accardo, vln; I Musici. Philips 6700 100 (2).
+ARG 11-76 p44 +NR 1-77 p6
++Gr 11-76 p823 ++RR 9-76 p75
+-HF 12-76 p118 ++SFC 9-26-76 p29
+HFN 9-76 p131 +St 1-77 p130
+MJ 1-77 p29

Concerto, violin, organ, harpsichord and strings, P 274, F major. cf Concerto, 4 flutes, 4 violins, 2 organs and strings, P 226, A major.

Concerto, violin, organ, harpsichord and strings, P 311, D minor. cf Concerto, 4 flutes, 4 violins, 2 organs and strings, P 226, A major.
Concerto 2 violins, op. 3, no. 8, A minor. cf Argo D69D3.
2844 Concerto, 2 violins and strings, P 436, C minor. Concerto, violoncello, P 282, D minor. L'Estro armonico, op. 3: Concerto, violin, no. 6, A minor; Concerto, 2 violins and strings, no. 8, A minor. Bohdan Warchal, Viliam Dobrucky, Anna Holblingova, Quido Holbling, vln; Juraj Alexander, vlc; Slovak Chamber Orchestra; Bohdan Warchal. Rediffusion Royale ROY 2006.
+-Gr 9-77 p438 ++RR 8-77 p64
Concerto, 2 violins, harpsichord and strings, P 222, A major. cf Concerto, flute, P 79, C major.
Concerto, 4 violins, op. 3, no. 10. cf Argo D69D3.
Concerti, violoncello, C minor, G minor, G major. cf Bruno Walter Society IGI 323.
Concerto, violoncello, P 282, D minor. cf Concerto, 2 violins and strings, P 436, C minor.
2845 Concerti, violoncello and strings, G major, A minor, G minor, A minor. Christine Walevska, vlc; Netherlands Chamber Orchestra; Kurt Redel. Philips 9500 144.
+ARG 9-77 p47 +RR 7-77 p62
+-Gr 8-77 p310 ++SFC 6-26-77 p46
+HFN 7-77 p123 ++St 10-77 p148
++NR 8-77 p6
Concerto, violoncello and strings, E minor. cf Concerto, flute, P 79, C major.
Concerto per ottavino, P 79. cf BIS LP 50.
Credo. cf Beatus vir.
2846 L'Estro armonico, op. 3. I Solisti Veneti; Claudio Scimone. Erato STU 70753/5 (3).
+-Gr 6-74 p65 +-MT 11-77 p924
+-Gr 3-77 p1410 +RR 12-73 p78
+HFN 4-77 p148 ++RR 3-77 p62
L'Estro armonico, op. 3: Concerto, violin, no. 6, A minor; Concerto, 2 violins and strings, no. 8, A minor. cf Concerto, 2 violins and strings, P 436, C minor.
L'Estro armonico, op. 3, no. 11, D minor. cf BACH: Cantata, no. 147, Jesu, joy of man's desiring.
L'Estro armonico, op. 3, no. 11, D minor: Largo. cf HMV RLS 723.
La fida ninfa. cf Golden Age GAR 1001.
2847 The four seasons (Concerti violin), op. 8, nos. 1-4. Itzhak Perlman, vln; LPO; Itzhak Perlman. Angel S 37053 Tape (c) 4XS 37053 (ct) 8XS 37053. (also HMV ASD 3293 Tape (c) TC ASD 3293)
+ARG 4-77 p27 +-NR 4-77 p3
+-Gr 2-77 p1290 +NYT 2-24-77 pD11
+-Gr 4-77 p1603 tape +RR 2-77 p65
+-HFN 3-77 p114 +-SFC 1-2-77 p34
+-HFN 6-77 p137 tape
2848 The four seasons, op. 8, nos. 1-4. New Tokyo Koto Ensemble. Angel S 37450 Tape (c) 4XS 37450 (ct) 8XS 37450.
+SFC 11-27-77 p66
2849 The four seasons, op. 8, nos. 1-4. AMF; Neville Marriner. Argo ZRG 654 Tape (c) ZRC 654.
+Gr 4-77 p1603 tape +NR 10-71 p7
+-HFN 1-72 p119 tape

2850 The four seasons, op. 8, nos. 1-4. Simon Standage, baroque violin; English Concert; Trevor Pinnock, hpd and cond. CRD CRD 1025.
+ARG 7-77 p36 +RR 10-76 p75
+Gr 11-76 p823 ++ST 1-77 p81
++NR 11-77 p7

2851 The four seasons, op. 8, nos. 1-4. Stuttgart Soloists; Marcel Couraud. Fontana Tape (c) 7327 003.
+Gr 2-77 p1325 tape ++Gr 4-77 p1603 tape

2852 The four seasons, op. 8, nos. 1-4. Astorre Ferrari, vln; Stuttgart Soloists; Marcel Couraud. Philips 6530 009.
++Gr 2-77 p1329

2853 The four seasons, op. 8, nos. 1-4. Felix Ayo, vln; Jeffrey Tate, hpd; Fritz Klingenstein, vlc; Berlin Chamber Orchestra; Vittorio Negri. Philips 9500 100 Tape (c) 7300 527.
+ARG 7-77 p35 +HFN 6-77 p139 tape
+-Gr 3-77 p1415 +NR 7-77 p7
+-Gr 4-77 p1603 tape ++RR 3-77 p62
+HFN 3-77 p114 +SFC 7-17-77 p42

2854 The four seasons (Il cimento dell'Armonia e dell'Invenzione), op. 8, nos. 1-4. Piero Toso, vln; I Solisti Veneti; Claudio Scimone. RCA AGL 1-2123.
+NR 7-77 p7 +-SFC 7-17-77 p42

2855 The four seasons, op. 8, nos. 1-4. James Galway, flt; Zagreb Soloists; James Galway. RCA LRL 1-2284 Tape (c) LRK 1-2284 (ct) LRS 1-2284. (also RCA RL 2-5034 Tape (c) RK 2-5034)
+FF 11-77 p55 +-NR 8-77 p6
+Gr 2-77 p1290 +RR 3-77 p63
+-Gr 5-77 p1743 tape +RR 7-77 p100 tape
+HFN 3-77 p114 +SFC 7-17-77 p42
+-HFN 6-77 p137 tape +-St 11-77 p158

2856 The four seasons, op. 8, nos. 1-4. Ralph Holmes, vln; Cantilena Chamber Players; Adrian Shepherd. RCA GL 2-5061.
+Gr 7-77 p188 +-HFN 8-77 p99 tape
+-HFN 7-77 p123 +-RR 7-77 p62

2857 The four seasons, op. 8, nos. 1-4. Slovak Chamber Orchestra; Bohdan Warchal. Royale ROY 2001. (Reissue from Supraphon SUAST 50767)
+Gr 8-77 p310 +-RR 6-77 p71
+-HFN 6-77 p135

2858 The four seasons, op. 8, nos. 1-4. ALBINONI: Adagio, G major. Giuliano Badini, vln; Sienna Sinfonia. Saga 5443.
+Gr 2-77 p1290 +-HFN 3-77 p114

The four seasons, op. 8, nos. 1-4. cf ALBINONI: Adagio, organ and strings, G minor.

Gloria. cf Beatus vir (CBS 76596).

Gloria. cf Beatus vir (Hungaroton SLPX 11695).

2859 Gloria, D major. Magnificat. (edit. Malipiero). Teresa Berganza, ms; Lucia Valenti Terrani, con; NPhO and Chorus; Riccardo Muti. HMV SQ ASD 3418. (also Angel S 37415)
+Gr 12-77 p1128 +RR 12-77 p83

2860 Invicti bellate. Longe mala umbrae terrores. Nisi dominus (Psalm no. 126). Teresa Berganza, con; ECO; Antonio Ros-Marba. HNH 4012. (also Pye NEL 2018)
+ARG 10-77 p46 +ON 10-77 p70
+HF 9-77 p108 ++SFC 12-18-77 p53
-NR 9-77 p11 ++St 9-77 p140

Invicti bellate. cf Beatus vir.
Kyrie. cf Beatus vir.
Lauda Jerusalem. cf Beatus vir.
2861 Longe mala umbrae terrores. Stabat mater. Livia Budai, ms; Ferenc Liszt Academy Chamber Orchestra; Sandor Frigyes. Hungaroton SLPX 11750.
+NR 4-77 p12 +St 2-77 p129
Longe mala umbrae terrores. cf Invicti bellate.
Magnificat. cf Gloria, D major.
Magnificat. cf BACH: Magnificat, S 243, D major.
Nisi dominus. cf Concerto, orchestra, G minor.
Nisi dominus (Psalm no. 126). cf Invicti bellate.
2862 Il Pastor Fido, op. 13. Jean-Pierre Rampal, flt; Robert Veyron-Lacroix, hpd. RCA FRL 1-5467 Tape (c) FRK 1-5467 (ct) FRS 1-5467.
++HF 4-77 p123 tape +-NR 3-77 p9
+MJ 2-77 p30 +St 4-77 p134
+-MU 3-77 p14
Sinfonia, no. 3, G major. cf GLUCK: Sinfonia, G major.
Sonata, op. 8, no. 2, C major. cf Arion ARN 90416.
Sonata, bassoon, B flat major. cf Works, selections (Odyssey Y 34614).
Sonata, oboe, C major. cf Works, selections (Odyssey Y 34614).
2863 Sonatas, violoncello and harpsichord, op. 14 (6). Yehuda Hanani, Christine Gummere, vlc; Lionel Party, hpd. Orion ORS 76249.
+-HF 12-77 p103 +NR 7-77 p7
++IN 10-77 p26 +St 6-77 p145
+MJ 5-77 p32
Sonata a due. cf Orion ORS 77269.
Sonata a 3, 2 trumpets and organ. cf COUPERIN: Concerto, trumpet and organ, no. 9: Ritratto dell'amore.
Songs: O di tua man mi svena; Dille ch'il viver mio; Se cerca se dice. cf HNH 4008.
Stabat mater. cf Concerto, orchestra, G minor.
Stabat mater. cf Longe mala umbrae terrores.
2864 Works, selections: Concerto, bassoon, flute, oboe and violin, C major. Concerto, bassoon, flute, oboe and violin, G minor. Concerto, flute and violin, D major. Concerto, bassoon, flute and violin, D minor. Sonata, oboe, C major. Sonata, bassoon, B flat major. Jean-Pierre Rampal, flt; Pierre Pierlot, ob; Robert Gendre, vln; Paul Hongne, bsn; Robert Veyron-Lacroix, hpd. Odyssey Y 34614.
++NR 11-77 p6

VIVIANI, Giovanni
Sonata, trumpet, no. 1, C major. cf DG 2530 792.

VOGELWEIDE, Walther von der
Palastinalied. cf Argo D40D3.

VOLKMANN, Robert
2865 Concerto, violoncello, op. 33. Konzertstuck, piano and orchestra, op. 42. Thomas Blees, vlc; Jerome Rose, pno; Hamburg Symphony Orchestra, Luxembourg Radio Orchestra; Alois Springer, Pierre Cao. Turnabout TVS 34576.
+HF 9-75 p96 +SFC 1-16-77 p43
+NR 7-75 p5
Konzertstuck, piano and orchestra, op. 42. cf Concerto, violoncello, op. 33.
Serenade, no. 2, op. 63, F major: Waltz. cf Citadel CT 6013.

VOLLINGER, William
More than conquerors. cf KAUFMAN: Reflections, clarinet and piano.
VON HAGEN
Funeral dirge on the death of General Washington. cf Vox SVBX 5350.
VON HESSEN, Moritz
Pavane. cf DG 2533 302.
VON RUGEN
Loybere risen. cf BIS LP 3.
VRANICKY, Anton
Marches in French style (3). cf DRUZECKY: Parthia, no. 3, D sharp major.
VULPIUS, Melchior
Die beste Zeit im Jahr. cf BIS LP 75.
WADE, John
Adeste fidelis. cf London OS 26437.
WAGNER, Richard
Adagio. cf BERNSTEIN: Sonata, clarinet and piano.
Adagio. cf Decca DDS 507/1-2.
2866 Arias: Der fliegende Hollander: Senta's ballad; Mogst du, mein Kind; Wie aus der Ferne. Gotterdammerung: Seit er von dir geschieden; Hier sitz ich zur Wacht. Die Meistersinger von Nurnberg: Wahn, Wahn; Morgenlich leuchtend. Parsifal: Ich sah das Kind; Das ist Karfreitagzuber, Herr. Das Rheingold: Weiche, Wotan, weiche. Siegfried: Ewig war ich. Tannhauser: Wie Todesahnung...O du, mein holder; Inbrunst im Herzen; Dich teure Halle. Tristan und Isolde Mild und leise. Die Walkure: Ein Schwert verhiess mir der Vater. Emmy Destinn, Frida Leider, Astrid Varnay, Birgit Nilsson, Leonie Rysanek, s; Sigrid Onegin, Karin Branzell, ms; Lauritz Melchior, Max Lorenz, Wolfgang Windgassen, t; Walter Soomer, Friedrich Schorr, Hans Hotter, bar; Paul Knupfer, Richard Mayr, Josef Greindl, bs; Orchestra, Munich Philharmonic Orchestra, Wurttemberg State Orchestra, Berlin State Opera Orchestra, Bavarian Radio Symphony Orchestra, Bayreuth Festival Orchestra; Bruno Seidler-Winkler, Ferdinand Leitner, Manfred Gurlitt, Arthur Rother, Robert Heger, Hermann Weigert, Karl Bohm. DG 2721 115 (2). (Reissues from 043064, 002416, 73940, 65598, 72863, LPM 19069, 18097, 66853, 72977, 30025, 19259, 67973, 19047, 19045, SLPM 139221/5)

+ARG 11-76 p10	+-ON 8-76 p39
+Gr 8-76 p337	+-RR 9-76 p35
+HFN 8-76 p88	+St 2-77 p136
+NR 1-77 p11	+STL 9-19-76 p36

Faust overture. cf BRAHMS: Tragic overture, op. 81.
2867 Der fliegende Hollander (The flying Dutchman). Janis Martin, s; Isola Jones, ms; Rene Kollo, Werner Krenn, t; Norman Bailey, bar; Martti Talvela, bs; CSO and Chorus; Georg Solti. London OSA 13119 (3) Tape (c) OSA 5-13119. (also Decca D24D3 Tape (c) K24K32.

+-ARG 12-77 p39	+MT 10-77 p827
+-FF 11-77 p56	+NR 9-77 p10
+Gr 5-77 p1730	+-NYT 7-10-77 pD13
+Gr 8-77 p349 tape	-ON 10-77 p70
-HF 8-77 p91	+-RR 5-77 p40
-HF 11-77 p138 tape	+RR 6-77 p98 tape

+HFN 5-77 p135 +-SFC 5-29-77 p42
++HFN 7-77 p127 tape +St 9-77 p90
+-MJ 11-77 p28

Der fliegende Hollander, excerpts. cf Works, selections (DG 2721 109).

Der fliegende Hollander, excerpts. cf Works, selections (DG 2721 110).

Der fliegende Hollander, excerpts. cf Works, selections (DG 2721 113).

The flying Dutchman, excerpt. cf HMV SLS 5073.

Der fliegende Hollander: Overture. cf Works, selections (HMV SLS 5075).

The flying Dutchman: Overture. cf KORNGOLD: Violanta: Prelude and carnival.

The flying Dutchman: Overture. cf Philips 6747 327.

2868 The flying Dutchman: Overture. Rienzi: Overture. Tannhauser: Overture and Venusberg music. Die Walkure: Ride of the Valkyries. NPhO, LSO, Netherlands Radio Philharmonic Orchestra; Carlos Paita, Leopold Stokowski, Erich Leinsdorf. Decca SPA 468 Tape (c) KCSP 468.

+HFN 4-77 p151 +-RR 4-77 p61
+-HFN 6-77 p139 tape

Der fliegende Hollander: Senta's ballad; Mogst du, mein Kind; Wie aus der Ferne. cf Arias (DG 2721 115).

Der fliegende Hollander: Trafft ihr das Schiff. cf Clubb 99-109.

Gotterdammerung, excerpts. cf Works, selections (DG 2721 109).

Gotterdammerung, excerpts. cf Works, selections (DG 2721 113).

2869 Gotterdammerung: Dawn and Siegfried's Rhine journey; Siegfried's funeral march; Brunnhilde's immolation and finale. LSO; Leopold Stokowski. RCA ARL 1-1317 Tape (c) ARK 1-1317 (ct) ARS 1-1317.

*ARG 3-77 p81 +MJ 4-77 p33
-Gr 7-77 p188 +-NR 2-77 p4
-Gr 8-77 p349 tape +-RR 7-77 p65
+HF 4-77 p121 tape +-SFC 1-30-77 p36
+-HFN 8-77 p95 ++St 4-77 p138
+HFN 8-77 p99 tape

2870 Gotterdammerung: Immolation scene. Siegfried: Love duet. Die Walkure: Brunnhilde's call. Kirsten Flagstad, s; La Scala Orchestra; Wilhelm Furtwangler. Everest SDBR 3414.

++NR 12-77 p9

Gotterdammerung: Seit er von dir geschieden; Hier sitz ich zur Wacht. cf Arias (DG 2721 115).

Gotterdammerung: Siegfried's Rhine journey; Siegfried's funeral march. cf Works, selections (HMV SLS 5075).

Gotterdammerung: Siegfried's Rhine journey; Siegfried's funeral march; Immolation of the Gods. cf Der Ring des Nibelungen (Decca SXL 6743).

Gotterdammerung: Zu neuen Taten. cf Club 99-109.

2871 Lohengrin. Elisabeth Grummer, s; Christa Ludwig, ms; Jess Thomas, t; Dietrich Fischer-Dieskau, bar; Gottlob Frick, Otto Wiener, bs; Vienna State Opera Chorus; VPO; Rudolf Kempe. HMV SQ SLS 5071 (5) Tape (c) TC SLS 5071. (Reissue from Angel SAN 121/5)

+Gr 12-76 p1057 +MT 3-77 p218
++HFN 12-76 p151 +RR 1-77 p42
++HFN 5-77 p138 tape +RR 5-77 p95 tape

2872 Lohengrin. Anja Silja, Astrid Varnay, s; Jess Thomas, t; Ramon Vinay, Tom Krause, bar; Franz Crass, bs; Bayreuth Festival Orchestra and Chorus; Wolfgang Sawallisch. Philips 6747 241 (4).
+Ha 3-77 p112 +NYT 3-14-76 pD15
/HF 6-76 p96 +ON 8-76 p39
+HFN 7-76 p99 +RR 7-76 p36
+MJ 5-76 p28 ++SFC 3-21-76 p28
+-MT 12-76 p1007 +St 7-76 p113
/NR 5-76 p10

Lohengrin, excerpts. cf Works, selections (DG 2721 109).
Lohengrin, excerpts. cf Works, selections (DG 2721 110).
Lohengrin, excerpts. cf Works, selections (DG 2721 113).
Lohengrin: Du armste kannst wohl nie ermessen. cf Seraphim IB 6105.
Lohengrin: Einsam in truben Tagen. cf Advent S 5023.
Lohengrin: Einsam in truben Tagen. cf BASF 22-22645-3.
Lohengrin: Elsa's dream. cf HMV RLS 719.
Lohengrin: Euch Luften. cf Club 99-109.
Lohengrin: Mein Herr und Gott; Prelude, Act 3; Treulich gefuhrt ziehet dahin; Das susse Lied verhallt; Hochstes Vertrau'n; In fernem Land, unnahbar euren Schritten; Mein lieber Schwan. cf Works, selections (Telefunken KT 11017).

2873 Lohengrin: Prelude, Act 1. Tannhauser: Overture and bacchanale. Tristan und Isolde: Prelude and Liebestod. BPhO; Herbert von Karajan. Angel S 37097 Tape (c) 4XS 37097 (ct) 8XS 37097. (also HMV (Q) ASD 3130 Tape (c) TC ASD 3130)
+Gr 12-75 p1054 ++NR 9-75 p2
+HF 12-75 p109 +-RR 12-75 p70
+-HFN 12-75 p169 +RR 2-77 p99 tape
+HFN 12-76 p155 tape

Lohengrin: Prelude, Acts 1 and 3. cf Works, selections (HMV SLS 5075).
Lohengrin: Prelude, Act 3. cf GLINKA: Russlan and Ludmilla: Overture.

2874 Die Meistersinger von Nurnberg. Hannelore Bode, s; Julia Hamari, ms; Adalbert Kraus, Martin Schomberg, Wolfgang Appel, Michel Senechal, Rene Kollo, Adolf Dallapozza, t; Norman Bailey, Bernd Weikl, bar; Kurt Moll, Martin Egel, Gerd Nienstedt, Helmut Berger-Tuna, Kurt Rydl, Rudolf Hartmann, bs; Gumpoldskirchener Spatzen; Vienna State Opera Chorus; VPO; Georg Solti. Decca D13D5 (5) Tape (c) K13K54. (also London 1512 Tape (c) 5-1512)
+-ARG 3-77 p8 +-OC 3-77 p53
+Gr 9-76 p468 +-ON 12-18-76 p76
+HF 2-77 p90 +-OR 6/7-77 p33
++HFN 9-76 p113 +-RR 9-76 p36
++HFN 12-76 p155 +-RR 1-77 p92 tape
+-MJ 7-77 p68 ++SFC 12-5-76 p58
+-MT 12-76 p1007 +St 3-77 p89
+-NR 6-77 p9

2875 Die Meistersinger von Nurnberg. Catarina Ligendza, s; Christa Ludwig, ms; Peter Maus, Loren Driscoll, Karl-Ernst Mercker, Martin Vantin, Placido Domingo, Horst Laubenthal, t; Dietrich Fischer-Dieskau, Klaus Lang, bar; Peter Lagger, Roland Hermann, Gerd Feldhoff, Ivan Sardi, bs; Berlin German Opera Orchestra and Chorus; Eugen Jochum. DG 2740 149. (also 2713 011)
+-ARG 3-77 p8 +-OC 3-77 p53
+-Gr 12-76 p1057 +ON 12-18-76 p76

+HF 2-77 p90 +OR 6/7-77 p33
+-HFN 1-77 p99 +-RR 12-76 p46
+-MJ 3-77 p46 +-SR 2-19-77 p42
+-MT 1-77 p47 +-St 3-77 p89
+NR 3-77 p11

2876 Die Meistersinger von Nurnberg (nearly complete recording). Maria Muller, s; Camilla Kallab, ms; Max Lorenz, Erich Zimmermann, Benno Arnold, Gerhard Witting, Gustaf Rodin, Karl Krollmann, t; Jaro Prohaska, Eugen Fuchs, bar; Josef Greindl, Fritz Krenn, Helmut Fehn, Herbert Gosebruch, Franz Sauer, Alfred Dome, Erich Pina, bs; Bayreuth Festival Orchestra and Chorus; Wilhelm Furtwangler. EMI Odeon 1C 181 01797/301 (5). (also Da Capo)
+Gr 12-76 p1057 +-RR 2-77 p39
+-HF 9-76 p100 +-St 11-76 p162
+-NYT 6-20-76 pD27 +STL 1-9-77 p35
+-ON 12-18-76 p76

2877 Die Meistersinger von Nurnberg, excerpts. Elfriede Marherr-Wagner, s; Robert Hutt, Carl Joken, t; Friedrich Schorr, Leo Schutzendorf, bar; Emmanuel List, bs; Orchestra; Leo Blech. Bruno Walter Society IGI 298 (2).
+NR 10-77 p9

Die Meistersinger von Nurnberg, excerpts. cf Works, selections (DG 2721 109).

Die Meistersinger von Nurnberg, excerpts. cf Works, selections (DG 2721 110).

Die Meistersinger von Nurnberg, excerpts. cf Works, selections (DG 2721 113).

Die Meistersinger von Nurnberg: Am stillen Herd; Was duftet doch der Flieder; Morgendlich leuchtend in rosigem. cf Works, selections (Telefunken KT 11017).

Die Meistersinger von Nurnberg: Gut'n Abend, Meister, O Sachs mein Freund. cf Seraphim IB 6105.

Die Meistersinger von Nurnberg: Overture. cf Bruno Walter Society RR 443.

Die Meistersinger von Nurnberg: Overture; Dance of the apprentices and entry of the masters. cf Works, selections (HMV SLS 5075).

Die Meistersinger von Nurnberg: Prelude. cf BARTOK: Concerto, orchestra.

2878 Die Meistersinger von Nurnberg: Preludes, Acts 1 and 3; Dance of the apprentices; Procession of the Meistersinger. Tannhauser: Overture and Venusberg music. PO; Eugene Ormandy. RCA ARL 1-1868 Tape (c) ARK 1-1868 (ct) ARS 1-1868.
+HF 7-77 p125 tape +MJ 1-77 p26
+NR 11-76 p4

2879 Die Meistersinger von Nurnberg: Prelude, Act 3; Dance of the apprentices; Entrance of the Meistersingers. Rienzi: Overture. Tristan und Isolde: Prelude and Liebestod. Die Walkure: Magic fire music. RPO; Leopold Stokowski. RCA ARL 1-0498 Tape (c) ARK 1-0498 (ct) ARL 1-0498.
+NR 11-77 p4 ++St 11-77 p163
+SFC 9-18-77 p42

Die Meistersinger von Nurnberg: Sankt Krispin, Lovet ihn. cf Decca SXL 6826.

Die Meistersinger von Nurnberg: Suite, Act 3. cf Pye Golden Hour GH 643.

Die Miestersinger von Nurnberg: Wahn, Wahn; Morgenlich leuchtend. cf Arias (DG 2721 115).

Die Meistersinger von Nurnberg: Was duftet doch der Flieder. cf Decca ECS 812.
Parsifal, excerpts. cf Works, selections (DG 2721 109).
Parsifal, excerpts. cf Works, selections (DG 2721 113).
Parsifal: Ich sah das Kind; Das ist Karfreitagzuber, Herr. cf Arias (DG 2721 115).
Parsifal: Nun achte wohl...Zum letzten Liebesmahle. cf Decca SXL 6826.
Parsifal: Prelude, Act 1. cf Works, selections (HMV SLS 5075).
Parsifal: Prelude, Act 1. cf BRUCKNER: Symphony, no. 7, E major.
Prize song. cf Discopaedia MOB 1018.

2880 Das Rheingold (sung in English). Valerie Masterson, Shelagh Squires, Lois McDonall, s; Helen Attfield, Anne Collins, ms; Katherine Pring, con; Robert Ferguson, Emile Belcourt, Gregory Dempsey, t; Derek Hammond-Stroud, Norman Bailey, Norman Welsby, bar; Robert Lloyd, Clifford Grant, bs; English Opera Orchestra; Reginald Goodall. HMV (Q) SLS 5032 (4). (also Angel SDC 3825)
+Gr 11-75 p903
+-HFN 12-75 p161
+MM 1-77 p29
+MT 7-76 p579
+NYT 3-14-76 pD15
+-ON 8-76 p39
++RR 12-75 p33
+SR 4-17-76 p51

Das Rheingold, excerpts. cf Works, selections (DG 2721 109).
Das Rheingold, excerpts. cf Works, selections (DG 2721 110).
Das Rheingold: Abendlich strahlt der Sonne Auge. cf Works, selections (Telefunken KT 11017).
Das Rheingold: Entrance of the Gods into Valhalla. cf Der Ring des Nibelungen (Decca SXL 6743).
Das Rheingold: Entry of the Gods into Valhalla. cf Works, selections (HMV SLS 5075).
Das Rheingold: Weiche, Wotan, weiche. cf Arias (DG 2721 115).

2881 Rienzi. Siv Wennberg, Janis Martin, Ingeborg Springer, s; Rene Kollo, Peter Schreier, t; Theo Adam, Nikolaus Hillebrand, Siegfried Vogel, Gunther Leib, bs; Leipzig Radio Chorus, Dresden State Opera Chorus; Dresden Staatskapelle; Heinrich Hollreiser. HMV SQ SLS 990 (5). (also Angel SX 3818)
+-Gr 11-76 p873
+-HF 3-77 p91
+-HFN 11-76 p151
+MT 1-77 p47
+-ON 2-19-77 p33
+-OR 9/10-77 p30
+-RR 11-76 p47
+SFC 11-14-76 p30
+St 4-77 p138

2882 Rienzi (rev. & ed. Scherchen), abridged. Raina Kabaiwanska; Giuseppe di Stefano, G. F. Cecchele, t; F. Piva, bs; Orchestra and Chorus; Hermann Scherchen. Rococo 1022 (2).
+-NR 2-77 p11

Rienzi: Gerechter Gott. cf Orion ORS 77271.
Rienzi: Overture. cf The flying Dutchman: Overture.
Rienzi: Overture. cf Die Meistersinger von Nurnberg: Prelude, Act 3; Dance of the apprentices; Entrance of the Meistersingers.
Rienzi: Overture. cf Works, selections (HMV SLS 5075).
Rienzi: Overture. cf HMV RLS 717.
Rienzi: Overture. cf Supraphon 110 1637.

2883 Der Ring des Nibelungen. Claire Watson, Birgit Nilsson, Regine Crespin, Joan Sutherland, s; Jean Madeira, ms; Christa Ludwig, Marga Hoffgen, con; Set Svanholm, Paul Kuen, Waldemar Kmentt, James King, Wolfgang Windgassen, Gerhard Stolze, t; Eberhard Wachter, Dietrich Fischer-Dieskau, bar; George London, Hans Hotter, bs-bar; Gustav Neidlinger, Walther Kreppel, Gottlob

Frick, bs; VPO; Georg Solti. Decca Tape (c) K2W29 (2), K3W30 (3), K3W31 (3), K4W32 (4).
++Gr 9-76 p494 tape +HFN 8-77 p97
++HFN 9-76 p133 tape +-RR 12-76 p110 tape

2884 Der Ring des Nibelungen. Birgit Nilsson, Kirsten Fladstad, Regine Crespin, s; Christa Ludwig, ms; Set Svanholm, Wolfgang Windgassen, t; George London, Hans Hotter, Dietrich Fischer-Dieskau, bar; Gottlob Frick, bs; VPO; Georg Solti. London Tape (c) RING S 5-1 (12). (also OSA 5-1309, 5-1508, 5-1509, 5-1604, Decca 100D19)
+-Gr 8-77 p345 +RR 8-77 p42
+HF 3-77 p123 tape +St 3-77 p144 tape
++HFN 8-77 p97

2885 Der Ring des Nibelungen. Walburga Wegner, Magda Gabory, Margarita Kenney, Kirsten Flagstad, Hilde Konetzni, Ilona Steingruber, Karen Marie Cerkal, Julia Moor, s; Elisabeth Hongen, Dagmar Schmedes, Margret Weth-Falke, Sieglinde Wagner, Polly Batic, con; Joachim Sattler, Gunther Treptow, Peter Markwort, Set Svanholm, Max Lorenz, t; Ferdinand Frantz, Angelo Mattiello, Josef Hermann, bar; Alois Pernerstorfer, Ludwig Weber, Albert Emmerich, bs; La Scala Orchestra and Chorus; Wilhelm Furt-Wangler. Murray Hill 940477 (11)
+Gr 10-77 p654 +-MT 3-77 p218
+HFN 8-76 p89 +-RR 10-76 p37

2886 Der Ring des Nibelungen, orchestral excerpts: Gotterdammerung: Siegfried's Rhine journey; Siegried's funeral march; Immolation of the Gods. Das Rheingold: Entrance of the Gods into Valhalla. Siegfried: Forest murmurs. Die Walkure: Ride of the Valkyries; Wotan's farewell and magic fire music. National Philharmonic Orchestra; Antal Dorati. Decca SXL 6743 Tape (c) KSXC 6743. (also London CS 6970 Tape (c) 5-6970)
+Gr 4-76 p1611 -RR 5-76 p59
++SFC 6-27-76 p29 -RR 12-76 p111 tape
++St 4-77 p138

Siegfried, excerpts. cf Works, selections (DG 2721 110).
Siegfried: Ewig war ich. cf Arias (DG 2721 115).
Siegfried: Ewig war ich. cf Club 99-109.
Siegfried: Forest murmurs. cf Der Ring des Nibelungen (Decca SXL 6743).
Siegfried: Forest murmurs. cf Works, selections (HMV SLS 5075).
Siegfried: Love duet. cf Gotterdammerung: Immolation scene.
Siegfried: Nothung, Nothung, neidliches Schwert; Schmiede, mein Hammer; Dass der Mein Vater nicht ist. cf Works, selections (Telefunken KT 11017).
Siegfried Idyll. cf BEETHOVEN: Symphony, no. 5, op. 67, C minor.
Siegfried Idyll (Rehearsal and performance). cf BEETHOVEN: Symphony, no. 5, op. 67, C minor (Rehearsal 1st and 2nd movements and performance).
Siegfried Idyll. cf BRUCKNER: Symphony, no. 7, E major.
Siegfried Idyll. cf FRANCK: Symphony, D minor.
Siegfried Idyll. cf KORNGOLD: Violanta: Prelude and carnival.
Songs: Der Engel. cf Bruno Walter Society BWS 729.
Songs: Traume. cf Seraphim 60274.

2887 Tannhauser. Birgit Nilsson, s; Wolfgang Windgassen, t; Dietrich Fischer-Dieskau, bar; Theo Adam, bs-bar; Berlin German Opera Orchestra; Otto Gerdes. DG 2711 008.
+-NYT 10-9-77 pD21

2888 Tannhauser. Helga Dernesch, s; Christa Ludwig, ms; Rene Kollo, t; Hans Braun, bs-bar; Hans Sotin, bs; VPO; George Solti. London OSA 1438.
+NYT 10-9-77 pD21

2889 Tannhauser, excerpts. Maria Reining, Margarete Baumer, s; Max Lorenz, Alfred Frey, Walther Ludwig, t; Karl Schmitt-Walter, bar; Ludwig Hoffmann, Walter Grossmann, bs; Berlin Opera Chorus; Berlin Radio Symphony Orchestra; Artur Rother. BASF 22 22119-9 (2).
+NR 4-77 p9

Tannhauser, excerpts. cf Works, selections (DG 2721 109).
Tannhauser, excerpts. cf Works, selections (DG 2721 113).
Tannhauser: Allmacht'ge Jungfrau. cf Club 99-109.
Tannhauser: Allmachtge Jungfrau, hor mein flehen. cf Seraphim IB 6105.
Tannhauser: Begluckt darf nun dich. cf Decca SXL 6826.
Tannhauser: Blick ich umher in diesem edlen Kreise; Inbrunst im Herzen. cf Works, selections (Telefunken KT 11017).
Tannhauser: Dich teure Halle. cf BASF 22 22645-3.
Tannhauser: Grand march. cf Rediffusion Royale 2015.
Tannhauser: Grand march. cf Virtuosi VR 7608.
Tannhauser: Overture. cf Bruno Walter Society RR 443.
Tannhauser: Overture. cf Classics for Pleasure CFP 40263.
Tannhauser: Overture and bacchanale. cf Lohengrin: Prelude, Act 1.

2890 Tannhauser: Overture and Venusberg music. Tristan und Isolde: Prelude and Liebestod; Prelude, Act 3. Minnesota Symphony Orchestra; Stanislaw Skrowaczewski. Turnabout QTV 34642.
+NR 2-77 p4

Tannhauser: Overture and Venusberg music. cf The flying Dutchman: Overture.
Tannhauser: Overture and Venusberg music. cf Die Meistersinger von Nurnberg: Preludes, Acts 1 and 3; Dance of the apprentices; Procession of the Meistersinger.
Tannhauser: Overture; Prelude, Act 3. cf Works, selections (HMV SLS 5075).
Tannhauser: Pilgrims chorus. cf Argo SPA 507.
Tannhauser: Pilgrims chorus. cf BBC REC 267.
Tannhauser: Prelude, Act 3. cf HMV RLS 717.
Tannhauser: Venusberg music. cf KORNGOLD: Violanta: Prelude and carnival.
Tannhauser: Wie Todesahnung...O du, mein holder Abendstern; Inbrunst im Herzen; Dich teure Halle. cf Arias (DG 2721 115).

2891 Tristan und Isolde. Margarete Klose, Helena Braun, s; G. Treptow, Albrecht Peter, t; Paul Schoffler, Ferdinand Frantz, bar; Bavarian State Opera Orchestra and Chorus; Hans Knappertsbusch. Bruno Walter Society IGI 345 (5).
+-NR 7-77 p9

2892 Tristan und Isolde. Birgit Nilsson, s; Regina Resnik, ms; Fritz Uhl, Ernst Kozub, Peter Klein, Waldemar Kmentt, t; Tom Krause, bar; Arnold van Mill, Theodor Kirschbichler, bs; Vienna Singverein, VPO; Georg Solti. Decca D41D5 (5) Tape (c) K41D53. (Reissue from SET 204/8)
+Gr 8-77 p345
+Gr 10-77 p706
+-HFN 8-77 p97
++HFN 10-77 p169 tape
+-RR 8-77 p43

Tristan und Isolde, excerpts. cf Works, selections (DG 2721 109).
Tristan und Isolde, excerpts. cf Works, selections (DG 2721 110).

Tristan und Isolde, excerpts. cf Works, selections (DG 2721 113).
Tristan und Isolde: Dein Werk. cf Club 99-109.

2893 Tristan und Isolde: Isolde's narrative and curse; Love duet; Liebestod. Helga Dernesch, s; Christa Ludwig, ms; Jon Vickers, Peter Schreier, Martin Vantin, t; Walter Berry, Bernd Weikl, bar; Karl Ridderbusch, bs; Berlin German Opera Chorus; BPhO; Herbert von Karajan. HMV ASD 3354 Tape (c) TC ASD 3354. (Reissue from SLS 963)

+-Gr 6-77 p102
+HFN 8-77 p95
++HFN 10-77 p171
+-RR 8-77 p43
++RR 11-77 p121 tape

Tristan und Isolde: Liebestod. cf International Piano Library IPL 109.
Tristan und Isolde: Liebestod. cf Seraphim 60274.
Tristan und Isolde: Mild und leise. cf Arias (DG 2721 115).
Tristan und Isolde: Mild und leise. cf BASF 22 22645-3.
Tristan und Isolde: Prelude, Act 1. cf BRUCKNER: Symphony, no. 7, E major.
Tristan und Isolde: Prelude, Act 1; Liebestod. cf BEETHOVEN: Symphony, no. 7, op. 92, A major.

2894 Tristan und Isolde: Prelude, Act 1; Weh, ach wehe dies zu dulden... Welcher Wahn...Kennst du der Mutter Kunste nicht; Act 2, Isolde, Tristan, Geliebter...O sink'hernieder...O ew'ge Nacht; Act 3, Liebestod. Martha Modl, s; Johanna Blatter, ms; Wolfgang Windgassen, t; Berlin Municpal Opera Orchestra; Artur Rother. Telefunken 6-48020 (2). (Reissue from originals, c1952-53)

+-HF 9-77 p113
+NR 10-77 p9
+SFC 11-13-77 p50

Tristan und Isolde: Prelude, Act 3. cf HMV RLS 717.
Tristan und Isolde: Prelude and Liebestod. cf Lohengrin: Prelude, Act 1.
Tristan und Isolde: Prelude and Liebestod. cf Die Meistersinger von Nurnberg: Prelude, Act 3; Dance of the apprentices; Entrance of the Meistersingers.
Tristan und Isolde: Prelude and Liebestod. cf Works, selections (HMV SLS 5075).
Tristan und Isolde: Prelude and Liebestod. cf Bruno Walter Society BWS 729.
Tristan und Isolde: Prelude and Liebestod; Prelude, Act 3. cf Tannhauser: Overture and Venusberg music.

2895 Die Walkure (The Valkyrie) (sung in English). Margaret Curphey, Rita Hunter, Katie Clarke, Helen Attfield, Anne Evans, s; Ann Howard, Elizabeth Connell, Sarah Walker, Shelagh Squires, Anne Collins, ms; Alberto Remedios, t; Norman Bailey, bs-bar; Clifford Grant, bs; English National Opera Orchestra; Reginald Goodall. HMV SLS 5063 (5) Tape (c) TC SLS 5063. (also Angel SX 3826 (Q) SELX 3826)

+-ARG 3-77 p37
++Gr 9-76 p473
+HF 8-77 p91
+HFN 9-76 p111
+-HFN 5-77 p138 tape
+MM 1-77 p29
+-MT 12-76 p1008
-NR 4-77 p10
+ON 2-19-77 p33
+RR 9-76 p41
+RR 5-77 p95 tape
+St 4-77 p140

Die Walkure, excerpts. cf Works, selections (DG 2721 109).
Die Walkure, excerpts. cf Works, selections (DG 2721 110).
Die Walkure: Brunnhilde's call. cf Gotterdammerung: Immolation scene.

Die Walkure: Du bist der Lenz. cf Seraphim IB 6105.
Die Walkure: Du bist der Lenz; Fort denn, eile; War es so schmahlich. cf Club 99-109.
Die Walkure: Ho-jo-to-ho. cf Seraphim 60274.
Die Walkure: Magic fire music. cf Die Meistersinger von Nurnberg: Prelude, Act 3; Dance of the apprentices; Entrance of the Meistersingers.
Die Walkure: Magic fire music. cf International Piano Library IPL 103.
Die Walkure: Der Manner Sippe. cf BASF 22-22645-3.
Die Walkure: Ride of the Valkyries. cf The flying Dutchman: Overture.
Die Walkure: Ride of the Valkyries. cf Works, selections (HMV SLS 5075).
Die Walkure: Ride of the Valkyries. cf BEETHOVEN: Symphony, no. 3, op. 55, E flat major.
Die Walkure: Ride of the Valkyries; Wotan's farewell and magic fire music. cf Der Ring des Nibelungen (Decca SXL 6743).
Die Walkure: Ein Schwert verhiess mir der Vater. cf Arias (DG 2721 115).
Die Walkure: Wintersturme wichen dem Wonnemond; Siegmund heiss ich. cf Works, selections (Telefunken KT 11017).
Die Walkure: Wotan's farewell and magic fire music. cf Decca ECS 812.
Wesendonk Lieder. cf BRAHMS: Alto rhapsody, op. 53.
Wesendonk Lieder: Der Engel; Stehe still; Im Treibhaus; Schmerzen; Traume. cf Club 99-108.

2896 Works, selections (Bayreuth Festival, 1900-1930): Der fliegende Hollander, excerpts. Gotterdammerung, excerpts. Lohengrin, excerpts. Die Meistersinger von Nurnberg, excerpts. Parsifal, excerpts. Das Rheingold, excerpts. Tannhauser, excerpts. Tristan und Isolde, excerpts. Die Walkure, excerpts. Emmy Destinn, Frida Leider, s; Sigrid Onegin, con; Heinrich Hensel, Lauritz Melchior, t; Friedrich Schorr, Walter Soomer, bar; Richard Mary, Paul Knupfer, bs; Bayreuth Festival Orchestra. DG 2721 109 (2).
+MT 4-77 p308 +-RR 12-76 p25

2897 Works, selections (Bayreuth Festival, 1930-1944): Die fliegende Hollander, excerpts. Lohengrin, excerpts. Die Meistersinger von Nurnberg, excerpts. Das Rheingold, excerpts. Siegfried, excerpts. Tristan und Isolde, excerpts. Die Walkure, excerpts. Elisabeth Ohms, Margarete Klose, Frida Leider, Maria Muller, s; Franz Volker, Max Lorenz, Lauritz Melchior, t; Heinrich Schlusnus, bar; Josef von Manowarda, bs; Bayreuth Festival Orchestra. DG 2721 110 (20).
+MT 4-77 p308 +-RR 12-76 p25

2898 Works, selections (orchestral): Der fliegende Hollander, excerpts. Gotterdammerung, excerpts. Lohengrin, excerpts. Die Meistersinger von Nurnberg, excerpts. Parsifal, excerpts. Tannhauser, excerpts. Tristan und Isolde, excerpts. Bayreuth Festival Orchestra; Hans Knappertsbusch, Richard Strauss, Karl Elmendorff, Wilhelm Furtwangler, Victor de Sabata, Eugen Jocuhm, Pierre Boulez, Karl Bohm. DG 2721 113 (2).
+-MT 6-77 p489 +-RR 12-76 p25

2899 Works, selections: Der fliegende Hollander: Overture. Gotterdammerung: Siegfried's Rhine journey; Siegfried's funeral march. Lohengrin: Preludes, Acts 1 and 3. Die Meistersinger von Nurnberg: Overture; Dance of the apprentices and Entry of the masters. Parsifal: Prelude, Act 1. Das Rheingold: Entry of the Gods into Valhalla. Rienzi: Overture. Siegfried; Forest murmurs. Tannhauser: Overture; Prelude, Act 3. Tristan und Isolde: Prelude and Liebestod. Die Walkure: Ride of the Valkyries. PhO; Otto Klemperer. HVM SLS 5075 (3) Tape (c) TC SLS 5075. (Reissues from Columbia SAX 2347/8, 2464)

+Gr 2-77 p1295 +-HFN 5-77 p139 tape
+HFN 4-77 p151 +-RR 6-77 p71

2900 Works, selections: Lohengrin: Mein Herr und Gott; Prelude, Act 3; Treulich gefuhrt ziehet dahin; Das susse Lied verhallt; Hochstes Vertrau'n; In fernem Land, unnahbar euren Schritten; Mein lieber Schwan. Die Meistersinger von Nurnberg: Am stillen Herd; Was duftet doch der Flieder; Morgendlich leuchtend in rosigem. Das Rheingold: Abendlich strahlt der Sonne Auge. Siegfried: Nothung, Nothing, neidliches Schwert; Schmiede, mein Hammer; Dass der Mein Vater nicht ist. Tannhauser: Blick ich umher in diesem edlen Kreise; Inbrunst im Herzen. Die Walkure: Wintersturme wichen dem Wonnemond; Siegmund heiss ich. Maria Muller, s; Margarethe Klose, con; Helge Roswange, t; Rudolf Bockelmann, bar; Max Lorenz, t; Franz Volker, cond. Telefunken KT 11017 (2). (Reissues from SKB 2050, GX 61020, SKB 2052, 2055, 2049, 2047/8, 1054, SK 1342, 1297, Polydor 67148, E 2091)

+-ARG 10-77 p10 +-HFN 3-73 p574
-Gr 4-73 p1933

WAISSELIUS, Matthaus

Deutscher Tanzt. cf DG 2533 302.
Fantasia. cf DG 2533 302.
Polonischer Tantz. cf Hungaroton SLPX 11721.

WALDTEUFEL, Emile

Acclamations, op. 223. cf Works, selections (HMV SQ ESD 7012).
Bella bocca, op. 163. cf Works, selections (HMV SQ ESD 7012).
Dans tes yeux. cf KELER BELA: Rakoczi overture, op. 76.
Espana, op. 236. cf Works, selections (HMV SQ ESD 7012).
Espana, op. 236. cf Seraphim S 60277.
L'Esprit francais, op. 182. cf Works, selections (HMV SQ ESD 7012).
Estudiantina, op. 191. cf Works, selections (HMV SQ ESD 7012).
Fleurs et baisers. cf KELER BELA: Rakoczi overture, op. 76.
Gouttes de rosee. cf KELER BELA: Rakoczi overture, op. 76.
Minuet, op. 168. cf Works, selections (HMV ESD 7012).
Les patineurs (The skaters), op. 183. cf Works, selections (HMV ESD 7012).
Les patineurs, op. 183. cf Mercury SRI 75098.
The skaters, op. 183. cf Seraphim S 60277.
The skaters, op. 183: Waltz. cf Angel S 37250.
The skaters, op. 183: Waltz. cf GOUNOD: Faust: Ballet music.
Prestissimo, op. 152. cf Works, selections (HMV ESD 7012).
Vision. cf KELER BELA: Rakoczi overture, op. 76.

2901 Works, selections: Acclamations, op. 223. Bella bocca, op. 163. Espana, op. 236. L'Esprit francais, op. 182. Estudiantina, op. 191. Les patineurs, op. 183. Prestissimo, op. 152. Minuite, op. 168. Monte Carlo Opera Orchestra; Willi Boskovsky. HMV SQ ESD 7012 Tape (c) TC ESD 7012. (also Angel S 37208 Tape (c) 4XS 37208)

-ARG 2-77 p44 ++HFN 9-77 p155
+Gr 11-76 p894 +NR 9-77 p2
++HF 2-77 p108 +RR 4-77 p94

+HFN 11-76 p169 +RR 11-76 p85
++HFN 5-77 p139 tape

WALKER, George
Sonata, piano, no. 3. cf BARBER: Excursions.

WALMISLEY, Thomas
Magnificat and nunc dimittis, D minor. cf Argo ZK 3.
Remember, O Lord. cf Abbey LPB 770.

WALOND, William
Voluntary, D minor. cf Cambridge CRS 2540.

WALTERS
Trumpets wild. cf Grosvenor GRS 1055.

WALTERS, Edmund
Iona. cf HMV ESD 7024.

WALTHER, Johann
Concerto, organ. cf HOWELLS: Psalm prelude.
Concerto by Albinoni. cf Crescent ARS 109.
Concerto by Meck. cf Crescent ARS 109.
Concerto del Sigr. Albinoni. cf BIS LP 27.
Jesu meine Freude, partita. cf BACH: Chorale preludes (Pelca PSR 40583).
Jesu meine Freude, partita. cf Wealden WS 159.
Was Gott tut. cf RC Records RCR 101.

WALTON, William

2902 Belshazzar's feast. Te deum. Benjamin Luxon, bar; Salisbury, Winchester and Chichester Cathedral Choirs, LPO and Chorus; Georg Solti; Ralph Downes, org. Decca SET 618. (also London S 26525)
+Gr 11-77 p879 +-RR 11-77 p110
+-HFN 11-77 p183

2903 Belshazzar's feast. Partita, orchestra. Donald Bell, bar; PhO and Chorus; William Walton. HMV SXLP 30236. (Reissue from Columbia SAX 2319)
+-Gr 3-77 p1443 +RR 3-77 p95
+HFN 4-77 p153

2904 Belshazzar's feast. Te deum. Sherrill Milnes, bar; Scottish National Orchestra and Chorus; Scottish Festival Brass Bands; Alexander Gibson. RCA RL 2-5105.
++Gr 11-77 p879 +-RR 11-77 p110

2905 Crown imperial, coronation march. Gloria. Orb and sceptre, coronation march. Te deum. Barbara Robotham, ms; Anthony Rolfe Johnson, t; Brian Rayner Cook, bar; Birmingham City Symphony Orchestra and Chorus; Worcester Cathedral Choristers; Louis Fremaux. HMV (SQ) ASD 3348 Tape (c) TC ASD 3348.
+Gr 6-77 p90 +RR 7-77 p93
+HFN 7-77 p123 ++RR 11-77 p121 tape
+HFN 10-77 p171 tape

Crown imperial. cf BAX: Coronation march.
Crown imperial. cf ELGAR: Pomp and circumstance marches, op. 39, nos. 1-5.
Crown imperial. cf Argo SPA 507.
Crown imperial. cf Decca SPA 500.
Crown imperial. cf Decca SB 715.
Crown imperial. cf Wealden WS 131.
Facade. cf HMV SLS 5073.

2906 Facade: Suites, nos. 1 and 2. The wise virgins: Ballet suite (after J. S. Bach). Birmingham City Symphony Orchestra; Louis Fremaux. HMV ASD 3317 Tape (c) TC ASD 3317.

++Audio 6-77 p134 +-HFN 6-77 p139 tape
+Gr 3-77 p1386 +-RR 3-77 p64
+-HFN 3-77 p114

Fanfare "Hamlet". cf RCA RL 2-5081.
Gloria. cf Crown imperial, coronation march.
Henry V suite: Touch her soft lips and part; Agincourt song. cf HMV ASD 3341.
Magnificat and nunc dimittis. cf Abbey LPB 770.
Orb and sceptre, coronation march. cf Crown imperial, coronation march.
Orb and sceptre. cf BAX: Coronation march.
Orb and sceptre. cf ELGAR: Pomp and circumstance marches, op. 39, nos. 1-5.
Partita, orchestra. cf Belshazzar's feast.
Portsmouth Point overture. cf HMV ESD 7011.
A queen's fanfare. cf Pye Nixa QS PCNHX 10.
A queen's fanfare. cf RCA RL 2-5081.
Sonata, violin. cf ELGAR: Sonata, violin, op. 82, E minor.
Te deum. cf Belshazzar's feast (Decca SET 618).
Te deum. cf Belshazzar's feast (RCA RL 2-5105).
Te deum. cf Crown imperial, coronation march.
Te deum. cf Vista VPS 1053.

2907 Troilus and Cressida. Janet Baker, s; Richard Cassilly, Gerald English, t; Benjamin Luxon, bar; ROHO and Chorus; Lawrence Foster. HMV SLS 997 (3).
+Gr 4-77 p1599 +MT 10-77 p827
+HFN 6-77 p115 +-RR 4-77 p20
+MM 11-77 p45

The wise virgins: Ballet suite. cf Facade: Suites, nos. 1 and 2.

WARD, John
Hope of my heart. cf BG HM 57/8.

WARLOCK, Peter
Adam lay y bounden. cf HELY-HUTCHINSON: Carol symphony.
Bethlehem down. cf HELY-HUTCHINSON: Carol symphony.
Capriol suite. cf ELGAR: Serenade, strings, op. 20, E minor.
Capriol suite. cf BIS LP 57.
Capriol suite: Branles. cf HMV SQ CSD 3781.
Motets. cf DELIUS: A late lark.

2908 Songs: After two years; As ever I saw; The bayly berith the bell away; The birds; The cricketers of Hambledon; The droll lover; Elore lo; Fair and true; The fox; The frostbound wood; Ha'nacker mill; Jillian of Berry; My own country; Passing by; Pretty ringtime; Robin goodfellow; Roister doister; Romance; Sigh no more ladies; Sleep; There is a lady sweet and kind; To the memory of a great singer; Twelve oxen; When as the rye reach to the chin; Yarmouth fair; Youth. Norman Bailey, bs-bar; Geoffrey Parsons, pno. L'Oiseau-Lyre DSLO 19 Tape (c) KDSLC 19.
+Gr 9-77 p469 -RR 9-77 p90
+-HFN 9-77 p150

Songs: Along the stream; Piggesnie. cf Argo ZK 28/9.
Songs: As ever I saw; To the memory of a great singer. cf Enigma VAR 1027.
Songs: The bayley beareth the bell away; Lullaby. cf HMV RLS 716.

WARREN, George
The Andes, March di bravoura. cf New World Records NW 257.

WATERS
Trumpets wild. cf Transatlantic XTRA 1169.

WATKINS, Michael Blake
Synthesis. cf BENNETT: Telegram.
WATSON
Songs: Anchored. cf Argo ZFB 95/6.
WATTS, John
Sonata, piano. cf BESTOR: Sonata, piano.
WAYDITCH, Gabriel (Gabor) von
2909 The Caliph's magician. Soloists; Budapest National Opera Orchestra; Andras Korodi. Musical Heritage Society MHS 3565/6 (2).
+-ON 9-77 p69
WEBBER, Amherst
Vieille chanson. cf HMV RLS 716.
WEBER, Ben
Quartet, strings, no. 2, op. 35. cf MONOD: Cantus contra cantum I.
WEBER, Carl Maria von
2910 Abu Hassan. Siegfried Gohler, Kurt Kachlicki, Gerd Biewer, August Hutten, Dorothea Garlin, speakers; Ingeborg Hallstein, s; Peter Schreier, t; Theo Adam, bs-bar; Gerhard Wustner Student Chorale, Dresden State Opera Chorus; Dresden Staatskapelle; Heinz Rogner. RCA LRL 1-5125.
+-Gr 8-76 p337 -MT 2-77 p135
+-HFN 8-76 p91 +RR 8-76 p25
Abu Hassan: Overture. cf Supraphon 110 1637.
Adagio and rondo. cf HMV SQ ASD 3283.
2911 Andante and rondo ungarese. Concerto, piano, no. 1, C major. Concerto, piano, no. 2, E flat major. Grand potpourri. Theme and variations on "A Schusserl und Reind'ri". Malcolm Frager, pno; Anner Bylsma, vlc; Rainer Moog, vla; North German Radio Orchestra; Marc Andreae. RCA PRL 2-9066 (2).
++Gr 2-77 p1295 +RR 2-77 p66
+-HFN 4-44 p149
2912 Concerto, bassoon, F major. Concerto, clarinet, no. 2, E flat major. Konzertstuck, F minor. Henri Helaerts, bsn; Griedrich Gulda, pno; Gervase de Peyer, clt; OSR, LSO, VPO; Ernest Ansermet, Colin Davis, Volkmar Andreae. Decca ECS 807. (Reissues from SXL 6375, LXT 5280, L'Oiseau-Lyre SOL 60035)
/Gr 12-77 p1097 +RR 12-77 p57
Concerto, clarinet, no. 2, E flat major. cf Concerto, bassoon, F major.
2913 Concerto, piano, no. 1, op. 11, C major. Concerto, piano, no. 2, E flat major. Symphony, no. 1, op. 19, C major. Symphony, no. 2, op. 51, C major. Malcolm Frager, pno; LSO, North German Radio Orchestra; Hans-Hubert Schonzeler, Marc Andreae. RCA CRL 2-2281 (2).
++ARG 11-77 p34 +NYT 12-4-77 pD18
+MJ 11-77 p30 ++SFC 9-25-77 p50
+NR 11-77 p2 +St 11-77 p158
Concerto, piano, no. 1, C major. cf Andante and rondo ungarese.
Concerto, piano, no. 2, E flat major. cf Andante and rondo ungarese.
Concerto, piano, no. 2, E flat major. cf Concerto, piano, no. 1, op. 11, C major.
2914 Die drei pintos (compl. Mahler). Lucia Popp, Jeanette Scovotti, s; Kari Lovaas, ms; Werner Hollweg, Heinz Kruse, t; Hermann Prey, bar; Kurt Moll, Franz Grundheber, bs; Netherlands Vocal Ensemble; Munich Philharmonic Orchestra; Gary Bertini. RCA PRL 3-9063 (3).

+ARG 7-77 p37 | +OC 9-77 p53
+-Gr 1-77 p1174 | +OC 12-77 p49
+HF 7-77 p110 | +ON 6-77 p44
+HFN 3-77 p96 | +-RR 1-77 p43
+MJ 7-77 p68 | +SR 7-9-77 p48
+NR 5-77 p11 | -St 4-77 p130

Duo concertant, op. 48. cf BRAHMS; Sonata, clarinet, no. 1, op. 120, no. 1, F minor.
Duo concertant, op. 48. cf BRAHMS: Sonata, clarinet, no. 2, op. 120, no. 2, E flat major.
Duo concertant, op. 48, E flat major. cf BRAHMS: Trio, clarinet, op. 114, A minor.
Duo concertant, op. 48, E flat major. cf GLINKA: Trio pathetique, D minor.
Euryanthe: Overture. cf BEETHOVEN: Concerto, piano, no. 1, op. 15, C major.
Der Freischutz: Nein langer trag ich nicht die Qualen. cf Telefunken AJ 6-42232.
Der Freischutz: Overture. cf Classics for Pleasure CFP 40263.
Der Freischutz: Und ob die Wolke. cf Decca D65D3.
Der Freischutz: Wie nahte mir der Schlummer...Leise leise fromme Weise. cf Seraphim IB 6105.
Grand potpourri. cf Andante and rondo ungarese.
Invitation to the dance, op. 63. cf Angel S 37250.
Konzertstuck, F minor. cf Concerto, bassoon, F major.
Oberon: Overture. cf BEETHOVEN: Symphony, no. 4, op. 60, B flat major.
Oberon: Overture. cf BRAHMS: Tragic overture, op. 81.
Oberon: Overture. cf MENDELSSOHN: Symphony, no. 4, op. 90, A major.
Oberon: Overture. cf Bruno Walter Society RR 443.
Quintet, clarinet, op. 34, B flat major. cf MOZART: Quintet, clarinet, K 581, A major.
Sonata, violoncello and piano, A major. cf ABC AB 67014.

2915 Songs: An sie, op. 80, no. 5; Elfenlied, op. 80, no. 3; Est sturmt auf der Flur, op. 30, no. 2; Die gefangenen Sanger, op. 47, no. 1; The four temperaments, op. 46; Die freien Sanger, op. 47, no. 2; Ich denke dein, op. 66, no. 3; Ich sah ein Roschen, op. 15, no. 5; Klage; Der kleine Fritz an seine jungen Freunde; Mein Schatzerl is hubsch, op. 64, no. 1; Meine Farben, op. 23, no. 3; Meine Lieder, meine Sange, op. 15, no. 1; Minnelied, op. 30, no. 4; Reigen, op. 30, no. 5; Sind es Schmerzen, sind es Freuden, op. 30, no. 6; Uber die Berge mit Ungestum, op. 25, no. 2; Unbefangenheit, op. 30, no. 3; Das Veilchen im Thale, op. 66, no. 1; Was zieht zu dienem Zauberkreise, op. 15, no. 4; Wiedersehen, op. 30, no. 1; Die Zeit, op. 13, no. 5. Martyn Hill, t; Christopher Hogwood, fortepiano. L'Oiseau-Lyre DSLO 523.

+Gr 1-77 p1168 | +RR 5-77 p85
+-HFN 1-77 p118

2916 Symphony, no. 1, op. 19, C major. Symphony, no. 2, op. 51, C major. Turandot: Overture and march. LSO; Hans-Hubert Schonzler. RCA LRL 1-5106.

++ARG 4-77 p10 | +-MM 12-76 p43
+-Gr 4-76 p1612 | +RR 4-76 p9
+HFN 4-76 p119 | +STL 4-11-76 p36
+MT 7-76 p580

2917 Symphony, no. 1, op. 19, C major. Symphony, no. 2, op. 51, C major. PCO; Dean Dixon. Supraphon 110 1635.
+ARG 5-77 p42 /NR 5-77 p2
+-Gr 2-77 p1295 +-RR 2-77 p66
+-HFN 2-77 p133 +St 11-77 p158

Symphony, no. 1, op. 19, C major. cf Concerto, piano, no. 1, op. 11, C major.

Symphony, no. 2, op. 51, C major. cf Concerto, piano, no. 1, op. 11, C major.

Symphony, no. 2, op. 51, C major. cf Symphony, no. 1, op. 19, C major(RCA 1-5106).

Symphony, no. 2, op. 21, C major. cf Symphony, no. 1, op. 19, C major (Supraphon 110 1635).

Theme and variations on "A Schusserl und Reind're". cf Andante and rondo ungarese.

Turandot: Overture and march. cf Symphony, no. 1, op. 19, C major.

WEBERN, Anton

Bagatelles, string quartet, op. 9 (6). cf HAUBENSTOCK-RAMATI: Quartet, strings, no. 1.

Movements, string quartet, op. 5 (5). cf HAUBENSTOCK-RAMATI: Quartet, strings, no. 1.

Pieces, orchestra, op. 10 (5). cf Westminster WGS 8338.

Quartet, strings, op. 28. cf HAUBENSTOCK-RAMATI: Quartet, strings, no. 1.

Sonata, violoncello. cf Laurel-Protone LP 13.

Variationen fur Klavier, op. 27. cf CP 3-5.

WEBSTER, Joseph

Songs: Sweet by and by; Willie's grave. cf New World Records NW 220.

WEBSTER, Michael

Pieces, solo clarinet. cf MARTINO: A set for clarinet.

WECK (15th century Germany)

Spanyoler Tanz and Hopper dancz. cf Nonesuch H 71326.

WEELKES, Thomas

As Vesta was from Latmos Hill. cf Argo ZK 25.

As Vesta was from Latmos Hill descending. cf DG 2533 347.

Hosanna to the Son of David. cf Abbey LPB 770.

Songs: All at once well met; The ape the money and the baboon; Cease sorrows now; O care thou wilt despatch me; On the plains fairy trains; Strike it up Tabor; Thule the period of cosmography; To shorten winter's sadness; Young cupid hath proclaimed. cf BG HM 57/8.

Songs: Cease sorrows now; Come sirrah, Jack ho; Since Robin Hood. cf Enigma VAR 1017.

Songs: Gloria in excelsis Deo; Hosanna to the son of David. cf HMV ESD 7050.

WEIGL, Karl

Night fantasies. cf RUDHYAR: Tetragram, nos. 1-3.

WEILL, Kurt

Berlin requiem. cf Works, selections (DG 2740 153).

Concerto, violin, wind orchestra and percussion, op. 12. cf Works, selections (DG 2740 153).

Death in the forest, op. 23. cf Works, selections (DG 2740 153).

2918 Die Dreigroschenoper (The three penny opera). New York Shakespeare Festival Recording. Columbia PZ 34326. (also Columbia X 798)
+-ARG 3-77 p46 +NR 2-77 p10
-MQ 7-77 p441 +-ON 4-2-77 p48

Die Dreigroschenoper: Suite, wind orchestra. cf Works, selections (DG 2740 153).
Happy end. cf Works, selections (DG 2740 153).
2919 Kleine Dreigroschenmusik. Mahagonny spongspiel. Soloists; Jerusalem Symphony, Music for Westchester Orchestra; Lukas Foss, Siegfried Landau. Turnabout TV 34675.
+Audio 11-77 p131
Kleine Dreigroschenmusik. cf KLEMPERER: Merry waltz.
Mahagonny songspiel. cf Kleine Dreigroschenmusik.
Mahagonny songspiel. cf Works, selections (DG 2740 153).
Protagonist, op. 14: Pantomime, no. 1. cf Works, selections (DG 2740 153).
Quodlibet, op. 9. cf KORNGOLD: Much ado about nothing, op. 11.
2920 The seven deadly sins. Lotte Lenya, s; Male Quartet and Orchestra; Wilhelm Brukener-Ruggeberg. CBS 73657. (Reissue from Philips ABL 3363)
+-Gr 9-77 p489 +RR 8-77 p80
+-HFN 8-77 p95
Vom Tod im Wald, op. 23. cf Works, selections (DG 2740 153).
2921 Works, selections: Berlin requiem. Concerto, violin, wind orchestra and percussion, op. 12. Die Dreigroschenoper: Suite, wind orchestra. Death in the forest, op. 23. Happy end. Mahagonny songspiel. Protagonist, op. 14: Pantomime, no. 1. Vom Tod im Wald, op. 23. Meriel Dickinson, Mary Thomas, ms; Philip Langridge, Ian Partridge, t; Benjamin Luxon, bar; Michael Rippon, bs; Nona Liddell, vln; London Sinfonietta; David Atherton. DG 2740 153 (3). (also DG 2709 064)
+ARG 5-77 p43 ++NR 6-77 p12
+Gr 11-76 p823 *ON 4-2-77 p48
++HF 5-77 p77 ++RR 11-76 p103
++HFN 11-76 p152 +SFC 4-3-77 p43
+MJ 5-77 p32 ++St 6-77 p130
+-MQ 7-77 p446 +-Te 3-77 p34

WEINBERG, Henry
Vox in Bama. cf Advance FGR 18.

WEINBERGER, Jaromir
Schwanda, the bagpiper: Polka. cf Angel S 37250.

WEINER, Lazar
Songs: A father and his son; A nign (arr.); Rhymes written in the sand; The story of the world; A tree stands on the road (arr.); What is the meaning of (arr.); Yidl and his fiddle. cf Olympic OLY 105.

WEINZWEIG
Red ear of corn: Barn dance. cf Citadel CT 6011.

WEISS, Robert
Ciacona. cf Telefunken AW 6-42155.
Prelude. cf Telefunken AW 6-42155.
Rigaudon. cf Telefunken AW 6-42155.

WEISS, Sylvius
Balletto. cf Classics for Pleasure CFP 40012.
Bourree. cf RCA ARL L-0864.
Chaconne, E flat major. cf d'ANGLETERRA: Carillon, G major.
Passacaglia. cf Angel S 36053.
Passacaglia. cf Philips 6833 159.
Sonata, lute, A minor. cf BACH: Prelude, fugue and allegro, S 998, E flat major.
Suite, C minor. cf d'ANGLETERRA: Carillon, G major.

Tombeau sur la mort de M. Comte de Logy. cf BACH: Suite, lute, S 997, C minor.

WEITZ, Guy

Grand choeur. cf Vista VPS 1047.

WELCHER, Dan

Concerto da camera, bassoon. cf LUKE: Concerto, bassoon.

WERDIN, Eberhard

Concertino, flute, guitar and strings. cf BIS LP 60.

WERLE, Lars

Nocturnal chase. cf BIS LP 34.

WERNER

Heidenroslein. cf Club 99-108.

WERNER, Lasse

Rituals in La. cf SEGERSTAM: Quartet, strings, no. 6.

WERNICK, Richard

A prayer for Jerusalem. cf MAYS: Invocations to the Svara mandala.

Songs of remembrance. cf MAXWELL DAVIES: Dark angels.

WESLEY, Samuel

In exitu Israel. cf Argo ZK 3.

Prelude, C minor. cf Vista VPS 1039.

WESLEY, Samuel S.

Ascribe unto the Lord. cf Abbey LPB 783.

Blessed be the God and Father. cf Abbey LPB 779.

Cast me not away. cf Abbey LPB 770.

2922 Songs (Anthems and hymns): Anthems, Blessed be God the Father; Brightest and best of the songs of the morning (Ephiphany); I am thine, o save me; The Lord is my shepherd; O Lord my God; O Lord, thou art my God; For this mortal must put on immortality; Psalm, no. 126, B flat major; Psalm, no. 127, F major; Wash me thoroughly from my wickedness. Hymns, The church's one foundation; O help us Lord; Each hour of need; O thou, who camest from above. Choral song and fugue. Largetto, F minor. Roy Massey, Robert Green, org; Hereford Cathedral Choir. RCA LRL 1-5129.

+Gr 9-76 p460 +-MT 1-77 p45
+HFN 9-76 p131 +RR 8-76 p77

Songs: Blessed be the God and Father; Thou wilt keep him in perfect peace; The wilderness. cf Argo ZK 3.

Thou wilt keep him. cf Pye Nixa QS PCNHX 10.

Thou wilt keep him. cf Vista VPS 1053.

Thou wilt keep him in perfect peace. cf Guild GRSP 701.

WHITE, C. A.

Trusting. cf New World Records NW 220.

WHITLOCK, Percy

Carol. cf Wealden WS 131.

Plymouth suite: Salix. cf Vista VPS 1051.

Plymouth suite: Toccata. cf Vista VPS 1035.

WHITTAKER, William

2923 Among the Northumbrian hills. I said in the noontide of my days, anthem. Quintet, winds. Amphion Wind Ensemble; London Soloists Vocal Ensemble; Ensemble; Roderick Spencer, org; John Bate. Viking VRSS 001.

+HFN 2-77 p133 ++RR 4-77 p84

I said in the noontide of my days, anthem. cf Among the Northumbrian hills.

Quintet, winds. cf Among the Northumbrian hills.

WIDDOES, Lawrence
From a time of snow. cf THOMAS: The death of Yukio Mishima.

WIDOR, Charles

2924 Symphonies, organ, nos. 1-5. Arthur Wills, Jane Parker-Smith, Graham Steed, org. RCA RL 2-5033 (4).
+Gr 10-77 p662 +RR 10-77 p80
+HFN 12-77 p183

2925 Symphony, organ, no. 5, op. 42, no. 1, F minor. Symphony, organ, no. 6, op. 42, no. 2, G major: Allegro. Symphony, organ, no. 8, op. 42, no. 4, B major: Prelude. David Sanger, org. Saga 5439 Tape (c) CA 5439.
+Gr 1-77 p1161 ++MT 8-77 p653
+HFN 2-77 p133 +-RR 2-77 p82
+-HFN 11-77 p187 tape

Symphony, organ, no. 5, op. 42, no. 1, F minor. cf GRISON: Toccata, F major.
Symphony, organ, no. 5, op. 42, no. 1, F minor: Toccata. cf Decca PFS 4416.
Symphony, organ, no. 5, op. 42, no. 1, F minor: Toccata. cf Odeon HQS 1356.
Symphony, organ, no. 6, op. 42, no. 2, G major: Allegro. cf Symphony, organ, no. 5, op. 42, no. 1, F minor.
Symphony, organ, no. 6, op. 42, no. 2, G major: Allegro. cf GUILMANT: Sonata, organ, no. 1, op. 42, D minor: Finale.
Symphony, organ, no. 6, op. 42, no. 2, G major: Allegro. cf Argo ZRG 864.
Symphony, organ, no. 6, op. 42, no. 2, G major: Allegro. cf Decca SDD 499.
Symphony, organ, no. 6, op. 42, no. 2, G major: Allegro. cf Calliope CAL 1922/4.
Symphony, organ, no. 6, op. 42, no. 2, G major: Allegro. cf Vista VPS 1060.
Symphony, organ, no. 8, op. 42, no. 4, B major: Prelude. cf Symphony, organ, no. 5, op. 42, no. 1, F minor.

WIEDOEFT, Rudy
Dans l'Orient. cf TCHAIKOVSKY: Danse hongroise.
Saxarella. cf TCHAIKOVSKY: Danse hongroise.
Saxophobia. cf TCHAIKOVSKY: Danse hongroise.
Valse Erica. cf TCHAIKOVSKY: Danse hongroise.
Valse Llewellyn. cf TCHAIKOVSKY: Danse hongroise.
Valse Marilyn. cf TCHAIKOVSKY: Danse hongroise.
Valse Mazanetta. cf TCHAIKOVSKY: Danse hongroise.
Valse Vanite. cf TCHAIKOVSKY: Danse hongroise.

WIENIAWSKI, Henryk
Caprice, A minor. cf HMV ASD 3346.
La carnaval Russe. cf Discopaedia MOB 1018.
Concerto, violin, no. 1, op. 14, F sharp minor. cf ERNST: Concerto, violin, op. 23, F sharp minor.
Concerto, violin, no. 2, op. 22, D minor: Romance. cf Discopaedia MB 1014.
Etudes caprices (3). cf PAGANINI: Sonatas, violin and guitar, op. 3, nos. 1-6.
Polonaise, op. 4, D major. cf ARENSKY: Concerto, violin, op. 54, A minor.
Polonaise, op. 4, D major. cf Enigma VAR 1025.
Polonaise brillante, no. 1, op. 4, D major. cf Discopaedia MB 1014.
Scherzo tarantelle, op. 16, G minor. cf Discopaedia MB 1013.

Scherzo tarantelle, op. 16, G minor. cf Discopaedia MB 1014.
Scherzo tarantelle, op. 16, G minor. cf Enigma VAR 1025.

WIEPRECHT, Wilhelm

2926 Marches for the Royal Prussian Army (14). Heeresmusikkops der Bundeswehr; Oberst Johannes Schade. Telefunken AW 6-42031.
+-RR 7-77 p65

WILBYE, John

The lady Oriana. cf Argo ZK 25.
The lady Oriana. cf DG 2533 347.
Madrigals (2). cf University of Iowa Press unnumbered.
Songs: Adieu sweet Amarillis; Thus saith my Cloris bright. cf Enigma VAR 1017.
Songs: Adieu sweet Amaryllis; Flora gave me fairest flowers; Lady when I behold; Oft have I vowed; Sweet honey-sucking bees. cf BG HM 57/8.

WILDER, Alec

2927 Quintets, nos. 4 and 5. Tidewater Brass Quintet. Golden Crest CRS 4156.
++MJ 3-77 p46
Suite, horn, tuba and piano, no. 2. cf HEIDEN: Variations, solo tuba and 9 horns.
Suite, piano. cf PROKOFIEV: Sonata, piano, no. 7, op. 83, B flat major.

WILHELM

Die Wacht am Rhein. cf Club 99-108.

WILLAERT, Adrian

E qui, la dira; Madonna qual certezza. cf L'Oiseau-Lyre 12BB 203/6.
Madrigal a 6, "Passa la nave". cf Telefunken AW 6-42033.
Ricercar a 3. cf Telefunken AW 6-42033.

WILLAN, Healy

Introduction, passacaglia and fugue, E flat minor. cf Vista VPS 1033.
O Lord, our governour. cf Vista VPS 1053.
WILLIAM BYRD AND HIS CONTEMPORARIES. cf RCA RL 2-5110.

WILLIAMS, William

Sonata, recorder, no. 4, A minor. cf Telefunken AW 6-42129.
Sonata in imitation of birds. cf HMV CSD 3781.

WILLIAMSON, Malcolm

The brilliant and the dark. cf HOPKINS: Riding to Canonbie.
Concerto, piano and strings, no. 2. cf HMV SLS 5080.
Elegy, J.F.K. cf Marc MC 5355.
Epitaphs for Edith Sitwell, nos. 1 and 2. cf HMV HQS 1376.
Ritual of admiration. cf BENNETT: Telegram.
Sinfonietta. cf GOOSENS: Concerto, oboe, op. 45.
Symphony, voices. cf BENNETT: Calendar.

WILLS, Arthur

Carillon on Orientis Partibus. cf Vista VPS 1030.
Piece, organ: Processional. cf Vista VPS 1033.

WILSON, Olly

Echoes. cf PECK: Automobile.
Sometimes. cf BLANK: Songs.

WILSON, Richard

Concert piece, violin and piano. cf McPHEE: Concerto, piano with wind octet.
Music for solo flute. cf McPHEE: Concerto, piano with wind octet.
THE WIND DEMON AND OTHER MID-NINETEENTH-CENTURY PIANO MUSIC. cf New World Records NW 257.

WINDSOR
Alpine echoes. cf Rediffusion 15-56.

WIREN, Dag
Serenade, strings, op. 11. cf HMV ESD 7001.
Serenade, strings, op. 11: March. cf Citadel CT 6013.

WISHART, Peter
Alleluya, a new work is come on hand. cf HMV CSD 3774.

WOLCOTT
Songs: Blessed is the spot; O thou, by whose name. cf L'Oiseau-Lyre DSLO 20.

WOLF, Hugo
Italian serenade. cf SCHUBERT: Quartet, strings, no. 14, D 810, D major.

2928 Italienisches Liederbuch. Edith Mathis, s; Peter Schreier, t; Karl Engel, pno. DG 2707 096 (2).
-Gr 8-77 p332 +-RR 8-77 p80
+HFN 8-77 p95

2929 Quartet, D minor. Keller Quartet. Oryx 1820.
+NR 9-77 p5

2930 Songs: (Goethe), nos. 1-4, 10-25, 28-36, 38, 39, 42-47, 49-51: Beherzignung II; Wanderers Nachtlied; (Heine) Madchen mit dem roten Mundchen; Du bist wie eine Blume; Wo wird einst; Wenn ich in deine Augen seh; Spatherbstnebel; Mit schwarzen Segeln; Wie des Mondes Abbild zittert; (Lenau): Frage nicht; Herbst; Abendbilder; Herstentschluss. Dietrich Fischer-Dieskau, bar; Daniel Barenboim, pno. DG 2709 066 (3).
+ARG 5-77 p44 ++NR 2-77 p12
+-HF 6-77 p100

2931 Songs: Andenken; Auf der Wanderung; Biterolf; Erwartung; Der Freund; Frohe Botschaft; Fruhlingsglocken; Gedichte von Michelangelo (3); Gesellenlied; Der Glucksritter; Ein Grab; Heimweh; Ich sagt es offen; In der Fremde, I, II, III; Ja, die Schonst; Keine gleicht von allen Schonen; Knabentod; Lieber alles; Liebesbotschaft; Liebesfruhling; Liebesgluck; Lied der transferierten Zettel; Liebchen wo bist du; Morgenstimmung; Der Musikant; Nach dem Abschiede; Nachruf; Die Nacht; Nachtgruss; Nachtzauber; Ruckkehr; Der Scholar; Der Schreckenberger; Der Schwalben Heimkehr; Seemanns Abschied; Skolie; Der Soldat I, II; Sonne der Schlummerlosen; Das Standchen; Standchen (Alles wiegt die stille Nacht); Standchen (Komm in der stille Nacht); Uber Nacht; Unfall; Verschwiegene Liebe; Der Verzweifelte Liebhaber; Wachterlied auf der Wartburg; Wohin mit der Freud; Zur Ruh zur Ruh. Dietrich Fischer-Dieskau, bar; Daniel Barenboim, pno. DG 2709 067 (3).
+-HF 12-77 p102

2932 Songs: Abendbilder; Anakreons Gab; Beherzignung; I and II; Blumengruss; Cophtisches Lied, I and II; Dank des Paria; Dies zu deuten bin erbotig; Du bist wie eine Blume; Ephiphanias; Erschaffen und Beleben; Frage nicht; Frech und Froh, I and II; Fruhling ubers Jahr; Ganymed; Genialisch Treiben; Gleich und gleich; Grenzen der Menschheit; Guttmann und Gutweib; Der Harfenspieler, 1-3; Hatt ich irgend wohl Bedenken; Herbst; Herbstentschluss; Komm Liebchen, komm; Koniglich Gebet; Locken haltet mich gefangen; Madchen mit dem roten Mundchen; Mit schwarzen Segeln; Nicht Gelegenheit macht Diebe; Der neue Amadis; Ob der Koran von Ewigkeit sei; Phanomen; Prometheus; Der Rattenfanger; Ritter Kurts Brautfahrt; St. Nepomuks Vorabend; Der

Sanger; Der Schafer; Solang man nuchtern ist; Spatherbstnebel; Spottlied; Trunken mussen wir alle sein; Wanderers Nachtlied; Was in der Schenke waren heute; Wenn ich dien gedenke; Wenn ich in deine Augen seh; Wie des Mondes Abbild zittert; Wie sollt ich heiter bleiben; Wo wird einst. Dietrich Fischer-Dieskau, bar; Daniel Barenboim, pno. DG 2740 156 (3).

+Gr 11-76 p861 ++SFC 2-27-77 p42
+HFN 11-76 p169 +St 4-77 p84
+RR 11-76 p104

2933 Songs: Alles endet, was entstehet; Andenken; Auf einer Wanderung; Biterolf; Ein Grab; Erwartung; Der Freund; Frohe Botschaft; Fruhlingsglocken; Fuhlt meine Seele; Gesellenlied; Der Glucksritter; Heimweh; In der Fremde, I, II, IV; Ja die Schonst; Keine gleicht von allen Schonen; Knabentod; Liebchen, wo bist du; Lieber alles; Libesbotschaft; Liebesfruhling; Liebesgluck; Lied das tranferierten Zettel; Morgenstimmung; Der Musikant; Nach dem Abschiede; Nachruf; Die Nacht; Nachtgruss; Nachtzauber; Ruckkehr; Der Scholar; Der Schreckenberger; Der Schwalben Heimkehr; Seemanns Abschied; Skolie; Der Soldat I, II; Sonne der Schlummerlosen; Das Standchen; Standchen (Alles wiegt die stille Nacht); Standchen (Komm in die stille Nacht); Uber Nacht; Unfall; Verschwiegene Liebe; Der Verzweifelte Liebhaber; Wachterlied auf der Wartburg; Wohin mit der Freud; Wohl denk ich oft; Zur Ruh zur Ruh. Dietrich Fischer-Diesdau, bar; Daniel Barenboim, pno. DG 2740 162 (3). (also DG 2709 067)

+-Gr 8-77 p331 +NR 12-77 p10
+HFN 8-77 p95 +RR 8-77 p81
+MJ 11-77 p30

Songs: Eichendorff Lieder: Waldmadchen; Verschwiegene Liebe. Goethe Lieder: Die Bekehrte; Epiphanias; Die Sprode. Morike Lieder: Schlafendes Jesuskind; Zum neuen Jahre. Spanish songbook: Ach des Knaben Augen; Die ihr schwebet; Nun wandre Maria. cf STRAUSS, R.: Songs (RCA ARL 1-1571).

Songs: Gesang Weylas; Neue Liebe. cf Bruno Walter Society BWS 729.

Songs: Schlafendes Jesuskind. cf Philips 9500 218.

WOLFF

La premiere lecon. cf Club 99-107.

WOLFF, Christian

Accompaniments. cf Lines.

2934 Lines. Accompaniments. Nathan Rubin, Thomas Halpin, vln; Nancy Ellis, vla; Judiyaba, vlc; Frederick Rzewski, pno. CRI SD 357.

+ARG 11-76 p44 -NR 11-76 p8
+-MJ 3-77 p46 +St 2-77 p124

WOLF-FERRARI, Ermanno

Il campiello: Intermezzo, Act 2; Ritornello. cf Works, selections (London STS 15362).

La dama boba: Overture. cf Works, selections (London STS 15362).

I Gioielli della Madonna (Jewels of the Madonna): Orchestra suite. cf Works, selections (London STS 15362).

I quattro Rusteghi: Prelude; Intermezzo, Act 2. cf Works, selections (London STS 15362).

School for fathers: Intermezzo. cf Citadel CT 6013.

2935 Il segreto di Susanna. Maria Chiara, s; Bernd Weikl, bar; ROHO; Lamberto Gardelli. Decca SET 617 Tape (c) KCET 617. (also London 1169 Tape (c) 5-1169)

++ARG 5-77 p45 +MT 7-77 p566
++Gr 11-76 p874 +NR 5-77 p10

++Gr 8-77 p349 tape
+HF 5-77 p94
++HF 10-77 p129 tape
+HFN 11-76 p169
++HFN 2-77 p135 tape
+MJ 7-77 p68
+NYT 3-16-77 pD15
+ON 2-26-77 p33
+OR 9/10-77 p28
++RR 11-76 p49
+St 6-77 p90

Il segreto di Susanna: Overture. cf Works, selections (london STS 15362).
Songs: Il campiello. cf Columbia M 34501.
Vanity fair. cf Citadel CT 6013.

2936 Works, selections: Il campiello: Intermezzo, Act 2; Ritornello. La dama boba: Overture. I gioielli della Madonna: Orchestral suite. I quattro Rusteghi: Prelude; Intermezzo, Act 2. Il segreto di Susanna: Overture. OSCCP; Nello Santi. London STS 15362. (Reissue) (also Decca SDD 452. Reissue)

+ARG 12-76 p44
+-HFN 1-76 p123
+MJ 4-77 p33
+-RR 1-76 p40

WOLKENSTEIN, Oswald
Songs: Mai; Unbek. cf BIS LP 75.

WOOD
Frescoes suite: Bandstand in Hyde Park. cf DJM Records DJM 22062.
London landmarks suite: The horseguards. cf DJM Records DJM 22062.

WOOD, Arthur
Dale dances. cf Grosvenor GRS 1048.

WOOD, Charles
Songs: Expectans expectavi. cf Abbey LPB 783.
Songs (anthems): Expectans expectavi; God omnipotent reigneth; Glory and honour and laud; Hail, gladdening light; Tis the day of resurrection; O thou the central orb. cf STANFORD: Songs (Argo ZRG 852).

WOOD, Gareth
Concerto, trumpet and brass band. cf Decca SB 328.

WOOD, Haydn
A brown bird singing. cf O'Oiseau-Lyre DSLO 20.
Paris suite: Montmartre. cf HMV SRS 5197.
Paris suite: Montmartre. cf Starline STS 197.

WOOD, Hugh
A garland for Dr. K. cf Universal Edition UE 15043.
Pieces, piano, op. 6 (3). cf BIRTWISTLE: Tragoedia.

WOOD, William
Sonata, violin and piano. cf THOMAS: Pricksong.

WOODBURY, I. B.
We are happy now, dear mother. cf New World Records NW 220.

WOODCOCK, Clement
Browning fantasy. cf BIS LP 57.
Browning fantasy. cf BIS LP 75

WOODCOCK, Robert
Concerto, flute and strings, D major. cf BOYCE: Concerto grosso, strings, E minor.
Concerto, oboe and strings, E flat major. cf BOYCE: Concerto grosso, strings, E minor.

WOODFIELD, Ray
The magic of Michel Legrand. cf EMI TWOX 1058.

WOODWARD
Songs: Ding don merrily on high; Unto us a son is born. cf HMV SQ CSD 3784.

THE WORCESTER FRAGMENTS. cf Nonesuch H 71308.

THE WORLD OF OPERA, vol. 2. cf Decca SPA 450.
WRANITZKY, Pavel (Paul)
German dances (10). cf DG 2533 182.
German dances (8). cf DG 2723 051.
Parthia, F major. cf Telefunken EK 6-35334.
Quodlibet. cf DG 2533 182.
Quodlibet. cf DG 2723 051.
WRIGHT, Maurice
Electronic composition. cf Odyssey Y 34139.
WUORINEN, Charles
Variations, bassoon, harp and timpani. cf BABBITT: Phenomena, soprano and piano.
WYLIE, Ruth
Psychogram. cf GOLDMAN: Sonata, violin and piano.
WYLKYNSON, Robert
Salve regina. cf Coimbra CCO 44.
WYNER, Yehudi
Intermedio. cf SOLLBERGER: Riding the wind I.
WYTON, Alec
Pieces. cf St. John the Divine Cathedral Church.
XENAKIS, Iannis
2937 Antikhthon. Auroura. Synaphai. NPhO; Geoffrey Douglas Madge, pno; Elgar Howarth. Decca HEAD 13.
+Gr 9-76 p435 +MM 4-77 p40
+Gr 9-77 p499 +MT 2-77 p135
++HFN 7-76 p101 +RR 7-76 p66
Aroura. cf Antikhthon.
Herma. cf CP 3-5.
Mikka. cf GLASS: Strung out.
Mikka "S". cf GLASS: Strung out.
Synaphai. cf Antikhthon.
YANNAY
Dawn and departure. cf Advance FGR 18.
YARDUMIAN, Richard
Passacaglia, recitative and fugue. cf GLAZUNOV: Concerto, piano, no. 1, op. 92, F minor.
YEFANOV
Battle of Liaoyang. cf HMV CSD 3782.
Victors triumph. cf HMV CSD 3782.
THE YIDDISH ART SONG. cf Olympic OLY 105.
YODER
Carolina clambake. cf EMI TWOX 1058.
YON, Pietro
Gesu bambino. cf London OS 26437.
Toccatina, flute. cf Vista VPS 1034.
YOSHIDA, Masao
Variations on a theme of "Sakura". cf RCA JRL 1-2315.
THE YOUNG CZIFFRA, paraphrases, transcriptions and improvisations. cf Connoisseur Society CS 2131.
THE YOUNG GILELS. cf Westminster WGM 8309.
YSAYE, Eugene
Mazurka, op. 11, no. 3, B minor. cf BARTOK: Hungarian folk songs.
Paganini variations. cf DG 2548 219.
Reve d'enfant, op. 14. cf Discopaedia MB 1016.
Sonata, violin and piano, no. 3, op. 27. cf BARTOK: Hungarian folk songs.

YUASA, Joji
Cosmos haptic. cf CP 3-5.
On the keyboard. cf CP 3-5.
ZABEL
La source. cf Musical Heritage Society MHS 3611.
ZACHOW, Friedrich
Prelude and fugue, G major. cf ABC ABCL 67008.
Suite, B minor. cf Toccata 53623.
ZADOR, Eugene
2938 Christopher Columbus. Robert Patterson, Dan Marek, Alice Garrott, Will Roy, Darrell Lauer, Harris Poor, Lawrence Farrar; Long Island Concert Choir; American Symphony Orchestra; Laszlo Halasz. Orion ORS 76251.
+ARG 10-77 p48
ZARZYCKI, Alexander
Mazurka, op. 26, G major. cf Discopaedia MB 1011.
ZEHLE, William
Wellington march. cf Grosvenor GRS 1048.
ZELEZNY, Lubomir
2939 Quintet, clarinet. Quintet, winds. Trio, piano. Jiri Stengl, clt; Semtana Trio, Prague Quartet, Prague Wind Quintet. Supraphon 111 1755.
/-ARG 6-77 p49 +NR 4-77 p5
Quintet, winds. cf Quintet, clarinet.
Trio, piano. cf Quintet, clarinet.
ZELLER, Karl
2940 Der Vogelhandler (The bird seller), excerpts. Anneliese Rothenberger, Renate Holm, s; Gisela Litz, con; Gerhard Unger, Adolf Dallapozza, t; Walter Berry, bs; VSOO and Chorus; Willi Boskovsky. Angel S 37165.
+HF 11-76 p134 ++SFC 6-27-76 p29
+NR 8-76 p10 +St 10-76 p140
+-ON 1-22-77 p33
Der Vogelhandler: Adam's song. cf KALMAN: Das Veilchen vom Montmartre: Heut nacht hab ich getraumt von dir.
Der Vogelhandler: Entr'acte, Act 3; Als begluht der Kirschenbaum. cf HMV ESD 7043.
ZESSO (15th century Italy)
E quando andaretu. cf Enigma VAR 1024.
E quando andarete al monte. cf Nonesuch H 71326.
ZIEHRER, Karl
Weaner Mad'ln, op. 338. cf Classics for Pleasure CFP 40213.
Wiener Burger, op. 419. cf Classics for Pleasure CFP 49213.
ZIMBALIST, Efrem
2941 Sonata, violin and piano, G minor. ZIMBALIST, Efrem Jr.: Sonata, violin and piano. Marilyn Thompson, John Sutherland, pno; Roy Malan, vln. Genesis GS 1070.
+SFC 3-6-77 p34
ZIMBALIST, Efrem Jr.
Sonata, violin and piano. cf ZIMBALIST, Efrem: Sonata, violin and piano, G minor.
ZINDARS, Earl
Trigon. cf BACH; Sonata, flute, E flat major.
ZIPOLI, Domenico
Suite, F major. cf ALBINONI: Concerto, 2 trumpets, C major.
ZUR, Menachem
Chants. cf Odyssey Y 34139.

Section II

MUSIC IN COLLECTIONS

ABBEY

LPB 712

2942 CAMPIAN: Shall I come, sweet love to thee. DANYEL: Dost thou withdraw thy grace. DOWLAND: In darkness let me dwell; Lady if you so spite me. HUMFREY: A hymn to God the Father. PURCELL: Evening hymn; The fatal hour; Man is for the woman made; The Queen's epicedium; Music for a while; There's nothing so fatal; Twas within a furlong of Edinburg town. ROSSETER: What then is love but mourning. ANON.: Like to the damask rose. Gerald English, t; David Lumsden, org, hpd; Jane Ryan, vla da gamba.

+HFN 12-77 p169 +RR 12-77 p76

LPB 719

2943 FRANCK: Piece heroique. IBERT: Pieces (3). HONEGGER: Choral. Fugue. MARTIN: Passacaille. MOTTU: Prelude et chorale. Josef Bucher, org.

+-Gr 11-77 p868 +RR 11-77 p92
+HFN 11-77 p177

LPB 752

2944 DAQUIN: Noel no. VII en trio et en dialogue. BACH: Trio sonata, no. 5, S 529, C major. BYRD: Ut Re me fa sol la. REGER: Fantasia and fugue on the chorale "Hallelujah, Gott zu loben bleibe meine Seelen Freund", op. 52, no. 3. STEWART: Prelude, organ and tape. Murray Somerville, org.

+Gr 10-76 p628 +RR 2-77 p77
+-HFN 11-76 p164

LPB 765

2945 BYRD: A gigg. COUPERIN, F.: L'Apotheose de Lully: Plainte des memes. CHEDEVILLE: Suite, no. 5, C major. HANDEL: Duo, F major. HOLBORNE: Consorts, 5 recorders. HOTTETERRE: Suite "La festin". JACOB: A consort of recorders. MORLEY: The frog galliard. NEWMAN: A pavyon. PURCELL: Chaconne, F major. SCHEIDT: Suite, C major. ANON.: Beata viscera; Coranto; Gentil prince. Dolmetsch Consort.

+Gr 9-77 p452 +RR 9-77 p77
+-HFN 9-77 p141

LPB 770

2946 BERKELEY: The Lord is my shepherd. BYRD: Praise our Lord, all ye gentiles. NAYLOR: O Lord, almighty God. RACHMANINOFF: Hymn to the cherubim. WALMISLEY: Remember, O Lord. WALTON: Magnificat and nunc dimittis. WEELKES: Hosanna to the Son of David. WESLEY, S. S.: Cast me not away. Chichester Cathedral Choir; John Birch.

+Gr 12-77 p1133 +-RR 12-77 p77
+HFN 12-77 p172

LPB 776

2947 BRUCKNER: Ecce sacerdos magnus. HAYDN: Insanae et vanae curae. LEIGHTON: God is ascended. SHEPHERD: Verbum caro factum est. STANFORD: Magnificat and nun dimittis, B flat major. TALLIS: O nata lux. TURNER: How far is it to Bethlehem. VAUGHAN WILLIAMS: For all the saints (arr. Bielby). TRAD.: O leave your sheep (arr. Leighton). Wakefield Cathedral Choir; Yorkshire Brass Ensemble; Jonathan Bielby, Peter Gould, org.
+Gr 12-77 p1133 +RR 10-77 p83

LPB 778

2948 ARNE: The lass with the delicate air. BOUGHTON: The immortal hour: Faery song. BRITTEN: The birds. MOZART: Le nozze di Figaro, K 492: Voi che sapete. SCHUBERT: Horch, horch, die Lerch; Der Musensohn, D 764; Lachen und Weinen, D 777; The shepherd on the rock, D 965. SPOHR: Zwiegesang. STANFORD: The monkey's carol. VAUGHAN WILLIAMS: Linden Lea. TRAD. (English): The tailor and mouse. TRAD. (Irish): The lark in the clear air (arr. Tate); Trottin' to the fair (arr. Stanford). Andrew Wicks, treble; Lee Stevenson, clt; John Birch, pno.
+Gr 8-77 p332 +MT 10-77 p838
+-HFN 7-77 p123 +RR 7-77 p89

LPB 779

2949 BACH: Cantata, no. 95, Content am I to leave thee. Cantata, no. 140, Zion hears her watchmen's voices. Cantata, no. 147, Jesu joy of man's desiring. BOURGEOIS: Dark'ning night the land doth cover (arr. Ley). BRAHMS: A German requiem, op. 45: How lovely is thy dwelling place. MUDD: Let thy merciful ears, O Lord. MUNDY: O Lord the maker of all things. STANFORD: Magnificat, G major. TALLIS: If ye love me. VAUGHAN WILLIAMS: Te deum, G major. WESLEY, S.S.: Blessed be the God and father. Noel Rawsthorne, org; Liverpool Cathedral Choir; Ronald Woan.
+-Gr 12-77 p1133 +MT 12-77 p1017
+-HFN 11-77 p177 +RR 10-77 p82

LPB 783

2950 BENNETT: Heare us, O heare us, Lord. FINZI: God is gone up with a triumphant shout. PALESTRINA: Dum complerentur dies Pentecostes. PARRY: Songs of farewell: My soul there is a country. SANDERS: Festival te deum. SUMSION: Magnificat, G major. Nunc dimittis. WESLEY, S.S.: Ascribe unto the Lord. WOOD: Expectans expectavi. Gloucester Cathedral Choir; John Sanders, Andrew Millington, org; John Sanders.
+Gr 11-77 p880 +RR 11-77 p103
+HFN 11-77 p171

ABC

L 67002

2951 Christmas music of the 15th and 16th centuries. AGRICOLA: Christum wir sollen loben schon; Gelobet seist du Jesu Christ. ECCARD: Ich steh an deiner Krippen hier. FULDA: Dies est laeticiae. GALLICULUS: In natali domini; Magnificat V toni. HERMAN: Lobt Gott ihr Christen. HOYOUL: Gelobet seist du Jesu Christ. KUGELMANN: Dies est laeticiae. NUCIUS: Freut euch ihr Auserwahlte. ANON.: En natus est Emanuel; Christum wir solen loben schon; In natali domini; Jure plaudant omnia. Capella Antiqua, Munich; Konrad Ruhland.

++HF 3-77 p108 +PRO 5-77 p25
+-MJ 2-77 p30 ++St 5-77 p54
+NR 11-77 p2

ABCL 67008 (2)

2952 AMMERBACH: Wer das Tochterlein haben will. BLITHEMAN: Eterne rerum. EBERLIN: Toccata sexta. Toccata e fuga tertia. FISCHER: Preludes and fugues, B minor, D major, E flat major, C minor. FROBERGER: Capriccio, no. 8. Ricercare, no. 1. FUX: Sonata quinta. KERLL: Canzona, G minor. Toccata con durezza e ligature. KREBS: Jesu meine Freude. Jesus meine Zuversicht. Von Gott will ich nicht lassen. MERULA: Un cromatico ovvero capriccio. MUFFAT: Fugue, G minor. NEWMAN: Pavan. PACHELBEL: Alle Menschen mussen sterben. Magnificat: Fugues, nos. 4, 5, 10, 13. Toccata and fugue, B flat major. PASQUINI: Canzone francese. Ricercare. STORACE: Ballo della battaglia aus. TAYLOR: Pavan and galliard. ZACHOW: Prelude and fugue, G major. ANON.: Gagliarda "Cathaccio". Gagliarda "Lodesana". Pavan and galliard. Gustav Leonhardt, org.

++HF 3-77 p108 +PRO 5-77 p25
+-MJ 2-77 p30 ++St 5-77 p54
+NR 4-77 p15

AB 67014

2953 DEBUSSY: Il pleure dans mon coeur. FAURE: Elegy, op. 24, C minor. Papillon, op. 77. Sicilienne, op. 78. RACHMANINOFF: Vocalise, op. 34, no. 14. TCHAIKOVSKY: Pezzo capriccioso, op. 62. Nocturne, op. 19, no. 4. TOCH: Impromptu, op. 90. WEBER: Sonata, violoncello and piano, A major. Jeffrey Solow, vlc; Doris Stevenson, pno.

+-ARG 11-77 p6

ADVANCE

FGR 18

2954 New choral music: The ineluctable modality. LONDON: Bjorne Enstabile's Christmas music. NEWELL: Ryo-nen. PECK: Questions from the electric chairman. POWELL: The Zelanski medley. WEINBERG: Vox in Bama. YANNAY: Dawn and departure. Instrumental Ensemble and Choir; Edwin London.

+-NR 10-77 p9

ADVENT

ASP 4007 (available from St. Peter's Episcopal Church, Tecumseh, Mich.)

2955 BACH: Chorale prelude, O Gott, du frommer Gott, S 767. Toccata and fugue, S 565, D minor. CARR: Variations on the Sicilian hymn. PAINE: Variations on the Austrian hymn. TAYLOR: Variations on Adeste fideles. THOMSON: Variations on Shall we gather at the river.

+MU 5-77 p17

S 5023

2956 Metropolitan opera 25th anniversary: BIZET: Carmen: Je dis que rien ne m'epouvante. KORNGOLD: Die tote Stadt, op. 12: Marietta's Lied. MOZART: Le nozze di Figaro, K 492: Dove sono. PUCCINI: Madama Butterfly: Un bel di. Suor Angelica: Senza mama. Tosca: Vissi d'arte. VERDI: Aida: Ritorna vincitor. Un ballo in maschera: Morro, ma prima in grazia. La forza del destino: Pace, pace mio Dio. WAGNER: Lohengrin: Einsam in

truben Tagen. Lucine Amara, s; Richard Woitach, pno.
+-NR 4-77 p10

AMBERLEE

ALM 602
2957 BYRD: The Earle of Salisbury pavan (arr. Lewin). DANIELSEN: Musical saunter. DVORAK: Humoresque, op. 101, no. 7, G sharp major (arr. Cinquavento). JANSEN: Fragments. LEWIN: The poacher. MacDOWELL: Woodland sketches, op. 51, no. 1: To a wild rose (arr. Husband). MOZART: Sonata, piano, no. 11, K 331, A major: Rondo alla turca. MUSSORGSKY: Fair at Sorochinsk: Gopak (arr. Lewin). POLDINI: Marionnettes: Poupee valsante (arr. Husband). PROKOFIEV: Symphony, no. 1, op. 25, D major: Gavotte (arr. Staines). TCHAIKOVSKY: The nutcracker, op. 71: Ballet suites (arr. Lovelock). TRAD.: Bobby Shafto (arr. Beck). Halle Wind Quintet, Norwegian Wind Quintet, Adelaide Wind Quintet.
+-Gr 3-77 p1461 +-RR 3-77 p73
+-HFN 4-77 p147

ANGEL

S 36053 Tape (c) 4XS 36053 (ct) 8XS 36053
2958 Christopher Parkening: Music of two centuries. COUPERIN (Parkening): Livres de clavecin, Bk II, Ordre no. 6: Les barricades mysterieuses. DEBUSSY (Marshall): Prelude, Bk I, no. 8: The girl with the flaxen hair. HANDEL (Parkening): Minuet, D major. Sarabande and variations. POULENC (Marshall): Pastourelle. RAVEL (Hyman): Empress of the pagodas. Pavane of the sleeping beauty. SATIE (Parkening): Gymnopedies, no. 1 and 3 (Marshall). SCARLATTI: Preambulo and allegro vivo. VISEE (Parkening): Giga. WEISS (Parkening): Passacaglia. TRAD. (Marshall): Afro-Cuban lullaby. Christopher Parkening, gtr.
++NR 11-77 p15 ++St 5-77 p103

S 36093
2959 GIULIANI: Grand overture, op. 61. MUDARRA: Fantasia. Gallarda. NARVAEZ: Variations on "Guardame las vacas". SANZ: Suite espanola. SCARLATTI, D.: Sonatas, guitar, L 83, A major; L 352, E minor; L 423, A minor; L 483, D major (arr. Romero). SOR: Variations on a theme by Mozart, op. 9. Angel Romero, gtr.
++NR 4-76 p14 +St 1-77 p71

S 36094 (also HMV HQS 1401)
2960 ALBENIZ: Espana, op. 165, no. 2: Tango. Cantos de Espana, op. 232: Cordoba. GRANADOS: Tonadillas al estilo antiguo: La Maja de Goya. RODRIGO: Fandango. TARREGA: Adelita. Estudio brillante. Maria. Marieta. Mazurka. Preludes, nos. 2, 5. TORROBA: Madronos. TURINA: Fandanguillo. Garrotin. Rafaga. Soleares. Angel Romero, gtr.
+Gr 12-77 p1117 +St 1-77 p71
*NR 7-76 p13

S 37219 (also HMV SQ HQS 1374)
2961 BACH: Partita, violin, no. 3, S 1006, E major: Prelude, gavotte, gigue. BIZET: L'Arlesienne: Minuetto, no. 1. KREISLER: Liebesfreud. Liebesleid. MENDELSSOHN: A midsummer night's dream, op. 61: Scherzo. MUSSORGSKY: Fair at Sorochinsk: Gopak. RACHMANINOFF: Daisies, op. 38, no. 3. Alfred de Musset, op. 21, no. 6, fragment. Humoresque, op. 10, no. 5 (revised version). Lil-

acs, op. 21, no. 5. Melodie, op. 3, no. 3, E major (revised version). Prelude, op. posth, D minor. RIMSKY-KORSAKOV: The tale of the Tsar Sultan: Flight of the bumblebee. SCHUBERT: Die schone Mullerin, op. 25, D 795: Wohin. TCHAIKOVSKY: Cradle song, op. 16, no. 1. Garrick Ohlsson, pno.

+-Gr 6-77 p75 +-NR 12-77 p5
+HF 11-77 p122 +RR 6-77 p83
+-HFN 7-77 p117 -SFC 1-24-77 p40

S 37250
2962 BIZET: L'Arlesienne: Suite, no. 2: Farandole. CHABRIER: Espana. Marche joyeuse. PONCHIELLI: La gioconda: Dance of the hours. TCHAIKOVSKY: Sleeping beauty, op. 66: Waltz. VERDI: Aida: Ballet music. WALDTEUFEL: The skaters, op. 183: Waltz. WEBER (arr. Berlioz): Invitation to the dance, op. 63. WEINBERGER: Schwanda, the bagpiper: Polka. PhO; Herbert von Karajan.

-NR 8-77 p3 ++St 6-77 p148

(Q) S 37304
2963 A Palm Court concert. BERGER: Amoureuse. DRDLA (arr. Dumont): Souvenir. DRIGO: Serenade. FARIS: Theme from Upstairs, Downstairs. GUNGL (arr. Winter): Casino dances. HERBERT (arr. Langley): The fortune teller, excerpts. L'Encore. HEUBERGER: Der Opernball: In chambre separee. LINCKE: Folies Bergere. The glow-worm. MONTI (arr. Baron): Czardas. MORET: Silver heels. PIEFKE (arr. Ascher-Mahl): Kutschke Polka. TIERNEY (arr. Lange): Irene: Alice blue gown. Albert White and His San Francisco Masters of Melody.

+ARG 2-77 p54 +St 2-77 p132
+NR 9-77 p2

S 37409 Tape (c) 4XS 37409
2964 BARBER: Adagio, strings. BUTTERWORTH: The banks of green willow. DEBUSSY: Prelude a l'apres-midi d'un faune. FALLA: The three-cornered hat: Dances (3). GLINKA: Russlan and Ludmilla: Overture. STRAUSS, J. II: Emperor waltz, op. 437. LSO; Andre Previn.

+HF 10-77 p128

S 37446
2965 BELLINI: I Puritani: O rendetemi la speme...Qui la voce...Vien diletto. CILEA: Adriana Lecouvreur: Poveri fiori. MASCAGNI: L'amico Fritz: Son pochi. MOZART: Le nozze di Figaro, K 492: E Susanna non viene...Dove sono. PUCCINI: Gianni Schicchi: O mio babbino caro. Madama Butterfly: Un bel di. La rondine: Chi il bel sogno il Doretto. Tosca: Vissi d'arte. Turandot: Signore, ascolta. VERDI: Otello: Ave Maria. La traviata: E strano...Ah fors'e lui...Follie Follie. Mirelli Freni, s; Orchestral accompaniments.

+NR 11-77 p13

ARGO

ZK 3 (Reissue from ZRG 5406)
2966 GOSS: If we believe that Jesus died. NARES: The souls of the righteous. WALMISLEY: Magnificat and nunc dimittis, D minor. WESLEY, S.: In exitu Israel. WESLEY, S.S.: Blessed be the God and Father; Thou wilt keep him in perfect peace; The wilderness. St. John's College Chapel Choir; Brian Runnett, org; George Guest.

++Gr 1-77 p1173 +-RR 1-77 p83
+HFN 1-77 p121

ZK 11
2967 FAURE: Requiem, op. 48: Pie Jesu. FRANCK: Panis Angelicus. HANDEL: Serse: Ombra mai fu. SCHUBERT: Salve regina. STANFORD: Biblical songs, op. 113, nos. 1-6. George Banks-Martin, treble; Simon Preston, org.
+Gr 11-77 p909 +-RR 11-77 p103
+HFN 11-77 p164

ZK 24 Tape (c) KZKC 24 (Reissue from ZRG 566)
2968 Music to entertain Henry VIII: BARBIREAU: En frolyk weson. BUSNOIS: Fortune esperee. CORNYSHE: Ah, Robin; Adieu, mes amours; Blow thy horn, hunter. DAGGERE: Downberry down. HENRY VIII, King: Pastime with good company; Taunder naken. ISAAC: La mi la sol. RICHAFORT: De mon triste deplaisir. ANON.: La morisque; Passo e mezzo; La rocha el fuso; La traditora; Vegnando da Bologna; Henry VIII's pavan; Hey trolly lolly lo; I am a jolly foster; Instrumental fancy; My Lady Carey's dumpe; Where be ye my love. Purcell Consort of Voices, Musica Reservata; Grayston Burgess, Michael Morrow.
+Gr 7-77 p211 +RR 7-77 p87
+HFN 8-77 p99 tape +RR 12-77 p97 tape
+HFN 10-77 p167

ZK 25 Tape (c) KZKC 25 (Reissue from ZRG 643)
2969 Music to entertain Elizabeth I: BENNET: All creatures now. EAST: Hence stars. FARMER: Fair nymphs. HOLMES: Thus bonny-boots. HUNT: Hark, did ye ever hear. MARSON: The nymphs and shepherds. MORLEY: Arise, awake; Hard by a crystal fountain. MUNDY: Lightly she whipped. PEELE: Descend ye sacred daughters; Long may she come. TOMKINS: The fauns and satyrs. WEELKES: As Vesta was from Latmos Hill. WILBYE: The lady Oriana. ANON.: Daphne; Strawberry leaves; A toye; La volta. Purcell Consort of Voices; London Cornett and Sackbut Ensemble; Elizabethen Consort of Viols; Grayston Burgess.
+Gr 7-77 p211 +RR 10-77 p97 tape
+HFN 10-77 p171 tape +RR 8-77 p76
+-HFN 10-77 p167

D26D4 (4) (Reissues from ZRG 860, 573, 680, 696)
2970 BUTTERWORTH: A Shropshire lad: Rhapsody. The banks of green willow. English idylls (2). BRITTEN: Variations on a theme by Frank Bridge, op. 10. ELGAR: Elegy, strings, op. 58. Introduction and allegro, op. 47. Serenade, strings, op. 20, E minor. Sospiri, op. 70. The Spanish lady, op. 89: Suite. TIPPETT: Concerto, 2 strings orchestras. Fantasia concertante on a theme by Corelli. Little music, string orchestra. VAUGHAN WILLIAMS: Fantasia on a theme by Thomas Tallis. Fantasia on "Greensleeves". The lark ascending. Variants on "Dives and Lazarus" (5). AMF; Neville Marriner.
+Gr 1-77 p1154 +RR 2-77 p45
+HFN 1-77 p121

ZK 28/9 (2) (Some reissues from ZRG 5418, 5439)
2971 BENNETT: Tom o'Bedlams song. BRIDGE: Goldenhair; Journey's end; Tis but a week; When you are old; So perverse. BUSCH: Come, o come, my life's delight; The echoing green; If thou wilt ease thine heart; The shepherd. BUSH: Voices of the prophets. DIEREN: Dream pedlary; Take, o take those lips away. DELIUS: To daffodils. GRAINGER: Bold William Taylor. IRELAND: Friendship in misfortune; The land of lost content; Love and friendship; The one hope; The trellis. MOERAN: The merry month of

May. RAINIER: Cycle for declamation. TIPPETT: Songs for Ariel. WARLOCK: Along the stream; Piggesnie. Peter Pears, t; Benjamin Britten, pno; Jean Dickson, vlc; Alan Bush, Viola Tunnard, pno.

+HFN 12-77 p185 +RR 12-77 p76

D40D3 (3) (Reissues from ZRG 673, 642, 728)

2972 BOLOGNA: Fenice fu. CONON DE BETHUNE: Ahi, amours. COUCY: Le noviaus tens. DIJON: Chanterei por mon corage. FINCK: Sauff aus und machs nit lang. FIRENZE: Da da a chi avareggia. GAUCELM FAIDIT: Foetz chausa es. ISAAC: Heliogierons nous. Innsbruck, ich muss dich lassen. Maudit seyt. La Mora. KOTTER: Kochersperger Spanieler. KREUTZENHOFF: Frisch und frolich wollen wir leben. LANDINI: La bionda treccia. Cara mie donna. Con bracchi assai. De dimni tu. Donna'l tuo partimento. Ecco la primavera. Giunta vaga bilta. Questa fanciulla amor. MARCABRU: Pax in nomine Domini. PIERO: Con dolce brama. RICHARD I, King of England: Ja nus hons pris. SENFL: Ach Elslein, liebes Elselein. Entlaubet ist der Walde. Das Glaut zu Speyer. Gottes Namen fahren wir. Ich weiss nit, was er ihr verhiess. Ich stund an einem Morgen. Meniger stellt nach Geld. Mit Lust tritt ich an diesen Tanz. Quis dabit oculis nostris. Was wird es doch. TERANO: Rosetta. THIBAUT DE CHAMPAGNE: Au tens plain de felonnie. VOGELWEIDE: Palastinalied. ANON: O nacio mane prima. Cum sint difficilia. Danse real. Istampitta Ghaetta. Je ne puis, Amors me tient, Veritatem. Lamento di Tristan. La Manfredina. O tocius Asie. Parti de mal. Quan je voy le duc. La quinte estampie real. Saltarelli (2). Sede Syon, in pulvere. La tierche estampie real. Trotto. La ultime estampie real. Welscher tantz. Early Music Consort; David Munrow.

+Gr 9-77 p455 +RR 9-77 p58
++HFN 9-77 p151

D69D3 (3) Tape (c) K69K33 (Reissues from ZRG 577, 820, 733/5, 585, 5444, 5400, 585, 836, 800/1, 644, 733/4, D3D4)

2973 ARNE: Concerto, harpsichord, no. 5, G minor. BACH: Concerto, flute and strings, G minor (arr. from S 1056). Concerto, oboe, violin and strings, D minor (arr. from S 1060). CORELLI: Concerto grosso, op. 6, no. 8, G minor. FASCH: Concerto, trumpet, D major. HANDEL: Concerto, oboe, no. 3, G minor. Concerto grosso, op. 3, no. 1. Concerto, organ, op. 7, no. 4, D minor. TELEMANN: Concerto, trumpet, D major. Concerto, viola, G major. VIVALDI: Concerto 2 trumpets, P 75. Concerto, violin, op. 4, no. 1, B flat major. Concerto, 4 violins, op. 3, no. 10. Concerto, 2 violins, op. 3, no. 8, A minor. George Malcolm, hpd and org; Carmel Kaine, Alan Loveday, Iona Brown, Roy Gillard, vln; William Bennett, flt; John Wilbraham, tpt; Tess Miller, Roger Lord, ob; Stephen Shingles, vla; Philip Jones, tpt; AMF; Neville Marriner.

+-Gr 9-77 p437 ++RR 9-77 p47
++HFN 9-77 p151 ++RR 12-77 p91 tape
++HFN 12-77 p187 tape

ZFB 95/6 (2)

2974 ADAMS: The holy city. BENNETT: The carol singers. BRAHE: Bless this house. BOND: A perfect day. CLARKE: The blind ploughman. DAVIS: God will watch over you. GLOVER: Rose of Tralee. GOULD: The curfew. HARRISON: Give me a ticket to heaven. HUHN: Invictus. KNIGHT: Rocked in the cradle of the deep. LAMB: The volunteer organist. LOHR: When Jack and I were children. MASCAGNI: Ave Maria (adapted from Intermezzo from Cavalleria Rusticana). MOSS: The floral dance. MURRAY: I'll walk beside you. PEPPER: Over the rolling sea. QUILTER: Now sleeps the crimson petal, op. 3, no. 2. RASBACH: Trees. SANDERSON: Friend o' mine. TOSTI: Parted. TOURS: Mother o' mine. WATSON: Anchored. ANON.: Mr. Shadowman (arr. E. Kaye). Benjamin Luxon, bar; David Willison, pno.

+Gr 10-76 p661 +RR 10-76 p96
+HFN 11-76 p160 +St 12-77 p95

SPA 507 (Reissue from ZRG 5448)

2975 CLARKE: Prince of Denmark's march (arr. Preston). ELGAR: Imperial march, op. 32 (arr. Martin). GUILMANT: March on a theme by Handel. HANDEL: Saul: Dead march (arr. Cunningham Woods). KARG-ELERT: March triomphale "Now that we all our God". PURCELL: Trumpet tune (arr. Trevor). SCHUMANN: Canons, op. 56, no. 5, B minor (arr. West). VIERNE: Symphony, no. 1, op. 14, D minor: Finale. WAGNER: Tannhauser: Pilgrims chorus (arr. Lemare). WALTON: Crown imperial (arr. Murrill). Simon Preston, org.

+Gr 11-77 p868 +-RR 10-77 p76
+HFN 10-77 p167

ZRG 806

2976 BACH: Passacaglia and fugue, S 582, C minor. BUXTEHUDE: Passacaglia, D minor. Prelude, fugue and chaconne, C major. BYRD: Ut, Re. CABANILLES: Passacalles du ler mode. CHAMBONNIERES: Chaconne, G major. FRESCOBALDI: Cento. Partite sopra passacagli. PACHELBEL: Chaconne, F minor. RAISON: Trio en passacaille. Peter Hurford, org.

++Gr 4-76 p1644 ++MT 9-77 p739
+HFN 3-76 p95 ++RR 3-76 p59

ZRG 823 Tape (c) KZRC 823

2977 AGRICOLA: Oublier veul. BYRD: Earle of Oxford's march (arr. Howarth). FARNABY: Giles Farnaby's dream. His rest. The new Sa-Hoo. The old spagnoletta. Tell me, Daphne. A toye. FRANCHOS: Trumpet intrada (ed. Herbert). GIBBONS: In nomine (arr. Howarth). Royal pavane (ed. Jones). LASSUS: Madrigal dell' eterna. PASSEREAU: Il est bel et bon (arr. Reeve). SUSATO: La bataille. Bergeret sans roch. Branle quatre branles. Mon amy, ronde. Ronde. La Mourisque. VECCHI: Saltarello. Philip Jones Brass Ensemble.

+Audio 12-76 p91 ++NR 10-77 p7
+Gr 6-76 p61 +RR 6-76 p63
+HFN 6-76 p97 +RR 10-76 p106 tape

ZRG 845 Tape (c) KZRC 845

2978 BARBER: Quartet, strings, op. 11, B minor: Adagio. COPLAND: Quiet city. COWELL: Hymn and fuguing tune, no. 10. CRESTON: A rumor. IVES: Symphony, no. 3. Celia Nicklin, ob, cor anglais; Michael Laird, tpt; AMF; Neville Marriner.

++Gr 7-76 p182 +-MT 2-77 p133
++Gr 11-76 p887 tape ++NR 6-76 p5
+-HF 10-76 p132 +NYT 7-4-76 pD1
+HFN 7-76 p84 +RR 7-76 p42
+-MJ 7-76 p57 ++St 10-76 p123

ZRG 851

2979 ARBAN: Etude characteristique (arr. Howarth). ARNOLD: Fantasy, trombone. DUFAY: Pasce tuos (arr. Howarth). BENNETT: Fanfare,

brass quintet. MAURER: Pieces, brass quintet (4). PREVIN: Outings for brass (4). SIMON: Quartet in form of a sonata. TCHAIKOVSKY: Sleeping beauty, op. 66: Waltz (arr. Fletcher). Philips Jones Brass Ensemble.
+Gr 7-77 p201 +RR 7-77 p68
+HFN 7-77 p111

ZRG 859
2980 BASSANO: Dic nobis Maria. CROCE: Ave virgo. GABRIELI, A.: Jubilate Deo; Maria stabat; Egredimini et videte; O Rex gloriae; Te deum patrem ingenitum. GABRIELI, G.: Cantata domine; Exultat iam Angelica; Deus, Deus meus; Beata es virgo. MONTEVERDI: Laudate dominum terzo. Richard Gowman, org; Magdalen College Choir; Bernard Rose.
+-Gr 4-77 p1588 +NR 9-77 p8
+HFN 3-77 p107 +RR 4-77 p83
+MT 9-77 p739

ZRG 864
2981 BULL: Dr. Bull's jewel. Dr. Bull's my selfe. DAQUIN: Noel suisse. GRIGNY: Tierce en taille. DUBOIS: Toccata. DUPRE: Fileuse. LANGLAIS: Dialogue sur lex mixtures. MULET: Tu es Petrus. SWEELINCK: Mein junges Leben hat ein End. TOMKINS: Worster braules. VIERNE: Impromptu. WIDOR: Symphony, organ, no. 6, op. 42, no. 2, G major: Allegro. Gillian Weir, orgn.
+Audio 9-77 p48 ++HFN 3-77 p109
+Gr 3-77 p1430 +RR 3-77 p82

ZRG 871
2982 BRAHMS: Geistliches Wiegenlied, op. 91, no. 2. BRUCKNER: Ave Maria; Graduale; Virga Jesse. ELGAR: Give unto the Lord, op. 67. FAURE: Ave verum. KALINNIKOV: I will love thee. RACHMANINOFF: Hymn of the cherubim (edit. Henderson). VERDI: Ave Maria; Pater noster. Christ Church Cathedral Choir; Colin Walsh, org; Simon Preston.
+-Gr 12-77 p1133 +RR 12-77 p81
+HFN 12-77 p177

ARION

ARN 34348
2983 Gregorian chant: Noel Provencal. Provence of the Abbey Saint-Victor Musicians.
+ARG 12-77 p55

ARN 90416
2984 CORELLI: Sonata, op. 5, no. 8, D minor. DE MONZA: Ecclesiastical concerto (realized by J. P. Mathieu). FRESCOBALDI: Canzoni, D major, G major, C major. MARCELLO: Sonata, A minor. PULITI: Concerti, nos. 4 and 5 (realized by. J. P. Mathieu). VIVALDI: Sonata, op. 8, no. 2, C major. Jean-Pierre Mathieu, trom and sackbut; Georges Delvallee, org.
+ARG 7-77 p39

ATLANTIC

SD 2-7000 (2) Tape (c) CS 2-7000 (ct) TP 2-7000
2985 BACH: Inventions, 2 part, S 775, D minor. EMERSON: Concerto, piano, no. 1. Fanfare for the common man (arr. from Copland). Pirates. LAKE: C'est la vie; Closer to believing; Hallowed be thy name; Lend your love to me tonight; Nobody loves you like

I do. PALMER: New Orleans. PALMER/EMERSON: L. A. nights; Tank. PALMER/SOUTH: Food for your soul. PROKOFIEV: Scythian suite, op. 20: The enemy god dances with the black spirit (arr.) Keith Emerson, pno, synthesizer; Greg Lake, gtr; Carl Palmer, drum; Greg Lake, vocals; LPO; Paris Opera Orchestra, Orchestra; John Mayer, Godfrey Salmon.
+-St 7-77 p116

BASF

22-22645-3 (2)
2986 BEETHOVEN: Fidelio, op. 72: Abscheulicher wo willst du hin. HALEVY: La Juive: Il va venir. MASCAGNI: Cavalleria rusticana: Voi lo sapete. MASSENET: Thais: Ah je suis seule...Dis-moi que je suis belle. PONCHIELLI: La gioconda: Suicidio. STRAUSS, R.: Elektra, op. 58: Allein weh ganz allein. Der Rosenkavalier, op. 59: Da geht er hin, Die Zeit die ist ein sonderbar Ding. Salome, op. 54: Ah du wolltest mich deinen Mund. VERDI: Aida: Ritorna vincitor. Don Carlo: O don fatale. Macbeth: La luce langue. Simon Boccanegra: Come in quest ora bruna. Il trovatore: D'amor sull ali rosee. WAGNER: Lohengrin: Einsam in truben Tagen. Tannhauser: Dich teure Halle. Tristan und Isolde: Mild und leise. Die Walkure: Der Manner Sippe. Astrid Varnay, s; Orchestral accompaniments.
+NR 4-77 p10

BBC

REC 267
2987 BIZET: Fair maid of Perth: Serenade (arr. Owen). DE RILLE: The martyrs of the arena. EVANS: Pant y Fedwen. EDWARDS (Hand): Take me home. GOUNOD: Faust: Soldiers chorus. HANDEL: Judas Maccabaeus: Sound an alarm. HARTSOUGH: Gwahoddiaa. WAGNER: Tannhauser: Pilgrims chorus. TRAD.: Dafydd y Garreg Wen; Jacob's ladder; Lily of the valley; The virgin Mary had a baby boy. Massed Welsh Male Voices; National Brass Band of Wales; John Peleg-Williams, Marilyn Phillips, pno; Donald Hendy.
+RR 4-77 p81

REH 290 (also RESL 48)
2988 Highlights from the last night of the Proms, 1974: ARNE: Rule Britannia (arr. Sargent). ELGAR: Pomp and circumstance march, op. 39, no. 1, D major. HOLST: The perfect fool, op. 39: Ballet music. PARRY: Jerusalem, op. 208 (orch. Elgar). VAUGHAN WILLIAMS: Serenade to music. Extracts from speech by Charles Groves. Norma Procter, con; BBC Singers; BBC Symphony Orchestra; Charles Groves.
+-Gr 11-77 p909 +-RR 11-77 p109

BG

HM 57/8 (2)
2989 The English madrigal school. BARTLETT: Of all the birds that I do know. BENNET: all creatures now are merry-minded; Weep o mine eyes. EDWARDS: In going to my naked bed; When griping griefs. JOHNSON: Defiled is my name; Benedicam Domino. MORLEY: He who comes here; Sweet nymph. SHEPHERD: O happy dames. TALLIS: Like as the doleful dove. VAUTOR: Mother I will have a

husband. WARD: Hope of my heart. WEELKES: All at once well met; The ape the monkey and the baboon; Cease sorrows, now; O care thou wilt despatch me; On the plains fairy trains; Strike it up Tabor; Thule the period of cosmography; To shorten winter's sadness; Young cupid hath proclaimed. WILBYE: Adieu sweet Amaryllis; Flora gave me fairest flowers; Lady when I behold; Oft have I vowed; Sweet honey-sucking bees. ANON.: The bitter sweet; The happy life; I smile to see how you devise. Deller Consort; Alfred Deller.
+NR 12-77 p7

BIS

LP 3
2990 DUNSTABLE: O rosa bella. ENCINA: Todos los bienes. GLOGAUER LIEDERBUCH: Zwe Lieder. HEINTZ: Da truncken sie. MUSET: Quant je voi. PAUMANN: Ellend du hast. PHALESE: Dances. SENFL: Lieder (5). SUSATO: Dances. VON RUGEN: Loybere risen. ANON.: Canciones (2). Estampie and trotto. Sumer is icumen in. Joculatores Upsaliensis.
/-Gr 7-75 p234 +-St 3-77 p148
+RR 6-75 p83

LP 22
2991 Music for lute and gamba. ABEL: Sonata, G major. BALLARD: Allemande. Courante. Prelude. Rocantins. CAIX d'HERVELOIS: Suite, A major. DOWLAND: Resolution. MORLEY: Fancy. Lamento. ORTIZ: Quinta pars. Recercadas primera y segunda. RAUH/NEUSIDLER: Ach Elslein. ROBINSON: The Queenes good night. Bengt Ericson, vla da gamba; Rolf La Fleur, lt.
+HFN 12-77 p177 ++St 8-76 p106

LP 27
2992 Old Swedish organs. BACH: Chorale preludes, S 636, S 711, S 729, S 731. BRAHMS: Choralvorspiele, op. 122, nos. 5, 8, 11. MENDELSSOHN: Sonata, organ, no. 2. REGER: Pieces, organ, op. 59, nos. 2, 5. RITTER: Sonatina. WALTHER: Concerto del Sigr. Albinoni. Rune Engso, org.
+Gr 3-77 p1454 +RR 2-77 p81
+HFN 2-77 p125

LP 30
2993 Works, for flute and guitar. CASTELNUOVO-TEDESCO: Sonatina, op. 205. GIULIANI: Sonata, flute and guitar, op. 85, A major. IBERT: Entr'acte. RAUTAVAARA: Sonata, flute and guitar. KOCH: Canto e danza. Gunilla von Bahr, flt; Diego Blanco, gtr.
+ARG 8-77 p38 +RR 6-76 p71
-Gr 6-76 p62

LP 32
2994 New Swedish music. ALLDAHL: Stem the blood flow. GEFORS: Songs about trusting (4). HERMANSON: Sound of a flute. Winter flute. HOLEWA: Concertino, no. 3, PERGAMENT: Sonatina. ROSELL: Poem in the dark. SANDSTROM: Close to. Gundrun Raccuja, ms; Rolf Leanderson, bar; Mats Persson, pno; Gunilla von Bahr, flt; Kjell-Inge Stevensson, clt; Christer Torge, hn; Martin Bergstrand, double bs; Jorgen Rorby, gtr; Helmut Sitar, Seppo Asikainen, perc; Musica Sveciae, Miklos Maros Ensemble.
+RR 7-77 p71

LP 34
2995 DEBUSSY: Chansons de Bilitis (3). HALLNAS: Songs (3). LIDHOLM:

Songs (6). MILHAUD: Catalogue de fleurs. NYSTROEM: Songs (3). RAVEL: Songs: Melodies populaires grecques (5). WERLE: Nocturnal chase. Marta Schele, s; Elsif Lunden-Bergfelt, pno.
+Gr 3-77 p1457 +St 3-77 p149
+-RR 2-77 p86

LP 37
2996 Jacqueline Delman, song recital. HEAD: A piper. MARTIN: Christmas songs (3). MESSIAEN: Poemes pour mi. PERGAMENT: Who plays in the night. SHOSTAKOVICH: Romances on words of Alexander Block, op. 127 (7). Jacqueline Delman, s; Emil Dekov, vln; Ake Olofsson, vlc; Gunilla von Bahr, flt; Lucia Negro, pno.
++St 2-77 p130

LP 45
2997 ALABIEFF (Alabiev): The Russian nightingale. ADAM: Variations on "Ah, vous dirai'je, Maman". BISHOP: Lo, here the gentle lark. BENEDICT: La capinera. DELL'ACQUA: Villanelle. DORROW: Dream; Pastourelles, pastoureux. MOZART: Il Re pastore, K 208: L' amero saro. ROUSSEL: Poemes de Ronsard (2). Dorothy Dorow, s; Gunilla von Bahr, flt; Lucia Negro, pno.
+RR 1-77 p76 +St 12-76 p153

LP 48
2998 BARSANTI: Sonata, recorder, C minor. EYCK: Come again. Fantasia in echo. FONTANA: Sonata prima. LINDE: Amarilli mia bella. SHINOHARA: Fragmente. TELEMANN: Sonata, recorder, E minor. Clas Pehrsson, rec; Bengt Ericson, vlc; Anders Ohrwall, hpd.
++HFN 11-77 p177 +-RR 11-77 p87

LP 50
2999 BASHMAKOV: Fantasia per flauti. MARTINU: Sonata, flute. MELLNAS: Fragments for family flute. PERGAMENT: Piece, 4 flutes. Sonata, flute, VIVALDI: Concerto per ottavino, P 79. Gunilla von Bahr, Robert von Bahr, flt; Kerstin von Hindart, pno; Instrumental Ensemble.
+-RR 3-77 p85

LP 57
3000 BOISMORTIER: Concerto, 5 recorders, D minor. BYRD: The leaves be green fantasy. HINDEMITH: Ploner Musiktag: Trio. LYNE: Epigrams (3). PERGAMENT: Little suite, 2 recorders. SCHEIDT: Cantus, 5 voices on "O Nachbar Roland". WARLOCK: Capriol suite. WOODCOCK: Browning fantasy. Musica Dolce.
+HFN 11-77 p177 +RR 11-77 p79

LP 59
3001 BRITTEN: Fanfare for St. Edmondsbury. GABRIELI, G.: Canzon septimitoni. HERMANSSON: Shadow play. HINDEMITH: Morgenmusik. KUBIK: Fanfare for the century. LOCKE: Music for His Majesty's sackbuts and cornetts. PREMRU: Tissington variations. SCHEIDT: Canzon a 5 and imitationem bergamas. Malmo Brass Ensemble.
++FF 9-77 p61 +RR 7-77 p49
+Gr 9-77 p500

LP 60
3002 CARULLI: Serenade, op. 109, no. 6, D major. FALLA: Cancion del pescador and farruca (arr. Diego Blanco). GOSSEC: Tambourin. DESPORTES: Pastorale joyeuse. HAHN: Suite in folk style, solo flute. PRAEGER: Introduction, theme and variations, op. 21, A minor. WERDIN: Concertino, flute, guitar and strings. Gunilla von Bahr, flt; Diego Blanco, gtr.
+ARG 8-77 p38 +-RR 3-77 p79

LP 62
3003 CRUSELL: Rondo, 2 clarinets and piano. DENISOV: Sonata, solo clarinet. LIDHOLM: Invention, clarinet and bass clarinet. MENDELSSOHN: Sonata, clarinet and piano. PENDERECKI: Miniatures, clarinet and piano (3). RAUTAVAARA: Sonata, clarinet and piano, op. 53. SHCHEDRIN: Basso ostinato, clarinet and piano. STRAVINSKY: Pieces, solo clarinet (3). Kjell-Inge Stevensson, Kjell Fageus, clt; Eva Knardahl, pno.
+RR 3-77 p79 ++St 7-77 p113

LP 75
3004 LANDINI: Ecco la primavera. MORLEY: Now is the gentle season. NEIDHART: Winder, diniu meil; Meie, din liehter schin; Meienzit. PEERSON: Fall of leafe. PRAETORIUS: Winter. SENFL: Laub, Gras u Blut. VULPIUS: Die beste Zeit im Jahr. WOODCOCK: Browning fantasy. WOLKENSTEIN: Mai; Unbek. GLOGAUER LIEDERBUCH: Dy katczen pfothe. ANON.: Lieder 13th-16th century. Joculatores Upsalienses; Sven Berger.
+-HFN 11-77 p177 +-RR 11-77 p103

BOSTON BRASS

BB 1001
3005 BOUTRY: Capriccio. BERGHMANS: La femme a barbe. DEFAYE: Danses (2). GUILMANT: Morceau symphonique, op. 88. ROPARTZ: Piece, E flat minor. SALZEDO: Piece concertante, op. 27. SAINT-SAENS: Cavatine, op. 144. Ronald Barron, trom; Fredrik Wanger, pno.
+FF 9-77 p58 +HF 8-76 p96

BRUNO WALTER SOCIETY

IGI 323 (2)
3006 BACH, C.P.E.: Concerto, violoncello, B flat major. BACH, J.C.: Concerto, violoncello, C minor. BOCCHERINI: Concerto, violoncello, D major. TARTINI: Concerto, violoncello, A major. VIVALDI: Concerti, violoncello, C minor, G minor, G major. Mstislav Rostropovich, vlc; Park Lane Ensemble; Mstislav Rostropovich.
+-NR 4-77 p4

RR 443 (2) (Reissues from Columbia and Telefunken
3007 BACH: Suite, orchestra, S 1068, D major. BEETHOVEN: Coriolan overture, op. 62. BRAHMS: Tragic overture, op. 81. RAVEL: Bolero. TCHAIKOVSKY: Overture, the year 1812, op. 49. WAGNER: Die Meistersinger von Nurnberg: Overture. Tannhauser: Overture. WEBER: Oberon: Overture. COA; Willem Mengelberg.
+-ARSC Vol VIII, no. 2-3, p91 +-NR 3-77 p1

BWS 729
3008 BEETHOVEN: Songs: Andenken; Klarchen's song, no. 1; Die Trammel geruhret; Wonne der Wehmut. BRAHMS: Wiegenlied. MOZART: Songs: Sehnsucht nach dem Fruhling, K 596; Das Veilchen, K 476; Warnung, K 433. MENDELSSOHN: Auf Flugeln des Gesanges, op. 34, no. 2. PUCCINI: Tosca: Vissi d'arte. SCHUBERT: Standchen. SCHUMANN: Songs: Auftrage, op. 77, no. 5; Volksliedchen, op. 51, no. 2. Liederkreis, op. 39: In der Fremde. STRAUSS, R.: Songs: Standchen, op. 17, no. 2; Traum durch die Dammerung, op. 29, no. 1; Zueignung, op. 10, no. 1. WAGNER: Tristan und Isolde: Prelude and Liebestod. Songs: Der Engel. WOLF: Songs: Gesang Weylas; Neue Liebe. Lotte Lehmann, s.
+-NR 5-77 p12

CALLIOPE

CAL 1848

3009 The play of Daniel. The Clerkes of Oxenford; David Wulstan.
+Gr 11-76 p844 +RR 1-77 p36

CAL 1922/4 (3)

3010 ALAIN: Suite pour orgue. Variations sur un theme de Jannequin, op. 78. BARIE: Toccata, op. 7. BONNAL: La vallee de Behorleguy. BOELLMANN: Suite Gothique: Toccata. DUPRE: Prelude and fugue, F minor. GIGOUT: Scherzo. GUILMANT: Sonata, organ, no. 1, op. 42, D minor: Finale. HURE: Communion sur un Noel. NIBELLE: Carillon orleanais. ROPARTZ: Prelude funebre. PIERNE: Prelude, G minor. TOURNEMIRE: Messe de l'assomption: Paraphrase carillon. VIERNE: Impromptu. Symphony, no. 3, op. 82, F minor: Toccata. WIDOR: Symphony, organ, no. 6, op. 42, no. 2, G major: Allegro. Andre Isoir, org.
+Gr 3-77 p1454 +RR 2-77 p83
+HFN 2-77 p125

CAMBRIDGE

CRS 2540

3011 A procession of voluntaries: ALCOCK: Voluntary, D major. BOYCE: Voluntaries, D major, G minor. GREENE: Voluntaries, G major, C minor. STANLEY: Voluntaries, D major, A minor. WALOND: Voluntary, D minor. Lawrence Moe, org.
+-ARG 5-77 p47 +MU 3-77 p13
+CL 11-76 p10 +NR 11-76 p11
+-HF 2-77 p108

CRS 2826

3012 BERTOLI: Sonata prima. BOISMORTIER: Rondeau, A minor. HANDEL: Sonata, flute, op. 1, no. 11, F major. LOEILLET: Sonata, recorder, G major. LAVIGNE: Sonata, recorder, C major. TELEMANN: Sonata, recorder, F major. Trio Primavera.
+ARG 2-77 p48 +NR 10-76 p7
+-HF 6-77 p102 +St 6-77 p148

CBS

30091 Tape (c) 40-30091

3013 BIZET: Carmen: March of the smugglers; Gypsy dance. GOUNOD: Faust: Ballet music. PONCHIELLI: La gioconda: Dance of the hours. SAINT-SAENS: Samson et Dalila: Bacchanale. VERDI: Aida: Ballet music. NYP; Leonard Bernstein.
+HFN 6-77 p139 tape +RR 3-77 p61
+-HFN 4-77 p153

61039 Tape (c) 40-61039

3014 BENJAMIN: Jamaican rumba. BRAHMS: Hungarian dance, no. 5, G minor. COPLAND: Rodeo: Hoe-down. DEBUSSY: Suite bergamasque: Clair de lune. DVORAK: Humoresque. FOSTER: Jeanie with the light brown hair. GERSHWIN: Porgy and Bess: Bess, you is my woman now. KREISLER: Liebesleid. RIMSKY-KORSAKOV: The tale of the Tsar Sultan: Flight of the bumblebee. SCHUBERT: Ave Maria, D 839. TCHAIKOVSKY: None but the lonely heart, op. 6, no. 6. TRAD.: Greensleeves. Isaac Stern, vln; Columbia Symphony Orchestra; Milton Katims.
+Gr 9-77 p517 +RR 9-77 p68
-HFN 9-77 p153 +RR 11-77 p120 tape
-HFN 10-77 p171 tape

61453
3015 BARTOLOZZI: Collage. MADERNA: Aulodia per Lothar. MATSUDAIRA: Somaksah. SINGER: Musica a 2. SINOPOLI: Numquid. VALDAMBRINI: Dioe. Lothar Faber, ob; Kate Wittlich, keyboards; Vincenzo Saldarelli, gtr; Francesco Valdambrini, pno.
+-Te 6-77 p41

61780
3016 BRAHMS: Hungarian dance, no. 6. COPLAND: Danzon cubano. Rodeo: Hoe-down. BERNSTEIN: Fancy free: Galop; Waltz; Danzon. DVORAK: Slavonic dance, op. 46, no. 1. FERNANDEZ: Reisado do pastoreio: Batuque. GLIERE: The red poppy, op. 70: Russian sailor's dance. GRIEG: Norwegian dance, op. 35, no. 2. GUARNIERI: Brazilian dance. MOZART: German dance, no. 3, K 605. TCHAIKOVSKY: The nutcracker, op. 71: Trepak. NYP; Leonard Bernstein.
+-Gr 6-77 p106 +RR 6-77 p35
+-HFN 6-77 p137

61781 Tape (c) 40-61781 (also Columbia M 34127)
3017 Age of gold: BORODIN: In the Steppes of Central Asia. Prince Igor: Dance of the Polovtsian maidens. GLIERE: The red poppy, op. 70: Sailors dance. IPPOLITOV-IVANOV: Caucasian sketches, op. 10: Procession of the Sardar. PROKOFIEV: Lieutenant Kije, op. 60: Troika. The love for three oranges, op. 33: March. RIMSKY-KORSAKOV: Snow maiden: Dance of the tumblers. SHOSTAKOVICH: The age of gold, op. 22. TCHAIKOVSKY: Eugene Onegin, op. 24: Waltz; Polonaise. The sleeping beauty, op. 66: Waltz. NYP; Leonard Bernstein.
+Gr 7-77 p226 +RR 7-77 p44
+HFN 8-77 p95 +-RR 8-77 p86 tape
+-HFN 8-77 p99 tape +SFC 10-31-76 p35
+NR 7-76 p3

72526 Tape (c) 40-72526 (also Columbia MS 6939)
3018 BACH: Prelude, fugue and allegro, S 998, E flat major. GIULIANI: Sonata, guitar, op. 15: 1st movement. MUDARRA: Diferencias sobre El Conde claros. Fantasia. PRAETORIUS: Terpsichore: Ballet. REUSNER: Suite, no. 2, C minor: Paduana. TORROBA: Aires de La Mancha. VILLA-LOBOS: Preludes, no. 2, E major; no. 4, E minor. John Williams, gtr.
+RR 7-75 p71 tape +St 1-77 p73

73589 (also Columbia M 34543)
3019 ALBENIZ: Iberia: Fete dieu a Seville. DEBUSSY: Estampes: Soiree dans Grenade. Suite bergamasque: Clair de lune. CHOPIN: Mazurka, op. 24, no. 4, B flat minor. Prelude, op. 28, no. 24, D minor. NOVACEK: Perpetuum mobile. RIMSKY-KORSAKOV: The tale of the Tsar Sultan: Flight of the bumblebee. Maid of Pskov: Overture. SHOSTAKOVICH: Prelude, op. 34, no. 14, E flat minor. TCHAIKOVSKY: Humoresque, op. 10. (all trans. Stokowski) NPhO; Leopold Stokowski.
+-FF 11-77 p63 ++RR 6-77 p67
+Gr 6-77 p106 ++SFC 9-4-77 p38
+-HFN 7-77 p121 +St 12-77 p158
++NR 11-77 p4

76522 Tape (c) 40-76522 (also Columbia M 34206)
3020 BERLIOZ: Beatrice et Benedict: Dieu, que viens-je d'entendre...Il m'en souvient. La damnation de Faust, op. 24: D'amour l'ardente flamme. GOUNOD: Romeo et Juliette: Depuis hier je cherche en vain. MASSENET: Cendrillon: Enfin, je suis ici. Werther: Va, laisse les couler mes larmes. MEYERBEER: Les Huguenots:

Nobles seigneurs, salut. OFFENBACH: La grande Duchesse de Gerolstein: Dites lui. La perichole: Ah, quel diner je viens de faire. THOMAS: Mignon: Connais-tu le pays. Frederica von Stade, ms; LPO; John Pritchard.

++Gr 7-76 p212 | +ON 1-29-77 p44
+-HF 2-77 p114 | +RR 7-76 p28
+HFN 7-76 p83 | +St 2-77 p83
++HFN 10-76 p185 tape | +STL 9-19-76 p36
+-NR 1-77 p10

76534
3021 Music from the time of the Popes at Avignon. DANDRIEU: Deploration sur la mort de Machaut. DUFAY: Gloria ad modam tubae. MACHAUT: Moult suis de bonne heure nee. PETRARQUE/BOLOGNE: Non al so amante. ANON.: Caccia de Zaccharias. Carole. Carole liturgique "Qui passus". Credo "Bonbarde". Impudenter. Kyrie. Saltarello. Paris Florilegium Musicum; Jean-Claude Malgoire.

+ARG 11-77 p41 | -MT 10-77 p827
+Gr 5-77 p1703 | +-RR 5-77 p83
+-HFN 5-77 p131

SQ 79200 (2) Tape (c) 40-47200 (also Columbia Tape (c) M2T 34256)
3022 BACH: Concerto, 2 violins and strings, S 1043, D minor. BEETHOVEN: Leonore overture, no. 3, op. 72. HANDEL: Messiah: Hallelujah chorus. RACHMANINOFF: Sonata, violoncello and piano, op. 19, G minor: Andante. SCHUMANN: Dicterliebe, op. 48. TCHAIKOVSKY: Trio, piano, op. 50, A minor: Pezzo elegiaco. Yehudi Menuhin, Isaac Stern, vln; Vladimir Horowitz, pno; Mstislav Rostropovich, vlc; Dietrich Fischer-Dieskau, bar; Leonard Bernstein, hpd; Oratorio Society; NYP; Leonard Bernstein.

+ARG 12-76 p26 | +-HFN 5-77 p139 tape
+-Audio 3-77 p92 | -MJ 4-77 p33
+-Gr 12-76 p1009 | +NR 1-77 p15
+HF 5-77 p101 tape | +-RR 12-76 p51
+-HFN 1-77 p104 | +-RR 4-77 p91 tape

Tape (c) 40-72728 (ct) 42-72728 (also Columbia MS 7195)
3023 BACH: Chaconne. BATCHELAR: Mounsiers almaine. DOWLAND: Queen Elizabeth, her galliard. Earl of Essex, his galliard. GIULINI: Variations on a theme by Handel, op. 107. PAGANINI: Caprice, op. 1, no. 24, A minor. SOR: Variations on a theme by Mozart, op. 9. John Williams, gtr.

++Gr 6-74 p115 tape | +St 1-77 p73

CITADEL

CT 6011
3024 ADASKIN: Algonquin symphony. CHAMPAGNE: Danse villageoise. CHOTEM: North country. JONES: Mirimachi ballad: The Jones boys. MacMILLAN: A Saint Malo. RIDOUT: Fall fair. WEINZWEIG: Red ear of corn: Barn dance. Toronto Symphony Orchestra; Victor Feldbrill, Walter Susskind.

+NR 4-77 p3

CT 6012
3025 The art of the alto saxophone. BONNEAU: Caprice en forme de valse. BOZZA: Aria. GLAZUNOV: Concerto, saxophone. HAYES: Concertino, saxophone. RARIG: Dance episode. REDDLE: Gypsy fantasy. VELLONES: Rhapsody for saxophone. Ralph Gari, alto sax; John Rarig, pno.

+NR 4-77 p16

CT 6013
3026 BIZET: L'Arlesienne suite, no. 1: Adagietto. DELA: Dans tous les cantons, Adagio. GIBBS: Fancy dress: Dusk. MENSELSSOHN: Songs without words, op. 67, no. 4: Spinning song (arr. Burt). RIMSKY-KORSAKOV: The tale of the Tsar Sultan: Flight of the bumblebee. VOLKMANN: Serenade, no. 2, op. 63, F major: Waltz. WIREN: Serenade, strings, op. 11: March. WOLF-FERRARI: School for fathers: Intermezzo. Vanity fair (Collins, arr. Burt). Hart House Orchestra; Boyd Neel.
+NR 3-77 p2

CLASSICAL CASSETTE COMPANY

CP 52 (also CCC AP 53, 54)
3027 Slavic orthodox liturgy programs. Bulgarian choirs under Georgi Robev, Mikhail Milkov.
+HF 10-77 p129 tape

CLASSICS FOR PLEASURE

CFP 40012 Tape (c) TC CFP 40012
3028 ALBENIZ: Suite espanola, no. 1: Granada. FRESCOBALDI: Air and variations (arr. Segovia). LAURO: Valse criollo. MALATS: Spanish serenade. RODRIGO: Concierto de Aranjuez, guitar and orchestra. WEISS: Balletto. John Zaradin, gtr; London Philomusica; Guy Barbier.
+Gr 8-72 p347 -RR 8-72 p64
-HFN 10-72 p1899 +-RR 11-77 p118 tape

CFP 40213 (Reissue from Columbia SCX 3279)
3029 GUNGL: Amoretten Tanze, op. 161. IVANOVICI: Waves of the Danube. LANNER: Die Schonbrunner, op. 200. LEHAR: Gold und Silber, op. 79. ZIEHRER: Weaner Mad'ln, op. 338. Wiener Burger, op. 419. Philharmonia Promenade Orchestra; Henry Krips.
+-Gr 3-77 p1462 +RR 3-77 p64

CFP 40236
3030 Viennese overtures: BEETHOVEN: The creatures of Prometheus, op. 43: Overture. HEUBERGER: Der Opernball: Overture. MOZART: The marriage of Figaro, K 492: Overture. SCHUBERT: Rosamunde, op. 26, D 797: Overture. STRAUSS, J. II: Die Fledermaus, op. 363: Overture. SUPPE: Poet and peasant: Overture. Halle Orchestra; James Loughran.
+Gr 3-77 p1461 +-RR 3-77 p61
+-HFN 3-77 p114

CFP 40263
3031 HUMPERDINCK: Hansel and Gretel: Overture. MENDELSSOHN: Ruy Blas overture, op. 95. NICOLAI: The merry wieves of Windsor: Overture. WAGNER: Tannhauser: Overture. WEBER: Der Freischutz: Overture. LPO; James Lockhart.
+-Gr 9-77 p517 +-RR 8-77 p65
+HFN 8-77 p81

CLEAR

TLC 2586 (3)
3032 GRIEG: Sonata, violin and piano, no. 2, op. 13, G minor. MANEN:

Concerto da camera. MENDELSSOHN: Concerto, violin, op. 64, E minor. RACHMANINOFF: Vocalise, op. 34, no. 14. SPALDING: Sonata, solo violin, E minor. Etchings, op. 5. TARTINI: Sonata, violin, G minor. TCHAIKOVSKY: Concerto, violin, op. 35, D major. Albert Spalding, Eddy Brown, Juan Manen, vln; Frieder Weismann, Andre Benoist, Clarence Adler, pno; PO, Orchestra; Eugene Ormandy, Herman Neuman.
+ARSC vol 9, no. 1, 1977 p85

CLUB

99-96
3033 BILLI: Canta il grillo. BIZET: Carmen: Habanera; Chanson boheme; Seguidille. BOITO: Mefistofele: Lontano; L'altra notte. DAVIDOV: Night, love, moon. FAURE: Le crucifix. GLINKA: Do not tempt me needlessly. GRODSKI: Sea gull's cry. MASSANET: Werther: Air des larmes; Tears. MEYER-HELMUND: In the morning I bring you violets. NAPRAVNIK: Dubrovsky: French duet; Masha's air. Harold: Cradle song. ROTOLI: Fior che langue. PUCCINI: Tosca: Vissi d'arte. RESPIGHI: Re Enzo: Stornellatrice. RUBINSTEIN: Night. TCHAIKOVSKY: Pique dame, op. 68: Air de Lisa. TOSTI: Penso. Medea Mei-Figner, s.
+ARSC vol 9, no. 1, 1977, p94

99-100
3034 BOITO: Mefistofele: L'altra notte. GIORDANO: Adriana Lecouvreur: Io sono l'umile ancella, Poveri fiori. Siberia: Nel suo amore, Non odi la il martis. Fedora: Morte di Fedora. LEONCAVALLO: Zaza: Dir che ci sono al mondo. MASCAGNI: Cavalleria rusticana: Voi lo sapete. MASCHERONI: Lorenza: Susanna al bagno. MASSENET: Manon: Ancor son io tutt attonita; Addio nostro piccolo desco. PONCHIELLI: La gioconda: Suicidio. PUCCINI: Tosca: Vissi d'arte; Quanto...Gia mi docon venal. La boheme: Mi chiamano Mimi, Donde lieta. THOMAS: Mignon: Non consoci il bel suol. TOSTI: Dopo. VERDI: Aida: Ritorna vincitor. VIRGILIO: Jana: Morte di Jane. Emma Carelli, s; Instrumental accompaniment.
*NR 8-76 p9 +-ON 2-26-77 p33

99-102
3035 FISCHER: In tiefem Keller. HANDEL: Acis and Galatea: O ruddier than the cherry. Israel in Egypt: The Lord is a man of war. Judas Maccabaeus: Arm, arm ye brave. Samson: Honour and arms. OFFENBACH: Genevieve de Brabant: Duo des gendarmes. MENDELSSOHN: Son and stranger: Heimkehr aus der Fremde. REEVE: I am a friar of orders gray. VERDI: I vespri siciliani: O tu palermo. Songs: Chu chin chow: Cobblers songs; The mighty deep, The windmill, My old shako; Fotsam and Jetsam; What's the matter with Rachmaninoff; Schubert's toy shop, Polonaise in the mall, Simon the bootlegger, Song of the air, Only a few of us left, The Alsatian and the Pekingese. Malcolm McEachern.
+-NR 4-77 p11

99-103
3036 BIZET: Carmen: Eccola...Ella vien...Habanera, Seguidilla, Chanson boheme, Vieni lassu sulla montagna, Air des cartes, Aragonaise, Su tu m'ami, Sei tu, Son io. DONIZETTI: La favorita: O mio Fernando. PONCHIELLI: La gioconda: Voce di donna. SAINT-SAENS: Samson et Dalila: Aprile foriero, S'apre per te il mio cor.

VERDI: Il trovatore: Stride la vampa. Gabriella Besanzoni, ms.
+-NR 3-77 p9

99-105
3037 BIXIO: The sun and the golden mountains. BOCCANEGRA: Teresita. CLUTSAM: Curly headed baby. CURTIS: Addio bel sogno, Non m'ami piu. DARGOMIZHSKY: Russalka: Some unknown power. DONIZETTI: La favorita: Spirto gentil. Lucia di Lammermoor: Tombe degli avi...Fra poco a me ricovero. IPPOLITOV-IVANOV: Treachery: Aria of Erekle. KHRENNIKOV: Much ado about nothing: Nightingale and the rose. LABRIOMA: Sailor's song. LEONCAVALLO: Au clair de la lune. MASSENET: Manon: Le reve. WERTHER: Pourquoi me reveiller. MATTEO: Bella donna. PETROV: Don't believe me child. TAGLIAFERRO: Mandolinata. Mikhail Alexandrovich.
-NR 4-77 p11

99-106 (Reissue from Fonotipia)
3038 BELLINI: Norma: Sgnombra e la sacra selva...Deh proteggi me, o Dio. COSTA: Serenata medioevale. DONIZETTI: La favorita: Quando le soglie...Ah l'altro ardor, Fernando Fernando...Che fino al ciel. FAINI: Amore e Maggio. GIORDANO: Andre Chenier: Vecchia madlon. GOUNOD: Quando canti "Quand tu chantes". MASSENET: Roi de Lahore: Viaggia o bella. MEYERBEER: Les Huguenots: Vaga donna; No no no giammai. PONCHIELLI: La gioconda: Voce di donna. PUCCINI: E l'uccelino ninna nanna. SAINT-SAENS: Samson et Dalila: Aprile foriero, S'apre per te il mio cor. VACCAI: Giullietta e Romeo: Tu dormi, sveglati. VERDI: Un ballo in maschera: Re del abisso. La forza del destino: Al suon del tamburo, Rataplan. Il trovatore: Ai nostri monti. Armida Parsi-Pettinella, ms.
+NR 3-77 p9

99-107
3039 ADAM: Le chalet: Vive le vin, l'amour, et le tabac. DELMET: Envoi de fleurs. DUPONT: Les boeufs; Les sapins. FLEGIER: Le cor. GOUNOD: Romeo et Juliette: Invocation, Dieu qui fis l'homme a ton image. Song: Primavera. HALEVY: Charles VI: Guerre aux tyrans. HEROLD: Le pre aux clercs: Le rendezvous de noble compagnie. MASSE: Galathee: Grand air. La mule de Pedro: Ma mule qui chaque semaine. MASSENET: Le Cid: Air de Don Diegue; Il a fait noblement. Song: Serenade du passant. MEHUL: Joseph: O toi le digne appui. MEYERBEER: Les Huguenots: Choral de Luther; Piff, paff. Robert le diable: Nonnes qui reposez. REYER: Sigurd: Chant du barde; Au nom de roi. SAINT-SAENS: Samson et Dalila: Gloire a Dagon. THOMAS: Mignon: Berceuse. Songe d'une nuit d'ete: Couplets de Falstaff, alors que tout s'apprete. WOLFF: La premiere lecon. Paul Aumonier, bs; Mary Boyer.
+ARG 7-77 p12

99-108
3040 FRANZ: Songs: Aus meinen grossen Schmerzen, op. 5, no. 1; Im Herbst, op. 17, no. 6. GILMOUR: Slumber song. MENDELSSOHN: Songs: Auf Flugeln des Gesanges, op. 34, no. 2. ROSSINI: Stabat mater: Inflammatus. SCHUBERT: Songs: Du bist die Ruh, D 776; Erlkonig, D 328; Gretchen am Spinnrade, D 118. SCHUMANN: Songs: Die Lotosblume, op. 25, no. 7; Widmung, op. 25, no. 1. SILCHER: Die Lorelei. STRAUSS, R.: Standchen, op. 17, no. 2. WAGNER: Wesendonck Lieder: Der Engel; Stehe still; Im Treibhaus; Schmerzen; Traume. WERNER: Heidenroslein. WILHELM: Die Wacht am Rhein. Johanna Gadski, s.
+-ARG 7-77 p19

99-109
3041 MASCAGNI: Cavalleria rusticana: Voi lo sapete. MOZART: Le nozze di Figaro, K 492: Porgi amor. Die Zauberflote, K 620: Ach, ich fuhl's. STRAUSS, R.: Salome, op. 54: Jochanaan, ich bin verliebt. THUILLE: Lobetanz: An allen Zweigen. VERDI: Aida: Ritorna vincitor; O patria mia. Un ballo in maschera: Ma dall arido. Il trovatore: D'amor sull ali rosee. WAGNER: Der fliegende Hollander: Trafft ihr das Schiff. Lohengrin: Euch Luften. Siegfried: Ewig war ich. Gotterdammerung: Zu neuen Taten. Tannhauser: Allmacht'ge Jungfrau. Tristan und Isolde: Dein Werk. Die Walkure: Du bist der Lenz; Fort denn, eile; War es so schmahlich. Johanna Gadski, s.
+ARG 7-77 p19

COIMBRA

CC 1
3042 The mass at Downham Market.
-RR 7-77 p86

CCO 34
3043 Faith of our fathers: Catholic hymns. Come, holy ghost, creator, come; Daily daily sing to Mary; Help Lord the souls that thou has made; Faith of our fathers; Hail queen of heaven; I'll sing a hymn to Mary; Jesus, my Lord, my God, my all; Look down O mother Mary; O purest of creatures; O salutaris hostia; Soul of my saviour; Sweet sacrament divine; Sweet saviour bless us ere we go; Tantum ergo sacramentum; To Christ, the prince of peace. St. Peter Singers; Arthur Gibson.
+Gr 9-77 p460

CCO 37
3044 Gregorian chant (Sunday compline): Veni creator spiritus; Compline; Pange lingua; Victimae paschali; Veni veni Emmanuel; Sacris solemnis; Veni sancte spiritus; Jesu nostra redemptio; Lucis creator; Te deum. St. Peter Singers, Male Voices; David Read.
+Gr 9-77 p469

CCO 44
3045 Tudor church music: BROWNE: Stabat iuxta Christi crucem. LAMBE: Salve regina. TAVERNER: Magnificat III. WYLKYNSON: Salve regina. BBC Singers; John Poole.
+-Gr 9-77 p469 +RR 8-77 p79

CCO 73
3046 Gregorian chant. Tyburn Convent Choir of Nuns; Male Voice Choir; David Read.
+-Gr 12-77 p1121

COLUMBIA

Melodiya M 33931
3047 DONIZETTI: La favorita: O mio Fernando. MUSSORGSKY: Khovanschina: Marfa's prophecy. RIMSKY-KORSAKOV: The Tsar's bride: Lyubasha's aria. Kashchei, the deathless: The night descends. SAINT-SAENS: Samson et Dalila: Mon coeur s'ouvre a ta voix; Amour, viens aider TCHAIKOVSKY: The maid of Orleans: Joan's aria. VERDI: Don Carlo: O don fatale. Elena Obraztsova, ms; Bolshoi Theatre Orchestra; Boris Khaikin, Odyssei Dimitriadi.

+-Audio 11-76 p106 +-ON 1-22-77 p33
+-FF 11-77 p60 +-SFC 4-4-76 p34
+-HF 7-76 p102 +SR 5-29-76 p52
+NR 5-76 p9 +-St 8-76 p106

M 34134
3048 Mormon Tabernacle Choir: A jubilant song. BERGER, J.: I lift up my eyes (Psalm 121). BRIGHT: Rainsong. CUNDICK: The west wind. BELLO JOIO: A jubilant song. GATES: Oh, my luve's like a red red rose. HANSON: Psalm, no. 150. LEAF: Let the whole creation cry. MECHEM: Make a joyful noise unto the Lord (Psalm 100). THOMPSON: Glory to God in the highest. Jo Ann Ottley, s; Robert Cundick, pno; Alexander Schreiner, org; Mormon Tabernacle Choir; Jerold D. Ottley.
+ARG 11-76 p49 +MU 6-77 p10
+CJ 1-77 p39 +NR 9-76 p8
+HF 11-76 p140 *ON 11-76 p98

34294 (3)
3049 Lily Pons: Coloratura assoluta. ALABIEFF: The nightingale. BACHELET: Chere nuit. BISHOP: Home, sweet home; Lo, here the gentle lar; Pretty mocking bird. DUPARC: L'Invitation au voyage. DAVID: La perle du Bresil: Charmant oiseau. DELIBES: Lakme: Pourquoi dans les grands bois; Bell song. DELL ACQUA: Villanelle. DONIZETTI: La fille du regiment: Il faut partir. Lucia di Lammermoor: Regnava nel silenzio...Quando rapita in estasi; Mad scene. FAURE: Les roses d'Ispahan, op. 39, no. 4. GRETRY: Zemire et Azor: La fauvette. JACOBSON: Chanson de Marie Antoinette. MEYERBEER: Dinorah: Ombre legere. MILHAUD: Chansons de Ronsard. MOORE: The last rose of summer. OFFENBACH: Les contes d'Hoffmann: Les oiseaux dans la charmille. PROCH: Theme and variations. RACHMANINOFF: Here beauty dwells; O cease thy singing, maiden fair, op. 4, no. 4; Vocalise, op. 34, no. 14. RIMSKY-KORSAKOV: Le coq d'or: Hymne au soleil. The rose and the nightingale. The legend of Sadko, op. 5: Song of India. SAINT-SAENS: Le timbre d'Argent: Le bonheur est une chose legere. VERDI: Rigoletto: Caro nome. La traviata: Ah, fors'e lui...Sempre libera. Lily Pons, s; Orchestras; Pietro Cimara, Andre Kostelantez, Maurice Abravanel.
+ARG 5-77 p49 +-NYT 4-24-77 pD27
+-MJ 9-77 p35 +ON 4-16-77 p37
+NR 4-77 p8 +St 5-77 p114

M 34501
3050 CATALANI: Vieni, deh, vien. LEONCAVALLO: Serenade francaise; Serenade napolitaine. MASCAGNI: M'ama, non m'ama; La luna; Serenata. PIZZETTI: I pastori. PUCCINI: Sole e amore; Menti all' avviso. RESPIGHI: Au milieu du jardin; Povero core; Razzolan; Soupir. TOSTI: Serenata; Malia. WOLF-FERRARI: Il campiello. Renata Scotto, s; John Atkins, pno.
-HF 6-77 p104 +-ON 2-26-77 p33
+-NR 7-77 p11 +SR 5-28-77 p42
+OC 6-77 p45 +-St 5-77 p120

CONNOISSEUR SOCIETY

CS 2131 (Some reissues from Angel 35610, 35528)
3051 The young Cziffra, paraphrases, transcriptions and improvisations. BRAHMS: Hungarian dance, no. 5, G minor. CZIFFRA: Fantaisie

Roumaine (after the gypsy style). RIMSKY-KORSAKOV: The tale of the Tsar Sultan: Flight of the bumblebee. ROSSINI: William Tell: Overture. La danza. STRAUSS, J. II: The beautiful blue Danube, op. 314. Die Fledermaus, op. 363: Reminiscences. Tritsch-Tratsch, op. 214. Gyorgy Cziffra, pno.
+ARG 11-77 p42 +-St 9-77 p146
+-NR 6-77 p13

CORONET

LPS 3036
3052 BACH, C.P.E.: Sonata, saxophone, A minor. DEBUSSY: Syrinx, flute. DEL BORGO: Canto, solo saxophone. GATES: Incantation and ritual. NODA: Improvisation, alto saxophone, no. 1. PERSICHETTI: Parable, op. 123. James Stoltie, alto saxophone.
+-NR 1-77 p15

COURT OPERA CLASSICS

CO 342
3053 ARDITI: Parla. DELL'ACQUA: Villanelle. GOUNOD: Romeo et Juliette: Je veux vivre dans ce reve. GRIEG: Peer Gynt, op. 46: Der Winger mag scheiden. OFFENBACH: Les contes d'Hoffmann: Les oiseux dans la charmille. ROSSINI: Il barbiere di Siviglia: Una voce poca fa. STRAUSS, Josef: Dorfschwalben aus Oesterreich, op. 164. STRAUSS, J. II: Fruhlingsstimmen, op. 410. VERDI: Rigoletto: E il sol dell'anima. La traviata: Ah fors'e lui...Follie. Melitta Heim, s; Instrumental accompaniment.
+-NR 8-76 p9 +-ON 4-2-77 p48

CO 347
3054 BELLINI: Norma: Dormono entrambi...Teneri, teneri figlia, In mia man alfin tu sei. CATALANI: Loreley: Non fui da um padre mai bendetta, Dove son, d'onde vengo...O forze recondite. La Wally: Ebben, ne andro lontana. GIORDANO: Siberia: Nel suo amore rianimata la coscienza. MEYERBEER: L'Africana: Di qui si vede il mar...Quai celesti concenti. PONCHIELLI: La gioconda: E un anatema...L'amo come il fulgor del creato, Cosi mantieni il patto. PUCCINI: Tosca: Amaro sol per te m'era il morire. VERDI: Aida: Ciel mio padre, A te grave cagion m'adduce, Aida, Pur ti riveggo. Ester Mazzoleni, s; Instrumental accompaniments.
+ON 2-26-77 p33 +-NR 8-76 p9

CP

3-5 (3)
3055 BERIO: Sequenza IV. BOULEZ: Sonata, piano, no. 1. BUSSOTTI: Pieces, piano, for David Tudor, no. 3. CAGE: Winter music. ICHIYANGANGI: Piano media. ISHII: Aphorismen II fur einen Pianisten. KONDO: Air I, amplified piano with trumpet. MATSUDAIRA: Allotropy for pianist. MESSIAEN: Mode de valeurs et d'intensities pour piano. MIZUNO: Tone for piano. SAEGUSA: Baire's theorem. SATOH: Calligraphy for piano. STOCKHAUSEN: Klavierstuck XI. TAKAHASHI: Meander. TAKEMITSU: Piano distance. Uninterrupted rests. WEBERN: Variations fur Klavier, op. 27. XENAKIS: Herma. YUASA: Cosmos haptic. On the keyboard. Aki Takahashi, pno.
+MJ 10-77 p28 +NR 8-77 p13

CRD 1019

3056 ALFONSO X, El Sabio: Rosa das rosas. DOWLAND: Captain Digorie piper's galliard. The King of Denmark's galliard. PRAETORIUS: Peasant dances. Terpsichore: Fire dance; Stepping dance; Windmills; Village dance; Sailor's dance; Fishermen's dance; Festive march. SOTHCOTT: Fanfare. TRAD. English: Good King Wenceslas, pavan. The dressed ship. Staines Morris. Here we come a-wassailing. Green garters. Fandango. God rest you merry gentlemen. I saw three ships. All hail to the days. TRAD. French: Branle de l'official. ANON. English: Edi beo thu. Ductia. As I lay. ANON. French: Alle psallite cum luya. ANON. Italian: La manfredina. Saltarello. St. George's Canzona; John Sothcott.

+HFN 12-75 p167 +St 1-77 p132
++RR 12-75 p93

Britannia BR 1077

3057 ARNE: Rule Britannia (arr. Sargent). BAX: Coronation march. BOOTH: Britons awake. Elizabeth. Salute to the Prince of Wales. BULLOCK: Fanfare for the coronation of Queen Elizabeth. CROFT: All people that on earth do dwell (arr. Vaughan Williams). DAVIES: R.A.F. march. ELGAR: Land of hope and glory. HANDEL: Zadok the priest. PARRY: Jerusalem, op. 208. STUART: Soldiers of the Queen. TRAD. (arr. Jacobs): God save the Queen. Coldstream Guards Regimental Band, London Festival Orchestra; Ambrosian Singers; Eric Rogers, R. A. Ridings.

+Gr 4-77 p1604

CRESCENT

ARS 109

3058 BACH, C.P.E.: Sonata, organ, no. 5, D major. BACH, J.S.: Fantasia, S 573, C major. DAQUIN: Noel suisse. KREBS: Chorale preludes: Allein Gott in der Hoh sei Ehr; Von Gott will ich nicht lassen; Jesu, meine Freude. WALTHER: Concerto by Albinoni. Concerto by Meck. Gerald Gifford, org.

+Gr 4-77 p1577 +RR 3-77 p83
+HFN 8-77 p81

BUR 1001

3059 BACH, J.C.: Sonata, harpsichord, op. 15, no. 3, D major. CHILCOT: Suite, no. 1. HANDEL: Fantasia, C major. HOLCOMBE: Three aires. LINLEY: Allegretto. SMITH: Lesson, op. 2, no. 7: Fugue. STORACE: Sonata, harpsichord, D major: 1st movement. Gerald Gifford, hpd.

+Gr 6-77 p76 +RR 3-77 p74

CRYSTAL

S 204

3060 BERNSTEIN: Fanfare for Bima. BOZZA: Sonatina, brass quintet. HUGGLER: Quintet, no. 1. RIETI: Incisioni. SPEER: Sonatas, G major/C major. Cambridge Brass Quintet.

+-HF 7-77 p118 ++NR 11-76 p7

S 206

3061 BACH: The art of the fugue, S 1080: Contrapunctus VII (trans. Posten). EAST: Peccavi (trans. Cran). FELD: Quintet. HARTLEY: Orpheus. McBETH: Brass. RENWICK: Dance. SPEER: Sonatas, brass (3) (trans. Fetter). Annapolis Brass Quintet.

+IN 9-77 p28 +NR 11-76 p7

S 221
3062 HEUSSENSTAMM: Tubafour, op. 30. PARKER: Au privave. PURCELL: Allegro and air. ROSS: Fancy dances, 3 bass tubas. SCHULLER: Moods for tuba quartet. STEVENS: Music, 4 tubas. New York Tuba Quartet.
+-FF 9-77 p71 +NR 9-77 p5

S 351
3063 AITKEN: Montages, solo bassoon. BOZZA: Sonatine, flute and bassoon. GABAYE: Sonatine, flute and bassoon. GERSTER: Bird in the spirit. GOODMAN: Jadis III. SMITH: Straws. Felix Skowronek, flt; Arthur Grossmann, bsn.
+IN 3-77 p28 +MJ 7-76 p57

S 361
3064 BUDD: New work, no. 5. CARTER: Canon for three. CHOU: Soliloquy of a Bhiksuni. HINDEMITH: Sonata, trumpet. LEWIS: Monophony VII. STRAVINSKY: Fanfare for a new theater. Thomas Stevens, tpt; Los Angeles Brass Society Members; Robert Henderson.
+IN 9-77 p28 +NR 2-77 p15

S 362
3065 The trumpet in contemporary chamber settings: CAMPO: Duet for equal trumpets. HOVHANESS: Sonata, trumpet and organ: 1st movement. PLOG: Fanfare, 2 trumpets. REYNOLDS: Music, 5 trumpets. SOUTHERS: Spheres. Anthony Plog, Russell Kidd, tpt; Ladd Thomas, org; Ken Wolfson, bsn; Sharon Davis, pno; Los Angeles Philharmonic Trumpet Section; Robert Henderson.
++FF 9-77 p68 +NR 9-77 p5

DECCA

JB 28 (Reissues from SXL 6040, 2198, 6495)
3066 STRAUSS, E.: Bahn frei, op. 45. STRAUSS, J. I: Radetzky march, op. 228. STRAUSS, J. II: Du und du, op. 367. Demolirer, op. 269. Die Fledermaus, op. 363, excerpt. Spanischer Marsch, op. 433. STRAUSS, Josef: Brennende Liebe, op. 129. Transaktionen, op. 184. STRAUSS, J. II/Josef: Pizzicato polka. STRAUSS, J. II/Josef/Eduard: Schutzenquadrille. VPO; Willi Boskovsky.
+RR 12-77 p55

D62D4 (4) Tape (c) D62K43 (Reissues from SXL 6058, 6353, 6655)
3067 BEETHOVEN: Concerto, piano, no. 5, op. 73, E flat major. BRAHMS: Concerto, piano, no. 2, op. 83, B flat major. MOZART: Concerto, piano, no. 20, K 466, D minor. RACHMANINOFF: Concerto, piano, no. 2, op. 18, C minor. TCHAIKOVSKY: Concerto, piano, no. 1, op. 23, B flat minor. Vladimir Ashkenazy, pno; LSO, MPO, CSO; Lorin Maazel, Kyril Kondrashin, Hans Schmidt-Isserstedt, Georg Solti, Zubin Mehta.
+-Gr 9-77 p443 +HFN 9-77 p151

D65D3 (3) (Reissues from SXL 2256, 6116, 2257, SET 232/4, 247/8, 454/5, 545/7, 256/8, 249/51, 412/5, 268/9, 212/4, 372/3, 259/61, 239/41, 320/2, 387/9)
3068 BELLINI: Beatrice di Tenda: Deh, so un urna. Norma: Casta diva. I puritani: Qui la voce. La sonnambula: Come per me sereno. BIZET: Carmen: Micaela's aria. DELIBES: Lakme: The bell song. DONIZETTI: La fille du regiment: Chacun le sait. Lucia di Lammermoor: Regnava nel silenzio. GOUNOD: Faust: The jewel song. HANDEL: Alcina: Tiranna, gelosia. Giulio Cesare: V'adoro

pupillo. Samson: Let the bright seraphim. MEYERBEER: Les Huguenots: O beau pays. MOZART: Don Giovanni, K 527: Or sai chi l'onore. Die Zauberflote, K 620: O zittre nicht. OFFENBACH: Les contes d'Hoffmann: Elle a fui, la tourterelle; Les oiseaux dans la charmille. PUCCINI: Tosca: Vissi d'arte. ROSSINI: Semiramide: Bel raggio. VERDI: Otello: Willow song; Ave Maria. Rigoletto: Caro nome. La traviata: E strano...Ah fors'e lui. WEBER: Der Freischutz: Und ob die Wolke. Joan Sutherland, s; ROHO, NSL, LSO, OSR, Maggio Musicale Fiorentino Orchestra, ECO, Santa Cecilia Orchestra, NPhO, Monte Carlo Opera Orchestra; Richard Bonynge, Francesco Molinari-Pradelli, Thomas Schippers, John Pritchard.

+Gr 9-77 p488 +RR 9-77 p39
+HFN 9-77 p153

SB 328 Tape (c) KBSC 328

3069 BALL: Morning rhapsody. FRIEDEMANN: Slavonic rhapsody, no. 1. GATES: If (arr.). LANGFORD: The seventies set. MOZART: The magic flute, K 620: Overture (arr. Rimmer). VERDI: Aida: Grand march (arr. D. Wright). VINTER: Portuguese party (arr. Barsotti). WOOD: Concerto, trumpet and brass band. Desford Colliery Band; James Watson, tpt; Robert Watson, hn; Albert Chappell.

+Gr 7-77 p229

SB 329

3070 ELGAR: Severn suite, op. 87. GEEHL: Romanza. HUMMEL: Theme and variations, op. 102. LANGFORD: North country fantasie. SAMPSON: High command. VINTER: Salute to youth. TRAD.: Crimond (arr. Hargreaves). Ever Ready (GB) Band; E. W. Cunningham, W. B. Hargreaves.

+Gr 10-77 p712

SPA 450

3071 BIZET: Carmen: Seguidilla. HANDEL: Rodelinda: Dove sei. MOZART: Le nozze di Figaro, K 492: Non piu andrai. PUCCINI: Madama Butterfly: Love duet. Tosca: Rcondita armonia. ROSSINI: Il barbiere di Siviglia: Una voce poco fa. VERDI: Aida: Celeste Aida. Nabucco: Va, pensiero...O chi piange. La traviata: Prelude, Act 1. Rigoletto: Caro nome. Marilyn Horne, Joan Sutherland, Renata Tebaldi, s; Bernadette Greevy, Regina Resnik, con; Placido Domingo, Bruno Prevedi, Carlo Bergonzi, t; Geraint Evans, Nicola Ghiaurov, Tom Krause, bar; Ambrosian Singers; Maggio Musicale Fiorentino Orchestra; John Pritchard and other conductors and orchestras.

+Gr 2-77 p1326 +RR 1-77 p34
+HFN 1-77 p121

SPA 473 Tape (c) KCSP 473

3072 BEETHOVEN: Sonata, piano, no. 8, op. 13, C minor: Adagio cantabile. BADARZEWSKA-BARANOWSKA: The maiden's prayer (arr. Cooper). CHOPIN: Chant polonais, op. 74, no. 5 (trans. Liszt). DEBUSSY: Prelude, Bk I, no. 8: La fille aux cheveux de lin. DOHNANYI: Rhapsody, op. 11, no. 3, C major. RACHMANINOFF: Prelude, op. 3, no. 2, C sharp minor. SINDING: Rustle of spring, op. 32, no. 3. Hidden melodies: I've got you under my skin (in the style of Mendelssohn). Three blind mice (Bach). Waltzing Matilda (Scarlatti). I saw three ships come sailing by (Schumann). When Johnny comes marching home (Schubert). For he's a jolly good fellow (Chopin). The Lambeth walk (Rachmaninoff). The Londonderry air (Brahms) Three blind mice (Debussy). Joseph Cooper, pno.

+Gr 12-76 p1072 +HFN 3-77 p119 tape
+HFN 1-77 p118 +RR 12-76 p87

SPA 491 Tape (c) KCSP 491 (Reissues)
3073 BIZET: The fair maid of Perth: Serenade. BRAHMS: Lullaby. FRANCK: Panis Angelicus. HOFSTETTER: Quartet, strings, F major: Serenade. MOZART: Concerto, piano, no. 21, K 467, C major: Andante. PUCCINI: La boheme: Che gelida manina. RIMSKY-KORSAKOV: Scheherazade, op. 35, excerpt. ROSSINI: The barger of Seville: Largo al factotum. STAINER: The crucifixion: God so loved the world. SUPPE: Poet and peasant: Overture. TCHAIKOVSKY: None but the lonely heart, op. 6, no. 6. VERDI: La traviata: Prelude, Act 1. Renata Tebaldi, s; Luciano Pavarotti, Kenneth McKellar, t; Sherrill Milnes, bar; Nicolai Ghiaurov, bs-bar; Radu Lupu, pno; St. John's College Choir; ROHO, OSR, LAPO; Georg Solti; Various other orchestras and conductors.
+HFN 1-77 p121 +-RR 1-77 p66
+HFN 3-77 p119 tape

SDD 499
3074 ALAIN: Litanies, op. 79. BACH: Toccata and fugue, S 565, D minor. LISZT: Prelude and fugue, on the name B-A-C-H, G 260. MESSIAEN: L'Ascension: Alleluias sereins; Transports de joie. REGER: Pieces, op. 59, no. 9: Benedictus. WIDOR: Symphony, organ, no. 6, op. 42, no. 2, G major: Allegro. Allan Wicks, org.
+Gr 9-76 p453 +MM 7-77 p37
+HFN 9-76 p126

SPA 500 Tape (c) KCSP 500 (also Argo SPA 500)
3075 BAX: Fanfare for the wedding of Princess Elizabeth, 1948. BLISS: Welcome the Queen. Antiphonal fanfare, 3 brass choirs. BLOW: God spake sometime in visions. BULLOCK: Fanfare for the coronation of Queen Elizabeth II. HANDEL: Zadok the priest. PARRY: I was glad. PURCELL: I was glad. VAUGHAN WILLIAMS: O taste and see. WALTON: Crown imperial. TRAD.: The national anthem (arr. Britten). Paul Esswood, Charles Brett, Timothy Brown, c-t; Ian Partridge, John Nixon, Ian Honeyman, t; Richard Brett, bar; Stafford Dean, Marcus Creed, Peter Allwood, Richard Wistreich, bs; St. John's College Chapel Choir, King's College Chapel Choir, St. Michaels College Choir, Westminster Abbey Choir; John Scott, James Lancelot, Simon Preston, org; Kenneth Heath, vlc; Thurston Dart, hpd; AMF, LSO and Chorus, Philips Jones Brass Ensemble; George Guest, David Willcocks, Lucian Nethsingha, William McKie, Arthur Bliss, Benjamin Britten.
+Gr 6-77 p95 +-HFN 8-77 p99 tape
+-Gr 8-77 p346 tape +RR 5-77 p83
+HFN 6-77 p137

DDS 507/1-2 (2)
3076 BACH, C.P.E.: Concerto, orchestra, D major: 2nd movement. BACH, J.S.: Cantata, no. 21: Sinfonia. Christmas oratorio, S 248: Sinfonia pastorale. BOCCHERINI: Quartet, strings, op. 61, no. 1, D major: Adagio. BULL: The chalet girl's Sunday. COATES: The three Elizabeths: Elizabeth of Glamis. ELGAR: The wand of youth suite, op. 1, no. 1, excerpts. FAURE: Shylock, op. 57: Nocturne. FALLA: El amor brujo: The magic circle. JONGEN: Legende, op. 89, no. 1. LARSSON: The disguised god: Lyric suite; Prelude. MOZART (attrib): Quintet, strings, K 46, B flat major: Adagio. NYSTROEM: The merchant of Venice: Nocturne. PUCCINI: Suor Angelica: Intermezzo. TCHAIKOVSKY: Suite, no. 4, op. 61, G major: Prayer. WAGNER (attrib): Adagio. Paris Symphony Orchestra Members; Jean-Paul Marty.

+Gr 7-77 p230 +RR 5-77 p55
+-HFN 5-77 p129

SDD 507 (also London OS 26537)

3077 The surprising soprano: ARDITI: L'Orologio. BEACH: The year's at the spring, op. 44, no. 1. BESLEY: Little fairy songs: The fairy children; Canterbury bells. BRAHMS: Serenata inutile. CHOPIN: Waltz, op. 64, no. 1, D flat major: Messaggero amoroso (arr. Buzzi-Peccia). FRASER (Simpson): A southern maid: Love's cigarette. LEHMANN: The cuckoo. LEONCAVALLO: Aprile. MILLOCKER: Drei Paar Schuhe: I und mei Bua. PUCCINI: Tosca: Vissi d'arte. RUBENS: The Maggie Teyte encore song "Ladies and Gentlemen". SCHUBERT: Der Erlkonig, D 328. TREVALSA: My treasure. TORRENS: How pansies grow. VASSEUR: La cruche cassee, chanson espagnol. VERDI: La traviata: Ah, fors'e lui. VISETTI: Diva, the Adelina Patti waltz. Michael Aspinall, s; Courtney Kenny, pno.

+Gr 2-77 p1331 +SFC 9-25-77 p50
+-HFN 2-77 p112 +St 10-77 p148
+RR 2-77 p90

SPA 519

3078 BRAHMS: Pieces, piano, op. 118, no. 3, G minor. GRAINGER: Country gardens. CHOPIN: Nocturne, op. 9, no. 2, E flat major. Polonaise, op. 53, A flat major. GRIEG: Lyric pieces, op. 43, no. 1: Butterfly. MacDOWELL: Woodland sketches, op. 51, no. 1: To a wild rose; no. 6: To a water lily. MAYERL: Marigold, op. 78. MOZART: Sonata, piano, no. 11, K 331, A minor: Rondo alla Turca. RACHMANINOFF: Polichinelle, op. 3, no. 4, F sharp minor. RAFF: La fileuse, op. 157, no. 2. Hidden melodies: Cheek to cheek (in the style of Mozart); Love walked in (Delius); Wouldn't it be loverly (Brahms); O what a beautiful morning (Grieg); The Vicar of Bray (Chopin); Crib, the melodies revealed. Joseph Cooper, pno.

+Gr 12-77 p1150 +RR 12-77 p73
+HFN 12-77 p168

SB 713

3079 Great Continental marches: BERLIOZ: The damnation of Faust, op. 24: Rakoczy. DONIZETTI: Daughter of the regiment. DUNKLER: Dutch Grenadiers. FREDERIKSEN: Copenhagen. FUCIK: Fearless and true. GANNE: Marche Lorraine. KRAL: Hoch Hapsburg. LEEMANS: March of the Belgian parachutists. LINCKE: Father Rhine. MARQUINA: Espani cani. RADECK: Fredericus Rex. TRAD.: Le baroudeur. Grenadier Guards Band; Peter Parkes.

++Gr 2-77 p1329

SB 715 Tape (c) KBSC 715

3080 BASHFORD: A Purcell suite: Rondo. A Windsor flourish. COATES: The three Elizabeths: Youth of Britain. ELGAR: Imperial march, op. 32. FLETCHER: The spirit of pageantry. KETELBY: With honour crowned. SULLIVAN: Henry VIII: Graceful dance (arr. Retford). WALTON: Crown imperial. TRAD.: The Agincourt song (arr. Bashford). Royal music of King James I: Almande, nos. 1, 5, 6. Grenadier Guards Band; Rodney Bashford.

+Gr 6-77 p106

ECS 811

3081 DONIZETTI: L'Elisir d'amore: Udite, udite, o rustici. LEONCAVALLO: I Pagliacci: Vesti la giubba. MASCAGNI: Cavalleria rusticana: Mamma, quel vino e generoso. MOZART: Don Giovanni, K 527: Madamina. Le nozze di Figaro, K 492: Non piu andrai. PUCCINI:

La boheme: Si, mi chiamano Mimi. Madama Butterfly: Un bel di. Tosca: Vissi d'arte. ROSSINI: La gazza ladra: Il mio piano e preparato. VERDI: Aida: Fu la sorte dell'armi. Un ballo in maschera: Ma dall'arido stelo divulsa; Morro, ma prima in grazia. Ljuba Welitsch, Renata Tebaldi, s; Ebe Stignani, ms; Mario del Monaco, t; Paul Schoeffler, bar; Fernando Corena, bs; Various orchestras and conductors.

+HFN 11-77 p185 +RR 12-77 p32

ECS 812

3082 BEETHOVEN: Fidelio, op. 72: Gott, Welch Dunkel hier. MOZART: Die Zauberflote, K 620: Dies Bildnis ist bezaubernd schon. STRAUSS, R.: Ariadne auf Naxos: Es gibt ein Reich. TCHAIKOVSKY: The queen of spades, op. 68: Ich muss am Fenster lehnen; Es geht auf Mitternacht. WAGNER: Die Meistersinger von Nurnberg: Was duftet doch der Flieder. Dei Walkure: Wotan's farewell and magic fire music. Ljuba Welitsch, Lisa della Casa, s; Anton Dermota, Julius Patzak, t; Paul Schoeffler, bs-bar; VPO; Karl Bohm, Rudolf Moralt, Hans Knappertsbusch, Heinrich Hollreiser.

+HFN 11-77 po85 +-RR 11-77 p43

PFS 4351 Tape (c) KPFC 4351 (Reissues from SDDN 436/8) (also London SPC 21130)

3083 BYRD: The Earl of Salisbury pavan (orch. Stokowski). Galliard (after Francis Tregian). CHOPIN: Mazurka, op. 17, no. 4, A minor (orch. Stokowski). CLARKE: Trumpet voluntary (arr. Stokowski). DUPARC: Extase (orch. Stokowski). DVORAK: Slavonic dance, op. 72, no. 2, E minor. ELGAR: Enigma variations, op. 36: Nimrod. RACHMANINOFF: Prelude, op. 3, no. 2, C sharp minor (orch. Stokowski). SCHUBERT: Moment musical, op. 94, no. 3, D 780, F minor (orch. Stokowski). TCHAIKOVSKY: Chant sans paroles, op. 40, no. 6, A minor (orch Stokowski). Howard Snell, tpt; David Gray, hn; CPhO, LSO; Leopold Stokowski.

*ARG 3-77 p51 +-RR 6-76 p52
+Gr 6-76 p95 +-SFC 1-30-77 p36
+HFN 7-76 p97 +-RR 1-77 p92 tape
+NR 8-77 p3

PFS 4387 (also London SPC 21156 Tape (c) SPC 5-21156)

3084 BACH: Cantata, no. 147, Jesu, joy of man's desiring (arr. Hess). BEN-HAIM: Pieces, piano, op. 34: Toccata. CHOPIN: Etude, op. 10, no. 12, C minor. DEBUSSY: Suite bergamasque: Clair de lune. LISZT: Hungarian rhapsody, no. 2, G 244, C sharp minor. Liebestraum, no. 3, G 541, A flat major. RACHMANINOFF: Prelude, op. 3, no. 2, C sharp minor. SATIE: Gymnopedie, no. 1. SCHUBERT: Moment musical, op. 94, no. 3, D 780, F minor. SCHUMANN: Kinderscenen, op. 15, no. 7: Traumerei. Ilana Vered, pno.

+Gr 2-77 p1308 +NR 9-77 p12
-HFN 4-77 p141 +-RR 2-77 p80
+-HF 11-77 p130

PFS 4416

3085 ALAIN: Litanies, op. 79. BACH: Chorale prelude: Wachet auf, ruft uns die Stimme, S 645. Toccata and fugue, S 565, D minor. BUCK Concert variations on the Star-spangled banner, op. 23. KARG-ELERT: Chorale improvisation, op. 65: Nun danket alle Gott, marche triomphale. VIERNE: Carillon de Westminster. WIDOR: Symphony, organ, no. 5, op. 42, no. 1, F minor: Toccata. Leslie Pearson, org.

+Gr 12-77 p1117 +-RR 12-77 p70

SXL 6781 Tape (c) KSXC 6781
3086 ADAM: Cantique Noel. BACH (Gounod): Ave Maria. BERLIOZ: Requiem, op. 5: Sanctus. BIZET: Agnus Dei (arr. Guiraud). FRANCK: Panis Angeliucs. GOUNOD: Ave Maria (Bach). MERCANDANTE: Le sette ultime parole di Nostro Signore sulla croce: Qual'ciglia candido (Parola quinta). SCHUBERT: Ave Maria, D 839. STRADELLA: Pieta Signore. ANON.: Adeste fideles. Luciano Pavarotti, t; NPhO; Wandsworth Boys Choir, London Voices; Kurt Herbert Adler.
+Gr 11-76 p888 +-RR 12-76 p87
+HFN 11-76 p165 -RR 3-77 p98 tape
+-HFN 3-77 p119 tape

SXL 6782 Tape (c) KSXC 6782 (Also London CS 7006 Tape (c) 5-7006)
3087 BEETHOVEN: The creatures of Prometheus, op. 43: Overture. BERLIOZ: Roman carnival, op. 9. BRAHMS: Academic festival overture, op. 80. GLINKA: Russlan and Ludmilla: Overture. ROSSINI: The thieving magpie: Overture. VERDI: The force of destiny: Overture. CO; Lorin Maazel.
++Gr 12-76 p1071 ++NR 5-77 p3
++HFN 1-77 p113 +RR 12-76 p68
++HFN 3-77 p119 tape

SXLR 6825 Tape (c) KSXCR 6825
3088 CATALANI: La wally: Ebben ne andro lontana. GIORDANO: Andrea Chenier: La mamma morta. MASCAGNI: Cavalleria rusticana: Voi lo sapete. PONCHIELLI: La gioconda: Suicidio. PUCCINI: Turandot: In questa reggia. VERDI: Macbeth: Nel di della vittoria ...Viene t'affretta...Or tutti sorgete. Il trovatore: Che piu t'arresti...Tacea la notte...Di tale amor. Montserrat Caballe, s; Barcelona Symphony Orchestra; Armando Gatto, Anton Guadagno.
+-HFN 5-77 p139 tape -RR 6-77 p98 tape
-RR 4-77 p40

SXL 6826 Tape (c) KSXC 6826 (Reissues from SET 514/7, 298/300, 209/11, 584/6, D13D5, SET 506/9, 550/4, 292/3)
3089 Grand opera choruses. BEETHOVEN: Fidelio, op. 72: O welche Lust. MUSSORGSKY: Boris Godunov: Vali suda (ed. Rimsky-Korsakov). PUCCINI: Madama Butterfly: Humming chorus. VERDI: Nabucco: Gli arredi festivi giu cadano infranti; Va, pensiero. Otello: Una vela una vela. WAGNER: Die Meistersinger von Nurnberg: Sankt Krispin, Lobet ihn. Parsifal: Nun achte wohl...Zum letzten Liebesmahle. Tannhauser: Begluckt darf nun dich. Vienna State Opera Chorus and Orchestra, VPO; Herbert von Karajan, Lamberto Gardelli, Georg Solti, Lorin Maazel.
+-Gr 3-77 p1450 +-HFN 5-77 p139 tape
+-HFN 3-77 p118 +RR 3-77 p34

SXL 6839 Tape (c) KSXC 6839 (Reissues from SET 503/5, D2D3, SET 372/3, 528/30, 484/6, 542/4, 510/2, 374/5, SXL 6534, SET 561/3) (also London OS 26510)
3090 DONIZETTI: L'Elisir d'amore: Quanto e bella; Una furtiva lagrima. La fille du regiment: Pour me rapprocher de Marie. Lucia di Lammermoor: Tu che a Dio spiegasti l'ali. Maria Stuarda: Ah, rimiro il bel sembiante. PUCCINI: Turandot: Non piangere, Liu. ROSSINI: Stabat mater: Cumus animam. VERDI: Un ballo in maschera: Le rivedra nell'estas; Di tu se fedele. Macbeth: O figli miell...La paterna mano. Requiem: Ingemisco. Rigoletto: Questa or quella; Ella mi fu rapita...Parmi veder le lagrime. Luciano Pavarotti, t; Roger Soyer, bar; Nicolai Ghiaurov, Leonardo Monreale, Nicolas Christou, bs; Helen Donath, Montserrat Caballe, s; ECO, Bologna Teatro Comunale Orchestra,

ROHO and Chorus, Santa Cecilia Orchestra, LSO, VPO; Richard Bonynge, Bruno Bartoletti, Lamberto Gardelli, Georg Solti, Istvan Kertesz, Zubin Mehta.
+Gr 5-77 p1738 ++NR 5-77 p10
+HFN 6-77 p137 +RR 5-77 p34
+HFN 7-77 p127 tape ++SFC 3-20-77 p34

DELOS

FY 001
3091 Grandes Heures Liturgiques. La Maitrise de Notre Dame Cathedral Choir; Various organists.
+NR 7-77 p14

FY 025
3092 ANDERSON: Canticle of praise: Te deum. BAKER: Far-West toccata. BERLINSKI: The burning bush. IVES: Variations on "America". PERSICHETTI: Drop, drop slow tears. ROBERTS: Prelude and trumpetings. George Baker, org.
+-MU 10-76 p16 ++NR 7-77 p14

DEL 25406
3093 BERIO: O King. DAVIDOVSKY: Synchronisms, no. 3. HARRIS: Ludis II. IVES: Largo, violin, clarinet and piano (1901). SCHWANTNER: Consortium I. In aeternum. Boston Musica Viva; Richard Pittman.
+NR 2-76 p6 +NYT 4-11-76 pD23
+NR 9-77 p6

DEUTSCHE GRAMMOPHON

2530 561
3094 BUSSOTTI: Ultima rara. CAROSO: Laura soave: Balletto, Gagliarda, Saltarello (Balleto). CASTELNUOVO-TEDESCO: La guarda cuydadosa. Tarantella. GIULIANI: Grande ouverture, op. 61. MURTULA: Tarantella. PAGANINI: Sonata, violin, op. 25, C major. RONCALLI: Suite, G major. ANON.: Suite, lute. Pieces, lute (5). Siegfried Behrend, gtr; Claudia Brodzinska Behrend, vocalist.
+Gr 11-75 p869 +RR 11-75 p75
+HFN 12-75 p155 ++SFC 2-8-76 p26
+NR 3-76 p12 +St 1-77 p72

2530 792
3095 STOELZEL: Concerto, trumpet, D major. TELEMANN: Concerto, trumpet and strings, D major. TORELLI: Concerto, trumpet and strings, D major. VIVALDI: Concerto, 2 trumpets and strings, C major. VIVIANI: Sonata, trumpet, no. 1, C major. Maurice Andre, tpt; Maurits Sillem, hpd; Hedwig Bilgram, org; ECO; Charles Mackerras.
+-Gr 9-77 p443 +RR 8-77 p64
++HFN 8-77 p91

2530 802
3096 Guitar music of the twentieth century: BALADA: Analogias. BROUWER: Parabola. KUCERA: Diario. MADERNA: Y despues. POULENC: Sarabande. RUIZ PIPO: Estancias. Narciso Yepes, gtr.
+Gr 4-77 p1577 +RR 4-77 p70
+HFN 4-77 p136

2533 173
3097 Italian Renaissance lute music. BARBETTA: Moresca detta le Canarie. CAPIROLA: Ricercare, nos. 1-2, 10, 13. NEGRI: Lo spagno-

letto. Il bianco fiore. MILANO: Fantasia. MOLINARO: Fantasias, nos. 1, 9-10. Ballo detto "Il Conte Orlando". Saltarello (2). PARMA: Aria del Gran Duca. Ballo del serenissimo Duca di Parma. La Cesarina. Corenta. Gagliarda Manfredina. La Mutia. La ne mente per la gola. SPINACINO: Ricercare. TERZI: Ballo tedesco a francese. Tre parti di gagliarde. Konrad Ragossnig, lt.

+Gr 4-75 p1840 +RR 3-75 p45
+NR 3-75 p15 +St 8-77 p59

2533 182 Tape (c) 3310 183
3098 Viennese dance music from the classical period: BEETHOVEN: Contradances, WoO14. EYBLER: Polonaise. GLUCK: Orfeo ed Euridice: Ballet. Don Juan: Allegretto. HAYDN: Minuets (2). MOZART: Landler, K 606 (6). Contredances, K 609 (5). SALIERI: Minuet. WRANITZKY: German dances (10). Quodlibet. Eduard Melkus Ensemble.

+Audio 4-76 p88 +HFN 9-75 p107
+Gr 8-75 p332 ++NR 11-75 p4
+HF 11-75 p130 ++RR 7-75 p37
+HF 2-77 p118 tape +St 8-76 p108

2533 284 Tape (c) 3300 284
3099 Ambrosian chant. Capella Musicale del Duomi di Milano; Luigi Benedetti, Luciano Migliavacca.

+-Gr 7-76 p199 +NR 4-76 p9
++HF 7-76 p96 ++RR 8-76 p76
+HF 1-77 p151 tape ++SFC 4-25-76 p30
+HFN 5-76 p91

2533 294
3100 BAKFARK: Fantasias (4). CATO: Praeludium, Galliardas I, II. DLUGORAJ: Carola Polonesa. Finale (2). Fantasia. Kowaly. Vilanella (2). POLAK: Praeludium. ANON.: Balletto Polacho. Konrad Ragossnig, lt.

++Gr 12-75 p1085 +RR 11-75 p75
+-HFN 12-75 p159 +St 8-77 p59
+NR 11-75 p16

2533 302
3101 Lute music of the Dutch Renaissance: ADRIAENSSEN: Branle Englese. Branle simple de Poictou. Courante. Fantasia. HOWET: Fantasie. JUDENKUNIG: Ellend bringt peyn. Hoff dantz. NEUSIDLER: Der Judentanz. Preambel. Welscher Tantz Wascha mesa. OCHSENKUHN: Innsbruck, ich muss dich lassen. SWEELINCK: Psalms, nos. 5, 23. HOVE: Galliarde. VALLET: Prelude Galliarde. Slaep, soete, slaep. VON HESSEN: Pavane. WAISSELIUS: Deutscher Tantz. Fantasia. ANON.: Der gestraifft Danntz. Konrad Ragossnig, Renaissance lute.

+Gr 7-76 p194 +RR 6-76 p72
+HFN 7-76 p91 ++SFC 8-22-76 p38
++NR 4-76 p14 +St 8-77 p59

2533 305
3102 Nigel Rogers, Canti amorosi. CACCINI: Amarilli mia bella; Belle rose porporine; Perfidissimo volto; Udite amante. CALESTANI: Damigella tutta bella. GAGLIANO: Valli profonde. d'INDIA: Cruda amarilli; Intenerite voi, lagrime mie. PERI: O durezza di ferro; Tra le donne; Bellissima regina. RASI: Indarno Febo. SARACINI: Da te parto; Deh, come invan chiedete; Giovinetta vezzosetta; Io moro; Quest'amore, quest'arsura. TURCO: Occhi belli. Nigel Rogers, t; Colin Tilney, hpd, positive organ;

Anthony Bailes, chitarrone; Jordi Saval, vla da gamba; Pere Ros, violone.

+Gr 4-76 p1650 | +NR 4-76 p10
+HFN 5-76 p97 | +RR 5-76 p68
+MJ 11-76 p45 | ++St 6-76 p72
++MT 1-77 p44

2533 310
3103 Gregorian chant: Chants for feasts of Mary; Propia Missarum; Antiphonae Mariae. Schola Cantorum Francesco Cordini; Fosco Corti.

+Gr 12-76 p1058 | +NR 4-77 p7
++HF 7-77 p118 | +RR 12-76 p91
+-HFN 12-76 p143

2533 320
3104 Gregorian chants for Palm Sunday. Choralschola Munsterschwarzach; Pater Godehard Joppich.

+ARG 8-77 p33

2533 323
3105 ADRIAENSSEN: Madonna mia pieta. Io vo gridando. DOWLAND: Mr. George Whitehead his almand. My Lord Willoughby's welcome home. GALILEI, V.: Fuga a l'unisono. HASSLER: Canzon. JOHNSON: Treble to a ground. MOLINARO: Ballo detto "Il Conte Orlando". Saltarello. PACOLINI: Passamezzo della battaglia. Saltarello della traditoria. Saltarello milanese. PICCININI: Canzona. ROBINSON: A toy. HOVE: Lieto godea. NON.: De la trumba. Drewries accords. Le rosignoll. Konrad Ragossnig, Jurg Hubscher, Dieter Kirsch, lt.

+ARG 2-77 p46 | +NR 7-77 p15
+Gr 2-77 p1302 | +RR 12-76 p74
++HFN 12-76 p145 | +St 8-77 p59

2533 347
3106 The triumphs of Oriana: BENNET: All creatures now are merry. CARLTON: Calm was the air. CAVENDISH: Come, gentle swains. COBBOLD: With wreaths of rose and laurel. EAST: Hence stars too dim of light. FARMER: Fair nymphs I heard one telling. GIBBONS, E.: Love live fair Oriana; Round about her charret. HILTON: Fair Oriana, beauty's queen. HOLMES: Thus bonny-boots the birthday celebrated. HUNT: Hark, did ye ever hear. JOHNSON, E.: Come blessed bird. JONES: Fair Oriana, seeming to wink. KIRBYE: With angels face. LISLEY: Fair Cytherea presents her doves. MARSON: The nymphs and shepherds danced. MILTON: Fair Orian, in the morn. MORLEY: Arise, awake, awake; Hard by a crystal fountain. MUNDY: Lightly she whipped o'er the dales. NICOLSON: Sing, shepherds all. NORCOME: With angel's face. TOMKINS: The fauns and satyrs tripping. WEELKES: As Vesta was from Latmos Hill descending. WILBYE: The lady Oriana. Pro Cantione Antiqua; Jennifer Smith, s; Leonie Mitchell, ms; Ian Partridge.

+Gr 8-77 p337 | *NR 12-77 p7
+HFN 8-77 p91 | +RR 8-77 p78

2535 254 Tape (c) 3335 254 (Reissues)
3107 BORODIN: Prince Igor: Polovtsian dances. FALLA: The three-cornered hat: Jota. KHACHATURIAN: Gayaneh: Sabre dance. RAVEL: Bolero. STRAUSS, R.: Salome, op. 54: Dance of the seven veils. STRAVINSKY: The rite of spring: Danse sacrale. Leningrad Philharmonic Orchestra, BPhO, BRSO, Monte Carlo Opera Orchestra; Gennady Rozhdestvensky, Herbert von Karajan, Karl Bohm, Lorin Maazel, Louis Fremaux, Michael Tilson Thomas.

+Gr 12-77 p1146 tape | +RR 11-77 p60
+HFN 11-77 p183

2548 148 Tape (c) 3348 148 (Reissue from 135017)
3108 BERLIOZ: Damnation of Faust, op. 24: Hungarian march. GRIEG: Peer Gynt, op. 46: In the hall of the mountain king. Sigurd Jorsalfar, op. 56: Homage march. MENDELSSOHN: A midsummer night's dream, op. 61: Wedding march; Fairies march. MILHAUD: Le carnaval d'Aix: Le Capitaine Cartuccia. MOZART: March, K 237, D major. PROKOFIEV: March, op. 99, B flat major. STRAUSS, J. I: Radetzky march, op. 228. TCHAIKOVSKY; Casse-Noisette, op. 71: March. Slavonic march, op. 31. Bavarian Radio Symphony Orchestra, Nordmark Symphony Orchestra, Bamberg Symphony Orchestra, Berlin Radio Symphony Orchestra, BPhO, Monte Carlo Opera Orchestra, Capella Coloniensis; Rafael Kubelik, Ferdinand Leitner, Heinrich Steiner, Richard Kraus, Ferenc Fricsay, Louis Fremaux.
+Gr 4-75 p1867 +-RR 4-75 p28
+HFN 3-77 p119 tape

2548 219 Tape (c) 3348 219
3109 Music of Venice: ALBINONI (Giazotti): Adagio, G minor. Sonata a 5, op. 5, no. 9, E minor. GEMINIANI: Concerto grosso, D minor (from Corelli's Sonata, violin, op. 5, no. 12). PACHELBEL: Canon, D major. ROSSINI: Sonata, 2 violins, violoncello and double bass, no. 3, C major. YSAYE: Paganini variations. Ensemble d'Archets Eugene Ysaye; Lola Bobesco, vln.
+-Gr 5-76 p1771 +HFN 3-77 p119 tape
++HFN 4-76 p109 +RR 4-76 p52

2584 004
3110 DUKAS: The sorcerer's apprentice. GINASTERA: Estancia: Danza final. KHACHATURIAN: Gayaneh: Sabre dance. MUSSORGSKY: A night on the bare mountain (arr. Rimsky-Korsakov). SAINT-SAENS: Danse macabre, op. 40. STRAVINSKY: The firebird: Danse infernale. BPO; Arthur Fiedler.
++Gr 6-76 p95 +-RR 6-76 p33
++NR 2-77 p3 ++St 2-77 p75
+HFN 5-76 p99

2710 019 (also 2723 054)
3111 Music of the Gothic era. ADAM DE LA HALLE: De ma dame vient. J'os bien a m'amie parler. BERNARD DE CLUNY: Pantheon abluiter. GILLES DE PUSIEUX: Ida capillorum. Rachel plorat filios. LEONINUS: Alleluia pascha nostrum. Gaude Maria virgo. Locus iste. Viderunt omnes. MACHAUT: Christe qui lux es. Hoquetus David. Lasse comment oublieray. Qui es promesses. PEROTIN LE GRAND: Sederunt principes. Viderunt omnes. PETRUS DE CRUCE: Aucun ont trouve. ROYLLART: Rex Karole. VITRY: Cum statua. Impudenter. ANON.: Alle psallite; Amor potest; Clap, clap par un matin; Degentis vita; Dominator Domine; En mai quant rosier; Febus mundo oriens; Hoquetus in saeculum I-VII; In mari miserie; Inter densas deserte meditans; Les l'ormel a la turelle; La mesnie fauveline; O matissima; El mois de mai; O Philippe Franci; On parole de batre; Quant je le voi; Quasi non ministerium; S'on me regards; Zelus familie. Early Music Consort; David Munrow.
++ARG 3-77 p43 +RR 11-76 p98
++HF 7-77 p120 ++St 5-77 p79
+NR 4-77 p12

2723 051 (6) (Reissues from 2533 111, 2533 150, 2533 172, 2533 303, 2533 303, 2533 182, 2533 134)
3112 Dance music from the 15th-19th centuries: ATTAINGNANT: La Brosse: Tripla, Tourdion. La gatta. La Magdalena. BACH, C.P.E.: Min-

uet and trio (2). Polonaises (5). Trio. BEETHOVEN: Modlinger Tanz, nos. 1-8. BESARD: Branle gay. BOUIN: La Montauban. CAROSO: Barriera. Celeste Giglio. CAROUBEL: Pavana de Spaigne. Courante (2). Volte (2). CHEDEVILLE: Musette. CORRETTE: Minuets, nos. 1 and 2. DALZA (Petrucci): Calata ala Spagnola. DESMARETS: Menuet. Passepied. DOWLAND: Mrs. Winter's jump. Queen Elizabeth her galliard. EYBLER: Polonaise. FISCHER: Bourree. Gigue. GERVAISE: Branle de Bourgogne. Branle de Champagne. GESUALDO: Gagiarda del Principi di Venosa. GIBBONS: Galliard. GLUCK: Don Juan: Allegretto. Orfeo ed Euridice: Ballet. GULIELMUS: Bassa danza a 2. HAYDN: Minuets (2). HAUSSMANN: Catkanei. Galliard. Paduan. Tantz. HOLBORNE: The funerals: Pavan. Heigh ho holiday: Coranto. Noel's galliard. HOTTETERRE: Bourree. LANNER: Ungarischer Galopp. LE ROY: Branle de Bourgogne. LOEILLET: Corente. Gigue. Sarabande. LULLY: Une noce de village: Derniere entree. MAINERIO: Schiarazula marazula. Tedesca: Saltarello. Ungaresca: Saltarello. MILAN: Pavanas 1 and 2. MOLINARO: Ballo detto "Il Conte Orlando": Saltarello. Saltarello. MOSCHELES: German dances. MOZART: Landler, K 606 (6). Contradances, K 609 (5). MUDARRA: Romanesca Guarda me las vacas. NEUSIDLER: Der Judentaz. Welscher Tanz Wascha mesa: Hupfauff. PAIX: Schiarazula Marazula. Ungaresca: Saltarello. PAMER: Waltz, E major. PHALESE: Passamezzo: Saltarello. Passamezzo d'Italye: Reprise, Gaillarde. POGLIETTI: Balletto. PRAETORIUS: Galliarde de la guerre. Galliarde de Monsieur Wustron. Reprinse. RAMEAU: Zoroastre: Dances (7). REUSNER: Suite (arr. Stanley). SALIERI: Menuetto. SANZ: Canarios. Espanoletas. Gallarda y villano. Passacalle de la Cavalleria de Napoles. SCHMID: Englischer Tanz. Tanz Du hast mich wollen nemmen. SCHUBERT: Landler, D 354 (4). Minuets, D 89 (5). SIMPSON: Alman. STARZER: Diane et Endimione: Dances. Gli Orazi e gli Zuriazi: Dances. Roger et Bradamante: Dances. SUSATO: Ronde. TORRE: Danza alta a 3. WRANITZKY: German dances (8). Quodlibet. ANON.: Bassa danza a 2. Bassa danza a 3. Cobbler's jigg. Gavotte. Greensleeves and pudding pyes. How can I keep my maiden head. Istampita Cominiciamento di gioia. Istampita Ghaetta. Kemp's jig. Lamenot di Tristano: Rotta. Linzer Tanz. Mascherade. Running footman. Saltarello. Trotto. Wiener Polka. Konrad Ragossnig, lt and gtr; Ulsamer Collegium, Eduard Melkus Ensemble; Josef Ulsamer.

+Gr 11-77 p853 +RR 10-77 p41
++HFN 11-77 p185 ++SFC 11-13-77 p50

DISCOPAEDIA

MB 1011

3113 Masters of the bow: Ossy Renardy. ERNST: Airs Hongrois varies, op. 22, A major. PAGANINI: Caprices, op. 1, nos. 1-24. Sonata, violin, op. posth., A major. SAINT-SAENS: Concerto, violin, no. 1, op. 20, A major. ZARZYCKI: Mazurka, op. 26, G major. Ossy Renardy, vln; Walter Robert, pno.

+NR 2-77 p31 +ST 5-76 p51

MB 1012

3114 Masters of the bow: Fritz Kreisler. ALBENIZ: Espana, op. 165. BACH: Partita, violin, no. 3, S 1006, E major: Preludio. BRANDL: Der Liebe Augustin: Du alter Stefansturm. COTTENET:

Chanson meditation. DOHNANYI: Ruralia Hungarica, op. 32. DRDLA: Souvenir. DVORAK: Songs, op. 55, no. 4. FALLA: Canciones populares espanolas, no. 6. FOSTER: The old folks at home. GARTNER: Aus Wien. GRIEG: Lyric pieces, op. 43, no. 6: To the spring. KREISLER: Caprice Viennois, op. 2. Leibesleid. Polichinelle. Schon Rosmarin. MASSENET: Thais: Meditation. SMETANA: Ma Vlast: Andantino. TCHAIKOVSKY: Chant sans paroles, op. 2: Souvenirs de Hapsal. Fritz Kreisler, vln.
++NR 2-77 p13 +ST 6-76 p151

MB 1013
3115 Masters of the bow: Yehudi Menuhin. BLOCH: Baal Shem: Ningun. BRAHMS: Hungarian dances, nos. 1, 7, 17. FIOCCO: Suite, no. 1, G major: Allegro. HANDEL: Dettingen Te Deum, op. 17. LECLAIR: Sonata, violin, op. 9, no. 3, D major. MONASTERIO: Sierra morena. MOZART: Concerto, violin, no. 3, K 216, G major: Adagio. NOVACEK: Caprices, op. 5, no. 4. RIES: La capricciosa. SAENGER: Concert miniatures, op. 130, no. 2. SAMAZEUIHL: Chant d'Espagne. SARASATE: Danzas espanolas, no. 6. SERRANO Y RUIZ: La cancion del olvido: Cancion de Marinella. SPOHR: Duet, no. 1, D minor. WIENIAWSKI: Scherzo tarantelle, op. 16, G minor. ANON.: La romanesca. Yehudi Menuhin, vln.
+NR 2-77 p13 +ST 6-76 p157

MB 1014
3116 Masters of the bow: Nathan Milstein. BRAHMS: Hungarian dance, no. 2, D minor. CHOPIN: Nocturne, op. posth., C sharp minor. GLUCK: Orphee ed Eurydice: Dance of the blessed spirits. KREISLER: Rondino on a theme by Beethoven. LISZT: Consolation, no. 3, G 172, D flat major. PAGANINI: Concerto, violin, no. 2, op. 7, B minor: Ronde a la clochette. PIZZETTI: Canti ad una Giovane Fidanzata: Affettuoso. RIMSKY-KORSAKOV: The tale of the Tsar Sultan: Flight of the bumblebee. SARASATE: Danzas espanolas: Romanza andaluza. SCHUMANN: Klavierstucke, op. 85: Abendlied. STAMITZ: Concerto, violin, B flat major: Adagio. SUK: Pieces, op. 17: Burleska. WIENIAWSKI: Concerto, violin, no. 2, op. 22, D minor: Romance. Polonaise brillante, no. 1, op. 4, D major. Scherzo tarantelle, op. 16. Nathan Milstein, vln.
+NR 3-77 p13 +ST 7-76 p237

MB 1015
3117 Masters of the bow: Georg Kulenkampff. ALBENIZ: Espana, op. 165, no. 2: Tango. BRAHMS: Concerto, violin, op. 77, D major: Adagio. BRUCH: Concerto, violin, no. 1, op. 26, G minor: Adagio. DVORAK: Humoresque, op. 101, no. 7, G flat major (arr. Kreisler). LULLY: Ballets du Roy. POLDINI: Marionnettes: Poupee valsante. SCHUBERT: Ave Maria, D 839. SCHUMANN: Klavierstucke, op. 85: Abendlied. SPOHR: Concerto, violin, no. 8, op. 47, A minor. SVENDSEN: Romance, op. 26, G major. TCHAIKOVSKY: Swan Lake, op. 20: Neapolitan dance. Georg Kulenkampff, vln.
+-NR 3-77 p13 +ST 7-76 p241

MB 1016
3118 Masters of the bow: Ruggiero Ricci. BACH: Sonata, violin, no. 2, S 1003, A minor. HINDEMITH: Sonata, violin, op. 31, no. 2. Sonata, violin, no. 3, E major. RACHMANINOFF: Vocalise, op. 34, no. 14. SARASATE: Introduction and tarantelle, op. 43. Danzas espanolas, op. 21: Habanera. Zigeunerweisen, op. 20, no. 1. YSAYE: Reve d'enfant, op. 14. Ruggiero Ricci, vln.
++NR 3-77 p13 +ST 8-77 p301

MOB 1018
3119 Masters of the bow: Paul Kochanski. BRAHMS (Joachim): Hungarian dance, no. 1, G minor. Sonata, violin and piano, no. 3, op. 108, D minor. KREISLER: La gitana. PIERNE: Serenade. RACHMANINOFF (Gutheil): Vocalise, op. 34, no. 14. RAFF: Cavatina. SARASATE: Danzas espanolas, op. 21, no. 1: Malaguena. TCHAIKOVSKY: Chant sans paroles. Melodie. WAGNER (Wilhelmj): Prize songs. WIENIAWSKI: La carnaval Russe. Paul Kochanski, vln; Joseph Kochanski, pno.
++ST 1-77 p755

MOB 1019
3120 Masters of the bow: Joseph Joachim, Marie Soldat-Roger, Deszo Szigeti. BACH: Partita, violin, no. 1, S 1002, B minor: Bourree (Joachim). Partita, violin, no. 3, S 1006, E major: Preludio (Soldat). Sonata, violin, no. 1, S 1001, G minor: Adagio (Joachim, Soldat). Suite, orchestra, S 1068, D major: Air on the G string (Soldat). BEETHOVEN: Romance, no. 2, op. 50, F major (Soldat). BRAHMS: Hungarian dances, nos. 1 and 2 (Joachim). HUBAY: Poeme Hongroise, nos. 1 and 2 (Szigeti). JOACHIM: Romance (Joachim). MOZART: Concerto, violin, no. 5, K 219, A major: 1st movement (Soldat). SCHUMANN: Abendlied, op. 107, no. 6 (Soldat). SPOHR: Concerto, violin, no. 9, op. 55, D minor: Adagio (Soldat). Joseph Joachim, Marie Soldat-Roger, Deszo Szigeti, vln; Otto Schulhoff, pno.
+-ST 1-77 p755

MOB 1020
3121 Masters of the bow: Henri Marteau and pupils, Goran Olsson-Follinger, Florizel von Reuter, Ferenc Aranyi. AULIN: Idyll. Polska (Olsson-Follinger). BACH: Partita, violin, no. 3, S 1006, E major (Marteau). Prelude, E major (Olsson-Follinger). Suite, orchestra, S 1068, D major: Air on the G string (Marteau). BOCCHERINI: Minuet (Marteau). BRAHMS (Joachim): Hungarian dance, no. 6 (Marteau). GODARD: Adagio pathetique (Marteau). HANDEL (Burmester): Arioso (Olsson-Follinger). Sonata, violin, op. 1, no. 5, G major: Minuet (Olsson-Follinger). HEGAR: Waltzes, nos. 2 and 4 (Marteau). KREISLER: Tambourin Chinois, op. 3 (Aranyi). MARTEAU: Cakewalk. Valse fantastic (Marteau). MOZART: Adagio, C major (Marteau). OLSSON-FOLLINGER: Mejlikan (Olsson-Follinger). Polska, A major (Marteau). PAGANINI: Moses fantasy on the G string (von Reuter). SARASATE: Carmen fantasy (Marteau). Danzas espanolas, op. 21, no. 2: Habanera (Marteau). SCHUBERT: Serenade (Marteau). Sonata, violin and piano, op. 162, D 574, A major: Finale (Marteau). Henri Marteau, Goran Olsson-Follinger, Florizel von Reuter, Ferenc Aranyi, vln; P. Vladigerov, pno.
+-ST 3-77 p917

DJM RECORDS

DJM 22026 Tape (c) DJM 42062
3122 BELTON: Down the mall. COATES: Knightsbridge. Oxford Street. ELY: Trafalgar Square. FARNON: Westminster waltz. KETELBEY: Cockney suite: Appy Ampstead. LEICESTER: Parliament Square. PARKES: Big Ben. PAYNE: London medley. STECK: Birdcage walk. WOOD: Frescoes suite: Bandstand in Hyde Park. London landmarks suite: The horseguards. Life Guards Band; Antony Richards.
+Gr 7-77 p226

EMI

TWOX 1058 Tape (c) TC TWOX 1058

3123 BARRET (Siebert): March of the cobblers. GOLDENBERG: Kojak theme. HERBERT: March of the toys (arr. Nestico). JONES: Ironside, theme. MANDEL: The shadow of your smile (arr. Woodfield). NEVILLE: Shrewsbury fair. RODGERS: Slaughter on Tenth Avenue. Victory at sea (arr. R. R. Bennett). SIMPSON: On the track (arr. Dawson). STEVENS: Policewoman, theme (arr. Woodfield). WOODFIELD: The magic of Michel Legrand. YODER: Carolina clambake. Michael Eastbrook, trom; Alan Webb, xylophone; Royal Marines School of Music Band; Paul Neville.

+Gr 7-77 p226

Italiana C 153 52425/31 (7)

3124 BEETHOVEN: Concerto, piano, no. 1, op. 15, C major. Concerto, piano, no. 4, op. 58, G major. Concerto, piano, no. 5, op. 73, E flat major. GRIEG: Concerto, piano, op. 16, A major. LISZT: Concerto, piano, no. 1, G 124, E flat major. MOZART: Concerto, piano, no. 9, K 271, E flat major. Concerto, piano, no. 20, K 466, D minor. Concerto, piano, no. 24, K 491, C minor. Concerto, piano, no. 25, K 503, C major. SCHUMANN: Concerto, piano, op. 54, A minor. Walter Gieseking, pno; PhO, Berlin State Opera Orchestra; Henry Wood, Hans Rosbaud, Herbert von Karajan.

+ARG 8-77 p3

ENIGMA

VAR 1015

3125 PUJOL VILARRUBI: Guajira. Tango. RODRIGO: Fandango. Pequenas sevillanas. Ya se van los pastores. SANZ: Espanoleta. Canarios. SOR: Fantasia elegiaca, op. 59. TARREGA: Capricho arabe. Gran jota. Maria. Sueno. TORROBA: Burgalesa. TRAD.: Brincan y bailan. Don Gato. La serrano. Ya se van la paloma (all arr. Bonell). Carlos Bonell, gtr.

+Gr 12-76 p1027 +RR 12-76 p84
+HFN 1-77 p107

VAR 1016

3126 Christmas carols from Westminster Cathedral: Angels we have heard on high; Away in the manger; The boar's head carol; Come to the manger; Ding dong merrily on high; The first Nowell; Good King Wenceslas; Hob and Colin; The holly and the ivy; Infant so gentle; O come all ye faithful; Personent hodie; While shepherds watched their flocks. (arr. Mawby, Wood, Holst, Malcolm, Getty). Westminster Cathedral Choir; Colin Mawby.

+Gr 12-77 p1156 +RR 11-77 p101
+HFN 12-77 p185

VAR 1017 Tape (c) TC VAR 1017

3127 Elizabethan and Jacobean madrigals: DOWLAND: Fine knacks for ladies. GIBBONS: The silver swan. MORLEY: April is my mistress face; Daemon and Phyllis; Fire, fire; I love, alas; Leave, alas, this tormenting; My bonny lass; Now is the month of Maying; O grief, even on the bud; Those dainty daffadillies; Though Philomela lost her love. MUNDY: Were I a king. VAUTOR: Sweet Suffolk owl. WEELKES: Cease sorrows now; Come sirrah, Jack ho; Since Robin Hood. WILBYE: Adieu sweet Amarillis; Thus saith my Cloris bright. The Scholars.

+-Gr 12-76 p1042 +-RR 8-77 p85 tape
+-HFN 1-77 p107 ++St 8-77 p130
+RR 1-77 p79

VAR 1020 Tape (c) TC VAR 1020
3128 BLONDEL: Quant je plus. CORNYSHE: Ah, Robin. ADAM DE LA HALLE: The play of Robin and Marion, excerpts. RAVENSCROFT: Now Robin lend to me thy bow. RICHARD I, King of England: Je nus hons pris. ANON/TRAD.: A la fontenella; Bonny, sweet Robin; English estampie; Greenwood/Dargason; Fanfare; The maid in the moon; My Robin; The nutting girl; The parson's farewell/ Goddeses; Redit aetas aurea; Robin Hood and the tanner; Sally's fancy/The maiden's blush; Shepherd's hey; Stingo; Tristan's lament; Two-voices estampie. St. George's Canzona; John Sothcott.
/-Gr 2-77 p1312 +RR 2-77 p90
-HFN 1-77 p117 +RR 8-77 p85

VAR 1023 Tape (c) TC VAR 1023
3129 BARTLETT: What thing is love. CAMPIAN: Beauty is but a painted hell; Come let us sound with melody; The cypress curtain of the night. DOWLAND: Can she excuse my wrongs; Come heavy sleep; Come away, come sweet love; In darkness let me dwell; Say love if ever thou didst find; Sweet stay awhile. FERRABOSCO: Come my Celia; If all these cupids; It was no policy of court; So beauty on the waters stood; Yes were the loves. JONES: Love is a babel; Love is a pretty frenzy; Now what is love. ROSSETER: What then is love. Wynford Evans, t; Carl Shavitz, lt; Peter Vel vla da gamba.
+Gr 7-77 p212 ++RR 7-77 p92
+HFN 7-77 p121 +RR 8-77 p85 tape

VAR 1024
3130 A tapestry of music for Christopher Columbus and his crew. ALMOROX: O dichoso. ATTAINGNANT: Basse dance. Galliard. Pavan. BINCHOIS: Files a marier. ENCINA: Congoxa mas. Si abra en este baldres. FOGLIANO: Tua volsi esser sempre mia. GABRIEL: De la dulce. ISAAC: E qui la dira. NOLA: Madonna nui sapima. OBRECHT: La stangetta. Tsaat een meskin. SCOTTO: O fallace speranza. TORRE: Pascua d'espiritu santo. TROMBONCINI: Io son l'ocello. ZESSO: E quando andaretu. ANON.: Al alva venid; El bisson; La Comarina; Dit le Bourguygnon; Der neue Bauernschwanz; La rocha el fuso; La spagna; Sola me dexastes; La traditora. St. George's Canzona; John Sothcott.
+-Gr 8-77 p332 +-RR 5-77 p85

VAR 1025
3131 BLOCH: Baal Shem: Ningun. BRAHMS: F-A-E sonata: Scherzo. GLUCK: Melodie (arr. Kreisler). PAGANINI: Cantabile, op. 17, D major. Caprices, op. 1, nos. 5, 13. Introduction and variations on "Di tanti palpiti", op. 13. PARADIS: Sicilienne. RAVEL: Habanera (arr. Catherine). RIMSKY-KORSAKOV: The tale of the Tsar Sultan: Flight of the bumblebee (arr. Hartmann). SARASATE: Danzas espanolas, op. 23, no. 2: Zapateado. STRAVINSKY: Russian maiden's song (arr. Dushkin). WIENIAWSKI: Polonaise, op. 4, D major. Scherzo tarantelle, op. 16, G minor. Maurice Hasson, vln; Ian Brown, pno.
+Gr 6-77 p69 ++RR 5-77 p76

VAR 1027 Tape (c) TC VAR 1027
3132 BRITTEN: Winter words, op. 53. BUSH: Echo's lament for Narcissus; The wonder of wonders. GIBBS: The fields are full; A song of shadows. GURNEY: Nine of the clock; Ploughman singing; Under the greenwood tree. HOLST: A little music; The floral bandit; The thought. QUILTER: Go lovely rose; O mistress mine. WARLOCK:

As ever I saw; To the memory of a great singer. Ian Partridge, t; Jennifer Partridge, pno.
+Gr 7-77 p1208 +RR 8-77 p73
++HFN 7-77 p112 +RR 8-77 p85 tape

EVEREST

SDBR 3402
3133 Gregorian chants. Trappist Monks Choir, Cisercian Abbey.
-NR 9-77 p10

FINNADAR

SR 9015
3134 New music for contrabass: ERB: Trio for two. JULIAN: Akasha. MINGUS: Goodbye, porkpie hat. SCHAFFER: Free form, no. 2: Evocazioni. Project. TURETZKY: Gamelan music. Bertram Turetzky, double bs; Assisting artists.
+MJ 10-77 p28 +NR 10-77 p16

FINNLEVY

SFLP 8569
3135 BACH: Chaconne, solo violin, D minor. MOZART: Rondo, violin, G major (arr. Kreisler). RAUTAVAARA: Varietude, solo violin, op. 82. SIBELIUS: Berceuse, op. 79, no. 6. SZYMANOWSKI: Notturno e tarantella, op. 28, nos. 1 and 2. TCHAIKOVSKY: Serenade melancolique, op. 26. Yuval Yaron, vln; Rena Stepelman, pno.
+-RR 10-77 p73

FOLKWAYS

FTS 32378
3136 Music for the Colonial Band. BILLINGS: Jargon. Judea. FELTON: Concerto, op. 2, no. 4, C major. Eighteenth century settings of The Star-Spangled Banner. FLAGG: Trumpet tune with fuguing section. HOLYOKE: Masonic processional march. SELBY: Anthem for Christmas day. A lesson. STANLEY: Concerto, D major. ANON.: Love in a village: O had I been by fate decreed, My Dolly was the fairest thing. Jonah: Save me O Lord. Colonial Band of Boston; David McKay.
+ARG 12-76 p12 +NR 10-77 p15

FREE LANCE

FLPS 675
3137 Arias: Kurt Baum. BIZET: Carmen. GIORDANO: Andrea Chenier. LEONCAVALLO: I Pagliacci. PONCHIELLA: La gioconda. PUCCINI: Manon Lescaut. Tosca. Turandot. VERDI: Aida. La forza del destino. Il trovatore. Kurt Baum, t.
+ON 1-15-77 p33

GASPARO

GS 103
3138 ARMA: Evolutions pour basson seul. ARNOLD: Fantasy, bassoon. BARATI: Triple exposure, solo violoncello. BIZET: Little duet.

HINDEMITH: Pieces, bassoon and violoncello (4). JACOB: Partita, bassoon. MOZART: Sonata, bassoon, K 292, B flat major. Otto Eifert, bsn; Roy Christensen, vlc.
++ARG 12-77 p48 ++NR 11-77 p10

GOLDEN AGE

GAR 1001

3139 BACH: Cantata, no. 211, Schweigt stille plaudert nicht (Coffee). BIZET: Carmen: Toreador song. BORODIN: Prince Igor: Igor's aria. CHOUHAJIAN: Garineh: Horhor's aria. HANDEL (attrib.): Dank sei dir, Herr. MOZART: Don Giovanni, K 527: Madamina; Finch'han dal vino. MURADIAN: Drunken with love. SPENDIARIAN: Oh, rose. VERDI: La traviata: Di provenza il mar. VIVALDI: La fida ninfa. Dicran Jamgochian, bar; Armenian Symphony Orchestra; Rafael Mangasarian.
+-St 12-77 p156

GOLDEN CREST

CRS 4148

3140 ADSON: Ayres (2). AICHINGER: Jubilate Deo. ANDERSON: Bugler's holiday. BEATLES: Eleanor Rigby; Penny Lane. BRADE: Pieces (2). JONES: Four movements, 5 brass. JOPLIN: Cascades. LE JEUNE: Le printemps, excerpt. MOURET: Fanfares (Rondeau). PEZEL: Pieces (3). PILSS: Scherzo. ANON.: Scarborough fair. Die Baneklsangerlieder: Sonate. New York Brass Choir.
+ARG 3-77 p40

RE 7064

3141 DONJON: Adagio nobile. Le chant du vent. Pan. Pipeaux. Offertoire. Rossignolet. HONEGGER: Romance. MONROE: Sketches, solo flute. MUSSORGSKY (Monroe): Hebrew song. PESSARD: Bolero. Ervin Monroe, flt; Fontaine Laing, pno.
-ARG 5-77 p50

RE 7068

3142 DARTER: Sonatina. ISRAEL: Dance variations. LINDENFELD: Combinations I: The last gold of perished stars. MORRILL: Studies, trumpet and computer. ROUSE: Subjectives VIII. SOKOL: Sonatina. Marice Stith, tpt; Brian Israel, pno and percussion.
++NR 7-77 p15

GROSVENOR

GRS 1043

3143 DRIGO: Les millions d'Arlequin: Serenade. GREEN: Sunset. HUME: BB and CF march. JENKINS: Life divine. LEAR: Shylock: Polka. McCUNN: The land of the mountain and the flood: Overture. SCHOLEFIELD: St. Clement, hymn tune. BOLTON (arr.): Bobby's tune (trans. E. Banks). Carlton Main Frickley Colliery Band; Robert Oughton.
++Gr 1-77 p1193

GRS 1048

3144 CROOKES: Way out West. DEL STAIGER: Napoli. GRIEG: Norwegian dance (arr. Ryan). KELLY: Divertimento. SONDHEIM: A little night music: Send in the clowns (arr. Bartram). SULLIVAN: H.M.S. Pinafore: Overture (arr. Martyn). WOOD, A.: Dale dances (3). ZEHLE: Wellington march. William Davis Construction

Band. John Berryman.
+Gr 1-77 p1193

GRS 1052
3145 DVORAK: Symphony, no. 9, op. 95, E minor: Largo (arr. Wright). FLETCHER: Epic symphony: Elegy. FOSTER: Jeanie with the light brown hair (arr. Howarth). HOLST: Fantasia on the Dargason. HUGGENS: Chorale and rock-out. PICON: Marche des bouffons. RIMMER: Rule Britannia. SOUSA: Manhattan Beach. TRAD.: Dashing away with a smoothing iron (arr. Wright). Cory Band; Bram Gay.
+-RR 12-77 p50

GRS 1055
3146 ALFORD: Colonel Bogey on parade. DEBUSSY: Children's corner suite: Golliwog's cakewalk (arr. Ball). FUCIK: Entry of the gladiators, op. 68 (arr. Langford). GERSHWIN: The little rhapsody in blue (arr. Thompson). KNIPPER: Cossack patrol (arr. Langford). MEYERBEER: Robert le diable, excerpt (arr. Moreton). NICHOLS/WILLIAMS: We've only just begun. STEPHENS: Copa cabana. VINTER: Entertainments: Elegy. WALTERS: Trumpets wild. Stalybridge Brass; Les Hine.
+-RR 10-77 p46

GUILD

GRSP 701
3147 Royal music from St. Pauls. BLISS: Fanfare. BYRD: O Lord, make thy servant Elizabeth. BRITTEN: Te deum, C major. GOSS: Praise my soul, the King of heaven. PARRY: I was glad. TOMKINS: The King shall rejoice. VAUGHAN WILLIAMS: Mass, G minor: Credo. All people that on earth do dwell; O taste and see. WESLEY, S.S.: Thou wilt keep him in perfect peace. TRAD. (arr. Jacob): The national anthem. St. Paul's Cathedral Choir; Barry Rose, org; Royal Military School of Music Trumpeters; Christopher Dearnley.
+Gr 6-77 p95

GRS 7008
3148 Music for a great cathedral: ATTWOOD: Turn thee again, O Lord, at the last. BATTEN: O sing joyfully. BATTISHILL: O Lord, look down from heaven. BOYCE: I have surely built thee an house. GOSS: Psalms, nos. 127, 128. GREENE: Lord, let me know mine end. MacPHERSON: A little organ book: Andante, G major. MENDELSSOHN: Above all praise and majesty. St. Paul: How lovely are the messengers. PARRY: Songs of farewell: My soul, there is a country. STAINER: Lord Jesus, think on me. St. Paul Cathedral Choir; Barry Rose, org; Christopher Dearnley.
+-Gr 6-77 p90 +-RR 6-77 p90
+HFN 6-77 p131

GRSP 7011
3149 ALCOCK: Introduction and passacaglia. DEARNLEY: Dominus regit me. Fanfare. FRANCK: Choral, no. 1, E major. GADE: Festligt praeludium "Lover den Herre". NIELSEN: Festpraeludium. SIBELIUS: Intrada, op. 111a. Christopher Dearnley, org.
++HFN 12-77 p176 +RR 11-77 p92

HARMONIA MUNDI

HMU 335

3150 Carmina Burana: Songs of drinking and eating; Songs of unhappy love; Bacche, bene venies; Virent prata hiemata; Nomen a sollempnibus; Alte clamat Epicurus; Vite perdite (2 versions); Vacillantis trutine; In taberna quando sumus; Iste mundus furibundus; Axe Phebus aureo; Dulce solum natalis patrie; Procurans odium; Sic mea fata canendo solo; Ich was ein chint so wolgetan. Clemenic Consort; Rene Clemencic.
+Gr 8-77 p337 +St 9-76 p129
++RR 4-77 p84

HMU 336

3151 Carmina Burana: Carmina moralis et divina; Carmina veris et amoris. Gustav Bauer, Ladislav Illavsky, Laszlo Kunz, bar; Franz Handlos, Karl Kastler, bs; Clemencic Consort; Rene Clemencic.
+Gr 8-77 p337

HMU 337

3152 Carmina Burana: Officium lusorum; Olim sudor Herculis; Virent prata hiemata. Clemencic Consort; Rene Clemencic.
+Gr 8-77 p337 +St 12-76 p152

HMU 338

3153 Carmina Burana: Carmina veris et amoris; Carmina moralia et divina. Clemencic Consort; Rene Clemencic.
+Gr 8-77 p337

HM 759

3154 Francis Chapelet: Organs of Spain. ARAJO: Batalha de sexto tono. d'ARAUXO: Tiento de medio registro de baxon de sexto tono. BERMUDO: Cantus del primero por mi bequadro. Conditor alme siderum. BRUNA: Variations on the Litany of the Virgin. CASANOVES: Paso, no. 7. LOPEZ DE VELASCO: Versos de quatro tono. MUDARRA: Gallarda. PERAZA: Tiento de medio registro alto de primero tono. ANON.: Batalla famosa; Himno sacris solemnis; Je vous; Pour un plaisir; Reveillez-vous; Versos varios. Francis Chapelet, org.
++St 2-77 p130

HMV

MLF 118

3155 BRAHE: Bless this house. BROWN: Love is where you find it. DONIZETTI: La fille du regiment; Chacun le sait, chacun le dit. GERSHWIN: Porgy and Bess: Summertime. HERMAN: Milk and honey, Shalom. LEHAR: The merry widow: Vilia. SONDHEIM: A little night music: Send in the clowns. STRAUSS, J. II: Die Fledermaus, op. 363: Czardas; Mein Herr Marquis. STRAUSS, O.: The chocolate soldier: My hero. STRAUSS, R.: Also sprach Zarathustra, op. 30: Opening. VERDI: Rigoletto: Caro nome. La traviata: Ah fors'e lui...Sempre libera. June Bronhill, s; Orchestra; Tommy Tycho.
+-Gr 10-77 p697

RLS 715 (3)

3156 ARTHUR: Today. BIZET: Les pecheurs de perles: Je crois entendre encore. BORODIN: Prince Igor: Vladimir's cavatina (orch. Rimsky-Korsakov/Glazunov). CAPUA: O sole mio. CARUSO: Dreams of long ago. CILEA: L'Arlesiana: E la solita storia. CURTIS: Torna a Surriento. FOSTER: Jeanie with the light brown hair. FRIML: The vagabond king: Only a rose. GEEHL: For you alone (2).

GIORDANO: Andrea Chenier: Come un bel di di maggio. Fedora: Amor ti vieta. GOUNOD: Faust: Salut, demeure chaste et pure. Romeo et Juliette: Ah leve toi, soleil (2). d'HARDELOT: Because. KALMAN: Die Czardasfurstin: Heut Nacht hab ich getraumt von dir. Das Veilchen vom Montmartre: Du Veilchen vom Montmartre. KORLING: Vita rosor. LAPARRA: L'Illustre Fregona: Melancolique tombe le soir. LEONCAVALLO: I Pagliacci: Recitar ...Vesti la giubba (2). MASCAGNI: Cavalleria rusticana: O Lola, bianca come fior di spino; Mamma que vino e generoso. MILLOCKER: Der Bettelstudent: Ich setz den Fall; Ich hab kein Gold. OFFENBACH: La belle Helene: Au mont Ida. PUCCINI: La boheme: Che gelida manina; O soave fanciulla. La fanciulla del West: Ch'ella mi creda libero. Manon Lescaut: Donna non vidi mai. Tosca: E lucevan le stelle; Recondita armonia (2). Turandot: Nessun dorma. RACHMANINOFF: In the silent night, op. 4, no. 3; Lilacs, op. 21, no. 5. SCHRADER: I de lyse naetter. SJOBERG: I drommen du ar mig nara; Tonerna. STRAUSS, J. II: Der Zigeunerbaron, op. 420: Wer uns getraut. TOSELLI: Serenata, op. 6, no. 1. TOSTI: Mattinata; Ideale. VERDI: Aida: Se quel guerrier io fossi...Celeste Aida. Rigoletto: Questa o quella; La donna e mobile (2). Il trovatore: Ah si ben mio, coll'essere io tuo; Di quella pira. TRAD.: Ack Varmeland, du skona; Allt under himmelens faste. Jussi Bjorling, t; Jens Warny Orchestra, Royal Opera House Orchestra, Stockholm, Concert Association Orchestra, Swedish Radio Orchestra, Orchestra; Nils Grevillius.

++Gr 12-77 p1141 ++RR 12-77 p30
+HFN 12-77 p165

RLS 716 (4) (Reissues)

3157 Opera, operetta and song recital. BERLIOZ: Les nuits d'ete, op. 7: Le spectre de la rose; Absence. CHAUSSON: Le colibri, op. 2, no. 7; Les papillons, op. 2, no. 3; Poeme de l'amour et la mer, op. 19; Les temps des lilas. DEBUSSY: Ballade des femmes de Paris; Chanson de Bilitis (3); Fetes galantes, I and II; Green. Pelleas et Melisande: Voici ce qu'il ecrit; Tu ne sais pas pourquoi. DUPARC: Chanson triste; Extase; L'invitation au voyage; Phidyle. FAURE: L'abasent, op. 5, no. 3; Apres un reve, op. 7, no. 1; Clair de lune, op. 46, no. 2; Dans les ruines d'une abbaye, op. 2, no. 1; Ici-bas, op, 8, no. 3; Nell, op. 18, no. 1; Les roses d'Ispahan, op. 39, no. 4; Le secret, op. 23, no. 3; Soir, op. 83, no. 2. FRANZ: O thank me not, op. 14, no. 1. GODARD: Chanson d'Estelle. GOETZE: Still as the night. HAHN: En sourdine; L'Heure exquise; L'Offrande; Si mes vers avaient des ailes. Mozart: Etre adore; L'adieu. d'HARDELOT: Because. KENNEDY (Frazer): Land of hearts desire. MASSENET: Elegie. MESSAGER: Monsieur Beaucaire: Philomel; I do not know; Lightly, lightly; What are the names. QUILTER: Now sleeps the crimson petal, op. 3, no. 2. PURCELL: King Arthur: Fairest isle. RAVEL: D'Anne jouant delespinette; D' Anne qui me jecta de la neige; Sheherazade. RHENE-BATON: Heures d'ete. RUSSELL: By appointment: White roses. STRAVINSKY: La rosee sainte. SZULC: Clair de lune, op. 81, no. 1. VAUGHAN WILLIAMS: Greensleeves. WARLOCK: The Bayley beareth the bell away; Lullaby. WEBBER: Vieille chanson. TRAD.: Coming thro the rye; Oft the stilly night; Vieille chanson de chasse. Maggie Teyte, s; Gerald Moore, Alfred Cortot, pno; Various orchestras and conductors.

+-ARSC vol VIII, no. 2-3 1976, p99 +HFN 11-76 p169
+RR 10-76 p94
+Gr 10-76 p657 +St 9-77 p54

HMV RLS 717 (3) (Reissues from Columbia LX 532/7, 712/3, 899/903, 918, 909, 861, 877/8, 897/8, 860/1, 868, 866, 898, CLX 2189/90, 2197/8, 2187/9, 2188, 2165/6, 2167/8, CAX 8184/7, 8717/26, 8733/6, 8737/9, 8742, DX 266, WAX 6050/1)

3158 BEETHOVEN: Leonore overture, no. 2, op. 72. Die Ruinen von Athens, op. 113: Overture. Symphony, no. 3, op. 55, E flat major. BERLIOZ: Les Troyens: Trojan march. BRAHMS: Symphony, no. 2, op. 73, D major. HANDEL: Alcina: Dream music (arr. Whittaker). LISZT: Les preludes, G 97. Mephisto waltz. STRAUSS, J. II: Voices of spring, op. 410. Wein, Weib und Gesange, op. 333. WAGNER: Rienzi: Overture. Tannhauser: Prelude, Act 3. Tristan und Isolde: Prelude, Act 3. VPO, LSO, LPO, OSCCP, British Symphony Orchestra; Felix Weingartner.

+-ARG 11-77 p35 +HFN 2-76 p113
+Gr 12-75 p1061 +RR 12-75 p37

RLS 719 (5)

3159 Dame Nelli Melba, The London recordings 1904-1926. ARDITI: Se saran rose. BACH (Gounod): Ave Maria (1904, 1906, 1913). BEMBERG: Les anges pleurent; Un ange est venu; Chant hindou; Chant venetien; Elaine: L'amour est pur; Nymphs et Sylvains; Sur le lac. BISHOP: Bid me discourse; Home, sweet home; Lo, hear the gentle lark. BIZET: Pastorale. CHAUSSON: Le temps des lilas. BARNARD: Come back to Erin. DONIZETTI: Lucia di Lammermoor: Mad scene. DUPARC: Chanson triste. FOSTER: Old folks at home. GOUNOD: Faust: Jewel song; Final trio. Romeo et Juliette: Waltz song. HAHN: Si mes vers avaient des ailes. HANDEL: Il pensieroso: Sweet bird (1904, 1926). d'HARDELOT: Three green bonnets. HENSCHEL: Spring. HUE: Soir paien. LALO: Le Roi d'Ys: Aubade. LIEURANCE: By the waters of Minnetonka. LOTTI: Pur dicesti. MASSENET: Le Cid: Pleurez mes yeux. Don Cesar de Bazan: Sevillana (2). MENDELSSOHN: O for the wings of a dove. MOZART: Le nozze di Figaro, K 492: Porgi amor. Il Re pastore, K 208: L'amero, saro costante. PUCCINI: La boheme: Addio di Mimi (2); On m'appelle Mimi; Si mia chiamano Mimi; Entrate...C'e Rodolfo; Donde lieta usci; Addio dolce svegliare alla mattina; Gavotta...Minuetto; Sono andate; Io Musetta... Oh come e bello e morbido. Tosca: Vissi d'arte. RIMSKY-KORSAKOV: The legend of Sadko, op. 5: Chanson hindoue. RONALD: Away on the hill; Down in the forest; Sounds of earth. SCOTT (Gatty) Goodnight. SCOTT: Annie Laurie. SZULC: Clair de lune, op. 81, no. 1. THOMAS: Hamlet: Mad scene (1904, 1910). TOSTI: Goodbye (two stanzas); Mattinata; La serenata. VERDI: La traviata: Ah fors'e lui; Sempre libera; Dite alla giovine. Otello: Piangea cantando; Ave Maria piena di grazia. Rigoletto: Caro nome; Quartet. WAGNER: Lohengrin: Elsa's dream. TRAD.: Auld lang syne; Coming thro the rye; God save the King; Swing low, sweet chariot (arr. Burleigh). Lord Stanley's address; Nellie Melba's farewell speech. Nelli Melba, s; Various other artists; Landon Ronald, pno; Orchestral accompaniment; Landon Ronald.

+Gr 11-76 p877 +RR 11-76 p23
+HFN 12-76 p133 +St 9-77 p54

RLS 723 (3)

3160 BACH: Sonata, violin, no. 2, S 1003, A minor: Andante. Suite, no. 3: Aria. BEETHOVEN: Variations on Mozart's "Bei Mannern" (7).

BOCCHERINI: Concerto, violoncello, B flat major (ed. Grutzmacher). Sonata, violoncello, no. 6, A major: Adagio and allegro. BRAHMS: Concerto, violin and violoncello, op. 102, A minor. DVORAK: Songs my mother taught me, op. 55. HAYDN: Sonata, no. 1, C major: Tempo di menuetto. LASERNA: Tonadillas. MENDELSSOHN: Songs without words, op. 109, D major. Trio, piano, no. 1, op. 49, D minor. RIMSKY-KORSAKOV: The tale of the Tsar Sultan: Flight of the bumblebee. SCHUMANN: Kinderscenen, op. 15, no. 7: Traumerei. Trio, piano, no. 1, op. 63, D minor. TARTINI: Concerto, violoncello, D major: Grave ed espressivo. VALENTI: Gavotte. VIVALDI: L'Estro armonico, op. 3, no. 11, D minor: Largo. Pablo Casals, vlc; Jacques Thibaud, vln; Alfred Cortot, Blas-Net, Otto Schulhof, pno; LSO, Casals Barcelona Orchestra; Landon Ronald, Alfred Cortot.

+ARSC vol 9, no. 1 1977, p87
+Gr 1-77 p1154
+HFN 1-77 p121
++RR 12-76 p52

SLS 863 (3) (also EMI OC 191-05410/2, also Seraphim SIC 6092)

3161 The art of courtly love. BINCHOIS: Amoreux suy. Bien puist. Files a marier. Je ne fai toujours. Jeloymors. Votre tres doulz regart. BORLET: He tres doulz roussignol. Ma tredol rosignol. CASERTA: Amour m'a le cuer mis. DANDRIEU: Armes, amours: O flour de flours. DUFAY: La belle se siet. Ce moys de may. Donnes l'assault. Helas mon dueil. Lamentatio Sanctae matris ecclesiae. Havre je suis. Par droit je puis. Vergine bella. FRANCISCUS (FRANCISQUE): Phiton, Phiton. GRIMACE: A l'arme, a l'arme. HASPROIS: Ma douce amour. LESCUREL: A vous douce debonaire. MACHAUT: Amours me fait desirer. Dame se vous m'estes. De bon espoir: Puis que la douce. De toutes flours. Douce dame jolie. Hareu, hareu: Helas, ou sera pris confors. Ma fin est mon commencement. Mes esperis se combat. Quant je sui mis. Quant j'ay l'espart. Quant Theseus: Ne quier veoir. Phyton, le merveilleus serpent. Se je souspir. Trop plus est belle: Biaute paree; Je ne sui. MERUCO: De home vray. MOLINS: Amis tous dous. PERUSIO: Andray soulet. Le gregnour bien. PYKINI: Plasanche or tost. SOLAGE: Fumeux, fume. Helas, je voy mon cuer. VAILLANT: Tres doulz amis: Ma dame; Cent mille fois. ANON.: La septime estampie real. Istampitta tre fontaine. Contre le temps. Restoes, restoes. Basse danses, I and II. Early Music Consort; David Munrow.

+-Gr 12-73 p1238
+HF 2-77 p108
+HFN 12-73 p2619
*NR 3-76 p9
+NYT 8-15-76 pD15
+PRO 3/4-76 p17
++RR 1-74 p60
++SFC 8-22-76 p38
+St 5-76 p78

SLS 988 (2) (also Angel SBZ 3810 Tape (c) 4X3S 3810)

3162 Instruments of the Middle Ages and Renaissance. (Includes 96 page booklet) Early Music Consort; David Munrow.

++Gr 6-76 p58
++HF 2-77 p108
++HF 9-77 p119
++HFN 6-76 p89
++NR 9-76 p15
+NYT 8-15-76 pD15
+RR 7-76 p71
++SFC 8-22-76 p38
++St 3-77 p148
+STL 5-9-76 p38

HQS 1376

3163 BLISS: Fanfares (3) (arr. Ramsey). BRITTEN: Prelude and fugue on a theme by Vittoria. ELGAR: Sonata, organ, no. 1, op. 28, G major. VAUGHAN WILLIAMS: Preludes on Welsh hymn tunes (3).

WILLIAMSON: Epitaphs for Edith Sitwell, nos. 1 and 2. Herbert Sumsion, Christopher Dearnley, Robert Joyce, Herrick Bunney, Allan Wicks, org.
+-Gr 8-77 p327 +-RR 7-77 p81
+-HFN 8-77 p93

SQ ASD 3283 Tape (c) TC ASD 3283

3164 BACH Toccata, C major: Adagio (arr. Siloti/Casals). DEBUSSY: Preludes, Bk I, no. 12: Minstrels. FAURE: Sicilienne, op. 78. HEKKING: Villageoise. KARJINSKY: Esquisse. MENDELSSOHN: Chant popularie (arr. de Hartmann). NIN: Granadina (arr. Kochanski). POPPER: Dance of the elves, op. 39 (ed Fournier). RACHMANINOFF: Vocalise, op. 34, no. 14. RIMSKY-KORSAKOV: The tale of the Tsar Sultan: Flight of the bumblebee. SCHUBERT: The bee, op. 13, no. 9 (arr. Casals). TORTELIER: Miniatures, 2 violoncellos (3). Valse, no. 1. WEBER: Adagio and rondo (arr. Piatigorsky). Paul Tortelier, vlc, Maud Tortelier, vlc; Maria de la Pau, pno.
++Gr 10-76 p661 +HFN 3-77 p119 tape
+HFN 10-76 p177 +RR 10-76 p85

HMV ASD 3302 Tape (c) TC ASD 3302 (Reissues from SLS 977, 962, 948, ASD 307/10, SAN 128/30, 159, 160/1, 117/9, 149/50, 252/4, 242/3, SLS 952, Columbia SAX 2442/3)

3165 Great tenors of today. BIZET: Carmen: La fleur que tu m'avais jetee (John Vickers). Les pecheurs de perles: Je crois entendre (Nicolai Gedda). GIORDANO: Andrea Chenier: Come un bel di (Franco Corelli). GOUNOD: Faust: Salut, demeure (Nicolai Gedda). MASCAGNI: L'Amico Fritz: Ed anche...Oh amore (Luciano Pavarotti). PUCCINI: Manon Lescaut: Donna non vidi mai (Placido Domingo). Tosca: E lucevan le stelle (Carlo Bergonzi). Turandot: Nessun dorma (Franco Corelli). SAINT-SAENS: Samson et Dalila: Arretez mes freres. VERDI: Aida: Se quel guerrier ...Celeste Aida (Placido Domingo). La forza del destino: O tu che in seno (Carlo Bergonzi). Otello: Niun mi tema (James McCracken). NPhO, Paris Opera-Comique Orchestra, Rome Opera Orchestra and Chorus, Paris Opera Orchestra, OSCCP, RPO, ROHO; Bruno Bartoletti, Riccardo Muti, Pierre Dervaux, Andre Cluytens, Gabriele Santini, Francesco Molinari-Pradelli, Rafael Fruhberg de Burgos, Georges Pretre, Lamberto Gardelli, John Barbirolli, Gianandrea Gavazzeni.
+-Gr 1-77 p1177 +HFN 5-77 p139 tape
+HFN 3-77 p118 +RR 1-77 p40

Melodiya ASD 3307 (also Angel 40270)

3166 Cossack folk songs: Be merry, Don Cossacks; By the forest; Bylina; A Cossack left for a far land; Dawn; Fellow Cossacks; Golden bee; In the meadow; Melt, you snowdrifts, it's time; My meadow my greensward; My native land; Oak grove; Oh, don't awaken me; Oh, in the garden; Oh, you oak grove; Oh, you frost, you bitter frost; One who truly loves his motherland; Wedding songs (6). Don Cossacks of Rostov; Anatoly Kvasov.
+Gr 2-77 p1330 +RR 1-77 p78
-NR 2-77 p8

ASD 3318

3167 ALBINONI: Concerto, oboe, op. 9, no. 2, D minor. BACH: Cantata, no. 147, Jesu, joy of man's desiring. BIBER: Sonata, trumpet, no. 4, C major. GABRIELI, D.: Sonata, trumpet, D major. GERVAISE: French Renaissance dances (7). LOEILLET: Sonata, trumpet, C major. PURCELL: The Queen's dolour. Maurice Andre, tpt; Alfred Mitterhofer, org.

+-Gr 4-77 p1569 ++RR 4-77 p70
+HFN 4-77 p147

SQ ASD 3338 Tape (c) TC ASD 3338
3168 Andre Previn's music night, vol. 2. BARBER: Adagio, strings. BUTTERWORTH: The banks of green willow. DEBUSSY: Prelude a l'apres-midi d'un faune. FALLA: The three-cornered hat: Suite, no. 2. GLINKA: Russlan and Ludmilla: Overture. STRAUSS, J. II: Emperor waltz, op. 437. LSO; Andre Previn.
+-Gr 5-77 p1704 +HFN 8-77 p99 tape
++HFN 5-77 p133 +-RR 5-77 p53

SQ ASD 3341 Tape (c) TC ASD 3341
3169 ALFORD: On the quarter deck. ARNE: The British Grenadiers (arr. Robinson). Rule Britannia (arr. Sargent). BLISS: Processional. COATES: The dambusters. DAVIES: R.A.F. march. ELGAR: Pomp and circumstance marches, op. 39, nos. 1, 4. HOLST: Songs without words, op. 22b: Marching song. VAUGHAN WILLIAMS: Coastal command: Dawn patrol (arr. Mathieson). WALTON: Henry V suite: Touch her soft lips and part; Agincourt song. TRAD.: Fantasia on British sea song: Hornpipe (arr. H. Wood). RPO, Royal Liverpool Philharmonic Orchestra; Charles Groves.
+Gr 6-77 p106 +HFN 7-77 p119
+Gr 8-77 p346 tape

ASD 3346 (also Angel S 37254)
3170 ALBENIZ: Espana, op. 165, no. 3: Malaguena. BRAHMS: Hungarian dance, F minor. CORELLI: Sarabande and allegretto. HEUBERGER: The opera ball: Midnight bells. MENDELSSOHN: Songs without words, op. 62, no. 1. MOZART: Serenade, no. 7, K 250, D major: Rondo. POLDINI: Marionettes: Dancing doll. TARTINI: Sonata, violin, G minor (The devil's trill). WIENIAWSKI: Caprice, A minor. TRAD.: Londonderry air. (all trans. Kreisler). Itzhak Perlman, vln; Samuel Sanders, pno.
+Gr 11-77 p909 ++RR 12-77 p72
+NR 11-77 p15 +St 12-77 p148

ASD 3357
3171 Improvisations. Yehudi Menuhin, vln; Ravi Shankar, sitar; Jean-Pierre Rampal, flt; Alla Rakha, tabla; Martine Geliot, hp; Nodu Mullick, tanpura.
+Gr 7-77 p197 ++RR 7-77 p67
++HFN 7-77 p115

SQ ASD 3375 Tape (c) TC ASD 3375 (also Angel S 37443)
3172 BACH: Cantata, no. 147, Jesu, joy of man's desiring (arr. Connah). GLUCK: Orfeo ed Euridice: Dance of the blessed spirits. HANDEL: Xerxes: Largo (arr. Connah). MOZART, L.: Cassation, orchestra and toys, G major: Toy symphony (attrib. Haydn). MOZART: Serenade, no. 13, K 525, G major. SCHUBERT: Rosamunde, op. 26, D 797: Entr'acte, B flat major. AMF; Neville Marriner.
+-Gr 10-77 p712 +NR 11-77 p3
+-Gr 12-77 p1145 tape ++RR 10-77 p57
+HFN 10-77 p148 +St 12-77 p157

SQ ASD 3393
3173 BUSATTI: Surrexit pastor bonus. DONATI: In te domine. GRANDI: O vos omnes; O beate benedicte. GUAMI: Canzona a 8. d'INDIA: Isti sunt duae olivae. LAPPI: Le negrona. MAINERIO: Dances (2). PORTA: Corda deo dabimus. PRIOLI: Canzona prima a 12. Early Music Consort; David Munrow.
+Gr 12-77 p1128

CSD 3774 Tape (c) TC CSD 3774
3174 BRITTEN: A hymn to the virgin. HADLEY: I sing of a maiden. HOWELLS: A spotless rose. KIRKPATRICK: Away in a manger. LEIGHTON: Lully, lulla, thou little tiny child. MENDELSSOHN: Hark, the herald angels sing. RAVENSCROFT: Remember, O thou man. WISHART: Alleluya, a new work is come on hand. TRAD.: Up, good Christen folk and listen. I saw three ships; O little town of Bethlehem; The first Nowell; O come all ye faithful (English). In dulci jubilo (German). Sans day carol (Cornish). Quelle est cette odeur agreable; Quittez pasteurs (French. Francis Grier, org; King's College Chapel Choir; Philips Ledger.
+HFN 12-76 p140 +RR 12-76 p96
+HFN 4-77 p155 tape

SQ CSD 3781 Tape (c) TC CSD 3781 (also Angel S 37263)
3175 DOWLAND: Dances (5). PAISIBLE: Sonata, 4 recorders. PURCELL: The history of Dioclesian: Chaconne. RICHARDSON: Beachcomber. RUBBRA: Meditazioni sopra "Coeurs desoles". VAUGHAN WILLIAMS: Greensleeves. Suite for pipes. WARLOCK: Capriol suite: Branles. WILLIAMS: Sonata in imitation of birds. ANON.: Greensleeves to a ground. Early Music Consort; David Munrow; George Malcolm, hpd.
++Gr 7-77 p198 +-RR 7-77 p65
+HFN 8-77 p83 ++RR 11-77 p120 tape
+HFN 10-77 p171 tape ++St 11-77 p161
+NR 10-77 p7

CSD 3782
3176 Russian marches. AGAPKIN: March of the Pechera Regiment. Slav girl's farewell. GLIERE: Red Army march. IPPOLITOV-IVANOV: Jubilee march. KHACHATURIAN: To the heroes. MIASKOVSKY: Army march. PROKOFIEV: March, op. 99, B flat major. SHOTAKOVICH: Festive march. STAROKADOMSKY: Victory march. YEFANOV: Battle of Liaoyang. Victors triumph. TRAD.: March of the Preobrazhensk Regiment. ANON.: March of the Kuban Cossacks Infantry Battalion. Todleben. USSR Defence Ministry Band; Nikolai Nazarov, Nikolai Sergeyev.
+Gr 10-77 p712 +RR 10-77 p46
+HFN 10-77 p153

SQ CSD 3784
3177 The joy of Christmas. GRANTHAM: When the crimson sun has set (arr. Gleathead). MENDELSSOHN: Hark, the herald angels sing (arr. and orch. Willcocks). NEALE: Good King Wenceslas (arr. and orch. Jacques). PLEKHCHEYEV (Tchaikovsky): The crown of roses. WOODWARD: Ding dong merrily on high; Unto us a son is born (arr. Wood, Willcocks). TRAD.: The birds; Christmas bells; Gloucestshire wassail; The holly and the ivy; I saw three ships; In dulci jubilo; Infant holy; The Sussex carol; The twelve days of Christmas; We wish you a merry Christmas. Geoffrey Mitchel Choir; ECO; Edward Heath.
+Gr 12-77 p1156 ++RR 12-77 p78
+HFN 12-77 p172

SLS 5049 (3) (also Seraphim SIC 6104)
3178 The art of the Netherlands. AGRICOLA: Comme femme. BARBIREAU: Songs: Ein frohlich wesen. BRUMEL: Songs: Du tout plongiet; Fors seulement, l'attente. Missa et ecce terrae motus: Gloria. Vray dieu d'amours. BUSNOIS: Songs: Fortuna desperata. COMPERE: O bone Jesu, motet. GHISELIN: Songs: Ghy syt die werste boven al (Verbonnet). HAYNE VON GHIZEGHEM: Songs: De tous

biens plaine; A la audienche. ISAAC, H.: Songs: Donna di dentro di tua casa; Missa la bassadanza: Agnus Dei; A la battaglia. JOSQUIN DES PRES: Songs: Allegez moy, doulce plaisant brunette; Adieu mes amours; El grillo e buon cantore; Guillaume se va; Scaramella va alla guerra. Motets: Benedicta es caelorum Regina; De profundis; Inviolata, integra et casta es, Maria. La Bernadina. La Spagna. Vive le roy. MOUTON: Nesciens Mater virgo virum. OBRECHT: Haec deum caeli laudemus nunc Dominum. OCKEGHEM: Songs: Prenez sur moi; Ma bouche rit. Motets: Intemerata Dei mater. RUE: Missa Ave santissima Maria: Sanctus. TINCTORIS: Missa 3 vocum: Kyrie. VERDELOT: Ave sanctissima Maria. ANON.: La guercia; Est-il conclu par un arret d'amour; Heth sold ein meisken garn om win; Lute dances (2) Mijm morken gaf mij een jonck wifjj; Andernaken. Early Music Consort; David Munrow.

+Gr 11-76 p861 ++RR 1-77 p76
+HF 8-77 p92 ++NR 11-77 p16
++HFN 1-77 p101

SLS 5057 (4) Tape (c) TC SLS 5057

3179 La divina: The art of Maria Callas. BELLINI: Norma: Sediziose voci...Casta diva. Il pirata: Oh, s'io potessi...Col sorriso d'innocenza. I puritani: O rendetemi la speme...Qui la voce. La sonnambula: Care compagne...Come per me sereno. CHERUBINI: Medea: Dei tuoi figli. CILEA: Adriana Lecouvreur: Respiro appena...Io son l'umile ancella. DELIBES: Lakme: Dov'e l'Indiana bruna. DONIZETTI: Lucia di Lammermoor: Oh giusto cielo; Ardon gl'incensi. MASCAGNI: Cavalleria rusticana: Voi lo sapete. MASSENET: Le Cid: Pleurez, mes yeux. Werther: Des cris joyeux. MEYERBEER: Dinorah: Ombra leggiera. PONCHIELLI: La gioconda: Suicidio. PUCCINI: La boheme: Donde lieta usci. Madama Butterfly: Con onor muore...Tu, tu, piccolo iddio. Manon Lescaut: Sola perduta, abbandonata. Turandot: In questa reggia. ROSSINI: Il barbiere di Siviglia: Una voce poca fa. SPONTINI: La vestale: Tu che invoco. THOMAS: Hamlet: A vos jeux...Partagez-vous mes fleurs...Et maintenant ecoutez ma chanson. VERDI: Attila: Oh nel fuggente nuvolo. Un ballo in maschera: Ecco l'orrido campo. Ernani: Surta e la notte. Don Carlo: Tu che le vanita. Macbeth: Nel di della vittoria. Nabucco: Ben io t'invenni. Rigoletto: Gualtier Malde...Caro nome. Il trovatore: Timor di me...D'amor sull ali rosee. I vespri siciliani: Merce, dilette amiche; Arrigo, ah parli a un core. Maria Callas, s; With various artists, orchestras and conductors.

+Gr 11-76 p878 +HFN 3-77 p119 tape
+HFN 1-77 p105 +RR 12-76 p42

SLS 5068 (5) Tape (c) TC SLS 5068 (Reissues from ASD 2938, 2465, 655, 2802, 334, 2363, 2361, 3262, 33CS 1140)

3180 BRUCH: Concerto, violin, no. 1, op. 26, G minor. ELGAR: Concerto, violoncello, op. 85, E minor. GRIEG: Concerto, piano, op. 16, A minor. HAYDN: Concerto, trumpet, E flat major. LISZT: Concerto, piano, no. 1, G 124, E flat major. MENDELSSOHN: Concerto, violin, op. 64, E minor. MOZART: Concerto, horn and strings, no. 4, K 495, E flat major. Concerto, piano, no. 21, K 467, C major. RACHMANINOFF: Concerto, piano, no. 2, op. 18, C minor. RODRIGO: Concierto de Aranjuez, guitar and orchestra. TCHAIKOVSKY: Concerto, piano, no. 1, op. 23, B flat minor. John Wilbraham, tpt; Dennis Brain, hn; Daniel Barenboim, John Ogdon,

Agustin Anievas, Horacio Gutierrez, pno; Jacqueline du Pre, vlc; Yehudi Menuhin, vln; Alirio Diaz, gtr; AMF, PhO, ECO, LSO, NPhO, Spanish National Orchestra; Neville Marriner, Herbert von Karajan, Daniel Barenboim, John Barbirolli, Paavo Berglund, Walter Susskind, Rafael Fruhbeck de Burgos, Moshe Atzmon, Andre Previn.
+-Gr 12-76 p1072 +RR 12-76 p56
+HFN 3-77 p119 tape

SLS 5073 (4)
3181 ELGAR: Enigma variations, op. 36, excerpts. GRIEG: Peer Gynt, op. 46, excerpts. HANDEL: Love in Bath: Suite. HUMPERDINCK: Hansel and Gretel, excerpts. MENDELSSOHN: A midsummer night's dream, op. 21/61: Overture. MOZART: Concerto, horn and strings, no. 4, K 495, E flat major. Serenade, no. 13, K 525, G major. RAVEL: Pavane pour une infante defunte. ROSSINI: William Tell: Overture. SIBELIUS: Karelia suite, op. 11. SMETANA: The bartered bride: Dance of the comedians. Ma Vlast: Vltava. STRAUSS, J. II: Waltzes (5). TCHAIKOVSKY: Romeo and Juliet, excerpts. Swan Lake: Waltz, Act 1. VERDI: Tosca: Vissi d'arte. WAGNER: The flying Dutchman, excerpt. WALTON: Facade. Various performers.
+-HFN 4-77 p153 +-RR 2-77 p44

SLS 5080 (4) (Reissues from CLP 1182, 1118, 1164, ASD 499, 621, EMD 5520)
3182 BLISS: Concerto, piano. BRITTEN: Concerto, piano, op. 13, D major. IRELAND: Concerto, piano, E flat major. RAWSTHORNE: Concerto, piano, no. 1. Concerto, piano, no. 2. RUBBRA: Concerto, piano, no. 2, op. 85, G major. TIPPETT: Concerto, piano. WILLIAMSON: Concerto, piano and strings, no. 2. Colin Horsley, Jacques Abram, Trevor Barnard, Moura Lympany, Denis Matthews, John Ogdon, Gwynneth Pryor, pno; RPO, PhO, BBC Symphony Orchestra, ECO Strings; Basil Cameron, Herbert Menges, Malcolm Sargent, Colin Davis, Yuval Zaliouk.
++Gr 6-77 p67 ++RR 8-77 p64
+-HFN 7-77 p125

SLS 5094 (4) (Reissues from ASD 272, ALP 1172, C 3610/5, C 3996/8, C 3761/2)
3183 BRAHMS: Concerto, piano, no. 1, op. 15, D minor. Concerto, piano, no. 2, op. 83, B flat major. GRIEG: Concerto, piano, op. 16, A minor. LISZT: Hungarian fantasia, G 123. SCHUMANN: Concerto, piano, op. 54, A minor. TCHAIKOVSKY: Concerto, piano, no. 1, op. 23, B flat minor. Solomon, pno; PhO; Herbert Menges, Rafael Kubelik, Issay Dobrowen, Walter Susskind.
++Gr 10-77 p653 +-HFN 11-77 p183

SLS 5104 (2) (Reissues from 33CX 1179/81, 1058/60, 1469/71, SAN 143/5, SAX 2316/7, 2410, 33CSX 12112/2, 1182/3, 1464/5, 1204, 1296/8, 1583/5, 1094/5, 1555/7, 1231, 1289/91, 1318/20, 1472/4, 1258/60, 1324/6, 1483/5, SAX 2503)
3184 BELLINI: Norma: Casta diva. I puritani: Son vergin vezzosa. La sonnambula: Ah non credea mirarti. BIZET: Carmen: L'amour est un oiseau rebelle. DONIZETTI: Lucia di Lammermoor: Regnava nel silenzio. GLUCK: Alceste: Divinites du Styx. LEONCAVALLO: I Pagliacci: Quel fiamma avea nel guardo. MASCAGNI: Cavalleria rusticana: Ineggiamo, il Signor non e morto. MASSENET: Manon: Adieu, notre petite table. PUCCINI: La boheme: Si mi chiamano Mimi. Gianni Schicchi: O mio babbino caro. Madama Butterfly: Un bel di vedremo. Manon Lescaut: Sola, perduta, abbandonata. Tosca: Vissi d'arte. Turandot: In questa reggia. ROSSINI: Il

barbiere di Siviglia: Una voce poco fa. Il Turco in Italia: Non si da follia maggiore. SAINT-SAENS: Samson et Dalila: Printemps que commence. VERDI: Aida: O patria mia. Un ballo in maschera: Morro, ma prima in grazia. La forza del destino: Pace, pace, mio Dio. Macbeth: La luce langue. Rigoletto: Caro nome. Il trovatore: Tacea la notte placida...Di tale amor. Maria Callas, s; La Scala Orchestra and Chorus, OSCCP, PhO, French National Radio Orchestra; Tullio Serafin, Antonino Votto, Georges Pretre, Herbert von Karajan, Victor de Sabata, Gianandrea Gavazzeni, Nicola Rescigno.

+Gr 12-77 p1138 +RR 12-77 p31

SRS 5197

3185 ANCLIFFE (arr. Lotter): Nights of gladness. BINGE: Elizabethan serenade. BYFIELD: A Cornish pastiche. Gabriel John. A pinch of salt. COATES: Springtime suite: Dance in the twilight. FARNON: Portrait of a flirt. HARTLEY: My love she's but a lassie yet (arr. Byfield). Rouge et noir. OSBORNE: Lullaby for Penelope. My northern hills (Younger, David). WOOD: Paris suite: Montmartre. Studio Two Concert Orchestra; Reginald Kilbey.

+RR 2-77 p66

SQ ESD 7001 Tape (c) TC ESD 7001

3186 DVORAK: Nocturne, op. 40, B major. GRIEG: Norwegian melodies, op. 63: In popular folk style; Cowkeeper's tune and country dance. NIELSEN: Little suite, op. 1, A minor. TCHAIKOVSKY: Elegie, G major. WIREN: Serenade, strings, op. 11. Bournemouth Sinfonietta; Kenneth Montgomery.

+Gr 7-76 p187 +RR 7-76 p61
+HFN 7-76 p87 +-RR 1-77 p91 tape
+HFN 2-77 p135 tape

ESD 7002 Tape (c) TC ESD 7002 (Reissue from CSD 1542)

3187 ARNE: Where the bee sucks. BOYCE: Heart of oak. DAVY: The Bay of Biscay. HORN: Cherry ripe. MORLEY: It was a lover and his lass. OXENFORD: The ash grove. STEVENS: Sigh no more, ladies. TRAD.: A hunting we will go. The bailiff's daughter of Islington. Charlie is my darling. Early one morning. John Peel. The keel row. The miller of Dee. Oh, the oak and the ash. Ye banks and braes. The vicar of Bray. Elizabeth Harwood, s; Owen Brannigan, bs; Hendon Grammar School Choir; Pro Arte Orchestra; Charles Mackerras.

++Gr 9-76 p483 +HFN 2-77 p135 tape
+HFN 7-76 p103 +RR 7-76 p74

ESD 7010 Tape (c) TC ESD 7010 (Reissue from Columbia 2 TWO 190)

3188 Overtures. HEROLD: Zampa. MENDELSSOHN: Ruy blas, op. 95. REZNICEK: Donna Diana. SUPPE: Light cavalry. THOMAS: Mignon. VERDI: Nabucco. Royal Liverpool Philharmonic Orchestra; Charles Groves.

++Gr 11-76 p888 +HFN 2-77 p135 tape
+HFN 11-76 p173 ++RR 11-76 p83

ESD 7011 Tape (c) TC ESD 7011 (Reissues from ASD 3131, 2960, 2784, 3002, 2894, 2990, 2784, SLS 864)

3189 ALBINONI: Adagio, G minor. BEETHOVEN: The creatures of Prometheus, op. 43: Overture. ENESCO: Roumanian rhapsody, op. 11, no. 1, A minor. HOLST: The planets, op. 32: Jupiter, the bringer of jollity. PROKOFIEV: Romeo and Juliet, op. 64, excerpts. TCHAIKOVSKY: Marche slav, op. 31. VAUGHAN WILLIAMS: Fantasia on "Greensleeves". WALTON: Portsmouth Point overture. LSO; Andre Previn.

+-Gr 11-76 p888 +RR 11-76 p50
+HFN 2-77 p135 tape

ESD 7024
3190 ANDERSON: Sleigh ride. GRUBER (arr. Walters): Stille Nacht. VAUGHAN WILLIAMS (arr. Walters): Wither's rocking carol. WALTERS: Iona. TRAD.: Babe of Bethlehem; The bells; Born in Bethlehem; The Chester carol; The cuckoo carol; Czech rocking carol; Deck the hall; Ding-dong-doh; Jingle bells; Little camel boy; Over the snow; Tua Bethlem Dref; The virgin Mary had a baby boy (arr. Walters, Wallace, Boughton). Robert Tear, t; Mair Jones, hp; Woodfall Junior School Choir; Royal Liverpool Philharmonic Orchestra and Chorus; Edmund Walters.
+-HFN 4-77 p155 tape +RR 1-77 p83

ESD 7043 Tape (c) TC ESD 7043 (*Reissues from CSD 3695, SLS 964, 1C 193 30 1945)
3191 DOSTAL: Clivia: Ich bin verliebt. Die Ungarische Hochzeit: Spiel mir das Lied von Gluck und Treu. KALMAN: Grafin Maritza: Lustige Zigeunerweisen; Hore ich Zigeunergeigen. KATTNIGG: Balkanliebe: Leise erklingen Glocken. KUNNEKE: Die Lockende Flamme: Lind ist die Nacht. LEHAR: Die lustige Witwe: Es lebt eine Vilja.* Giuditta: Meine Lippen sie kussen so heiss.* LINCKE: Frau Luna: Schlosser die im Monde liegen. BENATZKY: Im Weissen Rossl: Mein Liebeslied muss ein Walzer sein. STRAUSS, J. II: Casanova: Sul mare lucica (arr. Benatzky). Die Fledermaus, op. 363: Klange der Heimat. ZELLER: Der Vogelhandler: Entr'acte, Act 3; Als gebluht der Kirschenbaum.* Anneliese Rothenberger, s; Bavarian Radio Chorus; Graunke Symphony Orchestra, VSO; Robert Stolz, Willi Mattes, Willi Boskovsky.
+-Gr 9-77 p518 +HFN 10-77 p171 tape
+HFN 10-77 p167 ++RR 9-77 p40

ESD 7050
3192 Christmas music from King's. BYRD: Hodie beata Virgo Maria. Senex puerum portabat. GIBBONS: Hosanna to the Son of David. SWEELINCK: Hodie Christus natus est. VICTORIA: O magnum mysterium; Senex puerum portabat. WEELKES: Gloria in excelsis Deo; Hosanna to the Son of David. ANON.: Where riches is everlastingly. TRAD.: Angelus ad Virginem (1st and 2nd versions); The holly and the ivy; I sing of a maiden; My dancing day (arr. Poston); Nowell, Nowell, Nowell; That Lord that lay in Asse stall; Watts cradle song. King's College Chapel Choir; Douglas Whittaker, flt; Christopher van Kampen, vlc; Andrew Davis, org; David Willcocks.
+Gr 12-77 p1156 +RR 12-77 p76

HLM 7066 (Reissues from B 9323, C 3696, 3358, 3369, B 9541, C 3931, 3902, B 9213, 9222, C 3638)
3193 BIZET: The fair maid of Perth: Sweet echo, come tune thy lay. DELIBES: Lakme: Bell song. GOUNOD: Romeo and Juliet: Waltz song. MOZART: Die Entfuhrung aus dem Seraglio, K 384: I was heedless in my rapture. Lullaby. OFFENBACH: The tales of Hoffmann: Doll's song. PONCE: Star of love. ROSSINI: The barber of Seville: There is a voice within my heart. STRAUSS, J. II: Voices of spring, op. 410. THOMAS: Mignon: Behold, Titania. VERDI: Rigoletto: Gualtier Malde...Dearest name. La traviata: Tis wondrous...Ah was it he...What folly. Gwen Catley, s; Halle Orchestra, LSO, Orchestra; Gerald Moore, pno; Warwick Braithwaite, Hugo Rignold, Eric Robinson, Stanford Robinson.
+Gr 2-77 p1318 +-RR 1-77 p39
+HFN 1-77 p103

HLP 7109 (Reissues from C 3130, 3414, 3305, 3372, 3095, 3379, 3309, 3030, 3261, 3407, B 8747)

3194 BIZET: Carmen: Flower song. BOUGHTON: The immortal hour: The faery song. COLERIDGE-TAYLOR: Hiawatha's wedding feast: Onaway, awake beloved. GERMAN: Merrie England: English rose. GOUNOD: Faust: All hail thou dwelling pure and lowly. HANDEL: Xerxes: Grove so beautiful and stately...Shadows so sweet. Jeptha: Deeper and deeper still...Waft her angels. Semele: Where'er you walk. MENDELSSOHN: Elijah: Ye people rend your hearts. MOZART: Don Giovanni, K 527: Mine be her burden; Speak for me to my lady. PUCCINI: La boheme: Your tiny hand is frozen. SULLIVAN: The gondoliers: Take a pair of sparkling eyes. VERDI: Aida: What if tis I am chosen...Heavenly Aida. Webster Booth, t; LPO, Royal Liverpool Philharmonic Orchestra, Halle Orchestra, Orchestra; Wyn Reeves, Malcolm Sargent, Warwick Braithwaite, Leslie Howard, Clifford Greenwood.

+-Gr 5-77 p1737 +-RR 5-77 p34
+HFN 5-77 p133

Melodiya SXLP 30256

3195 GLIERE: Rondo, op. 43, no. 2. GLAZUNOV: Raymonda: Entr'acte. LIADOV: Prelude and pastorale. RACHMANINOFF: Serenade, op. 3, no. 5. SHOSTAKOVICH: Prelude and fugue, A flat major. SVETLANOV: Aria, string orchestra. TCHAIKOVSKY: Humoresque, op. 10, no. 2. Scherzo, op. 42, no. 2. Scherzo humoristique, op. 19, no. 2. The seasons, op. 37: Barcarolle; Autumn song. Valse sentimentale, op. 51, no. 6. Bolshoi Theatre Violinists Ensemble; Yuli Reyentovich.

+Gr 11-77 p909

SXLP 30259

3196 GLAZUNOV: Raymonda op. 57: Entr'acte. GLIERE: Rondo, op. 43, no. 2. LIADOV: Prelude and pastorale. RACHMANINOFF: Serenade, op. 3, no. 5. SHOSTAKOVICH: Prelude and fugue, op. 87, no. 17, A flat major. SVETLANOV: Aria, string orchestra. TCHAIKOVSKY: Humoresque, op. 10, no. 2. Scherzo, op. 42, no. 2. Scherzo humoristique, op. 19, no. 2. The seasons, op. 37: Barcarolle, Autumn song. Valse sentimentale, op. 51, no. 6. Bolshoi Theatre Violinists Ensemble; Yuli Reyentovich.

-RR 12-77 p56

HNH

4008

3197 Arias and songs of the Italian baroque. CALDARA: Vaghe luci. DURANTE: Danza, danza fanciulla. GIORDANI (attrib): Caro mio ben. PERGOLESI (attrib): Tre giorni son che nina. SCARLATTI, A.: Gia il sole dal gange; O cessate di piagarmi; Su venite a consiglio; Toglietemi la vita ancor; Le violette. VIVALDI: O di tua man mi svena; Dille ch'il viver mio; Se cerca se dice. Carlo Bergonzi, t; Felix Lavilla, pno.

+ARG 12-77 p46 ++NR 9-77 p11
+FF 11-77 p64 +-ON 10-77 p70
-HF 9-77 p108 +SFC 1-24-77 p40

HUNGAROTON

SLPX 11686

3198 Cimbalom recital. KURTAG: Duos, violin and cimbalom, op. 4. In memory of a winter twilight, op. 8. Splinters, op. 6c. LANG: Improvisation, cimbalom. PETROVICS: Nocturne. SARY: Sounds for cimbalom, excerpts. STRAVINSKY: Rag-time. SZOKOLAY: Lament and cultic dance. Marta Fabian, cimbalom; Alice Nemeth, s; Judit Hevesi, vln; Lorant Szucs, celesta, pno; Hedy Lubik, hp; Ferenc Petz, Vilmos Juptner, perc; Budapest Chamber Ensemble; Andras Mihaly.

++NR 1-77 p15

SLPX 11721

3199 Central European lute music, 16th and 17th centuries. BAKFARK: Fantasia (after "D'amour me plains" by Roger). CATO: Fantasia. Favorito. Villanella. CRAUS: Tantz, Hupff auff. Chorea, Auff und nider. Die trunken pinter. DLUGORAJ: Chorea polonica. Fantasia. Finale. Villanella polonica. HECKEL: Ein ungarischer Tantz, Proportz auff den ungarischen Tantz. NEUSIDLER: Ein guter Venezianer Tantz. Hie folget ein welscher Tantz Wascha Mesa, Der hupff auf. Der Juden Tantz, Der hupff auf zur Juden Tantz. Der polnisch Tantz, Der hupff auf. POLLONOIS: Courante. STOBAEUS: Alia Chorea polonica. WAISSELIUS: Polonischer Tantz. ANON.: Danza (ed. Chilesotti). Almande de Ungrie (ed. Phalese). Batori Tantz, Proportio. Tantz, Proportio (2). Paduana Hispanica. Lengyel tanc. Psalmus CXXX. Andras Kecskes, lt; Anges Meth, tabor, timbrel.

++NR 4-76 p14 ++SFC 8-22-76 p38
+RR 3-76 p56 +St 8-77 p59

SXLP 11808

3200 FARKAS: Passacaglia and postludium. KODALY: Praeludium. MAROS: Bagatelles. SULYOK: Te deum. SZONYI: Concerto, organ. Gabor Lehotka, org; HSO; Gyula Nemeth.

+HFN 2-77 p115 +RR 2-77 p80
+-NR 3-77 p14 +SFC 12-11-77 p61

SLPX 11811

3201 BOGAR: Three movements, brass quartet. HUZELLA: Miser catulle. KOCSAR: Sextet, brass. LANG: Cassazione. MIHALY: Little tower music. PATACHICH: Ritmi dispari. PETROVICS: Cassazione. Hungarian Brass Ensemble.

+-ARG 8-77 p39 +NR 9-77 p6
+FF 9-77 p63 +RR 5-77 p62

SLPX 11823

3202 BALAZS/Rossa: Salute to tomorrow. BALASZ/Pushkin: Toast. BARDOS/Lukin: Shepherd's pipe tune; Hymn to the sun. BARTOK: Letter to the home folks. FARKAS: Walking in woods and meadows. KODALY: St. Gregory procession. MORLEY: A deception. Choral Winners of the 3rd Competition Cycle, 1972/5.

+NR 10-77 p9

SLPX 11825

3203 CHAUSSON: Poeme, op. 25. FRANCK: Sonata, violin and piano, A major. KHACHATURIAN: Nocturne. SHCHEDRIN (Tzyganov): In the style of Albeniz. SHOSTAKOVITCH: Dances, op. 5 (3). STRAVINSKY: Petrouchka: Russian dance. Leila Rasonyi, vln; Gyorgy Miklos, pno.

+HFN 5-77 p123 +-RR 5-77 p70
+-NR 10-77 p7

IKON

IKO 4
3204 Orthodox Church music from Finland: Excerpts from the Moleben to the most holy mother of God and SS Sergius and Herman; Authentic bell ringing from Valamo Island; Bless the Lord o my soul; It is truly meet; Let my prayer arise; Before thy cross; Znamenny; When the tomb was opened; Thy resurrection, O Christ; The angel cried; O gladsome light; Praise ye the name of the Lord; Cherubic hymn; Beatitudes; The Lord's prayer; It is truly meet; Many years. Hymnodia Choir; Archbishop Paul of Karelia and All Finland.
+-Gr 4-77 p1588 +RR 11-76 p101

IKO 5
3205 Russian orthodox church music: The burial service of our Lord on Good Friday. Russian Orthodox Cathedral Choir; Michael Fortunatto.
+Gr 5-77 p1724 +RR 5-77 p79

INNER CITY

IC 1006
3206 BACH: Cantata, no. 147, Jesu, joy of man's desiring. BEETHOVEN: Bagatelle, A minor. Minuet, G major. Ode to joy. CHOPIN: Etude, op. 10, no. 3, E major. BRADY: Easy and hold on. Rosebud. RUBINSTEIN: Melody, F major. SCHUBERT: Serenade. TCHAIKOVSKY: Concerto, piano, no. 1, op. 23, B flat minor: Theme. Victor Brady, steel piano.
+St 6-77 p127

INTERNATIONAL PIANO LIBRARY

IPL 101
3207 BOCCHERINI (arr. Plante): Minuet. GLUCK (arr. Plante): Gavotte. BERLIOZ (arr. Redon) La damnation de Faust, op. 24: Mephisto's serenade. CHOPIN: Etudes, op. 10, nos. 4-5, 7; op. 25, nos. 1-2, 9, 11. MENDELSSOHN: Scherzo, op. 16, no. 2, E minor. Songs without words, op. 19, no. 3; op. 16, nos. 2, 6; op. 67, nos. 4, 6. SCHUMANN: Am springbrunne (arr. Debussy). Clavierstucke, op. 32, no. 2: Romanze. Romance, op. 28, no. 2, F sharp minor. Francis Plante, pno.
+-Gr 10-72 p764 +-RR 3-77 p26
+NR 6-73 p13

IPL 102
3208 BUSONI: Sonatina, no. 2. CASELLA: Contrasts, op. 31 (2). CHABRIER: Bourree fantasque. DUSSEK: La chasse. FIELD: Nocturne, no. 9, E minor. GODOWSKY: The gardens of Buitenzorg. JENSEN: Erotikon, op. 44: Eros. MacDOWELL: Woodland sketches, op. 51, no. 1: To a wild rose. MOSZKOWSKI: Valse, op. 34. PADEREWSKI: Legende, op. 16, no. 1, A flat major. RAFF: Rigaudon, op. 204, no. 3. RAVINA: Etude de style, op. 14, no. 1. REGER: Aus meinem Tagebuche, op. 82: Adagio and vivace. RUBINSTEIN: Prelude and fugue, op. 52, no. 2. Arthur Loesser, pno.
+ARG 2-77 p20 *MJ 2-77 p30
+ARSC vol 9, no. 1 1977 p78 +NR 6-73 p12
+NYT 2-20-77 pD19
+-Gr 10-72 p764 ++RR 3-77 p31
+Gr 3-77 p1453 ++St 12-76 p62

IPL 103
3209 BEETHOVEN (Rubinstein): The ruins of Athens, op. 113: Turkish march. CHOPIN: Nocturne, op. 15, no. 2, F sharp major. Polonaise, op. 40, no. 1, C minor. Scherzo, no. 1, op. 20, B minor, abbreviated. Waltz, op. 64, no. 2, C sharp minor. CHOPIN (Liszt): My joys. GLUCK (Brahms): Gavotte. HOFMANN: Mignonettes: Nocturne. LISZT: Hungarian rhapsody, no. 2, G 244, C sharp minor. Waldesrauschen, G 145. RACHMANINOFF: Preludes, op. 3, no. 2, C sharp minor; op. 23, no. 5, G minor. RUBINSTEIN: Melody, op. 3, no. 1, F major. SCARLATTI (Tausig): Pastorale and capriccio. WAGNER (Brassin): Die Walkure: Magic fire music. Josef Hofmann, pno.
+Gr 10-72 p764
++NR 8-73 p1e
++RR 3-77 p26

IPL 104 (also Desmar IPA 104)
3210 BACH: Prelude and fugue, S 545, C major. Partita, violin, no. 2, S 1004, D minor: Chaconne. Chorale prelude: Rejoice, beloved Christians, S 734 (arr. Busoni). BEETHOVEN: Eccossaises (arr. Busoni). BIZET: Carmen fantasy, op. 25 (arr. Busoni). BUSONI: Indian diary, Bk 1. CHOPIN: Etude, op. 10, no. 5, G flat major (2 versions). Etude, op. 25, no. 5, E minor. Nocturne, op. 15, no. 2, F sharp major. Prelude, op. 28, no. 7, A major. LISZT: Hungarian rhapsody, no. 13, G 244, A minor, abbreviated. Ferruccio Busoni, Michael von Zadora, Egon Petri, Edward Weiss, pno.
+ARG 2-77 p17
+ARSC vol 9, no. 1 1977, p78
+-Gr 10-72 p764
+-Gr 3-77 p1453
+HFN 1-77 p103
+MJ 2-77 p30
+NR 5-73 p12
+NYT 2-20-77 pD19
+RR 3-77 p26
+St 12-76 p62

IPL 108 (Reissues)
3211 BUSONI: Elegie, no. 4: Turandots Frauengemach. CHOPIN: Polonaise, op. 53, A flat major. DA MOTTA: Cantiga de amor, op. 9, no. 1. Chula (Danse Portugaise). Valse caprichosa. LISZT: Annees de pelerinage, 1st year, G 160: No. 7, Eglogue. Totentanz, G 126. MOZART (Busoni): Duettino concertante. SCHUBERT: Sonata, piano, no. 18, op. 78, D 894, G major: Minuetto. Wohin (arr. Liszt). Jose Vianna da Motta, Mlle de Castello Lopez, pno; Portuguese National Symphony Orchestra; Pedro de Freitas Branco.
+-Gr 9-74 p558
-HF 6-74 p108
+NR 5-73 p13
+RR 3-77 p26
+St 3-74 p121

IPL 109
3212 ALBENIZ: Improvisations (Albeniz). CHOPIN: Nocturne, op. 32, no. 1, B major (Larrocha). Waltz, op. 34, no. 2, A minor (Larrocha). Waltz, op. 64, no. 2, C sharp minor (Malats). GRANADOS: Spanish dances, nos. 7 and 10 (Granados). Goyescas: El Pelele (Granados). GRIEG: Norwegian dances, op. 35, no. 2 (Marshall). LISZT: Hungarian rhapsody, no. 13, G 244, A minor (Malats). MALATS: Serenata (Malats). SCARLATTI (Granados): Sonata, L 250 (K 190), B flat major (Granados). WAGNER (Liszt): Tristan und Isolde: Liebestod (Malats). Alicia de Larrocha, Isaac Albeniz, Joaquin Malats, Enrique Granados, Frank Marshall, pno.
+ARSC vol 9, no. 1 1977 p81
+-Gr 3-77 p1453
+-MJ 2-77 p30
++NR 3-77 p12
+RR 3-77 p26
+SR 5-28-77 p42
+St 12-76 p62

IPL 112 (1939 Schirmer recordings)

3213 BACH: The well-tempered clavier, Bk I, no. 3: Prelude and fugue, C sharp major. BRAHMS: Sonata, piano, no. 3, op. 5, F minor. Waltzes, op. 39, nos. 15 and 16, A flat major, C sharp minor. CHOPIN: Berceuse, op. 57, D flat major. COUPERIN: Livres de clavecin, Bk III, Ordre, no. 14: Le carillon de Cythere. DEBUSSY: Reverie. HANDEL: Suite, harpsichord, no. 5, E major (The harmonious blacksmith): Air and variations. MENDELSSOHN: Characeteristic pieces, op. 7, no. 4, A major. SCARLATTI: Sonata, keyboard, L 345, A major. SCHUBERT: Moment musical, op. 94, no. 3, D 780, F minor. SCHUMANN: Romance, op. 28, no. 1, B flat minor. Harold Bauer, pno.

++ARG 2-77 p15
+ARSC vol 9, no. 1 1977, p82
+Gr 3-77 p1453
+-MJ 2-77 p30
+NR 3-77 p12
+NYT 2-20-77 pD19
+RR 3-77 p26
+St 12-76 p62

IPL 113 (Reissues from 1930 British Columbia & 1925/26 Brunswick Recordings)

3214 CHAMINADE: The flatterer. Scarf dance. CHOPIN: Sonata, piano, no. 2, op. 35, B flat minor. DEBUSSY: Suite bergamasque: Clair de lune. RUBINSTEIN: Melody, op. 3, no. 1, F major. SCHUBERT (Godowsky): Die schone Mullerin, op. 25, D 795: Morgengruss. Winterreise, op. 89, D 911: Gute Nacht. SINDING: Rustle of spring, op. 32, no. 3. TCHAIKOVSKY: The months, op. 37: Barcarolle. Leopold Godowsky, pno.

++ARG 2-77 p15
+-ARSC vol 9, no. 1 1977 p83
+Gr 3-77 p1453
/MJ 2-77 p30
+-NR 4-77 p14
+NYT 2-20-77 pD19
+RR 3-77 p26
+St 12-76 p62

IPL 114 (Reissues from Columbia recordings, 1923-25, HMV 1927-29)

3215 CHOPIN: Etude, op. 10, no. 5, G flat major. Etude, op. 25, no. 1, A flat major. Nocturne, op. 48, no. 1, C minor. Ballade, no. 3, op. 47, A flat major. Preludes, op. 28, nos. 1, 7, 23. Waltz, op. 64, no. 3, A flat major. Waltz, op. 70, no. 1, G flat major. Waltz, op. posth., E minor. GLUCK (Brahms): Iphigenia: Gavotte. GLUCK (Sgambati): Orfeo ed Euridice: Melodie. LEVITZKI: Valse, A major. Valse de concert. MOSZKOWSKI: La jongleuse. LISZT: Etudes d'execution transcendente d'apres Paganini, no. 3, G 140, A flat major: La campanella. TCHAIKOVSKY: The months, op. 37: November. Mischa Levitzki, pno.

++ARG 2-77 p14
+Gr 3-77 p1453
+-MJ 2-77 p30
++NR 3-77 p12
+NYT 2-20-77 pD19
+RR 3-77 p26
+St 12-76 p62

IPL 117

3216 BEETHOVEN (Busoni): Ecossaises. BRAHMS: Hungarian dance, no. 1, G minor. Waltz, op. 39, no. 15, A flat major. Waltz, op. 39, no. 2, E major. CHOPIN: Nocturne, op. 15, no. 1, F major. Nocturne, op. 55, no. 1, F minor. Scherzo, no. 2, op. 31, B flat minor. Waltz, op. 64, no. 1, D flat major. DEBUSSY: Pelleas et Melisande: Mes longs cheveux. GRIEG: Sonata, piano, no. 7, E minor: Finale. LISZT: Hungarian rhapsody, no. 2, G 244, C sharp minor (abbreviated). Hungarian rhapsody, no. 15 G 244 (abbreviated). PAGANINI (Liszt): Concerto, violin, no. 2, op. 7, B minor (La campanella). RACHMANINOFF: Prelude, op.

23, no. 5, G minor. SCHUMANN (Tausig): Der contrabandiste, op. 74. SAINT-SAENS: Africa, op. 89: Improvised cadenca. TCHAIKOVSKY: Trepak, op. 72, no. 18. Mary Garden, s; Johannes Brahms, Ilona Eibenschutz, Edvard Grieg, Camille Saint-Saens, Claude Debussy, Arthur Friedheim, Percy Grainger, Josef Hofmann, Aleksander Michalowski, Vladimir de Pachmann, Ignace Jan Paderewski, Leopold Godowsky, Josef Lhevinne, pno.

+NR 12-77 p11

IPL 5001/2

3217 BEETHOVEN (Rubinstein): The ruins of Athens, op. 113: Turkish march. BRAHMS: Academic festival overture, op. 80. CHOPIN: Andante spianato and grande polonaise, op. 22, E flat major. Ballade, no. 1, op. 23, G minor. Berceuse, op. 57, D flat major. Etude, op. 25, no. 9, G flat major. Nocturne, op. 9, no. 2, E flat major. Nocturne, op. 15, no. 2, F sharp major. Waltz, op. 42, A flat major. Waltz, op. 64, no. 2, C sharp minor. HOFMANN: Chromaticon, piano and orchestra. MENDELSSOHN: Songs without words, op. 67, no. 4: Spinning song. MOSZKOWSKI: Caprice espagnole, op. 37. RACHMANINOFF: Prelude, op. 23, no. 5, G minor. RUBINSTEIN: Concerto, piano, no. 4, op. 70, D minor. Josef Hofmann, pno; Curtis Institute Symphony Orchestra; Fritz Reiner.

++Gr 9-76 p453
++NR 8-73 p14
++RR 3-77 p26
++St 12-71 p91

IPL 5007/8

3218 BEETHOVEN: Sonata, piano, no. 21, op. 53, C major. CHOPIN: Ballade no. 4, op. 52, F minor. Nocturne, op. 9, no. 3, B major. Polonaise, op. 26, no. 2, E flat minor. Waltz, op. 18, E flat major. Waltz, op. 64, no. 1, D flat major. SCHUBERT (Godowsky) Moment musical, op. 94, no. 3, D 780, F minor. SCHUMANN: Kreisleriana, op. 16. HOFMANN: Kaleideskop. Penguine. STOKOWSKI: Caprice oriental. Josef Hofmann, pno.

++ARG 2-77 p14
+-Gr 3-77 p1453
+-HF 3-76 p99
+MJ 2-77 p30
+NR 7-77 p12
+NYT 2-20-77 pD19
+RR 3-77 p26
+St 12-75 p130
+St 12-76 p62

KAIBALA

20B01

3219 BACH, J.C.: Rondo, B flat major (trans. Maros). BACH, J.S.: Fugue, A minor (trans. Weait). DANDRIEU: The fifers (trans. Weait). HANDEL: Rigaudon, bourree and march. LASSUS: Fantasias (trans Weiat). RAMEAU: Gavotte wiht 6 doubles (trans. Nakagawa). SENAILLE: Allegro spiritoso (trans. Rechtman). SCARLATTI: Sonata, L 465 (trans. Herder). Sonatas, L 104, C major; L 352, C minor; L 325, E minor; L 413, D minor; L 366, D minor; L 93, A minor; L 424, D major (trans. Noth). Hugo Noth, accord; Phoenix Woodwind Quintet, New York.

+NR 2-77 p8

40D03

3220 BRAHMS: Lullaby. GRIEG: Songs: From the fatherland; I love you; The first primrose; The dairy maid; The poet's heart; The swan. MEDINS: Caress. MOZART: Lullaby. SCHUBERT: Die schone Mullerin, op. 25, D 795: Impatience. Die Winterreise, op. 89, D 911: The post; The organ grinder. Serenade. Janis Zabers, t; Instrumental accompaniments.

+-NR 1-77 p11

KELSEY RECORDS

KEL 7601
3221 BARTOK: Rumanian folk dances (6) (arr. Szekeley). DINICU: Hora staccato (arr. Heifetz). HUBAY: Hejre Kati, op. 32. Hungarian poems (6). KREIN: Gipsy carnival (arr. Georgiadis). MONTI: Czardas. SARASATE: Zigeunerweisen, op. 20, no. 1. John Georgiadis, vln; Susan Georgiadis, pno.
+Gr 2-77 p1330 +RR 5-77 p75

LAUREL PROTONE

LP 13
3222 BOND: Sonata, violoncello. CASSADO: Requiebros. MANZIARLY: Dialogue. NIN: Spanish suite. PONCE: Preludes (3). VILLA-LOBOS: Song of the black swan. WEBERN: Sonata, violoncello. Gilberto Munguia, vlc; Thomas Hrynkiv, pno.
+MJ 3-77 p74 ++SFC 12-5-76 p58
+NR 11-76 p14

LISMOR

LILP 5078
3223 BAYCO: Royal Windsor march (arr. Richardson). BLISS: Gala fanfare. Things to come: Epilogue. Welcome the Queen (arr. Duthoit, Godfrey). COATES: The three Elizabeths: Youth of Britain. JACOB: Music for a festival: Overture and march. SOUSA: Fairest of the fair. TRAD.: God save the Queen (arr. Jacob). Music of the four countries (arr. Pinkney). Royal Artillery Alanbrooke Band; A. R. Pinkney.
+RR 8-77 p49

LONDON

CSA 2246 (2)
3224 BERNSTEIN: Candide: Overture. COPLAND: Appalachian spring. GERSHWIN: An American in Paris. IVES: Symphony, no. 2. Holidays: Decoration day. Variations on "America". SCHUMAN: Variations on "America" (after Charles Ives). LAPO; Zubin Mehta.
+-MJ 1-77 p26 ++SFC 8-8-76 p38
+NR 9-76 p2

CS 7015
3225 ALBENIZ; Recuerdos de viaje, op. 71: Rumores de la caleta (arr. Azpiazu). Zambra granadina (trans. Segovia). BACH: Prelude, A minor. Gavottes, nos. 1 and 2. Presto, A minor (arr. Krytiuk). CALLEJA: Cancion triste. DEBUSSY: Preludes, Bk 1, no. 8: La fille aux cheveux de lin (trans. Bream). GUIMARAES: Sounds of bells. SAGRERAS: El colibri. SAINZ DE LA MAZA: Campanas del Alba. SCARLATTI: Sonata, guitar, A minor (trans. Krytiuk). Sonata, A major (trans. Kima). TOMASI: Le muletier des Andes. Liona Boyd, gtr.
++NR 12-76 p14 ++St 1-77 p71
+SFC 9-19-76 p33

CS 7046 Tape (c) 5-7046
3226 BASSA: Minuet. CABANILLES: Batalla imperial. CABEZON: Pavana con su glosa. Diferencias sobre la Gallard milanesa. Diferencias sobre el Canto llano del caballero. DE LA TORRE: Alta. OLAGUE: Xacara. SOLER: Sonatas, B minor, D minor. Fandango. ANON.:

Minuet; Xacara. Jonathan Wood, hpd.
-HF 11-77 p132 ++SFC 6-26-77 p46
+NR 10-77 p14 +St 10-77 p150

OS 26435 Tape (c) OS 5-26435 (also Decca SXLR 6792)
3227 Music of Spain, Zarzuela arias. BARBIERI: El barberillo de lavapies; Cancion de paloma; Jugar con fuego; Romanza de la Duquesa. CABALLERO: Chateau Margaux: Romanza de Angelita. Gigantes y cabezudas: Romanza de Pilar. El senor Joaquin: Balada y alborada. CHAPI Y LORENTE: Las hijas del Zebedeo: Carceleras. La patria chica: Cancion de pastora. GIMENEZ: El barbero de Sevilla: Me llaman la primorosa. LUNA: El nino Judio: De Espana vengo. SERRANO Y RUIZ: El carro del sol: Cancion veneciana. Montserrat Caballe, s; Barcelona Orquesta Sinfonica; Eugenio Marco.
+ARG 11-76 p49 +ON 1-22-77 p33
+-Gr 11-76 p877 +RR 11-76 p40
+HFN 11-76 p165 +SR 11-13-76 p52
+NR 12-76 p11 ++St 11-76 p88

OS 26437
3228 ADAM: O holy night. BACH (Gounod): Ave Maria. BERLIOZ: Requiem, op. 5: Sanctus. BIZET: Agnus Dei. FRANCK: Panis Angelicus. MERCANDANTE: Le sette ultime parole di Nostro Signore sulla croce: Qual'ciglia candido (Parola quinta). SCHUBERT: Ave Maria, D 839; Mille cherubini in coro (arr. Melichar). STRADELLA: Pieta Signore. WADE: Adeste fidelis. YON: Gesu bambino. Luciano Pavarotti, t; Wandsworth Boys Choir, London Voices; National Philharmonic Orchestra; Kurt Herbert Adler.
-HF 8-77 p94 ++St 1-77 p131

OS 26493
3229 Gregorian anthology. Abbaye Saint-Pierre de Solesmes, Monks Choir; Joseph Gajard.
+-NR 9-77 p10

OS 26497 Tape (c) OS 5-26497 (also Decca SXLR 6825 Tape (c) KSXC 6825)
3230 CATALANI: La Wally: Ebben, ne andro lontana. GIORDANO: Andrea Chenier: La mamma morta. MASCAGNI: Cavalleria rusticana: Voi lo sapete. PONCHIELLI: La gioconda: Suicidio. PUCCINI: Turandot: In questa reggia. VERDI: Macbeth: Viente t'affretta...Or tutti sorgete. Il trovatore: Tacea la notte...Di tale amor. Montserrat Caballe, Cecilia Fondevila, s; Juan Pons, bs; Barcelona Orquesta Sinfonica; Armando Gatto, Anton Guadagno.
+-Gr 4-77 p1600 +-ON 6-77 p44
-HF 10-77 p126 +-OR 9/10-77 p30
+HFN 4-77 p135 +SFC 3-20-77 p34
+NR 7-77 p11 +-St 6-77 p145

OS 26499
3231 CILEA: Adriana Lecouvreur: Arias. DONIZETTI: L'Elisir d'amore: Arias. Don Pasquale: Arias. Lucia di Lammermoor: Arias. GIORDANO: Fedora: Arias. PUCCINI: La boheme: Arias. Manon Lescaut: Arias. Tosca: Arias. VERDI: Don Carlo: Arias. Luisa Miller: Arias. Rigoletto: Arias. Il trovatore: Arias. Giocomo Aragall, t; Barcelona Symphony Orchestra; Gianfranco Rivoli.
+-SFC 9-18-77 p42

SR 33221 (Some reissues from London 5482)
3232 BARBERIS (Galdieri): Munasterio e Santa Chiara. BIXIO (Neri, Ennio): Parlami d'amore Mariu. CESARINI: A la Barcillunisa; A la Vallelunghisa; Chiovu'Abballati; Cantu a Timuni; Firenze

sogno; Muttetti di la palieu; Nota de li Lavannari. CURTIS: A conzone e Napule; Ti voglio tanto bene. LAZZARO (Bruno): Chitarra romana. Giuseppe di Stefano, t; Instrumental accompaniments.
+-NR 5-77 p12 +-ON 2-26-77 p33

LOUISVILLE

LS 753/4
3233 BIRD: Carnival scene. CHADWICK: Euterpe. CONVERSE: Endymion's narrative, op. 10. Flivver ten million. FOOTE: Francesca da Rimini. ORNSTEIN: Nocturne and dance of the fates. Louisville Orchestra; Jorge Mester.
+HF 12-76 p120 +-NR 1-77 p2
+MJ 11-76 p44 +-NYT 7-3-77 pD11

LYRICHORD

LLST 7299
3234 BRAYSSING: Fantasia. CALVI: Suo corrent. COSENTINO: Misterios. FERANDIERE: Rondo. MADRIGUERA: Humorada. SEGOVIA: Dos anecdotas. SOR: Los adioses. TARREGA: Lagrima. VILLA-LOBOS: Preludio, no. 3. VISEE: Pascalle. Roberto Lara, gtr.
+-NR 8-77 p16

MARC

MC 5355
3235 ALBRIGHT: Pneuma. GUILLOU: Toccata. NOBLE: Scherzino. SALLINEN: Chaconne. SLAVICKY: Fresco. VILLASENOR: Paisaje. WILLIAMSON: Elegy, J. F. K. William Haller, org.
+MU 9-77 p12

MENC

76-11
3236 BERNSTEIN: West side story, excerpts. CRESTON: Celebration. FILLMORE: Americans we. GERSHWIN: Medley (arr. Bennett). GIANNINI: Preludium and allegro. GOULD: Fourth of July. KING: Garland entree. MIDDENDORF: Stand up for America. NELHYBEL: Trittico. PERSICHETTI: O cool is the valley. REED: Testament of an American. SCHUMAN: Chester overture. SOUSA: Stars and stripes forever. NBA High School Honors Band; William Revelli, Al Wright.
+IN 1-77 p20

MERCURY

SRI 75098
3237 DOHNANYI: Pierrette's veil: Waltz. LANNER: Die Schonbrunner waltz, op. 200. LEHAR: The merry widow: Waltz. STRAUSS, J. II: Artists life, op. 316. Roses from the south, op. 388. Voices of spring, op. 410. STRAUSS, Josef: Village swallows from Austria, op. 164. WALDTEUFEL: Les patineurs, op. 183. Minneapolis Symphony Orchestra, PH; Antal Dorati.
+NR 8-77 p6

MUSICAL HERITAGE SOCIETY

MHS 3340

3238 Twentieth century music for trumpet and organ. GENZMER: Sonata, trumpet and organ. JOLIVET: Arioso barocco. HOLLER: Choral variations on "Jesu meine Freude", op. 22, no. 2. TOMASI: Holy week at Cuzco. Maurice Andre, tpt; Hedwig Bilgram, org.
++MU 11-77 p17

MHS 3530

3239 Gregorian chant Easter processions. Deller Consort; Alfred Deller.
+MU 10-77 p8

MHS 3547

3240 ABBOTT: Alla caccia. DUKAS: Villanelle. HEIDEN: Sonata, horn and piano. KOHLERS: Sonata, horn and piano. MUSGRAVE: Music, horn and piano. NELHYBEL: Scherzo concertante. Charles Kavaloski, hn; Andrew Wolff, Evelyn Zuckerman, pno.
+FF 9-77 p69

MHS 3611

3241 Romantic music for harp ensemble and solo harp. DONIZETTI: Lucia paraphrase (Von Wurtzler). GLIERE: Concerto, harp, op. 74, E flat major: 3rd movement (Von Wurtzler). PAGANINI: Variations on etude, no. 6 (Von Wurtzler). RENIE: Concerto, harp in one movement (Von Wurtzler). SMETANA: Ma Vlast: The Moldau (Von Wurtzler). ZABEL: La source. New York Harp Ensemble; Aristid Von Wurtzler, solo harp and cond.
+ARG 11-77 p42

NATIONAL TRUST

NT 002

3242 Music for the Vyne. ARNE: Blow, blow thou winter wind; Come away death. BOYCE: Trio sonata, D major. FARNABY: The old spagnoletta. FORD: The pill to purge melancholy. HENRY VIII, King: Taunder naken. HOLBORNE: The image of melancholly. JENKINS: Newark siege. LAWES, H.: Sweet stay awhile. LAWES, W.: Gather ye rosebuds. MILAN: Toda mi vida os ame. Pavana. MUNDY: Robin. NICHOLSON: No more, good herdsman, of thy song. PURCELL, H.: Distressed innocence: Rondeau; Air; Minuet. ANON.: The Dalling alman; Come holy ghost; Dances: My robbin; Tickle my toe; Hollis berrie. Paul Elliott, t; King's Music.
++Gr 9-77 p504 +RR 8-76 p74
+-MT 10-76 p832

NEW ENGLAND CONSERVATORY

NEC 115

3243 CARTER: Elegy, string quartet. COWELL: Quartet euphometric. SCHULLER: Quartet, strings, no. 1. STRAVINSKY: Double canon. SWIFT: Quartet, strings, no. 4. Composers Quartet.
+-ARG 11-77 p40 +-St 7-77 p126

NEW WORLD RECORDS

NW 220

3244 BUCK: Rock of ages. DADMUN: The babe of Bethlehem. FISCHER: I love to tell the story. HICKS: The last hymn. LOWRY: Shall we know each other there. MELNOTTE: Angels visits. PRATT: Put my little shoes away. WEBSTER: Sweet by and by; Willie's grave. WHITE: Trusting. WOODBURY: We are happy now, dear mother. ANON.: Flee as a bird; Oh, you must be a lover of the Lord. Kathleen Battle, s; Rose Taylor, ms; Raymond Murcell, bar; Lawrence Skrobacs, pno and harmonium; Harmoneion Singers; Neely Bruce.

++St 7-77 p124

NW 243

3245 BARBER: Sure on this shining night. BOWLES: Once a lady was here; Song of an old woman. CHANLER: The children; Once upon a time; The rose; Moo is a cow; These, my Ophelia; Thomas Logge. CITKOWITZ: Chamber music: Songs (5). COPLAND: Song. DUKE: Luke Havergal; Miniver Cheevy; Richard Cory. HELPS: The running sun. SESSIONS: On the beach at Fontana. Bethany Beardslee, s; Donald Gramm, bar; Donald Hassard, Robert Helps, pno.

++St 8-77 p111

NW 257

3246 The wind demon and other mid-nineteenth-century piano music. BARTLETT: Grande polka de concert. BRISTOW: Dream land, op. 59. FRY: Adieu. GOTTSCHALK: Romance. GROBE: United States grand waltz, op. 43. HEINRICH: The Elssler dances: The laurel waltz. HOFFMAN: In memoriam L.M.G. Dixiana. HOPKINS: The wind demon, op. 11. MASON: A pastoral novellette. Silver spring, no. 6. WARREN: The Andes, march de bravoura. Ivan Davis, pno.

+ARG 11-77 p38

NONESUCH

H 71279

3247 Baroque masterpieces, trumpet and organ. BOYCE: Voluntary, no. 1, D major. GREENE: Voluntary, D major (Largo andante). KREBS: Wachet auf, ruft uns die Stimme (2). PEZEL: Sonata, trumpet and bassoon, C major. PRENTZL: Sonata, trumpet and bassoon, C major. PURCELL: Voluntary, organ, D minor. STANLEY: Suite of trumpet voluntaries, D major. Edward Tarr, Bengt Eklund, tpt; George Kent, org; Helmut Bocker, bsn.

++HF 8-73 p108
++MJ 10-73 p19
++NR 9-73 p13
++RR 6-77 p81
++St 10-73 p158

H 71308

3248 The Worcester fragments. Accademia Monteverdiana Chorus and Soloists; Denis Stevens.

++HF 10-75 p90
+HFN 8-75 p85
+MM 1-77 p31
++MT 9-75 p801
++NR 8-75 p9

H 71312

3249 Medieval German plainchant and polyphony: Advent/Christmas; Ad te levavi; Congaudeat turba; Puer natus est; Viderunt omnes; Johannes postquam senuit; Hodie progeditur. Passiontide: Gloria, laus et honor; Dominus Jesus; Christus factus est; Hely, hely. Easter: Confitemini...Laudate; Unicornis captivatu; Haec dies; Ad regnum...Noster cetus. Pentecost: Spiritus Domini; Factus est repente; Catholicorum concio. Parousia: Cum natus est Jesus; Gloria in excelsis. Schola Antiqua; R. John Blackley.

+-Gr 11-75 p888
+HFN 11-75 p163
+-MM 1-77 p31
+NR 10-75 p6
++RR 12-75 p90
+-St 7-76 p114

H 71326

3250 The pleasures of the royal courts. Courtly art of the trouvers: ADAM DE LA HALLE: Fines amouretes ai; Tant con je vivrai. MUSET: Quant je voy yver. ANON.: Ductia; La sexte estampie real; Souvent souspire mon cuer. Burgunidan Court of Philip the Good: DUFAY: Vergine bella. GULIELMUS: Falla con miseras. LEGRANT:Entre vous, noviaux maries. German Court of Emperor Maximilian I: ISAAC: Inssbruck, ich muss dich lassen. SENFL: Nun wollt ihr horen neue Mar. WECK: Spanyoler Tanz and Hopper dancz. Italian Music of the Medici Court: CARA: Non e tempo. FO: Tua voisi esser sempre mai. SCOTTO: O fallace speranza. ZESSO: E quando andarete al monte. ANON.: Polyphonic dances (6). Spanish Courts in the early Sixteenth Century: ALONSO: La tricotea Samartin. CABEZON: Diferencias sobre "La dama le demanda". ENCINA: Ay triste que vengo. ORTIZ: Recercadas primera y segunda. ANON.: Pase el agoa, ma Julieta; Rodrigo Martines. Early Music Consort; David Munrow.

+Gr 12-76 p1042
+HF 2-77 p108
+HFN 11-76 p165
+-MT 1-77 p47
+NR 4-77 p12
+NYT 8-15-76 pD15
+-RR 10-76 p91
++SFC 8-22-76 p38
+St 3-77 p148

H 71341

3251 BUCHTEL: Polka dots. CLARKE: Cousins. The maid of the mist, polka. Twilight dreams, waltz intermezzo. FILLMORE: Trombone family. GUMBERT: Cheerfulness. HANNEBERG: Triplets of the finest, concert polka. PRYOR: Blue bells of Scotland. Exposition echoes, polka. Thoughts of love, valse de concert. SMITH: The cascades, polka brilliant. Gerard Schwarz, Allan Dean, Mark Gould, cor; Ronald Barron, Norman Bolter, Douglas Ederman, trom; Kenneth Cooper, pno.

+FF 9-77 p12
+NYT 7-3-77 pD11
++St 12-77 p93

ODEON

HQS 1356

3252 Organ music from King's. BACH: Toccata and fugue, S 565, D minor. BRAHMS: Es ist ein Ros entsprungen, op. 122, no. 8. FRANCK: Chorale, no. 3, A minor. LISZT: Prelude and fugue on the name B-A-C-H, G 260. VAUGHAN WILLIAMS: Preludes on Welsh hymn tunes Rhosymedre. VIERNE: Berceuse. WIDOR: Symphony, organ, no. 5, op. 42, no. 1, F minor: Toccata. Philip Ledger, org.

+MU 12-77 p9

ODYSSEY

Y 33130 (Reissues)

3253 Bidu Sayao: French arias and songs. AUBER: Manon Lescaut: L'eclat de rire. CAMPRA: Les fetes venitiennes: Chanson du Papillon. CHOPIN: Melancoly, op. 74, no. 2. DEBUSSY: La damoiselle elui: Je voudrais qu'il fut deja pres de moi. L'Enfant prodigue: Lia's recitative and aria. Proses lyriques, no. 3: De fleurs. DUPARC: Chanson triste. HAHN: Si mes vers avaient des ailles. KOECHLIN: Si tu le veux. MORET: Le Nelumba. RAVEL: L'Enfant et les sortileges; Toi, le coeur de la rose. TRAD (arr. Crist) C'est mon ami. Bidu Sayao, s; Milne Charnley, pno; University

of Pennsylvania Womens Chorus; PO, Columbia Concert Orchestra; Paul Breisach, Eugene Ormandy.
+HF 4-77 p117 +ON 1-29-77 p44
+NR 5-77 p13 +St 4-77 p115

Y 33793 (Reissue from Edison)

3254 BACHELET: Chere nuit. CATALANI: La Wally: Ebben, ne andro lontana. CHOPIN: The maiden's wish, op. 74, no. 1 (in French). GIORDANO: Andrea Chenier: La mamma morta. GOMES: Salvator Rosa: Mia piccirella. LEONCAVALLO: I Pagliacci: Ballatella. MASCHERONI: Eternamente. PUCCINI: La boheme: Si, mi chiamano Mimi. SODERO: Crisantemi. TCHAIKOVSKY: Eugene Onegin, op. 24: Letter scene, excerpt (in Italian). VERDI: Il trovatore: Tacea la notte; D'amor sull ali rosee. Claudia Muzio, s; Albert Spalding, vln; Robert Gayler, pno; Orchestral accompaniment.
+ARG 2-77 p13 +ON 1-22-77 p33
+HF 3-77 p116 +SR 7-9-77 p48
++NR 1-77 p10

Y 34139

3255 CANN: Bonnylee. GRESSEL: Points in time. KRIEGER: Short piece. LANSKY: Mild und Leise. SEMEGEN: Electronic composition, no. 1. WRIGHT: Electronic composition. ZUR: Chants.
+HR 12-76 p126 -NR 2-77 p15

L'OISEAU-LYRE

DSLO 17

3256 BIRTWISTLE: Four interludes for a tragedy. DEBUSSY: Petite piece. MONTEVERDI: Il combattimento di Tancredi e Clorinda: Paraphrase on the madrigal by Monteverdi, op. 28 (arr. Goehr). SCHUMANN: Fantasiestucke, op. 73 (edit. Platt/Hacker). TRAD.: Mesomedes of Crete: Hymn to the sun. A Turkish Taksim (arr. Hacker). Alan Hacker, bassett clarinet, clarino, clt; Richard Burnett, fortepiano.
+-Gr 12-77 p1105 +-RR 12-77 p69
++HFN 12-77 p168

DSLO 20

3257 ADAMS: The holy city. BRAHE: Bless this house. HANDEL: Tolomeo: Silent worship (arr. Somervell). IRELAND: Sea fever. KEEL: Trade winds. MALOTTE: The Lord's prayer. MOSS: The floral dance. MURRAY: I'll walk beside you. RASBACH: Trees. SIBELIUS: Be still, my soul. SPEAKS: On the road to Mandalay. STANFORD: Songs of the sea, op. 9, no. 4: Drake's drum. SULLIVAN: The lost chord. VAUGHAN WILLIAMS: Songs of travel: The vagabond. Linden Lea. WOLCOTT: Blessed is the spot; O thou, by whose name. WOOD, H.: A brown bird singing. Norman Bailey, bs-bar; Geoffrey Parsons, pno.
+-Gr 5-77 p1750 +-RR 5-77 p78

12BB 203/6 (also Decca 12BB 203/6)

3258 Musicke of sundre kinds: Renaissance secular music, 1480-1620. ALISON: Dolorosa pavan. ATTAINGNANT: Content desir basse danse. AZZAIOLO: Quando le sera; Sentomi la formicula. BARBERIIS: Madonna, qual certezza. BOTTEGARI: Mi stare pone Totesche. CAURROY: Fantasia; Prince la France te veut. CAVENDISH: Wand' ring in this place. COMPERE: Virgo celesti. COSTELEY: Helas, helas, que de mal. DALZA: Recercar; Suite ferrarese; Tastar de corde. DUNSTABLE (Anon.): O rosa bella; Hastu mir. FAYRFAX: I love, loved; Thatt was my woo. FONTANA: Sonata, violin.

FORSTER: Vitrum nostrum gloriosum. FRESCOBALDI: Toccata. GABRIELI, A.: Canzona francese. GABRIELI, G.: Sanctus Dominus Deus. GESUALDO: Canzona francese; Mille volte il dir moro. GIBBONS: Now each flowery bank. GOMBERT: Caeciliam cantate. GUAMI: La brillantina. HECKEL: Mille regretz; Nach willen dein. HOFHEIMER: Nach willen dein. HUME: Musick and mirth. ISAAC, A.: Ne piu bella di queste; Palle, palle; Quis dabit pacem. ISAAC, H.: La la ho ho. JANNEQUIN: Les cris de Paris. LE JEUNE: Fiere cruelle. JOSQUIN DES PRES: Mille regretz. LASSUS: Morescas: Cathalina, apra finestra; Matona mia cara. MARENZIO: O voi che sospirate; Occhi lucenti. MERULO: Canzona francese. MODENA: Ricercare a 4. MONTEVERDI: Lamento d'Olimpia. MOUTON: La, la, la l'oysillon du bois. MUDARRA: Dulces exuyiae. NARVAEZ: Fantasia; Mille regretz. OBRECHT: Ic draghe die mutze clutze; Mijn morken gaf; Pater noster. ORTIZ: Dulce memoire; Recercada. PACOLINI: Padoana commun; Passamezzo commun. PORTER: Thus sang Orpheus. RONTANI: Nerinda bella. RORE: De la belle contrade. RUE: Pour ung jamais. SANDRIN: Doulce memoire. SERMISY: Content desir; La, je m'y plains. TERZI: Canzona francese. TIBURTINO: Ricercare a 3. TORRE: Adormas te, Senor; La Spagna. TROMBONCINO: Ave Maria; Hor ch'el ciel e la terra; Ostinato vo seguire. TYE: In nomine. VASQUEZ: Lindos ojos aveys, senora. VERDELOT: Madonna, qual certezza. WILLAERT (arr. Cabezon): E qui, la dira; Madonna qual certezza. ANON.: Belle, tenes mo; Calata; Celle qui m'a demande; Chui dicese e non l'amare; Der Katzenfote; Lady Wynkfyldes rownde; Las, je n'ecusse; L'e pur morto Feragu; Mignonne, allons; Pavana de la morte de la ragione; Pavana el tedescho; Pavane Venetiana; Der rather Schwanz; Le rossignol; Saltarello de la morte de la ragione; Se mai per maraveglia; Shooting the guns pavan; Sta notte; Suite regina; La triquotee. ANON/AZZAIOL Girometa. ANON/EDWARDS: Where griping griefs. ANON/ISAAC: Bruder Konrad. ANON/JUDENKUNIG: Christ der ist erstanden. ANON/PACOLINI: La bella Francheschina. ANON/SPINACINO: Je ne fay. ANON/WYATT: Blame not my lute. Consorte of Musicke; Anthony Rooley.

+-Gr 12-75 p1066 +NR 8-76 p15
+HF 1-77 p137 +NYT 8-15-76 pD15
++HFN 12-75 p159 +RR 12-75 p89
+MJ 11-76 p45 ++SFC 9-26-76 p29
+-MM 7-76 p37 ++STL 2-8-76 p36

SOL 343

3259 BEAUVARLET-CHARPENTIER: Fugue, G minor. FRANCK: Chorale, no. 2, B minor. JONGEN: Toccata. MANERI: Salve regina. ROBINSON: Improvisation on a submitted theme. McNeil Robinson, org.

+Gr 8-76 p323 +NR 12-76 p12
+HFN 7-76 p93 +RR 2-77 p78
+MU 3-77 p13

SOL 345 (Reissue from Cambrian SCLP 591)

3260 BELLINI: Almen se non poss'io. DONIZETTI: La conocchia. ROSSINI: L'Invito. SCHUBERT: Du bist die Ruh, D 776; Die Forelle, D 550; Gretchen am Spinnrade, D 118; Heidenroslein, D 257. VERDI: Ave Maria. TRAD.: Bugeilio'r Gwenith Gwyn; Dafydd y Garreg Wen; Wrth Fynd Efo Deio I Dowyn (arr. E. T. Davies); Y Bore Glas (arr. G. Williams); Y Deryn Pur. Margaret Price, s; James Lockhart, pno.

+-Gr 4-76 p1653 +RR 8-76 p63
+HFN 6-76 p102 ++St 10-77 p150

SOL 349
3261 BARRIOS: Danza paraguaya. FALLA: Homenaje a Debussy. GALILEI: Suite. LAURO: Vals venezolano. MUDARRA: Fantasia que contrahaza la harpa en la manera de Ludovico. RUIZ-PIPO: Cancion y danza. SAVIO: Batucada. Seroes. SOR: Variations on a theme by Mozart, op. 9. TARREGA: Recuerdos de la Alhambra. Lagrima. La alborada. VILLA-LOBOS: Preludes, no. 1, E minor; No. 5, D major. VIVALDI: Concerto, 2 mandolins: Adagio (arr. Walker). Timothy Walker, gtr.
+Gr 7-77 p202 +-RR 6-77 p81
+HFN 6-77 p123

DSLO 510
3262 The Cozens lute book. BATCHELAR: Pavan and galliard. DANYEL: Mistress Anne Grene her leaves be greene. DOWLAND: Fancy. Frogg galliard. Lachrimae pavan. HOLLIS: John Blundeville's last farewell. LAURENCINI: Fantasia. ROBINSON: Spanish pavan. ROMANI: Exercitium. SMYTHE: Galliards (2). ANON.: John come kiss me now. Mall Symms. Pavan. Preludium VI. Anthony Rooley, lt.
+Gr 10-77 p671 +RR 10-77 p76
+HFN 10-77 p153

OLYMPIC

OLY 105
3263 The Yiddish art song. ACHRON (arr.): In a little cottage. BINDER (arr.): Sabbath at the concluding meal. ENGEL (arr.): Dear father; Kaddish of Reb Levi-Itzchok of Barditchev; Listen. GELBERT: Keep moving on. GOLUB: Tanchum. MILNER: The hunter; In Cheyder. WEINER: A father and his son; A nigh (arr); Rhymes written in the sand; The story of the world; A tree stands on the road (arr.); What is the meaning of (arr.); Yidl and his fiddle. Leon Lishner, bs; Lazar Weiner, pno.
+NR 10-77 p11 +-St 2-77 p136

6118
3264 Bagpipe marches and music of Scotland: Balkan Hills; Barren rocks of Aden; Dovecote Park; Earl of Mansfield; Far o'er Struy; Green Hills of Tyrol; Hearken my love; Herding song; Highland laddie; Hills of Alva; Jimmy Tweedie's sealegs; Kitchener's army; Linen cap; Loch Broom Bay; Lord Byron; Lord James Murray; Miss Elspeth Campbell; Mrs. Lily Christie; My lodging's on the cold cold graound; My home; Peter McKenzie Warren; Rhodesian Regiment; Scotland, the brave; The shepherd's crook; Smith o' Chilliechassie; Smith's a gallant fireman; Sporting Jamie; 10th H.L.I. crossing the Rhine; Torosay castle; Walking the floor; Westering home. Shotts and Dykehead Caledonia Pipe Band; Tom McAllister, pipe major; Alex Duthart.
+NR 3-77 p15

6120
3265 German military marches: Bayrischer Defliermarsch; Castaldo Marsch; Fehrbeliner Reitenmarsch; Frei Weg; Des Grossen Kurfeursten Reitermarsch; Grosser Zapfenstreich; Helenenmarsch; Marsch des Yorckschen Korps; Mussinan Marsch; Petersburger Marsch; Waidmannsheil. German Air Force Band.
-NR 3-77.p15

6122
3266 Military Band of the Queen's Regiment: Arranger's holiday; Blue

tail fly; Expo march; Fiesta siesta; Regimental music of the 1st battalion; Regimental music of the 2nd battalion; Regimental music of the 3rd battalion; Regimental music of the 4th battalion; Rise and shine; Soldiers of the Queen; Trumpet fiesta.
+NR 3-77 p15

6131
3267 French military marches: Alsace Lorraine; Chant du Depart; Hymne de l'Infanterie de Marine; Marche consulaire; Marche de la Legion; Marche de la gendarmerie; Marche de la 2ieme D.B.; Marche des parachutistes; Marche du liere Regiment des Chasseurs Ardennais; Marche du liere Regiment des Carabiniers; La Marseillaise; Saint Cyr; Sambre et Meuse.
-ARG 2-77 p50 -NR 3-77 p15

ORION

ORS 75181
3268 BLOCH: Jewish life: 3 pieces. Meditation hebraique. BRUCH: Kol Nidrei, op. 47. CASALS: Song of the birds. CHOPIN: Sonata, violoncello, op. 65, G minor: Largo. DEBUSSY (Feuillard): Preludes, Bk I, no. 8: La fille aux cheveux de lin. FAURE: Apres un reve, op. 7, no. 1. Elegy, op. 24, C minor. RAVEL (Bazelaire): Piece en forme de habanera. Lillian Rehberg Goodman, vlc; Harold Bogin, pno.
+HF 10-75 p90 ++NR 8-75 p15
+IN 12-77 p26

ORS 75207
3269 HAINES: Sonata, harp. MEYER: Appalachian echoes. MONDELLO: Siciliana. SIEGMEISTER: American harp. STARER: Prelude. Pearl Chertok, hp.
+-MJ 3-77 p46 ++NR 11-76 p15

ORS 76212
3270 CORTES: Sonata, violin and piano. KRAFT: In memoriam Igor Stravinsky. LYNN: Vino. RODRIGUEZ: Variations, violin and piano. VAZZANA: Incontri. Harris Goldman, vln; Carolyn Brown, pno.
+ARG 8-77 p33 +NR 4-77 p5

ORS 76247
3271 BACON: The Burr frolic. FOOTE: Characteristic pieces (3). IVES: Variations on "America". KREISLER (arr. Montgomery): Liebesfreud. ROSSINI: Guglielmo Tell: Overture (arr. Gottschalk). SOUSA: The bride elect: Waltzes. Paul Hersh, David Montgomery, pno.
+Audio 1-77 p86 +-NR 9-77 p13

ORS 76255
3272 American organ music of three centuries. BARBER: Variations on a Shapenote hymn, op. 34. FARNAM: Toccata on "O fillii et filiae". KAY: Suite, organ, no. 1. MOLLER: Sinfonia. PARKER: Fugue, op. 36, no. 3, C minor. PAINE: Prelude, op. 19, no. 1, D flat major. SELBY: Voluntary, A major. SOWERBY: Prelude on "The King's majesty". Thomas Harmon, org.
++ARG 5-77 p47 +MU 4-77 p11
+CL 4-77 p10 +NR 2-77 p13
+HF 7-77 p120

ORS 77263
3273 GRAYSON: Promenade. KAGEL: Pandorasbox. NIGHTINGALE: Entente. NORDHEIM: Dinosaurus. NORGAARD: Arcana. Lou Anne Neill, hp; Stuart Fox, gtr; Tom Collier, perc; Mogen Ellegaard, Barbara Beisch, James Nightingale, accordion.
+FF 11-77 p65 +NR 12-77 p14

ORS 77269

3274 CHAGRIN: Pieces (2). CORDLE: Interlude. GALLIARD: Sonata, bassoon, A minor. PERLE: Inventions. TELEMANN: Sonata, bassoon, F minor. VIVALDI: Sonata a due. Andre Cordle, bsn; Sheryl Schrock, flt; Christine Daxelhofer, hpd.
+ARG 11-77 p43 +NR 7-77 p7

ORS 77271

3275 The art of Sonia Essin. ANIK: Changed. BRAHMS: Songs: Bei dir sind meine Gedanken; Gute Nacht; Madchenlied; Unbewegte Laue Luft. CHAUSSON: Le charme; Les papillons; Psyche. DONIZETTI: Anna Bolena: Deh non voler costringere. FALLA: Canciones populares espanolas: Jota; Nana; El pano moruno. GLUCK: Alceste: Divinities du Styx. HANDEL: Messiah: He shall feed his flock. JOHNSON: Song of the heart. RESPIGHI: Pioggia. TYSON: One little cloud. WAGNER: Rienzi: Gerechter Gott. TRAD.: Annie Laurie; I'm a'rollin; Sometimes I feel like a motherless child. Sonia Essin, s; Arpad Sandor, pno.
+NR 10-77 p12

PANDORA

PAN 2001

3276 BOZZA: Ballade, trombone. MICHALSKY: Concerto in re. MAHLER: Symphony, no, 3, D minor: 1st movement. PRYOR: Blue bells of Scotland. SIMONS: Atlantic zephyrs. Dennis Smith, trom.
+-FF 11-77 p61

PEARL

SHE 533

3277 DOPPLER: Valse di bravura. KOHLER: Fantasia on a theme by Chopin. PAGGI: Rimembranze Napolitane. REINECKE: Sonata, flute and piano, op. 167, E minor. SULLIVAN: Twilight, op. 12 (arr. Bennett). William Bennett, Trevor Wye, flt; Clifford Benson, pno.
+Gr 3-77 p1461 +RR 3-77 p83

PELCA

PSR 40571

3278 BACH: Chorale prelude: Kommst du nun, Jesu, S 650. CLARKE: Suite, D major. BRAGA: Batalha de 6 tom. KREBS: Prelude and fugue, F sharp major. Liebster Jesu wir sind hier. Wachet auf, ruft uns die Stimme. LUBECK: Prelude and fugue, F major. SCHEIBE: Trio, F major. TELEMANN: Suite, D major. Thomas Hartog, tpt; Martin Weyer, org.
++NR 1-77 p14

PSR 40599

3279 BACH: Prelude and fugue, S 545, C major. FRANCAIX: Suite Carmelite. MARTINI: Aria variata, thema and 4 variationen. REGER: Phantasie uber den choral "Straf mich nicht in deinem Zorn", op. 40, no. 2. STANLEY: Voluntary, D major. Gunter Eumann, org.
+NR 1-77 p14

PSR 40607

3280 BACH: Singet dem Herrn ein neues Lied, S 225. JOSQUIN DES PRES: Tu pauperum refugium. PENDERECKI: St. Luke passion: Miserere. REGER: Vater unser. SCHUTZ: Das ist je gewisslich wahr. Evangelische Jugendkantorei der Pfalz; Heinz Markus Gottsche.
+NR 7-77 p8

PETERS

PLE 013

3281 Gregorian chant: Midnight mass for Christmas eve; Mass for Christmas day. Abbey of St. Pierre de Solesmes Monks Choir; Dom Joseph Gajard.

+ARG 12-77 p55

PHILIPS

6392 023

3282 ADAM: O holy night (arr. Kelly). AITKEN: Maire my girl. BISHOP: My pretty Jane. FREIRE: ay, ay, ay. MENDELSSOHN: On wings of song (arr. Kelly). MURPHY: Connemara cradle song. SANTOS: O lonely moon. SCHUBERT: Standchen (arr. Kelly). SIMONS: Marta, rambling rose of the wildwood. STOLZ: Two hearts swing in three-quarter time. TOSELLI: Serenade (arr. Kelly). TRAD.: Loch Lomond. Frank Patterson, t; Irish Promenade Orchestra; Colman Pearce.

+-NR 10-77 p10

6500 926

3283 BIBER: Suite, 2 clarino trumpets. BLOW: Fugue, F major: Vers. BULL: Variations on the Dutch chorale "Laet ons met herten reijne". CAMPIAN: Never weather-beaten sail. DOWLAND: Lachrimae antiquae pavan. FANTINI: Sonata, 2 trumpets, B flat major. FRESCOBALDI: Capriccio sopra un soggetto. HANDEL: Concerto, trumpet, B flat major. MORLEY: La caccia, a 2. La sampogna. PURCELL: Trumpet tune (The cebell), D major. Trumpet tune and air. STANLEY: Trumpet voluntary, D major. ANON.: Hejnat Krakowska (2). Clarion Consort.

+ARG 3-77 p40
+-Gr 2-76 p1350
+HF 2-77 p112
+HFN 3-76 p107
++NR 7-77 p15
++RR 3-76 p52
++St 4-77 p142

6504 135

3284 Russian liturgical chants. Nicolai Gedda; Russian Cathedral Choir, Paris; Eugen Evetz.

+-Audio 11-77 p132
*MJ 5-77 p33

6580 105

3285 Gregorian chants: Resurrexi, Domine, probasti me. Victimae paschali laudes. Pascha nostrum. Post dies octo. Christus resurgens. Alleluia, lapis revolutus est. Te Deum. Requiem aeternam, Introitus and graduale. Absolve, Domine. Dies irae. Domine Jesu Christe. Lux aeterna. Libera me, Domine. In paradisum. Saint Maurice and Saint Maur Abbey Monks. Philips 6580 105.

+-Gr 11-75 p893
+HFN 8-75 p91
*MJ 5-77 p33
-MM 1-77 p29
+NR 9-77 p10

6580 114 Tape (c) 7317 135

3286 BACH, J.C.: Symphony, op. 3, no. 1, D major. BACH, J.S.: Brandenburg concerto, no. 3, S 1048, G major. MOZART: Concerto, piano, no. 24, K 491, C minor: 2nd movement. Symphony, no. 40, K 550, G minor: 1st movement. PURCELL: The Indian Queen: Trumpet overture. Yorkshire feast song. TORELLI: Sonata a 5, no. 7, D major. Don Smithers, Michael Laird, tpt; AMF; Neville Marriner.

+-Gr 6-76 p95
+HFN 7-76 p103
++NR 3-76 p15
+-RR 6-76 p35
+RR 1-77 p88 tape

6580 174 (Reissue from SAL 3574)
3287 BIZET: La jolie fille de Perth: Quand la flamme de l'amour. Les pecheurs de perles: L'orage s'est calme...O nadir, tendre ami de mon jeune age. GOUNOD: Romeo et Juliette: Mab, la reine des mensonges. GLUCK: Orfeo ed Euridice: Che puro ciel; Cara sposa ...Che faro senza Euridice. HANDEL: Partenope: Combattono il mio core. MASSENET: Manon: Les grands mots que voila...Epouse quelque brave fille. Thais: Voila donc la terrible cite. MEYERBEER: L'Africaine: Adamastor, roi des vagues profondes. MONTEVERDI: La favola d'Orfeo: Tu sei morta mia vita. MOZART: Don Giovanni, K 527: Deh viene alla finestra; Fin ch'han dal vino calda la testa. Le nozze di Figaro, K 492: Hai gia vinta la causa...Vedro mentr'io sospiro. THOMAS: Hamlet: O vin, dissipe ma tristesse. Gerard Souzay, bar; Lamoureux Orchestra; Serge Baudo.
+Gr 4-77 p1600 +RR 5-77 p36
+HFN 4-77 p153

6581 018
3288 ALBINONI: Concerto, trumpet and 6 clarinets, A major. CLARKE: Trumpet voluntary. PURCELL: Suite of dances. Trumpet tune and air. TELEMANN: Divertissement, 2 trumpets and strings, D major. STOELZEL: Concerto, trumpet, D major. VIVALDI: Concerto, trumpet, D major. Maurice Andre, tpt; Wind Ensemble, String Orchestra; Armand Birbaum.
-Gr 4-77 p1567 +-RR 3-77 p41
+-HFN 4-77 p132

5699 227 (Reissue)
3289 Irish songs. DAVIS: The West's awake. FERGUSON: The lark in the clear air. HUGHES: The stuttering lover. KICKHAM: She lived beside the Anner. McCALL: Kelly, the boy from Killane; Boolavogue. LOVER: The low-back'd car. MOORE: Believe me, if all those endearing young charms; The minstrel boy; The young May moon. TRAD.: An raibh tu ag an gCarrig; The bard of Armagh; Green bushes; The maid of the sweet boy. Frank Patterson, t; Orchestra; Thomas C. Kelly.
+Gr 6-73 p113 +RR 6-73 p89
+MJ 2-75 p31 +St 8-75 p107
+MJ 5-77 p33 +STL 7-8-73 p36

6747 327 (Reissues)
3290 Greatest music in the world. DUKAS: The sorcerer's apprentice. ELGAR: Pomp and circumstance march, op. 39, no. 1, D major. HANDEL: Messiah: Hallelujah chorus. HOLST: The planets, op. 32: Jupiter. MARTINI IL TEDESCO: Songs: Plaisir d'amour. MOZART: Sonata, piano, no. 11, K 331, A major: Rondo alla Turca. RACHMANINOFF: Prelude, op. 23, no. 5, G minor. RESPIGHI: Ancient airs and dances: Suite, no. 2: Bergamasca. RODRIGO: Concierto de Aranjuez, guitar and orchestra: Adagio. ROSSINI: Il Signor Bruschino: Overture. SCHUBERT: Impromptu, op. 90, no. 4, D 899, A flat major. TCHAIKOVSKY: Eugene Onegin, op. 24: Waltz. Mazeppa: Gopak. VERDI: La traviata, excerpt. WAGNER: The flying Dutchman: Overture. Cristina Deutekom, s; Angel Romero, gtr; Ingrid Haebler, Rafael Orozco, pno; Amusementorkest, Monte Carlo Opera Orchestra, AMF, LPO, COA, PH, LSO, Minneapolis Symphony Orchestra; Wolfgang Sawallisch, Bernard Haitink, Antal Dorati, Jean Fournet, Neville Marriner.
+-HFN 4-77 p153 +-RR 1-77 p54

6767 001 (4) (Reissues from 6500 523, 6500 660, 9500 023, 9500 307)
3291 BEETHOVEN: Ah, perfido, op. 65. Egmont, op. 84: Die Trommel geruhret; Freudvoll und leidvoll. Songs: No, non turbati. GLUCK: Alceste: Divinites du Styx. Armide: La perfide Renaud. Iphigenie en Aulide: Vous essayez en vain...Par la crainte; Adieu conservez dans votre ame. Iphigenie en Tauride: Non cet affreux devoir. Orfeo ed Euridice: Che puro ciel; Che faro senza Euridice. Paride ed Elena: Spiagge amate; Oh, del mio dolce amor; Le belle immagini; Dit te scordarmi. La rencontre imprevue: Bel inconnu; Je cherche a vous faire. HANDEL: Atalanta: Care selve. Ariodante: Dopo notte. Hercules: Where shall I fly. Joshua: O had I jubal's lyre. Lucrezia: Solo cantata. Rodelinda: Pompe vane di morte...Dove sei. Serse: Ombra mai fu. HAYDN: Arianna a Naxos: Berenice che fai. MOZART: Abendempfindung, K 523; Das Veilchen, K 476. La clemenza di Tito, K 621: Parto parto, ma tu ben mio; Deh per questo istante solo. SCHUBERT: Alfonso und Estrella, D 732: Konnt ich ewig hier verweilen. Rosamunde, op. 26, D 797: Der Vollmond strahlt. Lazarus, D 689: So schlummert auf Rosen. Songs: Zogernd Leise, D 920. Janet Baker, ms; Raymond Leppard, hpd, pno, fortepiano; Thea King, clt; ECO and Choir; Raymond Leppard.

+Gr 10-77 p696
+HFN 10-77 p167
+RR 9-77 p86

6833 159
3292 ALBENIZ: Suite espanola, no. 5: Asturias. BACH: Suite, lute, S 995, G minor. HANDEL: Sarabande, D major (trans. Lagoya). TARREGA: Recuerdos de la Alhambra. TORROBA: Nocturno. VILLA-LOBOS: Etude, guitar, no. 11. WEISS: Passacaglia (trans. Lagoya). Alexandre Lagoya, gtr.

+NR 12-75 p14
+St 1-77 p71

9500 203
3293 Jose Carreras, aria recital. BELLINI: Adelson e Salvini: Ecco, signor, la sposa. DONIZETTI: Il Duca d'Alba: Angelo casto e bel (completed by Matteo Salvi). Maria di Rohan: Alma soave e cara. MERCANDANTE: Il Giuramento: Bella adorata incognita; Compiuta e omai. PONCHIELLI: Il figliuol prodigo: Tenda natal. VERDI: Un ballo in maschera: Ma se m'en forza perderti. La forza del destino: O tu che in seno agli angeli. Jerusalem: O mes amis. I Lombardi: La mia letizia infondere. Luisa Miller: Quando le sere al placido. Jose Carreras, t; RPO; Roberto Benzi.

+ARG 3-77 p39
+-Gr 1-77 p1177
+-HF 3-77 p115
+HFN 1-77 p103
+MJ 2-77 p30
+NR 3-77 p10
+OC 6-77 p45
+-ON 1-15-77 p33
+OR 9/10-77 p30
+RR 1-77 p41
++St 2-77 p129

9500 218
3294 Frank Patterson: John McCormack favorites. CROUCH: Kathleen Mavourneen. FRANCK: Panis Angelicus. GLOVER: The rose of Tralee. GODARD: Jocelyn: Berceuse. HANDEL: Semele: Where'er you walk; O sleep, why dost thou leave me. MARSHALL: I hear you calling me. MARTINI IL TEDESCO: Plaisir d'amour. MOZART: Don Giovanni, K 527: Il mio tesoro. SCHUBERT: Ave Maria. WOLF: Schlafendes Jesuskind. TRAD.: Lna Ban; My lagan love. Frank Patterson, t; Orchestra; Christopher Seaman.

+NR 9-77 p10
+St 5-77 p120

POLYDOR

2489 542 Tape (c) 3150 560
3295 Fritz Wunderlich in Vienna. Songs: Denk dir die Welt war ein Blumenstrauss; Draussen in Sievering; Es steht ein alter Nussbaum drauss in Heiligenstadt; Herr Hofrat, erinnern Sie sich noch; Ich hab die schonen Maderin net erfunden; Ich kenn ein kleines Wegerl im Helenthal; Ich muss wieder einmal in Grinzing sein; Ich weiss auf der Wieden ein kleines Hotel; Im Prater bluh'n wider die Baume; In Wein gibt's manch winziges Gasserl; Wien, Wien, nur du allein; Wien wird bel Nacht erst schont. Fritz Wunderlich, t; Spilar Schrammel Ensemble,Vienna Volksoper Orchestra and Chorus; Robert Stolz.
+Gr 7-77 p230

PRELUDE

PRS 2505
3296 BISHOP: Home, sweet home. BOHM: Still as the night. BRAHMS: An eine Aeolscharfe, op. 19, no. 5; Meine Liebe ist grun, op. 63, no. 5; O wusst ich doch den Weg zuruck, op. 63; Der Tod, das ist die kuhle Nacht, op. 96, no. 1; Verzagen, op. 72, no. 4. COLERIDGE-TAYLOR: Big lady moon. HORN: Cherry ripe. LIDDELL: Abide with me. MAHLER: Des Knaben Wunderhorn: Das irdische Leben; Rheinlegendchen. Ruckert Lieder: Ich bin der Welt abhanden gekommen. MOLLOY: Love's old sweet song. NEVIN: The rosary. SULLIVAN: The lost chord. Norma Procter, con; Paul Hamburger, pno.
+Gr 12-76 p1042 +RR 2-77 p84
-HFN 12-76 p139

PRS 2512
3297 DEBUSSY: Suite bergamasque: Clair de lune. DELIUS: On hearing the first cuckoo in spring. Summer night on the river. FAURE: Masques et bergamasques, op. 112: Sicilienne. HONEGGER: Pastorale d'ete. IBERT: Escales: 2nd movement. MILHAUD: Symphony, no. 1. RAVEL: Pavane pour une infante defunte. NPhO; William Jackson.
+Gr 7-77 p188 +-RR 7-77 p55
+HFN 8-77 p87 +SFC 12-25-77 p42

PYE

Nixa PCNHX 6 Tape (c) ZCNPNH 6
3298 BEETHOVEN: Leonore overture, no. 3, op. 72. BERLIOZ: Roman carnival, op. 9. MOZART: Don Giovanni, K 527: Overture. ROSSINI: William Tell: Overture. SCHUBERT: Rosamunde, op. 26, D 797: Overture. National Philharmonic Orchestra; Leopold Stokowski.
+-Gr 1-77 p1184 +-RR 2-77 p96 tape
-HFN 12-76 p148

PCNH 9
3299 BLUMENFELD: Etude for the left hand, op. 36. GLAZUNOV: Etude, op. 31, no. 1, C major. MUSSORGSKY: Capriccio "In the Crimea". PROKOFIEV: Pieces, piano, op. 12: Rigaudon. RACHMANINOFF: Pieces, piano, op. 10, no. 3: Barcarolle. Polka de VR. RUBINSTEIN: Etude, op. 23, no. 2. SCRIABIN: Etude, op. 12, no. 12. STRAVINSKY: Petrouchka, excerpt. TCHAIKOVSKY: L'Espiegle,op. 72, no. 12. Romance. Un poco di Chopin, op. 72, no. 11. Peter Cooper, pno.
+HFN 5-77 p131

Nixa QS PCNHX 10

3300 Coronation music. GIBBONS: O clap your hands together. HANDEL: The king shall rejoice. Zadok the priest. PARRY: I was glad. STANFORD: Gloria in excelsis. VAUGHAN WILLIAMS: O taste and see; The old hundredth. WALTON: A queen's fanfare. WESLEY, S.S.: Thou wilt keep him. Winchester Cathedral Choir, Waynflete Singers, Winchester College Commoners, Bournemouth Sinfonietta; Martin Neary; James Lancelot, org.
+Gr 11-77 p880 -HFN 8-77 p85

Golden Hour GH 643

3301 BRAHMS: Academic festival overture, op. 80. DELIUS: On hearing the first cuckoo in spring. GRIEG: Peer Gynt, op. 46: Suite, no. 1: Morning. ELGAR: Enigma variations, op. 36: Nimrod. ROSSINI: William Tell: Ballet music (arr. Godfrey). SMETANA: The bartered bride: Overture. VAUGHAN WILLIAMS: Fantasia on "Greensleeves". WAGNER: Die Meistersinger von Nurnberg: Suite, Act 3. Halle Orchestra, Pro Arte Orchestra, LPO; John Barbirolli, Charles Mackerras, Adrian Boult.
+-RR 11-77 p57

QUINTESSENCE

PMC 7013

3302 Fiedler's favorite overtures. HEROLD: Zampa. NICOLAI: The merry wives of Windsor. OFFENBACH: Orpheus in the underworld. ROSSINI: William Tell. SUUPE: Light cavalry. THOMAS: Mignon. BPO; Arthur Fiedler.
+St 11-77 p146

RC RECORDS

RCR 101

3303 BACH: Chorale prelude: Liebster Jesu, S 731. FRANCK: Chorale, B minor. DUPRE: Magnificat, op. 18, no. 10. GRIGNY: Recit de tierce en taille. ROBERTS: Pastorale and aviary. ROUTH: Sonatina, op. 9. WALTHER: Was Gott tut. Richard Cunningham, org.
+MU 3-77 p13

RCA

LSC 9884

3304 ALFVEN: Sa tag mitt hjarta; Jag langtar dig. ALTHEN: Land du valsignade. KJORLING: Aftonstamning. PETERSON-BERGER: Bland skogens hoga furustammar; Nar jag for mig sjalv i morka skogen gar. SIBELIUS: Demanten pa marssnon; Sav sav susa. SJOBERG: Tonerna. SODERMAN: Kung Heimer och Aslog. STENHAMMAR: Sverige. Jussi Bjorling, t; Royal Theatre Orchestra; Nils Grevillius.
+HFN 7-77 p108 +RR 7-77 p84

ARL 1-0456 Tape (c) ARK 1-0456 (ct) ARS 1-0456

3305 ALBENIZ: Iberia: Evocacion. Bajo la palmera, op. 212. CARULLI: Serenade, op. 96. GIULIANI: Variazioni concertante, op. 130. GRANADOS: Danza espanola, op. 37, nos. 6 and 11. Julian Bream, John Williams, gtr.
+Gr 4-74 p1868 ++RR 4-74 p63
++MJ 2-75 p40 ++RR 12-74 p87 tape
+-NR 11-74 p12 +St 1-77 p73

ARL 1-0864 Tape (c) ARK 1-0864 (ct) ARK 1-0864
3306 ASENCIO: Dipso. BACH: Suite, solo violoncello, no. 1, S 1007, E major: 3 movements. BENDA: Sonatina, D major. Sonatina, D minor. PONCE: Prelude, E major. SCARLATTI: Sonatas, guitar (2). SOR: Andante, C minor. Minuets, C major, A major, C major. WEISS: Bourree. Andres Segovia, guitar.
+HF 11-76 p153 ++St 10-75 p119
+NR 10-75 p12 +St 1-77 p73

ARL 1-1323
3307 ALBENIZ: Capricho catalan (trans. Lorimer). BACH: Anna Magdalena notebook, S 508, excerpts. MOLLEDA: Variations on a theme. SAMAZEUILH: Serenade. SAN SEBASTIAN: Preludios vascos: Dolor (trans. Segovia). SOR: Introduction and variations on "Malbrough s'en va-t-en guerre". Sicilienne, D minor. Andres Segovia, gtr.
+NR 3-76 p13 +St 1-77 p73
+-SFC 2-29-76 p25

RK 1-1735
3308 ALBENIZ: Sonata, guitar (arr. Pujol). BERKELEY: Sonatina, op. 51. CIMAROSA (arr. Bream): Sonatas, guitar, C sharp minor, A major. FRESCOBALDI: Aria detto "La Frescobalda" (arr. Segovia). RAVEL: Pavane pour une infante defunte (arr. Bream). RODRIGO: En los trigales. ROUSSEL: Segovia, op. 29. SCARLATTI: Sonata, guitar, E minor (arr. Bream). Julian Bream, gtr.
+HFN 12-76 p155 tape +RR 1-77 p89 tape

CRM 1-1749 (Reissues from RCA originals, 1906-20)
3309 BIZET: Carmen: La fleur que tu m'avais jetee. DONIZETTI: L'Elisir d'amore: Una furtiva lagrima. FLOTOW: Martha: Ach, so fromm (in Italian). GOUNOD: Faust: Salut, demeure chaste et pure. HALEVY: La Juive: Rachel, quand du Seigneur. HANDEL: Serse: Ombra mai fu. LEONCAVALLO: I Pagliacci: Vesti la giubba. MEYERBEER: L'Africaine: O paradiso (in Italian). PONCHIELLI: La gioconda: Cielo e mar. PUCCINI: La boheme: Che gelida manina. Tosca: Recondita armonia; E lucevan le stelle. VERDI: Rigoletto: Questa o quella; La donna e mobile. Il trovatore: Di quella pira. Aida: Celeste Aida. Enrico Caruso, t; Orchestral accompaniment.
+ARSC vol VIII, no. 2-3, p106 +-NR 10-76 p12
+-NR 11-76 p9
+Gr 7-77 p216 +OC 6-77 p45
+-HF 11-76 p98 +ON 1-15-77 p33
+HFN 9-77 p139 +RR 8-77 p35
+MJ 11-76 p44

CRL 1-2064
3310 Arthur Fiedler: A legendary performer. FALLA: El amor brujo: Ritual fire dance. GADE: Jalousie (Tango Tzigane). GERSHWIN: Strike up the band (arr. Green). IPPOLITOV-IVANOV: Caucasian sketches, op. 10: Procession of the Sardar. KHACHATURIAN: Gayaneh: Sabre dance. LENNON-MCCARTNEY (arr. Hayman): I want to hold your hand. LOEWE (arr. Hayman): My fair lady: Selections. MASCAGNI: Cavalleria rusticana: Intermezzo. RAYE-PRINCE (arr. Hayman): The boogie-woogie bugle boy of Company B. SOUSA: Stars and stripes forever. STRAUSS, J. II: Wine, women and song, op. 333. TCHAIKOVSKY: The nutcracker, op. 71: Waltz of the flowers. BPO; Arthur Fiedler
+St 2-77 p75

JRL 1-2315
3311 BIZET (Borne): Carmen fantasy, op. 25. CHOPIN: Nocturne, op. 15, no. 2, F sharp major. Waltz, op. 64, no. 1, D flat major. DOPPLER: Fantaisie pastorale hongroise, op. 26. DEBUSSY: Suite bergamasque: Clair de lune. GENIN: Carnival of Venice. YOSHIDA: Variations on a theme of "Sakura". GLUCK: Orfeo ed Euridice: Minuet. KREISLER: Liebesfreud. Liebesleid. RAVEL: Habanera. Jean-Pierre Rampal, flt; Futaba Inoue, pno.
+-NR 11-77 p9 +St 12-77 p157

FRL 1-3504
3312 BACH: Brandenburg concerto, no. 3, S 1048, G major: Allegro. Cantata, no. 33: Aria. CIMAROSA: Melodie. CORELLI: Allemande. DEFAYE: Melancolie. Sur un air de Bach. Sur un air de Corelli. HANDEL: Allegro. Bourree. Concerto grosso. Water music: Aria. MARCELLO: Adieu a Venise. Final. Maurice Andre, tpt; Jean-Marc Pulfer, org; Guy Pedersen, double bs; Gus Wallez, drum.
+NR 9-77 p5

LFL 1-5094 Tape (c) RK 11719, LRK 1-5094 (ct) LRS 1-5094.
3313 BACH: Suite, orchestra, S 1067, B minor: Minuet; Badinerie. CHOPIN: Waltz, op. 64, no. 1, D flat major (arr. and orch. Gerhardt). DINICU: Hora staccato (arr. and orch. Gerhardt). DOPPLER: Fantaisie pastorale hongroise, op. 26. DRIGO: Les millions d'Arlequin: Serenade (arr. and orch. Gamley). GLUCK: Orfeo ed Euridice: Dance of the blessed spirits. GODARD: Pieces, op. 116: Waltz. MIYAGI: Haru no umi (arr. and orch. Gerhardt). PAGANINI: Moto perpetuo, op. 11, no. 2 (arr. and orch. Gerhardt). RIMSKY-KORSAKOV: The tale of the Tsar Sultan: Flight of the bumblebee (arr. and orch. Gerhardt). SAINT-SAENS: Ascanio: Ballet music, Adagio and variation. James Galway, flt; National Philharmonic Orchestra; Charles Gerhardt.
++Gr 11-75 p915 +MJ 11-76 p60
+Gr 1-76 p1244 ++NR 11-76 p16
+Gr 7-76 p230 tape +-RR 10-75 p38
+-HF 3-77 p123 tape +RR 1-77 p91 tape
+-HFN 12-75 p164

TVM 1-7201
3314 BIZET: Carmen: La fleur que tu m'avais jetee. DONIZETTI: Il Duca d'Alba: Angelo casto e bel. Don Sebastiano: Deserio in terra. FLOTOW: Marta: M'appari. HALEVY: La Juive: Rachel quand du Seigneur. GIORDANO: Andrea Chenier: Un di all'azzuro spazio. LEONCAVALLO: I Pagliacci: Recitar...Vesti la giubba; No Pagliaccio non son. MASCAGNI: Cavalleria rusticana: Addio alla madre. PONCHIELLI: La Gioconda: Cielo e mar. PUCCINI: La boheme: Che gelida manina; Testa adorata. Manon Lescaut: Donna non vidi mai. Tosca: Recondita armonia; E lucevan le stelle. Enrico Caruso, t; Instrumental accompaniments.
++NR 7-77 p9

TVM 1-7203
3315 BIZET: I pescatori di perle: Mi par di udir encor; Del templo al limitar. BOITO: Mefistofele: Dai campi dai prati. DONIZETTI: L'Elisir d'amore: Quanto e bella. Lucia di Lammermoor: Tombe degli avi miei; Tu che a Dio spiegasti l'all. GOUNOD: Faust: Salve dimora. LEONCAVALLO: I Pagliacci: Recitar...Vesti la giubba. MASCAGNI: Iris: Apri la tua finestra. MENDELSSOHN: Die Lorely, op. 98: Nel verde maggia. RIMSKY-KORSAKOV: The legend of Sadko, op. 5: Chanson Indoue. VERDI: Attila: Te sol quest'anima. La forza del destino: Solenne in quest'ora. La

traviata: De miei bollenti spiriti. Beniamino Gigli, t; Instrumental accompaniments.
++NR 7-77 p9

PRL 1-8020
3316 DAVIS: Little drummer boy. GRUBER: Stille Nacht. HANDEL: Joy to the world. HAYDN, M.: Anima nostra; In dulci jubilo. HERBECK: Angels we have heard on high; Kommet ihr Hirten; Pueri concinite. KODALY: Die Engel und die Hirten. MENDELSSOHN: Adeste fideles; Greensleeves; Hark, the herald angels sing. PRAETORIUS (arr.): Deck the halls; Es ist ein Ros entsprungen; Gloria, Gott in der Hoh. SCHULZ: Ihr Kinderlein kommet. Anton Neyder, pno; Johannes Sonnleitner, org; Vienna Boys Choir; Vienna Chamber Orchestra; Hans Gillesberger.
+ARG 12-76 p55 +NR 12-76 p1
+MU 2-77 p10

Tape (c) RK 2-5030
3317 The man with the golden flute, James Galway. BACH: Minuet and badinerie. BERKELEY: Sonatina. DEBUSSY: Little shepherd. Suite bergamasque: Clair de lune. Syrinx, flute. GLUCK: Orphee ed Euridice, excerpt. MOZART: Andante, K 315, C major. PAGANINI: Moto perpetuo, op. 11, no. 2. VIVALDI: Concerto, flute and bassoon, op. 10, no. 2, G minor. James Galway, flt.
+-HFN 5-77 p139 tape

PL 2-5046
3318 CAVIE: Variations on a theme by Lully. CLARKE: Trumpet voluntary (arr. Wright). DAMARE: Cleopatra. HOLST: Suite, no. 1, op. 28, no. 1, E flat major: March. MASCAGNI: Cavalleria rusticana: Intermezzo (arr. Wright). TCHAIKOVSKY: Sleeping beauty, op. 66: Waltz (arr. Newsome). THOMPSON: March: The Nybbs. TOMLINSON: Overture on famous English airs. TRAD.: The lark in the clear air; Sarie Marais (arr. Langford). Grimethorpe Colliery Band; Bryan Smith, hn; Tom Paulin, Peter Roberts, cor; Tom Paulin, tpt; Bryden Thomson.
++Gr 7-77 p226

GL 2-5062 Tape (c) GK 2-5062
3319 A garland for the Queen. BAX: What is it like to be young and fair. BERKELEY: Spring at the hour. BLISS: A garland for coronation morning. FINZI: White flowering days. HOWELLS: Inheritance. IRELAND: The hills. RAWSTHORNE: Canzonet. RUBBRA: Salutation, op. 82. TIPPETT: Dance, clarion air. VAUGHAN WILLIAMS: Silence and music. Exultate Singers; Garrett O'Brien.
+-Gr 6-77 p90 +-MT 12-77 p1017
+HFN 8-77 p85 +RR 9-77 p99 tape

RL 2-5081
3320 BLISS: Fanfare for a coming of age. Fanfare for a dignified occasion. Fanfare for the bride. Fanfare for heroes. Fanfare for the Lord Mayor of London. Fanfare, homage to Shakespeare. Interlude. Royal fanfare. Royal fanfare, no. 1: Sovereign's fanfare. Royal fanfares, nos. 5, 6. BENJAMIN: Fanfare for a festive occasion. Fanfares: For a state occasion; For a brilliant occasion; For a gala occasion. BRIAN: Festival fanfare. ELGAR: Civic fanfare. JACOB: Music for a festival: Interludes for trumpets and trombones, no. 1, Intrada; no. 2, Round of seven parts; no. 3, Interlude; no. 4, Saraband; no. 5, Madrigal. RUBBRA: Fanfare for Europe, op. 142. SIMPSON: Canzona for brass. TIPPETT: Fanfare, brass, no. 1. WALTON: Fanfare "Hamlet" (arr. Sargent). A queen's fanfare. TRAD.: National anthem (arr. Coe). Locke Consort of Brass; James Stobart.
++Gr 6-77 p67 ++HFN 8-77 p85

RL 2-5099

3321 Spanish music for harp. ALBENIZ: Suite espanola, no. 1: Granada. ANGLES: Aria, D minor. CABEZON: Pavane and variations. CHAVARRI: El viejo castillo moro. FALLA: El sombrero de tres picos: Danse du corregidor; Danse du meunier. Homenage a Debussy. MUDARRA: Tiento and fantasia. NEBRA: Sonata, no. 5. RIBAYAZ: Hachas. RUIZ-PIPO: Cancion y danza. Endecha. SOLER: Sonatas (2). TURINA Sacro-monte, op. 55, no. 5. David Watkins, hp.

+Gr 9-77 p460 +RR 10-77 p79
+HFN 12-77 p181

RL 2-5110 (2)

3322 William Byrd and his contemporaries. ALISON: Go from my windoe. Lady Francis Sidney's almayne. BACHILLER: The widoes mite. BYRD: Clarifica me, Pater. Come woeful Orpheus. Content is rich. Enemdemus in melius a 5. Fantasi a 3. Fantasia a 4. Haec dicit Dominus a 5. Libera me, Domine, et pone a 5. Praise our Lord, all ye gentiles. Wedded to will is witless. What pleasure have great princes. CUTTING: Galliard. DANYEL: Tyme cruell tyme. DOWLAND: Can she excuse. FERRABOSCO I: Pavan. HOLBORNE: The widowes myte. HUME: Touch me light; Tickle me quickly. JOHNSON, R.: The satyres masque. JONES: To sigh and to bee sad; Wither runneth my sweet hart. LUPO: Dance. MORLEY: Alman. See mine own sweet jewel. ANON.: The solemne pavin. Those eyes. David James, c-t; Paul Elliott, Leigh Nixon, t; Geoffrey Shaw, Paul Hillier, bar; Brian Etheridge, bs; Early Music Consort; James Tyler.

+-ARG 10-77 p677 +RR 11-77 p99

RL 2-5112

3323 English and French part-songs. BRITTEN: Hymn to St. Cecilia, op. 27. DEBUSSY: Chansons de Charles d'Orleans: Dieu qu'il la fait bon regarder; Quant j'ai ouy le tambourin; Yver, vous n'estes qu'un villian. ELGAR: The shower, op. 71, no. 1. POULENC: Un soir de neige. RAVEL: Nicolette; Trois beaux oiseaux de Paradis; Ronde. SAINT-SAENS: Calme des nuits; Les fleurs et les arbres. STANFORD: The blue bird, op. 119, no. 3. VAUGHAN WILLIAMS: The cloud-capp'd towers; Full fathom five; Over hill over dale. Swingle II.

++Gr 10-77 p685 +RR 11-77 p101

CRL 3-2026 (3)

3324 BORODIN: Quartet, strings, no. 2, D major: Nocturne (arr. Sargent). IPPOLITOV-IVANOV: Caucasian sketches, op. 10: Procession of the Sardar. PROKOFIEV: Love for three oranges, op. 33: March. Symphony, no. 1, op. 25, D major. RIMSKY-KORSAKOV: Le coq d'or: Bridal procession. SCRIABIN: Poeme de l'extase, op. 54. STRAVINSKY: L'Oiseau de feu: Suite. TCHAIKOVSKY: Capriccio italien, op. 45. Marche slav, op. 31. Romeo and Juliet: Overture. PO; Eugene Ormandy.

+MJ 2-77 p31 +NR 12-76 p5

GL 4-2125 Tape (c) GK 4-2125

3325 ALBENIZ: Suite espanola, no. 3: Sevillanas. BEETHOVEN: Bagatelle, op. 25, A minor. CHOPIN: Berceuse, op. 57, D flat major. Mazurka, op. 6, no. 2, C sharp minor. Waltz, op. 70, no. 1, G flat major. DE GROOT: Cloches dans le matin. FALLA: Love the magician: Ritual fire dance. GRANADOS: Spanish dance, op. 37: Andaluza. LISZT: Consolation, no. 3, G 172, D flat major. Liebestraum, no. 3, G 541, A flat major. MENDELSSOHN: Songs without words, op. 67, no. 4: Spinning song. POULENC: Pastourelle. RACHMANINOFF: Elegie, op. 3, no. 1. SATIE: Gymnopedies, nos. 1 and 3. Cor de Groot, pno.

+-Gr 7-77 p202
/-HFN 8-77 p89
+-HFN 8-77 p99 tape
+-RR 7-77 p83
-RR 12-77 p98 tape

CRM 5-1900 (5). (also RL 01900)

3326 BERLIOZ: Romeo and Juliet, op. 17: Queen Mab scherzo. DEBUSSY: Images pour orchestra: Iberia. La mer. MENDELSSOHN: A midsummer night's dream, op. 21/61: Incidental music. RESPIGHI: Feste Romane. SCHUBERT: Symphony, no. 9, D 944, C major. STRAUSS, R.: Tod und Verklarung, op. 24. TCHAIKOVSKY: Symphony, no. 6, op. 74, B minor. PO; Arturo Toscanini.

+ARG 2-77 p45
+-ARSC vol 9, no. 1 1977 p90
+-Gr 7-77 p188
+HF 1-77 p103
+HFN 9-77 p147
++NR 12-76 p2
++RR 7-77 p59
+-SFC 10-3-76 p33

CRM 6-2264

3327 BEETHOVEN: Sonata, violin and piano, no. 3, op. 12, E flat major. Sonata, violin and piano, no. 8, op. 30, no. 3, G major. Trio, strings, op. 3, no. 1, E flat major. Trio, strings, op. 9, no. 1, G major. Trio, strings, op. 9, no. 3, C minor. GRIEG: Sonata, violin and piano, no. 2, op. 13, G major. HANDEL: Passacaglia (trans. Halvorsen). Sonata, violin, op. 1, no. 15, E major. MOZART: Concerto, violin, no. 5, K 219, A major. Divertimento, string trio, K 563, E flat major. Duo, violin and viola, K 242, B flat major. Sonata, violin and piano, no. 17, K 296, C major. Sonata, violin and piano, no. 26, K 378, B flat major. Sonata, violin and piano, no. 32, K 454, B flat major. SINDING: Suite, op. 10, A minor. Jascha Heifetz, vln; Emanuel Feuermann, vlc; Gregor Piatigorsky, vlc; William Primrose, vla.

++ARG 12-77 p40
++MJ 10-77 p27
+NR 9-77 p7
+NYT 7-24-77 pD11

REDIFFUSION

15-56

3328 ALDER: Bi-Centenary USA march. ARNOLD: Little suite, no. 2. CATELINET: Our American cousins suite. HANMER: Stephen Foster fantasy. HEATH: Air and rondo. High spirits. RUBINSTEIN: Melody, F major (arr. Ball). SIEBERT: Hawaiian hoe-down. WINDSOR: Alpine echoes. Oxford City Youth Band; Terry Brotherhood.

+Gr 1-77 p1193

15-57

3329 ANDERSON: Fiddle-faddle. FERRARIS: Souvenir d'Ukraine. KARAS: Cafe Mozart waltz. Harry Lime theme. KENNY: Serenade for a gondolier. MASCAGNI: Cavalleria rusticana: Intermezzo. MASSENET: Thais: Meditation. PUCCINI: Manon Lescaut: Intermezzo. ROSSINI: Introduction and variations, clarinet and orchestra. SCOTT-WOOD: Shy serenade. TRAD.: Watching the wheat. Royal Artillery Orchestra; M. Pearce, clt; R. Pearce, hp; G. Starke, vln; R. Quinn, cond.

+-Gr 5-77 p1750

Royale 2015

3330 BACH: Cantata, no. 29: Sinfonia. Suite, orchestra, S 1068, D major: Air on the G string. DUSSEK: Andante, F major. HAYDN: St. Antoni chorale. MASSENET: Thais: Meditation. SAINT-SAENS: Suite algerienne, op. 60: Marche militaire Francaise. SCHUBERT: Marche militaire, no. 1, D 733, D major. Moment musical, op. 94, no. 3, D 780, F minor. SOUSA: Liberty bell. WAGNER: Tannhauser: Grand march. Carlo Curley, org.
+NR 12-77 p15

RICHARDSON

RRS 3

3331 Annapolis sounds. BILLINGS: Thus saith the high. BUXTEHUDE: Prelude and fugue, A minor. CLEMENT: Alleluia, the Lord is king. HASSLER: Cantata domino. HAYDN: Divertimento, B flat major (St. Antoni chorale). PHYLE: President's march. John Cooper, org; Aeolian Woodwind Quintet, Annapolis Brass Quintet, U. S. Naval Academy Glee Club; John Talley.
+MU 5-77 p18

RUBINI

GV 26

3332 ARDITI: Parla (in Italian). AUBER: Manon Lescaut: C'est l'histoire amoureuse (in French). CHOPIN: The nightingale's trill. CUI: The little cloud. DARGOMIZHSKY: On our street. DONIZETTI: Linda di Chamounix: O luce di quest'anima. FAURE: Crucifix. MEYERBEER: Les Huguenots: O beau pays; Beaute divine, enchanteresse. RUBINSTEIN: The demon: Sweet friends; The night is warm and quiet. VENZANO: O che assorta (in Italian). VERDI: Rigoletto: Ah veglia o donna; Si vendetta. La traviata: Un di felice. ANON. (Ukrainian folksongs, arr. Edlichko): My silken handkerchief; Storm breezes. Maria Michailova, s.
+-RR 8-77 p22

GV 29

3333 DONIZETTI: Lucia di Lammermoor: Tu che a Dio. GOUNOD: Faust: Salve dimora. MASCAGNI: Cavalleria rusticana: Siciliana; Brindisi. MASSENET: Manon: Ah dispar vision. Werther: Pourquoi me reveiller. PONCHIELLI: La gioconda: Cielo e mar. PUCCINI: La boheme: Che gelida manina. Tosca: Recondita armonia; E lucevan le stelle. VERDI: Rigoletto: Ella mi fu rapita... Parmi veder le lagrime. La traviata: Libiamo ne'lieti calici; Un di felice. Interview with Schipa. Tito Schipa, t.
+-Gr 9-77 p489 +-RR 8-77 p20

GV 34

3334 COCCIA: Per la patria: Bella Italia. DENZA: Culto; Occhi di fata. HEROLD: Zampa: Perche tremar. NOUGES: Quo Vadis: Amica l'ora. QUARANTA: O ma charmante. ROTOLI: La gondola nera; Mia sposa sara la mia bandiera. VERDI: Don Carlo: Per me giunto; Morro, ma lieta in corre. Il trovatore: Mira, d'acerbe; Vivra, contende. Mattia Battistini, bar.
+RR 8-77 p20

GV 38

3335 BRUNEAU: L'Attaque du Moulin: Adieux. BOUGAULT-DUCOUDRAY: Thamara: Reve de Noureddin. GLUCK: Armide: Plus j'observe ces lieux. GOUNOD: Romeo et Juliette: Madrigal; Juliette est vivante.

HALEVY: La Juive: Dieu que ma voix tremblante. LEROUX: Astarte: Adieu d'Hercule. MASSENET: Le mage: Ah, parais. MEYERBEER: L'Africaine: Pays marveillaux...Oh paradis. Les Huguenots: Tu l'as dit; Beaute divine enchanteresse. REYER: Sigurd: Esprit, gardiens; Duet de la fontaine. Le statue: Grand recitative. Agustarello Affre, t.
+RR 8-77 p20

GV 39
3336 ADAM: Le chalet: Vallons de l'Helvetie. BERLIOZ: La damnation de Faust, op. 24: Serenade. FLOTOW: Martha: Canzone del porter. GODARD: Embarquez-vous. GOUNOD: Faust: Serenade; Le veau d'or. Romeo et Juliette: Allons jeunes gens. MEYERBEER: L'Etoile du nord: O jours heureux. Robert le diable: Invocation. ROSSINI: Stabat mater: Pro peccatis. SCHUMANN: Les deux grenadiers. THOMAS: Le Caid: Air du Tambour-major (2 versions). VERDI: Don Carlo: Monoloque of Philip. Pol Plancon, bs.
+RR 8-77 p20

GV 43
3337 BOITO: Mefistofele: Dai campi, dai prati (in Swedish). BONINCONTRO: Kom ater. CASTBERG: Veste-blomme, enge-blomme. GRIEG: Mens jeg venter; En Svane. d'HARDELOT: Because (in Swedish). HOWARD: I wonder who's kissing her now (in Swedish). JORDAN: Baeken. LEONCAVALLO: I Pagliacci: Vesti la giubba; No Pagliaccio non son. MASCAGNI: Cavalleria rusticana: Sicliana (in Swedish); Addio alla madre. PUCCINI: La boheme: Che gelida manina. Tosca: Recondita armonia; E lucevan le stelle. SJOGREN: Evit dig til hjerlet trykke. VERDI: Rigoletto: La donna e mobile (in Swedish). Joseph Hislop, t.
+-RR 7-77 p39

GV 57
3338 BEETHOVEN: In questa tomba oscura (in Italian). BELLINI: Norma: Casta diva (in Italian). DAVID: La perle du Bresil: Charmant oiseau. DE LARA: Partir c'est mourir un peu. EMMETT: Dixie (in English). GODARD: La vivandiere: Viens avec nous petit. GORING (Thomas): Ma voisine. GOUNOD: Barcarolle; Serenade. HAHN: Dernier voeu; L'heure exquise. LULLY: Amadis de Gaule: Amour que veux-tu. MASCAGNI: Cavalleria rusticana: Voi lo sapete. MASSENET: Sappho: Pendant un an je jus ta femme. MOZART: Le nozze di Figaro, K 492: Voi che sapete. OFFENBACH: La perichole: O mon cher amant. ROSA: Quand on aime. Emma Calve, s.
+RR 8-77 p20

RS 301 (2)
3339 BILLI: E canta il grillo. BIZET: Carmen: Habanera; Chanson boheme; Seguidille. BOITO: Mefistofele: Lontano, lontano. DAVIDOV: Night, love and moon. FAURE: The crucifix. GLINKA: Do not tempt me needlessly. GRODSKI: The seagull's cry. HUMPERDINCK: Hansel und Gretel: Ein Mannlein steht im Walde. MASSENET: Werther: Air des larmes. MEYER-HELMUND: Violets. NAPRAVNIK: Dubrovsky: Never to see her. Harold: Lullaby. PUCCINI: Tosca: Vissi d'arte (2). RESPIGHI: Stornellatrice, 1st and 2nd edition. ROTOLI: Fior che langue. RUBINSTEIN: Night. TCHAIKOVSKY: At the ball, op. 38, no. 3. Queen of spades, op. 68: Twill soon be midnight. TOSTI: Penso. ANON.: Russian song. Interview with Medea Mei-Figner, Paris, 1949. Medea Mei-Figner, s.
+ARSC vol 9, no. 1 1977 p94
+Gr 9-77 p489
+HFN 3-76 p99
+RR 11-76 p24

SAGA

5384 Tape (c) CA 5384

3340 COUPERIN, F. Livres de clavecin, Bk II, Order no. 6: Les moissonneurs; Les barricades mysterieuses; Le moucheron. HANDEL: Suite harpsichord, no. 5, E major (The harmonious blacksmith). LOELLET. Suite, G minor. PARADISI: Sonata, harpsichord, no. 6, A major. SCARLATTI, D.: Sonata, harpsichord, no. 6, A major. David Sanger, hpd.
+-Gr 7-74 p243 /RR 8-74 p55
+-Gr 2-77 p1322 tape -RR 6-77 p96 tape

5412 Tape (c) CA 5412 (also Turnabout TV 34605)

3341 Music for 2 guitars. BARRIOS: Danza paraguaya. CRESPO: Nortena, homenaje a Julian Aguirre. LAURO: Suite venezelano: Vals. PONCE: Canciones populares mexicanas (3). Valse. VILLA-LOBOS: Choros, no. 1, E minor. Cirandinha: Therezinha de Jesus; Cirandinha, no. 10: A canoa virou. TRAD.: Boleras sevillanas. Buenos reyes. Cantar montanes. Cubana. De blanca tierra. Linda amiga. El pano moruno. El puerto. Salamanca. Tutu maramba. Villancico. Walter Feybli, Konrad Ragossnig, gtr.
+Gr 1-76 p1215 +HFN 2-77 p117
+Gr 2-77 p1322 tape +RR 3-76 p64
+HFN 12-75 p159 +RR 12-76 p108 tape
++HFN 10-76 p185 tape +RR 2-77 p82

5417 Tape (c) CA 5417

3342 BISHOP: Grand march, E major. CLARKE: Prince of Denmark's march. HAYDN: Feldpartita, B flat major. PEZEL: Suite, C major. PICK (attrib.): March and troop. Suite, B flat major. London Bach Ensemble; Trevor Sharpe.
+Gr 3-76 p1478 +-HFN 10-76 p185 tape
+-Gr 2-77 p1322 tape +-RR 12-75 p44
+HFN 1-76 p123 +-RR 12-76 p105 tape

5420

3343 CALVI: The Medici court. FOSCARINI: Il furioso. GALILEI: For the Duke of Bavaria. MELLII: For the Emperor Matthias. NEUSIDLER: The Burgher of Nuremberg. ROBINSON: Lessons for the lute. ANON.: At an English ale house. James Tyler, Renaissance lute, archlute, baroque guitar.
+Gr 4-76 p1643 +-SFC 8-22-76 p38
+MT 8-76 p666 +St 8-77 p59
+RR 3-76 p56

5421 Tape (c) CA 5421

3344 BEETHOVEN: Modlinger dances: Waltzes, nos. 3, 10, 11. DEBIASY: Dance. LANNER: Favorit-polka, op. 201. SCHUBERT: Ecossaises, D 529, nos. 1-3, 5; D 783, nos. 1 and 2. STELZMULLER: Dance. STRAUSS, E.: Unter der Enns, op. 121. STRAUSS, J. I: Lorely Rheinklange, op. 154. STRAUSS, J. II: Man lebt nur einmal, op. 167. Sinngedichte, op. 1. STRAUSS, Josef: Galoppin polka, op. 237. ANON.: Dance, D major. Schellerl-Tanz. Vienna Baroque Ensemble; Hans Totzauer.
+Gr 7-76 p221 +-RR 7-76 p53
++HFN 7-76 p85 +RR 11-77 p119 tape
+HFN 11-77 p187 tape

5425

3345 Music for recorder and harpsichord. BYRD: La volta. Wolsey's wilde. DOWLAND: Lachrimae: Antiquae pavan. Captain Digorie piper, his galliard. ECCLES: Sonata, recorder and harpsichord, F major. FARNABY: Tower hill. HANDEL: Sonata, flute, op. 1,

no. 9, B minor (transp.). CAIX d'HERVELOIS: Suite, op. 6, no. 3, G major. LOEILLET: Sonata, recorder and harpsichord, G major. ROSSETER: And would you see my mistress face. ANON.: Corranto. Faronels ground. Greensleeves to a ground. Nowells galliard. Paul's steeple. Michael Arno, rec; Adrian Bush, hpd.
+-Gr 1-77 p1160 +-RR 1-77 p72
+-HFN 2-77 p123

5438 Tape (c) CA 5438
3346 Renaissance works. ALISON: Sharp pavin. BERNIA: Toccata chromatica. BORRONO DA MILANO: Casteliono book: Pieces (3). CASTELLO: Sonata, mandora, theorbo and bass viol. CORBETTA: Suite. RORE: Contrapunto sopra "Non mi toglia il ben mio". DOWLAND: Fantasia. FERRABOSCO I: Spanish pavan. KAPSBERGER: Toccata. PICCININI: Toccata. VALLET: Suite. ANON.: Zouch, his march. Nigel Ryan, Jan Ryan, bass viol; Nigel North, theorbo, lute, cittern; Douglas Wootton, lt, mandora; James Tyler, lt, baroque guitar, mandora.
+Gr 2-77 p1302 +RR 3-77 p80
+HFN 5-77 p131 +RR 12-77 p97 tape
+HFN 11-77 p187 tape

5444
3347 Popular music from the time of Henry VIII. BARBIREAU: En frolyk weson. CORNYSH: Ay, Robin. Henry VIII, King: En vray amoure; O my heart. ANON.: Absent I am; And I were a maiden; Be peace ye make me spill my ale; Begone, sweit night; Consort piece XX; The duke of Somersettes domp; England be glad; Hey trolly lollo lo; I love unloved; Madame d'amours; My heartily service; O lusty May; Puzzle canon VI; This day day daws; Up I arose in verno tempore. Hilliard Ensemble, New London Consort Members.
+-Gr 7-77 p211 +RR 8-77 p77
+HFN 5-77 p133

5445
3348 English piano music. BAX: Burlesque. BOWER: Sonata, op. 35. BRIDGE: Sketches (3). GOOSSENS: Kaleidoscope, op. 18. IRELAND: Aubade. SCOTT: Water wagtail, op. 71, no. 3. Richard Deering, pno.
+-FF 9-77 p68 /MM 10-77 p44
+-Gr 5-77 p1717 +MT 11-77 p921
+-HFN 6-77 p123 +-RR 5-77 p68

5447
3349 BYRD: The Earl of Salisbury pavan and galliard. CAMPIAN: Oft have I sighed; There is a garden in her face. BULL: Les buffons. CORNYSH: Ah, Robin. JOHNSON, R.: Where the bee sucks. DOWLAND: Can she excuse my wrongs; Sorrow, stay. MORLEY: It was a lover and his lass. ANON.: All in a garden green; Good fellows must go learn to dance; Heven and erth; Jouissance; Packington's pound; The tree ravens; Walkins' ale; Wilson's wilde; The wind and the rain; Fortune my foe. London Camerata; Glenda Simpson, Barry Mason.
+Gr 6-77 p95 +-RR 7-77 p88
+-HFN 6-77 p131 ++SFC 11-6-77 p48

ST. JOHN THE DIVINE CATHEDRAL CHURCH

(Available from Cathedral Church, Cathedral Heights, NY)
3350 Seventy-five years of cathedral music: Saint John the Divine, New York. BACH: Chorale preludes: Ich ruf zu dir, S 639; Heut

triumphiret, S 630. Cantata, no. 147, Jesu, joy of man's desiring. COKE-JEPHCOTT: Bishop's promenade. Pieces, organ. Saint Anne with descant. Toccata on Saint Anne. FARROW: Pieces. HAYDN: Glorious things. IPPOLITOV-IVANOV: Bless the Lord, O my soul. PURCELL: Trumpet tune and air. VIERNE: Canzona. WYTON: Pieces. David Pizarro, org; St. John the Divine Choir; David Pizarro.
+MU 5-77 p18

SERAPHIM

SIB 6094 (2) (Reissues from Capitol CLSP 8399, 8458, CL SSAL 8385)
3351 BACH: Mein Jesu, was fur Seelenweh, S 487. Toccata and fugue, S 565, D minor. BARBER: Adagio, strings. DEBUSSY: Prelude a l'apres-midi d'un faune. Suite bergamasque: Clair de lune. DUKAS: La peri: Fanfare. GLUCK: Iphigenie en Aulide: Lento. Armide: Sicilienne. MUSSORGSKY (orch. Ravel): Pictures at an exhibition: Hut on fowl's legs, Great gate at Kiev. PURCELL: King Arthur: Hornpipe. SIBELIUS: Legens, op. 22: Swan of Tuonela. Finlandia, op. 26. STRAUSS, J. II: The beautiful blue Danube, op. 314. TCHAIKOVSKY: Symphony, no. 4, op. 36, F minor: Scherzo. TURINA: La oracion del torero, op. 34. Orchestra; Leopold Stokowski.
+NR 5-77 p3 +-RR 8-77 p57

IB 6105 (2)
3352 The art of Lotte Lehmann. BRAHMS: Die Mainacht, op. 43, no. 2. GIORDANO: Andrea Chenier: La mamma morta. GODARD: Jocelyn: Berceuse. GOUNOD: Faust: Je voudrais savoir...Il etait un Roi de Thule. KORNGOLD: Die tote Stadt, op. 12: Der erste der Lieb mich gelehrt. MASSENET: Manon: Il le faut...Adieu notre petite table. Werther: Letter scene. MOZART: Le nozze di Figaro, K 492: Deh viene, non tardar. OFFENBACH: Les contes d'Hoffmann: Elle a fui la tourterelle. PUCCINI: La boheme: Si mi chiamano Mimi. Madama Butterfly: Butterfly's entrance. Manon Lescaut: In quelle trine morbide. Tosca: Amaro sol per te. Turandot: Del primo pianto. SCHUBERT: An die Musik, D 547; Der Tod und das Madchen, D 531. SCHUMANN: Die Lotosblume, op. 25, no. 7; Der Nussbaum, op. 25, no. 3. STRAUSS, J. II: Die Fledermaus, op. 363: Mein Herr was dachten Sie von mir. STRAUSS, R.: Arabella, op. 79: Mein Elemer. Der Rosenkavalier, op. 59: Kann mich auch an ein Madel erinnern, Oh sei Er got Quinquin...Die Zeit die ist ein sonderbar Ding. Songs: Morgen, op. 27, no. 4. THOMAS: Mignon: Connais-tu le pays; Elle est la pres de lui...Elle est aimee. WAGNER: Lohengrin: Du armste kannst wohl nie ermessen. Die Meistersinger von Nurnberg: Gut'n Abend, Meister, O Sachs mein Freund. Tannhauser: Allmachtge Jungfrau, hor mein flehen. Die Walkure: Du bist der Lenz. WEBER: Der Freischutz: Wie nahte mir der Schlummer... Leise leise fromme Weise. Lotte Lehmann, s; Instrumental accompaniments.
+NR 11-77 p12 +St 11-77 p160
+ON 11-77 p70

60269
3353 Gregorian chant: Easter liturgy and Christmas cycle. Fathers of the Holy Spirit, Chevilly.
*MJ 5-77 p33

60274
3354 Great sopranos of the century. BOITO: Mefistofele: L'altra notte (Claudia Muzio). CANTELOUBE: Chants d'Auvergne: Lo fiolaire (Madeleine Grey). DEBUSSY: Green (Maggie Teyte). MEYERBEER: Dinorah: Ombra leggiera (Luisa Tetrazzini). PUCCINI: Madama Butterfly: Un bel di (Toti dal Monte). SCHUBERT: Ave Maria (Elisabeth Schumann). VALVERDE: Clavelitos (Conchita Supervia). VERDI: Aida: O patria mia (Eva Turner). Un ballo in maschera: Morro, ma prima in grazia (Maria Caniglia). La traviata: Ah, fors'e lui...Sempre libera (Nellie Melba). WAGNER: Traume (Kirsten Flagstad). Tristan und Isolde: Liebestod (Lotte Lehmann). Die Walkure: Ho-jo-to-ho (Frida Leider). Gerald Moore, pno; Various orchestras and conductors.
+NR 11-77 p12 +St 11-77 p162

S 60275
3355 DONIZETTI: Don Pasquale: Che interminabile andrivieni. Lucia di Lammermoor: Per poco fra le tenebre. MASCAGNI: Cavalleri rusticana: Regina coeli...Inneggiamo. PUCCINI: Turandot: Gira le cote. ROSSINI: Mose in Egitto: Dal duo stellato soglio. VERDI: Aida: Gloria all'egitto. Macbeth: Patria opressa. Nabucco: Va pensiero. Otello: Fuoco di gioia. Il trovatore: Vede, le fosche notturno. ROHO and Chorus; Lamberto Gardelli.
+NR 6-77 p10 -ON 3-12-77 p40

S 60277
3356 BIZET: L'Arlesienne: Suite, no. 2: Farandole. GLIERE: The red poppy, op. 70: Sailor's dance. GRIEG: Norwegian dance, op. 35, no. 2. KABALEVSKY: The comedians: Galop. KHACHATURIAN: Gayaneh: Sabre dance. MASSENET: Le Cid: Navarraise. RAVEL: Miroirs: Alborado del gracioso. RIMSKY-KORSAKOV: Capriccio espagnol, op. 34. TCHAIKOVSKY: Sleeping beauty, op. 66: Waltz, Act 1. WALDTEUFEL: The skaters, op. 183. Espana, op. 236. Hollywood Bowl Symphony Orchestra; Felix Slatkin.
+NR 6-77 p2 +-RR 8-77 p56

S 60280
3357 The art of Beniamino Gigli. DONAUDY: O del mio amato ben. FLOTOW: Martha: M'appari. LEONCAVALLO: I Pagliacci: Prologue. MEYERBEER: L'Africana: O paradiso. MOZART: Das Veilchen (La violetta), K 476. PUCCINI: La boheme: Che gelida manina. Tosca: Recondita armonia; O dolci mani. THOMAS: Mignon: Ah, non credevi tu. VERDI: Aida: Celeste Aida. I Lombardi: Qual volutta trascorrere. Attila: Te sol quest'anima. Il trovatore: Di quella pira. Elisabeth Rethberg, s; Beniamino Gigli, t; Ezio Pinza, bs; Various orchestras and conductors.
++NR 10-77 p8 +-St 9-77 p146
+RR 12-77 p30

M 60291
3358 BIZET: Carmen: L'amour est un oiseau rebelle; Pres des remparts de Seville; Les tringles des sistres tintaient. FALLA: Spanish popular songs (7). GRANADOS: Tonadillas. PUCCINI: La boheme: D'un pas leger. ROSSINI: La cenerentola: Nacqui all'affano... No no tergete il ciglio. THOMAS: Mignon: Connais-tu le pays. Conchita Supervia, s; Instrumental accompaniments.
+NR 12-77 p9

SERENUS

SRS 12067

3359 BURROWES: Auld lang syne. DUSSEK: My lodging is on the cold, cold ground. FIORILLO: Hanoverian air. HOWELL: Robin Adair. LATOUR: Variations on "God save the king". TRAD. (adapt. Leeves): Auld Robin Gray. Elizabeth Marshall, pno.
+-HF 12-77 p114 +NR 12-77 p12

1750 ARCH

S 1753

3360 DUFAY: Songs: C'Est bien raison de devoir essaucier; Je me complains piteusement; Invidia nimica; Malheureulx cuer que veux to faire; Par droit je puis bien complaindre. GRIMACE: A l'arme a l'arme. LANDINI: Se la nimica mie; Adiu adiu. PERUSIO: Andray soulet. VAILLANT: Par maintes foys. ANON.: Istampita Ghaetta. Music for a While.
++AR 2-76 p133 +NR 11-75 p11
++ARG 9-77 p28

SOCIETY FOR THE DISSEMINATION OF NATIONAL GREEK MUSIC

SDNM 101/2 107, 112

3361 Byzantine hymns of Christmas, Epiphany, Epitaphios and Easter, Service of the Akathistos hymn. Choir of the Society; Simon Karas.
+RR 5-77 p79

STARLINE

SRS 197 (Reissue from Columbia TWO 334)

3362 Elizabethan serenade. ANCLIFFE: Nights of gladness (arr. Lotter). BYFIELD: A Cornish pastiche. Gabriel John. A pinch of salt. BINGE: Elizabethan serenade. COATES: Springtime suite: Dance in the twilight. FARNON: Portrait of a flirt. HARTLEY: Rouge et noir. OSBORNE: Lullaby for Penelope. OSBORNE (Younger, David): My northern hills. WOOD: Paris suite: Montmartre. TRAD.: My love she's but a lassie yet (arr. Byfield). Studio Two Concert Orchestra; Reginald Kilbey.
+Gr 2-77 p1330

SUPRAPHON

110 1637

3363 FLOTOW: Allesandro stradella: Overture. KREUTZER: Das Nachtlager von Granada: Overture. NICOLAI: Die lustige Weiber von Windsor: Overture. REZNICEK: Donna Diana: Overture. WAGNER: Rienzi: Overture. WEBER: Abu Hassan: Overture. Brno State Philharmonic Orchestra; Zeljko Straka.
+Gr 1-77 p1184 +-RR 12-76 p68
+NR 11-76 p4 +SFC 3-20-77 p34

110 1890

3364 ALBINONI: Concerto, op. 5, no. 5, A major. CORELLI: Sarabande, gigue and badinerie. DURANTE: Concerto, strings, no. 5, A major. LEGRENZI: Sonata, strings, no. 6, E minor. MANFREDINI:

Concerto grosso, op. 3, no. 9, D major. Ostrava Janacek Chamber Orchestra; Zdenek Dejmek.
+FF 11-77 p60 ++RR 7-77 p44
+HFN 8-77 p76

111 1721/2 (2)
3365 AURIC: Adieu, New York. BURIAN: American suite. COPLAND: Piano blues (4). DEBUSSY: Children's corner suite: Golliwog's cakewalk. GERSHWIN: Preludes. HINDEMITH: Suite "1922", op. 26. MARTINU: Preludes. SATIE: Jack in the box. SCHULHOFF: Esquisses de jazze. Rag music. Peter Toperczer, Jan Vrana, Emil Leichner, Jan Marcol; Milos Mikula, pno.
+Gr 12-76 p1022 +NR 1-77 p14
+HFN 1-77 p109 -RR 1-77 p71

111 1867
3366 FISCHER: Suite, no. 6, F major. LULLY: Suite, C major. SCHEIDT: Suite, C major. TELEMANN: Suite, F major. VEJVANOVSKY: Sonata, no. 14 a 6 Campanarum. Musica de Camera, Praga.
+FF 11-77 p62 +NR 7-77 p7
+-Gr 8-77 p310 +RR 6-77 p58
+-HFN 7-77 p105

113 1323
3367 Concerto grosso for 7 voices. BEETHOVEN: Sonata, piano, no. 20, op. 49, no. 2, G major: Tempo di minuetto. CERNOHORSKY: Laudetur Jesus Christ, excerpt. HANDEL: Water music: Minuet. KOLOVRATEK: Parthia pastoralis, F major. PASCHA: Little Slovak suite: Andante cantabile, Con brio. Tota pulchra. PLIHAL: Fugue, F major. SCHMIDT: Cassation. SCHUBERT: Sonatina, violin and piano, no. 3, op. 137, D 408, G minor: Andante; Allegro giusto. VANHAL: Sonata, B flat major: Adadio cantabile; Rondo. ANON.: Psalm. The Linha Singers.
+-HFN 2-77 p115 -RR 1-77 p78
-NR 9-77 p11

SWEDISH SOCIETY

SLT 33205
3368 Guitarra espanola. ALBENIZ, M.: Sonata, guitar, D major. FALLA: Homenaje. RUIZ PIPO: Cancion y danza, no. 1. SOR: Minuet, op. 11, no. 6. Minuet, op. 22. TARREGA: Estudio brillante. Recuerdos de la Alhambra. TORROBA: Burgalesa. Nocturno. TURINA: Hommage a Tarrega. Diego Blanco, gtr.
+HFN 7-77 p113 +-RR 7-77 p81

SLT 33209
3369 BENEDICT: Greetings to America. BERG: Herdegossen. JOSPEHSON: Songs: Serenad; Sjung, sjung du underbara sang; Tro ej gladjen. LINDBLAD: Songs: Aftonen; Am aarensee; Manntro; En ung flickas morgonbetraktelse. MENDELSSOHN: Songs: Auf Flugeln des Gesanges, op. 34, no. 2; Fruhlingslied, op. 34, no. 3; Der Mond, op. 86, no. 5. SCHUMANN: Songs: An den Sonnenschein, op. 38, no. 4; Dicterliebe, op. 48: Im wunderschonen Monat Mai, Die Rose die Lilie; Roselein, op. 89, no. 6. THRANE (arr. Ahlstrom): Fajllvisan. Elisabeth Soderstrom, s; Jan Eyron, pno.
+-RR 12-77 p78

TELEFUNKEN

EK 6-35334 (2)

3370 BEETHOVEN: Rondino, E flat major. HAYDN: Divertimento, no. 15, F major. HOFFMEISTER: Parthia, E flat major. KRAMAR-KROMMER: Serenade, C minor. MOZART: The magic flute, K 620: Overture and airs. Serenade, E flat major. VANHAL: Parthia, B flat major. WRANITZKY: Parthia, F major. Consortium Classicum.
+RR 11-77 p60

6-41126 Tape (c) 4-41126

3371 Troubadour songs and instrumental works. BERNARD DE VENTADORN: Can vei la lauzeta mover. DIA: A chantar m'er so qu'eu no volria. GUIRAUT DE BORNELH: Leu chansonet a vil. VAQUEIRAS: Kalenda meia. VIDAL: Baron de mon dan covit. Early Music Quartet.
++HF 7-77 p125 tape

6-41928

3372 Music of the minstrels: Chose tassin; Ductia; Estampie (2); Chominciamento di gioia; La quinte, sexte, septime, ultime estampie real; Retrove; Saltarello; La tierche estampie royal.
+Gr 7-76 p205 +MT 12-77 p1017
+HF 7-76 p96 +NR 3-76 p10
+HFN 6-76 p81 +-RR 7-76 p67

AW 6-42033

3373 Recorder music of the Italian Renaissance. GABRIELI, A.: Ricercar del secondo tono. GABRIELI, G.: Canzone a 4, 5 and 6. LASSUS: Ricercar a 2 (2). MAINERIO: Primo libro de balli. PALESTRINA: Ricercar a 4. RORE: A la dolc'ombra. Pero piu ferm. ROSSI: Sinfonia grave a 5. Gagliarda "Norsina" a 5. SEGNI: Fantasien. WILLAERT: Madrigal a 6, "Passa la nave". Ricercar a 3. Vienna Recorder Ensemble.
+Gr 10-77 p654 +RR 10-77 p78
+HFN 10-77 p156

AW 6-42129

3374 DRAGHI: Trio sonata, G minor. HILTON: Fantasies, nos. 1-3. LOCKE: Suites, nos. 1 and 2. PURCELL: Chaconne "Two in one upon a ground". Symphonies, nos. 1 and 2. WILLIAMS: Sonata, recorder, no. 4, A minor. Quadro Hotteterre.
+Gr 9-77 p452 +RR 9-77 p78
+HFN 9-77 p141

AW 6-42155

3375 HAGEN: Sonata, lute, B major. KAPSBERGER: Toccata, no. 7. Canzona, no. 2. LULLY: La grotte de Versailles: Overture (trans. Visee). Les sourdines d'Armide. MARAIS: Les matelots: Air (trans. Visee). PICCININI: Corrente. Toccata, no. 10. WEISS: Ciacona. Prelude. Rigaudon. Toyohiko Soh, chitarrone, baroque lute.
+-Gr 9-77 p459 +-RR 10-77 p74
+-HFN 9-77 p136

AJ 6-42232

3376 FLOTOW: Martha: Ach, so fromm. LORTZING: Undine: Vater, Mutter. Zar und Zimmermann: Lebe wohl, mein flandrisch Madchen. MOZART: Don Giovanni, K 527: Nur ihrem Frieden; Folget der Heissgeliebten. Die Entfuhrung aus dem Serail, K 384: Hier soll ich dich denn sehen; Konstanze...O wie angstlich; Im Mohrenland. Die Zauberflote, K 620: Dies Bildnis. NICOLAI: Die lustige Weiber von Windsor: Horch, die Lerche. WEBER: Der Freischutz: Nein langer trag ich nicht die Qualen. Peter Anders, t; Berlin State Opera Orchestra; Hans Schmidt-Isserstedt.
+-HFN 12-77 p164 +-RR 12-77 p31

DT 6-48085 (2)
3377 BRAHMS: Songs: Des Abends kann ich nicht schlafen gehn; Ach, englische Schaferin; Ach konnt ich diesen Abend; All mein Gedanken, Gar lieblich hat sich gesellet; Erlaube mir, feins Madchen; Es ging ein Maidlein zarte; Fahr wohl, o Voglein; Feinsliebchen sollst nicht barfuss gehn; Guten Abend, mein tausinger Schatz; Mein Madel hat einen Rosenmund; Schwesterlein; Die Sonne scheint nicht mehr; Wach auf mein Herzens schone; Wie komm ich denn zur Tur herein; Wiegenlied. BRUCH: Waldpsalm, op. 38, no. 1. FROHLICH: Wem Gott will rechte Gunst erweisen. KUHN: Ausfahrt. LYRA: Wanderschaft. MENDELSSOHN: Abschied vom Walde; Fruhlingsgruss; Die Nachtigall, op. 59, no. 4. MOZART: Sehnsucht nach dem Fruhling, K 596. SCHULZ: Der Mond ist aufgegangen. SCHUMANN: Fruhlingsgruss; Gute Nacht. TRAD.: Alle Vogel sind schon da; Bitte; Bruderschaft; Es wollt ein Jagerlein jagen; Fangt euer Tagwerk frohlich an; Im Freien; Nach gruner Farb; Schone Fruhling, komm doch wieder; Die Voglein in dem Walde; Zwischen Berg und Tiefem Tal; Der weisse Hirsch. Peter Schreier, t; Rudolf Dunckel, pno; Dresden Kreuzchor; Rudolf Mauersberger.
+HFN 3-77 p105 +-RR 3-77 p88

TOCCATA

53623
3378 BACH, J.C.: Aria and 15 variations, A minor. BUXTEHUDE: Praeludium und Fuge, G minor. FISCHER: Musikalisches Blumenbuschlein: Suite, no. 6, D major. HANDEL: Suite, harpsichord, no. 3, D minor. ZACHOW: Suite, B minor. Bradford Tracy, hpd.
+MQ 10-77 p570

TR

1006
3379 Organ recital, Earl Barr. BACH: Chorale preludes: Nun komm, den Heiden Heiland, S 659. Prelude and fugue, S 532, D major. Prelude and fugue, S 550, G major. Toccata, D minor, excerpt. Fugue, S 578, G minor (Little), excerpt. KARG-ELERT: In dulci jubilo, op. 75, no. 2. KOERTSIER: Partita, English horn and organ. Earl Barr, org.
+-MU 4-77 p14

TRANSATLANTIC

XTRA 1169
3380 BROWNE: El cumbanchero. BULL (Byrd): Music from the Elizabethan court (arr. Howarth). MOZART: The magic flute, K 620: Overture (arr. Dent). SOUSA: Liberty bell. STEFFE: Glory, glory. SULLIVAN: Iolanthe: Overture (arr. Sargent). WATERS: Trumpets wild. TRAD. (arr. Langford): All through the night. Cory Band.
-RR 4-77 p60

TURNABOUT

TV 37086
3381 Medieval Paris, music of the city. CROIX: S'amours eust point de poer. l'ESCURIEL: Amours, cent mille merciz. ADAM DE LA HALLE:

De cueur pensieu; En mai, quant rosier. MUSET: Quant je voi yver retorner. PEROTIN LE GRAND: Alleluja. ANON.: Amours dont je sui espris; Ave virgo regia; Chaconnette; Danse real; Dieus, qui porroit; Ductia; Estampie royal; Hoquet; O Maria, virgo; Quant voi l'aloete; Un hocquet; Veris ad imperia. Praetorius Consort, Purcell Consort of Voices; Christopher Ball, Grayston Burgess.
+Gr 2-77 p1312 +MT 9-77 p739
+HFN 4-77 p138 +RR 3-77 p93

UNIVERSAL EDITION

UE 15043

3382 A garland for Dr. K. BEDFORD: A garland for Dr. K. BENNETT: Impromptu. BERIO: The modification and instrumentation of a famous hornpipe as merry and altogether sincere homage to Uncle Alfred. BIRTWISTLE: Some petals from the garland. BOULEZ: Pour le Dr. Kalmus. HALFFTER: Oda para felicitar a un amigo. HAUBENSTOCK-RAMATI: Rounds. POUSSEUR: Echoes Il de votre Faust. RANDS: Monotone. STOCKHAUSEN: Fur Dr. K. WOOD: A garland for Dr. K.
+-RR 12-77 p60

UNIVERSITY OF IOWA PRESS

Unnumbered

3383 BACH: Art of the fugue, S 1080: Contrapuncti (2). BOZZA: Sonatine, brass quintet. HUSA: Divertimento, brass quintet. LECLERC: Par monts et par vaux. MOURET: Rondeau. SMOKER: Brass in spirit. WILBYE: Madrigals (2). ANON.: Carmina, 16th century (4). Iowa Brass Quintet.
+-HF 7-77 p118 +NR 2-77 p7
+IN 5-77 p28

VANGUARD

SVD 71212

3384 Christmas Eve at the Cathedral of St. John the Divine: LVOV: To thy heavenly banquet (Communion anthem). PALESTRINA: Sanctus and benedictus (Soriano). SWEELINCK: Hodie Christus natus est. VICTORIA: O magnum mysterium. ANON.: Adeste fideles; The beatitudes; The first noel Hark, the herald angels sing; Ninefold alleluia and announcement of the holy gospel; The shepherd's carol; Trisagion; While shepherds watched their flocks. David Pizarro, org; Cathedral Choir and Cathedral Boys Choir; Richard Westernburg.
+Audio 5-77 p102 +NR 12-76 p1

VSD 71217

3385 A guide to Gregorian chant. Schola Antiqua; R. John Blackley.
++ARG 3-77 p45 +HF 7-77 p118
+-Audio 5-77 p100 ++NR 9-77 p10

VSD 71219/20

3386 Instruments of the Middle Ages and Renaissance: An illustrated guide to their ranges, timbres and special capabilities. Musica Reservata, London; Martin Bookspan, narrator.
+ARG 12-77 p16 +NR 12-77 p15

VIRTUOSI

VR 7608

3387 CUNDELL: Blackfriars, symphonic prelude. CLARKE, M.: Tyrolean tubas. DAVIES: Solemn melody (arr. Ord Hume). LALO: Le Roi d'Ys: Overture (arr. Wright). DVORAK: Rusalka: O silver moon (arr. Corbett). SAINT-SAENS: The swan (arr. Mott). VERDI: Nabucco: Chorus of the Hebrew slaves. WAGNER: Tannhauser: Grand march (arr. G. Hawkins). Virtuosi Brass band of Great Britain; Harry Mortimer.

++Gr 1-77 p1193

VISTA

VPS 1029

3388 DUPRE: Fileuse. DURUFLE: Suite, op. 5: Toccata, B minor. LANGLAIS: Triptyque. LITAIZE: Toccata sur le veni creator. SAINT-MARTIN: Toccata de la liberation. TOURNEMIRE: Chorale sur le victimae Paschali (arr. Durufle). Jane Parker-Smith, org.

+Gr 8-76 p320 ++MU 7-77 p14
+HFN 2-77 p125

VPS 1030

3389 BACH: Prelude and fugue, S 532, D major. COUPERIN: Chaconne, C major. FRANCK: Chorale, no. 2, B minor. LISZT: Prelude and fugue on the name B-A-C-H, G 260. MARCHAND: Basse de trompette. WILLS: Carillon on Orientis Partibus. Arthur Wills, org.

+Gr 8-76 p323 +-MU 7-77 p14
+HFN 2-77 p125 +RR 8-76 p67

VPS 1033

3390 COUPERIN: Dialogue sur les grands jeux. KARG-ELERT: Jerusalem, du hochgebaute Stadt, op. 65. MESSIAEN: La nativite du Seigneur: Dieu parmi nous. PEETERS: Koncertstuck, op. 52a. SALOME: Grand choeur. WILLAN: Introduction, passacaglia and fugue, E flat minor. WILLS: Piece, organ: Processional. Richard Galloway, David Harrison, Ronald Perrin, John Turner, Allan Wicks, Arthur Wills, org.

*Gr 3-77 p1444 +MU 8-77 p10
+-HF 7-77 p95 +-RR 4-77 p77

VPS 1034

3391 BENNETT: Alba. GUILMANT: Sonata, organ, no. 3, op. 56, C minor. LANG: Tuba tune, op. 15, D major. LEIGHTON: Festival fanfare. PARRY: Toccata and fugue (The wanderer). VIERNE: Pieces en style libre, op. 31, no. 5: Prelude. YON: Toccatina, flute. Jonathan Bielby, org.

+Gr 3-77 p1444 +MU 8-77 p10
+-HF 7-77 p95 +RR 4-77 p76

VPS 1035

3392 ANDRIESSEN: Theme and variations. BOSSI: Divertimento in forma de giga. HOLLINS: Concert overture, C minor. PARRY: Old 100th, chorale fantasia. Elegy, A flat major. REGER: Ein feste Burg, op. 67, no. 6. Introduction and passacaglia, D minor. SCHUMANN: Canon, op. 56, no. 5, B minor. SMART: Postlude, D major. WHITLOCK: Plymouth suite: Toccata. Roy Massey, org.

+Gr 10-77 p633 +MU 8-77 p10
+HFN 11-76 p164 +-RR 2-77 p80

VPS 1037

3393 BENNETT: Psalm, no. 71. BYRD: Motets: Sacerdotes domini; Veni sancte spiritus; Senex puerum portabat. DAVIES: God be in my

head. DUPRE: Prelude and fugue, op. 7, no. 1, B major. HOWELLS: Exultate Deo. A hymn for St. Cecilia. MARSHALL: Reveille pavan. MUDD: Let thy merciful ears, O Lord. SLATER: For life with all it yields. STANFORD: Evening service, C major. Lincoln Cathedral Choir; Philip Marshall, Roger Bryan, org; Philip Marshall.
+-Gr 2-77 p1312 +-RR 2-77 p87
+HFN 3-77 p101

VPS 1038
3394 BARBER: Die Natali, op. 38: Chorale prelude on "Silent night". Wondrous love: Variations on a Shapenote hymn, op. 34. BREMNER: Trumpet tune. COPLAND: Episode. BUCK: Concert variations on the "Star-Spangled Banner" op. 23. IVES: Variations on "America". Adeste fidelis in an organ prelude. SESSIONS: Chorale, no. 1, G major. Walter Hillsman, org.
+-Gr 3-77 p1444 +MU 8-77 p10
+-HF 7-77 p95 +RR 3-77 p74
+HFN 4-77 p141

VPS 1039
3395 CARLTON: Verse for 2 to play on one organ. LEIGHTON: Dialogues on the Scottish psalm-tune "Martyrs", op. 73. LUYTENS: Plenum IV, op. 100. MERKEL: Sonata, organ, op. 30, D minor. TOMKINS: Fancy for 2 to play. WESLEY, S.: Prelude, C minor (trans. of Bach's Fugue, S 552). Stephen and Nicholas Cleobury, org.
+Gr 3-77 p1444 -MU 7-77 p14
+-HF 7-77 p95 +-RR 3-77 p80
+-HFN 5-77 p131

VPS 1042
3396 GUILMANT: Sonata, organ, no. 5, op. 80, C minor. HOLLINS: Trumpet minuet. HARWOOD: Paean, op. 15, no. 2. KELLNER: Was Gott tut. SWEELINCK: Balletto del Granduca. Peter Goodman, org.
+Gr 3-77 p1444 +MU 7-77 p14
+-HF 7-77 p95 +RR 3-77 p82
+HFN 4-77 p141

VPS 1046
3397 COOK: Fanfare, organ. CRESTON: Toccata. DE MONFRED: In paradisum. REUBKE: Sonata on the 94th psalm. THALBEN-BALL: Tune, E major. TREDINNICK: Brief encounters. George Thalben-Ball, org.
+Gr 12-77 p118 +RR 11-77 p91
+HFN 12-77 p175

VPS 1047
3398 Organ music from the city of London. GIBBONS: Fantazia of foure parts. HEWITT-JONES: Fanfare. HOWELLS: Psalm prelude, op. 32, no. 1. MIDDLETON: Fantasie. PURCELL: Voluntary, G major. STANFORD: Postlude, D minor. STANLEY: Voluntary, op. 5, no. 1, C major. VAUGHAN WILLIAMS: Preludes on Welsh hymn tunes: Rhosymedre. WEITZ: Grand choeur. ANON.: Upon la mi re. Christopher Herrick, org.
+Gr 11-77 p867 +-RR 11-77 p93
+-HFN 12-77 p175

VPS 1051
3399 BRIDGE: First book of organ pieces. PARRY: Chorale prelude on "St. Thomas". STANFORD: Postlude, D minor. VIERNE: Symphony, no. 3, op. 28, F minor. WHITLOCK: Plymouth suite: Salix. Geoffrey Tristam, org.
+Gr 11-77 p868 +-RR 11-77 p92
+-HFN 12-77 p175

VPS 1053

3400 Coronation choral music 1953. DYSON: Be strong. HANDEL: Zadok the priest. HARRIS: Let my prayer come up. HOWELLS: Behold, O God our defender. JACOB: The national anthem (arr.). PARRY: I was glad. REDFORD: Rejoice in the Lord (attrib.). STANFORD: Gloria in excelsis. VAUGHAN WILLIAMS: The old hundredth; O taste and see. WALTON: Te deum. WESLEY, S.S.: Thou wilt keep him. WILLAN: O Lord, our governour. Exultate Singers, Dulwich College Preparatory School Choir; Timothy Farrell, org; Garrett O'Brien.

+Gr 8-77 p332 +-RR 8-77 p75
+HFN 8-77 p85

VPS 1060

3401 BACH: Chorale prelude: Wir glauben all an einen Gott, S 680. COUPERIN, F.: Plein jeu. Recit de tierce en taille. DAQUIN: Noel Suisse. HINE: A flute piece. HOWELLS: Master Tallis' testament. MUSHEL: Toccata. PIERNE: Pieces, op. 29 (3). VERSCHRAEGEN: Partita octavi toni super Veni Creator. WIDOR: Symphony, organ, no. 6, op. 42, no. 2, G major: Allegro. Donald Hunt, John Sanders, Roy Massey, org.

+-Gr 12-77 p1118 /-RR 11-77 p93

VOX

SVBX 5142 (3)

3402 Music of the French Baroque. COUPERIN, F.: La Sultane, D minor. La Steinquerque. COUPERIN, L.: Symphonies, violes (5). MARAIS: La gamma. Pieces de violes, Bk 2: Suite, B minor. RAMEAU: Pieces de clavecin en concert: Sonata, no. 4, G major. SAINTE-COLOMBE: Concerts, 2 bass violes (2). Oberlin Baroque Ensemble.

+HF 12-77 p115 +NR 11-77 p77

SVBX 5303 (3)

3403 Piano music in America, 1900-45. ANTHEIL: Sonata, no. 2. BARBER: Sonata, op. 26: Fuga. Nocturne, op. 33. CARPENTER: Impromptu. COPLAND: Variations. COWELL: Exultation. Invention. Aeolian harp. Advertisement. GERSHWIN: Preludes (3). GRIFFES: Roman sketches, op. 7, nos. 1, 3. HARRIS: Sonata, op. 1. IVES: The anti-abolitionist riots. Some southpaw pitching. Sonata, no. 2: The Alcotts. MacDOWELL: Woodland sketches, op. 51, nos. 3, 7-10. PISTON: Passacaglia. RIEGGER: New and old: The twelve tones, The tritone, Seven times seven, Tone clusters, Twelve upside down, Fourths and fourths. RUGGLES: Evocations. SCHUMAN: Three-score set. SESSIONS: From my diary. THOMSON: Sonata, no. 3. Ten etudes: Parallel chords, Ragtime bass. Roger Shields, pno.

+-ARG 5-77 p9 +NYT 7-3-77 pD11
+HF 9-77 p115 +SFC 4-10-77 p30
+NR 10-77 p13

SVBX 5307 (3)

3404 The avant garde woodwind quintet in the U.S.A. BARBER: Summer music, woodwind quintet, op. 31. BERGER: Quartet, C major. BERIO: Children's play, wind quartet, op. zoo. CARTER: Quintet, woodwinds. DAVIDOVSKY: Synchronisms, woodwind quintet and tape, no. 8. DRUCKMAN: Delizie contente che l'alme beate, woodwind quintet and tape. FINE: Partita, wind quintet. FOSS: The cave of the winds. HUSA: Preludes, flute, clarinet and bassoon (2). SCHULLER: Quintet, woodwinds. Dorian Quintet.

++ARG 12-77 p51 +NR 11-77 p8
+HF 12-77 p110

SVBX 5350 (3)
3405 America sings: The founding years, 1620-1800. Religious vocal music from the Ainsworth Psalter; The Bay Psalm Book; Tuft's Introduction; Walter's Grounds and Rules; Lyon's Urania. ANTES: How beautiful upon the mountains. BELKNAP: The season. BILLINGS: As the hart panteth; Consonance; Thus saith the high, the lofty one. BROWNSON: Salisbury. DENCKE: O, be glad, ye daughters of His people. HERBST: God was in Jesus. HOPKINSON: A toast; Beneath a weeping willow's shade; Come fair Rosina; Enraptur'd I gaze; My days have been so wond'rous free; My gen'rous heart disdains; My love is gone to sea; O'er the hills far away; See, down Maria's blushing hceek; The traveler benighted and lost. INGALLS: Northfield. LAW: Bunker Hill. KELLY: Last week I took a wife; The mischievous bee. LYON: Friendship. MORGAN: Amanda; Despair; Montgomery. PETER, J.F.: I will freely sacrifice to Thee; I will make an everlasting covenant. PETER. S.: Look ye, how my servants shall be feasting; O, there's a sight that rends my heart. READ: Down steers the bass; Russia. SELBY: O be joyful to the Lord; Ode for the New Year. STEVENSON: Behold I bring you glad tidings. SWAN: China. VON HAGEN: Funeral dirge on the death of General Washington. ANON.: Washington's march. Gregg Smith Singers and Orchestra; Gregg Smith.
+HF 7-76 p69 +SFC 4-10-77 p30
+MJ 11-76 p44 ++St 9-76 p117
+NR 10-76 p8

SVBX 5353 (3)
3406 America sings. BARBER: A stopwatch and an ordinance map. BERNSTEIN: The lark: Choruses. CARTER: Defense of Corinth; Musicians wrestle everywhere. COPLAND: Pieces, treble choir (2). COWELL: Luther's carol to his son. FINE: The choral New Yorker. FOSS: Behold I build an house. GERSHWIN: Madrigals (2). IVES: Election songs (2). PISTON: Psalm and prayer of David. RIEGGER: Who can revoke. SCHUMAN: Prelude. SEEGER: Chant. SESSIONS: Turn O libertad. TALMA: Let's touch the sky. THOMPSON, R.: Alleuia. THOMSON: Southern hymns (4). Gregg Smith Singers, Texas Boys Choir, Columbia University Men's Glee Club, Peabody Conservatory Concert Singers; Gregg Smith.
+NR 12-77 p8

SVBX 5483 (3)
3407 CHABRIER: Pieces,posth.: Ballabile. Bourree fantasque. Impromptu. CHAUSSON: Quelques danses. DEBUSSY: Children's corner suite. DUKAS: Variations, interlude and finale on a theme by Rameau. FAURE: Impromptu, no. 3, op. 34, A flat major. Impromptu, no. 5, op. 102, F sharp minor. Nocturne, op. 74, C sharp minor. FRANCK: Prelude, chorale and fugue. d'INDY: Chant des bruyeres. MILHAUD: Hymne de glorification. Romances. RAVEL: Serenade grotesque. ROUSSEL: Pieces, piano, op. 49. Suite, op. 14: Bourree. SAINT-SAENS: Etude, op. 135: Bourree. SATIE: Croquis et agaceries d'un gros bon homme en bois. Poudre d'or. Preludes veritables flasques pour un Chien. SEVERAC: Pippermitgit. Sous les Lauriers roses. Grant Johannesen, pno.
+NR 10-77 p12

WEALDEN

WS 131
3408 BACH: Fantasia, S 562, C minor. FAULKNER: Intrada. MESSIAEN: Le banquet celeste. MOZART: Fantasia, organ, K 608, F minor. REGER: Pieces, op. 59, no. 9: Benedictus. SWEELINCK: Unter der Linden grune. WALTON (arr. Murrill): Crown imperial. WHITLOCK: Carol. Robert Gower, org.
/-RR 1-77 p75

WS 159
3409 BACH: Passacaglia and fugue, S 582, C minor. JACKSON: Carillon. LANGLAIS: Suite francaise. REGER: Pieces, op. 80: Toccata and fugue. TOURNEMIRE: Cantilene improvisee. Te deum improvisee. VIERNE: Carillon de Westminster. WALTHER: Jesu meine Freude, partita. Nicholas Jackson, org.
+-RR 2-77 p74

WESTMINSTER

WG 1008
3410 Gregorian chant. Encalcat Abbey Monks Chorus.
+-Gr 2-77 p1321 +-RR 3-77 p92

WG 1012 (Reissues from XWN 18428, 18135, 18429, 18137)
3411 Lute and guitar recital. BACH: Suite, lute, S 996, E minor: Bourree. Prelude and fugue, S 554, D minor. DOWLAND: Melancholy galliard. Mrs. Vaux's gigue. Semper Dowland, semper dolens. SOR: Fantasia, no. 2: Largo. Sonata, guitar, op. 2: Rondo allegretto. TORROBA: Sonatine, A major. TURINA: Fandanguillo, op. 36. VILLA-LOBOS: Prelude, no. 1, E minor. Julian Bream, lt and gtr.
+Gr 4-77 p1578 +RR 4-77 p74

WGM 8309 (Reissues from 78 rpm originals)
3412 The young Gilels. CHOPIN: Ballade, no. 1, op. 23, G minor. LISZT: Hungarian rhapsody, no. 9, G 244, E flat major. Etudes d'execution transcendente d'apres Paganini, no. 5, G 140, E major. LULLY (Godowsky): Gigas. MENDELSSOHN: Songs without words, op. 38, no. 6. PROKOFIEV: The love for three oranges, op. 33: March. RAMEAU: Le Reppel des oiseaux. SCHUMANN: Der contrabandiste, op. 74 (arr. Liszt). Fantasiestucke, op. 12, no. 7: Traumeswirren. Toccata, op. 7, C major. Emil Gilels, pno.
+-ARSC 7-75 p90 +HF 11-75 p129
+-CL 10-77 p6

WGS 8338
3413 BELLINI: Little overture. IVES: The unanswered question. MOZART: Divertimento, no. 3, K 166, E flat major. ROGALSKI: Roumanian dances (2). SCHOECK: Concerto, horn, op. 65, D minor. WEBERN: Pieces, orchestra, op. 10 (5). Boris Afanasev, hn; Moscow Radio Ensemble, Bolshoi Theatre Orchestra; Gennady Rozhdestvensky.
+NR 5-77 p5

Section III

ANONYMOUS WORKS

A la fontenella. cf Enigma VAR 1020.
Absent I am. cf Saga 5444.
Ack Varmeland, du skona. cf HMV RLS 715.
Adeste fideles. cf Decca SXL 6781.
Adeste fideles. cf Vanguard SVD 71212.
Afro-Cuban lullaby. cf Angel S 36053.
The Agincourt song. cf Decca SB 715.
Al alva venid. cf Enigma VAR 1024.
All hail to the days. cf CRD CRD 1019.
All in a garden green. cf Saga 5447.
All through the night. cf Transatlantic XTRA 1169.
Alle psallite. cf DG 2710 019.
Alle psallite cum luya. cf CRD CRD 1019.
Alle Vogel sind schon da. cf Telefunken DT 6-48085.
Allemande prince. cf Argo ZK 24.
Allt under himmelens faste. cf HMV RLS 715.
Almande de Ungrie. cf Hungaroton SLPX 11721.
Amor potest. cf DG 2710 019.
Amours dont je sui espris. cf Turnabout TV 37086.
An raibh tu ag an gCarrig. cf Philips 6599 227.
And all in the morning (arr. Vaughan Williams). cf HELY-HUTCHINSON: Carol symphony.
And I were a maiden. cf Saga 5444.
Andernaken. cf HMV SLS 5049.
Angelus ad Virginem (1st and 2nd versions). cf HMV ESD 7050.
Annie Laurie. cf Orion ORS 77271.
Antyck. cf Argo ZK 24.
As I lay. cf CRD CRD 1019.
At an English ale house. cf Saga 5420.
Auld lang syne. cf HMV RLS 719.
Auld Robin Gray. cf Serenus SRS 12067.
Ave virgo regia. cf Turnabout TV 37086.
Babe of Bethlehem. cf HMV ESD 7024.
The bailiff's daughter of Islington. cf HMV ESD 7002.
Balletto Polacho. cf DG 2533 294.
The bard of Armagh. cf Philips 6599 227.
Le baroudeur. cf Decca SB 713.
Basse danza a 2. cf DG 2723 051.
Bassa danza a 3. cf DG 2723 051.
Basse danses I and II. cf HMV SLS 863.
Batalla famosa. cf Harmonia Mundi HM 759.
Batori Tantz, Proportio. cf Hungaroton SLPX 11721.
Be peace ye make me spill my ale. cf Saga 5444.
Beata viscera. cf Abbey LPB 765.

The beatitudes. cf Vanguard SVD 71212.
Begone, sweet night. cf Saga 5444.
La bella Franceschina. cf L'Oiseau-Lyre 12BB 203/6.
Belle, tenes mo. cf L'Oiseau-Lyre 12BB 203/6.
The bells. cf HMV ESD 7024.
Biance flour. cf Argo D40D3.
The birds. cf HMV SQ CSD 3784.
El bisson. cf Enigma VAR 1024.
Bitte. cf Telefunken DT 6-48085.
The bitter sweet. cf BG HM 57/9.
Blame not my lute. cf L'Oiseau-Lyre 12BB 203/7.
Bobby Shafto. cf Amberlee ALM 602.
Boleras sevillanas. cf Saga 5412.
Bonny, sweet Robin. cf Enigma VAR 1020.
Born in Bethlehem. cf HMV ESD 7024.
Branle de l'official. cf CRD CRD 1019.
Brincan y bailan. cf Enigma VAR 1015.
Bruder Konrad. cf L'Oiseau-Lyre 12BB 203/6.
Bruderschaft. cf Telefunken DT 6-48085.
Buenos reyes. cf Saga 5412.
Bugeilio'r Gwenith Gwyn. cf L'Oiseau-Lyre SOL 345.
Il buratto. cf Argo ZK 24.
Caccia de Zaccharias. cf CBS 76534.
Calata. cf L'Oiseau-Lyre 12BB 203/6.
Canciones (2). cf BIS LP 3.
Cantar montanes. cf Saga 5412.
Carmina, 16th century. cf University of Iowa Press unnumbered.
Carole. cf CBS 76534.
Carole liturgique "Qui passus". cf CBS 76534.
Celle qui m'a demande. cf L'Oiseau-Lyre 12BB 203/6.
C'est mon ami. cf Odyssey Y 33130.
Chaconette. cf Turnabout TV 37086.
Charlie is my darling. cf HMV ESD 7002.
The Chester carol. cf HMV ESD 7024.
Chevalier mult estes guariz. cf Argo D40D3.
Christ der ist erstanden. cf L'Oiseau-Lyre 12BB 203/6.
Christ ist erstanden. cf Argo D40D3.
Christmas bells. cf HMV SQ CSD 3784.
Christum wir solen loben schon. cf ABC L 67002.
Chui dicese e non l'amare. cf L'Oiseau-Lyre 12BB 203/6.
Clap, clap par un matin. cf DG 2710 019.
Cobbler's jigg. cf DG 2723 051.
La Comarina. cf Enigma VAR 1024.
Come holy ghost. cf National Trust NT 002.
Coming thro the rye. cf HMV RLS 716.
Coming thro the rye. cf HMV RLS 719.
Concerto, 2 oboes and strings, F major. cf BOYCE: Concerto grosso, strings, E minor.
Condio, O ancio mane prima. cf Argo D40D3.
Consort piece XX. cf Saga 5444.
Contre le temps. cf HMV SLS 863.
Coranto. cf Abbey LPB 765.
Corranto. cf Saga 5425.
Credo "Bonbarde". cf CBS 76534.
Crimond. cf Decca SB 329.
Cubana. cf Saga 5412.
The cuckoo carol. cf HMV ESD 7024.

Cum sint difficilia. cf Argo D40D3.
Czech rocking carol. cf HMV ESD 7024.
Dadme albricias. cf GABRIELI: O magnum mysterium.
Dafydd y Garreg Wen. cf BBC REC 267.
Dafydd y Garreg Wen. cf L'Oiseau-Lyre SOL 345.
The Dalling alman. cf National Trust NT 002.
Dance, D major. cf Saga 5421.
Dances: My robbin; Tickle my toe; Hollis berrie. cf National Trust 002.
Danse real. cf Argo D40D3.
Danse real. cf Turnabout TV 37086.
Danza. cf Hungaroton SLPX 11721.
Daphne. cf Argo ZK 25.
Dashing away with a smoothing iron. cf Grosvenor GRS 1052.
De blanca tierra. cf Saga 5412.
De la trumba. cf DG 2533 323.
Deck the hall. cf HMV ESD 7024.
Degentis vita. cf DG 2710 019.
Dieus, qui porroit. cf Turnabout TV 37086.
Ding-dong-doh. cf HMV ESD 7024.
Dit le Bourguygnon. cf Enigma VAR 1024.
Dominator Domine. cf DG 2710 019.
Don Gato. cf Enigma VAR 1015.
The dressed ship. cf CRD CRD 1019.
Drewries accords. cf DG 2533 323.
Ductia. cf CRD CRD 1019.
Ductia. cf Nonesuch H 71326.
Ductia. cf Turnabout TV 37086.
The Duke of Somersettes domp. cf Saga 5444.
E la Don, Don, Verges Maria. cf GABRIELI: O magnum mysterium.
Early one morning. cf HMV ESD 7002.
Edi beo thu. cf CRD CRD 1019.
En mai quant rosier. cf DG 2710 019.
En natus est Emanuel. cf ABC L 67002.
England be glad. cf Saga 5444.
English estampie. cf Enigma VAR 1020.
Es wollt ein Jagerlein jagen. cf Telefunken DT 6-48085.
Estampie and trotto. cf BIS LP 3.
Estampie royal. cf Turnabout TV 37086.
Est-il conclu par un arret d'amour. cf HMV SLS 5049.
Fandango. cf CRD CRD 1019.
Fanfare. cf Enigma VAR 1020.
Fangt euer Tagwerk frohlich an. cf Telefunken DT 6-48085.
Fantasia on British sea song: Hornpipe. cf HMV SQ ASD 3341.
Faronels ground. cf Saga 5425.
Febus mundo oriens. cf DG 2710 019.
The first noel. cf Vanguard SVD 71212.
The first Nowell. cf HMV CSD 3774.
Flee as a bird. cf New World NW 220.
Fortune my foe. cf Saga 5447.
Gagliarda "Cathaccio". cf ABC ABCL 67008.
Gagliarda "Lodesana". cf ABC ABCL 67008.
Gavotte. cf DG 2723 051.
Gentil prince. cf Abbey LPB 765.
Der gestraifft Danntz. cf DG 2533 302.
Girometa. cf L'Oiseau-Lyre 12BB 203/6.
Glogauer Liederbuch: Dy katczen pfothe. cf BIS LP 75.
Gloucestershire wassail. cf HMV SQ CSD 3784.

God rest you merry gentlemen. cf CRD CRD 1019.
God save the King. cf HMV RLS 719.
God save the Queen. cf CRD Britannia BR 1077.
God save the Queen. cf Lismor LILP 5078.
Good fellows must go learn to dance. cf Saga 5447.
Good King Wencelas, pavan. cf CRD CRD 1019.
Green bushes. cf Philips 6599 227.
Green garters. cf CRD CRD 1019.
Greensleeves. cf CBS 61039.
Greensleeves and pudding pyes. cf DG 2723 051.
Greensleeves to a ground. cf HMV SQ CSD 3781.
Greensleeves to a ground. cf Saga 5425.
Greenwood/Dargason: Fanfare. cf Enigma VAR 1020.
La guercia. cf HMV SLS 5049.
The happy life. cf BG HM 57/8.
Hark, the herald angels sing. cf Vanguard SVD 71212.
Hejnat Krakowska. cf Philips 6500 926.
Henry VIII's poem. cf Argo ZK 24.
Here we come a-wassailing. cf CRD CRD 1019.
Heth sold ein meisken garn om win. cf HMV SLS 5049.
Heven and erth. cf Saga 5447.
Hey trolly lolly lo. cf Argo ZK 24.
Hey trolly lolly lo. cf Saga 5444.
Himno sacris solemnis. cf Harmonia Mundi HM 759.
Hocquet. cf Turnabout TV 37086.
The holly and the ivy. cf HMV SQ CSD 3784.
The holly and the ivy. cf HMV ESD 7050.
Hoquetus in saeculum I-VII. cf DG 2710 019.
How can I keep my maiden head. cf DG 2723 051.
A hunting we will go. cf HMV ESD 7022.
I am a jolly foster. cf Argo ZK 24.
I love unloved. cf Saga 5444.
I saw three ships. cf CRD CRD 1019.
I saw three ships. cf HMV CSD 3774.
I saw three ships. cf HMV SQ CSD 3784.
I sing of a maiden. cf HMV ESD 7050.
I smile to see how you devise. cf BG HM 57/9.
Il me suffit. cf Argo ZK 24.
I'm a'rollin. cf Orion ORS 77271.
Im Freien. cf Telefunken DT 6-48085.
Impudenter. cf CBS 76534.
In natali domini. cf ABC L 67002.
In dulci jubilo. cf HMV CSD 3774.
In dulci jubilo. cf HMV SQ CSD 3784.
In mari miserie. cf DG 2710 019.
Infant holy. cf HMV SQ CSD 3784.
Instrumental fancy. cf Argo ZK 24.
Inter densas deserte meditans. cf DG 2710 019.
Istampita Cominiciamento di gioia. cf DG 2723 051.
Istampitta Ghaetta. cf Argo D40D3.
Istampita Ghaetta. cf DG 2723 051.
Istampita Ghaetta. cf 1750 Arch S 1753.
Istampitta tre fontaine. cf HMV SLS 863.
Jacob's ladder. cf BBC REC 267.
Je ne fay. cf L'Oiseau-Lyre 12BB 203/6.
Je ne puis, Amors me tient, Veritatem. cf Argo D40D3.
Je vous. cf Harmonia Mundi HM 759.

Jingle bells. cf HMV ESD 7024.
John come kiss me now. cf L'Oiseau-Lyre DSLO 510.
John Peel. cf HMV ESD 7002.
Jonah: Save me O Lord. cf Folkways FTS 32378.
Jouissance. cf Saga 5447.
Jure plaudant omnia. cf ABC L 67002.
Der Katzenfote. cf L'Oiseau-Lyre 12BB 203/6.
The keel row. cf HMV ESD 7002.
Kemp's jig. cf DG 2723 051.
Kyrie. cf CBS 76534.
Lady Wynkfyldes rownde. cf L'Oiseau-Lyre 12BB 203/6.
Lamento di Tristan. cf Argo D40D3.
Lamento di Tristano: Rotta. cf DG 2723 051.
The lark in the clear air. cf Abbey LPB 778.
The lark in the clear air. cf RCA PL 2-5046.
Las, je n'ecusse. cf L'Oiseau-Lyre 12BB 203/6.
L'e pur morto Feragu. cf L'Oiseau-Lyre 12BB 203/6.
Lengyel tanc. cf Hungaroton SLPX 11721.
Lieder 13th-16th century. cf BIS LP 75.
Like to the damask rose. cf Abbey LPB 712.
Lily of the valley. cf BBC REC 267.
Linda amiga. cf Saga 5412.
Linzer Tanz. cf DG 2723 051.
Little camel boy. cf HMV ESD 7024.
The liturgical year. cf ANTES: Chorales (3).
Lna Ban. cf Philips 9500 218.
Loch Lomand. cf Philips 6392 023.
Londonderry air. cf HMV ASD 3346.
Les l'ormel a la turelle. cf DG 2710 019.
Love in a village: O had I been by fate decreed, My Dolly was the fairest thing. cf Folkways FTS 32378.
Lute dances. cf HMV SLS 5049.
Madame d'amours. cf Saga 5444.
The maid in the moon. cf Enigma VAR 1020.
The maid of the sweet boy. cf Philips 6599 227.
Mall Symms. cf L'Oiseau-Lyre DSLO 510.
La Manfredina. cf Argo D40D3.
La Manfredina. cf CRD CRD 1019.
March of the Kuban Cossacks infantry batallion. cf HMV CSD 3782.
March of the Preobrazhensk regiment. cf HMV CSD 3782.
Mascherade. cf DG 2723 051.
La mesnie fauveline. cf DG 2710 019.
Mesomedes of Crete: Hymn to the sun. cf L'Oiseau-Lyre DSLO 17.
Mignonne, allons. cf L'Oiseau-Lyre 12BB 203/6.
Mijm morken gaf mij een jonck wifjj. cf HMV SLS 5049.
The miller of Dee. cf HMV ESD 7002.
Minuet. cf London CS 7046.
Mr. Shadowman. cf Argo ZFB 95/6.
El mois de mai. cf DG 2710 019.
Moravian chorale cycle. cf ANTES: Chorales.
Moravian funeral chorales. cf ANTES: Chorales.
La morisque. cf Argo ZK 24.
Music of the four countries. cf Lismor LILP 5078.
My dancing day. cf HMV ESD 7050.
My heartily service. cf Saga 5444.
My Lady Carey's dumpe. cf Argo ZK 24.
My lagan love. cf Philips 9500 218.

My love she's but a lassie yet. cf Starline SRS 197.
My Robin. cf Enigma VAR 1020.
My silken handkerchief. cf Rubini GV 26.
Nach gruner Farb. cf Telefunken DT 6-48085.
The national anthem. cf ELGAR: Coronation ode, op. 44.
The national anthem. cf ELGAR: Enigma variations, op. 36: Nimrod.
The national anthem. cf ELGAR: Pomp and circumstances marches, op. 39, nos. 1-5.
The national anthem. cf Decca SPA 500.
The national anthem. cf Guild GRSP 701.
The national anthem. cf RCA RL 2-5081.
Der neue Bauernschwanz. cf Enigma VAR 1024.
Ninefold alleluia and announcement of the holy gospel. cf Vanguard SVD 71212.
Nowell, Nowell, Nowell. cf HMV ESD 7050.
Nowells galliard. cf Saga 5425.
The nutting girl. cf Enigma VAR 1020.
O come all ye faithful. cf HMV CSD 3774.
O leave your sheep. cf Abbey LPB 776.
O little town of Bethlehem. cf HMV CSD 3774.
O lusty May. cf Saga 5444.
O Maria, virgo. cf Turnabout TV 37086.
O mitissima. cf DG 2710 019.
O Philippe Franci. cf DG 2710 019.
O tocius Asie. cf Argo D40D3.
Oft the stilly night. cf HMV RLS 716.
Oh, the oak and the ash. cf HMV ESD 7002.
Oh, you must be a lover of the Lord. cf New World NW 220.
On parole de batre. cf DG 2710 019.
Over the snow. cf HMV ESD 7024.
Packington's pound. cf Saga 5447.
Paduana Hispanica. cf Hungaroton SLPX 11721.
El pano moruno. cf Saga 5412.
The parson's farewell/Goddesses. cf Enigma VAR 1020.
Parti de mal. cf Argo D40D3.
Pase el agoa, ma Julieta. cf Nonesuch H 71326.
Passo e mezzo. cf Argo ZK 24.
Paul's steeple. cf Saga 5425.
Pavan. cf L'Oiseau-Lyre DSLO 510.
Pavan and galliard. cf ABC ABCL 67008.
Pavana de la morte de la ragione. cf L'Oiseau-Lyre 12BB 203/6.
Pavana el tedescho. cf L'Oiseau-Lyre 12BB 203/6.
Pavane Venetiana. cf L'Oiseau-Lyre 12BB 203/6.
Pieces, lute. cf DG 2530 561.
Polyphonic dances. cf Nonesuch H 71326.
Pour un plaisir. cf Harmonia Mundi HM 759.
Preludium VI. cf L'Oiseau-Lyre DSLO 510.
Psalm. cf Supraphon 113 1323.
Psalmus CXXX. cf Hungaroton SLPX 11721.
El puerto. cf Saga 5412.
Puzzle canon VI. cf Saga 5444.
Quan je voy le duc. cf Argo D40D3.
Quant je le voi. cf DG 2710 019.
Quant voi l'aloete. cf Turnabout TV 37086.
Quasi non ministerium. cf DG 2710 019.
Quelle est cette odeur agreable. cf HMV CSD 3774.
La quinte estampie real. cf Argo D40D3.

Quittez pasteurs. cf HMV CSD 3774.
Der rather Schwanz. cf L'Oiseau-Lyre 12BB 203/6.
Redit aetas aureau. cf Enigma VAR 1020.
Restoes, restoes. cf HMV SLS 863.
Reveillez-vous. cf Harmonia Mundi HM 759.
Riu, riu, chiu. cf GABRIELI: O magnum mysterium.
Robin Hood and the tanner. cf Enigma VAR 1020.
La rocha el fuso. cf Argo ZK 24.
La rocha el fuso. cf Enigma VAR 1024.
Rodrigo Martines. cf Nonesuch H 71326.
La romanesca. cf Discopaedia MB 1013.
Le rossignol. cf DG 2533 323.
Le rossignol. cf L'Oiseau-Lyre 12BB 203/6.
Royal music of King James I: Almande, nos. 1, 5, 6. cf Decca SB 715.
Running footman. cf DG 2723 051.
Russian song. cf Rubini RS 301.
Salamanca. cf Saga 5412.
Sally's fancy/The maiden's blush. cf Enigma VAR 1020.
Saltarelli (2). cf Argo D40D3.
Saltarello. cf CBS 76534.
Saltarello. cf DG 2723 051.
Saltarello (Italian). cf CRD CRD 1019.
Saltarello de la morte de la ragione. cf L'Oiseau-Lyre 12BB 203/6.
Sarie Marais. cf RCA PL 2-5046.
Scarborough fair. cf Golden Crest CRS 4148.
Schellerl-Tanz. cf Saga 5421.
Schone Fruhling, komm doch wieder. cf Telefunken DT 6-48085.
Se mai per maraveglia. cf L'Oiseau-Lyre 12BB 203/6.
Sede Syon, in pulvere. cf Argo D40D3.
La septime estampie real. cf HMV SLS 863.
La serrana. cf Enigma VAR 1015.
La sexte estampie real. cf Nonesuch H 71326.
The shepherd's carol. cf Vanguard SVD 71212.
Shepherd's hey. cf Enigma VAR 1020.
Shooting the guns pavan. cf L'Oiseau-Lyre 12BB 203/6.
Sola me dexastes. cf Enigma VAR 1024.
The solemne pavin. cf RCA RL 2-5110.
Sometimes I feel like a motherless child. cf Orion ORS 77271.
S'on me regards. cf DG 2710 019.
Song of the birds. cf COUPERIN: Pieces en concert.
Souvent souspire mon cuer. cf Nonesuch H 71326.
La spagna. cf Enigma VAR 1024.
Sta notte. cf L'Oiseau-Lyre 12BB 203/6.
Staines Morris. cf CRD CRD 1019.
Stingo. cf Enigma VAR 1020.
Stormy breezes. cf Rubini GV 26.
Strawberry leaves. cf Argo ZK 25.
Suite, lute. cf DG 2530 561.
Suite regina. cf L'Oiseau-Lyre 12BB 203/6.
Sumer is icumen in. cf BIS LP 3.
The Sussex carol. cf HMV SQ CSD 3784.
Swing low, sweet chariot. cf HMV RLS 719.
The tailor and the mouse. cf Abbey LPB 778.
Tantz, Proportio. cf Hungaroton SLPX 11721.
That Lord that lay in Asse stall. cf HMV ESD 7050.
This day day daws. cf Saga 5444.
Those eyes. cf RCA RL 2-5110.

La tierche estampie real. cf Argo D40D3.
Todleben. cf HMV CSD 3782.
A toye. cf Argo ZK 25.
La traditora. cf Argo ZK 24.
La traditora. cf Enigma VAR 1024.
The tree ravens. cf Saga 5447.
La triquottee. cf L'Oiseau-Lyre 12BB 203/6.
Trisagion. cf Vanguard SVD 71212.
Tristan's lament. cf Enigma VAR 1020.
Trottin' to the fair. cf Abbey LPB 778.
Trotto. cf Argo D40D3.
Trotto. cf DG 2723 051.
Tua Bethlem Dref. cf HMV ESD 7024.
A Turkish Taksim. cf L'Oiseau-Lyre DSLO 17.
Tutu maramba. cf Saga 5412.
The twelve days of Christmas. cf HMV SQ CSD 3784.
Two-voices estampie. cf Enigma VAR 1020.
La ultime estampie real. cf Argo D40D3.
Up, good Christen folk and listen. cf HMV CSD 3774.
Up I arose in verno tempore. cf Saga 5444.
Upon la mi re. cf Vista VPS 1047.
Vegnanado da Bologna. cf Argo ZK 24.
Veris ad imperia. cf Turnabout TV 37086.
Versos varios. cf Harmonia Mundi HM 759.
The vicar of Bray. cf HMV ESD 7002.
Vieille chanson de chasse. cf HMV RLS 716.
Villancico. cf Saga 5412.
The virgin Mary had a baby boy. cf BBC REC 267.
The virgin Mary had a baby boy. cf HMV ESD 7024.
Die Voglein in dem Walde. cf Telefunken DT 6-48085.
La volta. cf Argo ZK 25.
Washington's march. cf Vox SVBX 5350.
Wassail song (arr. Vaughan Williams). cf HELY-HUTCHINSON: Carol symphony.
Watching the wheat. cf Rediffusion 15-57.
Watkins' ale. cf Saga 5447.
Watts cradle song. cf HMV ESD 7050.
We wish you a merry Christmas. cf HMV SQ CSD 3784.
Der weisse Hirsch. cf Telefunken DT 6-48085.
Welscher Tantz. cf Argo D40D3.
Where be ye my love. cf Argo ZK 24.
Where griping griefs. cf L'Oiseau-Lyre 12BB 203/6.
Where riches is everlastingly. cf HMV ESD 7050.
While shepherds watched their flocks. cf Vanguard SVD 71212.
Wiener Polka. cf DG 2723 051.
Wilson's wilde. cf Saga 5447.
The wind and the rain. cf Saga 5447.
Wrth Fynd Efo Deio I Dowyn. cf L'Oiseau-Lyre SOL 345.
Xacara. cf London CS 7046.
Y Bore Glas. cf L'Oiseau-Lyre SOL 345.
Y Deryn Pur. cf L'Oiseau-Lyre SOL 345.
Ya se van la paloma. cf Enigma VAR 1015.
Ye banks and braes. cf HMV ESD 7002.
Zelus familie. cf DG 2710 019.
Zouch, his march. cf Saga 5438.
Zwischen Berg und tiefem Tal. cf Telefunken DT 6-48085.

Section IV

PERFORMER INDEX